The Oxford Handbook of Event-Related Potential Components

OXFORD LIBRARY OF PSYCHOLOGY

Editor-in-Chief PETER E. NATHAN

The Oxford Handbook of Event-Related Potential Components

Edited by

Steven J. Luck and Emily S. Kappenman

OXFORD
UNIVERSITY PRESS

OXFORD
UNIVERSITY PRESS

Oxford University Press, Inc., publishes works that further Oxford University's objective of excellence in research, scholarship, and education.

Oxford New York
Auckland Cape Town Dar es Salaam Hong Kong Karachi
Kuala Lumpur Madrid Melbourne Mexico City Nairobi
New Delhi Shanghai Taipei Toronto

With offices in
Argentina Austria Brazil Chile Czech Republic France Greece
Guatemala Hungary Italy Japan Poland Portugal Singapore
South Korea Switzerland Thailand Turkey Ukraine Vietnam

Published by Oxford University Press, Inc.
198 Madison Avenue, New York, New York 10016
www.oup.com

Oxford handbook of event-related potential components / edited by
Steven J. Luck and Emily S. Kappenman.
p. ; cm.
Handbook of event-related potential components
Event-related potential components
Includes bibliographical references.
ISBN 978-0-19-537414-8
1. Evoked potentials (Electrophysiology) I. Luck, Steven J. (Steven John), 1963-
II. Kappenman, Emily S. III. Title: Handbook of event-related potential components.
IV. Title: Event-related potential components.
[DNLM: 1. Evoked Potentials. WL 102]
QP376.5.O94 2011
616.8'047547—dc22
2010053196

9 8 7 6 5 4 3 2 1

Printed in China

SHORT CONTENTS

OXFORD LIBRARY OF PSYCHOLOGY

The *Oxford Library of Psychology*, a landmark series of handbooks, is published by Oxford University Press, one of the world's oldest and most highly respected publishers, with a tradition of publishing significant books in psychology. The ambitious goal of the *Oxford Library of Psychology* is nothing less than to span a vibrant, wide-ranging field and, in so doing, to fill a clear market need.

Encompassing a comprehensive set of handbooks, organized hierarchically, the *Library* incorporates volumes at different levels, each designed to meet a distinct need. At one level are a set of handbooks designed broadly to survey the major subfields of psychology; at another are numerous handbooks that cover important current focal research and scholarly areas of psychology in depth and detail. Planned as a reflection of the dynamism of psychology, the *Library* will grow and expand as psychology itself develops, thereby highlighting significant new research that will impact on the field. Adding to its accessibility and ease of use, the *Library* will be published in print and, later on, electronically.

The *Library* surveys psychology's principal subfields with a set of handbooks that capture the current status and future prospects of those major subdisciplines. This initial set includes handbooks of social and personality psychology, clinical psychology, counseling psychology, school psychology, educational psychology, industrial and organizational psychology, cognitive psychology, cognitive neuroscience, methods and measurements, history, neuropsychology, personality assessment, developmental psychology, and more. Each handbook undertakes to review one of psychology's major subdisciplines with breadth, comprehensiveness, and exemplary scholarship. In addition to these broadly-conceived volumes, the *Library* also includes a large number of handbooks designed to explore in depth more specialized areas of scholarship and research, such as stress, health and coping, anxiety and related disorders, cognitive development, or child and adolescent assessment. In contrast to the broad coverage of the subfield handbooks, each of these latter volumes focuses on an especially productive, more highly focused line of scholarship and research. Whether at the broadest or most specific level, however, all of the *Library* handbooks offer synthetic coverage that reviews and evaluates the relevant past and present research and anticipates research in the future. Each handbook in the *Library* includes introductory and concluding chapters written by its editor to provide a roadmap to the handbook's table of contents and to offer informed anticipations of significant future developments in that field.

An undertaking of this scope calls for handbook editors and chapter authors who are established scholars in the areas about which they write. Many of the nation's and world's most productive and best-respected psychologists have agreed to edit *Library* handbooks or write authoritative chapters in their areas of expertise.

For whom has the *Oxford Library of Psychology* been written? Because of its breadth, depth, and accessibility, the *Library* serves a diverse audience, including graduate students in psychology and their faculty mentors, scholars, researchers, and practitioners in psychology and related fields. Each will find in the *Library* the information they seek on the subfield or focal area of psychology in which they work or are interested.

Befitting its commitment to accessibility, each handbook includes a comprehensive index, as well as extensive references to help guide research. And because the *Library* was designed from its inception as an online as well as a print resource, its structure and contents will be readily and rationally searchable online. Further, once the *Library* is released online, the handbooks will be regularly and thoroughly updated.

In summary, the *Oxford Library of Psychology* will grow organically to provide a thoroughly informed perspective on the field of psychology, one that reflects both psychology's dynamism and its increasing interdisciplinarity. Once published electronically, the *Library* is also destined to become a uniquely valuable interactive tool, with extended search and browsing capabilities. As you begin to consult this handbook, we sincerely hope you will share our enthusiasm for the more than 500-year tradition of Oxford University Press for excellence, innovation, and quality, as exemplified by the *Oxford Library of Psychology.*

Peter E. Nathan
Editor-in-Chief
Oxford Library of Psychology

PREFACE

The primary goal of this handbook is to provide a comprehensive and thorough overview of research on ERP components for researchers in the mind and brain sciences. Our goal was to provide a single, comprehensive resource that researchers can use to learn about ERP components. We sought to provide breadth by covering a range of components and research domains and depth by devoting whole chapters to individual components and topics. We anticipate that most readers will initially peruse just a few chapters to learn about the components they regularly encounter, but that they will continue to come back to the volume as additional components become relevant throughout their evolving careers. The volume was designed to serve as a reference for ERP researchers of all levels of expertise, as well as researchers who use other techniques but wish to learn more about ERPs.

Origins of the Handbook

Over 100,000 studies using ERP components have been published, and the field continues to grow each year. Although these studies can be subdivided into a much smaller literature on each individual ERP component, finding and reviewing all of the relevant work on a single ERP component can be a daunting task. We see this every year at the ERP Boot Camp, an annual 10-day summer ERP workshop at UC Davis, where the participants often ask us where they can learn more about the *X* component (where *X* might be P300, N170, N400, P600, ERN, etc.). Our answer in the past was limited to something like, "Sorry, but there isn't a recent and comprehensive review. Here's a list of 40 journal articles you should read." This is an unsatisfactory answer, both for us and for the participants, because a thorough knowledge of the major ERP components is essential in the design and interpretation of ERP experiments. Furthermore, top ERP researchers rarely limit their research to a single ERP component, but rather use a range of ERP components depending on which is best suited for answering the question of interest in a given study. The ability to choose the most appropriate ERP component for answering a question and designing an experiment that properly uses that component requires extensive knowledge of ERP components.

We therefore assembled a volume in which leading ERP researchers wrote comprehensive overviews of the major ERP components in a format that is accessible to researchers with limited knowledge about ERPs, yet still useful to researchers with extensive ERP experience. Personally, we have found the drafts of the chapters to be extremely useful in our own research and writing, and we hope that others find the finished chapters equally useful.

Some readers may wonder whether an entire volume devoted solely to ERP components is warranted. In fact, an anonymous reviewer of our initial proposal for this volume raised this very issue. The review was thoughtful and detailed, but the gist can be summarized by the following excerpts:

> The concept of a component is a reasonable foundation for much ERP research, but much difficulty comes up in practice. There can be great difficulty in some circumstances in knowing when the one component of interest has been produced, as well as in assuring that other overlapping components do not have an undue influence on the measure in question . . .
>
> . . . it is often unwise to accept the assumption that each positive or negative deflection in the ERP waveform corresponds to a particular component. Deflections may be based on the summation of multiple components, which themselves are unknown, such that the amplitude and latency of the composite peak does not provide a valid characterization of any of the individual components.
>
> Accordingly, many ERP investigations in cognitive neuroscience no longer exemplify the strict component-centered approach, wherein a chief experimental goal was to understand an ERP component per se, and its associated cognitive process. Instead, difference-centered approaches have become prevalent, whereby experimental variables are manipulated based on theory-driven goals concerning specific neurocognitive functions.

We agree fully with this reviewer. As discussed in Chapter 1 of this volume, ERP components can be difficult to define and challenging to isolate. It is often unclear which component has changed across conditions or across groups, and it can be difficult to be certain that the same underlying component is being observed in different studies. As a result, ERP studies that have had a large impact outside the community of ERP researchers have often used designs based on difference waves that do not depend on identifying specific ERP components (see the discussion of *component-independent experimental designs* in Chapter 1).

Why, then, should we devote a 22-chapter volume to ERP components? The answer is that sophisticated ERP researchers typically design their experiments to take advantage of their knowledge of the properties of specific components, even if the conclusions of that experiment do not depend on isolating a specific component. Knowing about ERP components can help researchers determine what questions can be addressed with ERPs, because a psychological or neural process can be studied only if variations in that process lead to a measurable change in the ERP. And knowing what factors influence the amplitude and latency of an ERP component can help in creating an optimal design that generates robust ERP effects. For example, before attempting to use ERPs to study grammatical abnormalities in children with autism, one must first determine whether any ERP components exist that are sensitive to grammatical variables. Once these components are known, experiments can be designed that examine grammatical variables that have a clear ERP signature using parameters that optimize the robustness of the ERP signal. Consequently, a volume focusing on ERP components, in which researchers can gain breadth of knowledge about the available components as well as depth of knowledge about the properties of individual components, is much needed.

Organization of the Handbook

A terrific lineup of ERP researchers (including the discoverers of several ERP components) contributed chapters to this handbook, and we were extremely pleased by the comprehensive and insightful nature of the chapters they wrote. Authors were asked to address specific questions relating to: (1) defining and identifying the components; (2) isolating the components; (3) describing the neural and psychological processes reflected by the components; (4) which brain systems contribute to the generation of the components; and (5) the major variables affecting the amplitude and latency of the components.

The handbook is organized into four main sections. The first section contains three chapters devoted to broader issues surrounding ERP components, including the definition and isolation of ERP components in general, the relationship between EEG oscillations and ERP components, and the formal mathematical approach embodied by Independent Component Analysis. This section is extremely important because, although the concept of an ERP component is broadly used, the concept is rarely defined or discussed in the contemporary ERP literature. Furthermore, knowledge about EEG oscillations and Independent Component Analysis is essential to researchers wishing to evaluate new and evolving work using these approaches. We hope these chapters will be helpful to beginning ERP researchers who are just learning about the concept of an ERP component and to more advanced researchers who are familiar with the general concept but would like to develop a more sophisticated perspective. Each chapter in this section will be most useful if read in its entirety.

The second section contains a set of chapters focused on individual components that are often examined in isolation, including the basic sensory ERP components, the N170 component, the mismatch negativity, the P300 component, the slow anticipatory components, the lateralized readiness potential, and the error-related negativity. The chapters in this section are useful to anyone interested in learning more about these specific components. Some readers may wish to read these chapters in their entirety, and others may take advantage of the subheadings to read specific portions of interest.

The third section is composed of chapters that review multiple components related to a specific psychological domain, including chapters on attention, visual perception, working memory, long-term memory, language, and emotion. These chapters should be useful as a reference for researchers who conduct ERP research within these domains and also for researchers who use other techniques but wish to be able to better understand and evaluate ERP research in their field. Moreover, researchers who wish to apply ERPs to the study of a new domain may find these chapters useful as a guideline to adapting ERPs for use in a field. For example, emotion may not seem amenable to ERP methods at first glance, and understanding the process of applying ERPs to study emotion may provide a useful model for adapting ERPs to other domains. It should also be noted that there is some overlap between the components covered in this section and the components covered in the section on individual ERP components. In this way, the present volume provides both an in-depth methodology-focused account of individual components and a contextualized description of the use of the components to answer broader questions about the mind and brain.

The final section focuses on how ERP components vary across different groups of individuals, including chapters focused on infant and child development, aging, schizophrenia, depression, neurological disorders, and nonhuman primates. These chapters review a range of components, which will include many of the components covered in the second section of the volume (readers should consult the individual component chapters for more detailed discussions of the components). These chapters will be especially useful to ERP researchers who focus on these groups of individuals, as well as researchers who use other techniques but wish to evaluate ERP research within their domain. In addition, researchers who wish to apply ERP research to study a population not addressed in the present section may find the chapters useful as a general guideline to the application of ERPs in the study of special populations.

ABOUT THE EDITORS

Steven J. Luck

Steven J. Luck received a Ph.D. in neuroscience in 1993 from the University of California, San Diego, under the mentorship of Steve Hillyard. Dr. Luck spent 12 years on the psychology faculty at the University of Iowa before moving to the University of California, Davis, where today he is Director of the Center for Mind & Brain. His work includes basic science research on attention and working memory and translational research on schizophrenia.

Emily S. Kappenman

Emily S. Kappenman is a Ph.D. candidate at the University of California, Davis, working with Steve Luck. She began conducting ERP experiments in the laboratory of Bill Hetrick at Indiana University in 2002. Her research uses ERPs to examine neurocognitive processes in normal adults and in patients with psychopathology.

CONTRIBUTORS

Marcel Bastiaansen
Max Planck Institute for Psycholinguistics
Donders Institute
Centre for Cognitive Neuroimaging
Radboud University
Nijmegen, the Netherlands

Koen B. E. Böcker
Alan Turing Institute Almere
Almere, the Netherlands

Megan Boudewyn
Department of Psychology
University of California, Davis
Davis, CA

Colleen A. Brenner
Department of Psychology
University of British Columbia
Vancouver, British Columbia

Gerard E. Bruder
Division of Cognitive Neuroscience
New York State Psychiatric Institute
Department of Psychiatry
Columbia University College of Physicians and Surgeons
New York, NY

Cornelis H. M. Brunia
Department of Psychology
Tilburg University
AMC Department of Neurology
University of Amsterdam
Amsterdam, the Netherlands

C. Christine Camblin
Department of Psychology
Washington University
St. Louis, MO

Joshua Carp
Department of Psychology
University of Michigan
Ann Arbor, MI

Donna Coch
Department of Education
Reading Brains Lab
Dartmouth College
Hanover, NH

Dan Foti
Department of Psychology
Stony Brook University
Stony Brook, NY

David Friedman
Cognitive Electrophysiology Laboratory
Department of Psychiatry
New York State Psychiatric Institute
Columbia University Medical Center
New York, NY

William J. Gehring
Department of Psychology
University of Michigan
Ann Arbor, MI

Margaret M. Gullick
Department of Psychological and Brain Sciences
Dartmouth College
Hanover, NH

Greg Hajcak
Department of Psychology
Stony Brook University
Stony Brook, NY

Corentin Jacques
Face Categorization Lab
University of Louvain
Louvain-la-Neuve,Belgium
Department of Psychology
Stanford University, USA

Ole Jensen
Donders Institute, Centre for Cognitive Neuroimaging
Radboud University
Nijmegen, the Netherlands

Emily S. Kappenman
Center for Mind & Brain
University of California, Davis
Davis, CA

Jürgen Kayser
Division of Cognitive Neuroscience
New York State Psychiatric Institute
Department of Psychiatry
Columbia University
New York, NY

Kairi Kreegipuu
Cognitive Brain Research Unit
Department of Psychology,
University of Helsinki
Helsinki, Finland

Kerry Ledoux
Cognitive Neurology/Neuropsychology
Johns Hopkins University
Baltimore, MD

Yanni Liu
Department of Psychology
University of Michigan
Ann Arbor, MI

Steven J. Luck
Center for Mind & Brain
University of California, Davis
Davis, CA

Annmarie MacNamara
Department of Psychology
Stony Brook University
Stony Brook, NY

Scott Makeig
Swartz Center for Computational Neuroscience
Institute for Neural Computation
University of California, San Diego
La Jolla, CA

Ali Mazaheri
Donders Institute, Centre for Cognitive Neuroimaging
Nijmegen, the Netherlands

Jeff O. Miller
Department of Psychology
University of Otago
Dunedin, New Zealand

Risto Näätänen
Institute of Psychology
University of Tartu
Tartu, Estonia
Centre of Integrative Neuroscience (CFIN)
University of Aarhus
Aarhus, Denmark
Cognitive Brain Research Unit, Cognitive Science
Institute of Behavioural Sciences
University of Helsinki
Helsinki, Finland

Margaret A. Niznikiewicz
Department of Psychiatry
VA Medical Center Boston
Harvard Medical School
Beth Israel Deaconess Hospital
Boston, MA

Brian F. O'Donnell
Department of Psychological and Brain Sciences
Indiana University
Department of Psychiatry
Indiana University School of Medicine
Bloomington, IN

Julie Onton
Swartz Center for Computational Neuroscience
Institute for Neural Computation
University of California San Diego
La Jolla, CA

Joseph M. Orr
Department of Psychology
University of Michigan
Ann Arbor, MI

Veronica B. Perez
Department of Psychology
University of Oregon
Eugene, OR

John Polich
The Scripps Research Institute
La Jolla, CA

Hillel Pratt
Technion–Israel Institute of Technology
Haifa, Israel

Charan Ranganath
Center for Neuroscience
Department of Psychology
University of California, Davis
Davis, CA
Bruno Rossion
Face Categorization Lab
University of Louvain
Louvain-la-Neuve, Belgium
Dean F. Salisbury
Cognitive Neuroscience Laboratory
McLean Hospital
Department of Psychiatry
Harvard Medical School
Belmont, MA
Fren T. Y. Smulders
Faculty of Psychology and Neuroscience
Department of Cognitive Neuroscience
Maastricht University
Maastricht, The Netherlands
Tamara Y. Swaab
Department of Psychology
University of California, Davis
Davis, CA
Craig E. Tenke
Division of Cognitive Neuroscience
New York State Psychiatric Institute
Department of Psychiatry
Columbia University
New York, NY
Geert J. M. van Boxtel
Department of Psychology
Tilburg University
Department of Electrical Engineering
Eindhoven University of Technology
Eindhoven, The Netherlands
Rolf Verleger
Department of Neurology
University of Lübeck
Lübeck, Germany
Edward K. Vogel
Department of Psychology
University of Oregon
Eugene, OR
Jenifer L. Vohs
Department of Psychological and
Brain Sciences
Indiana University
Department of Psychiatry
Indiana University School of Medicine
Bloomington, IN
Anna Weinberg
Department of Psychology
Stony Brook University
Stony Brook, NY
Edward L. Wilding
Cardiff University Brain Research
Imaging Centre
School of Psychology
Cardiff University
Cardiff, Wales, UK
Geoffrey F. Woodman
Vanderbilt Vision Research Center
Center for Integrative and Cognitive
Neuroscience
Vanderbilt University
Nashville, TN

CONTENTS

ABBREVIATIONS

ABR auditory brainstem response
ACC anterior cingulate cortex
AD Alzheimer's disease
AD/HD attention deficit/hyperactivity disorder
ALS amyotrophic lateral sclerosis
BCG ballistocardiographic
BEM boundary element method
BIC binaural interaction component
BIS behavioral inhibition system
BP Bereitschaftspotential
BOLD blood oxygen-level dependent
CA cerebellar atrophy
CC closed-class
CDA contralateral decay activity
CNV contingent negative variation
CRN correct-response negativity
CSD current source density
DA dopamine
Dm difference due to memory
DTI diffusion tensor imaging
ECG electrocardiography
EEG electroencephalography
ELAN early left anterior negativity
EM episodic memory
EMG electromyography
EOG electro-oculogram
EPN early posterior negativity
EPSP excitatory postsynaptic potential
ERD event-related desynchronization
ERMF event-related magnetic field
ERN error-related negativity
ERP event-related potential
ERSP event-related spectral perturbation
FFA fusiform face area
fMRI functional magnetic resonance imaging
FRN feedback-related negativity
HD Huntington's disease
IAPS International Affective Picture System
IC independent component
ICA independent component analysis
IPS intraparietal sulcus
ISI interstimulus interval
ITC intertrial coherence
ITI intertrial interval
KR knowledge of results
LAN left anterior negativity
LC-NE locus coeruleus–norepinephrine
LFP local field potential
LH left hemisphere
LIP lateral intraparietal
LPC lateral positive complex
LPP late positive potential
LRP lateralized readiness potential
LTP long-term potentiation
LVF left visual field
MCI mild cognitive impairment
MDD major depressive disorder
MEG magnetoencephalography
MFN medial-frontal negativity
MLR midlatency evoked response
MMN mismatch negativity
MMNm neuromagnetic mismatch negativity
MOT multiple object tracking
MP motor potential
NARI noradrenaline reuptake inhibitor
Nc negative central
Nd negative difference
NE norepinephrine
NMDA *N*-methyl-D-aspartate

NDRI	noradrenaline/dopamine reuptake inhibitor
NS	negative shift
NSW	negative slow wave
OC	open-class
OCD	obsessive-compulsive disorder
OFA	occipital face area
OFC	orbitofrontal cortex
OR	object-relative
PAD	probable Alzheimer's disease
PAM	postauricular muscle
PCA	principal component analysis
PD	Parkinson's disease
Pe	error positivity
PET	positron emission tomography
PFC	prefrontal cortex
PMC	premotor complex
PMP	premovement positivity
PPI	phase preservation index
PRO	predictor of response outcome
PRP	psychological refractory period
PSP	postsynaptic potential
PSP	progressive supranuclear palsy
pSTS	posterior superior temporal sulcus
PSW	positive slow wave
RH	right hemisphere
RL-ERN	reinforcement-learning theory of event-related negativity
RSVP	rapid serial visual presentation
RT	reaction time
RVF	right visual field
SAT	speed-accuracy trade-off
SCR	skin conductance response
SEF	supplementary eye field
SEP	somatosensory evoked potential
SLI	specific language impairment
SMA	supplementary motor area
SNR	signal-to-noise ratio
SNS	spatially nonstereotyped
SOA	stimulus onset asynchrony
SPCN	sustained posterior contralateral negativity
SPN	stimulus-preceding negativity
SR	subject-relative
S–R	stimulus–response
SSRI	selective serotonin reuptake inhibitor
TMS	transcranial magnetic stimulation
TTI	target-to-target interval
VEP	visual evoked potential
VHF	visual half field
VPP	vertex positive potential
VWM	visual working memory
WM	working memory
WT	wavelet transform

PART 1

Conceptual Bases

CHAPTER

1 ERP Components: The Ups and Downs of Brainwave Recordings

Emily S. Kappenman *and* Steven J. Luck

Abstract

This chapter provides a framework for understanding, interpreting, and using event-related potential (ERP) components in the broad domain of mind, brain, and behavior sciences. The first section defines the term *ERP component*, describing the neural events that give rise to ERP components and explaining how multiple components sum together to form the observed ERP waveform. The next section describes the problems involved in isolating individual ERP components from the observed waveform, which is often much more difficult than researchers realize. This is followed by a discussion of the challenges involved in linking an ERP component with a specific neural or psychological process and then using this link to answer broader questions about the mind and brain. The chapter concludes with a discussion of what types of questions are most easily answered with ERPs and the approaches that have proven effective in overcoming the challenges of the technique.

Keywords: event-related potential, ERP component, peaks, waves, reverse inference

The goal of this chapter is to provide a framework for understanding, interpreting, and using event-related potential (ERP) components in the broad domain of mind, brain, and behavior sciences. Researchers in other areas such as political science, economics, law, and medicine may also find this overview useful as a guide to a broad understanding of ERP components. Event-related potentials have been used for decades to uncover aspects of the sensory, cognitive, and motor processes that underlie human thought and behavior. The excellent temporal resolution of the technique provides a narration of neural processes as they unfold millisecond by millisecond, adding whole pages to the story of the mind that behavioral and imaging techniques leave blank. However, the ERP technique is not without limitations. As reflected in the title of this chapter, there are both advantages and limitations of the ERP technique, and we will explore both the ups and the downs of ERPs in this chapter.

The first section of the chapter is aimed at defining the term *ERP component*, describing the neural events that give rise to ERP components and explaining how multiple components sum together to form the observed ERP waveform. The next section describes the problems involved in isolating individual ERP components from the observed waveform, which is often much more difficult than researchers realize. This is followed by a discussion of the challenges involved in linking an ERP component with a specific neural or psychological process and then using this link to answer broader questions about the mind and brain. These challenges may seem insurmountable, but researchers have developed experimental and analytic approaches that can overcome them in many cases. The key to using ERPs effectively is to understand what questions can be answered by ERP experiments and how the limitations of the technique can be avoided. Indeed, despite its limitations, the ERP technique is often the best one for answering certain types of questions. The chapter therefore ends with a discussion of what types of questions are most easily answered with ERPs and the approaches that have proven effective in overcoming the challenges of the technique.

Although a number of the issues we address are discussed elsewhere in the literature (e.g., see Luck, 2005), this chapter provides a comprehensive and concise overview of the nature and use of ERP components from a vantage point that is readily accessible to researchers from a wide range of backgrounds. Readers who have no familiarity at all with the ERP technique may wish to first read the more basic introduction provided by Luck (in press).

The Nature of ERP Components

What Is an ERP Component?

The ERP waveform appears on the scalp as a series of positive and negative peaks[1] that vary in polarity, amplitude, and duration as the waveform unfolds over time. However, the actual waveform is continuous, with no sudden transitions between one peak and the next, and division of the ERP waveform into discrete peaks is somewhat arbitrary. Indeed, this peak-centered view of the ERP waveform may reflect an intrinsic predisposition of the human visual system to use *minima of curvature* (places where orientation reverses direction) to define the parts of complex real-world objects (Hoffman & Richards, 1984). Although the peaks are visually salient, there is no a priori reason to believe that each peak reflects a specific brain process. However, early ERP researchers tended to make this assumption, and this has had a major influence on the terminology and analytical techniques used in ERP research. Sophisticated ERP researchers have recognized for decades that the peaks are somewhat arbitrary, and they make a distinction between *peaks* (local voltage maxima) and *components* (discrete intracranial sources of voltage that reflect specific neurocognitive processes, defined further below). Nonetheless, it is still common for researchers to assume that a peak in the observed ERP waveform is equal (or approximately equal) to an underlying ERP component. Perhaps the most important goal of this chapter will be to encourage readers to look beyond the visually salient peaks to the underlying components; it is the underlying components rather than the peaks that directly reflect the neural and psychological processes we wish to study.

To clarify the relationships among peaks and components, it is important to begin with some clear definitions. We can define the observed ERP waveform as *a depiction of the changes in scalp-recorded voltage over time that reflect the sensory, cognitive, affective, and motor processes elicited by a stimulus.* We can define an ERP peak as *a reliable local positive or negative maximum in the observed ERP waveform* (the term *reliable* allows us to disregard local maxima that result from high-frequency noise).

The term *ERP component* is more challenging to define. This term gets bandied about in the literature very frequently, but it is rarely defined or conceptualized beyond the peaks in the observed ERP waveform. In some sense, the term *ERP component* is analogous to the concept of attention: Just as "everyone knows what attention is" (James, 1890, p. 381), everyone knows what an ERP component is (at least everyone in the ERP world). Moreover, despite the fact that attention researchers all believe they know what attention is, they vary substantially in how they use the term *attention* (Luck & Vecera, 2002), and ERP researchers similarly vary in how they use the term *component.* Therefore, just as it is difficult to elicit agreement on the term *attention* in a room full of attention experts, it is no easy task to find a simple, concise, and widely accepted definition of the term *ERP component.* Furthermore, there is an important distinction between how these terms have evolved: although attention researchers frequently debate the fundamental nature of attention, ERP researchers rarely discuss the nature of ERP components.

There are, of course, counterexamples to this sweeping generalization about the nature of ERP components. For example, Manny Donchin has written extensively and explicitly throughout his career about ERP components and their existence beyond the peaks in the observed ERP waveform (e.g., see Donchin & Heffley, 1978). More recently, Luck (2005) provided a comprehensive discussion of the distinction between components and peaks. The concept of a component has also been discussed in the context of mathematical techniques for isolating components, such as principal component analysis (Donchin & Heffley, 1978) and independent component analysis (see Chapter 3, this volume). However, this important issue is often ignored in the ERP literature and warrants continued discussion.

In a general sense, we can define the term *ERP component* as *a scalp-recorded voltage change that reflects a specific neural or psychological process.* Although most researchers understand and use words such as *reflect* and *process,* such terms themselves refer to loose concepts without clear definitions. Consequently, it will be necessary to fill out the details of this definition over the course of this chapter. However, this concise definition does provide a reasonable approximation of the way the term *ERP component* is usually used by ERP researchers. We will illustrate the relationship between the ERP waveform and the underlying ERP components in the following sections, first discussing the neural events that give rise to the observed ERP waveform and the process of isolating the ERP waveform from other electrical activity. We will then illustrate the differences between the peaks in the ERP waveform and the underlying ERP components through the use of simulated waveforms.

Where Do ERP Components Come From?

Event-related potentials are voltage fluctuations in the ongoing electroencephalogram (EEG) that are time-locked to an event, such as the onset of a stimulus or the execution of a manual response. Electroencephalographic research began long before laboratory computers were available, and early researchers were able to observe only large ERPs that were visible on single trials (Davis, 1939) prior to the advent of computer averaging in the early 1960s (Galambos & Sheatz, 1962). However, most ERPs are rather small in comparison with the ongoing EEG activity and usually become visible only when multiple EEG epochs are combined together to form an average ERP waveform. This averaging process proved extremely beneficial to the field of ERPs and was the first occurrence in which signal averaging "revealed the existence of novel, previously unknown, neural processes" (Donchin et al., 1978, p. 349).

To understand the intricate mixture of signals we record on the surface of the scalp, we must first understand where and how these signals arise neurally. Although it is difficult to know with certainty how scalp-recorded voltage changes originate at the neural level, the following represents the best estimate based on our understanding of both biophysics and the properties of neural communication.

The changes in scalp-recorded voltage that give rise to the ERP waveform reflect the summation of postsynaptic potentials (PSPs) that occur simultaneously in large numbers of cortical pyramidal cells that are orientated in a similar manner with respect to the scalp (see Luck, 2005, chap. 1). These PSPs are a result of changes in electrical potential that occur when ion channels open or close in response to neurotransmitters binding with receptors on the postsynaptic cell membrane, which leads to the flow of ions into or out of the cell. When a PSP occurs at one end of a cortical pyramidal neuron, the result can be considered an electrical *dipole*, with positive on one end and negative on the other end. When PSPs occur simultaneously in many neurons that are spatially aligned, such that their dipoles point in the same direction, the dipoles sum together to form a large dipole known as an *equivalent current dipole*. If a sufficiently large number of spatially aligned neurons are simultaneously active, the equivalent current dipole is large enough to be reliably recorded on the surface of the scalp. This requires the simultaneous activation of thousands of neurons, due in part to the many layers of tissue that separate the scalp electrodes from the neurons. This is most likely to occur in groups of pyramidal cells in cerebral cortex, which are lined up together perpendicularly with respect to the cortical surface and are often active in unison. In other words, ERPs are almost always the result of PSPs in large groups of cortical pyramidal cells. It should be noted that, except in a few unusual cases, scalp ERPs do not reflect action potentials. Thus, ERPs represent the inputs to a group of neurons rather than the outputs of those neurons. Also, due to the necessity for such large numbers of spatially aligned neurons to be simultaneously active in scalp recordings, much of the neural activity in the brain that gives rise to cognition and behavior is not visible to an electrode placed on the scalp.[2]

For a given equivalent current dipole or neural generator source, the specific distribution of positive and negative voltages recorded on the scalp is determined by the position of the dipole in the head and its orientation with respect to the scalp (although it should be noted that the choice of reference electrode can also play a factor in the voltage distribution; see Luck, 2005, chap. 3). In other words, each equivalent current dipole will produce both positive and negative voltages on the head, with a band of zero separating the positive and negative voltage halves. This voltage reversal on the opposite side of the equivalent current dipole is often not very noticeable, because electrodes are not generally placed over the entire head, but the reversal is easily observed for some components (such as the N170; see Chapter 5, this volume). The positive or negative polarity of an ERP component at a given electrode site is related to several factors, including the orientation of the equivalent current dipole with respect to the electrode, and it is not usually possible to link the polarity to the type of neural processing (such as inhibition versus excitation). For a more detailed discussion of the factors that affect the polarity of an ERP, see Luck (2005, chap. 1).

Because electrical potential travels close to the speed of light, the transmission through the brain, meninges, skull, and scalp is essentially instantaneous. In other words, the voltages measured on the scalp at a particular time reflect synaptic activity at that particular instant, with no measurable delay. Thus, ERPs provide a direct and instantaneous millisecond-resolution measure of activity related to neurotransmission.

Summation of Components in the Observed ERP Waveform

It is important to note that although the ERP waveform at a particular instant reflects synaptic activity at that moment, it does not reflect *only* the neural activity that *began* at that particular instant. Specifically, the PSPs that give rise to ERPs last on the order of tens or even hundreds of milliseconds.[3] Therefore, as new mental processes are unfolding, the previous neural activations persist. In other words, multiple groups of neurons are active simultaneously in different regions in the brain. If we think of this neural activity in terms of dipoles, this means that multiple equivalent current dipoles are active simultaneously. In fact, source localization studies have shown that as many as 10 separate equivalent current dipoles may be active at a given time (Di Russo et al., 2002; Picton et al., 1999). If we return to our conception of ERP components, in which we define an ERP component as a signature of an individual neural process, each equivalent current dipole is essentially a separate ERP component. In other words, when we say that multiple equivalent current dipoles are active simultaneously, this really means that multiple ERP components are generated simultaneously.

In some cases, neurons engaged in one mental process may be distributed in different areas of the brain, such as the simultaneous processing of a single auditory signal in both the left and right temporal lobes. This would essentially lead to two equivalent current dipoles. Should we consider these two dipoles as two separate ERP components or as a single ERP component? They are typically treated as parts of a single component under the assumption that both hemispheres are engaging in essentially the same mental process. However, this is a fine detail of the definition of an ERP component, with little practical significance for the use of ERP components. Furthermore, resolution of this issue would require a precise definition of what is meant by *mental process* in terms of the behavior of neurons, both individually and as a group. That is, how do we determine whether the same mental process is occurring in two individual neurons, and on a larger scale, in groups of neurons? This is a complex issue that remains to be resolved by future research.

The combination of multiple ERP components on the scalp leads to the *superposition problem*, which is depicted in Figure 1.1. When multiple ERP components are simultaneously active, the recorded voltage at the scalp is based on the sum of the voltages from all the individual components. This is a simple additive process. That is, if you knew the true waveform for each individual component, you could add all the component waveforms together to get the ERP waveform at each electrode site (scaling each component by a weighting factor that reflects the contribution of the component to the voltage measured at a specific electrode site). Unfortunately, the true waveform for each component is not known in real recordings, and it is quite difficult to reduce the sum of the components in the observed data to the individual components. However, understanding with simulated data how the voltage recorded at a particular electrode site reflects the various internal generator sources can help us understand the properties and intricacies of the ERP signals.

The propagation of voltage from a single generator site to a particular electrode site depends on the position and orientation of the ERP generator source with respect to the electrode, along with the conductance of the brain, skull, and scalp. This can be quantified with a weighting factor: The contribution of a given generator to the voltage recorded from a given electrode site is simply the waveform at the generator multiplied by the weighting factor (see Figure 1.1). There will be a separate weighting factor specifying the relationship between each electrode site and each internal neural generator source. Together, the set of weighting values between each source and each electrode site provides a *mixing matrix* that defines how the different components mix together at each site. Some mathematical techniques for recovering the underlying components work by computing an *unmixing matrix* that reverses this process, passing the observed data through the unmixing matrix to compute the component waveforms (see Chapter 3, this volume).

When multiple ERP components are simultaneously generated in different brain areas, the voltages from these components sum together. The voltage recorded at each site will therefore be the sum of each of the internally generated ERP components, with each scaled by the weight between that electrode site and each of the generator locations. The value at

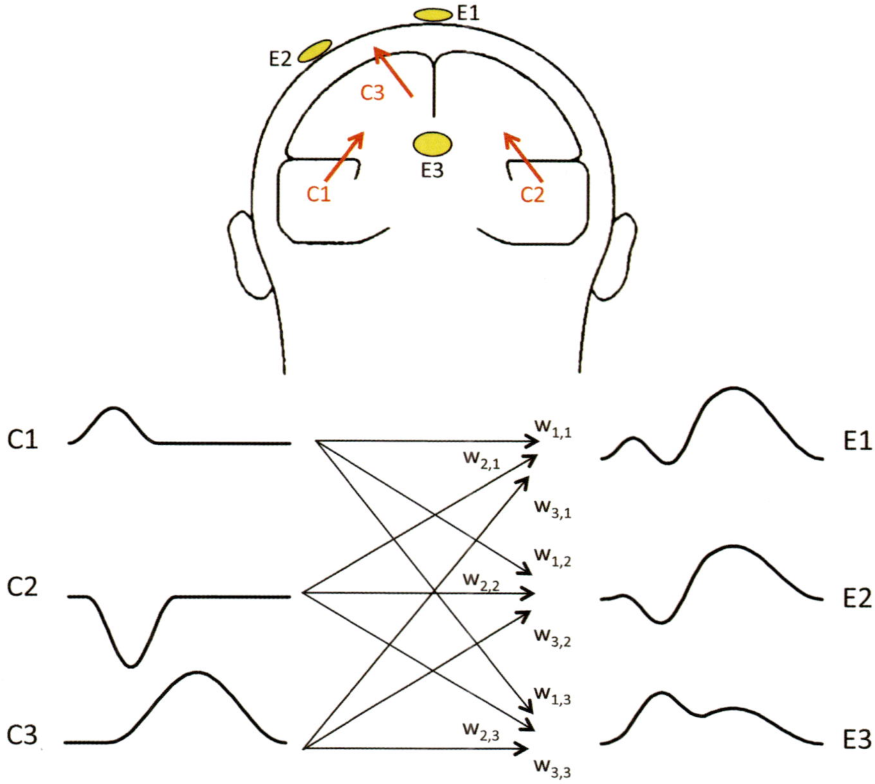

Fig. 1.1. Relation between the underlying component waveforms and the observed scalp waveforms. In this example, three components are present (C1, C2, C3), each of which has a waveform (shown at the bottom left) and a generator location (represented by the arrows in the head). The contribution of each component waveform to the observed waveform at a given electrode site is determined by a weighting factor that reflects the location and orientation of the generator relative to that electrode, along with the conductivity of the tissues that form the head. The observed waveform at a given electrode site (shown at the bottom right) is equal to the sum of each of the component waveforms, multiplied by the weighting factor between each component and that electrode site. The weights are indicated by the *w*'s on the arrows between the component waveforms and the observed waveforms (e.g., $w_{2,3}$ represents the weighting factor between component 2 and electrode 3).

a given electrode site at a particular moment in time is equivalent to the magnitude of each component at that time, scaled by the appropriate weighting factor and then summed together. Consequently, the ERP waveform at each electrode site contains information about all of the neural generators in the brain, not just the generator sources located close to the electrode (although nearby sources will usually have a greater weight).

The inability to relate the ERP waveform at a particular electrode site to the neural tissue directly below the electrode site is made even more severe by the properties of the head. Specifically, as electrical activity travels from the brain to the surface of the scalp, the activity must pass through layers of skull and scalp. Although these constituents of the head are sufficiently conductive to allow the electrical activity generated in the brain to appear on the surface of the head, they are not perfect conductors, and the high resistance of the skull relative to the low resistance of the underlying brain and overlying scalp causes the voltage to spread laterally as it travels. The signals are therefore blurred together by the head, which further distorts the relationship between the voltage at a particular electrode site and the cortex directly under that site.

Of course, anyone who has seen the ERP waveforms from multiple electrode sites knows that differences exist in the shape and size of the ERP waveform across electrode sites. In other words, although the waveform at each electrode site reflects neural signals from all over the brain, the summated signals are not identical at each site. It is tempting to use the scalp distribution information to estimate the location of the neural generator source by, for example, determining at which electrode site the signal is largest. However, the superposition of multiple components and the blurring of the voltages across the head make it impossible to determine the locations of the generator sources solely from the observed waveforms. In fact, an infinite number of internal generator configurations could produce any observed distribution of ERP activity over the scalp (see Luck, 2005, chap. 7). Thus, there is no technique that can determine, with certainty, the locations of the sources and the waveform at each source without bringing in difficult-to-verify assumptions or other sources of evidence.

To summarize, the ERP waveform reflects ongoing synaptic activity related to mental processing as it unfolds millisecond by millisecond. However, because scalp-recorded signals require the simultaneous activation of large groups of spatially similar oriented neurons, only a portion of the neural activity that occurs in response to a stimulus will be measurable from electrodes on the surface of the scalp. Furthermore, the ERP waveform at a given electrode site reflects the contribution of many simultaneously active ERP components that overlap in time, and it is difficult to mathematically unmix the observed waveforms and determine the original component waveforms.

Other Approaches to Defining ERP Components

In this section, we will consider the relationship between the definition of the term *ERP component* that we have proposed in this chapter and the way that components are defined by four other approaches: *source localization*, *principal component analysis* (PCA; see Donchin & Heffley, 1978), *independent component analysis* (ICA; see Chapter 3, this volume), and *time-frequency analysis* (see Chapter 2, this volume). We will concentrate on the spatial variants of PCA and ICA, in which components are defined on the basis of scalp distribution information (see Spencer et al., 2001, for a discussion of temporal and spatiotemporal PCA).

We will begin by considering the source localization, ICA, and PCA approaches. In these three approaches, a component is defined solely by its scalp distribution, which is assumed to remain stable over the course of a single experimental session (this is a reasonable assumption given that brain geometry is unlikely to undergo major changes within a few hours). As mentioned in the previous section, these techniques provide an *unmixing matrix* that reflects the estimated scalp distributions of the individual components; the waveform for each component is computed by passing the observed waveforms through this matrix. That is, rather than passing the component waveforms through the weights shown in Figure 1.1 to obtain the observed waveforms at each electrode (moving from left to right in the figure), these techniques pass the waveforms observed at each electrode site through an unmixing matrix to obtain the component waveforms (moving from right to left). Unfortunately, there is no unique solution to the problem of determining the underlying component waveforms from the observed scalp waveforms, and these three techniques use different assumptions to pick a single solution to this problem (without any guarantee that the correct solution will be found).

In source localization techniques, a component is equivalent to a neural generator source. These techniques use biophysical assumptions about the flow of current through the conductive tissues of the head to define the scalp distribution of each component (and thereby compute a unique unmixing matrix). To obtain a unique solution, these techniques must also rely on additional assumptions, such as a specific number of discrete dipoles or maximal smoothness in the distribution of current flow over the cortical surface. That is, these techniques find the set of single-component scalp distributions that can sum together to provide the best fit to the observed scalp distribution as it varies over time while also being consistent with a variety of assumptions (for a review and critique, see Luck, 2005, chap. 7). Thus, source localization techniques define a component as activity arising from a region of cortex, which is similar to our definition of an ERP component as reflecting a specific brain process (on the assumption that most brain processes occur in discrete areas[4]). However, our definition of the term *ERP component* goes further, because more than one brain process may occur in a given region of cortex. Moreover, source localization approaches differ considerably from the traditional approach to defining components in the procedures used to discover and define individual components. Whereas source localization techniques use a variety of assumptions to select a set of scalp distributions that together provide a quantitative account of the data from a given experiment, traditional approaches to defining components are based on using experimental manipulations to test hypotheses about the link between a voltage deflection and an underlying neural or psychological process (as discussed further in a later section).

Principal component analysis and ICA make no biophysical assumptions, but instead use the statistical properties of the data to derive the scalp distributions of the components. That is, the observed scalp distribution changes from moment to moment and from condition to condition as the underlying components wax and wane, and the statistical relationships between the values observed at the different electrode sites are used to determine the scalp distributions of the individual components. In PCA, for example, two electrode sites will tend to contribute strongly to the same component if they tend to covary in voltage. Principal component analysis is designed to find an unmixing matrix in which a small number of components—each with its own scalp distribution—can sum together to explain most of the variations in the observed scalp distribution. It reduces a large and complex set of observed scalp distributions (for each time point, condition, etc.) to a small number of component scalp distributions. In contrast, ICA is designed to find an unmixing matrix that maximizes the independence of each component so that every individual component represents the largest possible amount of information. The scalp distributions of the components in ICA may be correlated with each other (as would be expected for two independent but nearby neural sources), but the strength of activation of each component varies independently of the strength of the other components over time points and over conditions. Whereas PCA attempts to lump as much information as possible into a small number of components, ICA attempts to split apart the information into different components (for a detailed comparison, see Chapter 3, this volume).

Because it is a "lumping" technique, spatial PCA by itself is unlikely to produce components that are related to individual neural and psychological processes. However, the essence of ICA corresponds well with a reasonable assumption about these processes. Specifically, for something to count as a unique process, it must be dissociable from other processes. This is largely identical to saying that the process must sometimes vary independently of other processes, and this is exactly the type of independence that ICA uses to define components. Thus, although ICA uses a mathematical approach rather than a hypothesis-testing approach to derive the components, it shares much with the definition of the term *ERP component* that we have proposed in this chapter. Moreover, the components isolated by ICA often have a scalp distribution that matches what would be expected for a single dipole, even though the technique makes no biophysical assumptions about dipoles (see, e.g., Figure 3.9 in Chapter 3, this volume).

There are, however, some practical problems associated with linking ICA components to ERP components as we have defined them here. First, ICA is applied to single-subject data, and it can be difficult to determine the correspondence between the ICA components obtained for the different subjects. The same problem arises when comparing components across experiments. Second, the ICA computational approach requires that the number of ICA components is always equal to the number of electrodes, and this means that multiple true components may be lumped together into a single ICA component or that a single true component may be distributed across multiple ICA components. It remains to be seen how well the traditional approach and the ICA approach to defining and isolating components can be combined.

The time-frequency approach is very different from the source localization, ICA, and PCA approaches (although it can be combined with them). In the time-frequency approach, the EEG is decomposed into the sum of a set of oscillations, and the power in each frequency band is estimated at each moment in time (with varying degrees of temporal precision; see Chapter 2, this volume, for details). The results of this approach can be related to conventional ERP components in two main ways.

First, if the oscillations vary randomly in phase from trial to trial, they will ordinarily disappear when the single-trial EEG epochs are averaged together; oscillations of this sort are completely invisible in conventional averaged ERP waveforms (for an exception, see Mazaheri & Jensen, 2008). In such cases, oscillations within a given frequency band are often considered as being analogous to ERP components, reflecting a specific neural or psychological process. However, many different processes might lead to oscillations in a given frequency band, so it is problematic to assume that power in a given frequency band in one experiment reflects the same process reflected by power in that same frequency band in another experiment (e.g., theta-band activity in one experiment may reflect very different processes than theta-band activity in a different experiment). Assuming that a given band reflects a specific process would be analogous to assuming that any positive deflection in the P3 latency range reflects a single process.

A second possibility is that a stimulus might perturb the phase of an ongoing oscillation, causing the phase to become consistent across trials during the period immediately after the stimulus. In such cases, the phase consistency across trials will allow the oscillation to survive the averaging process (see Figure 2.2 in Chapter 2, this volume). When this happens, a component in an average ERP may actually consist of a portion of an ongoing oscillation rather than reflecting a discrete voltage deflection that is elicited by the stimulus.

Challenges in Isolating ERP Components

We have defined the term *ERP component* as scalp-recorded neural activity that is associated with a particular neural or psychological process. It is the nature of the underlying process that we are seeking to uncover with ERP research; however, as discussed in the preceding section, the ERP waveform that we can record contains a mixture of many different ERP components. Deconstructing the ERP waveform into its ERP components is no trivial task. An infinite number of combinations of underlying components could sum together to give rise to a given ERP waveform. This section is devoted to illustrating the difficulty in assessing changes in a component from the observable ERP waveform. To illustrate these points, we will use simulated data for which the underlying ERP components are known and modifiable. This section is primarily aimed at pointing out the limitations of ERP component research. Although this section may make ERP research seem dismal, you should not become disheartened with ERPs. The final section of this chapter will provide some tools that have been successful for using ERPs to answer questions about the mind, brain, and behavior.

ERP Peaks ≠ ERP Components

As discussed earlier, the ERP waveform looks like a succession of distinct and easily separable peaks, but these peaks do not map onto distinct ERP components in a simple one-to-one manner. The neural activation associated with each distinct mental process persists for tens or hundreds of milliseconds, which means that the ERP signature from one process will overlap with the ERP signature for subsequent processes either in part or in whole. Even if these neural processes occur in separate parts of the brain, the ERP waveform at a given electrode site will be the weighted sum of all of the underlying components. In other words, each peak in the waveform is usually determined by more than one, and often several, separate ERP components.

Much ERP research has centered on evaluating differences in the size or timing of an ERP component across conditions or across groups of subjects. Such changes can speak volumes about differences in neural processing. However, the problem of overlapping components makes it difficult to ascertain whether a change in a peak in the observed ERP waveform is due to a change in one component, a change in a different component, or changes in a combination of multiple components. In the language literature, for example, it is not always clear whether a putative increase in N400 amplitude might actually be a decrease in P3 amplitude, and a great deal of work was needed to determine that the P600 component elicited by syntactic anomalies was different from the P3 wave (see Chapter 15, this volume).

Figure 1.2 illustrates some of the measurement problems that arise due to the overlap of ERP components. In this simulated example, the observed waveform shown in Figure 1.2A is the sum of the three underlying components shown in Figure 1.2B. In other words, Figure 1.2B is the observed ERP waveform and Figure 1.2A shows the underlying components (which we cannot observe directly in real experiments). Looking at the observed waveform, the ERP appears to consist of a positive component from 0 to 90 ms, a negative component from 90 to 180 ms, and a positive component from 180 to 450 ms. However, the underlying components are much longer in duration, with the first positive component active from 0 to 200 ms, the negative component active from 50 to 325 ms, and the second positive component active from 100 to 450 ms. Thus, one cannot easily determine the duration of an underlying component from the duration of the peak in the observed waveform. The difficulty of assessing component duration from the ERP waveform is a problem in experimental contexts as well, particularly when a smaller component is preceded or followed by a much larger component. For example, it is difficult to assess the duration of the N2 component when it is followed closely by the much larger P3 component. Although it is often the case that evaluating the length of a peak in the waveform minimizes the apparent duration of a component, the waveform can also make a component seem longer in duration than it is in actuality. For example, the late positive potential (LPP) in the emotion literature appears as a single component that is hundreds of milliseconds in duration; however, the LPP may actually be composed of several distinct shorter-duration components (see Chapter 16, this volume). Therefore, the duration of peaks in the ERP waveform is often quite different from the duration of the underlying components.

Changes in the timing or size of components across experimental conditions or groups of subjects

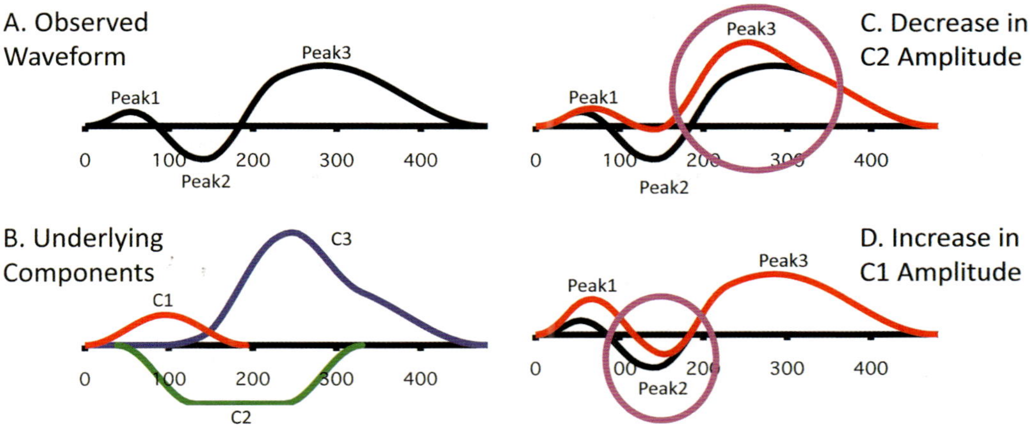

Fig. 1.2. Example of how the peaks in an observed waveform can misrepresent the underlying components. Panel A shows the observed waveform, and Panel B shows the underlying components that sum together to produce the observed waveform. Note that Peak 1 is much earlier than the peak of component C1, and the shape of Peak 2 is very different from the shape of component C2. Panel C shows the original waveform overlaid with a waveform in which the amplitude of component C2 has been decreased. Note that this change in C2 causes an increase in the amplitude of Peak 3, even though component C3 does not differ between these waveforms. Panel D shows the original waveform overlaid with a waveform in which the amplitude of component C1 has been increased. Note that this changes the amplitude and latency of Peak 2, even though component C2 does not differ between these waveforms.

can also be difficult to assess from the ERP waveform. Figure 1.2C shows the effect of an experimental manipulation that decreases the amplitude of the negative component. In addition to decreasing the measured amplitude of the negative peak in the observed waveform, this manipulation greatly increases the amplitude of the second positive peak (even though the manipulation did not change the amplitude of the second positive component). This is one clear example of how changes in the amplitude of one component (the negative component) can result in an amplitude change in a subsequent part of the waveform (the second positive peak). Based on a superficial evaluation of the waveform, these changes would lead to the erroneous conclusion that the difference between conditions was the result of modulations in two underlying ERP components; however, in this case, both peak modulations were caused by a change in a single underlying ERP component. Therefore, researchers may draw substantially incorrect conclusions if they assume that a change in the size of a peak reflects a change in the size of a particular component.

Similarly, Figure 1.2D shows the effect of a manipulation that increases the amplitude of the first positive component. In addition to increasing the measured amplitude of the first positive peak in the observed waveform, this manipulation decreased the measured amplitude of the negative peak. The manipulation of the amplitude of the first positive component also increased the apparent latency of the negative peak, even though no latency shift occurred for any of the underlying components. In other words, a change in the amplitude of one component can in some cases masquerade as a shift in the latency of a different component. Therefore, it is often difficult to determine whether a specific type of modulation of the ERP waveform is related to the same type of change in the underlying components. In other words, measured shifts in peak latency can sometimes be caused merely by changes in component amplitude, and measured changes in peak amplitude can sometimes result from shifts in component latency.

Although we have shown a few cases of the difficulty in linking changes in the ERP waveform with changes in particular underlying ERP components, this is by no means an exhaustive description of the ways in which changes in underlying components can affect the observed ERP waveform. We encourage anyone interested in exploring these effects to create simulated data and see how modulations in the underlying components affect various parts of the ERP waveform (this is easy to do in a spreadsheet program, such as Excel). Furthermore, it should be noted that the simulation shown in Figure 1.2 may actually underestimate the severity of the problem of measuring amplitudes and latencies from the ERP waveform, because modeling efforts suggest that 6–10 generators may be active within a given 150 ms period (Di Russo et al., 2002; Picton et al., 1999), in contrast to the 3 neural generators used in the simulation shown in Figure 1.2. On the other hand, considerable information about the underlying component structure can often be obtained by examining the waveforms from multiple electrode sites, because different components will be weighted differently at each electrode.

Variability in ERPs

Amplitudes and latencies are almost always measured from the average of multiple EEG segments but separately for each individual subject. In other words, all of the trials in a condition are averaged together for a given subject, and the amplitude and latency measures are computed for each subject from this average waveform. Each subject then contributes a value to the statistical test for differences across conditions or groups, with the variance across subjects contributing to the ability to detect a significant experimental effect. This process of signal averaging is incredibly important and integral to the utilization of ERPs; averaging across multiple EEG epochs reveals ERPs that are not visible on single trials, and data from multiple subjects provide a measure of variance that is important to assessing statistically significant changes. However, it is important to understand distortions that can be introduced by the averaging process.

The process of averaging across multiple trials to form an average ERP waveform relies on several assumptions, the most important of which is that the timing of the signal of interest is the same on each trial. However, this is often not the case. Specifically, just as the behavioral reaction time varies substantially from trial to trial in an experiment, the timing of the underlying neural processes that give rise to the ERP components may also vary from trial to trial. The variability in the timing of a component across trials is known as *latency jitter*, and it can actually be quite problematic to the interpretation of an average waveform. When latency jitter is present for a component, as depicted in Figure 1.3, the average ERP waveform will contain a "smeared-out" version of the component. Specifically, the average ERP waveform will reflect both the *earliest* onset and *latest* offset times of the component, as opposed to reflecting the *average* onset and offset times. In addition, latency jitter can greatly reduce the measured peak amplitude (discussed more fully later in the chapter). Furthermore, although this variation in timing across trials is informative about the nature of the process reflected by the component, it can make the comparison of the size and timing of a component across conditions or across groups of subjects more difficult. Specifically, greater variability in the timing of a component across conditions may be incorrectly interpreted as a change in the size or duration of the component. For example, a comparison of the two conditions depicted in Figures 1.3A and 1.3B might lead to the erroneous conclusion of a smaller component in condition A than in condition B, even though the only difference between the conditions lies in the variability in the component timing across trials. Therefore, understanding how latency jitter

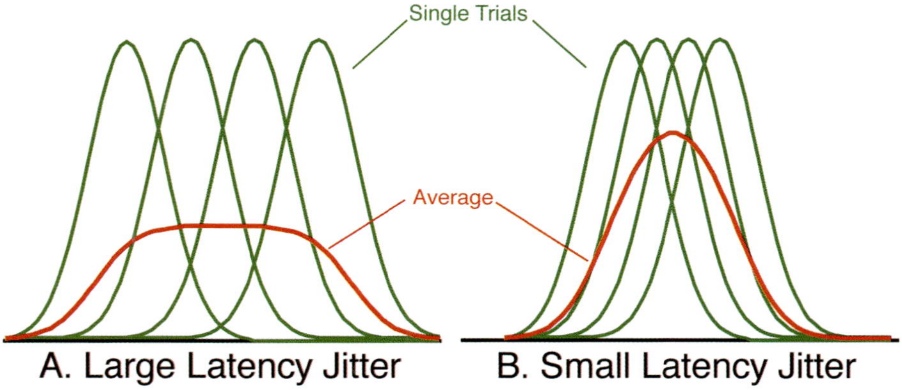

Fig. 1.3. Example of how differences in latency jitter (the amount of variability in component latency across trials) influence the average waveform. The green waveforms are the single-trial data, and the red waveforms are the averages across trials. The jitter in single-trial latency is greater in (A) than in (B), leading to a broader averaged waveform with a lower peak amplitude in (A) than in (B). That is, even though the amplitude of the single-trial waveforms is equivalent in (A) and (B), the peak amplitude of the averages differs between (A) and (B). In addition, the onset time and offset time of each average reflect the earliest onset times and latest offset times of the single trials rather than the average of the single-trial onset and offset times.

can impact the average waveform can be useful in interpreting experimental effects.

Measures of amplitude and latency are almost always taken from individual subject waveforms. By contrast, most ERP papers show the grand average ERP waveform across subjects, as opposed to each of the individual subject waveforms. Therefore, the characterizations we can make of the components in a particular experiment are generally taken from an average representation of all the subject waveforms in the study. It is tempting to think that the grand average ERP waveform would reflect the average of all of the individual subject waveforms that make up the average; however, just as the average of multiple EEG segments within a subject reflects the range of the epochs, a grand average across subjects actually reflects the *earliest* onset and *latest* offset times and not the average of the onsets and offsets of the components. In other words, if there is substantial variability in the timing of the components across subjects, the grand average ERP will reflect that variability.

One of the most salient factors when measuring the amplitudes and latencies of ERP components from the individual subject waveforms is the quite substantial variation in shape across waveforms. For example, consider the waveforms in Figure 1.4A. The bottom waveform is the grand average across subjects, and the other waveforms reflect 8 randomly selected subjects from the 20 individuals who contributed to the grand average. The highlighted portion of the figure corresponds to the time period one might select to measure the P2 wave, because it covers almost the entire duration of the wave in the grand average. However, the activity within this time window varies considerably across the individual subject waveforms. For some of the subjects, the first positive wave peaks prior to the

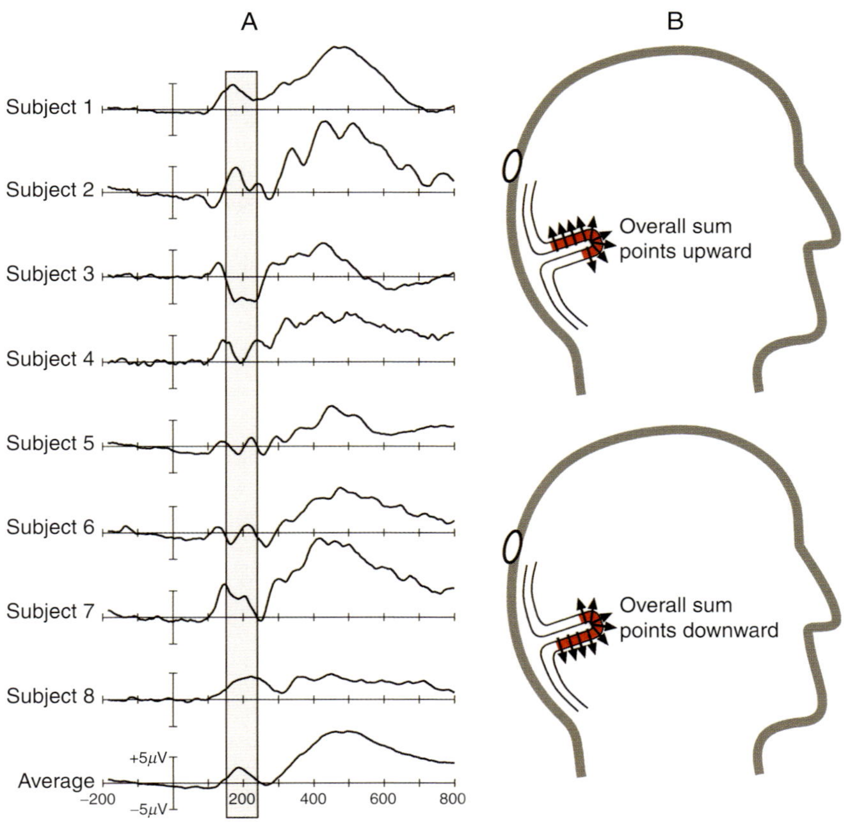

Fig. 1.4. (A) Single-subject ERP waveforms from 8 of 20 subjects in an oddball paradigm, along with the grand average of all 20 subjects (data from the study of Luck et al., 2009). (B) Example of how small differences between two subjects in the position of an active area of cortex within a sulcus could lead to opposite polarities at the electrode shown on the surface of the head. Each arrow represents the equivalent current dipole in a small patch of cortex, with positive at the arrowhead end and negative at the opposite end. Many of these dipoles will cancel each other, and the surface voltage will reflect the activity in the noncanceling dipoles.

beginning of the window (e.g., subjects 3, 4, and 7), and one subject's waveform is entirely negative during this window (subject 3).

The between-subject variations in the ERP waveform can be quite disconcerting when measuring a component from the single-subject waveforms. It is very unlikely that the same process reaches maximal activity at 145 ms in one healthy adult (e.g., subject 7) and at 220 ms in another (e.g., subject 6), so it does not seem appropriate to use a window that is broad enough to include peaks at such different latencies. And it is hard to understand how the negative deflection exhibited by subject 3 could represent the same functional brain activity as the positive deflection exhibited by subjects 1 and 2 in this same interval. However, as discussed above, peaks in the ERP waveform do not correspond directly to the underlying components. So, how problematic are these individual-subject waveform differences?

To understand whether the differences among individual-subject waveforms adversely affect our characterization of the components, we must first understand the source of the differences. For later periods of the waveform that reflect higher cognitive processes, differences in size and shape may reflect differences in the strategies subjects engage in during cognitive processing. Therefore, individual differences in the size and shape of the waveform may reflect actual processing differences. However, for the sensory processing that occurs within ~200 ms after the onset of the stimulus, it is unlikely that differences in waveform size and shape reflect differences in strategy or processing, at least in healthy subjects. Instead, the waveform differences most likely arise from differences across subjects in nonfunctional "nuisance" factors such as skull thickness and cortical folding patterns.

Just as fingerprints are unique to each individual, so is the intricate pattern of sulci and gyri in the human brain. Such changes in folding pattern could easily lead to differences in ERP waveforms across subjects like those illustrated in Figure 1.4A. For example, Figure 1.4B shows how a relatively small difference between two subjects in the location of an active strip of cortex within a sulcus could lead to opposite polarities for those two subjects at a given electrode site. More of the active region is on one side of the sulcus for one subject and more is on the opposite side of the sulcus for the other subject, leading to an overall equivalent current dipole pointing upward for one subject and pointing downward for the other subject. Consequently, the overall activity at a given scalp electrode will be positive for one subject and negative for the other.

Although this variability can be problematic for studies designed to assess individual differences, there is considerable similarity in the grand average ERP waveforms from different experiments that utilize similar tasks. This gives us some confidence that reliable conclusions can be drawn by comparing reasonably sized groups of subjects, even if the individual subjects within a group vary considerably in waveform shape. For example, there is great similarity across P3 oddball studies in grand average ERP waveforms, despite the fact that these waveforms are made up of different underlying individual-subject waveforms. Consider the ERP waveforms in Figure 1.5. The top left panel shows all 20 individual-subject waveforms from a P3 oddball task, subdivided at random into two separate groups of 10 subjects each. There is enormous variability between subjects in the amplitude and shape of the ERP waveform, with much larger P3 waves in some subjects than in others. However, as can be seen at the bottom of Figure 1.5, the grand averages across these two subgroups of subjects are quite similar in amplitude, timing, and shape, despite the large differences in the underlying individual-subject ERP waveforms that make up those grand averages. In other words, the individual-subject differences do not alter the overall experimental effect when the sample size is sufficient. However, it is important to remember that some measurement techniques may be more affected by this between-subjects variance than other techniques. We will address the issue of measurement later in the chapter. It should also be noted that statistical techniques can be applied that allow measurements to be made from grand averages rather than from single-subject waveforms, capitalizing on the stability of the grand averages (Kiesel et al., 2008; Miller et al., 1998, 2009).

How to Identify and Define an ERP Component

Given how difficult it is to isolate a specific ERP component from the ERP waveform, you may be wondering how we even know that a specific ERP component exists. For example, how do we know that there is an N1 wave, a P3 wave, an N400, and so on? Of course, there is a voltage deflection in a broad time range corresponding to each of these components, but as we have already seen, there are usually multiple components active simultaneously in a given time range. So, how do we know that a voltage deflection is caused by a specific ERP

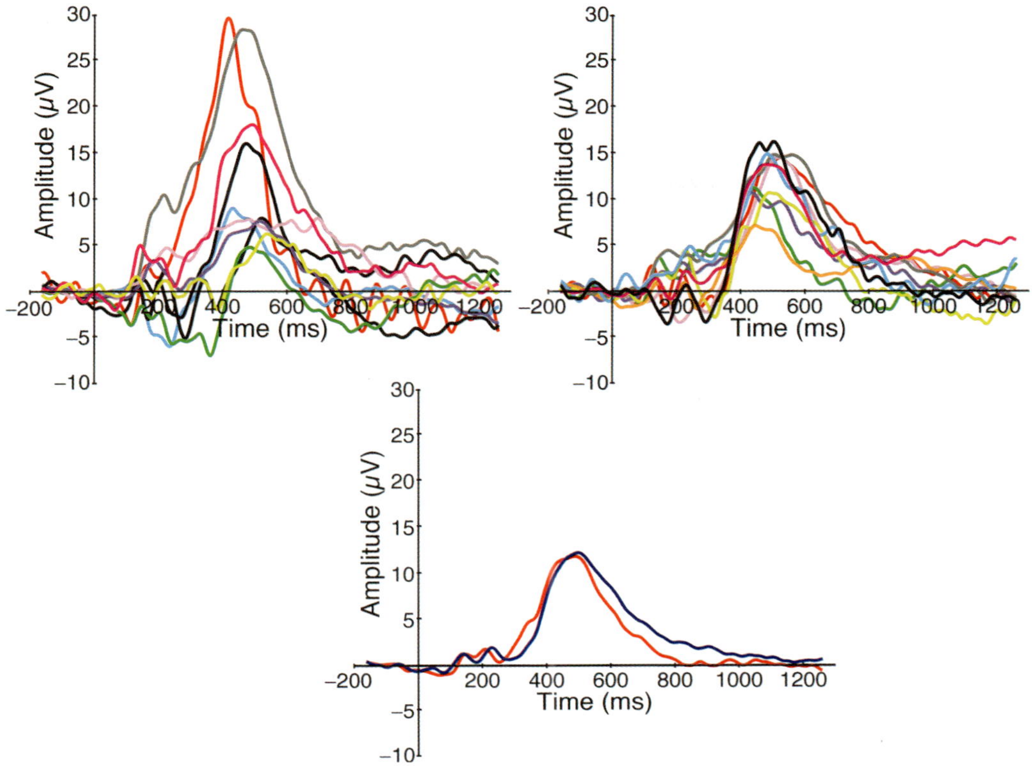

Fig. 1.5. Example of the similarity of grand average waveforms despite substantial differences among the single-subject waveforms. Waveforms from 20 subjects in an oddball experiment were randomly divided into two groups of 10. The single-subject waveforms for each group are shown at the top left and top right. Note the large variability in the amplitude and shape of the waveforms. The grand averages of these two subgroups of 10 subjects are shown at the bottom. Despite the large differences among the individual subjects, the grand averages from the two subgroups are quite similar.

component in one study, and how do we know that that same ERP component is active in subsequent studies? In other words, how do we *operationally identify and define* an ERP component?

Event-related potential components are often defined in terms of a combination of polarity, latency, and scalp distribution. This method of defining ERP components is evident from the common naming scheme in which ERP components are named in terms of polarity and latency (given either in milliseconds or as the ordinal position in the waveform). However, as we will see below, these dimensions describe the observed peaks and do not provide a stable and precise means of defining the underlying ERP components. That is, the factors of polarity, timing, and scalp distribution can vary from context to context, rendering them unstable representations of a component. We will explore each of these factors in turn and will end with some strategies for defining and isolating ERP components.

As discussed above, the timing of a neural process can vary across trials, subjects, and experiments. And because an ERP component is a scalp-recorded signature of a neural process, it stands to reason that the timing of an ERP component will vary across these same contexts. We can see timing variability quite clearly in studies of the P3 component, which can vary across conditions by hundreds of milliseconds, sometimes occurring before the manual response and sometimes appearing after the response. This is one reason that the moniker P300 is often shortened to P3, to eliminate the association with the time value of 300 ms. Although the timing of most ERP components is not nearly as variable as that of the P3, timing variability does occur for all ERP components. Visual sensory components, for example, increase in latency as stimulus brightness decreases for the simple reason that the amount of time required for information to reach cortex increases as brightness decreases. In addition, most components change in latency across early development (see Chapter 17, this volume) and across aging (see Chapter 18, this volume). Examining the variation in time windows over which the components are measured in different studies makes the variation in component latencies across experiments quite obvious. Therefore, although a specific latency is often denoted by the name of an ERP component, this latency is approximate and often specific to the context in which the component was first identified, and latency cannot be used as a direct means of determining whether a component in a given study is the same as a component observed in previous studies, especially if the subjects, stimuli, or task differ considerably across studies.

Many ERP component names also make reference to the polarity of the component, but polarity may vary for a single component. For example, the C1 wave reverses in polarity for stimuli in the upper visual field compared with stimuli presented in the lower visual field owing to the cortical folding pattern of primary visual cortex. Both the positive- and negative-polarity C1 waves reflect the same underlying process and are therefore the same ERP component by any reasonable definition. Although other ERP components do not reverse polarity so dramatically, differences in cortical folding pattern across subjects might occasionally lead to polarity differences from one subject to the next at a given electrode site (see, e.g., subject 3 in Figure 1.4A). Furthermore, as discussed above, all ERP components are positive on one end of the dipole and negative on the other end, and all ERP components therefore reverse polarity at some place on the head.

If the polarity and timing information cannot be used to identify a component, what about the scalp distribution? Scalp distribution is often used to distinguish between components that have the same polarity and similar latencies, such as the "frontally distributed P3a" versus the "centroparietal P3b." In these cases, researchers often refer to a *family* of components (e.g., the N2 family of components) consisting of a set of *subcomponents* (e.g., the N2a, N2b, N2c, and N2pc subcomponents). Each subcomponent is actually a full-blown component, reflecting a different functionally and anatomically defined process, and the different subcomponents within a family are united only by their common polarity and similar timing.

Although adding the scalp distribution information can help to define a component, it will be ineffective if multiple subcomponents have similar scalp distributions (e.g., it seems likely that multiple different brain processes will produce a positive voltage deflection over the frontal lobes between 300 and 600 ms). Moreover, the scalp distribution for a single ERP component may vary across experimental contexts. For example, one subcomponent of the auditory N1 family arises from tonotopically mapped auditory cortex, and its scalp distribution therefore changes according to the pitch of the stimulus (Bertrand et al., 1991). Moreover, the scalp distribution in any given time range is influenced by all the components active in that range, which makes it difficult to determine the true distribution of a single component in a given experiment (unless that component has been isolated using one of the approaches described later in this chapter). Furthermore, the apparent scalp distribution can vary widely, depending on the choice of reference electrode (see Luck, 2005, chap. 3).

One additional variable that is often used to identify and define ERP components is their sensitivity to experimental manipulations or factors (see Donchin et al., 1978, for a thorough discussion). That is, what are the tasks, stimuli, timing parameters, and other factors that allow the component to be observed, and how do changes in these various factors modulate the timing, amplitude, and scalp distribution of the component? For example, the N2pc is observed for a target stimulus surrounded by distractors but not for a target stimulus presented in isolation (see Chapter 12, this volume). This dependence of the N2pc on the presence of distracting information in the display has played a large role in shaping various theories of the component. Furthermore, the N2pc has been shown to be largely unaffected by the probability of the target item (see Chapter 12, this volume), in contrast to the P3b. Therefore, sensitivity to experimental factors can help to identify the nature of a component and to distinguish among different components. However, just as discussed above with the variables of polarity, timing, and scalp distribution, sensitivity to experimental factors is not by itself a sufficient method for defining a component. For example, multiple ERP components may be sensitive to the same experimental manipulation, such as the similar dependence of P2 and P3b amplitude on the probability of the target stimulus. Furthermore, it is difficult to determine if an experimental manipulation has modulated the strength or timing of a specific component, or rather has resulted in a change in task strategy that has affected some other overlapping component. That is, it is difficult to assess whether an experimental manipulation has made an impact on a specific component, and it is also difficult to determine whether the experimental manipulation changed the strength, location, or timing of the neural process.

From these considerations, it should be clear that it is not appropriate to formally define an *ERP component* in terms of a combination of polarity, timing, scalp distribution, and sensitivity to experimental manipulations. These variables may be associated with a given component, but they do not define the component. We have instead argued that the term *ERP component* is best defined in terms of the scalp-recorded activity generated by a specific neural or psychological process, which in turn produces the polarity, latency, and scalp distribution of the component (which vary as that process varies), along with the sensitivity of the component to experimental manipulations. Unfortunately, our preferred definition is not very useful as an *operational* definition (i.e., a definition that describes the operations necessary to determine whether a specific voltage deflection reflects a specific component), because it is not usually possible to determine from the observed waveforms the voltage that is attributable to a specific known process.

Thus, in practice, the best way to identify a specific component is to take a *converging evidence* approach that intelligently combines various factors (including but not limited to polarity, latency, scalp distribution, and sensitivity to experimental manipulations) that would be expected to be true of a given process in a given context. For example, imagine that an oddball task was used in a study of elderly individuals, and a large positive voltage with a parietal maximum was observed to peak at 500 ms for the oddball stimuli, with a much smaller voltage observed for the standard stimuli. Four pieces of evidence converge on the conclusion that this voltage consists predominantly of the P3b component: (1) the voltage is positive at sites where the P3b is typically positive; (2) the latency is what we would expect given that cognition is typically slowed in elderly individuals; (3) the scalp distribution is consistent with previous studies of the P3b; and (4) the voltage shows the typical dependence on target probability. Now consider an example in which 5-year-old children are asked to passively view pictures of same-race faces and pictures of different-race faces, and a greater positive voltage is observed for the different-race faces with a peak latency of 325 ms. Imagine also that the voltage for both same-race and different-race faces was largest at parietal electrode sites, but the difference in voltage between same-race and different-race faces was largest at central sites. Is this a P3b component? A superficial analysis might lead to the conclusion that a larger P3b component was observed for the different-race faces, because the voltage was positive, peaked near 300 ms, and was maximal at parietal electrode sites. However, 325 ms would be an unusually early latency for a visual P3b component, especially in 5-year-old children. Moreover, even if a P3b were present in this latency range, the difference between conditions had a more central scalp distribution than is typical for the P3b component. Thus, it would be unlikely that this experimental manipulation primarily influenced P3b amplitude.

When this converging evidence approach is taken, it is important to consider both the strength of the evidence that a given component has a specific property and the degree to which other components might have that same property (this is essentially an application of Bayes's theorem). For example, although the N400 component is almost always present between 300 and 600 ms (see Chapter 15, this volume), many other components are also active in this latency range, so the finding that a given voltage deflection occurs in this latency range is not strong evidence that the deflection is an N400 component. In contrast, the lateralized readiness potential (LRP; see Chapter 9, this volume) and the N2pc component (see Chapter 12, this volume) have distinctive lateralized scalp distributions that are not present for many other components; the presence of this distinctive scalp distribution therefore provides strong (although not infallible) evidence that an LRP or N2pc was present.

With this approach, one is never completely certain that a specific component has been identified, and the strength of a conclusion will depend on both the number of pieces of converging evidence and the strength of each piece. Although it may be disappointing that one can never be certain that a specific component has been identified, this kind of uncertainty is common in all fields of science. Moreover, as discussed in the latter part of this chapter, it is sometimes possible to use *component-independent experimental designs* in which the conclusions of a study do not depend on identifying a specific ERP component.

Linking Components with Processes: The Problems of Forward and Reverse Inference

Up to this point, we have assumed that we already know what neural or psychological process is reflected by a given ERP component. In this section, we consider how one might create this link (which we call the *problem of forward inference*) and how one might use this information to draw conclusions in new experiments (which Poldrack, 2006, called the *problem of reverse inference* in the context of neuroimaging).

The Problem of Forward Inference

It is more difficult than one might think to demonstrate that a given ERP component (or any other physiological measure) reflects a specific neural or psychological process. The challenge arises from the fact that we are looking for a neural measure of a given process because we do not fully understand the process and wish to use the neural measure to study the process. Because we do not fully understand the process, it is difficult to design unambiguous tests of the hypothesis that a given component reflects this process. For example, imagine that component A is hypothesized to reflect the encoding of information in verbal working memory. We could test this hypothesis by comparing the ERPs in a condition in which subjects are asked to encode words in working memory and a condition in which they passively view the same words. However, it is possible that working memory encoding is fairly automatic and would occur in both conditions; thus, the absence of a difference in component A between conditions might not be strong evidence against the hypothesis that this component reflects working memory updating. Moreover, if component A is found to differ between conditions, this could reflect some other process that differs between these conditions (see Shulman, 1996, for an interesting discussion of a related set of issues in the context of neuroimaging).

This problem could potentially be solved with a bootstrapping approach (the term *bootstrapping* refers to "pulling oneself up by one's bootstraps"). In this approach, one begins by trying the most obvious and unassailable manipulations of a given process to see if the component is present under the conditions in which everyone would agree that the process should be present. If the hypothesis survives multiple tests of this nature, it is tentatively accepted. The component is then used to test new hypotheses about the process it is thought to reflect. If these experiments yield results that are broadly consistent with evidence from other approaches, then confidence in the link between the component and the process continues to grow. If discrepancies arise, then researchers must reappraise the link between the component and the process.

As an example, consider the N2pc component (for a detailed discussion, see Chapter 12, this volume). Luck and Hillyard (1994) proposed that this component reflects an attentional filtering process that is used to suppress inputs from distractor objects surrounding a potential target. This was initially tested with the most obvious possible manipulations, such as removing the distractors to see if the N2pc component would disappear. A second set of experiments tested more refined manipulations based on findings from monkey single-unit experiments (Luck et al., 1997). The results of these experiments were consistent with the proposed link between N2pc and attentional filtering, and subsequent experiments assumed that this link was true and used it to test hypotheses about attention. For example, one study asked whether the same putative filtering mechanism was used by targets defined by different types of features (Girelli & Luck, 1997), and another series of experiments asked whether this mechanism was applied in parallel or in serial (Woodman & Luck, 1999, 2003b). However, later evidence demonstrated that the N2pc does not reflect filtering of the distractors per se, instead reflecting operations that must be applied to the attended object itself when distractors are present (Hickey et al., 2009). This is a modest change in the process thought to be reflected by the N2pc, but it was enough to slightly change the conclusions that can be drawn from the previous studies.

The Problem of Reverse Inference

Once the problem of forward inference has been solved and a given component has been linked with some certainty to a given process, it is desirable to use this component as a measure of the presence, magnitude, and timing of that process in new experiments. This leads to the problem of reverse inference: If a component is present at a particular time, can we conclude that the process was present at that time? In Poldrack's (2006) analysis of this problem in the context of neuroimaging, the question is framed as follows: If brain activity has previously been observed in area X when process P is active, can we use the presence of activity in area X in a new experiment as evidence that process P was active in that experiment? As an example, Poldrack cited experiments using differences in activity in the dorsal striatum across conditions, which had previously been associated with reward processing, as evidence that reward mechanisms were differentially active in these conditions.

However, one must be cautious about using reverse inference. Reverse inference is actually a case of the well-known logical error of *affirming the consequent*. If the presence of P (e.g., reward) leads to the occurrence of X (activity in the striatum), this does not mean that the occurrence of X necessarily entails the presence of P. For example, sleeping (P) causes the eyes to close (X), but eye closure (X) does not necessarily mean that someone is asleep (P). Reverse inference is valid only when it is possible to say that X occurs if *and only if* P occurs (i.e., X never occurs without P). In functional magnetic resonance imaging (fMRI) this standard is difficult to meet, because it is likely that the thousands of neurons in a given voxel and the millions of neurons within a cortical area are involved in multiple processes (e.g., the same neurons in visual cortex that are involved in perception are also involved in working memory). Consequently, it is not usually possible to assert that activity in a given voxel occurs if and only if a single process occurred.

Fortunately, an if-and-only-if condition is not as difficult to achieve for ERP components, because scalp ERPs represent a subset of the activity occurring within a given brain area. As described earlier, ERPs reflect the synchronous activity of cortical pyramidal cells, and many processes that occur within a given brain region will not lead to an ERP signature on the scalp. Consequently, whereas almost any process within a given brain region will change metabolic activity and therefore change the blood oxygen-dependent (BOLD) activity, only a subset of processes within a given region will produce a measurable ERP on the scalp. This makes ERP components more likely than BOLD responses to be tied to a specific process, and makes it less likely that a change in a given ERP component reflects different processes in different experiments. In other words, it is more plausible that a specific ERP component will be present if and only if a given process is present than that a BOLD response in a specific voxel will be present if and only if a given process is present.

For example, the evidence to date indicates that the N2pc component is present if and only if attention is allocated to an object in the presence of distractors. Of course, future research may demonstrate that the N2pc component can sometimes be elicited under conditions that do not involve this attention process, but it is at least plausible that this component might be present if and only if this attention process occurs. For example, when Luck and Ford (1998) found that an N2pc was present for conjunction targets and not for feature targets, they were reasonably justified in using reverse inference to draw the conclusion that a specific mechanism of attention was allocated to the conjunction targets and not to the feature targets. In contrast, there is no area of the brain in which one could reasonably assume that the presence of an increased BOLD signal necessarily reflected the allocation of attention.

Two main problems must be solved for reverse inference to be used with a given ERP component to draw strong conclusions. First, it is necessary to conduct a comprehensive set of experiments testing the hypothesis that the component of interest is present if and only if the corresponding process occurs. This is the problem of forward inference, and it is made difficult by the fact that we do not usually know enough about the process that a component hypothetically reflects to know whether this process is present or absent in a given experimental condition. Second, once the problem of forward inference has been solved, new experiments that attempt to use reverse inference must solve the problem of component identification. That is, one must be able to demonstrate that voltage deflections observed in the new experiments represent the same component observed in the earlier studies that established the link between the component and the process.

These two challenges are sufficiently difficult that it may never be possible to use reverse inference with complete certainty. However, as Poldrack (2006) discussed in the context of neuroimaging, one can use a Bayesian approach to draw probabilistic inferences on the basis of reverse inference. This involves assessing the probability that the ERP component would be present even if the corresponding process was not active and the probability that the corresponding process would be active without eliciting the ERP component. These probabilities are difficult to calculate, so this Bayesian approach is usually used informally. For example, we do not know the probability that an N2pc component would be present without the allocation of attention, and we do not know the probability that the allocation of attention may occur without an N2pc component. Thus, we cannot provide a precise probability for the claim that the variety of attention indexed by the N2pc component is needed for conjunction targets but not for feature targets (based on the presence of an N2pc for the former but not the latter). However, given that several experiments support the contention that N2pc is observed if and only if this particular mechanism of attention is present, and given that the N2pc can be isolated quite well from other components because of its distinctive contralateral scalp distribution, we can say something informal such as "The finding that an N2pc was present for conjunction targets but not feature targets provides good evidence that the attentional processes that were present in prior N2pc experiments are needed for the detection of conjunction targets but not for the detection of feature targets."

Interestingly, the logic of reverse inference may sometimes allow stronger conclusions to be drawn from the absence of an ERP component than from its presence. If we can say that a given physiological measure X is always present when process P occurs—without the if-and-only-if restriction—then we can use the *modus tollens* argument from classical logic. This argument says that if we know that the presence of P entails X, then the absence of X entails the absence of P. That is, if previous experiments demonstrate that process P always leads to physiological measure X, then the absence of physiological measure X in a new experiment can be used to deduce that process P was not present. For example, Vogel and colleagues (1998) assumed that working memory encoding leads to the occurrence of a P3 wave (for supporting evidence, see Chapter 7, this volume). They found that this component was absent under conditions that led to an "attentional blink," and from this they concluded that no working memory encoding occurred for stimuli presented during the attentional blink. This is a logically valid conclusion. However, its truth depends on the validity of the initial assumption that working memory encoding leads to a P3 wave, which is not certain. Nevertheless, this general approach is less problematic than the typical use of reverse inference, which is based both on the assumption that a component is present when the corresponding process occurs and on the further assumption that the component is absent when the process does not occur. Of course, it is important to ensure that the absence of a voltage deflection in a given condition truly reflects the absence of the component of interest rather than cancellation by an opposite-polarity component, latency jitter, poor signal quality, low statistical power, and so on.

Solving and Avoiding the Problems Associated with ERP Components

We have now seen how difficult it can be to associate changes in the observed ERP waveform with changes in an underlying ERP component. You may find yourself rightfully wondering, so what is this technique good for? In this section, we explore methods and strategies that have proven successful in using ERP components to answer questions about the mind and brain.

Event-related potentials provide a unique window into ongoing processing in the brain, serving as a continuous play-by-play of processing as it unfolds over time. It is this high temporal resolution of ERPs that makes them so desirable as a measure of brain processing. With ERPs, we can see processing before, during, and after the execution of behavioral responses, providing us with additional insights that cannot be gained with behavioral measures alone. However, the limitations of the ERP technique discussed in the previous sections mean that ERPs are only well suited for answering certain types of questions. Understanding the types of questions that can be readily answered with ERPs is essential for the successful application of the technique, and the remainder of the chapter will focus on describing several types of questions that ERPs have proven useful in answering.

The domains covered here may not encompass every current or potential use of ERPs; for example, ERPs may be useful as potential biomarkers in mental illness (Javitt et al., 2008; Luck et al., 2011). However, the topics covered here provide a broad overview of the ways in which ERPs have been most commonly used to make scientific progress. These can be broadly divided into four domains, which we will explore in turn below: (1) determining which cognitive or neural process differs across conditions or across groups (e.g., perception, attention, response selection); (2) determining whether and when the brain has completed some set of processes; (3) uncovering new mental processes and subdividing known processes; and (4) covert monitoring of processing in situations in which overt behavior is difficult to measure or interpret (e.g., coma, infancy). We will examine each of these areas, providing specific examples of how ERPs have been used to expand our understanding in each domain.

Using Specific Components to Index Specific Processes

One of the most notable and widely used applications of ERPs is to determine which specific neural or psychological process is affected by the factors of interest in the experiment. In other words, does a particular manipulation affect process A or process B or alternatively, do two groups of individuals differ in process A or process B? Using ERPs in this manner usually requires that (1) the precise neural or psychological process indexed by a component is known and understood and that (2) the component can be successfully isolated from the surrounding and overlapping ERP components. As discussed earlier in the chapter, both of these requirements are difficult to meet; therefore, this branch of research typically relies on a number of assumptions concerning the specific nature of the ERP component of interest. These assumptions about the nature of the component are usually based on a wealth of previous research on the component and ideally include both studies in which the experimental manipulations that alter the component are explored and studies that are specifically aimed at elucidating the functional nature of the component (termed *ERPology* by Luck, 2005). We will first give an example of using components in this process-dependent manner to make the main issues facing researchers in this domain concrete, followed by some tips on how to successfully isolate and measure an ERP component.

Imagine that we wanted to understand why schizophrenia patients show prolonged reaction times (RTs) across a wide variety of behavioral tasks, an effect that has been observed for decades (see the review by Nuechterlein, 1977). In other words, which stage or stages of processing are slowed in schizophrenia patients, producing the slowing of behavioral RTs? We can address this question by examining whether particular ERP components are affected in the patient group compared to healthy controls. That is, is the scalp-recorded signature of a particular cognitive process delayed in latency or decreased in amplitude in the patients compared to the controls? This general approach has been used in studies of schizophrenia to examine abnormalities in many components, including the mismatch negativity (MMN), the P1 wave, the N2pc component, the P3 wave, the lateralized readiness potential, and the error-related negativity (Bates et al., 2002; Butler et al., 2007; Javitt, 2000; Jeon & Polich, 2003; Luck et al., 2006, 2009; see Chapter 19, this volume, for a review).

This approach—as typically applied—requires that previous experiments have already linked a component to a process, and it requires determining that a newly observed difference between patients and controls reflects a change in this specific component and not some other component (see the earlier sections on forward and reverse inference). For example, the N2pc component was used to assess whether prolonged behavioral RTs are accompanied by delays in the allocation of covert visual spatial attention in schizophrenia patients (Luck et al., 2006), which relied on previous work demonstrating that the N2pc is a scalp-recorded signature of covert shifts of visual attention and on the ability to isolate the N2pc from the surrounding ERP activity (which was achieved by using contralateral-minus-ipsilateral difference waves, as described in more detail below). Additionally, we can use ERPs to assess whether multiple stages of processing are affected in a patient group. For example, is the RT slowing exhibited by schizophrenia patients caused by a generalized slowing of all cognitive and neural processing or a combination of some subset of processes?

METHODS FOR ISOLATING AN ERP COMPONENT

As described above, the ability to use ERP components as indexes of specific processes (reverse inference) depends on the ability to successfully isolate the component of interest from the surrounding ERP components. This is not an easy task. It may even seem impossible. However, there are a number of tricks that can be used to isolate a particular ERP component of interest from all of the other ongoing activity. Although the specific methods will depend on the specific task, ERP component, question of interest, and so on, the following strategies have proven successful in a number of different contexts.

One strategy is to focus the experimental design on ERP components that are large compared to the surrounding components. For example, the P3 wave is often >10 microvolts, making it easy to distinguish from the much smaller surrounding and overlapping ERP components. A second strategy is to focus the task design such that only one or two ERP components differ across conditions. When the design focuses on a small number of ERP components, it is easier to avoid significant component overlap, making the measurement of a specific component much easier. A third strategy involves subtracting out overlapping ERP components by creating difference waves between conditions or between electrode sites. For example, the lateralized readiness potential (LRP) is a difference wave created by subtracting the voltage at sites ipsilateral to the response hand from the activity at sites contralateral to the response hand. This subtraction process effectively isolates only the activity related to response selection, subtracting away the many other processes that do not differ between the contralateral and ipsilateral hemispheres; indeed, any brain activity that differs between the contralateral and ipsilateral electrode sites (relative to the hand that responds) must be generated during or after the process that determines which hand should respond (see Chapter 9, this volume). Similarly, by computing a rare-minus-frequent difference wave in an oddball paradigm, it is possible to isolate probability-sensitive ERP components such as the P3 wave (see, e.g., Luck et al., 2009; Vogel et al., 1998).

Although difference waveforms can be an effective tool in isolating specific ERP components, they are not a panacea. First, a difference waveform is effective in isolating a specific ERP component only when all or most other components do no not vary across the two conditions used in the subtraction. Second, when a difference wave varies in amplitude across groups or across conditions, it is difficult to know which of the two waveforms used in the subtraction actually varies. For example, the LRP is decreased in schizophrenia patients relative to control subjects (Luck et al., 2009), but this could reflect less activation over the contralateral hemisphere or more activation over the ipsilateral hemisphere. Third, activity in a difference wave could reflect latency differences between the two original waveforms rather than a difference in amplitude.

An additional class of strategies uses scalp distribution information to isolate components. A simple version of this strategy is simply to measure a given component at an electrode site where this component is relatively large and other components are relatively small. A somewhat more sophisticated approach is to use a *vector filter*, which combines the data across all scalp sites in a manner that reflects the scalp distribution of a given component (see, e.g., Gehring et al., 1992). Event-related potential source localization techniques go one step further, providing a source waveform for each estimated generator site. In addition, ICA and PCA can use scalp distribution information to isolate the time course of each component.

When evaluating these different approaches, it is important to remember that, just as every researcher has his or her own individual limitations, each technique used to isolate ERP components is limited in its own special way. No technique—despite what its proponents may shout loudly from the research pulpit—is without its shortcomings, flaws, and limitations. Successfully using any of the techniques at our disposal requires that we know and understand the limitations of the method. Before using source localization, ICA, or even simple difference waves, one must be careful to fully understand how the technique works and when it might fail.

METHODS FOR MEASURING AN ERP COMPONENT

Once a component has been successfully isolated from the overlapping activity, some quantitative assessment of the component must be made in order to compare it across conditions or across groups of subjects. The most widely used quantitative characterizations of ERP components include amplitude and latency assessments. Despite the inherent difference between peaks and components described above, it is common for ERP researchers to quantify ERP results by measuring the amplitude and latency of the peaks. Peak amplitude and peak latency measures are generally computed by choosing a time window surrounding a peak in the waveform and finding the most positive point in that time window (or the most negative point for a negative-going peak). The amplitude at this point is used as a representation of the magnitude of the component, and the latency of this point is used as a representation of the timing of the component. Historically, peak measures were employed because, as Donchin and Heffley (1978) so aptly stated, "it requires nothing but an x-y plotter, a ruler, and enough time" (p. 557). These were often all that a typical ERP researcher had at his or her disposal in the early days of ERP research, but researchers today have computers capable of performing much more advanced algorithms than those that a ruler can accomplish, and we are no longer limited to such simple measurement techniques.

Is there anything special about the amplitude or timing of the peaks in the observed ERP waveforms? As Figure 1.2 illustrates, the amplitude and timing of the peaks in the observed waveform may be quite different from the amplitude and timing of the underlying components that sum together to produce the observed waveform. And as Figure 1.3 illustrates, factors such as latency variability can strongly influence peak amplitude. Moreover, it seems simplistic to assume that a process that extends over hundreds of milliseconds can be quantified by the value of a single time point. In addition, when the values are measured at multiple electrode sites, it makes no sense to use the peak at each electrode site to measure a single component: The peak will occur at a different time at each electrode site, but a given component necessarily has the same time course at each electrode site (because of the instantaneous transmission of voltage). Peak measures have other shortcomings as well (summarized in Luck, 2005, chap. 6), and there is a clear trend away from peak measures among sophisticated ERP researchers.

How, then, can one better quantify the magnitude and timing of an ERP component? The first step is usually to isolate the component by computing some kind of difference wave that subtracts away most of the other components. As an example, consider the MMN data shown in Figure 1.6. In this experiment, subjects were presented with a frequently occurring *standard* pitch or a rare *deviant* pitch every 1000 ms (see Chapter 6, this volume, for details). When the deviant pitch was sufficiently different from the standard pitch, the ERP waveform was more negative for the deviant pitch than for the standard pitch from approximately 100 to 200 ms poststimulus. If we attempted to quantify the magnitude of this effect by measuring the amplitude of the most negative peak between 100 and 200 ms, we would face two serious problems. First, because the overall waveform contains a P2 peak during this interval, there is no negative peak to be measured in many of the waveforms shown in Figure 1.6 (especially in the waveforms elicited by the standards). Second, even if we could find a negative peak, the voltage at this peak would reflect a combination of this P2 wave, the MMN, and any other components that were active during this period.[5] Thus, it is better to quantify the magnitude of the

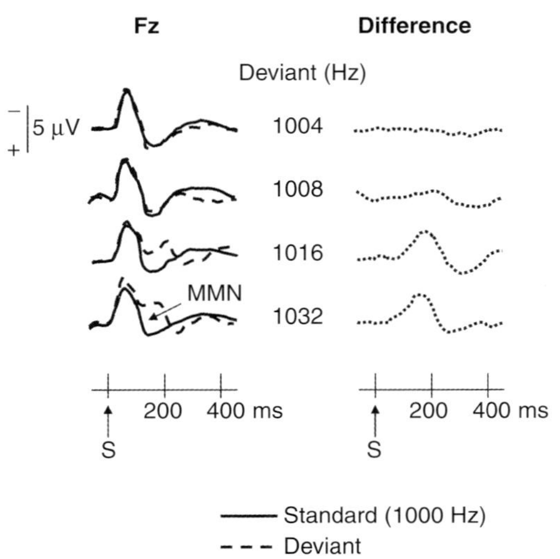

Fig. 1.6. Example of the use of difference waves in the context of MMN. The left side shows the waveform elicited by a 1000 Hz standard tone that occurred on 80% of trials, overlaid with deviant stimuli that differed in pitch from the standard by varying amounts and occurred on 20% of trials. The right side shows the deviant-minus-standard difference waves. Note that this is the same as Figure 6.1 in Chapter 6, this volume.

MMN from the deviant-minus-standard difference wave.

By measuring amplitude or latency from a difference wave, the contributions of the overlapping peaks are reduced or eliminated. Of course, this will work well only if the other components are equivalent across the two waveforms that are used for the subtraction so that they are eliminated in the difference wave. One could use a peak amplitude measure to quantify the amplitude of a component in the difference, and this would certainly be an improvement over measuring peak amplitude from the two original waveforms used in the subtraction. However, there is still no particular reason to choose this one point as a reflection of the magnitude of the underlying process. If one is interested in the overall magnitude of a brain response, it is usually more reasonable to measure the area under the curve or the mean voltage over the duration of the component (these are nearly equivalent: mean is simply area divided by duration). An important exception arises, however, when one is trying to measure the amplitude of a component that varies in latency across conditions or across groups; in this case, it may be necessary to use a method that finds the peak and then measures the amplitude at (or around) this peak.

Peak latency is also a poor measure in most cases, because the latency of the peak is not usually a particularly interesting time point. Quantifying the latency of an ERP component by finding the peak is analogous to quantifying RT by finding the mode of the RT distribution for each subject. Instead, it is sometimes possible to quantify the midpoint of a component by finding the time point that divides the area under the curve into two equal portions. This is called the *50% area latency* measure, and it is closely related to median RT (see Luck, 2005, chap. 6). In addition, theories of cognitive processes often make predictions about the onset or duration of a process rather than the midpoint. Kiesel and colleagues (2008) provide an excellent comparison of the different methods that can be used for the onset of a component, and these methods can be easily extended to measure the offset and duration of a component.

Assessing the Time Course of Processing

The temporal resolution of ERPs makes them an excellent tool for determining the time course of a neural or psychological process. The simplest way to do this is to measure the latency of a given peak in two different conditions or two different groups and use this as a measure of the amount of time required for this process to occur in the two conditions or two groups. However, this approach is not usually very powerful, because it does not isolate a specific component and because it uses the peak as a measure of timing. A more powerful approach is to compare the waveforms from two conditions or from two groups of subjects to ascertain the point in time at which the waveforms begin to diverge. For example, ERPs have been used in the emotion literature to determine when, after the onset of a stimulus, processing differs between emotion-evoking and neutral stimuli (see Chapter 16, this volume). There are advantages and limitations to using ERPs in this manner, and we will explore both of these through some examples below.

Let's consider the emotion example mentioned above, in which we wish to know by what point in time processing related to the emotional content of a stimulus has begun. In other words, by what point in time has the brain distinguished between emotional and nonemotional stimuli? We can answer this question by comparing the ERP waveforms elicited by neutral stimuli (e.g., a picture of a landscape) and emotion-eliciting stimuli (e.g., a picture of a mutilation). We can use the time point at which the waveforms begin to diverge as a measure of when the brain has distinguished between the neutral and emotional stimuli. That is, the waveforms between an emotional and a nonemotional condition cannot diverge until the brain has begun to distinguish the emotional content of the stimulus (provided that all other factors, including physical stimulus factors, are matched between the conditions). The advantage of this approach is that, although specific ERP components may differ between the conditions, the conclusions about timing do not rely on isolating a specific ERP component. That is, the presence of a difference between conditions at a given time indicates that the brain has distinguished between the two conditions by this time, regardless of which component was responsible for this difference. This approach is one case of what are called *component-independent experimental designs* (see Luck, 2005, chap. 2).

Because this method does not require isolating a specific component or linking a component with a specific process, it generally requires fewer assumptions than using ERPs in a component-dependent manner. However, there are some limitations to this approach. For example, it is important to note that this method provides an upper bound on the timing of an effect. Because many processes may be invisible in scalp ERP recordings, the brain might make a distinction between two stimuli long before the first point at which the scalp-recorded signals differ. Therefore, one can use ERPs to say that a particular effect has occurred *by* a particular time point, but one cannot use ERPs to conclude that an effect did not begin *until* a particular time. In our emotional content example, one could conclude that the brain has begun to process information related to emotional information by the point at which the waveforms diverge. However, one could not say that emotional processing did not begin until that time point, because the effect could have begun earlier in brain areas that did not give rise to a scalp-recorded ERP. Generally speaking, this technique is valuable in providing evidence that an effect happens early in the processing stream, but it cannot be used to prove that an effect does not happen until late in the processing stream.

The limitations in the conclusions that can be drawn about timing from ERPs may seem debilitating to the technique, but using ERPs in this manner has answered many important questions about cognitive and neural processing. For example, ERPs were able to end a long-standing debate in the attention literature about whether attention operates at an early stage or a late stage of processing (for reviews, see Hillyard et al., 1998; Luck et al., 2000). It is difficult to determine from behavioral studies whether the effects of attention on response speed and accuracy arose from changes in perceptual processing or changes in a postperceptual stage of processing. However, because ERPs provide a continuous measure of processing between the stimulus and the response, they can indicate whether the attention effects begin early or late in the processing stream. That is, the *locus of selection* can be assessed directly by asking whether the ERP waveforms for attended and ignored stimuli diverge early in time (e.g., within the first 100 ms after stimulus onset) or late in time (e.g., more than a few hundred milliseconds after stimulus onset). Research using this approach has shown that—at least under some conditions—attention influences sensory processing within the first 50 ms after stimulus onset for auditory stimuli and within the first 100 ms after stimulus onset for visual stimuli (see Chapter 11, this volume). These ERP results provided key evidence in favor of early selection models of attention, helping to answer a fundamental question that could not be easily addressed using behavioral techniques.

This time-based approach is often combined with the process-specific approach described in the previous section, in which the effects are linked with specific components. For example, researchers have argued that the early ERP attention effects consist of modulations of specific sensory-evoked ERP components (see, e.g., Di Russo et al., 2003; Woldorff et al., 1993). This has been difficult to establish with complete certainty because of the many difficulties associated with trying to identify specific components, as discussed earlier in the chapter. However, the *converging evidence* approach described earlier in this chapter has been used to provide substantial support for the hypothesis that attention influences specific ERP components. Even more important, the simple fact that the waveforms for attended and unattended stimuli diverge at an early time provides very strong evidence that attention can influence perceptual processing.

MEASURING PROCESSES THAT OCCUR PRIOR TO A COMPONENT

A related approach uses an ERP component to assess the processes that must have occurred *prior* to the ERP component. The advantage of this approach is that it does not require that we first determine a solid link between an ERP component and a specific process (i.e., we do not need to solve the forward inference problem). Instead, we can use simple assumptions about the processes that must have occurred prior to the ERP component to draw inferences about these processes.

As an example, consider the N400 component, which countless studies have shown is larger for words that mismatch the current semantic context than for words that match this context (reviewed by Kutas, 1997). For example, the word *nurse* will elicit a larger N400 if it is preceded by an unrelated word such as *cup* than if it is preceded by a related word such as *doctor*. Substantial controversy surrounds the question of exactly what process the N400 component represents (see Chapter 15, this volume). However, it is safe to assume that this difference in N400 between words that match and mismatch a semantic context could not occur unless the words were perceived. Thus, if we see that a given word elicits a larger N400 when the preceding word was related than if it was unrelated, then we can be certain that the words were perceived. This logic has been used to show that, under certain conditions, attention does not influence sensory processing and that words are fully perceived even when unattended (Luck et al., 1996; Vogel et al., 1998, 2005). That is, although attention influences sensory processing under some conditions, modulating the early sensory-evoked components, under other conditions attention only influences postperceptual processes that follow word identification (see the reviews by Luck & Hillyard, 1999; Luck & Vecera, 2002). Under these latter conditions, attention has no impact on the difference in N400 amplitude for words that match versus mismatch the current semantic context.

As a second example, consider the P3b component, which every ERP researcher knows is larger for infrequent target stimuli than for frequently occurring standard stimuli. However, an important implication of this probability dependence often goes unnoticed. Specifically, the onset of the difference in P3b amplitude between rare and frequent stimuli cannot occur until the brain has at least begun to determine whether the eliciting stimulus belongs to the rare category or the frequent category. This implication was spelled out very clearly by Kutas and colleagues (1977), who framed it in terms of the then-popular idea that the P3b component was elicited by surprising stimuli: "before a stimulus can surprise it must be identified. As P300 commonly appears as a discriminative response to specific stimuli within a series, its elicitation must be preceded by an adequate evaluation of the stimulus at some level of processing" (p. 792–793). This idea is commonly described by saying that the latency of the P3 wave reflects *stimulus evaluation time*, but this is a somewhat vague description. It is much more precise—and powerful—to say that the onset of the difference between the waveforms elicited by the rare and frequent stimuli reflects a time at which the brain has begun to determine whether the stimulus belongs to the rare or the frequent category. That is, the waveforms between these two conditions could not differ until the brain has determined whether the stimulus belongs to the rare or the frequent category, indicating that by that point the brain has begun to categorize the stimuli.

We have applied this more precise framing of P3b latency to understanding why behavioral RTs are slowed in patients with schizophrenia (Luck et al., 2009). Each stimulus in this experiment was a digit or a letter, with one category rare ($p = .2$) and the other frequent ($p = .8$). Subjects were asked to press a button with one hand for digits and a button with the other hand for letters, and patient RTs were approximately 60 ms slower than control RTs. As shown in Figure 1.7, the voltage in the P3 latency range was larger for control subjects than for patients, for both the rare and frequent stimulus categories, but the latency of the P3 peak was similar across groups. However, given that many different processes

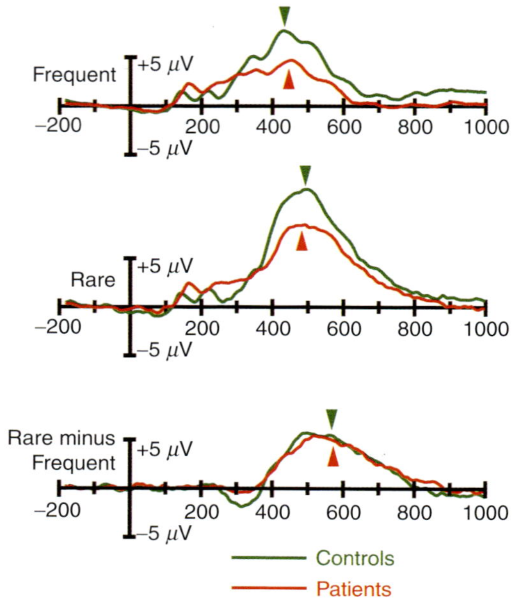

Fig. 1.7. Grand average ERPs recorded at the Pz electrode site from schizophrenia patients and control subjects (from the study of Luck et al., 2009). The patient and control waveforms are overlaid for the frequent stimuli, the standard stimuli, and the rare-minus-frequent difference wave. Triangles show mean P3 latency for each group, quantified as peak latency for the rare and frequent stimuli and 50% area latency (the point that divides the area under the curve into two equal portions) for the difference wave.

presumably overlap during the P3 time range, it is difficult to draw firm conclusions on the basis of the time of the peak voltage in this time range. More precise conclusions can be drawn by examining the rare-minus-frequent difference waves in each group (Figure 1.7, bottom). These difference waves reflect the differential processing of the rare and frequent stimulus categories, and any nonzero voltages in these difference waves must be a consequence of a preceding process that determined the category to which a stimulus belonged. The only difference between patients and controls in these difference waves was a reduction in amplitude in the time range of the N2 wave in patients. The difference waves were nearly identical across groups in the P3 time range, and the midpoint of the deflection in this wave (the time that divided the area under the curve into equal halves) was nearly identical across groups. Thus, no delay was observed in the brain's differential responding to rare versus frequent stimuli in the patients compared to the controls, despite a 60 ms slowing of the behavioral response in patients. This suggests that the slowing of behavioral responses was not caused by a slowing of the processes that lead up to the categorization of the stimuli, but was instead caused by postcategorization slowing. This conclusion was further supported by a reduction in the amplitude and a slowing of the latency of the lateralized readiness potential in the patients compared to the controls.

It is important to note that, in both of these examples, conclusions were drawn about the processes that logically must have preceded the component being measured rather than the process the component directly reflected. That is, the N400 was used to assess the perceptual processes that must occur before the brain can distinguish between semantically related and unrelated words, and the P3b was used to assess the perceptual and categorization processes that must occur before the brain can determine whether a stimulus belongs to a rare or a frequent category. An important advantage of this approach is that we do not need to know with certainty what process produces a given ERP component. Instead, we can make very straightforward assumptions about what processes must occur for a component to differ across conditions. In many cases, it does not actually matter which component differs across conditions; the mere presence of a difference indicates that the brain has made a specific discrimination by a given point in time. Thus, this is another example of a component-independent approach. This does not mean that components are irrelevant in the design of the experiment. Instead, it means that the conclusions do not depend on whether a specific component has been identified in the results.

Uncovering and Subdividing Mental Processes

Event-related potentials have also been useful in identifying new, previously unknown mental processes and subdividing known processes into multiple separate subprocesses. From behavioral measures, it is difficult to ascertain how many mental processes intervene between the occurrence of a stimulus and the execution of a behavior. However, ERPs provide a continuous measure of processing before, during, and after the execution of the behavior. Therefore, it is possible with ERPs to identify processes that were previously unknown.

For example, error-related negativity (ERN; see Chapter 10, this volume) occurs after the execution of a response and therefore reflects a process that behavioral measures cannot directly measure. Although previous studies had pointed to the existence of processes related to detecting and correcting errors (e.g., Laming, 1979; Rabbitt, 1966), no one had hypothesized a process with the timing of the ERN. The ERN helped to focus research on the processes occurring within 100 ms of an error response, which has led not only to numerous studies of processes related to error detection, but also to a large literature on response-conflict monitoring.

Similarly, ERPs can be used to determine whether a given behavioral effect is the result of a change in a single process or of multiple separable subprocesses. Almost every experimental manipulation that produces a behavioral effect leads to differences between conditions in multiple ERP components, and this naturally leads to the idea that the behavioral effect reflects changes in more than one process. Consider, for example, manipulations of attention. It is parsimonious to assume that any experiment in which behavioral responses are faster or more accurate for attended stimuli than for unattended stimuli reflects the operation of a single mechanism of attention, and most behavior-inspired theories of attention have taken a monolithic view of attention. However, ERP studies have demonstrated that different manipulations of attention influence different ERP components, demonstrating that different mechanisms of attention operate to produce the observed behavioral effects under different conditions (see Chapter 11, this volume). These ERP studies have inspired behavioral studies demonstrating that the details of the behavioral attention effects are indeed best explained by the existence of multiple mechanisms of attention (see, e.g., Vogel et al., 2005). Thus, the ability to monitor multiple processes with ERPs makes it possible to provide empirical evidence against simplistic explanations of behavior that invoke a single mechanism.

Covert Monitoring

A final ERP approach involves using ERPs as a means of "covertly monitoring" processing in situations in which behavioral output is uninformative, inapplicable, or unavailable. There are three general situations in which this approach is applied: (1) assessing processing in individuals who cannot or will not make a behavioral response (e.g., infants, coma patients); (2) assessing processing under conditions in which requiring a behavioral response might invalidate the task (e.g., monitoring the processing of unattended stimuli); and (3) assessing processes that might not be evident in behavior (e.g., the processing of subliminal stimuli). In this section, we will provide examples of all three of these situations.

Behavioral methods used with infants almost always take advantage of the fact that infants tend to orient toward some types of stimuli (e.g., complex, dynamic, or novel stimuli) more than other types of stimuli (Brennan et al., 1966). And if they exhibit greater looking times toward one category of stimuli than another, then this is evidence that they were able to distinguish between these categories (Spelke, 1985). The categories can be simple sensory categories (e.g., the presence versus absence of a fine pattern) or complex conceptual categories (e.g., animal versus nonanimal). However, it is always possible that infants are able to make a particular discrimination even if they fail to exhibit any behavioral orienting on the basis of this discrimination. Moreover, these techniques are difficult to use prior to about 4 months of age owing to poor motor control. Event-related potentials can be useful in these situations to determine whether the brain has made a given discrimination.

For many years, ERPs have been used in this way to determine whether newborn infants might be suffering from hearing loss. Specifically, a rapid sequence of clicks is presented, and the amplitude and latency of the early brainstem evoked responses are used to determine whether the sensory response is abnormal (Stapells, 1989). The auditory MMN component has also been widely used to assess the ability of infants to make more complex perceptual discriminations, such as distinctions between phonemes (see Chapter 6, this volume). Other components have been used to assess higher-level aspects of visual processing in infancy, such as face perception, and even higher-level cognitive discriminations (see Chapter 17, this volume). It is generally easier to assess lower-level sensory processes than higher-level cognitive processes with ERPs, because the sensory processes can typically be assessed without any kind of task. Higher-level processes are typically task-dependent, and it is difficult to teach infants a task that will elicit these processes reliably. One can sometimes take advantage of spontaneous differences in processing between, for example, rare and frequent stimulus categories, but these spontaneous differences may habituate before enough trials have been acquired to obtain reliable average ERP waveforms.

Event-related potentials can also be used in individuals who are unable to make behavioral responses due to a medical condition. In amyotrophic lateral sclerosis, for example, ERPs have been used to create brain–computer interfaces that allow patients to communicate with their families and caregivers

(Silvoni et al., 2009). Another recent example comes from coma research, where ERPs have been used to predict which patients are likely to recover (Fischer et al., 2004). There are also cases in which an individual might refuse to make a valid behavioral response, such as a suspect in a in crime, and ERPs have been used to assess whether people have knowledge of an event that they are not admitting (e.g., Farwell & Donchin, 1991).

Another type of covert monitoring approach is used when the requirement to make a behavioral response might interfere with the processing of a task. The most obvious example of this arises in attention research, in which ERPs have been widely used to compare the processing of attended and unattended stimuli (see Chapter 11, this volume). Requiring a behavioral response for an unattended stimulus presumably creates an incentive to attend to the stimulus, which is problematic for the study of attention. However, because ERPs can be recorded just as easily for unattended stimuli as for attended stimuli, they can be used to assess the processing of stimuli for which there is absolutely no incentive to attend.

This approach has also been used extensively in language research (see Chapter 15, this volume). In studies of sentence comprehension, it is difficult to assess the processing of each individual word by means of behavioral measures, because this would require interrupting the sentence for a response. Eye movement measures have often been used for this purpose in studies of reading, because the eye movements are a naturally occurring part of the reading process. However, the eye movements are still discrete events that occur some time after the eyes have landed on a given word, and they are applicable primarily in the context of written language comprehension rather than spoken language comprehension.

The third variety of covert monitoring involves asking questions about processes that might not be evident in behavior. That is, the brain may engage in a given process and reach a specific result without that result reaching awareness or triggering a behavioral response. The most obvious case of this arises in research on perception without awareness. By using ERPs, it is possible to determine how much information has been extracted from a stimulus that fails to reach awareness. For example, research has shown that a specific type of masking (*object substitution masking*) does not eliminate the orienting of attention to a target stimulus, as indexed by the N2pc component (Woodman & Luck, 2003a), but it does impair the processes needed to generate an N400 difference between words that match versus mismatch a semantic context (Reiss & Hoffman, 2006). This pattern of results indicates that this variety of masking operates after early perceptual processing but prior to semantic analysis. Similarly, stimuli that are associated with a given response will activate the preparation of that response, as indexed by the LRP, even if the subject is unaware of the stimulus and does not actually execute the response (Dehaene et al., 1998).

Conclusions

The ERP technique provides a unique and highly informative perspective on brain processing, but like all techniques it suffers from challenges, difficulties, and limitations. The goal of this chapter was to chronicle both the positive and negative sides of ERPs, exploring issues that are often unaddressed in the literature while providing a detailed set of strategies that allow the technique to be optimally employed. We hope that these recommendations allow the reader to understand and avoid the down sides of ERP research while also adopting our view that the positives of ERP research outweigh the negatives.

Acknowledgment

We thank Bill Gehring and Greg Hajcak for helpful editorial comments and suggestions.

Notes

1 The negative peaks of the waveform are sometimes referred to as *troughs*; however, there is nothing special about whether the activity is positive or negative in polarity. Therefore, we will refer to both the positive and negative deflections in the waveform as *peaks*.

2 The necessity for summation across large groups of neurons to observe a scalp ERP has implications that are often neglected by researchers. First, the magnitude of an ERP will depend on both the size of the individual postsynaptic potentials and the number of neurons that are active. Second, many neurons that are simultaneously active within a given cortical region may actually be doing very different things, and an ERP component may therefore reflect a mixture of different neural responses. Bill Gehring suggested to us that recording the ERP waveform is analogous to measuring the number of cars crossing the San Francisco Bay Bridge at a given time of day: These cars may have nothing in common except that many of their drivers are heading home for dinner. Thus, ERPs may be useful for answering broad questions about neural activity (analogous to asking when most people end their workday in San Francisco) and not as useful for answering narrow questions (analogous to asking where individual cars are going or what their occupants are doing).

3 Analogous effects can be seen for neural firing rates; for example, the duration of a change in the firing rate of a typical neuron in visual cortex following a brief stimulus is typically at least 100 ms. This is presumably a result of PSPs that last at least 100 ms.

4 The claim that brain processes involve individual brain areas requires us to be a bit more specific about what we mean by the term *process*, because much brain activity involves the interaction of multiple brain areas. We are using the term *process* to mean an elementary computation that might plausibly occur within a single brain area (e.g., spatial filtering based on lateral inhibition within an area) rather than a multistep computation that likely involves the coordinated operation of multiple brain areas (e.g., retrieval of an item from memory).

5 It should be noted that these two problems are not this extreme in all cases. For example, if one measures the amplitude or latency of the P3 peak when this component is much larger than all of the other components, then these measures will not be greatly distorted by the overlapping components. However, other shortcomings of peak measures still apply in this situation, and small differences between groups or conditions could easily reflect differences in the overlapping components.

References

Bates, A. T., Kiehl, K. A., Laurens, K. R., & Liddle, P. F. (2002). Error-related negativity and correct-response negativity in schizophrenia. *Clinical Neurophysiology, 113*, 1454–1463.

Bertrand, O., Perrin, F., & Pernier, J. (1991). Evidence for a tonotopic organization of the auditory cortex with auditory evoked potentials. *Acta Otolaryngologica, 491*, 116–123.

Brennan, W., Ames, E. W., & Moore, R. W. (1966). Age differences in infants' attention to patterns of different complexities. *Science, 151*, 354–356.

Butler, P. D., Martinez, A., Foxe, J. J., Kim, D., Zemon, V., Silipo, G., Mahoney, J., Shpaner, M., Jalbrzikowski, M., & Javitt, D. C. (2007). Subcortical visual dysfunction in schizophrenia drives secondary cortical impairments. *Brain, 130*, 417–430.

Davis, P. A. (1939). Effects of acoustic stimuli on the waking human brain. *Journal of Neurophysiology, 2*, 494–499.

Dehaene, S., Naccache, L., Le Clec'H, G., Koechlin, E., Mueller, M., Dehaene-Lambertz, G., van de Moortele, P. F., & Le Bihan, D. (1998). Imaging unconscious semantic priming. *Nature, 395*, 597–600.

Di Russo, F., Martinez, A., & Hillyard, S. A. (2003). Source analysis of event-related cortical activity during visuo-spatial attention. *Cerebral Cortex, 13*, 486–499.

Di Russo, F., Martinez, A., Sereno, M. I., Pitzalis, S., & Hillyard, S. A. (2002). Cortical sources of the early components of the visual evoked potential. *Human Brain Mapping, 15*, 95–111.

Donchin, E., & Heffley, E. F., III. (1978). Multivariate analysis of event-related potential data: A tutorial review. In D. Otto (Ed.), *Multidisciplinary perspectives in event-related brain potential research* (pp. 555–572). Washington, DC: U.S. Government Printing Office.

Donchin, E., Ritter, W., & McCallum, W. C. (1978). Cognitive psychophysiology: The endogenous components of the ERP. In E. Callaway, P. Tueting, & S. H. Koslow (Eds.), *Event-related brain potentials in man* (pp. 349–441). New York: Academic Press.

Farwell, L. A., & Donchin, E. (1991). The truth will out: Interrogative polygraphy ("lie detection") with event-related brain potentials. *Psychophysiology, 28*, 531–547.

Fischer, C., Luaute, J., Adeleine, P., & Morlet, D. (2004). Predictive value of sensory and cognitive evoked potentials for awakening from coma. *Neurology, 63*, 669–673.

Galambos, R., & Sheatz, G. C. (1962). An electroencephalographic study of classical conditioning. *American Journal of Physiology, 203*, 173–184.

Gehring, W. J., Gratton, G., Coles, M., & Donchin, E. (1992). Probability effects on stimulus evaluation and response processes. *Journal of Experimental Psychology: Human Perception and Performance, 18*, 198–216.

Girelli, M., & Luck, S. J. (1997). Are the same attentional mechanisms used to detect visual search targets defined by color, orientation, and motion? *Journal of Cognitive Neuroscience, 9*, 238–253.

Hickey, C., Di Lollo, V., & McDonald, J. J. (2009). Electrophysiological indices of target and distractor processing in visual search. *Journal of Cognitive Neuroscience, 21*, 760–775.

Hillyard, S. A., Vogel, E. K., & Luck, S. J. (1998). Sensory gain control (amplification) as a mechanism of selective attention: Electrophysiological and neuroimaging evidence. *Philosophical Transactions of the Royal Society: Biological Sciences, 353*, 1257–1270.

Hoffman, D. D., & Richards, W. A. (1984). Parts of recognition. *Cognition, 18*, 65–96.

James, W. (1890). *The principles of psychology*. New York: Holt.

Javitt, D. C. (2000). Intracortical mechanisms of mismatch negativity dysfunction in schizophrenia. *Audiology and Neurotology, 5*, 207–215.

Javitt, D. C., Spencer, K. M., Thaker, G. K., Winterer, G., & Hajos, M. (2008). Neurophysiological biomarkers for drug development in schizophrenia. *Nature Reviews Drug Discovery, 7*, 68–83.

Jeon, Y. W., & Polich, J. (2003). Meta-analysis of P300 and schizophrenia: Patients, paradigms, and practical implications. *Psychophysiology, 40*, 684–701.

Kiesel, A., Miller, J., Jolicoeur, P., & Brisson, B. (2008). Measurement of ERP latency differences: A comparison of single-participant and jackknife-based scoring methods. *Psychophysiology, 45*, 250–274.

Kutas, M. (1997). Views on how the electrical activity that the brain generates reflects the functions of different language structures. *Psychophysiology, 34*, 383–398.

Kutas, M., McCarthy, G., & Donchin, E. (1977). Augmenting mental chronometry: The P300 as a measure of stimulus evaluation time. *Science, 197*, 792–795.

Laming, D. (1979). Choice reaction performance following an error. *Acta Psychologica, 43*, 199–224.

Luck, S. J. (2005). *An introduction to the event-related potential technique*. Cambridge, MA: MIT Press.

Luck, S. J. (in press). Event-related potentials. In D. L. Long (Ed.), *APA handbook of research methods in psychology*. Washington, DC: American Psychological Association.

Luck, S. J., & Ford, M. A. (1998). On the role of selective attention in visual perception. *Proceedings of the National Academy of Sciences, USA, 95*, 825–830.

Luck, S. J., Fuller, R. L., Braun, E. L., Robinson, B., Summerfelt, A., & Gold, J. M. (2006). The speed of visual attention in schizophrenia: Electrophysiological and behavioral evidence. *Schizophrenia Research, 85*, 174–195.

Luck, S. J., Girelli, M., McDermott, M. T., & Ford, M. A. (1997). Bridging the gap between monkey neurophysiology and human perception: An ambiguity resolution theory of visual selective attention. *Cognitive Psychology*, *33*, 64–87.

Luck, S. J., & Hillyard, S. A. (1994). Spatial filtering during visual search: Evidence from human electrophysiology. *Journal of Experimental Psychology: Human Perception and Performance*, *20*, 1000–1014.

Luck, S. J., & Hillyard, S. A. (1999). The operation of selective attention at multiple stages of processing: Evidence from human and monkey electrophysiology. In M. S. Gazzaniga (Ed.), *The new cognitive neurosciences* (pp. 687–700). Cambridge, MA: MIT Press.

Luck, S. J., Kappenman, E. S., Fuller, R. L., Robinson, B., Summerfelt, A., & Gold, J. M. (2009). Impaired response selection in schizophrenia: Evidence from the P3 wave and the lateralized readiness potential. *Psychophysiology*, *46*, 776–786.

Luck, S. J., Mathalon, D. H., O'Donnell, B. F., Hämäläinen, M. S., Spencer, K. M., Javitt, D. C., & Uhlhaas, P. J. (2011). A roadmap for the development and validation of ERP biomarkers in schizophrenia research. *Biological Psychiatry*, *70*, 28–34.

Luck, S. J., & Vecera, S. P. (2002). Attention. In S. Yantis (Ed.), *Stevens' handbook of experimental psychology: Vol. 1: Sensation and perception* (3rd ed., pp. 235–286). New York: Wiley.

Luck, S. J., Vogel, E. K., & Shapiro, K. L. (1996). Word meanings can be accessed but not reported during the attentional blink. *Nature*, *382*, 616–618.

Luck, S. J., Woodman, G. F., & Vogel, E. K. (2000). Event-related potential studies of attention. *Trends in Cognitive Sciences*, *4*, 432–440.

Mazaheri, A., & Jensen, O. (2008). Asymmetric amplitude modulations of brain oscillations generate slow evoked responses. *Journal of Neuroscience*, *28*, 7781–7787.

Miller, J., Patterson, T., & Ulrich, R. (1998). Jackknife-based method for measuring LRP onset latency differences. *Psychophysiology*, *35*, 99–115.

Miller, J., Ulrich, R., & Schwarz, W. (2009). Why jackknifing yields good latency estimates. *Psychophysiology*, *46*, 300–312.

Nuechterlein, K. H. (1977). Reaction time and attention in schizophrenia: A critical evaluation of the data and theories. *Schizophrenia Bulletin*, *3*, 373–428.

Picton, T. W., Alain, C., Woods, D. L., John, M. S., Scherg, M., Valdes-Sosa, P., Bosch-Bayard, J., & Trujillo, N. J. (1999). Intracerebral sources of human auditory-evoked potentials. *Audiology & Neurotology*, *4*, 64–79.

Poldrack, R. (2006). Can cognitive processes be inferred from neuroimaging data? *Trends in Cognitive Sciences*, *10*, 59–63.

Rabbitt, P. M. (1966). Errors and error correction in choice-response tasks. *Journal of Experimental Psychology*, *71*, 264–272.

Reiss, J. E., & Hoffman, J. E. (2006). Object substitution masking interferes with semantic processing: Evidence from event-related potentials. *Psychological Science*, *17*, 1015–1020.

Shulman, R. G. (1996). Interview with Robert G. Shulman. *Journal of Cognitive Neuroscience*, *8*, 474–480.

Silvoni, S., Volpato, C., Cavinato, M., Marchetti, M., Priftis, K., Merico, A., Tonin, P., Koutsikos, K., Beverina, F., & Piccione, F. (2009). P300-based brain–computer interface Communication: Evaluation and follow-up in amyotrophic lateral sclerosis. *Frontiers in Neuroscience*, *3*, 60.

Spelke, E. (1985). Preferential-looking methods as tools for the study of cognition in infancy. In G. Gottlieb & N. Krasnegor (Eds.), *Measurement of audition and vision in the first year of postnatal life: A methodological overview* (pp. 323–363). Norwood, NJ: Ablex.

Spencer, K. M., Dien, J., & Donchin, E. (2001). Spatiotemporal analysis of the late ERP responses to deviant stimuli. *Psychophysiology*, *38*, 343–358.

Stapells, D. R. (1989). Auditory brainstem response assessment of infants and children. *Seminars in Hearing*, *10*, 229–251.

Vogel, E. K., Luck, S. J., & Shapiro, K. L. (1998). Electrophysiological evidence for a postperceptual locus of suppression during the attentional blink. *Journal of Experimental Psychology: Human Perception and Performance*, *24*, 1656–1674.

Vogel, E. K., Woodman, G. F., & Luck, S. J. (2005). Pushing around the locus of selection: Evidence for the flexible-selection hypothesis. *Journal of Cognitive Neuroscience*, *17*, 1907–1922.

Woldorff, M. G., Gallen, C. C., Hampson, S. A., Hillyard, S. A., Pantev, C., Sobel, D., & Bloom, F. E. (1993). Modulation of early sensory processing in human auditory cortex during auditory selective attention. *Proceedings of the National Academy of Sciences USA*, *90*, 8722–8726.

Woodman, G. F., & Luck, S. J. (1999). Electrophysiological measurement of rapid shifts of attention during visual search. *Nature*, *400*, 867–869.

Woodman, G. F., & Luck, S. J. (2003a). Dissociations among attention, perception, and awareness during object-substitution masking. *Psychological Science*, *14*, 605–611.

Woodman, G. F., & Luck, S. J. (2003b). Serial deployment of attention during visual search. *Journal of Experimental Psychology: Human Perception and Performance*, *29*, 121–138.

CHAPTER

2 Beyond ERPs: Oscillatory Neuronal Dynamics

Marcel Bastiaansen, Ali Mazaheri, *and* Ole Jensen

Abstract

The event-related potential (ERP) approach has provided a wealth of fine-grained information about the time course and the neural basis of cognitive processing events. However, in the 1980s and 1990s, an increasing number of researchers began to realize that an ERP only represents a certain part of the event-related electroencephalographic (EEG) signal. This chapter focuses on another aspect of event-related EEG activity: oscillatory EEG activity. There exists a meaningful relationship between oscillatory neuronal dynamics, on the one hand, and a wide range of cognitive processes, on the other hand. Given that the analysis of oscillatory dynamics extracts information from the EEG/magnetoencephalographic (EEG/MEG) signal that is largely lost with the traditional time-locked averaging of single trials used in the ERP approach, studying the dynamic oscillatory patterns in the EEG/MEG is at least a useful addition to the traditional ERP approach.

Keywords: ERP, oscillatory EEG activity, EEG oscillations, cognitive processes, oscillatory dynamics

Introduction

The Discovery of Rhythmic EEG Oscillations

Since the initial report on the existence of electrical activity stemming from the brain in 1875 by Richard Caton (Caton, 1875), researchers have been intrigued for more than half a century by the striking ubiquitousness of rhythmic, oscillatory activity in electroencephalographic (EEG) recordings. In the early reports on EEG, a particularly counterintuitive finding was that the observed oscillations seemed to disappear upon sensory stimulation. Adolf Beck, one of the pioneers of EEG research, struggled with the interpretation of this observation, as evidenced in the following passage from his Ph.D. thesis:

> An important phenomenon which occurred in nearly all experiments . . . was the arrest of the spontaneous oscillations of the action current. The explanation of this phenomenon is not too easy. I would interpret it as an expression of the arrest of the active state at a certain point and a suppression of the changes which occurred spontaneously in the active state. In a word, one can explain it by inhibition. It is nothing new to us that the excitation of some centres causes inhibition of the active state of others.
>
> *Beck*, 1891 (Polish thesis; for the English translation, see Beck, 1973)

Some 20 years later, when the recording equipment was sufficiently powerful, part of the riddle was solved through the identification of two types of rhythmic activity in the EEG of the dog (Práwdicz-Neminski, 1913). The relatively high-amplitude, low-frequency oscillations described by Caton and Beck did not fully disappear upon sensory stimulation, but shifted to a different pattern of low-amplitude, high-frequency oscillations. These distinct EEG patterns were initially termed *waves of the first order* and *waves of the second order*. Later, these waves were called *A-waves* and *B-waves*, and today they are known as *alpha waves* and *beta waves*. In 1929 the famous paper by Hans Berger was published (Berger, 1929), which constitutes the first report on human scalp EEG. Berger carefully described the conditions under which the alpha and beta rhythms appear in humans, and noted the inverse relation between the amplitude and frequency of EEG rhythms. He is most famous for describing the phenomenon of *alpha blocking* (also known as the *Berger effect*), not only upon the opening of the eyes, but also upon the execution of a cognitive task (such as performing arithmetic operations).

From EEG to ERPs

With the development, between the 1930s and 1960s of better recording devices, more advanced signal analysis techniques, and more computational power, EEG became an increasingly popular tool to study the brain in action. As a matter of fact, for almost half a century, that is, from the early 1960s to the present, EEG has been by far the most widely used experimental technique to investigate the relationship between cognitive functions and brain activity (although functional magnetic resonance imaging [fMRI] might have taken over this position in the last couple of years). Interestingly, however, EEG researchers have moved from studying oscillatory EEG phenomena to studying event-related potentials (ERPs). In fact, since the early 1960s, the ERP approach has been the dominant approach to studying the relation between EEG and cognition. As far as we can see, the main motivation for this shift in focus has been the insight that a large part of the recorded EEG is not related to the processing of the experimental event in question (a stimulus, a response, or a given cognitive process). Therefore, a procedure is needed to extract the truly event-related EEG (the signal) from this so-called background EEG (the noise). An obvious way of extracting the signal from the noise is to average the EEG across a number of epochs that are time-locked to the experimental event. In practice, this means that one typically repeats a given experimental paradigm a number of times (say, 30 times), and then one averages the 30 EEG recordings that are recorded time-locked to the experimental event. The noise (which is assumed to be randomly distributed across trials) diminishes each time a trial is added to the average, while the signal (which is assumed to be stationary across trials), gradually emerges out of the noise as more trials are added to the average.

Phase-Locked versus Non-Phase-Locked or Evoked versus Induced EEG Responses

The ERP approach has provided a wealth of fine-grained information about the time course and the neural basis of cognitive processing events. The success of this approach is evidenced, for instance, by the book you are now reading. However, in the 1980s and 1990s, an increasing number of researchers began to realize that an ERP only represents *a certain part* of the event-related EEG signal (we will return to this in a minute). Therefore, recent years have seen a renewed interest in another aspect of event-related EEG activity, on which we focus in this chapter. This aspect is related to the event-related fluctuations in rhythmic, oscillatory EEG activity. Indeed, more than 100 years after the initial discovery of EEG oscillations, interest in these phenomena has returned, mainly as a result of the view that they might provide a window on the dynamics of the coupling and uncoupling of functional networks involved in cognitive processing (see, e.g., Singer, 1993, 1999; Varela et al., 2001), as we explain in more detail later.

In studying oscillatory EEG responses, it is important to realize that any change in oscillatory activity that is related to an experimental event is time-locked to this event (otherwise, it wouldn't be event-related) but not necessarily phase-locked to the event.[1] The reason is that oscillations are ongoing phenomena that also exist in the absence of any experimental task. As a result, the phase of the oscillation at the time of occurrence of the event is variable. Therefore, such event-related changes in oscillatory EEG activity are termed *non-phase-locked responses*. Note that, although such an oscillatory response may be meaningfully related to the event in question (e.g., the decrease in amplitude of posterior alpha-band activity when a subject opens his or her eyes—the famous Berger effect), straightforward averaging of a number of trials in which this response occurs (i.e., using the ERP approach) does not work. As such, non-phase-locked responses are not stationary (because the phase of the response varies from trial to trial), and therefore they will be severely reduced (and will fully cancel in the long run) as a result of averaging across trials. The different results of averaging for phase-locked event-related EEG responses (i.e., ERPs) and non-phase-locked responses are illustrated in Figure 2.1.

We would like to note in passing that because event-related oscillatory responses are essentially modulations of ongoing activity, they are often termed *induced responses*. In contrast, the transient

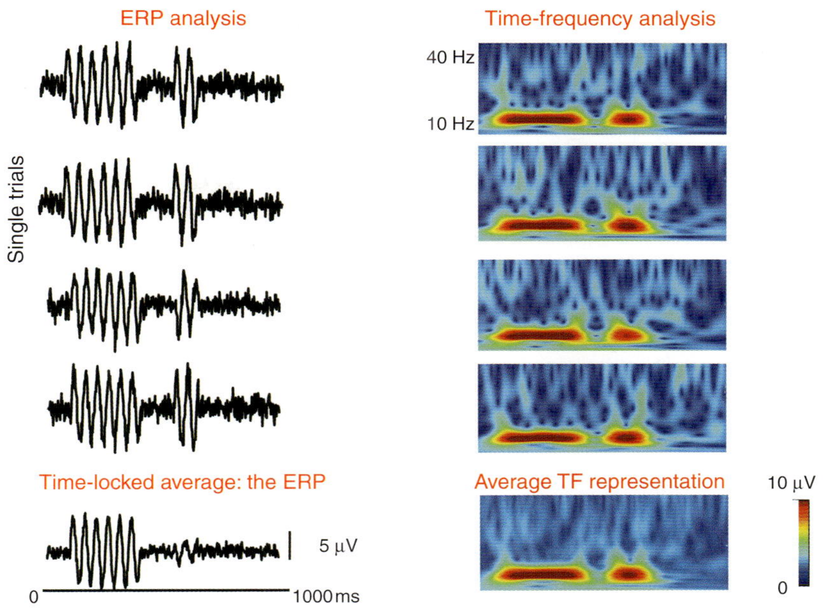

Fig. 2.1. Simulated EEG data illustrating the difference between phase-locked (evoked) activity and non-phase-locked (induced) activity. The left-hand side of the figure presents single-trial EEG time courses that show two consecutive event-related responses (an amplitude increase at 10 Hz). The first response is phase-locked with respect to the reference time point ($t = 0$), and as a result, this evoked response is adequately represented in the average ERP. The second response is time-locked, but not phase-locked to $t = 0$, and as a result, this induced response is largely lost in the average ERP. The right-hand side of the figure shows time-frequency (TF) representations of each single trial, with red colors coding for the amplitude increase at 10 Hz. Crucially, the average TF representation contains both the phase-locked and the non-phase-locked responses.

time- and phase-locked activity that is represented by the ERP is often termed *evoked activity*, because this activity is initially nonexistent and is only driven (evoked) by the experimental event. However, in the remainder of this chapter, we will use the more neutrally descriptive (and signal analysis–oriented) terms *phase-locked activity* and *non-phase-locked activity*.

Oscillations, Synchronization, and Functional Networks

The distinction between phase-locked and non-phase-locked responses, or between evoked and induced EEG, for that matter, is useful in the context of signal analysis. That is, for non-phase-locked EEG responses, one needs to use different analysis tools (see the section on power and coherence changes below) than for phase-locked responses, where simple averaging across trials is sufficient. However, this raises the question of whether there is also a more fundamental difference between the two types of responses.

One thing that has become very clear on the basis of positron emission tomography (PET) and fMRI studies is that a one-to-one mapping between a brain area and a specific component of a cognitive function is very often far too simplistic. Imaging studies often report activations of the same area during different tasks or cognitive processes. This indicates that individual cortical areas can be recruited dynamically in more than one functional network (Mesulam, 1998). This raises the question of how, for a given function (e.g., visual perception), the dynamic recruitment of the participating cortical and subcortical areas takes place. A related question is how different types of information (e.g., information about the color, shape, and movement of an object), which are stored in different parts of the network, are integrated in order to form a coherent neural representation of the object.

An answer to both of these questions may be found in the patterns of synchronization and desynchronization of neuronal activity. Over the last 15 years, evidence has accumulated that (de)synchronization is related to the coupling and uncoupling of functional networks in the brain (see, e.g., Pfurtscheller & Berghold, 1989; Pfurtscheller & Lopes da Silva, 1999; Pfurtscheller & Neuper, 1997; Singer, 1993, 1999; Varela et al., 2001, and many more). The idea is that synchronous, repetitive firing of neurons facilitates the activation of functional networks because it increases the probability that neurons entrain one another in synchronous firing (e.g., Konig & Schillen, 1991). In addition, elements pertaining to one and the same functional network are identifiable as such by the fact that they fire synchronously at a given frequency. This frequency specificity allows one and the same neuron (or neuronal pool) to participate at different times in different representations. Hence, synchronous oscillations in a wide range of frequencies are considered to play a crucial role in linking areas that are part of the same functional network. Importantly, in addition to recruiting all the relevant network elements, oscillatory neuronal synchrony serves to bind together the information represented in the different elements, as was elegantly demonstrated in a seminal paper by Gray and colleagues (1989).

As there is always a substantial level of synchrony in neuronal firing, oscillatory activity dominates raw EEG recordings. Experimental events will occur at random phases of the ongoing oscillations and will modulate these oscillations, thus resulting in non-phase-locked responses. To cut a long story short, in contrast to phase-locked responses (ERPs), non-phase-locked responses predominantly reflect the extent to which the underlying neuronal activity synchronizes. Since, as we have argued, synchronization and desynchronization are indicative of the coupling and uncoupling, respectively, of functional networks, it follows that *event-related, non-phase-locked oscillatory EEG responses provide us with a window on the functional network dynamics in the brain.*

Power and Coherence as Measures of Functional Network Dynamics

As noted, the ERP approach is not adequate for extracting non-phase-locked responses from the raw EEG. Therefore, methods other than traditional ERP analysis are needed to optimally capture (de) synchronization phenomena. The question is, then, which analysis procedures would be adequate for studying changes in oscillatory synchrony?

In this context, it is sensible to realize that the neocortex is organized such that at any place in the cortex, there are very dense connections to adjacent cortical areas. This facilitates neuronal synchrony at a local scale. In addition, there are long-range connections, albeit much sparser, to more distant cortical areas, which enable neuronal synchrony across larger distances. It therefore seems reasonable to assume that distributed functional networks consist of a number of local "nodes" that operate in synchrony with each other through long-range synchronization. As a result, it makes sense also at the level of signal analysis to try to distinguish between local synchrony, that is, synchronization *within* a node of a functional network, on the one hand, and long-range synchrony, that is, synchronization *between* different nodes of a network. Let us see how that works out.

Local synchronous activation of a large number of neurons will, by virtue of the spatial summation of postsynaptic potentials, result in an increase in amplitude of the resulting field potential at the recording site. It follows that an increase in local synchronous oscillatory firing of a neuronal population will lead to an increase in amplitude of the scalp-recorded EEG oscillations. Such amplitude (or power, i.e., squared amplitude) changes should be studied at the single-trial level (although the average amplitude, or power change, can subsequently be computed across trials; see Figure 2.1). Different methods for studying event-related power changes have been reported in the literature. Widely used are the more traditional event-related band power analyses like event-related desynchronization (ERD; Pfurtscheller & Aranibar, 1979; Pfurtscheller & Lopes da Silva, 1999) or induced band power (IBP; Klimesch et al., 1998), the more fashionable single-trial wavelet analysis (Tallon-Baudry et al., 1998), or the more recently proposed and more sophisticated multitaper analysis (Mitra & Pesaran, 1999).

It should be noted here that the mere presence of power in a given frequency band does not necessarily imply the presence of an oscillation at that frequency. Transients in the EEG signals, such as ERPs, muscle artifacts, and eye-movement artifacts will yield power in specific frequency bands without the presence of any oscillations. For instance, performing a time-frequency analysis on an averaged ERP (say, a succession of N1/P2/N400 components) will yield power in the alpha (for the N1/P2 complex) and theta (for the N400) frequency ranges, but these ERPs have nothing to do with oscillations. In general, a good indication of the presence of oscillatory activity in a time-frequency decomposition is the presence of narrowband power increases that last for at least one or two (for low frequencies) or more (for high frequencies) periods of the oscillation (e.g., 400 ms for a 5 Hz oscillation, or 100–200 ms for a 40 Hz oscillation). Brief broadband responses (e.g., a 40–100 Hz increase in power that lasts only for a few tens of milliseconds) are usually indicative of the presence of muscle artifacts, stimulus-related artifacts, or other nonbrain transients. Moreover, regardless of the exact time-frequency method used, one should realize that the time resolution of a time-frequency decomposition is inherently poor, that is, on the order of one to two periods of the slowest oscillation under consideration (cf. Knosche & Bastiaansen, 2002, for more details).

Activity in distant neuronal populations is, to a large extent, not picked up by the same electrode, but by different electrodes. Therefore, the above-mentioned principle of spatial summation, and hence power increase as an indication of local synchrony, does not apply. Here the relevant information lies in the phase relationship between the field potentials recorded from or above different areas. An increase in synchrony will logically lead to increased stability of the phase difference of two oscillatory field potentials, which is typically termed *coherence* (see, e.g., Varela et al., 2001, for extensive arguments). This can be quantified using methods such as event-related coherence analysis (Andrew & Pfurtscheller, 1996; Nunez et al., 1997) or phase-locking statistics (Lachaux et al., 1999). In sum, the power of a given frequency band can be used to assess synchronization changes in local neuronal ensembles (i.e., within nodes of the functional network), while the coherence between electrode sites in a given frequency band yields information about synchronization changes between two or more such local ensembles (i.e., between different nodes).

However, although power and coherence are by far the most widely studied aspects of event-related modulations of oscillatory activity, it is important to realize that there are other aspects (and related measures) of such modulations that may be equally meaningful in better understanding the neural basis of cognitive functions. An excellent overview of potentially relevant phenomena was given by Makeig and colleagues (2004). For instance, there have been several reports of event-related phase resetting (the phase realignment, across trials, of oscillatory activity with respect to an experimental event; see, e.g., Braeutigam et al., 2001; Rizzuto et al., 2003). Interestingly, because such phase resetting will by definition result in an event-related, phase-locked EEG response (our familiar ERP), it has been proposed (Makeig et al., 2002) that such a phase-resetting mechanism might constitute the basis for many (if not all) ERP-like phenomena. In the next section we discuss this idea in more detail, and we present experimental evidence that speaks against this idea.

Are ERPs Really Just Oscillations? Phase Resetting versus the Additive Model

Over the past few years, the role of the EEG in cerebral processing has been extensively reconsidered (Makeig et al., 2004; Mazaheri & Jensen, 2006), and ongoing cerebral activity can no longer be thought of as just relatively random background noise that must be removed in order to see the event-related responses. However, to date, researchers studying electrophysiological brain activity have concentrated either on oscillatory dynamics or on additive effects. Moreover, the relationship between ERPs and oscillatory brain activity has remained elusive. There are currently two different theories that attempt to account for the relationship between stimulus-evoked components and the ongoing EEG activity.

One view, often referred to as the *additive model*, proposes that the ERPs and EEG are different neural events that are either in the strictest sense completely independent of one another or in milder version of this view, are potentially modulated by one another. According to the additive model of ERP generation, the stimulus "evokes" an additive, phase-locked response in each trial, and by averaging, which removes the spontaneous oscillatory activity that is not phase-locked to the onset of the stimulus, this leaves behind the (time- and phase-locked) evoked components. Because the ERP should be seen in the frequency domain as a transient change in amplitude, the additive model is also often referred to as the *amplitude-modulation theory* (Makinen et al., 2005; Mazaheri & Jensen, 2006; Mazaheri & Picton, 2005; Shah et al., 2004).

An opposing view, referred to as the *phase-resetting view*, which of late has started to gain some prominence, claims that the EEG and ERP are the same neuronal event. In the phase-resetting view, the ERP is generated because of stimulus-evoked phase perturbations in the ongoing EEG. According to the phase-resetting view, upon the onset of a stimulus, the phases of the ongoing background oscillations become aligned (phase-reset or partially phase-reset) to the stimulus (Gruber & Muller, 2005; Klimesch et al., 2004; Makeig et al., 2002; Penny et al., 2002). By averaging the stimulus-locked trials, the phase-locked oscillatory activity emerges as the evoked component in the average. Figure 2.2 shows schematically how this works. Since alpha oscillations are the predominant oscillatory activity in the EEG/MEG, it is believed that the phase resetting of these oscillations is particularly relevant for producing the ERP (Gruber et al., 2005; Klimesch et al., 2004;

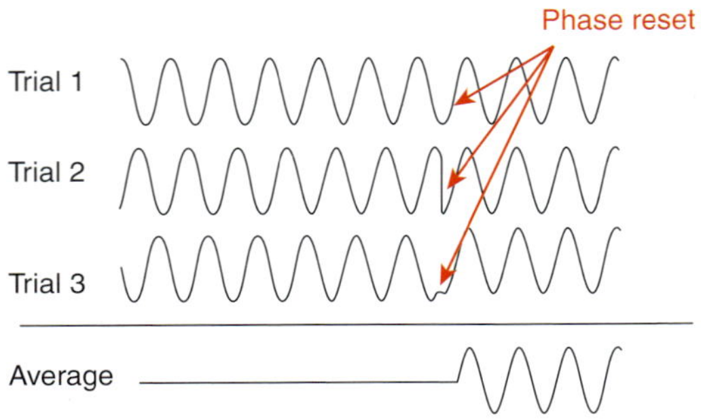

Fig. 2.2. Simulated data illustrating the principle of phase resetting. Three single trials are shown whose phases are not aligned initially. Red arrows indicate the point in time at which an event-induced phase reset occurs. The bottom trace shows what the average ERP would look like if a sufficient number of such trials (in practice >30 trials) are averaged.

Makeig et al., 2002). We will now critically discuss the validity of arguments for and against phase resetting as a mechanism for ERP generation.

A fundamental feature of the phase-resetting hypothesis is that following the presentation of a stimulus, the phases of ongoing EEG rhythms are shifted to lock to the stimulus. From this, it follows that during prestimulus intervals, the distribution of the phase at each EEG frequency would be random, whereas upon stimulus presentation, the phases would be set (or reset) to specific values (for each frequency). The resetting of the phases causes an ERP waveform to appear in the average (Makeig et al., 2002; Penny et al., 2002). The most commonly cited evidence for the phase-resetting hypothesis is the fact that the trial-to-trial phase coherence increases after the onset of the stimulus, suggesting that the phases of the oscillations were perturbed by the stimulus (Gruber et al., 2005; Klimesch et al., 2004; Makeig et al., 2002; Penny et al., 2002). However, as has recently been demonstrated by many studies, an increase in trial-to-trial phase coherence can in no way disambiguate between the additive theory and the phase resetting hypothesis (Mazaheri & Picton, 2005 Yeung, 2004; Yeung et al., 2007)

This is because adding an ERP waveform (with set phases for each of its component frequencies) to a randomly phased EEG causes the recorded phases to move toward the phases of the ERP waveform. The addition of a signal to an unchanging background can therefore look much the same as a phase locking of the background activity. Thus, phase synchronization during the ERP is clearly not proof that the ERP is generated by phase resetting of the EEG (see Figure 2.3)

Other support for phase resetting comes from studies reporting no net power change in the EEG at the time of the evoked response (accompanied by an increase in phase synchronization across trials). However, this too unfortunately cannot disambiguate between the two theories, since a decrease in the amplitude of the EEG (particularly in the alpha band) at the same time as the evoked potential could

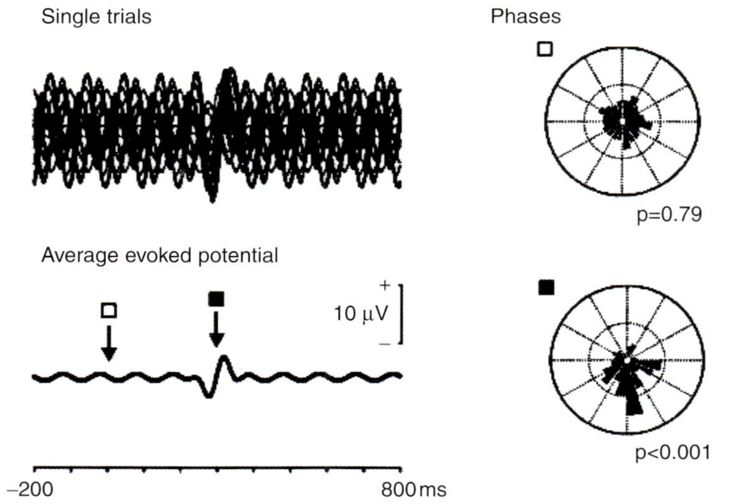

Fig. 2.3. The upper left side of the figure shows 20 superimposed trials of a model evoked potential consisting of a single cycle of activity added to ongoing activity of the same frequency with variable phase and amplitude. Below is the average evoked potential over 100 trials. At the upper right are the polar plots showing the phase distributions and the frequency of the evoked potential (and background activity) during baseline and at the middle of the evoked potential. Adapted with permission from Mazaheri and Picton (2005).

mask out any transient evoked increases (Hanslmayr et al., 2007). In sum, arguments in favor of the phase-resetting hypothesis have remained unconvincing so far.

Now we turn to the arguments favoring the additive model of ERP generation. Intracortical animal recordings have shown that the visual-evoked ERP can occur when there is little background EEG activity (Shah et al., 2004). This is contradictory to the phase-resetting model since one key requirement of the phase modulation theory is the existence of ongoing EEG rhythms whose phases can be modulated. However, although arguably unlikely, it is possible that the visual-evoked ERP of these recordings might have triggered phase locking of EEG rhythms that were not recorded in the multielectrodes but that might have been visible in the scalp recording. To make matters more complicated, human intracortical recordings of 7–16 Hz rhythms during cognitive processing show clear phase resetting with no consistent increase in amplitude (Rizzuto et al., 2003).

Using a modeling approach, Makinen et al. (2005) suggested that the amplitude variance across trials could be used as a parameter for distinguishing between the two theories of ERP generation. The authors argued that if the oscillations in the EEG are reset as a result of a stimulus, then a drop in the variance of the amplitude should be seen at the time of the evoked response. They not only found that the amplitude of the event-related responses was independent of the amplitude of prestimulus oscillatory activity, but also that the amplitude variance increased across trials (Makinen et al., 2005). Klimesch et al. (2006a) were able to counter the claim of amplitude variance serving as an indicator of a phase reset by integrating event-related decreases in alpha amplitude in the model and finding that the variance was indeed unaffected in a phase-reset (Klimesch et al., 2006a). Thus, amplitude variance of the oscillatory activity in the EEG cannot be used to argue against either of the models.

Using a different approach, Mazaheri and Jensen (2006) developed a phase-preservation index (PPI; see Figure 2.4) that estimates the stability of the instantaneous poststimulus phase compared with the prestimulus phase (Mazaheri & Jensen, 2006). The PPI is a variant of the conventional phase-locking measures in that instead of looking at instantaneous phase locking across trials, it looks at the uniformity of difference between a reference phase and an instantaneous phase. The authors reasoned that to fully disambiguate the two theories, it is crucial to examine the phase of the ongoing oscillations

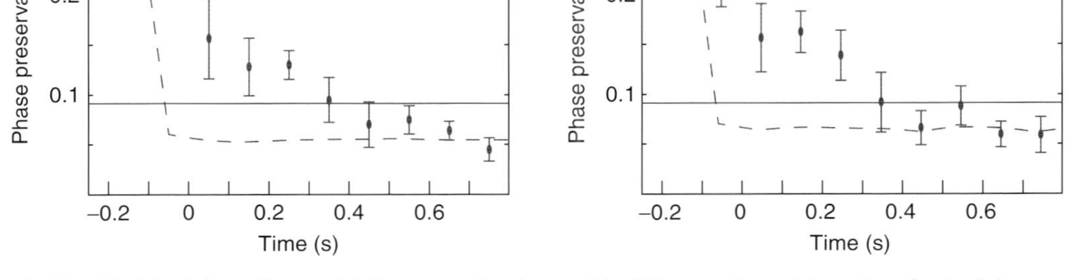

Fig. 2.4. The PPI of the alpha oscillations. (A) The temporal evolution of the PPI averaged over eight subjects for the alpha frequency identified in individual subjects. Error bars indicate the SEM. The reference phase was determined at –0.25 s. The PPI decays slowly, showing that the poststimulus phases are preserved with respect to the prestimulus phase up to approximately 0.3 s poststimulus. The PPI values above the red line are considered statistically significant yes, p should be lowercase ($P < .01$;). The dashed line indicates the PPI for trials shuffled in time (temporally uncorrelated). (B) The PPI for the unstimulated trials. The PPI values between the stimulated and unstimulated trials were not significantly different across time. (*t* test, $p < .05$). Reproduced with permission from Mazaheri, A., & Jensen, O. (2006). Posterior alpha activity is not phase-reset by visual stimuli. *Proceedings of the National Academy of Sciences USA, 103*(8), 2948–2952. Copyright (2006) National Academy of Sciences, U.S.A.

before the stimulus and see if it is preserved *after* the event-related response. If the phase of the oscillatory activity was perturbed by the stimulus, than there should be no poststimulus–prestimulus phase relationship. The authors found that ongoing alpha oscillations (8–13 Hz) after visual stimuli preserved their phase relationship to the oscillations prior to the stimuli, indicating that no phase resetting had taken place. However, it must be noted that even though the PPI is able to assess the stability of phase across time, a breakdown in the PPI (strong evidence *against* phase resetting) cannot be used as evidence for phase resetting. This is because, as with the conventional phase-locking measures, an additive signal to the ongoing oscillations can bias the measurement of the instantaneous phase and cause it to become uncoupled from the reference phase.

If the evoked potentials are phase-reset EEG oscillations, one would then expect a strong relationship between the amplitude of the prestimulus oscillations and the amplitude of the evoked potential. Using this premise, Becker and colleagues (2008) found that high alpha amplitudes were not able to change the amplitude of visual ERPs and in fact decreased the amount of phase locking across trials.

Thus, it can safely be said that to date, no method has been able to find unequivocal evidence for either of the two theories of ERP generation. Makeig and his colleagues (2004) have proposed that the full brain response can be mapped onto an "event-related brain dynamic state" with the dimensions of frequency, power, and phase synchronization. This would mean that any cognitive event can both activate a specific pattern of response and reset the phase.

One idea that has been put forward is that the EEG and evoked responses are generated by overlapping generators (Mazaheri & Picton, 2005; Shah et al., 2004). Termed the *shared generator hypothesis*, the idea is that EEG rhythms share neuronal generators such that neurons in a particular region of cortex are involved both in the generation of EEG rhythms and in stimulus-evoked ERPs. A stimulus may thus cause several changes. First, the stimulus may activate cells that are either relatively quiescent in a precisely time-locked way (yielding a phase-locked response) or variably time-locked (producing a non-phase-locked response). Second, the stimulus may cause neurons that are already active in producing certain ongoing oscillations to entrain themselves to the stimulus (phase resetting). This entrainment may result in a power decrease in the frequency of that oscillation and a subsequent power increase in lower or higher frequencies at the time of the phase entrainment.

After all of this discussion, one might be tempted to ask whether trying to establishing a relationship between ERPs and oscillatory dynamics is of any use. One reason for trying to establish such a relationship is that neuronal activity measured using electrophysiology exhibits spontaneous variability on time scales of hundreds of milliseconds to seconds, suggesting that the functional state of cortical networks is continuously changing. Yet, the ERP alone in principle cannot directly provide any insight into this variability since it reflects the brain's phase-locked response *after* the stimulus. A link between ERP and EEG can provide a unified temporal account (prestimulus and poststimulus) of how information is processed by the brain. Finally, other imaging modalities can benefit from further insight into this relationship of ERPs. Event-related potentials can provide a critical link between the hemodynamic response indexed by fMRI and the temporal dynamics of underlying neuronal activity (Murray et al., 2002). If ERPs are just a by-product of phase perpetuation of the EEG, this link would have to be expanded and thoroughly revised. It has been suggested that blood oxygen-level dependent (BOLD) responses are likely related to single-trial EEG amplitudes but not to the phase concentration across trials. In order to gain a deeper understanding of the relation between event-related EEG and fMRI, separation of the amplitude and phase contributions would be a vital and crucial endeavor (Fell, 2007).

A Selective Literature Review of Oscillatory Neuronal Dynamics in Human Cognition

Since the early 1990s, there has been an explosion of research efforts directed at the relationship between oscillatory neuronal dynamics, on the one hand, and cognitive processes, on the other hand. Research has been performed with a variety of experimental tools and a wide range of experimental paradigms ranging, to give a few examples, from attentional modulations in spike-field coherence estimates in single-cell recordings of monkey visual cortex (e.g., Fries et al., 2001), to subdural recordings of oscillatory EEG activity during spatial navigation tasks in epileptic patients (e.g., Kahana et al., 1999), to scalp EEG measures of power and coherence during the processing of semantic errors in sentences in healthy adults (Hagoort et al., 2004). Clearly, the amount of work that has been published is far too large to be reviewed in this chapter (see, e.g., Klimesch & Neuper, 2006, for a much broader sample of research activities). The literature review presented here is therefore necessarily limited. In the remainder of this chapter, we therefore selectively discuss some of the work that has explored the relationship between oscillatory responses in the scalp EEG or MEG of healthy human adults and cognitive processes such as memory encoding/retrieval, working memory, and language comprehension, with an emphasis on our own work.

Working Memory and Long-Term Memory

OSCILLATORY DYNAMICS IN THE GAMMA BAND REFLECT WORKING MEMORY MAINTENANCE

The primary role of the working memory system is to maintain a given stimulus representation after the stimulus is no longer present. For visual stimuli, it has been proposed that this is achieved by the visuospatial sketchpad, which serves to maintain the actual memory trace (Baddeley, 1992). Recent findings in humans point to memory representations being maintained by sustained oscillatory activity in the gamma band using both MEG and EEG. This has among others been demonstrated in the visual and auditory domains (Kaiser et al., 2003; Tallon-Baudry et al., 1998). In both of these working memory paradigms, sustained gamma activity was observed in comparison to a control task without a memory component. In these studies, attention demands are a confound since the working memory task is more demanding than the control task. To circumvent this problem, recent studies have aimed at selectively engaging different posterior areas in terms of stimulus specificity. A recent study (see Figure 2.5) relied on subjects maintaining either the identity or the orientation of presented faces (Jokisch & Jensen, 2007). The aim was to engage, respectively, the ventral stream and the dorsal stream. When subjects were maintaining the face orientation, induced gamma activity was observed over posterior midline areas. When subjects were maintaining the face identity, posterior midline gamma activity was not observed. Importantly, the face identity condition was the more demanding of the two conditions. This means that the sustained gamma activity is specific to the working memory conditions with the spatial component (face orientation) and is not explained by general increases in task demands. Source modeling revealed that the posterior gamma activity was produced in occipital areas.

These findings are consistent with a recent MEG study on working memory of saccades (Medendorp et al., 2006). Subjects were cued to prepare a saccade during a 2 s interval either to the left or the right. Gamma activity was observed over the hemisphere contralateral to the saccade direction during the retention period. In summary, these findings suggest that sustained gamma activity produced in the visual system is partly responsible for carrying the working memory traces during maintenance.

INHIBITION OF VISUAL INPUTS DURING WORKING MEMORY MAINTENANCE

Successful working memory maintenance relies on the ability to retain the relevant representations. Failure to do so might be due to interference by other incoming stimuli. Thus, the working memory system might benefit from inhibiting irrelevant sensory input during retention. Recent findings suggest that posterior alpha activity might serve this purpose. Posterior alpha activity has traditionally been interpreted as an idling rhythm (Pfurtscheller et al., 1996). This is motivated by the finding that alpha activity emerges strongly when subjects are resting and have their eyes closed. The idling hypothesis has been challenged by findings demonstrating an increase in alpha power during working memory maintenance (Klimesch et al., 1999; Krause et al., 1996). In particular, the alpha activity was shown to increase parametrically with the working memory load (Jensen et al., 2002). The sources accounting for the parametric increase were localized to the parieto-occipital sulcus, an area typically not associated with working memory maintenance (Tuladhar et al., 2007). On the basis of these findings, it has been proposed that posterior alpha activity serves to inhibit incoming visual stimuli that might interfere with memory maintenance (Jensen et al., 2002; Klimesch et al.,). This interpretation is supported by the study of Jensen and Jokisch (2007), mentioned above, in which subjects had to maintain the identity or orientation of presented faces. The alpha activity over dorsal areas was strongest during the face identify condition. In support of the inhibition hypothesis, this shows that when the ventral stream is engaged during the retention period, alpha activity in the dorsal stream is increased. This case is further supported by the task on delayed saccades also mentioned above (Medendorp et al., 2006). Alpha activity in the hemisphere contralateral to the saccade was dramatically stronger compared with the ipsilateral hemisphere. The inhibition hypothesis has gained indirect support from studies combining EEG and fMRI methods demonstrating a negative correlation between the BOLD signal and alpha

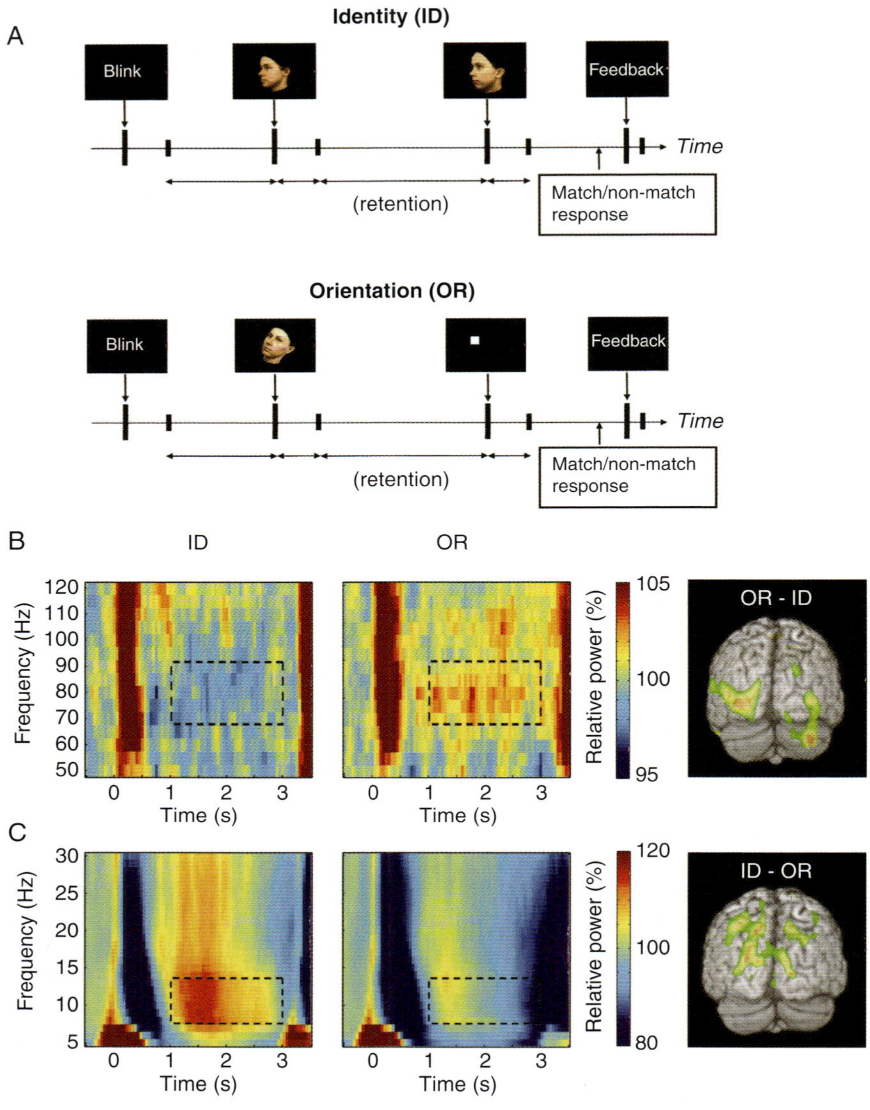

Fig. 2.5. A working memory task showing sustained alpha and gamma activity. (A) Subjects were requested to maintain either the identity (ID) or orientation (OR) of a presented face. After a 3 s retention period, a sample item was presented and the subject had to indicate a match or nonmatch by means of a button press. (B) Only in the OR condition, which engaged the dorsal stream, was sustained gamma activity (60–90 s) present centrally during the retention interval. Sources accounting for the gamma activity were identified within occipital cortex. (C) Only in the ID condition, which engaged the ventral stream, was sustained alpha activity present centrally during the retention interval. Sources accounting for the alpha activity were identified around the parieto-occipital sulcus. Reproduced with permission from Jokisch and Jensen (2007).

activity produced around the parieto-occipital sulcus (Goldman et al., 2002; Laufs et al., 2003). Importantly, this negative correlation was found to increase with memory load (Scheeringa et al., in preparation). It should be mentioned that oscillatory activity in the alpha band has been proposed to play an active role in cognitive processing as well (Palva & Palva, 2007). This notion is based mainly on studies focusing on the phase relationship between different alpha generators recorded in subjects performing tasks engaging the working memory system (Palva et al., 2005). In conclusion, the power of posterior alpha activity has been proposed to reflect functional inhibition of visual input during working memory maintenance. The main argument is based on findings demonstrating that alpha activity increases in areas not engaged in a given task. This functional inhibition serves to suppress interfering input, thus devoting resources to areas engaged in the actual working memory operations.

OSCILLATORY GAMMA ACTIVITY REFLECTS SUCCESSFUL LONG-TERM MEMORY OPERATIONS

In contrast to working memory, long-term memory representations remain stable over longer time periods. There is a strong consensus that synaptic plasticity is responsible for encoding of long-term memories. Long-term potentiation is considered the physiological mechanism responsible for synaptic plasticity. Experimentally, it has been demonstrated that the induction of long-term potentiation is modulated by the phase of ongoing gamma oscillations (Wespatat et al., 2004). This provides a direct link between gamma oscillations and long-term memory formation. Extensive theoretical work has explored how memory representations can be encoded synaptically in network models of mutually coupled neurons. What these models have in common is that synaptic weights can only be decoded by actually reactivating the initial representations in the network. This activation might be reflected by oscillatory activity in the gamma band. In sum, these considerations motivate the study of oscillatory gamma activity during long-term memory operations. Several EEG, intracranial EEG, and MEG studies have demonstrated that induced gamma activity reflects successful long-term memory operations (Gruber et al., 2004; Osipova et al., 2006; Sederberg et al., 2006). In the study by Osipova et al., 240 pictures of landscapes and buildings were shown during the encoding session. In a subsequent recognition session, these pictures mixed with 240 new pictures were shown. Strong gamma activity produced in occipital areas was observed during the encoding session. Importantly, the gamma activity predicted the encoding of items that later were correctly recalled. During the recall session, gamma activity produced in the same areas correlated with successful long-term memory retrieval. In conclusion, gamma activity reflects successful long-term memory encoding. This might be explained by a stronger feedforward drive due to synchronization in the gamma band promoting synaptic plasticity. The gamma activity during long-term memory recall could reflect a recall process in which the initially encoded memory representations are retrieved.

In conclusion, the findings point to oscillatory brain activity in various frequency bands playing an important role in human memory representations. Activity in the gamma band is likely to reflect neuronal activation of the actual representations. Thus, the sustained gamma activity could reflect maintenance of neuronal representations during working memory and/or the activation of the memory representations during recall. Gamma activity during long-term memory encoding is likely to promote synaptic plasticity. By contrast, posterior alpha activity seems to reflect functional inhibition of brain areas not engaged in a given memory task. Future work manipulating oscillatory activity by pharmacological means or transcranial magnetic stimulation is required to further establish the direct causal role of oscillatory brain responses in human memory.

Language Comprehension

In this section, we concentrate on the rapid dynamics of the neural processes underlying language comprehension. However, before turning to this subject, we will briefly delineate at a very general level what is thought to be the cognitive architecture of language comprehension.

It is generally agreed that during language comprehension, incoming sounds or orthographic patterns trigger a cascade of memory retrieval operations that make available the phonological, syntactic, and semantic properties of individual words. Once available, these different ingredients have to be integrated (unified) at the sentence and/or discourse levels into a meaningful whole in order to yield a coherent interpretation of the linguistic input (see Hagoort, 2005, for a more detailed elaboration of this framework). Thus, two different cognitive processes, namely memory retrieval operations and unification operations, play a crucial role during language comprehension.

Power and coherence changes in oscillatory neuronal responses during language comprehension have been observed in four different frequency bands: theta (4–7 Hz), alpha (8–12 Hz), lower beta (13–18 Hz), and gamma (above 30 Hz). In (relatively loose) agreement with the cognitive architecture of language comprehension, the observed effects can roughly be subdivided into effects related to memory retrieval operations during language comprehension and effects related to unification of linguistic information.

OSCILLATORY NEURONAL DYNAMICS RELATED TO THE RETRIEVAL OF LEXICAL INFORMATION

A number of studies suggest that oscillatory neuronal dynamics in the theta frequency range are involved in the retrieval of lexical-semantic information. In an initial study (Bastiaansen et al., 2005), we examined EEG power changes in a range of frequencies from 1 to 30 Hz while subjects read a short story. Power changes were averaged selectively for open-class (OC) words (e.g., nouns, verbs, and adjectives), which carry most of the semantic information in a sentence, and for closed-class (CC) words (e.g., articles, determiners, and prepositions), which carry much less semantic information but rather serve as "syntactic glue" at the sentence level. As Figure 2.6 shows, OC and CC words both elicited a power increase in the theta frequency range, together with decreases in the alpha and beta frequency ranges, roughly within 100–600 ms after word onset. The OC words elicited stronger power changes in general. Interestingly, however, while the scalp topography of the alpha and beta responses was qualitatively similar, in the theta frequency range we observed a qualitative difference between OC and CC words. Although both types elicited a theta power increase over left occipital and midfrontal areas, the OC words additionally elicited a power increase over left temporal areas that was not observed for the CC words. A regression analysis showed that this effect was not dependent upon word length or word frequency. This topographical pattern connects well to existing hemodynamic data. The left occipital power increase may be related to complex visual processing either in Broca's area 18/19 (see, e.g., Indefrey et al., 1997; Petersen et al., 1988) or in the fusiform gyrus (the putative visual word form area; cf. Cohen et al., 2000; McCandliss et al., 2003). Most interestingly, however, with respect to the differential (OC-specific) theta response over the left temporal cortex, it has been shown that left temporal areas (more precisely, the left posterior superior and/or middle temporal gyrus) are involved in lexical retrieval (see, e.g., Indefrey, 2004; Indefrey & Cutler, 2005). Therefore, the qualitative difference between OC and CC words led us to hypothesize that theta-band synchronization of neuronal activity is related to lexical-semantic retrieval.

A subsequent experiment (Bastiaansen et al., 2008 in press) aimed at further testing this hypothesis. Subjects performed a lexical decision task in which they had to decide whether or not a visually presented string of letters constituted a real word.

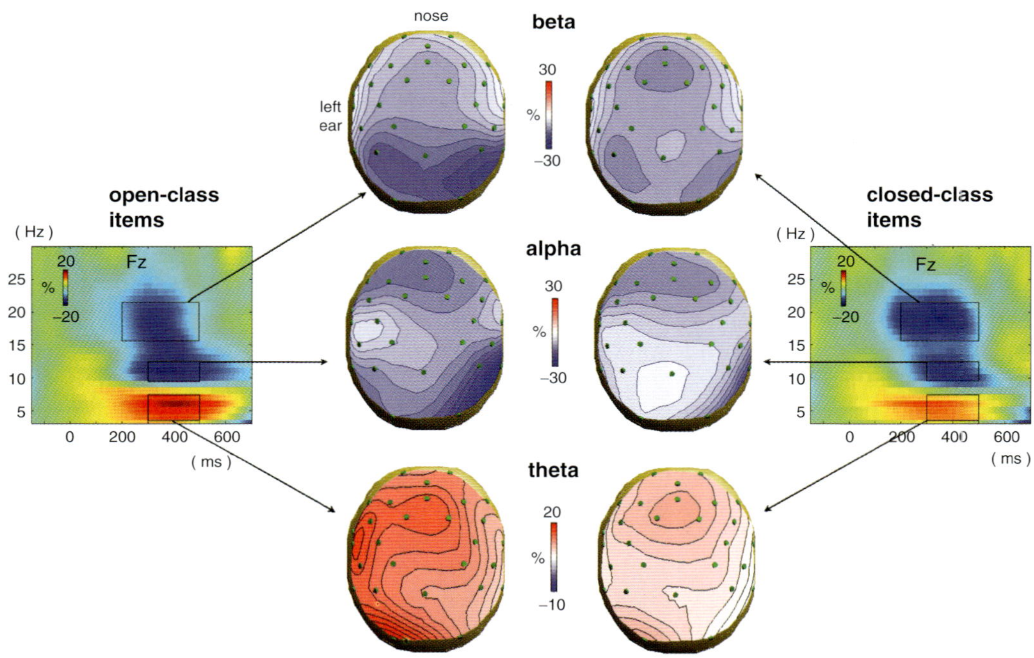

Fig. 2.6 Results of a time-frequency analysis of power changes for open-class words and closed-class words separately. The left- and right-hand parts of the figure show a time-frequency representation of the percentage power change at one frontal channel (Fz). Words are presented at $t = 0$. The middle part of the figure shows the topographical distribution of three time-frequency components, indicated by the black tiles in the time-frequency representations. Note the qualitative difference in scalp topography of the theta power increase.

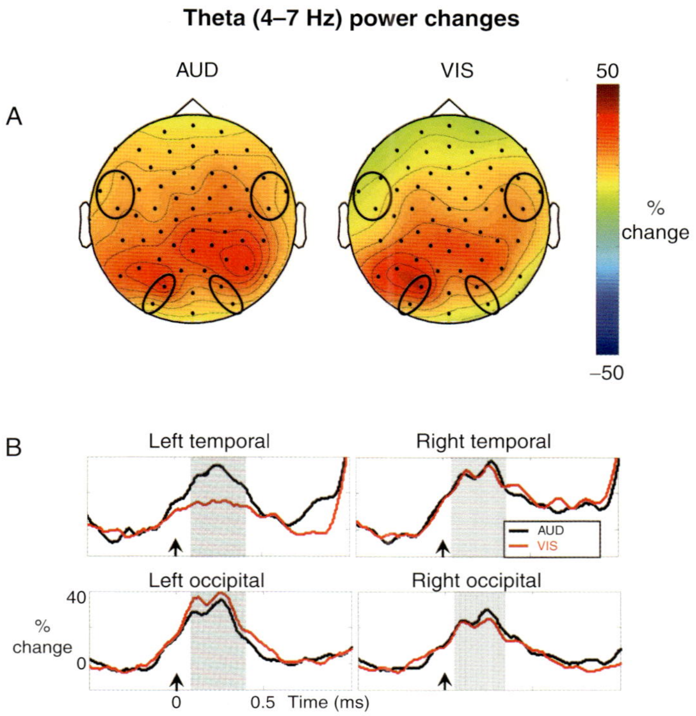

Fig. 2.7. Topographical distribution (A) and region-of-interest time courses (B) of percentage theta power changes induced by visually presented words with auditory semantic properties (AUD) and visual semantic properties (VIS). The regions of interest are indicated by the black ellipses in (A). Shaded areas in (B) indicate the time interval used for the statistical analyses.

Nonwords could be either pseudowords (phonologically legal but nonsensical letter strings) or consonant strings. Real words could either be nouns with visual semantic properties (that is, referring to colors and shapes) or nouns with auditory semantic properties (that is, referring to sounds).

Again, both sets of words elicited an increase in theta power and decreases in the alpha and beta frequency ranges. And again, qualitatively different responses were recorded only in the theta band, where the following double dissociation was found (see Figure 2.7): Words with auditory semantic properties showed larger theta power increases in electrodes overlying the left auditory cortex than in electrodes overlying the left visual cortex, while the opposite pattern was found for words with visual semantic properties. Note that, as both sets of words were presented visually and were matched for length and frequency, they differed only in terms of their semantic properties. We therefore concluded that the results of this study were consistent with the hypothesis that neuronal synchronization in the theta frequency range is involved in the retrieval of lexical-semantic information. In addition, our data are in agreement with the hypothesis that spatially distributed functional networks form the basis of semantic representations and that the topographies of these networks reflect the semantic properties of individual items (for similar hypotheses, see Martin & Chao, 2001; Pulvermueller, 1999, 2001).

While the above studies suggest a strong involvement of theta-band synchronization in retrieving lexical-semantic information, the neuronal dynamics of language comprehension–related retrieval operations are certainly not restricted to the theta frequency band. Other studies have related semantic memory operations to power changes in the alpha frequency band (for review, see Klimesch, 1999). In addition, a few studies have reported effects of memory retrieval in higher frequency bands (beta/gamma; Pulvermueller et al., 1999; Weiss & Mueller, 2003).

OSCILLATORY DYNAMICS RELATED TO UNIFICATION OPERATIONS IN LANGUAGE COMPREHENSION

As sentences extend over relatively long time periods, maintaining the working memory (WM) trace of the linguistic input is a prerequisite to performing unification operations. In addition to taking part in retrieval operations, theta-band changes in neuronal synchrony appear to play a role in such WM processes.

For example, theta power (Bastiaansen et al., 2002a) and coherence (Weiss & Mueller, 2003) linearly increase over the course of correct sentences. In addition, theta power is larger over the frontocentral midline following words constituting syntactic (Bastiaansen et al., 2002b) and semantic (Hagoort et al., 2004; Hald et al., 2006) violations in sentences. These effects were interpreted to be related to the larger demands that they impose on verbal WM (Hald et al., 2006). Also, theta coherence is greater following WM-demanding object-relative versus subject-relative clauses (Weiss et al., 2005). A clear relationship between theta-band oscillations and WM has also been found using intracranial recordings in humans (Raghavachari et al., 2001; Rizzuto et al., 2003).

There is, however, more to unification than just maintaining the input in WM. Unification requires the active manipulation of phonological, syntactic, semantic, and likely also pragmatic information, resulting in a message-level understanding of the linguistic input. So far, experimental research into the oscillatory neuronal dynamics of unification operations has concentrated on semantic and syntactic unification only.

Semantic Unification

Recently, we have addressed the oscillatory correlates of semantic unification in the following way (Hagoort et al., 2004; Hald et al., 2006). Subjects read the following three versions of sentences such as "The Dutch trains are *yellow/white/sour* and very crowded." In fact, Dutch trains are yellow, and therefore the first version of this sentence is the correct one. However, the linguistic meaning aspects of the alternative color term *white* apply just as well to trains as does the predicate *yellow.* It is world knowledge about trains in Holland that make the second version of this sentence false (the world knowledge violation condition). The situation is different for the third version. The core meaning of *sour* is related to taste and food. Under standard interpretation conditions, a predicate requires an argument whose semantic features overlap with those of its predicate. For vehicles such as trains, this is clearly not the case, since semantic features related to taste and food do not apply to the materials that trains are made of. Thus, for semantic-internal reasons, the third sentence is an outright semantic violation condition.

The results are presented in Figure 2.8. In the correct sentence condition, where normal semantic unification takes place, we observed a small increase in gamma power relative to baseline in response to the critical word (e.g., *yellow*). This gamma power increase was much stronger in the world knowledge violation condition, where semantic unification is difficult but not impossible. In the semantic violation condition, however, where semantic unification is impossible for the reasons mentioned above, the gamma power increase was absent.

The parametric relation between semantic unification and gamma power described above suggests that gamma oscillations are functionally related to semantic unification operations. Several other studies support this notion. First, gamma coherence is larger for semantically congruous than for incongruous sentence endings (Weiss et al., 2003). In a different approach, van Berkum and colleagues (2004) studied the effects of referential success, referential ambiguity, and referential failure in a sentence. They found a drop in gamma power following the ambiguous and failing reference conditions, which render semantic unification problematic.

Taken together, the above studies lead us to hypothesize that neuronal synchronization in the gamma frequency range is in some way related to the neuronal implementation of semantic unification.

Syntactic Unification

The neuronal dynamics of syntactic unification have been addressed in a number of different experimental paradigms. For instance, Haarmann and colleagues (2002) used sentences in which EEG coherence was examined in the interval between the object and the main verb of a sentence (the so-called filler-gap interval). Although this filler-gap interval places a relatively large demand on verbal WM, the gap filling that the reader has to perform online can be seen as a syntactic unification operation. The authors found increased coherence over many scalp areas in the sentences containing a filler-gap interval compared to non-filler sentences in the lower beta frequency range (roughly 15–18 Hz). This increased coherence is indicative of an increase in long-range neuronal synchronization during the syntactically more demanding filler sentences. A study by the Weiss group (2005) computed EEG coherence during sentences in which the subject of the main clause was also the subject of the relative clause (so-called subject-relative [SR] sentences). This was contrasted with coherence during sentences in which the subject of the main verb was the object of the relative clause (so-called object-relative [OR] sentences). Subject-relative sentences are syntactically relatively simple and occur frequently in everyday

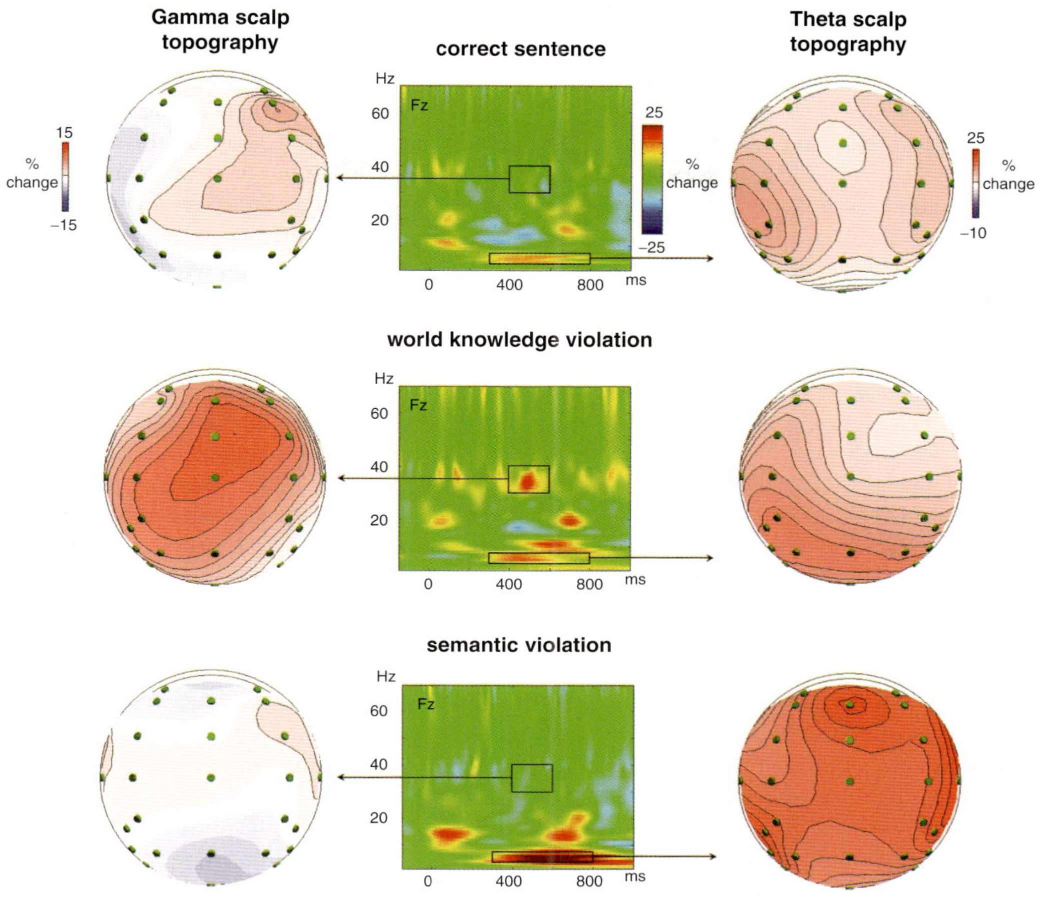

Fig. 2.8. Time-frequency analysis of the power changes elicited by the critical words in correct sentences, sentences containing violations of world knowledge, and sentences containing semantic violations. The middle part shows the time-frequency representations; the left- and right-hand parts show the topographical distribution of gamma and theta power changes, respectively. Note the small but significant increase in gamma power over lateral frontal areas in the correct condition, the strong gamma power increase following a world knowledge violation, and the absence of gamma power increase after a semantic violation.

language, whereas OR sentences are less frequent, syntactically more demanding, and place a larger load on verbal WM. The authors showed that increased theta and gamma coherence was associated with the higher WM load. The effects of syntactic complexity (which become most evident in the time period immediately following the relative clause), however, were expressed in the lower beta frequency range (13–18 Hz). In a similar type of experiment with Dutch subjects and stimuli, we are currently investigating power changes in the MEG between OR and SR sentences. Preliminary results from this experiment suggest that OR sentences are accompanied by a power decrease in the 13–18 Hz frequency range. Similar results were obtained in a recent study in which syntactic violations occurred in sentences, where violations lead to a relative decrease in power in the same frequency range (Bastiaansen et al., submitted).

Overall, the experimental data suggest that the two components of language comprehension, namely, the retrieval of lexical information from the mental lexicon and the subsequent unification of semantic and syntactic information, yield distinct patterns of synchronization in the brain's language network: Retrieval operations are associated with neuronal dynamics in the theta and alpha frequency ranges, whereas unification operations are associated with neuronal synchronization in the beta and gamma frequency ranges. As such, the general pattern of results suggests that at the level of the dynamic neuronal mechanisms by which the brain operates during language comprehension, domain-general processes are operative (i.e., the observed mechanisms are not exclusively engaged by linguistic processing, but are also observed in other cognitive domains). This is in contrast to the representational level, where there is relative domain specificity for language comprehension (i.e., there are brain areas that are partly dedicated to the storage and manipulation of linguistic information, as evidenced by hemodynamic studies).

Thus, it seems that it is possible to capture the dynamics of the brain's language network by a careful analysis of the event-related changes in power and coherence of EEG and MEG data in a wide range of frequencies, in combination with subtle experimental manipulations in a range of language comprehension tasks.

Before ending this chapter with some conclusions, we will address one issue that often leads to misconceptions among researchers who are not familiar with time-frequency analyses of oscillatory EEG/MEG phenomena. This issue relates to the different frequency bands in which experimental effects are reported. One may be tempted to conceive of effects occurring in the same frequency band as being functionally or neurophysiologically related effects. For instance, one may think that theta-band power increases reported in one set of studies (on, say, WM) may be related to theta increases in another set of studies (on, say, error detection). We would like to emphasize here that such inferences are not warranted. Frequency-band similarity (or dissimilarity) between different event-related responses only means that in both cases neurons are synchronizing at the same (or at a different) rate. This may be due to relatively low-level neurophysiological parameters (axonal conductance times, exact connectivity patterns, number of intervening synapses etc.) and, as such, can hardly be viewed as functionally relevant for the cognitive neuroscientist. Having said this, we should also note that in some special cases there are good reasons to relate oscillation frequency to specific anatomical pathways. For instance, alpha-band responses are often linked to thalamo-cortical interactions (see, e.g., Lopes da Silva, 1991; Steriade et al., 1990), whereas theta-band responses are sometimes related to cortico-hippocampal circuits (see, e.g., Bastiaansen & Hagoort, 2003; Miller, 1991).

Conclusion

In our opinion, the above experimental data illustrate that there exists a meaningful relationship between oscillatory neuronal dynamics, on the one hand, and a wide range of cognitive processes, on the other hand. Given that the analysis of oscillatory dynamics extracts information from the EEG/MEG signal that is largely lost with the traditional time-locked averaging of single trials used in the ERP approach, studying the dynamic oscillatory patterns in the EEG/MEG thus proves to be at least a useful addition to the traditional ERP approach.

The analysis methodologies involved in studying oscillatory neuronal dynamics are optimally suited to zoom in on the patterns of synchronization and desynchronization of neuronal activity; thus, they provide the necessary means to empirically address issues related to the coupling and uncoupling of functional networks during cognitive processing. Interestingly, there is a clear parallel between the development and rising popularity of EEG/MEG measures of network dynamics, on the one hand, and modern fMRI data analysis techniques that focus on network dynamics through measures of functional and effective connectivity (such as studying low-frequency synchronization of BOLD responses, diffusion tensor imaging, or structural equation modeling), on the other hand. These parallel developments are particularly exciting because they essentially demonstrate the growing awareness among cognitive neuroscientists that functional network dynamics are at the core of human cognition.

Note

1 The amplitude of an oscillation is, roughly speaking, the size of its (positive or negative) peak deflection relative to some baseline—that is, how big the oscillation is. The phase of an oscillation is, roughly speaking, the slope (or direction) of the signal at a given one point in time, which is equivalent to the left–right shift of the oscillation.

References

Andrew, C., & Pfurtscheller, G. (1996). Event-related coherence as a tool for studying dynamic interaction of brain regions. *Electroencephalography and Clinical Neurophysiology, 98*, 144–148.

Baddeley, A. (1992). Working memory. *Science, 255*(5044), 556–559.

Bastiaansen, M. C. M., & Hagoort, P. (2003). Event-induced theta responses as a window on the dynamics of memory. *Cortex, 39*(4–5), 967–992.

Bastiaansen, M. C. M., Magyari, L., & Hagoort, P. (submitted). Dissociation between syntactic and working memory processes in oscillatory dynamics during online sentence comprehension. Manuscript submitted for publication.

Bastiaansen, M., Magyari, L., & Hagoort, P. (2010). Syntactic unification operations are reflected in oscillatory dynamics during on-line sentence comprehension. *J Cogn Neurosci, 22*(7), 1333–1347.

Bastiaansen, M. C., Oostenveld, R., Jensen, O., & Hagoort, P. (2008). I see what you mean: theta power increases are involved in the retrieval of lexical semantic information. *Brain and Language, 106*(1), 15–28.

Bastiaansen, M. C. M., van Berkum, J. J., & Hagoort, P. (2002a). Event-related theta power increases in the human EEG during online sentence processing. *Neuroscience Letters, 323*(1), 13–16.

Bastiaansen, M. C. M., van Berkum, J. J., & Hagoort, P. (2002b). Syntactic processing modulates the theta rhythm of the human EEG. *Neuroimage, 17*(3), 1479–1492.

Bastiaansen, M. C. M., Van der Linden, M., ter Keurs, M., Dijkstra, T., & Hagoort, P. (2005). Theta responses are involved in lexico-semantic retrieval during language processing. *Journal of Cognitive Neuroscience, 17*(3), 530–541.

Beck, A. (1891). *Oznaczenie lokalizacyi W Mozgu I rdzeniu za pomoca zjawisk elektrycznych.* Kraków: University of Jagiellonski University.

Beck, A. (1973). The determination of localizations in the brain and spinal cord with the aid of electrical phenomena. *Acta Neurobiologiae Experimentalis, Suppl 3*, 1–55.

Becker, R., Ritter, P., & Villringer, A. (2008). Influence of ongoing alpha rhythm on the visual evoked potential. *Neuroimage, 39*(2), 707–716.

Berger, H. (1929). Über das Elektroencephalogramm des Menschen. *Archiv Psychiatrie Nerven.-Krankheith., 87*, 527–570.

Braeutigam, S., Bailey, A. J., & Swithenby, S. J. (2001). Phase-locked gamma band responses to semantic violation stimuli. *Brain Research Cognition and Brain Research, 10*(3), 365–377.

Caton, R. (1875). The electric currents of the brain. *British Medical Journal, 2*, 278.

Cohen, L., Dehaene, S., Naccache, L., Lehericy, S., Dehaene-Lambertz, G., Henaff, M. A., et al. (2000). The visual word form area: Spatial and temporal characterization of an initial stage of reading in normal subjects and posterior split-brain patients. *Brain, 123*(Pt 2), 291–307.

Fell, J. (2007). Cognitive neurophysiology: beyond averaging. *Neuroimage, 37*(4), 1069–1072.

Fries, P., Reynolds, J. H., Rorie, A. E., & Desimone, R. (2001). Modulation of oscillatory neuronal synchronization by selective visual attention. *Science, 291*(5508), 1560–1563.

Goldman, R. I., Stern, J. M., Engel, J., Jr., & Cohen, M. S. (2002). Simultaneous EEG and fMRI of the alpha rhythm. *Neuroreport, 13*(18), 2487–2492.

Gray, C. M., Konig, P., Engel, A. K., & Singer, W. (1989). Oscillatory responses in cat visual cortex exhibit inter-columnar synchronization which reflects global stimulus properties. *Nature, 338*(6213), 334–337.

Gruber, T., & Muller, M. M. (2005). Oscillatory brain activity dissociates between associative stimulus content in a repetition priming task in the human EEG. *Cerebral Cortex, 15*(1), 109–116.

Gruber, T., Tsivilis, D., Montaldi, D., & Muller, M. M. (2004). Induced gamma band responses: An early marker of memory encoding and retrieval. *Neuroreport, 15*(11), 1837–1841.

Gruber, W. R., Klimesch, W., Sauseng, P., & Doppelmayr, M. (2005). Alpha phase synchronization predicts P1 and N1 latency and amplitude size. *Cerebral Cortex, 15*(4), 371–377.

Haarmann, H. J., Cameron, K. A., & Ruchkin, D. S. (2002). Neural synchronization mediates on-line sentence processing: EEG coherence evidence from filler-gap constructions. *Psychophysiology, 39*, 820–825.

Hagoort, P. (2005). On Broca, brain, and binding: A new framework. *Trends in Cognitive Sciences, 9*(9), 416–423.

Hagoort, P., Hald, L., Bastiaansen, M., & Petersson, K. M. (2004). Integration of word meaning and world knowledge in language comprehension. *Science, 304*(5669), 438–441.

Hald, L. A., Bastiaansen, M. C., & Hagoort, P. (2006). EEG theta and gamma responses to semantic violations in online sentence processing. *Brain and Language, 96*(1), 90–105.

Hanslmayr, S., Klimesch, W., Sauseng, P., Gruber, W., Doppelmayr, M., Freunberger, R., et al. (2007). Alpha phase reset contributes to the generation of ERPs. *Cerebral Cortex, 17*(1), 1–8.

Indefrey, P. (2004). Hirnaktivierungen bei syntaktischer Sprachverarbeitung: eine Meta-Analyse. In H. M. Müller & G. Rickheit (Eds.), *Neurokognition der Sprache* (pp. 31–50). Tübingen: Stauffenburg Verlag.

Indefrey, P., & Cutler, A. (2004). Prelexical and lexical processing in listening. In M. Gazzaniga (Ed.), *The cognitive neurosciences III* (pp. 759–774). Cambridge, MA: MIT Press.

Indefrey, P., Kleinschmidt, A., Merboldt, K. D., Kruger, G., Brown, C., Hagoort, P., et al. (1997). Equivalent responses to lexical and nonlexical visual stimuli in occipital cortex: A functional magnetic resonance imaging study. *Neuroimage, 5*(1), 78–81.

Jensen, O., Gelfand, J., Kounios, J., & Lisman, J. E. (2002). Oscillations in the alpha band (9–12 Hz) increase with memory load during retention in a short-term memory task. *Cerebral Cortex, 12*(8), 877–882.

Jokisch, D., & Jensen, O. (2007). Modulation of gamma and alpha activity during a working memory task engaging the dorsal or ventral stream. *Journal of Neuroscience, 27*(12), 3244–3251.

Kahana, M. J., Sekuler, R., Caplan, J. B., Kirschen, M., & Madsen, J. R. (1999). Human theta oscillations exhibit task dependence during virtual maze navigation. *Nature, 399*, 781–784.

Kaiser, J., Ripper, B., Birbaumer, N., & Lutzenberger, W. (2003). Dynamics of gamma-band activity in human magnetoencephalogram during auditory pattern working memory. *Neuroimage, 20*(2), 816–827.

Klimesch, W. (1999). EEG alpha and theta oscillations reflect cognitive and memory performance: A review and analysis. *Brain Research Reviews, 29*, 169–195.

Klimesch, W., Doppelmayr, M., Schwaiger, J., Auinger, P., & Winkler, T. (1999). "Paradoxical" alpha synchronization in a memory task. *Brain Research and Cognitive Brain Research, 7*(4), 493–501.

Klimesch, W., Hanslmayr, S., Sauseng, P., & Gruber, W. R. (2006a). Distinguishing the evoked response from phase reset: A comment to Makinen et al. *Neuroimage, 29*(3), 808–811.

Klimesch, W., & Neuper, C. (Eds.). (2006). *Event-related dynamics of brain oscillations.* (Vol. 159). Amsterdam: Elsevier.

Klimesch, W., Russegger, H., Doppelmayr, M., & Pachinger, T. (1998). A method for the calculation of induced band power: Implications for the significance of brain oscillations. *Electroencephalography and Clinical Neurophysiology, 108*, 123–130.

Klimesch, W., Sauseng, P., & Hanslmayr, S. (2006b). EEG alpha oscillations: The inhibition-timing hypothesis. *Brain Research Brain Research Reviews*, Volume 53, Issue 1, January 2007, Pages 63–88.

Klimesch, W., Schack, B., Schabus, M., Doppelmayr, M., Gruber, W., & Sauseng, P. (2004). Phase-locked alpha and theta oscillations generate the P1-N1 complex and are related to memory performance. *Brain Research Cognition and Brain Research, 19*(3), 302–316.

Knosche, T. R., & Bastiaansen, M. C. M. (2002). On the time resolution of event-related desynchronization: A simulation study. *Clinical Neurophysiology, 113*(5), 754–763.

Konig, P., & Schillen, T. B. (1991). Stimulus-dependent assembly formation of oscillatory responses: I. Synchronization. *Neural Computation, 3,* 155–166.

Krause, C. M., Lang, A. H., Laine, M., Kuusisto, M., & Porn, B. (1996). Event-related EEG desynchronization and synchronization during an auditory memory task. *Electroencephalography and Clinical Neurophysiology, 98*(4), 319–326.

Lachaux, J. P., Rodriguez, E., Martinerie, J., & Varela, F. J. (1999). Measuring phase synchrony in brain signals. *Human Brain Mapping, 8*(4), 194–208.

Laufs, H., Kleinschmidt, A., Beyerle, A., Eger, E., Salek-Haddadi, A., Preibisch, C., et al. (2003). EEG-correlated fMRI of human alpha activity. *Neuroimage, 19*(4), 1463–1476.

Lopes da Silva, F. H. (1991). Neural mechanisms underlying brain waves: From neural membranes to networks. *Electroencephalography and Clinical Neurophysiology, 79,* 81–93.

Makeig, S., Debener, S., Onton, J., & Delorme, A. (2004). Mining event-related brain dynamics. *Trends in Cognitive Sciences, 8*(5), 204–210.

Makeig, S., Westerfield, M., Jung, T. P., Enghoff, S., Townsend, J., Courchesne, E., et al. (2002). Dynamic brain sources of visual evoked responses. *Science, 295*(5555), 690–694.

Makinen, V., Tiitinen, H., & May, P. (2005). Auditory event-related responses are generated independently of ongoing brain activity. *Neuroimage, 24*(4), 961–968.

Martin, A., & Chao, L. L. (2001). Semantic memory and the brain: structure and processes. *Current Opinion in Neurobiology, 11*(2), 194–201.

Mazaheri, A., & Jensen, O. (2006). Posterior alpha activity is not phase-reset by visual stimuli. *Proceedings of the National Academy of Sciences USA, 103*(8), 2948–2952.

Mazaheri, A., & Picton, T. W. (2005). EEG spectral dynamics during discrimination of auditory and visual targets. *Brain Research Cognition and Brain Research, 24*(1), 81–96.

McCandliss, B. D., Cohen, L., & Dehaene, S. (2003). The visual word form area: Expertise for reading in the fusiform gyrus. *Trends in Cognitive Sciences, 7*(7), 293–299.

Medendorp, W. P., Kramer, G. F., Jensen, O., Oostenveld, R., Schoffelen, J. M., & Fries, P. (2006). Oscillatory activity in human parietal and occipital cortex shows hemispheric lateralization and memory effects in a delayed double-step saccade task. *Cerebral Cortex.*

Medendorp WP, Kramer GF, Jensen O, Oostenveld R, Schoffelen JM, Fries P (2007) Oscillatory activity in human parietal and occipital cortex shows hemispheric lateralization and memory effects in a delayed double-step saccade task. *Cereb Cortex* 17:2364–2374.

Mesulam, M. M. (1998). From sensation to cognition. *Brain, 121*(Pt 6), 1013–1052.

Miller, R. (1991). *Cortico-hippocampal interplay and the representation of contexts in the brain.* Berlin: Springer.

Mitra, P. P., & Pesaran, B. (1999). Analysis of dynamic brain imaging data. *Biophysiology Journal, 76*(2), 691–708.

Murray, M. M., Wylie, G. R., Higgins, B. A., Javitt, D. C., Schroeder, C. E., & Foxe, J. J. (2002). The spatiotemporal dynamics of illusory contour processing: Combined high-density electrical mapping, source analysis, and functional magnetic resonance imaging. *Journal of Neuroscience, 22*(12), 5055–5073.

Nunez, P. L., Srinivasan, R., Westdorp, A. F., Wijesinghe, R. S., Tucker, D. M., Silberstein, R. B., et al. (1997). EEG coherency. I: Statistics, reference electrode, volume conduction, Laplacians, cortical imaging, and interpretation at multiple scales. *Electroencephalography and Clinical Neurophysiology, 103*(5), 499–515.

Osipova, D., Takashima, A., Oostenveld, R., Fernandez, G., Maris, E., & Jensen, O. (2006). Theta and gamma oscillations predict encoding and retrieval of declarative memory. *Journal of Neuroscience, 26*(28), 7523–7531.

Palva, J. M., Palva, S., & Kaila, K. (2005). Phase synchrony among neuronal oscillations in the human cortex. *Journal of Neuroscience, 25*(15), 3962–3972.

Palva, S., & Palva, J. M. (2007). New vistas for alpha-frequency band oscillations. *Trends in Neuroscience, 30*(4), 150–158.

Penny, W. D., Kiebel, S. J., Kilner, J. M., & Rugg, M. D. (2002). Event-related brain dynamics. *Trends in Neuroscience, 25*(8), 387–389.

Petersen, S. E., Fox, P. T., Posner, M. I., Mintun, M., & Raichle, M. E. (1988). Positron emission tomographic studies of the cortical anatomy of single-word processing. *Nature, 331*(6157), 585–589.

Pfurtscheller, G., & Aranibar, A. (1979). Evaluation of event-related desynchronization (ERD) preceding and following voluntary self-paced movements. *Electroencephalography and Clinical Neurophysiology, 46,* 138–146.

Pfurtscheller, G., & Berghold, A. (1989). Patterns of cortical activation during planning of voluntary movement. *Electroencephalography and Clinical Neurophysiology, 72,* 250–258.

Pfurtscheller, G., & Lopes da Silva, F. H. E. (1999). *Event-related desynchronization: Handbook of electroencephalography and clinical neurophysiology, revised series.* (Vol. 6). Amsterdam: Elsevier.

Pfurtscheller, G., & Neuper, C. (1997). Motor imagery activates primary sensorimotor area in humans. *Neuroscience Letters, 239,* 65–68.

Pfurtscheller, G., Stancak, A., Jr., & Neuper, C. (1996). Event-related synchronization (ERS) in the alpha band—an electrophysiological correlate of cortical idling: A review. *International Journal of Psychophysiology, 24*(1–2), 39–46.

Prawdicz-Neminski, W. W. (1913). Ein Versuch der Registrierung der elektrischen Gehirnerscheinungen. *Centralblatt für Physiologie, 27,* 951–960.

Pulvermueller, F. (1999). Words in the brain's language. *Behavioural Brain Science, 22,* 253–336.

Pulvermueller, F. (2001). Brain reflections of words and their meaning. *Trends in Cognitive Sciences, 5*(12), 517–524.

Pulvermueller, F., Lutzenberger, W., & Preissl, H. (1999). Nouns and verbs in the intact brain: Evidence from event-related potentials and high-frequency cortical responses. *Cerebral Cortex, 9*(5), 497–506.

Raghavachari, S., Kahana, M. J., Rizzuto, D. S., Caplan, J. B., Kirschen, M. P., Bourgeois, B., et al. (2001). Gating of human theta oscillations by a working memory task. *Journal of Neuroscience, 21*(9), 3175–3183.

Rizzuto, D. S., Madsen, J. R., Bromfield, E. B., Schulze-Bonhage, A., Seelig, D., Aschenbrenner-Scheibe, R., et al. (2003). Reset of human neocortical oscillations during a working memory task. *Proceedings of the National Academy of Sciences USA, 100*(13), 7931–7936.

Scheeringa, R., Bastiaansen, M. C. M., Petersson, K. M., Oostenveld, R., Norris, D. G., & Hagoort, P. (in preparation). BOLD correlates of a WM-related alpha network: A simultaneous EEG/fMRI study.

Scheeringa, R., Petersson, K. M., Oostenveld, R., Norris, D. G., Hagoort, P., & Bastiaansen, M. C. (2009). Trial-by-trial coupling between EEG and BOLD identifies networks related to alpha and theta EEG power increases during working memory maintenance. *Neuroimage, 44*(3), 1224–1238.

Sederberg, P. B., Gauthier, L. V., Terushkin, V., Miller, J. F., Barnathan, J. A., & Kahana, M. J. (2006). Oscillatory correlates of the primacy effect in episodic memory. *Neuroimage, 32*(3), 1422–1431.

Shah, A. S., Bressler, S. L., Knuth, K. H., Ding, M., Mehta, A. D., Ulbert, I., et al. (2004). Neural dynamics and the fundamental mechanisms of event-related brain potentials. *Cerebral Cortex, 14*(5), 476–483.

Singer, W. (1993). Synchronization of cortical activity and its putative role in information processing and learning. *Annual Review of Physiology, 55*, 349–374.

Singer, W. (1999). Neuronal synchrony: A versatile code for the definition of relations? *Neuron, 24*(1), 49–65, 111–125.

Steriade, M., Gloor, P., Llinás, R. R., Lopes da Silva, F. H., & Mesulam, M. M. (1990). Basic mechanisms of cerebral rhythmic activities. *Electroencephalography and Clinical Neurophysiology, 76*, 481–508.

Tallon-Baudry, C., Bertrand, O., Peronnet, F., & Pernier, J. (1998). Induced gamma band activity during the delay of a visual short-term memory task in humans. *Journal of Neuroscience, 18*, 4244–4254.

Tuladhar, A. M., Huurne, N. T., Schoffelen, J. M., Maris, E., Oostenveld, R., & Jensen, O. (2007). Parieto-occipital sources account for the increase in alpha activity with working memory load. *Human Brain Mapping, 28*(8), 785–792.

van Berkum, J. J., Zwitserlood, P., Bastiaansen, M., Brown, C., & Hagoort, P. (2004). *So who's "he" anyway? Differential ERP and ERSP effects of referential success, ambiguity and failure during spoken language comprehension.* Presented at the annual meeting of the Cognitive Neuroscience Society, San Francisco.

Varela, F., Lachaux, J. P., Rodriguez, E., & Martinerie, J. (2001). The brainweb: Phase synchronization and large-scale integration. *Nature Review of Neuroscience, 2*(4), 229–239.

Weiss, S., & Mueller, H. M. (2003). The contribution of EEG coherence to the investigation of language. *Brain and Language, 85*(2), 325–343.

Weiss, S., Mueller, H. M., Schack, B., King, J. W., Kutas, M., & Rappelsberger, P. (2005). Increased neuronal communication accompanying sentence comprehension. *International Journal of Psychophysiology, 57*(2), 129–141.

Weiss, S., Rappelsberger, P., Schack, B., & Mueller, H. M. (2003). Koharenz- und Phasenuntersuchungen und ihre Bedeutung fur die Untersuchung von Sprachprozessen. In G. R. H. M. Mueller (Ed.), *Neurokognition der Sprache* (pp. 211–258). Tubingen: Stauffenburg.

Wespatat, V., Tennigkeit, F., & Singer, W. (2004). Phase sensitivity of synaptic modifications in oscillating cells of rat visual cortex. *Journal of Neuroscience, 24*(41), 9067–9075.

Yeung, N., Bogacz, R., Holroyd, C. B., Nieuwenhuis, S., & Cohen, J. D. (2007). Theta phase resetting and the error-related negativity. *Psychophysiology, 44*(1), 39–49.

CHAPTER

3 ERP Features and EEG Dynamics: An ICA Perspective

Scott Makeig *and* Julie Onton

Abstract

This chapter considers the relationship between ongoing electroencephalographic (EEG) activity as recorded in event-related paradigms and trial averages time locked to some class of experimental events, known as *event-related potentials* (ERPs). It first discusses the concept of the ERP as averaging potentials generated by spatially coherent activity within a number of cortical EEG source areas as well as nonbrain sources typically treated as data artifacts. Event-related potential image (ERP-image) plotting is used to visualize variability in EEG dynamics across trials associated with events of interest using an example data set. The concept and use of independent component analysis (ICA) is then introduced to undo the effects of source signal mixing at scalp electrodes and to identify EEG sources contributing to the averaged ERP. After presenting some basic time-frequency measures useful for studying trial-to-trial variability, the chapter takes another look at trial-to-trial variability, now focusing on the contributions of selected independent component processes to the recorded scalp signals.

Keywords: EEG activity, event-related potentials, ERP, independent component analysis

Prefatory Remarks

Following the advent of averaging computers in the early 1960s, event-related potential (ERP) averaging became the first functional brain imaging method to open a window into human brain processing of first sensory and then cognitive events, and the first to demonstrate statistically reliable differences in this processing depending on the *contextual significance* of these events—or their unexpected absence. Yet the same response averaging methods, now easily performed on any personal computer, may have encouraged the separation of electrophysiological brain research into two camps. For nearly half a century now, researchers mainly in psychology departments have recorded human scalp electroencephalographic (EEG) data and studied the features of human average ERPs time-locked to events and behavior, while researchers mainly in physiology departments have measured averaged event-related changes in the number of spikes emitted by single neurons in animals, captured from high-pass filtered local field potential (LFP) recordings from microelectrodes. Since the spatial scales of these phenomena are so different, these two groups have had little to say to one another. In fact, relationships between these quite different average brain response measures can be learned only by studying the spatiotemporal complexities of the whole EEG and LFP signals from which they are respectively extracted, and by understanding not only their average behavior, but also the complexities of their moment-to-moment dynamics.

Many questions remain about the nature and variability of these electrophysiological signals, including the functional relationships of their dynamics to behavior and experience. This investigation is now beginning, with much more remaining to be discovered about the distributed macroscopic electromagnetic brain dynamics that allow our brains to support us to optimize our behavior and brain activity *to meet the challenge of each moment.* From this point of view, key open questions for those interested in understanding the nature and origins of average scalp ERPs are how to identify the brain sources of EEG and EEG-derived ERP dynamics, their locations, and their dynamic interrelationships. To adequately address these questions, new analysis methods are required and are becoming available.

Our research over the past dozen years has convinced us, and an increasing number of other researchers, that using independent component analysis (ICA) to find spatial filters for *information sources* in scalp-recorded EEG (and other) data, combined in particular with trial-by-trial visualization and time/frequency analysis methods, form a powerful approach to identifying the *complex spatiotemporal dynamics* that underlie both ERP averages and the continually unfolding and varying brain field potential phenomena for which they provide a partial measure. In essence, ICA is a method for training or learning spatial filters that, when applied to data collected from many scalp locations, each focus on one source of maximally distinct information in the data. Characterizing the information content of data rather than its variance (the most common goal of previous signal processing methods) is a powerful new approach to analysis of complex signals that is becoming ever more important for data mining of all sorts. Applied to EEG data, ICA tackles "head on" the major confounding factor that has limited the development of EEG-based brain imaging methods, namely, the broad, instantaneous spread of EEG source potentials through the brain, skull, and scalp and the complicated additive mixtures of these widely broadcast signals that are recorded by each scalp electrode.

EEG Sources and Source Projections

In this chapter, we consider the relationship between ongoing EEG activity as recorded in event-related paradigms and trial averages time-locked to some class of experimental events, known as *event-related potentials* (ERPs). We first discuss the concept of ERPs as averaging potentials generated by spatially coherent activity within a number of cortical EEG source areas as well as nonbrain sources typically treated as data artifacts. We use 'ERP-image' plotting to visualize the variability in EEG dynamics across trials associated with events of interest using an example data set. We then introduce the concept and use of ICA to undo the effects of source signal mixing at scalp electrodes and to identify EEG sources contributing to the averaged ERP. We note that *independent components* are brain or nonbrain processes more or less active throughout the data set, and thus represent a quite different use of the term *component* than in the title and elsewhere in this volume. After introducing some basic time/frequency measures useful for studying trial-to-trial variability, we take another look at trial-to-trial variability, now focusing on the contributions of selected independent component processes to the recorded scalp signals in the same data set. We hope the chapter will help the reader interested in event-related EEG analysis to think carefully about trial-to-trial EEG stability and variability. The latter, we suggest, may largely reflect not 'ERP noise' but instead the brain's carefully constructed distributed response to the highly individual, complex, and context-defined challenges posed by unfolding events.

What Is an EEG Source?

A fundamental fact about electrophysiological signals recorded at any spatial scale is that they reflect and index emergent partial coherence (in both time and space) of electrophysiological events occurring at smaller scales. Brain electrophysiological signals recorded by relatively large and/or distant electrodes can be viewed as phenomena emerging from the possibly 1 quadrillion synaptic events that occur in the human brain each second. These events, in turn, arise within the still more vast complexity of brain molecular and submolecular dynamics. The synchronies and near-synchronies, in time and space, of synaptic and nonsynaptic neural field dynamics precipitate not only neural spikes, but also other intracellular and extracellular field phenomena—both those measurable only at the *near field* (e.g., within the range of a neural arbor) and those recorded only at the *far field* (in particular, as far from the brain electrically as the human scalp). The emergence of spatiotemporal field synchrony or near-synchrony across an area of cortex is conceptually akin to the emergence of a galaxy in the plasma of space. Both are spontaneously emergent dynamic phenomena large enough to be detected and measured at a distance—via EEG electrodes and powerful telescopes, respectively.

The emergence of synchronous or near-synchronous local field activity across some portion of the cortical mantle requires that cells in the synchronized cortical area be physically coupled in some manner. A basic fact of cortical connectivity is that corticocortical connections between cells are highly weighted toward local (e.g., shorter than 0.5 mm) connections, particularly those coupling nearby inhibitory cells whose fast gap-junction connections support the spread of near-synchronous field dynamics through local cortical areas (Murre & Sturdy, 1995; Stettler et al., 2002). Also important for sustaining rhythmic EEG activity are thalamocortical connections that are predominantly (though not exclusively) organized in a radial one-to-one manner (Frost & Caviness, 1980). EEG activity is therefore likely to arise as emergent *mesoscopic* patterns (Freeman, 2000) of local field synchrony or near-synchrony in compact thalamocortical networks. Potentials arising from vertical field gradients associated with pyramidal cells arrayed orthogonal to the cortical surface produce the local field potentials recorded on the cortical surface (Luck, 2005; Nunez, 2005). Synchronous (or near-synchronous) field activity across a cortical patch produces the far-field potentials that are conveyed by volume conduction to scalp electrodes. Both scalp and direct cortical recordings agree that in nearly all cognitive states, such locally coherent field activities arise within many parts of human cortex, often with distinctive dynamic signatures in different areas. Direct observations in animals report that cortical EEG signals are indeed associated with subcentimeter-size cortical patches whose spatial patterns resemble "phase cones" (like "pond ripples"; Freeman & Barrie, 2000) or repeatedly spreading "avalanche" events (Beggs & Plenz, 2003), though more adequate multiresolution recording and modeling are needed to better define their spatiotemporal geometry and dynamics.

In this chapter, we will use the term *EEG source* to mean a compact cortical patch (or occasionally, connected patches) within which temporally coherent (or partially coherent) local field activity emerges, thereby producing a far-field potential contributing appreciably to the EEG signals recorded on the scalp. While noncortical brain sources may contribute to the recorded EEG (e.g., auditory brain stem potentials), their contributions are typically small compared to those of cortical potentials; we will therefore assume in this chapter that resolvable EEG signals of interest are of cortical origin. We will use the phrase *source activity* to refer to the varying far-field potential arising within an EEG source area and volume-conducted to the scalp electrodes. Recorded EEG signals are then, in this view, the sum of EEG source activities, contributions of nonbrain sources such as scalp muscle, eye movement, and cardiac artifacts, plus (ideally small) electrode and environmental noise.

Note that the activity contributed by a cortical source to the recorded EEG typically does *not* comprise all the local field activity within the cortical source domain, since potentials recorded with small cortical electrodes at different points in a cortical source domain may only be *weakly* coherent with the far-field activity that is partially coherent across the domain and is therefore *not* projected to the scalp electrodes. That is, only the portion of the local field activity in a source domain that is synchronous across the domain will contribute appreciably to the net source potentials recorded by scalp electrodes. Thus, cortical electrophysiology is by its nature multiscale; its properties differ, depending on the size of the recording electrodes and their distance from the source areas in ways that are currently far from adequately observed or modeled. Scalp EEG recordings predominantly capture the sum of locally coherent source activities originating within a number of cortical source domains plus projections of nonbrain (or 'artifactual') signal sources.

Roles of EEG Source Activities

The primary function of our brain is to organize and control our behavior "in the moment" so as to optimize its outcome. For many neurobiologists, field potential recordings have been considered of possible interest at best only as passive, indirect, and quite poor statistical indices of changes in neural spike rates, their primary measure of interest. In fact, however, the variations in electrical potential recorded by EEG or LFP electrodes better reflect variations in concurrent dendritic synaptic input to neurons, input that may or may not provoke action potentials. Action potential generation is provoked by receipt of sufficient dendritic input within a brief (several-millisecond) time window. The emergence of synchronized local field potentials across a cortical area may therefore reflect changes in the occurrence of joint spiking events across groups of associated neurons in that area. Some recent experimental results also suggest that local field potentials may also actively affect spike timing and the degree of synchrony between neurons within a partially synchronized source domain, biasing their joint spike timing toward (or away from) a concentration into brief, potent volleys (Francis et al., 2003; Froehlich & McCormick, 2010, Radman et al., 2006; Voronin et al., 1999, Womelsdorf & Fries, 2007). By this means, small statistical adjustments in joint spike timing of neurons with common axonal targets effected by spatially synchronized local field potentials might be associated with large changes in effective neural communication, and thence with behavior and behavioral outcome (Fries et al., 2007).

According to recent reports, the timing and phase of extracellular fields may also enhance or weaken the effects of concurrent input on future cell and areal responsivity, by affecting the amount of long-term synaptic potentiation (LTP) produced by that input (Dan & Poo, 2004). Thus, locally synchronous (or near-synchronous) field activity arising within compact cortical source areas may not only weakly index neural dynamics on spatial scales larger than a single neuron, but may also play a more direct *and active* role in organizing the distributed brain dynamics that support experience, behavior, and changes in psychophysiological state. The spatiotemporal dynamics of cortical field synchrony, and their relationship to neural spike timing, have been relatively little studied (Bollimunta et al., 2008; Logothetis et al., 2001) and, contrary to standard presumption, much more is likely to be discovered about the relationships between extracellular fields and intracellular potentials in living brains.

SPATIAL SOURCE VARIABILITY

The concept that an EEG source represents the emergence of synchronous field activity across a cortical patch is undoubtedly a simplification of the actual more complex, multiscale dynamics that produce near-synchronous activity within an EEG source domain. Although the concept that EEG is produced by synchronous field activity in small cortical domains is supported qualitatively by functional magnetic resonance imaging (fMRI) results that are generally dominated by roughly centimeter-scale or smaller pockets of enhanced activity, and by a few reports of direct field potential grid recordings (Bullock, 1983; Freeman & Barrie, 2000), when cortical activity in animals is viewed at a smaller (submillimeter) scale using optical imaging, smaller-scale moving waves of electromagnetic activity are observed (Arieli et al., 1995). However, simple calculation of the phase difference between the edges and the center of a "pond ripple" pattern at EEG frequencies, based on the estimated (square-centimeter-scale) domain sizes and observed traveling wave velocities (1–2 m/s), suggests that the spatial wavelength of the radiating "ripples" is considerably larger than the size of the domain, meaning that the topographic scalp projection of a cortical "phase cone" is close to that of totally synchronous activity across the patch, as in our simplified EEG source model.[1]

However, larger-scale traveling or meandering waves at slow (1–3 Hz) delta or infraslow (<1 Hz) EEG frequencies have also been observed in epilepsy, sleep, and migraine (Massimini et al., 2004), and (near 12-Hz) sleep spindles may also "wander" over cortex in concert with spatially varying activity in coupled regions of the thalamic reticular nucleus (Rulkov et al., 2004). Sufficiently detailed recordings from high-density multiresolution arrays are not yet available to allow models of the relationship between these moving activity patterns and seemingly stable EEG source dynamics in waking life.

TEMPORAL SOURCE VARIABILITY

A hallmark of EEG is that its temporal dynamics are highly nonstationary and exhibit continuous changes on all time scales (Linkenkaer-Hansen et al., 2001). Changing EEG dynamics index changes in and between local synchronies that are driven or affected by a variety of mechanisms, including sensory information as well as broadly projecting brainstem-based arousal or "value" systems identified by their central neurotransmitters—dopamine, acetylcholine, serotonin, neurepinephrine, and so on. These *neuromodulatory* systems based in brainstem areas project to widespread cortical areas and are very likely an important source of temporal variability in the spatiotemporal coherence that produces far-field signals recorded at the scalp, variability that gives flexibility and individuality to our distributed brain responses so as to respond most appropriately to the particular challenge of the moment. Ranganath and Rainer (2003) have reviewed what is known and still unknown about these systems and their interactions with cortical field potentials.

Volume Conduction and Source Mixing

Experimental neurobiology suggests that the spontaneous emergence of partial coherence of complex rhythmic temporal patterns of local field activity across compact cortical "phase cone" or "avalanche" domains a few millimeters or larger in diameter produces the scalp EEG and therefore ERP signals. The differences between distinct parts of neurons within the partially synchronized and nearly aligned pyramidal cell domain sum coherently both in local recordings and as measured at any distance after passing by *volume conduction* through intervening conductive media including cortical gray and white matter, cerebrospinal fluid (CSF), skull, and skin (Akalin-Acar & Gencer, 2004; Gencer & Akalin-Acar, 2005). The very broad cortical source field patterns (each generally resembling the double-lobed pattern that iron filings take when placed around a bar magnet) are attenuated by the partial resistance of these media, and their propagation patterns are spatially distorted at tissue boundaries where conductance changes. The broadly projecting, spatially distorted, and severely attenuated signals are then summed within conductive EEG electrodes attached to the scalp (Nunez, 1977).[2]

FORWARD AND INVERSE MODELING

A first question for EEG/ERP researchers, therefore, is or should be how to separate the recorded EEG activities recorded at all the scalp channels into a set of activities originating within different spatial source domains. Finding the appropriate spatial filters is, unfortunately, technically difficult, and were any arbitrary three-dimensional (3-D) arrangement of source configurations physiologically possible, any number of them could be found that would produce the same scalp potentials (Grave de Peralta-Menendez & Gonzalez-Andino, 1998). A biophysical solution to this so-called inverse problem must begin with construction of a forward head model specifying (1) where in the brain the electromagnetic sources may be expected to appear, and (2) in which orientations, and (3) how their electromagnetic fields propagate through the subject's head to the recording electrodes (Akalin-Acar & Gencer, 2004). Fortunately, the well-grounded assumption that the brain EEG sources are cortical source patches whose field patterns are oriented near-perpendicular to the local cortical surface allows more physiologically plausible estimates since these assumptions allow a fair MR image-derived model of the shape, location, and local orientation of subject cortex (Fischl et al., 2004; Akalin Acar & Makeig, 2009). Constructing individualized cortical models for EEG analysis requires extensive computation and expensive MR head images; thus, the process is still rarely carried out for routine EEG/ERP experiments. While adequate head models are needed to develop EEG into a 3-D functional brain imaging modality, different analysis goals may require various degrees of anatomical accuracy, and use of standard head models may suffice for many analysis purposes.

Given an accurate forward head model, the inverse problem is still underdetermined if multiple sources contribute to the observed scalp potential distribution whose sources are to be determined. In contrast, the solution is much simpler, and its answer better determined, if the scalp maps whose source projections they sum are *simple maps* representing the activity of only one source. Both EEG and magnetoencephalography (MEG) researchers have long attempted to consider scalp maps of amplitude peaks in average ERPs to be simple maps. That is, they have attempted to use ERP averaging, a purely temporal filtering method, as a means of, in effect, *spatial filtering*, attempting to eliminate from each channel the projections of EEG sources not directly affected by the time-locking events.

In many cases, taking the difference between two average ERPs time-locked to related sets of experimental events may further restrict the number of strongly contributing brain source areas. Unfortunately, trial averaging or differencing is rarely completely effective for this purpose, since meaningful sensory (as well as purely cognitive) events rapidly perturb the statistics of many cortical EEG source areas as well as subcortical areas (Halgren et al., 1998; Schroeder et al., 1998). Therefore, except for very early sensory brainstem and cortical potentials, ERP maps sum activities that arise, typically, in many source domains. This means that scalp maps of ERP or ERP difference-wave peaks are rarely *simple maps* representing potentials projected to the scalp from a single source. To optimally estimate the source distribution responsible for EEG or ERP data, it is desirable to find a better way to isolate simple maps representing the projection of single sources contributing to the data, a subject we will return to in "Separating EEG Sources using ICA"

ERP Trial Averaging and Trial Variability

Figure 3.1A shows a single-subject ERP for all 238 scalp channels averaged over 500 data epochs time-locked to onsets (a latency 0) of an infrequently presented visual target disc in a visuospatial selective attention task (Makeig et al., 1999a, 1999b, 2002, 2004b). The ERP traces for all 238 channels are overlaid on the same plot axis. Interpolated scalp maps show the ERP scalp distribution at four indicated latencies. The bottom panel shows the ERP average of the same 500 epochs, but now time-locked to the subject's button press in each trial. In both panels, the data were averaged after removing artifacts produced by eye movements, eye blinks, electrocardiographic (ECG) activity, and electromyographic (EMG) activity from scalp and neck muscles using ICA, as explained later in the chapter[3,4] The late positive complex (LPC or "P300") feature of this averaged response to the infrequent attended visual targets is attenuated by the highpass filtering above 1 Hz applied to the data; it is most evident in the motor response-locked data average (Figure 3.1B). We will use these data throughout this chapter to explore the relations between the average ERPs, as shown in this figure, and the single EEG data epochs that were averaged to produce them.

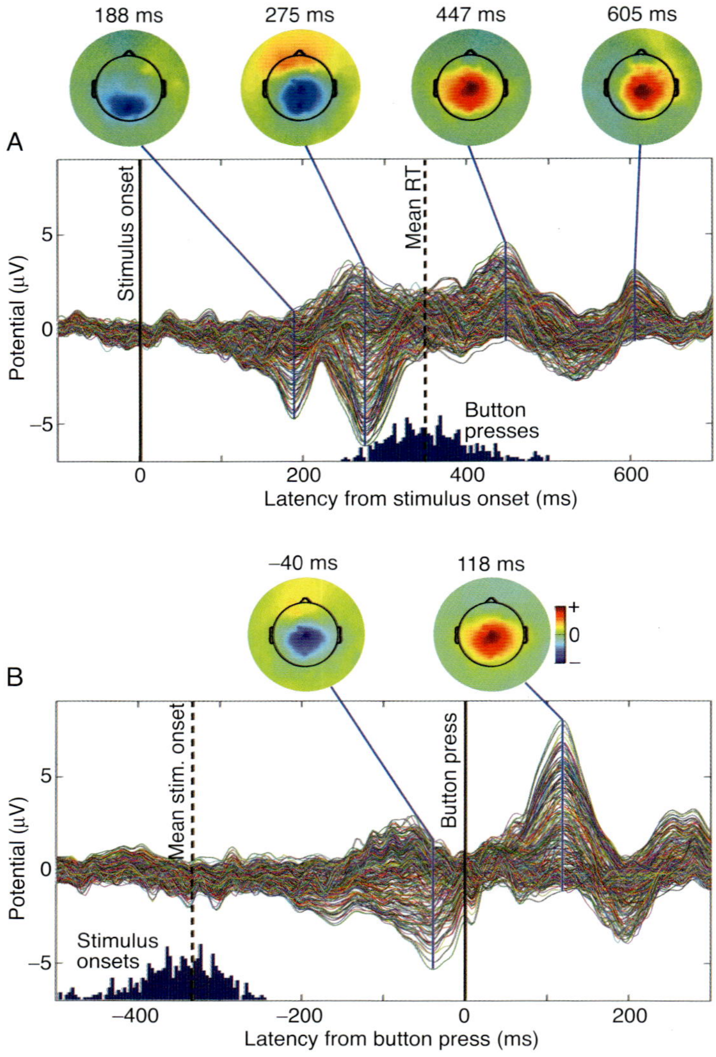

Fig. 3.1. Event-related potential traces from each of 238 scalp channels averaged over 500 EEG epochs in a single subject, time-locked to (A) continually anticipated but infrequently presented visual target stimuli in a visuospatial selective attention task and (B) immediately following speeded subject button presses cued by target presentation.[4] Solid lines cutting vertically through the ERP channel trace bundles lead to cartoon heads showing the interpolated scalp potential distribution at the moment indicated.

The Trial-Averaging Model

Over the past near half-century, the predominant method for reducing the complexity of event-related EEG data collected in sensory and cognitive paradigms has been to form ERP averages of trial records time-locked to sets of experimental events assumed by the experimenter to generate the same or essentially similar brain responses. To gain a realistic understanding of the features of ERP trial averages and their relationship to the underlying brain dynamics, it is important to understand both the strengths and limitations of ERP averaging. The physiological model underlying ERP averaging is that cortical processing of sensory (or other) event information follows a fixed spatiotemporal sequence of source activities, and this processing produces a fixed sequence of deviations in scalp potentials whose distribution reflects the locations of their cortical generators. However, these traces of the cortical processing sequence are obscured in single response epochs by typically much larger ongoing EEG activities generated in many brain areas, as well as by artifacts generated in nonbrain structures. Crucially, these ongoing activities are assumed to be unaffected by the time-locking events of interest.

Thus, in the ERP averaging model, EEG epochs are assumed to sum (1) a temporally consistent *event-related* activity sequence (the evoked response) plus (2) *event-unrelated* ongoing or spontaneous EEG activities (not contributing to the ERP). Under these circumstances, averaging a sufficient number of event-locked epochs subtracts, cancels, or spatially filters out the unrelated brain activities, leaving a single average response epoch dominated by the consistent event-related activity sequence, recorded on the scalp as the flowing ERP field "movie." The amplitudes of EEG (or other nonbrain) activities unaffected by the time-locking events that remain in the average will be approximately $1/N^{1/2}$ of the amplitude of those activities in the single trials, where N is the number of epochs averaged. Thus, achieving a faithful representation of the actual (and typically relatively small) evoked response sequence usually requires averaging a relatively large number of event-related epochs known or assumed to contain the same time-locked ERP sequence.

To understand how ERP averaging leads to a reduction in event-*un*related EEG activity, we first need to define the phase of an EEG source signal. The simplest definition of the term *phase* might be the sign or polarity of the recorded potential at a given time point, either positive or negative, in relation to some baseline potential (typically established by averaging the potentials recorded in some period of the recording assumed to be unaffected by the events of interest). A more specific meaning of the *phase* of a signal at some time point and frequency refers to the phase of a best-fitting brief, tapered sinusoid at the given frequency centered on that time point. Thus, the phase of an EEG source or scalp signal, at any given time point and frequency, is defined by the relation of its value at that time point to its values in an enclosing window of time points. Note that at a given time point, a signal has a different phase at each frequency. Also, since EEG signals are relatively smooth, EEG phase differences between neighboring time points cannot vary freely but must change smoothly.[5]

Event-related potential averaging can remove the contributions of those source activities unrelated to the time-locking events by means of *phase cancellation*, which works as follows. If a given source signal is unaffected by the time-locking events, and if the timing of the experimental events is not based on the ongoing EEG signals, their phase at each latency and frequency will differ randomly across trials. Mathematically, the sum of random-phase signals at a given frequency tends to become smaller and smaller (at that frequency) as the number of summed trials increases. We can see this most easily by considering the signs of the signals (+ or -) instead of their phases. If the signs of a set of signal epoch values at a given latency are random, then in the average of those epochs the positive-phase and negative-phase values in different epochs will partially cancel each other, and the magnitude of the average epoch at that latency will be smaller than the average of the same values were they all of the same sign.

Similarly, if at a given analysis frequency (e.g., 9 Hz) the single-trial EEG signals have a random phase distribution when measured in a time window centered at some latency (say, 200 ms following the time-locking event), then the vectors that can be used to represent their amplitudes and phases at that frequency will be more or less evenly distributed around the phase circle, and the expected length of the vector average of these vectors will become smaller as the number of trial vectors averaged increases, assuming that the exact timing of experimental events cannot be predicted by the brain based on previous events. On average, the length of the average phase vector will decrease as the square root of the number of trials averaged. In this way, trial averaging filters out all features of the data that are not wholly or at least partially *phase-locked* to the time-locking events at any frequency and latency.

It is important to understand how event-related phase locking and time locking differ. For example, imagine a set of EEG trial epochs, time-locked to a particular type of event, that each contain a burst of 10 Hz alpha-band activity centered 500 ms after the time-locking event. Further, imagine that these alpha bursts, while undeniably *time*-locked to the experimental events of interest, may exhibit any phase at 500 ms (ascending, descending, etc.). These bursts, therefore, are not *phase*-locked to the events. An ERP average of enough such epochs would therefore contain little trace of 10 Hz activity at 500 ms, even though this is a striking feature of the single-trial data. This is because trial averaging filters out all activity that is not *both time-locked and phase-locked* to the time-locking event (see the discussion of time locking and phase locking below). Thus, scalp ERPs do *not* capture all of the consistent event-related dynamics in the averaged EEG epochs, but only those dynamic processes that affect the phase distribution of their signals at some analysis frequencies and trial latencies. We will consider this question again below (see also Chapter 2, this volume).

However, if a given brain source contributes a fixed activity sequence to a given scalp channel signal in every trial, at each analysis frequency and epoch latency the phase of its contributions to the scalp signals will be consistent across trials, and the ERP average at that scalp channel will contain *all* of that source's activity sequence, without diminution. If the phase of its single-trial activity at a given frequency and latency is variable and only weakly consistent, relative to a true random phase process, then the source will contribute only weakly to the scalp ERP. If the source activity has truly random phase at the given frequency and latency, then its ERP contribution will be minimal and further decreasing towards zero as more trials are averaged.

If all the EEG sources that project to a scalp channel have fixed evoked activity sequences, then their collective contribution to the channel in each trial will be the sum of all their source activities, and the average ERP at that channel will be the average of the summed source activities at each trial latency. Thus, source mixing that occurs at the scalp electrode could decrease (or increase) the apparent ERP magnitude through the same process of phase cancellation. But again, only those signals that are phase-locked to the time-locking events from trial to trial will be retained in the average. And conversely, for activity at a given frequency and latency to be removed from the signal by averaging, only the signal *phase* in the single trials need be random.

Limitations of Event-Related Averaging

The mean of any distribution is simply one statistical measure of the distribution—a statistic that, if provided apart from other statistics, may be informative and/or misleading. For example, telling a New Guinean unacquainted with Americans that the average adult American height is 5 ft.6 in. (1.68 m) might give him or her an adequate concept of the distribution of American adult heights—assuming the shapes and the widths of the (near-normal) height distributions in the two cultures are not dissimilar. However, sending Martian scientists the arithmetically equally correct information that the average human is half male and half female might well engender quite incorrect ideas about human biology and society. The problem here is that human sexual physiology has not one but two quite distinct modes (female and male), information that is not captured in or conveyed by the average. Thus, the average of a distribution may or may not in itself provide or suggest a useful and realistic model of the underlying distribution or its features. This may be even more problematic for time series averages that sum disparate activities of many distinct brain and nonbrain sources whose detailed features are of primary interest, including their spatial and temporal trial-to-trial variability.

SPATIAL VARIABILITY OF EVENT-RELATED ACTIVITY

Note how the scalp topographies of the ERPs in both panels of Figure 3.1 differ slightly before and after the button press. If ERP spatial variations were generated within or directly under the scalp itself, such changes would reflect potential changes occurring directly below the most strongly affected electrodes. Such an interpretation, while having naive appeal, is, however, contrary to the anatomical and biophysical facts about volume-conducted cortical field potentials that actually produce scalp EEG signals, as summarized above. The very broad "point-spread" pattern of potentials propagating out by volume conduction from each cortical source area means that each source contributes to some extent to the signals recorded at nearly all of the scalp electrodes, and contributes appreciably to many of them.

Further, if each source area is spatially fixed (or nearly so), by itself it cannot produce a moving topographic pattern of field activity on the scalp; it can only produce proportional and *simultaneous* changes across all the electrodes in its projection pattern. Therefore, changes in the scalp map of average ERP data (as in Figure 3.1) must reflect the sums of time-varying potentials projected in the broad and highly overlapping scalp patterns from several spatially fixed EEG source activities, each contributing to the ERP in spatially and temporally overlapping time windows. This view is compatible with fMRI results showing that cortical activations and deactivations mainly occur within compact cortical domains—though direct high-resolution, multiscale observations of electrocortical activity that could constitute "ground truth" tests of this assumption are not yet available.

Two main points are noted here. First, basic biophysical knowledge is *not* consistent with an interpretation of ERP potentials as exclusively (or even principally) reflecting activity generated directly below each electrode, a fact it is easy to lose sight of when focusing on the details of single-channel ERP or EEG waveforms. Second, although, when animated, changes in high-density ERP time series appear to flow across the scalp, features of most ERPs recorded in cognitive experiments are much more likely produced by sums of time-varying activities of relatively small, spatially fixed cortical generator domains, each with a broad point-spread pattern of projection to the scalp surface. Looking ahead a bit: Although the exact size distribution of these domains is not yet known, fairly precise indications of their centers can be obtained by methods (e.g., ICA) that spatially filter the scalp EEG data to focus on single sources.

ERPS AS SPATIAL FILTERS?

Early ERP analysis attempted to deal with the difficulty in interpretation posed by volume conduction and scalp mixing by assuming that event-related averaging provides sufficient spatial filtering of the many source signals reaching the scalp electrodes so that activity from only one affected source area contributes to each ERP amplitude peak. That is, some early ERP researchers hoped that the sequence of peaks comprising ERP waveforms would each spatially filter out all activities not generated in a single cortical area.

However, subsequent research clearly suggested that soon after sensory signals arrive in cortex following meaningful events, multiple EEG sources begin to contribute to ERP averages. It has now been shown that in animals, coordination of activities between early visual areas (Grinvald et al., 1994) and between primary visual and auditory cortex (Foxe & Schroeder, 2005) begins as early as 30 ms after stimulus onsets, and invasive recordings from epileptic patients for clinical purposes show that by at most 150 ms after presentation of meaningful visual stimuli, the phase statistics of local field processes are altered in many parts of the brain, both cortical and subcortical (Klopp et al., 2000). Thus, the somewhat different scalp distributions in the single-channel ERP scalp field maps of Figure 3.1 in fact represent differently weighted mixtures of mean event-related activities, time-locked to subject button presses, from several to many cortical sources having broad and strongly overlapping scalp projections. Also, direct cortical recordings from both animals and humans show that, via activity cycles through thalamocortical and other network connections involving significant delays, single cortical areas often produce multipeak complexes instead of single peaks in response to single stimulus presentation events (Swadlow & Gusev, 2000), adding to the spatial overlapping of source activities in ERP waveforms.

Thus, response averaging is not an efficient method for spatial filtering of event-related EEG data since it typically does not produce a sequence of simple maps each reflecting the projection to the scalp of a single cortical source. Computing the ERP *difference wave* at each scalp channel between ERPs in two contrasting conditions may more effectively isolate condition differences in the activity of a single generator or set of generators, although in general, the effectiveness of this approach cannot be guaranteed. Finding more effective methods for spatial filtering of EEG data into EEG source activities is therefore of urgent importance to human electrophysiology research.

TEMPORAL VARIABILITY OF EVENT-RELATED ACTIVITY

Another problem with modelling event-related EEG dynamics based on ERP averages alone is that ERP averaging collapses and thus conceals both the orderly (event-, task-, or context-related) as well as the disorderly (event-, task-, and context-unrelated) trial-to-trial variability in the recorded EEG scalp and source signals—giving no way for the user to determine the relative proportions or types of these two classes of signal variations that are present in the single trials. A model of brain activity built solely on an ERP average of scalp activity in event-related epochs must fail to include many aspects of the brain processes that produce it. If the single-trial EEG epochs each sum activity and time-varying activities from multiple spatial sources whose dynamics are tightly linked to multiple task or context-related factors, focusing solely on their average may discourage study of their orderly trial-to-trial variations.

Averaging itself simplifies not only the spatial pattern, but also the temporal patterns of the signals averaged, retaining only that portion of the signals (typically a small portion) that is both time-locked and phase-locked to the time-locking signals (as explained above and in Chapter 2, this volume). Far too often, trial-to-trial variability of EEG signals is simply dismissed by researchers (either explicitly or implicitly) as representing irrelevant brain "noise"—without sufficient consideration or evidence for this assumption. To consider this point more carefully, let us examine some aspects of the trial-to-trial temporal variability in scalp EEG activity before and after target presentations in the selective visual attention task session for which Figure 3.1 showed the ERP averages.

A first step toward understanding trial-by-trial variability in EEG epochs time-locked to some class of time-locking events is to find useful ways to visualize their variability. Makeig and Jung have developed a method of sorting the order of single trials by some criterion and then plotting them as horizontal color-coded lines in a rectangular image they called the *ERP image* (Jung et al., 2001; Makeig et al., 1999b). In ERP-image plots, single-trial traces are drawn as horizontal colored lines, with color (instead of vertical placement) encoding potential. The colored lines can be fused into a rectangular image and smoothed, if desired, with a vertical moving average to bring out trends. Crucially, the order in which the trials are sorted (bottom to top) need not follow the actual time order of the trials, but can be based on any other criterion. Unlike the average ERP, which is fixed, the ERP images resulting from different trial-sorting orders can differ dramatically from each other, each bringing out one or more aspects of trial-to-trial variability in the data.

Panel A of Figure 3.2 uses this method to visualize nearly 500 single trials time-locked to visual target stimulus presentations at a scalp site near the vertex (Cz)—the same stimulus-locked trials averaged to form the ERP shown in Figure 3.1A, with the (vertical) order of trials here sorted in their original time order. For easier visualization of event-related patterns across trials, in each panel of Figure 3.2 we have smoothed the single-trial data (vertically) with a 20-trial moving window. The solid vertical line in panel A marks the onset of the visual target, and the vertical dashed line marks the subject's median button press latency. The average ERP (shown below the panel) appears to have only a small and temporally diffuse late positive complex (LPC or, for historical reasons, "P300") feature, here peaking near 400 ms after stimulus onset.

Panel B of Figure 3.2 shows the same data trials, but here sorted by the latency of the subject's button press (dashed trace), again smoothing with a 20-trial moving average. We now see that in most trials, a (much more distinct) LPC follows the subject's button press by about 120 ms, with LPC amplitude smaller in long-reaction-time (RT) trials (near the top of the ERP image panel). This panel shows which features of the poststimulus ERP are primarily time-locked to the stimulus onset itself (e.g., the negative blue preresponse N2 peak) and which to the subject's behavioral response (the following LPC and two sparser ensuing positive wave fronts). The trial-to-trial latency variability of the LPC cannot be labeled as irrelevant trial-to-trial noise, and cannot be deduced from the stimulus-locked trial average (the blue trace below panels A and B) in which trial-to-trial variability time-locked to the button press rather than the stimulus onset is temporally "smeared out."

Panel C of Figure 3.2 shows the same trials, here sorted by mean potential in the N2 response ERP latency window indicated by the dashed lines. We see that only in about the bottom half of the trials is the mean potential in this window actually negative. But this includes about a (bottom) third of the trials in which the negativity is relatively large, thereby "outweighing" the contributions of the trials in which the single-trial value is positive, thus producing a negative peak in the average ERP (again shown below the ERP-image panel).

The curving post-RT positive (red-orange) wave fronts in the ERP-image plot in panel B reveal that the evoked LPC can be more accurately represented by an ERP average of the same data epochs time-aligned to the subject's button press rather than to stimulus onset—as in panel D. In particular, the average motor response–aligned ERP (below panel D) better reflects the abrupt onset, slope, and duration of the post–motor LPC in the single trials than the stimulus-aligned ERP (below panel A). Like panel B, panel D shows that the post–button press positivity is generally stronger in (lowermost) trials with short button press latencies and is weak or even absent in (uppermost) trials with long response latencies. Also, the slightly curving (red) response column in panel D suggests that the time locking of the LPC peak to the button press is only relatively constant, the LPC appearing slightly wider and centered slightly later in shortest-latency trials relative to other trials.

These scalp data, measuring the potential difference between an "active" electrode near the vertex (Cz) and a "reference" electrode on the right mastoid (see the cartoon head above panel A), actually sum broadly spreading projections of field activities projecting by volume conduction from several to many cortical sources. The mean power spectrum for these trials (above panel A) contains a strong alpha-band peak at 10 Hz. Panel E shows the same response-aligned trials as in panel D, now sorted according to the best-fitting 10 Hz phase in the indicated three-cycle (300 ms) wide trial sorting window centered on the LPC. In this ERP-image view of the data, the central positive (red) LPC peak of the activity in the single trials forms a red curving wave front near the center of the sorting window. The peak latency difference (from bottommost to topmost trials in the image), clearly visible in panel E, is hidden in panel D using a different trial-sorting order.

Panel F shows the same response-aligned trial data, but now sorted by alpha phase in the preresponse (but poststimulus) period between the dotted lines. Here, we see that ongoing alpha activity is

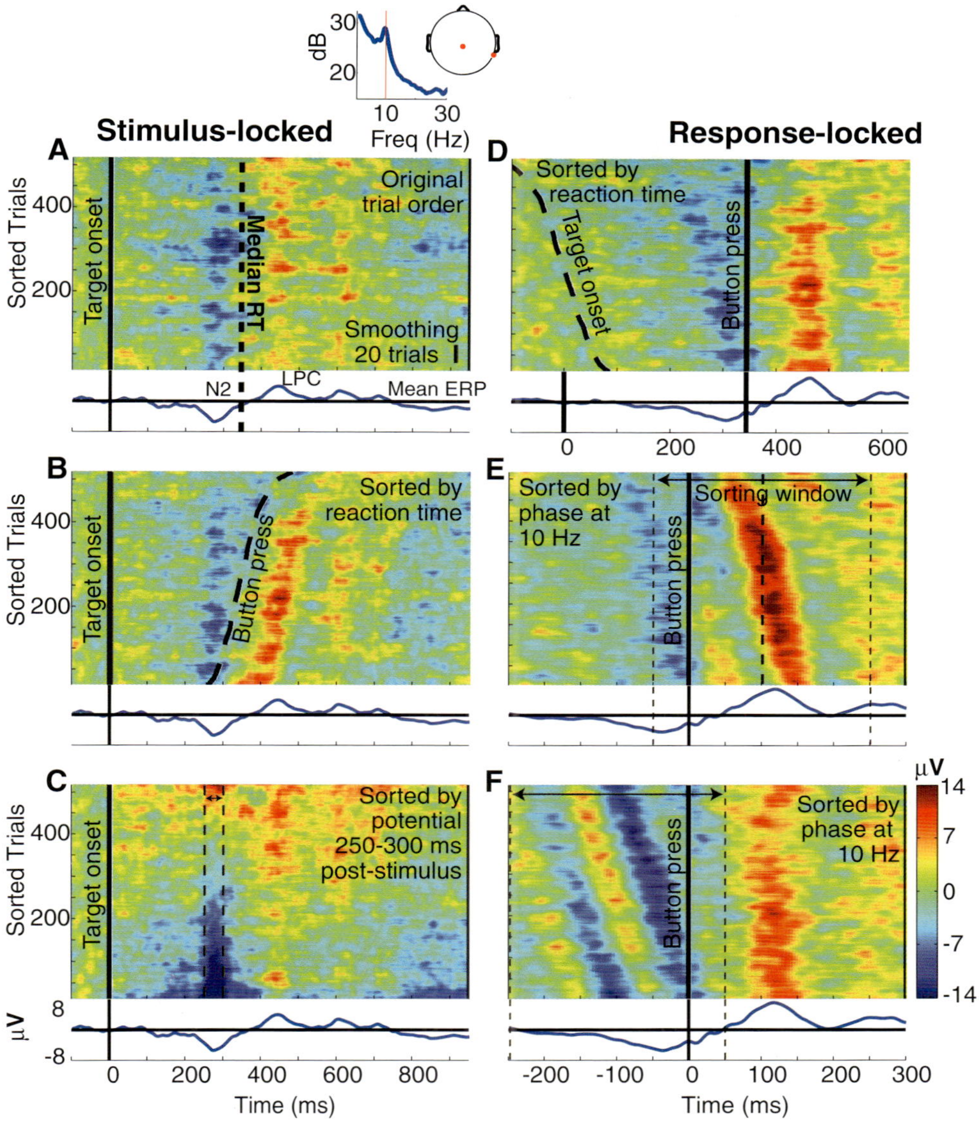

Fig. 3.2. Six ERP-image plots of nearly 500 visual target response epochs (same data as in Figure 3.1) at a scalp electrode near the vertex (referred to right mastoid). In each panel, trial potentials are sorted (from bottom to top) in the order indicated, then smoothed (vertically) with a 20-trial moving window, and finally color-coded (see the color bar at the lower right). In the left panels (A–C), the trials are aligned to the moment of stimulus onset in each trial. In the right panels (D–F), they are aligned to the moment of the subject button-press response. The trace below each image shows the ERP average of the trials. Stimulus-locked ERP peaks N2 and LPC (late positive complex) are labeled in (A). The trial-sorting criteria are indicated in each ERP-image panel. Horizontal arrows show the value (C) and phase-sorting (E, F) windows. The ERP images illustrate the wide variety of trial-to-trial differences in the data.

random phase before the button press (as reflected in the perfectly diagonal positive and negative wave fronts in the sorting window), but the ensuing LPC positivity peaking in this view about 120 ms after the button press is nearly vertical in this view, and therefore appears to be independent of the phase of the preceding alpha-band activity (which might well have a different set of sources than those generating the LPC). Note that the LPC positivity is again wider than the tighter peak obtained in the alpha-sorted view of the same data in panel E, though panels D–F) all visualize aspects of the same data and have the same motor response-locked ERP.

What conclusions can we draw from these six quite different ways of plotting the same data? Note that each panel highlights a different way of separating the ERP trial average into a set of single-trial activities, followed by moving-average smoothing to bring out trial-to-trial trends in the trial-sorted data images. Each panel represents, in a way, a decomposition of the single-trial data highlighting some foreground features and smoothing other types of trial-to-trial variability to form a kind of background noise (as it were). Which of these decompositions in this sense—if any of them—is the most physiologically realistic or "correct" decomposition? Arithmetically, there is no difference between them; in each case the same trial data, aligned as in panels A–C or D–F, have the same ERP average, no matter how the trials are sorted—just as five pennies are equally the sum of 3 + 1 + 1 or 1 + 3 + 1 coin subsets.

The standard model underlying ERP analysis is that the EEG data essentially sum (1) contributions to the ERP occurring in each trial and (2) other (undefined) variable activities unaffected by the time-locking events and therefore not contributing to the ERP. Is this standard (ERP + background) decomposition of the single-trial signals more physiologically realistic, in any sense, than the different implied decompositions of the data into the quite different features that are highlighted (plus those obscured) in these six quite different ERP-image panels? In particular, do any of these views parse the data into physiologically distinct source contributions?

From these ERP-image representations of the data, we can at least see some ways in which an average ERP of the trials is simply one statistical measure of them, a measure that does not reveal their orderly (though complex) temporal variations or multiple spatial sources. In each trial, the subject performed the same task—attempting to produce quick button press responses to target stimuli while withholding responses to nontargets. Panel B clarifies at least one aspect of trial-to-trial EEG variability directly related to subject behavior (e.g., their manual response latency). What other trial-to-trial differences, either in stimulus features (here, in target location) and/or in the trial context (here, the history of preceding stimuli and manual responses), may have altered the nature of the "challenge" posed to the subject's brain—and thus affected aspects of trial-to-trial EEG variations? Neither the ERP averages nor these six ERP-image panels answer these questions. There are many other possible decompositions into putative underlying EEG source activities that examination of its trial average ERP cannot rule in or out as reflecting physiologically valid distinctions among spatial sources and their event-related temporal patterns.

At the least, these panels illustrate the fact that trial-to-trial variations in EEG data are not simply EEG noise. Rather, they include several types of orderly trial-to-trial variability linked to several EEG source and/or behavioral and task parameters. Panels C–F, in particular, pose an interesting question. Is the LPC activity at this channel, time-aligned to the subject's button presses, dominated by positive-phase alpha activity (as panel E suggests) or by a broader fixed-latency, central LPC peak (as in panel F)—or perhaps by both types of activity arising in different cortical domains and both projecting to the vertex to differing relative extents in different trials?

A simple ERP model of the trial-to-trial variability in EEG data represents the non-ERP portion of the data as summing contributions of brain source activities that are unaffected by the time-locking events. In an extreme version of this model, the amplitude of the ERP might be assumed to be identical in every trial, with any trial differences reflecting additional task-irrelevant EEG (or nonbrain artifact) activity. But Figure 3.2 suggests that such a strict version of the model is unlikely to adequately capture the trial-to-trial variability in the event-related dynamics in these trials.

Panel A of Figure 3.3 shows the same trial data as in Figure 3.2, but now sorted by mean potential in the indicated postresponse LPC data window. Panels B and C visualize an ERP-based decomposition of the data in panel A, all three panels using the same trial-sorting order (sorting by LPC amplitude). Panel B shows the estimated contribution of the mean ERP to each trial, as determined by finding the best least-square fit of the mean ERP to each single-trial epoch but not allowing negative ERP trial weights in the few lowest trials. Panel C shows the remaining (non-ERP) data for each trial. Thus, the sum of the values in panels B and C are the whole-trial data as shown in panel A. At least two latency windows in panel C (90–150 ms and 200–300 ms) exhibit systematic differences from a random trial distribution, indicating that the trial-to-trial variability of potentials in those windows is partially independent of the overall amount of ERP-like activity in the trial. But is this attempted decomposition of the single-channel data (in panel A) into ERP and non-ERP data portions (in panels B and C) a physiologically valid separation between completely motor response-locked (ERP) and motor response-independent (non-ERP) cortical source processes in the data?

How can we begin to find answers to this question? To model the nature of the highly

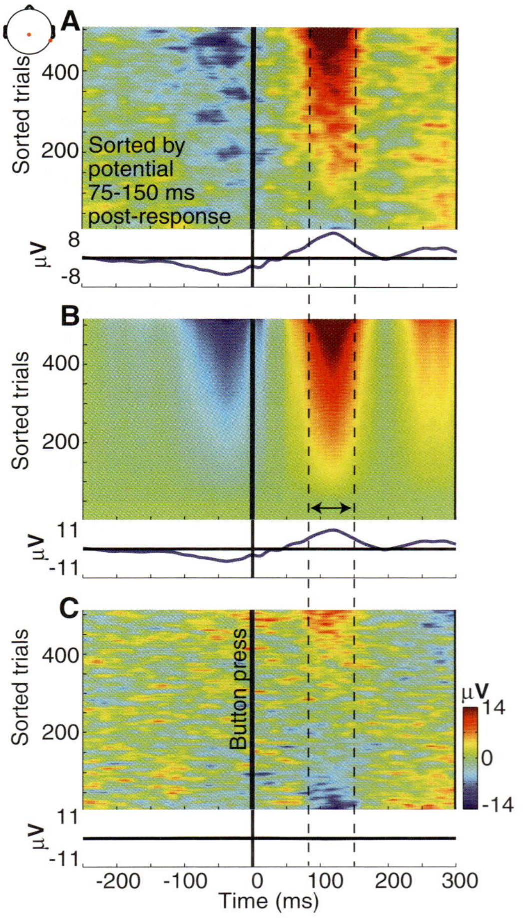

Fig. 3.3. Event-related potential-image plots separating the ERP image in (A), where the trials are sorted by mean potential between the two dashed lines surrounding the postresponse ERP peak into the sum of (B) an ERP-image of the best-fitting (nonnegative) mean-ERP contribution to each trial and (C) the remaining unexplained portion of each trial. As in Figure 3.2, the mean of the single-trial data in each panel is shown below the panel. The ERP has been largely (though not completely) removed from the lower-panel data. This decomposition is compatible with the assumption that the single-trial data sum a single ERP response of variable amplitude (B) plus unrelated EEG activity (C). However, many EEG sources may make separate contributions to the data and to the ERP average. Regression of the whole average ERP trace on the single-data trials does not take into account the spatiotemporal variability of the independent sources of trial-to-trial variation—for example, the partial sorting by value remaining within the sorting window and near 300 ms. (The vertical smoothing window width is 20 trials.)

variable signals occurring during these recorded data epochs, ideally we should first find a way to separate the whole-scalp EEG data into a set of functionally and physiologically distinct source activities.

Separating EEG Sources Using ICA

In 1995 the first author and colleagues at Salk Institute (La Jolla, California) performed the first decomposition of multichannel EEG data into ir maximally independent components (Makeig, 1996) using a then-new and elegant "infomax" algorithm (Bell & Sejnowski, 1995) that followed insights, a few years earlier, that weighted sums of independent source signals should be separable "blindly" into the individual source signals *without* advance knowledge of the nature of the source processes, as had been thought necessary (Comon, 1994; Jutten & Herault, 1991). The prototypical example of this problem is the "cocktail party problem," in which an array of microphones records mixtures of the voices of several people talking at once at a cocktail party. Individually, the recordings sound like indecipherable "cocktail party noise." The blind source separation problem is to determine how to combine the recorded signals so as to separate out each speaker's voice, "blind" to any knowledge of the nature or properties of individual sources.

At root, the insight allowing the solution to this problem is that the individual speakers' voices are the only sources of independent *information* in the recorded data. By adapting randomly weighted sums of the recorded signals in such a way as to make the weighted-sum signals more and more *temporally independent* of each other, the unmixing process must finally produce the individual voice signals. In signal processing terms, the joint microphone data are separated into ir maximally independent signal *components*, which must be the original voice sources since they are the only possible independent sources in the recorded mixtures.

When the process proceeds without relying on any knowledge of the qualities of the individual source signals, the unmixing process is called *blind source separation*. Using temporal independence to separate the source signals is a form of blind source separation called *independent component analysis* (ICA). Intuitively, two time series are maximally independent when their waveforms are maximally distinct from each other. More technically, independent time series have no *mutual information*, meaning that knowing the value of one process at a given time gives no information at all (even partial or probabilistic) about the concurrent value of the other process.

Reading about the ICA solution to the cocktail problem in the influential paper of Bell and Sejnowski before its publication in 1995, the first author suspected that the same approach should be applicable to EEG data. The results of the first EEG decomposition (Makeig, 1996) were highly promising, and subsequent work over the next dozen years or more has confirmed the ability of ICA to identify both temporally and functionally independent source signals in multichannel EEG or other electrophysiological data. In effect, ICA creates a set of spatial filters. Each *independent component* (IC) filter cancels out the contributions of all but one of the distinct source signals that contribute to the multichannel data. Thus, ICA can be thought of as a method of data-driven spatial filtering or beamforming (Iyer et al., 1990) that learns spatial filters that each focus on a distinct physical source of EEG signals—separating out distinct brain generator processes as well as different nonbrain ('artifact') signals.

More formally, a linear decomposition of a (channels by time points) signal matrix is its representation as any weighted sum of component signal matrices of the same size (the same numbers of channels and time points). Figure 3.4 schematically visualizes the simple matrix algebraic formulation of the linear signal decomposition used in ICA. The scalp data channel signals are formed into a matrix (top center). Independent component analysis decomposition finds an *unmixing* matrix (W) that, when multiplied by the data matrix, *decomposes* the data (downward-pointing arrow) into a matrix of IC signals, called the *IC activations* (lower right), of the same size as the input scalp data. Multiplying the IC activations matrix by the inverse of the unmixing matrix W (lower middle) reconstitutes or *back-projects* the original scalp data channels (upward-pointing arrow).

The inverse of the unmixing matrix, W^{-1}, is the component *mixing* matrix (lower center) whose columns give the relative strengths and polarities of the projections of one component source signal to each of the scalp channels. In Figure 3.4, the values in the columns of the mixing matrix are color coded and interpolated onto cartoon heads to visualize the topographic projection patterns or *scalp maps* associated with each of the sources.

The component scalp maps found by ICA decomposition are not constrained to have any particular

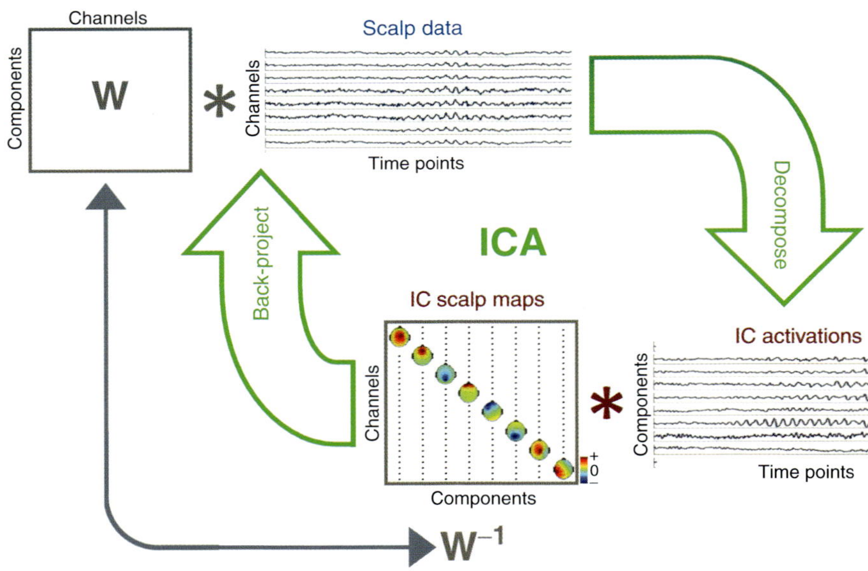

Fig. 3.4. Schematic flowchart for ICA data decomposition and back-projection. Independent component analysis applied to a matrix of EEG scalp data (upper middle) finds an "unmixing" matrix of weights (W, upper left) that, when multiplied by the (channels by time points) scalp data matrix, gives a matrix of IC activities or activations (lower right). This is the process of ICA decomposition (downward arrow) of the data into maximally temporally independent processes, each with its distinct time series and scalp map. The process of back-projection (upward arrow) recaptures the original scalp data by multiplying the IC activations matrix (lower right) by the matrix of IC scalp maps (lower center) whose columns give the relative projection weights from each component to each scalp channel. The IC scalp map or "mixing" matrix (W^{-1}, lower center) is the inverse of the mixing matrix (W, upper left). In simple matrix algebra form, if the indicated scalp data matrix is X and the component activations matrix is U, then algebraically $WX = U$ and $X = W^{-1}U$. Here, W is a matrix of spatial filters learned by ICA from the EEG scalp data that, when applied to the data, finds the activity projections of the underlying EEG source processes and the IC activations (lower right). This general schematic holds for all complete linear decomposition methods returning as many components as there are data channels.

relationship to each other (unlike in PCA decomposition). They may be highly (though not perfectly) correlated. They may also have any (simple or complex) spatial pattern, although in practice, scalp maps for components truly accounting for a distinct source process (contributing independent temporal information to the data) must reflect the relative projections of the source process to the individual scalp channels. Thus, a source comprised of spatially coherent local field activity across a cortical patch must have a scalp map that nearly perfectly matches that of a single tiny battery (dipole) placed in "the electrical center of mass" (as it were) of the source patch and called its *equivalent dipole* (Scherg, 1990). However, IC scalp maps may again have any form, depending on the pattern of the source projection to the scalp electrodes and on the degree of dominance of a (maximally) independent component by a single signal source.

What Is an IC?

Before going further, we must first discuss a basic terminological confound. Independent component analysis (like PCA and other linear decompositions) uses the term *component* to mean something quite different than its use elsewhere in this volume (including its title): It is used as a contraction of the term *ERP component feature*—some identifiable feature in an ERP waveform (typically associated with a single peak). By broader definition, an ERP component may be any functionally distinct feature or portion of an ERP waveform, that is, a feature with a functionally distinct relationship to experimental parameters and/or an ERP feature generated in a particular brain region (see Chapter 1, this volume).

In this chapter, however, we will use the term *component process* (or *component* for short) to mean some portion of an entire multichannel recorded data set separated from the remaining recorded data by linear decomposition. To minimize confusion, we will substitute a terminological equivalent, *ERP peak* or *ERP peak feature*, for the more usual term *ERP component*. Thus again, in this chapter, *components* will not refer to ERP peaks or other features but to *EEG source processes*, each accounting for some portion of the continuous EEG activity (at all time points) forming a multichannel data set. Each data set component naturally then also accounts for some portion (large, small, or negligible) of any ERP average of epochs drawn from the data set.

As shown in Figure 3.4, an independent component (process) of an EEG data set (or IC for short) comprises *both* a fixed scalp map *and* a time series that gives its relative amplitude (or *activation*) and polarity (positive or negative) at each time point. The scalp map shows the relative weights or projection strengths (and polarities) of the projection from the component process to each electrode location. The component activation time series gives the relative amplitude and polarity of the component's activity at each time point. Because we define an EEG source as being spatially stable, a component scalp map remains constant over time. The back-projection of each component process to each scalp channel is the product of the component activation time series with the scalp map weight for that channel. The IC back-projection to all the channels is the portion or component of the scalp data (at all channels) contributed by the component process. The original channel signals are the sums of the back-projected activities of all the independent components. That is, the scalp data are the collection of all the summed back-projections of all the independent components to all the channels.

In simple matrix algebra form, if the scalp data matrix is X and the component activations matrix is U, then algebraically $WX = U$, where W is the unmixing matrix of spatial filters learned by ICA decomposition of the EEG scalp data. This equation simply says that applying spatial filters W to the data X (by simple matrix multiplication) gives the activation time courses of the IC processes. The converse process that reconstitutes the data from the ICs is, algebraically, $X = W^{-1}U$, where W^{-1} (the matrix inverse of W) is the component mixing matrix. These same equations can be used to represent any linear decomposition method, though other methods may use different names for the matrices.

ICs of EEG Data

It is important to understand that each scalp EEG recording channel is itself in effect a spatially filtered measure of the varying scalp potential field, recording only the time-varying potential difference between two scalp electrodes, the *active* electrode and one or more *reference* electrodes—at which brain and nonbrain signals are potentially just as "active" as the so-called active electrode. Independent component analysis attempts to replace these scalp channel electrode-difference filters with IC filters using other linear electrode combinations chosen so as to pass individual EEG source signals while rejecting all other sources. The degree of source fidelity ICA can achieve depends on the number of data channels versus the number of active sources—as well as on the length and quality of the data.

Before considering in detail the assumptions underlying ICA and giving heuristic guidelines for how to apply it, let us first show model examples of independent EEG components or ICs. Independent components of EEG data can be roughly separated into three types: ICs accounting for brain and nonbrain (artifact) processes, respectively, and small ICs whose maps and activities appear noisy and are poorly if at all replicated from session to session. This last category can be considered a noisy part of the EEG signals that ICA is unable to resolve into components dominated by a single source (although not every small IC fits this description). Between these three IC categories, there may be ICs in "gray areas" whose assignment to one of these three categories is difficult. Here, let us first consider ICs clearly accounting for activity from particular nonbrain (artifact) sources.

INDEPENDENT NONBRAIN COMPONENT PROCESSES: NOISE OR SIGNALS?

Characteristically, ICA separates several important classes of nonbrain EEG artifact activity from the rest of the EEG signal into separate sources including eye blinks, eye movement potentials, EMG and ECG signals, line noise, and single-channel noise (Jung et al., 2000a, 2000b). This important benefit of ICA decomposition of EEG data was apparent from the first attempt to apply it (Makeig, 1996). Thus, ICA has found initial use in many EEG laboratories simply as a method for removing eye blinks and other artifacts from data. For data sets heavily contaminated by eye blinks or other artifacts—for instance, data collected from young children—the ability to analyze brain activity in data trials including eye movement artifacts can mean the difference between analyzing and rejecting the subject data altogether.

Unlike regression-based methods for artifact removal, ICA artifact separation allows artifact subtraction (often called artifact *correction*) without requiring a separate (*pure*) reference channel for each signal. In practice, regression methods risk eliminating brain signals that also project to the (impure) reference channel (e.g., frontal brain sources also project to an electrooculographic [EOG] channel near the eyes). Figure 3.5 shows scalp maps, spectra, and ERP-image plots (above their trial-average ERPs) for four typical independent artifact source components separated by ICA from the visual selective attention task session considered in earlier figures. The highly distinct activity features separated from the data by ICA make the qualitative implications of temporal independence clear. The recovered component waveforms are the most *temporally distinct* portions of the recorded data. Separation by ICA of nonbrain source processes allows detailed analysis of the separated source process time courses. Note, for example, that the subject refrained from blinking for over 1 s following target stimulus presentations (upper panels in Figure 3.5).

When, as here, the electrode montage includes both head and neck sites, scalp maps of head muscle components exhibit a characteristic polarity reversal at the insertion point of the muscle into the skull, with the direction of the dipole following the direction of the muscle fibers. Note the scalp map associated with an EMG component signal (Figure 3.5, lower panel). Note also the abrupt changes in EMG activity level in the component ERP-image plot near trials 180, 300, and 370 (lower left), a common occurrence for scalp muscle activity recorded during experiments in which the subject is sitting comfortably while attempting to minimize head and eye movements. These marked changes in activity level of this muscle were likely neither willfully controlled nor noted by the subject. By so clearly separating nonbrain processes contributing to EEG data, ICA allows these activities to be analyzed as concurrently recorded biological (or other) signals instead of simply being rejected as nonbrain artifacts.

SPATIALLY STEREOTYPED VERSUS NONSTEREOTYPED ARTIFACTS

It is important, however, to understand the distinction between spatially stereotyped nonbrain signal sources, such as eye blinks and scalp muscle activities that always project with the same topographic pattern to the scalp channels, and nonstereotyped nonbrain signal phenomena that have varying spatial scalp projections. Consider, for example, the case of an unruly subject who vigorously scratches his or her scalp for a second or two during the EEG recording. This quickly produces a series of hundreds of EEG data points (i.e., EEG scalp maps) whose topographic patterns do not match each other or appear elsewhere in the data. The one-time-only appearance of each of these scalp maps is in effect temporally independent of all other data sources, possibly hugely increasing the number of temporally independent sources ICA needs to separate into a finite number of component activities. Further, during this period, the changes in electrode contact with the skin may alter the spatial pattern with which the other brain and nonbrain signal sources project to the electrode array, violating the

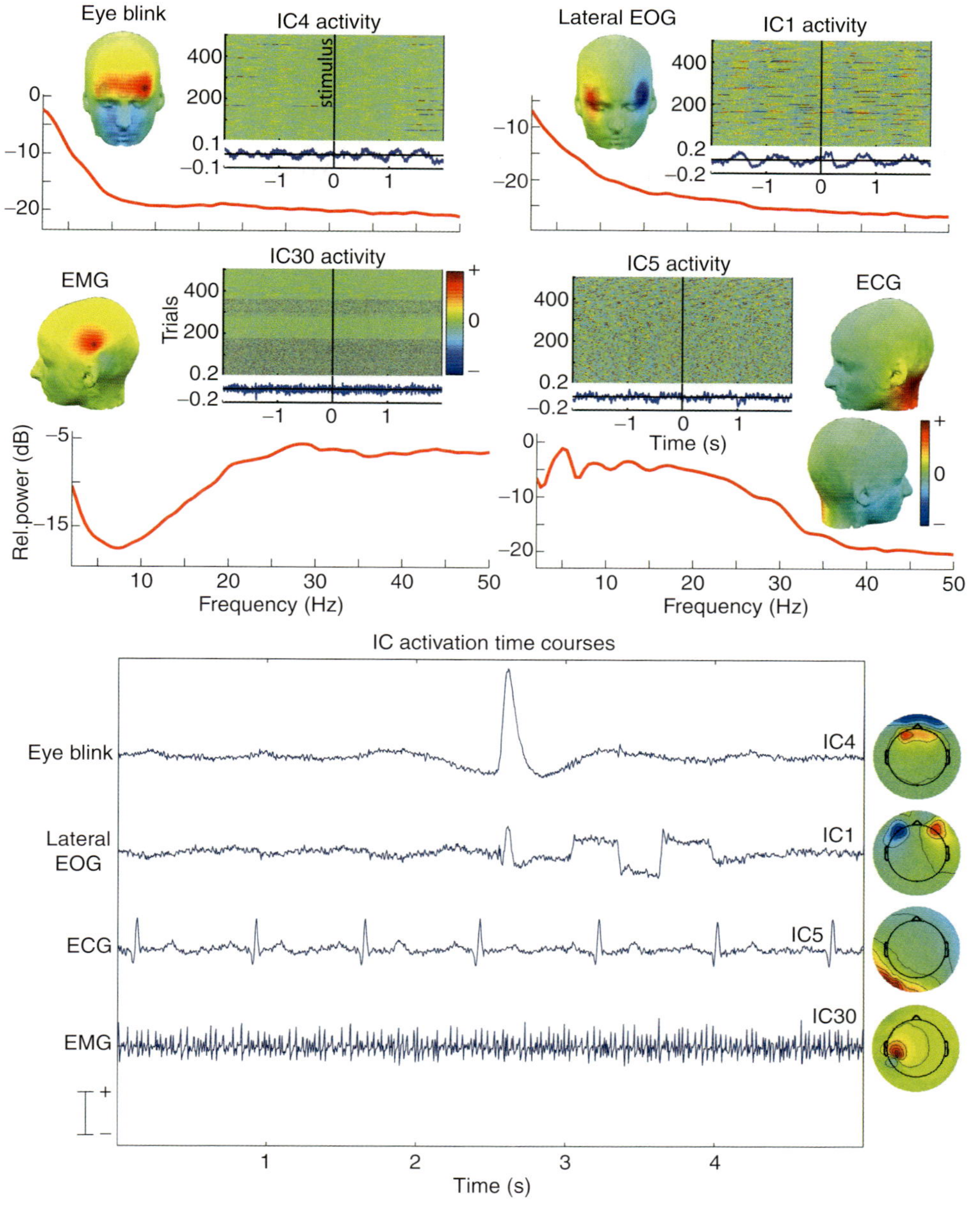

Fig. 3.5. Typical component properties of four nonbrain independent component processes accounting, respectively, for eye blinks, lateral eye movements, left postauricular EMG activity, and ECG activity from the 238-channel EEG recording studied in Figures 3.2 and 3.3. The upper axes show the interpolated component scalp map, activity ERP image, average ERP (below the ERP images), and mean power spectrum. The lower axis the maximally independent activities of the four processes during a 5 s period. Note the characteristic activity elements, which are also seen in the ERP-image representations.

ICA assumption that these spatial projection patterns are stable throughout the data.

Thus, including a stretch of data dominated by this or any other spatially nonstereotyped (SNS) artifact in the data given to ICA for decomposition can only limit the success of the decomposition in identifying physiologically distinct EEG source processes. Such SNS periods may be identified by eye while scrolling through the data, by the use of simple heuristics (Delorme et al., 2007a), by similar observations of a preliminary ICA decomposition of the whole data, or even automatically during ICA training by computing the probability of each data point fitting the current ICA model and rejecting highly improbable data points from further training.

Cases intermediate between spatially stereotyped and nonstereotyped artifacts are phenomena with spatially stereotyped but nonstationary scalp patterns, for example slow blinks (Onton & Makeig, 2006), ballistocardiographic (BCG) cardiac artifacts recorded within a high-field magnetic resonance scanner (Debener et al., 2008), and slow waves in sleep (Massimini et al., 2004). In such cases, ICA typically finds a small set of *maximally* independent components that each captures one or more time periods of the repeating activity pattern, thereby separating it from other source activities.

INDEPENDENT BRAIN COMPONENT PROCESSES

Using ICA solely to remove nonbrain source processes, while valuable, does not exploit the power of ICA to separate the activities of individual *brain* sources that contribute to the scalp data. Some might ask that since no part of the brain acts *wholly* independently of the rest of the brain, how can ICA decomposition extract physiologically meaningful component signals? The answer to this question is that ICA finds the *maximally* independent components for a data set, even if traces of dependence remain between them. This dependence might be transient—for example, the sometimes strong similarity of occasional evoked activity in otherwise maximally independent left and right lateral occipital processes produced by central visual stimulus presentation (Makeig et al., 2002). Or the dependence might be limited—for example, only reflected in weakly coherent, low-amplitude, high-frequency activity.

Even in cases in which two ICs have remnant mutual dependence, that is, when their joint activities could be said to form "a dependent two-dimensional subspace" of the data, ICA should still separate the activity of this subspace from the activities of other single IC processes. For example, a moving scalp artifact produced by slow eye blinks might be separated into two or more ICs, each accounting for one phase of the moving blink potential. In this case, the spatiotemporally overlapping component activities would differ from one another, not allowing their parsing as a single IC. Though the ICs are not completely independent of each other, their time courses might still be independent of any other IC time course in the data and sufficiently different from each other to require more than one IC (Meinecke et al., 2002).

Applied to high-density EEG data of adequate length and quality, ICA decomposition typically produces from one to three dozen components with low mutual information and scalp maps highly compatible with an origin in a single cortical patch (or occasionally in a bilaterally symmetric pair of cortical patches), as in our definition of an EEG source above. As an example, Figure 3.6 shows 3-D scalp maps and equivalent dipole locations in an individualized subject boundary element method (BEM) head model for six independent brain components separated by ICA from the same recorded data as in earlier figures. Beside each component scalp map, the residual scalp map variance (r.v.) not explained by the best-fit single (or dual bilateral) equivalent dipole model is shown. The component scalp maps are nearly dipolar, that is, nearly matching the computed projection of a single equivalent dipole (or bilateral dipole pair), whose locations are here within an individualized BEM head model built from a subject MR image (Oostenveld & Oostendorp, 2002).

Again, the computed IC single *equivalent dipole* locations cannot represent the spatial distribution of the cortical generator domains. Instead, they represent the computed positions (in the BEM head model) of vanishingly small oriented dipoles whose scalp projection patterns match most closely the actual IC component maps (across all electrodes). In general, an equivalent dipole for a cortical patch source is typically deeper in the brain than the cortical patch itself (Scherg, 1990). Recent advances in distributed inverse source localization methods suggest that it may soon be possible to estimate, using subject MR images, the patch (or patches) of subject cortex that most likely constitute each IC source domain. Such a goal is likely not reachable by the alternative strategy of first computing an ERP average and then finding the inverse source distributions of one or more ERP scalp maps, since the ERP scalp map at any point in time is typically a weighted mixture of contributions from several cortical source areas.

ICA Assumptions

As illustrated in Figures 3.5 and 3.6, ICA decomposition has proven to be highly successful for studying EEG data. Why? An important part of the answer must be that there is an approximate fit, at least, between ICA assumptions and the physiological nature of EEG sources themselves (Makeig et al., 2004a; Onton & Makeig, 2006; Onton et al., 2006). Basically, ICA "blindly" separates the scalp data given it into component processes whose

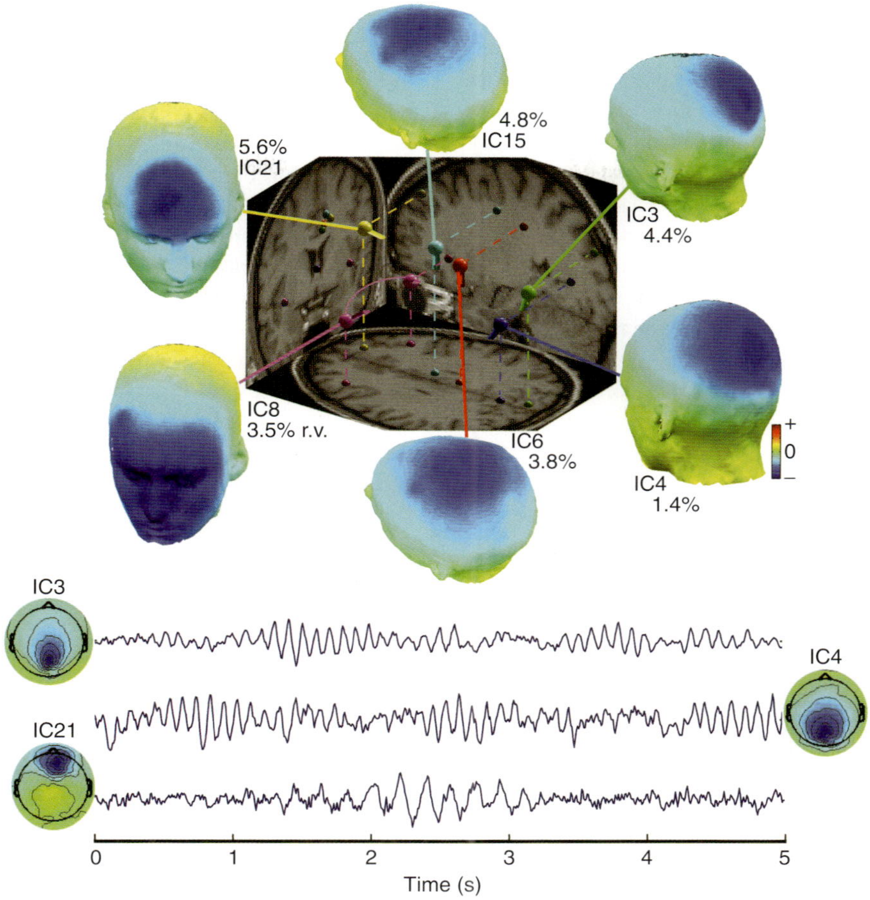

Fig. 3.6. Equivalent dipoles for six maximally independent brain source components. Near each IC index (ranked in order of variance contributed to the scalp data), the residual variance (r.v.) of the equivalent dipole model across the 238-channel component scalp map is indicated, based on fitting the measured 3-D electrode locations to an individualized three-shell boundary element method (BEM) head model. All the residual variances are low (<6%), indicating that the component maps are compatible with an origin in a single cortical patch (or, IC8, in dual bilateral cortical patches). The equivalent dipoles (center) are likely situated somewhat closer to the cortical surface than the locations of the equivalent model dipoles. The 5 s activation periods shown in the lower panel give representative (not concurrent) examples of bursts of frontal midline (IC21) theta (and higher-frequency) activity and posterior (IC3, IC4) alpha source activities for three of the components.

spatial and temporal properties are not known in advance, based on the following five assumptions:

1. The component source locations (and thereby their topographic projection patterns to the scalp sensors) are fixed throughout the data.
2. The projected component source activities are summed linearly at the sensors.
3. There are no differential delays involved in projecting the source signals to the different sensors.
4. The probability distributions of the IC source activity values are not precisely Gaussian.
5. The component source activity waveforms are (maximally) temporally independent of one another.

The last (independence) assumption can be translated informally as saying that the component source activity time patterns are *maximally distinct* from one another, an assumption partially supported by intracranial recordings from neighboring areas (Destexhe et al., 1999). More technically, a set of signals are temporally independent, in the sense used for ICA, if knowing the activity (microvolt) values of any subset of the signals at a given time point gives no clue about the activity values of any subset of remaining sources at the same time point. Thus, each component source signal is, in a particular sense, *an independent source of information* in the data, contributing a temporal pattern that is in no way determinable from the values (at the same time point) of the other component source signals.

The spread of information-based signal processing into nearly every signal processing application area in the last decade (Jutten & Karhunen, 2004) derives primarily from the basic interest of investigators in all research areas in identifying the sources of information that contribute to their multidimensional data. However, the value of ICA for decomposing any signal is determined by the degree to which the ICA assumptions fit the manner in which the data are actually generated and recorded. For EEG signals, the assumptions of simple summation at the electrodes (2) and lack of differential delay (3) are met precisely. The non-Gaussian distribution assumption (4) is plausible for EEG sources generted by nonlinear cortical dynamics as well as for nonbrain artifact sources including cardiac signals, line noise, muscle signals, eye blinks, eye movements, and so on that are not themselves sums of smaller *un*correlated signals.

As mentioned earlier, the ICA spatial source stationarity assumption (1) is consistent with indirect evidence from fMRI and other brain imaging methods, and the independence assumption (5) is consistent with the very sparse long-range corticocortical coupling and the predominantly radial thalamocortical connectivity profile. However, both of these assumptions have limitations. In particular, optical dye recordings in animals of local field potentials at the millimeter and smaller scale reveal moving wave patterns (Arieli et al., 1995), and comparison of ICA solutions across a group of subjects participating in the same task suggests that the spatially stable EEG sources separated from the data by ICA depend in part on the task the subject is performing. Thus, further research is needed on methods of identifying spatial lability in EEG source data (Anemuller et al., 2003) and of identifying changes in the spatial distribution of the sources as the subject's task, strategy, or preoccupation changes (Lee, 2000). However, given a hypothetical switch between two sites of EEG signal generation as the subject alternately performs two tasks, ICA should in theory return two components each showing the task-related activity only during one performance condition.

DUAL-DIPOLE IC PROCESSES

From the viewpoint of ICA decomposition, an EEG source is nothing more than an independent time course of spatially stationary information in the data, whatever its scalp projection pattern. The scalp projections (and hence, scalp maps) of ICA components are thus constrained only by the projection patterns of the actual physical sources of the data. Cortical (or other) source signals arising in separate cortical patches may be partially or wholly synchronized if the separate patches are physically linked by dense white matter tracts (such as corpus callosum) or are identically stimulated. In this case, ICA decomposition will (rightly) return a component summing the scalp projections of the two (physical) source patches. For example, a single IC typically accounts for eye blink artifacts from the two eyes, whose synchronized small upward movements during the blink induce electrical activity accounted for by an equivalent dipole located in each eye.

Similarly, ICA may return one or more brain components whose scalp maps sum the projections of two equivalent dipoles, usually with bilaterally near-symmetrical locations and scalp projections, compatible with patches connected by corpus callosum. Theoretically, cortical activities on either end of any dense white matter tract might synchronize and their activities might be combined into a single IC, though for us this has not yet been conclusively demonstrated. It makes no sense to say that ICA fails to separate "sources" in this case, unless one for example (re)defines the term *EEG source* to mean activity in a single cortical patch. However, in practice, the number of dual-dipolar ICs is relatively small (except for most ICs accounting for eye movements, thankfully).

ICA AMBIGUITY

Discussions of the polarity and amplitude ambiguity inherent in IC activations in some early ICA papers have been confusing to some readers. In fact, this ambiguity is present only when the IC activations and IC scalp maps are considered separately. We might say that the sign and scaling of the (back-projected) component in the data are split (arbitrarily) between its activation and scalp map. Since $-1 \times -1 = 1$, inverting the signs of both an IC activation and its scalp map will not change their product, or the back-projection of the IC into the original data, which will retain its original polarity. These ambiguities should be kept in mind when examining or comparing IC activations or scalp maps.

However, the microvolt scaling of the back-projected IC scalp activity is precisely the product of the scalp map values with the activation time series. Thus, ICA decomposition does not lose this information, as is sometimes mistakenly suggested. Also, while IC potentials at the cortical surface are also proportional to the IC activation, accurate source location and electrical head models are needed to determine the actual IC strength on or in the cortex, since this depends on the resistance between scalp and cortex, which in turn varies across heads and source locations.

Note that ICA does not itself sort the components into any fixed order. Thus, decompositions of similar data, even data from the first and second halves of the same recording session, are not guaranteed to return ICs in the same order. The ICs from different data sets need to be compared with each other using one or more measures of their time courses and/or scalp maps—for example, their power spectra and equivalent dipole locations.

NUMBER OF DATA CHANNELS

How many data channels should be used for ICA filtering to be successful? The most computationally efficient and robust ICA methods, such as infomax ICA, neither increase nor decrease the dimensionality of the data; they find the same number of components as there are data channels and are therefore called *complete* decomposition methods.[6] How many independent sources contribute to EEG data? It is highly likely that there are always more (brain and nonbrain) sources with distinct (e.g., near-independent) time courses and unique scalp maps than any possible number of recording channels, since synchronized cortical field activity likely occurs, at least transiently, at more than one spatial scale, and to some extent uncorrelated noise is generated at each of the electrodes. Most such source activities will be small to negligible, but their presence guarantees that the number of degrees of freedom of the recorded data will never be less than the number of data channels.

Data contributions from numbers of sources beyond the number of available component degrees of freedom (i.e., beyond the number of data channels) will be mixed into some or all of the resulting components, thereby adding a kind of noise to the results of the decomposition. The noise inherent in ICA decomposition of EEG data is evidenced by the indeterminate scalp maps of the very smallest ICs in a high-dimensional data decomposition, ICs that may not prove stable under repeated decomposition and whose scalp maps are often far from dipolar (i.e., resembling the projection of a single dipolar source). Because of the need for ICA to mix all of the EEG sources into the available number of components, decomposing data with a larger number of (clean signal) channels may be preferable when there are enough data to decompose them (see the following discussion). But decomposing a smaller number of channels will likely prove beneficial as well.

DATA REQUIREMENTS

Successful ICA decomposition requires an adequate amount of data. We may say, metaphorically, that the independence of many source signals cannot be "expressed" in brief mixtures of them. To express their independence (or, less metaphorically, for an ICA algorithm to recognize it), a considerable amount of data is typically required. Thus, successful ICA decomposition typically profits from being applied to a large amount of data, typically *the entire collection of continuous data or extracted and then concatenated single trials from an event-related EEG/ERP task session*. The most frequent mistake researchers make in attempting to apply ICA to their data is to attempt to apply ICA decomposition to too few data points. For data with large numbers of channels (64 or more), we have observed that it is better to decompose a number of time points at least 20 or more times the number of channels *squared*. This is only a heuristic standard, not a strict minimum, and using this much data does not in itself guarantee an optimal decomposition. For very dense scalp arrays, this standard could require an unreasonable amount of data. For example, to decompose 256-channel data 20×256^2 time points at 256 points/second would require over 80 min of recording time and occupy nearly 1.5 GB, though by this same standard for 64-channel data, a 22-min recording occupying about 0.35 GB would suffice.

We are not as sure about the influence of sampling rate on ICA decomposition. Doubling the sampling rate during a recording period shortened by half might not produce as effective a decomposition, since the higher frequencies captured in the data acquired with a higher sampling rate would be small, relative to lower-frequency activity, and might have a lower source signal-to-noise ratio. See Onton and Makeig (2006) for further discussion.

Optimally, the data should be from a period in which the subject is predominantly in the same state (e.g., awake and attentive) and performing the same type of task or tasks. Although standard ICA methods are theoretically able to separate data into sources that are principally active only during different periods in the data set, a promising newer mode of ICA decomposition, AMICA, allows learning multiple sets of ICs wherein each time point is associated with one decomposition (Palmer, 2008).

Since most ICA decompositions do not use relationships among time points to perform the source separation (infomax ICA actually reshuffles the order of the time points in each training step), it makes no difference whether the data are from contiguous time periods or from separate data epochs. For example, Makeig et al. (2002) reported a 31-channel decomposition of data only from the N1 response-peak period following presentations of visual stimuli. This was possible because of the large number of such stimuli (2500) viewed by each subject and the relatively small number of channels recorded (31).

Finally, it should be noted that ICA is reference free, since any rereferencing of the data that preserves its dimensionality does not change its information content or its sources. After rereferencing, the IC scalp maps will change but IC activation dynamics and equivalent model dipole locations should not change except as a result of normal statistical variability, which is typically small for ICs with highly dipolar scalp maps.

ICA VERSUS PCA

Another well-known method of linear decomposition of multichannel data, *principal component analysis* (PCA), transforms multichannel data into a sum of uncorrelated principal components so named because they each, in sequence, account for the most possible (or principal) variance in the remaining uncorrelated (or orthogonal) portion of the signal data not accounted for by the preceding principal components. By contrast, ICs produced by ICA have no natural order—though it is common to sort them by descending variance of the (back-projected) scalp data they each account for. Again, for either PCA or ICA, the whole-scalp data are the sum of the individual component contributions. The simple system of Figure 3.4 thus applies to PCA as well, though for PCA, W^{-1} is called the *eigenvector matrix* and W is its inverse. Also, in PCA, both the eigenvector matrix and the activations or factor weights matrices are normalized, and an intervening diagonal matrix E, the *eigenvalues* matrix, is used to hold the relative scaling of the components (i.e., $X = W^{-1}EU$), while the columns of the mixing matrix as well as the rows of the activations matrices are each normalized (to have unity root-mean-square values).

The important difference between ICA and PCA is in their quite different goals or objectives. We may say that PCA attempts to *lump* together maximum signal variance from however many sources into as few principal components as possible, whereas ICA attempts to *split* the signal into its separate information sources, regardless of their variance. This makes PCA useful for compressing the number of dimensions in the data while preserving as much data variance as possible. However, elimination of low-variance principal components for the purpose of dimension reduction most likely deletes portions of nearly *all* the source activities, not just the smaller ones. When the amount of data is not sufficient to successfully decompose all available channels, another possibility is to perform ICA decomposition of data from some channel subset. The relative values of these two approaches (principal subspace versus channel subspace) are difficult to evaluate in advance.

The "maximum successive variance" objective of PCA also forces both the principal component activities and the scalp projections (scalp maps) to be mutually uncorrelated (orthogonal). Since the scalp projections of brain (and nonbrain) sources are rarely themselves orthogonal, this property forces all but the first very few principal component scalp maps to resemble checkerboards that cannot reasonably represent the activity of single EEG sources. In general, principal component maps do not resemble the projection of a single EEG source unless one source (often, eye blinks) or two sources with near-orthogonal maps (e.g., lateral and vertical eye movements) dominate the signal variance.

For this reason, some ERP researchers advocate the use of post-PCA component rotation methods developed for earlier factor analysis approaches, such as Varimax or Promax (Dien et al., 2005). These may help focus the scalp maps of the very first components to emphasize a few large source activities (such as eye blink artifacts and lateral eye movements), but both simulations and actual decompositions show that their power to accomplish this for many brain and nonbrain sources pales in comparison to ICA methods when properly applied to sufficient data (Makeig et al., 1999a, 2002).

Independence among source waveforms, however, is a much stronger assumption than the simple absence of correlations between source pair signals. Substituting the stronger assumption of independence between component activities instead of requiring them only to be uncorrelated allows ICA to return ICs having any (nonidentical) scalp maps. Every IC scalp map is then free to represent the projection of a single brain or nonbrain signal source, whereas PCA component maps are constrained to be uncorrelated; therefore, most of them have a checkerboard appearance not compatible with a single cortical (or other) source projection.

Theoretically, exact independence is such a strict requirement that it can never be established for EEG signals with finite length. Therefore, ICA algorithms may at best produce components with *maximal* independence by ensuring that components continually *approach* independence as the ICA algorithm iteratively applied to the data. The degree of IC independence achieved may differ for different data sets and also for different ICA algorithms applied to the same data set. Our discussion (above) of the nature of brain EEG sources implies that the more independent the recovered IC activities, the more dipolar (or occasionally, bilateral dual-dipolar) the IC scalp maps of brain components, a trend supported by recent tests (A. Delorme, unpublished data).

IC Contributions to Single Trials and ERPs

By definition and design, IC processes contribute nearly independent temporal variability to sets of single-trial epochs. Each IC represents an independent EEG process whose continuous activity variations in the single trials are available for inspection and analysis. In particular, brain-based ICs with near-dipolar scalp maps may each be presumed to index the near-synchronous field activity arising in a single patch of cortical neuropile (or occasionally, simultaneously in two bilaterally symmetric and likely tightly coupled cortical patches). Examining the trial-to-trial variability in the IC activities relative to a set of time-locking events may allow a more detailed understanding of event-related brain dynamics than examination of raw scalp channel data themselves, since the effects of source (and artifact) mixing by volume conduction have been removed or strongly reduced by ICA.

As an example of this, Figure 3.7 shows ERP-image plots of the activities of six independent components in the same response-locked single-trial data as earlier figures. In each ERP image, IC activity is scaled (in microvolts) as it projects to the near-vertex (Cz) electrode. Locations of the equivalent IC dipoles are shown in the central panel. Trials are ordered exactly as in Figure 3.3, by the amplitude of the LPC peak 120 ms after the button press (0 ms) at the vertex. Note the quite high degree of overlap of the dipolar scalp maps of these midline components that contribute temporally *independent* contributions to the recorded EEG signals.

The sum of the signals projected by these six component processes is shown in the middle right (gray) panel. Note that the trial order used here, which sorts the trials by ERP amplitude at the selected channel (as in Figure 3.3B), only partially sorts the postresponse amplitude of each IC activation. This is shown by the uneven gradations of the post–motor response positivity in the component ERP-image panels and in the sum of their contributions at the same near-vertex channel (middle right panel). This implies that contributions to the activity fitting the mean ERP template in Figure 3.3B sum *varying spatial combinations* of these and other brain source processes in the different trials.

Note also in Figure 3.7 the *different peak latencies* of the LPC peak for ICs 2, 6, and 15 (top row). The trial ordering selected in Figure 3.3 based on the amplitude of the average ERP ignores these single-trial and component process differences. It sorts trials according to the amplitude of the *average summed contributions* of these and other independent sources, rather than on the varying amplitudes and latencies of the individual source processes.

The summed and similarly smoothed (smaller) contributions of all the other 232 nonartifact components to the same channel in the single trials are shown in the lower right (gray) panel. The channel ERP image in Figure 3.3A is thus the sum of the two right (gray) panels, as well as the sum of the two ERP-image panels (Figures 3.3 B and 3.3C). Neither of the "remainder" ERP images (Figure 3.3C or Figure 3.7, lower right) suggests a satisfactory modeling of the LPC as summing just two factors—that is, neither the "six-ICs" model imaged in Figure 3.7 (lower right) nor the "invariant-ERP" model in Figure 3.3 (B and C). Using the ICA model, however, we may examine, for example, whether the particular trial-to-trial differences in the LPC window of each identified IC may indicate that its response varies with some dimension of the varying trial context (e.g., each trial's particular cognitive and behavioral demands and demand history).

Figure 3.8 shows that a portion of the IC trial-to-trial variability highlighted in Figure 3.7 is indeed linked in orderly ways to behavioral trial differences. It shows ERP-image plots for the same six ICs as in Figure 3.7, again scaled as they contribute to the central scalp channel (center right) but here sorted by subject reaction time and then smoothed with a wider (50-trial) moving window to more clearly visualize trends. Note that the wider averaging window reduces the overall amplitude of the imaged data through phase cancellation of trial-to-trial IC variability in neighboring trials (compare the color microvolt scale limits here with those in Figure 3.7).

The middle right panel of Figure 3.8 shows the summed contribution of the six ICs to the whole-channel signal, and the lower right panel again shows

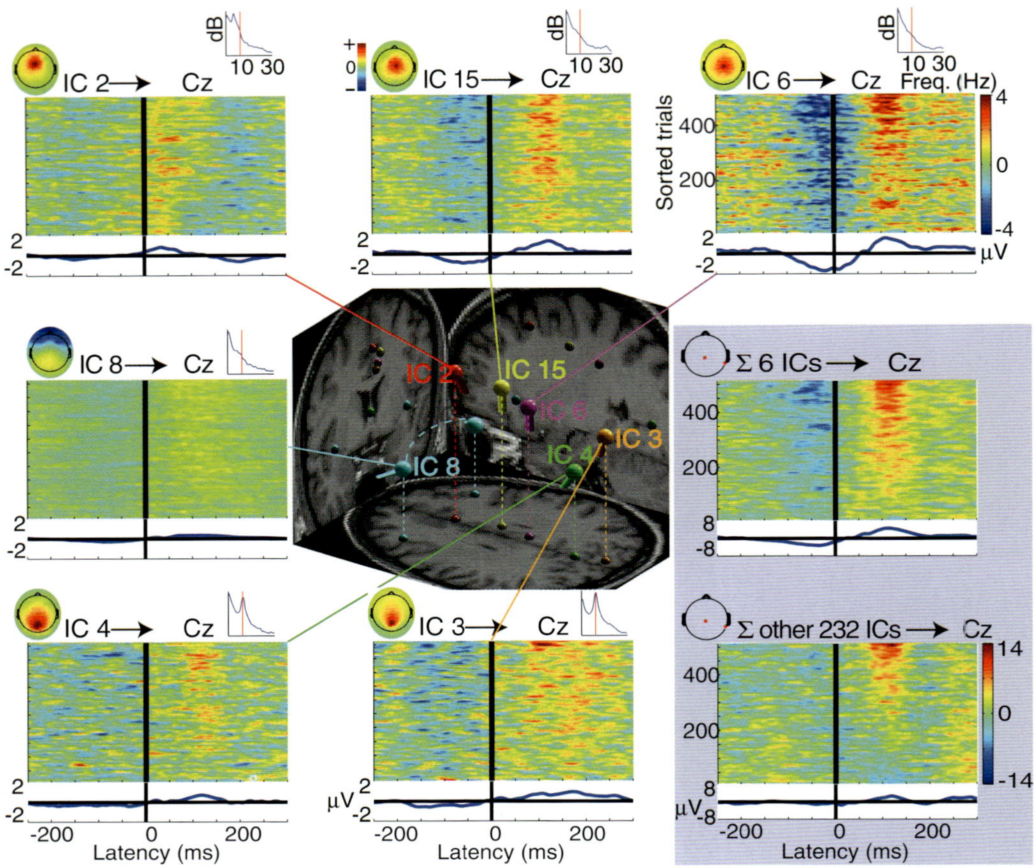

Fig. 3.7. Event-related potential images for six midline ICs and one bilateral brain IC in the same session as earlier figures (compare Figure 3.6), with trials sorted in the same order as in Figure 3.3 and scaled as they contribute to the near-vertex channel signal imaged in Figure 3.3. There, the trials were sorted by the amplitude of the LPC peak centered 120 ms after the button press (at latency 0 ms) at the near-vertex channel shown on the head cartoon (middle right panel). The sum of the signals projected by these six component processes is shown in the same panel (note the difference in color scale). The summed contributions of the other 232 nonartifact components to the same channel are shown in the lower right panel. The channel ERP-image plot (Figure 3.3A) is thus the sum of the two ERP-image panels with gray backgrounds at the right of this figure. Separately, the contributions to the ERP of each of the other components summed in the lower right panel are smaller than those of the six components whose contributions are shown in the other panels.

the remainder of the whole-channel signals they do not account for. This view reveals that a portion of the trial variability evidenced in Figure 3.7 is tightly linked to differences in subject reaction time. For anterior sources IC2 and IC8, the negativity preceding the button press is time-locked to stimulus onsets, while the subsequent LPC is mainly time-locked to the subject button press. For central midline sources IC15 and IC6, the negativity onset and offset are time-locked to the stimulus and button press, respectively. Posterior sources IC3 and IC4 appear to exhibit partial phase resetting of their alpha activities following stimulus presentations. The single-scalp channel data and ERP sum all these (and doubtless other) event-related source process contributions.

It may be worth the reader's effort to reexamine carefully the trial variability in different dimensions visualized for scalp channel data in Figures 3.2 and 3.3 and for IC data in Figures 3.7 and 3.8.

IC Clustering

To compare, group, or further average ERPs across subjects and/or sessions, channel data are typically identified by the labeled (Cz, Pz, etc.) or measured (x, y, z) channel positions on the scalp. Though equating of equivalent scalp locations across sessions and subjects is adequate for many purposes, it ignores

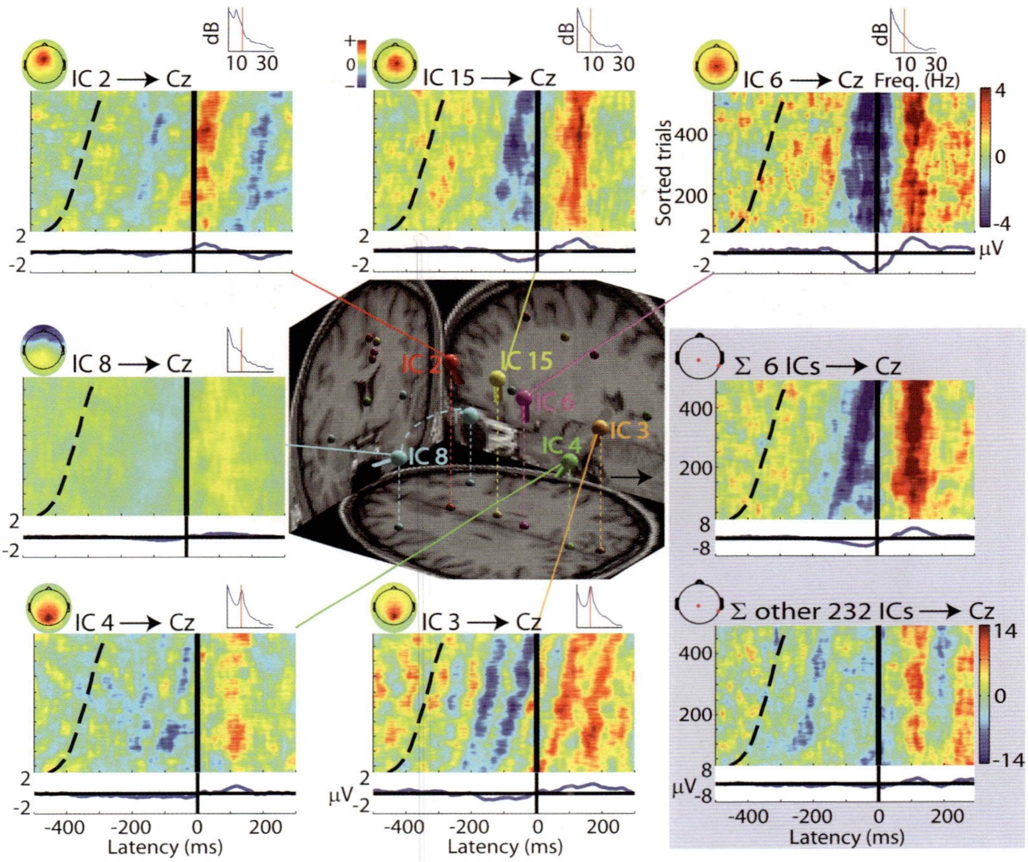

Fig. 3.8. Event-related potential images for the same six ICs as in Figure 3.7, again scaled as they contribute to the central scalp channel (center right cartoon head) but here sorted by subject reaction time and then smoothed with a broader (50-trial) moving window (note the color scales). The middle right panel shows the summed contribution of the six ICs to the whole channel signal, again (lower right) with the difference between the whole signals and the sum of these six source contributions. This view reveals that a portion of the trial variability evidenced in Figure 3.7 is tightly linked to differences in subject reaction time. For anterior sources IC2 and IC8, the negativity preceding the button press is time-locked to stimulus onsets, while the subsequent LPC is mainly time-locked to the subject button press. For central midline sources IC15 and IC6, the negativity onset and offset are time-locked to the stimulus and button press, respectively. Posterior sources IC3 and IC4 appear to exhibit partial phase resetting of their alpha activities by stimulus presentations. The single scalp channel signal sums all these (and other) event-related source dynamics.

the variety of individual cortical configuration differences, particularly in the positions and orientations of cortical sulci, that may orient anatomically equivalent EEG source projections toward different scalp areas in different subjects. In this case, functionally equivalent sources may have quite different scalp maps, and therefore electrodes at analogous locations will record different weighted mixtures of source activities. Thus, for example, signals from "my Cz" and "your Cz" may not be equivalent, even if our brains have equivalent cortical areas that function identically. This produces unavoidable and rarely considered variability in scalp recordings that are compared or averaged across subjects.

Since under favorable circumstances ICA can separate scalp-recorded signals into the volume-conducted activities of maximally independent brain sources, it may be more accurate to group, compare, and characterize *functionally* equivalent clusters of ICs across subjects and/or sessions. Finding these IC equivalence classes is the challenge of *IC clustering* across subjects and/or sessions. Independent component clusters may be selected on the basis of their equivalent dipole locations, ERPs, and/or other measures.[7]

Figure 3.9 shows a sample application of IC clustering to a grand mean ERP averaging data from 12 subjects who participated in a visual

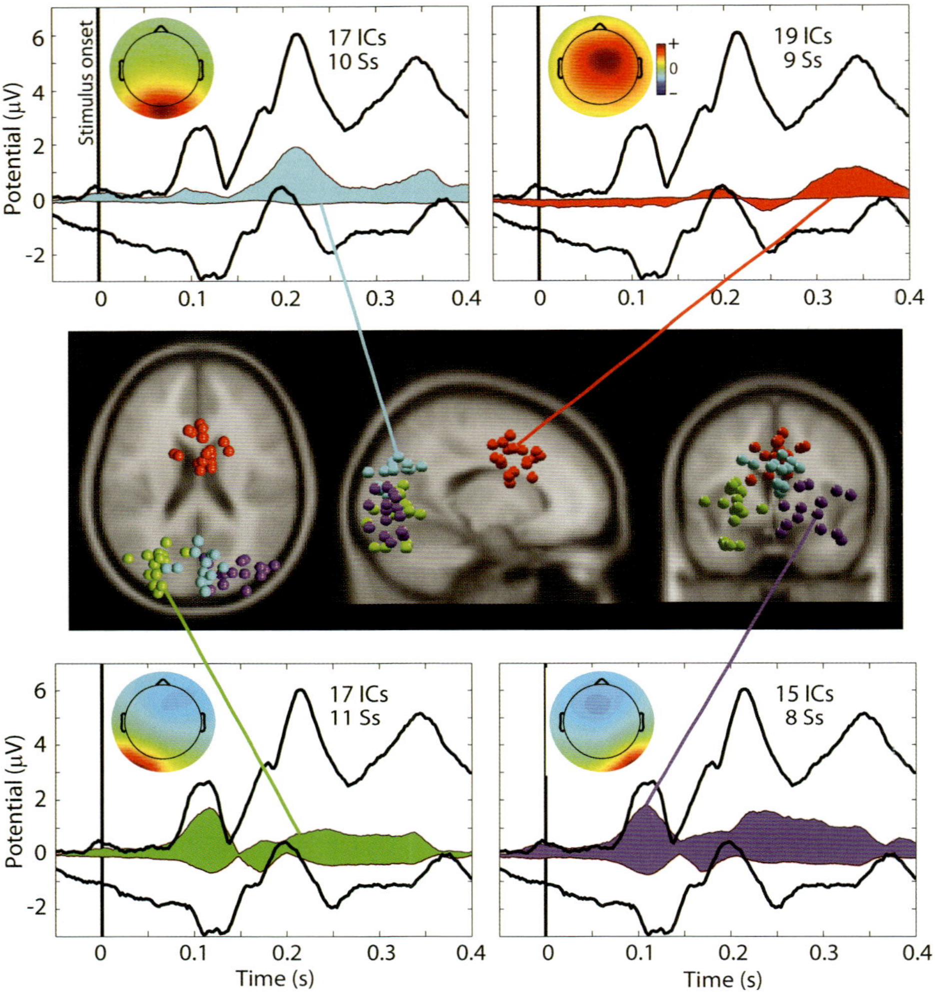

Fig. 3.9. Equivalent model dipole locations, mean scalp maps, and cluster projection envelopes of 4 (of 22) IC cluster contributions to a visual stimulus ERP (grand mean over 12 subjects) in a visual attention-shift experiment. The ERP envelopes show only the most positive and most negative channel values at each response latency. Here, the envelope (see text) of the grand-average back-projection of the indicated IC cluster is color-filled. The outer black traces are the envelope of the whole grand-mean ERP after removing the contributions of component clusters and outlier components accounting for eye, muscle, and other nonbrain artifacts. The bottom two panels show clusters accounting for most of the P1 peak in the grand-mean ERP. The upper two panels indicate the portions of the grand-mean ERP accounted for by a central posterior cluster (blue, with maximal contribution to the peak labeled P2) and a midline cluster (red, with maximal contribution to the later peak labeled P3).

attention-shift experiment. Throughout the experiment, subjects made speeded manual choice responses to indicate in which dimension (shape or color) the lateral target stimulus (presented at 0 ms) differed from a simultaneously presented neutral background stimulus. In the 12 subjects' ICA-decomposed data, we identified 22 clusters of similarly located and similarly reacting ICs by comparing equivalent dipole locations, mean power spectra, and event-related spectral perturbations (see section IV) in three stimulus conditions. Figure 3.9 focuses on a grand-mean ERP time locked to the stimulus presentation (at latency 0 s) in one condition. The central panel shows IC equivalent dipole locations for 4 of 22 identified IC clusters.

The black traces in the four top and bottom plot panels show the *envelope* of the grand mean ERP (i.e., its maximum and minimum channel values at each latency). The four top and bottom panels show the cluster-mean scalp maps and the boundaries of the colored regions, the *envelopes* of those portions of the grand-mean ERP accounted for by each of the four clusters.[8] Envelope plotting allows the ERP contributions of one or more ICs or IC clusters to be visually compared with the envelope of the whole-scalp ERP.

The lower panels show two lateral occipital IC clusters (see the green and purple IC dipoles) that accounted for nearly all the bilateral positive peak near 110 ms in the ERP, plus a later sustained ridge-like feature. The upper two panels show the portions of the grand mean ERP accounted for by a central posterior cluster (blue) whose maximum ERP contribution was to the positive peak near 220 ms and a midline cluster (red) that contributed maximally to a later positive peak near 350 ms.

Note that although the model dipoles are represented, for visual convenience, as small balls, the actual uncertainty about their individual locations is rather larger, as are the distributions of cortical territory across which synchronized local field activity (in our model) produce the far-field potentials recorded by the scalp electrodes. Sources of dipole location error in Figure 3.9 include possible differences in recorded electrode positions relative to each other and the scalp, errors in coregistering the electrodes to the head model, and differences in head shape, as well as possible differences in head tissue conductivity parameters. Although the equivalent model dipole locations shown in the middle panel are relatively tightly grouped, their spread may also reflect differences in the locations of functionally equivalent cortical areas across subjects, since similarities between activity measures were also considered in assigning components to clusters.

Independent component clustering is required to compare ICA decompositions from more than one subject or session. It can be used to understand the locations and dynamics of IC processes contributing to average ERPs as well as to the unaveraged single trials. Independent component clustering provides an involved but, under favorable circumstances, we believe a more adequate answer to the inverse problem of estimating the distributed sources of ERP scalp maps and the relationship of the source dynamics to experimental events and conditions. In particular, IC clustering gives a more adequate solution than simply attempting to model the distributed cortical sources of ERP scalp maps themselves. In addition, IC clustering allows testing for differences within and/or between subject groups reflected in the presence or absence of ICs in one or more clusters and/or on details of the clustered IC locations or activities.

Time/Frequency Analysis of Event-Related EEG Data

Time-Locked But Not Phase-Locked: Event-Related Spectral Perturbations

To understand the relationship of ERP features to the event-related dynamics of the entire EEG signals from which they are derived, it is convenient to use time/frequency analysis that models the single-trial data as summing an ever-changing collection of sinusoidal bursts across a wide frequency range. Note that producing this representation of the data does not mean that the EEG is necessarily composed of such bursts, or that the burst shape or *window* employed in the analysis is necessarily a physiologically accurate template. Rather, as Joseph Fourier first showed for heat flows along a copper tube, frequency analysis, and later nonstationary time-frequency analysis, can be used to represent *any* temporal activity pattern, not limited to those portions of the recorded signals that do indeed resemble single time-frequency basis elements (e.g., symmetric and smoothly tapered bursts at a single frequency). However, the frequent appearance of periodicities at multiple frequencies is a clear and remarkable feature of EEG records, and this property of the signals shows quite clear and spatially distinct changes accompany changes in arousal and attention, making time-frequency analysis clearly useful for EEG analysis.

Rather than averaging the recorded ("time-domain") event-related data epochs directly, one may average their time-frequency transforms (see also Chapter 2, this volume). Averaging time-frequency power or log power values in a regular grid of time-frequency windows gives an event-related spectrogram that is nearly always dominated by relatively large low-frequency activities. Normalizing the result, therefore, by subtracting the mean log power spectrum within some defined baseline period (prestimulus or otherwise, as relevant to the analysis) allows a color-coded time-frequency image of mean log spectral differences that we call the *event-related spectral perturbation* (ERSP) image (Makeig, 1993). Basing the ERSP on changes in log power implicitly assumes a multiplicative model in which EEG spectral changes represent the multiplication or division of the baseline power at each frequency in each latency window relative to the time-locking events.

Determining either the amplitude or the phase of activity at a particular time-frequency point involves matching the data in a window surrounding the given time point to the oscillatory basis element (typically a tapered sinusoidal burst or *wavelet*). To measure low frequencies, this window must be relatively long, limiting the frequency range considered for short data epochs. Also, event-related changes in spectral power may last longer than significant features in the ERP. For these reasons, our own typical time-frequency analyses use epochs including at least 1 s before the time-locking event and continuing to 2 s or more following it, allowing a frequency decomposition based on a three-cycle tapered sinusoidal wavelet down to 3 Hz.

The mean ERSP of a set of event-related data epochs can index event-related dynamics that leave no trace at all in the ERP average of the same epochs, as first shown for alpha-band activity by Pfurtscheller and Aranibar (1977). Thus, the ERSP transform of the average ERP for a set of data epochs, while of possible interest to compute, may bear little or no resemblance to the average ERSP for the same collection of epochs. For one thing, significant ERSP features may long outlast the reliable ERP features. For example, Figure 3.10 (top panel) shows a mean ERSP time-frequency image for a left-frontal IC (IC2) time-locked to button presses following target stimuli (from the same session shown in Figures 3.1–3.8). Regions of nonsignificant difference from baseline (here $p < .001$, uncorrected for multiple comparison) are masked with light green. The ERSP image reveals that mean alpha-band power just below 10 Hz increases weakly following the button press, while mean low-beta activity (15–20 Hz) in two frequency ranges increases most markedly after 400 ms. Activity at the 6 Hz baseline spectral peak (see the top left side-facing blue baseline spectrum) does not change, though activity below 5 Hz increases weakly around the button press, and then decreases beginning 200 ms after the button press.

Spectral power in the average ERP is often referred to as the spectrum of activity *evoked* by events, while changes in spectral power appearing in the ERSP are dubbed changes *induced* by events. However, this terminological distinction should not suggest that the two are necessarily physiologically distinct. To see this, we need to consider changes in phase statistics associated with experimental events.

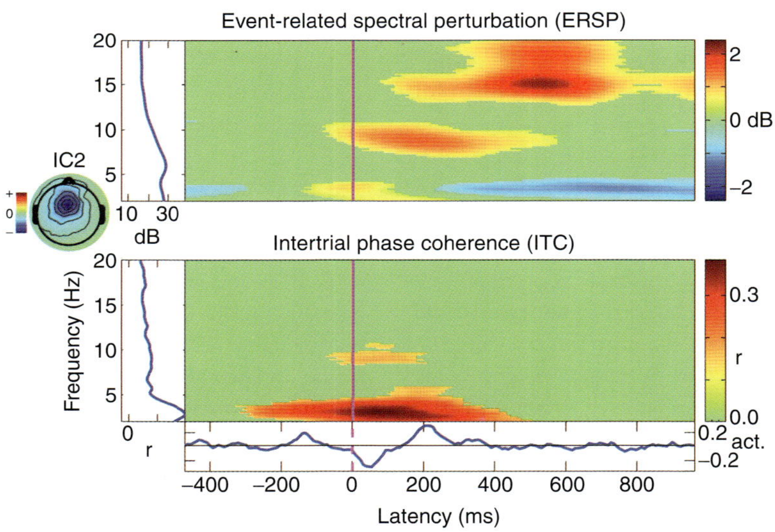

Fig. 3.10. Event-related time/frequency analysis of the set of IC trials shown in Figures 3.7 and 3.8 (upper left). Mean event-related spectral perturbation (ERSP, top) and intertrial coherence (ITC, (bottom) time-frequency images for a leftward-pointing midfrontal IC (IC2) time-locked to button presses following target stimuli. Regions of nonsignificant difference from baseline ($p < .001$, uncorrected) are masked with light green. The top (ERSP) image reveals that mean alpha-band power at 10 Hz increases weakly following the button press. On average, low-beta activity (15–20 Hz) activities first decrease slightly, then increase after 400 ms. Activity at the baseline spectral peak (6 Hz; see the top left baseline spectrum plot) does not change, though activity below 5 Hz is maximal at the button press. The bottom (ITC) image shows that 4 Hz activity becomes partially but significantly phase-locked around the button press, meaning that the portion of the component ERP (lower trace) near 4 Hz is statistically significant (compare the ERP trace below), as is its weak 10 Hz "scalloping" between –50 and 200 ms. Component activation units ("act.") are proportional to scalp microvoltage. The statistically significant changes in mean spectral power in the beta band, shown in the upper panel, are not associated with significant ERP features and therefore represent changes in component activity time-aligned but not phase-aligned to the button presses.

Phase Locking Across Trials: Intertrial Coherence

The ERSP disregards completely the consistency or inconsistency of the *phase* of the activity at each frequency and latency in a set of event-related epochs. *Intertrial coherence* (ITC) or, more precisely, *intertrial phase coherence*, introduced as the "phase-locking factor" by Tallon-Baudry et al. (1996), measures the degree of consistency, across trials, of the phase of the best-fitting time-frequency basis element at each latency-frequency point. Phase consistency is measured on a scale from 0 (no consistency; phase across trials is random and uniform around the phase circle) to 1 (phase is perfectly consistent across trials). The ITC for any *finite* set of randomly selected data epochs will typically not be 0. Therefore, it is important to compute a baseline threshold for the appearance of significantly nonrandom phase coherence. An ITC reliability threshold for a set of trial data can be found using either parametric or nonparametric statistical methods (Delorme & Makeig, 2004; Mardia, 1972).

It is important to note that the ITC and *ERSP* images for a given set of event-locked data epochs may have few or even no common features. For example, in Figure 3.10 the postmotor response increases in alpha-band and then in beta-band power in the frontal midline IC spectrum are not mirrored by significant changes in ITC at the same latencies and frequencies.

However, there is an intimate relationship between the ITC and the *ERP*. In particular, the occurrence of a significant ERP peak or another feature requires significant ITC. In this sense, a significant ERP value at any time point reflects and requires significant ITC values at one or more frequencies at that time point (except in odd, improbable cases). Note also that the ITC for any frequency may be significant even at latencies at which the mean potential ERP value is 0. For example, if the 0 value in the ERP occurs during the zero crossing of an alpha oscillation in each trial, then the ITC at that alpha frequency might be highly significant, although the ERP at that time point might have a value of 0. The ITC may also be significant, at a particular latency, at more than one frequency. If so, this will be reflected in the shape of the ERP waveform surrounding the latency in question.

For example, in the ITC image in Figure 3.10 (lower panel), activity near 4 Hz becomes partially but significantly phase-locked around the button press event (0 ms), meaning that the portion of the component ERP (bottom trace) at 4 Hz is statistically significant. As well, the ITC becomes (barely) significant at 10 Hz, a fact reflected in the weak 10 Hz "scalloping" in the ERP waveform between –200 and +300 ms. No other ITC frequencies, and therefore no other ERP frequencies, are significantly different from chance. The statistically significant ERSP changes in mean spectral power in the beta and low-gamma bands, shown in the upper panel, are not associated with significant ITC features and therefore represent component activity that is phase inconsistent, that is, not phase-aligned (or phase-coherent) across trials, and so does not contribute significantly to the trial-average ERP.

ERPs and Partial Phase Resetting

The relatively low peak ITC values in Figure 3.10 (~0.4) are not unusual for longer-latency ERP features. An alternative model, first proposed for selected event-related data as early as 1974 by Sayers et al. (1974), is known as the *phase-resetting* or *partial phase-resetting model. Phase resetting* refers to a phenomenon seen both in mathematical models and in biological systems in which the phase of an ongoing periodicity (e.g., the cardiac or circadian cycle) is reset to a fixed value relative to the delivered perturbing stimulus. For example, brief exposure to strong light delivered to a dark-adapted rat (or human) at almost any phase of the wake-sleep cycle will tend to reset the cycle to a fixed phase value (Czeisler et al., 1986; Honma et al., 1987; Tass, 1999; Winfree, 1980). At the frequency of the ongoing, spontaneous rhythm, an ITC measure time-locked to comparable events delivered at random time points throughout the session will become significant as the phase of the rhythm in some or most of the trials is reset to a fixed value. If the phase of the rhythmic activity then tends to continue to advance in a regular manner from its initial reset value, the ITC time-locked to the events of interest will remain significant for some number of cycles until, across trials, natural variability randomly separates the advancing phase values.

The term *phase resetting* has been applied to EEG dynamics in a less formal sense, since in most cases there is no constant, ongoing rhythm for experimental events to perturb. Rather, in many cases, the signal contains only intermittent bursts of alpha or other frequency spindles of various lengths. The term *phase resetting*, therefore, can be formally applied in some statistical sense to mean that the phase statistics (as measured, for example, by the ITC) are transiently perturbed following events of interest. If, whenever rhythmic activity at a given frequency is present, its phase distribution following the time-locking events becomes nonuniform, ITC will increase and may tend to remain significant for as long as the rhythmic activity is present.

Figure 3.11 shows ERP-image plots for two sets of visual-stimulus locked trial data from two ICs captured in different subjects under rather different task conditions. In panel A (left), the stimulus is a briefly flashed disk presented at a central, visually unattended target square located above a central fixation cross during a visually selective attention task. The IC shown here has a bilateral equivalent dipole model in or near primary visual cortex and produces abundant alpha-band activity (see the power spectrum), likely reflecting the "alpha flooding" of visual cortical areas sensitive to the foveal fixation region when the subject places his or her visual attention elsewhere in the visual field (Worden et al., 2000). This alpha activity appears to be partially phase-reset (ITC ~ 0.4) for nearly 500 ms (five alpha cycles) following stimulus presentation. In the ERP image, trials are sorted by alpha phase in a three-cycle window ending 50 ms after stimulus onset. The possibility of partial phase resetting is suggested by the bending and then near-vertical alignment of the positive and negative wave fronts beginning near 100 ms in the ERP image when the ITC becomes significant.

Note that the visual evidence presented by this ERP image, including the finding of a significant ITC (lower ITC trace), is *not* in itself sufficient evidence to prove that these data truly fit a phase-resetting model (for more discussion, see Chapter 2, this volume). Nor does it necessarily rule out a true-ERP model for the data—for example, a model in which the same ERP (upper ERP trace) resembling

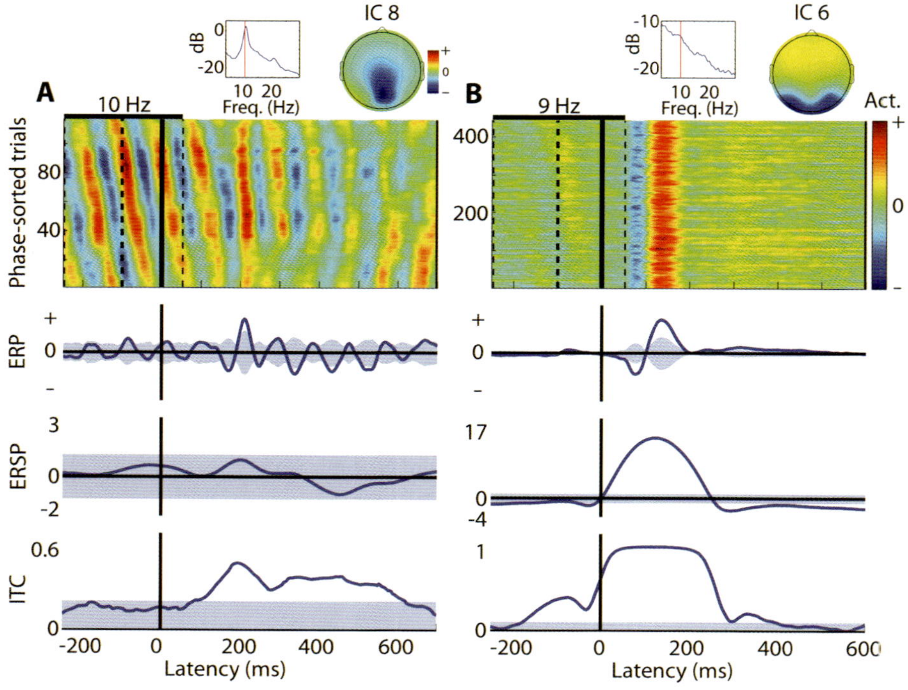

Fig. 3.11. Event-related potentials and partial phase resetting. Panel B shows an ERP-image plot visualizing the responses in over 400 single trials of a bilateral occipital IC process following presentation of a letter at the central fixation point of a subject participating in a working memory task. The bilateral component (IC6, sixth by variance expressed in the data) produces a response largely resembling a one-cycle sinusoid at 9 Hz. The mean ERP trace (below the ERP image) plots its mean time course, time-locked to stimulus onset. The ERSP trace (below that) shows that the mean level of 9 Hz energy in the data during this period is 15 dB or more higher than in the prestimulus (or ensuing) period. The ITC trace (below that) confirms that during this ERP feature the phase of the entire 9 Hz activity in the trials is highly consistent (ITC approaching 1). The blue backgrounds show $p < .01$ probability limits, demonstrating that all three measures are highly significantly different from baseline in this period. Panel A shows a quite different set of over 100 data trials for a medial (or medial bilateral) occipital IC process in the five-box task of Figure 3.1, in which stimuli were presented above and to the left of a central fixation cross while the subject retained fixation. The large amount of alpha-band activity produced by this IC process under these circumstances likely reflects "alpha flooding" of relevant visual cortex when visual attention is forced by the task to remain elsewhere in the visual field (Worden et al., 2000).

an alpha burst is simply *added* to ongoing alpha and other EEG activity in every trial (as in Figure 3.3) and the ongoing alpha activity is in turn reduced in amplitude just enough to make total mean alpha power at each latency constant, as observed here (middle ERSP trace). However, keep in mind that these data are the result of spatial filtering by ICA of a single independent source, very likely focused on a single source area (or closely spaced medial bilateral areas), given the highly dipolar form of the IC scalp map. It thus seems to us physiologically implausible that, following these visual events, this same cortical source area produces ongoing random-phase alpha activity *plus* a fixed but wholly *unrelated* alpha-burst ERP. Several groups have recently proposed measures to further test phase-resetting models on data such as these (Hanslmayr et al., 2007; Martinez-Montes et al., 2008; Mazaheri & Jensen, 2006). Ultimately, the issue will likely be settled by fitting concurrent scalp and intracranial EEG recordings to generative models of cortical field dynamics, a process begun by groups studying human brain responses during cortical recording (Wang et al., 2005).

In panel B of Figure 3.11 (right), on the other hand, the time-locking stimulus is a letter presented *at fixation* in a letter working memory task. The spectrum of the bilateral lateral-occipital IC (inset) has only a weak alpha band peak, and no sign of prolonged alpha-band phase resetting following the highly stereotyped (ITC >0.8) component IC stimulus-evoked response (which contributes strongly to the P1-N1-P2 features of the full scalp ERP, not shown). At the frequency best fitting the ERP complex (9 Hz), mean single-trial amplitude during the ERP is nearly six times (over 15 dB) higher than the mean amplitude of activity at the same frequency in the prestimulus baseline. Phase-sorting the single trials at 9 Hz in a window ending 50 ms after stimulus onset (as in Figure 3.11A) shows that the phase of the weak alpha activity present in single trials during the baseline period has no obvious effect on the latencies of the subsequent evoked-response activity in the same trials. For this IC stimulus response, therefore, a partial phase-resetting model seems unnatural and a true-ERP model adequate. However, even here, one may ask whether, for example, the frequency peak of the ERP (9 Hz) may not also be a peak of the spontaneous (baseline) spectrum of this cortical area.

Many authors have attempted to draw a clear distinction between evoked and induced event-related activities, defining evoked activity as being activity completely time-locked and phase-locked to the stimulus (ITC = 1) and thereby composing the ERP, while the remainder of the single-trial activity, having no phase locking to the time-locking events (ITC = 0), is defined as induced (Galambos, 1992). While this distinction may be useful for some purposes, drawing this terminological distinction does not mean that this decomposition of the EEG signal into evoked (ERP) activity plus induced (other EEG) activity has any natural physiological basis. Think of a stack of five pennies: Does this stack really sum two groups of two and three or of groups of four and one? In fact, the stack of pennies retains no trace of how it was constructed and thus cannot be said to be any more 3 + 2 than 4 + 1, no matter how it was originally constructed. The same applies to the model of event-related EEG data illustrated in Figure 3.3: EEG data = ERP + Other, a model that, as ICA decomposition and Figures 3.7 and 3.8 suggest, disregards the varying single-trial contributions of spatially separable data information sources, some clearly linked to trial-by-trial behavioral differences.

Figures 3.7 and 3.8 suggest that scalp ERPs sum channel activity arising from different mixtures of spatial source processes in different trials. But how should we think of the average response of a single IC? Assuming that an IC activation does index locally synchronous or near-synchronous field activity of a single patch of cortex, can the IC activity producing the IC ERP activity (strictly time-locked to the set of evoking events) be physiologically distinct from other (non-phase-locked) EEG activity originating at the same moments in presumably the same cortical patch?

Linear summation in cortex, even of direct sensory input and ongoing cortical dynamics, appears physiologically implausible without strong nonlinear interactions. Fiser and colleagues (2004) have noted that even at prototypical sensory cortex—the input layer of primary visual cortex (in ferrets)—only a small percentage of the synapses deliver information directly from the eyes via the lateral geniculate nucleus (LGN). In accord with this fact, they report that "at all ages including the mature animal, correlations in spontaneous neural firing [during natural vision] were only slightly modified by visual stimulation, irrespective of the sensory input. These results suggest that in both the developing and mature visual cortex, sensory evoked neural activity represents the modulation and triggering of ongoing circuit dynamics by input signals, rather than directly reflecting the structure of the input signal itself" (p. 573). If this is the case even for V1, it should not be less so for cortical areas that are not primary sensory areas. Clearly, deeper understanding of the EEG dynamic changes associated with sensory and other events will require more detailed observation and modeling of brain dynamics at multiple spatial scales. In terms of EEG research, more detailed observations and modeling are needed of trial-by-trial differences in oscillatory activity and its relationship to its transformation by experimental events.

Event-Related Coherence

Cognitive events—moments at which we apperceive the *significance* of some sensory event and mentally grasp its immediate consequences for our attention and behavioral planning—must involve and/or produce complex and distributed changes in EEG dynamics. Furthermore, some mechanism of information transfer between brain regions must exist that is dynamically dependent both on the nature of the stimulus and on its relation to the subject's expectations and intentions. A possible mechanism for this transfer may be indexed by transient temporal coupling between pairs of sources relative to experimental events. One measure of this coupling is event-related coherence (ERC; Delorme & Makeig, 2003).

The preponderance of coherence of all sorts observed between pairs of scalp channel signals is accounted for by ICA as deriving from common IC projections to both scalp channels. A change in amplitude of a single IC, relative to other ICs that project to the same channel pair, may produce a change in their measured (zero-lag) scalp channel coherence *without* any actual coherence changes occurring at the cortical source level. By maximally reducing the effects of volume conduction on the data, ICA decomposition allows a more principled study of transient or intermittent coherence between IC source activities.

Recently, we used ICA decomposition of target response data from the same five-box task used here for illustrative purposes to show that brief and weakly spatially coherent theta wave complexes arise in frontal midline, somatomotor, and parietal cortex in many subjects following significant events (Makeig et al., 2004b), often beginning in frontal polar cortex (Delorme et al., 2007b). But how can the activities of independent components be (occasionally) phase coherent? As described previously, ICA decomposition actually derives *maximally* independent components—which allows the discovered IC activity patterns to exhibit occasional transient dependence—for example, in the five-box data at one frequency in at most a fifth of the trials. In this and related cases, we found the partial coherences to have nonzero phase lags, and to remain when each component ERP was (artificially) regressed out of each single trial activity and coherence was computed only on the remainder. Event-related coherence is another measure that cannot be deduced from ERP waveforms alone, and cannot be confidently interpreted when computed for pairs of scalp-channel signals. Independent component analysis preserves only those coherences that represent transient coupling of the frequency-domain activities of two EEG sources with a fixed latency difference.

A close-up example of similar "phase reorganization" in human brain was recently provided by the study by Wang and colleagues of event-related local field activity in multichannel "thumbtack" electrodes pushed through a small piece of intact cortex in anterior cingulate before its clinically required removal in a brain operation (Wang et al., 2005). They reported that theta band activity was generated in superficial layers of anterior cingulate cortex (ACC) both before and after presentations of a variety of task-relevant stimuli, while after presentations, phase locking between ACC and other brain areas increased transiently.

Meeting the Challenge of the Moment

For us to survive and thrive, at each moment our brain must integrate its awareness of its present situation and environment, including existing plans for action and/or inaction, with its emerging sensory experience and mnemonic associations. It must optimally engage or revise its attentional distribution, action plans, and physiological body state in a way that is adequate to *meet the challenge of the moment.*

This volume summarizes the results of nearly 50 years of scientific experience in studying the shapes and sizes of average ERP responses of scalp EEG signals to sensory or other events, responses that depend in large part on the *significance* of the events to the subject and on the *context* in which they occur. Electroencephalography is the oldest and most noninvasive functional brain imaging modality; it is also the least expensive and most portable. The continuing promise of EEG brain imaging is that the highly labile dynamics of EEG scalp fields, signaling changes in local field synchrony within and between cortical areas, can provide detailed indices of changes in human attentional, intentional, and affective state, both post hoc and even, to an increasing extent, online, with potentially important applications to basic scientific research, to clinical and workplace monitoring, and to other fields of human interest and endeavor.

In this chapter, we have discussed the origins in *local cortical synchrony* of both EEG signals and ERP waveforms derived from them. We have defined the concept of an EEG source, based on both EEG analysis and physiological evidence, and have demonstrated the utility of ICA for separating multichannel EEG recordings into a set of temporally and functionally independent brain and nonbrain source processes. Finally, we have shown a simple example of using ICA decomposition to study the sources that contribute to (as well as those that contaminate) ERPs and their activities in single-trial EEG data. We have given examples of using ERP-image plotting to visualize the dependence of EEG responses in single trials on behavioral, EEG, or other parameters; have introduced time-frequency analysis in the form of ITC to show that the activity captured in average ERPs reflects trial-to-trial phase consistency; and have introduced the concept that *some* ERP features may reflect reorganization (or *perturbation*) of the exact timing or phase statistics of *ongoing* activity in the same cortical areas, as long suggested by investigators familiar with dynamic modeling methods used in engineering.

We believe the increasingly urgent challenge for the field of ERP and more general EEG research is to discover the brain source dynamics that produce the characteristic features of evoked responses and to model the trial-by-trial (and condition-by-condition) differences in EEG (and ERP) dynamics associated with the large variety of events that unfold continually in our daily lives within an ever-evolving situational context—events that pose a wide variety of challenges to which our brains respond effectively.

We believe this to be an exciting time to study human electrophysiology, an era in which noninvasive EEG recording is moving toward fulfilling its promise of becoming a true functional brain imaging modality. Current knowledge and understanding of EEG dynamics are likely to advance steadily as new analysis tools developed for this purpose become more widely applied. One result should be a deeper and fuller understanding of the nature and significance of ERP features.

Acknowledgment

Thanks to Arnaud Delorme and David Groppe for helpful discussions, and to our many colleagues from the Swartz Center for Computational Neurosciences, UCSD and Computational Neurobiology Laboratory, Salk Institute, for their invaluable support, insight, and companionship on our long and continuing quest to better understand human event-related brain dynamics.

Notes

1 For example, at 20 Hz and traveling at 1 m/s, a radiating pond ripple would reach the edge of a 1 cm source domain in 5 ms, one-tenth of a 20 Hz cycle. Therefore, there would only be a $2\pi/10 = 36$ degree phase lag between the center and the edge of the patch, and the spatiotemporal pattern of potentials at scalp electrodes would be highly correlated with the pattern produced by completely synchronous 20 Hz activity across the same 1 cm domain.

2 Though less commonly appreciated, intracranial electrodes also record volume-conducted signals from distant sources along with local field activity produced just under the electrode (Lee, 2005).

3 These EEG data were collected synchronously from 250 scalp electrodes plus 4 infraocular and 2 ECG electrodes with an active reference (Biosemi, Amsterdam) at a sampling rate of 256 Hz and 24-bit A/D resolution. Onsets and offsets of target discs, as well subject button presses, were recorded in a simultaneously acquired event channel. The recording montage covered most of the skull, forehead, and lateral face surface, omitting the chin and fleshy cheek areas. Locations of the electrodes relative to skull landmarks for each subject were recorded (Polhemus, Inc.). Electrodes with grossly abnormal activity patterns were removed from the data, leaving 238 channels. After rereferencing to digitally linked mastoids, the data were digitally filtered to emphasize frequencies above 1 Hz. Data periods containing broadly distributed, high-amplitude muscle noise and other irregular artifacts were identified by tests for high kurtosis or low-probability activity and removed from analysis. Occurrence of eye blink, other eye movement, or isolated muscle noise artifact was not a criterion for rejection. Remaining data time points were then concatenated and submitted to decomposition by extended infomax ICA using the binica function available in the EEGLAB toolbox (http://sccn.ucsd.edu/eeglab). Decompositions used extended-mode infomax ICA (Makeig et al., 1997) with default training parameters. Extended infomax was used to allow recovery of any components with subgaussian activity distributions, including 60 Hz line noise contamination. Independent components clearly and predominantly accounting for eye movement, muscle, cardiac, single-channel, or other artifactual activity were removed from the ERP data. Both the target stimulus-locked and motor response-locked epochs analyzed in the figures were referred to a mean baseline in a 500 ms period before target stimulus onsets.

4 Data figures in this chapter were produced using software tools from the freely available EEGLAB Matlab software environment (http://sccn.ucsd.edu/eeglab). The single-subject 256-channel data set from which we derived most of the figures was recorded and first studied by Delorme et al. (2007b) and is available for download in raw and in EEGLAB formats from the EEGLAB web site (above).

5 In particular, the phase of a digitally recorded signal cannot be defined above its Nyquist frequency (half of its sampling rate) and is ambiguous at its Nyquist frequency.

6 Methods that find more components are available but require narrower source assumptions and more computation time.
7 EEGLAB includes Matlab-based tools for applying, evaluating, and exploring component clustering.
8 Data used for the IC cluster figure were collected by Klaus Gramann at the University of Munich from 12 subjects performing a visual feature discrimination task. The EEG was recorded continuously at a sampling rate of 500 Hz using 64 Ag/AgCl electrodes mounted on an elastic cap. The EEG signals were amplified using a 0.1–100 Hz bandpass filter and filtered offline using a 1–40 Hz bandpass. All electrodes were recorded referenced to Cz and then rereferenced offline to linked mastoids. Average ERPs in an 800 ms epoch were computed relative to a 200 ms prestimulus baseline. Independent component analysis decomposition used extended infomax. Independent components were clustered across subjects using EEGLAB clustering functions based on their respective dynamics under three target stimulus-difference conditions (whether or not the target had a different color, different shape, or both than the accompanying standard stimuli). Only ICs whose equivalent dipole projection to the scalp had a residual variance from the IC scalp map below 15% and an equivalent dipole location within the brain volume were considered for clustering. These ICs were separated into 22 clusters based on their equivalent dipole locations, ERSPs and ITCs in the 500 ms following target onsets.

References

Akalin-Acar, Z., & Gencer, N. G. (2004). An advanced boundary element method (BEM) implementation for the forward problem of electromagnetic source imaging. *Physics of Medical Biology, 49*, 5011–5028.

Anemuller, J., Sejnowski, T. J., & Makeig, S. (2003). Complex independent component analysis of frequency-domain electroencephalographic data. *Neural Networks, 16*, 1311–1323.

Arieli, A., Shoham, D., Hildesheim, R., & Grinvald, A. (1995). Coherent spatiotemporal patterns of ongoing activity revealed by real-time optical imaging coupled with single-unit recording in the cat visual cortex. *Journal of Neurophysiology, 73*, 2072–2093.

Beggs, J. M., & Plenz, D. (2003). Neuronal avalanches in neocortical circuits. *Journal of Neuroscience, 23*, 11167–11177.

Bell, A. J., & Sejnowski, T. J. (1995). An information-maximization approach to blind separation and blind deconvolution. *Neural Computation, 7*, 1129–1159.

Bollimunta, A., Chen, Y., Schroeder, C. E., & Ding, M. (2008). Neuronal mechanisms of cortical alpha oscillations in awake-behaving macaques. *Journal of Neuroscience, 28*, 9976–9988.

Bullock, T. H. (1983). Electrical signs of activity in assemblies of neurons: Compound field potentials as objects of study in their own right. *Acta Morphologica Hungarica, 31*, 39–62.

Comon, P. (1994). Independent component analysis, a new concept. *Signal Processing, 36*, 287–314.

Czeisler, C. A., Allan, J. S., Strogatz, S. H., Ronda, J. M., Sanchez, R., Rios, C. D., Freitag, W. O., Richardson, G. S., & Kronauer, R. E. (1986). Bright light resets the human circadian pacemaker independent of the timing of the sleep-wake cycle. *Science, 233*, 667–671.

Dan, Y., & Poo, M. M. (2004). Spike timing-dependent plasticity of neural circuits. *Neuron, 44*, 23–30.

Debener, S., Mullinger, K.J., Niazy, R. K., & Bowtell, R. W. (2008). Properties of the ballistocardiogram artefact as revealed by EEG recordings at 1.5, 3 and 7 T static magnetic field strength. *International Journal of Psychophysiology, 67*, 189–199.

Delorme, A., & Makeig, S. (2003). EEG changes accompanying learned regulation of 12-Hz EEG activity. *IEEE Transactions in Neural System Rehabilitation Engineering, 11*, 133–137.

Delorme, A., & Makeig, S. (2004). EEGLAB: An open source toolbox for analysis of single-trial EEG dynamics including independent component analysis. *Journal of Neuroscience Methods, 134*, 9–21.

Delorme, A., Sejnowski, T., & Makeig, S. (2007a). Enhanced detection of artifacts in EEG data using higher-order statistics and independent component analysis. *Neuroimage, 34*, 1443–1449.

Delorme, A., Westerfield, M., & Makeig, S. (2007b). Medial prefrontal theta bursts precede rapid motor responses during visual selective attention. *Journal of Neuroscience, 27*, 11949–11959.

Destexhe, A., Contreras, D., & Steriade M. (1999). Spatiotemporal analysis of local field potentials and unit discharges in cat cerebral cortex during natural wake and sleep states. *Journal of Neuroscience, 19*, 4595–4608.

Dien, J., Beal, D. J., & Berg, P. (2005). Optimizing principal components analysis of event-related potentials: Matrix type, factor loading weighting, extraction, and rotations. *Clinical Neurophysiology, 116*, 1808–1825.

Fischl, B., Salat, D. H., van der Kouwe, A. J., Makris, N., Segonne, F., Quinn, B. T., & Dale, A. M. (2004). Sequence-independent segmentation of magnetic resonance images. *Neuroimage, 23*(Suppl 1), S69–S84.

Fiser, J., Chiu, C., & Weliky, M. (2004). Small modulation of ongoing cortical dynamics by sensory input during natural vision. *Nature, 431*, 573–578.

Foxe, J. J., & Schroeder, C. E. (2005). The case for feedforward multisensory convergence during early cortical processing. *Neuroreport, 16*, 419–423.

Francis, J. T., Gluckman, B. J., & Schiff, S. (2003). Sensitivity of neurons to weak electric fields. *Journal of Neuroscience, 23*, 7255–7261.

Freeman, W. J. (2000). Mesoscopic neurodynamics: From neuron to brain. *Journal of Physiology (Paris), 94*, 303–322.

Freeman, W. J., & Barrie, J. M. (2000). Analysis of spatial patterns of phase in neocortical gamma EEGs in rabbit. *Journal of Neurophysiology, 84*, 1266–1278.

Fries, P., Nikolic, D., & Singer, W. (2007). The gamma cycle. *Trends in Neuroscience, 30*, 309–316.

Frost, D. O., & Caviness, V. S., Jr. (1980). Radial organization of thalamic projections to the neocortex in the mouse. *Journal of Comparative Neurology, 194*, 369–393.

Galambos, R. (1992). A comparison of certain gamma band (40-Hz) brain rhythms in cat and man. In E. B. T. H. Basar (Ed.), Induced rhythms in the brain (pp. 201–216). Boston: Birkhauser.

Gencer, N. G., & Akalin-Acar, Z. (2005). Use of the isolated problem approach for multi-compartment BEM models of electro-magnetic source imaging. *Physics of Medical Biology, 50*, 3007–3022.

Grave de Peralta-Menendez, R., & Gonzalez-Andino, S. Ll (1998). A critical analysis of linear inverse solutions to the neuroelectromagnetic inverse problem. *IEEE Transactions in Biomedical Engineering, 45*, 440–448.

Grinvald, A., Lieke, E. E., Frostig, R. D., & Hildesheim, R. (1994). Cortical point-spread function and long-range

lateral interactions revealed by real-time optical imaging of macaque monkey primary visual cortex. *Journal of Neuroscience, 14*, 2545–2568.
Halgren, E., Marinkovic, K., & Chauvel, P. (1998). Generators of the late cognitive potentials in auditory and visual oddball tasks. *Electroencephalography and Clinical Neurophysiology, 106*, 156–164.
Hanslmayr, S., Klimesch, W., Sauseng, P., Gruber, W., Doppelmayr, M., Freunberger, R., Pecherstorfer, T., & Birbaumer, N. (2007). Alpha phase reset contributes to the generation of ERPs. *Cerebral Cortex, 17*, 1–8.
Honma, K., Honma, S., & Wada, T. (1987). Phase-dependent shift of free-running human circadian rhythms in response to a single bright light pulse. *Experientia, 43*, 1205–1207.
Iyer, V. K., Ploysongsang, Y., & Ramamoorthy, P. A. (1990). Adaptive filtering in biological signal processing. *Critical Reviews in Biomedical Engineering, 17*, 531–584.
Jung, T. P., Makeig, S., Humphries, C., Lee, T. W., McKeown, M. J., Iragui, V., & Sejnowski, T. J. (2000a). Removing electroencephalographic artifacts by blind source separation. *Psychophysiology, 37*, 163–178.
Jung, T. P., Makeig, S., Westerfield, M., Townsend, J., Courchesne, E., & Sejnowski. T. J. (2000b). Removal of eye activity artifacts from visual event-related potentials in normal and clinical subjects. *Clinical Neurophysiology, 111*, 1745–1758.
Jung, T. P., Makeig, S., Westerfield, M., Townsend, J., Courchesne, E., & Sejnowski, T. J. (2001). Analysis and visualization of single-trial event-related potentials. *Human Brain Mapping, 14*, 166–185.
Jutten, C., & Herault, J. (1991). Blind separation of sources. 1. An adaptive algorithm based on neuromimetic architecture. *Signal Processing, 24*, 1–10.
Jutten, C., & Karhunen, J. (2004). Advances in blind source separation (BSS) and independent component analysis (ICA) for nonlinear mixtures. *International Journal of Neural Systems, 14*, 267–292.
Klopp, J., Marinkovic, K., Chauvel, P., Nenov, V., & Halgren, E. (2000). Early widespread cortical distribution of coherent fusiform face selective activity. *Human Brain Mapping, 11*, 286–293.
Lee, I., Worrell, G., & Makeig, S. (2005). Relationships between concurrently recorded scalp and intracranial electrical signals in humans. *Human Brain Mapping Abstracts*. Toronto, Canada.
Lee, T. W., & Lewicki, M. S. (2000). The generalized Gaussian mixture model using ICA. *International workshop on ICA*, Helsinki, Finland, 239–244.
Linkenkaer-Hansen, K., Nikouline, V. V., Palva, J. M., & Ilmoniemi, R. J. (2001). Long-range temporal correlations and scaling behavior in human brain oscillations. *Journal of Neuroscience, 21*, 1370–1377.
Logothetis, N. K., Pauls, J., Augath, M., Trinath, T., & Oeltermann, A. (2001). Neurophysiological investigation of the basis of the fMRI signal. *Nature, 412*, 150–157.
Luck, S. J. (2005). An introduction to event-related potentials. In (Ed.), *An introduction to the event-related potential technique* (pp. 1–50). Cambridge, MA: MIT Press.
Makeig, S. (1993). Auditory event-related dynamics of the EEG spectrum and effects of exposure to tones. *Electroencephalography and Clinical Neurophysiology, 86*, 283–293.
Makeig, S., Bell, A. J., Jung, T. P., & Sejnowski, T. J. (1996). Independent component analysis of electroencephalographic data. *Advances in Neural Information Processing Systems, 8*, 145–151.
Makeig, S., Debener, S., Onton, J., & Delorme, A. (2004a). Mining event-related brain dynamics. *Trends in Cognitive Science, 8*, 204–210.
Makeig, S., Delorme, A., Westerfield, M., Jung, T. P., Townsend, J., Courchesne, E., & Sejnowski, T. J. (2004b). Electroencephalographic brain dynamics following manually responded visual targets. *Public Library of Science, 2*, e176.
Makeig, S., Jung, T. P., Bell, A. J., Ghahremani, D., & Sejnowski, T. J. (1997). Blind separation of auditory event-related brain responses into independent components. *Proceedings of the National Academy of Sciences USA, 94*, 10979–10984.
Makeig, S., Westerfield, M., Jung, T. P., Covington, J., Townsend, J., Sejnowski, T. J., & Courchesne, E. (1999a). Functionally independent components of the late positive event-related potential during visual spatial attention. *Journal of Neuroscience, 19*, 2665–2680.
Makeig, S., Westerfield, M., Jung, T. P., Enghoff, S., Townsend, J., Courchesne, E., & Sejnowski, T. J. (2002). Dynamic brain sources of visual evoked responses. *Science, 295*, 690–694.
Makeig, S., Westerfield, M., Townsend, J., Jung, T. P., Courchesne, E., & Sejnowski, T. J. (1999b). Functionally independent components of early event-related potentials in a visual spatial attention task. *Philosophical Transactions of the Royal Society of London B, Biological Science, 354*, 1135–1144.
Mardia, K. V. (1972). *Statistics of directional data*. New York: Academic Press.
Martinez-Montes, E., Cuspineda-Bravo, E. R., El-Deredy, W., Sanchez-Bornot, J. M., Lage-Castellanos, A., & Valdes-Sosa, P. A. (2008). Exploring event-related brain dynamics with tests on complex valued time-frequency representations. *Statistical Medicine, 27*, 2922–2947.
Massimini, M., Huber, R., Ferrarelli, F., Hill, S., & Tononi, G. (2004). The sleep slow oscillation as a traveling wave. *Journal of Neuroscience, 24*, 6862–6870.
Mazaheri, A., & Jensen, O. (2006). Posterior alpha activity is not phase-reset by visual stimuli. *Proceedings of the National Academy of Sciences USA, 103*, 2948–2952.
Meinecke, F., Ziehe, A., Kawanabe, M., & Muller, K. R. (2002). A resampling approach to estimate the stability of one-dimensional or multidimensional independent components. *IEEE Transactions in Biomedical Engineering, 49*, 1514–1525.
Murre, J. M., & Sturdy, D. P. (1995). The connectivity of the brain: Multi-level quantitative analysis. *Biology and Cybernetics, 73*, 529–545.
Nunez, P. (1977). The dipole layer as a model for scalp potentials. *TIT Journal of Life Sciences, 7*, 65–72.
Nunez, P., & Srinivasan, R. (2005). *Electric fields of the brain: The neurophysics of EEG*. Oxford: Oxford University Press.
Onton, J., & Makeig, S. (2005). Independent component analysis (ICA) source locations vary according to task demands. *Organization for Human Brain Mapping Abstracts*. Toronto, Canada.
Onton, J., & Makeig, S. (2006). Information-based modeling of event-related brain dynamics. *Progress in Brain Research, 159*, 99–120.
Onton, J., Westerfield, M., Townsend, J., & Makeig, S. (2006). Imaging human EEG dynamics using independent component analysis. *Neuroscience and Biobehavior Reviews, 30*, 808–822.
Oostenveld, R., & Oostendorp, T. F. (2002). Validating the boundary element method for forward and inverse EEG

computations in the presence of a hole in the skull. *Human Brain Mapping, 17*, 179–192.

Palmer, J. A., Makeig, S., Kreutz-Delgado, K., & Rao, B. D. (2008). Newton method for the ICA mixture model. *IEEE International Conference on Acoustics, Speech and Signal Processing*, 1805–1808.

Pfurtscheller, G., & Aranibar, A. (1977). Event-related cortical desynchronization detected by power measurements of scalp EEG. *Electroencephalography and Clinical Neurophysiology, 42*, 817–826.

Radman, T., Parra, L., & Bikson, M. (2006). Amplification of small electric fields by neurons: Implications for spike timing. *Conference Proceedings of the IEEE Engineering Medical Biology Society, 1*, 4949–4952.

Ranganath, C., & Rainer, G. (2003). Neural mechanisms for detecting and remembering novel events. *Nature Reviews in Neuroscience, 4*, 193–202.

Rulkov, N. F., Timofeev, I., & Bazhenov, M. (2004). Oscillations in large-scale cortical networks: Map-based model. *Journal of Computational Neuroscience, 17*, 203–223.

Sayers, B. M., Beagley, H. A., & Henshall, W. R. (1974). The mechanism of auditory evoked EEG responses. *Nature, 247*, 481–483.

Scherg, M. (1990). Fundamentals of dipole source potential analysis. *Advances in Audiology, 6*, 40–69.

Schroeder, C. E., Mehta, A. D., & Givre, S. J. (1998). A spatiotemporal profile of visual system activation revealed by current source density analysis in the awake macaque. *Cerebral Cortex, 8*, 575–592.

Stettler, D. D., Das, A., Bennett, J., & Gilbert, C. D. (2002). Lateral connectivity and contextual interactions in macaque primary visual cortex. *Neuron, 36*, 739–750.

Swadlow, H. A., & Gusev, A. G. (2000). The influence of single VB thalamocortical impulses on barrel columns of rabbit somatosensory cortex. *Journal of Neurophysiology, 83*, 2802–2813.

Tallon-Baudry, C., Bertrand, O., Delpuech, C., & Pernier, J. (1996). Stimulus specificity of phase-locked and non-phase-locked 40 Hz visual responses in human. *Journal of Neuroscience, 16*, 4240–4249.

Tass, P. (1999). *Phase resetting in medicine and biology: Stochastic modelling and data analysis*. Springer.

Voronin, L. L., Volgushev, M., Sokolov, M., Kasyanov, A., Chistiakova, M., & Reymann, K. G. (1999). Evidence for an ephaptic feedback in cortical synapses: Postsynaptic hyperpolarization alters the number of response failures and quantal content. *Neuroscience, 92*, 399–405.

Wang, C., Ulbert, I., Schomer, D. L., Marinkovic, K., & Halgren, E. (2005). Responses of human anterior cingulate cortex microdomains to error detection, conflict monitoring, stimulus-response mapping, familiarity, and orienting. *Journal of Neuroscience, 25*, 604–613.

Winfree, A. T. (1980). *The geometry of biological time* (2nd ed., p. 808). New York: Springer.

Worden, M. S., Foxe, J. J., Wang, N., & Simpson, G. V. (2000). Anticipatory biasing of visuospatial attention indexed by retinotopically specific alpha-band electroencephalography increases over occipital cortex. *Journal of Neuroscience, 20*, RC63.

PART 2

Commonly Studied ERP Components

CHAPTER

4 Sensory ERP Components

Hillel Pratt

Abstract

This chapter provides an overview of the main sensory evoked potentials of the auditory, somatosensory, and visual modalities. The short-latency sensory auditory, somatosensory, and visual evoked potentials are widely used in clinical practice, and their role in the context of event-related potential (ERP) studies is mostly as controls for the integrity of the sensory input to the central nervous system. Their description in this chapter therefore includes recording methods, waveform descriptions, and generators. The auditory middle-latency and long-latency components reflect cortical activity that often includes elements of cognitive processing of the sensory input. Their descriptions are therefore more detailed.

Keywords: event-related potential, auditory evoked potentials, somatosensory evoked potentials, visual evoked potentials

The recording of event-related potentials (ERPs) typically employs sensory stimuli to evoke brain responses. The cognitive processing of external sensory stimuli is preceded by stimulus transduction in sensory organs and conduction of neural signals along the sensory pathways. This exogenous sensory activity is the input that triggers the cognitive processes manifesting in the endogenous ERP component. Moreover, in some cases, attention or pathology can modulate the sensory inputs and their evoked activity. Therefore, monitoring the early sensory components can be helpful in assessment of attention, vigilance, and gating of sensory inputs in normal subjects and patients with pathologies that affect sensory processing (e.g., in schizophrenia). Sensory components are therefore of interest in studies of psychiatric and neurological patients, because they can be used to screen subjects to determine whether the disorder impacts sensory input to the subsequent cognitive processing. Recording the activity of sensory systems is thus used to assess the efficacy of the stimuli used in ERP experiments and to monitor the early processing preceding cognitive activity. Some of this activity is recorded in ERP waveforms as the initial exogenous components preceding and, at times, overlapping endogenous components. Other sensory components require specific recording settings that may be different from those of the cognitive ERPs. Sensory components of ERPs and sensory evoked potentials in general are, therefore, crucial for defining and understanding the subsequent cognitive ERPs.

This chapter provides an overview of the main sensory evoked potentials of the auditory, somatosenory, and visual modalities. The short-latency sensory auditory, somatosensory, and visual evoked potentials are widely used in clinical practice, and their role in the context of ERP studies is mostly as controls for the integrity of the sensory input to the central nervous system. Their description in this chapter therefore includes recording methods, waveform descriptions, and generators. The auditory middle-latency and long-latency components reflect cortical activity that often includes elements of cognitive processing of the sensory input. Their descriptions are therefore more detailed.

Because ERP studies typically use abrupt and transient stimuli, only sensory evoked potentials to such stimuli will be reviewed; sensory potentials to continuous steady-state stimuli will not be elaborated. More information on steady-state potentials may be found in reviews on their recording and analysis (Cacace & McFarland, 2007; Krishnan, 2007; Regan, 1989) as well as on their clinical uses (Picton, 2007).

Auditory Evoked Potentials

The exact timing of sounds is important in sound localization because it utilizes the difference in time of arrival of sound at the two ears. Therefore, the auditory system is particularly sensitive to the timing of stimuli, and this timing is maintained to some degree along the pathway at all of its levels. This temporal accuracy makes for excellent time locking of neural activity in response to the auditory stimulus that evoked it. Consequently, auditory evoked potentials include components spanning the full length of the auditory pathway, from the sensory organ, the cochlea, through the auditory nerve and brainstem, by way of the thalamus to the cortex. The respective evoked potentials include the auditory nerve and brainstem responses (ABRs), recorded in the initial 10 ms following stimulus onset; the auditory middle-latency evoked responses (MLRs), defined between 10 and 60 ms after stimulus onset; and the long-latency components defined between 60 and 200 ms after sound onset. The long-latency components are such relative to the earlier sensory components, but are shorter in latency than the cognitive components of ERPs such as the P300.

As the evoked activity ascends the auditory pathway, the number of activated neurons at each consecutive level increases because of axonal divergence. Consequently, amplitudes of components increase from ABRs (tenths of a microvolt) through MLRs (just under a microvolt) to the long-latency components (a few microvolts). In addition, the synchrony of neural populations at consecutive levels decreases as a result of serial synaptic transmissions along the pathway. As a result of this progressive desynchronization, the waveforms of ABRs, which include fast voltage oscillations every millisecond, are followed by much slower oscillations in the MLRs that include peaks that are approximately 25 ms apart, which in turn are followed by the long-latency peaks that are approximately 100 ms apart. These waveform differences are the reason for the differences in filter settings used for the recording of ABRs, MLRs, and long- latency potentials, with lower frequency bands the later the potentials. As a result of all this, the ABRs appear as the initial small deflection on the MLRs waveform, which, in turn appear as the first small component of the long-latency auditory evoked potentials (Figure 4.1.)

Auditory Nerve and Brainstem Potentials

RECORDING AND WAVEFORM

Auditory brainstem potentials include the transient-evoked ABRs (Pratt et al., 1999) and the steady-state frequency following potentials. Transient-evoked ABRs are optimally recorded from the scalp in response to high-intensity clicks presented at a rate of about 10/s. The potential difference between an electrode on the top of the head (C_z or F_z according to the 10–20 system; Jasper, 1958) and an electrode in the vicinity of the stimulated ear (mastoid or earlobe) is amplified in a bandpass of about 30–3000 Hz and averaged across a few thousand sweeps of 10–15 ms. To avoid waveform distortions and to enhance latency resolution, the sampling rate should not be lower than 20 kHz (50 μs/address or less). Under such stimulus and recording conditions, the normal waveform (Figure 4.2A) includes a series of five to seven voltage oscillations approximately 1 ms apart during the first 6–10 ms after stimulus onset. Amplitudes of the components are in the order of tenths of a microvolt, and components are affected by auditory sensitivity in the range of 1000–4000 Hz and are independent of the subject's state of arousal or the effects of drugs. Components are labeled by Roman numerals, beginning with the vertex positive peak at approximately 1.5 ms after stimulus onset, which is followed by a prominent negative trough. Peak I and its associated trough (I′) are followed by four fast oscillations, II, III, IV, and V, about 1 ms apart, which appear to be superimposed on a slowly rising positivity (the *pedestal*), with peak V typically identified as the most positive in the complex. The components following V exhibit large intersubject variability as well as large within-subject session-to-session variability.

The ascending auditory pathway in the brainstem includes monaural neurons that are activated only by right ear or only by left ear stimulation and binaural neurons that can be activated by either ear. With binaural stimulation, monaural neurons sensitive to left and right ear stimulation are activated in addition to the binaurally sensitive neurons. Arithmetic summation of the monaural responses to left and right ear stimulation obviously results in a larger response than to each ear alone. This sum is also larger than the response to binaural stimulation

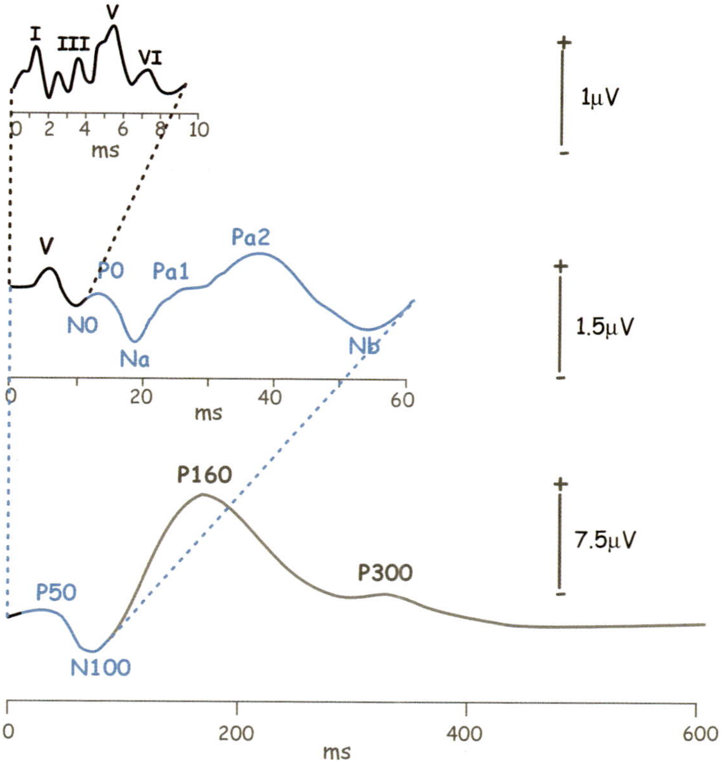

Fig. 4.1. Auditory brainstem responses, MLRs, and long-latency evoked potentials and their different time scales and amplitudes. Corresponding activity is plotted in the corresponding colors. Note that the different time scales and filter settings result in the entire ABR waveform being represented by the initial positive component of the MLRs, while the entire MLR waveform manifests in the initial long-latency component P50.

because it includes the neurons activated by either ear twice:—once in response to left ear stimulation and once in response to right ear stimulation. This inclusion of the same cells twice in the summed potentials is called *occlusion*. Binaural responses are therefore larger in amplitude than monaural responses but smaller than the sum of the monaural responses (Dobie & Norton, 1980). Auditory brainstem evoked potentials binaural interaction components (BICs) can be derived by subtracting the potentials to binaural stimulation from the sum of the potentials evoked by monaural left and right stimulation (Figure 4.2B). Binaural interaction components have been derived from humans (Debruyne, 1984; Decker & Howe, 1981; Dobie & Norton, 1980; Hosford et al., 1979; Levine, 1981; McPherson & Starr, 1993; Robinson & Rudge, 1981; Wrege & Starr, 1981) and correlated with psychophysical measures of sound lateralization (Furst et al., 1985, 1990; Polyakov & Pratt 1996; Pratt & Polyakov, 1996). The extent to which BICs reflect neural events that are uniquely binaural, rather than occlusion of monaural and binaural activity, has not been determined. However, the method of their derivation reflects the nonlinear interaction of activity from binaural compared to monaural stimulation.

GENERATORS

The generators of ABRs have been the subject of much research and controversy and are still not entirely agreed upon. However, for auditory function assessment, the following relationships between ABR peaks and neuroanatomy are adequate. The first peak in the sequence, peak I, is the only one on which there is general agreement regarding its generator. Peak I is the only one to survive section of the auditory nerve central to the internal auditory canal, placing its generator in the cochlea. It is synchronous with the compound action potential recorded from the cochlea, it is affected by masking, and it does not change polarity between condensation and rarefaction clicks (as the hair-cell microphonic potentials do), compatible with susceptibility to refractoriness and the all-or-none nature of neural action potentials from the cochlea. The polarity

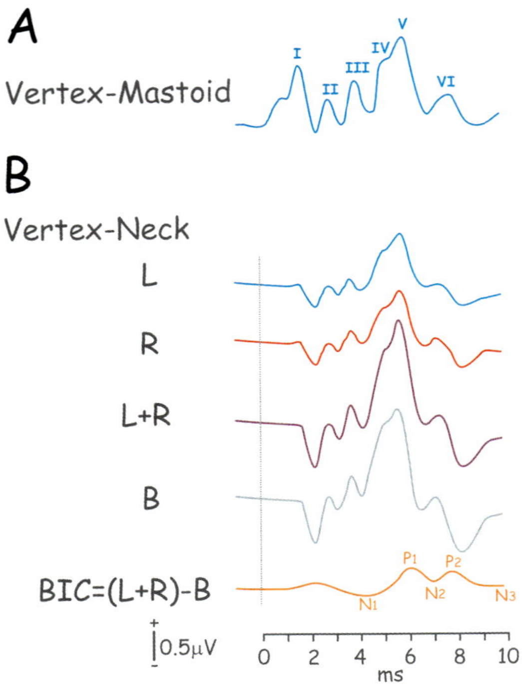

Fig. 4.2. (A) Auditory brainstem responses recorded between the vertex and the mastoid ipsilateral to the stimulated ear. Vertex-positive peaks are marked by Roman numerals in the order of their appearance. (B) Derivation of the auditory brainstem BIC. The potentials to binaural stimulation (B) are subtracted from the sum (L+R) of the potentials to left ear (L) and right ear (R) stimulation. Binaural interaction components include negative and positive peaks denoted by their polarity and order of appearance.

reversal of peak I around the mastoid reinforces its origin in activity within the temporal bone, diagonally directed toward the center of the head, compatible with auditory nerve action potential.

The generators of the second peak, II, are among the most controversial. Correlations with intraoperative intracranial recordings from humans (Hashimoto et al., 1981; Moller et al., 1988) and from experimental animals (Starr & Zaaroorr, 1990) indicate that it is synchronous with proximal auditory nerve activity. However, it does not reverse polarity around the mastoid, and it is not entirely obliterated with auditory nerve section. Some attribute it to overlapping activity from the auditory nerve and from the cochlear nucleus (Zaaroor & Starr, 1991). Others suggest that it is generated by the electric field distortions due to the changing geometry and conductance of the volume conductor as the nerve passes through the internal auditory canal and then through the posterior fossa to its merging into the brainstem (Martin et al., 1995). Regardless of the fine details, there is general agreement that peak II is generated in the vicinity of the auditory nerve's entry into the brainstem.

Peak III is generally agreed to be generated in the brainstem, but there is disagreement on the exact generator set because intracranial recordings from humans and correlations with animal experiments have not been conclusive. Suggested generators span the lower brainstem between the cochlear nucleus, through the trapezoid body to the superior olivary complex. Because the anatomical distance between these structures is small, for practical clinical purpose the generators may be attributed to the lower pons.

The fourth component is not always identified in human subjects, and it is generally absent in animals. When identified, peak IV is usually partially merged with peak V, creating a bifid IV–V complex. It is generally agreed that the fourth peak of animals corresponds to the human V. The exact generators of this complex are still under debate, but all evidence points to the upper pons, between the superior olivary complex, through the lateral lemniscus, with a possible contribution from the inferior colliculus. For practical purposes, the IV–V complex can be attributed to the upper pons and its junction with the midbrain.

Results of recording binaural interaction components from patients with brain lesions (Pratt et al., 1998) indicate that the component occurring at the time of peak IV is dependent on the integrity of the ventral-caudal trapezoid body, that the ventral trapezoid body contributes to the component at the time of peak V (P1), and that the rostral lateral lemniscus is a contributing generator of the binaural interaction component occurring at the time of peak VI (P2).

WAVEFORM MEASURES

Auditory brainstem potential audiometric measures include the lowest stimulus intensity at which a response is detected (detection threshold). Neuro-otological measures of ABRs include peak latencies and amplitudes, as well as interpeak latency differences and amplitude ratios (Table 4.1). In general, peak latencies measure the time lapse between stimulus onset and the time of highest synchronous activity in the generator set of the measured component. With auditory evoked potentials, this time includes the sound conduction time from the transducer (typically an earphone or speaker), through mechanical transmission in the middle and inner ears, through receptor transduction and auditory nerve activation, as well as transmission along the auditory pathway to the generators of the surface recorded activity. Peak latencies are therefore affected by all these processes,

Table 4.1 Typical adult normative upper limits for BAEP measures to 10/sec, 70 dBnHL clicks. Values are representative typical values and should not be used as norms for any particular setting. The norms for each laboratory should be calculated based on records from subjects recorded on the same system in the same setting. Neonates and children may have different normative values. Latency measures are in msec, Interaural differences are in the same units as their monaural counterparts in the same line, amplitude ratios are in decimal fractions

		MEAN	UPPER LIMIT	INTERAURAL DIFFERENCE
ABSOLUTE LATENCIES	I	1.75	2.2	0.3
	II	2.8	3.3	0.4
	III	3.9	4.5	0.4
	IV	5.1	5.9	0.6
	V	5.7	6.4	0.5
INTERPEAK LATENCY DIFFERENCE	V-I	4.0	4.5	0.5
	III-I	2.1	2.5	0.5
	V-III	1.9	2.4	0.5
AMPLITUDE RATIO	V/I	1.3	Lower limit 0.5	

which should all be taken into consideration in the interpretation of latency data. The multitude of factors contributing to peak latency has led to the definition of interpeak latency difference measures. By deriving the difference in latency between two peaks, factors affecting the latency of both peaks are canceled out, and the latency measure gains in specificity. Thus, for example, the V–I interpeak latency difference reflects transmission time from the cochlea to the ponto-midbrain junction and is usually unaffected by more peripheral factors such as conductive hearing loss. With ABRs, the most widely used interpeak latency differences are V–I, III–I, and V–III, reflecting transmission along the auditory pathway between the cochlea and the ponto-midbrain junction, between the cochlea and the ponto-medullary junction, and along the pons, respectively.

The functionally intact auditory pathway is typically symmetrical, with equal transmission times in response to left ear and right ear stimulation. Thus, a significant interaural difference in respective measures between responses to left and right ear stimulation may be indicative of a unilateral functional abnormality. Peak amplitudes are affected by a variety of recording factors such as electrode impedance, the precise location of the electrodes relative to the generators of the potentials, and the effective stimulus intensity reaching the inner ear. Amplitude measures, therefore, exhibit very large variability across subjects as well as between sessions with the same subject. Amplitude ratios between peaks in the same record quantify the amount of synchronous activity at different levels of the pathway, canceling out sources of variance that are common to both peaks, such as electrode impedance and relative positions of electrodes and generators. The most widely used amplitude ratio is the one between peaks V and I, reflecting the amount of synchronous activity at the upper pons normalized for the cochlear output that gave rise to it.

FACTORS AFFECTING THE POTENTIALS

Auditory brainstem potentials are affected by a variety of nonpathological factors, including the subject's age, body temperature, and gender, as well as stimulus factors such as frequency composition, intensity, presentation rate, and envelope. Auditory brainstem potentials reflect the firing of neural elements that are distributed along the auditory pathway and summate on the scalp. Therefore, the more synchronous the neural activation, the larger the peak amplitudes. Fast-rising stimulus envelopes result in more synchronous neural activation. Because high-frequency stimuli have a fast rise time, they are more effective in synchronous neural activation. Moreover, because high frequencies are represented in the more basal portions of the cochlea, where longitudinal mechanical coupling along the basilar membrane is more pronounced, the responses across a wider frequency range are synchronous. The mechanical traveling wave reaches the basal cochlea earlier; therefore, potentials evoked by high frequencies have shorter latencies. In summary, high-frequency stimuli and stimulus envelopes with a short rise time evoke ABRs with larger amplitudes and shorter latencies. Increasing stimulus intensity results in earlier activation of nerve fibers, usually because the cochlear postsynaptic potentials rise higher and faster and reach threshold earlier. Thus, ABR peak latencies shorten with increasing stimulus intensity. With an increasing stimulus rate, synaptic efficacy is reduced, typically because recycling of the transmitter cannot keep up with the rate of its depletion. Thus, the synaptic potentials are smaller, resulting in longer delays until the threshold of the next-order neurons is reached. Consequently, the cumulative effect along the chain of synapses connecting successive generator sets of ABRs progressively prolongs peak latencies with increasing stimulus rate: The later peaks' latencies are more affected by high rates than the earlier peaks. As a result, interpeak latency differences are prolonged with increasing stimulus rate.

The effect of age may be divided into two periods: maturation and adulthood. During maturation, which can be followed in premature neonates, through full-term infants and up to 2 years of age, significant changes in component amplitudes, latency, and interpeak latency differences take place. In premature neonates, ABRs are not detectable before 26 weeks postconception. After 26 weeks, components II, IV, and VI are not well defined and peak latencies and interpeak latency differences are prolonged. Interpeak latency differences shorten with maturation at a rate of 0.1–0.2 ms/week until week 40 and at a slower rate until the age of 2 years, when they reach adult values. During childhood, the definition of components II, IV, and VI improves. Peak amplitudes are typically larger than in adults, particularly component I, which in infants is often larger than V. As a result, the amplitude ratio between V and I, which is typically larger than 1 in adults, is smaller than 1 in children. Consequently, the V/I amplitude ratio must be used with caution, taking into account the profound effects of age on this measure. During adulthood, the age-related changes in amplitude and latency measures are so small as to not warrant age-adjusted norms for clinical use. The effects of increasing stimulus rate on interpeak latency differences are large in neonates, decrease during childhood, and reach stable adult values at adolescence.

The effect on ABRs of lowering the body temperature is prolongation of peak latencies such that peak V latency increases by 0.2 ms/ °C down to a core temperature of 27°C. Below 27°C component definition becomes difficult. Some effects of drugs and alcohol, as well as circadian rhythms in ABR measures, have been attributed to changes in body temperature.

The effect of gender on ABRs is apparent only in adults, where latencies and interpeak latency differences are shorter and amplitudes are larger in women. Attempts have been made to relate these differences to gender differences in body and brain size and/or to differences in body temperature, water content, and adipose tissue. The differences are small and well within normal variability of each gender, but some insist on separate normative adult data for men and women.

Middle-Latency Potentials

The auditory MLRs include the transient-evoked potentials and the 40 Hz steady-state potentials. The transient-evoked potentials are generated by thalamocortical and cortical activity, as well as by ascending subcortical activity in the extralemniscal auditory pathway between 10 and 50 ms after the onset of a brief auditory stimulus such as a tone burst of a few milliseconds' duration or a click (Figure 4.3). Early components of MLRs are susceptible to muscle activity and are also affected by

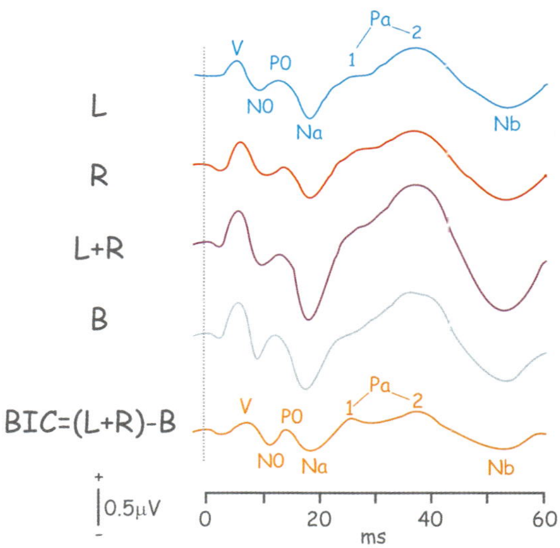

Fig. 4.3. Auditory middle-latency evoked potentials and their binaural interaction components. The potentials include four vertex positive and three negative peaks, the first of which is brainstem component V, followed by negative and positive alternations beginning with N0, through P0, Na, the double-peaked Pa, and Nb. These components are followed by peak Pb (not shown), which is similar in polarity and latency to ERP peak P50. The derivation of the MLR BIC is presented as well. L marks potentials to left ear stimulation, R to right ear stimulation, and B to binaural stimulation.

the subject's wakefulness and age, effects that increase with increased stimulus rate.

Clinical application of MLRs is limited by their sensitivity to subjects' age and to sedation. When these limitations are addressed, MLRs can be useful to estimate frequency-specific auditory sensitivity up to the cortical level, to assess auditory pathway functional integrity, and to localize lesions at the thalamocortical and primary auditory cortex levels. In addition, MLRs are useful for research on the functional role of auditory cortex in auditory processing.

WAVEFORM

The normal transient-evoked MLR waveform is recorded using a bandpass of 3–1000 Hz and includes a series of about three voltage oscillations between 10 and 50 ms after stimulus onset. Waveforms begin with brainstem component V and end with the long-latency potentials' P1. The MLR components include four vertex-positive and three vertex-negative peaks: V, N 0, P 0, Na, Pa, Nb, and Pb (Figure 4.3). The waveform and amplitudes of components may vary widely, depending on the subject's age (above or below 10 years), state of arousal (awake or asleep), stimulus rate (10/s or above or much lower), and acquisition filter settings.

The most consistent MLR component, Pa, has a peak latency of about 25 ms and an amplitude of about 0.5–1 μV in adults in response to a high-intensity (60–70 dBnHL) stimulus (click or tone pip). The following prominent positive peak, Pb, is less consistent and may be absent even in normally hearing and neurologically intact adults. It follows Pa by about 25 ms. The positive peak P 0, preceding Pa, is even more variable and is prone to contamination with myogenic potentials. These myogenic artifacts are recorded from muscles that participate in reflexes that move the auricles toward the source of sound. Although such auricular movements to sound are prominent in animals, they are undetected in humans, but the muscle activity involved in them is recordable from humans as the postauricular muscle (PAM) response to sound. These muscle potentials are sensitive to neck tension, and their artifact can be reduced by tilting the subject's head backward, releasing nuchal muscle tension.

Components V and N0 are actually the brainstem component V and the subsequent negative deflection (also known as SN10–slow negativity at 10 ms), as recorded with MLR filters. The negative peak preceding Pa is Na, which typically appears at 15–20 ms and may be affected in waveform and peak latency by possible contamination of PAM potentials, but for the most part it is neural in origin. The negative peak between Pa and Pb is termed Nb, and its detectability and scalp distribution closely resemble those of Pa.

As explained in the section on ABR binaural interaction, the difference waveform between the algebraic sum of the responses to monaural stimulation of the right and left ears, and the response to binaural stimulation, is called the *binaural interaction components* (BICs) (Figure 4.2). Binaural interaction components of the human MLR are at a latency of about 20–40 ms (Debruyne, 1984; Dobie & Norton, 1980; McPherson et al., 1989; Polyakov & Pratt, 1995, 2003; Woods & Clayworth, 1985;). Initially, substantial BICs were indicated in component Pa (Dobie & Norton, 1980; Woods & Clayworth, 1985) and later in the preceding and following Na and Nb as well (McPherson et al., 1989; Polyakov & Pratt, 1995). Binaural interaction component amplitudes amount to nearly 50% of the sum of the monaural responses, a value significantly greater than that reported for the brainstem binaural interaction potentials. This suggests that binaural processes constitute a major function of the central auditory pathway rostral to the pons.

GENERATORS

The generators of MLR have long been suggested to be auditory cortex (primary and/or secondary) with possible upper brainstem and thalamic contributions (Dieber et al., 1988; Kraus et al., 1982; Ozdamar et al., 1982; Scherg & von Kramon, 1986). Evidence in support of these hypotheses has been obtained from research involving both animal and human subjects (Arezzo et al., 1975; Buchwald et al., 1981; Celesia, 1976; Kileny et al., 1987; Lee et al., 1984; Parving et al., 1980; Woods et al., 1987).

The Na component is currently believed to originate subcortically from the midbrain, thalamus, or thalamocortical radiations. This hypothesis is based on the results of depth recording of Na (Hashimoto, 1982) and on scalp topography analyses of the Na component (Dieber et al., 1988; Kraus & McGee, 1988).

Studies on the generators of Pa (Celesia, 1976; Kraus et al, 1982; Lee et al, 1984; Ozdamar & Kraus, 1983; Picton et al., 1974) demonstrated that the vertex-recorded Pa component originated bilaterally from vertically oriented dipoles within the primary auditory cortices, with contributions from the mesencephalic reticular formation (Kraus et al., 1982) and the medial geniculate body (McGee et al., 1992). The vertex-recorded MLR has been suggested to have different physiological origins than the Pa component that is recorded over the temporal scalp (Kraus & McGee, 1988). In addition, a preserved "Pa" component has been reported despite the presence of bilateral lesions that affect the primary auditory cortex (Kileny et al., 1987; Parving et al., 1980; Rosati et al., 1982); Woods et al., 1987).

The sleep sensitive component of Pa may reflect subcortical ascending reticular activity, and Pa has a dual set of generators with both primary auditory cortex and ascending subcortical contributions. Accordingly, equivalent dipole studies of the human MLR and its BICs (Polyakov & Pratt, 1994, 1995) have suggested that Pa and its BIC are composites of activity from primary auditory cortex as well as from subcortical nonspecific ascending structures, manifesting in two subcomponents—Pa1 and Pa2, respectively (Figure 4.3).

On the other hand, the generators of component "Nb" are still not clear because no anatomical localization of lesion can be inferred from its alteration. Previous studies (Yoshiyura et al., 1995) estimated the locations of MLR sources using magnetoencephalography and magnetic resonance images. They demonstrated that no significant spatial difference was observed between the Pa and Nb sources. Thus, either these two successive components reflect activities of the same neural population or the sources of these two responses are too close to each other to be distinguished clearly. Studies using functional magnetic resonance imaging (fMRI) similarly find that brain areas are most responsive to lower stimulation rates at later stages in the ascending auditory system from the brainstem to secondary cortical fields (Giraud et al., 2000; Harms & Melcher, 2002). These findings lend further support to the suggestion that MLRs, with their relative resilience to high stimulus rates, are from primary auditory cortex and perhaps subcortical contributions. They also indicate that the activity reflected in the MLRs relates to rapid changes in stimulus timing and envelopes, suggesting a role in sound localization, speech processing, and auditory object definition. Factors affecting the potentials will now be discussed.

Middle-latency response components are most affected by auditory sensitivity to the stimulus used to evoke them (Maurizi et al., 1984; Oates & Stappells, 1997; Ozdamar et al., 1982; Scherg & Volk, 1983) and by the subject's age (Amenedo & Diaz, 1998; Chambers, 1992; Chambers & Griffiths, 1991; Pasman et al., 1992; Rogers et al., 1989; Rotteveel et al., 1987; Stapells et al., 1988; Suzuki & Hirabayashi, 1987; Suzuki et al., 1983; Tucker & Ruth, 1996; Woods & Clayworth, 1986; Yamada et al., 2003). Specifically, amplitudes of MLR components increase with increasing stimulus intensity, from near-threshold intensities up to about 70 dB above threshold. Latencies decrease with increasing stimulus intensity across a range of only 40–50 dB above behavioral threshold and then stabilize.

Young children and infants may not present MLRs, even when their auditory and neurological functions are intact, because of their higher sensitivity to stimulus rate and the lability of their waveforms to filter settings (see below). When stimulus rates are slow enough (1–2/s) and the recording high-pass is low enough (e.g., 10 Hz), an MLR may be recorded from neonates and infants. The latency of Pa is then about twice the normal adult value, and its waveform is broader than in adults (Fifer & Sierra-Irizarry, 1988). The amplitude of Pa increases from infancy until late childhood and then decreases through adolescence and advancing age. The most consistently recorded MLR component in children is Na, which may be detected even in the absence of Pa (Kraus et al., 1985). In general, MLRs from children under the age of about 10 years should be interpreted with caution.

Many of the factors affecting MLRs, particularly stimulus rate and filter setting, interact in their effects with the subject's age, especially in neonates and children (Jerger et al., 1987; Kraus et al., 1987; Stapells et al., 1988). A stimulus rate of 10/s, which is optimal for adults, may be much too fast for recording reliable MLRs from young children. The detectability of MLRs with this stimulus rate declines with decreasing age from adolescence to infancy. Thus, to record MLRs from infants, a stimulus rate as slow as 1/s may be necessary. The recording filter high-pass setting also affects the detectability of MLRs at different ages, such that higher detection rates are obtained using 15 Hz compared to a 3 Hz high-pass cutoff (Kraus et al., 1987).

Unlike the earlier cochlear and brainstem potentials, MLRs are also affected by the subject's attention (Kadobayashi & Toyoshima, 1984) and state of arousal (Aceto et al., 2003; Davies et al., 1996; Dieber et al, 1989; Drummond, 2000; Erwin & Buchwald, 1986; Firsching et al., 1987; Osterhammel et al., 1985; Suzuki et al., 1992; Telles et al., 1993), as well as by some drugs (Antognini & Wang, 1999; Jaaskelainen et al., 1999; Japaridze et al., 1993; Kudoh & Matsuki, 1999; Masahiro et al., 1989; Plourde & Villemure, 1996; Prosser & Arslan, 1985; Smith & Kraus, 1987; Tatsumiet et al., 1995).

Waveform measures will now be considered. Middle-latency response measures include the stimulus intensity at which a response is just detected, as well as peak latencies. Amplitudes are too variable to be clinically useful, and in order to reduce variability, peak-to-peak measures between positive peaks and the following negative trough are sometimes used. Table 4.2 presents the average and standard deviation values of peak latencies and peak-to-peak amplitudes of the major MLR peaks, from a normative population of normal-hearing young adults, in response to monaural, 3.3/s clicks at an intensity of 65 dBnHL. These values should not be used as normative values, even when using the exact same stimulus and recording parameters, because factors such as ambient acoustic noise, recording filter settings, and click acoustic spectra, to name only a few, may vary among laboratories.

As Table 4.2 shows, peak latencies have a relatively narrow normal range. Consequently, the normative ranges of peak latencies of successive peaks of the same polarity (e.g., Na and Nb) do not overlap and allow unambiguous identification of peaks. This is even true for the two constituent peaks of Pa. In contrast, the normal variability of amplitude measures is much higher, and even with the least variable peak-to-peak measures, the normative range of amplitudes may span the range between an amplitude of zero and as high as twice the normative average. Therefore, amplitudes are rarely used as diagnostic measures, whereas latency prolongations or absence of components are considered clinically significant.

Long-Latency Potentials

GENERAL DESCRIPTION AND RECORDING

The long-latency auditory evoked potentials (Figure 4.4) include the initial components of an auditory ERP recorded with a typical bandpass of 0.1–100 Hz. They comprise an initial scalp-positive peak at about 50 ms (P50) followed by a prominent scalp-negative peak at approximately 100 ms (N100), followed by a series of alternating scalp-positive and negative peaks (P160, N200, P300, and more) that are not specific to auditory stimuli and reflect processing beyond sensory perception. This chapter focuses on sensory potentials and will therefore be limited to the P50 and N100 components of the auditory long-latency evoked potentials.

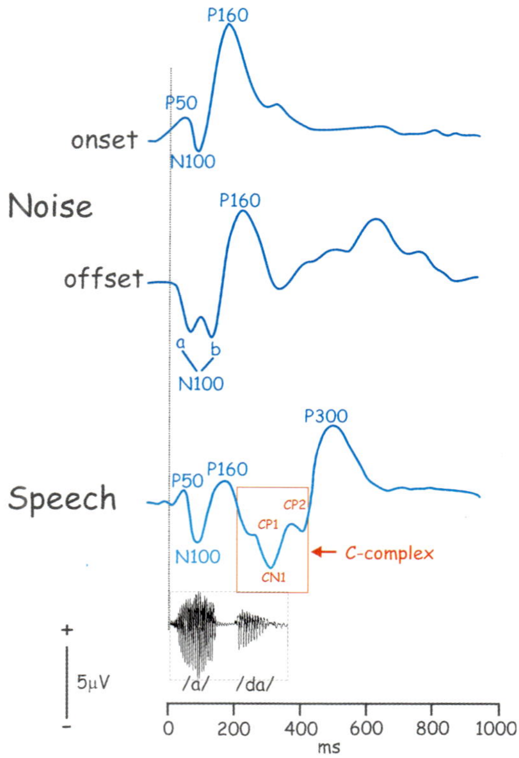

Fig. 4.4. Auditory long-latency evoked potentials to noise onset and offset (top) and to a meaningless target word/ada/ (bottom). The components include a sequence of P50, N100 (which is bifid to stimulus offset), and P160. When the stimulus is task relevant (as in the word *stimulus* in this case), these components are followed by a P300. The acoustic waveform of the word is presented in the bottom trace. Noise and word onsets as well as noise offset are marked by the first vertical line. The potentials to the acoustic change of the second syllable in the ongoing sound are delineated by the inset. The components of this change complex (C-complex) are designated with a C prefix: CP1, CN1, and CP2.

Table 4.2 Normative latencies and amplitudes of click-evoked auditory middle-latency components

	Peak Latency (ms)							Peak-to-Peak Amplitude (μV)		
	V	No	Po	Na	Pa1	Pa2	Nb	V–No	Po–Na	Pa2–Nb
Mean	5.46	8.96	12.24	16.88	23.28	34.16	47.16	0.45	0.37	0.69
SD	±0.46	±1.16	±1.46	±1.18	±1.54	±1.98	±2.42	±0.13	±0.10	±0.25

P50

The P50 component is the earliest (around 50 ms), the smallest in amplitude, the most variable, and consequently the least studied of the auditory ERPs. Early reports on long-latency evoked potentials typically reported its presence, as part of the "P1–N1–P2 complex" but rarely elaborated on parametric effects on its characteristics (e.g., Knight et al., 1980; Naatanen & Picton 1987). Some early reports detailed its latency and amplitude values, stating that it was not affected by factors such as age (Barnet et al., 1975) or attention (Picton & Hillyard, 1974). More recently, reports have described extensive changes in P50 with maturation, beginning with its domination of the P50–N100–P160 complex in young children (Ceponiene et al., 2002; Sharma et al., 1997) to its small amplitude in adults. The normal maturation of P50 (Sharma et al., 1997) has been used to determine a period of about 3.5 years during which the human central auditory system remains maximally plastic and therefore optimal for cochlear implantation (Sharma et al., 2002). The latency of P50 was used as the indicator of auditory system maturation and of the effects of deprivation due to deafness on auditory function (Eggermont et al., 1997). Maturation of P50 evoked by pairs of clicks has also indicated that the neural circuits underlying sensory gating are functional very early in postnatal development (Kisley et al., 2003b).

The generators of P50 have been attributed to the primary auditory cortex at Heschl's gyrus (Huotilainen et al., 1988; Liegeois-Chauvel et al., 1994; Ponton et al., 2002; Pool et al., 1989; Reite et al., 1988; Wood & Woolpaw, 1982), with the earlier work describing it as part of the middle-latency potentials' Pb. However, more recent work suggests that these are distinct components, with P50 having more complex generators that also include the hippocampus, the planum temporale, and the lateral temporal cortex (Howard et al., 2000; Liegeois-Chauvel et al., 1994, 1999) and neocortical areas (Grunwald et al., 2003; Kisley et al., 2003b).

A number of studies reported P50 sensitivity to reticular formation nonspecific cholinergic activation (Buchwald et al., 1991) and consequently to levels of arousal (de Lugt et al., 1996; Erwin & Buchwald, 1986), sensory activation, and a variety of disorders. Sensory gating represents the nervous system's ability to inhibit responding to irrelevant environmental stimuli. A common method of assessing gating is measuring response suppression when pairs of clicks with an interval of 250, 500, or 1000 ms are presented. The earliest component to show such suppression is P50. Sensory gating (Skinner et al., 1999) and habituation (Pitman et al., 1999) of P50 was found impaired in subjects with posttraumatic stress disorder (PTSD) compared to controls, indicating dysregulation of sensory processing in PTSD. Such decreased gating was also observed in normal adolescents compared to normal older subjects (Rasco et al., 2000). The amplitude of P50 was found to be attenuated in autism (Buchwald et al., 1988, 1992), Alzheimer's disease (Buchwald et al., 1989; Fein et al., 1994; Green et al., 1992; O'Mahony et al., 1994), Huntington's disease (Uc et al., 2003), attention deficit hyperactivity disorder (ADHD; Kemner et al., 1996), and narcolepsy (Boop et al., 1994), suggesting decreased reticular arousal by sound. P50 was reported to be diminished and prolonged or absent in Parkinson's disease, improving following posterior ansa-pallidotomy, except in one patient who showed mild worsening attributed to postoperative sleepiness (Mohamed et al., 1996). Increased P50 amplitudes have been reported in mild cognitive impairment (Golob et al., 2007; Irimajiri et al., 2005) and in human immunodeficiency virus 1 (HIV-1) infection, correlating with indices of disease progression (Schroeder et al., 1994), but also in normal elderly subjects (Smith et al., 1980). Diminished habituation of P50 was reported in schizophrenia (Erwin et al., 1994; Kisley et al., 2003a; Schall et al., 1997; see Chapter 19, this volume), autism (Buchwald et al., 1992). Parkinson's disease (Teo et al., 1998; see Chapter 21, this volume), and major depression (Franks et al., 1983; see Chapter 20, this volume). Adults with sensory hypersensitivity without additional health or mental problems have been reported to have less robust P50 suppression (Kisley et al., 2004), along with "overinclusion" of irrelevant sounds in their focus of attention. The amplitude of P50 in an auditory task was reported to be significantly increased in irritable bowel syndrome (IBS) patients compared to controls, compatible with a generalized preattentive increase in central nervous system reactivity in this disorder (Berman et al., 2002).

The auditory neuromagnetic P50 field has been reported to be present and indistinguishable in its sources from N100 in response to stimulus onset but absent from offset responses (Hari et al., 1987; Pantev et al., 1996). This neuromagnetic response, P40m, peaking about 40 ms after stimulus onset, preceded a prominent deflection in the opposite direction at about 100 ms (N100m). Both deflections could be explained by cortical activity within the Sylvian fissure (Hari et al., 1987). Striking similarities were found between the N100m of the on- and off-responses in their latency, in the estimated sources in the supratemporal plane, and in their amplitude dependence on stimulus rate (Hari et al., 1987). However, a weaker source strength was reported for the off-response (Pantev et al., 1996) and only the on-response was preceded by P40m (Hari et al., 1987; Pantev et al., 1996). This suggested that P50 and N100 are not causally related (Hari et al., 1987). Moreover, while N100 seems to reflect cortical activity related to any abrupt change in the auditory environment (Hari et al., 1987), P50 reflects a distinct process evoked by stimulus onset.

P50 in response to stimulus onset thus appears to have a complex set of generators that are sensitive to reticular brain activation, and to pathologies affecting sensory, cognitive, and motor processing. In response to sound offset, P50 is less prominent or absent. The aspects of auditory processing associated with P50 to noise onset that are absent in response to noise offset appear to be preattentive orienting to a new sound in the scene (Pratt et al., 2008).

N100

Component N100 (~100 ms from stimulus onset) of the auditory ERPs (Figure 4.4) has been suggested to signal the detection of acoustic change in the environment (Hyde, 1997). Change is defined as a deviation from a preceding constant state, and such deviation can be effected by a variety of sound attributes. N100 has accordingly been shown to include, in addition to the transient onset and offset responses, three processes at latencies around N100 that are associated with changes in the auditory environment: (1) the *M-process* in response to deviation (*mismatch*) from the preceding auditory sequence of stimuli (Jones et al., 1998, Vaz Pato & Jones, 1999); (2) the *C-process* in response to acoustic *change* within an ongoing stream (Jones, 2003; Jones & Perez, 2001); and (3) the *F-process* associated with *fusion* of acoustic elements to form a new auditory object (Laufer & Pratt, 2003a, 2005). The M-process's mismatch negativity (MMN) is the subject of a separate chapter (see Chapter 6, this volume).

ONSET RESPONSE

The N100 potentials evoked by short transient stimuli, such as clicks or tones that contain both onsets and frequency change, and by onsets of noise, which involve only sound onset and no frequency change (Figure 4.4, top), are similar and single-peaked (Pratt et al., 2005). The single-peaked N100 has been shown to include three temporally overlapping constituents. The dominant contribution to N100 is fronto-central, with a peak latency of 100 ms and slightly greater activity at the scalp over the hemisphere contralateral to stimulation. As originally suggested (Vaughan & Ritter, 1970), and later confirmed using a regional dipole model (Scherg & von Cramon, 1986a, 1986b), it has bilateral and tangential neural generators in the auditory cortices on the superior temporal lobe. The second constituent of N100 is maximally recorded at midtemporal electrodes and is therefore called the *T-complex* (Wolpaw & Penry, 1977). It consists of a positive peak around 100 ms (Ta) and a negative peak around 150 ms (Tb), and is thought to originate in the auditory association cortex of the superior temporal gyrus with a radial orientation (Ponton et al., 2002; Scherg & von Cramon, 1985, 1986a). The T-complex is much larger and slightly earlier over the hemisphere contralateral to stimulation compared to the ipsilateral hemisphere (Ponton et al., 2002; Wolpaw & Penry, 1977). The third constituent of N100 has a variable latency, around 100 ms, and may overlap and continue beyond the major constituent of N100 (Hari et al., 1982; Velasco et al., 1985). This contribution may be associated with widespread transient arousal to increase sensory and motor responses to sound. It is most easily recorded at stimulus intensities greater than 60 dB sound pressure level (SPL) presented at intervals longer than 4 to 5 s (Näätänen & Picton, 1987).

Dipole source estimation indicated two distinct source activities underlying N100 with latencies of 100 and 145 ms, both symmetrically localized in Heschl's gyrus of both hemispheres (Scherg & Picton, 1991). The later of the two components could be the negative Tb of the T-complex. Another source localization study (Ponton et al., 2002) found a symmetrical source for the N100–P160 complex in Heschl's gyrus with no frontal sources, because the procedure limited source locations to two regional dipoles. A magnetic source study (Lütkenhöner & Steinsträter, 1998) located the major constituent of N100 in the planum temporale. Another study of the electric N100 fields (Picton et al., 1999) found that the dominant sources were in the supratemporal plane and the lateral temporal lobe, with some activity possibly originating in the anterior cingulate cortex or supplementary motor areas of the prefrontal cortex.

OFFSET RESPONSE

Recent studies on potentials to sound offset without the confounds of frequency change used noise offset

(Michalewski et al., 2005; Pratt et al., 2005, 2007) and noted a double-peaked N100 to sound offset (Figure 4.4, second from the top) called the *N-complex*. The first peak (N100a) occurred at ~100 ms, was frontal in distribution, and was similar to N1 of short transients and sound onset. The following peak (N100b) occurred at ~150 ms with a central/temporal scalp distribution, with distinct sources and a distinct time course of their activity. Cortical activity associated with the N-complex was located (Pratt et al., 2005, 2007) bilaterally in temporoparietal regions, with left hemisphere prominence of the first peak (N100a) and right hemisphere prominence of the second peak (N100b).

A double-peaked N100 has also been described in response to speech signals with long voice onset times (VOTs). The first N100 peak coincided with the release burst and the second with the onset of voicing (Sharma & Dorman, 2000). The same pattern of responses was obtained in intracranial recordings from the cortical surface of animals (Steinschneider et al., 1994) and humans (Liégeois-Chauvel et al., 1999; Steinschneider et al., 1999). The transition from a single-peaked to a bifid N100 that coincides with the transition from voiced to voiceless stops (in English) has led to the suggestion that the perception of voicing is neurophysiologically based and is not language specific (Liégeois-Chauvel et al., 1999; Simos et al., 1997, Steinschneider et al., 1994, 1999). Other studies showed that changes in N100 morphology do not correlate with behavioral voicing perception (Sharma & Dorman, 2000; Sharma et al., 2000). However, subjects with different native languages having different VOTs all exhibited the same double-peaked N100 to offsets of nonspeech white noise (Michalewski et al., 2005; Pratt et al, 2005). Thus, the bifid N100 to nonspeech stimuli, which was the same regardless of the native language, suggests that the double peaking reflects a more general aspect of sound processing that is independent of language.

Whereas the first constituent of N100 to sound offset (N100a) appears to be similar to the onset-evoked N100, the second constituent, N100b, is unique to sound offset, is later than N100a, and is associated with a specific change in the stimulus from sound to silence: —termination of an ongoing stimulus. Comparing N100b with the other ERP components associated with stimulus change that are at the same latency (fusion, change, or mismatch), indicates a number of differences. Sound offset does not involve fusion of acoustic elements; thus, the F-process is an unlikely contributor to N100b. Comparing scalp distribution and source currents and their time courses reveals differences between N100b and the other contenders: The change potential negative peak is associated with prominent left auditory cortex activity (Laufer, 2003), as opposed to right temporal activity during N100b (Pratt et al., 2005). The early portions of the best-studied mismatch component, MMN, do, indeed reflect the M-process and are located in the temporal lobe (Giard et al., 1990; Jemel et al., 2002). Moreover, its single-unit correlates have been described in the auditory cortex of experimental animals (Ulanovski et al., 2003). Mismatch negativity lateralization has been shown to be affected by the stimulus type involved: Tones in different spatial locations have been reported to evoke an MMN that is lateralized to the right (Deouell et al., 1998), as did frequency deviance (Liebenthal et al., 2003), while verbal stimuli lateralize the early MMN to the left (Naatanen et al., 1997). White noise, being a nonspeech stimulus, could thus be expected to evoke an M-complex that is lateralized to the right, as N100b actually is.

However, N100b is not evoked by the sound, but rather by its ending. Whereas the M- or C-processes reflect the response to an actual acoustic stimulus—a deviant stimulus showing a mismatch in a series of stimuli (M-complex, MMN) or the continuation of the same stimulus after it has changed (C-complex)—N100b reflects the perception of "negation" or termination of an ongoing stimulus's transition to silence. No effects of attention on the constituents of N100 were observed, and N100b was therefore suggested to reflect preattentive perception of the cessation of an ongoing sound.

CHANGE COMPLEX

The change-complex potentials were termed *C (change) potentials* to distinguish potentials elicited by acoustic change per se (Figure 4.4, bottom) from the "onset-type" N100/P160 in response to stimulus onset (Jones et al., 1998). Specifically, the C-potentials are auditory evoked components to groupings of separate sound components, usually tones and notes (Alain & Woods, 1994; Jones & Perez, 2001; Jones et al., 1998; Makela et al., 1988; Martin & Boothroyd, 1999), to form an auditory stream. In a series of studies on fusion of rapid stimulus alternations (Jones et al., 2000a, 2000b; Vaz Pato & Jones, 1999), continuously synthesized musical instrument notes oscillated between two pitches at a rapid rate (8–16 notes/s, i.e., every 63–125 ms) and the "change-type" N100 and P160 potentials associated with each individual change were abolished.

The vertex-maximal C-potentials were found to be associated with both sequential and spectral streaming (e.g., Jones, 2003; Jones & Perez, 2001, 2002; Jones et al., 1998), as well as transition from a periodic to an aperiodic sound (Martin & Boothroyd, 1999; Ostroff et al., 1998). C-potentials were also elicited by infrequently interspersed changes of timbre occurring after a stream of rapid pitch changes (Jones et al., 1998). These results indicate that the C-potentials reflect a mismatch between the incoming acoustic change and the preceding stream. Since the change-potentials were mainly affected by the time interval between pitch changes and not by the magnitude of the pitch change (Jones & Perez, 2001), the C-potentials were concluded to reflect the analysis of the spectral composition of sounds (*spectral profile analysis*) that may also be important in maintaining the perceptual constancy of timbre (Jones & Perez, 2001, 2002).

FUSION COMPLEX

The grouping of the frequency components of sounds emanating from the same source into coherent auditory objects (fusion) is performed by the auditory cortex (Creutzfeldt et al., 1980; Gehr et al., 2000; Leppelsack, 1978; Newman & Wollberg, 1973; Pantev et al., 1995; Wang et al., 1995). The nonlinear summation of brain activity evoked by each of the fusion elements reflects their interaction in the fusion process. The net fusion effect (the nonlinear element) can therefore be derived by waveform subtraction of the sum of the responses to each of the stimuli when presented alone (Figure 4.5., middle) from the response to the fused stimulus (Figure 4.5., top). The resultant difference waveform (Figure 4.5., bottom) is termed the *F(fusion)-complex* (Laufer & Pratt, 2003). The resulting difference waveform components, N1, P2, and N2b obtained by the subtraction procedure are termed FN1, FP2, and FN2b, where *F* stands for *fusion.*

The main features of the F-complex components are a negative potential (FN1) with a parieto-central distribution, peaking with a mean latency of around 100 ms, depending on the spatial source of the sounds (right ear advantage); a positive potential (FP2) maximal at electrodes Fz and Cz, peaking at around a mean latency of about 180 ms, depending on the spatial source of the sounds (right ear advantage); and a subsequent negative potential (FN2b) maximal at Pz, peaking at approximately 260 ms in all fusion conditions. At the latencies of FN1 and FP2, activity tends to be enhanced in the right temporal cortex.

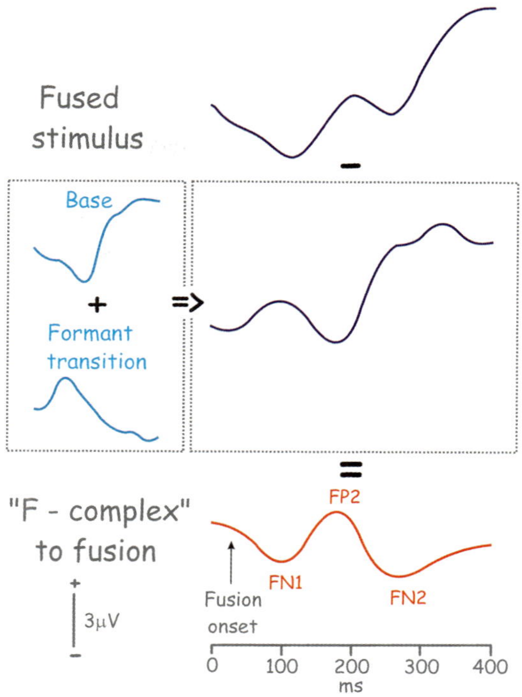

Fig. 4.5. Derivation and components of the fusion complex (F-complex). The F-complex is derived by waveform subtraction of the sum (inset) of the responses to each of the fused stimuli when presented alone (in this example, specific frequency bands of a speech sound) from the response to the fused stimulus (the speech sound). The resultant difference waveform is termed the *F (fusion)-complex*, and its components are termed FN1, FP2, and FN2b, where *F* stands for *fusion*.

Somatosensory Evoked Potentials

Somatosensory evoked potentials (SEPs) are sensory potentials evoked mainly in large-diameter fibers of peripheral nerves at spinal, subcortical, and cortical levels of the somatosensory system (Chiappa, 1990; Mauguiere et al., 1999). The somatosensory system is a mediocre timekeeper because natural stimuli interact with its sensory receptors after undergoing transformations in body tissues (e.g., skin mechanical distortions, heat conduction through the skin). Therefore, to ensure synchronous and time-locked neural activity, SEPs are typically evoked in response to electrical stimulation of peripheral nerves (Jones, 1993), bypassing the interactions of natural stimuli in tissues. Peripheral nerves most often used are the median nerve of the upper extremity (at the wrist) and posterior tibial nerves of the lower extremity (at the ankle). Typical filter settings have a high-pass setting of less than 3 Hz and a low-pass setting of over 2000 Hz.

Median nerve SEPs are typically recorded on an analysis time base of 30–50 ms, while lower extremity SEPs require a longer time base of 60–100 ms. Somatosensory evoked potentials can also be elicited from almost any other peripheral nerve, as well as by natural stimulation such as mechanical taps on skin (Pratt et al., 1979) and tendons (Cohen et al., 1985) and by heat and pain pulses delivered by a laser beam to the skin (Carmon et al., 1978; Garcia-Larrea et al., 2003). Peripheral nerve and spinal components are best recorded from noncephalic electrodes, while brainstem and cortical potentials are recorded from the scalp.

Figures 4.6. and 4.7 present SEPs to upper and lower extremity stimulation, and Table 4.3 lists typical values of SEP latencies and amplitudes to stimulation of peripheral nerves in the upper extremity (A) and lower extremity (B) of normal subjects. Latency values vary in normal subjects according to body size (height) and age because of their dependence on conduction along peripheral nerves.

Peripheral Nerve Potentials

UPPER EXTREMITY

Peripheral nerve potentials from the upper extremity are recorded from electrodes placed over the nerve (e.g., at the elbow or axilla) or over the brachial plexus at Erb's point (Figure 4.6). The Erb's point potential is called N9, and when it appears double-peaked (mostly in children), the first peak is chosen to represent this component. It reflects the compound action potential of the brachial plexus trunks.

LOWER EXTREMITY

Peripheral nerve potentials from the lower limb are recorded from the skin above the nerve in the case of the posterior tibial nerve—from the popliteal fossa behind the knee, referenced to the medial surface of the knee (Figure 4.7). The knee potential, N8, is the principal negative peak arising from the tibial nerve or its parent sciatic nerve.

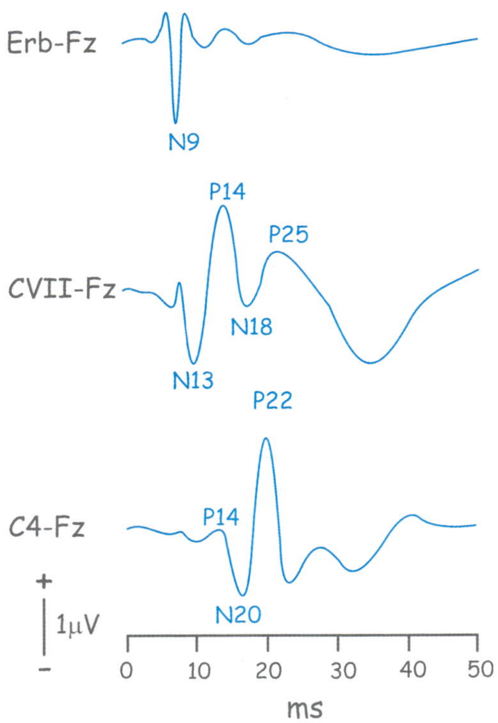

Fig. 4.6. Somatosensory evoked potentials to upper extremity electrical stimulation of the left median nerve at the wrist. Components are named by their polarity and latency. Peripheral nerve potentials are recorded from Erb's point referenced to the shoulder or forehead (Fz), spinal neck potentials are recorded from over cervical spinous process 7 (CVII) referenced to the forehead, and cortical potentials are recorded between an electrode on the scalp contralateral to the stimulated nerve (C4 in this example) referenced to the forehead.

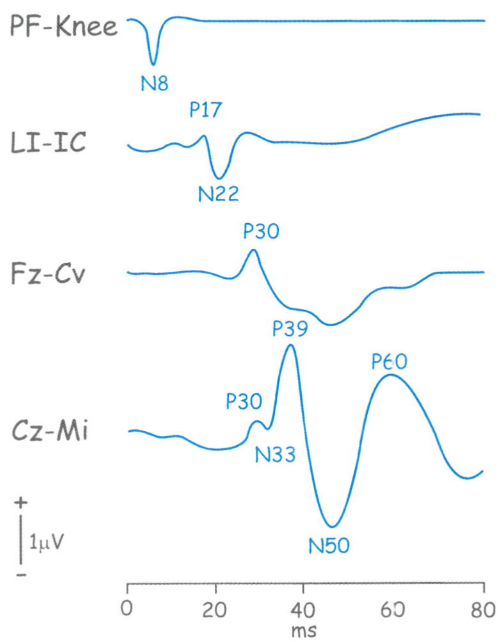

Fig. 4.7. Somatosensory evoked potentials to lower extremity electrical stimulation of the posterior tibial nerve and the ankle. Components are named by their polarity and latency. Peripheral nerve potentials are recorded between an electrode at the popliteal fossa behind the knee (PF) referenced to the middle of the knee (Knee), and lumbar spinal potentials are recorded between the middle of the back over the first lumbar vertebra (LI) referenced to the iliac crest of the hip (IC). Brainstem potentials are recorded between forehead (Fz) and the fifth cervical vertebra (Cv), while cortical potentials are recorded between vertex (Cz) and the mastoid ipslateral to the stimulated limb (Mi).

Table 4.3 Typical values of latency of the major somatosensory components to upper and lower extremity stimulation

A. Median nerve stimulation in normal young adults 1.70 m tall		
	Latencies (ms)	
Peak	Average	Upper limit
N9	9.8	11.5
N13	13.3	14.5
P14	14.3	16.7
N20	19.8	23.0
	Amplitudes (μV)	
Peak	Average	Lower limit
N9	4.8	1.0
N13	2.3	0.5
N20–P25	3.2	0.8
B. Posterior tibial nerve stimulation in normal young adults 1.7 m tall		
	Latencies (ms)	
Peak	Average	Upper limit
N8	8.5	10.5
N22	21.8	25.2
P39	38.0	43.9
	Amplitudes (μV)	
Peak	Average	Lower limit
N22	1.1	0.3
P30	0.6	—
P39N50	1.8	0.5

Spinal Potentials

UPPER EXTREMITY

The cervical potentials can be recorded from an electrode placed on the back of the neck around the spinous process of the fifth cervical vertebra, referenced to a supraglottal reference. A small negative potential, N11, is often recorded but, except for children, is difficult to differentiate from the major cervical component, N13. N11 is attributed to the ascending volley in dorsal columns at the cervical level. The cervical N13 potential is recorded as a negative component at the back of the neck (Figure 4.6) and as a positive P13 component at the anterior neck. The likely generator of the cervical N13/P13 is the postsynaptic potential evoked in the dorsal cervical gray matter by the afferent volley from the dorsal roots. This root volley manifests in the preceding N9/P9 deflection.

LOWER EXTREMITY

Low back potentials are recorded from the skin overlying the spinous processes of lumbar or lower thoracic vertebrae, referenced to the iliac crest contralateral to the stimulated nerve. The most prominent component recorded from lower thoracic and upper lumbar levels is N22, which is preceded by a small positive P17 component (Figure 4.7). N22 is the major component, attributed to postsynaptic potentials in the dorsal gray matter of the lumbosacral cord, and it may be accompanied by an earlier N19 that reflects ascending volleys in spinal somatosensory pathways. P17 is thought to originate in lumbosacral plexus trunks.

Brainstem Potentials

UPPER EXTREMITY

Scalp-recorded somatosensory potentials include far-field peripheral nerve and spinal potentials (N9, P11, P13), followed by a P14, which is attributed to the lower brainstem, close to the cervico-medullary junction. In some subjects, P14 and P13 are barely distinguishable (Figure 4.6); in others, P14 is barely an inflection on the negative-going slope following the peak of P13.

Immediately following P14 in response to upper extremity stimulation is a prolonged scalp negativity peaking at 18 ms, called N18 and attributed to the brainstem between the foramen magnum and thalamus. It can best be recorded from parietal regions ipsilateral to stimulation with a noncephalic reference to minimize interference with the subsequent cortical potentials.

LOWER EXTREMITY

Midline scalp electrodes with a noncephalic reference record a P30 potential that is most prominent frontally. It is likely to be the lower extremity homologue of the median nerve P14, which is generated at the cervico-medullary junction. A small N33 potential preceding the cortical potentials is most likely the lower extremity equivalent of the median nerve N18 from the brainstem level between the foramen magnum and the thalamus.

Cortical Potentials

UPPER EXTREMITY

The first cortical potential is N20 (Figure 4.6), most prominent at the parietal scalp, showing a polarity reversal across the central sulcus and often showing several small peaks. It is attributed to the primary somatosensory cortex in the posterior wall of the central sulcus. The following positive P22 or P25 is most prominent at the parietal scalp contralateral to the stimulated nerve and is also attributed to the primary somatosensory cortex.

LOWER EXTREMITY

The first cortical potential is P39, most prominent at the postcentral midline scalp and consistently followed by N50 and P60, forming together the *W-complex* of cortical components (Figure 4.7.). P39 is generated at the postcentral somatosensory area of the leg, in the interhemispheric fissure, with a dipolar distribution of positivity in the ipsilateral parietal region and negativity in the frontocentral areas contralateral to stimulation.

Visual Evoked Potentials

Visual evoked potentials (VEPs) are the electric manifestation of cortical and subcortical activation of the visual pathway (Celesia & Brigell, 1999). Light interacts with the neural elements by way of a photochemical transduction process and electrotonic conduction through the retina. As a result, the time locking of neural activity in the visual pathway is poor, its timing is sluggish, and its ability to follow fast transitions is limited. Consequently, visual ERP components are almost exclusively limited to slow retinal potentials (the electroretinigram [ERG]), which are not described in this chapter, and cortical potentials.

Although the temporal sensitivity of the visual system is mediocre at best, its sensitivity to luminance changes is high and its sensitivity to pattern change is excellent. Therefore, the stimuli used to evoke VEPs involve either unpatterned luminance change or patterned change without luminance change. This chapter describes VEPs to luminance change and pattern change first, followed by VEPs to the stimuli that are usually used in ERP experiments: visual displays, such as pictures or written text, that involve both pattern and luminance changes.

Because of the low synchrony of ascending activity in the visual pathway, recording passband is typically 1–250 Hz or somewhat wider. Subcortical potentials include the ERG, recorded from the vicinity of the eye, and the short-latency "oscillatory" potentials recorded from the anterior, central, and posterior scalp, usually in response to flashes and using a higher bandpass (up to a low-pass of 1000 Hz). Cortical VEPs are best recorded from occipital electrodes referenced to frontal or central electrodes, where they manifest in opposite polarities to the occipital potentials.

Luminance Change Evoked Potentials

Stimuli

Unpatterned stimuli may consist of an increase or decrease of luminance, but most often they are short (less than 5 ms) flashes that are effectively an increase, immediately followed by a decrease of luminance. The short duration of the flashes, coupled with the low temporal sensitivity of the visual system, results in summation of the onset and offset responses to the flash. Flashes are typically generated in gas discharge tubes, stroboscopes, or light-emitting diodes. Stroboscopic flashes are used with diffusers that are visibly white and may be fitted with colored filters to enhance the color selectivity of the response.

Subcortical Potentials

Short-latency VEPs to flash (Figure 4.8, top), as early as 28 ms after the flash, have been reported since the early days of scalp-recorded human brain responses (Allison et al., 1977; Ciganek, 1961; Cobb & Dawson, 1960; Vaughan, 1966). Oscillatory potentials beginning at 12 ms and lasting until 90 ms after the evoking flash have also been reported (Cracco & Cracco, 1978). The origin of these potentials has been debated: Some (Sigfried & Lukas, 1981) have shown that the waves recorded at 50 ms between vertex and inion behaved differently than the retinal potentials, suggesting a nonretinal origin. Others (Pratt et al., 1982, 1995) have shown, based on amplitude and polarity changes periocularly and across the scalp, that the potentials before 40 ms originated within the eyeball and those between 40 and 70 ms originated in the optic nerves and tracts, while the later potentials were generated more centrally. These interpretations were later corroborated by findings from brain-dead patients and the effects of stimulus rate on normal subjects (Perez-Arroyo & Chiappa, 1985), as well as from depth recordings from the optic nerve during surgery (Moller et al., 1987). The scalp distribution and color sensitivity of retinal potentials led others (Whittaker & Siegfried, 1983) to conclude that components between 35 and 70 ms were generated occipitally, while components with latencies beyond 70 ms were generated in more distant visual and associative cortex.

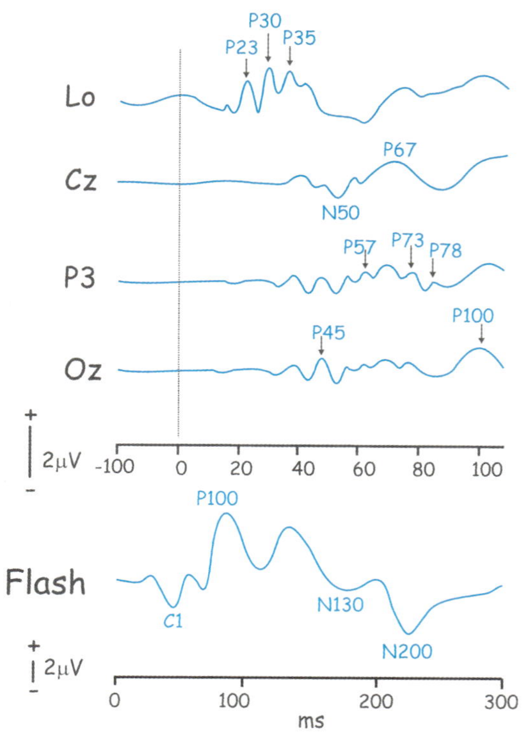

Fig. 4.8. Visual evoked potentials to flash. The short-latency components are presented at the top and the long-latency waveform at the bottom. The short-latency components earlier than 45 ms are best recorded from periocular electrodes (e.g., Lo—below the left orbit, top trace); the intermediate components, attributed to the optic nerves and tracts, are optimally recorded from the vertex (Cz) and parietal sites (e.g., P3); and the components after 70 ms are attributed to the cortex and are best recorded from occipital electrodes (Oz). The long-latency components (bottom) are best recorded from vertex and include an initial negative C1 followed by a P100 that is sometimes notched by N130 and N200.

Cortical Potentials

The cortical potentials evoked by luminance change (Figure 4.8, bottom) include a series of negative and positive components, beginning with a negative peak at around 50–70 ms (C1), through a positive peak at around 100 ms (P100), which is often notched by a negative peak at about 130 ms, followed by a negative peak at approximately 200 ms (N200). Due to the large intersubject variability in the potentials to flashes, a universal nomenclature of components is difficult. Current source estimations and animal experiments indicate that the 70 ms negativity is generated by primary visual cortex excitation, the 100 ms positivity may be due to inhibitory processes resulting from thalamic input to visual cortex, and the notch around 130 ms reflects intracortical activity, with later peaks having extrastriatal origin (Kraut et al., 1985). Results from cats indicate that regardless of the stimulus type (e.g., patterned, unpatterned, motion), every stimulus type activates similar portions of visual cortex (Mitzdorf, 1986).

Pattern Change Evoked Potentials

The stimulus of choice for clinical assessment of the visual pathways is a checkerboard pattern in which the black and white checks reverse at the rate of stimulation. The checks reverse without a change in total luminance, hence the term *pattern reversal* or *pattern shift* for this type of stimulus. The main advantage of this stimulus compared to other forms of stimulation that also include luminance change is the small intersubject and intrasubject test-to-test variability and the higher sensitivity to impaired conduction along the visual pathways. The most prominent VEP component to pattern reversal stimulation (Figure 4.9) is P100, with a typical average latency of just under 100 ms and a normative range of 90–110 ms. P100 is usually between two negative peaks at approximately 70 and 135 ms (N70 and N135).

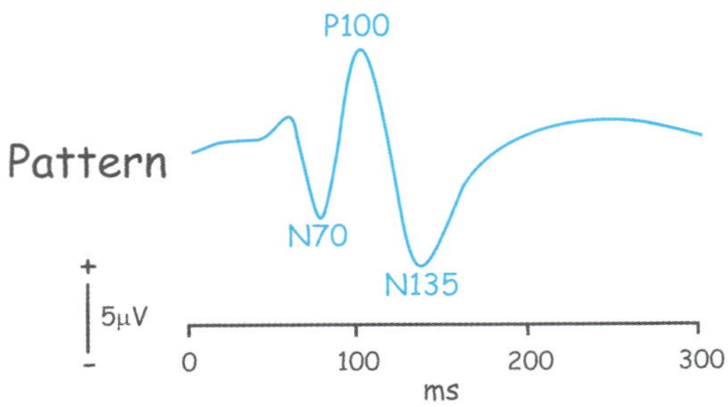

Fig. 4.9. Pattern reversal VEPs include cortical peaks N70, P100, and N135, best recorded from occipital electrodes referenced to anterior scalp.

The combined contrast reversal and illusory movement of boundaries are particularly effective in activating visual cortex neurons; hence, the VEPs recorded are closely related to the underlying cortical activation, being sensitive to retinotopic organization and partial field stimulation. Thus, the initial peak (N70 or C1) reverses in polarity across a midline row of electrodes, between stimulation of the upper and lower halves of the visual field (Jeffreys, 1977), compatible with cortical mapping of the visual field. Interestingly, when potentials are recorded from a horizontal row of occipital electrodes in response to left and right hemifields, a reversal is also observed: The electrodes ipsilateral to the stimulated hemifield record a typical NPN (N70–P100–N135) pattern, similar to the potentials to whole-field stimulation, while the contralateral electrodes record a PNP complex. This polarity reversal is unexpected, as it is the contralateral electrodes that should record the typical waveform and the ipsilateral electrodes that should be reversed. This paradox was explained by the orientation of the generator neurons, which "are largely situated on the medial and posteromedial surface of the visual cortex where the neurons are transversely oriented. Thus, electrodes over the hemisphere ipsilateral to the field stimulated are optimally placed to record the response from these generator areas, whereas electrodes over the contralateral hemisphere" record the activity of these neurons from the opposite side, resulting in polarity reversal of the potentials (Barrett et al., 1976). Asymmetrical activation of visual cortex because of partial field stimulation or because of lesions is accompanied by a shift of the optimal recording site from midline, 5 cm above the inion, to more lateral sites, typically about 5 cm lateral and 5 cm above the inion.

In addition to lesions affecting the visual pathway, which typically result in latency prolongation and distorted scalp distribution of P100, various nonpathological factors can affect VEPs to pattern reversal. Latency decreases as a function of increasing log unit brightness of the pattern and its contrast. The color of the pattern does not seem to affect pattern reversal VEPs as long as all other parameters (brightness, contrast, check size) are the same. Visual field size affects P100 amplitude (but not latency), which diminishes with diminishing field size. The central 8° of the visual field contribute most of the response, but the peripheral field, up to 32°, also contributes. Central areas evoke larger amplitudes with smaller check sizes (17 min of arc) and peripheral areas optimally evoke with larger checks (60′), congruent with visual acuity in these areas. The reversal rate affects both the latency and the amplitude of P100, which diminishes and is later at the higher rates between 1 and 4/s, and is fused to a steady-state VEP at 8–10/s.

The generators of the pattern reversal VEPs are attributed to visual cortex, but different studies suggest different areas as responsible for the N70–P100 sequence. The source of N70 has been suggested to be area 17 (Jeffreys, 1977; Jeffreys & Axford, 1972, Lesevre, 1982) or area 18 (Halliday & Michael, 1970; Maier et al., 1986). P100 has been attributed to areas 18 and 19 (Halliday & Michael, 1970; Lesevre, 1982) or to areas 17 and 18 (Maier et al., 1986).

Evoked Potentials to Pictures and Written Text

The study of VEPs to luminance change or pattern reversal is typically used for clinical assessment of the integrity of the visual pathways. In ERP studies, stimuli consist of both luminance change and patterns with meaning, such as pictures or written text, and the VEP waveforms are therefore more complex and may include additional early components.

The components most often observed in cognitive experiments, which usually involve stimulus onsets, include C1, P1, and N1. C1 is the initial (ca. 75 ms) negative peak in the VEP, attributed to the visual cortex, with occipital prominence and polarity inversion from back to front along the midline. It is followed by P1 (ca. 100 ms), which is attributed to thalamocortical inhibitory inputs to visual cortex and a subsequent negativity at approximately 135–150 ms (N1), attributed to extrastriate activity. Because of the composite pattern and onset nature of such stimuli, the waveform, and in fact each of its components, are a composite of temporally overlapping activations involving striate and extrastriate cortex.

The contributions of magnocellular or parvocellular neural inputs to these components may be biased by the stimuli used to evoke them. Stimuli emphasizing the magnocellular system include low-contrast isolated checks and low spatial frequency stimuli; the magnocellular inputs can be emphasized using high spatial frequency sinusoidal gratings (Butler et al., 2007). High-contrast stimuli activate both magno- and parvocellular inputs to the cortex. Biasing toward parvocellular or magnocellular activation can also be achieved by manipulation of chromatic and achromatic contrast, respectively (Schechter et al., 2005). C1 has been shown to be primarily evoked by parvocellular and mixed parvo- and magnocellular inputs. P1 was elicited primarily by magnocellular and mixed magno- and parvocellular inputs, whereas N1 was elicited primarily by parvocellular and mixed parvo- and magnocellular stimuli (Schechter et al., 2005).

Schizophrenic patients showed a significant reduction in amplitude and an increase in latency of the C1 component, as well as reductions in P1 amplitudes (Schechter et al., 2005). Furthermore, a large decrease in the P1 component in response to magnocellular-biased isolated check stimuli was seen in patients compared with controls, and C1 and N1 were reduced in amplitude to magnocellular-biased low spatial frequency stimuli but were intact to parvocellular-biased high spatial frequency stimuli (Butler et al., 2007). In addition to schizophrenia, where P1 but not C1 has been reported to be significantly affected (Butler et al., 2007; Foxe et al., 2001); Schechter et al., 2005), hemineglect was found to affect N1 but not C1 or P1 (Di Russo et al., 2008). Thus, of the early VEPs, only C1 is entirely exogenous and preattentive.

The generators of C1, P1, and N1 have been estimated using topographic maps of voltage and current source density in conjunction with dipole modeling. The topography of C1 was found to be consistent with a generator in striate cortex, and this component was unaffected by attention. In contrast, the P1 and N1 components (ca. 95 and 170 ms) exhibited current density foci overlying lateral extrastriate cortex (Gomez Gonzalez et al., 1994) and inferior occipital-temporal cortex (Hopf et al., 2002). P1 and N1 were larger for attended stimuli than for unattended stimuli (Gomez Gonzalez et al., 1994). Effects of spatial attention on P1 and N1, but not on C1, have corroborated the results using semantic processing of written words (McCarthy & Nobre, 1993). In addition to spatial attention, both the latency and amplitude of P1 have been shown to be affected by attention to nonspatial visual features (Taylor, 2002). The voltage topographies during the latency range of theses components could be modeled with a five-dipole configuration consisting of a single striate dipole and left-right pairs of dipoles located in lateral extrastriate and inferior occipito-temporal areas (Gomez Gonzalez et al., 1994). Such multiple generators for each of the early VEP components were also suggested by data on flow of activation from V1 to frontal cortex in humans (Foxe & Simpson, 2002).

The most intriguing visually evoked early component to meaningful patterns is the N170 recorded from the posterior scalp and typically associated with face processing (Bentin et al., 1996). More recent findings indicate that it is associated with expert visual scrutiny of visual objects in general (Bigman & Pratt, 2004). In visual object processing a cascaded categorization and identification process has been proposed, beginning with expert object definition (e.g., faces, cars) reflected by N170, leading to the final identification of a particular familiar item in that category manifesting in N250 (Scott et al., 2006; Tanaka et al., 2006). In the specific case of face processing, faces evoked an equally prominent N170 that was significantly larger than the ERPs to non-face categories, regardless of whether they were ignored, whether they had an equal status with other categories or were the targets in a task (Carmel & Bentin, 2002). Thus, the visual N170 (and N250) do not require that the subject attend to the features presented.

References

Aceto, P., Valente, A., Gorgoglione, M., Adducci, E., & De Cosmo, G. (2003). Relationship between awareness and middle latency auditory evoked responses during surgical anaesthesia. *British Journal of Anaesthesiology, 90*, 630–635.

Alain, C., & Woods, D. L. (1994). Signal clustering modulates auditory cortical activity in humans. *Perceptual Psychophysics, 56*, 501–516.

Allison, T., Matsumiya, Y., Goff, G. D., & Goff, W. R. (1977). The scalp topography of of human visual evoked potentials. *Electroencephalography and Clinical Neurophysiology, 42*, 185–197.

Amenedo, E., & Diaz, F. (1998). Effects of aging on middle-latency auditory evoked potentials: A cross-sectional study. *Biological Psychiatry, 43*, 210–219.

Antognini, J. F., & Wang, X. W. (1999). Isoflurane indirectly depresses middle latency auditory evoked potentials by action in the spinal cord in the goat. *Canadian Journal of Anaesthesiology, 46*, 692–695.

Arezzo, J., Pickoff, A., & Vaughan, H. G. (1975). The sources and intracerebral distribution of auditory evoked potentials in the alert rhesus monkey. *Brain Research, 90*, 57–73.

Barnet, A. B., Ohlrich, E. S., Weiss, I. P., & Shanks, B. (1975). Auditory evoked potentials during sleep in normal children from ten days to three years of age. *Electroencephalography and Clinical Neurophysiology, 39*, 29–41.

Barrett, G., Blumhardt, L., Halliday, A. M., Halliday, E., & Kriss, A. (1976). A paradox in the lateralization of the visual evoked response. *Nature, 261*, 253–255.

Bentin, S., Allison, T., Puce, A., Perez, E., & McCarthy, G. (1996). Electrophysiological studies of face perception in humans. *Journal of Cognition and Neuroscience, 8*, 551–565.

Berman, S. M., Naliboff, B. D., Chang, L., Fitzgerald, L., Antolin, T., Camplone, A., & Mayer, E. A. (2002). Enhanced preattentive central nervous system reactivity in irritable bowel syndrome. *American Journal of Gastroenterology, 97*, 2791–2797.

Bigman, Z., & Pratt, H. (2004). Time course and nature of stimulus evaluation in category induction as revealed by visual event-related potentials. *Biological Psychology, 66*, 99–128.

Boop, F. A., Garcia-Rill, E., Dykman, R., & Skinner, R. D. (1994). The P1: Insights into attention and arousal. *Pediatric Neurosurgery, 20*, 57–62.

Buchwald, J. S., Erwin, R. J., Schwfel, J., & Tanguay, P. (1988). Abnormal P1 potentials in autistic subjects. *Neuroscience Abstracts, 14*, 771.

Buchwald, J. S., Erwin, R. J., Van Lancker, D., & Cummings, J. L. (1989). Midlatency auditory evoked responses: Differential abnormalities of P1 in Alzheimer's disease. *Electroencephalography and Clinical Neurophysiology, 74*, 378–384.

Buchwald, J. S., Erwin, R., Van Lancker, D., Guthrie, D., Schwafel, J., & Tanguay, P. (1992). Midlatency auditory evoked responses: P1 abnormalities in adult autistic subjects. *Electroencephalography and Clinical Neurophysiology, 84*, 164–171.

Buchwald, J. S., Hinman, C., Norman, R. J., Huang, C. M., & Brown, K. A. (1981). Middle- and long-latency auditory evoked responses recorded from the vertex of normal and chronically lesioned cats. *Brain Research, 205*, 91–109.

Buchwald, J. S., Rubinstein, E. H., Schwafel, J., & Stranburg, R. J. (1991). Midlatency auditory evoked responses: Differential effects of a cholinergic agonist and antagonist. *Electroencephalography and Clinical Neurophysiology, 80*, 303–309.

Butler, P. D., Martinez, A., Foxe, J. J., Kim, D., Zemon, V., Silipo, G., Mahoney, J., Shpaner, M., Jalbrzikowski, M., & Javitt, D. C. (2007). Subcortical visual dysfunction in schizophrenia drives secondary cortical impairments. *Brain, 130*, 417–430.

Cacace, A. T., & McFarland, D. J. (2007). Frequency domain analysis of event-related potentials and oscillations. Chapter 6 R.F. Burkard, M. Don & J.J. Eggermont (Eds.), *Auditory evoked potentials: Basic principles and clinical applications* (pp. 124–139). Philadelphia: Lippincott Williams & Wilkins.

Carmel, D., & Bentin, S. (2002). Domain specificity versus expertise: Factors influencing distinct processing of faces. *Cognition, 83*, 1–29.

Carmon, A., Dotan, Y., & Sarne, Y. (1978) Correlation of subjective pain experience with cerebral evoked responses to noxious thermal stimulation. *Experimental Brain Research, 33*, 445–453.

Celesia, G. (1976). Organization of auditory cortical areas in man. *Brain, 99*, 403–414.

Celesia, G. G., & Brigell, M. G. (1999). Recommended standards for pattern electroretinograms and visual evoked potentials. In G. Deuschl & A. Eisen (Eds.), *Recommendations for the practice of clinical neurophysiology*. (pp. 53–67). Amsterdam: Elsevier.

Ceponiene, R., Rinne, T., & Naatanen, R. (2002). Maturation of cortical sound processing as indexed by event-related potentials. *Clinical Neurophysiology, 113*, 870–882.

Chambers, R. D. (1992). Differential age effects for components of the adult auditory middle latency response. *Hearing Research, 58*, 123–131.

Chambers, R. D., & Griffiths, S. K. (1991). Effects of age on the adult auditory middle latency response. *Hearing Research, 51*, 1–10.

Chiappa, K. H. (1990). Short latency somatosensory evoked potentials: Methodology. In K. H. Chiappa (Ed.), *Evoked potentials in clinical medicine* (pp. 307–370). New York: Raven Press.

Ciganek, I (1961). The EEG response (evoked potential) to light stimulus in man. *Electroencephalography and Clinical Neurophysiology, 13*, 165–172.

Cobb, W. A., & Dawson, G. D. (1960). The latency and form in man of the occipital potentials evoked by bright flashes. *Journal of Physiology, 152*, 108–121.

Cohen, L. G., Starr, A., & Pratt, H. (1985). Cerebral somatosensory potentials evoked by muscle stretch, cutaneous taps and electrical stimulation of peripheral nerves in lower extremity in humans. *Brain, 108*, 103–121.

Cracco, R. Q., & Cracco, J. B. (1978). Visual evoked potentials in man: Early oscillatory potentials. *Electroencephalography and Clinical Neurophysiology, 45*, 731–739.

Creutzfeldt, O., Hellweg, F. C., & Schreiner, C. (1980). Thalamocortical transformation of responses to complex auditory stimuli. *Experimental Brain Research, 39*, 87–104.

Davies, F. W., Mantzaridis, H., Kenny, G. N., & Fisher, A. C. (1996). Middle latency auditory evoked potentials during repeated transitions from consciousness to unconsciousness. *Anaesthesia, 51*, 107–113.

Debruyne, F. (1984). Binaural interaction in early, middle and late auditory evoked responses. *Scandinavian Audiology, 13*, 293–296.

Decker, T. N., & Howe, S. W. (1981). Auditory tract asymmetry in brainstem electrical responses during binaural stimulation. *Journal of the Acoustical Society of America, 9*. 1084–1090.

Deiber, M. P., Ibañez, V., Bastuji, H., Fischer, C., & Mauguiere, F. (1989). Changes of middle latency auditory evoked potentials during natural sleep in humans. *Neurology, 39*, 806–813.

Deiber, M. P., Ibañez, V., Fischer, C., Perrin, F., & Mauguiere, F. (1988). Sequential mapping favours the hypothesis of distinct generators for Na and Pa middle latency auditory evoked potentials. *Electroencephalography and Clinical Neurophysiology, 71*, 187–197.

de Lugt, D. R., Loewy, D. H., & Campbell, K. B. (1996). The effect of sleep onset on event related potentials with rapid rates of stimulus presentation. *Electroencephalography and Clinical Neurophysiology, 98*, 484–492.

Deouell, L. Y., Bentin, S., & Giard, M. H. (1998). Mismatch negativity in dichotic listening: Evidence for interhemispheric differences and multiple generators. *Psychophysiology, 35*, 355–365.

Dobie, R. A., & Norton, S. J. (1980). Binaural interaction in human auditory evoked potentials. *Electroencephalography and Clinical Neurophysiology, 49*, 303–313.

Drummond, J. C. (2000). Monitoring depth of anesthesia: With emphasis on the application of the bispectral index and the middle latency auditory evoked response to the prevention of recall. *Anesthesiology, 93*, 876–882.

Eggermont, J. J., Ponton, C. W., Don, M., Waring, M. D., & Kwong, B, (1997). Maturational delays in cortical evoked potentials in cochlear implant users. *Acta Otolaryngology, 117*, 161–163.

Erwin, R. J., & Buchwald, J. S. (1986). Midlatency auditory evoked responses: Differential effects of sleep in the human. *Electroencephalography and Clinical Neurophysiology, 65*, 383–392.

Erwin, R. J., Shtasel, D., & Gur, R. E, (1994). Effects of medication history on midlatency auditory evoked responses in schizophrenia. *Schizophrenia Research, 11*, 251–258.

Fein, G., Biggins, C., & van-Dyke, C. (1994). The auditory P50 response is normal in Alzheimer's disease when measured via a paired click paradigm. *Electroencephalography and Clinical Neurophysiology, 92*, 536–545.

Fifer, R. C., & Sierra-Irizarry, B. (1988). Clinical applications of the auditory middle latency response. *American Journal of Otology, 9*(Suppl), 47–56.

Firsching, R., Luther, J., Eidelberg, E., Brown, W. E., Jr., Story, J. L., & Boop, F. A. (1987). 40 Hz—middle latency auditory evoked response in comatose patients. *Electroencephalography and Clinical Neurophysiology, 67*, 213–216.

Foxe, J. J., Doniger, G. M., & Javitt, D. C. (2001). Early visual, processing deficits in schizophrenia: Impaired PI generation revealed by high-density electrical mapping. *Neuroreport, 12*, 3815–3820.

Foxe, J. J., & Simpson, G. V. (2002). Flow of activation from V1 to frontal cortex in humans—a framework for defining "early" visual processing. *Experimental Brain Research, 142*, 139–150.

Franks, R. D., Adler, L. E., Waldo, M. C., Alpert, J., & Freedman, R. (1983). Neurophysiological studies of sensory gating in mania: Comparison with schizophrenia. *Biological Psychiatry, 18*, 989–1005.

Furst, M., Eyal, S., & Korczyn, A. D. (1990). Prediction of binaural click lateralization by brainstem auditory evoked potentials. *Hearing Research, 49*, 347–359.

Furst, M., Levine, R. A., & McGaffigan, P. M. (1985). Click lateralization is related to the β component of the dichotic brainstem auditory evoked potentials of human subjects. *Journal of the Acoustical Society of America, 78*, 1644–1651.

Garcia-Larrea, L., Frot, M., & Valeriani, M. (2003). Brain generators of laser-evoked potentials: From dipoles to functional significance. *Neurophysiologie Clinique/Clinical Neurophysiology, 33*, 279–292.

Gehr, D. D., Komiya, H., & Eggermont, J. J. (2000). Neuronal responses in cat primary auditory cortex to natural and altered species-specific calls. *Ear and Hearing, 150*, 27–42.

Giard, M. H., Perrin, F., Pernier, J., & Bouchet, P. (1990). Brain generators implicated in the processing of auditory stimulus deviance: A topographic event-related potential study. *Psychophysiology, 27*, 627–640.

Giraud, A. L., Lorenzi, C., Ashburner, J., Wable, J., Johnsrude, I., Frackowiak, R., & Kleinschmidt, A. (2000). Representation of the temporal envelope of sounds in the human brain. *Journal of Neurophysiology, 84*, 1588–1598.

Golob, E. J., Irimajiri, R., & Starr, A. (2007). Auditory cortical activity in amnestic mild cognitive impairment: Relationship to subtype and conversion to dementia. *Brain, 130*, 740–752.

Gomez Gonzalez, C. M., Clark, V. P., Fan, S., Luck, S. J., & Hillyard, S. A. (1994). Sources of attention-sensitive visual event-related potentials. *Brain Topography, 7*, 41–51.

Green, J. B., Flagg, L., Freed, D. M., & Schwankhaus, J. D. (1992). The middle latency auditory evoked potential may be abnormal in dementia. *Neurology, 42*, 1034–1036.

Grunwald, T., Boutros, N. N., Pezer, N., von-Oertzen, J., Fernandez, G., Schaller, C., & Elger, C. E. (2003). Neuronal substrates of sensory gating within the human brain. *Biological Psychiatry, 53*, 511–519.

Halliday, A. M., & Michael, W. F. (1970). Changes in the pattern-evoked responses in man associated with the vertical and horizontal meridians of the visual field. *Journal of Physiology, 208*, 499–513.

Hari, K., Kaila, K., Katila, T., Tuomisto, T., & Varpula, T. (1982). Interstimulus interval dependence of the auditory vertex response and its magnetic counterpart: Implications for their neural generation. *Electroencephalography and Clinical Neurophysiology, 54*, 561–569.

Hari, R., Pelizzone, M., Makela, J. P., Hallstrom, J., Leinonen, L., & Lounasmaa O. V. (1987). Neuromagnetic responses of the human auditory cortex to on- and offsets of noise bursts. *Audiology, 26*, 31–43.

Harms, M. P., & Melcher, J. R. (2002). Sound repetition rate in the human auditory pathway: Representations in the waveshape and amplitude of fMRI activation. *Journal of Neurophysiology, 88*, 1433–1450.

Hashimoto, I. (1982). Auditory evoked potentials from the human midbrain: Slow brain stem responses. *Electroencephalography and Clinical Neurophysiology, 53*, 652–657.

Hashimoto, I., Ishiyama, Y., Yoshimoto, T., & Nemoto, S. (1981). Brainstem auditory evoked potentials recorded directly from human brainstem and thalamus. *Brain, 104*, 841–859.

Hopf, J. M., Vogel, E., Woodman, G., Heinze, H. J., & Luck, S. J. (2002). Localizing visual discrimination processes in time and space. *Journal of Neurophysiology, 88*, 2088–2095.

Hosford, H. L., Fullerton, B. C., & Levine, R. A. (1979). Binaural interaction in human and cat brain stem evoked responses. *Journal of the Acoustical Society of America, 65*, s86.

Howard, M. A., Volkov, I. O., Mirsky, R., Garell, P. C., Noh, M. D., Granner, M., Damasio, H., Steinschneider, M., Reale, R. A., Hind, J. E., & Brugge, J. F. (2000). Auditory cortex on the human posterior superior temporal gyrus. *Journal of Comparative Neurology, 416*, 79–92.

Huotilainen, M., Winkler, I., Alho, K., Escera, C., Virtanen, J., Ilmoniemi, R. J., Jaaskelainen, I. P., Pekkonen, E., & Naatanen, R (1998). Combined mapping of human auditory EEG and MEG responses. *Electroencephalography and Clinical Neurophysiology, 108*, 370–379.

Hyde, M. (1997). The N1 response and its applications. *Audiology and Neuro-Otology, 2*, 281–307.

Irimajiri, R., Golob, E. J., & Starr, A. (2005). Auditory brain-stem, middle- and long-latency evoked potentials in mild cognitive impairment. *Clinical Neurophysiology, 116*, 1918–1929.

Jaaskelainen, I. P., Hirvonen, J., Huttunen, J., Kaakkola, S., & Pekkonen, E. (1999). Scopolamine enhances middle-latency auditory evoked magnetic fields. *Neuroscience Letters, 259*, 41–44.

Japaridze, G., Kvernadze, D., Geladze, T., & Kevanishvili, Z. (1993). Effects of carbamazepine on auditory brainstem response, middle-latency response, and slow cortical potential in epileptic patients. *Epilepsia, 34*, 1105–1109.

Jasper, H. H. (1958). The ten twenty electrode system of the international federation. *Electroencephalography and Clinical Neurophysiology, 10*, 371–375.

Jeffreys, D. A. (1977). The physiological significance of pattern visual evoked potentials. In J. E. Desmedt (Ed.), *Visual evoked potentials in man* (pp. 134–167). Oxford: Clarendon Press.

Jeffreys, D. A., & Axford, J. G. (1972). Source location of pattern specific components of human visual evoked potentials. *Experimental Brain Research, 16*, 1–21.

Jemel, B., Achenbach, C., Muller, B. W., Ropcke, B., & Oades, R. D. (2002). Mismatch negativity results from bilateral asymmetric dipole sources in the frontal and temporal lobes. *Brain Topography, 15*, 13–27.

Jerger, J., Chmiel, R., Glaze, D., & Frost, J. D. (1987). Rate and filter dependence of the middle-latency response in infants. *Audiology, 26*, 269–283.

Jones, S. J. (1993). Somatosensory evoked potentials. I. Methodology, generators and special techniques, II Clinical observations and applications. In A. M. Halliday (Ed.), *Evoked potentials in clinical testing* (pp. 383–466). Edinburgh: Churchill Livingstone.

Jones, S. J. (2003). Sensitivity of human auditory evoked potentials to the harmonicity of complex tones: Evidence for dissociated cortical processes of spectral and periodicity analysis. *Experimental Brain Research, 150*, 506–514.

Jones, S. J., Longe, O., & Vaz Pato, M. (1998). Auditory evoked potentials to abrupt pitch and timbre change of complex tones. Electrophysiological evidence of "streaming"? *Electroencephalography and Clinical Neurophysiology, 108*, 131–142.

Jones, S. J., & Perez, N. (2001). The auditory "C-process" analyzing the spectral envelope of complex sounds. *Clinical Neurophysiology, 112*, 965–975.

Jones, S. J., & Perez, N. (2002). The auditory C-process of spectral profile analysis. *Clinical Neurophysiology, 113*, 1558–1565.

Jones, S. J., Vaz Pato, M., & Sprague, L. (2000a). Spectro-temporal analysis of complex tones: Two cortical processes dependent on retention of sounds in the long auditory store. *Clinical Neurophysiology, 111*, 1569–1576.

Jones, S. J., Vaz Pato, M., Sprague, L., Stokes, M., Munday, R., & Haque, N. (2000b). Auditory evoked potentials to spectro-temporal modulation of complex tones in normal subjects and patients with severe brain injury. *Brain, 123*, 1007–1016.

Kadobayashi, I., & Toyoshima, A. (1984). Effects of attention on auditory evoked middle latency potentials. *Folia Psychiatric Neurology Japan, 38*, 459–463.

Kemner, C., Verbaten, M. N., Koelega, H. S., Buitelaar, J. K., van der Gaag, R. J., Camfferman, G., & van Engeland, H. (1996). Event-related brain potentials in children with attention-deficit and hyperactivity disorder: Effects of stimulus deviancy and task relevance in the visual and auditory modality. *Biological Psychiatry, 40*, 522–534.

Kileny, P., Paccioretti, D., & Wilson, A. F. (1987). Effects of cortical lesions on middle-latency auditory evoked responses (MLR). *Electroencephalography and Clinical Neurophysiology, 66*, 108–120.

Kisley, M. A., Noecker, T. L., & Guinther, P. M. (2004). Comparison of sensory gating to mismatch negativity and self-reported perceptual phenomena in healthy adults. *Psychophysiology, 41*, 604–612.

Kisley, M. A., Olincy, A., Robbins, E., Polk, S. D., Adler, L. E., Waldo, M. C., & Freedman, R. (2003a). Sensory gating impairment associated with schizophrenia persists into REM sleep. *Psychophysiology, 40*, 29–38.

Kisley, M. A., Polk, S. D., Ross, R. G., Levisohn, P. M., & Freedman, R. (2003b). Early postnatal development of sensory gating. *Neuroreport, 14*, 693–697.

Knight, R. T., Hillyard, S. A., Woods, D. L., & Neville, H. J. (1980). The effects of frontal and temporal-parietal lesions on the auditory evoked potential in man. *Electroencephalography and Clinical Neurophysiology, 50*, 112–124.

Kraus, N., & McGee, T. (1988). Color imaging of the human middle latency response. *Ear and Hearing, 9*, 159–167.

Kraus, N., Ozdamar, O., Hier, D., & Stein, L. (1982). Auditory middle latency responses (MLRs) in patients with cortical lesions. *Electroencephalography and Clinical Neurophysiology, 54*, 275–287.

Kraus, N., Reed, N., Smith, D. I, Stein, L., & Cartee, C. (1987). High-pass filter settings affect the detectability of MLRs in humans. *Electroencephalography and Clinical Neurophysiology, 68*, 234–236.

Kraus, N., Smith, D., Reed, N. L., Stein, L. K. & Cartee, C. (1985). Auditory middle latency responses in children: Effects of age and diagnostic category. *Electroencephalography and Clinical Neurophysiology, 62*, 343–351.

Kraut, M. A., Arezzo, J. C., & Vaughan, H. G., Jr. (1985). Intracortical generators of the flash VEP in monkeys. *Electroencephalography and Clinical Neurophysiology, 62*, 300–312.

Krishnan, A. (2007). Frequency-following response. In R.F. Burkard, M. Don & J.J. Eggermont (Eds.), *Auditory evoked potentials: Basic principles and clinical applications* (pp. 313–335). Philadelphia: Lippincott Williams & Wilkins.

Kudoh, A., & Matsuki, A. (1999). Middle latency auditory evoked potentials during total intravenous anesthesia with droperidol, ketamine and fentanyl. *Clinical Electroencephalography, 30*, 79–83.

Laufer, I. (2003). *The electrophysiology of auditory object formation by fusion of a phonemic element with a preceding stream.* Unpublished doctoral dissertation, Technion–Israel Institute of Technology.

Laufer, I., & Pratt, H. (2003). The electrophysiological net response (F-complex) to spatial fusion of speech elements forming an auditory object. *Clinical Neurophysiology, 114*, 818–834.

Laufer, I., & Pratt, H. (2005). The "F-complex" and MMN tap different aspects of deviance. *Clinical Neurophysiology, 116*, 336–352.

Lee, Y. S., Lueders, H., Dinner, D. S., Lesser, R. P., Hahn, J., & Klemm, G. (1984). Recording of auditory evoked potentials in man using chronic subdural electrodes. *Brain, 107*, 115–131.

Leppelsack, H.-J. (1978). Unit responses to species-specific sounds in the auditory forebrain center of birds. *Federation Proceedings 37*, 2336–2341.

Leserve, N. (1982). Chronotopographical analysis of the human evoked potential in relation to the visual field. *Annals of the New York Academy of Sciences, 388*, 156–182.

Levine, R. A. (1981). Binaural interaction in brainstem potentials of human subjects. *Annals of Neurology, 9*, 384–393.

Liebenthal, E., Ellingson, M. L., Spanaki, M. V., Prieto, T. E., Ropella, K. M., & Binder, J. R. (2003). Simultaneous ERP and fMRI of the auditory cortex in a passive oddball paradigm. *Neuroimage, 19,* 1395–1404.

Liégeois-Chauvel, C., De Graaf, J. B., Laguitton, V., & Chauvel, P. (1999). Specialization of left auditory cortex for speech perception in man depends on temporal coding. *Cerebral Cortex, 9,* 484–496.

Liegeois-Chauvel, C., Musolino, A., Badier, J. M., Marquis, P., & Chauvel, P. (1994). Evoked potentials recorded from the auditory cortex in man: Evaluation and topography of the middle latency components. *Electroencephalography and Clinical Neurophysiology, 92,* 204–214.

Lutkenhöner, B., & Steinsträter, O. (1998). High-precision neuromagnetic study of the functional organization of the human auditory cortex. *Audiology and Neuro-Otology, 3,* 191–213.

Maier, J., Dangelie, G., Spekreijse, H., & van Dijk, B. W. (1986). Principal component analysis for source locations of VEPs in man. *Vision Research, 27,* 165–177.

Makela, L. P., Han, R., & Leinonen, L. (1988). Magnetic responses of the human auditory cortex to noise/square wave transitions. *Electroencephalography and Clinical Neurophysiology, 69,* 423–430.

Martin, B. A., & Boothroyd, A. (1999). Cortical, auditory, event-related potentials in response to periodic and aperiodic stimuli with the same spectral envelope. *Ear and Hearing, 20,* 33–44.

Martin, B. A., & Boothroyd, A. (2000). Cortical, auditory, evoked potentials in response to change of spectrum and amplitude. *Journal of the Acoustical Society of America, 107,* 2155–2161.

Martin, W. H., Pratt, H., & Schwegler, J. W. (1995). The origin of the human auditory brain-stem response wave II. *Electroencephalography and Clinical Neurophysiology, 96,* 357–370.

Masahiro, K., Tomoaki, T., & Noboru, S. (1989). The effects of 30% nitrous oxide on auditory evoked middle latency responses. *SAAD Digest,* 7, 206–209.

Mauguiere, F., Allison, T., Babiloni, C., Buchner, H., Eisen, A. A., Goodin, D. S., Jones, S. J., Kakigi, R., Matsuoka, S., Nuwer, M., Rossini, P. M., & Shibasaki, H. (1999). Somatosensory evoked potentials. In G. Deuschl & A. Eisen (Eds.), *Recommendations for the practice of clinical neurophysiology* (pp. 79–90). Amsterdam: Elsevier.

Maurizi, M., Ottaviani, F., Paludetti, G., Rosignoli, M., Almadori, G., & Tassoni, A. (1984). Middle-latency auditory components in response to clicks and low- and middle-frequency tone pips (0.5–1 kHz). *Audiology, 23,* 569–580.

McCarthy, G., & Nobre, A. C. (1993). Modulation of semantic processing by spatial selective attention. *Electroencephalography and Clinical Neurophysiology, 88,* 210–219.

McGee, T., Kraus, N., Littman, T., & Nicol, T. (1992). Contributions of medial geniculate body subdivisions to the middle latency response. *Hearing Research, 61,* 147–154.

McPherson, D. L., & Starr, A. (1993). Binaural interaction in auditory evoked potentials: Brainstem, middle- and long-latency components. *Hearing Research, 66,* 91–98.

McPherson, D. L., Tures, C., & Starr, A. (1989). Binaural interaction of the auditory brain-stem potentials and middle latency auditory evoked potentials in infants and adults. *Electroencephalography and Clinical Neurophysiology, 74,* 124–130.

Michalewski, H. J., Starr, A., Nguyen, T. T., Kong, Y.-Y., & Zeng, F.-G. (2005). Auditory temporal processes in normal-hearing individuals and in patients with auditory neuropathy. *Clinical Neurophysiology, 116,* 669–680.

Mitzdorf, U. (1986). The physiological causes of VEP: Current source density analysis of electrically and visually evoked potentials. In R. Q. Cracco & I. Bodis-Wollner (Eds.), *Evoked potentials* (pp. 141–154). New York: A. R. Liss.

Mohamed, A. S., Iacono, R. P., & Yamada, S. (1996). Normalization of middle latency auditory P1 potential following posterior ansa-pallidotomy in idiopathic Parkinson's disease. *Neurology Research, 18,* 516–520.

Moller, A. R., Burgess, J. E., & Sekhar, L. N. (1987). Recording compound action potentials from the optic nerve in man and monkeys. *Electroencephalography and Clinical Neurophysiology, 67,* 549–555.

Moller, A. R., Jannetta, P. J., & Sekhar, L. N. (1988). Contributions from the auditory nerve to the brain-stem auditory evoked potentials (BAEPs): Results of intracranial recording in man. *Electroencephalography and Clinical Neurophysiology, 71,* 198–211.

Näätänen, R., Lehtokoski, A., Lennes, M., Cheour, M., Huotilainen, M., Iivonen, A., Vainio, M., Alku, P., Ilmoniemi, R. J., Luuk, A., Allik, J., Sinkkonen, J., & Alho, K. (1997). Language-specific phoneme representations revealed by electric and magnetic brain responses. *Nature, 385,* 432–434.

Näätänen, R., & Picton, T. (1987). The N1 wave of the human electric and magnetic response to sound: A review and an analysis of the component structure. *Psychophysiology, 24,* 375–425.

Newman, J. D., & Wollberg, Z. (1973). Multiple coding of species-specific vocalizations in the auditory cortex of squirrel monkeys. *Brain Research, 54,* 287–304.

Oates, P., & Stapells, D. R. (1997). Frequency specificity of the human auditory brainstem and middle latency responses to brief tones. II. Derived response analyses. *Journal of the Acoustical Society of America, 102,* 3609–3619.

O'Mahony, D., Rowan, M., Feely, J., Walsh, J. B., & Coakley, D. (1994). Primary auditory pathway and reticular activating system dysfunction in Alzheimer's disease. *Neurology, 44,* 2089–2094.

Osterhammel, P. A., Shallop, J. K., & Terkildsen, K. (1985). The effect of sleep on the auditory brainstem response (ABR) and the middle latency response (MLR). *Scandinavian Audiology, 14,* 47–50.

Ostroff, M. J., Martin, B. A., & Boothroyd, A. (1998). Cortical evoked responses to acoustic change within a syllable. *Ear and Hearing, 19,* 290–297.

Ozdamar, O., & Kraus, N. (1983). Auditory middle-latency responses in humans. *Audiology, 22,* 34–49.

Ozdamar, O., Kraus, N., & Curry, F. (1982). Auditory brain stem and middle latency responses in a patient with cortical deafness. *Electroencephalography and Clinical Neurophysiology, 53,* 224–230.

Pantev, C., Eulitz, C., Hampson, S., Ross, B., & Roberts, L. E. (1996). The auditory evoked "off" response: Sources and comparison with the "on" and the "sustained" responses. *Ear and Hearing, 17,* 255–265.

Pantev, C., Hoke, M., Lehnertz, K., Lukenhoner, B., Anogianakis, G., & Wittkowski, W. (1995). Tonotopic organization of different areas of the human auditory cortex revealed by simultaneous magnetic and electric recordings. *Electroencephalography and Clinical Neurophysiology, 94,* 26–40.

Parving, A., Salomon, G., Elberling, C., Larsen, B., & Lassen, N. A. (1980). Middle components of the auditory evoked response in bilateral temporal lobe lesions. *Scandinavian Audiology, 9,* 161–167.

Pasman, J. W., Rotteveel, J. J., de Graaf, R., Stegeman, D. F., & Visco, Y. M. (1992). The effect of preterm birth on brainstem, middle latency and cortical auditory evoked responses (BMC AERs). *Early Human Development, 31,* 113–129.

Perez-Aroyo, M., & Chiappa, K. H. (1985). Early visual evoked potentials in normal subjects and brain-dead patients. *Electroencephalography and Clinical Neurophysiology, 61,* S38.

Picton, T. W. (2007). Audiometry using auditory steady-state responses. In R.F. Burkard, M. Don & J.J. Eggermont (Eds.), *Auditory evoked potentials: Basic principles and clinical applications* (pp. 441–462). Philadelphia: Lippincott Williams & Wilkins.

Picton, T. W., Alain, C., Woods, D. L., John, M. S., Scherg, M., Valdes-Sosa, P., Bosch-Bayard, J., & Trujillo, N. J. (1999). Intracerebral sources of human auditory-evoked potentials. *Audiology and Neuro-Otology, 4,* 64–79.

Picton, T. W., & Hillyard, S. A. (1974). Human auditory evoked potentials: II. Effects of attention. *Electroencephalography and Clinical Neurophysiology, 36,* 191–199.

Picton, T. W., Hillyard, S. A., Krausz, H. I., & Galambos, R (1974). Human auditory evoked potentials. I. Evaluation of the components. *Electroencephalography and Clinical Neurophysiology, 36,* 179–190.

Pitman, R. K., Orr, S. P., Shalev, A. Y., Metzger, L. J., & Mellman, T. A. (1999). Psychophysiological alterations in post-traumatic stress disorder. *Seminars in Clinical Neuropsychiatry, 4,* 234–241.

Plourde, G., & Villemure, C. (1996). Comparison of the effects of enflurane/N_2O on the 40-Hz auditory steady-state response versus the auditory middle-latency response. *Anesthesia and Analgesia, 82,* 75–83.

Polyakov, A., & Pratt, H. (1994). Three-channel Lissajous' trajectory of human middle latency auditory evoked potentials. *Ear and Hearing, 15,* 390–399.

Polyakov, A., & Pratt, H. (1995). Three-channel Lissajous' trajectory of the binaural interaction components of human auditory middle-latency evoked potentials. *Hearing Research, 82,* 205–215.

Polyakov, A., & Pratt, H. (1996). Evidence for spatio-topic organization of binaural processing in the human brainstem. *Hearing Research, 94,* 107–115.

Polyakov, A., & Pratt, H. (2003). Electrophysiological correlates of azimuth and elevation cues for sound localization in human middle latency auditory evoked potentials. *Ear and Hearing, 24,* 143–155.

Ponton, C. W., Eggermont, J. J., Khosla, D., Kwong, B., & Don, M. (2002). Maturation of human central auditory system activity: Separating auditory evoked potentials by dipole source modeling. *Clinical Neurophysiology, 113,* 407–420.

Pool, K. D., Finitzo, T., Hong, C. T., Rogers, J., & Pickett, R. B. (1989). Infarction of the superior temporal gyrus: A description of auditory evoked potential latency and amplitude topology. *Ear and Hearing, 10,* 144–152.

Pratt, H., Aminoff, M., Nuwer, M. R., & Starr, A. (1999). Short-latency auditory evoked potentials. In G. Deuschl & A. Eisen (Eds.), *Recommendations for the practice of clinical neurophysiology* (pp. 69–78). Amsterdam: Elsevier.

Pratt, H., Amlie, R. N., & Starr, A. (1979). Short latency mechanically evoked somatosensory potentials in humans. *Electroencephalography and Clinical Neurophysiology, 47,* 524–531.

Pratt, H., Bleich, N., & Berliner, E. (1982). Short latency visual evoked potentials in man. *Electroencephalography and Clinical Neurophysiology, 54,* 55–62.

Pratt, H., Bleich, N., & Martin, W. H. (1995). Short-latency visual evoked potentials to flashes from light-emitting diodes. *Electroencephalography and Clinical Neurophysiology, 96,* 502–508.

Pratt, H., Bleich, N., & Mittelman, N. (2005). The composite N1 component to gaps in noise. *Clinical Neurophysiology, 116,* 2648–2663.

Pratt, H., & Polyakov, A. (1996). Electrophysiological evidence of a sound localizing binaural subsystem in the human auditory brainstem. *Journal of Basic Clinical and Physiological Pharmacology, 7,* 235–244.

Pratt, H., Polyakov, A., Aharonson, V., Korczyn, A. D., Tadmor, R., Fullerton, B. C., Levine, R. A., & Furst, M. (1998). Effects of localized pontine lesions on auditory brainstem evoked potentials and binaural processing in humans. *Electroencephalography and Clinical Neurophysiology, 108,* 511–520.

Pratt, H., Starr, A., Michalewski, H. J., Bleich, N., & Mittelman, N. (2007). The N1 complex to gaps in noise: Effects of preceding noise duration and intensity. *Clinical Neurophysiology, 118,* 1078–1087.

Pratt, H., Starr, A., Michalewski, H. J., Bleich, N., & Mittelman, N. (2008). The auditory P50 component to onset and offset of sound. *Clinical Neurophysiology, 119,* 376–387.

Prosser, S., & Arslan, E. (1985). Does general anaesthesia affect the child's auditory middle latency response (MLR)? *Scandinavian Audiology, 14,* 105–107.

Rasco, L., Skinner, R. D., & Garcia-Rill, E. (2000). Effect of age on sensory gating of the sleep state-dependent P1/P50 midlatency auditory evoked potential. *Sleep Research Online, 3,* 97–105.

Regan, D. (1989). *Human brain electrophysiology. Evoked potentials and evoked magnetic fields in science and medicine* (pp. 70–111, 271–275, 294–295, 380–390). New York: Elsevier.

Reite, M., Teale, P., Zimmerman, J., Davis, K., & Whalen, J. (1988). Source location of a 50-msec-latency auditory evoked field component. *Electroencephalography and Clinical Neurophysiology, 70,* 490–498.

Robinson, K., & Rudge, P. (1981). Wave form analysis of the brain stem auditory evoked potential. *Electroencephalography and Clinical Neurophysiology, 52,* 583–594.

Rogers, S. H., Edwards, D. A., Henderson-Smart, D. J., & Pettigrew, A. G. (1989). Middle latency auditory evoked responses in normal term infants: A longitudinal study. *Neuropediatrics, 20,* 59–63.

Rosati, G., Bastiani, P. D., Paolino, E., Prosser, A., Arslan, E., & Artioli, M. (1982). Clinical and audiological findings in a case of auditory agnosia. *Journal of Neurology, 227,* 21–27.

Rotteveel, J. J., Stegeman, D. F., de Graaf, R., Colon, E. J., & Visco, Y. M. (1987). The maturation of the central auditory conduction in preterm infants until three months post term. III. The middle latency auditory evoked response (MLR). *Hearing Research, 27,* 245–256.

Schall, U., Schon, A., Zerbin, D., Bender, S., Eggers, C., & Oades, R. D. (1997). A left temporal lobe impairment of auditory information processing in schizophrenia: An event-related potential study. *Neuroscience Letters, 229,* 25–28.

Schechter, I., Butler, P. D., Zemon, V. M., Revheim, N., Saperstein, A. M., Jalbrzikowski, M., Pasternak R., Silipo, G., & Javitt, D. C. (2005). Impairments in generation of early-stage transient visual evoked potentials to magno- and parvocellular-selective stimuli in schizophrenia. *Clinical Neurophysiology, 116*, 2204–2015.

Scherg, M., & Picton, T. W. (1991). Separation and identification of event-related potential components by brain electric source analysis. *Electroencephalography and Clinical Neurophysiology, Suppl. 42*, 24–37.

Scherg, M., & Volk, S. A. (1983). Frequency specificity of simultaneously recorded early and middle latency auditory evoked potentials. *Electroencephalography and Clinical Neurophysiology, 56*, 443–452.

Scherg, M., & von Cramon, D. (1985). Two bilateral sources of the late AEP as identified by a spatio-temporal dipole model. *Electroencephalography and Clinical Neurophysiology, 62*, 32–44.

Scherg, M., & von Cramon, D. (1986a). Evoked dipole source potentials of the human auditory cortex. *Electroencephalography and Clinical Neurophysiology, 65*, 344–360.

Scherg, M., & von Cramon, D. (1986b). Psychoacoustic and electrophysiologic correlates of central hearing disorders in man. *European Archives of Psychiatry and Neurological Sciences, 236*, 56–60.

Schroeder, M. M., Handelsman, L., Torres, L., Dorfman, D., Rinaldi, P., Jacobson, J., Wiener, J., & Ritter, W. (1994). Early and late cognitive event-related potentials mark stages of HIV-1 infection in the drug-user risk group. *Biological Psychiatry, 35*, 54–69.

Scott, L. S., Tanaka, J. W., Sheinberg, D. L., & Curran, T. (2006). A reevaluation of the electrophysiological correlates of expert object processing. *Journal of Cognitive Neuroscience, 18*, 1453–1465.

Sharma, A., & Dorman, M. (2000). Neurophysiologic correlates of cross-language phonetic perception. *Journal of the Acoustical Society of America, 107*, 2697–2703.

Sharma, A., Dorman, M. F., & Spahr, A. J. (2002). A sensitive period for the development of the central auditory system in children with cochlear implants: Implications for age of implantation. *Ear and Hearing, 23*, 532–539.

Sharma, A., Kraus, N., McGee, T. J., & Nicol, T. G. (1997). Developmental changes in P1 and N1 central auditory responses elicited by consonant-vowel-consonant syllables. *Electroencephalography and Clinical Neurophysiology, 104*, 540–545.

Sharma, A., Marsh, C. M., & Dorman, M. F, (2000). Relationship between N1 evoked potential morphology and the perception of voicing. *Journal of the Acoustical Society of America, 108*, 3030–3035.

Siegfried, J. B., & Lukas, J. (1981). Early wavelets in the VECP. *Investigations in Ophthalmology and Visual Science, 20*, 125–129.

Simos, P. G., Molfese, D. L., & Brenden, R. A. (1997). Behavioral and electrophysiological indices of voicing-cue discrimination: Laterality patterns and development. *Brain and Language, 57*, 122–150.

Skinner, R. D. M., Rasco, B. A., Fitzgerald, J., Karson, C. N., Matthew, M., Williams, D. K., & Garcia-Rill, E. (1999). Reduced sensory gating of the P1 potential in rape victims and combat veterans with posttraumatic stress disorder. *Depression and Anxiety, 9*, 122–130.

Smith, D. B., Michalewski, H. J., Brent, G. A., & Thompson, L. W. (1980). Auditory averaged evoked potentials and aging: Factors of stimulus, task and topography. *Biological Psychology, 11*, 135–151

Smith, D. I., & Kraus, N. (1987). Effects of chloral hydrate, pentobarbital, ketamine, and curare on the auditory middle latency response. *American Journal of Otolaryngology, 8*, 241–248.

Stapells, D. R., Galambos, R., & Costello, J. A. (1988). Inconsistency of auditory middle latency and steady-state responses in infants. *Electroencephalography and Clin Neurophysiology, 71*, 289–295.

Starr, A., & Zaaroor, M. (1990). Eighth nerve contributions to cat auditory brainstem responses (ABR). *Hearing Research, 48*, 151–160.

Steinschneider, M., Schroeder, C. E., Arezzo, J. C., & Vaughan, H. G. (1994). Speech-evoked activity in primary auditory cortex: Effects of voice onset time. *Electroencephalography and Clinical Neurophysiology, 92*, 30–43.

Steinschneider, M., Volkoc, I., Noh, M., Garell, P., & Howard, M. T. (1999). Temporal encoding of the voice onset time phonetic parameter by field potentials recorded directly from human auditory cortex. *Journal of Neurophysiology, 82*, 2346–2357.

Suzuki, T., & Hirabayashi, M. (1987). Age-related morphological changes in auditory middle-latency response. *Audiology, 26*, 312–320.

Suzuki, T., Hirabayashi, M., & Kobayashi, K. (1983). Auditory middle responses in young children. *British Journal of Audiology, 17*, 5–9.

Suzuki, T., Kobayashi, K., Aoki, K., & Umegaki, Y. (1992). Effect of sleep on binaural interaction in auditory brainstem response and middle latency response. *Audiology, 31*, 25–30.

Tanaka, J. W., Curran, T., Porterfield, A. L., & Collins, D. (2006). Activation of preexisting and acquired face representations: The N250 event-related potential as an index of face familiarity. *Journal of Cognitive Neuroscience, 18*, 1488–1497.

Tatsumi, K., Hirai, K., Furuya, H., & Okuda, T. (1995). Effects of sevoflurane on the middle latency auditory evoked response and the electroencephalographic power spectrum. *Anesthesia and Analgesia, 80*, 940–943.

Taylor, M. J. (2002). Non-spatial attentional effects on P1. *Clinical Neurophysiology, 113*, 1903–1908.

Telles, S., Joseph, C., Venkatesh, S., & Desiraju, T. (1993). Alterations of auditory middle latency evoked potentials during yogic consciously regulated breathing and attentive state of mind. *International Journal of Psychophysiology, 14*, 189–198.

Teo, C., Rasco, L., Skinner, R. D., & Garcia-Rill, E. (1998). Disinhibition of the sleep state–dependent p1 potential in Parkinson's disease improvement after pallidotomy. *Sleep Research Online, 1*, 62–70.

Tucker, D. A., & Ruth, R. A. (1996). Effects of age, signal level, and signal rate on the auditory middle latency response. *Journal of the American Academy of Audiology, 7*, 83–91.

Uc, E. Y., Skinner, R. D., Rodnitzky, R. L., & Garcia-Rill, E (2003). The midlatency auditory evoked potential P50 is abnormal in Huntington's disease. *Journal of Neurological Science, 212*, 1–5.

Ulanovsky, N., Las, L., & Nelken, I. (2003). Processing of low-probability sounds by cortical neurons. *Nature Neuroscience, 6*, 391–398.

Vaughan, H. G, (1966). The perceptual and physiologic significance of visual evoked responses recorded from the scalp of man. In Burian H.M., Jacobsen J.H. (Eds.), *Clinical electroretinography (Suppl. Vison Research)*, (pp. 203–223). Oxford: Pergamon Press.

Vaughan, H. G., & Ritter, W. (1970). The sources of auditory evoked responses recorded from the human scalp. *Electroencephalography and Clinical Neurophysiology, 28*, 360–367.

Vaz Pato, M., & Jones, S. J. (1999). Cortical processing of complex tone stimuli: Mismatch negativity at the end of a period of rapid pitch modulation. *Cognition and Brain Research, 7*, 295–306.

Velasco, M., Velasco, F., & Olvera, A. (1985). Subcortical correlates of the somatic, auditory and visual vertex activities in man. I. Bipolar EEG responses and electrical stimulation. *Electroencephalography and Clinical Neurophysiology, 61*, 519–529.

Wang, X., Merzenich, M. M., Beitel, R., & Schreiner, C. E. (1995). Representation of a species-specific vocalization in the primary auditory cortex of the common marmoset: Temporal and spectral characteristics. *Journal of Neurophysiology, 74*, 2685–2706.

Whittaker, S. G., & Siegfried, J. B. (1983). Origin of wavelets in the visual evoked potential. *Electroencephalography and Clinical Neurophysiology, 55*, 91–101.

Wolpaw, J. R., & Penry, J. K. (1977). Hemispheric differences in the auditory evoked response. *Electroencephalography and Clinical Neurophysiology, 43*, 99–102.

Woods, D. L., & Clayworth, C. C. (1985). Click spatial position influences middle latency auditory evoked potentials (MAEPs) in humans. *Electroencephalography and Clinical Neurophysiology, 60*, 122–129.

Woods, D. L., & Clayworth, C. C. (1986). Age-related changes in human middle latency auditory evoked potentials. *Electroencephalography and Clinical Neurophysiology, 65*, 297–303.

Woods, D. L., Clayworth, C. C., Knight, R. T., Simpson, G. V., & Naeser, M. A. (1987). Generators of middle- and long-latency auditory evoked potentials: Implications from studies of patients with bitemporal lesions. *Electroencephalography and Clinical Neurophysiology, 68*, 132–148.

Wood, C. C., & Wolpaw, J. R. (1982). Scalp distribution of human auditory evoked potentials: II. Evidence for multiple sources and involvement of auditory cortex. *Electroencephalography and Clinical Neurophysiology, 54*, 25–38.

Wrege, K. S., & Starr, A. (1981). Binaural interaction in human auditory brainstem evoked potentials. *Archives of Neurology, 38*, 572–580.

Yamada, T., Nakamura, A., Horibe, K., Washimi, Y., Bundo, M., Kato, T., Ito, K., Kachi, T., & Sobue, G. (2003). Asymmetrical enhancement of middle-latency auditory evoked fields with aging. *Neuroscience Letters, 337*, 21–24.

Yoshiura, T., Ueno, S., Iramina, K., & Masuda, K. (1995). Source localization of middle latency auditory evoked magnetic fields. *Brain Research, 703*, 139–144.

Zaaroor, M., & Starrs A. (1991) Auditory brain-stem evoked potentials in cat after kainic acid induced neuronal loss. II. Cochlear nucleus. *Electroencephalography and Clinical Neurophysiology, 80*, 436–445.

CHAPTER

5

The N170: Understanding the Time Course of Face Perception in the Human Brain

Bruno Rossion *and* Corentin Jacques

Abstract

This chapter reviews the contribution of electromagnetic measures, mostly event-related potentials (ERPs), to our understanding of the time course of face processing in the normal adult brain, with a focus on the 100–200 ms time window after stimulus onset, that is, during the occipitotemporal component termed the N170. It first describes the N170 component, how it can be defined, and its relationship to the vertex positive potential (VPP) response to faces that was reported prior to the N170 in the literature. It then addresses the question of the origin of the largest N170 to faces in terms of electroencephalographic (EEG) signal, neural sources, and functional processes that lead to this effect. It also discusses the controversial issue of whether the N170 reflects underlying processes that can be at least partly recruited for processing nonface objects following extensive visual experience with these objects. The chapter summarizes the evidence showing that the N170 reflects both the initial basic-level categorization of the stimulus as a face through the activation of neural face representations and the coding of individual face representations. It then briefly discusses why the N170 may be a critical time window for other types of face categorizations before summarizing the chapter and addressing the question of how the N170 can be taken as a tool to clarify the dynamics and the nature of early face processes in future research.

Keywords: event-related potential, N170, face perception, perceptual processing, occipito-temporal component

This chapter reviews the functional properties of a human visual event-related potential (ERP) component, the N170, which has been associated with the perceptual processing of faces. A face can be detected in a visual scene extremely rapidly (e.g., Fei-Fei et al., 2007; Lewis & Edmonds, 2003; Rousselet et al., 2003), and a familiar person can be identified from his or her face in a few hundred milliseconds (e.g., Young et al., 1985). Various kinds of information can also be extracted quickly and efficiently from the face in order to categorize the person's gender, facial expression, ethnic origin, direction of gaze, and so on. (Bruce & Young, 1998). Since the early 1970s, with the huge increase in the amount of empirical work on face processing (Ellis, 1986), experimental psychologists and psychophysicists have aimed at clarifying the nature of the facial cues that are diagnostic to process faces (e.g., Haig, 1985), how these cues are integrated into global face representations (e.g., Sergent et al., 1984; Young et al., 1987), and how face processes and representations can be distinguished and organized in an information processing framework (e.g., Bruce & Young, 1986).

However, while behavioral studies have access only to the input and output of the system, the diagnostic information for face categorization is dynamically processed in the human brain in the period between the onset of the visual stimulation and a behavioral response several hundreds of milliseconds later. Clarifying the exact time course of face processes is a major goal of cognitive neuroscience. Because of their high temporal resolution, noninvasive electromagnetic measures, mostly ERPs but also event-related magnetic fields (ERMFs—in magnetoencephalography [MEG]) recorded on the human scalp, can greatly contribute to reach this objective. Event-related potentials to simple visual stimulation with a few electrodes were extensively investigated and described in the 1960s and 1970s (see Regan, 1989), but it is only since the late 1980s that ERPs to complex visual stimuli, in particular faces, have been systematically studied (Bötzel & Grüsser, 1989; Jeffreys, 1989).

The goal of this chapter is to assess the contribution of electromagnetic measures, mostly ERPs, to our understanding of the time course of face processing in the normal adult brain, with a focus on the 100–200 ms time window after stimulus onset, that is, during the occipitotemporal component termed the N170. Currently, more than hundreds of studies refer to the N170 component, without any published review. Reviewing all the findings and issues raised by these studies is clearly beyond the scope of this chapter. Therefore, for sake of clarity and in accordance with the objective of this volume, we have chosen to concentrate on summarizing and discussing *basic* issues regarding the N170 in the normal adult human brain. We will first describe the N170 component, how it can be defined and its relationship to the vertex positive potential (VPP) response to faces that was reported prior to the N170 in the literature (see "The Early ERP Studies of Face Processing and the N170 Face Effect"). Then we will address the question of the origin of the largest N170 to faces in terms of EEG signal, neural sources, and functional processes that lead to this effect (see the section "Why Is the N170 Larger to Faces?"). We will also discuss the controversial issue of whether the N170 reflects underlying processes that can be at least partly recruited for processing nonface objects following extensive visual experience with these objects (see the section "Are Early Face Processes Flexible?"). In the fourth section ("The N170: A Tool to Disentangle and Clarify the Time Course of Face Processes"), we will summarize the evidence showing that the N170 reflects both the initial basic-level categorization of the stimulus as a face through the activation of neural face representations *and* the coding of *individual* face representations. We will then briefly discuss why the N170 may be a critical time window for other types of face categorizations before summarizing this chapter (see the section "Summary, Questions to Clarify, and Future Directions") and addressing the question of how the N170 can be taken as a tool to clarify the dynamics and the nature of early face processes in future research.

The Early ERP Studies of Face Processing and the N170 Face Effect

The VPP as the Vertex Positive Counterpart of the N170

The first systematic ERP studies of face processing (e.g., Bötzel & Grüsser, 1989; Jeffreys, 1989; Jeffreys et al., 1992; Seek & Grüsser, 1992) reported a large positive potential peaking at the vertex between 140 and 180 ms following the presentation of a face stimulus (Figure 5.1a), termed the *vertex positive potential* (VPP) (following Jeffreys et al., 1989). In reviewing the response properties of the VPP, Jeffreys (1996) emphasized its larger amplitude in response to faces than other visual object categories and noted that the VPP presented a negative counterpart at bilateral occipitotemporal sites, suggesting sites of origin in areas of the temporal cortex (Figure 5.1a). However, the investigation of the VPP was emphasized in these initial studies because few if any electrodes were located on posterior lateral temporal regions of the scalp. Moreover, most of these studies used a mastoid reference located closely to the electrode sites picking up the occipitotemporal side of the dipolar activity. As a result, the amplitude of the occipitotemporal negativity was attenuated and the VPP increased, as can be demonstrated even with a large array of electrodes (see Joyce & Rossion, 2005; Figure 5.1b).

In subsequent studies of face stimulation, the use of a different reference (e.g., common average, nose; Bentin et al., 1996; Bötzel et al., 1995; George et al., 1996) to analyze the ERPs, and the availability or EEG recording systems with a larger number of electrodes covering the whole scalp favored the investigation of the occipitotemporal negative counterpart of the VPP, peaking at about 160–170 ms with a larger amplitude in the right hemisphere (Bötzel et al., 1995; George et al., 1996). This negativity was termed the N170 by Bentin and colleagues (1996).

Over the past 15 years, hundreds of ERP studies of face processing referring to the N170 component have been published. Magnetoencephalographic scalp recordings revealed a "M170" component with response properties similar to those of the N170 (e.g., Liu et al., 2000; Halgren et al., 2000; Sams et al., 1997). The advantage of focusing on the N170 rather than on the VPP is twofold. First, the electrodes recording the N170 on the scalp are closer to the neural generators of the component. Second, studying the N170 rather than the VPP allows

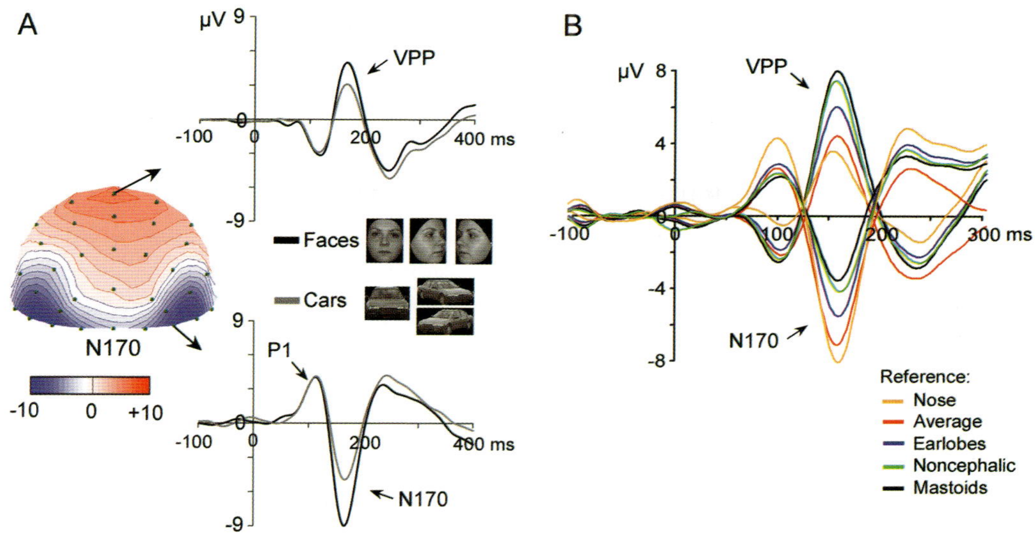

Fig. 5.1. (A) A typical N170 recorded from posterior lateral electrode sites following the presentation of faces and nonface objects (pictures of cars). It peaks at about 160–170 ms following stimulus onset and is most prominent at the lowest occipitotemporal electrode sites, usually maximal on channels P8(T6) or PO8, or on lower channels in this area if available. It is associated with a temporally coincident positivity on the vertex (Cz), the vertex positive potential (VPP). The VPP shows the same response properties as the N170 and largely reflects the projection of the occipitotemporal dipolar sources to the vertex. The data displayed are grand averages of 20 subjects presented with full-front and three-quarter profile pictures averaged together (180 trials/condition/subject, common average reference; adapted with permission from Rossion & Jacques, 2008). (B) The inversion of polarity between the N170 and VPP. The relative amplitude of the two "faces" of the component shows an inversely proportional relationship depending on the location of the reference electrode (adapted with permission from Joyce & Rossion, 2005).

investigation of the hemispheric lateralization of face processes (Joyce & Rossion, 2005).

The N1, the N170, and the N170 Face Effect

In the jargon of ERP researchers, the N170 corresponds to the visual N1 component: It is the first negative deflection on posterior scalp regions, following early posterior visual components C1[1] (peak ~70 ms) and P1 (peak ~100 ms), which can be observed in response to any visual stimulus. This N1 has a peak latency of 130–200 ms (see, e.g., Clark et al., 1995; Vogel & Luck, 2000; Chapter 4, this volume). However, the N1 is particularly large in response to pictures of faces and peaks on average at about 160–170 ms for these stimuli. Furthermore, the N170 marks the earliest, strongest, and most reliable difference in amplitude on the scalp between faces and nonface objects (e.g., Bentin et al., 1996; Bötzel et al., 1995; Rossion et al., 2000).

Together with its peak latency and its occipitotemporal topography, this larger amplitude to faces than to any other object category is what *defines* the N170 in the literature (Figure 5.1a). While some ERP researchers have referred to an N1 component in response to objects versus an N170 in response to faces (Carmel & Bentin, 2002; Itier & Taylor, 2004a), we believe that it is most appropriate to use the same label for the ERP component elicited by faces and objects. In short, the posterior lateral N1 component recorded to any visual stimulation varies in amplitude for different stimuli (see Rossion et al., 2000) and is particularly enhanced in response to faces. An important question is, of course, whether this face effect is due to stronger activation of the *same* neural sources that are recruited for nonface visual stimuli or to the addition of *other* sources specific to faces. As we will see in the next section, ERP researchers can address this issue only indirectly. Moreover, for sake of clarity in this field, the answer to this question should not change the terminology given to the basic ERP component that is used as a marker of high-level visual processes: It should either be the N1 or the N170 for all visual stimuli. Because of peak latency variability, the term *N1* may seem more appropriate. However, the term *N170* has become widely used in the face processing literature for the past 15 years. For this reason, we will refer to the *N170* (for both faces and nonface objects) and to the *N170 face effect* (the largest amplitude to faces) in the remainer of the chapter.

Why Is the N170 Larger to Faces?

While the N170 is systematically larger in amplitude for pictures of faces than for other object categories tested, there are also substantial N170 amplitude differences among nonface object categories (Itier & Taylor, 2004a; Rossion et al., 2000). In particular, pictures of highly familiar objects such as cars elicit a quite large N170 component, yet systematically smaller than the N170 to faces (Rossion & Jacques, 2008). From the published literature, it is currently impossible to quantify the magnitude of the N170 face effect because the amplitude of the N170 varies substantially among participants, and the categories of stimuli compared to faces, as well as the tasks that are used (passive viewing, one-back detection task, orientation judgment), differ greatly among studies. In fact, it is impossible to identify a "typical" ERP paradigm used to assess face and object differences at the level of the N170 in the literature, unlike what is done in functional magnetic imaging resonance (fMRI) studies to localize the areas of the visual cortex responding preferentially to faces (e.g., the so-called *fusiform face area* [FFA]; Kanwisher et al., 1997). Moreover, it is difficult to quantify the magnitude of the N170 face effect because there are other methodological parameters that can greatly influence this effect (e.g., the location of the reference electrode; see Joyce & Rossion, 2005; Figure 5.1) and that also vary substantially among studies. Yet, unless one reduces the N170 component amplitude through extremely severe low-pass filtering (Schweinberger et al., 2004) or measures its amplitude at the wrong electrode sites (e.g., medial occipital; see Rossion & Jacques, 2008), the N170 is systematically and substantially larger in response to faces than to nonface visual stimuli.

Why is the N170 larger for faces? Interpreting a differential amplitude of a scalp ERP component between two conditions is not straightforward, and in the case of the N170 face effect there are several issues to consider. The first issue refers to the origin of the effect in terms of EEG signal. This will be examined in the next section.

The N170: Time-Locked Increase in EEG Amplitude Rather Than Intertrial Phase Realignment

According to the traditional view of the generation of ERP components, the N170 originates from a massive synchronized increase in postsynaptic neural activity time-locked and phase-locked to stimulation onset, superimposed on background electrophysiological activity unrelated to the stimulation. In this framework, the N170 face effect simply reflects a *larger* increase in neural activity to faces compared to objects. This leads to a larger increase in EEG amplitude at a constant latency and polarity on the scalp for faces. Alternatively, the N170 face effect may be simply due to face stimuli eliciting an electrophysiological response at a more consistent latency from trial to trial compared to objects. This smaller intertrial latency jitter in response to faces would correspond either to a lower variance in the peak latency of the N170 from trial to trial or to a more precise phase resetting of ongoing EEG oscillations (i.e., preceding the stimulus; see Chapter 2, this volume). This phenomenon would also lead to a larger N170 after averaging in the time domain (Sayers & Beagley, 1974) without necessarily being associated with an increased recruitment of neural sources compared to nonface objects. This is an interesting idea because it has often been claimed that, compared to many object categories, faces form a particularly visually homogeneous category (Damasio et al., 1982), thus potentially leading to a better alignment of visual responses to members of the face class than to nonface objects. Moreover, it has been proposed that the visual N1 component to simple stimuli can indeed be largely generated by such a phase resetting of EEG ongoing oscillations in the alpha range (Makeig et al., 2002; but see Mazaheri & Jensen, 2005; Sauseng et al., 2007).

However, there is currently no evidence in favor of the phase-resetting model as accounting, even partly, for the N170 component and for the N170 face effect. That is, the largest N170 to faces is associated with a massive increase of power in the 5 to 15 Hz band time-locked to stimulus onset (Rousselet et al., 2007), which would not be observed in a case of pure phase resetting. Moreover, the supposedly larger visual homogeneity between exemplars of the face category than the nonface category (Damasio et al., 1982) is also irrelevant with respect to the N170 face effect: Most ERP studies have compared ERPs in response to faces and to members of the *same* nonface object class (e.g., cars), with exemplars of the nonface object class being highly similar (see Rossion & Jacques, 2008).

Even though the N170 face effect is largely due to a time-locked larger increase in EEG amplitude for faces, comparing faces to objects with various shapes, textures, and colors may possibly increase the N170 face effect artificially and create latency differences between categories in the averaged N170 response. For instance, when homogeneous pictures of faces are compared to pictures of nonface objects with various shapes, there appears to be an increase in the peak *latency* of the averaged N170 as well as a *widening* of the component for nonface objects (e.g., Itier & Taylor, 2004a), two phenomena that could be due to an increase in latency jitter between trials (Regan, 1989; Figure 5.2). This caveat can be circumvented by comparing the N170 to face and nonface stimuli of similar visual homogeneity, with the ERP response being averaged separately for each category, as in the majority of N170 studies (Rossion & Jacques, 2008). In these conditions, when visual stimuli are segmented from the background scene, the N170 is consistently larger in amplitude to faces than objects, but the component is not wider for nonface objects (Figure 5.1).

Do We Need a "Face Localizer" Approach in N170 Studies?

A question of interest is whether it would be possible, or even worthwhile, to design a typical face localizer paradigm that should be used across all or most studies to identify the N170 face effect, similar to fMRI studies prelocalizing the "face areas." The answer to this question is probably negative for several reasons. First, a face localizer paradigm as it is currently used in most fMRI studies is inappropriate, as it compares a set of visually homogeneous faces to various kinds of object categories (see Rossion & Jacques, 2008). Moreover, in traditional fMRI face processing studies, it is our experience that a one-back matching task commonly used in

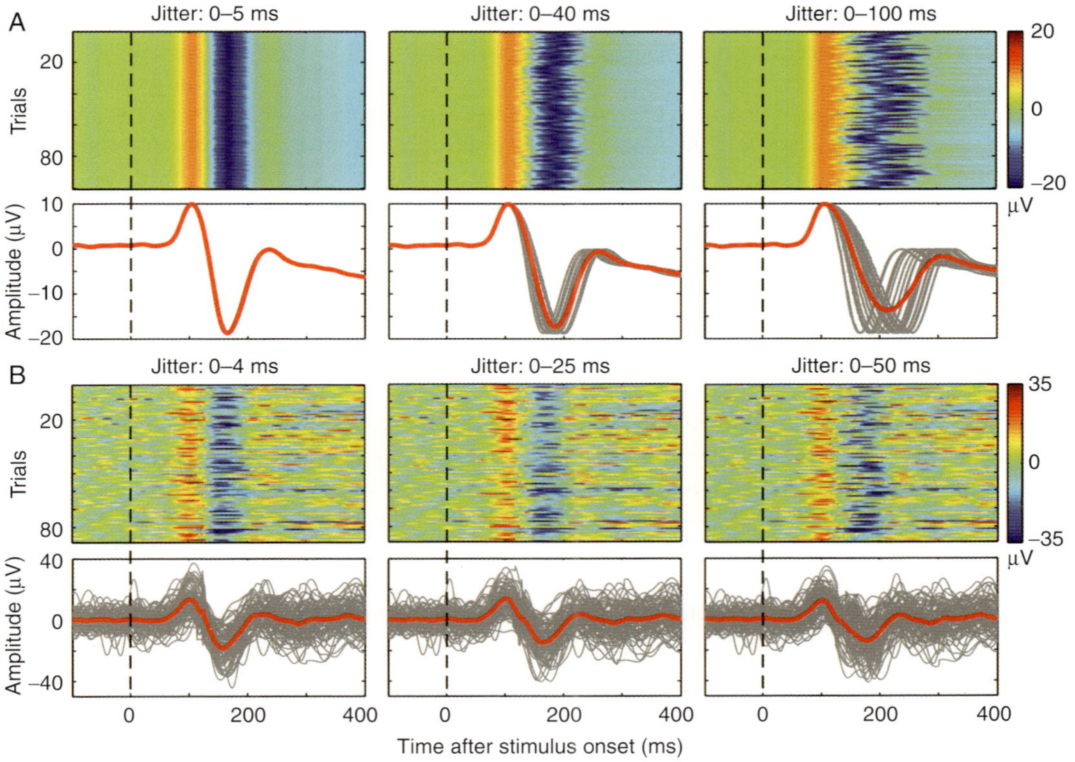

Fig. 5.2. Effect of increasing intertrial time jitter in the peak latency of the N170. (A) Simulation of jitter in the slope between the P1 peak and the N170 peak based on a grand averaged ERP response to face stimuli. The jitter was generated by randomly stretching the ERP from 100 to 160 ms after stimulus onset in the range of 0–5 ms (left), 0–40 ms (middle), and 0–100 ms (right). The upper row shows ERP images (trials x time, color-coded for amplitude; number of trials generated = 100) of the jitter simulation. The lower row shows a subset of individual trials (thin gray traces) as well as an average of 100 simulated trials (thick red trace). (B) Simulation of intertrial jitter using real EEG data. The simulated data were generated by adding sections of a grand averaged ERP response corresponding to the P1, N170, and P2 to 90 individual real EEG epochs containing no ERP. The ERP section corresponding to the N170 component was randomly jittered in latency (range, 0–4, 0–25, and 0–50 ms) and in amplitude before it was added to the EEG background. Note in both simulations the reduction of amplitude, the latency increase, and the smearing of the N170 as time jitter increases. Upper row: ERP images of the 90 simulated trials. Lower row: the 90 individual trials (thin gray traces) and the average of these trials (thick red trace).

so-called localizer paradigms (e.g., Kanwisher et al., 1997) is unbalanced: It is harder for faces than objects. Second, one would have to identify a specific category of stimuli that could be adequately compared to faces in terms of complexity, symmetry, familiarity, and so on. This issue has plagued the behavioral face processing literature for a long time, and experimenters generally admit that there is no such perfect control stimulus. Consequently, various stimuli are used in different studies (cars, houses, birds, chairs, etc.). Third, while regions such as the FFA can be disclosed in the individual human brain only by using a statistical criterion to estimate a differential level of activation to faces and objects, the N170 can be readily identified as a large voltage change (with respect to the reference electrode) in a single condition (i.e., face stimulation), without the need to make a statistical comparison with a nonface object stimulation. Moreover, considering the limited spatial resolution of scalp ERPs/ERMs, there is currently no evidence that the topographical distribution of the N170/M170 in response to faces is different than the topography of the N170/M170 face effect (Rossion et al., 2003). In addition, most ERP studies are interested in testing hypotheses about the time course of faces processes using the N170 as a tool, and simply need to identify the component in response to different face stimulations without asking direct questions about the face specificity of the effects. Finally, limiting analysis to spatiotemporal regions specific to faces may hide potentially interesting effects occurring outside of face-specific spatiotemporal windows as identified by the localizer. This is particularly problematic when experimental effects that are specific to faces occur outside such spatiotemporal windows.

For all these reasons, we believe that using a face localizer approach, which may be useful but not without its own problems in fMRI studies (see Friston et al., 2006; Saxe et al., 2006), is unnecessary for ERP studies of face processing. Yet, if such a typical paradigm to identify the N170 effect had to be used in order to address questions concerning only face-specific processes during the N170 time window, several issues are worth considering. First, the ERP response to pictures of faces should be compared to the ERP response obtained by averaging EEG segments elicited by pictures of the same visually homogeneous object category, such as cars, rather than mixing different object classes together. If time is not too constrained, several object classes can be used, provided that that the ERP averages are determined separately for each object class (e.g., Rossion et al., 2000). Second, in order to remove potential ERP effects due to low-level visual differences between faces and nonface objects, these stimuli should also be presented as phase-scrambled versions, controlling for the global luminance, contrast, and power spectra of the images (e.g., scrambled faces and cars; see Figure 5.3). The interaction between shape stimuli and their scrambled counterpart (faces–scrambled faces; cars–scrambled cars) should reveal the spatiotemporal time window that is most sensitive to faces. Third, all conditions should be randomized within each block of trials in order to avoid differential repetition effects and attentional confounds. Fourth, participants should perform an active task (e.g., one-back matching) to maintain their attention level quite high throughout the face localizer experiment, with a task that is equally difficult for all categories of stimuli. Finally, given the high temporal resolution of the method, identifying the N170 face effect on the scalp requires a spatio*temporal* definition: Which exact time window, for each electrode and in each individual participant of a study, shows a statistically larger response to faces? While this approach to defining the N170 face effect in an independent localizer may possibly be interesting in some specific cases, we still believe that the outcome of an experiment that relies on this approach would be largely identical to that achieved with a classical approach, that is, merely identifying the channels showing the largest N170 response to faces based on topographical maps and in keeping with the literature to test for an effect of interest.

The Sources of the N170 Face Effect and the Issue of Multiple Components

Is the N170 face effect due to faces eliciting a stronger activation of the sources that generate the N170 to both faces and nonfaces (a *quantitative* effect) or to the addition of one or several specific cortical source(s) for faces (a *qualitative* effect)?

The N170 takes place during a quite long time window (~130–200 ms) at a latency that is well beyond the average onset activation in the primary visual cortex (~50 ms in humans; e.g., Clark et al., 1995; Foxe & Simpson, 2002) and that is compatible with the activation in interlocked time courses of dozens of visual areas in the human brain located on the latero-medial, ventro-dorsal, and antero-posterior axes of the occipital, temporal, and parietal lobes (e.g., Foxe & Simpson, 2002; Vanni et al., 2004). Thus, it is reasonable to assume that the N170/VPP complex on the scalp is due to a configuration of bilateral equivalent dipoles reflecting the vectorial sum of multiple neural sources overlapping in time.

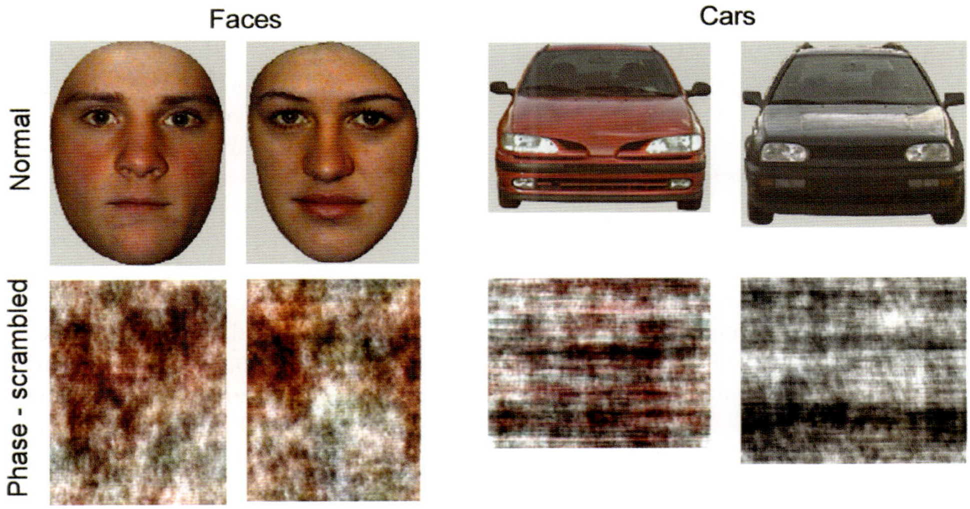

Fig. 5.3. Examples of stimuli that could be used in order to define properly the time window of the N170 face-specific increase in amplitude. Top left, pictures of faces; top right, luminance-matched pictures of a highly familiar category (cars). Pictures of faces and cars differ in terms of color variations and power spectra. Below, these differences can be taken into account by presenting phase-scrambled versions of the above stimuli. The interaction [(Faces–Scrambled faces)–(Cars–Scrambled Cars)] computed point by point should reveal the differences between faces and nonface stimuli that cannot be accounted for by low-level variations.

Source localization of the N170 to faces using constrained dipolar fit methods (e.g., Scherg & Berg, 1991) reported equivalent bilateral dipole solutions in the lateral occipitotemporal cortex or slightly more medially in the posterior part of the fusiform gyrus (Bötzel et al., 1995; Deffke et al., 2007; Pizzagalli et al., 2002; Rossion et al., 2003a; Schweinberger et al., 2002b; Shibata et al., 2002; Figure 5.3). The M170 has generally been localized in the very same region (Deffke et al., 2007; Halgren et al., 2000; Swithenby et al., 1998; Tanskanen et al., 2005; Tarkainen et al., 2002; Watanabe et al., 1999). This localization would rather correspond to the region of the so-called *occipital face area* (OFA; in BA19) than of the FFA (in BA37) identified in fMRI studies by contrasting pictures of faces and objects (see Haxby et al., 2000; Figure 5.4). However, some studies have also reported a more anterior location of the M170 source in the middle fusiform gyrus, more compatible with an FFA localization (MEG: Linkenkaer-Hansen et al., 1998; Sams et al., 1997; EEG: Mnatsakanian & Tarkka, 2004, as well as two posterior sources in the lingual gyrus; Taylor et al., 2001). Given that the FFA and the OFA are located only about 2 cm apart in the posterior-anterior axis along the ventral visual stream (Figure 5.4), the low

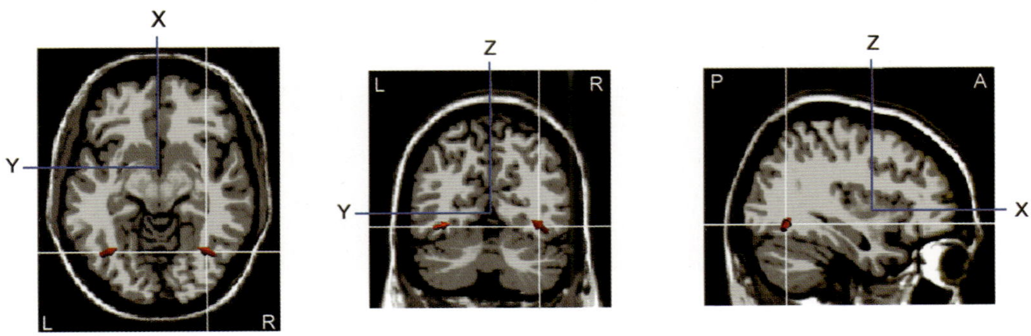

Fig. 5.4. Example of the dipole source localization of the N170 in response to faces (adpated with permission from Rossion et al., 2003) in the posterior fusiform gyrus/lateral occipitotemporal cortex, with their orientation. Many EEG/MEG studies have reported roughly similar localizations of the equivalent dipoles of the N170 (Bötzel et al., 1995; Deffke et al., 2007; Halgren et al., 2000; Pizzagalli et al., 2002; Schweinberger et al., 2002b; Shibata et al., 2002; Swithenby et al., 1998; Tanskanen et al., 2005; Tarkainen et al., 2002; Watanabe et al., 1999). Axes: X = antero-posterior, Y = left-right, Z = dorsal-ventral.

resolution of the EEG source localization, together with the constrains of the dipole fit procedure, may explain this slight difference in source localization.

However, distributed source localization methods without a priori assumptions about the number of sources (e.g., LAURA: Grave de Peralta Menendez et al., 2001; LORETA: Pascual-Marqui et al., 2002) have provided different and contrasting results. Itier and Taylor (2004b) reported a dominant source of the N170 to faces in the posterior part of the superior temporal sulcus (pSTS; see also Watanabe et al., 2003), while Herrmann et al. (2005b) reported main sources in the anterior part of the fusiform gyrus (BA 20), together with multiple activations in a parieto-temporal-occipital network of areas. Henson et al. (2007) used a distributed source localization method with constraints on the number of dipoles and their orientation to test the respective weights of these sources. They also reported dominant sources of the differential M170 for faces and scrambled faces quite anteriorly in the fusiform gyrus, with a strong right hemispheric dominance.

Taken together, these results indicate the presence of multiple cortical sources accounting for the N170 component to faces, with dominant sources in the lateral part of the posterior fusiform gyrus and in the anterior/middle fusiform gyrus. Unfortunately, very few studies have reported the sources of the N170 face effect (differential amplitude for faces and objects) or of the N170 in response to objects. Rossion et al. (2003) found equivalent source localization for faces and cars in the posterior fusiform gyrus, but with different strengths and orientations. Itier and Taylor (2004b) reported that faces recruit an additional pSTS source compared to multiple nonface categories, but the sources also varied among nonface categories and were very similar to faces for some categories (e.g., road signs).

To summarize, in light of the current evidence, the question of whether the sources of the N170 to faces and objects differ (i.e., whether the N170 face effect is due to the addition of specific sources or not) is currently unresolved by EEG/MEG studies. In the human brain, fMRI studies have identified several visual areas—the FFA, OFA, and pSTS—that respond more strongly to faces than to other object categories (Haxby et al., 2000; Figure 5.5). However, it is unclear if any of these areas respond *selectively* to faces. High-resolution fMRI has revealed that the FFA is a heterogeneous functional region made up of a high proportion of clusters of the size of several cortical columns, responding selectively to faces, mixed together with clusters responding nonspecifically to any category (Grill-Spector et al., 2006). These face-selective clusters may be the generators of local field potentials such as the intracranial N200 recorded on the surface of the ventral occipitotemporal cortex (Allison et al., 1999) or the much larger P160 response to faces than abstract visual patterns recently reported with intracerebral electrodes implanted in the posterior fusiform gyrus (Barbeau et al., 2008). Similarly, face selectivity is observed in the monkey brain at the level of single neurons (Gross et al., 1972; Perrett et al., 1992) grouped in columns (Wang et al., 1996), which could also be clustered to form larger patches of face selectivity *below* the level of organization of a whole visual area (Tsao et al., 2006).

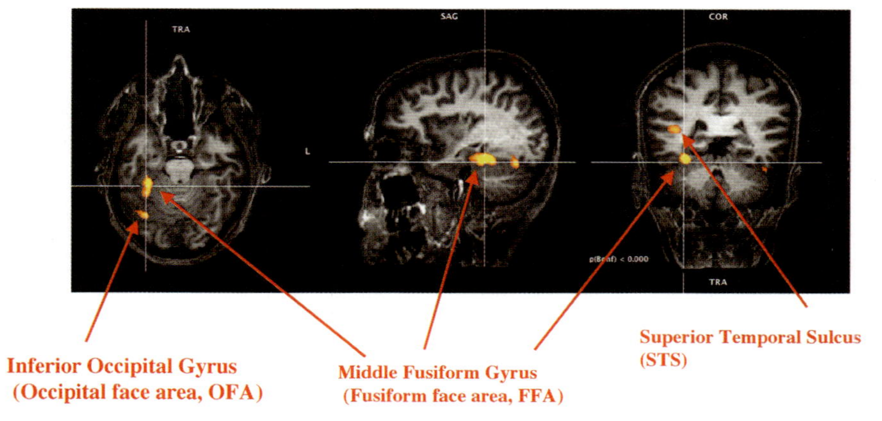

Fig. 5.5. The three functional areas responding more strongly to faces than to nonface visual stimuli in the human brain as identified in fMRI (adapted with permission from Haxby et al., 2000). They are illustrated here in the right hemisphere, in a single normal brain, during a functional face localizer contrast (faces vs. objects).

Considering this evidence, the most reasonable account of the N170 face effect on the scalp is that in addition to the contribution of general sources in visual areas responding to object shapes (e.g., the lateral occipital complex), faces recruit a few additional sources in these face areas (i.e., face-selective clusters) between 100 and 200 ms. These sources would contribute heavily to the N170 face effect observed on the scalp.

What Drives the N170 Face Effect?

Does the N170 face effect truly reflect the perception of a face stimulus or rather the low-level properties differing between faces and nonface object categories? In most ERP studies of face processing, low-level properties (e.g., size, luminance, contrast, spatial frequency spectrum) of the categories of stimuli compared, known to influence the amplitude of early visual potentials (see Regan, 1989), are usually not tightly controlled. In general, these factors may also influence the N170 parameters, and thus potentially affect the differential amplitude of this component for faces and nonface object categories. In some studies, however, low-level properties have been controlled as much as possible between faces and the control object category compared (e.g., houses in Rousselet et al., 2005, 2007). In these conditions, the N170 is still much larger in response to faces.

In any case, many observations in the literature indicate that the N170 face effect reflects high-level processes, that is, the *perception* of the stimulus as a face. Two clear illustrations are the larger N170 to the identical two-tone "Mooney" images when they are presented in an upright orientation—and thus are generally perceived as faces—than when they are presented upside down (George et al., 2005; Jeffreys, 1993; Figure 5.6A). In a similar vein, the famous paintings of the sixteenth-century Italian artist Arcimboldo, in which a face is made up of nonface objects (usually organic elements), elicit a clear N170, which decreases substantially when the picture's orientation is reversed and the face is no longer perceived (Figure 5.6B). In other cases, whenever a stimulus contains enough information (either in the local elements, or in their global configuration, or both) to be interpreted as a face by the visual system, the N170 is large in amplitude. This is true for face photographs obviously, but also for schematic faces, faces with features rearranged, inverted faces, faces cut in half, isolated eyes, faces with contrast inverted, faces without eyes, and so on (see Figure 5.7; e.g., Bentin et al., 1996; Eimer, 1998;

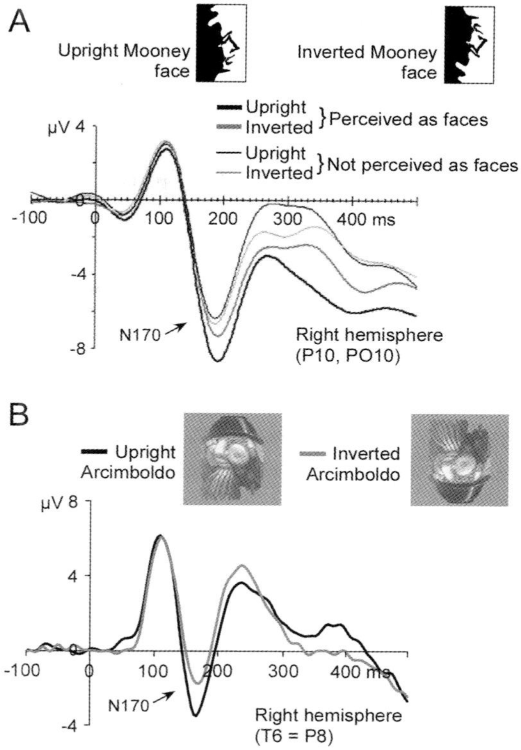

Fig. 5.6. (A) When two-tone ("Mooney") images are presented upright, they usually lead to the perception of a face stimulus, yielding a larger N170 than when the same pictures are presented inverted and do not lead to the perception of a face (figure adapted with permission from George et al., 2005). (B) The same effect is observed for pictures of the paintings of Giuseppe Arcimboldo (1527–1593), where the face stimulus is perceived as emerging from the organization of nonface features such as fruits and vegetables (figure adapted with permission from Rossion & Jacques, 2008).

George et al., 1996; Itier & Taylor, 2002; Rossion et al., 1999b; Sagiv & Bentin, 2001).

However, when a transformation removes most of the diagnostic information used to perceive the stimulus as a face, the N170 is extremely small in amplitude (e.g., an isolated nose or mouth: [Bentin et al., 1996]; superimposed random noise in frequency bands critical for face perception [Tanskanen et al., 2005]). This reduction is also observed when single-stimulus transformations that usually do not reduce the N170 amplitude nevertheless lead to such a reduction when their *combination* affects face perception. For instance, while masking the facial elements through noise *or* inverting the face (i.e., masking the global configuration) may not lead to an N170 amplitude decrease, *combining* the two transformations makes the stimulus difficult to perceive as a face, leading to a substantial N170 amplitude decrease (Schneider et al., 2007).

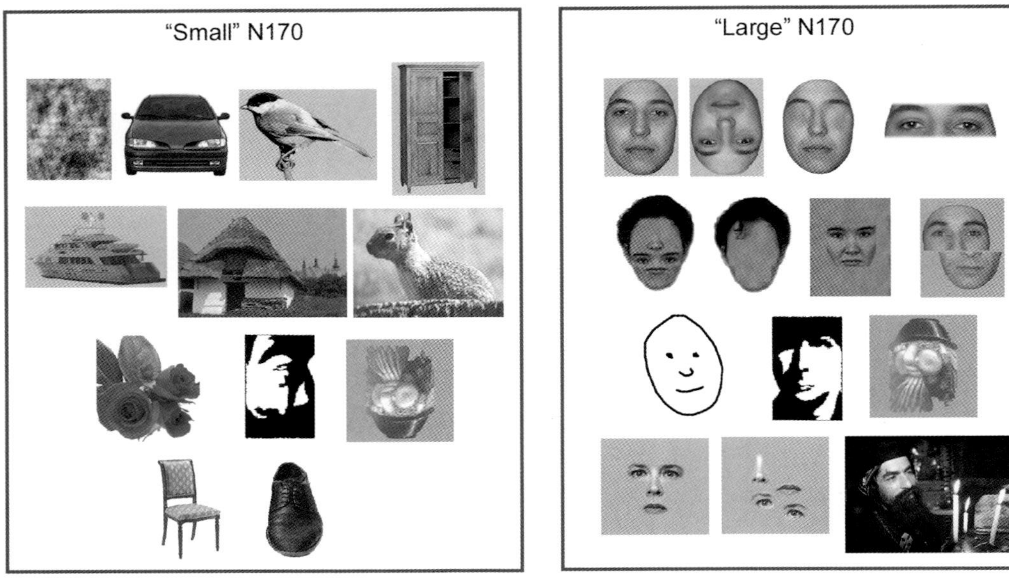

Fig. 5.7. The N170 amplitude is large in response to stimuli that are perceived as a face, across various formats (right), and in comparison to equally complex visual stimuli that can be matched for low-level visual properties (left). When the stimulus is transformed by isolating, masking, or removing facial features, or when the organization of the features is disrupted, the N170 remains large in amplitude or can even be increased as long as the stimulus is perceived as a face. This observation suggests that the N170 onset marks the access to face representations in the human brain. This access is generally slightly delayed in latency (10–20 ms) when the face stimulus is transformed, either at the level of local features (e.g., removing the eyes) or at the level of their first-order organization (e.g., inverting the position of the features).

All of these instances indicate that what drives the increased N170 response is that there must be enough information in the visual stimulus, either as local elements or in their organization, to activate face representations and allow the stimulus to be perceived as a face. Pushing this idea to the limit, Bentin and colleagues also found that the very same simple stimuli, originally not perceived as faces or facial elements, elicited a face-like N170 response only after they conceptually primed study participants' awareness to the physiognomic value of the stimuli (Bentin & Golland, 2002; Bentin et al., 2002). These observations reinforce the view that the N170 face effect reflects the perception of a face and appears to be largely driven by the early activation of neural representations of faces in high-level visual cortex.

Are Early Face Processes Flexible? The N170 and Visual Expertise

Whether neural mechanisms tuned optimally for face perception are strictly modular (domain-specific), or whether they are flexible and potentially recruited for nonface objects following visual expertise, has long been debated (e.g., Diamond & Carey, 1986; Ellis & Young, 1989; Kanwisher, 2000; Tarr & Gauthier, 2000). The N170 face effect is an interesting phenomenon for this debate because it allows testing of the hypothesis that early *perceptual* processes devoted to faces can also be recruited for nonface objects of visual expertise. Supporting this hypothesis, two ERP studies have reported an N170 amplitude increase in bird and dog experts (Tanaka & Curran, 2001) as well as in fingerprint experts (Busey & Vanderkolk, 2005) when they are presented with members of their categories of expertise. One limitation of these studies is that it is unclear whether this amplitude modulation really taps into face processes. This question was addressed directly using an ERP concurrent stimulation paradigm (Jacques & Rossion, 2004). When observers fixate a face stimulus remaining on the screen, the N170 response to another face stimulus presented at a different location is substantially reduced (with respect to a control condition in which the first stimulus is a phase-scrambled face; Jacques & Rossion, 2004). This strong effect is usually taken as evidence for competition between overlapping neural representations and processes. Similarly, when observers fixate a centrally presented object of expertise, the N170 time-locked

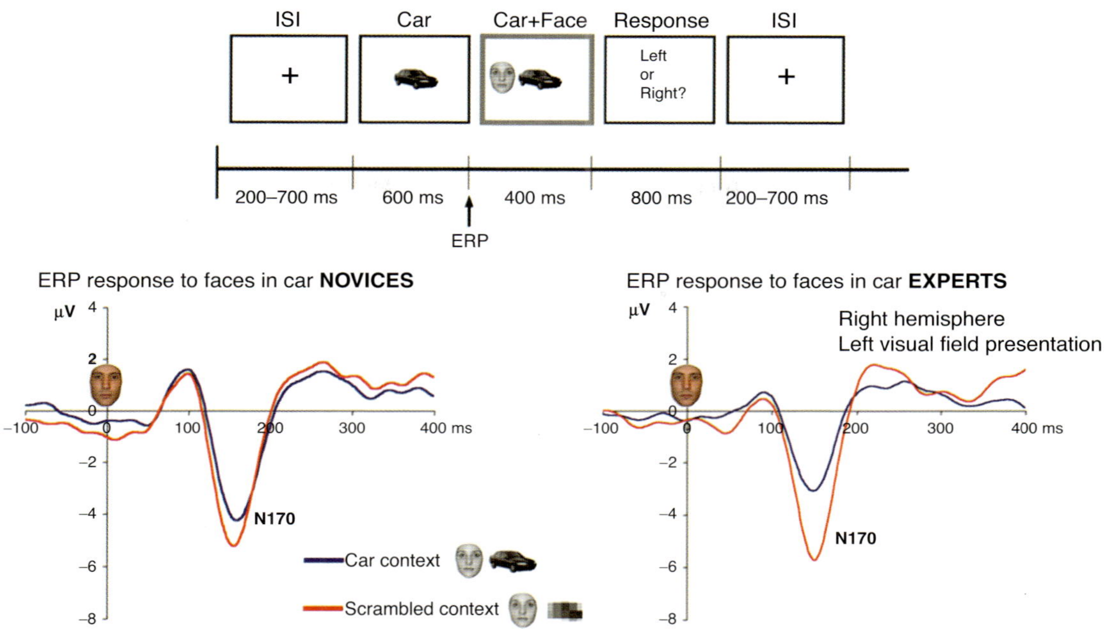

Fig. 5.8. A concurrent stimulation paradigm can be used to show that faces and nonface objects share common perceptual processes during the N170 time window. The ERPs are recorded in response to a face stimulus presented laterally while car experts and car novices fixate either a car or a control stimulus presented in the center of the screen. Relative to novices, the N170 in response to lateralized faces (average of three occipitotemporal electrodes in the right hemisphere) is massively reduced when car experts fixate the picture of a car but not when they fixate the control stimulus. Figure adapted with permission from Rossion et al. (2007).

to a lateralized face stimulus is substantially reduced in amplitude between 130 and 200 ms (Figure 5.8; Rossion et al., 2004, 2007). This sensory competition effect is much larger for experts than for novices and is not found when participants fixate a control nonface stimulus. It is observed for nonface objects learned either in the laboratory (*Greebles*; Rossion et al., 2004) or in real-life conditions (*Cars* in car experts; Rossion et al., 2007). These observations suggest that experts recruit face processes when they fixate nonface objects of expertise, such that the face stimulus that follows can no longer activate the same processes, leading to a reduced N170. Supporting this claim, the degree of visual expertise measured independently through a behavioral task is strongly correlated with the amount of amplitude reduction of the face N170 in the concurrent stimulation paradigm (Rossion et al., 2007). These effects are substantial; they are measured on the N170 elicited by faces, not objects; and they are larger in the right hemisphere in agreement with fMRI localization of visual expertise effects (Gauthier et al., 2000) and the general right hemispheric advantage for processing faces (e.g., Sergent et al., 1992). Furthermore, there is no evidence that these N170 modulations could be due to an increase of central attention to the nonface object of expertise in experts: When manipulated, spatial attention modulates the N170 amplitude to the lateralized face stimulus in an orthogonal (i.e., additive) way to the competition effect and also affects the preceding P1 component (which is unaffected by visual expertise; Jacques & Rossion, 2007a). In summary, by virtue of the excellent temporal resolution offered by ERP recordings and the spatial sampling of the whole system, these observations demonstrate that visual competition between faces and objects of expertise takes place as early as 130 ms in the human brain, during a limited time window, in occipitotemporal areas. However, it remains unclear whether this sensory competition effect results from the recruitment of the exact same neural sources (i.e., clusters of neuronal columns; see the section "The Source of the N170 Face Effect and the Issue of Multiple Components") for faces and nonface objects of expertise or from increased competitive interactions between distinct populations of cells located in the same area through local lateral inhibitory connections (Allison et al., 2002; Wang et al., 2000). Irrespective of this question, the perceptual mechanisms reflected by the N170 do not appear to be rigidly dedicated to visual stimuli with a facial configuration. They are particularly tuned to faces but remain flexible enough so that they can be partly recruited for some nonface objects following the development of a visual experience at processing these objects.

The N170: A Tool to Disentangle and Clarify the Time Course of Face Processes

This section will address the question of what kinds of face processes take place during the N170 time window, and their putative relations to earlier and later face processes as identified in ERPs.

Basic-Level Face Categorization at the Level of the N170

Because of the evidence reviewed above (in the section "What Drives the N170 Face Effect?"), ERP researchers generally acknowledge that the basic-level categorization of the stimulus as a face, or the detection of a face in a visual scene, takes place during the VPP/N170 time window (Bentin et al., 1996; Jeffreys, 1996; Rousselet et al., 2004). This basic-level face categorization stage has been associated with the *structural encoding stage* described in an influential information processing model in the face processing literature (Bruce & Young, 1986). However, as pointed out earlier (Rossion & Gauthier, 2002), this is conceptually incorrect, since the structural encoding stage defined by Bruce and Young (1986) does not refer to a face detection stage, but to the activation of an initial *individual* face representation, common for both familiar and unfamiliar faces, irrespective of the format of presentation of the stimulus (variable in size, viewpoint, etc.).[2] The question of whether individual faces are coded during the time window of the N170 will be addressed below (see the section "The Coding of Individual Face Representations during the N170 Time Window").

Based on the larger N170 amplitude to faces than nonface objects and its correlation with the perception of a face per se (a face percept), it is legitimate to consider that the process of face detection is indeed taking place during the N170 time window. The N170 face effect usually starts at about 130 ms after stimulus onset, although the onset time has rarely been measured precisely (see Rousselet et al., 2005, for an exception) or even discussed. Moreover, the N170 face effect is found for segmented faces or for faces inserted in visual scenes (Rousselet et al., 2004a; 2004b), and appears to be insensitive to large variations of face stimulation in size, position (to some extent), or face viewpoint (Jeffreys, 1996; Rousselet et al., 2005). However, there are two important issues to consider when claiming that the N170 marks the onset of the categorization of the stimulus as a face based on an access to face representations.

DEGRADING FACE STIMULATION DELAYS THE N170

The first issue is that basic-level categorization of a face, or face detection, is partially affected by certain stimulus transformations such as face inversion (Lewis & Edmonds, 2003; 2005; Purcell & Stewart, 1988; Rousselet et al., 2003), which nevertheless do *not* decrease the N170 face effect. In fact, the N170 face effect may even be *larger* following stimulus inversion (Figure 5.9A), because this manipulation

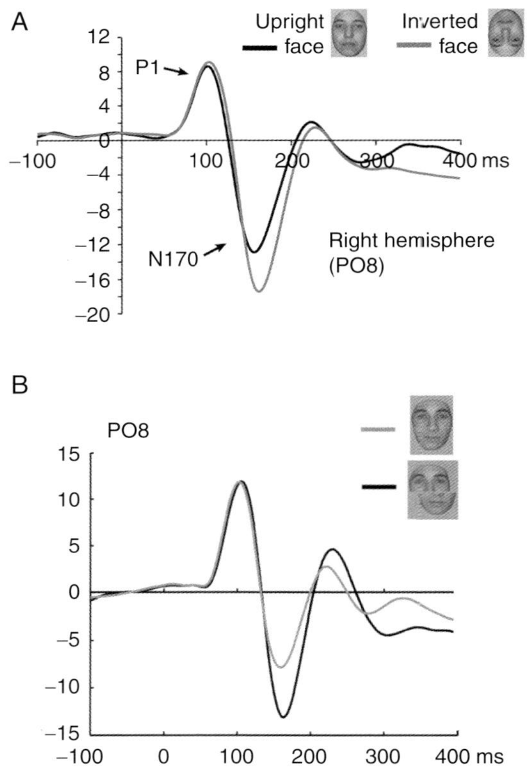

Fig. 5.9. (A) As demonstrated in numerous studies, inversion of a segmented face stimulus, which leads to a massive decrement in individual recognition performance, causes a substantial increase in N170 amplitude (e.g., Eimer, 2000b; Rossion et al., 1999b). Figure adapted with permission from Rossion and Jacques (2008). There are currently no satisfactory accounts of this paradoxical increase in amplitude, which is at odds with the slight reduction of neural activity for inverted faces as recorded in fMRI (e.g., Kanwisher et al., 1998), single neurons (Perrett et al., 1998), or local field potentials recorded on the cortical surface (Allison et al., 1999). (B) Breaking the face stimulus into two parts, even slightly, also leads to an amplitude increase in the N170 (Letourneau & Mitchell 2008; ERP waveforms of this figure from Jacques & Rossion, unpublished data).

increases the amplitude of the N170 to faces while leaving the N170 to nonface objects of identical or similar amplitude (e.g., Rossion et al., 2000). The same paradoxical N170 increase is found for other manipulations that may affect the categorization of the stimulus as a face, such as isolating the eyes (e.g., Bentin et al., 1996; Taylor et al., 2001), changing the features' positions (George et al., 1996), inverting the contrast of the face (Itier & Taylor, 2002), or laterally offsetting the bottom part of the face (see Figure 5.9b; Letourneau & Mitchell, 2008). Moreover, as noted above, the N170 remains very large for faces without eyes, for instance (Eimer, 1998; Itier et al., 2007) or when a small amount of visual noise is added to the image (e.g., Jemel et al., 2003c; Schneider et al., 2007). How can these observations be reconciled with the idea that the N170 largely reflects the initial activation of face representations associated with the categorization of the stimulus as a face? One critical element to consider here is that all these transformations of the face stimulus, which increase the N170 amplitude or leave it unaffected, do not *prevent* the stimulus from being categorized as a face. As mentioned above (in the section "What Drives the N170 Face Effect?"), if sufficient elements are present, either as features or as a first-order configuration, so that the stimulus is perceived as a face, the N170 will be large in amplitude. However, removing or degrading some elements of the face will generally *slow down* the activation of the representation, an effect that is reflected in the delay (10–20 ms) of the N170 following inversion (e.g., Bentin et al., 1996; Rossion et al., 1999b, 2000; Figure 5.9) as well as for the above-mentioned stimulus transformations (e.g., Bentin et al., 1996; Eimer, 1998; George et al., 1996; Itier & Taylor, 2002; Itier et al., 2007). One possibility is thus that the delay of the N170 due to stimulus transformations such as inversion merely reflects a delay in the activation of face representations or a slower accumulation of evidence at the level of the neuronal population coding for faces (see Perrett et al., 1998). Recent evidence suggests in fact that both mechanisms may be at play, because the latency delay measured at the N170 peak for inverted faces is correlated with variations of the ERP signal as early as 120–130 ms (N170 onset) but is maximal at the N170 peak (Jacques & Rossion, 2007b).

In summary, despite stimulus transformations that slow down face detection, whenever a face representation is activated, it is associated with a large N170 response.

EARLY FACE DETECTION (P1/M1) IS BASED ON LOW-LEVEL VISUAL FEATURES

A second issue to consider is whether the time window of the N170 is too late to reflect the basic categorization of a face stimulus, a process that is extremely fast. Indeed, both ERP and forced-choice saccadic eye movement studies indicate that categorization of animal or human faces in pictures of visual scenes, for instance, can take place within 110–150 ms following stimulus onset, *including the perceptual decision* (Crouzet, Kirchner, & Thorpe, 2010; Thorpe et al., 1996, VanRullen & Thorpe, 2001). How can this finding be reconciled with the idea that the N170 marks the onset of basic-level categorization of faces? One possibility is that a stimulus may be detected in a visual scene and categorized as a face above chance level before the onset of the N170, but that this fast categorization is not based on the activation of face representations. Rather, it could be based on an accumulation of evidence from low-level cues that are statistically more frequently associated with faces (e.g., roundness, specific color distribution in the visual scene, local contrast, distribution of energy in different frequency bands) and can lead to fast basic-level face categorization.

Along these lines, several studies have reported a larger P1 (or M1 in MEG) in response to faces than to objects (e.g., Eimer, 1998, 2000a; Goffaux et al., 2003; Herrmann et al., 2005a, 2005b; Itier & Taylor, 2004a; Liu et al., 2002) at electrodes near the medial occipital pole. This P1/M1 face effect is not consistently observed (e.g., Boutsen et al., 2006; Rossion et al., 2003; Rousselet et al., 2005, 2007) and is not as large as the N170 face effect (e.g., Goffaux et al., 2003; Itier & Taylor, 2004a; Liu et al., 2002; see Figure 5.10). It has sometimes been associated in the literature with high-level face processes, such as basic-level face categorization (Herrmann et al., 2005a), the perception of facial parts (Liu et al., 2002), or holistic/configural face processing (Halit et al., 2000; Latinus & Taylor, 2005).

However, several elements suggest that the P1/M1 face effect does not reflect face perception per se, but rather is related to differences between faces and nonface stimuli in terms of intrinsic low-level visual information (see also Rossion & Caharel, 2011). First, the visual P1/M1 is an early component, peaking at around 100 ms following stimulus onset, and thought to originate mainly from striate and lateral extrastriate visual areas (Clark et al., 1995;

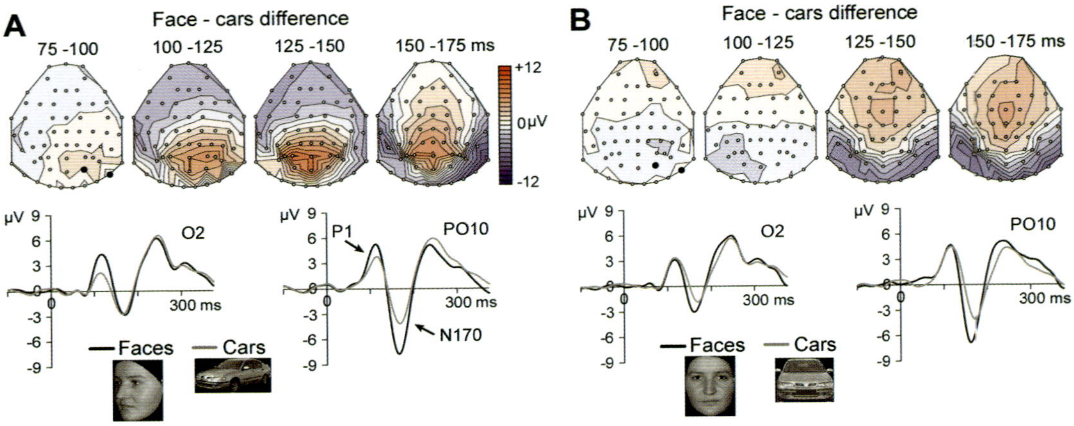

Fig. 5.10. (A) Grand average ERP to three-quarter views of faces and cars. The upper row shows scalp topographies of the difference between faces and cars (from 75 to 175 ms after stimulus onset). The lower row depicts raw ERPs at two posterior electrodes (location indicated in black on the left scalp topography). Note the large amplitude difference at the level of the P1 and N170 components both on ERP waveforms and on scalp topographies. (B) Grand average ERP to front views of faces and cars. Details are identical to those in (A). When front-view symmetrical stimuli are presented, ERPs to faces and cars no longer differ at the level of the P1 component, while the N170 is still much larger for faces. This is visible on ERP waveforms and scalp topographies. The fact that the N170 (but not the P1) is larger for faces irrespective of the viewpoint indicates that this effect is robust even for symmetrical full-front images of the two categories (as for Figure 5.1, unpublished data). Figure adapted with permission from Rossion and Jacques (2008).

Di Russo et al., 2002; Halgren et al., 2000; Tanskanen et al., 2005; Tarkiainen et al., 2002), even though some studies have reported a contribution of the posterior fusiform gyrus (Hermann et al., 2005b; Liu & Ioannides, 2006). P1 amplitude is known to be sensitive to many low-level visual features such as luminance, color, contrast, or spatial frequencies of the stimulus (see Regan, 1989). Supporting this view, the early M1 difference between photographs of faces and other categories can be reversed in amplitude (Halgren et al., 2000) and can be canceled when surface information (color and texture) of the face stimuli is removed (while the M170 face effect resists these low-level transformations). In the same vein, studies in which the face and object stimuli are well controlled for low-level features do not report P1 face effects (e.g., Rousselet et al., 2005). More intriguingly, while the N170 face effect appears to generalize across views of the stimuli, the P1 face effect may disappear when symmetrical full-front pictures of faces and cars are presented, indicating that it is not always reliable (Rossion & Jacques, 2008; Figure 5.10).

Second, given the early onset latency of the P1 (~80 ms) compared to the mean onset latency of face-selective neurons (100 ms in the monkey brain e.g., [Kiani et al., 2005], probably slightly later in the human brain, see Schroeder et al., 2004), it is unlikely that this P1/M1 face effect reflects the activation of facial representations. Similarly, intracranial recordings have so far demonstrated earlier face-preferential or face-specific responses clearly after 100 ms, that is, N200s in the ventral occipitotemporal cortex and lateral middle temporal gyrus (Allison et al., 1999) and P160 in the posterior fusiform gyrus (Barbeau et al., 2008; see also Halgren et al., 1994). Third and finally, strong support for the dissociation between a low-level and a high-level origin of the P1 and N170 face effects, respectively, has been reported by studies varying parametrically the amount of visual noise or the noise spatial frequency added to a face image. While the P1/M1 is not correlated with the amount of face information in the image manipulated parametrically through random noise, the N170/M170 amplitude increases with visibility of the face (Jemel et al., 2003c; Tarkiainen et al., 2002). Most interestingly, an elegant MEG experiment (Tanskanen et al., 2005) dissociated the M1 and M170 effects by masking face stimuli with narrow-band spatial frequency noise. When the noise was presented in the frequency bands optimal for face perception (11–16 cycles per image), the face could not be perceived adequately, but the occipital M1 was maximal in amplitude. In contrast, the M1 was minimal and the M170 was maximal at the lowest and highest noise spatial frequencies, in parallel with the clear perception of the face (Figure 5.11). These results underline the two

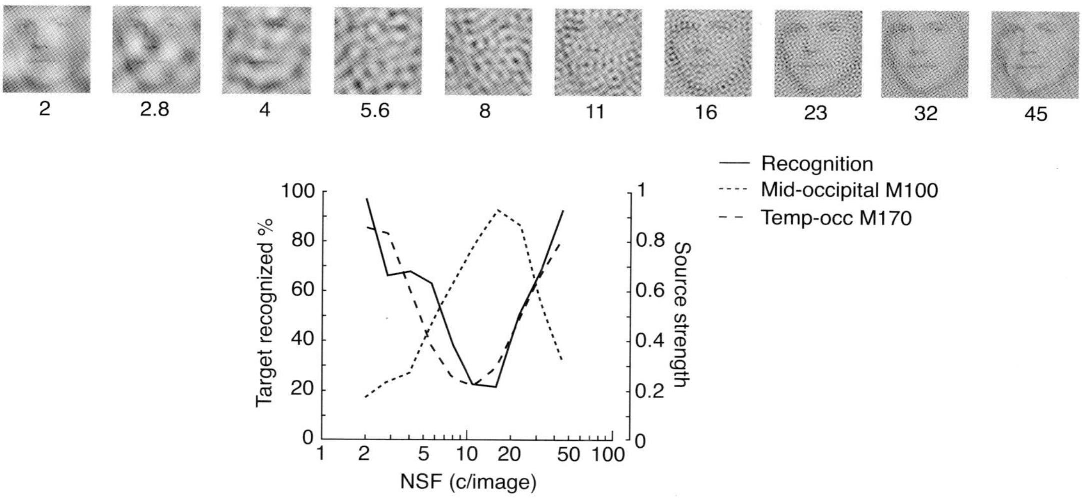

Fig. 5.11. By adding noise at different frequency bands on face stimuli (top row), Tanskanen and collaborators (2005) showed a dissociation between the M100 (M1) and M170 amplitude response properties. The M170 is larger when the noise is in the highest and lowest spatial frequency bands, that is, when it does not affect the perception of the stimulus as a face (see the full line). In contrast, the M100 is larger when the energy is higher in the mid-frequency bands, masking the perception of the stimulus as a face. This shows not only that the M170 amplitude is a function of the perception of the stimulus as a face, but also that the face-sensitivity effects found on the M100 may be due to the specific power spectrum of face stimuli, with more information in the mid-frequency bands. Figure adapted with permission from Tanskanen et al. (2005).

important points of this section. First, the M170 response is sensitive to the visibility of a face and is closely related to face perception. Second, the M1 is not sensitive to the perception of a face per se, but its response is largest to the middle spatial frequencies that are critical for face perception. This strongly suggests that at the level of the M1, the face effect reported by some studies is meaningful and reflects the early accumulation of evidence to categorize the stimulus as a face. However, this categorization is based on low-level visual information, such as the spatial frequency spectrum or color of the stimuli, rather than on the activation of face representations per se (see also Rossion & Caharel, 2011 for recent direct evidence). Therefore, the N170 time window appears to offer the most reliable time frame to investigate the nature of face perceptual mechanisms in the human brain, even if earlier face sensitivity can be observed.

The Coding of Individual Face Representations during the N170 Time Window

While detecting a face in the visual scene is a complex and biologically relevant task, in most circumstances our face processing system has to go beyond this initial categorization and extract an individual face representation in order to be able to determine if that person has been seen previously. How fast does the system extract a representation that is detailed enough to individualize a face?

In humans, this question has been mostly investigated by measuring the ERP responses to repeated individual faces. The rationale behind repetition studies is that the time point at which the ERP signal diverges for repeated and unrepeated faces indicates the speed at which the system is sensitive to the differences among individual faces.

A series of studies have used a *delayed repetition* paradigm that includes a phase during which a number of faces are learned (either only visually or by association with a name and/or brief semantic information such as an occupation; e.g., Curran & Hancock, 2007; Joyce & Kutas, 2005; Paller et al., 2000; Yovel & Paller, 2004). Other studies have compared the ERP response to the first presentation of faces with the ERP response to the same faces presented in a subsequent block of trials (e.g., Henson et al., 2003; Schweinberger et al., 2002a; Tanaka et al., 2006). All of these studies thus include a variable number of intervening face stimuli between the first and subsequent face presentations. Notably, *none* of these delayed face repetition studies has reported a modulation of the N170 when comparing repeated to unrepeated faces. Rather, the most robust finding in these studies is that repeated faces elicit a larger N400 potential compared to unrepeated/new faces in a time window ranging from around 300 to 500 ms. This effect has been termed the *ERP repetition effect* or the *old/new ERP effect* (e.g., see Paller et al., 2000).

From these observations, one might conclude that the coding of individual faces occurs no sooner than 300–400 ms after stimulus onset, that is, about 200–300 ms after the stimulus has been categorized as a face (i.e., at the onset of the N170, around 130 ms). However, this conclusion is at odds with the speed at which individual faces can be discriminated behaviorally (Figure 5.12), as well as the known temporal dynamics of face information encoding by face-selective neurons in the nonhuman primate inferotemporal cortex. These neurons have an average onset latency of about 100 ms and accumulate information about both global face category and face identity simultaneously (Tovee & Rolls, 1995), with information about individual faces being significantly represented in the neurons' responses not more than 40–50 ms after information about the global category is represented (Matsumoto et al., 2005; Sugase et al., 1999).

This discrepancy suggests that the use of a delayed repetition paradigm in ERPs may not provide reliable information about how sensory/early visual representations are modulated by repetition. That is, these representations may not hold the trace of a previously presented stimulus for a prolonged time interval.

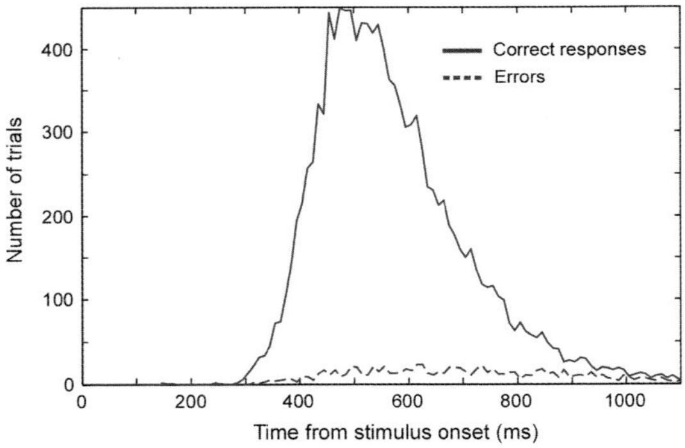

Fig. 5.12. Distribution of response times in an individual face discrimination same/different task (data from Jacques & Rossion, 2007b) in which faces were presented at 12 orientations in the picture plane (0–330°). The histograms represent the distribution of correct and incorrect responses pooled across all face orientations. The number of trials in successive 10 ms time bins is plotted as a function of time from stimulus onset. Note that the earliest correct responses start at around 300 ms after stimulus onset. If we consider that a minimum of 100 ms is needed to generate a motor command (see VanRullen & Thorpe, 2001), this response time distribution points to a coding of individual faces before 200 ms.

When using *immediate* face repetition, some studies found that the N170 is slightly reduced in response to a face preceded by the same individual face compared to a face preceded by a different face (e.g., Campanella et al., 2000; Guillaume & Tibergien, 2001; Heisz et al., 2006; Itier & Taylor, 2002; Jemel et al., 2003b, 2005). However, other studies did not find any N170 amplitude difference between repeated and unrepeated faces (e.g., Huddy et al., 2003; Mnatsakanian & Tarkka, 2004; Schweinberger et al., 1995, 2002b, 2004). The factors accounting for this discrepancy between studies are difficult to identify. Moreover, a common criticism of these effects is that they may reflect general repetition effects that could also be due to image-based elements and not to the repetition of an individual face per se.

Recently, two ERP paradigms were used to address these limitations and investigate systematically the time course of individual face coding. First, in a long adaptation paradigm (~3000 ms duration for the adapter) with a short interstimulus interval (100–300 ms) between the adapting face and the target face, the N170 amplitude was substantially reduced when the test face was of the same identity as the adapting face, starting at around 160 ms (Jacques et al., 2007; Figure 5.13). This effect was found despite the use of different photographs and a change of size between the adapter and the target face (Jacques et al., 2007). Importantly, when the identical face stimuli were presented upside down (Figure 5.13), the difference between same and different faces was not found on the N170 but was delayed by about 30 ms (i.e., starting at ~190 ms). Moreover, a recent ERP adaptation experiment with similar timing parameters (i.e., presentation duration and interstimulus interval) indicates that the adaptation effect to face identity on the N170 is robust enough to generalize at least partly across a 30 degrees viewpoint change between adapting and target faces (Caharel et al., 2009a). These observations further rule out an interpretation of this N170 individual face adaptation effect as being due to simple physical differences rather than to perceived differences between individual faces. This strong and replicable effect of visual adaptation found for individual faces on the N170 (Caharel et al., 2009a; 2009b; Jacques et al., 2007; Jacques & Rossion, 2009) stands in contrast with inconsistent effects found in previous face identity repetition studies. Several factors may account for this discrepancy,

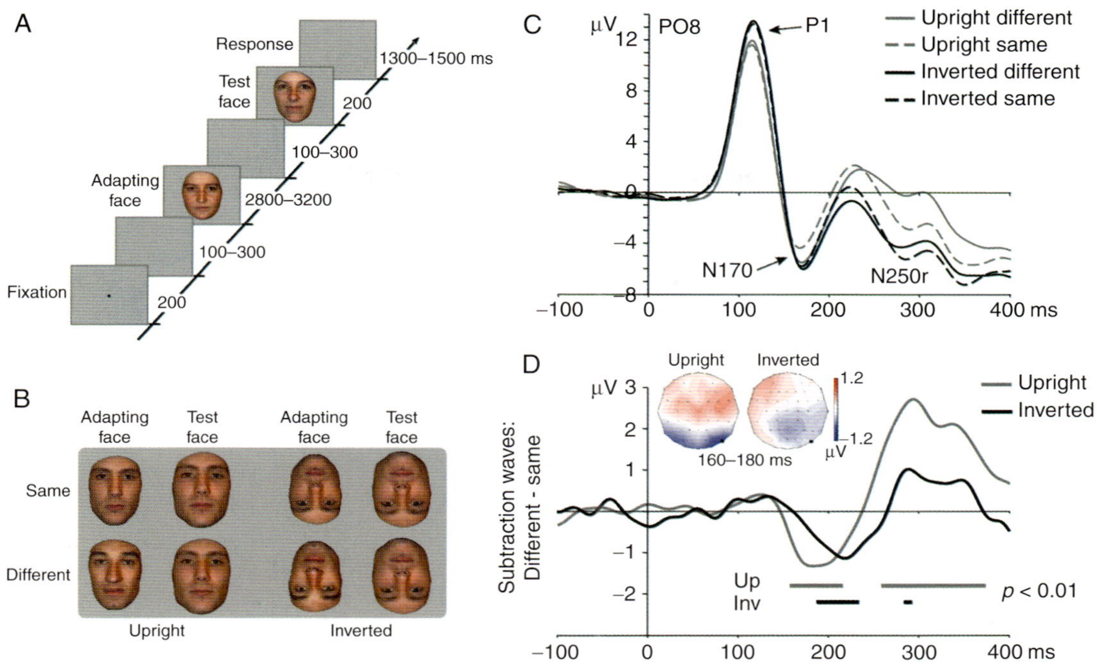

Fig. 5.13. (A) An individual face adaptation ERP paradigm (Jacques et al., 2007) with a long duration of the adaptor and a short interstimulus interval was used during a delayed matching task. (B) The four conditions of the experiment, crossing adaptation (adapting and test faces = same or different) with orientation (upright or inverted). (C) The ERP response to the test face. There is a strong interaction between adaptation and orientation starting at the N170 level, with a reduction of amplitude for the upright same faces only. (D) Subtraction waveforms (different–same faces) on electrode PO8 and topographical maps, showing that the identity adaptation effect starts at around 160 ms for upright faces but takes place about 30 ms later (after the N170 component) for the inverted faces and is much weaker. Significant differences ($p < .01$) between waveforms recorded in the same versus different conditions are indicated as horizontal gray bars for upright faces (Up) and black bars for inverted faces (Inv). Note also the large effect of identity adaptation following the N170 at the level of the N250 component (see, e.g., Schweinberger et al., 2004). Figure adapted with permission from Jacques et al. (2007).

in particular the longer duration of the first stimulus (adapter), which is necessary to elicit behavioral face adaptation effects (see, e.g., Leopold et al., 2005) and has been also used successfully in an ERP-adaptation study at the level of the face category (Kovacs et al., 2006). Another element to consider is the short interstimulus interval between the adapter and the target face (~200 ms) used by Jacques, Rossion and colleagues in their studies compared to the longer intervals (usually >1 s) used in previous immediate repetition studies.

Second, converging evidence of individual face coding at the level of the N170 is found when a continuous face identity reversal paradigm is used (Jacques & Rossion, 2006). Here, instead of recording the N170 in response to a "flashed" face (i.e., presented after a blank screen period), the ERP is recorded to an individual face that follows immediately the presentation of another face (i.e., pattern-reversal stimulation, or face identity reversal here). In these conditions of identity reversal stimulation (~2 Hz), early visual components preceding the N170 are abolished and a "pure" N170 response can be isolated. This stimulation mode allows measurement of the ERP response reflecting the difference between two individual faces (Figure 5.14). Using morphed stimuli in a categorical face perception design, it was found that the isolated N170 response was larger when the two faces reversing identity were located on different sides of the identity boundary, compared to when they were located on the same side of the identity boundary (Jacques & Rossion, 2006), again ruling out a low-level visual account of these observations.

To summarize, both ERP adaptation and face identity reversal stimulation indicate that the system can discriminate between individual face representations as early as 160 ms during the late N170 time window. These observations suggest that the N170 should not only be described as a face detection

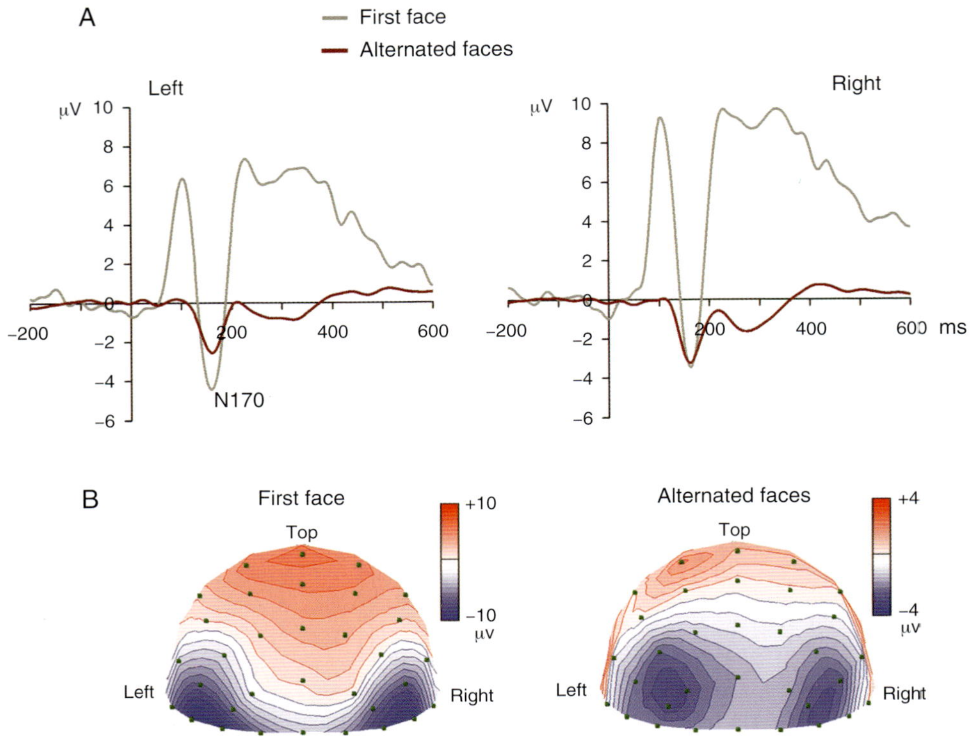

Fig. 5.14. (A) An ERP to the first face of a block of trials (preceded by a blank screen) superimposed on the ERP in response to the face identity reversal (electrodes PO7 and PO8). Note that the onset of the response to a face identity reversal is at the N170 level (130 ms), that is, the P1 component is no longer present. (B) The scalp topographies (back view of the head) depict the distribution of the ERP response at 160 ms following the onset of the first face of each block (left) or the alternated face (right). Figure adapted with permission from Jacques and Rossion (2006).

stage. Rather, it is a time window during which multiple face processes take place, including both face detection *and* the coding of individual faces. Or, to put it differently, once a face representation has been activated, it can be rapidly refined into an individual representation *within* the N170 time window. Of course, this is not to say that the whole process of the individual face representation is completed during the N170 time window. For instance, an additional individual face repetition effect is usually observed starting at around 220–250 ms after stimulus onset. This repetition effect generally arises in the form of a more negative ERP for repeated compared to unrepeated faces around 250 ms over temporal scalp regions, and has accordingly been termed the *N250* or *N250r* (e.g., Begleiter et al., 1995; Schweinberger et al., 1995; Tanaka et al., 2006). Thus, even though the processing of an individual face starts during the N170 time window, information continues to be further accumulated during the later time window, as reflected by the repetition effects occurring at later time points (see Jacques et al., 2007, for a time-point-by-time-point analysis of individual face repetition effects).

Are Long-Term Face Representations Activated during the N170 Time Window?

When are long-term stored representations of individual faces activated? Most studies have found that the N170 does not discriminate between unfamiliar faces and famous faces (e.g., politicians, celebrities; Bentin & Deouell, 2000; Eimer, 2000a; Henson et al., 2003; Jemel et al., 2003a, 2003b, 2005; Schweinberger et al., 2002a) or learned faces (Rossion et al, 1999a). In contrast, famous faces usually elicit an enhanced N400 component (i.e., between 300 and 500 ms; Bentin & Deouell, 2000; Eimer, 2000a; Jemel et al., 2003a, 2003b) over central or frontocentral electrodes and an increased positivity between 500 and 700 ms over central or centroparietal sites (Bentin & Deouell, 2000; Eimer, 2000a; Henson et al., 2003) when compared with unfamiliar faces.

However, a series of recent studies have found a larger N170 for *personally* familiar faces (the subject's own face, mother's face, friends' faces) or very famous faces compared to unknown faces (Caharel et al., 2002, 2006). Similarly, an MEG study (Kloth et al., 2006), found a larger M170 to personally familiar faces (lecturers and fellow university students) compared to unfamiliar faces. The larger amplitude for familiar faces starts at around the peak of the N170 (160–170 ms) and is maximal in the descending slope of the N170, similar to the timing of individual face adaptation effects on the N170 (Jacques et al., 2007). Furthermore, the N170 difference between familiar and unfamiliar faces is no longer present for faces presented upside down (Caharel et al., 2006), in agreement with the behavioral face inversion effect and with N170 adaptation findings (Jacques et al., 2007).

Part of the discrepancy between the studies that found or did not find an effect of face familiarity on the N170 is due to the comparison of unfamiliar with *personally* familiar faces in the latter group (Caharel et al., 2002, 2006; Kloth et al., 2006), whereas the former used famous faces, for which there may be large individual differences in the degree of familiarity of the participants with each face. Specifically, the visual coding of personally familiar faces, which would be associated with more robust representations (Tong & Nakayama, 1999), may be facilitated by the extensive visual experience that observers have with these faces, hence yielding a differential N170 response when compared to unfamiliar faces (Caharel et al., 2002).

Alternatively, these familiarity effects on the N170 might reflect a top-down modulation from stored face representations, as suggested by the finding of a strong familiarity effect on the N170 (comparing famous to unfamiliar faces) only when faces had been previously presented (i.e., a priming paradigm; Jemel et al., 2003b). More precisely, the (prolonged) activation of stored robust face representations, due either to the large number of repetitions of familiar faces (Caharel et al., 2002) or to the use of semantically related familiar faces (e.g., friends, family members, fellow students; Caharel et al., 2002, 2006; Kloth et al., 2006), may have biased the visual encoding of individual faces taking place at the N170. It is therefore currently unclear whether this N170 familiarity effect arises due to face familiarity per se (i.e., the information about face familiarity is contained in the individual face representations extracted during the N170) or to top-down modulations.

The N170 and Other Face Categorizations

This review has largely focused on how the N170 reflects the coding of a face in order to detect faces and process their identity. However, faces are extremely complex stimuli, carrying a large number of cues that are important for social interactions. From a face, we are able to extract information allowing us to recognize the facial expression and the mood of the person, and to categorize the face's sex, infer its race, or infer its apparent age (Bruce & Young, 1998). Furthermore, primates can also detect rapidly and automatically the direction of gaze to determine where the person is looking (see, e.g., Emery, 2000). The extraction of the cues leading to these categorizations of the face stimulus is notoriously fast and efficient, yet little is known about its time course. Most EEG/MEG studies that have addressed the issue of the speed and time course of face categorization besides identity processing have contrasted the perception of different stimuli (e.g., male and female faces, faces with different expressions, directed and averted gaze). With the exception of studies contrasting different eye-gaze directions (e.g., Conty et al., 2007; Taylor et al., 2001), the majority of these studies have failed to report modulations at the level of the N170 (for expression, see, e.g., Eimer & Holmes, 2002; for age and gender, see, e.g., Mouchetang-Rostaing & Giard, 2003; for race, see, e.g., Caldara et al., 2004). Some studies have reported amplitude modulations of the N170 for different facial expressions, in particular a larger N170 to fearful faces than neutral faces (e.g., Batty & Taylor, 2003; Blau et al., 2007). However, it is unclear if these effects are due to low-level features (e.g., increased contrast between dark and white areas of the face in fearful expressions) or to the differentiation of facial expressions per se. Other modulations of the N170 with facial expression are largely inconsistent among studies. Our view on this issue is in line with the evidence reviewed above: Since the N170 marks the early access to both global and fine facial information, there is no reason to believe that the extraction of cues to categorize rapidly and efficiently a face according to its gender, age, race, or facial expression would not also take place predominantly within that time window. However, there is no reason to expect that the raw N170 amplitude, which reflects the global activation of the system, would differ reliably in response to various face stimuli (e.g., two faces with different expressions) that activate largely overlapping populations of neurons. Rather, the sensitivity of the component to the differential information contained in these stimuli needs to be assessed through ERP adaptation paradigms (i.e., changing expression between adapter and target) or continuous changes of expression, for instance (as performed in some eye gaze studies; e.g., Conty et al., 2007; Watanabe et al., 2001). Future ERP/ERMF research using such paradigms will then have to disentangle these different effects in both time and space in order to clarify the exact time course of face categorization processes.

Summary, Questions to Clarify, and Future Directions

Summary

The N170 is a visual component that is much larger in response to pictures of faces than to any kind of comparable visual stimulation (i.e., the N170 face effect) and has accordingly been studied as a marker of perceptual face processes (see the section "The Early ERP Studies of Face Processing and the N170 Face Effect"). As indicated in the introduction, there are currently more than 100 published ERP/ERMF studies focusing on the N170/M170 component, and an extensive review of these studies, their findings, and their implications was clearly beyond the scope of this chapter. In concentrating on answering basic questions regarding the N170 in this chapter, we had to omit a number of important issues, such as whether the component can be modulated by spatial and selective attention (Eimer, 2000c; Jacques & Rossion, 2007a), how it is affected in clinical populations (prosopagnosia, autism, etc.; e.g., Dawson et al., 2005; Eimer & McCarthy, 1999), what its developmental course is (de Haan et al., 2002; Kuefner et al., 2010; Taylor et al., 1999), and clever experiments using the particular tuning of the N170 to faces to understand the mechanisms of visual working memory maintenance, for instance (Sreenivasan et al., 2007). The interest in these studies directly depends first on how the basic issues that were addressed in this chapter are clarified and understood among the scientific community of N170 researchers.

Here, in summary, we have seen that the N170 corresponds to a time-locked increase of EEG amplitude (see the section "The N170: Time-Locked Increase in EMG Amplitude Rather Than Intertrial Phase Realignment") and originates most likely from multiple cortical sources, with dominant bilateral sources in the posterior fusiform/lateral occipitotemporal complex as well as the middle fusiform gyrus (see the section "The Sources of the N170 Face Effect and the Issue of Multiple Components"). The N170 is larger in the right hemisphere in response to faces. Even though the N170 face effect appears to reflect the important contribution of processes taking place in visual areas activated preferentially for face stimuli, these processes remain plastic enough in the adult visual system to be recruited partly for nonface objects following extensive expertise training (see the section "Are Early Face Processes Flexible?").

There is widespread evidence that the N170 onset (~130 ms) reflects the earliest activation of face representations in the occipitotemporal cortex: The N170 is large in amplitude as long as the stimulus is perceived as a face, even if either the local features or their global configuration is disrupted (see the section "What Drives the N170 Face Effect?"). As long as the stimulus remains interpretable as a face, the N170 presents a large amplitude. However, it can be delayed when the access to face representations is slowed down following manipulations such as removing diagnostic features, presenting the features in isolation, or breaking the first-order face organization through scrambling or inversion (see the section "Basic-Level Face Categorization at the Level of the N170"). Contrary to this categorization of the stimulus as a face based on the nature of the features and their configuration, earlier face-sensitive effects at the level of the P1 appear to reflect largely low-level visual differences between faces and nonface objects such as their differential power spectra or color distribution (see Rossion & Caharel, 2011). Future studies should go beyond a mere debate between ERP components by performing point-by-point correlation between electrophysiological and behavioral responses. When performing such analyses (e.g., Jacques & Rossion, 2007b; Philiastides & Sajda, 2006), the exact time point at which the effects of stimulus and task manipulations arise can be identified. Such analyses usually reveal significant effects *after* the P1 component in the downward slope of the N170 (e.g., Jacques & Rossion, 2007b; Rousselet et al., 2007).

The first access to individual face representations takes place during the late N170 time window, in agreement with the speed of individual face coding in the monkey brain (see the section "The Coding of Individual Face Representations during the N170 Time Window"). However, the question of whether personally familiar face representations are already activated at that latency requires stronger evidence (see the section "Are Long-Term Face Representations Activated during the N170 Time Window?").

Although the processing of an individual face starts during the N170 time window, further information continues to be accumulated during the later time window, as reflected by repetition effects occurring at later time points (e.g., the N250 component; see, e.g., Schweinberger et al., 1995; 2002b). It is our view that other finer-level face categorizations of the stimulus (e.g., expression, gender) are also performed during the N170 time window at various latencies (see the section "The N170 and Other Face Categorizations"), but decisive evidence requires the use of more sensitive paradigms than the comparison of different stimuli presented in isolation.

The Road Ahead: Caveats and Recommendations for N170 Research

Even though we know that both basic-level (coarse) and fine face categorization processes take place during the N170 time window, the *nature* of the face representations and the time course of their activations during the N170 time window remain largely unclear. For instance, one may ask whether certain facial features or properties have more weight in the face representations or are activated earlier than others. More generally, during the N170 time window, are faces processed through the extraction of local facial parts that are then integrated into a global representation (i.e;? *local to global*) or rather from an initial coarse global face picture to a finer-grained representation (i.e., *global to local*)?

The majority of studies that have addressed this question of the nature of face representations and its dynamics have proceeded by comparing the N170 amplitude in response to single face stimuli that are transformed or degraded. This is done either in a hypothesis-driven way in most studies (e.g., revealing, masking, or modifying only certain predetermined features of the face; e.g., Bentin et al., 1996; Eimer, 1998; Itier et al., 2007) or in an unbiased sampling of small portions of the stimulus (e.g., Schyns et al., 2003). The nature of the information coded is then inferred from the modulation of N170 amplitude with the stimulation, much as the response properties of single neurons of the monkey infero-temporal cortex are inferred from their spiking rate following degradation of complex visual stimuli (e.g., Tanaka, 1996). The ERP studies adopting this approach have mainly confirmed the well-known finding that the eyes are a dominant feature of the face (e.g., Haig, 1985), eliciting a conspicuous N170 even when presented in isolation (Bentin et al., 1996). This large N170 to isolated eyes has been sometimes interpreted as evidence that there is a distinct source devoted to the eyes of the face contributing heavily to the N170 (e.g., Bentin et al., 1996; Itier et al., 2007) or that the representation of the face during the N170 reflects mainly the local information about the eyes (Schyns et al., 2003). Yet, this interpretation is problematic because the N170 amplitude is large in response to a face defined only through its first-order configuration, even without any eyes or other features (e.g., the Arcimboldo paintings; see the section "What Drives the N170 Face Effect?"; Figure 5.6). Moreover, as long as the stimulus is still perceived as a face, removing the eyes from a face photograph does not attenuate the N170 amplitude at all (Eimer, 1998; Itier et al., 2007). Hence, one cannot infer from the larger N170 to isolated eyes that the eyes are perceptually processed in a distinct population of neurons or that they represent a critical feature to elicit the N170.

In a similar vein, the interpretation of an *earlier* representation of the local eye region than any other features based on the N170 amplitude during the presentation of randomly selected facial information through small apertures (Schyns et al., 2003, 2007) could be mistaken. Indeed, this effect may be simply due to a *quantitative* difference, that is, the fact that the N170 amplitude is large in response to isolated eyes but not to other isolated internal features (Bentin et al., 1996). However, when a full-face stimulus is presented, there is currently no evidence that information on the eyes is processed before information on the other features of the face. This example illustrates how the raw N170 amplitude in response to a single stimulus, while being informative about the faceness of that stimulus, cannot be directly taken as reflecting the nature of the representation at that latency. One reason for this limitation is that, as we have seen (see the section "What Drives the N170 Face Effect"?), once there is enough evidence in the stimulus to activate a face representation (a process that can be facilitated by the viewer's expectations), a large N170 is evoked. The eyes may have more weight in the activation of the representation of a face, in particular when the stimulus is not segmented or masked by noise (Paras et al., 2007), but this does not mean that the eyes of the face are processed in a separate neural source, that they are critical, or that their representation is activated first in time when features are presented altogether. Moreover, the N170 amplitude is certainly *not proportional* to the strength of activation of the representation, being in fact larger to degraded or transformed face stimuli

(Figure 5.9). A second reason for this limitation is that contrary to the response of a single neuron, the N170 amplitude reflects the *global* contribution of multiple cortical sources that are activated in interlocked time courses and can compensate for or counteract each other. For these reasons, directly inferring the nature of the coding from the raw N170 amplitude to single stimuli that reveal partial information of a face appears to be misleading.

As noted earlier in this review, and similarly to what is currently being done in many fMRI studies (Grill-Spector & Malach, 2001), a more adequate and sensitive approach to characterizing the nature of the face representations during the N170 time window may be to rely on modulations of the N170 amplitude following face adaptation (*ERP adaptation* within the face domain; e.g., Ewbank et al., 2008; Harris & Nakayama, 2007; Caharel et al., 2009b; Jacques et al., 2007; Figure 5.13) or reversal of certain features of the face in a continuous stimulation paradigm (see Jacques & Rossion, 2006; Figure 5.14). Here the interest is no longer in clarifying what the component reflects but rather in using it as a *tool* to clarify the nature of early face representations and processes. For instance, presenting a face stimulus after an adapter and modifying separately or in combination various properties, such as the overall shape of the face, its pigmentation, particular facial features, distances between features, and so on, should potentially reveal *if*, *how strongly*, and *when* exactly these cues are coded during the N170 time window (or later) (for a recent example see Caharel et al., 2009b). Another advantage of this approach is that it isolates specific effects during the processing of *whole faces* rather than degraded or transformed stimuli. Given that the literature reviewed in this chapter clearly indicates that the N170 is a critical time window for investigating human face processing, an approach that treats the N170 component as a tool to investigate the nature of face processes and their time course during natural stimulation of whole faces appears to us to be one of the most promising in this field.

Notes

1 The C1 reverses polarity with the presentation of the stimulus in the upper/lower visual field (associated with a negative/positive polarity, respectively) due to reversal of the orientation of the sources with respect to the calcarine sulcus (see, e.g., Clark et al., 1995; but see Ales et al., 2010).

2 According to Bruce and Young (1986, p. 307), the structural encoding stage is considered as a level "which capture those aspects of the structure of a face essential to distinguish it from other faces" and thus supposedly reflects individual face coding.

References

Ales, J. M., Yates, J. L., Norcia, A. M. (2010). V1 is not uniquely identified by polarity reversals of responses to upper and lower visual field stimuli. *Neuroimage*, 52, 1401–1409.

Allison, T., Puce, A., & McCarthy, G. (2002). Category-sensitive excitatory and inhibitory processes in human extrastriate cortex. *Journal of Neurophysiology, 88*, 2864–2868.

Allison, T., Puce, A., Spencer, D. D., & McCarthy, G. (1999). Electrophysiological studies of human face perception. I: Potentials generated in occipitotemporal cortex by face and non-face stimuli. *Cerebral Cortex, 9*, 415–430.

Barbeau, E. J, Taylor, M. J., Regis, J., Marquis, P., Chauvel, P., & Liégeois-Chauvel, C. (2008). Spatiotemporal dynamics of face recognition. *Cerebral Cortex, 18*, 997–1009.

Batty, M., & Taylor, M. J. (2003). Early processing of the six basic facial emotional expressions. *Cognitive Brain Research, 17*, 613–620.

Begleiter, H., Porjesz, B., & Wang, W. (1995). Event-related brain potentials differentiate priming and recognition to familiar and unfamiliar faces. *Electroencephalography and Clinical Neurophysiology, 94*, 41–49.

Bentin, S., & Deouell, L. Y. (2000). Structural encoding and identification in face processing: ERP evidence for separate mechanisms. *Cognitive Neuropsychology, 17*, 35–54.

Bentin, S., & Golland, Y. (2002). Meaningful processing of meaningless stimuli: The influence of perceptual experience on early visual processing of faces. *Cognition, 86*, B1–B14.

Bentin, S., McCarthy, G., Perez, E., Puce, A., & Allison, T. (1996). Electrophysiological studies of face perception in humans. *Journal of Cognitive Neuroscience, 8*, 551–565.

Bentin, S., Sagiv, N., Mecklinger, A., Friederici, A., & von Cramon, Y. D. (2002). Priming visual face-processing mechanisms: Electrophysiological evidence. *Psychological Science, 13*, 190–193.

Blau, V. C., Maurer, U., Tottenham, N., & McCandliss, B. D. (2007). The face-specific N170 component is moduated by emotional facial expression. *Behavioral and Brain Function, 3*, 7.

Bötzel, K., & Grusser, O. J. (1989). Electric brain potentials evoked by pictures of faces and non-faces—a search for face-specific EEG-potentials. *Experimental Brain Research, 77*, 349–360.

Bötzel, K., Schulze, S., & Stodieck, S. R. G. (1995). Scalp topography and analysis of intracranial sources of face-evoked potentials. *Experimental Brain Research, 104*, 135–143.

Boutsen, L., Humphreys, G. W., Praamstra, P., & Warbrick, T. (2006). Comparing neural correlates of configural processing in faces and objects: An ERP study of the Thatcher illusion. *Neuroimage, 32*, 352–367.

Bruce, V., & Young, A. (1986). Understanding face recognition. *British Journal of Psychology, 77*, 305–327.

Bruce, V., & Young, A. W. (1998). *In the eye of the beholder: The science of face perception*. Oxford: Oxford University Press.

Busey, T. A., & Vanderkolk, J. R. (2005). Behavioral and electrophysiological evidence for configural processing in fingerprint experts. *Vision Research, 45*, 431–448.

Caharel, S., d'Arripe, O., Ramon, M., Jacques, C., & Rossion, B. (2009a). Early adaptation to unfamiliar faces across viewpoint changes in the right hemisphere: evidence from the N170 ERP component. *Neuropsychologia*, 47, 639–643.

Caharel, S., Jiang, F., Blanz, V., Rossion, B. (2009b). Recognizing an individual face: 3D shape contributes earlier than 2D surface reflectance information. *NeuroImage, 47*, 1809–1818.

Caharel, S., & Rossion, B. (in press). ERP evidence for the speed of face categorization in the human brain: disentangling the contribution of low-level visual cues from face perception. *Vision Research*.

Caharel, S., Fiori, N., Bernard, C., Lalonde, R., & Rebai, M. (2006). The effects of inversion and eye displacements of familiar and unknown faces on early and late-stage ERPs. *International Journal of Psychophysiology, 62*, 141–151.

Caharel, S., Poiroux, S., Bernard, C., Thibaut, F., Lalonde, R., & Rebai, M. (2002). ERPs associated with familiarity and degree of familiarity during face recognition. *International Journal of Neuroscience, 112*, 1499–1512.

Caldara, R., Rossion, B., Bovet, P., & Hauert, C.A. (2004). Event-related potentials and time course of the "other-race" face classification advantage. *Neuroreport, 15*, 905–910.

Campanella, S., Hanoteau, C., Depy, D., Rossion, B., Bruyer, R., Crommelinck, M., & Guerit, J. M. (2000). Right N170 modulation in a face discrimination task: An account for categorical perception of familiar faces. *Psychophysiology, 37*, 796–806.

Carmel, D., & Bentin, S. (2002). Domain specificity versus expertise: Factors influencing distinct processing of faces. *Cognition, 83*, 1–29.

Clark, V. P., Fan, S., & Hillyard, S. A. (1995). Identification of early visual evoked potential generators by retinotopic and topographic analyses. *Human Brain Mapping, 2*, 170–187.

Conty, L., N'Diaye, K., Tijus, C., & George, N. (2007). When eye creates the contact! ERP evidence for early dissociation between direct and averted gaze motion processing. *Neuropsychologia, 45*, 3024–3037.

Crouzet, S.M., Kirchner, H., Thorpe, S.J., (2010). Fast saccades toward faces: Face detection in just 100 ms. J Vis 10(4): 16, 1–17.

Curran, T., & Hancock, J. (2007). The FN400 indexes familiarity-based recognition of faces. *Neuroimage, 36*, 464–471.

Damasio, A. R., Damasio, H., & Van Hoesen, G. W. (1982). Prosopagnosia: Anatomic basis and behavioral mechanisms. *Neurology, 32*, 331–341.

Dawson, G., Webb, S. J., & McPartland, J. (2005). Understanding the nature of face processing impairment in autism: Insights from behavioral and electrophysiological studies. *Developmental Neuropsychology, 27*, 403–424.

Deffke, I., Sander, T., Heidenreich, J., Sommer, W., Curio, G., Trahms, L., & Lueschow, A. (2007). MEG/EEG sources of the 170-ms response to faces are co-localized in the fusiform gyrus. *Neuroimage, 35*, 1495–1501.

de Haan, M., Pascalis, O., & Johnson, M. H. (2002). Specialization of neural mechanisms underlying face recognition in human infants. *Journal of Cognitive Neuroscience, 14*, 199–209.

Di Russo, F., Martinez, A., Sereno, M. I., Pitzalis, S., & Hillyard, S. A. (2002). Cortical sources of the early components of the visual evoked potential. *Human Brain Mapping, 15*, 95–111.

Diamond, R., & Carey, S. (1986). Why faces are and are not special: An effect of expertise. *Journal of Experimental Psychology: General, 115*, 107–117.

Eimer, M. (1998). Does the face-specific N170 component reflect the activity of a specialized eye processor? *Neuroreport, 9*, 2945–2948.

Eimer, M. (2000a). Event-related brain potentials distinguish processing stages involved in face perception and recognition. *Clinical Neurophysiology, 111*, 694–705.

Eimer, M. (2000b). Effects of face inversion on the structural encoding and recognition of faces: Evidence from event-related brain potentials. *Cognitive Brain Research, 10*, 145–158.

Eimer, M. (2000c). Attentional modulations of event-related brain potentials sensitive to faces. *Cognitive Neuropsychology, 17*, 103–116.

Eimer M., & Holmes, A. (2002). An ERP study on the time course of emotional face processing. *Neuroreport, 13*, 427–431.

Eimer, M., & McCarthy, R. A. (1999). Prosopagnosia and structural encoding of faces: Evidence from event-related potentials. *Neuroreport, 10*, 255–259.

Ellis, H. D., & Young, A. W. (1989). Are faces special? In A. W. Young & H. D. Ellis (Eds.), *Handbook of research on face processing* (pp. 1–26). Amsterdam: Elsevier Science.

Emery, N. J. (2000). The eyes have it: The neuroethology, function and evolution of social gaze. *Neuroscience and Biobehavior, 24*, 581–604.

Ewbank, M. P., Smith, W. A, Hancock, E. R., & Andrews, T. J. (2008). The M170 reflects a viewpoint-dependent representation for both familiar and unfamiliar faces. *Cerebral Cortex, 18*, 364–370.

Fei-Fei, L., Iyer, A., Koch, C., & Perona, P. (2007). What do we perceive in a glance of a real-world scene? *Journal of Vision, 7*, 1–29.

Foxe, J. J., & Simpson, G. V. (2002). Flow of activation from V1 to frontal cortex in humans A framework for defining "early" visual processing. *Experimental Brain Research, 142*, 139–150.

Friston, K. J., Rotshtein, P., Geng, J. J., Sterzer, P., & Henson, R. N. (2006). A critique of functional localisers. *Neuroimage, 30*, 1077–1087.

Gauthier, I., Tarr, M. J., Moylan, J., Skudlarski, P., Gore, J. C., & Anderson, A. W. (2000). The fusiform "face area" is part of a network that processes faces at the individual level. *Journal of Cognitive Neuroscience, 12*, 495–504.

George, N., Evans, J., Fiori, N., Davidoff, J., & Renault, B. (1996). Brain events related to normal and moderately scrambled faces. *Cognitive Brain Research, 4*, 65–76.

George, N., Jemel, B., Fiori, N., Chaby, L., & Renault, B. (2005). Electrophysiological correlates of facial decision: Insights from upright and upside-down Mooney-face perception. *Cognitive Brain Research, 24*, 663–673.

Goffaux, V., Gauthier, I., & Rossion, B. (2003). Spatial scale contribution to early visual differences between face and object processing. *Cognitive Brain Research, 16*, 416–424.

Grave de Peralta Menendez, R., Gonzalez Andino, S., Lantz, G., Michel, C. M., & Landis, T. (2001). Noninvasive localization of electromagnetic epileptic activity. I. Method descriptions and simulations. *Brain Topography, 14*, 131–137.

Grill-Spector, K., & Malach, R. (2001). fMR-adaptation: A tool for studying the functional properties of human cortical neurons. *Acta Psychologica, 107*, 293–321.

Grill-Spector, K., Sayres, R., & Ress, D. (2006). High-resolution imaging reveals highly selective nonface clusters in the fusiform face area. *Nature Neuroscience, 9*, 1177–1185.

Gross, C. G., Rocha-Miranda, C. E., & Bender, D. B. (1972). Visual properties of neurons in inferotemporal cortex of the macaque. *Journal of Neurophysiology, 35*, 96–111.

Guillaume, F., & Tiberghien, G. (2001). An event-related potential study of contextual modifications in a face recognition task. *Neuroreport, 12*, 1209–1216.
Haig, N. D. (1985). How faces differ—a new comparative technique. *Perception, 14*, 601–615.
Halgren, E., Baudena, P., Heit, G., Clarke, J. M., Marinkovic, K., & Clarke, M. (1994). Spatio-temporal stages in face and word processing. I. Depth-recorded potentials in the human occipital, temporal and parietal lobes. *Journal of Physiology, 88*, 1–50.
Halgren, E., Raij, T., Marinkovic, K., Jousmaki, V., & Hari, R. (2000). Cognitive response profile of the human fusiform face area as determined by MEG. *Cerebral Cortex, 10*, 69–81.
Halit, H., de Haan, M., & Johnson, M. H. (2000). Modulation of event-related potentials by prototypical and atypical faces. *Neuroreport, 11*, 1871–1875.
Harris, A., & Nakayama, K. (2007). Rapid face-selective adaptation of an early extrastriate component in MEG. *Cerebral Cortex, 17*, 63–70.
Haxby, J. V., Hoffman, E. A., & Gobbini, I. A. (2000). The distributed neural system for face perception. *Trends in Cognitive Science, 4*, 223–233.
Heisz, J. J., Watter, S., & Shedden, J. A. (2006). Automatic face identity encoding at the N170. *Vision Research, 46*, 4604–4614.
Henson, R. N., Goshen-Gottstein, Y., Ganel, T., Otten, L. J., Quayle, A., & Rugg, M. D. (2003). Electrophysiological and haemodynamic correlates of face perception, recognition and priming. *Cerebral Cortex, 13*, 793–805.
Henson, R. N., Mattout, J., Singh, K. D., Barnes, G. R., Hillebrand, A., & Friston, K. (2007). Population-level inferences for distributed MEG source localisation under multiple constraints: Application to face-evoked fields. *Neuroimage, 38*, 422–438.
Herrmann, M. J., Ehlis, A. C., Ellgring, H., & Fallgatter, A. J. (2005a). Early stages (P100) of face perception in humans as measured with event-related potentials (ERPs). *Journal of Neural Transmission, 112*, 1073–1081.
Herrmann, M. J., Ehlis, A. C., Muehlberger, A., & Fallgatter, A. J. (2005b). Source localization of early stages of face processing. *Brain Topography, 18*, 77–85.
Huddy, V., Schweinberger, S. R., Jentzsch, I., & Burton, A. M. (2003). Matching faces for semantic information and names: An event-related brain potentials study. *Cognitive Brain Research, 17*, 314–326.
Itier, R. J., Alain, C., Sedore, K., & McIntosh, A. R. (2007). Early face processing specificity: It's in the eyes! *Journal of Cognitive Neuroscience, 19*, 1815–1826.
Itier, R. J., & Taylor, M. J. (2002). Inversion and contrast polarity reversal affect both encoding and recognition processes of unfamiliar faces: A repetition study using ERPs. *Neuroimage, 15*, 353–372.
Itier, R. J., & Taylor, M. J. (2004a). N170 or N1? Spatiotemporal differences between object and face processing using ERPs. *Cerebral Cortex, 14*, 132–142.
Itier, R. J., & Taylor, M. J. (2004b). Source analysis of the N170 to faces and objects. *Neuroreport, 15*, 1261–1265.
Jacques, C., d'Arripe, O., & Rossion, B. (2007). The time course of the inversion effect during individual face discrimination. *Journal of Vision, 7*(8), 3, 1–9.
Jacques, C., & Rossion, B. (2004). Concurrent processing reveals competition between visual representations of faces. *Neuroreport, 15*, 2417–2421.
Jacques, C., & Rossion, B. (2006). The speed of individual face categorization. *Psychological Science, 17*, 485–492.
Jacques, C., & Rossion, B. (2007a). Electrophysiological evidence for temporal dissociation between spatial attention and sensory competition during human face processing. *Cerebral Cortex, 17*, 1055–1065.
Jacques, C., & Rossion, B. (2007b). Early electrophysiological responses to multiple face orientations correlate with individual discrimination performance in humans. *Neuroimage, 36*, 863–876.
Jacques, C., Rossion, B. (2009). The initial representation of individual faces in the right occipito-temporal cortex is holistic: electrophysiological evidence from the composite face illusion. *Journal of Vision, 9*(6):8, 1–16, http://journalofvision.org/9/6/8/
Jeffreys, D. A. (1989). A face-responsive potential recorded from the human scalp. *Experimental Brain Research, 78*, 193–202.
Jeffreys, D. A. (1993). The influence of stimulus orientation on the vertex positive scalp potential evoked by faces. *Experimental Brain Research, 96*, 163–172.
Jeffreys, D. A. (1996). Evoked potential studies of face and object processing. *Visual Cognition, 3*, 1–38.
Jeffreys, D. A., Tukmachi, E. S. A., & Rockley, G. (1992). Evoked-potential evidence for human brain mechanisms that respond to single, fixated faces. *Experimental Brain Research, 91*, 351–362.
Jemel, B., Calabria, M., Delvenne, J. F., Crommelinck, M., & Bruyer, R. (2003a). Differential involvement of episodic and face representations in ERP repetition effects. *Neuroreport, 14*, 525–530.
Jemel, B., Pisani, M., Calabria, M., Crommelinck, M., & Bruyer, R. (2003b). Is the N170 for faces cognitively penetrable? Evidence from repetition priming of Mooney faces of familiar and unfamiliar persons. *Cognitive Brain Research, 17*, 431–446.
Jemel, B., Pisani, M., Rousselle, L., Crommelinck, M., & Bruyer, R. (2005). Exploring the functional architecture of person recognition system with event-related potentials in a within- and cross-domain self-priming of faces. *Neuropsychologia, 43*, 2024–2040.
Jemel, B., Schuller, A. M., Cheref-Khan, Y., Goffaux, V., Crommelinck, M., & Bruyer, R. (2003c). Stepwise emergence of the face-sensitive N170 event-related potential component. *Neuroreport, 14*, 2035–2039.
Joyce, C., & Kutas, M. (2005). Event-related potential correlates of long-term memory for briefly presented faces. *Journal of Cognitive Neuroscience, 17*, 757–767.
Joyce, C., & Rossion, B. (2005). The face-sensitive N170 and VPP components manifest the same brain processes: The effect of reference electrode site. *Clinical Neurophysiology, 116*, 2613–2631.
Kanwisher, N. (2000). Domain-specificity in face perception. *Nature Neuroscience, 3*, 758–763.
Kanwisher, N., McDermott, J & Chun, M.M. (1997). The fusiform face area a module in human extrastriate cortex specialized for face perception. *Journal of Neuroscience, 17*, 4302–4311.
Kanwisher, N., Tong, F., & Nakayama, K. (1998). The effect of face inversion on the human fusiform face area. *Cognition, 68*, B1–B11.
Kiani, R., Esteky, H., & Tanaka, K. (2005). Differences in onset latency of macaque inferotemporal neural responses to primate and non-primate faces. *Journal of Neurophysiology, 94*, 1587–1596.

Kirchner, H., & Thorpe, S. J. (2006). Ultra-rapid object detection with saccadic eye movements: Visual processing speed revisited. *Vision Research, 46*, 1762–1776.

Kloth, N., Dobel, C., Schweinberger, S. R., Zwitserlood, P., Bolte, J., & Junghofer, M. (2006). Effects of personal familiarity on early neuromagnetic correlates of face perception. *European Journal of Neuroscience, 24*, 3317–3321.

Kovacs, G., Zimmer, M., Banko, E., Harza, I., Antal, A., & Vidnyanszky, Z. (2006). Electrophysiological correlates of visual adaptation to faces and body parts in humans. *Cerebral Cortex, 16*, 742–753.

Kuefner, D., de Heering, A., Jacques, C., Palmero-Soler, E., Rossion, B. (2010). Early visually evoked electrophysiological responses over the human brain (P1, N170) show stable patterns of face-sensitivity from 4 years to adulthood. *Frontiers in Human Neuroscience. 3:67*. doi: 10.3389/neuro.09.067.2009

Latinus, M., & Taylor, M. J. (2005). Holistic processing of faces: Learning effects with Mooney faces. *Journal of Cognitive Neuroscience, 17*, 1316–1327.

Leopold, D. A., Rhodes, G., Müller, K. M., & Jeffery, L. (2005). The dynamics of visual adaptation to faces. *Proceedings of the Royal Society B , 272*, 897–904.

Letourneau, S. M., & Mitchell, T. V. (2008). Behavioral and ERP measures of holistic face processing in a composite task. *Brain and Cognition, 67*, 234–245.

Lewis, M. B., & Edmonds, A. J. (2003). Face detection: Mapping human performance. *Perception, 32*, 903–920.

Linkenkaer-Hansen, K., Palva, J. M., Sams, M., Hietanen, J. K., Aronen, H. J., & Ilmoniemi, R. J. (1998). Face-selective processing in human extrastriate cortex around 120 ms after stimulus onset revealed by magneto- and electroencephalography. *Neuroscience Letters, 253*, 147–150.

Liu, J., Harris, A., & Kanwisher, N. (2002). Stages of processing in face perception: An MEG study. *Nature Neuroscience, 5*, 910–916.

Liu, J., Higuchi, M., Marantz, A., & Kanwisher, N. (2000). The selectivity of the occipitotemporal M170 for faces. *Neuroreport, 11*, 337–341.

Liu, L., & Ioannides, A. A. (2006). Spatiotemporal dynamics and connectivity pattern differences between centrally and peripherally presented faces. *Neuroimage, 31*, 1726–1740.

Makeig, S., Westerfield, M., Jung, T. P., Enghoff, S., Townsend, J., Courchesne, E., & Sejnowski, T. J. (2002). Dynamic brain sources of visual evoked responses. *Science, 295*, 690–694.

Matsumoto, N., Okada, M., Sugase-Miyamoto, Y., Yamane, S., & Kawano, K. (2005). Population dynamics of face-responsive neurons in the inferior temporal cortex. *Cerebral Cortex, 15*, 1103–1112.

Mazaheri, A., & Jensen, O. (2005). Posterior alpha activity is not phase-reset by visual stimuli. *Proceedings of the National Academy of Sciences USA, 109*, 2948–2952.

Mnatsakanian, E. V., & Tarkka, I. M. (2004). Influence of the background on the recognition and comparison of targets: Behavioral and ERP study. *International Journal of Psychophysiology, 54*, 156–156.

Mouchetant-Rostaing, Y, & Giard, M. H. (2003). Electrophysiological correlates of age and gender perception on human faces. *Journal of Cognitive Neuroscience, 15*, 900–910.

Paller, K. A., Gonsalves, B., Grabowecky, M., Bozic, V. S., & Yamada, S. (2000). Electrophysiological correlates of recollecting faces of known and unknown individuals. *Neuroimage, 11*, 98–110.

Paras, C., McDermott, K., Webster, S., & Webster, M. (2007). Stimulus requirements for perceiving a face: An analysis based on "totem poles." *Journal of Vision, 7*, 497.

Perrett, D. I., Hietanen, J. K., Oram, M. W., & Benson, P. J. (1992). Organization and functions of cells responsive to faces in the temporal cortex. *Philosophical Transactions of the Royal Society B, 335*, 23–30.

Perrett, D. I., Oram, M. W., & Ashbridge, E. (1998). Evidence accumulation in cell populations responsive to faces: An account of generalisation of recognition without mental transformations. *Cognition, 67*, 111–145.

Philiastides, M. G., & Sajda, P. (2006). Temporal characterization of the neural correlates of perceptual decision making in the human brain. *Cerebral Cortex, 16*, 509–518.

Pizzagalli, D. A., Lehmann, D., Hendrick, A. M., Regard, M., Pascual-Marqui, R. D., & Davidson, R. J. (2002). Affective judgments of faces modulate early activity (approximately 160 ms) within the fusiform gyri. *Neuroimage, 16*, 663–677.

Purcell D. G., & Stewart A. L. (1988). The face-detection effect: configuration enhances detection. *Perception & Psychophysics, 43*, 35–66.

Regan, D. (1989). *Human brain electrophysiology: Evoked potentials and evoked magnetic fields in science and medicine.* New York: Elsevier.

Rossion, B., & Caharel, S. (2011). ERP evidence for the speed of face categorization in the human brain: Disentangling the contribution of low-level visual cues from face perception. *Vision Research, 51*, 1297–1311.

Rossion, B., Campanella, S., Gomez, C. M., Delinte, A., Debatisse, D., Liard, L., Dubois, S., Bruyer, R., Crommelinck, M., & Guerit, J. M. (1999a). Task modulation of brain activity related to familiar and unfamiliar face processing: An ERP study. *Clinical Neurophysiology, 110*, 449–462.

Rossion, B., Collins, D., Goffaux, V., & Curran, T. (2007). Long-term expertise with artificial objects increases visual competition with early face categorization processes. *Journal of Cognitive Neuroscience, 19*, 543–555.

Rossion, B., Delvenne, J. F., Debatisse, D., Goffaux, V., Bruyer, R., Crommelinck, M., & Guerit, J. M. (1999b). Spatio-temporal localization of the face inversion effect: An event-related potentials study. *Biological Psychology, 50*, 173–189.

Rossion, B., & Gauthier, I. (2002). How does the brain process upright and inverted faces? *Behavioral and Cognitive Neuroscience Review, 1*, 62–74.

Rossion, B., Gauthier, I., Tarr, M. J., Despland, P., Bruyer, R., Linotte, S., & Crommelinck, M. (2000). The N170 occipito-temporal component is delayed and enhanced to inverted faces but not to inverted objects: An electrophysiological account of face-specific processes in the human brain. *Neuroreport, 11*, 69–74.

Rossion, B., & Jacques, C. (2008). Does physical interstimulus variance account for early electrophysiological face preferential responses in the human brain? Ten lessons on the N170. *Neuroimage, 39*, 1959–1979.

Rossion, B., Joyce, C. A., Cottrell, G. W., & Tarr, M. J. (2003). Early lateralization and orientation tuning for face, word, and object processing in the visual cortex. *Neuroimage, 20*, 1609–1624.

Rossion, B., Kung, C. C., & Tarr, M. J. (2004). Visual expertise with nonface objects leads to competition with the early perceptual processing of faces in the human occipitotemporal cortex. *Proceedings of the National Academy of Sciences USA, 101*, 14521–14526.

Rousselet, G. A., Husk, J. S., Bennett, P. J., & Sekuler, A. B. (2005). Spatial scaling factors explain eccentricity effects on face ERPs. *Journal of Vision, 5*, 755–763.

Rousselet, G. A., Husk, J. S., Bennett, P. J., & Sekuler, A. B. (2007). Single-trial EEG dynamics of object and face visual processing. *Neuroimage, 36*, 843–862.

Rousselet, G. A., Mace, M. J. M., & Fabre-Thorpe, M. (2003). Is it an animal? Is it a human face? Fast processing in upright and inverted natural scenes. *Journal of Vision, 3*, 440–455.

Rousselet, G. A., Mace, M. J. M., & Fabre-Thorpe, M. (2004a). Animal and human faces in natural scenes: How specific to human faces is the N170 ERP component? *Journal of Vision, 4*, 13–21.

Rousselet, G. A., Mace, M. J. M., & Fabre-Thorpe, M. (2004b). Spatiotemporal analyses of the N170 for human faces, animal faces and objects in natural scenes. *Neuroreport, 15*, 2607–2611.

Sagiv, N., & Bentin, S. (2001). Structural encoding of human and schematic faces: Holistic and part-based processes. *Journal of Cognitive Neuroscience, 13*, 937–951.

Sams, M., Hietanen, J. K., Hari, R., Ilmoniemi, R. J., & Lounasmaa, O. V. (1997). Face-specific responses from the human inferior occipito-temporal cortex. *Neuroscience, 77*, 49–55.

Sauseng, P., Klimesch, W., Gruber, W. R., Hanslmayr, S., Freunberger, R., & Doppelmayr, M., 2007. Are event-related potential components generated by phase resetting of brain oscillations? A critical discussion. *Neuroscience, 146*, 1435–1444.

Saxe, R., Brett, M., & Kanwisher, N. (2006). Divide and conquer: A defense of functional localizers. *Neuroimage, 30*, 1088–1096.

Sayers, B. M., & Beagley, H. A. (1974). Objective evaluation of auditory evoked EEG responses. *Nature, 251*, 608–609.

Scherg, M., & Berg, P. (1991). Use of prior knowledge in brain electromagnetic source analysis. *Brain Topography, 4*, 143–150.

Schneider, B. L., DeLong, J. E., & Busey, T. A. (2007). Added noise affects the neural correlates of upright and inverted faces differently. *Journal of Vision, 7*, 4.

Schroeder, C. E., Molholm, S., Lakatos, P., Ritter, W., & Foxe, J. J. (2004). Human–simian correspondence in the early cortical processing of multisensory cues. *Cognitive Processes, 5*, 140–151.

Schweinberger, S. R., Huddy, V., & Burton, A. M. (2004). N250r: A face-selective brain response to stimulus repetitions. *Neuroreport, 15*, 1501–1505.

Schweinberger, S. R., Pfutze, E. M., & Sommer, W. (1995). Repetition priming and associative priming of face recognition—evidence from event-related potentials. *Journal of Experimental Psycholoy: Learning, Memory and Cognition, 21*, 722–736.

Schweinberger, S. R., Pickering, E. C., Burton, A. M., & Kaufmann, J. M. (2002a). Human brain potential correlates of repetition priming in face and name recognition. *Neuropsychologia, 40*, 2057–2073.

Schweinberger, S. R., Pickering, E. C., Jentzsch, I., Burton, A. M., & Kaufmann, J. M. (2002b). Event-related brain potential evidence for a response of inferior temporal cortex to familiar face repetitions. *Cognitive Brain Research, 14*, 398–409.

Schyns, P. G., Jentzsch, I., Johnson, M., Schweinberger, S. R., & Gosselin, F. (2003). A principled method for determining the functionality of brain responses. *Neuroreport, 14*, 1665–1669.

Schyns, P. G., Petro, L. S., & Smith, M. L. (2007). Dynamics of visual information integration in the brain for categorizing facial expressions. *Current Biology, 17*, 1580–1585.

Seeck, M., & Grüsser, O. J. (1992). Category-related components in visual evoked-potentials— photographs of faces, persons, flowers and tools as stimuli. *Experimental Brain Research, 92*, 338–349.

Sergent, J. (1984). An investigation into component and configural processes underlying face perception. *British Journal of Psychology, 75*, 221–242.

Sergent, J., Ohta, S., & Macdonald, B. (1992). Functional neuroanatomy of face and object processing—a positron emission tomography study. *Brain, 115*, 15–36.

Shibata, T., Nishijo, H., Tamura, R., Miyamoto, K., Eifuku, S., Endo, S., & Ono, T. (2002). Generators of visual evoked potentials for faces and eyes in the human brain as determined by dipole localization. *Brain Topography, 15*, 51–63.

Sreenivasan, K. K., Katz, J., & Jha, A. P. (2007). Temporal characteristics of top-down modulations during working memory maintenance: An event-related potential study of the N170 component. *Journal of Cognitive Neuroscience, 19*, 1836–1844.

Sugase, Y., Yamane, S., Ueno, S., & Kawano, K. (1999). Global and fine information coded by single neurons in the temporal visual cortex. *Nature, 400*, 869–873.

Swithenby, S. J., Bailey, A. J., Brautigam, S., Josephs, O. E., Jousmaki, V., & Tesche, C. D. (1998). Neural processing of human faces: A magnetoencephalographic study. *Experimental Brain Research, 118*, 501–510.

Tanaka, J. W., & Curran, T. (2001). A neural basis for expert object recognition. *Psychological Science, 12*, 43–47.

Tanaka, J. W., Curran, T., Porterfield, A. L., & Collins, D. (2006). Activation of preexisting and acquired face representations: The N250 event-related potential as an index of face familiarity. *Journal of Cognitive Neuroscience, 18*, 1488–1497.

Tanaka, K. (1996). Inferotemporal cortex and object vision. *Annual Review of Neuroscience, 19*, 109–139.

Tanskanen, T., Nasanen, R., Montez, T., Paallysaho, J., & Hari, R. (2005). Face recognition and cortical responses show similar sensitivity to noise spatial frequency. *Cerebral Cortex, 15*, 526–534.

Tarkiainen, A., Cornelissen, P. L., & Salmelin, R. (2002). Dynamics of visual feature analysis and of object-level processing in face versus letter-string perception. *Brain, 125*, 1125–1136.

Tarr, M. J., & Gauthier, I. (2000). FFA: A flexible fusiform area for subordinate-level visual processing automatized by expertise. *Nature Neuroscience, 3*, 764–769.

Taylor, M. J., Itier, R. J., Allison, T., & Edmonds, G. E. (2001). Direction of gaze effects on early face processing: Eyes-only versus full faces. *Cognitive Brain Research, 10*, 333–340.

Taylor, M. J., McCarthy, G., Saliba, E., & Degiovanni, E. (1999). ERP evidence of developmental changes in processing of faces. *Clinical Neurophysiology, 110*, 910–915.

Thorpe, S., Fize, D., & Marlot, C. (1996). Speed of processing in the human visual system. *Nature, 381*, 520–522.

Tong, F., & Nakayama, K. (1999). Robust representations for faces: Evidence from visual search. *Journal of Experimental Psychology: Human Perception and Performance*, 25, 1016–1035.

Tovee, M., & Rolls, E. T. (1995). Information encoding in short firing rate epochs by single neurons in the primate temporal visual cortex. *Visual Cognition, 2*, 35–58.

Tsao, D. Y., Freiwald, W. A., Tootell, R. B. H., & Livingstone, M. S. (2006). A cortical region consisting entirely of face-selective cells. *Science, 311*, 670–674.

Vanni, S., Warnking, J., Dojat, M., Delon-Martin, C., Bullier, J., & Segebarth, C. (2004). Sequence of pattern onset responses in the human visual areas: An fMRI constrained VEP source analysis. *Neuroimage, 21*, 801–817.

VanRullen, R., & Thorpe, S. J. (2001). The time course of visual processing: From early perception to decision-making. *Journal of Cognitive Neuroscience, 13*, 454–461.

Vogel, E. K., & Luck, S. J. (2000). The visual N1 component as an index of a discrimination process. *Psychophysiology, 37*, 190–203.

Wang, G., Tanaka, K. & Tanifuji, M. (1996). Optical imaging of functional organization in the monkey inferotemporal cortex. *Science, 272*, 1665–1668.

Wang, Y., Fujita, I., & Murayama, Y. (2000). Neuronal mechanisms of selectivity for object features revealed by blocking inhibition in inferotemporal cortex. *Nature Neuroscience, 3*, 807–813.

Watanabe, S., Kakigi, R., Koyama, S., & Kirino, E. (1999). Human face perception traced by magneto- and electro-encephalography. *Cognitive Brain Research, 8*, 125–142.

Watanabe, S., Kakigi, R., & Puce, A. (2001). Occipitotemporal activity elicited by viewing eye movements: A magnetoencephalographic study. *Neuroimage, 13*, 351–363.

Watanabe, S., Kakigi, R., & Puce, A. (2003). The spatiotemporal dynamics of the face inversion effect: A magneto- and electro-encephalographic study. *Neuroscience, 116*, 879–895.

Young, A. W., Hay, D. C., McWeeny, K. H., Ellis, A. W., & Barry, C. (1985). Familiarity decisions for faces presented to the left and right cerebral hemispheres. *Brain and Cognition, 4*, 439–450.

Young, A. W., Hellawell, D., & Hay, D. C. (1987). Configurational information in face perception. *Perception, 16*, 747–759.

Yovel, G., & Paller, K. A. (2004). The neural basis of the butcher-on-the-bus phenomenon: When a face seems familiar but is not remembered. *Neuroimage, 21*, 789–800.

CHAPTER

6 The Mismatch Negativity (MMN)

Risto Näätänen *and* Kairi Kreegipuu

Abstract

The auditory mismatch negativity (MMN) is a change-specific component of the auditory event-related brain potential (ERP) that is elicited even in the absence of attention and can be used as an objective index of sound-discrimination accuracy and auditory sensory memory. The MMN enables one to reach a new level of understanding of the brain processes forming the biological substrate of central auditory perception and the different forms of auditory memory. A review of MMN studies indicates that the central auditory system performs complex cognitive operations, such as generalization leading to simple concept formation (e.g., a rising pair irrespective of the specific frequency values), rule extraction, and the anticipation of the next stimulus at the preattentive level. These findings demonstrate the presence of a cognitive change-detection mechanism in the auditory cortex.

Keywords: mismatch negativity (MMN), auditory event-related potential, sound discrimination, auditory sensory memory

The Mismatch Negativity (MMN): An Introduction

The auditory mismatch negativity (MMN; Näätänen, 1979; Näätänen & Michie, 1979; Näätänen et al., 1978; for a review see Näätänen et al., 2007) is a change-specific component of the auditory event-related brain potential (ERP) that is elicited even in the absence of attention and can be used as an objective index of sound-discrimination accuracy and auditory sensory memory. The MMN has therefore opened an unprecedented window to central auditory processing and the underlying neurophysiology, affected in a large number of different clinical conditions. The MMN enables one to reach a new level of understanding of the brain processes forming the biological substrate of central auditory perception and the different forms of auditory memory (for a review, see Näätänen & Kreegipuu, 2010). Furthermore, the MMN sheds new light on brain mechanisms controlling the access of auditory sensory input to conscious perception and higher forms of memory. In addition, it can be recorded with a multitude of brain-research methods, including magnetoencephalography (MEG; Hari et al., 1984), optical imaging (OI; Rinne et al., 1999; Tse et al., 2006), functional magnetic resonance imaging (fMRI; Celsis et al., 1999; Molholm et al., 2005; Opitz et al., 2002), positron emission tomography (PET; Müller et al., 2002; Tervaniemi et al., 2000), and intracranial recording techniques (Halgren et al., 1995; Kropotov et al., 1995, 2000).

The "traditional" MMN is generated by the brain's automatic response to any change in auditory stimulation exceeding a certain limit roughly corresponding to the behavioral discrimination threshold (see Figure 6.1). In a typical experimental situation, the subject is reading a book or watching (silenced) videos while a sequence of sound stimuli (e.g., tones of 1000 Hz) are presented, for example, at 1 s stimulus onset asynchronies (SOA) with an instruction to ignore the stimuli. These "standard" stimuli are occasionally interspersed with "deviant" stimuli (e.g., tones of 1100 Hz). The standard stimuli elicit a typical ERP with the predominant N1 (peaking at about 100 ms from stimulus onset) and P2 (180–200 ms) components, whereas in ERPs to deviant stimuli there is an additional negative component, the MMN. The MMN typically has a frontocentral scalp distribution, commences at the peak of the N1 wave, and overlaps the P2 component, typically lasting for approximately 100 to 250 ms poststimulus. The MMN is also elicited by other kinds of change, such as those in intensity (Näätänen et al., 1988), duration (Näätänen et al., 1989, 2004b), locus of origin (Paavilainen et al., 1989), and timbre (Toiviainen et al., 1998). In addition, phonetic changes elicit the MMN (Dehaene-Lambertz et al., 1997). Moreover, the MMN is elicited by different kinds of abstract changes in auditory stimulation such as grammar violations in mother-tongue sentences. In addition, an analogous response occurs in other sensory modalities. In the somatosensory modality, for example, Kekoni et al. (1997) obtained an MMN to infrequent changes on the locus of stimulation on the skin. This MMN reached its maximal amplitude over the portion of somatosensory cortex that represents the hand, contralateral to the point of stimulation on the skin. Moreover, Akatsuka and colleagues, using the somatosensory MMN, objectively determined the temporal (Akatsuka et al., 2005) and spatial (Akatsuka et al., 2007) resolution of the somatosensory modality. An analogous response has also been found in vision (Alho et al., 1992; Heslenfeld, 2003; Kremlàček et al., 2006; Tales et al., 1999) and olfaction (Krauel et al., 1999).

Importantly, the MMN is elicited irrespective of the subject or patient's direction of attention (Näätänen, 1979; Näätänen et al., 1978) even though strong focusing of attention to a concurrent auditory stimulus stream may attenuate its amplitude (Woldorff et al., 1991). Hence, no behavioral task is needed; such tasks are used instead to direct the subject's attention away from the MMN-eliciting stimulus sequence in order to prevent the elicitation of attention-dependent ERP components.

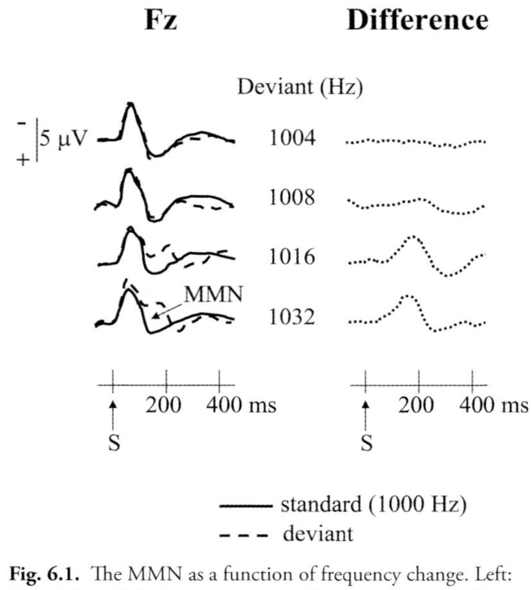

Fig. 6.1. The MMN as a function of frequency change. Left: frontal (Fz) ERPs (averaged across subjects reading a book) to randomized 1000 Hz standard (80%, solid line) and deviant (20%, broken line) stimuli of different frequencies, as indicated in the center. Right: difference waves obtained by subtracting the standard-stimulus ERP from the deviant-stimulus ERP for the different deviant stimuli. Data from Sams et al. (1985).

How Is MMN Defined and Isolated from Other Overlapping Components?

The MMN is often visualized in a difference wave obtained by subtracting the ERP to frequent, standard stimuli from that to deviant stimuli. In these difference waves, the MMN response is seen as a negative displacement usually peaking 150–250 ms from the onset of the deviant stimulus (or, for duration deviants, from deviance onset), in particular at the frontocentral and central scalp electrodes (relative to a mastoid or nose reference electrode). One must, however, take into account the possible differences in the obligatory ERPs between standards and deviants caused by physical stimulus differences between the two stimuli and those in the refractoriness of the neural populations activated by the two stimuli because of the probability difference. Therefore, the difference wave is often formed by subtracting the ERP to a given stimulus when it is presented as a standard in one stimulus block from that elicited by the same stimulus when it is presented as a deviant in another block. These differences in the obligatory components are, however, usually rather small in amplitude and mainly involve the N1 time zone. Therefore, post-N1 measurements usually provide quite reliable estimates of the "genuine" MMN, provided that deviants do not elicit the N2b (Näätänen et al., 1982; Renault & Lesévre, 1978, 1979; for a review, see Näätänen & Gaillard, 1983) that is typically elicited by deviants in the attended input stream irrespective of their target/nontarget status. Hence, N2b contamination can be avoided by drawing the subject's attention away from auditory stimuli to some interesting primary task, such as watching exciting videos; therefore, as the MMN is still elicited, it seems that MMN measurement accuracy benefits by withdrawing attention. Another way to avoid N2b contamination is to measure the MMN amplitude from the mastoid recordings (when the nose is used as a reference) where the MMN reverses its polarity and the N2b is missing. Furthermore, MEG recordings permit one to measure the MMNm (the MEG equivalent of the MMN) in the absence of N2b contamination, as there is no N2b equivalent in the MEG.

The prerequisite of MMN elicitation is that before the occurrence of the deviant stimulus, the central auditory system has been able to form a representation of the repetitive aspects of auditory stimulation (for a review, see Näätänen & Winkler, 1999). An MMN is then elicited by a stimulus that violates this representation. Most studies have employed simple paradigms (e.g., a deviant tone of 1100 Hz against the background of repetitive 1000 Hz standard tones). The MMN can, however, also be elicited by changes in complex stimuli such as speech sounds (Dehaene-Lambertz, 1997; Näätänen et al., 1997) and even by stimuli that deviate from an abstract rule followed by the ongoing auditory stimulation, such as a tone repetition in a sequence of descending tones with no constant standard stimulus (Tervaniemi et al., 1994).

The MMN Subcomponents and Generators

The MMN gets a contribution from at least two intracranial processes: (1) a bilateral supratemporal process generating the supratemporal MMN subcomponent (and the polarity-reversed "MMN" in nose-referenced mastoid recordings) and (2) a predominantly right-hemispheric frontal process generating the frontal MMN subcomponent (Baldeweg et al., 1999; Deouell, 2007; Giard et al., 1990; Näätänen et al., 1978; Rinne et al., 2000). The supratemporal component is, presumably, associated with preperceptual change detection, occurring even in the comatose state when the patient will later regain consciousness (Fischer et al., 2006). The frontal component, in turn, appears to be related to the initiation (*call*; Öhman, 1979) of involuntary attention switch caused by an auditory change (Escera et al., 1998; Giard et al., 1990; Näätänen & Michie, 1979; Näätänen et al., 1978; Rinne et al., 2000; Schröger, 1996). The MMN generators reflect the nature of the stimulus (e.g., they usually are left lateralized for language stimuli; Näätänen et al., 1997; Pulvermüller et al., 2003; Shestakova et al., 2002) and are located anteriorly to those of the N1 (Rosburg, 2003; Rosburg et al., 2004a, 2004b). The separability of the two MMN subcomponents can be demonstrated through their differential sensitivity to experimental manipulation. For example, ethanol dramatically reduced the frontally recorded MMN, which is a composite of the auditory- and frontal-cortex processes, whereas it did not affect the polarity-reversed mastoid recordings indexing (most of) the supratemporal component (Jääskeläinen et al., 1996b). Interestingly, in a following study (Jääskeläinen et al., 1996a) ethanol also abolished the hit rate decrease in a visual target–detection task caused by blocking the path of auditory distraction to frontal mechanisms of involuntary attention switching.

Additional evidence for different generators behind the MMN is provided by the fact that different types of auditory change activate at least partially different neural populations in the auditory cortex. This was previously suggested by Paavilainen et al. (1991), who found different polarity-reversal ratios for the frequency, duration, and intensity MMNs. Subsequently, studies using dipole modeling of the MMN and MMNm sources reported differences of a few millimeters in location and/or in orientation between the sources of the MMN responses to intensity, frequency, interstimulus interval (ISI), or duration changes (Deouell & Bentin, 1998; Deouell et al., 1998; Doeller et al., 2003; Frodl-Bauch et al., 1997; Giard et al., 1995; Rosburg, 2003). Furthermore, a recent fMRI study (Molholm et al., 2005) found that frequency and duration changes activate different areas in both the supratemporal and frontal cortices.

Memory Dependence of MMN Elicitation

The MMN depends on the presence of a short-term memory trace in the auditory cortex representing the repetitive aspects of the preceding auditory events and lasting for a few seconds. Therefore, rather than eliciting the MMN, a sound presented after a long silence elicits enhanced obligatory P1, N1, and P2 responses (Korzyukov et al., 1999). Hence, the MMN seen in the difference wave delineated by subtracting the standard-stimulus ERP from that to the deviant stimulus is not, as already discussed, simply due to a larger exogenous response to an infrequent than frequent stimulus even though the early part of the difference wave may, depending on the magnitude and nature of the stimulus difference, be modulated by the enhancements of the generator processes of the temporally overlapping exogenous components N1 and P2 (for a review, see Näätänen et al., 2005).

The memory-trace interpretation is, in particular, supported by fact that an MMN is elicited by a decrement of the ISI (Ford & Hillyard, 1981; Nordby et al., 1988; Rüsseler et al., 2001). This large, long-duration MMN cannot be explained by the activation of an additional pool of afferent neurons, because a shorter ISI should lead to a reduced sensory response. In the same vein, such a difference in the activation of neurons cannot account for the MMN elicited by stimulus omission in a stimulus sequence presented at a very fast constant rate (Rüsseler et al., 2001; Yabe et al., 1997, 1998). Nor can it account for the fact that the MMN peak latency strongly depends on the magnitude of the difference between the deviant and standard stimuli (Näätänen et al., 1989; Tiitinen et al., 1994; see Figure 6.2). The sensory information carried by the sensory-memory traces underlying the MMN generation indeed corresponds to sound perception and memory (and thus provides the central sound representation) rather than just to the acoustic elements composing the stimulus (Näätänen & Winkler, 1999).

As already mentioned, memory traces of relatively short duration are involved in MMN generation. In young subjects, these traces usually last for about 5–10 s, judging from the ISIs that still permit MMN elicitation (Böttscher-Gandor & Ullsperger, 1992; Näätänen et al., 1987; Sams et al., 1993), but the trace duration gets shorter with aging (Jääskeläinen et al., 1999; Pekkonen et al., 1996). Furthermore, this age-dependent shortening of the memory trace is increased by chronic alcoholism (Polo et al., 1999) and, in particular, by neurodegenerative brain diseases such as Alzheimer's disease (Pekkonen et al., 1994).

Deviant-Stimulus Probability

The MMN amplitude is decreased with an increasing deviant-stimulus probability (Haenschel et al., 2005; Ritter et al., 1992; Sabri & Campbell, 2001; Sato et al., 2000). This is partially due to the fact that the standard stimulus is often replaced by deviant stimuli (being then unable to contribute to memory trace strength) but with shorter deviant-stimulus intervals, these stimuli may also develop a trace of their own, which in turn might inhibit the MMN to the initial standard (Ritter et al., 1992; Rosburg, 2004; Rosburg et al., 2004a, 2004b; Sams et al., 1984). In fact, Näätänen (1984) explained the MMN phenomenon in terms of the input from the eliciting deviant stimulus "starting" to develop the representation of its own in the auditory sensory-memory system already engaged by the representation of a different stimulus (the standard).

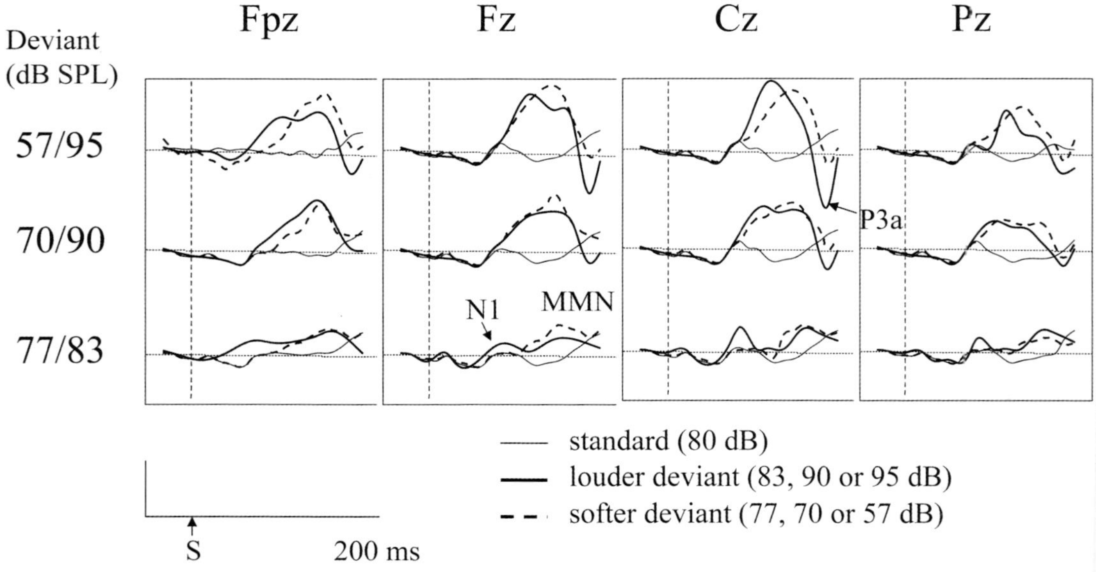

Fig. 6.2. The MMN to intensity decrement and increment. Frontal pole (Fpz), frontal (Fz), central (Cz), and parietal (Pz) difference waves obtained by subtracting the ERP elicited by the 80 dB standard tone (thin line) from those elicited by deviant tones of six different intensity levels presented in separate blocks (grand averaged data). Solid lines represent louder deviants and broken lines softer deviants. The difference from the standard signal was large (57 or 95 dB, upper panel), moderate (70 or 90 dB, middle panel), or subtle (77 or 83 dB, lower panel). It can be seen that whereas the N1 amplitude is increased with an increasing deviant-tone intensity, the MMN amplitude is increased and the MMN peak latency is decreased with the increasing magnitude of deviation from the standard in either direction. Data from Näätänen (1992).

Automaticity of MMN Elicitation

As already mentioned, MMN generation is an automatic brain process in the sense that its occurrence does not depend on attention (Alain et al., 1994; Alho et al., 1989; Näätänen & Michie, 1979; see also Muller-Gass et al., 2005), even though strong focusing of attention to a concurrent stimulus stream may attenuate its amplitude (Alain & Izenberg, 2003; Alain & Woods, 1994, 1997; Muller-Gass et al., 2005; Näätänen et al., 1993a; Otten et al., 2000; Trejo et al., 1995; Woldorff et al., 1991, 1998). Arnott and Alain (2002) and Woldorff and colleagues (1991) even claimed that the MMN was totally abolished by the very strong focusing of attention on another auditory input stream, but the data presented suggest that a small residual MMN may nevertheless have remained (for a discussion, see Näätänen, 1991; see also Alain & Izenberg, 2003). Futhermore, in a number of studies, attention to another auditory stimulus stream had no influence on MMN amplitude (e.g., Alho et al., 1989; Kaukoranta et al., 1989; Lounasmaa et al., 1989; Näätänen et al., 1978, 1982, 1989; Paavilainen et al., 1993). In addition, numerous studies also varied the visual task load but usually found no attentional modulation of MMN amplitude (e.g., Alho et al., 1992, 1994; Dittmann-Balcar et al., 1999; Dyson et al., 2005; Harmony et al., 2000; Kathmann et al., 1999; Müller et al., 2002; Muller-Gass et al., 2005, 2006; Otten et al., 2000; Takegata et al., 2005; Winkler et al., 2005; see, however, Kramer et al., 1995; Woods et al., 1992; Yucel et al., 2005a, 2005b). Even opposite results were reported: Zhang et al. (2006) found that with an increasing visual attentional load, the MMN amplitude for a frequency change in irrelevant background stimulation became larger (whereas the amplitude of the subsequent P3a became smaller, indicating decreased involuntary attention switching with increased visual task load).

Involuntary Attention Switch to Auditory Change (Passive Attention)

It is assumed that the activation of the auditory change-detection mechanism reflected by the MMN may also trigger the switching of attention to potentially important events in the unattended auditory environment (Giard et al., 1990; Näätänen & Michie, 1979). Results of numerous studies (e.g., Alho et al., 1997; Escera et al., 1998; Schröger, 1996) indeed showed that MMN-eliciting sound changes in irrelevant auditory background stimulation decrease task performance and, further, that they also elicit a subsequent P3a response thought to be associated with the actual orienting of attention to a deviant in an unattended sound sequence or in some irrelevant feature of attended sounds (see Escera et al., 1998, 2001; Ford et al., 1976; Squires et al., 1975; for reviews, see Friedman et al., 2001, and Ranganath & Rainer, 2003). Furthermore, the MMN may also be followed by autonomic nervous system (ANS) responses associated with the involuntary orienting of attention such as heart rate deceleration and the skin conductance response (Lyytinen et al., 1992; Sokolov et al., 2002).

Discrimination Accuracy

In the foregoing, it was suggested that the MMN represents a preattentive feature-specific code of stimulus change and, further, that it might provide an objective index of the discrimination accuracy for the different acoustic feature dimensions. Consistent with this, in general, MMN sensitivity to small stimulus changes corresponds quite well to the behavioral discrimination thresholds; this holds with both normal subjects and clinical populations. For example, Lang et al. (1990) found that the accuracy of the behavioral discrimination of a frequency difference between two successively presented tone stimuli strongly correlated with the MMN amplitude (recorded in a separate, passive session) in high school pupils 17 years of age (see Figure 6.3).

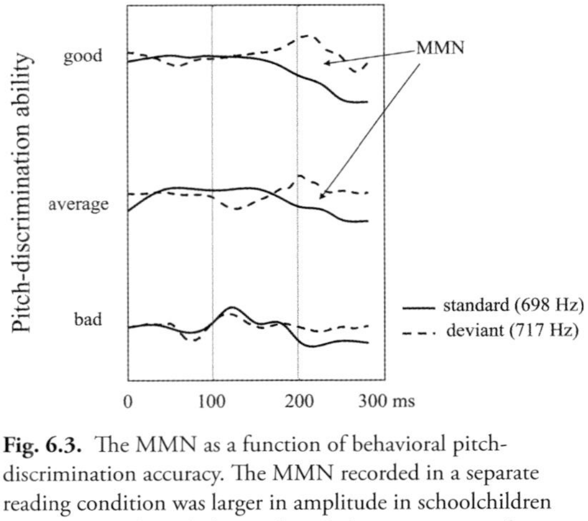

Fig. 6.3. The MMN as a function of behavioral pitch-discrimination accuracy. The MMN recorded in a separate reading condition was larger in amplitude in schoolchildren who were good in a behavioral pitch discrimination task (Seashore's test of musicality) than in those who were average or bad in this task. Consequently, MMN as an objective index of sound *discrimination* opens a new dimension in evoked-response audiometry (ERA, the objective determination of auditory processing capabilities by using ERPs), which previously was able to index sound *detection* only. Data from Lang et al. (1990).

With clinical populations, a close relationship between speech perception and the MMN amplitude was found in cochlear implant patients in several studies (Groenen et al., 1996; Kelly et al., 2005; Kraus et al., 1995; Roman et al., 2005). Moreover, Aaltonen et al. (1993) observed that in aphasic patients, the MMN, or its absence, could provide specific information with regard to the perceptual deterioration caused by a brain lesion. Two of their patients, who had a posterior left-hemispheric lesion, had a normal MMN to the frequency change of a simple tone, whereas a vowel change elicited no MMN. In contrast, in two other patients, who had anterior left-hemispheric lesions, both changes elicited an MMN. Furthermore, in studying schoolchildren who were either good or bad in discriminating /ba/ and /da/ syllables, Kraus et al. (1996) found a distinct MMN for these syllables in the children with good behavioral discrimination only. Importantly, the children with speech discrimination difficulties were the ones who also had learning problems, which suggests a role for these discrimination difficulties in the emergence of learning or other problems at school. For reviews on the MMN as an index of discrimination accuracy, see Näätänen and Alho (1995, 1997). Näätänen and Alho (1995) proposed that the MMN provides the best available neurophysiological measure of automatic central processing in audition. This is particularly true for infants, who often cannot give reliable behavioral responses. The MMN is the ontologically earliest cognitive response that can be recorded from the human brain. It can be recorded even from newborns, providing a potential early screening methodology for different auditory abnormalities and for securing the presence of all different central auditory processing prerequisites for language learning. Surprisingly, MMN can be recorded even from a fetus in the womb (through the mother's stomach wall by using a magnetometer (Draganova et al., 2005; Huotilainen et al., 2005), with the earliest MMN response being obtainable at the 33rd week from conception (Draganova et al., 2007).

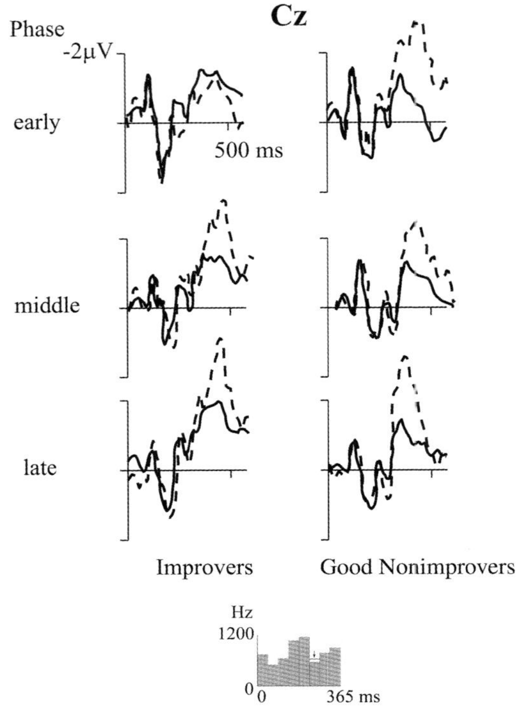

Fig. 6.4. The MMN as a function of training. On the left side: vertex (Cz) ERPs of subjects reading a book to standard (solid lines) and deviant (broken lines) stimulus patterns during the early, middle, and late MMN recording blocks of the session, each preceded by a behavioral discrimination task (grand-averaged data). Subjects' performance in the discrimination test was initially weak but was considerably improved after the second phase and especially after the third phase. This improvement was accompanied by MMN emergence. The stimulus pattern is schematically illustrated at the bottom. The only difference between the standard (the sixth segment, 565 Hz) and deviant (650 Hz) patterns is indicated by the arrow. On the right side: the five subjects who were good (near 100%) in discriminating deviants even at the early phase had an MMN even at this early phase of the session. Data from Näätänen et al. (1993b).

Learning/Training Effects

The MMN can also reflect improvement in discrimination performance as a result of training (e.g., Näätänen et al., 1993b; Winkler et al., 1999). Several studies showed that with the MMN, one can monitor progress in sound discrimination. For example, in an early study, Näätänen et al. (1993b) used a complex spectrotemporal stimulus pattern as the standard stimulus. They found that subjects who were able to detect a slightly deviant pattern in a behavioral discrimination task showed an MMN to this deviant stimulus in a subsequent passive condition (Figure 6.4). In contrast, no MMN was elicited in those subjects who were unable to behaviorally discriminate the stimuli in the preceding discrimination condition. However, after they learned to discriminate the stimuli during the course of the session, the MMN was elicited by the deviant patterns in the subsequent passive conditions.

Subsequently, Atienza et al. (2001), using identical stimuli, showed that this MMN training effect could even be recorded in REM sleep 3 days after the training. However, for the MMN to become a useful tool for everyday clinical practice, one should be able to measure it reliably in single subjects and patients. To this end, more time-effective paradigms (see Näätänen et al., 2004a Pakarinen et al., 2007) and improved signal-analysis methods (e.g., Ha et al., 2003; Marco-Pallares et al., 2005; McGee et al., 1997; Ponton et al., 1997) have been recently developed.

For MMN replicability, see Escera et al. (1999), Escera and Grau (1996), Frodl-Bauch et al. (1997), Joutsiniemi et al. (1998), Kathmann et al. (1999), Kujala et al. (2001), and Tervaniemi et al. (1999). In general, MMN replicability is quite good at the group level, but at the individual level there is ample room for improvement before the MMN provides a reliable clinical tool.

Speech Stimuli

As already mentioned, the MMN (and MMNm) is also elicited when speech sounds are presented in a passive oddball paradigm (Aaltonen et al., 1987; Cheour et al., 1998; Dehaene-Lambertz & Baillet, 1998). Furthermore, a number of studies using speech sounds found that with the MMN, one can probe the permanent language-specific speech-sound memory traces. Näätänen et al. (1997; see also Dehaene-Lambertz, 1997) found that an infrequent vowel deviant presented in a sequence of native-language vowel standards elicited a larger-amplitude MMN when it was a typical exemplar of a vowel category of the subject's native language (Finnish) than when it was not a typical vowel in this language (e.g., "/õ/" in the Estonian language, which does not exist in Finnish). The subsequent MEG recordings showed that the vowel-related MMN enhancement originated from the left posterior auditory cortex, which suggested this cortex as the locus of the language-specific vowel traces (Figure 6.5). In contrast, the concomitant acoustic change necessarily accompanying the vowel change elicited a bilateral auditory-cortex MMN subcomponent.

According to the authors, these long-term or permanent traces serve as recognition patterns activated by the corresponding speech sounds, enabling one to correctly perceive them. Further, these traces provide reference information for pronunciation.

In addition, using a similar Finnish-Estonian cross-linguistic design, Cheour et al. (1998) obtained

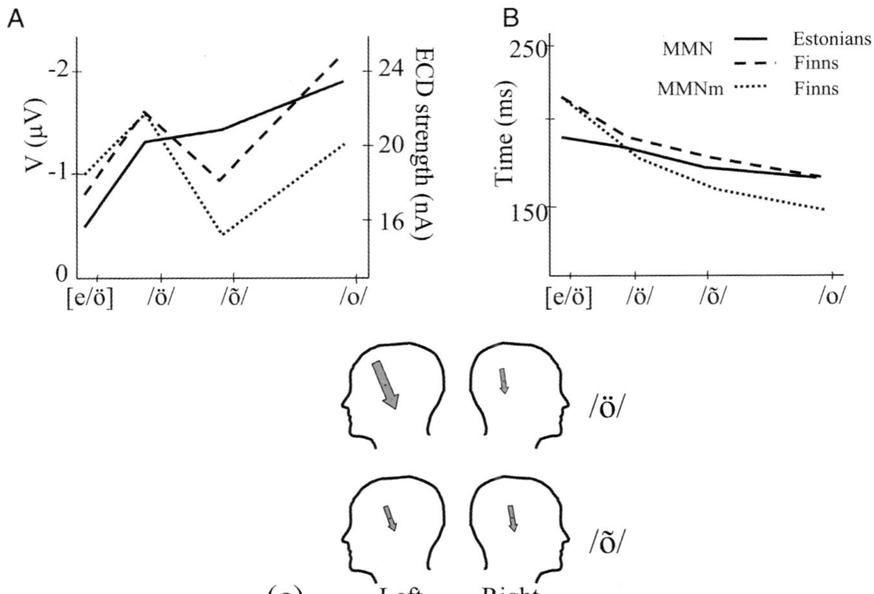

Fig. 6.5. (A) The MMN to mother and foreign tongue deviants. The MMN peak amplitude at the mid-frontal recording site (Fz) in Finns (broken line) and Estonians (solid line) as a function of the deviant stimulus, arranged according to the increasing second-formant (F2) difference from the standard stimulus. The dotted line indicates the strength of the equivalent current dipole (ECD; average of nine Finnish subjects) modeling the left auditory cortex MMNm for the different deviant stimuli (right *y*-axis). (B) The MMN (solid line for Estonians, broken line for Finns) and MMNm (dotted line) peak latency as a function of the deviant stimulus. (C) The left- and right-hemisphere MMNms of one typical Finnish subject for deviants /ö/ and /õ/ presented in contour maps. The arrows represent ECDs indicating activity in the auditory cortex; the black dots in these arrows show the centers of gravity of the MMNm. It can be seen that the prototype /ö/ elicits a much larger MMNm in the left than in the right hemisphere, whereas the nonprototype /õ/ responses in both hemispheres are small and quite similar in amplitude. Data from Näätänen et al. (1997).

results suggesting that the language-specific speech-sound memory traces develop between 6 and 12 months of age (see also Rivera-Gaxiola et al., 2005).

Most of the MMN studies with speech sounds used only one (acoustically constant) exemplar of each phonetic category included. Therefore, the MMN obtained could, at least in most studies, be due either to phonetic-category or mere acoustic change. Therefore, Shestakova et al. (2002) used 150 different male-voice exemplars of each of the three vowel (*a*, *u*, *i*) categories involved. These vowels were presented in short sequences of each category with continuously varying exemplars (standards), with no break before the onset of the sequence of another vowel. Hence, the first stimulus of each sequence served as a deviant stimulus. In almost all subjects, an MMNm response was obtained above the left hemisphere only; hence, the acoustic MMN component was abolished.

Furthermore, with the MMN, one can also probe the memory representations of higher-order linguistic phenomena. For example, MMN evidence for memory traces of mother-tongue syllables was obtained by Alho et al. (1998) and Shtyrov et al. (1998). Subsequently, Korpilahti et al. (2001) and Pulvermüller et al. (2001) found that even the memory traces of mother-tongue words can be probed with the MMN.

Moreover, MMN data (Pulvermüller & Shtyrov, 2003; Shtyrov et al., 2003) suggest the automatic processing of grammar. These data demonstrated, according to the authors, for the first time that the brain detects grammatical violations even when subjects are instructed to direct their attention away from the language input; in other words, early syntax processing in the human brain may take place outside the focus of attention.

Musical Stimuli

Studies on the MMN have also contributed to our understanding of music perception. Tervaniemi and Brattico (2004), in their review, concluded that by using the MMN, one can look into the separate submodules of music perception, "examining with optimal time resolution the dynamic stages of information processing, as well as their automaticity and possible top-down modulation (which may be determined, for example, by implicit or explicit knowledge of musical sounds)" (p. 15). In the following studies, a finding of fundamental importance was that the MMN was elicited even by timbre change (Tervaniemi et al., 1997; Toiviainen et al., 1998). This and related findings highlight, according to Tervaniemi and Brattico (2004), the ability of the auditory cortex to differentiate sounds on the basis of a multidimensional sound attribute such as timbre, which in acoustical terms is quite complex but forms a common cue in our daily auditory scene to differentiate sounds even with the same spatial origin and time course (see Bregman, 1990).

Further, several studies found that musicians were superior to nonmusicians in processing musical sound material. For example, Vuust et al. (2005) observed that expert jazz musicians had a larger and earlier MMNm response to subtle deviations in rhythm than did musically inexperienced nonmusicians. Moreover, in the musicians, the MMNm for these deviations was left-lateralized, whereas that of the nonmusicians was right-lateralized. Vuust et al (2005) regarded this left lateralization as reflecting the functional adaptation of the musicians' brains to a communication task that in these musicians was, according to the authors, much like that of language when they played jazz together, with subtle rhythmic deviations forming signals of communication.

For further MMN studies showing superior auditory preattentive processing in musicians, see, for example, Brattico et al. (2002), Fujioka et al. (2005), Koelsch et al. (1999, 2002), and Rüsseler et al. (2001).

Abstract-Feature MMNs

As already mentioned, the preattentive auditory analysis reflected by the MMN is not restricted to physical, or "first-order," stimulus features but rather includes even more complex invariance, such as the *relationships* between various physical stimulus features. In these *abstract-feature* MMN studies, there is no physically identical, repetitive standard stimulus but rather a class of several physically different standard stimuli united by some common rule that they all obey.

In the first of these studies, Saarinen et al. (1992) presented subjects with a sequence of tone pairs with their positions in the frequency scale randomly varying over a wide range; therefore, there was no physically identical, repetitive standard stimulus. Instead, the standard pairs all were ascending pairs (i.e., the second tone of a pair being higher in frequency than the first tone), whereas the deviant pairs were descending pairs. Nevertheless, the MMN was elicited by the direction-deviant pairs in an ignore condition, showing that the preattentively formed sensory representations encoded a common invariant feature (rising pair) from a set of individually varying physical events. Subsequently, Korzyukov et al. (2003), using a similar paradigm, localized the source of the abstract-feature MMN with EEG and MEG recordings at the auditory cortex. For studies showing abstract-feature MMNs in newborns, see Ruusuvirta et al. (2003) and Carral et al. (2005). Moreover, Paavilainen et al. (1999) found that in addition to the direction of frequency and intensity change, even invariant frequency *ratios* (musical intervals) can be automatically derived from acoustically varying stimulation.

In addition, Tervaniemi et al. (1994) demonstrated that the MMN mechanism extrapolates traces representing the forthcoming stimuli on the basis of the regularities or trends detected in the immediate auditory past. Their stimuli consisted of a long sequence of steadily descending tones, occasionally interrupted by an ascending tone or a tone repetition. It was found that these deviant events elicited an MMN (Figure 6.6). Again, all standard stimuli were physically different, and the deviant events were composed of physically identical stimuli that had occurred in the immediate auditory past. Hence, it appeared that the impinging stimuli were automatically compared with an extrapolated trace, one automatically developed on the basis of the trend detected in the

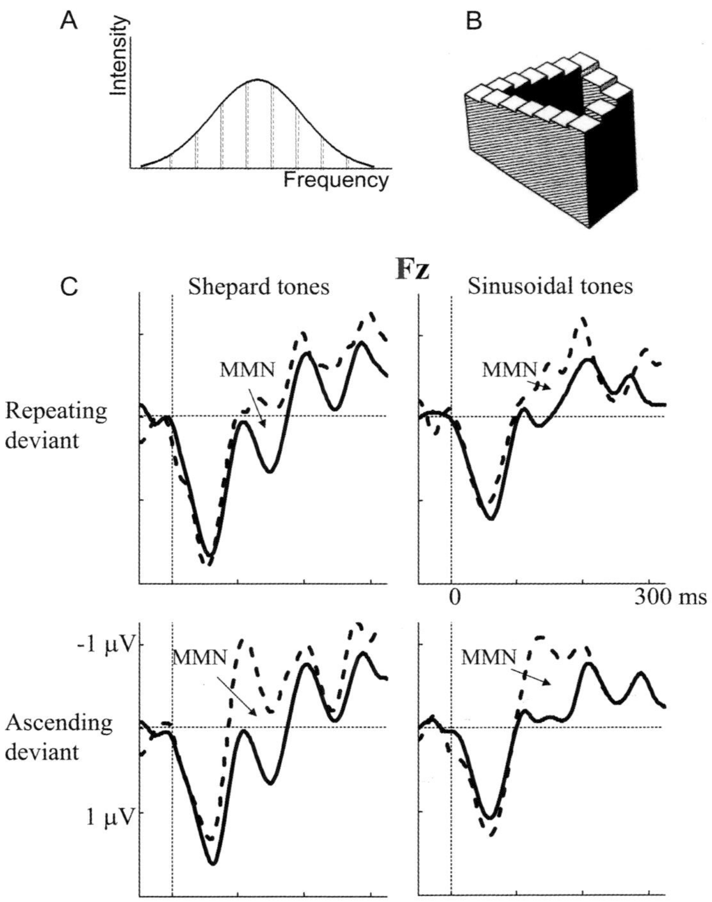

Fig. 6.6. The MMN to trend violation (descending standards). (A) The spectrum of an individual Shepard sound that, when presented in ascending or descending 12-sound sequences with one semitone steps, causes a pitch to ascend or descend in an endless manner. One Shepard sound consists of 10 frequency components, one octave apart, with a bell-shaped spectrum. During the presentation of a Shepard-sounds tone series, the tone-height (equivalent to the sense of octave) perception is made to disappear by manipulating the sound spectrum. (B) A visual analogy of the Shepard illusion, the endlessly ascending or descending stairs. (C) The ERPs recorded at the frontal electrode (Fz) from reading subjects at the time of the auditory stimulation to Shepard (left) and sinusoidal (right) tones, with the solid line presenting responses to the standard stimulus and the broken line to the deviant stimuli. The left column shows a regularly descending Shepard tone sequence randomly replaced by a repeating (top) or an ascending (bottom) tone (deviants). The right column shows a regularly descending sinusoidal tone sequence with occasional repetitive (top) or ascending (bottom) deviants. The dotted line at time 0 indicates the deviant-stimulus onset, and the statistically significant area of the MMN is indicated by the arrow. Data from Tervaniemi et al. (1994).

sequence of the preceding stimuli. For further corroborating data, see Paavilainen et al. (2007), who used sound stimuli that varied in duration and frequency, either being short (50 ms) or long (150 ms) and low (1000 Hz) or high (1500 Hz). All four stimuli (short-low, short-high, long-low, long-high) were presented at $p = .25$ at an ISI of 300 ms. Further, the duration of each stimulus predicted the frequency of the *next* stimulus as follows: (1) present stimulus short, next stimulus low; (2) present long, next high. Occasional deviant events (e.g., the present stimulus short, the next high) broke these rules. In this design, all four different stimuli could appear either as a standard or a deviant event, depending on the duration of the preceding stimulus. Nevertheless, deviant events elicited, in an ignore condition, an MMN, one peaking at 150–200 ms and reversing its polarity at the mastoids, which suggested a source in the auditory cortex.

In the subsequent attend conditions, subjects were asked (without being given an explanation of the rules obeyed by the stimulus sequence) to press a button to any stimuli they found somehow "strange" or "deviant." An MMN was again elicited, although subjects could detect only about 15% of the deviant events and none of them could verbally express the rules in the later interviews. This suggests that the neural mechanism modeling the auditory environment may automatically learn the covariation between the features of the successive events and make predictions about the properties of the forthcoming stimuli. Further, if the predictions are not fulfilled, then the MMN is generated.

Conclusion

These MMN studies indicate that the central auditory system performs complex cognitive operations, such as generalization leading to simple concept formation (e.g., a rising pair irrespective of the specific frequency values), rule extraction, and the anticipation of the next stimulus at the preattentive level. These findings appear to demonstrate the presence of "primitive sensory intelligence" in the auditory cortex (for a review, see Näätänen et al., 2001). The information extracted by the sensory-memory mechanisms often seems to be in an implicit form, not directly available to conscious processes and difficult to express verbally. These results are, consequently, consistent with the framework suggested by Winkler et al. (1996), according to which the main function of the MMN process is to adjust a neural model to the various regularities of the auditory environment. This would enable the central auditory system to process a large part of the subsequent input automatically, that is, without requiring the limited resources of the controlled processing system.

References

Aaltonen, O., Niemi, P., Nyrke, T., & Tuhkanen, M. (1987). Event-related brain potentials and the perception of a phonetic continuum. *Biological Psychology, 24*, 197–207.

Aaltonen, O., Tuomainen, J., Laine, M., & Niemi, P. (1993). Cortical differences in tonal versus vowel processing as revealed by an ERP component called mismatch negativity (MMN). *Brain and Language, 44*, 139–152.

Akatsuka, K., Wasaka, T., Nakata, H., Inui, K., Hoshiyama, M., & Kakigi, R. (2005). Mismatch responses related to temporal discrimination of somatosensory stimulation. *Clinical Neurophysiology, 116*, 1930–1937.

Akatsuka, K., Wasaka, T., Nakata, H., Kida, T., Hoshiyama, M., Tamura, Y., et al. (2007). Objective examination for two-point stimulation using a somatosensory oddball paradigm: An MEG study. *Clinical Neurophysiology, 118*, 403–411.

Alain, C., & Izenberg, A. (2003). Effects of attentional load on auditory scene analysis. *Journal of Cognitive Neuroscience, 15*, 1063–1073.

Alain, C., & Woods, D. L. (1994). Signal clustering modulates auditory cortical activity in humans. *Perception & Psychophysics, 56*, 501–516.

Alain, C., & Woods, D. L. (1997). Attention modulates auditory pattern memory as indexed by event-related brain potentials. *Psychophysiology, 34*, 534–546.

Alain, C., Woods, D. L., & Ogawa, K. H. (1994). Brain indices of automatic pattern processing. *Neuroreport, 5*, 140–144.

Alho, K., Connolly, J. F., Cheour, M., Lehtokoski, A., Huotilainen, M., Virtanen, J., Aulanko, R., & Ilmoniemi, R. J. (1998). Hemispheric lateralization in preattentive processing of speech sounds. *Neuroscience Letters, 258*, 9–12.

Alho, K., Escera, C., Díaz, R., Yago, E., & Serra, J. M. (1997). Effects of involuntary auditory attention on visual task performance and brain activity. *Neuroreport, 8*, 3233–3237.

Alho, K., Sams, M., Paavilainen, P., Reinikainen, K., & Näätänen, R. (1989). Event-related brain potentials reflecting processing of relevant and irrelevant stimuli during selective listening. *Psychophysiology, 26*, 514–528.

Alho, K., Woods, D. L., & Algazi, A. (1994). Processing of auditory stimuli during auditory and visual attention as revealed by event-related potentials. *Psychophysiology, 31*, 469–479.

Alho, K., Woods, D. L., Algazi, A., & Näätänen, R. (1992). Intermodal selective attention II: Effects of attentional load on processing auditory and visual stimuli in central space. *Electroencephalography and Clinical Neurophysiology, 82*, 356–368.

Arnott, S. R., & Alain, C. (2002). Stepping out of the spotlight: MMN attenuation as a function of distance from the attended location. *Neuroreport, 13*, 2209–2212.

Atienza, M., & Cantero, J. L. (2001). Complex sound processing during human REM sleep by recovering information from long-term memory as revealed by the mismatch negativity (MMN). *Brain Research, 901*, 151–160.

Baldeweg, T., Williams, J. D., & Gruzelier, J. H. (1999). Differential changes in frontal and sub-temporal components of mismatch negativity. *International Journal of Psychophysiology, 33*, 143–148.

Böttcher-Gandor, C., & Ullsperger, P. (1992). Mismatch negativity in event-related potentials of auditory stimuli as a function of varying interstimulus interval. *Psychophysiology, 29*, 546–550.

Brattico, E., Näätänen, R., & Tervaniemi, M. (2002). Context effects on pitch perception in musicians and non-musicians: Evidence from ERP recordings. *Music Perception, 19*, 199–222.

Bregman, A. S. *Auditory scene analysis: The perceptual organization of sound.* (1990). Cambridge, MA: Bradford Books.

Carral, V., Huotilainen, M., Ruusuvirta, T., Fellman, V., Näätänen, R., & Escera, C. (2005). A kind of auditory 'primitive intelligence' already present at birth. *European Journal of Neuroscience, 21*, 3201–3204.

Celsis, P., Boulanouar, K., Doyon, B., Ranjeva, J. P., Berry, I., Nespoulous, J. L., et al. (1999). Differential fMRI responses in the left posterior superior temporal gyrus and left supramarginal gyrus to habituation and change detection in syllables and tones. *Neuroimage, 9*, 135–144.

Cheour, M., Ceponiene, R., Lehtokoski, A., Luuk, A., Allik, J., Alho, K., et al. (1998). Development of language-specific phoneme representations in the infant brain. *Nature Neuroscience, 1*, 351–353.

Dehaene-Lambertz, G. (1997). Electrophysiological correlates of categorical phoneme perception in adults. *Neuroreport, 8*, 919–924.

Dehaene-Lambertz, G., & Baillet, S. (1998). A phonological representation in the infant brain. *Neuroreport, 9*, 1885–1888.

Deouell, L. (2007). The frontal generator of the mismatch negativity revisited. *Journal of Psychophysiology, 21*, 188–203.

Deouell, L., & Bentin, S. (1998). Variable cerebral responses to equally distinct deviance in four auditory dimensions: A mismatch negativity study. *Psychophysiology, 35*, 745–754.

Deouell, L., Bentin, S., & Giard, M. H. (1998). Mismatch negativity in dichotic listening: Evidence for interhemispheric differences and multiple generators. *Psychophysiology, 35*, 355–365.

Dittmann-Balcar, A., Thienel, R., & Schall, U. (1999). Attention-dependent allocation of auditory processing resources as measured by mismatch negativity. *Neuroreport, 10*, 3749–3753.

Doeller, C. F., Opitz, B., Mecklinger, A., Krick, C., Reith, W., & Schröger, E. (2003). Prefrontal cortex involvement in preattentive auditory deviance detection: Neuroimaging and electrophysiological evidence. *Neuroimage, 20*, 1270–1282.

Draganova, R., Eswaran, H., Murphy, P., Huotilainen, M., Lowerey, C., & Preissl, H. (2005). Sound frequency change detection in fetuses and newborns, a magnetoencephalographic study. *Neuroimage, 28*, 354–361.

Draganova, R., Eswaran, H., Murphy, P., Lowery, C., & Preissl, H. (2007). Serial magnetoencephalographic study of fetal and newborn auditory discriminative evoked responses. *Early Human Development, 83*, 199–207.

Dyson, B. J., Alain, C., & He, Y. (2005). Effects of visual attentional loads on low-level auditory scene analysis. *Cognitive, Affective, & Behavioral Neuroscience, 5*, 319–338.

Escera, C., Alho, K., Winkler, I., & Näätänen, R. (1998). Neural mechanisms of involuntary attention switching to novelty and change in the acoustic environment. *Journal of Cognitive Neuroscience, 10*, 590–604.

Escera, C., & Grau, C. (1996). Short-term replicability of the mismatch negativity. *Electroencephalography and Clinical Neurophysiology, 100*, 549–554.

Escera, C., Yago, E., & Alho, K. (2001). Electrical responses reveal the temporal dynamics of brain events during involuntary attention switching. *European Journal of Neuroscience, 14*, 877–883.

Escera, C., Yago, E., Polo, M. D., & Grau, C. (1999). The individual replicability of mismatch negativity at short and long interstimulus intervals. *Clinical Neurophysiology, 110*, 1–6.

Fischer, C., Luauté, J., Némoz, C., Morlet, D., Kirkorian, G., & Mauguière, F. (2006). Improved prediction of awakening or nonawakening from severe anoxic coma using tree-based classification analysis. *Critical Care Medicine, 34*, 1–5.

Ford, J. M., & Hillyard, S. A. (1981). Event-related potentials (ERPs) to interruptions of a steady rhythm. *Psychophysiology, 18*, 322–330.

Ford, J. M., Roth, W. T., & Kopell, B. S. (1976). Attention effects on auditory evoked potentials to infrequent events. *Biological Psychology, 4*, 65–77.

Friedman, D., Cycowicz, Y. M., & Gaeta, H. (2001). The novelty P3: An event-related brain potential (ERP) sign of the brain's evaluation of novelty. *Neuroscience and Biobehavioral Reviews, 25*, 355–373.

Frodl-Bauch, T., Kathmann, N., Möller, H.-J., & Hegerl, U. (1997). Dipole localization and test-retest reliability of frequency and duration mismatch negativity generator processes. *Brain Topography, 10*, 3–8.

Fujioka, T., Trainor, L. J., Ross, B., Kakigi, R., & Pantev, C. (2005). Automatic encoding of polyphonic melodies in musicians and nonmusicians. *Journal of Cognitive Neuroscience, 17*, 1578–1592.

Giard, M. H., Lavikainen, J., Reinikainen, K., Perrin, F., Bertrand, O., Thévenet, M., et al. (1995). Separate representation of stimulus frequency, intensity, and duration in auditory sensory memory. *Journal of Cognitive Neuroscience, 7*, 133–143.

Giard, M. H., Perrin, F., Pernier, J., & Bouchet, P. (1990). Brain generators implicated in processing of auditory stimulus deviance: A topographic event-related potential study. *Psychophysiology, 27*, 627–640.

Groenen, P., Snik, A., & van der Broek, P. (1996). On the clinical relevance of mismatch negativity: Results from subjects with normal hearing and cochlear implant users. *Audiology and Neuro-Otology, 1*, 112–124.

Ha, K., Youn, T., Kong, S.W., Park, H. J., Ha, T. H., Kim, M. S., & Kwon, J. S. (2003). Optimized individual mismatch negativity source localization using a realistic head model and the Talairach coordinate system. *Brain Topography, 15*, 233–238.

Haenschel, C., Vernon, D. J., Dwiyedi, P., Gruzelier, J. H., & Baldeweg, T. (2005). Event-related brain potential correlates of human auditory sensory memory trace formation. *Journal of Neuroscience, 25*, 10494–10501.

Halgren, E., Baudena, P., Clarke, J. M., Heit, G., Liegeois, C., Chauvel, P., et al. (1995). Intracerebral potentials to rare target and distractor auditory and visual stimuli: I. Superior temporal plane and parietal lobe. *Electroencephalography and Clinical Neurophysiology, 94*, 191–220.

Hari, R., Hämäläinen, M., Ilmoniemi, R., Kaukoranta, E., Reinikainen, K., Salminen, J., et al. (1984). Responses of the primary auditory cortex to pitch changes in a sequence of tone pips: Neuromagnetic recordings in man. *Neuroscience Letters, 50*, 127–132.

Harmony, T., Bernal, J., Fernández, T., Silva-Pereyra, J., Fernández-Bouzas, A., Marosi, E., et al. (2000). Primary task demands modulate P3a amplitude. *Cognitive Brain Research, 9*, 53–60.

Heslenfeld, D. J. Visual mismatch negativity. (2003). In J. Polich (Ed.), *Detection of change: Event-related potential and fMRI findings* (pp. 41-59). Boston: Kluwer Academic.

Huotilainen, M., Kujala, A., Hotakainen, M., Parkkonen, L., Taulu, S., Simola, J., et al. (2005). Short-term memory functions of the human fetus recorded with magnetoencephalography. *Neuroreport, 16*, 81–84.

Jääskeläinen, I. P., Alho, K., Escera, C., Winkler, I., Sillanaukee, P., & Näätänen, R. (1996a). Effects of ethanol and auditory distraction on forced choice reaction time. *Alcohol, 13*, 153–156.

Jääskeläinen, I. P., Hautamäki, M., Näätänen, R., & Ilmoniemi, R. J. (1999). Temporal span of auditory sensory memory and mismatch negativity: revisited. *Neuroreport, 10*, 1305–1308.

Jääskeläinen, I. P., Pekkonen, E., Hirvonen, J., Sillanaukee, P., & Näätänen, R. (1996b). Mismatch negativity subcomponents and ethyl alcohol. *Biological Psychology, 43*, 13–25.

Joutsiniemi, S.-L., Ilvonen, T., Sinkkonen, J., Huotilainen, M., Tervaniemi, M., Lehtokoski, A., et al. (1998). The mismatch negativity for duration decrement of auditory stimuli in healthy subjects. *Electroencephalography and Clinical Neurophysiology, 108*, 154–159.

Kathmann, N., Frodl-Bauch, T., & Hegerl, U. (1999). Stability of the mismatch negativity under different stimulus and attention conditions. *Clinical Neurophysiology, 110*, 317–323.

Kaukoranta, E., Sams, M., Hari, R., Hämäläinen, M., & Näätänen, R. (1989). Reactions of human auditory cortex to changes in tone duration. *Hearing Research, 41*, 15–22.

Kekoni, J., Hämäläinen, H., Saarinen, M., Gröhn, J., Reinikainen, K., Lehtokoski, A., et al. (1997). Rate effect and mismatch responses in the somatosensory system: ERP-recordings in humans. *Biological Psychology, 46*, 125–142.

Kelly, A. S., Purdy, S. C., & Thorne, P. R. (2005). Electrophysiological and speech perception measures of auditory processing in experienced adult cochlear implant users. *Clinical Neurophysiology, 116*, 1235–1246.

Koelsch, S., Schmidt, B.-H., & Kansok, J. (2002). Effects of musical expertise on the early right anterior negativity: An event-related brain potential study. *Psychophysiology, 39*, 657–663.

Koelsch, S., Schröger, E., & Tervaniemi, M. (1999). Superior pre-attentive auditory processing in musicians. *Neuroreport, 10*, 1309–1313.

Korpilahti, P., Krause, C. M., Holopainen, I., & Lang, A. H. (2001). Early and late mismatch negativity elicited by words and speech-like stimuli in children. *Brain and Language, 76*, 332–339.

Korzyukov, O., Alho, K., Kujala, A., Gumenyuk, V., Ilmoniemi, R. J., Virtanen, J., et al. (1999). Electromagnetic responses of the human auditory cortex generated by sensory-memory based processing of tone-frequency changes. *Neuroscience Letters, 276*, 169–172.

Korzyukov, O., Winkler, I., Gumenyuk, V., & Alho, K. (2003). Processing abstract auditory features in the human auditory cortex. *Neuroimage, 20*, 2245–2258.

Kramer, A. F., Trejo, L. J., & Humphrey, D. (1995). Assessment of mental workload with task-irrelevant auditory probes. *Biological Psychology, 40*, 83–100.

Krauel, K., Schott, P., Sojka, B., Pause, B. M., & Ferstl, R. (1999). Is there a mismatch negativity analogue in the olfactory event-related potential? *Journal of Psychophysiology, 13*, 49–55.

Kraus, N., McGee, T., Carrell, T. D., King, C., Tremblay, K., & Nicol, T. (1995). Central auditory plasticity associated with speech discrimination training. *Journal of Cognitive Neuroscience, 7*, 25–32.

Kraus, N., McGee, T. J., Carrell, T. D., Zecker, S. G., Nicol, T. G., & Koch, D. B. (1996). Auditory neurophysiologic response and discrimination deficits in children with learning problems. *Science, 273*, 971–973.

Kremlàček, J., Kuba, M., Kubovà, Z., & Langrovà, J. (2006). Visual mismatch negativity elicited by magnocellular system activation. *Vision Research, 46*, 485–490.

Kropotov, J. D., Alho, K., Näätänen, R., Ponomarev, V. A., Kropotova, O. V., Anichkov, A. D., et al. (2000). Human auditory-cortex mechanisms of preattentive sound discrimination. *Neuroscience Letters, 280*, 87–90.

Kropotov, J. D., Näätänen, R., Sevostianov, A. V., Alho, K., Reinikainen, K., & Kropotova, O. V. (1995). Mismatch negativity to auditory stimulus change recorded directly from the human temporal cortex. *Psychophysiology, 32*, 418–422.

Kujala, T., Kallio, J., Tervaniemi, M., & Näätänen, R. (2001). The mismatch negativity as an index of temporal processing in audition. *Clinical Neurophysiology, 112*, 1712–1719.

Lang, H. A., Nyrke, T., Ek, M., Aaltonen, O., Raimo, I., & Näätänen, R. (1990). Pitch discrimination performance and auditory event-related potentials. In C. H. M. Brunia, A. W. K. Gaillard, A. Kok, G. Mulder, & M. M.Verbaten (Eds.), *Psychophysiological brain research* (Vol. 1, pp. 294–298). Tilburg, the Netherlands: Tilburg University Press.

Lounasmaa, O. V., Hari, R., Joutsiniemi, S.-L., & Hämäläinen, M. (1989). Multi-SQUID recordings of human cerebral magnetic fields may give information about memory processes. *Europhysics Letters, 9*, 603–608.

Lyytinen, H., Blomberg, A. P., & Näätänen, R. (1992). Event-related potentials and autonomic responses to a change in unattended auditory stimuli. *Psychophysiology, 29*, 523–534.

Marco-Pallares, J., Grau, C., & Ruffini, G. (2005). Combined ICA-LORETA analysis of mismatch negativity. *Neuroimage, 25*, 471–477.

McGee, T., Kraus, N., & Nicol, T. (1997). Is it really a mismatch negativity? An assessment of methods for determining response validity in individual subjects. *Electroencephalography and Clinical Neurophysiology, 104*, 359–368.

Molholm, S., Martinez, A., Ritter, W., Javitt, D. C., & Foxe, J. J. (2005). The neural circuitry of pre-attentive auditory change-detection: An fMRI study of pitch and duration mismatch negativity generators. *Cerebral Cortex, 15*, 545–551.

Müller, B. W., Jüptner, M., Jentzen, W., & Müller, S. P. (2002). Cortical activation to auditory mismatch elicited by frequency deviant and complex novel sounds: A PET study. *Neuroimage, 17*, 231–239.

Muller-Gass, A., Stelmack, R. M., & Campbell, K. B. (2005). "... and were instructed to read a self-selected book while ignoring the auditory stimuli": The effects of task demands on the mismatch negativity. *Clinical Neurophysiology, 116*, 2142–2152.

Muller-Gass, A., Stelmack, R. M., & Campbell, K. B. (2006). The effect of visual task difficulty and attentional direction on the detection of acoustic change as indexed by the mismatch negativity. *Brain Research, 1078*, 112–130.

Näätänen, R. (1979). Orienting and evoked potentials. In H. D. Kimmel, E. H. van Olst, & J. F. Orlebeke (Eds), *The orienting reflex in humans* (pp. 66–75). Hillsdale, NJ: Erlbaum.

Näätänen, R. (1984). In search of a short-duration memory trace of a stimulus in the human brain. In L. Pulkkinen & P. Lyytinen (Eds), *Human action and personality: Essays in honour of Martti Takala.* (pp. 22–36). Jyväskylä Studies in Education, Psychology and Social Research, Vol. 54. Jyväskylä, Finland: University of Jyväskylä.

Näätänen, R. (1991). Mismatch negativity (MMN) outside strong attentional focus: A commentary on Woldorff et al. *Psychophysiology, 28*, 478–484.

Näätänen, R. (1992). *Attention and brain function*. Erlbaum, Hillsdale, NJ: Erlbaum.

Näätänen, R., & Alho, K. (1995). Mismatch negativity—a unique measure of sensory processing in audition. *International Journal of Neuroscience, 80*, 317–337.

Näätänen, R., & Alho, K. (1997). Mismatch negativity—the measure for central sound representation accuracy. *Audiology and Neuro-Otology, 2*, 341–353.

Näätänen, R., Gaillard, A. W. K., & Mäntysalo, S. (1978). Early selective-attention effect on evoked potential reinterpreted. *Acta Psychologica, 42*, 313–329.

Näätänen, R., & Gaillard, A. W. K. (1983). The orienting reflex and the N2 deflection of the event-related potential (ERP). In A. W. K. Gaillard & W. Ritter (Eds.), *Tutorials in event related potential research: Endogenous components* (pp. 119–141). Amsterdam: North Holland.

Näätänen, R., Jacobsen, T., & Winkler, I. (2005). Memory based or afferent processes in mismatch negativity (MMN): A review of the evidence. *Psychophysiology, 42*, 25–32.

Näätänen, R., & Kreegipuu, K. (2010). The mismatch negativity (MMN) as an index of different forms of memory in audition. In L. Bäckman & L. Nyberg (Eds.), *Memory, ageing, and brain* (pp. *287–299*). London: Psychology Press.

Näätänen, R., Lehtokoski, A., Lennes, M., Cheour, M., Huotilainen, M., Iivonen, A., et al. (1997). Language-specific phoneme representations revealed by electric and magnetic brain responses. *Nature, 385*, 432–434.

Näätänen, R., & Michie, P. T. (1979). Early selective attention effects on the evoked potential: A critical review and reinterpretation. *Biological Psychology, 8*, 81–136.

Näätänen, R., Paavilainen, P., Alho, K., Reinikainen, K., & Sams, M. (1987). Interstimulus interval and the mismatch negativity. In C. Barber & T. Blum (Eds.), *Evoked potentials III*, The Third International Evoked Potentials Symposium (pp. 392–397). Boston: Butterworths.

Näätänen, R., Paavilainen, P., & Reinikainen, K. (1989). Do event-related potentials to infrequent decrements in duration of auditory stimuli demonstrate a memory trace in man? *Neuroscience Letters, 107*, 347–352.

Näätänen, R., Paavilainen, P., Rinne, T., & Alho, K. (2007). The mismatch negativity (MMN) in basic research of central auditory processing: A review. *Clinical Neurophysiology, 118*, 2544–2590.

Näätänen, R., Pakarinen, S., Rinne, T., & Takegata, R. (2004a). The mismatch negativity (MMN)-towards the optimal paradigm. *Clinical Neurophysiology, 115*, 140–144.

Näätänen, R., Paavilainen, P., Tiitinen, H., Jiang, D., & Alho, K. (1993a). Attention and mismatch negativity. *Psychophysiology, 30*, 436–450.

Näätänen, R., Schröger, E., Karakas, S., Tervaniemi, M., & Paavilainen, P. (1993b). Development of a memory trace for a complex sound in the human brain. *Neuroreport, 4*, 503–506.

Näätänen, R., Simpson, M., & Loveless, N. E. (1982). Stimulus deviance and evoked potentials. *Biological Psychology, 14*, 53–98.

Näätänen, R., Syssoeva, O., & Takegata, R. (2004b). Automatic time perception in the human brain for intervals ranging from milliseconds to seconds. *Psychophysiology, 41*, 660–663.

Näätänen, R., Tervaniemi, M., Sussman, E., Paavilainen, P., & Winkler, I. (2001). "Primitive intelligence" in the auditory cortex. *Trends in Neuroscience, 24*, 283–288.

Näätänen, R., & Winkler, I. (1999). The concept of auditory stimulus representation in neuroscience. *Psychology Bulletin, 125*, 826–859.

Nordby, H., Roth, W. T., & Pfefferbaum, A. (1988). Event-related potentials to breaks in sequence of alternating pitches or interstimulus intervals. *Psychophysiology, 25*, 262–268.

Öhman, A. (1979). The orienting response, attention and learning: An information processing perspective. In H. D. Kimmel, E. H. van Olst, & J. F. Orlebeke (Eds), *The orienting reflex in humans* (pp. 443–471). Hillsdale, NJ: Erlbaum.

Opitz, B., Rinne, T., Mecklinger, A., von Cramon, D. Y., & Schröger, E. (2002). Differential contribution of frontal and temporal cortices to auditory change detection: fMRI and ERP results. *Neuroimage, 15*, 165–174.

Otten, L. J., Alain, C., & Picton, T. (2000). Effects of visual attentional load on auditory processing. *Neuroreport, 11*, 875–880.

Paavilainen, P., Alho, K., Reinikainen, K., Sams, M., & Näätänen, R. (1991). Right hemisphere dominance of different mismatch negativities. *Electroencephalography and Clinical Neurophysiology, 78*, 466–479.

Paavilainen, P., Arajärvi, P., & Takegata, R. (2007). Preattentive detection of nonsalient contingencies between auditory features. *Neuroreport, 18*, 159–163.

Paavilainen, P., Jaramillo, M., Näätänen, R., & Winkler, I. (1999). Neuronal populations in the human brain extracting invariant relationships from acoustic variance. *Neuroscience Letters, 265*, 179–182.

Paavilainen, P., Karlsson, M.-L., Reinikainen, K., & Näätänen, R. (1989). Mismatch negativity to change in the spatial location of an auditory stimulus. *Electroencephalography and Clinical Neurophysiology, 73*, 129–141.

Paavilainen, P., Tiitinen, H., Alho, K., & Näätänen, R. (1993). Mismatch negativity to slight pitch changes outside strong attentional focus. *Biological Psychology, 37*, 23–41.

Pakarinen, S., Takegata, R., Rinne, T., Huotilainen, M., & Näätänen, R. (2007). Measurement of extensive auditory discrimination profiles using mismatch negativity (MMN) of the auditory event-related potential. *Clinical Neurophysiology, 118*, 177–185.

Pekkonen, E., Jousimäki, V., Könönen, M., Reinikainen, K., & Partanen, J. (1994). Auditory sensory memory impairment in Alzheimer's disease: An event-related potential study. *Neuroreport, 5*, 2537–2540.

Pekkonen, E., Rinne, T., Reinikainen, K., Kujala, T., Alho, K.,& Näätänen, R. (1996). Aging effects on auditory processing: An event-related potential study. *Experimental Aging Research, 22*, 171–184.

Polo, M. D., Escera, C., Gual, A., & Grau, C. (1999). Mismatch negativity and auditory sensory memory in chronic alcoholics. *Alcohol: Clinical and Experimental Research, 23*, 1744–1750.

Ponton, C. W., Don, M., Eggermont, J. J., & Kwong, B. (1997). Integrated mismatch negativity (MMN): A noise-free representation of evoked responses allowing single-point distribution-free statistical tests. *Electroencephalography and Clinical Neurophysiology, 104*, 143–150.

Pulvermüller, F., Kujala, T., Shtyrov, Y., Simola, J., Tiitinen, H., Alku, P., et al. Memory traces for words as revealed by the mismatch negativity. *Neuroimage, 2001*, 14: 607–616.

Pulvermüller, F., & Shtyrov, Y. (2003). Automatic processing of grammar in the human brain as revealed by the mismatch negativity. *Neuroimage, 20*, 159–172.

Pulvermüller, F., Shtyrov, Y., & Ilmoniemi, R. (2003). Spatiotemporal dynamics of neural language processing: An MEG study using minimum-norm current estimates. *Neuroimage, 20*, 1020–1025.

Ranganath, C., & Rainer, G. (2003). Neural mechanisms for detecting and remembering novel events. *Nature Reviews in Neuroscience, 4*, 193–202.

Renault, B., & Lesévre, N. (1978). Topographical study of the emitted potential obtained after the omission of an expected visual stimulus. In D. Otto D (Ed.), *Multidiscliplinary perspectives in event-related brain potential research*, EPA Publication 600/9–77-043 (p. 202–208). Washington, DC: U.S. Government Printing Office.

Renault, B., & Lesévre, N. (1979). A trial-by-trial study of the visual omission response in reaction time situations. In D. Lehmann & E. Callaway (Eds.), *Human evoked potentials* (pp. 317–329). New York: Plenum Press.

Rinne, T., Alho, K., Ilmoniemi, R. J., Virtanen, J., & Näätänen, R. (2000). Separate time behaviors of the temporal and frontal mismatch negativity sources. *Neuroimage, 12*, 14–19.

Rinne, T., Gratton, G., Fabiani, M., Cowan, N., Maclin, E., Stinard, A., et al. (1999). Scalp-recorded optical signals make sound processing in the auditory cortex visible. *Neuroimage, 10*, 620–624.

Ritter, W., Paavilainen, P., Lavikainen, J., Reinikainen, K., Alho, K., Sams, M., et al. (1992). Event-related potentials to repetition and change of auditory stimuli. *Electroencephalography and Clinical Neurophysiology, 83*, 306–321.

Rivera-Gaxiola, M., Silva-Pereyra, J., & Kuhl, PK. (2005). Brain potentials to native and non-native speech contrasts in 7- and 11-month-old American infants. *Developmental Science, 8*, 162–172.

Roman, S., Canévet, G., Marquis, P., Triglia, J.-M.,& Liégeois-Chauvel, C. (2005). Relationship between auditory perception skills and mismatch negativity recorded in free field in cochlear-implant users. *Hearing Research, 201*, 10–20.

Rosburg, T. (2003). Left hemispheric dipole locations of the neuromagnetic mismatch negativity to frequency, intensity and duration deviants. *Cognitive Brain Research, 16*, 83–90.

Rosburg, T. (2004). Effects of tone repetition on auditory evoked neuromagnetic fields. *Clinical Neurophysiology, 115*, 898–905.

Rosburg, T., Haueisen, J., & Kreitschman-Andermahr, I. (2004a). The dipole location shift within the auditory evoked neuromagnetic field components N100m and mismatch negativity (MMNm). *Clinical Neurophysiology, 115*, 906–913.

Rosburg, T., Marinou, V., Haueisen, J., Smesny, S., & Sauer, H. (2004b). Effects of lorazepam on the neuromagnetic mismatch negativity (MMNm) and auditory evoked field component N100m. *Neuropsychopharmacology, 29*, 1723–1733.

Rüsseler, J., Altenmüller, E., Nager, W., Kohlmetz, C., & Münte, TF. (2001). Event related brain potentials to sound omissions differ in musicians and non-musicians. *Neuroscience Letters, 308*, 33–36.

Ruusuvirta, T., Huotilainen, M., Fellman, V., & Näätänen, R. (2003). The newborn human brain binds sound features together. *Neuroreport, 14*, 2117–2119.

Saarinen, J., Paavilainen, P., Schröger, E., Tervaniemi, M., & Näätänen, R. (1992). Representation of abstract stimulus attributes in human brain. *Neuroreport, 3*, 1149–1151.

Sabri, M., & Campbell, K. B. (2001). Effects of sequential and temporal probability of deviant occurrence on mismatch negativity. *Cognitive Brain Research, 12*, 171–180.

Sams, M., Alho, K., & Näätänen, R. (1984). Short-term habituation and dishabituaton of the mismatch negativity of the ERP. *Psychophysiology, 21*, 434–441.

Sams, M., Hari, R., Rif, J., & Knuuttila, J. (1993). The human auditory sensory memory trace persists about 10 s: Neuromagnetic evidence. *Journal of Cognitive Neuroscience, 5*, 363–370.

Sams, M., Paavilainen, P., Alho, K., & Näätänen, R. (1985). Auditory frequency discrimination and event-related potentials. *Electroencephalography and Clinical Neurophysiology, 62*, 437–448.

Sato, Y., Yabe, H., Hiruma, T., Sutoh, T., Shinozaki, N., Nashida, T., et al. (2000). The effect of deviant stimulus probability on the human mismatch process. *Neuroreport, 11*, 3703–3708.

Schröger, E., Tervaniemi, M., Wolff, C., & Näätänen, R. (1996). Preattentive periodicity detection in auditory patterns as governed by time and intensity information. *Cognitive Brain Research, 4*, 145–148.

Shestakova, A., Brattico, E., Huotilainen, M., Galunov, V., Soloviev, A., Sams, M., et al. (2002). Abstract phoneme representations in the left temporal cortex: Magnetic mismatch negativity study. *Neuroreport, 13*, 1813–1816.

Shtyrov, Y., Kujala, T., Ahveninen, J., Tervaniemi, M., Alku, P., Ilmoniemi, R. J., et al. (1998). Background acoustic noise and the hemispheric lateralization of speech processing in the human brain: Magnetic mismatch negativity study. *Neuroscience Letters, 251*, 141–144.

Shtyrov, Y., Pulvermüller, F., Näätänen, R., & Ilmoniemi, R. J. (2003). Grammar processing outside the focus of attention: An MEG study. *Journal of Cognitive Neuroscience, 15*, 1195–1206.

Sokolov, E. N., Spinks, J. A., Näätänen, R., & Lyytinen, H. (2002). *The orienting response in information processing*. Mahwah, NJ: Erlbaum.

Squires, N. K., Squires, K. C., & Hillyard, S. A. (1975). Two varieties of long-latency positive waves evoked by unpredictable auditory stimuli in man. *Electroencephalography and Clinical Neurophysiology, 38*, 387–401.

Takegata, R., Brattico, E., Tervaniemi, M., Varyagina, O., Näätänen, R., & Winkler, I. (2005). Preattentive representation of feature conjunctions for simultaneous, spatially distributed auditory objects. *Cognitive Brain Research, 25*, 169–179.

Tales, A., Newton, P., Troscianko, T., & Butler, S. (1999). Mismatch negativity in the visual modality. *Neuroreport, 110*, 3363–3367.

Tervaniemi, M., & Brattico, E. (2004). From sounds to music. Towards understanding the neurocognition of musical sound perception. *Journal of Consciousness Studies, 11*, 9–27.

Tervaniemi, M., Kujala, A., Alho, K., Virtanen, J., Ilmoniemi, R. J., & Näätänen, R. (1999). Functional specialization of the human auditory cortex in processing phonetic and musical sounds: A magnetoencephalographic (MEG) study. *Neuroimage, 10,* 330–336.

Tervaniemi, M., Maury, S., & Näätänen, R. (1994). Neural representations of abstract stimulus features in the human brain as reflected by the mismatch negativity. *Neuroreport, 5,* 844–846.

Tervaniemi, M., Medvedev, S. V., Alho, K., Pakhomov, S. V., Roudas, M. S., van Zuijen, T. L., et al. (2000). Lateralized automatic auditory processing of phonetic versus musical information: A PET study. *Human Brain Mapping, 10,* 74–79.

Tervaniemi, M., Winkler, I., & Näätänen, R. (1997). Pre-attentive categorization of sounds by timbre as revealed by event-related potentials. *Neuroreport, 8,* 2571–2574.

Tiitinen, H., May, P., Reinikainen, K., & Näätänen, R. (1994). Attentive novelty detection in humans is governed by pre-attentive sensory memory. *Nature, 370,* 90–92.

Toiviainen, P., Tervaniemi, M., Louhivuori, J., Saher, M., Huotilainen, M., & Näätänen, R. (1998). Timbre similarity: Convergence of neural, behavioral, and computational approaches. *Music Perception, 16,* 223–241.

Trejo, L. J., Ryan-Jones, D. L., & Kramer, A. F. (1995). Attentional modulation of the mismatch negativity elicited by frequency differences between binaurally presented tone bursts. *Psychophysiology, 32,* 319–328.

Tse, C.-Y., Tien, K. R., & Penney, T. B. (2006). Event-related optical imaging reveals the temporal and frontal cortex activation in pre-attentive change detection. *Neuroimage, 29,* 314–320.

Vuust, P., Pallesen, K. J., Bailey, C., van Zuijen, T. L., Gjedde, A., Roepstorff, A., et al. (2005). To musicians, the message is in the meter: Pre-attentive neuronal responses to incongruent rhythm are left-lateralized in musicians. *Neuroimage, 24,* 560–564.

Winkler, I., Czigler, I., Sussman, E., Horváth, J., & Balázs, L. (2005). Preattentive binding of auditory and visual stimulus features. *Journal of Cognitive Neuroscience, 17,* 320–339.

Winkler, I., Karmos, G., & Näätänen, R. (1996). Adaptive modeling of the unattended acoustic environment reflected in the mismatch negativity event-related potential. *Brain Research, 742,* 239–252.

Winkler, I., Kujala, T., Tiitinen, H., Sivonen, P., Alku, P., Lehtokoski, A., et al. (1999). Brain responses reveal the learning of foreign language phonemes. *Psychophysiology, 36,* 638–642.

Woldorff, M. G., Hackley, S. A., & Hillyard, S. A. (1991). The effects of channel-selective attention on the mismatch negativity wave elicited by deviant tones. *Psychophysiology, 28,* 30–42.

Woldorff, M. G., Hillyard, S. A., Gallen, C. C., Hampson, S. R., & Bloom. F. E. (1998). Magnetoencephalographic recordings demonstrate attentional modulation of mismatch-related neural activity in human auditory cortex. *Psychophysiology, 35,* 283–292.

Woods, D. L., Alho, K., & Algazi, A. (1992). Intermodal selective attention I: Effects on event-related potentials to lateralized auditory and visual stimuli. *Electroencephalography and Clinical Neurophysiology, 82,* 341–355.

Yabe, H., Tervaniemi, M., Reinikainen, K., & Näätänen, R. (1997). Temporal window of integration revealed by MMN to sound omission. *Neuroreport, 8,* 1971–1974.

Yabe, H., Tervaniemi, M., Sinkkonen, J., Huotilainen, M., Ilmoniemi, R. J., & Näätänen, R. (1998). Temporal window of integration of auditory information in the human brain. *Psychophysiology, 35,* 615–619.

Yucel, G., Petty, C., McCarthy, G., & Belger, A. (2005a). Visual task complexity modulates the brain's response to unattended auditory novelty. *Neuroreport, 16,* 1031–1036.

Yucel, G., Petty, C., McCarthy, G., & Belger, A. (2005b). Graded visual attention modulates brain responses evoked by task-irrelevant auditory pitch changes. *Journal of Cognitive Neuroscience, 17,* 1819–1828.

Zhang, P., Chen, X., Yuan, P., Zhang, D., & He, S. (2006). The effect of visuospatial attentional load on the processing of irrelevant acoustic distractors. *Neuroimage, 33,* 715–724.

CHAPTER

7 Neuropsychology of P300

John Polich

Abstract

The discovery of the P300 event-related potential (ERP) stimulated the use of brain recording methods to assess human cognition. This chapter reviews the background and develops an integrated interpretation of P300. First, empirical issues and a theoretical overview are presented. Second, applied uses of P300 are reviewed, with normative and clinical studies highlighted. Third, the neuropsychological background and neurophysiological foundations of the P3a and P3b subcomponents are outlined. Fourth, neuropharmacological processes related to these constituent potentials are sketched to suggest how neurotransmitter systems may contribute to P300 production. Fifth, the P3a and P3b are proposed to result from neuroinhibition that is engaged when incoming stimuli garner attentional processes to facilitate memory encoding.

Keywords: P300, event-related potential, neuroinhibition, memory, cognition

Introduction

Discovery of the P300 event-related potential (ERP) stimulated the use of brain recording methods to assess human cognition. Early theoretical explanations of this waveform were unitary and functional in nature, but the phenomenon is now considered to reflect information-processing cascade associated with attentional and memory mechanisms. A convergence of approaches is beginning to limn the basic circuitry and transmitter systems related to P300 generation, with theoretical inroads being made into how this brainwave is associated with the experience of mental events.

The present chapter reviews the background and develops an integrated interpretation of P300. It is organized into sections: First, empirical issues and a theoretical overview are presented. Second, applied uses of P300 are reviewed, with normative and clinical studies highlighted. Third, the neuropsychological background and neurophysiological foundations of the P3a and P3b subcomponents are outlined. Fourth, neuropharmacological processes related to these constituent potentials are sketched to suggest how neurotransmitter systems may contribute to P300 production. Fifth, the P3a and P3b are proposed to result from neuroinhibition that is engaged when incoming stimuli garner attentional processes to facilitate memory encoding.

To maintain nomenclature consistency, the term *P300* is used to refer to the canonical ERP component, which also has been called the *P3* and sometimes the *late positive component* (LPC). The terms *P3a* and *P3b* denote distinct subcomponents of the P300 as defined below. The contemporary background to P300 is covered, with early findings reviewed (Johnson, 1986; Picton, 1992; Pritchard, 1981) and additional summaries of related topics available elsewhere (Cycowicz, 2000; DeBoer et al., 2006; Jeon & Polich, 2003; Kramer & Parasuraman, 2008; Reinvang, 1999; Verleger, 2003).

Characteristics and Theory

A Short History

The P300 was first reported by Sutton and colleagues (1965). Its origins stemmed in part from the confluence of information theory on psychological research and increased technological capability for signal averaging of human neuroelectric measures. The original studies manipulated stimulus information to assess how electric brain patterns varied among conditions (Sutton, 1979; see Bashore & van der Molen, 1991, for a superb review). Subsequent results elucidated stimulus probability and task relevance effects often obtained from *oddball* paradigm data (Donchin et al., 1978).

Figure 7.1 illustrates variants of the oddball task. The single-stimulus procedure presents an infrequent target with no other stimuli. The two-stimulus task is the traditional oddball and presents an infrequent target in a background of frequent standard stimuli. The three-stimulus task presents an infrequent target in a background of frequently occurring standard stimuli and infrequently occurring distracter stimuli. In each case, the subject is instructed to respond mentally or physically to the target stimulus and typically not to respond otherwise. For example, a common procedure is the *auditory oddball*, in which two tones that typically vary in pitch (500 vs. 1000 Hz) and probability of occurrence (.80 vs. .20) are presented randomly with an interstimulus interval of 1 to 2 s. Discriminating the infrequent *target* from the more frequently occurring *standard* produces a robust P300. Some task situations require responses to both the target and the standard, which will elicit P300 components from both stimulus types that vary in amplitude (see below).

The P300 component is measured by assessing its amplitude and latency. Amplitude (µV) usually is defined as the difference between the mean

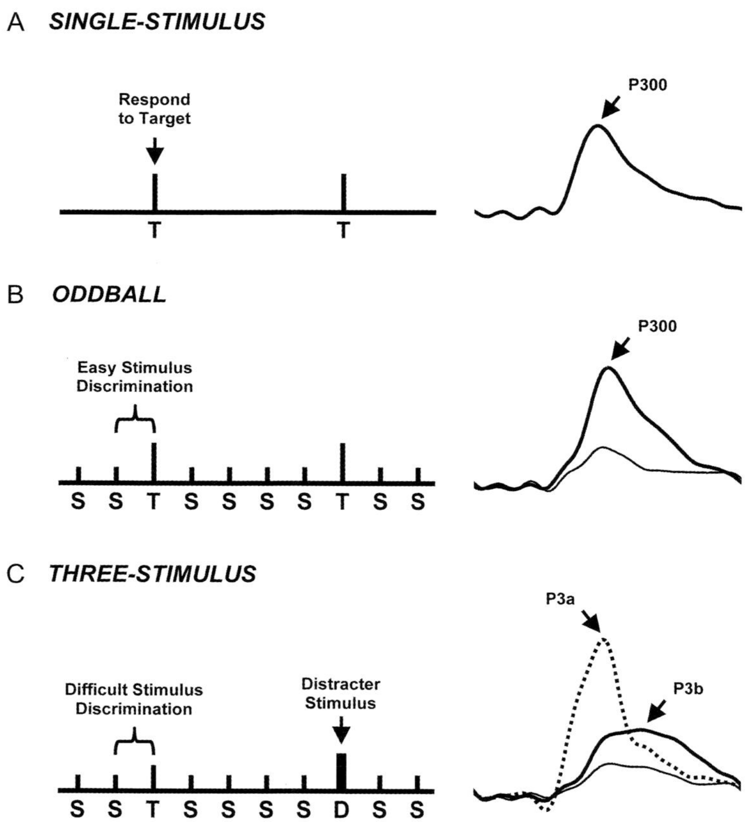

Fig. 7.1. Schematic illustration of the single-stimulus, oddball, and three-stimulus paradigms, with the elicited ERPs from the stimuli of each task at the right. (A) The single-stimulus task presents an infrequent target ("T") without any other stimuli. (B) The oddball task presents two different stimuli in a random sequence, with one occurring less frequently than the other (T = target, S = standard). (C) The three-stimulus task is similar to the oddball task but has a difficult target/standard discrimination and a compelling distracter ("D") stimulus that occurs infrequently. In each task, the subject is instructed to respond only to the target and to refrain from responding otherwise. The distracter elicits a P3a, and the target elicits a P3b (after Polich, 2007).

prestimulus baseline voltage and the largest positive-going peak of the ERP waveform within a time window determined by stimulus modality, task conditions, subject age, and other factors. Latency (ms) is typically defined as the time from stimulus onset to the point of maximum positive amplitude within this same time window. P300 scalp distribution is characterized as the amplitude change over the midline electrodes (Fz, Cz, Pz) that increases from the frontal to the parietal electrode sites for target stimuli (Johnson, 1993).

Increases in P300 amplitude or size are usually interpreted as reflecting greater neuronal activity in some fashion that is modulated by internal biological and external cognitive factors, as outlined below. These factors also often influence component peak latency, which is interpreted as indexing stimulus evaluation time and is relatively short when task processing is easy and long when task processing is difficult. Biological as well as task variables contribute to peak timing, and these also are summarized below. The amplitude and latency distribution over the scalp topography, wherein amplitude becomes larger and latency generally shorter from frontal to parietal electrode sites for typical P300 components from an oddball task, is used as a means to identify the potential and to differentiate the several subtypes that are discussed below. Thus, component size, its timing, and its shape across the scalp provide the definitional characteristics of the P300 and its variants.

Context Updating Theory

Figure 7.2 schematically illustrates a theoretical account of the oddball task, which posits that the P300 indexes brain activities underlying revision of the mental representation induced by incoming stimuli (Donchin, 1981). After initial sensory input, an attention-driven comparison evaluates the representation of the previous event in working memory. If no stimulus attribute change is detected, the current mental model or *schema* of the stimulus context is maintained, and only sensory potentials are evoked. When a new stimulus attribute is detected, the "updating" of the neural stimulus representation in working memory occurs and P300 is produced.

Despite the simplicity of the task situation and the reliability of obtaining ERPs using oddball paradigms, a clear understanding of how and why the brain produces the P300 remains elusive. Task demands determine how the context is refurbished, but updating operations occur as the intervening nontarget events engage attention and the current neural representation is modified (Donchin et al., 1986). The context-updating hypothesis is the major theoretical account of P300, although other positions have emerged (e.g., Mecklinger & Ullsperger, 1993, 1995; Nieuwenhuis et al., 2005). However, P300 amplitude from an oddball task does demonstrate habituation and dishabituation effects, which strongly imply that this ERP component indexes fundamental attention and memory-related operations (Kok, 1997; Polich, 1989a; Rushby et al., 2005).

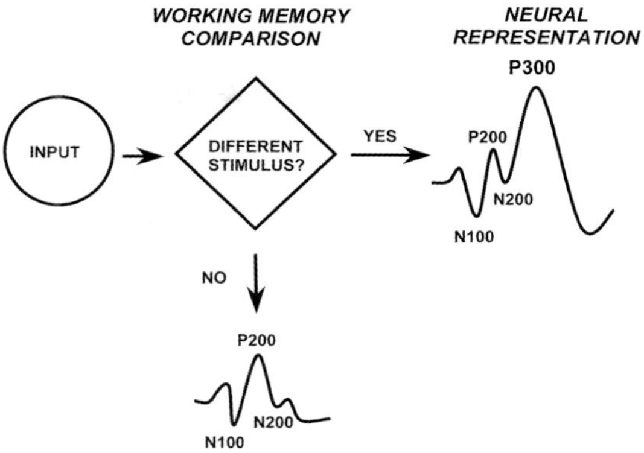

Fig. 7.2. Schematic illustration of the P300 context-updating model. Stimuli enter the system and a memory comparison process is engaged that ascertains whether the current stimulus is the same as the previous stimulus or not (e.g., in the oddball task, whether a standard or a target stimulus was presented). If the incoming stimulus is the same, the neural model of the stimulus environment is unchanged, and only sensory potentials are evoked (N100, P200, N200). If the incoming stimulus is not the same and the subject allocates attentional resources to the target, the neural representation of the stimulus environment is updated, such that a P300 potential (P3b) is elicited in addition to the sensory potentials (after Polich, 2007).

The P300 context-updating hypothesis was derived in large measure from manipulating target stimulus probability in the oddball task. Discriminating a target from a standard stimulus produces a robust P300 that increases in amplitude as the global and local sequence probability for the target stimulus decreases (Duncan-Johnson & Donchin, 1977; Squires et al., 1976). These findings implied that P300 originates at least in part from working memory comparisons and that conscious awareness is affected by stimulus sequence effects (Leuthold & Sommer, 1993; Sommer et al., 1990).

P300 Amplitude

Resource Allocation and P300

P300 amplitude also is sensitive to the amount of attentional resources engaged during dual-task performance. In a dual-task design, a primary task that varies cognitive demands is performed while the subject also is engaged in a secondary task of mentally counting target oddball stimuli. As primary task difficulty is increased, target stimulus P300 amplitude from the oddball task decreases, regardless of modality or the motor requirements of the primary task (Isreal et al., 1980; Wickens, Kramer et al., 1983). When primary task conditions are easy, P300 amplitude from the secondary task is relatively large, whereas during difficult primary tasks, P300 amplitude is small since more processing resources are used for task performance (Kramer et al., 1985).

Target-to-Target Interval

P300 amplitude for a target in an oddball task increases as the number of consecutive standards preceding the target increases. Hence, P300 amplitude to a target preceded by a target is relatively small, whereas P300 amplitude to a target preceded by several standards is relatively large. This component amplitude sensitivity to the preceding stimulus structure has been interpreted as reflecting the local sequential probability (Sommer et al., 1990; Squires et al., 1976), which is governed by the overall global target stimulus probability (Duncan-Johnson & Donchin, 1977). A key determinant of these effects, however, is the rate at which the stimuli were presented.

Figure 7.3 illustrates the influence of time on P300 as a function of target-to-target interval (TTI), which is determined by the sequences of nontarget stimuli (Gonsalvez & Polich, 2002). This phenomenon is an important empirical restriction on P300 probability and sequence effects, as TTI governs how quickly resources can be directed to target stimuli (Pashler, 1994). Short intervals produce smaller P300 components than longer intervals, with TTIs of 6–8 s or greater eliminating probability effects (Polich, 1990). These limitations may originate from temporal memory trace development that governs the representational event quality underlying P300 generation (Gonsalvez et al., 2007). Thus, context updating may be optimally controlling P300 amplitude when relatively rapid stimulus changes occur, with resource allocation ultimately determining component size when TTI is long.

Such an interpretation is consonant with P300 findings from single-stimulus paradigms in which only the target stimulus occurs randomly and variably in time (Polich et al., 1994). This task produces P300 components comparable to the traditional two-stimulus oddball paradigm across modality, probability, and response parameters (Katayama & Polich, 1996; Mertens & Polich, 1997).

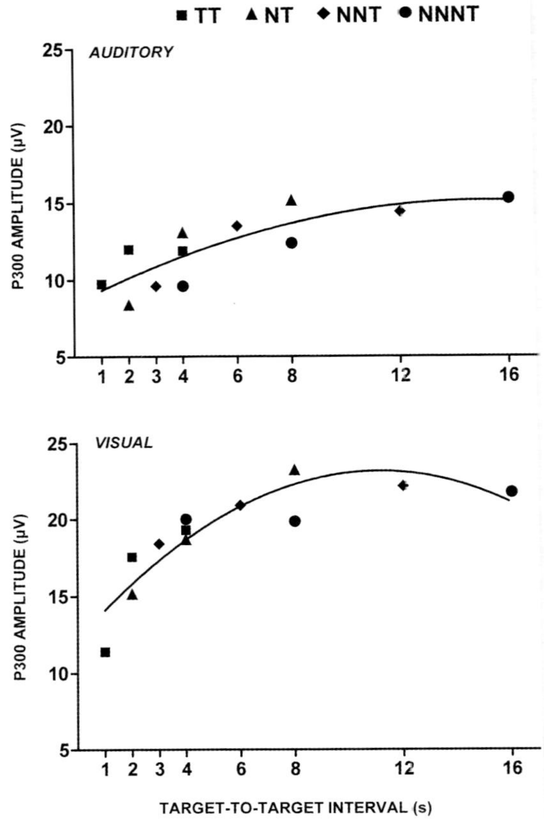

Fig. 7.3. P300 amplitude as a function of target-to-target interval (TTI) time, which varies across stimulus sequences by the number of nontarget ("N") items between each occurrence of a target ("T"). The subject is instructed to respond to the target stimulus. P300 amplitude increases independently of sequence and global target probability. The regression lines reflect the curvilinear best fit for a second-order polynomial. Similar results have been found for the single-stimulus paradigm when only target stimuli are presented (after Polich, 2007).

Furthermore, topographic localization methods have demonstrated similar waveforms, amplitude distributions, and dipole coordinates across tasks (Croft et al., 2003; Tarkka & Stokic, 1998). Even when target stimulus probability is unitary, the time between events is the primary determinant of P300 amplitude.

Memory and P300

The context updating origin of P300 theory was derived in part from consideration of the orienting response (Donchin et al., 1986; Sokolov, 1977), with attention allocation findings leading to using P300 as an index of recall memory (Karis et al., 1984). Lists of words were presented sequentially and ERPs recorded. In some lists, one of the words was presented in a different font size than the other words so that stimulus distinctiveness would enhance encoding to facilitate recall memory. Stimuli that were made distinctive were more likely to be recalled and elicited larger P300 components during encoding than those that were not recalled. However, P300 size was directly affected by the rehearsal method, such that component amplitude was larger for items subsequently recalled if participants used a rote rather than a semantic strategy (Fabiani et al., 1986). When participants employed an elaborative strategy such as developing an internal story line or related images of the stimulus words to facilitate memorization, recalled and unrecalled words produced P300 components that were unaffected by stimulus distinctiveness.

These findings implied that memory storage operations from rote memorization reflected incidental rather than effortful processing, suggesting that P300 amplitude changes were related to initial encoding (cf. Fabiani et al., 1990; Paller et al., 1988). Indeed, subsequent studies have found that stimuli presented in a nonmemory response task yield representations in working memory (words, objects) that elicit larger P300 components when presented in a recognition paradigm to their recurrence than stimuli not formerly presented (Curran & Cleary, 2003; Guo et al., 2006; McEvoy et al., 2001). Memory-related changes in P300 amplitude, therefore, appear to index the updating processes associated with encoding operations that facilitate subsequent access of the remembered.

Serial Position Memory and P300

Event-related potential memory effects often are obtained with recognition rather than recall procedures (see Chapter 14, this volume). After the participant memorizes a list of items, a probe stimulus is presented with the instruction to indicate whether that item is from the memory set. A robust finding is that probe stimuli from the last serial position produce shorter response times and larger P300 components from recognition than those from the beginning or middle of the list (Crites et al., 1998; Patterson et al., 1991). P300 amplitude therefore reflects the strength of memory formed during encoding and storage processes, which vary across serial position in recognition tasks.

Event-related potentials and serial position at the time of encoding also have been studied, with participants instructed to use an elaborative learning strategy to chunk the words in groups of three or five items (Rushby et al., 2002). Comparisons made between the recalled and unrecalled words for the primacy, middle, and recency positions found that early ERP components were modulated by serial position: Primacy recall was associated with increased P1 and P2 amplitudes, whereas recall at the middle positions only demonstrated increased P1 amplitude. Even though behavioral recall reflected both primacy and recency effects, ERPs were not modulated by recency during encoding.

These mechanisms are likely to be operating when assays of serial position are used to elicit ERPs, with relatively large P300 amplitudes obtained for recency items rather than earlier probes (Chao & Knight, 1996). Indeed, P300 amplitude for probe stimuli increases linearly from the initial to the ending list items, suggesting that recognition primacy and recency effects reflect different processing mechanisms (Golob & Starr, 2004; Wiswede et al., 2007). Studies involving ERPs have demonstrated a strong relationship between encoding and subsequent retrieval (Friedman & Trott, 2000; Guo et al., 2006). Depth of encoding produces stronger retrieval performance for items in the initial compared to subsequent serial positions, as these receive greater attentional processing through rehearsal (Curran, 2004), and the strength of attention declines across successive trials (Farrell & Lewandowsky, 2002; Page & Norris, 1998).

Figure 7.4 summarizes findings that illustrate this theoretical position (Azizian & Polich, 2007). Participants were instructed to remember lists of 12 common nouns each presented once every 1.5 s, for 20 different word lists. A recall signal followed the last word in each list to indicate that all remembered items should be written on paper. Figure 7.4A illustrates the recall performance data, which evinced the classic serial position curve. Figure 7.4B illustrates the mean amplitude area for the late positive component (500–750 ms) from each of the midline electrodes. For the primacy items, area was greater for the recalled compared to the unrecalled words. Figure 7.4C illustrates the amplitude scalp distributions for recalled and unrecalled items for each serial position block. P300 amplitude appears to index the operation of an attentional gradient that modulates stimulus encoding for memory storage. Given that the attentional gradient strength is reflected by P300, primacy items that were correctly recalled produced larger amplitudes relative to recency items that had the same amplitude for both recalled and unrecalled items. Hence, words studied with full attention are recognized more confidently and have larger P300 amplitudes (Fabiani et al., 1986; Karis et al., 1984). The encoding strength therefore underlies successful memory storage and

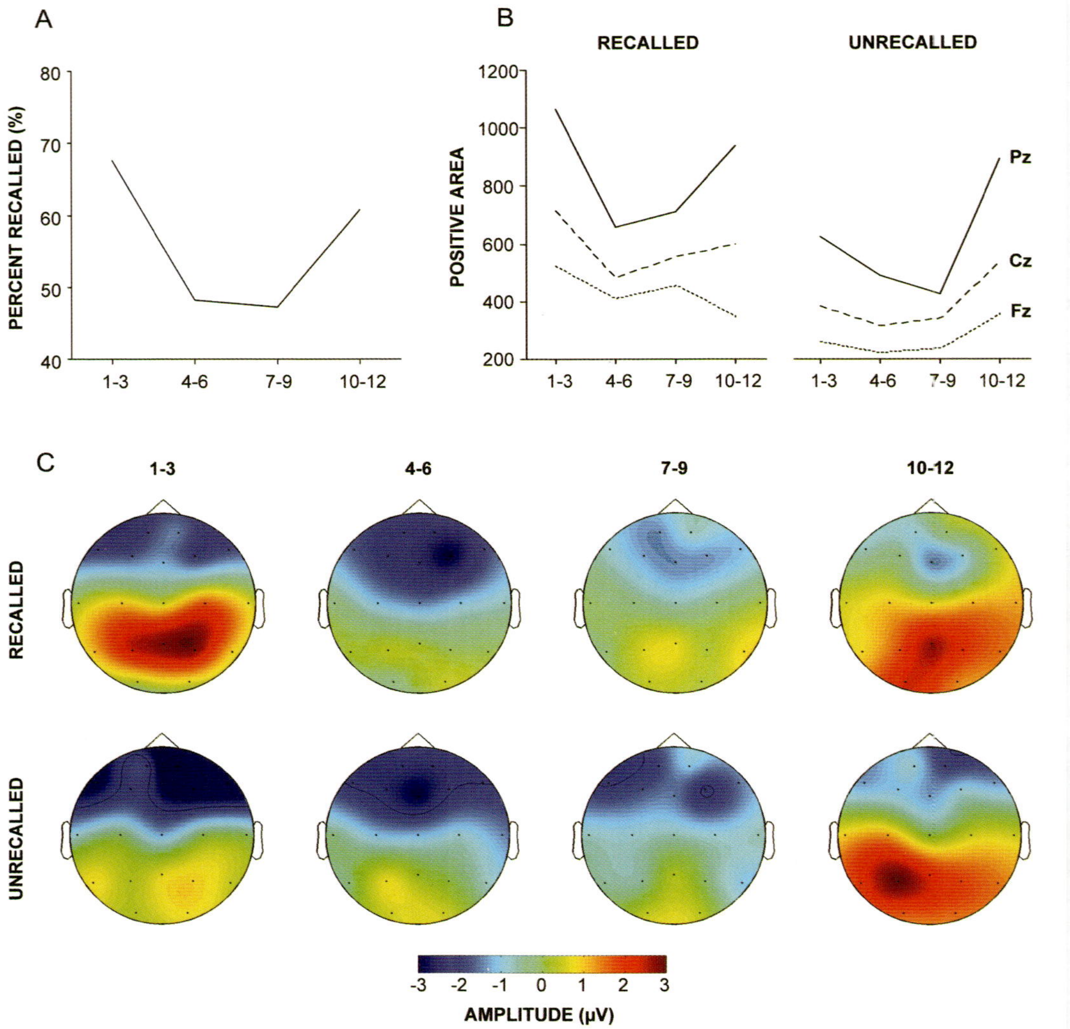

Fig. 7.4. (A) Percentage of correctly recalled word items from each serial position block (comprised of word trials 1–3, 3–6, 7–9, and 10–12). (B) Mean waveform amplitude (500–750 ms) as a function of serial position block for correctly recalled and unrecalled word items from the Fz, Cz, Pz electrodes (N = 30). (C) Topographic amplitude (500–750 ms) distributions from each serial position block for recalled and unrecalled stimulus words (after Azizian & Polich, 2007).

retrieval (Azizian & Polich, 2007; Wiswede et al., 2007).

Memory Encoding and P300

Increases in memory load reduce P300 amplitude in a manner that suggests that fewer attentional resources are engaged because of increased task demands to process these items (Gomer et al., 1976; Kok, 2001; Wijers et al., 1989). Moreover, retrieval does not produce amplitude variation of the early ERP components, which is inconsistent with findings that primacy recall is reflected by increased P1 and P2 amplitudes (Rushby et al., 2002). Larger late positive amplitudes occur for the recency items from both recalled and unrecalled items, most likely because the temporal cue for recall indicated that the memory set would end. This finding is similar to the enhanced positivity observed for cue stimuli that index cognitive shifts induced by tasks such as the Wisconsin Card Sorting Test (Barceló et al., 2002; Watson et al., 2006).

As noted, the primacy ERP outcomes for recalled and unrecalled items were distinguishable, but the recency effects were similar across both recalled and unrecalled items. The discrepancy may indicate that the enhanced positivity for primacy items reflects serial position encoding and memory storage, whereas recency items were limited to encoding. The difference between recall memory for items in the primacy and recency positions could stem from interference and memory overload (Doyle & Rugg, 1992; Otten et al., 2006). Although the P300 may not be a direct reflection of storage processes, memory for items that elicit a P300 is stronger than memory for items that do not elicit a P300 even for task conditions that require delayed retrieval. Thus, the close relationship of P300 amplitude and delayed retrieval supports the view that this potential is one of the series of processes that reflect the encoding of events for storage in memory.

P300 Latency

P300 latency is thought to index classification speed, which is proportional to the time required to detect and process a target item (Kutas et al., 1977; Magliero et al., 1984). P300 peak latency changes over the scalp and is shorter over frontal areas and longer over parietal areas (Conroy & Polich, 2007; Polich et al., 1997). Stimulus and task requirements contribute to the association between P300 latency and response time, but the strength or sensitivity of the relationship between latency and response time varies across stimulus-response compatibility and Stroop choice-response tasks (Duncan-Johnson & Kopell, 1981; McCarthy & Donchin., 1981). Semantic-based compatibility tasks produce a larger P300 latency/response time difference compared to spatial compatibility tasks. Furthermore, P300 latency has been used as a metric for timing mental events related to other ERP components (Renault et al., 1982; Ritter et al., 1983). Inferences based on component timing therefore need to consider the processes underlying response activation (Pfefferbaum et al., 1986; Ragot, 1984). Indeed, since P300 timing is sensitive to both stimulus- and response-related variables when responding is fast (cf. Ilan & Polich, 1999; Verleger, 1997), this conclusion has suggested that P300 may originate from the neural events that link stimulus perception and event response (Verleger et al., 2005).

Individual differences for P300 latency are correlated with mental speed, such that shorter latencies are related to superior cognitive performance (Pelosi et al., 1992a; Polich et al., 1983). The neuropsychological tests that produce the strongest correlation between P300 latency and cognitive capability assess how rapidly subjects can allocate attentional resources (Houlihan et al., 1998; Reinvang, 1999). P300 latency decreases as children develop (Howard & Polich, 1985; Polich et al., 1990b) and increases with normal aging (Fjell &Walhovd, 2001; Polich, 1996). Component peak latency also becomes longer as dementia level increases (O'Donnell et al., 1992; Polich et al., 1986, 1990a), although how brain insult or disease prolongs ERP timing is unclear (Bashore & Ridderinkhof, 2002; Potter & Barrett, 1999 Rossini et al., 2007).

Applied P300

Individual Differences

Test-retest correlation coefficients for oddball task P300 amplitude range from .50 to .80 and for peak latency from 0.40 to 0.70 (Fabiani et al., 1987; Segalowitz & Barnes, 1993; Walhovd & Fjell, 2002). Latency jitter from individual trial variability can contribute to amplitude effects, but it is relatively minimal for oddball tasks (Cohen & Polich, 1997; Michalewski et al., 1988). Individual ERP differences are influenced by ultradian rhythms that affect measurement stability (Lin & Polich, 1999; Ravden & Polich, 1999). Neuroelectric signals are genetically transmitted, with strong EEG spectral power similarities observed among biological family members (Almasy et al., 1999; Eischen et al., 1995). Electroencephalographic (EEG) spectral characteristics are highly similar for identical twins (van Beijsterveldt & van Baal, 2002), and P300 is virtually identical for pairs of monozygotic twins, less so for dizygotic twins, and different for unrelated controls (Katsanis et al., 1997; O'Connor et al., 1994; Polich & Burns, 1987). P300 heritability also is evidenced by biologically related family members who demonstrate significant interfamily member correlations for ERP measures (Eischen & Polich, 1994; Polich & Bloom, 1999). Specific loci on the human genome appear to determine ERP characteristics (Begleiter et al., 1998), which suggests that the P300 may be used as a biomarker for some disease phenotypes (e.g., Carlson et al., 1999; Jeon & Polich, 2003; Porjesz et al., 2005).

The genetic underpinnings for P300 are supported by findings for personality attributes such as introversion/extraversion, sensation seeking, and impulsivity (Gurrera et al., 2001; Polich & Martin, 1992; Stelmack & Houlihan, 1994). Although the relationship among ERP measures and personality is still murky, a correlation between individual differences for personality-related arousal levels and P300 is generally observed: Low-arousal individuals have smaller amplitudes compared to high-arousal individuals who have larger components (Brocke, 2004; De Pascalis, 2004). However, ERP personality variation measures also are modulated by biological factors (Polich & Kok, 1995), paradigm differences (DiTraglia & Polich, 1991; Stenberg, 1992, 1994), and psychopathology (Iacono et al., 2002, 2003; Justus et al., 2001). Furthermore, individual differences for attentional resource capabilities also are affected by variability for neurotransmitter function (Hill et al., 1998, 1999; Polich & Criado, 2006).

Biological Factors

P300 measures reflect naturally occurring tonic and situational phasic arousal (Kok, 1990; Pribram & McGuinness, 1975). Event-related potential trait and state arousal effects originate from biological factors that engage attention and working memory processes, such that the resources underlying P300 generation are altered (Kok, 1997, 2001). Tonic changes are relatively slow manifestations of energetic state fluctuations, whereas phasic arousal indicates the organism's energetic reaction to specific events to affect the availability of attentional resources and therefore the strength of P300. These biological determinants occur spontaneously, can be induced by the environment, or stem from constitutional differences (Polich & Kok, 1995).

Table 7.1 summarizes P300 biological effects. Event-related potential studies typically employ within-subject designs, so that most of these variables do not change within a recording session, and their influence can be distributed by standard counterbalancing methods. However, between-subjects designs that compare groups should measure biological determinants of P300 to minimize intergroup variability and statistical error. Biological influences can be reduced by ensuring that ERPs are recorded similarly with respect to temporal variables (circadian, ultradian, seasonal, etc.) and bodily functions (fatigue, food, temperature, etc.), and by matching patients on constitutional factors (age, gender, handedness, etc). Consideration of neuropsychological differences also provides control over cognitive capabilities (Colet et al., 1993; Kujala & Näätänen, 2003; Orlebeke et al., 1989; Pelosi et al., 1992a, 1992b; Stelmack & Houlihan, 1994).

Figure 7.5 illustrates an example of how individual biological variation can influence P300 (Polich & Ochoa, 2004). The topographic amplitude mappings from a visual target in an oddball task are presented for young adult groups that vary in the risk for alcoholism (defined by a high incidence of familial alcoholism) and smoking status (defined by consistent tobacco smoking for more than 6 months). P300 amplitude is smaller for individuals with a high compared to a low risk for alcoholism but is decreased even more for those who are currently smoking tobacco. Indeed, tobacco smoking accounted for twice the P300 amplitude variance relative to alcoholism risk background, which may imply that the groups differ in dopaminergic systems (Anokhin et al., 2000); Hill et al., 2001). Thus, control of such variability is an important consideration for evaluating applied P300 and other ERP component studies.

Normal Aging and P300

Figure 7.6 illustrates P300 amplitude and latency data elicited by auditory and visual stimuli in an oddball task from a large study of normal adults with equal numbers of each gender (Polich, 1997a). With increased adult age, P300 amplitude declines and peak latency increases. The impetus for assessing clinical ERPs stemmed from the need to evaluate older demented subjects (Bashore, 1990; Polich & Luckritz, 1995). Behavioral studies had found that compared to younger subjects, healthy elderly subjects demonstrate a slowdown of information processing and a decline of short-term memory (Bashore & Ridderinkhof, 2002). However, this goal was difficult to implement empirically because normative age-related ERP variability is greater than that found for sensory evoked potentials (Fein & Turetsky, 1989; Polich & Starr, 1984).

Table 7.2 summarizes the findings of a meta-analysis of P300 normative studies of aging that was performed to define moderator variables as sources of interstudy variability based on obvious classification schemes (male vs. female, number of rare stimuli, etc.) and those derived from the variable's low-to-high (stimulus intensity, duration, etc.) values (Polich, 1996). The findings suggest that P300 latency variability from normal aging could be reduced by systematic evaluation of sample and task characteristics. For example, both auditory and visual modalities should be used in separate conditions, with task conditions designed to produce different degrees of difficulty and elicit various P300 subcomponents (P3a, P3b, go/no-go, etc.), as outlined below, to provide neuroelectric assays of different mental abilities and brain areas. Single-stimulus or passive paradigms may be used for children or patients who cannot follow task instructions (Knight, 1987; Walhovd & Fell, 2008).

Clinical P300 Applications

The P300 component has been used to study neurological and psychiatric cognitive impairment. However, its utility has been limited because normative data and standard clinical methods have not been established, although progress is being made

Table 7.1 P300 amplitude and latency biological determinants (after Polich and Kok, 1995)

FACTOR	AMPLITUDE	LATENCY	COMMENT
Natural			
Circadian	Indirect	Indirect	Circadian body changes affect P300 measures
Body Temperature	No	Yes	Increased temperature, decreased latency
Heart Rate	No	Yes	Faster heart rate, decreased latency
Food Intake	Yes	Some	Amplitude increases, latency shorter
Activity Time	Yes	Some	Food interacts with activity preference time
Ultradian	Some	Yes	90 min latency cycles
Seasonal	Yes	No	Seasons with light, increased amplitude
Menstrual	No	No	Neutral stimuli, no effects
Induced			
Exercise	Indirect	Direct	Affects overall arousal level
Tonic	Yes	Yes	Increases amplitude, decreases latency
Chronic	No	Yes	Decreased latency, variable results across studies
Fatigue	Yes	Yes	Decreased amplitude, increased latency
Drugs (Common)	Yes	Yes	Specific drug, arousal level, tonic/chronic use
Caffeine	Some	Yes	Amplitude increases if fatigued, latency decreases
Nicotine	Small	Yes	Weak amplitude effects, latency decreases
Alcohol			
Acute	Yes	Yes	Amplitude decreases, latency increases
Chronic	No	No	Social drinking: No permanent long-term effects
Alcoholism Risk	Yes	No	At-risk: smaller amplitudes with visual tasks
Constitutional			
Age	Yes	Yes	Modality, task, response parameters important
Children	Yes	Yes	Amplitude increases, latency decreases
Adults	Yes	Yes	Amplitude decreases, latency increases
Intelligence	Yes	Yes	Amplitude from complex tasks smaller for more intelligent, latency shorter for perceptual/speeded classification tasks for more intelligent
Handedness	Yes	Yes	Amplitude: left > right for frontal/central sites Latency: left < right for frontal/central sites
Gender	Small	Small	Amplitude: female > male, latency: female < male
Personality	Yes	No	Amplitude: introverts < extroverts
Genetic	Yes	Yes	Amplitude and latency genetically determined

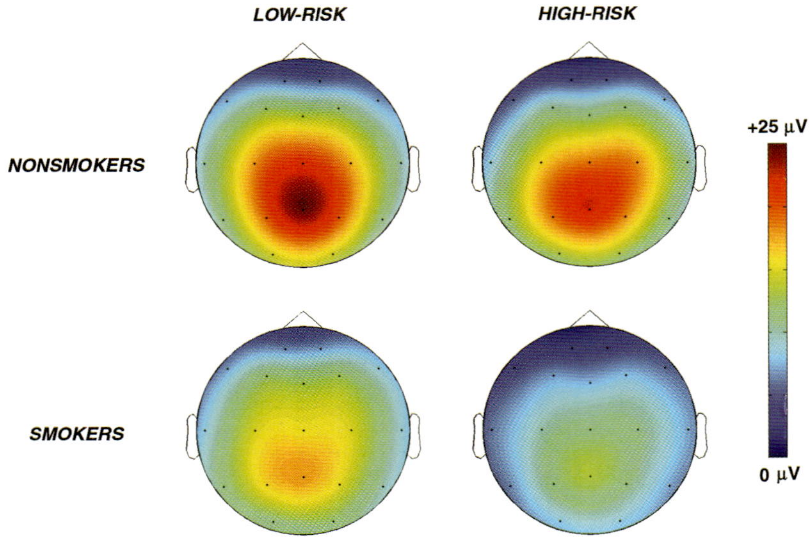

Fig. 7.5. Mean P3b amplitude from a visual target stimulus for young adults at low risk and high risk for alcoholism who were nonsmokers or regular cigarette smokers (n = 20/group, F = M). All subjects were assessed in their normal tobacco use state (after Polich & Ochoa, 2004).

(Duncan et al., 2008). The suggestion that the P300 might be used to assess cognitive function originated from studies of dementing illness, since peak latency from patients was prolonged compared to age-matched normal subjects (Goodin et al., 1978; Polich et al., 1986). Most brain disorders affect the fundamental cognitive operations of attention allocation and immediate memory, and therefore influence P300 amplitude or latency. A systematic evaluation of normative P300 measures compared to routine biomedical assays found them to be quite similar (Polich & Herbst, 2000). P300 latency variability is actually more homogeneous than many standard clinical panels. Despite a lack of clinical specificity for P300, the development of this objective and relatively inexpensive means to assess cognitive efficiency should be highly useful in clinical settings (Jeon & Polich, 2003; Katada et al., 2004; Polich, 2004a).

Clinical ERP studies should include measures of (1) patient disease characteristics and medication; (2) sample characteristics such as female and male proportions, education, age, and other demographic data; and (3) food intake time, body temperature, physical/mental fatigue level, and related factors, so that these variables can be controlled directly or used as covariates in the statistical analysis. Further, stimulus, task, and measurement methods need to be established empirically (e.g., Polich, 1997b; Salisbury et al., 2004), with both P300 amplitude and latency reported for at least the midline electrodes (Fz, Cz, Pz). As P300 timing varies systematically across the scalp, latency values need to be evaluated from individual electrodes to take into account the temporal changes of scalp topography (Fjell & Walhovd, 2001; Polich et al., 1997).

P300 and Alzheimer's Disease

A variety of oddball paradigms has been used to assess Alzheimer disease (AD). Early studies often employed a mental counting task, whereas latter reports used a button-press response that permits averaging of only those trials to which subjects respond correctly. The sample populations, stimuli, methods, and recording conditions varied appreciably across studies, but the majority of studies found differences between AD patients and controls: P300 amplitude was smaller and latency longer for AD patients compared to unaffected controls (Jeong, 2004; Olichney & Hillert, 2004; Polich, 2004a).

Figure 7.7 illustrates grand averages from patients in the early stages of AD compared to controls in a study that manipulated oddball difficulty and stimulus modality (Polich & Corey-Bloom, 2005). The strongest differences between the AD and control groups were obtained with the relatively easy oddball tasks—especially for P300 amplitude. Component amplitude produced a more consistent difference between AD and control subjects, whereas peak latency was affected by the task and modality factors in an additive rather than interactive fashion. Indeed, peak latency effects were not robust overall and were minimal for the hard task. These outcomes

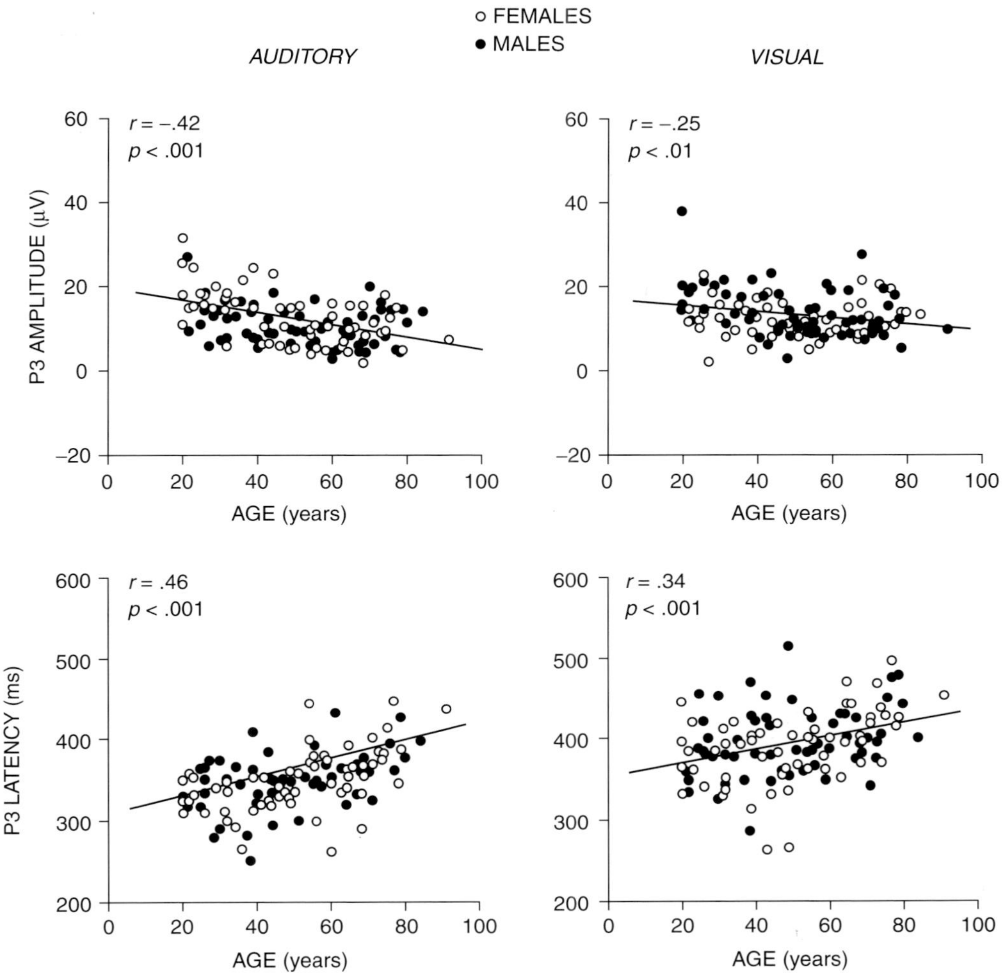

Fig. 7.6. Scattergrams of P300 (Pz) amplitude (upper) and latency (lower) from a large (N = 120) normative aging study for auditory and visual stimuli with aging (after Polich, 1997a).

indicate that P300 measures for AD are relatively stable regardless of task or modality and can discriminate between AD and controls at the group level (Rossini et al., 2007).

Passive and Single-Stimulus Tasks

Some populations, such as young children and cognitively impaired subjects, may necessitate the use of passive paradigms to avoid active task responding (Bennington & Polich, 1999; McIsaac & Polich, 1992; Polich, 1989b; Polich & McIsaac, 1994). The single-stimulus paradigm, as opposed to the oddball task, provides several advantages for recording from clinical populations. It (1) produces similar scalp topographies for both auditory and visual stimuli; (2) yields the same outcomes when stimulus probability and interstimulus interval are varied; and (3) demonstrates the same effects for auditory intensity and frequency variation (Cass & Polich, 1997; Katayama & Polich, 1996; Polich & Heine, 1996; Polich & Margala, 1997). Moreover, a single-stimulus paradigm presented as a passive task produces strong P300 components relative to those from a passive oddball task (Mertens & Polich, 1997). Sufficient normative data for these methods are not yet available, but application of single-stimulus techniques to clinical populations should prove fruitful for subjects who require very simple or passive task conditions.

Neuropsychology of P3a and P3b

An infrequent distinct tone presented in a series of frequent tones without a task can produce a positive-going waveform having a central/parietal maximum

Table 7.2 Summary of moderator variable effects on normative aging P300 latency studies (after Polich, 1996).

Moderator Variable	Comment
Sample Characteristics	
Size *(N)*	Large ($n > 80$) samples more variable than small ($n < 40$) samples
Density (mean *n*/decade)	Equal and moderately unequal densities most effective
Female/Male (% males)	Equal numbers of female and male subjects best
Stimulus Factors	
Modality (stimulus type)	Auditory stimuli produce best general measure
Number of Stimuli (*n*)	Two-stimulus oddball better than three-stimulus paradigms
Probability (target)	.20 better than <.20 conditions
Target-Standard Difference	Medium/large stimulus differences better than small
Stimulus Intensity	Medium/high intensity levels better than low
Duration (ms)	Short/medium durations better than long
Task Conditions	
Response Type	Count > press but produces more performance errors
Task Difficulty	Medium > hard > easy levels

amplitude distribution and relatively short peak latency. This component was dubbed the *P3a* to distinguish it from the task-relevant *P3b* potential elicited during target stimulus processing (Snyder & Hillyard, 1976; Squires et al., 1975). P3a from an auditory oddball task can be readily observed in about 10%–15% of normal young adults (Polich, 1988); visual stimuli without a task also can produce a P3a-like potential using passive viewing (Jeon & Polich, 2001).

Several ERPs appear related to the P3a, which are elicited by distracter stimuli inserted into the target/standard sequence (see Figure 7.1C). When perceptually novel distracters (dog barks, color forms, etc.) occur in a series of more typical stimuli (tones, letters of the alphabet, etc.), a frontal/central P300 can be elicited with a relatively short peak latency that rapidly habituates (Courchesne et al., 1975; Knight, 1984). This potential has been called the *novelty P300* and is interpreted as reflecting frontal and hippocampal activity (Knight, 1996). As novelty P300 amplitude decreases with repeated stimulus presentations, it may be more directly related to the orienting response than the P3b (Kok, 2001; Rushby et al., 2005).

If nonnovel repeated stimuli (tones, letters, etc.) are used as distracters that do not require a response in a three-stimulus oddball, a no-go P300 is elicited (Kok, 1986; Pfefferbaum et al., 1985). The P300 from this type of distracter has a maximum amplitude over the central/parietal areas (Falkenstein et al., 1999; Katayama & Polich, 1996). The scalp distribution for the no-go is more central than the target P300, and this distribution has linked the no-go to response inhibition mechanisms (Azizian et al., 2006; Falkenstein et al., 2000).

P3a and Stimulus Context

The P3a, novelty, and no-go P300 findings suggest that the type of nontarget distracter and task demands determine amplitude portraiture (Donchin et al., 1997; Friedman et al., 2001). The experimental paradigms used fostered the idea that these components were distinct entities, but this view changed when task difficulty was manipulated in the three-stimulus oddball paradigm. Katayama and Polich (1998) varied perceptual discrimination difficulty between the auditory target and the standard stimulus to manipulate the amount of focal attention engaged, with the same physical tone repeatedly used as a distracter stimulus. For the easy task, P300 amplitude from the distracter and target was largest over the parietal electrodes; for the difficult task, P300 from the distracter was larger than that from the target over the frontal/central electrodes. Subsequent studies using the same difficulty manipulations found

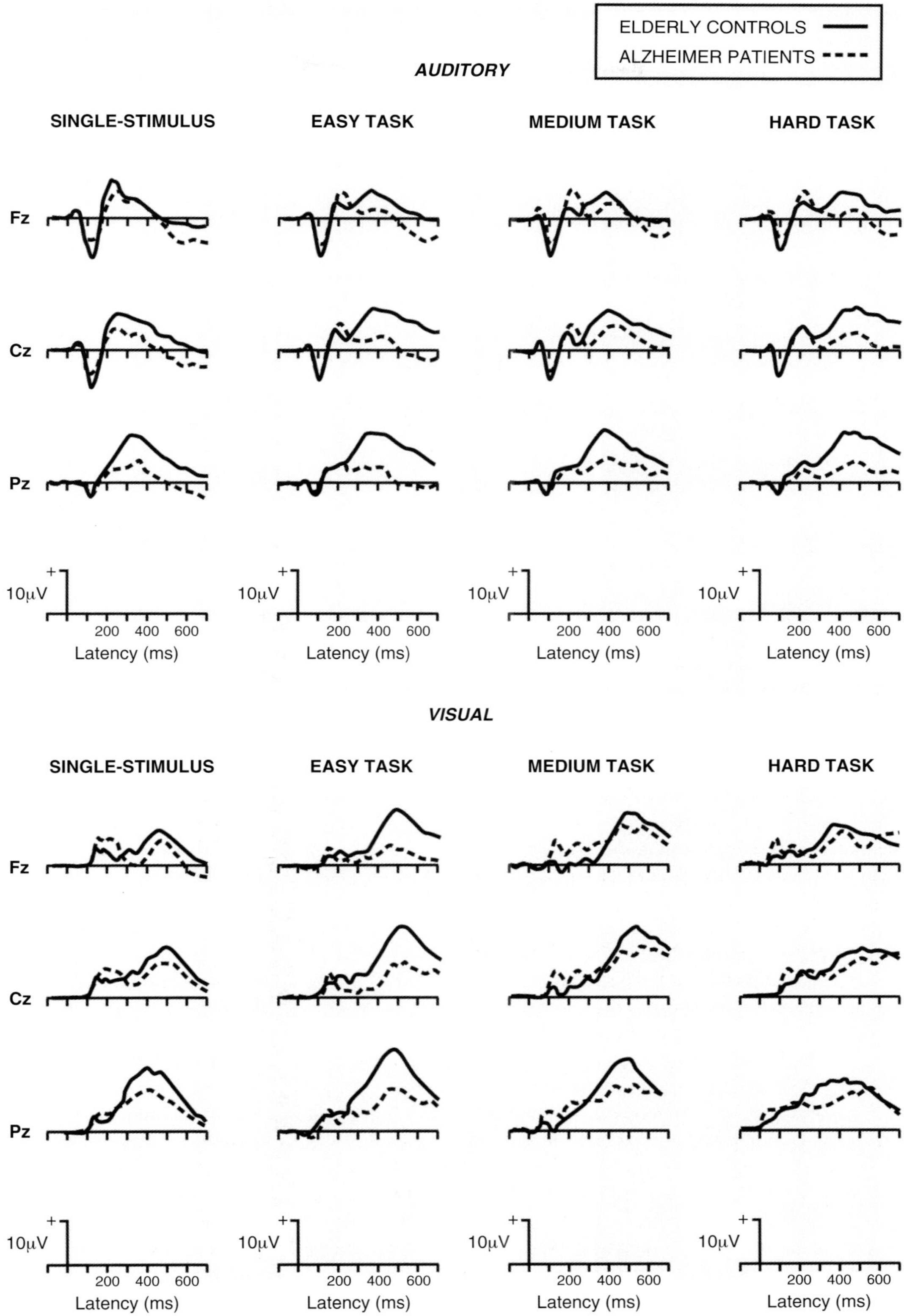

Fig. 7.7. Grand averaged ERPs for patients with early Alzheimer's disease and unaffected matched controls (n = 16/group) from single-stimulus, easy oddball, medium oddball, and hard oddball tasks that were presented in the auditory (upper) and visual (lower) modalities (after Polich and Corey-Bloom, 2005).

similar outcomes for both auditory and visual stimuli (Comerchero & Polich, 1998, 1999). Further, if standard/target discrimination difficulty is varied, easy tasks produce P300 topography similar to no-go P300 potentials, whereas difficult tasks produce P3a potentials similar to novelty P300 (cf. Dien et al., 2004; Hagen et al., 2006).

An auditory oddball target presented with novel distracters in a three-stimulus task using a large electrode array and principal components analysis (PCA) demonstrated that the novelty P3 was a different component from the classic P300—that is, P3b (Spencer et al., 1999). Replication of the original nonnovel auditory P3a and novelty P300 tasks found no differences between the two components using PCA (Simons et al., 2001). Replication of the original visual novelty P300 tasks compared novel nonrepeating abstract color stimuli and nonnovel repetitive blue-square distracters (Polich & Comerchero, 2003). The easy task yielded a central maximum distribution for the novel stimuli and a central/parietal maximum P300 potential for the nonnovel stimuli—that is, the same topography as the no-go P300. The hard task produced central maximum topographies for both the novel and nonnovel distracters. Similar results were obtained for the auditory task with white noise bursts, novel sounds, and a high-frequency tone used as distracters: White noise and novel stimuli generated P3a components that were larger over the central electrodes compared to the tone distracters (Combs & Polich, 2006). Thus, the findings suggest that the P3a, novelty, and no-go P300 are most likely variants of the same ERP that varies in scalp topography as a function of attentional and task demands.

Theoretical Perspective

Figure 7.8 illustrates P3a and P3b data from 120 normal young adults using a difficult target/standard discrimination task in a three-stimulus paradigm (Conroy & Polich, 2007). The distracter was a large black/white checkerboard square, and the target was a blue circle that was slightly larger than the standard (not shown) blue circle. Figure 7.8A illustrates the grand averages for the P3a from the checkerboard distracter (thick line) and P3b from the target circle (thin line); the vertical lines indicate the peak latencies and mean response time. Figure 7.8B illustrates the topographic distributions of the mean amplitude and latency for the distracter (top) and target (bottom) stimuli. P3a has a central maximum, whereas P3b has a parietal maximum. Peak latency for both potentials was shorter over the frontal and longer over the parietal electrode sites. Figure 7.8C illustrates the topographic distributions of the correlation coefficients between peak latency and response time to target stimuli at each electrode position across subjects. P3a demonstrated weak correlations, but P3b was significantly correlated over the parietal areas.

These results demonstrate that P3a and P3b have distinct topographic amplitude distributions. Their scalp latency distributions are similar but covary differently with response time. The topographic variation may be induced by different stimulus/task contexts to produce overlapping neural activation patterns that are functionally distinct. This perspective is consistent with the view that novelty processing is modulated by contextual and familiarity effects. Nonrepeating stimulus events define novel items, whereas repeating stimulus events engage top-down processing, so novelty P300 and P3a may differ with respect to how attentional processes are engaged for distracter stimuli (Chong et al., 2008; Knight & Nakada, 1998; Ranganath & Rainer, 2003). As distracter stimuli do not receive an overt response, topographic differences among potentials necessarily reflect stimulus-driven attributes that can be manipulated by discrimination task demands. Given the relatively simple paradigms assessed to date, it is reasonable to infer that stimulus evaluation engages focal attention (P3a) to facilitate context representational maintenance (P3b), which is associated with memory operations (Hartikainen & Knight, 2003; Kok, 2001; Polich, 2003).

Neural Origins of P3a and P3b

P300 neural generators are imprecisely delineated, although appreciable progress has been made in the last 25 years (Linden, 2005; Molnár, 1994; Soltani & Knight, 2000). Patients with frontal lobe lesions demonstrated diminution of P3a amplitude, whereas the same patients demonstrated a parietal maximum for the P3b (Knight, 1984; Knight et al., 1995). Moreover, patients with focal hippocampal lesions yielded reduced P3a amplitude from novel distracters but normal P3b components from targets, so that frontal lobe and hippocampal integrity are therefore necessary for P3a generation (Knight, 1990, 1996, 1997).

Initial studies of the hippocampal formation using depth electrodes in humans suggested that at least some portion of the P300 (P3b) is generated in the medial temporal lobe (Halgren et al., 1980; McCarthy et al., 1989). However, subsequent reports of scalp recordings from individuals after

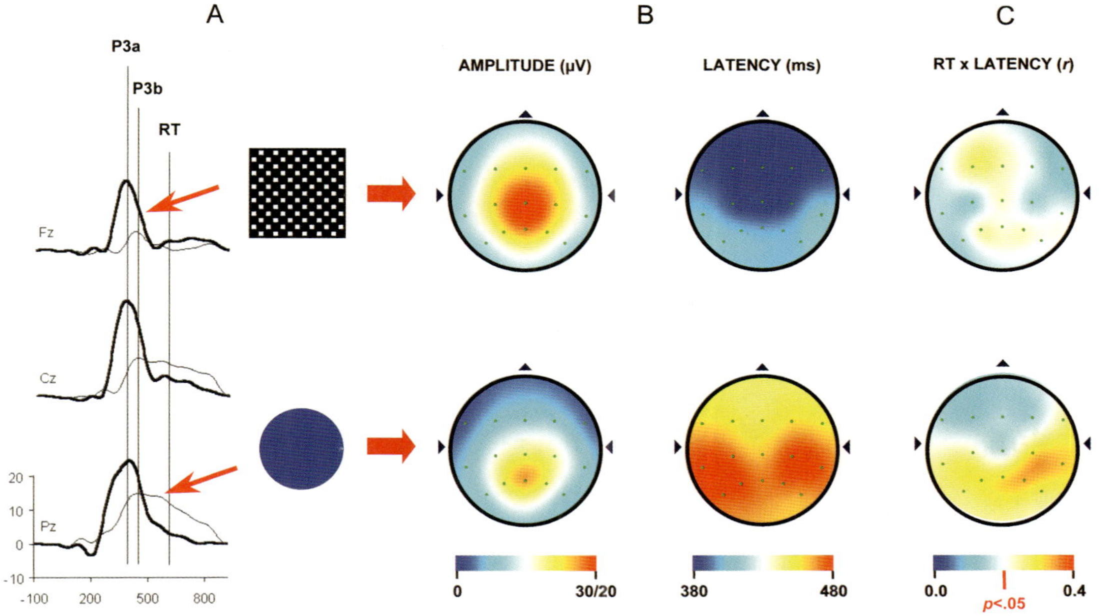

Fig. 7.8. (A) Grand averages of the P3a and P3b (classic P300) components from midline electrodes with mean response time (RT) from a three-stimulus oddball task ($N = 120$). Subjects were instructed to press a button whenever an infrequent target (5.0 cm diameter) circle was detected in a series of standard (4.5 cm diameter, not shown) stimuli. Infrequently presented distracter checkerboard patterns (18 cm^2) were employed to elicit the P3a. (B) Topographic distributions for mean P3a (upper) and P3b (lower) amplitude and latency. Note the distinct amplitude patterns for each subcomponent; amplitude scales (30/20) refer to μV for P3a and P3b, respectively. (C) Correlation topographic distributions between P3a (top) and P3b (lower) peak component latency and RT. P3a latency and RT were weakly associated, whereas P3b and RT were strongly correlated over parietal areas (after Conroy & Polich, 2007).

temporal lobectomy (Johnson, 1988, 1993; Smith & Halgren, 1989), from monkeys with experimental excisions (Paller et al., 1988, 1992), and from patients with medial temporal lobe damage found that the hippocampal formation does not contribute directly to P300 generation (Onofrj et al., 1992; Rugg et al., 1991).

Figure 7.9A provides an ERP example from patients ($n = 5$) with bilateral hippocampal lesions and matched normal controls ($n = 20$) using auditory and visual stimuli in an oddball task (Polich & Squire, 1993). Similar morphological effects were obtained for both groups, and no reliable P300 amplitude or latency differences were found. Figure 7.9B illustrates additional observations from some patients and controls whose magnetic resonance imaging (MRI) hippocampal size measures were available. P3b amplitude is positively correlated with hippocampal size relative to temporal lobe size. This outcome implies that larger hippocampal size is associated with larger P300 amplitudes (Polich, 2004b). Furthermore, other studies of lesion patients have found that the integrity of the temporal–parietal lobe junction is involved with either transmission or generation processes subsequent to hippocampal activity and contributes to component recordings at the scalp (Knight et al., 1989; Verleger et al., 1994; Yamaguchi & Knight, 1992). These findings suggest that P3a and P3b are produced by a neural circuit pathway between frontal and temporal/parietal brain areas (Polich, 2003; Soltani & Knight, 2000).

Figure 7.10 presents a neuropsychological model for P3a and P3b based on these results. Discrimination between target and standard stimuli in an oddball paradigm is hypothesized to initiate frontal lobe activity that is sensitive to the attentional demands induced by task performance (Pardo et al., 1991; Posner & Dehaene, 1994). Functional MRI (fMRI) and ERP findings have demonstrated increased frontal lobe activity for the detection of rare or physically alerting stimuli (McCarthy et al., 1997; Potts et al., 1996). P3a may be generated when such stimuli are processed if sufficient attentional focus is engaged. P3b appears to occur when subsequent attentional resource activations promote memory operations in temporal-parietal areas (Brázdil et al., 2001, 2003; Squire &

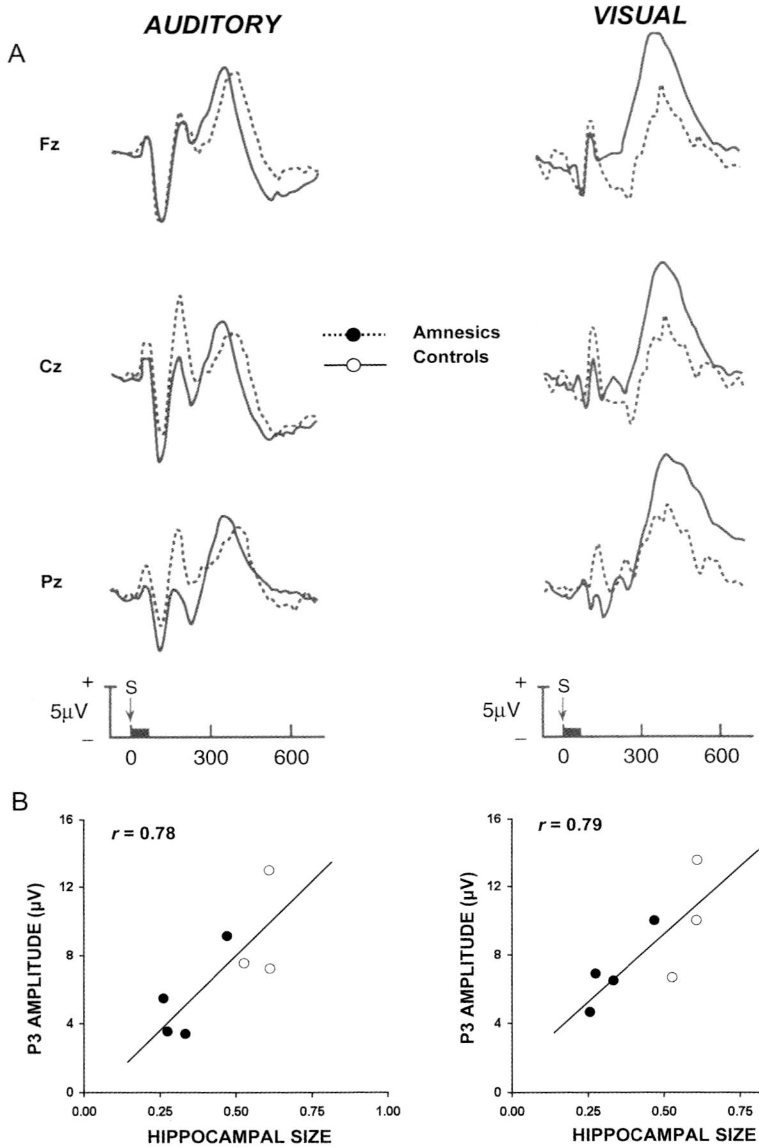

Fig. 7.9. (A) Grand averages of bilateral hippocampal lesion amnesic patients (n = 5) and matched normal controls (n = 20) from auditory and visual oddball target stimuli at the Fz, Cz, and Pz midline electrodes. (B) P3 target stimulus amplitude as a function of normalized relative hippocampus size from available patient and control MRI images (after Polich & Squire, 1993).

Kandel, 1999). Indeed, cellular recording studies in primates indicate that information induced by changes in frontal activation during a matching-to-sample task is shunted to inferotemporal structures that index task context updating for stimulus presentations (Desimone et al., 1995). It is therefore reasonable to suppose that P3a and P3b generation stem from frontal and temporal/parietal activations, respectively (Ebmeier et al., 1995; Kirino et al., 2000).

This view is consonant with the neurocognitive assumptions that incoming stimuli invoke top-down attention switching that is engaged to capture new information, and that response organization and production are guided by bottom-up memory-driven operations (Debener et al., 2002; Goldstein et al., 2002). Neuroimaging studies further imply that a frontal attention mechanism governs neural responsivity to novelty (Daffner et al., 2000a, 2000b, 2000c; Suwazono et al., 2000), thereby engaging top-down control (Bledowski et al., 2004a; Kiehl et al., 2005; Opitz et al., 1999). Thus, stimulus and task structure determine distracter and target evaluation to shape ERP topography and timing.

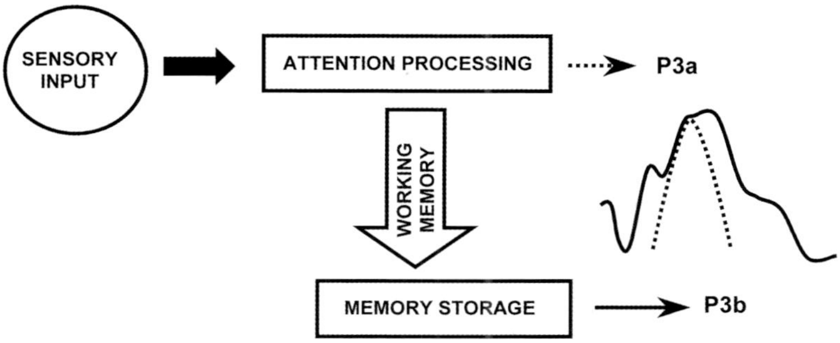

Fig. 7.10. Schematic model of P300 cognitive activity. Sensory input is hypothesized to engage attention-driven working memory processes that produce P3a; memory storage operations associated with temporal-parietal lobe activation produce P3b (after Polich, 2003).

Neuropharmacology of P300

Dual-Transmitter Hypothesis

The exact neurotransmitter systems underlying P300 generation are unclear, although various mechanisms have been implicated (Frodl-Bauch et al., 1999; Hansenne, 2000). Available data suggest that since P3a is related to frontal focal attention and working memory, it is likely mediated by dopaminergic activity. Since P3b is related to temporal-parietal processes, it is associated with norepinephrine activity (cf. Braver & Cohen, 2002; Nieuwenhuis et al., 2005; Pineda, 1995).

Several lines of evidence imply catecholaminergic mediation of frontal P300 (P3a) generation: (1) Parkinson's disease patients who have decreased levels of dopamine demonstrate deficient P300 measures (Hansch et al., 1982; Stanzione et al., 1991). (2) The dopamine antagonist sulpiride increases P300 in low-amplitude subjects and decreases it in high-amplitude subjects (Takeshita & Ogura, 1994). (3) Pharmacological studies have found dopaminergic mediation of P300 amplitude and latency (Hansenne et al., 1995; Wang et al., 2000). (4) Children at elevated risk for alcoholism demonstrate dopamine-related genetic differences such that their amplitude deficits reflect externalizing disorders (Hicks et al., 2007; Hill et al., 1998).

A comprehensive review of the limited P300 neuropharmacology literature suggests that the locus coeruleus–norepinephrine (LC-NE) system underlies parietal P300 (P3b) generation for a target detection task (Nieuwenhuis et al., 2005). Since this wide-ranging evidence stems from ERPs elicited in rat, cat, and monkey populations, differences in paradigm and task performance need to be considered in evaluating the findings. However, the suggestion that LC-NE contributes to P300 generation is consonant with attention resource allocation and arousal-related effects in humans (Intriligator & Polich, 1995; Kok, 2001). The topographic LC-NE activation of temporal-parietal areas also is in agreement with overall P300 characteristics (Aston-Jones & Cohen, 2005; Pineda et al., 1989).

Illustration of P3a and P3b Neuropharmacology

Figure 7.11 illustrates clinical examples of P300-defined neurotransmitter effects (Poceta et al., 2006). P3a and P3b amplitude data were obtained using a three-stimulus paradigm to compare unaffected controls, patients with restless legs syndrome, and patients with Parkinson's disease. Restless legs syndrome originates from dopaminergic deficits, and Parkinson's disease patients exhibit even greater dopaminergic deficits (Trenkwalder & Winkelmann, 2003). P3a amplitude is largest for the controls, decreased for restless legs syndrome, and almost eliminated for the Parkinson's disease patients. P3b amplitude for the controls and restless legs patients is comparable but greatly reduced for the Parkinson's disease patients. Thus, the P3a and some portion of the P3b are affected by dopaminergic activity (Polich & Criado, 2006).

P300 and Neuroinhibition

The P300 is produced by a distributed network of brain processes associated with attention and memory operations. However, providing a singular overarching explanation for this neuroelectric phenomenon has proven difficult, primarily because the P300 is observed in any task that requires stimulus discrimination—a fundamental psychological event that determines many aspects of cognition. Empirical differentiation of the P3a and P3b has

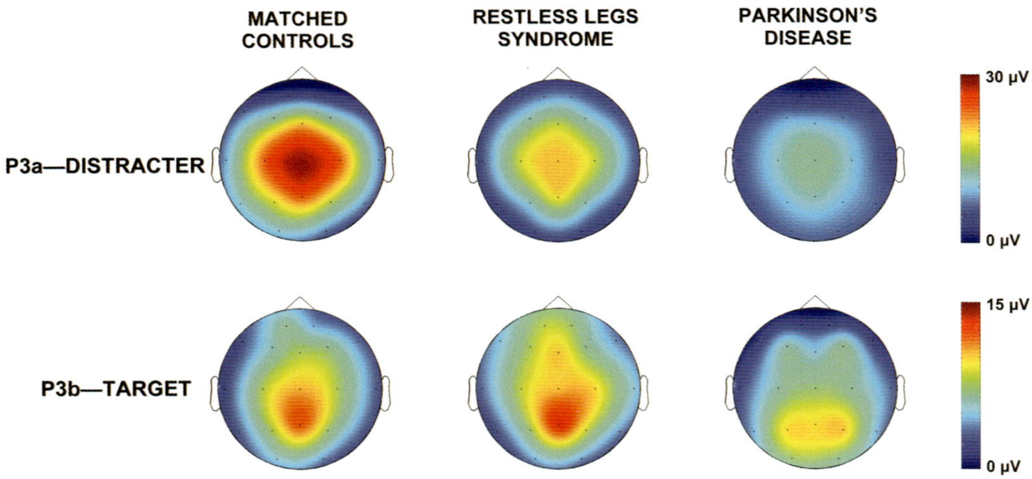

Fig. 7.11. Topographic amplitude distributions from the distracter (upper) and target (lower) stimuli in a three-stimulus task for unaffected controls, restless legs syndrome, and Parkinson's disease patients (n = 7/group). P3a amplitudes illustrate increasing dopaminergic deficits from left to right; P3b amplitudes demonstrated little difference between controls and restless legs syndrome patients, with Parkinson's disease patients demonstrating appreciably smaller amplitudes (after Polich, 2007).

begun to elucidate the interaction between initial and subsequent P300 processes, but the resulting component's ontology is still unclear.

Generation of a neuroelectric event linked to attention and memory might be caused by brain mechanisms engaged to inhibit extraneous brain activation (Polich, 2007). The implication of this hypothesis is that the P300 observed at the scalp could reflect rapid inhibition of neural activity to facilitate the transmission of information from frontal (P3a) to temporal-parietal (P3b) locations. P300 signals may originate from the initial need to enhance focal attention to isolate the task-relevant contents of working memory during stimulus detection (cf. Klimesch et al., 2007). Reduction of extraneous "neural chatter" would focus incoming stimulus information and sharpen memory encoding. This speculation is supported broadly by neurophysiological results, cognitive findings, and ERP differences among personality types.

An inhibition hypothesis is consistent with functional descriptions of P300. (1) Infrequent low-probability stimuli can be biologically important, so it is adaptive to inhibit unrelated activity to promote processing efficiency, thereby yielding large P300 amplitudes. (2) Difficult discrimination and dual-processing tasks that induce a high cognitive demand delimit the availability of attentional resources to produce relatively small P300 components because of strong inhibition induced by the stimulus/task condition. (3) Arousal modulates the level of neural inhibition engaged, as it governs the amount of attention resources available for task performance to affect P300 measures. Since the endogenous arousal level is considered a major personality characteristic, individual ERP profile differences could be accounted for by this scheme: low arousal and high inhibition, high arousal and low inhibition. (4) The relationship between P300 peak latency and cognitive capability indexes the celerity with which extraneous processes are inhibited—a highly advantageous quality associated with intelligence. (5) Declines in P300 amplitude and lengthening of latency with aging and dementing illness stem in part from breakdowns in cortical processes underlying inhibitory signals. (6) The neurotransmitter systems suggested for P3a and P3b are congenial with an inhibition hypothesis, as these neurochemical effects influence inhibitory signals. Thus, a neural inhibition mechanism could produce P300 when attention is engaged to process stimuli into memory.

Figure 7.12 presents a schematic outline of how these processes might occur. Attention-demanding stimuli elicit a P3a when the contents of working memory change. These events are hypothesized to initiate neural activity toward the areas associated with P3b production and subsequent memory storage. If the resultant ERP waveforms originate from inhibitory signals initiated by stimulus and task processing, a major question is how P3a information is transmitted (black arrows) to structures used for P3b generation. Communication between frontal attention and working memory events to

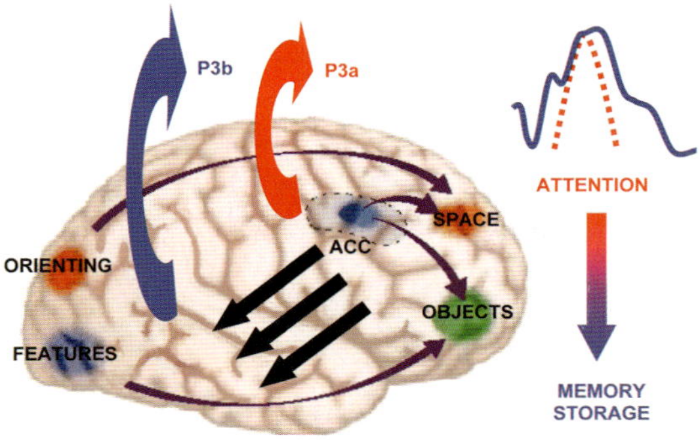

Fig. 7.12. Schematic representation of brain activation patterns underlying P3a and P3b generation. The model suggests that stimulus information is maintained in frontal lobe working memory, perhaps monitored by the anterior cingulate. When focal attention for the standard stimulus is disrupted by the detection of a distracter or a target (stimuli that garner attention automatically or purposefully from task demands), the P3a is generated by the change in frontal lobe activation pattern. This attention-driven neural signal then engages memory storage operations, and the P3b is generated via temporal-parietal cortical structures. As suggested by the red and blue waveforms, every "P300" is composed of P3a and P3b subcomponents, and the resulting scalp topographies are determined by the specific stimulus and task conditions that elicit them (after Polich, 2007).

temporal-parietal locations has been documented in primates and findings from intracranial as well as neuroimaging recordings in humans (cf. Brázdil et al., 2001, 2003; Desimone et al., 1995; Halgren et al., 1998; Simons & Spiers, 2003). A neuroelectric connection between frontal and temporal-parietal structures could modulate P3a and P3b activity at the scalp.

Figure 7.13 illustrates fMRI activation from a three-stimulus oddball task using nonnovel distracter stimuli and a difficult target/standard discrimination task (Bledowski et al., 2004b). The fMRI hemodynamic responses indicate that strong frontal lobe activation occurs for distracter stimulus processing, whereas the target stimulus produces both frontal and temporal-parietal activation. Whether the fMRI signals reflect the hypothesized sequelae of rapid neuroelectric inhibitory mechanisms is unknown, but the similarity of the magnetic and electric signal locales is suggestive. Indeed, structural images of gray matter volume indicate that P3a amplitude variation is positively correlated with frontal lobe area size, whereas P3b amplitude variation is positively correlated with parietal lobe area size (Ford et al., 1994). Such individual differences in cortical area may contribute to normative P3a and P3b amplitude and latency variability (Conroy & Polich, 2007; Polich, 1988). These functional and structural imaging data mimic the neuroelectric morphological signatures of P3a and P3b.

Figure 7.13 also indicates greater activation for the right compared to the left hemisphere. Given a frontal-parietal right-hemisphere attentional network system, wider activation for right-hemisphere processing would be expected (Posner & Dehaene, 1994), and larger P300 amplitude is found over the right compared to the left frontal/central areas in an oddball task (Alexander et al., 1995, 1996). Communication between the frontal and temporal-parietal areas may be propagated across the corpus callosum and contribute to ERP morphology

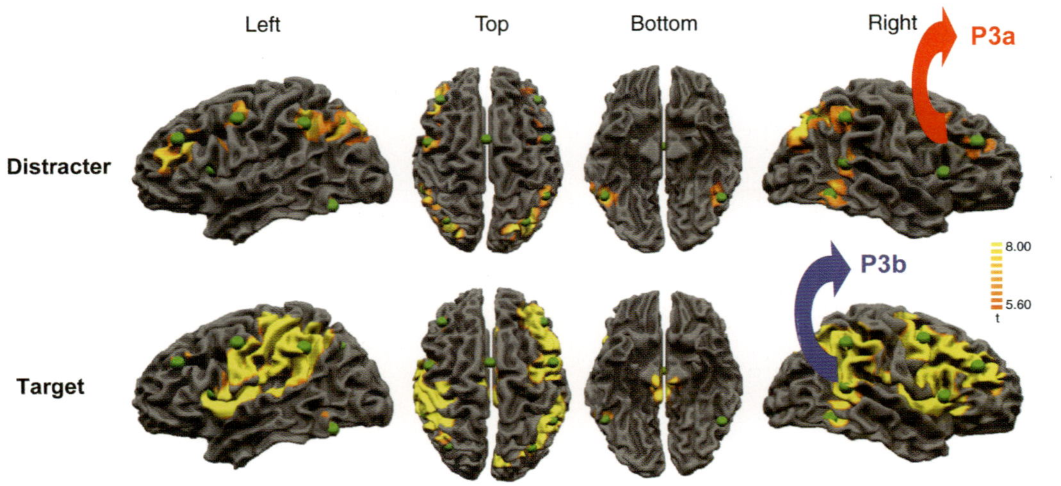

Fig. 7.13. Brain activation patterns from a visual three-stimulus oddball task modeled after that described in Figure 7.8. Arrows and labels for the P3a and P3b have been added on the right for emphasis. The green spheres reflect dipole generator sources and appeared on the original figure (after Polich, 2007).

(Barceló et al., 2000; Baudena et al., 1995; Satomi et al., 1995). Interhemispheric transmissions are affected by callosal pathway size, as left-handers have larger corpus callosal fiber tracks than right-handers (Driesen & Raz, 1995; Witelson, 1992), and P300 components with larger amplitudes and shorter peak latencies are obtained for left- compared to right-handers (Alexander & Polich, 1995, 1997; Polich & Hoffman, 1998). Taken together, these findings suggest that the hemispheric attentional pathways, as indexed by fMRI hemodynamic and P300 neuroelectric patterns, are similar in their location and function.

Neuroelectric Determinants of P3a and P3b

Conventional ERP analyses are performed in the time domain by assessing the amplitudes and latencies of prominent peaks and associating the resulting measures with information processing variables. The distinct time scales for the brainstem, middle latency, and endogenous potentials indicate that frequency is an important parameter for ERP interpretation (Hillyard & Picton, 1986; Makeig et al., 2002). Procedures such as PCA and independent component analysis (ICA) are based on mathematical transformations of ERP data and can be employed to separate functionally distinct events that occur simultaneously (Kayser & Tenke, 2003, 2005; Makeig et al., 2004; see also Chapters 1 and 3, this volume). Analysis of ERP data in the frequency domain also has revealed that neuroelectric rhythms in specific oscillatory ranges are functionally related to cognition and behavior (Başar et al., 2001; McEvoy et al., 2000; Pfurtscheller & Lopes da Silva, 1999; see also Chapter 2, this volume).

Wavelet transform (WT) analysis is an efficient time-frequency decomposition method that has been applied to ERP signals (Ademoglu et al., 1998; Daubechies, 1990; Kolev et al., 1997). The major advantage of the WT approach is its multiresolution property that employs shorter time windows for higher frequencies and longer time windows for lower frequencies—an attribute that closely matches the structural properties of ERP signals. The variable time-frequency localization method therefore takes into consideration overlapping components and provides efficient analysis of the transient nonstationary ERP signals (Demiralp et al., 1999; Chapter 2, this volume).

Figure 7.14A illustrates an application of the WT to assess nonnovel and novel distracter stimuli

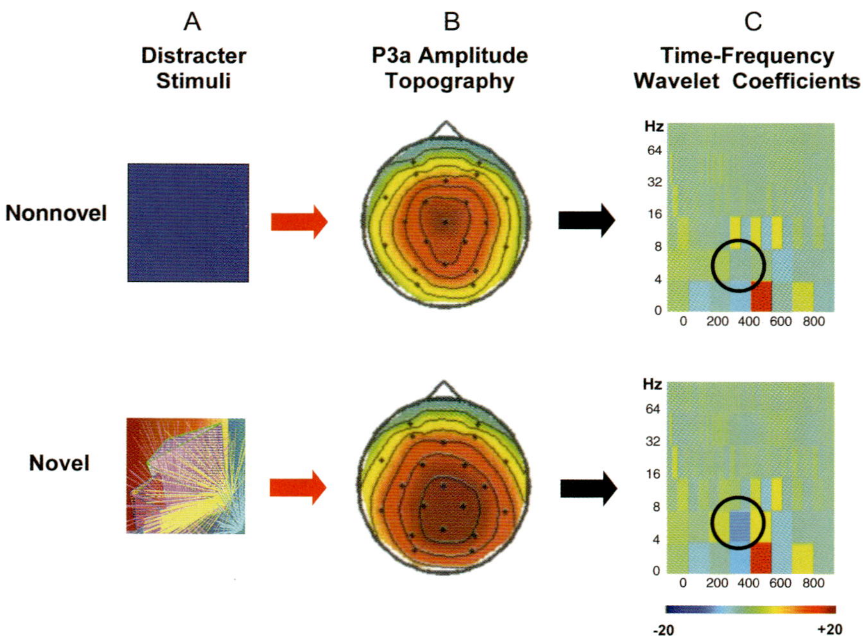

Fig. 7.14. (A) Illustrations of nonnovel blue-square distracter and novel distracter stimuli (18 cm²) used in a three-stimulus oddball task. The nonnovel distracter blue squares were always the same, whereas the novel distracters varied in form, contrast, and color across trials. (B) Topographic distributions of the grand average P3a components (μV) from the blue-square and novel distracter stimuli. (C) Time-frequency wavelet analysis representation of the blue square and novel distracter stimuli. The amplitudes of the wavelet coefficients are defined with brighter colors indicating greater spectral power, and the black circles encompass theta activity (after Demiralp et al., 2001).

when P3a is generated by a difficult oddball task (Demiralp et al., 2001). Figure 7.14B illustrates the amplitude topography distributions for each distracter type, as is typically found for perceptually demanding discrimination tasks. Figure 7.14C characterizes the WT coefficients in spectral density plots, which are highly similar for both distracter types. However, as indicated by the somewhat different shades of blue within the circles on each plot, the novel visual stimuli produced significantly more theta band activity during the P3a time interval than did the nonnovel stimuli. This outcome implies that theta band frequencies may be engaged more for novel compared to nonnovel repetitive stimuli. Furthermore, spectral analysis of the P3b ERP waveform from an auditory oddball task has demonstrated that cognitive variables strongly affect theta band activity (Spencer & Polich, 1999). If P3a originates from initial processes fostered by attention to a distracting stimulus and P3b indexes subsequent memory encoding, theta frequency modulations could underlie inhibitory control of these processes (Klimesch et al., 2007). Indeed, these results may reflect attentional mechanisms that modulate slow alpha activity (Klimesch, 1997; Klimesch et al., 1993), just as episodic memory operations alter fast alpha (10–12 Hz) activity (Hanslmayr et al., 2007; Klimesch, 2003).

Event-Related Desynchronization and P300

Phasic alpha event-related desynchronization (ERD) can be induced by sensory stimulation across modalities. Alpha ERD also occurs during task conditions requiring attention and memory (Pfurtscheller &

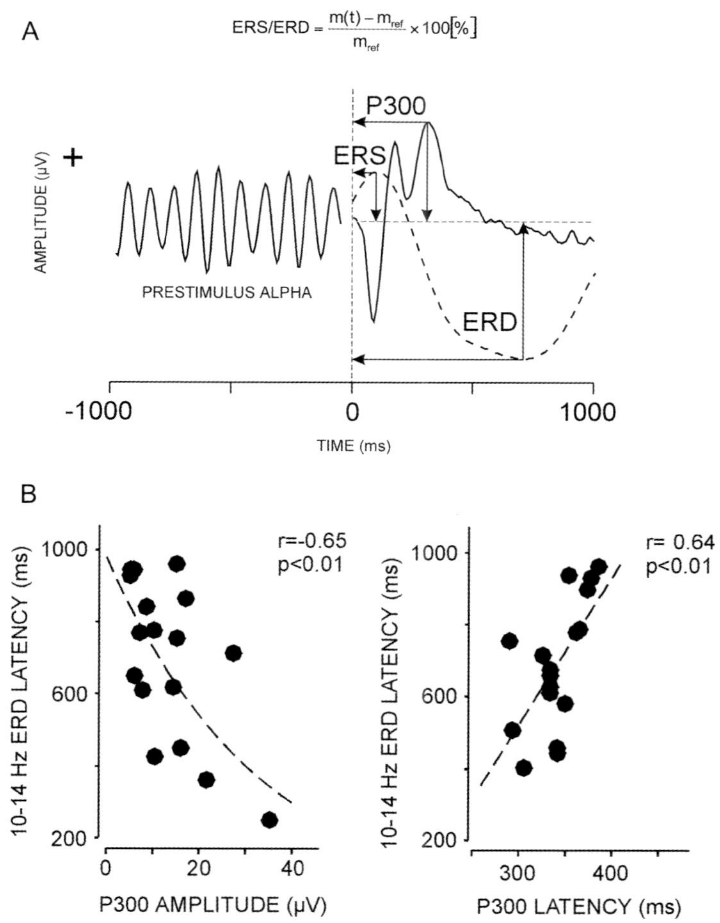

Fig. 7.15. (A) Illustration of event-related synchronization (ERS) and event-related desynchronization (ERD) from stimulus-related alpha activity calculations obtained during an ERP auditory oddball task. (B) The ERD latency has been plotted against P300 amplitude (left) and latency (right) from normal young adults (after Yordanova et al., 2001).

Klimesch, 1992; Sergeant et al., 1987). Event-related desynchronization originates from a reduction of fast non-phase-locked oscillations, whereas P300 is composed of phase-locked delta and theta-range synchronized oscillations (Başar-Eroglu et al., Başar, 1992; Yordanova & Kolev, 1998a, 1998b). Task-induced alpha frequencies have specific functions: Slow alpha ERD (8–10 Hz) is related to attention engagement, and fast alpha ERD (10–12 Hz) is related to memory operations (Schack et al., 2005). Therefore, ERD indexes variation in alpha activity power and frequency associated with specific cognitive processes.

Figure 7.15 summarizes how ERD is calculated and illustrates the relationship of alpha frequency deactivation to P300 amplitude and latency from an auditory oddball task (Yordanova et al., 2001). Measures of ERD were obtained from poststimulus late alpha power. Strong relationships between ERD and P300 (P3b) were found: (1) Onset of ERD was negatively associated with P300 amplitude—the shorter the onset time, the larger the component amplitude. (2) Onset of ERD was positively associated with P300 latency—the shorter the onset time, the shorter the component latency. (3) Homogeneous variability for each measurement pairing was observed across subjects. Assuming that ERD is related to neural inhibition by virtue of its desynchronization origins, P300 amplitude and latency could therefore stem from alpha-band deactivation. Whether P3a is regulated by this process is unknown, but P3b is clearly related to desynchronization of late alpha EEG.

Conclusion

The present chapter traced the development of P300, with neuropsychological and neurophysiological mechanisms underlying its function outlined. Neuroimaging and neuropharmacological studies provided a framework on how the P3a and P3b subcomponents could be produced. The empirical and theoretical background suggests that the P300 may stem from neural inhibitory activity that enhances the attentional focus to promote memory storage. The proposed model posits that "the P300" is comprised of a P3a that results from an early attention-related process stemming from a working memory representational change, and P3b occurs when the attention-driven stimulus signal is transmitted to temporal and parietal structures. Thus, the P300 waveform may reflect neural inhibition that occurs when stimulus and task demands engage fundamental cognitive mechanisms.

The activity and desynchronization of the theta and alpha frequency bands appear related to the neuroelectric architecture of this phenomenon. It is notable in this context that theta and alpha frequencies have been strongly linked to meditation effects—a self-induced neuroelectric inhibition (Cahn & Polich, 2006). These activation patterns also may reflect the effects of gamma band signals (Canolty et al., 2006). The P300 therefore may be the neuronal consequence of stimulus events important enough to inhibit concomitant brain activity and index processes that are generated by brain mechanisms underlying various types of attention and memory operations. If these activities determine P3a and P3b generation, understanding the origin of neuroelectric information and its transmission are important next steps in discerning the meaning of P300. As the relationships between neurotransmitter function and the concomitant neuroelectric signals recorded at the scalp are clarified, articulating how these variables interact will fulfill the cognitive promise that the P300 inspired when it was discovered over 40 years ago.

Acknowledgment

This work was supported by NIH Grant RO1-DA018262. I thank Kelly Courtney and Brian Lopez for their perspicacious editorial comments.

References

Ademoglu, A., Demiralp, T., Yordanova, J., Kolev, V., & Devrim, M. (1998). Decomposition of event-related brain potentials into multicomponents using wavelet transform. *Applied Signal Processing, 5*, 142–151.

Alexander, J. E., Bauer, L. O., Kuperman, S., Rohrbaugh, J., Morzorati, S., O'Connor, S. J., Rohrbaugh, J., Porjesz, B., Begleiter, H., & Polich, J. (1996). Hemispheric differences for P300 amplitude from an auditory oddball task. *International Journal of Psychophysiology, 21*, 189–196.

Alexander, J. E., & Polich, J. (1995). P300 differences between sinistrals and dextrals. *Cognitive Brain Research, 2*, 277–282.

Alexander, J. E., & Polich, J. (1997). Handedness and P300 from auditory stimuli. *Brain and Cognition, 35*, 259–270.

Alexander, J. E., Porjesz, B., Bauer, L. O., Kuperman, S., Mozorati, S., O'Connor, S. J., Rohrbaugh, J., Begleiter, H., & Polich, J. (1995). P300 amplitude hemispheric asymmetries from a visual discrimination task. *Psychophysiology, 32*, 467–475.

Almasy, L., Porjesz, B., Blangero, J., Chorlian, D. B., O'Connor, S. J., Kuperman, S., Rohrbaugh, J., Bauer, L. O., Reich, T., Polich, J., & Begleiter, H. (1999). Heritability of event-related brain potentials in alcoholic families. *Human Genetics, 88*, 383–390.

Anokhin, A. P., Vedeniapin, A. B., Sirevaag, E. J., Bauer, L. O., O'Connor, S. J., Kuperman, S., Porjesz, B. Reich, T., Begleiter, H., Polich, J., & Rohrbaugh, J. W. (2000). The P300 brain potential is reduced in smokers. *Psychopharmacology, 149*, 409–413.

Aston-Jones, G., & Cohen, J. D. (2005). An integrative theory of locus coeruleus–norepinephrine function: Adaptive gain and optimal performance. *Annual Review of Neuroscience, 28*, 403–450.

Azizian, A., Freitas, A. L., Watson, T. D., & Squires, N. K. (2006). Electrophysiological correlates of categorization: P300 amplitude as an index of target similarity. *Biological Psychology, 71*, 278–288.

Azizian, A., & Polich, J. (2007). Evidence for attentional gradient in the serial position memory curve from event-related potentials. *Journal of Cognitive Neuroscience, 19*, 2071–2081.

Barceló, F., Periáñez, J. A., & Knight, R. T. (2002). Think differently: A brain orienting response to task novelty. *Neuroreport, 13*, 1887–1892.

Barceló, F., Suwazono, S., & Knight, R. T. (2000). Prefrontal modulation of visual processing in humans. *Nature Neuroscience, 3*, 399–403.

Başar, E., Başar-Eroglu, C., Karakaş, S., & Schürmann, M. (2001). Gamma, alpha, delta, and theta oscillations govern cognitive processes. *International Journal of Psychophysiology, 39*, 241–248.

Başar-Eroglu, C., Başar, E., Demiralp, T., & Schürmann, M. (1992). P300-response: Possible psychophysiological correlates in delta and theta frequency channels. A review. *International Journal of Psychophysiology, 13*, 161–179.

Bashore, T. R. (1990). Age-related changes in mental processing revealed by analyses of event-related brain potentials. In J. Rohrbaugh, R. Parasuraman, & R. Johnson (Eds.), *Event-related brain potentials: Basic issues and applications* (pp. 242–278). New York: Oxford University Press.

Bashore, T. R., & Ridderinkhof, K. R. (2002). Older age, traumatic brain injury, and cognitive slowing: Some convergent and divergent findings. *Psychological Bulletin*, 2002, 151–198.

Bashore, T. R., & van der Molen, M. W. (1991). Discovery of the P300: A tribute. *Biological Psychology, 32*, 155–171.

Baudena, P., Halgren, E., Heit, G., & Clarke, J. (1995). Intracerebral potentials to rare target and distractor auditory and visual stimuli. III. Frontal cortex. *Electroencephalography and Clinical Neurophysiology, 94*, 251–264.

Begleiter, H., Porjesz, B., Reich, T., Edenberg, H., Goate, A., Blangero, J., Almasy, L., Foroud, T., Van Eerdewegh, P., Polich, J., Rohrbaugh, J., Kuperman, S., Bauer, L., O'Connor, S., Chorlian, D., Li, T., Conneally, P., Hesselbrock, V., Rice, J., Schuckit, M., Cloninger, R., Nurnberger, J., Jr., Crowe, R., & Bloom, F. E. (1998). Quantitative trait loci analysis of human event-related brain potentials: P3 voltage. *Electroencephalography and Clinical Neurophysiology, 108*, 244–250.

Bennington, J. Y., & Polich, J. (1999). Comparison of P300 from passive and active tasks for auditory and visual stimuli. *International Journal of Psychophysiology, 34*, 171–177.

Bledowski, C., Prvulovic, D., Goebel, R., Zanella, F., & Linden, D. (2004a). Attention systems in target and distractor processing: A combined ERP and fMRI study. *Neuroimage, 22*, 530–540.

Bledowski, C., Prvulovic, D., Hoechstetter, K., Scherg, M., Wibral, M., Goebel, R., & Linden, D. (2004b). Localizing P300 generators in visual target and distractor processing: A combined event-related potential and functional magnetic resonance imaging study. *Journal of Neuroscience, 24*, 9353–9360.

Braver, T. S., & Cohen, J. D. (2002). On the control of control: The role of dopamine in regulating prefrontal function and working memory. In S. Monsell & J. Driver (Eds.), *Control of cognitive processes: Attention and performance XVIII* (pp. 713–737). Cambridge, MA: MIT Press.

Brázdil, M., Rektor, I., Daniel, P., Dufek, M., & Jurak, P. (2001). Intracerebral event-related potentials to subthreshold target stimuli. *Clinical Neurophysiology, 112*, 650–661.

Brázdil, M., Roman, R., Daniel, P., & Rektor, I. (2003). Intracerebral somatosensory event-related potentials: Effect of response type (button pressing versus mental counting) on P3-like potentials within the human brain. *Clinical Neurophysiology, 114*, 1489–1496.

Brocke, B. (2004). The multilevel approach in sensation seeking: Potentials and findings of a four-level research program. In R. M. Stelmack (Ed.), *On the psychobiology of personality* (pp. 267–293). Amsterdam: Elsevier.

Cahn, B. R., & Polich, J. (2006). Meditation states and traits: EEG, ERP, and neuroimaging studies. *Psychological Bulletin, 132*, 180–211.

Canolty, R. T., Edwards, E., Dalal, S. S., Soltani, M., Nagarajan, S. S., Kirsch, H. E., Berger, M. S., Barbaro, N. M., & Knight, R. T. (2006). High gamma power is phase-locked to theta oscillations in human neocortex. *Science, 313*, 1626–1628.

Carlson, S. R., Katsanis, J., Iacono, W. G., & Mertz, A. K. (1999). Substance dependence and externalizing psychopathology in adolescent boys with small, average, or large P300 event-related potential amplitude. *Psychophysiology, 36*, 583–590.

Cass, M., & Polich, J. (1997). P300 from an auditory single-stimulus paradigm: Intensity and tone frequency effects. *Biological Psychology, 46*, 51–65.

Chao, L. L., & Knight, R. T. (1996). Prefrontal and posterior cortical activation during auditory working memory. *Cognitive Brain Research, 4*, 27–37.

Chong, H., Riis, J., McGinnis, S. M., Williams, D. M., Holcomb, P. J., & Daffner, K. R. (2008). To ignore or explore: Top-down modulation of novelty processing. *Journal of Cognitive Neuroscience, 20*, 120–134.

Cohen, J., & Polich, J. (1997). On the number of trials needed for P300. *International Journal of Psychophysiology, 25*, 249–255.

Colet, A. V., Piera, P. J. F., & Pueyo, A. A. (1993). Initial stages of information processing and inspection time: Electrophysiological correlates. *Personality and Individual Differences, 14*, 733–738.

Combs, L. A., & Polich, J. (2006). P3a from white noise stimuli. *Clinical Neurophysiology, 117*, 1106–1112.

Comerchero, M. D., & Polich, J. (1998). P3a, perceptual distinctiveness, and stimulus modality. *Cognitive Brain Research, 7*, 41–48.

Comerchero, M. D., & Polich, J. (1999). P3a and P3b from typical auditory and visual stimuli. *Clinical Neurophysiology, 110*, 24–30.

Conroy, M. A., & Polich, J. (2007). Normative variation of P3a and P3b from a large sample ($N = 120$): Gender, topography, and response time. *Journal of Psychophysiology, 21*, 22–32.

Courchesne, E., Hillyard, S. A., & Galambos, R. (1975). Stimulus novelty, task relevance, and the visual evoked potential in man. *Electroencephalography and Clinical Neurophysiology, 39*, 131–143.

Crites, S. L., Jr., Devine, J. V., Lozano, D. I., & Moreno S. (1998). Event-related potentials and serial position effects in a visual probe recognition task. *Psychophysiology, 35*, 293–304.

Croft, R. J., Gonsalvez, C. J., Gabriel, C., & Barry, R. J. (2003). Target-to-target interval versus probability effects on P300 in one- and two-tone tasks. *Psychophysiology, 40*, 322–328.

Curran, T. (2004). Effects of attention and confidence on the hypothesized ERP correlates of recollection and familiarity. *Neuropsychologia, 42*, 1088–1106.

Curran, T., & Cleary, A. M. (2003). Using ERPs to dissociate recollection from familiarity in picture recognition. *Brain Research Cognitive Brain Research, 15*, 191–205.

Cycowicz, Y. M. (2000). Memory development and event-related brain potentials in children. *Biological Psychology, 54*, 145–174.

Daffner, K., Mesulam, M., Holcomb, P., Calvo, V., Acar, D., Chabrerie, A., Kikinis, R., Jolesz, F., Rentz, D., & Seinto, L. (2000a). Disruption of attention to novel events after frontal lobe injury in humans. *Journal of Neurology, Neurosurgery, and Psychiatry, 68*, 18–24.

Daffner, K., Mesulam, M., Seinto, L., Acar, D., Calvo, V., Faust, R., Chabrerie, A., Kennedy, B., & Holcomb, P. (2000b). The central role of the prefrontal cortex in directing attention to novel events. *Brain, 123*, 927–939.

Daffner, K., Mesulam, M., Seinto, L., Calvo, V., Faust, R., & Holcomb, P. (2000c). An electrophysiological index of stimulus unfamiliarity. *Psychophysiology, 37*, 737–747.

Daubechies, I. (1990). The wavelet transform, time-frequency localization and signal analysis. *IEEE Transactions on Information Theory, 36*, 961–1005.

Debener, S., Kranczioch, C., Herrman, C., & Engel, A. (2002). Auditory novelty oddball allows reliable distinction of top-down and bottom-up processes of attention. *International Journal of Psychophysiology, 46*, 77–84.

DeBoer, T., Scott, L. S., & Nelson, C. A. (2006). ERPs in developmental populations. In T. C. Handy (Ed.), *Event-related potentials: A methods handbook* (pp. 263–297). Cambridge, MA: MIT Press.

Demiralp, T., Ademoglu, A., Comerchero, M., & Polich, J. (2001). Wavelet analysis of P3a and P3b. *Brain Topography, 13*, 251–267.

Demiralp, T., Yordanova, J., Kolev, V., Ademoglu, A., & Devrim, M. (1999). Time-frequency analysis of single-sweep event-related potentials by means of fast wavelet transform. *Brain and Language, 66*, 129–145.

De Pascalis, V. (2004). On the psychophysiology of extraversion. In R. M. Stelmack (Ed.), *On the psychobiology of personality* (pp. 297–329). Amsterdam: Elsevier.

Desimone, R., Miller, E. K., Chelazzi, L., & Lueschow, A. (1995). Multiple memory systems in the visual cortex. In M. S. Gazzaniga (Ed.), *The cognitive neurosciences* (pp. 475–486). Cambridge, Mass: MIT Press.

Dien, J., Spencer, K. M., & Donchin, E. (2004). Parsing the late positive complex: Mental chronometry and the ERP components that inhabit the neighborhood of the P300. *Psychophysiology, 41*, 665–678.

DiTraglia, G., & Polich, J. (1991). P300 and introverted/extroverted personality types. *Psychophysiology, 28*, 177–184.

Donchin, E. (1981). Surprise!... Surprise? *Psychophysiology, 18*, 493–513.

Donchin, E., Karis, D., Bashore, T. R., Coles, M. G. H., & Gratton, G. (1986). Cognitive psychophysiology and human information processing. In M. G. H. Coles, E. Donchin, & S. W. Porges (Eds.), *Psychophysiology: Systems, processes, and applications* (pp. 244–267). New York: Guilford Press.

Donchin, E., Ritter, W., & McCallum, C. (1978). Cognitive psychophysiology: The endogenous components of the ERP. In E. Callaway, P. Tueting, & S. Koslow (Eds.), *Brain-event-related potentials in man* (pp. 349–411). New York: Academic Press.

Donchin, E., Spencer, K. M., & Dien, J. (1997). The varieties of deviant experience: ERP manifestations of deviance processors. In G. J. M. van Boxtel & K. B. E. Bocker (Eds.), *Brain and behavior: Past, present, and future of deviance detectors in the brain* (pp. 67–91). Tilburg: The Netherlands University Press.

Doyle, M. C., & Rugg, M. D. (1992). Event-related potentials and recognition memory for low- and high-frequency words. *Journal of Cognitive Neuroscience, 4*, 69–79.

Driesen, N., & Raz, N. (1995). The influence of sex, age and handedness of corpus callosum morphology: A meta-analysis. *Psychobiology, 23*, 240–247.

Duncan, C. C., Barry, R. J., Connolly, J. F., Fischer, C., Michie, P. T., Näätänen, R., Polich, J., Reinvang, I., & Van Petten, C. (2008). Event-related potentials in clinical research: Guidelines for eliciting, recording, and quantifying mismatch negativity, P300, and N400. *Clinical Neurophysiology, 11*, 1883-1908.

Duncan-Johnson, C. C., & Donchin, E. (1977). On quantifying surprise: The variation of event-related potentials with subjective probability. *Psychophysiology, 14*, 456–467.

Duncan-Johnson, C. C., & Kopell, B. S. (1981). The Stroop effect: Brain potentials localize the source of interference. *Science, 20*, 938–940.

Ebmeier, K. P., Steele, J. D., MacKenzie, D. M., O'Carroll, R. E., Kydd, R. R., Glabus, M. F., Backwood, D. H. R., Rugg, M. D., & Goodwin, G. M. (1995). Cognitive brain potentials and regional cerebral blood flow equivalents during two- and three-sound auditory "oddball tasks." *Electroencephalography and Clinical Neurophysiology, 95*, 434–443.

Eischen, S. E., Luckritz, J. Y., & Polich, J. (1995). Spectral analysis of EEG from families. *Biological Psychology, 41*, 61–68.

Eischen, S. E., & Polich, J. (1994). P300 from families. *Electroencephalography and Clinical Neurophysiology, 92*, 369–372.

Fabiani, M., Gratton, G., Karis, D., & Donchin, E. (1987). The definition, identification, and reliability of measurement of the P300 component of the event-related brain potential. In P. K. Ackles, J. R. Jennings, & M. G. H. Coles (Eds.), *Advances in psychophysiology, Vol. 2* (pp. 1–78). Greenwich, CT: JAI Press.

Fabiani, M., Karis, D., & Donchin, E. (1986). P300 and recall in an incidental memory paradigm. *Psychophysiology, 23*, 298–308.

Fabiani, M., Karis, D., & Donchin, E. (1990). Effects of mneumonic strategy manipulation in a von Restorff paradigm. *Electroencephalography and Clinical Neurophysiology 75*, 22–35.

Falkenstein, M., Hoormann, J., Christ, S., & Hohnsbein, J. (2000) ERP components on reaction errors and their functional significance: A tutorial. *Biological Psychology, 51*, 87–107.

Falkenstein, M., Hoormann, J., & Hohnsbein, J. (1999). ERP components in go/no-go tasks and their relation to inhibition. *Acta Psychologica, 101*, 267–291.

Farrell, S., & Lewandowsky, S. (2002). An endogenous distributed model of ordering in serial recall. *Psychonomic Bulletin & Review, 9*, 59–79.

Fein, G., & Turetsky, B. (1989). P300 latency variability in normal elderly: Effects of paradigm and measurement technique. *Electroencephalography and Clinical Neurophysiology, 72*, 384–394.

Fjell, A. M., & Walhovd, K. B. (2001). P300 and neuropsychological tests as measures of aging: Scalp topography and cognitive changes. *Brain Topography, 14*, 25–40.

Ford, J., Sullivan, E., Marsh, L., White, P., Lim, K., & Pfefferbaum, A. (1994). The relationship between P300 amplitude and regional gray matter volumes depends on the attention system engaged. *Electroencephalography and Clinical Neurophysiology, 90,* 214–228.

Friedman, D., Cycowicz, Y. M., & Gaeta, H. (2001). The novelty P3: An event-related brain potential (ERP) sign of the brain's evaluation of novelty. *Neuroscience and Biobehavioral Review, 25,* 355–373.

Friedman, D., & Trott, C. (2000). An event-related potential study of encoding in young and older adults. *Neuropsychologia, 38,* 542–557.

Frodl-Bauch, T., Bottlender, R., & Hegerl, U. (1999). Neurochemical substrates and neuroanatomical generators of the event-related P300. *Neuropsychobiology, 40,* 86–94.

Goldstein, A., Spencer, K. M., & Donchin, E. (2002). The influence of stimulus deviance and novelty on the P300 and novelty P3. *Psychophysiology, 39,* 781–790.

Golob, E. J., & Starr, A. (2004). Serial position effects in auditory event-related potentials during working memory retrieval. *Journal Cognitive Neuroscience, 16,* 40–52.

Gomer, F. E., Spicuzza, R. J., & O'Donnell, R. D. (1976). Evoked potential correlates of visual item recognition during memory scanning tasks. *Physiological Psychology, 4,* 61–65.

Gonsalvez, C. J., Barry, R. J., Rushby, J. A., & Polich, J. (2007). Target-to-target interval, intensity, and P300 from an auditory single-stimulus task. *Psychophysiology, 44,* 245–250.

Gonsalvez, C. J., & Polich, J. (2002). P300 amplitude is determined by target-to-target interval. *Psychophysiology, 39,* 388–396.

Goodin, D., Squires, K., Henderson, B., & Starr, A. (1978). Age-related variations in evoked potentials to auditory stimuli in normal human subjects. *Electroencephalography and Clinical Neurophysiology, 44,* 447–458.

Guo, C., Duan, L., Li, W., & Paller, K. A. (2006). Distinguishing source memory and item memory: Brain potentials at encoding and retrieval. *Brain Research, 1118,* 142–154.

Gurrera, R. J., O'Donnell, B. F., Nestor, P. G., Gainski, J., & McCarley, R. W. (2001). The P3 auditory event-related brain potential indexes major personality traits. *Biological Psychiatry, 49,* 922–929.

Hagen, G. F., Gatherwright, J. R., Lopez, B. A., & Polich, J. (2006). P3a from visual stimuli: Task difficulty effects. *International Journal of Psychophysiology, 59,* 8–14.

Halgren, E., Marinkovic, K., & Chauvel, P. (1998). Generators of the late cognitive potentials in auditory and visual oddball tasks. *Electroencephalography and Clinical Neurophysiology, 106,* 156–164.

Halgren, E., Squires, N., Wilson, C., Rohbraugh, J., Bab, T., & Crandall, P. (1980). Endogenous potentials in the human hippocampal formation and amygdala by infrequent events. *Science, 210,* 803–805.

Hansch, E., Syndulko, K., Cohen, S., Goldberg, Z., Potvin, A., & Tourtellotte, W. (1982). Cognition in Parkinson disease: An event-related potential perspective. *Annals of Neurology, 11,* 599–607.

Hansenne, M. (2000). The P300 cognitive event-related potential. II. Individual variability. *Neurophysiologie Clinique/Clinical Neurophysiology, 30,* 211–231.

Hansenne, M., Pitchot, W., Gonzalez-Moreno, A., Papart, P., Timsit-Berthier, M., & Ansseau, M. (1995). Catecholaminergic function and P300 amplitude in major depressive disorder (P300 and catecholamines). *Electroencephalography and Clinical Neurophysiology, 96,* 194–196.

Hanslmayr, S., Klimesch, W., Sauseng, P., Gruber, W., Doppelmayr, M., Freunberger, R., Pecherstorfer, T., & Birbaumer, N. (2007). Alpha phase reset contributes to the generation of ERPs. *Cerebral Cortex, 17,* 1–8.

Hartikainen, K., & Knight, R. T. (2003). Lateral and orbital prefrontal cortex contributions to attention. In J. Polich (Ed.), *Detection of change: Event-related potential and fMRI findings* (pp. 99–116). Boston: Kluwer Academic Press.

Hicks, B. M., Bernat, E., Malone, S. M., Iacono, W. G., Patrick, C. J., Krueger, R., & McGue, M. (2007). Genes mediate the association between P3 amplitude and externalizing disorders. *Psychophysiology, 44,* 98–105.

Hill, S., Locke, J., Zezza, N., Kaplan, B., Neiswanger, K., Steinhauer, S., Wipprecht, G., & Xu, J. (1998). Genetic association between reduced P300 amplitude and the DRD2 dopamine receptor A1 allele in children at high risk for alcoholism. *Biological Psychiatry, 43,* 40–51.

Hill, S. Y., De Bellis, M., Keshavan, M., Lowers, L., Shen, S., Hall, J., & Pitts, T. (2001). Right amygdala volume in adolescent and young adult offspring from families at high risk for developing alcoholism. *Biological Psychiatry, 49,* 894–905.

Hill, S. Y., Shen, S., Locke, J., Steinhauer, S., Konicky, C., Lowers, L., & Connolly, J. (1999). Developmental delay in P300 production in children at high risk for developing alcohol-related disorders. *Biological Psychiatry, 46,* 970–981.

Hillyard, S. A., & Picton, T. W. (1986). Electrophysiology of cognition. In S. R. Geiger (Ed.), *Handbook of physiology: A critical, comprehensive presentation of physiological knowledge and concepts* (pp. 519–584). New York: American Physiology Association.

Houlihan, M., Stelmack, R., & Campbell, K. (1998). P300 and cognitive ability: Assessing the roles of processing speed, perceptual processing demands, and task difficulty. *Intelligence, 26,* 9–25.

Howard, L., & Polich, J. (1985). P300 latency and memory span development. *Developmental Psychology, 21,* 283–289.

Iacono, W. G., Carlson, S. R., Malone, S. M., & McGue, M. (2002). P3 event-related potential amplitude and the risk for disinhibitory disorders in adolescent boys. *Archives of General Psychiatry, 59,* 750–757.

Iacono, W. G., Carlson, S. R., Malone, S. M., & McGue, M. (2003). Substance use disorders, externalizing psychopathology, and P300 event-related potential amplitude. *International Journal of Psychophysiology, 48,* 147–178.

Ilan, A. B., & Polich, J. (1999). P300 and response time from a manual Stroop task. *Clinical Neurophysiology, 110,* 367–373.

Intriligator, J. M., & Polich, J. (1995). On the relationship between EEG and ERP variability. *International Journal of Psychophysiology, 20,* 59–74.

Isreal, J. B., Chesney, G. L., Wickens, C. D., & Donchin, E. (1980). P300 and tracking difficulty: Evidence for multiple resources in dual-task performance. *Psychophysiology, 17,* 259–273.

Jeon, Y. W., & Polich, J. (2001). P3a from a passive visual paradigm. *Clinical Neurophysiology, 112,* 2202–2208.

Jeon, Y. W., & Polich, J. (2003). Meta-analysis of P300 schizophrenia studies: Patients, paradigms, and practical implications. *Psychophysiology, 40,* 684–701.

Jeong, J. (2004). EEG dynamics in patients with Alzheimer's disease. *Clinical Neurophysiology, 115,* 1490–1505.

Johnson, R. (1986). A triarchic model of P300 amplitude. *Psychophysiology, 23,* 367–384.

Johnson, R. (1988). The amplitude of the P300 component of the event-related potential: Review and synthesis. In P. Ackles, J. R. Jennings, & M. G. H. Coles (Eds.), *Advances in psychophysiology: A research annual*, Vol. *3* (pp. 69–137). Greenwich, CT: JAI Press.

Johnson, R. (1993). On the neural generators of the P300 component of the event-related potential. *Psychophysiology, 30*, 90–97.

Justus, A. N., Finn, P. R., & Steinmetz, J. E. (2001). P300, disinhibited personality, and early-onset alcohol problems. *Alcoholism: Clinical and Experimental Research, 25*, 1457–1466.

Karis, D., Fabiani, M., & Donchin, E. (1984). "P300" and memory: Individual differences in the von Restorff effects. *Cognitive Psychology, 16*, 117–216.

Katada, E., Sato, K., Ojika, K., & Ueda, R. (2004). Cognitive event-related potentials: Useful clinical information in Alzheimer disease. *Current Alzheimer Research, 1*, 63–69.

Katayama, J., & Polich, J. (1996). P300 from one-, two-, and three-stimulus auditory paradigms. *International Journal of Psychophysiology, 23*, 33–40.

Katayama, J., & Polich, J. (1998). Stimulus context determines P3a and P3b. *Psychophysiology, 35*, 23–33.

Katsanis, J. G., Iacono, W. G., McGue, M. K., & Carlson, S. R. (1997). P300 event-related potential heritability in monozygotic and dizygotic twins. *Psychophysiology, 34*, 47–58.

Kayser, J., & Tenke, C. E. (2003). Optimizing PCA methodology for ERP component identification and measurement: Theoretical rationale and empirical evaluation. *Clinical Neurophysiology, 114*, 2307–2325.

Kayser, J., & Tenke, C. E. (2005). Trusting in or breaking with convention: Towards a renaissance of principal components analysis in electrophysiology. *Clinical Neurophysiology, 116*, 1747–1753.

Kiehl, K. A., Stevens, M. C., Laurens, K. R., Pearlson, G., Calhoun, V. D., & Liddle, P. F. (2005). An adaptive reflexive processing model of neurocognitive function: Supporting evidence from a large scale (n = 100) fMRI study of an auditory oddball task. *Neuroimage, 25*, 899–915.

Kirino, E., Belger, A., Goldman-Rakic, P., & McCarthy, G. (2000). Prefrontal activation evoked by infrequent target and novel stimuli in a visual target detection task: An event-related functional magnetic resonance study. *Journal of Neuroscience, 20*, 6612–6618.

Klimesch, W. (1997). EEG-alpha rhythms and memory processes. *International Journal of Psychophysiology, 26*, 319–340.

Klimesch, W. (2003). Interindividual differences in oscillatory EEG activity and cognitive performance. In I. Reinvang, M. W. Greenlee, & M. Herrmann (Eds.), *The cognitive neuroscience of individual differences* (pp. 87–99). Oldenburg, Saxony, Germany: BIS.

Klimesch, W., Sauseng, P., & Hanslmayr, S. (2007). EEG alpha oscillations: The inhibition-timing hypothesis. *Brain Research Reviews, 53*, 63–88.

Klimesch, W., Schimke, H., & Pfurtscheller, G. (1993). Alpha frequency, cognitive load and memory performance. *Brain Topography, 5*, 241–251.

Knight, R. T. (1984). Decreased response to novel stimuli after prefrontal lesions in man. *Electroencephalography and Clinical Neurophysiology, 59*, 9–20.

Knight, R. T. (1987). Aging decreases auditory event-related potentials to unexpected stimuli in humans. *Neurobiology of Aging, 8*, 109–113.

Knight, R. T. (1990). Neural mechanisms of event-related potentials from human lesion studies. In J. Rohrbaugh, R. Parasuraman, & R. Johnson (Eds.), *Event-related brain potentials: Basic issues and applications* (pp. 3–18). New York: Oxford University Press.

Knight, R. T. (1996). Contribution of human hippocampal region to novelty detection. *Nature, 383*, 256–259.

Knight, R. T. (1997). Distributed cortical network for visual attention. *Journal of Cognitive Neuroscience, 9*, 75–91.

Knight, R. T., Grabowecky, M., & Scabini, D. (1995). Role of human prefrontal cortex in attention control. *Advances in Neurology, 66*, 21–34.

Knight, R. T., & Nakada, T. (1998). Cortico-limbic circuits and novelty: A review of EEG and blood flow data. *Review of Neuroscience, 9*, 57–70.

Knight, R. T., Scabini, D., Woods, D., & Clayworth, C. (1989). Contributions of temporal parietal junction to the human auditory P3. *Brain Research, 502*, 109–116.

Kok, A. (1986). Effects of degradation of visual stimulation on components of the event-related potential (ERP) in go/nogo reaction tasks. *Biological Psychology, 23*, 21–38.

Kok, A. (1990). Internal and external control: A two-factor model of amplitude change of event-related potentials. *Biological Psychology, 74*, 203–236.

Kok, A. (1997). Event-related potential (ERP) reflections of mental resources: A review and synthesis. *Biological Psychology, 45*, 19–56.

Kok, A. (2001). On the utility of P3 amplitude as a measure of processing capacity. *Psychophysiology, 38*, 557–577.

Kolev, V., Demiralp, T., Yordanova, J., Ademoglu, A., & Isoglu-Alkac, U. (1997). Time-frequency analysis reveals multiple functional components during oddball P300. *Neuroreport, 8*, 2061–2065.

Kramer, A. F., & Parasuraman, R. (2008). Neuroergonomics: Application of neuroscience to human factors. In J. Cacioppo, L. Tassinary, & G. Berntson (Eds.), *Handbook of psychophysiology* (3rd ed., pp. 704–722). New York: Cambridge University Press.

Kramer, A. F., Wickens, C. D., & Donchin, E. (1985). Processing of stimulus properties: Evidence for dual-task integrality. *Journal of Experimental Psychology: Human Perception and Performance, 11*, 393–408.

Kujala, A., & Näätänen, R. (2003). Auditory environment and change detection as indexed by the mismatch negativity (MMN). In J. Polich (Ed.), *Detection of change: Event-related potential and fMRI findings* (pp. 1–22). Boston: Kluwer Academic Press.

Kutas, M., McCarthy, G., & Donchin, E. (1977). Augmenting mental chronometry: The P300 as a measure of stimulus evaluation. *Science, 197*, 792–795.

Leuthold, H., & Sommer, W. (1993). Stimulus presentation rate dissociates sequential effects in event-related potentials and reaction times. *Psychophysiology, 30*, 510–517.

Lin, E., & Polich, J. (1999). P300 habituation patterns: Individual differences and ultradian rhythms. *Perceptual and Motor Skills, 88*, 1111–1125.

Linden, D. (2005). The P300: Where in the brain is it produced and what does it tell us? *The Neuroscientist, 6*, 563–576.

Magliero, A., Bashore, T., Coles, M. G. H., & Donchin, E. (1984). On the dependence of P300 latency on stimulus evaluation processes. *Psychophysiology, 21*, 171.

Makeig, S., Debener, S., Onton, J., & Delorme, A. (2004). Mining event-related brain dynamics. *Trends in Cognitive Science, 8*, 204–210.

Makeig, S., Westerfield, M., Jung, T. P., Enghoff, J., Townsend, J., Courchesne, E., & Sejnowski, T. J. (2002). Dynamic brain sources of visual evoked responses. *Science, 295*, 1466–1466.

McCarthy, G., & Donchin, E. (1981). A metric for thought: A comparison of P300 latency and reaction time. *Science, 211*, 77–80.

McCarthy, G., Luby, M., Gore, J., & Goldman-Rakic. P. (1997). Infrequent events transiently activate human prefrontal and parietal cortex as measured by functional MRI. *Journal of Neurophysiology, 77*, 1630–1634.

McCarthy, G., Wood, C. C., Williamson, P. D., & Spencer, D. (1989). Task-dependent field potentials in human hippocampal formation. *Journal of Neuroscience, 9*, 4235–4268.

McEvoy, L. K., Smith, M. E., & Gevins, A. (2000). Test-retest reliability of cognitive EEG. *Clinical Neurophysiology, 111*, 457–463.

McEvoy, L. K., Smith, M. E., & Gevins, A. (2001). Neurophysiological signals of working memory in normal aging. *Cognitive Brain Research, 11*, 363–376.

McIsaac, H., & Polich, J. (1992). Comparison of infant and adult P300 from auditory stimuli. *Journal of Experimental Child Psychology, 53*, 115–128.

Mecklinger, A., & Ullsperger, P. (1993). P3 varies with stimulus categorization rather than probability. *Electroencephalography and Clinical Neurophysiology, 86*, 395–407.

Mecklinger, A., & Ullsperger, P. (1995). What makes a category a category? ERP correlates of stimulus-to-category assignments. *Electroencephalography and Clinical Neurophysiology* (*Supplement*), *44*, 255–260.

Mertens, R., & Polich, J. (1997). P300 from a single-stimulus paradigm: Passive versus active tasks and stimulus modality. *Electroencephalography and Clinical Neurophysiology, 104*, 488–497.

Michalewski, H. J., Prasher, D., & Starr, A. (1988). Latency variability and temporal relationships of the auditory event-related potential (N1, P2, N2, and P3) in normal subjects. *Electroencephalography and Clinical Neurophysiology, 65*, 59–71.

Molnár, M. (1994). On the origin of the P300 event-related potential component. *International Journal of Psychophysiology, 17*, 129–144.

Nieuwenhuis, S., Aston-Jones, G., & Cohen, J. (2005). Decision making, the P3, and the locus coeruleus–norepinephrine system. *Psychological Bulletin, 131*, 510–532.

O'Connor, S. O., Morzorati, S., & Li, T.-K. (1994). Heritable features of the auditory oddball event-related potential: Peaks, latencies, morphology, and topography. *Electroencephalography and Clinical Neurophysiology, 59*, 238–248.

O'Donnell, B., Friedman, S., Swearer, J., & Drachman, D. (1992). Active and passive P3 latency and psychometric performance: Influence of age and individual differences. *International Journal of Psychophysiology, 12*, 187–195.

Olichney, J. M., & Hillert, D. G. (2004). Clinical applications of cognitive event-related potentials in Alzheimer's disease. *Physical Medicine and Rehabilitation Clinics, 15*, 205–233.

Onofrj, M., Fulgente, T., Nobiolio, D., Malatesta, G., Bazzano, S., Colamartino, P., & Gambi, D. (1992). P300 recordings in patients with bilateral temporal lobe lesions. *Neurology, 42*, 1762–1767.

Opitz, B., Mecklinger, A., Von Cramon, D. Y., & Kruggel, F. (1999). Combining electrophysiological and hemodynamic measures of the auditory oddball. *Psychophysiology, 36*, 142–147.

Orlebeke, J. F., Kok, A., & Zeillemaker, C. W. (1989). Disinhibition and the processing of auditory stimulus intensity: An ERP-study. *Personality and Individual Differences, 10*, 445–451.

Otten, L. J., Quayle, A. H., Akram, S., Ditewig, T. A., & Rugg, M. D. (2006). Brain activity before an event predicts later recall. *Nature Neuroscience, 9*, 489–491.

Page, M. P. A., & Norris, D. (1998). The primacy model: A new model of immediate serial recall. *Psychological Review, 105*, 761–781.

Paller, K. A., McCarthy, G., Roessler, E., Allison, T., & Wood, C. C. (1992). Potentials evoked in human and monkey medial temporal lobe during auditory and visual oddball paradigms. *Electroencephalography and Clinical Neurophysiology, 84*, 269–279.

Paller, K. A., McCarthy, G., & Wood, C. C. (1988). ERPs predictive of subsequent recall and recognition performance. *Biological Psychology, 26*, 269–276.

Paller, K. A., Zola-Morgan, S., Squire, L. R., & Hillyard, S. A. (1988). P3-like brain waves in normal monkeys and in monkeys with medial temporal lesions. *Behavioral Neuroscience, 102*, 714–725.

Pardo, J. V., Fox, P., & Raichle, M. (1991). Localization of human system for sustained attention by positron emission tomography. *Nature, 349*, 61–64.

Pashler, H. (1994). Dual-task interference in simple tasks: Data and theory. *Psychological Bulletin, 116*, 220–244.

Patterson, J. V., Pratt, H., & Starr, A. (1991). Event-related potential correlates of the serial position effect in short-term memory. *Electroencephalography and Clinical Neurophysiology, 78*, 424–437.

Pelosi, L., Holly, M., Slade, T., Hayward, M., Barrett, G., & Blumhardt, L. (1992a). Event-related potential (ERP) correlates of performance on intelligence tests. *Electroencephalography and Clinical Neurophysiology, 84*, 515–520.

Pelosi, L., Holly, M., Slade, T., Hayward, M., Barrett, G., & Blumhardt, L. (1992b). Wave form variations in auditory event-related potentials evoked by a memory-scanning task and their relationship with tests of intellectual function. *Electroencephalography and Clinical Neurophysiology, 84*, 344–352.

Pfefferbaum, A., Christensen, C., Ford, J., & Kopell, B. (1986). Apparent response incompatibility effects on P3 latency depend on the task. *Electroencephalography and Clinical Neurophysiology, 64*, 424–437.

Pfefferbaum, A., Ford, J. M., Weller, B. J., & Kopell, B. S. (1985). ERPs to response production and inhibition. *Electroencephalography and Clinical Neurophysiology, 60*, 423–434.

Pfurtscheller, G., & Klimesch, W. (1992). Functional topography during a visuoverbal judgment task studied with event-related desynchronization mapping. *Journal of Clinical Neurophysiology, 9*, 120–131.

Pfurtscheller, G., & Lopes da Silva, F. H. (1999). Event-related EEG/MEG synchronization and desynchronization: Basic principles. *Clinical Neurophysiology, 110*, 1842–1857.

Picton, T. W. (1992). The P300 wave of the human event-related potential. *Journal of Clinical Neurophysiology, 9*, 456–479.

Pineda, J. A. (1995). Are neurotransmitter systems of subcortical origin relevant to the electrogenesis of cortical ERPs? *Electroencephalography and Clinical Neurophysiology* (*Supplement*), *44*, 143–150.

Pineda, J. A., Foote, S. L., & Neville, H. J. (1989). Effects of locus coeruleus lesions on auditory, long-latency, event-related potentials in monkey. *Journal of Neuroscience, 9*, 81–93.

Poceta, S. J., Houser, M., & Polich, J. (2006). Event-related potentials in restless legs syndrome and Parkinson's disease [abstract]. *Sleep, 28*, A274.

Polich, J. (1988). Bifurcated P300 peaks: P3a and P3b revisited? *Journal of Clinical Neurophysiology*, 5, 287–294.

Polich, J. (1989a). Habituation of P300 from auditory stimuli. *Psychobiology, 17*, 19–28.

Polich, J. (1989b). P300 from a passive auditory paradigm. *Electroencephalography and Clinical Neurophysiology, 74*, 312–320.

Polich, J. (1990). P300, probability, and interstimulus interval. *Psychophysiology, 27*, 396–403.

Polich, J. (1996). Meta-analysis of P300 normative aging studies. *Psychophysiology, 33*, 334–353.

Polich, J. (1997a). EEG and ERP assessment of normal aging. *Electroencephalography and Clinical Neurophysiology, 104*, 228–243.

Polich, J. (1997b). On the relationship between EEG and P300: Individual differences, aging, and ultradian rhythms. *International Journal of Psychophysiology, 26*, 299–317.

Polich, J. (2003). Overview of P3a and P3b. In J. Polich (Ed.), *Detection of change: Event-related potential and fMRI findings* (pp. 83–98). Boston: Kluwer Academic Press.

Polich, J. (2004a). Clinical application of the P300 event-related potential. *Physical Medicine and Rehabilitation Clinics, 15*, 133–161.

Polich, J. (2004b). Neuropsychology of P3a and P3b: A theoretical overview. In N. C. Moore & K. Arikan (Eds.), *Brainwaves and mind: Recent developments* (pp. 15–29). Wheaton, IL: Kjellberg.

Polich, J. (2007). Updating P300: An integrative theory of P3a and P3b. *Clinical Neurophysiology, 118*, 2128–2148.

Polich, J., Alexander, J. E., Bauer, L. O., Kuperman, S., Rohrbaugh, J., Mozarati, S., O'Connor, S. J., Porjesz, B., & Begleiter, H. (1997). P300 topography of amplitude/latency correlations. *Brain Topography, 9*, 275–282.

Polich, J., & Bloom, F. E. (1999). P300, alcoholism heritability, stimulus modality. *Alcohol, 17*, 149–156.

Polich, J., & Burns, T. (1987). P300 from identical twins. *Neuropsychologia, 25*, 299–304.

Polich, J., & Comerchero, M. D. (2003). P3a from visual stimuli: Typicality, task, and topography. *Brain Topography, 15*, 141–152.

Polich, J., & Corey-Bloom, J. (2005). Alzheimer's disease and P300: Review and evaluation of task and modality. *Current Alzheimer's Research, 2*, 515–526.

Polich, J., & Criado, J. R. (2006). Neuropsychology and neuropharmacology of P3a and P3b. *International Journal of Psychophysiology, 60*, 172–185.

Polich, J., Ehlers, C. L., Otis, S., Mandell, A. J., & Bloom, F. E. (1986). P300 latency reflects cognitive decline in dementing illness. *Electroencephalography and Clinical Neurophysiology, 63*, 138–114.

Polich, J., Eischen, S. E., & Collins, G. E. (1994). P300 from a single auditory stimulus. *Electroencephalography and Clinical Neurophysiology, 92*, 253–261.

Polich, J., & Heine, M. R. D. (1996). P300 topography and modality effects from a single stimulus paradigm. *Psychophysiology, 33*, 747–752.

Polich, J., & Herbst, K. L. (2000). P300 as a clinical assay: Rationale, evaluation, and findings. *International Journal of Psychophysiology, 38*, 3–19.

Polich, J., & Hoffman, L. D. (1998). P300 and handedness: Possible corpus callosal contributions to ERPs. *Psychophysiology, 35*, 497–507.

Polich, J., Howard, L., & Starr, A. (1983). P300 latency correlates with digit span. *Psychophysiology, 20*, 665–669.

Polich, J., & Kok, A. (1995). Cognitive and biological determinants of P300: An integrative review. *Biological Psychology, 41*, 103–146.

Polich, J., Ladish, C., & Bloom, F. E. (1990a). P300 assessment of early Alzheimer's disease. *Electroencephalography and Clinical Neurophysiology, 77*, 179–189.

Polich, J., Ladish, C., & Burns, T. (1990b). Normal variation of P300 in children: Age, memory span, and head size. *International Journal of Psychophysiology, 9*, 237–248.

Polich, J., & Luckritz, J. Y. (1995). EEG and ERPs in young and elderly subjects. In G. Karmos, M. Molnár, V. Csépe, I. Czigler, & J. E. Desmedt (Eds.), *Perspectives in event-related potential research, EEG Supplement 44* (pp. 358–368). Amsterdam: Elsevier.

Polich, J., & Margala, C. (1997). P300 and probability: Comparison of oddball and single-stimulus paradigms. *International Journal of Psychophysiology, 25*, 169–176.

Polich, J., & Martin, S. (1992). P300, cognitive capability, and personality: A correlational study of university undergraduates. *Personality and Individual Differences, 13*, 533–543.

Polich, J., & McIsaac, H. (1994). Comparison of auditory P300 habituation from active and passive conditions. *International Journal of Psychophysiology, 17*, 25–34.

Polich, J., & Ochoa, J. C. (2004). Alcoholism risk, tobacco smoking, and P300. *Clinical Neurophysiology, 115*, 1374–1383.

Polich, J., & Squire, L. (1993). P300 from amnesic patients with bilateral hippocampal lesions. *Electroencephalography and Clinical Neurophysiology, 86*, 408–417.

Polich, J., & Starr, A. (1984). Evoked potentials in aging. In M. L. Alpert (Ed.), *The clinical neurology of aging* (pp. 149–177). New York: Oxford University Press.

Porjesz, B., Rangaswamy, M., Kamarajan, C., Jones, K. A., Padmanabhapillai, A., & Begleiter, H. (2005). The utility of neurophysiological markers in the study of alcoholism. *Clinical Neurophysiology, 116*, 993–1018.

Posner, M. I., & Dehaene, S. (1994). Attentional networks. *Trends in Neuroscience, 17*, 75–79.

Potter, D., & Barrett, K. (1999). Assessment of mild head injury with ERPs and neuropsychological tasks. *Journal of Psychophysiology, 13*, 173–189.

Potts, G., Liotti, M., Tucker, D., & Posner, M. I. (1996). Frontal and inferior temporal cortical activity in visual target detection: Evidence from high spatially sampled event-related potentials. *Brain Topography, 9*, 3–14.

Pribram, K. H., & McGuinness, D. (1975). Arousal, activation, and effort in the control of attention. *Psychological Review, 82*, 116–149.

Pritchard, W. S. (1981). Psychophysiology of the P300. *Psychological Bulletin, 89*, 506–540.

Ragot, R. (1984). Perceptual and motor space representation: An event-related potential study. *Psychophysiology, 2*, 159–170.

Ranganath, C., & Rainer, G. (2003). Neural mechanisms for detecting and remembering novel events. *Nature Review of Neuroscience, 4*, 193–202.

Ravden, D., & Polich, J. (1999). On P300 measurement stability: Habituation, intra-trial block variation, and ultradian rhythms. *Biological Psychology, 51*, 59–76.

Reinvang, I. (1999). Cognitive event-related potentials in neuropsychological assessment. *Neuropsychology Review, 9*, 231–248.

Renault, B., Ragot, R., Lesèvre, N., & Remond, A. (1982). Onset and offset of brain events as indexes of mental chronometry. *Science, 215*, 1413–1415.

Ritter, W., Simson, R., Vaughan, H., & Macht, M. (1983). Manipulation of event-related potential manifestations of information processing stages. *Science, 218*, 909–911.

Rossini, P. M., Rossi, S., Babiloni, C., & Polich, J. (2007). Clinical neurophysiology of aging brain: From normal aging to neurodegeneration. *Progress in Neurobiology, 83*, 375–400.

Rugg, M., Pickles, C., Potter, D., & Roberts, R. (1991). Normal P300 following extensive damage to the left medial temporal lobe. *Journal of Neurology, Neurosurgery, and Psychiatry, 54*, 217–222.

Rushby, J. A., Barry, R. J., & Doherty, R. J. (2005). Separation of the components of the late positive complex in an ERP dishabituation paradigm. *Clinical Neurophysiology, 116*, 2363–2380.

Rushby, J. A., Barry, R. J., & Johnstone, S. S. (2002). Event-related potential correlates of serial-position effects during an elaborative memory test. *International Journal of Psychophysiology, 46*, 13–27.

Salisbury, D. F., Griggs, C. B., Shenton, M. E., & McCarley, R. W. (2004). The nogo P300 "anteriorization" effect and response inhibition. *Clinical Neurophysiology, 115*, 1550–1558.

Satomi, K., Horai, T., Kinoshita, Y., & Wakazono, A. (1995). Hemispheric asymmetry of event-related potentials in a patient with callosal disconnection syndrome: A comparison of auditory, visual, and somatosensory modalities. *Electroencephalography and Clinical Neurophysiology, 94*, 440–449.

Schack, B., Klimesch, W., & Sauseng, P. (2005). Phase synchronization between theta and upper alpha oscillations in a working memory task. *International Journal of Psychophysiology, 57*, 105–114.

Segalowitz, S., & Barnes, K. (1993). The reliability of ERP components in the auditory oddball paradigm. *Psychophysiology, 30*, 451–459.

Sergeant, J., Geuze, R., & van Winsum, W. (1987). Event-related desynchronization and P300. *Psychophysiology, 24*, 272–277.

Simons, J. S., & Spiers, H. J. (2003). Prefrontal and medial temporal lobe interactions in long-term memory. *Nature Review Neuroscience, 4*, 637–648.

Simons, R. F., Graham, F. K., Miles, M. A., & Chen, X. (2001). On the relationship of P3a and the novelty-P3. *Biological Psychology, 56*, 207–218.

Smith, M. E., & Halgren, E. (1989). Dissociation of recognition memory components following temporal lobe lesions. *Journal of Experimental Psychology: General, 15*, 50–60.

Snyder, E., & Hillyard, S. (1976). Long-latency evoked potentials to irrelevant, deviant stimuli. *Behavioral Biology, 16*, 319–331.

Sokolov, E. (1977). Brain functions: Neuronal mechanisms of learning and memory. *Annual Review of Psychology, 28*, 85–112.

Soltani, M., & Knight, R. (2000). Neural origins of the P300. *Critical Review in Neurobiology, 14*, 199–224.

Sommer, W., Matt, J., & Leuthold, H. (1990). Consciousness of attention and expectancy as reflected in event-related potentials and reaction times. *Journal of Experimental Psychology, 16*, 902–915.

Spencer, K. M., Dien, J., & Donchin, E. (1999). A componential analysis of the ERP elicited by novel events using a dense electrode array. *Psychophysiology, 36*, 409–414.

Spencer, K., & Polich, J. (1999). Post-stimulus EEG spectral analysis and P300: Attention, task, and probability. *Psychophysiology, 36*, 220–232.

Squire, L. R., & Kandel, E. R. (1999). *Memory from mind to molecules*. New York: Scientific American Library.

Squires, K., Wickens, C., Squires, N. K., & Donchin, E. (1976). The effect of stimulus sequence on the waveform of the cortical event-related potential. *Science, 193*, 1142–1146.

Squires, N. K., Squires, K., & Hillyard, S. (1975). Two varieties of long-latency positive waves evoked by unpredictable auditory stimuli in man. *Electroencephalography and Clinical Neurophysiology, 38*, 387–401.

Stanzione, P., Fattapposta, F., & Guiunti, P. (1991). P300 variations in Parkinsonian patients before and during dopaminergic monotherapy: A suggested dopamine component in P300. *Electroencephalography and Clinical Neurophysiology, 80*, 446–453.

Stelmack, R. M., & Houlihan, M. (1994). Event-related potentials, personality and intelligence: Concepts, issues, and evidence. In D. H. Saklofske & M. Zaidner (Eds.), *International handbook of personality and intelligence* (pp. 349–365). New York: Plenum Press.

Stenberg, G. (1992). Personality and the EEG: Arousal and emotional arousability. *Personality and Individual Differences, 13*, 1097–1113.

Stenberg, G. (1994). Extraversion and the P300 in a visual classification task. *Personality and Individual Differences, 16*, 543–560.

Sutton, S. (1979). P300—Thirteen years later. In H. Begleiter (Ed.), *Evoked brain potentials and behavior* (pp. 107–125). New York: Plenum Press.

Sutton, S., Braren, M., Zubin, J., & John, E. R. (1965). Evoked potential correlates of stimulus uncertainty. *Science, 150*, 1187–1188.

Suwazono, S., Machado, L., & Knight, R. T. (2000). Predictive value of novel stimuli modifies visual event-related potentials and behavior. *Clinical Neurophysiology, 111*, 29–39.

Takeshita, S., & Ogura, C. (1994). Effect of the dopamine D2 antagonist sulpiride on event-related potentials and its relation to the law of the initial value. *International Journal of Psychophysiology, 16*, 99–106.

Tarkka, I. M., & Stokic, D. S. (1998). Source localization of P300 from oddball, single stimulus, and omitted-stimulus paradigms. *Brain Topography, 11*, 141–151.

Trenkwalder, C., & Winkelmann, J. (2003). Pathophysiology of the restless legs syndrome. In S. Chokroverty, W. A. Hening, & A. S. Walters (Eds.), *Sleep and movement disorders* (pp. 322–332). Philadelphia: Butterworth/Heinemann.

van Beijsterveldt, C. E., & van Baal, G. C. (2002). Twin and family studies of the human electroencephalogram: A review and meta-analysis. *Biological Psychology, 61*, 111–138.

Verleger, R. (1997). On the utility of P3 latency as an index of mental chronometry. *Psychophysiology, 34*, 131–156.

Verleger R. (2003). Event-related EEG potential research in neurological patients. In A. Zani & A. M. Proverbio (Eds.), *The cognitive electrophysiology of mind and brain* (pp. 309–341). Amsterdam: Academic Press.

Verleger, R., Heide, W., Butt, C., & Kömpf, D. (1994). Reduction of P3b in patients with temporo-parietal lesions. *Cognitive Brain Research, 2*, 103–116.

Verleger, R., Jaśkowskis, P., & Wascher, E. (2005). Evidence for an integrative role of P3b in linking reaction to perception. *Journal of Psychophysiology, 19*, 150–150.

Walhovd, K. B., & Fjell, A. M. (2002). One-year test-retest reliability of auditory ERPs in young and old adults. *International Journal of Psychophysiology, 46*, 29–40.

Walhovd, K. B., Rosquist, H., & Fjell, A. M. (2008). P300 amplitude age reductions are not caused by latency jitter. *Psychophysiology, 45*, 545–553.

Wang, L., Kuroiwa, Y., Li, M., Kamitani, T., Wang, J., Takahashi, T., Suzuki, Y., Ikegami, T., & Matsubara, S. (2000). The correlation between P300 alterations and regional cerebral blood flow in non-demented Parkinson's disease. *Neuroscience Letters, 282*, 133–136.

Watson, T. D., Azizian, A., & Squires, N. K. (2006). Event-related potential correlates of extradimensional and intradimensional set-shifts in a modified Wisconsin Card Sorting Test. *Brain Research, 1092*, 138–151.

Wickens, C., Kramer, A., Vanesse, L., & Donchin, E. (1983). The performance of concurrent tasks: A psychophysiological analysis of the reciprocity of information processing resources. *Science, 221*, 1080–1082.

Wijers, A. A., Mulder, G., Okita, T., & Mulder, L. J. (1989). Event-related potentials during memory search and selective attention to letter size and conjunctions of letter size and color. *Psychophysiology, 26*, 529–547.

Wiswede, D., Rüsseler, J., & Münte, T.F. (2007). Serial position effects in free memory recall—An ERP study. *Biological Psychology, 75*, 185–193.

Witelson, S. F. (1992). Cognitive neuroanatomy: A new era. *Neurology, 42*, 709–713.

Yamaguchi, S., & Knight, R. T. (1992). Effects of temporal-parietal lesions on the somatosensory P3 to lower limb stimulation. *Electroencephalography and Clinical Neurophysiology, 84*, 139–148.

Yordanova, J., & Kolev, V. (1998a). A single-sweep analysis of the theta frequency band during an auditory oddball task. *Psychophysiology, 35*, 116–126.

Yordanova, J., & Kolev, V. (1998b). Event-related alpha oscillations are functionally associated with P300 during information processing. *Neuroreport, 9*, 3159–3164.

Yordanova, J., Kolev, V., & Polich, J. (2001). P300 and alpha event-related desynchronization (ERD). *Psychophysiology, 38*, 143–152.

CHAPTER

8

Negative Slow Waves as Indices of Anticipation: The Bereitschaftspotential, the Contingent Negative Variation, and the Stimulus-Preceding Negativity

Cornelis H. M. Brunia, Geert J. M. van Boxtel, *and* Koen B. E. Böcker

Abstract

Preparing a movement or waiting for a stimulus that will show up in a few seconds is accompanied by a slow negative wave in the electroencephalogram (EEG). It is the result of a summation of a large number of postsynaptic potentials (EPSPs) in the cell columns of the cortical brain areas that will—in one way or another—be involved in the processing of the future event. Three different types of anticipatory slow waves can be distinguished based on the experimental context in which they can be elicited: the *Bereitschaftspotential* (BP), the *contingent negative variation* (CNV), and the *stimulus-preceding negativity* (SPN). This chapter discusses these potentials in detail. First, it considers which components might be distinguished and which brain areas might be involved in their emergence. It then answers the question of which neurotransmitter systems underlie their appearance and what functions they might have.

Keywords: anticipatory slow waves, *Bereitschaftspotential*, contingent negative variation, stimulus preceding negativity, postsynaptic potentials

The electrical currents of the grey matter appear to have a relation to its function. When any part of the grey matter is in a state of functional activity, its electric current usually exhibits negative variation

–(*R. Caton*, 1875).

Introduction

Preparing a movement or waiting for a stimulus that will show up in a few seconds is accompanied by a slow negative wave in the electroencephalogram (EEG). It is the result of a summation of a large number of (presumably excitatory) postsynaptic potentials (EPSPs) in the cell columns of the cortical brain areas that will—in one way or another—be involved in the processing of the future event. Three different types of anticipatory slow waves can be distinguished based on the experimental context in which they are elicited: the *Bereitschaftspotential* (BP), the contingent negative variation (CNV), and the stimulus-preceding negativity (SPN). Figure 8.1 presents schematically the timing of stimulus presentation and movement onset and the way the averaging is achieved to obtain these slow waves.

The BP is a slow negative-going wave that is recorded prior to the execution of a voluntary movement (see Figure 8.2). In a typical case, subjects are asked to make a series of self-paced button-press responses (e.g., approximately one response every 5 s). No stimuli are needed; the impending voluntary movement is a sufficient condition for the emergence of the BP. Approximately 30–40 trials are typically needed to produce, after averaging, a robust slow wave (see the Appendix for a discussion of technical issues related to recording and analyzing the BP and other slow potentials).

The CNV is a slow negative-going wave that shows up if a warning stimulus (S1) announces that, within a few seconds, an imperative stimulus (S2)

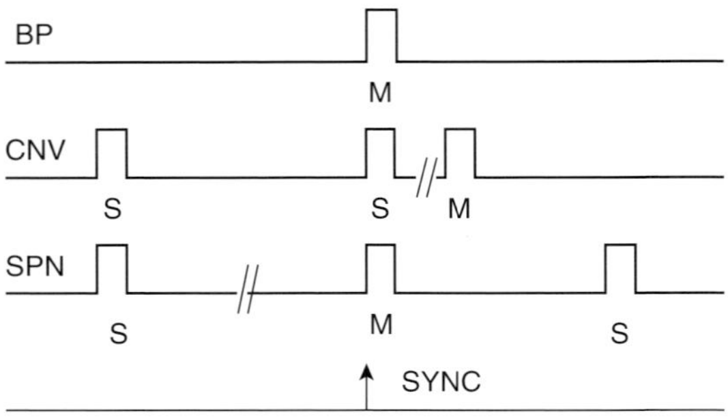

Fig. 8.1. Timing diagram displaying the three experimental situations described in the text. Sync indicates the point in time used for aligning the single trials in averaging.

will arrive, asking for a quick response. The instruction to respond seems to be crucial. Again, a number of trials are needed to produce a clearly recognizable slow wave (see Figure 8.3).

The SPN is a slow negative-going wave that shows up if subjects are waiting for a stimulus that will arrive in a few seconds and will provide significant information. No movement is involved during the waiting interval. In a typical case, subjects are warned by a stimulus that they have to press a button 3 s later. Two seconds after the movement, a feedback stimulus is presented, which informs the subjects about the correctness of their timing. The SPN is observed during the period preceding this feedback stimulus. As with the other slow potentials, an averaging procedure is necessary to make the SPN visible (see Figure 8.4).

We will discuss these potentials in detail later on. First, we will see which components might be distinguished and which brain areas might be involved in their emergence. Next, we will answer the question of which neurotransmitter systems underlie their appearance and what functions they might have.

All three of these potentials are slowly increasing negative shifts that continue to increase up to some significant event. In the case of the BP (also called the *readiness potential* [RP]), the event is a voluntary movement. In the case of the CNV, the event is the imperative stimulus and the speeded-up response to that stimulus; in the case of the SPN, the event is a stimulus carrying some relevant information. It is often thought that the processes underlying the negative shift activate the brain areas prior to the events, presumably because of efficiency considerations. Thus, in the context of the CNV, an analogy with athletes getting ready for the Olympics 100 m race is frequently made, in which a warning stimulus precedes the starting gun: "ready . . . go!" The negative shift before the second stimulus is thought to allow for a more efficient processing of that stimulus and for a quicker start of the response. Even in non-CNV situations, slow brain potentials preceding a particular event are often thought to somehow speed up the processing after that event.

However, such an interpretation runs into various problems. For instance, in the context of the BP, it is not clear what exactly needs to be speeded up if the movements that are asked to be produced are truly voluntary. It is true that the amplitude of the BP increases with shorter time delays between the movements (Deecke & Kornhuber, 1977), but there seems no logic in the finding that preparatory activity should start more than 1 s before the

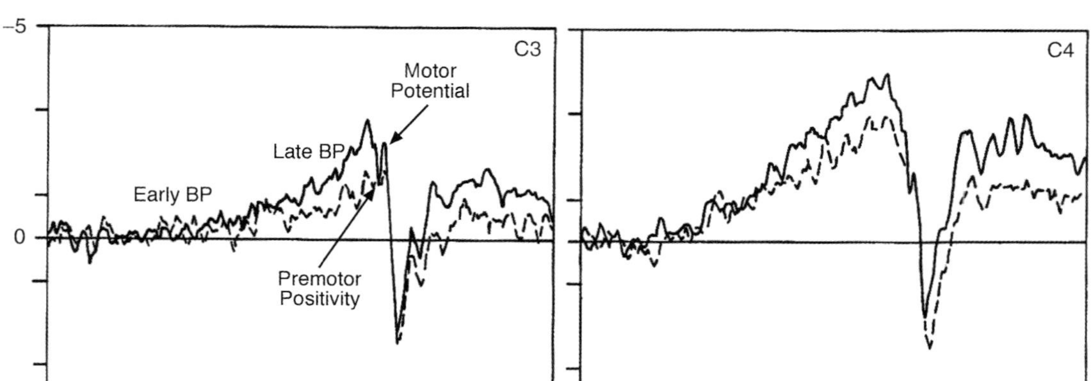

Fig. 8.2. Left: *Bereitschaftspotential* (BP) recorded prior to a right finger movement (solid line) is larger over the left hemisphere (C3) than that recorded prior to a left finger movement (dotted line). Right: A BP recorded prior to a right foot movement (solid line) is larger over the right hemisphere (C4) than that recorded prior to a left foot movement (dotted line). Time zero is movement onset.

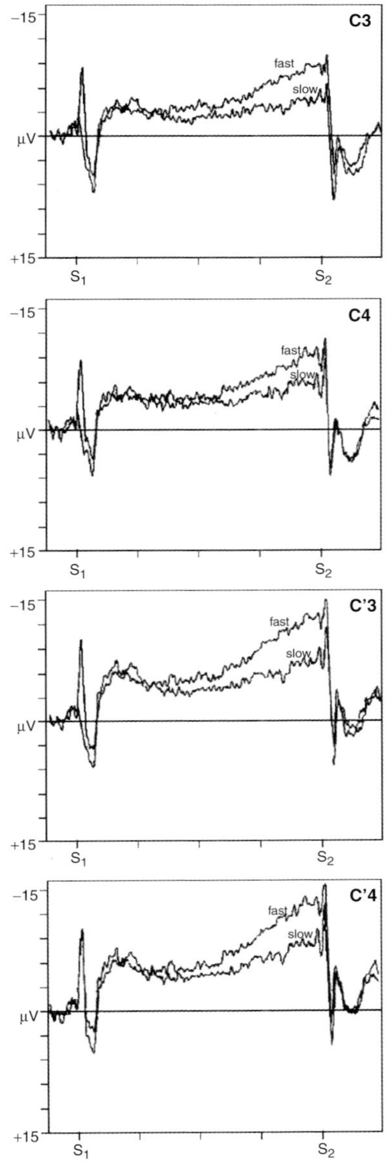

Fig. 8.3. Contingent negative variation (CNV) recorded during a 4 s foreperiod prior to a right foot movement. C'3 and C'4 are located in between CZ and C3 or C4, respectively. Note the larger amplitudes over the right hemisphere ipsilateral to the foot movement side. The early wave has its peak at about 850 ms following the warning signal (WS = S1). The late wave reaches its maximal amplitude shortly after the imperative stimulus (R2 = S2), on which the response had to be given. In this example, motor preparation is clearly manifest in the late wave. If the 10 fastest and the 10 slowest trials of each subject are averaged separately, larger amplitudes are found prior to fast responses. The largest amplitudes are found near the midline, in agreement with a source in the midline. Reprinted with permission from Brunia and Vingerhoets, 1980.

movement is actually made. Secondly, in the context of the SPN, relations between SPN amplitude preceding a stimulus conveying feedback about past performance and the amplitude or latency of postfeedback potentials have, to our knowledge, never been found. In addition, one might wonder why a processing speed increase would be needed in a situation in which a feedback stimulus is presented at the very end of a trial, after which several seconds of resting activity usually follow before the next trial starts.

In the following sections, we shall discuss the functional significance of these potentials shifts in more detail.

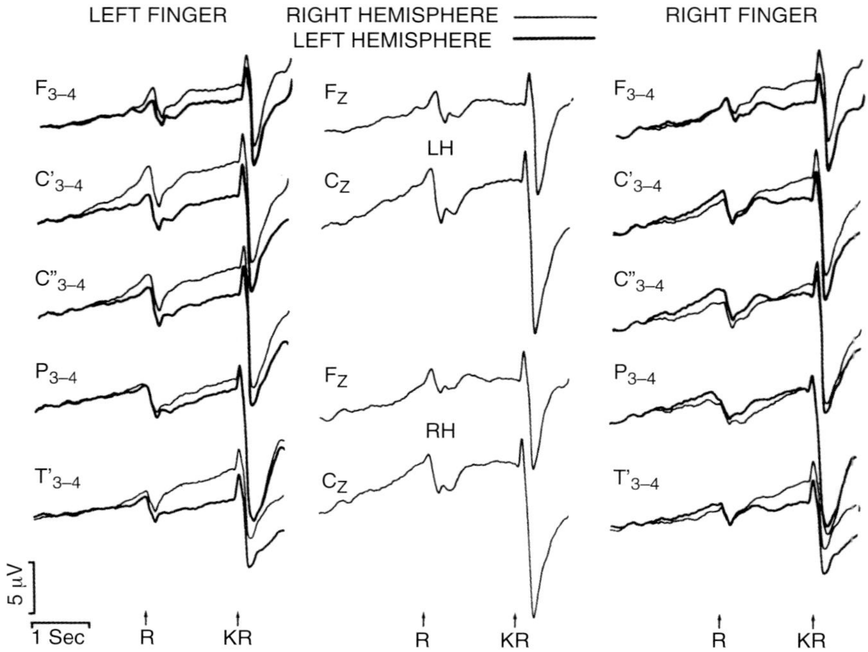

Fig. 8.4. Example of the classical stimulus-preceding negativity (SPN), recorded prior to auditory stimuli providing KR about the correctness of the timing of a preceding finger press. Subjects were instructed to make a movement with either the right or left index finger at intervals of 20–22 s. The KR stimulus indicated whether the movement was made too early, on time, or too late. The button was pressed at time point R, and the KR stimulus was presented 2 s later. Brain activity was recorded over the frontal, precentral (C'3–4 = 1 cm anterior to C3–4, respectively), postcentral (C'3–4 = 1 cm posterior to C3–4, respectively), temporal (T'3–4 = 1 cm anterior to T3–4, respectively), and parietal areas. Over the central areas, BP amplitudes were larger over the hemisphere contralateral to the movement side. Just prior to the presentation of the KR stimulus, SPN amplitudes were larger over the right hemisphere, suggesting that the right hemisphere was more involved in the attention for this stimulus. Reprinted with permission from Brunia (1988). In later experiments we used a warning stimulus 3 s after which the movement had to be made (see also Figure 8.1). Two seconds later, the KR stimulus was presented.

The BP

If a simple movement is going to be made, a slow negative wave can be derived from the EEG, showing systematically increasing amplitude up to the moment the relevant muscles are activated (see Figure 8.2). This potential is known as the *Bereitschaftspotential* (BP) or *readiness potential* (RP; Kornhuber & Deecke, 1965). For a discussion of the BP beyond the scope of this chapter, the reader is referred to the book by Jahanshahi and Hallett (2003) and to the review paper by Shibasaki and Hallett (2006).

Components

Usually, at least two components are distinguished in the negative potential that precedes voluntary movement (see Figure 8.2, left). The slowly increasing early negative BP starts between 2000 and 1500 ms prior to movement onset, depending on the rate at which the movements occur. It shows a steeper increase in amplitude 400 to 500 ms preceding the movement: the late BP or the *negative shift* (NS; Shibasaki et al., 1980). A further component is the so-called motor potential (MP), which, after a small premovement positivity (PMP), shows a very sharp increase in negativity from about 80 ms prior to movement onset. The anticipatory movement-preceding slow waves are followed by a strong positive wave, usually considered a manifestation of "reafferent" activity.

Deecke and colleagues (1976) suggested that the early BP might be related to activation of the supplementary motor area (SMA) because of its largest amplitude in the midline. They related the late BP and the MP, because of their contralateral predominance, to activation of the primary motor cortex (MI). The PMP (also known as the *P-50*) seems to show up at the side ipsilateral to the movement. Whether the PMP reflects a specific neural process (Ball et al., 1999; Shibasaki & Kato, 1975) or is simply a trough between two subsequent partially overlapping negative components (Böcker et al., 1994; Shibasaki & Hallett, 2006) is not yet clear.

Are Self-Paced and Externally Triggered Movements Different?

This question refers to the functional interpretation of the BP. In the early papers, it was observed that the BP started very early in relation to the reaction time of a simple externally triggered movement and did not seem necessary for executing the movement. The best comparison of movement-preceding negativities prior to self-initiated movements and similar externally triggered movements has been made by Jahanshahi and colleagues (1995). These authors presented imperative stimuli (tones) at the same frequency as each subject's rate of voluntary movements. Since the tones in the externally triggered movements cause brain activity that might confound the movement-preceding negativity, the authors presented the same tones after the self-initiated movements in the BP recording, the instruction being to ignore them. This tone-related activity was later subtracted from the recordings in the externally triggered movements after careful rearrangement. The self-initiated movements were accompanied by larger movement-related negativities over the midline than the stimulus-triggered movements.

In a positron emission tomography (PET) recording in that same study, Jahanshahi and colleagues (1995) found an activation of SMA, the bilateral anterior cingulated cortex (ACC), right dorsolateral prefrontal cortex (DLPFC), left putamen, left thalamus, left sensorimotor cortex, bilateral premotor cortex, bilateral insular cortex, and bilateral parietal cortex (area 40). The only difference between the self-initiated movements and the externally triggered movements was the activation of the right DLPFC in the self-initiated movements, which was interpreted to be related to non routine decision making. Although this was a nicely controlled study, part of the differences might have been related to attention being paid to the tones in the externally paced condition and movement preparation due to the predictable response rate (which was instructed to be about once per 3 s).

In a follow-up study (Jenkins et al., 2000), variable-rate movements were studied (self-initiated and yoked, one per 2–7 s) instead of fixed-rate movements. In the variable-rate, and thus unpredictable, condition, no BP whatsoever was discernible. The associated PET results showed that the rostral SMA was uniquely active in the self-initiated (variable-rate) movement condition, and caudal SMA, adjacent ACC and bilateral DLPFC were more active during the same condition compared to the (unpredictable) externally triggered movement condition. This suggests that the rostral SMA, or pre-SMA, plays a primary role in movement preparation.

Brain Areas

Because the EEG reflects activity from cortical neurons, part of the sources of the BP have a cortical origin, and these will be discussed first. Yet, subcortical structures contribute to the emergence of the BP as well, as we will see later.

CORTICAL SOURCES

Cortical sources of movement-preceding activity have been demonstrated (1) by direct measurement of intracerebral activity in humans (Rektor, 2003), (2) by epicortical recordings in humans (Ikeda et al., 1993), (3) by source localization studies in humans using surface EEG (Böcker et al., 1994; Knösche et al., 1996; Praamstra et al., 1996), (4) by functional magnetic resonance imaging (fMRI) studies in humans (Cunnington et al., 2003), (5) by PET studies in humans (Jahanshahi et al., 1995; Jenkins et al., 2000, as discussed above), and (6) by studies in monkeys using subdural or intracortical electrodes for recording either the BP itself (Sasaki & Gemba, 1981) or the unit activity underlying this slow wave (Tanji & Mushiake, 1996). Apart from the monkey studies, most of the data recorded by means of these techniques are summarized in Jahanshahi and Hallett (2003). The following cortical areas are active during the generation of the BP: the primary motor cortex, (MI), the primary somatosensory cortex (SI), the premotor cortex (PMC), and the four motor areas Dum and Strick (1991) distinguished in the mesial frontal cortex. These mesial motor areas are the pre-SMA, the SMA proper, and the rostral and caudal cingulate motor areas, CMAr and CMAc, respectively. The last three contribute to the descending corticospinal motor tracts (Dum & Strick, 1996) and might influence the interneuronal networks around the motoneurons in the spinal cord as a function of motor preparation. Because the pre-SMA does not make such a contribution, this suggests a special role for this area, which can be considered a supra- or executive motor area without connections to MI or the spinal cord, in contrast to the SMA proper.

SMA AND MI: A DIFFERENT TIME COURSE OF ACTIVATION?

Above, we suggested an executive role for the pre-SMA in self-initiated movement, which would predict that this area is activated prior to MI, as suggested by Deecke et al. for the whole SMA in 1976, although the distinction between both areas was not known at that time (Deecke et al., 1976). One would expect that epidural electrodes would provide the most pertinent information on this subject, because they seem a natural way to deblur the scalp recordings while preserving the time resolution and waveforms of these recordings. However, Ikeda and colleagues (1993) concluded that the onset of the epidural BP is generally equal in pre-SMA, SMA, and M1-S1.

This might pose a problem for the interpretation of electrical source analysis approaches. Using time-varying but locally fixed dipoles, we (Böcker et al., 1994) concluded that the main generators of BP are in bilateral motor areas at the dorsal brain surface. However, we realized that current dipoles in the medial wall of the SMA may cancel each other by their both pointing to the midline. Under normal circumstances, such source configurations cannot be identified accurately from scalp recordings alone, especially if the time series of the BP from the different areas overlap to the extent suggested by Ikeda et al. (1993). Lang et al. (1991) investigated a patient with an infarct in the right SMA, thus avoiding the possibility of dipole cancellation. They found a clear BP over the left SMA prior to a simple movement, demonstrating an active focus related to movement preparation in the SMA. Using a more advanced source analysis technique (distributed source modeling based on realistic head models derived from individual MRI scans co-registered with f-MRI), Ball et al. (1999) studied the premovement activity in a number of brain areas. They concluded that a simple flexion movement is preceded first by activation of higher motor centers (the pre-SMA, the cingulate motor area and the inferior parietal lobe), followed by activation of the SMA proper and the MI. By scanning a single slice with 100 ms time resolution, Weilke and coworkers (2001) were able to replicate this result based on imaging information alone.

Recording of unit activity in nonhuman primates also suggested a special role for the pre-SMA. Many neurons in the pre-SMA become active prior to movement execution (Matsuzaka et al., 1992), suggesting that this area is involved in the preparation of movements. More specifically, one of the functions of the pre-SMA might be the organization of temporal sequences (Shima & Tanji, 2000; Tanji & Mushiake, 1996). These authors showed that the pre-SMA contains neurons that fire during the waiting period prior to a specific memorized movement sequence (e.g., turn, pull, and push) but are silent prior to a different memorized sequence (e.g., turn, push and pull). The premovement activity in the pre-SMA thus seems, among other things, to suggest the participation of this area in a working memory (slave) circuit.

Summarizing these studies, we reach the same conclusion as Shibasaki and Hallett (2006). Preceding self-paced movements, the following brain areas are activated subsequently during the early BP: the pre-SMA and the SMA proper; the bilateral PMC and MI become activated subsequently. During the late BP (NS'), MI and the PMC are mainly active contralateral to the movement side. Yet, in view of the epidural recordings, it remains possible that unit activity, hemodynamic responses, and postsynaptic potentials giving rise to the scalp-recorded BP show different time courses.

SUBCORTICAL SOURCES

It is likely that the cortical sources of the BP are triggered from subcortical areas like the cerebellar dentate nucleus, the thalamus, and the basal ganglia. In particular, the dentate-thalamocortical pathway seems crucial for the emergence of the BP, as Ikeda and co-workers have demonstrated clearly (see Ikeda & Shibasaki, 2003, for a review). Subcortical structures play a role in networks that are responsible for planning and executing voluntary and automatic movements, response selection, and the inhibition of alternative motor programs. In five out of seven patients undergoing deep brain stimulation for treatment of essential tremor (n = 6) or myoclonus dystonia (n = 1), Paradiso and colleagues (2004) recorded BPs from the cortex and underlying thalamus. The depth electrode had four contacts, and with monopolar recordings all showed positive-going potentials, meaning that much of the potential was volume conducted from another part of the cortex. However, bipolar voltages were also detected between the successive contacts of the quadripolar depth electrode, and these voltages reflected locally generated activity. Thus, the authors were able to record bipolar premovement potentials with the same time delay to movement onset as the cortically recorded BP. The similar time course makes it impossible to decide in what area the activity started. Paradiso et al. (2004) reported maximal amplitudes in the ventral lateral nucleus of the thalamus in four patients in whom the location of the electrodes was confirmed by postoperative MRI. In four thalamic recordings a phase reversal was reported, which is a sound argument in favor of a local source.

It is of theoretical interest that the thalamic premovement potentials were present before movements on either side. Since we don't know where the activity started, we may conclude that either one cortical area projects to either thalamus or that both motor cortices from one thalamus are activated. The latter might play a role in the emergence of the BP ipsilateral to the movement, instead of or together with an activation of ipsilateral motor areas via the corpus callosum.

Subcortical and intracortical BP recordings have been made in therapy-resistant epileptic patients in whom it might be necessary to remove the focus of pathological activity from the brain. To determine the exact location of the epileptic focus, depth electrodes can be used, which remain in place for some time. Rektor and coworkers (for an overview, see Rektor, 2003) have made BP recordings from a number of different areas. Apart from the MI, the SMA, and the anterior and posterior cingulate gyrus, they were able to record the BP in the thalamus and the basal ganglia. More specifically, they found the BP in the pallidum, the putamen, and the head of the caudate nucleus. Taken together, these data suggest that the different cortical and subcortical areas from which the BP can be recorded may be considered nodes in preparatory networks. The exact functional role of these nodes needs further research. With scalp-recorded brain potentials, it will be possible to study the roles of the cortical nodes more precisely. This will reflect the contributions of subcortical nodes insofar as they influence the cortical ones, for the distance from the sensors and the absence of massive parallel fibers in subcortical nodes prevent their direct contribution to scalp-recorded potentials.

Psychopharmacology

Most of the information about the psychopharmacology of the BP comes from studies of Parkinson's disease (PD). The clinical symptom of impaired movement initiation leads to the prediction that the (early) phase of the BP should be affected. Indeed, the weight of the evidence (reviewed in Colebatch, 2007; Verleger, 2004; Chapter 21, this volume) indicates that the early phase of the BP is reduced in PD patients compared to normal controls, whereas the late BP, or NS, is relatively spared. Furthermore, the effect is larger for more complex and free movements compared to simple and externally triggered movements. This is generally interpreted as evidence that the output from the basal ganglia to the (pre) SMA is disturbed, whereas the cerebellar input to the motor cortex is relatively preserved. This suggests the involvement of dopamine in the (subcortical) generation of the early phase of the BP. The initial study on this subject also showed that the early BP increases with L-DOPA administration in PD patients as well as normal controls (Dick et al., 1987). In the same vein, Fattaposta and colleagues (2002) found a normalization of BP after acute L-DOPA administration. In their study, the BP was larger in the PD group compared to the control group for reasons that might be related to the exact task used or increased motivation in the patients.

These conclusions are supported and extended by findings on the BP in schizophrenia. In general, both the early and late BP are reduced in these patients compared with controls (as reviewed by Westphal, 2003). This is especially the case in patients with negative symptoms (Fuller et al., 1999). It is now generally assumed that negative symptoms stem from deficient activity in the mesiocortical dopamine system. In cases of dopaminergic supersensitivity, that is, tardive dyskinesia, a very large BP is present (Adler et al., 1989).

Functional Significance

Research on the functional significance of the BP has been greatly limited by the fact that only very basic manipulations can be applied to voluntary movements. Participants can be asked to produce movements with either hand, with various fingers (either in isolation or in a sequence), with high or low force, slowly or rapidly, but all of these manipulations are quite basic in nature. And of course, many of these manipulations have been reported in the literature.

In normal subjects, effects on the early symmetrical part of the BP are usually found not on its amplitude but rather on its onset latency, that is, the time between the start of the early BP and the impending movement. Thus, the early BP starts later at high as opposed to low movement rates (Kornhuber & Deecke, 1965), for simple as opposed to complex movements (Lang et al., 1991), and for hand movements as opposed to foot movements (Brunia & Van den Bosch, 1984). These are very general variables that may be tentatively related to aspects of the selection of the appropriate strategy for the movements.

It has been shown above that brain circuits involving the basal ganglia are important for the generation of the early BP, and indeed, the amplitude of the early BP has been found to be attenuated in patients with PD (Dick et al., 1989). The finding that these patients have difficulty with response selection (e.g., Van den Wildenberg et al., 2006), and other evidence linking the basal ganglia with selection and implicit memory (e.g., Mishkin et al., 1997), supports the above suggestion that the early BP might be related to the selection of the appropriate motor strategy from memory. This might be called an *internal system*, and its anatomical basis is in the medial (basal ganglia–SMA–medial premotor) structures.

Effects on the asymmetrical late part of the BP have been found for extremity, force, and time parameters, as well as for external factors such as online feedback. When movements with both hands are compared, the familiar contralateral dominance of the late BP was found (Kornhuber & Deecke, 1965). Together with anatomical knowledge from the motor system, this finding constituted the late BP as movement-related. When foot movements were compared to hand movements (e.g., Brunia, 1980), ipsilateral but not contralateral dominance was found. The surface potentials were hypothesized to still originate from the contralateral hemisphere. This was confirmed by magnetoencephalography (MEG) recordings of Hari et al. (1983) and by EEG source localization studies by Böcker et al. (1994). Variations of response force and timing have effects on the late BP but not on the early BP. Thus, greater force produces an enhanced late BP (e.g., Becker & Kristeva, 1980), as do movements that reach the same force level quickly as opposed to slowly (ballistic versus ramp movements; e.g., Grünewald & Grünewald-Zuberbier, 1983). These findings suggest that one of the functions of the late BP is related to the specification of the appropriate parameters for the impending movement.

When movements are executed that contain aspects of feedback, such as in tracking, the late BP is smaller and its scalp distribution is more posterior (Deecke et al., 1984). In patients withPD, the late BP is usually enhanced relative to controls (Dick et al., 1989). Both findings seem to suggest a dependence of the late BP on external factors. The idea in PD is that these patients have to rely more on external factors that control for the successful initiation and performance of movements. The late BP has been shown above to be dependent on circuits involving the dentate nucleus in the cerebellum and the primary motor cortex. Input to this system involves motor and sensory association cortices (Allen & Tsukahara, 1974). This is consistent with dependence of the late BP on external (e.g., sensory) factors.

Summarizing, the late asymmetrical part of the BP appears to be an external system, and its anatomical basis is in the lateral (cerebellum–motor cortex) structures. The primary function of the system is thought to be the specification of the precise movement parameters for the impending motor act, depending on the actual environmental context.

It should be noted that there seems to be no reason why the hypothesized functions of the early and late BP would need to take such a long time for relatively simple motor acts. On the other hand, there also seems to be no reason why such processes could not take all the time required when the motor acts are truly voluntary.

Finally, the BP paradigm has been studied to elucidate the nature of voluntary decisions. Libet and colleagues (1983) had participants execute voluntary movements while viewing a clock and report afterward the moment at which they decided to perform the movement. As usual, the BP started about 500 ms prior to the movements, whereas subjects reported making the decision 200 ms prior to the movements, or 300 ms after the onset of the BP. This finding and its replications started a great debate on the nature of voluntary decisions and consciousness. One of the associated findings was that, patients with parietal cortex damage reported the voluntary decision even later, at the moment of the movement itself (Sirigu et al., 2004), indicating a role for this brain area in motor awareness.

The Contingent Negative Variation

If a warning stimulus (S1) announces the future arrival of an imperative stimulus (S2), reaction time is faster than without S1. Grey Walter and his colleagues discovered that during the interval between S1 and S2 a slow negative wave can be recorded in the EEG, which they called the *contingent negative variation* (CNV), in line with the quote of Caton at the beginning of this chapter. The slow wave did not show up if just the two stimuli were presented, without a task. The instruction to respond seemed to be the crucial factor.

Components

Initially, an interstimulus interval of 1 s was used, and it soon became clear that a number of psychological factors played a role in the emergence of the CNV. Moreover, researchers realized that there might be a connection between the CNV and the BP, which was discovered at around the same time. After all, both slow waves show up prior to a button press. Yet, in their first publication, Grey Walter et al. (1964) mentioned that the CNV could be recorded as long as subjects pressed the button and remained attentive. This was an early indication that the preparation of the movement was not the only determining factor; attention was involved as well. Järviletho and Frühstorfer (1970) were the first to suggest that there was a frontally dominant activity related to the properties of S1 and a central activity related to the response preparation. Conner and Lang (1969) started to use experiments with a longer foreperiod, resulting in the emergence of two slow waves, one early and one late.

It appeared that the two components described by Järviletho and Frühstorfer (1970) were reflected in the early and late waves, respectively. The late wave resembled the BP more than did the CNV, recorded during a 1 s interval, suggesting that the late wave and the BP might be identical (see, e.g., Figure 8.3).

Rohrbaugh and colleagues (1976) recorded brain activity after the presentation of a tone and prior to a voluntary button press. Next, they showed the coupling of both results in a CNV-like picture, suggesting that the CNV is just a summation of brain activity elicited after the tone and brain activity recorded prior to the movement. If that was correct, the inescapable conclusion would be that the CNV late wave and the BP are identical. However, if the BP and the CNV were recorded in the same trials, the amplitude of the CNV late wave was larger than that of the BP, although the potential distribution for finger movements and the paradoxical potential distribution for foot movements (Brunia, 1980) were similar for both slow waves (Brunia & Vingerhoets, 1981). The larger amplitude of the CNV late wave suggested that other processes are involved, while the distribution itself was an argument for the motor interpretation. In line with the earlier paper of Rohrbaugh et al. (1976), Rohrbaugh and Gaillard (1983) presented a number of arguments in favor of the identity of the BP and the CNV late wave. Yet, it has become clear that the motor interpretation, however important, is only part of the story.

What and When?

In a classic fixed forewarned simple reaction time task, the subject knows what to do and when. These (relative) certainties can be manipulated. In a choice RT task there is event uncertainty: It is not known what the response should be unless a cue is presented giving complete or partial information about the expected response (Van Boxtel & Brunia, 1994). In that case, smaller late-wave amplitudes are recorded. If foreperiods of different lengths are presented, there is time uncertainty, with comparable consequences (for a discussion of this topic, see Brunia, 2003). Event and time uncertainties have electrophysiological and behavioral consequences that have to be kept in mind when considering the results of PET and fMRI studies.

Brain Areas

We have seen that even for a very simple movement a number of brain structures have to be activated, and that for self-initiated and externally triggered movements, apart from the DLPFC, the same structures are activated, albeit in a different strength (Jahanshahi et al. 1995). The premovement activity in the externally triggered movement takes place between stimulus presentation and motor act, while in the CNV there is time for anticipatory attention to the imperative stimulus and for motor preparation before the S2 is presented. The latter might include the areas activated in the externally triggered movements, but the attention for the upcoming stimulus is a different process in which different brain areas are involved. Taking our interpretation of the anticipatory slow wave as the point of departure, components of the CNV will be present over all areas that are, in one way or another, involved in the upcoming task. Therefore, the topographical distribution of the CNV (especially the late wave) is far more complex than that of the BP, since it depends upon all aspects of the task to be performed. This is also the reason why Lang et al. (1984) found a different slow wave distribution when movements had to be carried out under the guidance of somaesthetic or visual guidance. What at that time was called *directed attention* potential could indeed, at least partly, be considered an index of anticipatory attention and monitoring activity. The distribution of slow potentials recorded during task performance reflects *area-specific* activity (see Birbaumer et al., 1992) that starts with anticipatory slow waves as soon as it is known when a certain area has to be activated and for what reason (for a short discussion of slow potentials during task performance, see Brunia, 2003). Investigating long-term memory, Rösler and co-workers (Heil et al., 1997; Rösler et al., 1995) found different slow potential distributions when verbal, spatial, or color information had to be memorized. Another source of a differing potential distribution is the emotional value of S2. Because we believe that the functional role of the brain areas involved in anticipatory attention for emotional stimuli can be studied better in motor-free conditions, we will discuss that subject in the section on the SPN.

Psychopharmacology

The most explicit pharmacological model for the CNV states that its amplitude is determined by the activity of cholinergic neurons. These neurons are, in turn, under the control of the excitatory catecholamines dopamine (DA) and norepinephrine (NE) and the inhibitory GABA-ergic activity (Hansenne et al., 2000); Papart et al., 1997; Timsit-Berthier, 1991). Effects of DA, NE, and GABA have been demonstrated by acute drug manipulations. However, Rockstroh and colleagues (1991) had suggested that the GABA-ergic dampening effects were not very specific, but were shared between benzodiazepines and, for example, carbamazepine (an anticonvulsant that probably acts on voltage rather than receptor-gated ion channels). The relationship between DA and CNV amplitude is probably an inverted U shape (Tecce, 1991), like other dose-effect relations of DA. Furthermore, CNV has been found to be decreased in patients with PD (Verleger, 2004) and with schizophrenia (e.g., Verleger et al., 1999). With respect to NE, it has been reported that beta-adrenergic antagonists decreased migraines in patients showing high baseline-level CNV amplitudes (Schoenen et al., 1986). The treatment also normalized their CNV amplitudes (Maertens et al., 1986),

However, the idea that cholinergic neurons are the final common pathway for CNV generation has not received much support lately. Nicotine, a cholinergic agonist, increased CNV only in some of the males studied (Daurignac et al., 2001; Houlihan et al., 2002), depending on personality characteristics (Cook et al., 1996). Yet, the cholinergic generation of slow potentials has been suggested to depend on muscarinergic and not nicotinergic receptors. Nevertheless, we suggest that it may be more fruitful to integrate the pharmacological data on CNV generation in a slightly different and more general model. The EEG/ERP and slow potential recordings are probably dominated by postsynaptic potentials at the pyramidal neurons. Currently, the general model for transmitter control of the pyramidal cells states that they get a great deal of inhibitory input, generating inhibitory postsynaptic potentials, all over the cell membrane from GABA-ergic interneurons. This is balanced by excitatory input from corticocortical glutamatergic connections that generate excitatory postsynaptic potentials at the apical dendrite. The monoaminergic (DA, NE, and serotonin) and acetylcholinergic input stems from subcortical (brainstem, basal forebrain, and tegmental) nuclei and has synapses at the level of the basilar dendrites of the pyramidal cells. The latter inputs probably control the firing rate of the pyramidal cell by adaptive filtering of the amino acids GABA and glutamate inputs on the neuron.

In this model, serotonin could also influence CNV recordings, yet only a few studies have addressed such an effect. Papart (1996) selected a group of depressed patients with high CNV amplitude and showed that they responded better on selective serotonin reuptake inhibitors (SSRIs) than a group with low CNV amplitudes. Whereas the clinical effect of SSRIs is thought to depend on an increase in serotonin activity, following down-regulation of presynaptic 5HT1A receptors (involved in negative feedback), a single study with flesinoxan, a full 5HT1A agonist, suggested that this receptor is not involved in CNV modulation (Hansenne et al., 2000). Finally, a study by Ashton and colleagues (1994) showed that suicide-prone depressed subjects with low serotonin levels are characterized by low CNV amplitude. In sum, serotonin does affect CNV amplitudes, but the existing data are equivocal given that that low serotonin levels were associated with either high or low CNV amplitudes in two different samples of depressed patients.

Two recommendations follow from this short overview of the psychopharmacology of the CNV. First, it would be useful to compare the psychopharmacology of CNV and BP more directly by recording them in the same session (cf. Brunia & Vingerhoets, 1981). This would also give indications about the similarities and differences between those slow potentials. Second, CNV researchers should use interstimulus intervals that make it possible to distinguish between CNV early and late waves. Given the distinction between phasic stimulus-related arousal and tonic response-related activation (McGuinness & Pribram, 1980), it could be hypothesized that the CNV early wave is related to noradrenergic arousal and the late wave to dopaminergic activation.

Functional Significance

If the slow brain potentials indeed reflect the preparation of brain areas required for task execution after the relevant event, as suggested above, then it should come as no surprise that motor areas will be activated when the task focuses on motor execution. In other words, if the manipulations in a CNV task concern the response extremity (mostly the hand versus the foot), force and time parameters, or the speed of the motor response, it can be expected that the late CNV wave resembles the late BP wave to a great extent (see Figure 8.3). This is what Rohrbaugh and Gaillard (1983) concluded based on the many experiments that had manipulated such factors at the time of their writing.

Thus, if a typical CNV experiment is focused on the comparison between hand and foot movements, then the same seemingly paradoxical results will be found for the late CNV as for the late BP (Brunia, 1980). Similarly, if such an experiment concerns manipulations of response force or velocity, the late CNV will be found, just like the late BP, to be greatest preceding more forceful responses (Low & McSherry, 1968) and preceding responses with a more rapid force increase (Van Boxtel et al., 1993). Also, the emphasis on response speed in many CNV experiments results in greater late CNV amplitudes for fast compared to slow responses (e.g., Rohrbaugh et al., 1976), suggesting that the late CNV should be interpreted in terms of motor variables.

Taken together, these lines of evidence strongly suggest that if a fast response is required after the RS in a CNV task, then motor-related activity is present before S2. Figure 8.3 shows that larger amplitudes of the late wave are found preceding fast responses than preceding slow responses. This activity resembles the late BP, but because of the experimental paradigm, it will be denoted as late CNV. It follows that the functional significance of the late CNV in such cases is much the same as that of the late BP, that is, the specification of the precise response parameters based on the environmental context (the motor task in this case).

As an aside, it would be interesting to investigate the similarity of the early CNV and the early BP. Both have more or less the same scalp distribution (midline frontocentral), and although the early CNV is usually viewed as a reaction to S1 (Rohrbaugh & Gaillard, 1983), there is also some evidence of motor effects on the early CNV (e.g., Van Boxtel et al., 1993), which possibly have the same functional significance as the early BP, that is, the selection of a motor strategy from memory-based on the characteristics of the S1).

If the task is more perceptual than motor, or even only perceptual, then the late CNV is much smaller compared to the one in which a fast motor response is required (e.g., Gaillard & Perdok, 1980). However, it is also true that it varies as a function of more perceptual variables. For instance, its scalp distribution is shifted to more posterior areas known to be involved in sensory processing (Gaillard & Perdok, 1980; Lang et al., 1978). The same is true when no motor response at all is required, such as in the experiments reported by Simons and colleagues (1979), who observed a late CNV preceding interesting slides, or those of Ruchkin and coworkers (1986), in which subjects had to judge the similarity of S1 and S2. If tasks are used in which timing (Macar & Bonnet, 1997) or mental effort (Van Boxtel, 1994) are key ingredients, the scalp distribution of the late CNV is more frontal than central even when a fast motor response is required.

The emerging picture of the functional significance of the late CNV is one in which both motor and non-motor negativities are present. The amplitude and scalp distribution of the late CNV are mainly determined by the task requirements at S2, which often, but not always, includes a motor response.

The disadvantage of the warned reaction time paradigm is that attention for the upcoming stimulus and preparation for the movement take place simultaneously. Thus, the CNV late wave reflects the confounding of both processes. In order to separate in time the movement preparation from the anticipatory attention, we did a series of experiments with a time estimation paradigm. It began the investigation of the next slow wave.

The SPN

Although we started our research of the SPN with the time estimation paradigm, it later became clear that this slow wave precedes different stimuli in various tasks and with a different potential scalp distribution. An overview of that research can be found in van Boxtel and Böcker (2004). These authors distinguished four types of stimuli prior to which an SPN can be recorded: (1) stimuli providing knowledge of results (KR);(2) stimuli conveying instructions for a task at hand; (3) probe stimuli against which the result of an actual performance should be matched; and (4) affective stimuli.

Just as the CNV late wave is largest when some motor activity is demanded, the SPN is largest prior to KR stimuli and prior to affective stimuli. We will discuss the latter two in more detail and point to van Boxtel and Böcker (2004) for the SPN preceding instruction and probe stimuli.

In the time estimation task, subjects have to press a button 2 s after an instruction stimulus and 2 s after the button-press feedback is presented about the correctness of their performance by means of a KR stimulus. As the KR stimulus, we used a vertical line if the subject pressed the button within the correct time window, a horizontal line (minus sign) if the button was pressed too early, and a plus sign if it was pressed too late. Two different slow waves were recorded with a different potential distribution. The first was the BP, prior to the button press, and the second was the SPN, prior to the KR stimulus presentation. The BP is larger over the hemisphere contralateral to the movement side, while the SPN is mostly larger over the right hemisphere (see Figure 8.4).

We tried to relate the results of investigations into the SPN, both prior to KR stimuli and prior to affective stimuli, to involved brain areas. We discuss these results in the next section.

Brain Areas

The SPN preceding KR stimuli in a time estimation task was studied by Brunia & Damen (1988), Damen and Brunia (1987, 1994), Chwilla and Brunia (1991), and Kotani and Aihara (1999). To evaluate whether attention was crucial, Chwilla and Brunia (1991) presented the time estimation task described above in three conditions: real feedback as usual, false feedback in which the symbols presented had no relation to the actual performance, and no feedback. The SPN showed up only in the true feedback condition, suggesting that attention for a stimulus that provides useful information is a necessary condition for the appearance of the SPN. Although the amplitude of the pre-KR SPN depends on the amount of information conveyed by the KR stimulus, we will see later that it reflects a motivational function too.

The pre-KR SPN shows a frontal plateau, and a right parietal maximum. At the latter scalp position, the negative slope increases up to the presentation of the KR. This differentiation of waveforms suggests multiple sources. In a dipole modeling study, Böcker and colleagues (1994) suggested that a bilateral frontotemporal dipole could explain most of the variance of the activity recorded between the button press and the presentation of the KR stimulus. They supposed that the insula Reili could be one of the locations activated when waiting for a KR stimulus. In that study there was no indication of the right-hemisphere preponderance that had been found so far in all other studies. Comparing the true and false feedback conditions of the study of Chwilla and Brunia (1991) in a PET study, Brunia and coworkers (2000) found an activation in the PFC (area 45), the temporoparietal insula Reili, and the parietal cortex. These data suggest that waiting for information about a past performance goes along with activation of a network in the right hemisphere, in which the PFC, insula Reili, and parietal cortex participate. This was later confirmed in an fMRI study (Tsukamoto et al., 2006), where again the right hemisphere was found to be activated.

This right-hemisphere preponderance of the KR circuit is independent of the kind of stimuli used. Modality-specific effects are manifest as relatively greater amplitude over the frontal areas prior to auditory KR stimuli and in the occipital areas prior to visual KR stimuli (Brunia & Van Boxtel, 2004; Kotani et al., 1999). Because the KR network is independent of modality effects, we interpret the modality-specific activity as a separate SPN, reflecting the anticipation of the future arrival of the visual or auditory information in the sensory areas. After being processed, this sensory information could contribute to the activity in the KR circuitry. A further argument for this interpretation is the use of verbal KR stimuli, which had no influence on the right-hemisphere preponderance of the KR circuit either (Brunia & Van Boxtel, 2004). Probably due to an insufficient number of electrodes, we found in that study no language-specific SPN, which, according to our hypotheses, should be found in future investigations.

In a recent study, Ohgami et al. (2006) demonstrated that reward and punishment related to KR induce differential lateralization of the SPN. Whereas the pre-KR SPN in a punishment condition showed the classical right-hemisphere preponderance, the pre-KR SPN in a reward condition did not. While in this study a mixture of KR and emotional stimuli was used, the latter are more specifically addressed in the following studies.

The other robust SPN is recorded in anticipation of emotional stimuli (Böcker et al., 2001; Van Boxtel & Böcker, 2004; see also Simons et al., 1979). Emotional stimuli are, by definition, those that evoke affective, motivational, and physiological responses. They activate the limbic circuits. With positive stimuli the dopaminergic nucleus accumbens and the reinforcement circuit are predominantly involved, whereas negative stimuli activate primarily the amygdala. Furthermore, stimuli that induce approach motivation lead to left-hemisphere activation, as assessed by frontal alpha desynchronization, and those that induce avoidance motivation activate the right frontal lobe (Davidson & Irwin, 1999).

Stimuli that occur predictably and that clearly evoke affective and physiological reactions, ranging from electric shocks (Böcker et al., 2001; Irwin et al., 1966) to pictures of opposite-sex nudes (e.g., Howard et al., 1992; Simons et al., 1979), are indeed preceded by negative slow waves. This class of SPNs is generally characterized by a right anterior distribution. Whether this lateralization depends on the polarity of the affect needs more formal testing using extended electrode configurations. Using the same experimental setup as

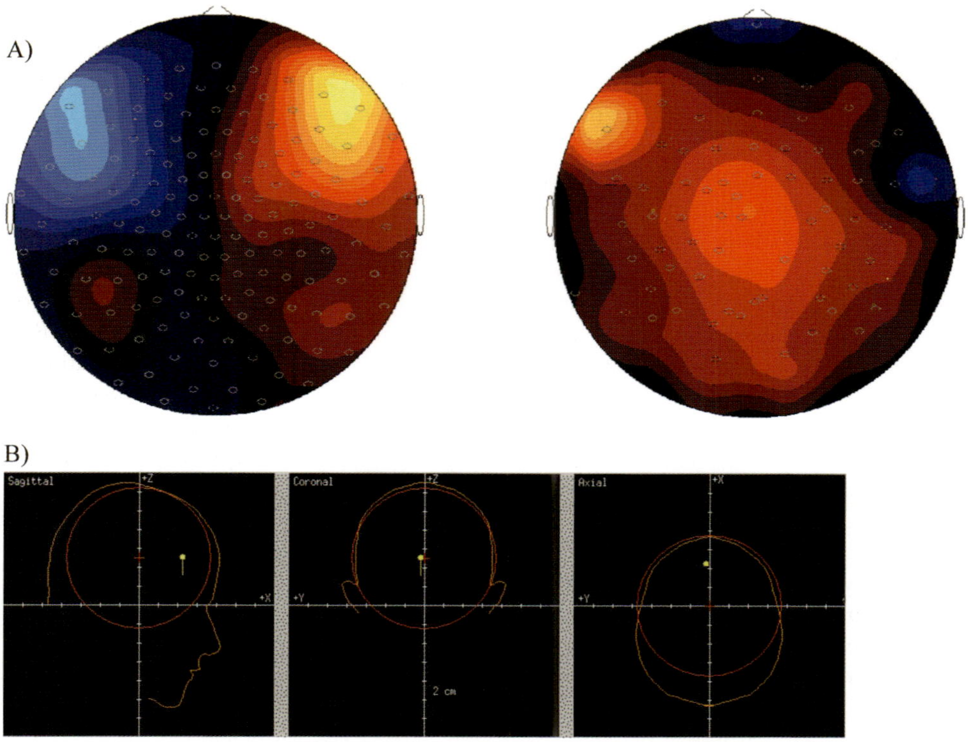

Fig. 8.5. The SPN during a threat of shock (n = 8). (A) Magnetic (left; red = influx, blue = efflux) and electrical (right; red = negative, blue = positive) field distribution at 1.1 s after cue onset, with possible shock (on 6% of trials) 100–300 ms later. (B) Instantaneous dipole explaining 82% of the magnetic distribution in (A).

Baas and colleagues (2002), we recently showed that 82% of the variance in the magnetic counterpart of the pre-shock SPN could be explained by a dipolar source in the vicinity of the ACC, whereas this was not the case in the appropriate control condition (see Figure 8.5). In similar paradigms in neuroimaging studies, the ACC was indeed activated, together with the insular cortex (Büchel et al., 1998).

Psychopharmacology

As the SPN is far less studied than the BP and CNV, not much is known about its psychopharmacology. It has been shown that the (preinstruction) SPN is not as clearly reduced in schizophrenia as the CNV (Reuter et al., 2006). Furthermore, it has been shown that the prereward SPN is reduced in patients with PD (Mattox et al., 2006). The latter finding confirms the firmly established view that DA reflects reward anticipation. Preceding negative emotional stimuli, such as shocks, GABA and serotonin might also play a role. Hypothetically, the catecholamines might be involved in the generation of nonmotivational SPN that is more a manifestation of anticipatory cognitive processing and attention.

Functional Significance

The SPN concept was introduced by our group mainly to describe the differences between the BP and the late CNV (see Brunia, 1988, for a review). That is, we conceived of it as a purely nonmotor slow brain potential (the "true" CNV). The functional significance of the SPN as we interpreted it in the initial years was roughly as sketched above—preparatory activity aimed at speeding up brain processes after the relevant stimulus. Initially, we interpreted the motor–nonmotor distinction fairly dichotomously; there was the motor BP and the nonmotor SPN, and the combination of the two resulted in the CNV. Needless to say, this strict interpretation soon ran into problems. Adding the BP to the SPN did not result in a scalp distribution that matched that of the CNV (Böcker, 1994), nor could the functional significance of the BP and SPN be combined into a CNV (Van Boxtel, 1994).

Van Boxtel and Böcker (2004) tentatively suggested three overlapping components that could underlie the SPN observed in these situations, of which the functional significance could be best described as (1) a general anticipation of the impending stimulus; (2) anticipation of the information content of the stimulus; or (3) emotional anticipation. These three components were thought to be differentially active in various tasks in which the SPN could be recorded. For instance, in the classic feedback studies (e.g., Damen & Brunia, 1987), the SPN is likely to be mainly influenced by the first and third factors, giving rise to the frequently observed scalp distribution of bilateral parietal and right frontal activity. The functional significance in that situation would be a combination of general stimulus anticipation and emotional anticipation.

It can be expected that this description will be refined or summarized differently as SPN research continues and the types of stimuli used increase. This expectation is based on the overall picture that we have attempted to lay out in this chapter: that the function of the slow brain potentials is to prepare the brain for the upcoming event or action. Therefore, if other events are being used, different types of anticipation or preparation will be invoked and the potentials observed at the scalp will change as a result, probably most clearly in their scalp distribution.

In a very broad sense, the SPN can also be viewed as an index of attention. Such a view might constitute a possible umbrella use of the SPN in a variety of situations. For instance, Donkers and colleagues (2005) and Donkers and Van Boxtel (2005) described a "slot machine" experiment in which a series of digits were presented at intervals of 1 s. A reward or a loss occurred when three identical digits were presented in succession. The authors wanted to study what happened to the feedback-related negativity (FRN) after a stimulus was presented that discontinued a series of identical digits. However, they also wanted to assess whether the participants actually paid attention to the stimuli. This was important because the experiment did not include a task or a response that could be used to evaluate the participants' attention to the task. Instead, the authors analyzed the SPN and observed that it resembled the prefeedback SPN in all respects. Hence, they concluded that the participants paid attention to the stimuli. It should be noted that temporal predictability seems to be a necessary condition for preparatory allocation of attention, for attention has limited capacity and consumes energy (Requin et al., 1991). Hence, SPN is an index of attention in a limited number of environmental contexts and might in turn also reflect timing aspects.

To summarize, the functional significance of the SPN can best be described by different forms of stimulus anticipation, that is, basic anticipation of the stimulus, anticipation of the information content of that stimulus, and emotional anticipation. In a broader context, the term *attention* might be used in practical applications of the SPN.

Conclusion

Anticipation involves realizing in advance what has to be done sometime later without causing premature actions. The time scale we are interested in is on the order of several seconds. Within this period electrophysiological changes in the central nervous system take place, related to anticipatory attention or motor preparation. Although perception and movement are different behavioral domains, the underlying processes are largely comparable. The cortical and subcortical structures activated in advance might be differently localized, depending upon what the future behavior will be; the basal mechanisms are the same. The location of the structures involved is directly related to the task at hand, and the basal mechanisms are related to the anticipatory processes.

The slow potentials, recorded over a certain cortical area, are in general a summation of the EPSPs in the underlying cell columns of that area. Different neurotransmitters play a role in the balance between excitation and inhibition. The distribution of the amplitudes of the anticipatory slow waves may reflect which areas will be involved in the future task. Therefore, it is understandable that the number of cortical areas contributing to the BP is smaller than the number of cortical areas contributing to the CNV, where, apart from the movement requirements, perceptual processes are also involved. While the BP is largely a movement-related potential, the CNV late wave is both movement-related and perception-related. Apart from that, cognitive processes (memory, attention) play a role in the BP but a larger one in the CNV late wave. The SPN is a nonmotor anticipatory slow wave that is especially large preceding affective-motivational stimuli such as KR.

Although the three slow waves discussed in this chapter reflect anticipatory processes, and although we are beginning to understand some of these processes, there is still a lot to discover before we can conclude that we know what is going on during anticipation.

Appendix

There are a few technical requirements that researchers should take into account when recording and analyzing slow brain potentials.

First, during recording, an appropriate time constant should be used. The time constant is the value in seconds that it takes for a voltage to drop a value of 0.368. That is, if the true amplitude of a certain signal is set to 100%, then the time constant is the number of seconds that it takes for the recorded signal to drop to 63% of its true value (a drop of 37%). The time constant is only relevant for AC-coupled recording equipment; DC recordings have an infinite time constant. The time constant is functionally equivalent to a high-pass filter, which attenuates the low frequencies. The relation between the time constant and the –3 dB cutoff point of a high-pass filter is given by

$$F \bullet T = \frac{1}{2 \bullet \pi}$$

where F is the –3 dB low cutoff frequency in Hertz (Hz) and T is the time constant in seconds. It follows that $F = 1/2\pi T$ and $T = 1/2\pi F$.

Nowadays, DC recording equipment is readily available and the time constant is less of an issue. For AC recordings, we recommend that the time constant be at least three to four times as long as the longest interstimulus interval. For instance, in a CNV experiment with an interval of 3 s between S1 and S2, the time constant should be at least 9–12 s. For typical BP recordings, a time constant of 5 s should suffice. Obviously, longer time constants allow a more precise approximation of the true underlying signal. We often used a time constant of 30 s. Finally, the electrodes are an integral part of the AC-coupling. Silver–silver chloride electrodes are a prerequisite for obtaining system time constants of several seconds.

The low-pass filter can be set quite low, even as low as 5–10 Hz, when the sole interest is in slow brain potentials. At least it can be set well below the 50 or 60 Hz mains interference frequency. We often used a high cutoff frequency of 30 Hz because we often want to analyze some other potentials, such as N2 or P3. Of course, it is well known that the sampling frequency should be at least twice as high as the high cutoff value of the low-pass filter in order to avoid aliasing.

A second issue is directly related to the long time constant used during recording. A disadvantage of DC recordings or AC recordings with a long time constant is that not only is the signal of interest amplified, but many ultraslow artifacts as well. For instance, activity of the sweat glands at frontal and prefrontal electrodes may cause slow potentials drifts, and small movement of the wires connecting the electrodes to the amplifiers may cause impedance changes that, after amplification, also look like slow potential drifts. It is important to filter for these kinds of artifacts, but the problem is that they are in the same frequency range as the signal of interest. We have found the following procedures useful:

1. Define a window that constrains the minimum and maximum voltages in a segment (trial). For instance, if a window width of 80 μV is used, then the difference between the minimum and maximum values within that segment (trial) must not exceed 80 μV. If it does, the trial should be discarded. This is a fairly standard min-max criterion. For slow brain potentials, apply this criterion after having low-pass filtered the data with 5 Hz or so. The width of the window depends on the exact recording circumstances and generally is selected between 50 and 200 μV. For slow brain potentials low-pass filtered for artifact rejection, small values between 50 and 100 μV should be used.

2. Low-pass filter the data with 2 Hz and divide the segment (trial) into four epochs of equal length (baseline not included). Calculate the mean amplitude in each of these epochs. This yields four values for each segment (trial), and these four values should not differ by more than a certain voltage (25 μV is a useful criterion for many applications). If they do, the trial should be discarded. This procedure will have the effect of discarding trials with very-low-frequency artifacts.

3. An important issue in separating signal from noise is that the noise does not have a fixed time relationship with the events in the experiment, while the signal does. Therefore, remaining artifacts can be expected to be attenuated when multiple trials are averaged with synchronization to the events, provided that enough trials are taken into the average. At least 30 to 40 trials should be included in the average (the more the better).

These procedures should be applied in addition to standard artifact rejection or correction techniques, for instance for eliminating eye movements and blinks. Note that the low-pass filtering applied in these procedures is only done for selecting trials without artifacts. Once these trials have been selected, averaging is done with less strictly filtered data. In our experience, these procedures work better than outlier analysis based on percentages or on a mean plus or minus standard-deviations criteria. However, we have never systematically evaluated more modern artifact rejection techniques based on PCA or ICA, for instance.

A third issue relates to the statistical evaluation of the potentials. Amplitude scoring can be done using area measures (taking the mean over a certain interval) or by PCA, but never by peak picking. The reason is that the potentials of interest are slow enough to allow for area measures, while peak picking has the risk of picking noise instead of signal. When area measures are taken, be sure to use an area of sufficient width in order to estimate the slow potential optimally. Depending on the interstimulus interval and filter settings used in the experiment, a mean of 50 to 100 ms should be taken. In some of our CNV experiments with a 4 s foreperiod, we even used area measurements over 200 ms. Note that taking area measures is functionally equivalent to low-pass filtering. Therefore, the more strongly the data have been filtered, the shorter the area can be, and vice versa.

Acknowledgement

The authors are grateful to Hilde Huizenga, Jan de Munck and the VUMC MEG Centre, Amsterdam, for their cooperation in recording and analyzing the data presented in Figure 8.5.

References

Adler, L. E., Pecevich, M., & Nagamoto, H. (1989). Bereitschaftspotential in tardive-dyskinesia. *Movement Disorders, 4*, 105–112.

Allen, G. I., & Tsukahara, N. (1974). Cerebrocerebellar communication systems. *Physiological Reviews, 54*, 957–1006.

Ashton, C. H., Marshall, E. F., Hassanyeh, F., Marsh, V. R., & Wrighthonari, S. (1994). Biological correlates of deliberate self-harm behavior—A study of electroencephalographic, biochemical and psychological variables in parasuicide. *Acta Psychiatrica Scandinavica, 90*, 316–323.

Baas, J. M. P., Kenemans, J. L., Böcker, K. B. E., & Verbaten, M. N. (2002). Threat-induced cortical processing and startle potentiation. *Neuroreport, 13*, 133–137.

Ball, T., Schreiber, A., Feige, B., Wagner, M., Lücking, C. H., & Kristeva-Feige, R. (1999). The role of higher-order motor areas in voluntary movement as revealed by high-resolution EEG and fMRI. *Neuroimage, 10*, 682–694.

Becker, W., & Kristeva, R. (1980). Cerebral potentials prior to various force deployments. In H. H. Kornhuber & L. Deecke (Eds.), *Progress in brain research: Vol. 54. Motivation, motor and sensory processes of the brain: Electrical potentials, behavior and clinical use* (pp. 189–194). Amsterdam: Elsevier.

Birbaumer, N., Roberts, L. E., Lutzenberger, W., Rockstroh, B., & Elbert, T. (1992). Area-specific self-regulation of cortical slow potentials on the sagittal midline and its effects on behavior. *Electroencephalography and Clinical Neurophysiology, 84*, 353–361.

Böcker, K. B. E., Brunia, C. H. M., & Cluitmans, P. J. M. (1994). A spatio-temporal dipole model of the readiness potential in humans. I. Finger movement. *Electroencephalography and Clinical Neurophysiology, 91*, 275–285.

Böcker, K. B. E., Baas, J. M. P., Kenemans, J. L., & Verbaten, M. N. (2001). Stimulus-preceding negativity induced by fear: A manifestation of affective anticipation. *International Journal of Psychophysiology, 43*, 97–90.

Brunia, C. H. M. (1980). What is wrong with legs in motor preparation? In H. H. Kornhuber & L. Deecke (Eds.), *Progress in brain research: Vol. 54. Motivation, motor and sensory processes of the brain* (pp. 1–811). Amsterdam: Elsevier.

Brunia, C. H. M. (1988). Movement and stimulus preceding negativity. *Biological Psychology, 26*, 165–178.

Brunia, C. H. M. & Damen, E. J. P. (1988). Distribution of slow potentials related to motor preparation and stimulus anticipation in a time estimation task. *Electroencephalography and Clinical Neurophysiology, 69*, 234–243.

Brunia, C. H. M. (2003). CNV and SPN: Indices of anticipatory behavior. In M. Jahanshahi & M. Hallet (Eds.), *The Bereitschaftspotential: Movement-related cortical potentials* (pp. 207–227). New York: Kluwer Academic/Plenum.

Brunia, C. H. M., & Van Boxtel, G. J. M. (2004). Anticipatory attention to verbal and non-verbal is reflected in a modality-specific SPN. *Experimental Brain Research, 156*, 231–239.

Brunia, C. H. M. & van den Bosch, W. E. J. (1984). Movement-related slow potentials I. A contrast between finger and foot movements in right-handed subjects. *Electroencephalography and Clinical Neurophysiology, 57*, 515–527.

Brunia, C. H. M. (1980). CNV and EMG preceding a plantar flexion of the foot. *Biological Psychology, 11*, 181–191.

Brunia, C. H. M., & Vingerhoets, A. J. J. M. (1981). Opposite hemisphere differences in movement related potentials preceding foot and finger flexions. *Biological Psychology, 13*, 261–269.

Brunia, C., De Jong, B., Van den Berg-Lenssen, M., & Paans, A. (2000). Visual feedback about time estimation is related to a right hemisphere activation measured by PET. *Experimental Brain Research, 130*, 328–337.

Büchel, C., Morris, J., Dolan, R. J., & Friston, K. J. (1998). Brain systems mediating aversive conditioning: An event-related fMRI study. *Neuron, 20*, 947–957.

Caton, R. (1875). The electric currents of the brain. *British Medical Journal, 2*, 278.

Chwilla, D. J., & Brunia, C. H. M. (1991). Event-related potentials to different feedback stimuli. *Psychophysiology, 28*, 123–132.

Colebatch, J. G. (2007). Bereitschaftspotential and movement-related potentials: Origin, significance, and application in disorders of human movement. *Movement Disorders, 22*, 601–610.

Connor, W. H., & Lang, P., J. (1969). Cortical slow wave and cardiac rate responses in stimulus orientation and reaction time conditions. *Journal of Experimental Psychology, 82*, 310–320.

Cook, M. R., Gerkovich, M. M., Hoffman, S. J., McClernon, F. J., & OConnell, K. A. (1996). Effects of smoking and telic/paratelic dominance on the contingent negative variation (CNV). *International Journal of Psychophysiology, 23*, 101–110.

Cunnington, R., Windischberger, C., Deecke, L., & Moser, E. (2003). The preparation and readiness for voluntary movement: A high-field event-related fMRI study of the Bereitschafts-BOLD response. *Neuroimage, 20*, 404–412.

Damen, E. J. P., & Brunia, C. H. M. (1987). Changes in heart rate and slow brain potentials related to motor preparation and stimulus anticipation in a time estimation task. *Psychophysiology, 24*, 700–713.

Damen, E. J. P., & Brunia, C. H. M. (1994). Is a stimulus conveying task-relevant information a sufficient condition to elicit a stimulus-preceding negativity? *Psychophysiology, 31*, 129–139.

Daurignac, E., Perez-Diaz, F., Grillon, C., & Jouvent, R. (2001). Gender and activation level in smokers. *European Psychiatry, 16*, 313–316.

Davidson, R. J., & Irwin, W. (1999). The functional neuroanatomy of emotion and affective style. *Trends in Cognitive Sciences, 3*, 11–21.

Deecke, L., Grözinger, B., & Kornhuber, H. H. (1976). Voluntary finger movement in man: Cerebral potentials and theory. *Biological Cybernetics, 23*, 99–119.

Deecke, L., Heise, B., Kornhuber, H. H., Lang, M., & Lang, W. (1984). Brain potentials associated with voluntary manual tracking: Bereitschaftspotential, conditioned premotion positivity, directed attention potential, and relaxation potential. In R. Karrer, J. Cohen, & P. Tueting (Eds.), *Brain and information: Event-related potentials*. (pp.1–768). New York: Annals of the New York Academy of Sciences.

Deecke, L., & Kornhuber, H. H. (1977). Cerebral potentials and the initiation of voluntary movement. In J. E. Desmedt (Ed.), *Progress in clinical neurophysiology: Vol. 1. Attention, voluntary contraction and event-related cerebral potentials* (pp. 132–150). Basel: Karger.

Dick, J. P. R., Cantello, R., Buruma, O., Gioux, M., Benecke, R., Day, B. L., et al. (1987). The Bereitschaftspotential, L-DOPA and Parkinson's disease. *Electroencephalography and Clinical Neurophysiology, 66*, 263–274.

Dick, J. P. R., Rothwell, J. C., Day, B. L., Cantello, R., Buruma, O., Gioux, M., Benecke, R., Berardelli, A., Thompson, P. D., & Marsden, C. D. (1989). The Bereitschaftspotential is abnormal in Parkinson's disease. *Brain, 112*, 233–244.

Donkers, F. C. L., Nieuwenhuis, S., & Van Boxtel, G. J. M. (2005). Mediofrontal negativities in the absence of responding. *Cognitive Brain Research, 25*, 777–787.

Donkers, F. C. L. & Van Boxtel, G. J. M. (2005). Mediofrontal negativities to averted gains and losses in the slot-machine task: A further investigation. *Journal of Psychophysiology, 19*, 256–262.

Dum, R. P., & Strick, P. L. (1991). The origin of corticospinal projections from the premotor areas in the frontal lobe. *Journal of Neuroscience, 11*, 667–689.

Dum, R. P., & Strick, P. L. (1996). Spinal cord termination of the medial wall motor areas in macaque monkeys. *Journal of Neuroscience, 16*, 6513–6525.

Fattaposta, F., Pierelli, F., My, F., Mostarda, M., Del Monte, S., Parisi, L., et al. (2002). L-dopa effects on preprogramming and control activity in a skilled motor act in Parkinson's disease. *Clinical Neurophysiology, 113*, 243–253.

Fuller, R., Nathaniel-James, D., & Jahanshahi, M. (1999). Movement-related potentials prior to self-initiated movements are impaired in patients with schizophrenia and negative signs. *Experimental Brain Research, 126*, 545–555.

Gaillard, A. W. K., & Perdok, J. (1980). Slow brain potentials in the CNV-paradigm. *Acta Psychologica, 44*, 147–163.

Grünewald, G., & Grünewald-Zuberbier, E. (1983). Cerebral potentials during voluntary ramp movements in aiming tasks. In A. W. K. Gaillard & W. Ritter (Eds.), *Tutorials in ERP research: Endogenous components* (pp. 311–327). Amsterdam: North-Holland.

Hansenne, M., Pitchot, W., Pinto, E., Papart, P., & Ansseau, M. (2000). Serotonergic-1a activity and contingent negative variation. *Biological Psychology, 52*, 259–265.

Hari, R., Antervo, A., Katila, T., Poutanen, T., Seppänen, M., Tuomisto, T., & Varpula, T. (1983). Cerebral magnetic fields associated with voluntary limb movements in man. *Nuovo Cimento, 2*, 484–494.

Heil, M., Rösler, F., & Henninghausen, E. (1997). Topography of brain electrical activity dissociates the retrieval of spatial versus verbal information from episodic long-term memory in humans. *Neuroscience Letters, 222*, 45–48.

Houlihan, M. E., Pritchard, W. S., Guy, T. D., & Robinson, J. H. (2002). Smoking/nicotine affects the magnitude and onset of lateralized readiness potentials. *Journal of Psychophysiology, 16*, 37–45.

Howard, R., Longmore, F., & Mason, P. (1992). Contingent negative variation as an indicator of sexual object preference: Revisited. *International Journal of Psychophysiology, 13*, 185–188.

Ikeda, A., Lüders, H. O., Burgess, R. C., & Shibasaki, H. (1993). Movement-related potentials associated with single and repetitive movements recorded from human supplementary motor area. *Electroencephalography and Clinical Neurophysiology, 89*, 269–277.

Ikeda, A., & Shibasaki, H. (2003). Generator mechanisms of the Bereitschaftspotential as studied by epicortical recording in patients with intractable partial epilepsy. In M. Jahanshahi & M. Hallet (Eds.), *The Bereitschaftspotential: Movement-related cortical potent*ials (pp. 45–59). New York: Kluwer Academic/Plenum.

Irwin, D. A., Knott, J. R., McAdam, D. W., & Rebert, C. S. (1966). Motivational determinants of the "contingent negative variation." *Electroencephalography and Clinical Neurophysiology, 21*, 538–543.

Jahanshahi, M., & Hallet, M. (2003). *The Bereitschaftspotential: Movement-related cortical potentials*. New York: Kluwer Academic/Plenum.

Jahanshahi, M., Jenkins, I. H., Brown, R. G., Marsden, C. D., Passingham, R. E., & Brooks, D. J. (1995). Self-initiated versus externally triggered movements. I. An investigation using measurement of blood flow with PET and movement-related potentials in normal and Parkinson's disease subjects. *Brain, 118*, 913–933.

Järvilehto, T., & Frühstorfer, H. (1970). Differentiation between slow cortical potentials associated with motor and mental acts in man. *Experimental Brain Research, 11*, 309–317.

Jenkins, I. H., Jahanshahi, M., Jueptner, M., Passingham, R. E., & Brooks, D. J. (2000). Self-initiated versus externally triggered movements. II. The effect of movement predictability on regional cerebral blood flow. *Brain, 123*, 1216–1228.

Knösche, T., Praamstra, P., Stegeman, D., & Peters, M. (1996). Linear estimation discriminates midline sources and a motor cortex contribution to the readiness potential. *Electroencephalography and Clinical Neurophysiology, 99*, 183–190.

Kornhuber, H. H., & Deecke, L. (1965). Hirnpotentialänderungen bei Willkürbewegungen und passiven Bewegungen des Menschen: Bereitschaftspotential und reafferente Potentiale. *Pflügers Archiv, 284*, 1–17.

Kotani, Y., & Aihara, Y. (1999). The effect of stimulus discriminability on stimulus-preceding negativities prior to instructive and feedback stimuli. *Biological Psychology, 50*, 1–18.

Lang, P. J., Öhman, A., & Simons, R. F. (1978). The psychology of anticipation. In J. Requin (Ed.) *Attention and performance*, Vol. 7 (pp. 469–485). Amsterdam: North-Holland.

Lang, W., Cheyne, D., Kristeva, R., Lindinger, G., & Deecke, L. (1991). Functional localisation of motor processes in the human cortex. In C. H. M. Brunia, G. Mulder, & M. N. Verbaten (Eds.), *Event-related brain research (EEG—Supplement 42)* (pp. 97–115). Amsterdam: Elsevier.

Lang, W., Lang, M., Kornhuber A., Deecke, L., & Kornhuber, H. H. (1984). Brain potentials related to voluntary hand tracking: Motivation and attention. *Human Neurobiology, 3*, 235–240.

Libet, B., Gleason, C. A., Wright, E. W., & Pearl, D. K. (1983). Time of conscious intention to act in relation to onset of cerebral activity (readiness-potential). *Brain, 106*, 623–642.

Low, M. D., & McSherry, J. W. (1968). Further observations of psychological factors involved in CNV genesis. *Electroencephalography and Clinical Neurophysiology, 25*, 203–207.

Macar, F., & Bonnet, M. (1997). Event-related potentials during temporal information processing. In G. J. M. van Boxtel & K. B. E. Böcker (Eds.), *Brain and behavior: Past, present, and future* (pp. 49–66) Tilburg, the Netherlands: Tilburg University Press.

Maertens, d. N., Timsit-Berthier, M., Timsit, M., & Schoenen, J. (1986). Contingent negative variation in headache. *Annals of Neurology, 19*, 78–80.

Mattox, S. T., Valle-Inclan, F., & Hackley, S. A. (2006). Psychophysiological evidence for impaired reward anticipation in Parkinson's disease. *Clinical Neurophysiology, 117*, 2144–2153.

Matzusaka, Y., Aizawa, H., & Tanji, J. (1992). A motor area rostral to the supplementary motor area (presuplementary motor area) in the monkey: Neuronal activity during a learned motor task. *Journal of Neurophysiology, 68*, 653–662.

McGuiness, D., & Pribram, K. H. (1980). The neuropsychology of attention: emotional and motivational controls. In M. C. Wittrock (Ed.), *The brain and psychology* (pp. 95–139). New York: Academic Press.

Mishkin, M., Suzuki, W. A., Gadian, D. G., & Vargha-Khadem, F. (1997). Hierarchical organization of cognitive memory. *Philosophical Transactions of the Royal Society of London, Biological Sciences, 352*, 1461–1467.

Ohgami, Y., Kotani, Y., Tsukamoto, T., Omura, K., Inoue, Y., Aihara, Y., et al. (2006). Effects of monetary reward and punishment on stimulus-preceding negativity. *Psychophysiology, 43*, 227–236.

Papart, P. (1996). CNV in mood disorders. In C. Ogura, Y. Koga, & M. Shimokoshi (Eds.), *Recent advances in event-related potential research* (pp. 897–900). Amsterdam: Elsevier.

Papart, P., Ansseau, M., & Timsit-Berthier, M. (1997). Influence of diazepam on Contingent Negative Variation. *Human Psychopharmacology–Clinical and Experimental, 12*, 95–98.

Paradiso, G., Cunic, D., Saint-Cyr, J., Hoque, T., Lozano, A. M., Lang, A., & Chen, R. (2004). Involvement of human thalamus in the preparation of self-paced movement. *Brain, 127*, 2717–2731.

Praamstra, P., Stegeman, D. F., Horstink, M. W. I. M., & Cools, A. R. (1996). Dipole source analysis suggests selective modulation of the supplementary motor area contribution to the readiness potential. *Electroencephalography and Clinical Neurophysiology, 98*, 468–477.

Rektor, I. (2003). Intracerebral recordings of the Bereitschaftspotential and related potentials in cortical and subcortical structures in human subjects. In M. Jahanshahi & M. Hallett (Eds.), *The Bereitschaftspotential: Movement-related cortical potentials* (pp. 61–77). New York: Kluwer Academic, Plenum.

Requin, J., Brener, J., & Ring, C. (1991). Preparation for action. In J. R. Jennings & M. G. H. Coles (Eds.), *Handbook of cognitive psychology: Central and autonomic nervous system approaches* (pp. 357–448). New York: Wiley.

Reuter, B., Herzog, E., Endrass, T., & Kathmann, N. (2006). Brain potentials indicate poor preparation for action in schizophrenia. *Psychophysiology, 43*, 604–611.

Rockstroh, B., Elbert, T., Lutzenberger, W., & Altenmüller, E. (1991). Effects of the anticonvulsant benzodiazepine clonazepam on event-related brain potentials in humans. *Electroencephalography and Clinical Neurophysiology, 78*, 142–149.

Rohrbaugh, J. W., & Gaillard, A. W. K. (1983). Sensory and motor aspects of the contingent negative variation. In A. W. K. Gaillard & W. Ritter (Eds.), *Tutorials in ERP research: Endogenous components* (pp. 269–310). Amsterdam: North-Holland.

Rohrbaugh, J. W., Syndulko, K., & Lindsley, D. B. (1976). Brain wave components of the contingent negative variation in humans. *Science, 191*, 1055–1057.

Rösler, F., Heil, M., & Henninghausen, E. (1995). Distinct cortical activation patterns during long-term memory retrieval of verbal, spatial, and color information. *Journal of Cognitive Neuroscience, 7*, 51–65.

Ruchkin, D. S., Sutton, S., Mahaffey, D., & Glaser, J. (1986). Terminal CNV in the absence of motor response. *Electroencephalography and Clinical Neurophysiology, 63*, 445–463.

Sasaki, K., & Gemba, H. (1981). Cortical field potentials preceding self-paced and visually initiated hand movements in one and the same monkey and influences of cerebellar hemispherectomy upon the potentials. *Neuroscience Letters, 25*, 287–292.

Schoenen, J., Maertens, d. N., Timsit-Berthier, M., & Timsit, M. (1986). Contingent negative variation and efficacy of beta-blocking agents in migraine. *Cephalalgia, 6*, 229–233.

Shibasaki, M., Barrett, G., Halliday, E., & Halliday, A. M. (1980). Components of the movement-related cortical potential and their scalp topography. *Electroencephalography and Clinical Neurophysiology, 49*, 213–226.

Shibasaki, M., & Hallett, M. (2006). What is the Bereitschaftspotential? *Clinical Neurophysiology, 117*, 2341–2356.

Shibasaki, M., & Kato, M. (1975). Movement-associated cortical potentials with unilateral and bilateral simultaneous hand movement. *Journal of Neurology, 208*, 1432–1459.

Shima, K., & Tanji, J. (2000). Neuronal activity in the supplementary and presupplementary motor areas for temporal organization of multiple movements. *Journal of Neurophysiology, 84*, 2148–2160.

Simons, R. F., Öhman, A., & Lang, P. J. (1979). Anticipation and response set: Cortical, cardiac, and electrodermal correlates. *Psychophysiology, 16*, 222–233.

Sirigu, A., Daprati, E., Ciancia, S., Giraux, P., Nighoghossian, N., Posada, A., et al. (2004). Altered awareness of voluntary action after damage to the parietal cortex. *Nature Neuroscience, 7*, 80–84.

Tanji, J., & Mushiake, H. (1996). Comparison of neuronal activity in the supplementary motor area and primary motor cortex. *Cognitive Brain Research, 3,* 143–150.

Tecce, J. J. (1991). Dopamine and CNV: Studies of drugs, disease and nutrition. *Electroencephalography and Clinical Neurophysiology, Suppl. 42,* 153–164.

Timsit-Berthier, M. (1991). Contingent negative variation (CNV) in psychopharmacology. *Electroencephalography and Clinical Neurophysiology, Suppl. 42,* 142–152.

Tsukamoto, T., Kotani, Y., Yoshimi, O., Omura, K., Inoue, Y., & Aihara, Y. (2006). Activation of insular cortex and subcortical regions related to feedback stimuli in a time estimation task: An fMRI study. *Neuroscience Letters, 399,* 39–44.

Van Boxtel, G. J. M. (1994). *Non-motor components of slow brain potentials.* Unpublished doctoral thesis, Tilburg University, the Netherlands.

Van Boxtel, G. J. M., & Böcker, K. B. E. (2004). Cortical measures of anticipation. *Journal of Psychophysiology, 18,* 61–76.

Van Boxtel, G. J. M., & Brunia, C. H. M. (1994). Motor and non-motor aspects of slow brain potentials. *Biological Psychology, 38,* 37–51.

Van Boxtel, G. J. M., Van den Boogaart, B., & Brunia, C. H. M. (1993). The Contingent Negative Variation in a choice reaction time task. *Journal of Psychophysiology, 7,* 11–23.

Van den Wildenberg, W. P. M., Van Boxtel, G. J. M., Van der Molen, M. W., Bosch, D. A., Speelman, J. D., & Brunia, C. H. M. (2006). Stimulation of the subthalamic region facilitates the selection and inhibition of motor responses in Parkinson's disease. *Journal of Cognitive Neuroscience, 18,* 626–636.

Verleger, R. (2004). Malfunctions of central control of movement studied with slow brain potentials in neurological patients. *Journal of Psychophysiology, 18,* 105–120.

Verleger, R., Wascher, E., Arolt, V., Daase, C., Strohm, A., & Kompf, D. (1999). Slow EEG potentials (contingent negative variation and post-imperative negative variation) in schizophrenia: Their association to the present state and to Parkinsonian medication effects. *Clinical Neurophysiology, 110,* 1175–1192.

Walter, W. G., Cooper, R., Aldridge, V. J., McCallum, W. C., & Winter, A. L. (1964). Contingent Negative Variation: An electric sign of sensori-motor association and expectancy in the human brain. *Nature, 203,* 380–384.

Weilke, F., Spiegel, S., Boecker, H., Von Einsiedel, H. G., Conrad, B., Schwaiger, M., & Erhard, P. (2001). Time-resolved fMRI of activation patterns in M1 and SMA during complex voluntary movement. *Journal of Neurophysiology, 85,* 1858–1863.

Westphal, K. P. (2003). The Bereitschaftspotential in schizophrenia and depression. In M. Jahanshahi & M. Hallet (Eds.), *The Bereitschaftspotential: Movement-related cortical potentials* (pp. 155–173). New York: Kluwer Academic/Plenum.

CHAPTER

9 The Lateralized Readiness Potential

Fren T. Y. Smulders *and* Jeff O. Miller

Abstract

When we respond with one hand, a negative potential can be observed over the motor cortex contralateral to the responding hand. This lateralized readiness potential (LRP) starts even before the response is emitted, and its onset has been taken as a measure of the time at which the brain began preparing to make the response. The LRP has also been used to measure the preliminary activation of a response that is never actually produced. The LRP is a valuable measure of relatively central response activation that is now widely used in many areas of psychology involving reaction time tasks. This chapter discusses the definition of LRP, methods of isolating LRP from other components, brain systems generating the LRP, uses of LRP, and analyses of LRP.

Keywords: lateralized readiness potential, reaction time, central response activation

When we respond with one hand, a negative potential can be observed over the motor cortex contralateral to the responding hand. This lateralized readiness potential (LRP) starts even before the response is emitted, so its timing reflects the time at which the brain has begun to prepare to make the response. The LRP is a valuable measure of relatively central response activation that is now widely used in many areas of psychology involving reaction time (RT) tasks. Introduced barely 20 years ago, it came into prominence very rapidly both because (1) the lateralization-based nature of its derivation allowed an unusually clear specification of its functional significance right from the start and (2) a measure of central response activation is extremely useful in addressing many fundamental questions about cognitive mechanisms. Because it was the first lateralization-based event-related potential (ERP) component to be used widely, LRP researchers often had to chart new experimental paradigms, control conditions, and methods of inference, and many of these were later adapted to more recent lateralization-based components (see, e.g., Chapters 12 and 13, this volume). Thus, the LRP is not only an important ERP component in its own right, but also an important advance in ERP methodology that served as a model forerunner for the study of other lateralization-based components.

LRP Definition

Voluntary movements with one hand are preceded by a negative wave called the *Bereitschaftspotential* (readiness potential; Deecke et al., 1976) that can be measured on both sides of the scalp using an earlobe reference. A short time before movement onset, this negative wave displays a preponderance over the hemisphere that is contralateral to the responding hand. Kutas and Donchin (1980) found the same lateralization well in advance of movement if the subject was informed which hand would be required to respond. They concluded that it reflects preparation to execute specific motor acts. Later, it was realized that the motor lateralizations could be isolated from lateralizations related to other structural and functional asymmetries between hemispheres using a double subtraction technique (De Jong et al., 1988; Gratton et al., 1988; Smid et al., 1987).

Figure 9.1 shows how the double subtraction technique isolates the LRP from other components, lateralized as well as unlateralized. First, potentials are recorded from both hemispheres at electrode locations close to the cortical areas that control left and right hand finger movements (C3′ and C4′, close to the standard locations C3 and C4) for both left and right hand responses in a number of trials (for the figure, artificial data were generated for reasons of clarity). Second, for each electrode and hand, potentials at each time point are averaged across trials, resulting in ERPs like those in Figures 9.1A and 9.1B. Third, one of the following computations is made:

$$(C3'(t) - C4'(t))\text{right hand} - (C3'(t) - C4'(t))\text{left hand} \quad \text{(De Jong et al., 1988)} \quad (1)$$

or

$$\left[(C4'(t) - C3'(t))\text{left hand} + (C3'(t) - C4'(t))\text{right hand}\right] \Big/ 2 \quad \text{(Coles, 1989)} \quad (2)$$

with $C3'(t)$ and $C4'(t)$ denoting the digitized scalp potentials at C3′ and C4′ for multiple time points.

If a right hand response is activated, the electrode over the contralateral hemisphere (C3′)—but not that over the ipsilateral hemisphere (C4′)—will become more negative; hence, the result of the $C3'(t) - C4'(t)$ subtraction will become more negative (Figure 9.1C). Likewise, the result of the $C3'(t) - C4'(t)$ subtraction will become more positive if the left hand is activated (Figure 9.1C). Their subtraction in equation (1) effectively sums the two effects (Figure 9.1E). In equation (2), the potential at the electrode that is ipsilateral to the responding hand is subtracted from the (more negative) potential at the contralateral electrode for each hand, and the results are averaged across hands. Thus, both computations are equally sensitive to one response hand becoming more activated than the other in preparation for a unimanual response. The only difference is in the division by 2 in equation (2), a linear scaling factor to which analysis techniques will be insensitive.[1] Secondly, any lateralized activity that is unrelated to the responding hand (i.e., lateralized activity that is the same regardless of which hand responds, like component II in Figure 9.1) is canceled out because it appears once with a plus sign and once with a minus sign in both computations. Thus, in the computation of LRP, not only are nonlateralized potentials (e.g., component I in Figure 9.1) eliminated by the double subtraction of homologous electrodes over the two hemispheres, but also any lateralized potentials that are not related to the responding hand are eliminated (e.g., component II), as well as any differences between hands that are not lateralized (e.g., component III). Clearly, any activation related to bimanual preparation is canceled in the subtractions as well. The result of equation (1) was originally called *corrected motor asymmetry* (CMA) by De Jong and coworkers, but the name *lateralized readiness potential*, introduced by Gratton et al. (1988), has become widely accepted.

Isolating LRP from Other Components

Because of the logic of its derivation, the LRP is widely interpreted as a manifestation of hand-specific response activation. This interpretation is plausible regardless of whether it is evoked during a precue interval after some information about the response has been given or during the short interval between an imperative stimulus and a speeded response. Because it is recorded at the same time and at the same electrode locations where components like N2, P300, and other slow waves are also active, it will overlap these components in an additive manner. However, the double subtraction technique ensures the removal of any of these components from the LRP. As a result, the waveform used to isolate the LRP is usually effectively zero for about the first 200 ms after the stimulus that carries the information about which response hand will be required.

Still, the logic of its derivation does not completely rule out the influence of nonmotoric contributions. To appreciate this, consider that the LRP is actually an effect of the experimental factor "hand," with levels "left" and "right," on the difference between C3′ and C4′. If the factor "hand" is confounded with some other factor that leads to lateralization, the LRP need not be purely an index of response activation. Three examples will be given.

First, if stimuli are presented to the left or the right of the participant's midline and the response is made with the hand on the same side as the stimulus, lateralized sensory potentials will contaminate the LRP. For instance, a stimulus presented in the right visual field will evoke a larger N160 over the left than the right hemisphere (Rugg et al., 1984). Since the activation of the right hand response is also associated with increased left hemisphere negativity, the two lateralizations will sum.[2] Thus, in this situation, the formulas for computation of LRP confound

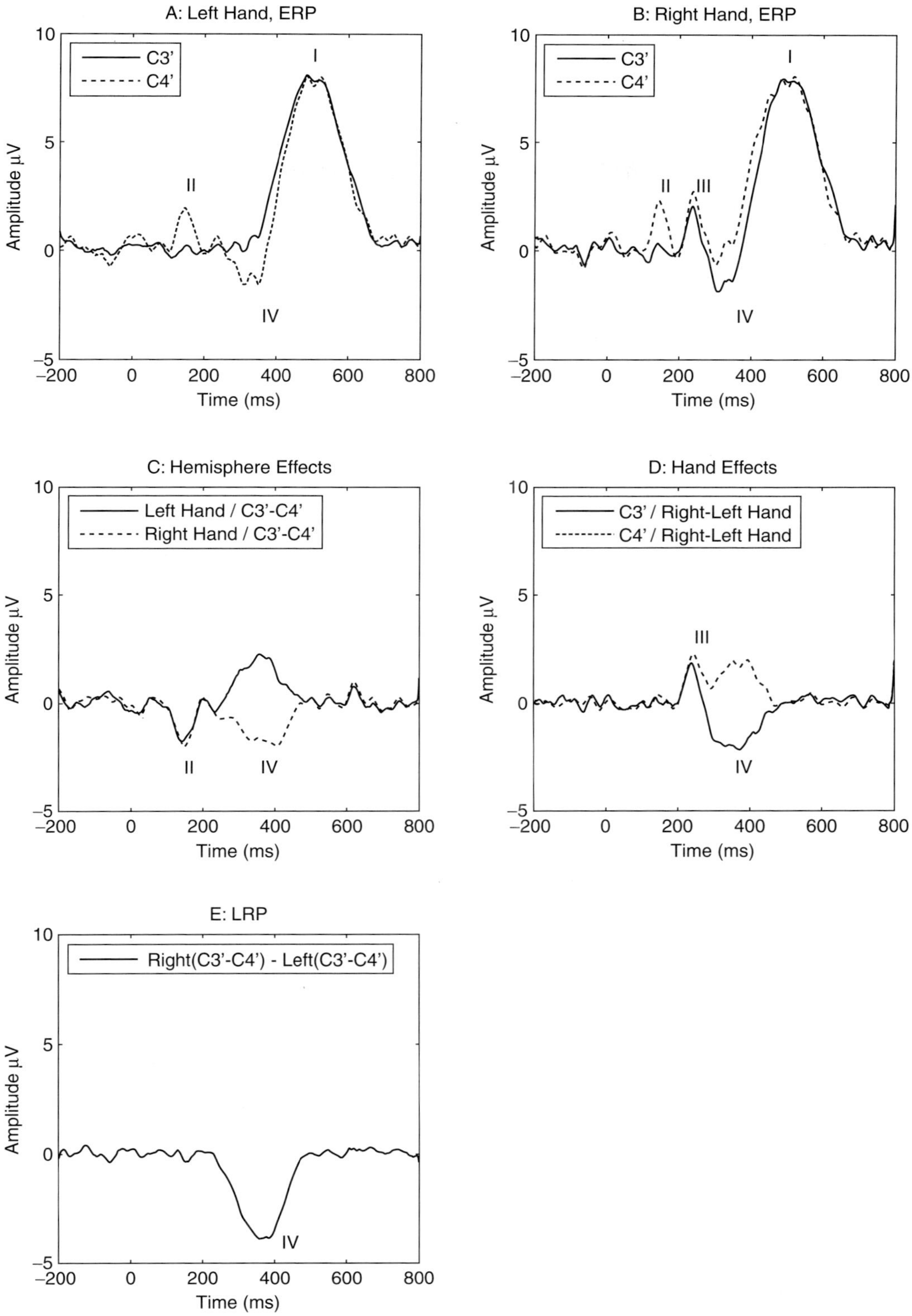

Fig. 9.1. The isolation of the LRP from other components by means of the double subtraction technique. Four artificially generated ERP components and some random noise were added. The largest, component I, was present at both electrode locations C3 and C4 for both the left and right hands. Its peak latency and amplitude were 500 ms and 8 μV, respectively. Component II (150 ms, 2 μV)

(Continued)

activity contralateral to the response with activity contralateral to the stimulus, preventing interpretation of the LRP as a pure measure of motor processing. If the direction of attentional shifts covaries with the response hand, similar considerations apply (for reviews, see Eimer, 1998; Praamstra, 2007). This problem has led investigators of spatial stimulus–response compatibility and the Simon effect to arrange both stimuli and responses along a vertical dimension (e.g., De Jong et al., 1994). With this arrangement, spatial correspondence (or lack thereof) between stimulus position and response position is retained, and behavioral data reflect an effect of such correspondence. However, since all stimulation occurs along the midline, sensory activity is not lateralized and cannot contaminate the LRP.

Second, horizontal eye movements may contaminate the LRP. For instance, if participants move their eyes to the right, the left hemisphere will be more negative than the right hemisphere, because the eye's retina is more negative than the cornea (Luck, 2005, chap. 4). As long as eye movements do not systematically covary with left versus right hand responses, contaminating lateralizations will be suppressed by the averaging. But if the covariation is systematic—for instance, when participants tend to look at the hand that is about to respond—the LRP will be contaminated. Fortunately, the electro-oculogram (EOG) provides a good check for this type of contamination. If the bipolar horizontal EOG derivation of left hand trials is subtracted from that of right hand trials, the remaining signal is a good "source monitor" of the artifactual contribution to the LRP. Many researchers (e.g., Miller & Hackley, 1992) have thus plotted and analyzed this horizontal electro-oculogram (HEOG) derivation in a manner analogous to LRP, that is, subtracting averages of left- and right hand trials. Normally, about 20% of the HEOG amplitude would be expected to propagate to central electrodes (Elbert et al., 1985). If at most small lateralizations are observed in the EOG derivation, the possibility of substantial contamination of the LRP by eye movements can be eliminated.

Finally, close to the overt response, a lateralized contribution from reafferent activity can also be expected. In a number of studies, such lateralization was found to start almost directly after the beginning of the movement (Praamstra et al., 1996; Wascher & Wauschkuhn, 1996). Pope and colleagues (2007), using a high-density electrode array, found that the reafferent dipole source could be separated from movement-related activation and peaked 70 ms later. Lateralized reafferent activity is clearly motor-related, and yet it does not reflect response activation per se, because it instead involves sensorimotor feedback produced after the response is initiated. Most investigators have ignored this possible contaminator of LRP because—as will be seen later—the LRP has been used mostly to study the processes leading up to response initiation, so subsequent reafferent activity happens too late to contaminate the LRP in the time range of interest.

Brain Systems Generating the LRP

It is assumed that the primary motor cortex plays a large role in the generation of the LRP (e.g., Coles, 1989; De Jong et al., 1988). This assumption is supported by evidence from studies using depth electrodes and magnetic field recordings (for reviews, see Eimer, 1998; Miller & Hackley, 1992). Also, the finding that the lateralization is reversed for foot movements (Brunia & Vingerhoets, 1980; Leuthold & Jentzsch, 2002) is evidence for localization within M1, given that the foot area is located medially within the longitudinal fissure, and its current dipole is oriented toward the hemisphere ipsilateral to the movement. The somatotopic mapping of other areas, such as the supplementary motor area, does not lead to the prediction of opposite polarity lateralizations for hand versus foot movements.

As for any ERP component, the anatomical structures that are responsible for the generation of LRP can be examined through an analysis of its scalp distribution. An early multielectrode study was carried out by Sommer and colleagues (1994). They recorded LRP from positions 2, 4, and 6 cm lateral to Cz, as well as two more anterior and

Fig. 9.1. (Continued)
was a positivity over the right hemisphere that did not depend on the responding hand. Component III (250 ms, 2 uV) was a positivity for right hand responses that did not depend on the hemisphere. Finally, the LRP was added as a negative component IV (-4 uV, 360 ms) to the contralateral hemisphere for left and right hand responses, with a strong temporal overlap with components I and III. (A,B) Resulting ERPs at C3 and C4 for the left and right hand, respectively, showing the sum of all components. (C) The result of subtraction of the waveforms at C3 and C4. Components I and III are effectively removed through subtraction. (D) The result of subtraction of the waveforms for left and right hands. Components I and II are effectively removed through subtraction. (E) Components I, II, and III have been removed through double subtraction, leaving component IV, the LRP.

posterior positions. They found comparable LRPs at most locations, except at the most medial and anterior ones, where LRP was smaller but still robustly different from zero. These results suggest that electrode positioning is not very critical (though see Oostenveld et al., 2003, for the importance of symmetry). Possibly related to this, LRP researchers have not come to complete agreement about the exact scalp locations where LRP should be measured. Although C3′ and C4′ are most often reported, there is variability in the actual locations these labels refer to. While Kutas and Donchin (1980) affixed their electrodes 4 cm lateral to Cz, and labeled these locations C3′ and C4′, others have used locations labeled by Jasper (1958) as C1/C2 (~3.4 cm lateral to Cz; Van der Lubbe et al., 2001), C3/C4 (~6.8 cm lateral; Hackley & Valle-Inclán, 1999; Smulders Kok et al., 1995), or slightly more anterior (1 cm with respect to C3/C4; De Jong et al., 1988), or more anterior and superior (1 cm from C3/C4 in both directions; e.g., Miller & Hackley, 1992), or still other positions. Apparently, good results have been obtained with electrodes at quite a wide range of scalp locations.

Using a high electrode density, Pope et al. (2007) found the largest LRP at, or immediately posterior to, C3/C4, depending on the type of movement (isometric vs. isotonic, respectively). Before fitting a dipole model to the scalp distribution of LRP, several researchers have adopted the procedure used by Praamstra et al. (1996). First, the LRP was computed for every homologous pair of lateral electrodes, with the two electrodes of each pair replacing C3′ and C4′ in equations (1) and (2). With this derivation, however, lateralized movement-related activity cannot be determined independently for each hemisphere. Effectively, the resulting LRP waveforms cover only half of the scalp, providing an incomplete electrode distribution for dipole analysis. Second, Praamstra et al. (1996) projected the movement-related activity (i.e., the LRP) to one (arbitrary) hemisphere and then replicated it at the homologous electrode sites of the other hemisphere—but with polarity inverted. This results in an antisymmetric scalp distribution, implying that data can be fitted by (pairs of) dipoles constrained to symmetrical locations in each hemisphere and to antisymmetric orientations. Using dipole modeling of this type, it has been found that somewhat different sources generate LRP preceding actual movements as compared with LRP during a preparatory interval during which participants have only partial information about the upcoming response and are awaiting complete information (see the paradigm of Rosenbaum, 1980, discussed below). For LRPs that precede actual response execution, two dipoles are typically needed to adequately model the scalp distribution, with one dipole describing movement-related activity, presumably in the primary motor area, peaking at the moment of the response, and another dipole describing reafferent activity, peaking after the response (Leuthold & Jentzsch, 2001, 2002; Pope et al., 2007; Praamstra et al., 1996). Leuthold and Jentzsch (2002) studied lateralized activity during a time interval following the presentation of information about which hand was required to respond later. For this preparatory activity, they found evidence for a source configuration that differed from execution-related activity: There was a combination of a source in the more anterior primary motor area and a source within the parietal lobe.

Uses of the LRP

The LRP has attracted considerable interest from researchers both because its functional significance is fairly clear from its derivation and localization, and also because it can be used in a relatively direct fashion to address a number of intriguing experimental questions. It has been employed not only by researchers working on the psychophysiology and neuroscience of motor control, but also by those studying various functional questions within cognitive psychology. These uses may be broadly categorized into two main types: assessing motor preparation and partitioning the reaction time interval.[3]

Assessing Motor Preparation

Given that we know from its derivation that the LRP indexes response activation, this component has, of course, played a crucial role in studies of movement preparation and organization. Many studies have used a variant of the movement precuing paradigm developed by Rosenbaum (1980), in which participants are given an advance signal to prepare one hand to make the upcoming response. A clear LRP develops after such a signal is presented (e.g., Leuthold et al., 1996), and this LRP is maintained fairly steadily for at least 1–2 s while the participant waits to produce the prepared response (e.g., Hackley & Miller, 1995). The size of the LRP observed during this interval depends on exactly which characteristics of the upcoming movement are known in advance (e.g., hand, finger, direction and extent of movement, required force), consistent with partially hierarchical models of motor preparation (e.g., Ulrich et al., 2003). It also depends on the complexity of the response being prepared (e.g., Hackley & Miller, 1995) but not on the forcefulness of the upcoming movement (e.g., Sommer et al., 1994) or the task for which that movement needs to be produced (Miller & Low, 2001).

The LRP has also proven useful in studying the nature of deficits in motor preparation in certain types of patient groups. For example, preparatory LRP similar to that in normals is also observed in patients with Parkinson's disease (e.g., Praamstra et al., 1996), despite the substantial difficulties these patients have with movement execution. Yet, it appears that these patients need more time than normal individuals to establish movement preparation in the first place, especially when preparing complex movements (e.g., Low et al., 2002).

Interestingly, although the LRP is clearly associated with movement preparation, strong LRPs can also be observed in the absence of movement. In the stop-signal paradigm (e.g., Logan, 1994), for example, participants perform a choice RT task but are in some trials given a special countermanding signal telling them to inhibit the normal response to the choice stimulus. It turns out that participants still produce a strong LRP to the choice reaction stimulus even in trials where they successfully inhibit the response following the countermanding signal (e.g., De Jong et al., 1990). This shows that the LRP can be used to detect partial preparation of movement, even when no actual movement is produced. As is discussed next, this feature of the LRP has made it a very attractive tool for addressing a number of theoretical questions involving partial response preparation.

Because the LRP can be used to assess partial movement preparation dissociated from overt movement, it can be a very powerful tool for checking whether online response preparation actually occurs in a variety of choice RT tasks where there is theoretical dispute about whether any such preparation takes place. In the flankers paradigm, for example, the stimulus is typically a center letter surrounded by two or more identical flanker letters (e.g., Eriksen & Eriksen, 1974). Participants are instructed to focus their attention on the center letter, responding with the left or right hand depending on its identity, and to ignore the flanker letters. Despite these instructions, people cannot completely ignore the flanker letters. Responses are slowed, relative to a neutral condition, in a so-called "incompatible" condition in which the identities of these letters are associated with the response hand opposite to that indicated by the relevant center letter. One plausible account of this slowing is that it arises from response conflict generated by the flanker letters, and the LRP provides a very convenient experimental window through which to look for such conflict. Using the LRP, Gratton and colleagues (1988) were able to show that the incorrect hand is briefly prepared in the incompatible condition and that the slowing of behavioral responses in this condition was directly associated with that incorrect-hand preparation. Thus, the LRP provided direct evidence for the response-conflict explanation of the flanker incompatibility effect. Analogous studies have shown response-level effects produced by spatial cues (Eimer, 1995), and even by masked stimuli of which participants were not consciously aware (e.g., Eimer & Schlaghecken, 1998). Such effects are all the more impressive because they must be fairly large and consistent across trials to show up in an across-participants average waveform (*grand average*), where they would be diluted or reversed by measurements from trials in which there was no conflict.

A similar brief lateralization in the "incorrect" direction has also been observed in studies of the Simon effect (Simon & Rudell, 1967). In a horizontal version of the Simon task, colored squares are presented to the left or right of fixation. The color determines the response hand, and its location should be ignored. Nevertheless, responses are faster when the location corresponds to the side of the responding hand than when these do not correspond. The LRP has been studied in this task because it might give evidence of an early automatic activation of the response hand that is on the same side as the stimulus. However, Praamstra (2007) disputed the inference from LRP to response preparation in this horizontal version of the task because of the likely involvement of sensory or attentional lateralizations (e.g., N2pc, N2cc; Praamstra, 2006). As noted above, when stimuli and responses are arranged along a vertical dimension, the contribution of sensory lateralized potentials can be ruled out. Using such an arrangement, several studies have found evidence of a brief activation of the incorrect response based on the location of the stimulus (De Jong et al., 1994; Stürmer et al., 2002; Valle-Inclán, 1996, Experiment 3), whereas others have not found it (Vallesi et al., 2005; Wiegand & Wascher, 2005).

Assessing partial response preparation with the LRP has also contributed to cognitive models of RT by helping to establish the point at which preparation of a response can begin. Many studies of this topic have used a hybrid task combining choice RT and go/no-go decisions, with some stimuli requiring left or right hand responses and other stimuli requiring the no-go response (e.g., Miller & Hackley, 1992; Osman et al., 1992). In most versions of the task, it is relatively easy to see which hand might have to respond, but it takes longer to make the go/no-go decision, and the crucial question is whether an LRP develops in no-go trials. For example, Miller and Hackley (1992, Experiment 2) required some participants to respond with the left hand when presented with a large *S*, to respond with the right hand when presented with a large *T*, and to withhold the response when presented with a small *S* or *T*, counterbalancing across participants the assignments of letter to hand and of size to go/no-go. The stimuli were constructed so that the *S/T* discrimination would be much easier than the large/small distinction, thereby ensuring that participants would know which hand might have to respond before they knew whether the response would actually be required. Critically, a clear LRP was observed in no-go trials, indicating that partial response preparation could occur before stimulus analysis was complete. Numerous other studies have also documented that partial LRPs can occur before stimulus analysis is complete (e.g., Band & Miller, 1997, Experiment 2; Hackley & Valle-Inclán, 1999; Heil et al., 1998; Osman et al., 1992; Smid et al., 1996, Experiment 1; Smid et al., 1991, 1992), although there are also quite a number of circumstances in which such partial LRPs do not occur, consistent with models in which stimulus analysis must finish before response preparation can begin (e.g., Band & Miller, 1997, Experiment 1; Ilan & Miller, 1998; Low & Miller, 1999; Miller & Hackley, 1992, Experiment 4; Smid et al., 1996, Experiment 2; Smid & Heinze, 1997).

Partitioning the RT Interval

In many studies cited in the preceding section, the main interest was in the size or polarity of the LRP during a particular time interval. An alternative approach uses the moment the LRP starts as a measure of the timing of hand-specific motor preparation. Given the nature of the LRP derivation, the lateralization cannot possibly begin until the brain has begun to determine which response is signaled by a given stimulus. For this reason, the time from stimulus presentation to the onset of the LRP (S→LRP) can be used as a measure of the duration of stimulus processing prior to the point at which response activation begins. Likewise, the interval between LRP onset and the overt response (LRP→R) can be regarded as an index of the duration of response activation and peripheral motor processes. Thus, these two temporal intervals can be regarded as decomposing the overall RT interval into two subintervals preceding and following the onset of response activation. For the first interval, the averaging across trials in equations (1) and (2) is typically done for time points (t) relative to the stimulus; for the second interval, time points are defined with respect to the moment of the overt response or of the onset of muscle activity (EMG) over the forearms (e.g., Osman & Moore, 1993). These two types of averaging are referred to as *stimulus-locked* and *response-locked*, respectively.

Broadly, the utility of partitioning the RT interval in this way is twofold. First, partitioning can provide information about the temporal locus of experimental effects. As will be illustrated later with several examples, it is often theoretically important to determine whether an experimental manipulation has its effect on relatively early sensory and perceptual processes or on rather late response activation and motor processes. Partitioning studies can address this question directly by revealing whether a manipulation has its effect before or after the onset of LRP. Second, partitioning studies also provide information about the functional locus of the LRP within the information processing structure. In principle, as an index of selective response activation, the LRP might be generated as soon as the response selection has begun, although it might also start only at some later point during response selection or subsequent motor stages. By manipulating task variables, the point of LRP onset may be defined more precisely (e.g., Masaki et al., 2004).

Table 9.1 lists a selection of studies that have used LRP onset latency to partition RT. Each of the experimental manipulations listed could have an effect on the S→LRP interval (a "+" in the S-LRP column), the LRP→R interval ("+" in the LRP-R column), or both. Sometimes a large effect on one interval was combined with a smaller effect on the other, indicated by "++" and "+," respectively. Especially interesting are selective effects (Sternberg, 2001), where an effect is present in one interval (+) but absent in another ("0").[4]

The first thing to note about the table is the abundant evidence for experimental effects on the LRP→R interval. Variables acting on the LRP→R interval include the precueing of movement parameters (e.g., Leuthold et al., 1996); Osman et al., 1995), response complexity (Smulders et al., 1995),

Table 9.1 Effects on the S-LRP interval, the LRP-R interval, and RT for various manipulations

Topic/Reference	Manipulation	S-LRP	LRP-R	dRT	dS-LRP	dLRP-R	Loss (ms)	Loss %
Precueing								
Osman et al. (1995)	Finger precued vs. uncued	+	+	84	29	45	10	12
Muller-Gethmann et al. (2000)	Exp.1, force precued vs. uncued	+	0	95	73	8	14	15
	Exp.1, direction precued vs. uncued	+	+	102	59	25	18	18
	Exp. 2, force precued vs. uncued	+	0	107	84	–6	29	27
	Exp. 2, direction precued vs. uncued	+	+	105	67	15	23	22
Leuthold et al. (1996)	Exp.1, direction precued vs. uncued	0	+	44	15	117	–88	–200
	Exp. 2, direction precued vs. uncued	0	+	115	17	93	5	4
Leuthold (2003)	Exp. 1, high prep. group, valid direction precue vs. uncued	+	+	56	38	26	–8	–14
	Exp.1, high prep. group, invalid direction precue vs. valid	+	0	40	50	–17	7	18
	Exp. 2, hand precued vs. uncued[a]	+	0	91	83	1	7	8
	Exp. 2, direction precued vs. uncued[a]	+	0	112	111	–19	20	18
Speed–accuracy trade-off								
Rinkenauer et al. (2004)	Exp.1, line length discrimination, low vs. high stress	+	+	113	65	34	14	12
	Exp.1, line length discrimination, low vs. moderate stress	+	+	66	29	26	11	17
	Exp.1, line length discrimination, moderate vs. high stress	+	0	47	36	8	3	6
	Exp. 2, lexical retrieval, low vs. high stress	+	+	198	106	31	61	31
	Exp. 2, lexical retrieval, low vs. moderate stress	+	+	114	52	24	38	33
	Exp. 2, lexical retrieval, moderate vs. high stress	+	0	84	54	7	23	27
	Exp. 3, Eriksen flanker task, low vs. high stress[b]	+	+	79	60	16	3	4

	Exp. 3, Eriksen flanker task, low vs. moderate stress	+	+	44	21	17	6	14
	Exp. 3, Eriksen flanker task, moderate vs. high stress	+	0	35	39	–1	–3	–8
Osman et al. (2000)	SAT macro-trade-off (speed-accuracy instruction)[c]	0	+	67	8	22	37	55
	SAT micro-trade-off (random variation)	+	0	131	57	–2	76	58
Van der Lubbe et al. (2001)	Choice-by-location, time pressure	0	+	39	18	36	–15	–38
	Simon task, corresponding, time pressure[d]	0	+	32	–4	25	11	34
	Simon task, noncorresponding, time pressure	0	+	26	7	17	2	8
Temporal preparation/accessories								
Muller-Gethmann et al. (2003)	Exp.1, visual S1, auditory S2, foreperiod (FP) 50 vs. 400 ms[e]	++	+	55	45	30	–20	–36
	Exp.1, visual S1, auditory S2, FP 400 vs. 3200 ms[e]	++	+	78	38	5	35	45
	Exp. 2, auditory S1, visual S2, FP 50 vs. 400 ms[e]	+	0	20	10	7	3	15
	Exp. 2, auditory S1, visual S2, FP 400 vs. 3200 ms[e]	+	0	80	47	–3	36	45
Hackley et al. (2007).	FP 600 vs. 3000 ms	++	+	83	60	11	12	14
Tandonnet et al. (2003)	FP 500, 2500 ms	n.a.	0[f]	20	n.a.	–6		
Hackley & Valle-Inclan (1998)	Accessory presence vs. absence	+	0	20	>0	n.s.		
Hackley and Valle-Inclan (1999)	Go trials, accessory presence vs. absence	+	0	34	36	n.s.		
Individual differences								
Rammsayer and Stahl (in 2007)	Individual differences in intelligence related to two-choice RT/LRP	0	+[g]	n.a.	n.a.	n.a.		
Rammsayer and Stahl (2004)	Extraverts (faster) vs. introverts	0	+	8	–11	19	0	0
Aging								
Band and Kok (2000)	Aging, 21 vs. 67 years, on average	n.a.	+	533	n.a.	199		

(*Continued*)

Table 9.1 (Continued)

Topic/Reference	Manipulation	S-LRP	LRP-R	dRT	dS-LRP	dLRP-R	Loss (ms)	Loss %
Psychopharmaca								
Rammsayer and Stahl (2006)	Dopamine agonist pergolide in two-choice task (slowing RT)	+	0	8	–43	5	46	575
Smid et al. (1997)	Hypoglycemia vs. baseline	+	n.a.	27	72	n.a.		
Psych. refractory period								
Osman and Moore (1993)	Exp.1, stimulus-onset asynchrony (SOA) short vs. long	+	0	279	119	40	120	43
	Exp. 2, SOA short vs. long	+	0	247	179	4	64	26
Jentzsch et al. (2007)	Psychological refractory period (PRP) task, SOA short vs. long	+	0	70	38	n.s.		
	PRP task, stimulus contrast	+	0	26	39	n.s.		
	Residual PRP effect, SOA RT1 <400	+	0	39	35	n.s.		
	Residual PRP effect, fast vs. slow RT1 bin	+	0	50	35	n.s.		
Sommer et al. (2001)	Choice RT task, SOA short vs. long	+	0	380	235	55	90	24
Additive factor method								
Smulders et al. (1995)	Stimulus degradation[h]	+	0	35	31	–2	6	17
	Response complexity (1 vs. 3 taps)[h]	0	+	25	6	22	–3	–12
Masaki et al. (2004)	Symbolic S–R compatibility[i]	+	0	103	50	–33	86	84
	Time-to-peak force, fast vs. slow[i]	0	+	28	21	59	–52	–188
Low et al. (2002)	Parkinson vs. age-matched controls	+	+	111	66	n.a.		
	Simple response, Parkinson vs. controls	+	0	42	n.a.	n.s.		
	Complex response, Parkinson vs. controls	+	+	180	n.a.	121		
	Stimulus discriminability (mixed)	+	+	149	44	45	60	40
	Response complexity (1 vs. 3 taps)	0	+	172	n.s.	169		

Donders's subtraction method								
Smid et al. (2000)	Choice: go/no-go vs. go/no-go (for fast RTs)	0	+	35	–16	48	3	7
	Choice: go/no-go vs. go/no-go (for slow RTs)	+	+	58	60	36	–39	–67
Miller and Low (2001)	Task type: simple RT vs. go/no-go	+	~0	48	16	22	10	21
	Go/no-go vs. choice RT	+	~0	46	43	1	2	4
Mordkoff et al. (2007)	Simple RT vs. go/no-go	n.a.	+	39	n.a.	97		
	Go/no-go vs. choice RT	n.a.	0	49	n.a.	4		
Miscellaneous								
Miller et al. (1999)	Choice RT, visual intensity (5 vs. 11 cd/m²)	+	0	28	33	0	–5	–18
	Choice RT, auditory intensity (46 vs. 80 dB)	?[j]	?[j]	18	8	0	10	56
Mordkoff et al. (1996)	Redundant vs. faster single target	+	0	30	32	n.s.		
Miller (2007)	Redundant vs. average single target	0	+	28	11	11	6	21
Keus et al. (2005)	Numerical magnitude–hand compatibility effect (SNARC)	+	0	14	n.a.	n.a.		
Smulders et al. (1996)	Intact, correct vs. error	+	–	38	45	–10	3	8
	Degraded, correct vs. error	+	–	32	50	–20	2	6
	Intact, fast vs. intermediate (micro-SAT)	++	+	50	45	5	0	0
	Intact, intermediate vs. slow (micro-SAT)	++	+	55	45	10	0	0
	Degraded, fast vs. intermediate (micro-SAT)	+	0	50	60	0	–10	–20
	Degraded, intermediate vs. slow (micro-SAT)	++	+	60	40	5	15	25
	Intact vs. degraded	+	0	45	40	5	0	0
Miller and Ulrich (1998)	2 vs. 6 response alternatives (Exp. 1)	+	+	226	55	79	92	41

(Continued)

Table 9.1 (Continued)

Topic/Reference	Manipulation	S-LRP	LRP-R	dRT	dS-LRP	dLRP-R	Loss (ms)	Loss %
	2 vs. 4 response alternatives (Exp. 2, total)	+	0	61	38	n.s.		
	2 vs. 4 response alternatives (Exp. 2, within hand)	–	+	46	-68	70	44	96
	Probability (Exp. 3)	?	?	33	2	30	1	3
	Index vs. middle finger (Exp. 3)	0	+	n.a.	n.s.	>0		
	S–R compatibility (within hand, order 1, Exp. 4)	0	+	124	n.s.	150		
	S–R compatibility (within hand, order 2, Exp.4)	0	+	31	n.s.	14		

Note: The columns S-LRP and LRP-R reflect effects of the manipulation on the S→LRP and LRP→R intervals, respectively, using the following symbols: "+": an effect in the same direction as the RT effect (an increase in the interval along with an increase in RT); "++" combined with "+" in the same row: a larger effect; "-" an effect opposite to the RT effect (a decrease in the interval along with an increase in RT); "0": no (or no significant) effect. Other symbols: "~" : about; "?": undetermined; "n.a.": not available; "n.s.": nonsignificant. dRT is the size of the effect on RT in milliseconds. dS-LRP and dLRP-R are the sizes of effect on these intervals in milliseconds. Here, negative values indicate effects opposite to the direction of the RT effect (RT is increased, but the interval is decreased). "Loss" is the difference between the sum of effects on the S→LRP and LRP→R intervals, on the one hand, and the RT effect,on the other hand. The column "Loss %" is the same, but expressed as a percentage of the RT effect.

[a] A hand precue strongly affected the LRP baseline.

[b] The effect on the S-LRP interval almost disappeared after a correction for guessing was applied.

[c] This was an Eriksen flanker task: all LRP onsets possibly determined by automatic activation effects.

[d] This was a Simon task: all LRP onsets possibly determined by automatic activation effects.

[e] More FPs were used; values were estimated from the plot.

[f] The authors concluded that late processes were affected, based on ipsilateral and contralateral activation and inhibition, not based on LRP.

[g] A correlational design; the effect denotes a positive correlation between the LRP-R interval and intelligence, which was again positively correlated with response speed.

[h] These were main effects, i.e., averaging across levels of other factor(s) that had additive effects.

[i] These were main effects on electromyographic onset latency, i.e., averaging across levels of other factor(s) that had additive effects.

[j] LRP was undetermined. The locus of intensity effects was concluded to precede motor processes, based on their sizable effects on the latencies of N100 and P300

and time pressure or speed stress (e.g., Van der Lubbe et al., 2001). As shown in the table, the authors often arrived at the conclusion that the effects on the LRP→R interval were selective. The observation that it is possible for at least some cognitive manipulations to exert an effect on the LRP→R interval is important, because it shows that that LRP does not merely covary with the overt response, adding importance to studies that found the entire RT effect to be located in the S→LRP interval (Osman et al., 1995).

The results from partitioning studies have been very helpful in validating and refining models that had been developed on the basis of purely behavioral measures. As a first example, many stage models of RT (e.g., Sanders, 1980, 1983, 1998; Sternberg, 1969, 2001) localize the effects of stimulus manipulations (e.g., intensity, stimulus quality) solely on an early perceptual stage and the effects of response manipulations solely on a late motor stage. Consistent with these models, Miller and colleagues (1999) found that visual stimulus intensity has a selective effect—at least when manipulated across a moderate range—influencing only the S→LRP interval and not the LRP→R interval. Similarly, Smulders et al. (1995) found selective effects of stimulus quality on the S→LRP interval, and they also found selective effects of response complexity on the LRP→R interval. In this study, the effects of stimulus degradation and response complexity on mean RT were additive, consistent with previous suggestions that two factors having additive effects on RT most likely affect two different processing stages (cf. Sternberg's 1969 additive factor method). A similar combination of RT additivity and selective effects on LRP intervals was reported by Masaki and coauthors (2004). They also found that a motor manipulation (i.e., instructed variation in the time to reach peak response force) selectively affected the LRP→R interval; in addition, they found that symbolic stimulus–response (S–R) compatibility selectively affected the S→LRP interval. The results concerning S–R compatibility were generalized by Keus and colleagues (2005), who obtained evidence that a type of S–R compatibility between numerical magnitude and the response hand also affected the S→LRP interval. Together, these and several other studies support the important conclusion that LRP probably begins after response selection (which is affected by S–R compatibility) but before motor programming (which is affected by response complexity and time-to-peak force). Finally, in a study with patients suffering from Parkinson's disease, Low and coauthors (2002) manipulated stimulus discriminability and response complexity. These factors had additive effects on RT for both patients and controls, even though the effects of response complexity (but not discriminability) were greatly enhanced for the patients. Both the effect of response complexity and its enhancement in the patient group were selective for the LRP→R interval, and this pattern of effects can be explained by the same two-stage model in both groups ("stage robustness," Sanders, 1980). In contrast with Smulders et al., however, the effects of stimulus discriminability were not completely selective for the S→LRP interval in the study of Low et al.

A second example of the usefulness of the partitioning method comes from studies testing the movement parameter specification notion of Rosenbaum (1980). Rosenbaum found that giving participants advance information about some parameters of a soon-to-be-required movement (i.e., precueing those parameters) reduces RT in response to a later imperative stimulus, and he concluded that the advance information reduces RT because the motor system can set up the prespecified parameters in advance, thus reducing the time needed for motor programming after the imperative stimulus is eventually presented. According to that view, precueing movement parameters should reduce the LRP→R interval. As shown in Table 9.1, there is good evidence that the LRP→R interval is indeed reduced by precueing the response finger or the direction of movement (flexion versus extension). In contrast, however, precueing response force has a selective effect on the S→LRP interval. In short, Rosenbaum's predictions about the effects of movement precuing have been confirmed for some movement parameters but not others.

Third, Donders (1868/1969) proposed that the duration of a processing stage can be computed simply by subtracting the reaction times obtained in two tasks that differ only in the presence versus absence of that stage. A crucially important assumption of this subtraction method is that a stage may be inserted in or deleted from the reaction process without influencing the duration of the stages that the two tasks have in common—the so-called assumption of "pure insertion." This assumption was tested in three studies listed in Table 9.1 using variants of the original tasks employed by Donders. Smid and colleagues (2000) inserted a selection between left-hand and right-hand responses into a go/no-go task and used the LRP to investigate whether the processing stages following the onset of hand-specific response activation indeed remained constant. Contrary to the assumption of pure insertion, they found that the LRP→R interval was larger when the response hand had to be selected after stimulus onset than when it was known in advance. Using rather different tasks, Miller and Low (2001) preserved the insertion of stimulus discrimination and response selection, as in Donders' original tasks, but attempted to increase task similarity by minimizing task differences in response preparation. Preparation immediately prior to the imperative stimulus was measured using LRP, contingent negative variation (CNV), and muscle activity, and it was found to be equivalent across tasks. In contrast to the results of Smid et al., the S→LRP interval but not the LRP→R interval was significantly increased by inserting the extra stages. Finally, Mordkoff and colleagues (2007) used a balance between response probabilities different from that used by Miller and Low, and—like Smid et al.—they found that the LRP→R interval was increased when discrimination was added to the task. Overall, this rather mixed pattern of results suggests that it is possible to keep motor time constant across certain versions of the tasks employed by Donders, but that this can only be done by carefully equating response preparation across tasks using methods like those of Miller and Low (2001). In the absence of special precautions designed to equate response preparation, the assumption of motor invariance made by the subtraction method appears to be violated.

The above three examples illustrate cases in which partitioning studies have provided at least a degree of support for the models they were intended to test, but some other partitioning studies have found rather surprising results that contrasted with the models' predictions. As a case in point, Table 9.1 lists three studies of the speed-accuracy trade-off (SAT), the strategic trading of speed for accuracy that participants exhibit when they are placed under varying levels of speed stress. According to most models of SAT, participants adjust premotor processes in order to trade off speed and accuracy. For instance, accumulation models of SAT assume that gradually accumulated evidence about the stimulus must reach a threshold before a response is initiated and that the motor processing after response initiation is relatively constant (cf. Luce, 1986). In such models, participants raise or lower the threshold in order to produce faster and more error-prone responses or slower, more accurate ones, and these adjustments influence only the duration of premotor processes. Yet, in the first two studies of SAT effects using the partitioning approach, speed stress influenced only the LRP→R interval, not the S→LRP interval (Osman et al., 2000; Van der Lubbe et al., 2001). Rinkenauer and colleagues (2004) manipulated speed stress in an easy line-discrimination task and a slower lexical decision task, choosing these tasks to maximize the contribution of deliberate, controlled stimulus analysis processes and thereby maximize the chances of premotor adjustments to speed stress. In both tasks, they again found effects on the LRP→R interval, but there were also effects on the S→LRP interval. Rinkenauer et al. concluded that speed stress affects the duration of both decision and postdecision processes and that none of the extant SAT models was sufficient to explain the complete pattern of results.

Partitioning studies have also produced unexpected results in studies of foreperiod effects. The foreperiod is the time interval between a warning signal and an imperative stimulus, and it has long been known that short foreperiods yield faster responses than long ones when foreperiod length is fixed within a block (Klemmer, 1956). Some evidence (see Müller-Gethmann et al., 2003, and Sanders, 1998, for reviews) suggests that a short foreperiod enables more motor preparation than a long one, so the RT decrease associated with short foreperiods has generally been explained in terms of a late motor speedup (Sanders, 1983, 1998). This explanation suggests that foreperiod manipulations should act on the LRP→R interval, not the S→LRP interval. As listed in Table 9.1, this prediction has been tested in three studies including six experiments. In all cases where the S→LRP interval was examined, it was substantially affected by the manipulation of foreperiod, whereas the LRP→R interval was affected in only three cases, and even then it was affected to a relatively small degree. This evidence of relatively early foreperiod effects contradicts the view that foreperiod manipulations affect primarily or uniquely a late motor stage, especially in conjunction with the evidence reviewed above that manipulations of motor programming do exert an effect on the LRP→R interval. In addition, several recent findings support the idea of foreperiod effects on perceptual processes, consistent with the view that foreperiod effects arise before selective response activation (Rolke & Hofmann, 2007; Smulders & Meijer, 2008).

Finally, partitioning studies have provided clear-cut diagnostics for cases in which prior results had not provided a strong rationale for associating an experimental effect with either early or late processes.

One case concerns the effects of *accessory* stimuli in a choice RT task. An accessory stimulus is a task-irrelevant stimulus (e.g., a tone) that is presented together with a relevant stimulus (e.g., a visual stimulus). The effects of accessory stimuli on RT have been investigated with LRP in two listed studies. Accessory stimuli are known to reduce choice RT, but there exist several conflicting theoretical accounts of this effect. Some theorists have argued that alerting by an accessory stimulus speeds a relatively early perceptual or decision-level process (e.g., Posner, 1978), whereas others argue that it speeds late motor processes (e.g., Sanders, 1980). Hackley and Valle-Inclán (1998, 1999) attempted to decide between these alternative theories by checking whether accessory stimuli reduce the S→LRP interval or the LRP→R interval. In their first study, they found an effect only on the S→LRP interval, favoring the former interpretation in terms of speeding at a perceptual or decision level. In the second study, they managed to shift a portion of response selection to the LRP→R interval by allowing hand selection to be based on easy-to-discriminate letter identity (*S* vs. *T*), while finger selection was based on slower color discrimination and S–R mapping. Again, accessory stimuli affected only the S→LRP interval. Since accessory stimuli led to a notable increase in the amplitude of a no-go LRP in the same task, the authors concluded that the selection of the responding hand constituted at least one of the loci of the accessory stimulus effect.

In summary, studies using LRP onset to partition the RT interval have led to both expected and unexpected results, thereby making a substantial contribution to the study of mental chronometry. The time interval between stimulus and LRP onset was found to be affected by stimulus intensity, stimulus degradation, S–R compatibility (but see Miller & Ulrich, 1998, Experiment 4), speed stress, and the presence of discrimination and response selection stages, whereas the interval between LRP onset and the response was affected by response complexity, time-to-peak force, speed stress, and prior information about movement parameters, although the last was not always selective. A number of these results contradicted or decided between previous theories, especially the effects of foreperiod (S→LRP interval), accessory stimuli (S→LRP interval), and speed stress (LRP→R interval). In these cases, LRP results clearly argued for a reinterpretation of the mechanisms causing RT effects. In addition, partitioning studies have helped to validate and refine ideas about the functional significance of the LRP observed during the short interval between an imperative stimulus and a speeded response to it. As an index of selective response activation, the LRP could in principle start being generated either as soon as response selection has begun or else at some later point during the response selection or motor stages, as discussed earlier. Effects observed in partitioning studies—especially studies varying S–R compatibility and the necessity of response selection (both of which affect the S→LRP interval), as well as studies varying motor programming requirements (which affect the LRP→R interval)—collectively indicate that the LRP starts after response selection but before motor programming.

Scoring and Testing

Analyses of LRP have been restricted to averages across trials, as in equations (1) and (2), with typical averages of between 50 and 100 trials for each hand. No attempts have been made to score LRP for single trials for at least two reasons. First, LRP peak amplitude is typically only a few microvolts, and any ERP component that small would be very difficult to detect within the ongoing electroencephalogram (EEG). Second and perhaps even more importantly, the definition of the LRP as a preponderance of contralateral activity, averaging across left and right hand responses, seems to require consideration of at least two sets of trials (i.e., one for the left hand, one for the right hand) in order to isolate an LRP.

Motor Preparation and Incorrect Lateralizations

When the LRP is used to detect the preparation of a motor response, whether by the correct response hand or by the incorrect one, the focus is generally on LRP amplitude. Using equation (1) or (2), an amplitude reliably different from baseline activity indicates that motor preparation did take place, with a negative or positive amplitude indicating that the correct or incorrect response hand was prepared, respectively. Thus, motor preparation during a predefined interval and more qualitative phenomena like incorrect-hand preparation (the *Gratton dip*) can be tested for significance using standard methods. Typically, the amplitude is measured as the mean amplitude of the LRP waveform within a certain time window. Each amplitude can then be tested against zero or against the amplitude in another condition (e.g., Gratton et al, 1988). If it is desirable to describe the time course of these differences in more detail, the time windows can be made accordingly smaller. In this case, as a protection against Type I errors associated with multiple testing, effects lasting less than some duration may not be interpreted (e.g., Smid et al., 2000).

LRP Onset Latency in Partitioning Studies

For the purpose of partitioning RT using the onset latency of the LRP, it would theoretically be best to find for each subject an estimate of the mean across trials of the moment in time at which the LRP starts to deviate from zero. This estimate could then be compared to mean RT in the same set of trials. Since finding this moment in time in each trial is not feasible, researchers have developed various alternative strategies. As mentioned earlier, the interval between the stimulus and LRP onset is usually estimated by an analysis of the stimulus-locked average LRP, and the interval between LRP onset and the response is estimated by an analysis of the response-locked average LRP (Osman & Moore, 1993). Nearly all of the studies listed in Table 9.1 have followed this method, although a few have reported only one interval. Sternberg (2001, p. 226) suggested that researchers should also consider the complement of these measures, that is, the interval between stimulus-locked LRP onset and RT, and between stimulus onset and the onset of response-locked LRP, but to our knowledge these measures have not yet been reported by other researchers.

Unfortunately, the measurement of LRP onset latency is a difficult technical problem. In addition to its small peak amplitude, LRP generally has a rather small initial slope, making it extremely difficult to determine, for instance, when the LRP first reaches a level beyond what would be expected due to noise. Quite a few methods have been used to measure LRP onset latencies (for reviews, see Miller et al., 1998; Mordkoff & Gianaros, 2000). One way to measure the latency of the grand-average LRP is to do a statistical test of its deviance from zero for each of multiple time points, and search for the point in time at which the average starts to deviate significantly and consistently from zero (De Jong et al., 1988; Smid et al., 1997). Other methods yield LRP onsets for each participant in every experimental condition, including (1) searching for a time point at which the participant's average LRP first exceeds a criterion based on the noise level during a baseline preceding the stimulus (Osman et. al., 1992) and (2) fitting a model consisting of various line segments to the entire LRP waveform from stimulus onset to LRP peak (Mordkoff & Gianaros, 2000; Schwarzenau et al., 1998). If reliable LRP onset latency estimates can be obtained for each participant in each condition, these estimates can then be analyzed further with analysis of variance (ANOVA) or another standard statistical technique. In many experiments, however, the signal-to-noise ratio of the LRP is so low that there is no clear LRP at all for at least one participant in one condition, precluding such scoring of single-participant LRP onset latencies. Fortunately, an alternative jackknife-based method seems to work quite well (Miller et al., 1998; Ulrich & Miller, 2001). The jackknifing method was originally based on the realization that LRP onset latency in an across-participants average waveform (grand average) is probably quite accurate but, being a single number for each condition, does not yield an estimate of variation in onset latency among participants, precluding tests of statistical significance. With jackknifing, within each experimental condition, every participant's average waveform is replaced by the average across the other $n-1$ participants, called a *subaverage*. Obviously, the subaverage waveforms, each being based on $n-1$ participants, have a substantially larger signal-to-noise ratio than the individual participants' waveforms, facilitating reliable determination of LRP latencies in the subaverages. Importantly, the variability among the subaverages still reflects the variability among the individual LRPs. In each subaverage, the moment at which a certain criterion is exceeded by the waveform can be used as the subaverage latency score. There are appropriate statistical techniques for analyzing the subaverage scores. The jackknifing method can also be employed with correlational designs (Stahl & Gibbons, 2004) and with ERP components other than LRP (Kiesel et al., 2008). Simulations suggest that jackknifing provides a powerful method for analyzing LRP onset latencies, and these simulations seem to be backed up by practical experiences in many labs, judging from the fact that the jackknifing approach was used in the majority of studies listed in Table 9.1 that were published after its introduction.

Conclusions and Future Directions

Approximately 20 years have passed since the introduction of the LRP as an index of hand-specific motor activation. As long as contamination of the LRP by other lateralized effects is ruled out through careful experimental design, there seems to be no reason to doubt this strong interpretation of LRP, and indeed, opposing evidence seems absent in the literature. A more detailed account of the cognitive processes indexed by LRP will naturally require assumptions about the precise nature of the processing architecture. On the one hand, the presence or polarity of LRP in some experimental conditions has been used to detect the early communication of partial stimulus information to the motor system, and some have suggested that the gradual rise of the LRP might reflect the gradual buildup of stimulus information that is immediately passed on to the motor system (Coles et al., 1988; Smid et al., 1991), although a gradual buildup can also result from averaging across trials with discrete steps at different time points (Meyer et al., 1988). On the other hand, a similar use of LRP provided support for asynchronous discrete coding of stimulus features (Miller, 1982). Also, in partitioning studies, the LRP has provided evidence for the selective influence of experimental factors on specific portions of the RT interval, which supports the concept of information processing in discrete sequential stages (Sternberg, 1969, 2001). In a stage model, the early part of LRP (i.e., its onset) has now been localized between the end of response selection and the start of motor programming or motor preparation. The functional significance of later parts of the LRP remains unclear, but these most likely reflect later motor programming and execution stages. The LRP has had a substantial impact not only on psychophysiological research but also on cognitive psychology in general, as can be seen by examining the journals represented by the articles in Table 9.1.

There are at least two promising avenues for further scientific progress using the LRP. One is based on the idea of extending the partitioning technique to include one or more additional ERP-based markers of temporal intervals, thereby partitioning RT in at least three intervals. For example, Hackley and colleagues (2007) examined foreperiod effects in a paradigm allowing them to extract not only the LRP but also the N2pc. They found that foreperiod length affected the latency of stimulus-locked LRP but not the latency of N2pc, indicating that this manipulation affects processes between the onsets of these two components. A similar approach has been used to study time pressure effects by Van der Lubbe and coauthors (2001). Clearly, the combination of additional time markers with LRP onset latency has great potential for more narrowly determining the temporal locus of experimental effects.

Another promising avenue for further development of the LRP concerns its precise relation to RT. Ideally, researchers would like to compare directly—that is, in milliseconds—the effect of an experimental manipulation on RT with its effects on the S→LRP and LRP→R intervals. Given that stimulus- and response-locked LRPs partition the RT interval into two mutually exclusive and exhaustive subintervals, any overall effect on RT would logically be predicted to equal the sum of the effects on these two subintervals (Miller et al., 1999). Indeed, many studies have already provided details about the actual size of the experimental effects, in milliseconds, on stimulus-locked and response-locked LRP latency and on RT, and the columns "dS-LRP" and "dLRP-R" of Table 9.1 were derived from these reports. These two columns represent the sizes of the experimental effects on the S→LRP and LRP→R intervals, with positive numbers indicating effects in the same direction as the RT effect, and their sum would logically be expected to equal the RT effect. The column "Loss" specifies the extent to which the combined effects on the S→LRP and LRP→R intervals fall short in explaining the RT effect—that is, the number of "lost" milliseconds. This is expressed as a percentage of the RT effect in the column "Loss%." As can be seen, the loss varies considerably across experiments, reaching extremes of –200% and 575%. The extreme negative value is the result of a 44 ms increase in RT, along with a 15 ms increase in the S→LRP interval and a 117 ms increase in the LRP→R interval (Leuthold et al., 1996), although a mere 4% loss was obtained in a second experiment using a similar paradigm. The extreme positive value was obtained in a study reporting a nonsignificant increase in RT of 8 ms, along with a 43 ms decrease in the S→LRP interval and a further 5 ms increase in the LRP→R interval (Rammsayer & Stahl, 2006).[5] Frequently, the correspondence is good, with a loss between –20% and 20%. Still, considering that all studies met at least reasonable methodological standards, many deviations from 0% are too large to attribute to mere statistical error. Instead, it seems likely that these deviations stem at least partly from limitations in our current methods for scoring LRP onset latencies, at least for purposes of comparison with effects on mean RT, so there is clearly room for further improvement in the comparison of effects on RT versus LRP onset latencies. Indeed, some of the problems that need to be addressed are already well known. For one, the onset of an averaged waveform tends to be determined by those trials in which the onset is relatively early (Luck, 2005; Meyer et al., 1988). For stimulus-locked LRP, these will be the trials with a relatively short S→LRP interval, but for response-locked averages, the onset will be determined by trials with a relatively long LRP→R interval. Given that all trials contribute equally to mean RT, the unequal influence of different trials on LRP onset latencies could clearly create differences between the RT and LRP measures. Indeed, the latencies of stimulus-locked and response-locked LRP are probably biased differently from one another, because most experimental manipulations affect the shape of the RT distribution as well as its location (Luce, 1986; Wagenmakers & Brown, 2007).

Acknowledgment

This work was prepared while the first author was a visiting researcher at the Department of Psychology of the University of Otago. We thank Saul Sternberg for his comments on a previous version.

Notes

1 Still other computations have been used by others; for example, Miller et al. (1992) computed (C3′(*t*) – C4′(*t*))left hand – (C3′(*t*) – C4′(*t*))right hand, generating a positive-going potential. Again, this yields an LRP that differs from the other derivations only by a linear scaling.

2 Oostenveld and colleagues (2001) quantified the overlap between LRP and lateralized components generated in other brain areas using a realistically shaped volume conduction model.

3 The term *bisecting* is sometimes used to convey the same meaning (e.g., Osman et al., 1995), but its first meaning is "to divide into two *equal* parts." The term *partitioning* avoids the suggestion that the parts should be equal.

4 This table does not present an exhaustive summary of the literature using this technique, nor does it even include every experimental manipulation that has been studied, but we have attempted to include a wide and representative set of studies. In addition, various studies were excluded because they employed a paradigm in which a partial LRP could obviously be generated by partial information extracted from the stimulus—for example, if one of several stimulus features allowed for hand selection, but the overt response had to await the analysis of the other feature(s). The latter category includes studies of Eriksen's flanker compatibility effect and most studies on spatial compatibility and the Simon effect (as discussed above). We did include some studies in which a precue specified the response hand or the response hand was held constant throughout a block of trials, allowing for selective preparation before the imperative stimulus. In these cases, a movement-related LRP could be dissociated from preparation-related LRP by a steep rise just prior to the manual response that was used for partitioning RT.

5 We recognize that the percentage increases to infinity as the RT effect approaches zero. However, for most reported studies, RT effects were substantial.

References

Band, G. P. H., & Miller, J. (1997). Mental rotation interferes with response preparation. *Journal of Experimental Psychology: Human Perception and Performance, 23*, 319–338.

Band, G. P. H., & Kok, A. (2000). Age effects on response monitoring in a mental-rotation task. *Biological Psychology, 51*(2–3), 201–221.

Brunia, C. H. M., & Vingerhoets, A. (1980). Cnv and Emg preceding a plantar flexion of the foot. *Biological Psychology, 11*, 181–191.

Coles, M. G. (1989). Modern mind-brain reading: Psychophysiology, physiology, and cognition. *Psychophysiology, 26*, 251–269.

Coles, M. G. H., Gratton, G., & Donchin, E. (1988) Detecting early communication: Using measures of movement-related potentials to illuminate human information processing. *Biological Psychology, 26*, 69–89.

De Jong, R., Coles, M. G. H., Logan, G. G., & Gratton, G. (1990). In search of the point of no return: The control of response processes. *Journal of Experimental Psychology: Human Perception and Performance, 16*, 164–182.

De Jong, R., Liang, C.-C., & Lauber, E. (1994). Conditional and unconditional automaticity: A dual-process model of effects of spatial stimulus–response correspondence. *Journal of Experimental Psychology: Human Perception and Performance, 20*, 731–750.

De Jong, R., Wierda, M., Mulder, G., & Mulder, L. J. M. (1988). Use of partial stimulus information in response processing. *Journal of Experimental Psychology: Human Perception and Performance, 14*, 682–692.

Deecke, L., Grozinger, B., & Kornhuber, H. H. (1976). Voluntary finger movement in man: Cerebral potentials and theory. *Biological Cybernetics, 23*, 99–119.

Donders, F. C. (1868/1969). Over de snelheid van psychische processen [On the speed of mental processes] (W. Koster, Trans.). In W. G. Koster (Ed.), *Attention and performance II* (pp. 412–431). Amsterdam: North Holland.

Eimer, M. (1995). Stimulus–response compatibility and automatic response activation: Evidence from psychophysiological studies. *Journal of Experimental Psychology: Human Perception and Performance, 21*, 837–854.

Eimer, M. (1998). The lateralized readiness potential as an on-line measure of central response activation processes. *Behavior Research Methods, Instruments and Computers, 30*, 146–156.

Eimer, M., & Schlaghecken, F. (1998). Effects of masked stimuli on motor activation: Behavioral and electrophysiological evidence. *Journal of Experimental Psychology: Human Perception and Performance, 24*, 1737–1747.

Elbert, T., Lutzenberger, W., Rockstroh, B., & Birbaumer, N. (1985). Removal of ocular artifacts from the EEG—A biophysical approach to the EOG. *Electroencephalography and Clinical Neurophysiology, 60*, 455–463.

Eriksen, B. A., & Eriksen, C. W. (1974). Effects of noise letters upon identification of a target letter in a nonsearch task. *Perception and Psychophysics, 16*, 143–149.

Gratton, G., Coles, M. G. H., Sirevaag, E., Eriksen, C. W., & Donchin, E. (1988). Pre- and post-stimulus activation of response channels: A psychophysiological analysis. *Journal of Experimental Psychology: Human Perception and Performance, 14*, 331–344.

Hackley, S. A., & Miller, J. (1995). Response complexity and precue interval effects on the lateralized readiness potential. *Psychophysiology, 32*, 230–241.

Hackley, S. A., Schankin, A., Wohlschlaeger, A., & Wascher, E. (2007). Localization of temporal preparation effects via trisected reaction time. *Psychophysiology, 44*, 334–338.

Hackley, S. A., & Valle-Inclán, F. (1998). Automatic alerting does not speed late motoric processes in a reaction-time task. *Nature, 391*, 786–788.

Hackley, S. A., & Valle-Inclán, F. (1999). Accessory stimulus effects on response selection: Does arousal speed decision making? *Journal of Cognitive Neuroscience, 11*, 321–329.

Heil, M., Rauch, M., & Henninghausen, E. (1998). Response preparation begins before mental rotation is finished: Evidence from event-related brain potentials. *Acta Psychologica, 99*, 217–232.

Ilan, A. B., & Miller, J. (1998). On the temporal relations between memory scanning and response preparation. *Journal of Experimental Psychology: Human Perception and Performance, 24*, 1501–1520.

Jasper, H. H. (1958). The ten twenty electrode system of the international federation. *Electroencephalography and Clinical Neurophysiology, 10*, 371–375.

Jentzsch, I., Leuthold, H., & Ulrich, R. (2007). Decomposing sources of response slowing in the PRP paradigm. *Journal of Experimental Psychology: Human Perception and Performance, 33*, 610–626.

Keus, I. M., Jenks, K. M., & Schwarz, W. (2005). Psychophysiological evidence that the SNARC effect has its functional locus in a response selection stage. *Cognitive Brain Research, 24*, 48–56.

Kiesel, A., Miller, J., Jolicaeur, P., & Brisson, B. (2008). Measurement of ERP latency differences: A comparison of single-participant and jackknife-based scoring methods. *Psychophysiology, 45*, 250–274.

Klemmer, E. T. (1956). Time uncertainty in simple reaction-time. *Journal of Experimental Psychology, 51*, 179–184.

Kutas, M., & Donchin, E. (1980). Preparation to respond as manifested by movement-related brain potentials. *Brain Research, 202*, 95–115.

Leuthold, H. (2003). Programming of expected and unexpected movements: Effects on the onset of the lateralized readiness potential. *Acta Psychologica, 114*, 83–100.

Leuthold, H., & Jentzsch, I. (2001). Neural correlates of advance movement preparation: A dipole source analysis approach. *Cognitive Brain Research, 12*(2), 207–224.

Leuthold, H., & Jentzsch, I. (2002). Distinguishing neural sources of movement preparation and execution—An electrophysiological analysis. *Biological Psychology, 60*, 173–198.

Leuthold, H., Sommer, W., & Ulrich, R. (1996). Partial advance information and response preparation: Inferences from the lateralized readiness potential. *Journal of Experimental Psychology: General, 125*, 307–323.

Logan, G. D. (1994). On the ability to inhibit thought and action: A users' guide to the stop signal paradigm. In D. Dagenbach & T. H. Carr (Eds.), *Inhibitory processes in attention, memory, and language* (pp. 189–239). New York: Academic Press.

Low, K. A., & Miller, J. (1999). The usefulness of partial information: Effects of go probability in the choice/Nogo task. *Psychophysiology, 36*, 288–297.

Low, K. A., Miller, J., & Vierck, E. (2002). Response slowing in Parkinson's disease: A psychophysiological analysis of premotor and motor processes. *Brain, 125*, 1980–1994.

Luce, R. D. (1986). *Response times: Their role in inferring elementary mental organization*. Oxford: Oxford University Press.

Luck, S. J. (2005). *An introduction to the event-related potential technique*. Cambridge: MIT.

Masaki, H., Wild-Wall, N., Sangals, J., & Sommer, W. (2004). The functional locus of the lateralized readiness potential. *Psychophysiology, 41*, 220–230.

Meyer, D. E., Osman, A. M., Irwin, D. E., & Yantis, S. (1988). Modern mental chronometry. *Biological Psychology, 26*, 3–67.

Miller, J. O. (1982). Discrete versus continuous stage models of human information processing: In search of partial output. *Journal of Experimental Psychology: Human Perception and Performance, 8*, 273–296.

Miller, J. (1998). Effects of stimulus–response probability on choice reaction time: Evidence from the lateralized readiness potential. *Journal of Experimental Psychology: Human Perception and Performance, 24*, 1521–1534.

Miller, J. (2007). Contralateral and ipsilateral motor activation in visual simple reaction time: A test of the hemispheric coactivation model. *Experimental Brain Research, 176*, 539–558.

Miller, J., & Hackley, S. A. (1992). Electrophysiological evidence for temporal overlap among contingent mental processes. *Journal of Experimental Psychology: General, 121*, 195–209.

Miller, J. O., & Low, K. (2001). Motor processes in simple, go/no-go, and choice reaction time tasks: A psychophysiological analysis. *Journal of Experimental Psychology: Human Perception and Performance, 27*, 266–289.

Miller, J., Patterson, T., & Ulrich, R. (1998). Jackknife-based method for measuring LRP onset latency differences. *Psychophysiology, 35*, 99–115.

Miller, J., & Ulrich, R. (1998). Locus of the effect of the number of alternative responses: Evidence from the lateralized readiness potential. *Journal of Experimental Psychology: Human Perception and Performance, 24*, 1215–1231.

Miller, J., Ulrich, R., & Rinkenauer, G. (1999). Effects of stimulus intensity on the lateralized readiness potential. *Journal of Experimental Psychology: Human Perception and Performance, 25*, 1454–1471.

Mordkoff, J. T., Aderhold, L., Chen, P., Sullivan, B., Von der Heide, R., & Halterman, R. (2007). *Response-locked LRPs from the complication procedure: Evidence against pure insertion*. Presented at the 48th annual meeting of the Psychonomic Society, November 15–18, Long Beach, CA.

Mordkoff, J. T., & Gianaros, P. J. (2000). Detecting the onset of the lateralized readiness potential: A comparison of available methods and procedures. *Psychophysiology, 37*, 347–360.

Mordkoff, J. T., Miller, J., & Roch, A.C. (1996). Absence of coactivation in the motor component: Evidence from psychophysiological measures of target detection. *Journal of Experimental Psychology: Human Perception and Performance, 22*, 25–41.

Müller-Gethmann, H., Rinkenauer, G., Stahl, J., & Ulrich, R. (2000). Preparation of response force and movement direction: Onset effects on the lateralized readiness potential. *Psychophysiology, 37*, 507–514.

Müller-Gethmann, H., Ulrich, R., & Rinkenauer, G. (2003). Locus of the effect of temporal preparation: evidence from the lateralized readiness potential. *Psychophysiology, 40*(4), 597–611.

Oostenveld, R., Praamstra, P., Stegeman, D. F., & van Oosterom, A. (2001). Overlap of attention and movement-related activity in lateralized event-related brain potentials. *Clinical Neurophysiology, 112*, 477–484.

Oostenveld, R., Stegeman, D. F., Praamstra, P., & van Oosterom, A. (2003). Brain symmetry and topographic analysis of lateralized event-related potentials. *Clinical Neurophysiology, 114*, 1194–1202.

Osman, A., Bashore, T. R., Coles, M. G., Donchin, E., & Meyer, D. E. (1992). On the transmission of partial information: Inferences from movement-related brain potentials. *Journal of Experimental Psychology: Human Perception and Performance, 18*, 217–232.

Osman, A., Lou, L., Müller-Gethmann, H., Rinkenauer, G., Mattes, S., & Ulrich, R. (2000). Mechanisms of speed–accuracy tradeoff: Evidence from covert motor processes. *Biological Psychology, 51*, 173–199.

Osman, A., & Moore, C. M. (1993). The locus of dual-task interference: Psychological refractory effects on movement-related brain potentials. *Journal of Experimental Psychology: Human Perception and Performance, 19*, 1292–1312.

Osman, A., Moore, C. M., & Ulrich, R. (1995). Bisecting RT with lateralized readiness potentials: Precue effects after LRP onset. *Acta Psychologica, 90*, 111–127.

Pope, P. A., Holton, A., Hassan, S., Kourtis, D., & Praamstra, P. (2007). Cortical control of muscle relaxation: A lateralized readiness potential (LRP) investigation. *Clinical Neurophysiology, 118*, 1044–1052.

Posner, M. I. (1978). *Chronometric explorations of mind.* Hillsdale, NJ: Erlbaum.

Praamstra, P. (2006). Prior information of stimulus location: Effects on ERP measures of visual selection and response selection. *Brain Research, 1072*, 153–160.

Praamstra, P. (2007). Do's and don'ts with lateralized event-related brain potentials. *Journal of Experimental Psychology: Human Perception and Performance, 33*, 497–502.

Praamstra, P., Stegeman, D. F., Horstink, M. W. I. M., & Cools, A. R. (1996). Dipole source analysis suggests selective modulation of the supplementary motor area contribution to the readiness potential. *Electroencephalography and Clinical Neurophysiology, 98*, 468–477.

Rammsayer, T., & Stahl, J. (2004). Extraversion-related differences in response organization: Evidence from lateralized readiness potentials. *Biological Psychology, 66*, 35–49.

Rammsayer, T., & Stahl, J. (2006). Sensorimotor effects of pergolide, a dopamine agonist, in healthy subjects: A lateralized readiness potential study. *Psychopharmacology, 187*, 36–46.

Rammsayer, T. H., & Stahl, J. (2007). Identification of sensorimotor components accounting for individual variability in Zahlen-Verbindungs-Test (ZVT) performance. *Intelligence, 35*, 623–630.

Rinkenauer, G., Osman, A., Ulrich, R., Müller-Gethmann, H., & Mattes, S. (2004). On the locus of speed–accuracy tradeoff in reaction time: Inferences from the lateralized readiness potential. *Journal of Experimental Psychology: General, 133*, 261–282.

Rolke, B., & Hofmann, P. (2007). Temporal uncertainty degrades perceptual processing. *Psychonomic Bulletin and Review, 14*, 522–526.

Rosenbaum, D. A. (1980). Human movement initiation—specification of arm, direction, and extent. *Journal of Experimental Psychology: General, 109*, 444–474.

Rugg, M. D., Lines, C. R., & Milner, A. D. (1984). Visual evoked potentials to lateralized visual stimuli and the measurement of interhemispheric transmission time. *Neuropsychologia, 22*, 215–225.

Sanders, A. F. (1980). Stage analysis of reaction processes. In G. E. Stelmach & J. Requin (Eds.), *Tutorials in motor behavior* (pp. 331–354). Amsterdam: North Holland.

Sanders, A. F. (1983). Towards a model of stress and human performance. *Acta Psychologica, 53*, 61–97.

Sanders, A. F. (1998). *Elements of human performance: Reaction processes and attention in human skill.* Hillsdale, NJ: Erlbaum.

Schwarzenau, P., Falkenstein, M., Hoormann, J., & Hohnsbein, J. (1998). A new method for the estimation of the onset of the lateralized readiness potential (LRP). *Behavior Research Methods Instruments and Computers, 30*, 110–117.

Simon, J. R., & Rudell, A. P. (1967). Auditory S–R compatibility—Effect of an irrelevant cue on information processing. *Journal of Applied Psychology, 51*, 300–304.

Smid, H. G., Böcker, K. B., van Touw, D. A., Mulder, G., & Brunia, C. H. (1996). A psychophysiological investigation of the selection and the use of partial stimulus information in response choice. *Journal of Experimental Psychology: Human Perception and Performance, 22*, 3–24.

Smid, H. G., Fiedler, R., & Heinze, H. J. (2000). An electrophysiological study of the insertion of overt response choice. *Journal of Experimental Psychology: Human Perception and Performance, 26*, 1053–1071.

Smid, H. G. O. M., & Heinze, H. J. (1997). An electrophysiological study of the selection of the color and shape of alphanumeric characters in response choice. *Biological Psychology, 44*, 161–185.

Smid, H. G. O. M., Lamain, W., Hogeboom, M. M., Mulder, G., & Mulder, L. J. M. (1991). Psychophysiological evidence for continuous information transmission between visual search and response processes. *Journal of Experimental Psychology: Human Perception and Performance, 17*, 696–714.

Smid, H. G. O. M., Mulder, G., & Mulder, L. J. M. (1987). The continuous flow model revisited: Perceptual and central motor aspects. In J. J. R., Parasuraman & J. W. Rohrbaugh (Eds.), *Current trends in event-related potential research EEG supplement 40;* (pp. 270–278), Amsterdam, Elsevier.

Smid, H. G., Mulder, G., Mulder, L. J., & Brands, G. J. (1992). A psychophysiological study of the use of partial information in stimulus-response translation. *Journal of Experimental Psychology Human Perception and Performance, 18*(4), 1101–1119.

Smid, H., Trumper, B. G., Pottag, G., Wagner, K., Lobmann, R., Scheich, H., et al. (1997). Differentiation of hypoglycaemia induced cognitive impairments—An electrophysiological approach. *Brain, 120*, 1041–1056.

Smulders, F. T. Y., Kenemans, J. L., & Kok, A. (1996). Effects of task variables on measures of the mean onset latency of LRP depend on the scoring method. *Psychophysiology, 33*, 194–205.

Smulders, F. T. Y., Kok, A., Kenemans, J. L., & Bashore, T. R. (1995). The temporal selectivity of additive factor effects on the reaction process revealed in ERP component latencies. *Acta Psychologica, 90*, 97–109.

Smulders, F. T. Y., & Meijer, E. H. (2008). Extraversion and performance: A cognitive-energetical approach. *Personality and Individual Differences, 44*, 475–486.

Sommer, W., Leuthold, H., & Schubert, T. (2001). Multiple bottlenecks in information processing? An electrophysiological examination. *Psychonomic Bulletin and Review, 8*, 81–88.

Sommer, W., Leuthold, H., & Ulrich, R. (1994). The lateralized readiness potential preceding brief isometric force pulses of different peak force and rate of force production. *Psychophysiology, 31*, 503–512.

Stahl, J., & Gibbons, H. (2004). The application of jackknife-based onset detection of lateralized readiness potential in correlative approaches. *Psychophysiology, 41*, 845–860.
Sternberg, S. (1969). The discovery of processing stages: Extensions of Donders' method. *Acta Psychologica, 30*, 276–315.
Sternberg, S. (2001). Separate modifiability, mental modules, and the use of pure and composite measures to reveal them. *Acta Psychologica, 106*, 147–246.
Stürmer, B., Leuthold, H., Soetens, E., Schroeter, H., & Sommer, W. (2002). Control over location-based response activation in the Simon task: Behavioral and electrophysiological evidence. *Journal of Experimental Psychology: Human Perception and Performance, 28*, 1345–1363.
Tandonnet, C., Burle, B., Vidal, F., & Hasbroucq, T. (2003). The influence of time preparation on motor processes assessed by surface Laplacian estimation. *Clinical Neurophysiology, 114*, 2376–2384.
Ulrich, R., & Miller, J. (2001). Using the jackknife-based scoring method for measuring LRP onset effects in factorial designs. *Psychophysiology, 38*, 816–827.
Valle-Inclán, F. (1996). The locus of interference in the Simon effect: An ERP study. *Biological Psychology, 43*, 147–162.
Vallesi, A., Mapelli, D., Schiff, S., Amodio, P., & Umilta, C. (2005). Horizontal and vertical Simon effect: Different underlying mechanisms? *Cognition, 96*, B33–B43.
Van der Lubbe, R. H. J., Jaskowśki, P., Wauschkuhn, B., & Verleger, R. (2001). Influence of time pressure in a simple response task, a choice-by-location task, and the Simon task. *Journal of Psychophysiology, 15*, 241–255.
Wagenmakers, E. J., & Brown, S. (2007). On the linear relation between the mean and the standard deviation of a response time distribution. *Psychological Review, 114*, 830–841.
Wascher, E., & Wauschkuhn, B. (1996). The interaction of stimulus- and response-related processes measured by event-related lateralizations of the EEG. *Electroencephalography and Clinical Neurophysiology, 99*, 149–162.
Wiegand, K., & Wascher, E. (2005). Dynamic aspects of stimulus-response correspondence: Evidence for two mechanisms involved in the Simon effect. *Journal of Experimental Psychology: Human Perception and Performance, 31*, 453–464.
Wild-Wall, N., Sangals, J., Sommer, W., & Leuthold, H. (2003). Are fingers special? Evidence about movement preparation from event-related brain potentials. *Psychophysiology, 40*, 7–16.

CHAPTER

10 The Error-Related Negativity (ERN/Ne)

William J. Gehring, Yanni Liu, Joseph M. Orr, *and* Joshua Carp

Abstract

We review two decades of research on the error-related negativity (ERN or Ne), a component of the event-related brain potential that accompanies errors in speeded performance. Theories of the ERN must contend with a wealth of experimental data, both in healthy subjects and in individuals with neurological and psychiatric conditions. Data regarding a number of other components, including the error positivity, feedback-related negativity, correct response negativity, and theta oscillations are thought by many to also constrain ERN theorizing. We attempt to characterize the past highlights and current trajectory of theorizing, computational modeling, and empirical research. We consider how the way in which ERN research is conducted affects its success, and we discuss some promising trends for the future. Although two decades have resulted in impressive theories and data, the ERN community awaits breakthrough developments by new investigators.

Keywords: error-related negativity, ERN, error negativity, Ne, feedback-related negativity, FRN, error detection, response conflict, reinforcement learning, anterior cingulate cortex.

Introduction

It has been 20 years since the first reports of an event-related brain potential component associated with error commission in choice reaction time performance (Falkenstein et al., 1989, 1990, 1991; Gehring et al., 1990, 1993). The anniversary provides a good opportunity to review the state of research on this component, which is known as the *error negativity* (Ne) or *error-related negativity* (ERN). Here we refer to it as the *ERN*. Over the two-decade span, a large number of studies have followed, and ERN research has proven influential to scientists in diverse fields both inside and outside the ERP research community.

In this chapter, we present a somewhat selective overview of ERN research. Excellent reviews of major experimental findings already exist (Falkenstein, 2004a; Falkenstein et al., 2000; Holroyd et al., 2004b; Nieuwenhuis et al., 2004a; Olvet & Hajcak, 2008; Overbeek et al., 2005; Ullsperger, 2006) and collections of papers concerning the ERN and related topics appear regularly in special issues of journals (e.g., Elton et al., 2000; Falkenstein, 2004b, 2005; Kok et al., 2006; Mars et al., 2008). Instead of exhaustively reviewing the ERN literature, our goal here is to step back and characterize the big picture—what we can safely conclude about the functional significance of the ERN, how the way in which ERN research is conducted has affected this knowledge, and how considering these things can suggest some new directions for future research. Our review focuses primarily on the classic response-locked ERN, referring to other components only insofar as they shed light on the ERN.

In evaluating the progress of over 20 years of research on the ERN, it is instructive to consider the comparable 20-year span in the investigation of the P300 component (see Chapter 7, this volume). Sutton and Ruchkin (1984) looked back on the P300 research that had taken place since the initial report of the P300 by Sutton and colleagues (1965). They did not see 20 years of increasing clarity regarding the significance and neural origins of the P300. Rather, the primary message of their chapter was that 20 years of research had made the situation remarkably complicated. In particular, they noted problems with "issues related to the increase in the number of components that have been identified and to the problem of deciding which components are being dealt with in a particular experiment" (p. 1). Our survey of ERN research suggests that Sutton and Ruchkin's remarks on the P300 of 1984 could easily apply to the ERN of today. Still, although the ERN is more complicated than we thought 20 years ago, the level of theoretical and methodological sophistication has grown at a remarkable pace, and the degree to which ERN researchers have influenced—and been influenced by—the larger community of cognitive neuroscientists is unusual in the history of ERP research.

The ERN

The ERN was first observed in speeded choice reaction time tasks (Falkenstein et al., 1989, 1991; Gehring et al., 1990, 1993; see Renault et al., 1980, for what is probably the earliest appearance of this component in published data), where it appeared in response-locked waveforms as a difference between error trials and correct trials. Figure 10.1 shows the ERN waveforms reported in Falkenstein et al. (1991) and Gehring et al. (1993); note that negative is plotted downward in Panel A of the figure and upward in Panel B. The onset of the ERN occurs at or shortly before the moment of the erroneous button press and peaks around 100 ms later. The precise latency depends on the time-locking event: the ERN will appear later in a waveform time-locked to the onset of electromyogram (EMG) activity than in a waveform from the same set of trials time-locked to the button-press switch closure. The lateness of the ERN seen in Figure 10.1B compared to that in Figure 10.1A may be in part because the data in Figure 10.1B were time-locked to EMG onset and those in Figure 10.1A were time-locked to the button-press switch closure. It is also likely that some of the variability in ERN latency in the literature is related

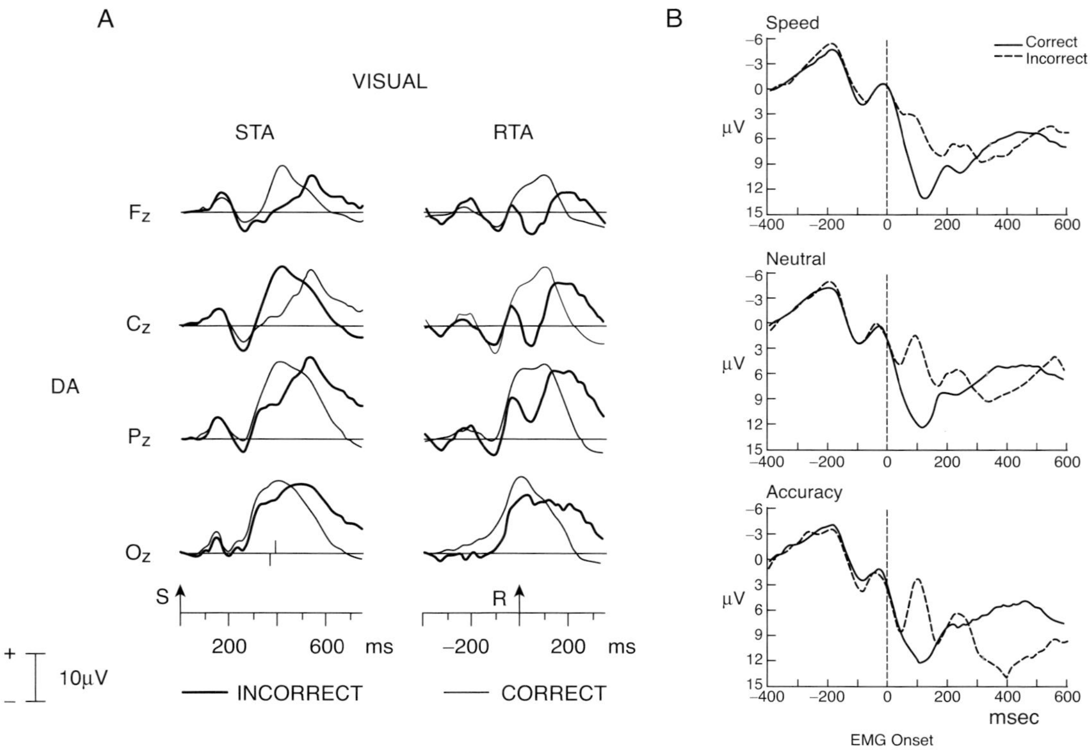

Fig. 10.1. Initial reports of the error-related negativity (ERN/Ne). (A) Stimulus-locked (STA) and response-locked (RTA) grand average ERP waveforms for correct and error trials evoked during a visual discrimination task with divided attention blocks (DA). Response-locked data used the button-press switch closure as the time-locking event. Modified from Falkenstein et al. (1991), Figure 2, reprinted with permission from Elsevier. (B) Response-locked grand average ERP waveforms for correct and error trials evoked during a flanker paradigm. The time-locking event was the onset of electromyogram (EMG) activity associated with a squeeze response. From Gehring et al. (1993), Figure 1, reprinted with permission of John Wiley & Sons, Inc.

to variability across different response devices in the time it takes for a button press to travel from a resting position to switch closure. Studies have found the scalp distribution of the ERN to be maximal at midline frontocentral scalp locations, most typically the 10–20 location FCz. The ERN occurs on error trials in a wide variety of speeded-response tasks (Falkenstein et al., 1995; Gehring et al., 1995) involving visual (Falkenstein et al., 1991; Gehring et al., 1993), auditory (Falkenstein et al., 1991), and tactile (Forster & Pavone, 2008) stimuli, and unimanual (Falkenstein et al., 1991; Gehring et al., 1993), bimanual (Murata & Katayama, 2005), foot (Forster & Pavone, 2008; Gehring & Fencsik, 2001; Holroyd et al., 1998), oculomotor (Endrass et al., 2005; Nieuwenhuis et al., 2001), and vocal (Masaki et al., 2001) responses. It may even be elicited by auditory, visual, and somatosensory error feedback stimuli (Holroyd & Coles, 2002; Miltner et al., 1997) and by losses in gambling tasks (Gehring & Willoughby, 2002).

The ERN has attracted a great deal of interest, both within the ERP research community and in cognitive neuroscience more generally. Much of this interest arose because of evidence that the ERN is generated in the anterior cingulate cortex (ACC) and because of the burgeoning interest in the role of the ACC in those *cognitive control* functions that enable the brain to adapt behavior to changing task demands and environmental circumstances (Botvinick et al., 2001; Ridderinkhof et al., 2004). Cognitive control functions include processes that detect when control is needed—as when performance breaks down—and processes that implement control through changes in attentional focus and other strategic adjustments. Because an error is a salient marker that performance has broken down, the ERN is generally thought to reflect a process involved in evaluating the need for, or in implementing, control. As we describe in this chapter, the quest to determine the precise nature of that process has spawned a rich and fascinating debate involving evidence that includes neuroimaging, neurological and psychiatric patient studies, animal neurophysiology, neuropharmacology, and computational modeling.

The ACC

Gehring et al. (1993) suggested that the ACC and the adjacent supplementary motor area (SMA) were likely candidates for the neural generator of the ERN. A short time later, Dehaene and colleagues (1994), using Brain Electromagnetic Source Analysis (BESA; Scherg, 1990), showed that an equivalent dipole within the ACC accounted well for the midline-frontal scalp distribution of the ERN (Figure 10.2). Several subsequent BESA modeling efforts have supported an ACC locus (e.g., Holroyd et al., 1998; Mathewson et al., 2005; van Veen & Carter, 2002).

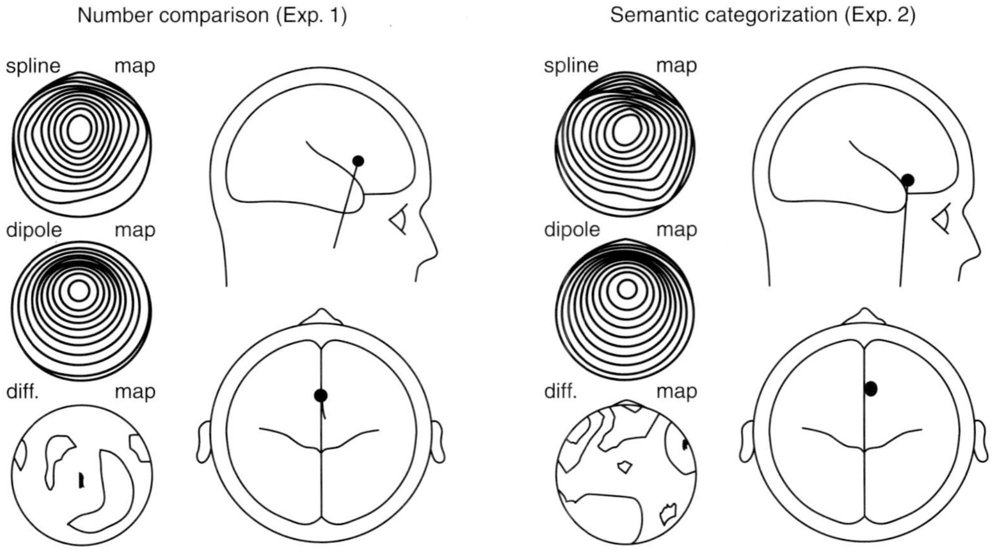

Fig. 10.2. Dehaene and colleagues (1994), using Brian Electromagnetic Source Analysis (BESA) to analyze data recorded with a dense electrode array, showed that an equivalent dipole placed roughly in the ACC accounted well for the midline-frontal scalp distribution of the ERN. From Dehaene et al. (1994), Figure 2, reprinted with permission of John Wiley & Sons, Inc.

Consistent with this ERP evidence, several functional magnetic resonance imaging (fMRI) studies showed error-related blood oxygen-level dependent (BOLD) signal increases in the ACC (Figure 10.3; Carter et al., 1998; Kiehl et al., 2000; Menon et al., 2001; Ullsperger & von Cramon, 2001; for reviews, see Hester et al., 2004; Ridderinkhof et al., 2004; Taylor et al., 2007). For example, Carter and colleagues showed that a region of the ACC responsive to conflict was also involved in processing errors in a modified version of the continuous performance task (see Figure 10.3A). In addition, one magnetoencephalographic study in humans has identified an ACC source (Miltner et al., 2003; but see Stemmer et al., 2004b). The ACC encompasses several subdivisions along its rostral-caudal extent, and the precise locus of the ERN generator within this region is usually argued to be in a dorsal region of the ACC (e.g., Debener et al., 2005; Garavan et al., 2003; Holroyd & Coles, 2002; Holroyd et al., 2004b; Ridderinkhof et al., 2004; Yeung et al., 2004b).

Neurophysiological studies in nonhuman primates support the existence of ACC activity related to errors. Early studies found ACC single-unit activity related to the absence of an earned reward (Niki & Watanabe, 1979), and a subsequent study using local field potentials (LFPs) found ACC activity during intermediate stages of learning in a task where monkeys had to learn the appropriate response to a visual cue (Gemba et al., 1986). A series of recent studies has used an oculomotor stop-signal task that is similar to the speeded tasks typically used to elicit the ERN. In the stop-signal task, an error is a trial where subjects are directed by a stop signal not to respond, yet they fail to withhold the response. Both single-unit (Ito et al., 2003) and LFP (Emeric et al., 2008) recordings in monkeys

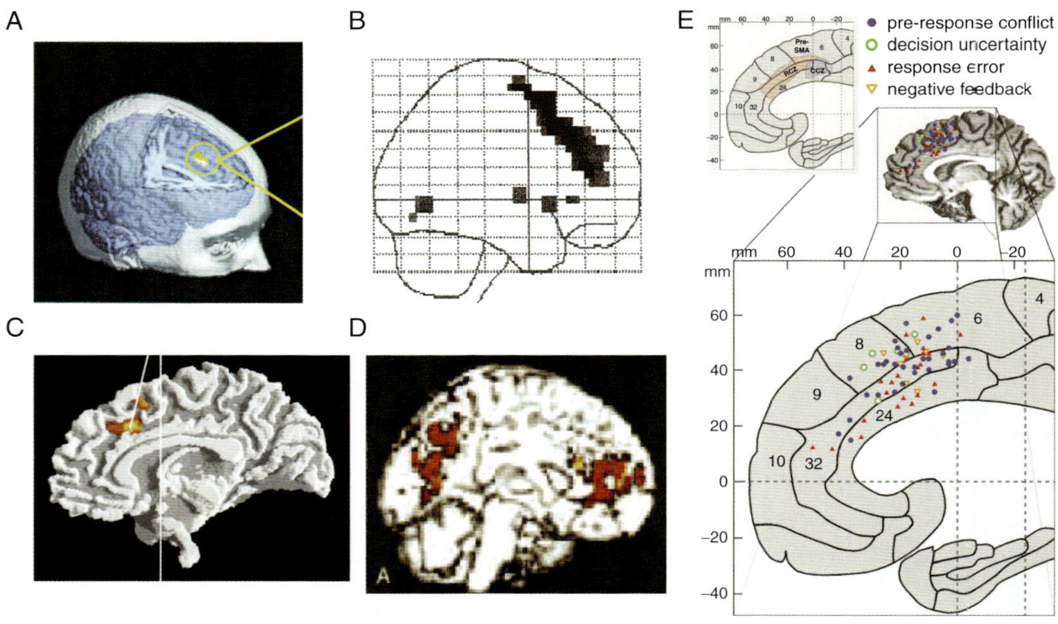

Fig. 10.3. Error-related fMRI activations. Studies showing the medial frontal loci of fMRI BOLD response activations associated with errors. (A) Carter and colleagues showed that a region of the ACC that responded to processing conflict also was involved in processing errors. The fMRI activations were elicited by a modified continuous performance task. From Carter et al. (1998), Figure 2, reprinted with permission from AAAS. (B) In a go/no-go task, errors of commission were associated with BOLD activation in the medial frontal cortex consisting of voxels in the caudal ACC, rostral ACC, and medial frontal gyrus. From Kiehl et al. (2000), Figure 1A, reprinted with permission of John Wiley & Sons, Inc. (C) Errors in a flanker task were associated with fMRI activations in the cingulate motor area (CMA). From Ullsperger and von Cramon (2001), Figure 2, reprinted with permission from Elsevier. (D) In a go/no-go task, errors of commission were associated with fMRI activations in the ACC and medial prefrontal cortex, as well as the insula and precuneus/posterior cingulate. From Menon et al. (2001), Figure 1A, reprinted with permission of John Wiley & Sons, Inc. (E) Ridderinkhof and colleagues (2004) presented a meta-analysis of fMRI studies finding activations in the medial frontal cortex related to performance monitoring. Error-related activations were found in various regions of the medial frontal cortex, including the rostral cingulate zone (RCZ), the caudal cingulate zone (CCZ), and the pre-SMA. From Ridderinkhof et al. (2004), Figure 1, reprinted with permission from AAAS.

have shown ACC activity related to errors (see Figure 10.4). Interestingly, analogs of the ERN may even be evident in rats when they make errors, in medial-frontal cortex locations homologous to the human ACC (Smith et al., 2009).

If the ERN is generated in the ACC, lesions there should reduce ERN amplitude. There are only a few studies of humans that have tested this prediction, in part because well-circumscribed ACC lesions are not common. Swick and Turken (2002) described an individual with a focal left-hemisphere lesion encompassing the rostral to middorsal ACC who failed to show an ERP difference between error and correct trials, which could be consistent with a reduction in the ERN. Nevertheless, the individual showed a large negativity on both error and correct trials, which could indicate that the structures necessary for generating the ERN were intact but receiving faulty input. Stemmer and colleagues (2004a) reported that three of five individuals with ACC damage showed no ERN, but that the other two showed an ERN in at least one of the two flanker

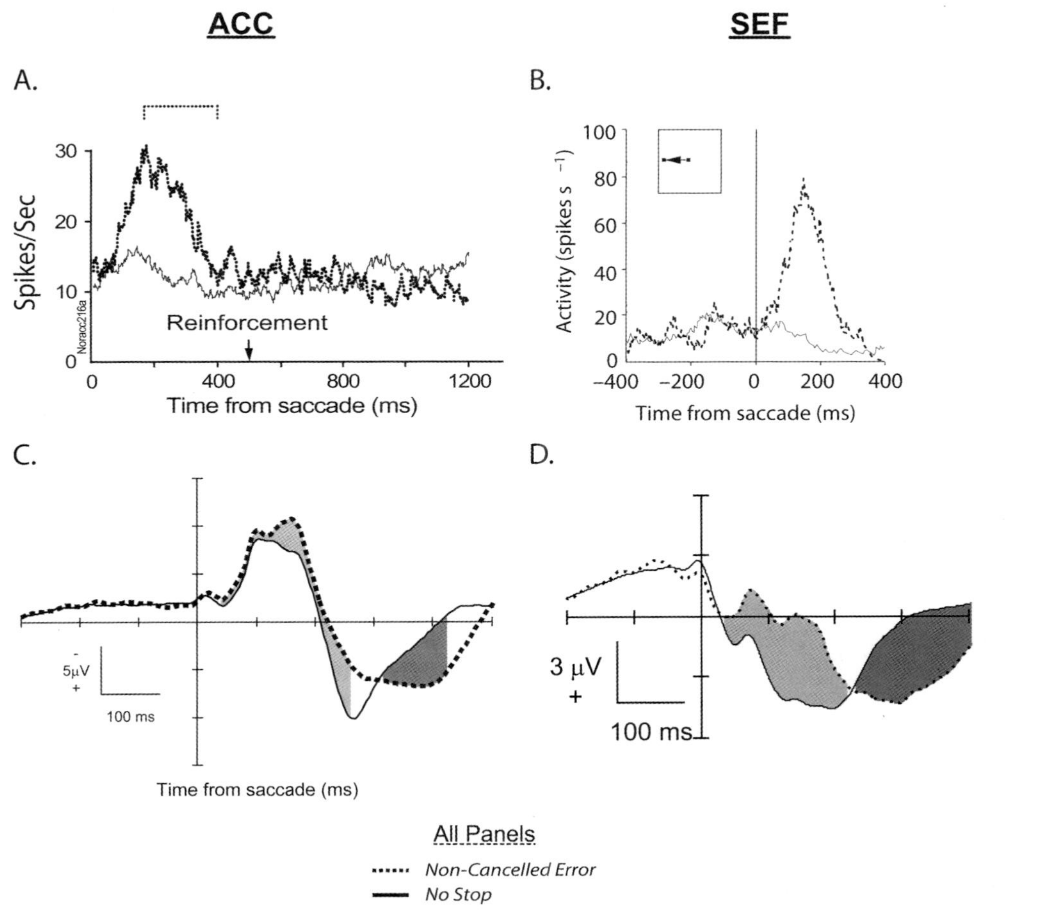

Fig. 10.4. Single-unit (top) and local field potentials (bottom) recordings in macaque monkeys comparing activity for correct no-stop signal trials and erroneous noncanceled trials. These figures demonstrate error activity that manifested earlier in the supplementary eye field (SEF) than in the anterior cingulate cortex (ACC). (A) The average spike rate in an ACC neuron peaked at about 200 ms after the initiation of an erroneous noncanceled saccade. The bracket reflects the range of saccade initiation times for errors. From Ito et al. (2003), reprinted with permission from AAAS. (B) The average spike rate in an SEF neuron peaked at about 180 ms after the initiation of an erroneous noncanceled saccade. From Stuphorn et al. (2000), reprinted with permission from Macmillan Publishers Ltd. Error activity peaks earlier in the SEF (panel B) than in the ACC (panel A). (C) Local field potentials averaged across multiple sites from the ACC revealed a greater negativity for errors that peaked at about 180 ms after saccade onset and a greater positivity for errors that peaked at about 400 ms after saccade onset. From Emeric et al. (2008), reprinted with permission from the American Physiological Society. (D) Local field potentials averaged across sites in the SEF revealed a greater negativity for errors that peaked at about 80 ms after saccade onset and a greater positivity for errors that peaked at about 300 ms after saccade onset. From Emeric et al. (2010), reprinted with permission from the authors. Both the early error negativity and the later error positivity peaked earlier in the SEF (panel D) than in the ACC (panel C).

task variants used in that study. Precise lesion loci were not reported, but Stemmer et al. suggested that the individuals with no ERN had damage in the rostral ACC and those with spared ERNs tended to have damage in the most ventral (subgenual) ACC (see also Stemmer et al., 2000). However, performance measures differed between patients and controls in these studies. Thus, although the studies report results consistent with an ACC source of the ERN, the inconsistency of lesion locations and effects on the ERN, as well as behavioral differences between groups, point to the need for further work.

Although much of the evidence discussed thus far is consistent with a dorsal ACC generator, a number of considerations suggest that areas outside of the dorsal ACC—such as the rostral ACC and the pre–supplementary motor area (pre-SMA)—should also be considered as candidate neural generators for the ERN. Some fMRI studies of errors (Kiehl et al., 2000; Menon et al., 2001) showed error-related ACC activity in the rostral anterior cingulate. Indeed, across the literature, there is substantial variability in fMRI findings: a meta-analysis of fMRI studies by Ridderinkhof and coworkers (2004) found peaks for error-related activity throughout the medial frontal cortex, including areas adjacent to the dorsal ACC, such as the rostral ACC and pre-SMA (see Figure 10.3). Another reason for questioning the dorsal ACC assumption is that it is still an open question whether the fMRI BOLD response and the ERN reflect the same neural activity. The most direct test of this assumption was reported by Debener et al. (2005), who used simultaneous recording of EEG and fMRI to show that a single-trial measure of the ERN predicted the ACC BOLD response on the same trial. Nevertheless, their single-trial measure of the ERN was sensitive to the positive deflection following the ERN in addition to the ERN itself (see the discussion of the early error positivity below). Also, Mathalon and colleagues (2003) measured the ERN and BOLD response in separate sessions and found the ERN to be correlated with a more rostral ACC region in addition to the dorsal ACC (whose activation extended into neighboring Brodmann's area 8). Even reports of dorsal ACC activity should not be taken at face value; Nachev and coworkers (2008) argue that many of the dorsal ACC fMRI activations reported in the literature are more accurately characterized as pre-SMA activations.

The use of BESA equivalent-dipole models that provides much of the support for a dorsal ACC source calls for additional caution. First, some evidence using BESA has pointed to other possible generators, particularly the SMA and a more rostral/ventral part of the ACC (Dehaene et al., 1994; Luu et al., 2003). Moreover, excessive reliance on BESA is itself a cause for concern, because of BESA's limited accuracy in localizing deep sources such as the ACC and because of its inability to distinguish a single deep source from a more widespread and superficial distribution of cortical activity. Even for the most experienced and careful BESA modelers, the number of sources reached by the technique can never deviate from the number of dipoles stipulated in advance by the modeler (see Luck, 2005). In addition, even with the freedom to select any number of dipoles, ERN investigators have reported models that are rather unimpressive in their ability to explain the observed data: in some cases, the unexplained variance in fitting the observed data to the data predicted by the model approaches or even exceeds 10%, making it plausible that alternative models would be more appropriate.

Other methods support the possibility that sources other than (or in addition to) the dorsal ACC contribute to the ERN, including the rostral ACC and the pre-SMA. Studies of nonhuman primates, for example, showed error-related single-unit activity (Stuphorn et al., 2000) and LFPs (Emeric et al., 2010) in the supplementary eye field (SEF) in addition to those in the ACC. The SEF may serve the same function for oculomotor movements that the pre-SMA or SMA does for manual movements (Schall & Boucher, 2007; Stuphorn et al., 2000). Similarly, in a study of humans, Herrmann and colleagues (2004) reported an analysis employing low-resolution electromagnetic tomography (LORETA) that showed a pre-SMA source for the ERN (and, interestingly, a dorsal ACC source for the Pe, which could have influenced the results of Debener et al., 2005, described above). Vidal and coworkers (2000) suggested that the enhancement of the ERN by a surface Laplacian analysis is more consistent with a superficial source such as the SMA than with a deeper source. In the study by Miltner et al. (2003), at least four of the six subjects showing a magnetic equivalent of the ERN showed sources more consistent with a rostral than with a dorsal ACC source. Brazdil and colleagues, using intracranial recordings in humans, identified multiple sources of ERN-like activity, including the rostral ACC and pre-SMA (Brazdil et al., 2002, 2005). One study using transcranial magnetic stimulation (TMS) found that medial frontal stimulation of the pre-SMA led to an attenuation of the ERN (Rollnik et al., 2004).

The orbitofrontal cortex (OFC) has emerged as another possible contributor to the ERN, although the evidence is both limited and mixed. Intracranial ERN-like activity has been observed in the OFC (Brazdil et al., 2002, 2005). Turken and Swick (2008) found that lesions in the OFC reduced the amplitude of the ERN. The lesions in that study extended to the rostral portion of the ACC that was associated with a reduced ERN in the study of Stemmer et al. (2004a) described earlier, so it is possible that the more rostral ACC damage contributed to the effect. Ullsperger and coworkers (2002) failed to find reduced ERNs in patients who had damage to the frontopolar cortex that included the OFC. Still, the plausibility of an OFC source is further supported by neurophysiological recordings in monkeys that show neurons with some similarity to those observed in the ACC (e.g., Thorpe et al., 1983).

Whatever the source of the ERN, connectivity between that structure and the lateral prefrontal cortex (PFC) appears to be critical for generating the ERN. Gehring and Knight (2000) reported that a group of individuals with focal lateral PFC lesions showed ERN activity equivalent to that of controls on error trials, but what appeared to be an ERN on correct trials (see discussion of the correct-response negativity below), resulting in no difference between error and correct trials. Other studies found reduced ERNs caused by lateral prefrontal damage (Ullsperger & von Cramon, 2006b; Ullsperger et al., 2002). Consistent with these findings, lesions from sickle-cell disease—thought to interrupt the communication between the lateral and medial frontal cortices—also reduced the ERN (Hogan et al., 2006; Figure 10.5). Other evidence of communication between the generator of the ERN and the lateral PFC comes from measures of phase coherence between medial and lateral frontal sites (Cavanagh et al., 2009) and a correlational functional connectivity analysis based on LORETA (Holmes & Pizzagalli, 2008).

It may be the case that the source of the ERN uses information supplied by the PFC to distinguish errors and correct responses (Gehring & Knight, 2000) or that the ERN is inhibited on correct trials. Alternatively, damage to the PFC may increase the response conflict on correct trials (Cohen et al., 2000). It is also possible that the ERN detects conflict or errors and signals the PFC to engage cognitive control processes (Kerns et al., 2004), but if the direction of communication runs from the medial to the lateral PFC, some kind of bidirectional communication might be necessary to explain how lateral PFC lesions would diminish the size of the ERN (see, e.g., Banich, 2009). Note, however, that not all evidence points to a role for the PFC in generating the ERN; the TMS study mentioned above found no effect of lateral PFC stimulation on the ERN (Rollnik et al., 2004).

More recent evidence suggests that the generator of the ERN may communicate with a number of regions in addition to the lateral PFC. Cohen (2011) correlated measures of theta-band activity likely to be sensitive to the ERN (see below) and white matter tract strength based on diffusion tensor imaging. The amount of error-related activity was related to the strength of white matter connections

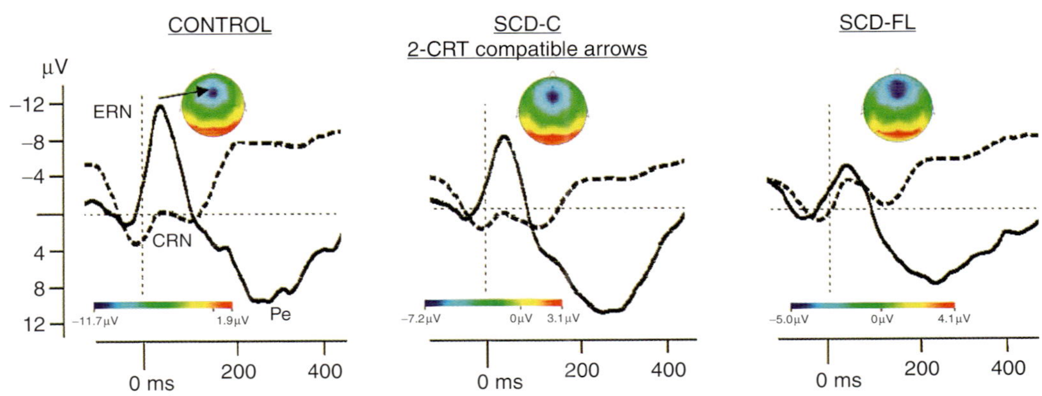

Fig. 10.5. Error-related negativity waveforms from healthy controls, control subjects with sickle-cell disease and an absence of brain lesions (SCD-C), and patients with frontal lesions due to sickle-cell disease (SCD-FL). The lesions in frontal white matter decreased the amplitude of the ERN, supporting a prefrontal contribution to the generation of the ERN. From Hogan et al. (2006), Figure 3, reprinted with permission from Oxford University Press.

linking the medial frontal cortex to the motor cortex, inferior frontal cortex and ventral striarum. Moreover, subjects showing a greater degree of theta-band phase synchrony between FCz and other scalp electrodes showed stronger connections linking the medial frontal cortex to the corpus callosum and to white matter tracts leading to the superior frontal gyrus.

Many investigators assume that the ERN arises from the dorsal ACC, yet the inconsistency of the evidence supporting this hypothesis, coupled with other evidence consistent with alternative sources, raises the intriguing possibility that one or more regions outside the dorsal ACC contribute to the ERN. The ERN could arise because of multiple sources, such as the SMA/pre-SMA, along with a deeper source such as the rostral ACC or the dorsal ACC (Falkenstein, 2004a). It is even possible that the ERN arises from a single region other than the dorsal ACC, such as the rostral ACC, the pre-SMA, or the SMA (Vidal et al., 2000). In any case, it seems prudent for ERN research to consider some of these alternatives rather than to proceed with the assumption that the dorsal ACC is the only region that generates the ERN.

Related Components

Before reviewing the major theories and models of the functional significance of the ERN, we turn to other ERP components that are relevant to these theories. Some are also elicited by errors; others may share a common neural generator.

ERROR POSITIVITY

In the response-locked error-trial waveform, a positivity known as the *error positivity* (Pe) usually follows the ERN (Falkenstein et al., 1990; for a comprehensive review, see Overbeek et al., 2005). The Pe often has a centroparietal topography with a maximum amplitude between 200 and 400 ms after an erroneous response (e.g., see Figures 10.1, 10.5, 10.6, 10.15b, 10.18, and 10.19). The Pe, unlike the ERN, is consistently larger for errors that the subject reports than for errors that go unreported (Endrass et al., 2005, 2007; Nieuwenhuis et al., 2001; O'Connell et al., 2007; Vidal et al., 2000) and is related to the increase in skin conductance response (SCR) following error responses relative to that following correct responses (Hajcak et al., 2003b). Falkenstein (2004a) and Overbeek et al. (2005) review evidence for three hypotheses of the functional significance of the Pe: that it reflects an affective response to the error, that it is involved in awareness of the error, and that it is involved in adapting response strategies following an error.

Nevertheless, definitive conclusions await a better understanding of the component structure of the post-error positivities, an understanding that is just beginning to emerge. There is wide variability in the reported Pe scalp distributions (Arbel & Donchin, 2009; Overbeek et al., 2005), and there is some disagreement about what constitutes a Pe. For example, several investigators suggested that the Pe may be a delayed parietal P300 (P3b) associated with detecting or evaluating an error (Davies et al., 2001; Leuthold & Sommer, 1999; Overbeek et al., 2005), an assertion supported by the similar response of the P300 and the centroparietal Pe to variations in the intertrial interval (Ridderinkhof et al., 2009). It is likely, however, that there are two components: an early fronto-central Pe and a later, more posterior component (Arbel & Donchin, 2009; Ruchsow et al., 2005b; van Veen & Carter, 2002). The anterior Pe may have a neural generator overlapping or near that of the ERN in the medial frontal cortex (Herrmann et al., 2004; van Veen & Carter, 2002). The early Pe and the ERN are also similar in that both are unrelated to error awareness in an antisaccade task (Endrass et al., 2007). As we discuss below, these similarities have led to the suggestion that the ERN and early the Pe are both parts of a single oscillatory potential. As for the later, more posterior positivity, it could be generated in the rostral ACC and reflect error awareness (Endrass et al., 2007) or a subjective affective response (van Veen & Carter, 2002). It could also be a P300 (P3b) associated with the error (Arbel & Donchin, 2009).

N200/N450

Larger stimulus-locked N200s appear on incongruent trials than on congruent trials in conflict tasks like the Eriksen flanker task (Gehring et al., 1992; Kopp et al., 1996) and in other conflict conditions (Nieuwenhuis et al., 2003). In conflict tasks using verbal stimuli, such as the Stroop task, an N450 component on incongruent trials may be a delayed instance of the N200 (Liotti et al., 2000; West, 2003). As we discuss later, the conflict monitoring theory claims that the congruence effect on the N200/N450 and the ERN reflect the same component. It is plausible that the N200 and the ERN might be related, because the scalp distribution and time course of the ERN bear some similarity to those of the classic N200 component (Simson et al., 1976; Squires et al., 1976; for reviews, see Folstein & Van Petten, 2008; Pritchard et al., 1991). Studies relating the ERN to reinforcement learning show some additional parallels: the N200-like component elicited by error feedback (see below) and the appearance of an N200 in response to stimuli that violate expectancies acquired during sequence learning (Eimer et al., 1996; Ferdinand et al., 2008; Russeler et al., 2003).

CORRECT-RESPONSE NEGATIVITY

Although the ERN is usually much larger on error trials than on correct trials, a negativity often appears on correct trials at the same latency in the response-locked waveform as the ERN (Figure 10.6; Ford, 1999; Gehring & Knight, 2000; Luu et al., 2000b; Scheffers & Coles, 2000; Vidal et al., 2000). The correct-response negativity (CRN) is usually smaller than the ERN, but the two components show a similar scalp distribution (Vidal et al., 2000). In some published waveforms the CRN is quite striking, especially in individual subject data (e.g., Swick & Turken, 2002). Observations of a CRN have raised the question of just how specific the ERN is to errors (Vidal et al., 2000). We will discuss the CRN in more detail later in the chapter.

FEEDBACK-RELATED NEGATIVITY

Miltner and colleagues (1997) reported that error-feedback stimuli (such as tones informing the subject that an error has occurred) elicit activity that resembles the ERN (Figure 10.7; Badgaiyan & Posner, 1998; Gehring & Willoughby, 2002; Holroyd & Coles, 2002; for a review, see Nieuwenhuis et al., 2004a). Whereas the classic ERN is time-locked to the error response, the feedback-related negativity (FRN) occurs approximately 250–300 ms following a feedback stimulus. In the Miltner et al. task, for example, subjects made a time estimation judgment using a button-press response, and the task was difficult enough that subjects did not know the accuracy of their judgment until the feedback stimulus occurred 600 ms later. In gambling tasks, subjects make a choice and later receive random gain or loss feedback (e.g., Gehring & Willoughby, 2002). In some ways, the FRN seems similar to the ERN: it is a negative-going component with roughly the same time course and a frontocentral scalp distribution. The FRN can be modeled as one or two equivalent dipoles in roughly the same location observed in studies of the classic ERN (e.g., Gehring & Willoughby, 2002; Miltner et al., 1997). If the FRN represents the same component as the ERN, the FRN would be evidence that the ERN is a general-purpose system for error detection (Miltner et al., 1997). We discuss this component in more detail in the section on the reinforcement-learning theory of the ERN.

THETA OSCILLATIONS

In numerous reports of the ERN, one gets the visual impression of multiple negative peaks, with a small

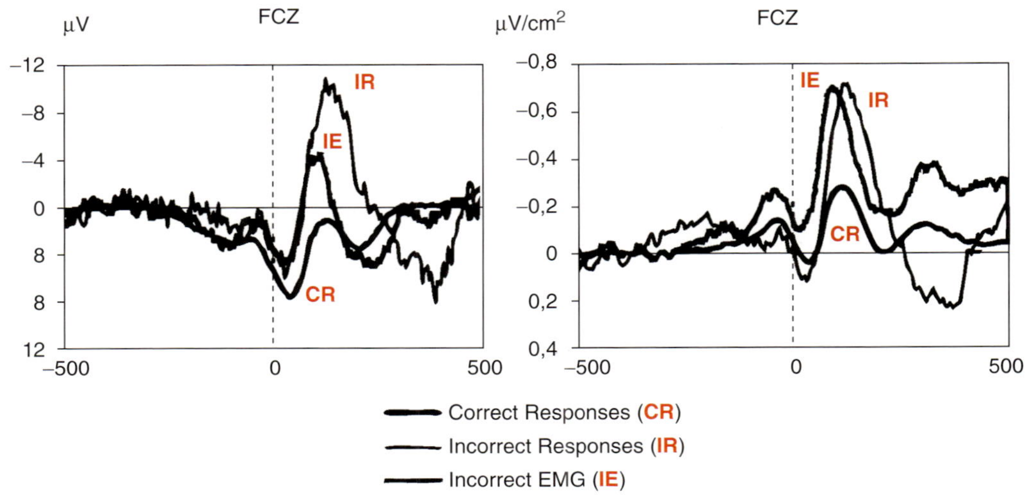

Fig. 10.6. The CRN. Grand average EMG-locked monopolar activity (left) and Laplacian transformed activity (right) elicited during a go/no-go task. The authors compared activity for correct responses, incorrect responses, and partial errors (where incorrect EMG activity was observed but no incorrect response was made). The authors observed ERN-like activity on all three trial types; the ERN was greater for partial errors than for correct responses and greatest for incorrect responses. Modified from Vidal et al. (2000), Figure 2, reprinted with permission from Elsevier.

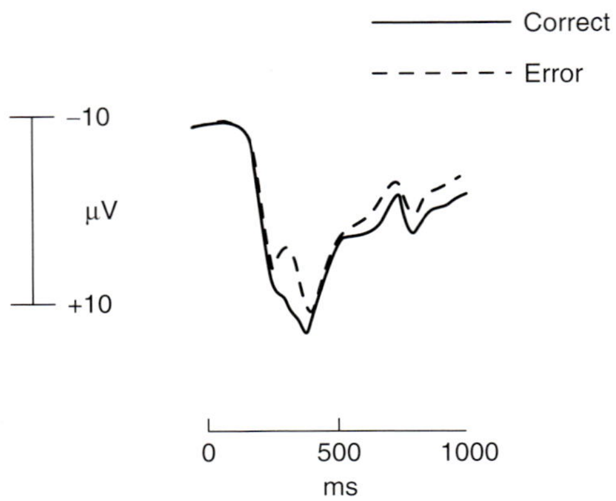

Fig. 10.7. Grand average ERP waveforms evoked by visually presented correct and error feedback stimuli in a time estimation task. This report suggested that the negative-going component elicited by negative feedback is the same as the classic response-locked ERN. Adapted from Miltner et al. (1997), Figure 3, reprinted with permission from MIT Press.

negativity often following the ERN (e.g., Figures 10.1b, 10.6). This peak does not always follow the ERN in average waveforms, and indeed, it may be a correction-related negativity associated with the corrections following an error response, as Fiehler and colleagues (2005) claim. However, Luu and Tucker (2001) proposed an alternative possibility: that the ERN is just one peak in a theta-frequency (4–7 Hz) oscillation. Studies that have applied time-frequency analysis to the ERN have confirmed that the ERN does indeed appear within that frequency range, but thus far, such techniques cannot reveal definitively whether the ERN actually consists of single or multiple peaks (Bernat et al., 2005; Cavanagh et al., 2009; Gehring & Willoughby, 2004; Hall et al., 2007; Trujillo & Allen, 2007; see Figure 10.8). Other studies have identified a delta (1.5–3.5 Hz) contribution to the ERN in addition to the theta activity (Yordanova et al., 2004). These investigators argue that the delta contribution reflects error-specific processing, whereas the theta activity is more generally related to response monitoring because of its presence in both the CRN and the ERN (Yordanova et al., 2004). This argument suggests that the separate time-frequency analysis of both errors and correct trials will be necessary for a complete picture of the time-frequency content of the ERN, because not all time-frequency studies have included correct-trial data.

The notion that the ERN is part of a theta oscillation suggests an alternative interpretation of the early, anterior Pe discussed earlier (e.g., van Veen & Carter, 2002): the ERN and the early, anterior Pe could both be part of a single oscillatory potential. The biphasic nature of intracranial field potentials recorded in monkeys would be consistent with the sources of the ERN and early Pe being close to each other (Emeric et al., 2008). Arbel and Donchin (2009) suggested that the early Pe was not part of the same oscillation that causes the ERN, however, because they found that the ERN and the frontocentral Pe responded differently to speed-accuracy instructions. Still, it is plausible that experimental variables could influence an oscillatory generator differently at different points in the oscillation if

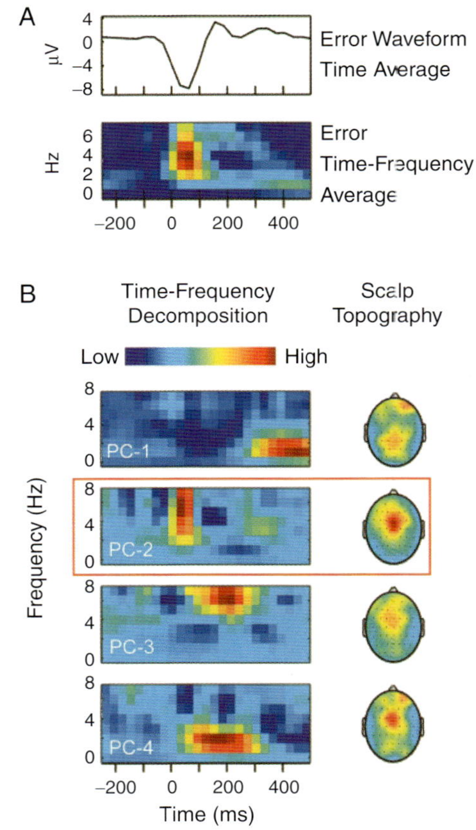

Fig. 10.8. Grand average error waveform (top) elicited by a flanker task and the average time-frequency data showing a peak in the theta range. Time-frequency surfaces show components as colored blobs in the time-frequency plane. Multiple components can occur and overlap in the time-frequency plane much as they do in the time-domain waveforms; thus, such surfaces require techniques to separate one component from another. Here, the authors used principal components analysis to extract the ERN (component 2) from the other overlapping activity. Modified from Hall et al. (2007), Figure 3, reprinted with permission of John Wiley & Sons, Inc.

time-varying information flows continuously to the generator.

A related issue is whether the ERN represents theta activity that is more phase-aligned on error trials than on correct trials. Phase alignment on error trials alone could produce an ERN in the average error waveform merely because the single-trial peak happens to occur more consistently at the same latency on error trials, even if there is no actual change in amplitude. Luu et al. (2004) supported this idea through an analysis of single-trial data filtered for theta-frequency activity (see also Trujillo & Allen, 2007). Yeung and colleagues (2004a, 2007a), however, suggested that such analyses must be viewed with caution. For example, digital filters can cause ringing artifacts that appear as oscillations, and an increase in a single discrete ERN-like event on error trials could cause data to appear as though phase alignment was greater on those trials than on correct trials. (Illusory oscillations could also result from Morlet wavelets, such as those used by Trujillo and Allen [2007].) Nevertheless, the Yeung et al. studies do not rule out either the theta-oscillation or phase-alignment hypotheses, and ringing artifacts may not be a satisfactory explanation for the oscillatory appearance of the ERN (Cavanagh et al., 2009; Trujillo & Allen, 2007). Some evidence indicates that phase alignment and amplitude enhancement both play a role (Cavanagh et al., 2009). Finally, further complicating the picture, a study of the FRN showed evidence for phase-amplitude coupling across frequency bands, raising the intriguing possibility that the computations underlying the ERN and FRN may be more accurately characterized by examining the interactions among activities in multiple frequency bands (Cohen et al., 2009).

Flies in the ERN Ointment

The previous section has already introduced some of the thorny issues in the effort to understand the ERN, and this is as good a place as any to make those issues explicit. First, the range of electrophysiological phenomena to be covered by a theory of the ERN (or by a chapter about the ERN) is changing and is a matter of active debate: Does the ERN include feedback-related components? Does the ERN include the ERN-like peak that happens on correct trials? Does the ERN comprise not only the negative-going peak following the error, but also the sharp fronto-central positivity that sometimes accompanies the ERN? Is the ERN really a multiple-peak oscillation?

Second, as will become clear below, studying a phenomenon related to errors presents unique difficulties: unlike an experiment that can manipulate a critical independent variable—whether a stimulus is attended or unattended, for example—a study of errors cannot specify in advance whether a particular response will be correct. Thus, the analysis of errors must always be to some degree correlational and post hoc. This presents practical problems (designing an experiment in which subjects will produce enough errors for analysis) as well as more conceptual ones (errors have multiple possible causes, and the experimenter cannot know with certainty why a particular error occurred).

Third, the important distinction for the purposes of theory is not what is correct or an error in the eyes of the experimenter, but rather what is deemed correct or an error by the brain of the subject. These are not identical, and some confusion in the literature arises from the assumption that they are. The CRN is an example; it is possible that the CRN occurs because the brain labels a response as an error that is not an error according to the experimenter (Coles et al., 2001). Another example is found in the conflict literature (see below), where it is sometimes assumed that as long as no overt error (like a button press) occurred, there was no error for the brain to detect (see, e.g., Carter et al., 1998), overlooking the fact that the brain might detect errors at a level of response activation lower than that required to produce an overt error (see Gratton et al., 1988; Murthy et al., 2007).

And finally, what is an error at one level of analysis might be correct at another; thus, the brain's response to an error might depend on consequences at some level of analysis other than the one specified by the instructions to the subject. The subject's negative affective reaction to an error might not be determined by the small monetary penalty designated by the experimenter, but instead by the fear of the experimenter's disapproval or the desire to perform better than other subjects or in accord with the experimenter's exhortations. Hence, if changing the penalty from 10 cents to 25 cents fails to affect a component, the reason might be that the component is sensitive to a different, perhaps social, incentive—not that the component is insensitive to incentives.

Functional Significance of the ERN—Major Theories

A spirited debate has emerged regarding the computation represented by the ERN. In this section, we introduce several influential theories of ERN and describe some of the data supporting them.

Error Detection/Comparator Theory

For the first few years of ERN research, the dominant view of the ERN was that it reflected a computation involved in error detection. Specifically, this error-detection theory posited that the ERN reflects a process that compares the output of the motor system (as represented by an efference copy of the movement command) to the best estimate of the correct response at the time of ERN occurrence (Falkenstein et al., 1991; Gehring et al., 1993). The short latency of the ERN and its close temporal association with the response suggest that the ERN is triggered when a comparator receives the efference copy corresponding to the response that is being executed at that moment (Coles et al., 2001). The notion that the ACC might be a comparator was first proposed by the motor physiologist Vernon Brooks (Brooks, 1986). The schematic diagram (Figure 10.9) published by Coles et al. (2001) illustrates this idea: in speeded-response tasks, an error usually occurs because the subject responds before stimulus evaluation is complete. As the response is executed, stimulus processing continues. A comparison process computes the difference between the representation of the correct response (derived from continuing stimulus processing) and a representation of the current, ongoing response (the efference copy). A discrepancy between these two representations gives rise to a mismatch or error signal.

There are several possible ways the ERN could originate from such a system. The ERN could reflect the output of the comparison process (Coles et al., 2001; Falkenstein et al., 1990) or the comparison process itself (Falkenstein et al., 2000; Vidal et al., 2000). Still another possibility is that the ERN reflects a process that uses information in the error signal to prevent or correct the error or to make some kind of strategic adjustment (Gehring et al., 1993; Holroyd & Coles, 2002). Another variant of the error-detection view emphasizes stimulus representations: the premature response is associated with an anticipated stimulus, and it is the mismatch between the anticipated and actual stimuli that causes the ERN (Bernstein et al., 1995; Schmidt & Gordon, 1977).

Conflict-Monitoring Theory

The conflict-monitoring theory of the ACC was originally proposed as an alternative to the error-detection theory (Carter et al., 1998). Proponents of the conflict-monitoring theory argued that the error-detection model was computationally implausible (Botvinick et al., 2001; Carter et al., 1998; Yeung et al., 2004b). In their view, a comparator would have to have information that specified which of the representations being compared was the correct one. For the comparator to have that information, it would have to be able to access information outside of the series of processes responsible for task performance. In particular, the system generating the ERN would have to know the intended (correct) action (Carter et al., 1998). And if the brain

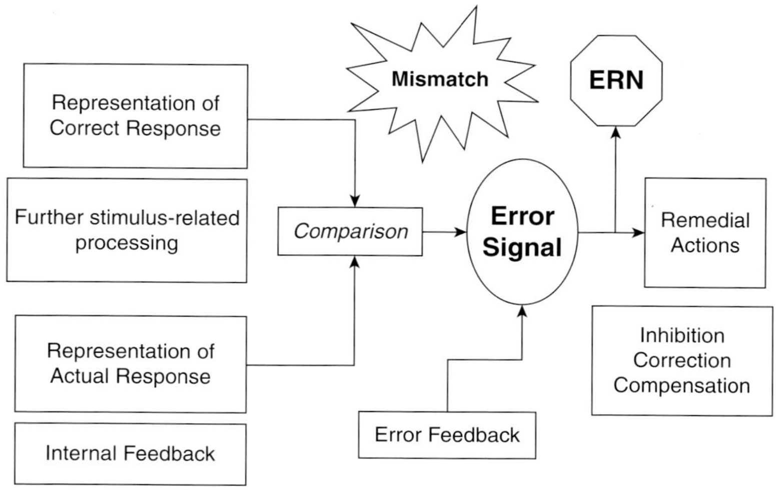

Fig. 10.9. Schematic diagram illustrating the error-detection theory of the ERN, which posits a comparison process that computes the difference between a representation of the correct response (derived from stimulus processing) and a representation of the current, ongoing response (the efference copy). A discrepancy between these two representations gives rise to a mismatch or error signal, which underlies the ERN. From Coles et al. (2001), Figure 1, reprinted with permission from Elsevier.

knows which response is correct, the argument goes, then why isn't the brain executing the correct response?

According to the conflict model, *response conflict*—defined as concurrent activation of multiple competing responses—can account for the ERN without postulating an all-knowing homunculus. In typical choice tasks, the activation of more than one response might signal that something is awry. In this way, response conflict can track performance accuracy without the system "knowing" which response is correct. Conflict signals the need for increased control: following high-conflict trials, the task set is strengthened, leading to improved performance on postconflict trials. Computational implementations of this idea (see Figure 10.10) model response selection using parallel distributed processing networks based on the one originally proposed by Cohen and colleagues (1990). These models define conflict as the Hopfield energy of the response units—in a simple two-choice task, twice the product of the activation of two response units, weighted by the strength of the inhibitory connections among responses.

The conflict model has been used to model performance in a wide variety of tasks (Botvinick et al., 2001), and Yeung et al. (2004b) showed that the simulated conflict signal mimicked the time course of the ERN and its sensitivity to a variety of experimental manipulations, including flanker

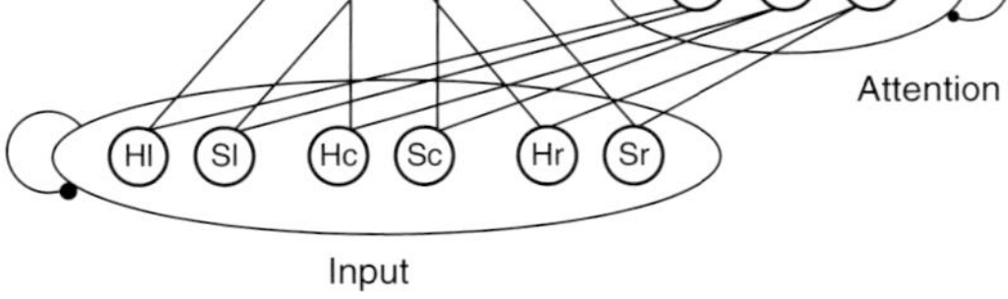

Fig. 10.10. Conflict-monitoring model for the Eriksen flanker task. Stimulus processing units corresponding to the target and flanking letters *H* and *S* activate the corresponding responses. Conflict is generated in the response layer by the coactivation of mutually inhibitory response channels (e.g. "*H*" and "*S*"). The conflict-monitoring process detects the presence of response conflict and signals adjustments in attention that will reduce conflict from the flanking letters on subsequent trials. From Botvinick et al. (2001), Figure 7, reprinted with permission from the American Psychological Association.

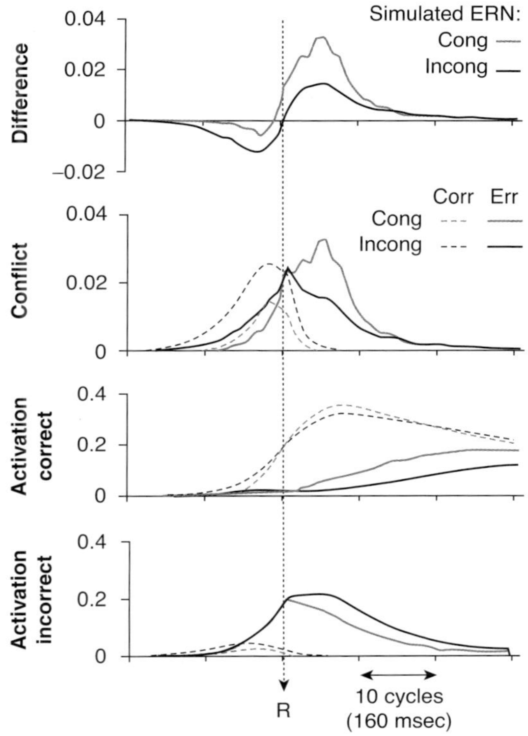

Fig. 10.11. Conflict-monitoring model simulation of response-locked ERN activity for congruent and incongruent trials in a flanker task, shown for correct and error responses. Conflict is the product of activation in the correct channel and the incorrect channel. On congruent error trials, the correct channel is highly activated following the erroneous response, leading to a large simulated ERN. In contrast, conflict is greatest prior to the response on correct incongruent trials, because incorrect response activation from the flanking letters subsides before the correct response. From Yeung et al. (2004b), Figure 4, reprinted with permission from the American Psychological Association.

congruency (see Figure 10.11). For example, the enhanced ERN observed in cases where accuracy is emphasized over speed (see the discussion below and Figure 10.1) can be produced in the conflict model by focusing attention more strongly on the central letter and adopting a more conservative response criterion (Yeung et al., 2004b).

Reinforcement Learning Theory of the ERN

Holroyd and Coles proposed the reinforcement-learning theory of the ERN (abbreviated RL-ERN; Figure 10.12), which is also instantiated as a computational model (Holroyd & Coles, 2002; Holroyd et al., 2005). According to this theory, a monitoring mechanism in the basal ganglia produces an error signal when events occur that are worse than expected, where the expectation has developed

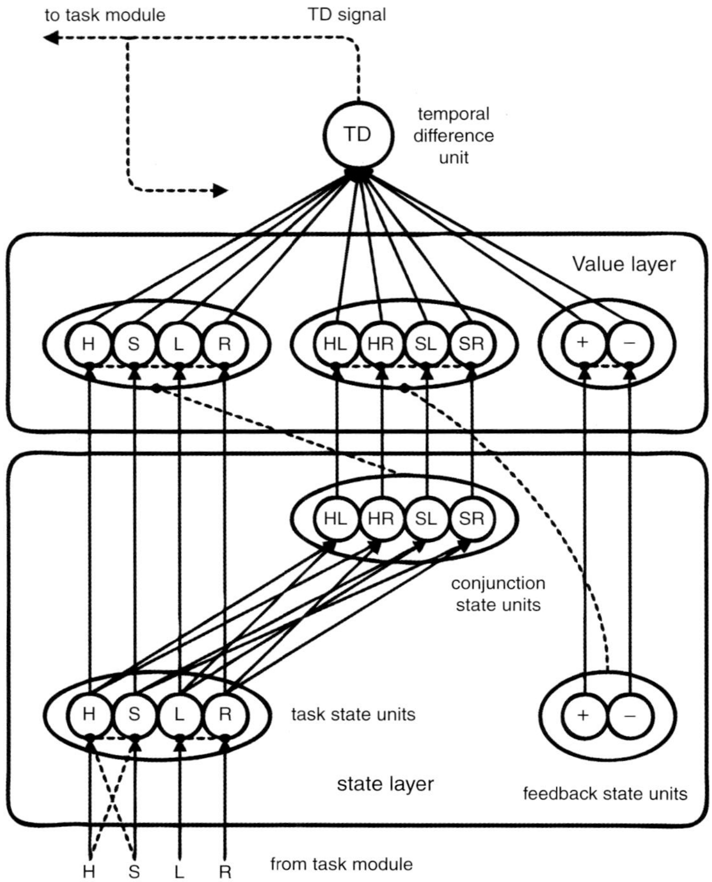

Fig. 10.12. Diagram of the RL-ERN model. A task module (not shown) produces overt behavior in response to external stimulus input, simulating choice reaction time performance. The monitor module (shown) evaluates the output of the task module according to the current context, with additional input from external feedback, and assigns values (good or bad) to the current behavior. In the case of a bad outcome, the monitor module sends an error signal to the task module in order to improve performance. From Holroyd et al. (2005), Figure 8, reprinted with permission from the American Psychological Association.

according to the history of prior reinforcements associated with a response. The midbrain dopamine system conveys this error signal to the ACC, where the signal is used to improve task performance by changing the manner in which control over the motor system is allocated to various competing systems in the brain. The theory has roots both in artificial intelligence models of reinforcement learning (Sutton & Barto, 1988) and in the literature relating dopamine to reinforcement learning (Schultz, 2002). An important prediction of the model is that the monitoring mechanism responds to the earliest information that something has gone wrong. Hence, before a task is learned and early in learning, errors occur because the system does not yet represent the contingencies between stimuli, responses, and reward; thus, the system determines whether performance is good or bad via external feedback. After learning establishes associations between reward values and stimulus–response conjunctions, errors can be detected immediately when a response occurs, without the need to wait for external feedback (Holroyd & Coles, 2002).

A recent extension to the RL-ERN framework is the prediction of response outcome (PRO) theory of Alexander and Brown (2010). The PRO theory says that the medial prefrontal cortex predicts the outcomes of an action based on past experience, and it compares the predicted response outcomes to the outcomes that actually occur. A major difference from the RL-ERN model is that the prediction of response outcomes in the PRO theory does not distinguish between good and bad outcomes. In fact, undesirable outcomes (errors) can be the most likely outcomes, in which case the ERN will occur on correct trials, because it is elicited by the unexpected outcome (see also Oliveira et al., 2007). An additional difference from the RL-ERN model is that the outcome prediction implemented in the PRO theory is not dependent on a dopaminergic error signal.

Affect/Motivation

As most experimenters know, subjects often become aware of their errors, and the awareness often manifests itself in language unfit for polite company. Some investigators have hypothesized that the ERN represents just such an affective response to errors. Luu and colleagues (2003), for example, suggested that it reflects distress associated with the violation of expectancy caused by the error (see also Tucker et al., 1999). They argued that the ERN reflects part of a limbic circuit for action regulation, a function that, in their view, encompasses both online control of action and learning from action outcomes. Activity in the circuit is coordinated by theta-frequency (4–7 Hz) oscillations, and the ERN reflects one portion of this theta activity (Luu et al., 2004). Luu and Pederson (2004) suggested that this view is not necessarily inconsistent with the conflict-monitoring or error-detection accounts, as the detection of errors, conflict, or a loss of reward could cause an affective response (see also Yeung, 2004). In contrast to the RL-ERN and conflict theories, the affective/motivational theory has not been formalized as a quantitative model.

Core Empirical Phenomena

Although new theories of the ERN (e.g., Jocham & Ullsperger, 2009) and of the ACC (e.g., Anderson et al., 2008; Brown & Braver, 2005) continue to emerge, the four perspectives outlined above have dominated the literature. This section is concerned with the empirical evidence relevant to these theories. First, we review some of the major empirical findings that are not closely aligned to any particular theory of the ERN.

Speed/Accuracy Emphasis (or Error Probability)

The speed–accuracy trade-off describes the fact that subjects can respond quickly, making many errors, or slowly, avoiding errors (Pachella, 1974). Studies that have instructed subjects to perform under varying levels of speed-versus-accuracy emphasis have tended to show that speed emphasis decreases the amplitude of the ERN relative to accuracy emphasis (Falkenstein et al., 1990, 1995; Gehring et al., 1993; see also Ganushchak & Schiller, 2006; Hajcak et al., 2003b; Ullsperger & Szymanowski, 2004). The Gehring et al. study compared error responses of exactly the same latency, establishing that the slowing of responses in the accuracy condition was not responsible for the effect. However, by definition, a speed/accuracy manipulation is confounded with error probability, so it is possible that such findings could be explained by the principle that unexpected outcomes elicit more activity than expected ones (Alexander & Brown, 2010; Oliveira et al., 2007).

Individuals can differ in their speed and accuracy, but studies relating the ERN to individual differences in performance are less consistent in their findings. Some studies have reported that high-accuracy subjects showed larger ERNs (e.g., Hajcak et al., 2003b; Pieters et al., 2007). Others failed to find such a relationship (Falkenstein et al., 2000; Mathewson et al., 2005). A negative finding was also reported in one attempt to relate the ERN to within-subject, block-by-block fluctuations in accuracy (Ullsperger & Szymanowski, 2004). Falkenstein and colleagues (2000) have suggested that it is time pressure per se, rather than error rate, that accounts for the findings of studies that manipulate speed/accuracy emphasis, because subjects in their study who differed in error rate showed equivalent ERNs (Falkenstein et al., 2000; see also Ullsperger & Szymanowski, 2004). Note, however, that analyses relating the ERN to differences in performance across subjects or across task blocks may not be conclusive tests of the hypothesis that speed/accuracy emphasis modulates ERN amplitude. Speed and accuracy are determined by many factors other than strategic trade-offs, including fatigue and endogenous lapses of attention (e.g., Weissman et al., 2006).

Each of the major theoretical approaches can accommodate the effects of speed/accuracy instructions on the ERN. Yeung et al. (2004b), using the conflict-monitoring framework, showed that varying response thresholds and attentional focus affected simulated response conflict in a manner consistent with empirical observations of speed/accuracy effects on the ERN. Error detection accounts of the ERN can also explain these observations. Gehring et al. (1993) argued that instructions emphasizing accuracy over speed enhanced the error-monitoring process underlying the ERN. Similarly, Falkenstein et al. (2000) suggested that accuracy emphasis strengthens the representation of the correct response, yielding a stronger mismatch signal on error trials. The enhanced ERN related to the unexpectedness of low-probability errors would be consistent with the RL-ERN and PRO accounts. Finally, emotional accounts of the ERN could argue that emphasizing accuracy affects the ERN by making subjects experience errors as more aversive.

Error Detection and Correction

A series of studies conducted by Rabbitt and Laming beginning in the 1960s showed that subjects attempt to prevent error commission and, failing that, try to correct their errors and avoid subsequent mistakes (Laming, 1968; Rabbitt, 1966, 1967, 1968, 1981; Rabbitt et al., 1978). The theories outlined above predict a relationship between the ERN and compensatory behavior. Error-detection accounts of the ERN suggested that corrective behavior would be more likely or greater in magnitude when the ERN accompanying an error is larger (Falkenstein et al., 1995; Gehring et al., 1993). The RL-ERN model says that the ERN represents a process that reallocates control among various motor controllers, suggesting that the size of the ERN is related to changes in response strategy (Holroyd & Coles, 2002). The conflict-monitoring model makes similar predictions, because the conflict-monitoring system can drive changes in response strategy and attentional focus (Botvinick et al., 2001).

ERROR CORRECTION

When an error occurs, subjects often make the correct response soon afterward (e.g., Rabbitt, 1966). Findings concerning the relationship between the ERN and such immediate error corrections are inconsistent. Some studies have found that corrected errors are accompanied by larger ERNs than uncorrected errors (Falkenstein et al., 1995, 1996; Gehring et al., 1993; Rodríguez-Fornells et al., 2002), but others have failed to find such a difference (Fiehler et al., 2005). As for ERN latency, several studies have reported later ERNs when errors are not corrected than when they are corrected, which could mean that the ERN must occur quickly if it is to assist in error correction (Falkenstein et al., 1996; Fiehler et al., 2005; Hoffmann & Falkenstein, 2010). However, Rodríguez-Fornells et al. (2002) failed to find a latency difference. One might also predict that the error trials with the largest ERNs would also be corrected most quickly; this prediction has been confirmed (Gentsch et al., 2009; Rodríguez-Fornells et al., 2002) as well as disconfirmed (Fiehler et al., 2005). Group differences are also inconsistent with a link between the ERN and immediate error corrections. Whereas older adults typically show reduced ERNs relative to younger controls (e.g., Falkenstein et al., 2001b; Gehring & Knight, 2000; Mathewson et al., 2005; Nieuwenhuis et al., 2002), they can show this difference yet still correct their errors equally often (Falkenstein et al., 2001b; Gehring & Knight, 2000).

Most studies record button-press responses, limiting the analysis to the latency of the switch closure and its accuracy. It is more useful when other movement parameters, such as force or velocity, are also recorded. When response force data are available, it is usually observed that errors are less forceful than correct responses, suggesting that subjects may be inhibiting the error as it is being executed (Carbonnell & Falkenstein, 2006; Gehring et al., 1993; Rabbitt et al., 1978). The existence of partial errors, where there is electromyographic (EMG) activity that does not result in a button press, also supports this idea (Burle et al., 2008; Coles et al., 1985). The ERN could reflect activity involved in inhibiting the error in two ways: First, the error signal could give rise to the inhibition (in which case a larger error signal presumably would call for more inhibition); second, the activity itself could have an inhibitory effect. Here again the evidence is mixed. Gehring et al. (1993), using a single-trial measure of ERN amplitude, showed that larger ERNs tended to be associated with less forceful responses. In a go/no-go task, however, Scheffers et al. (1996) found no such relationship. Carbonnell and Falkenstein (2006) did not find that ERN amplitude differed between partial and full errors, but they found that ERN latency was shorter for partial errors than for full errors (see also Vidal et al., 2000; Figure 10.6). Endrass and colleagues (2008) also found that partial errors were accompanied by ERNs that were earlier, but in their study those ERNs were smaller than those associated with full responses. These error-inhibition analyses suffer from some ambiguities: responses might be less forceful or incomplete because of mutual inhibition between the error and the correct response (Ohtsuki, 1981) rather than an inhibitory process undertaken to stop the error. Thus far, the literature has failed to distinguish between these alternatives. Nevertheless, the latency analyses are especially intriguing, because they could indicate that an ERN—if it represents an attempt to stop the error response—has to be early to be effective. If the process intervenes early enough, the response can be stopped, but if it is too late, the response will proceed to completion.

The ERN occurs after lateralized readiness potential (LRP; see Chapter 9, this volume) activity associated with activating the error response, and in some reports following the onset of LRP activity associated with the error correction (Burle et al., 2008; Rodríguez-Fornells et al., 2002). Some have suggested that this latency makes the ERN too late to reflect activity that inhibits or corrects the error (Rodríguez-Fornells et al., 2002). Burle et al. (2008) conducted perhaps the most detailed study of the timing of the ERN relative to measures of EMG, LRP, and behavior (see Figure 10.13). In their study, the onset of the ERN did not vary in latency when the error corrections were fast compared with when they were slow. However, the duration of the ERN was longer, and the peak was higher (and later) when corrections were late. The ERN itself appeared in the interval between the error and the error correction. As Burle et al. suggest, the results seem as if the ERN was interrupted by the error correction: the ERN began at a set time following the error but was then interrupted by the beginning of the error correction. The authors hypothesized that the ERN represents an alarm signal that lasts until remediation of the error begins.

Nevertheless, it is not hard to come up with alternative hypotheses that are also consistent with the time course of the ERN reported by Burle et al. (2008). For example, the ERN could represent a process necessary for error correction. Late error corrections could be late because they required more of this processing (that is, a greater-amplitude and longer-duration process). In order to execute a fast error correction, it may be necessary to suppress a still-active erroneous motor program to permit the correct response to occur. One computational model includes such a process: in the executive-process interactive control (EPIC) cognitive architecture (Meyer & Kieras, 1997), a motor memory buffer retains recently programmed movement features. Those features remain until a different movement is required. At that point, the movement features must be deleted from the buffer so that the new movement can be programmed. Random variability in the activation of the movement features to be deleted would be enough to produce variability in the time needed to delete them, but there might be other reasons as well. Could the ERN reflect some process that is needed to enable the error correction by getting the just-executed motor program out of the way? It would be difficult to disprove such a hypothesis using ERN latency measures without showing trials where the error correction was completed—and not simply initiated—prior to occurrence of the ERN.

A plausible alternative hypothesis would be compatible with an amended version of the conflict-monitoring theory in which the conflict signal is used as a control signal that resolves conflict (e.g., Seymour & Schumacher, 2009). Some error corrections may have

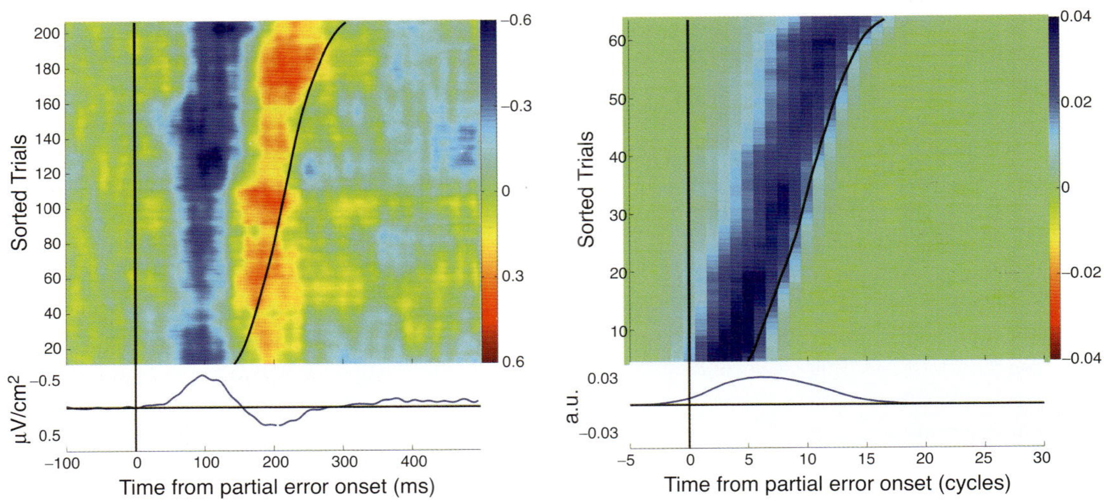

Fig. 10.13. Analysis of single-trial experimental ERN data (left) and simulated data (right) derived from the conflict-monitoring computational model. The trials all consisted of partial error responses followed by correct responses. The surfaces represent those trials time-locked to the onset of the partial error, sorted according to the length of the interval between the partial error and the subsequent correct response (indicated by the black line). In the experimental data, the ERN appears as a blue patch occurring on all trials around 100 ms following the partial error; the amplitude and duration grow larger as the interval between the partial error and the correct response lengthens. In the simulated data, the conflict model predicts that the time of maximum conflict will correspond most closely with the time of the correct response. Thus, the experimental data disconfirm the prediction of the conflict model. From Burle et al. (2008), Figure 3, reprinted with permission from MIT Press.

been late because there was more conflict on those trials, perhaps because of random variability in error-response activation. The ERN could represent a process that resolves that conflict (rather than simply detects it); thus, the size and duration of the ERN would be proportional to the conflict—still present following the error—that must be resolved for an error correction to be initiated.

Consistent with these two possibilities (which might be termed the *error-clearing* and *conflict-resolution hypotheses*, respectively), in the data of Burle et al. (2008) the late corrections were also associated with a greater amount of error EMG activity, such that the amount and duration of the ERN corresponded to the amount and duration of EMG activity associated with the eliciting error. Interestingly, the peak of the ERN in their study corresponded roughly to the offset of the error EMG activity. It is clear that ERN research would do well to probe the generality of the findings of Burle et al., applying their approach to a broader range of tasks and subject strategies.

There is an additional complication that must be considered in relating the ERN to measures of error correction: if the ERN represents a process involved in preventing or suppressing an error, then error trials represent those trials where the ERN occurred but failed. For example, the notion that the ERN, occurring after the error, is too late to reflect suppression of the error response itself overlooks the possibility that errors might be precisely those cases in which ERN activity occurred too late to be effective. Successfully inhibited error trials will (by definition) not occur. Thus, the typical ERN average could suffer from a selection bias: in this case, there could be another set of trials on which error activity occurs but the ERN occurs quickly enough to inhibit the error. Those trials will ultimately appear as correct trials; therefore, the resulting estimate of ERN latency will not include those trials (see Isoda & Hikosaka, 2007, for similar ideas applied to single-unit recordings from the pre-SMA). This ERN would not appear as a classic CRN (following the onset of the response); rather, it would occur prior to the correct response, possibly not even appearing in the average waveform because of latency variability (although the N2 activity on conflict trials identified by Yeung et al., 2004b, would be consistent with such a process).

A similar principle makes it difficult to come up with an airtight prediction about the relationship between the ERN amplitude and the likelihood of an error correction. If the process reflected by the ERN is involved in clearing or deleting portions of the error motor program to enable correction of the error, the appearance of an error correction will depend on whether the process represented by the ERN is successful. If a lot of activity is required on some trials because clearing out the error motor program is difficult on those trials, it is possible that the process could fail to delete the error program more frequently on those trials, creating an inverse relationship between the size of the ERN and the likelihood of error correction. Running counter to this tendency, however, could be a (possibly coexisting) dependence of the ERN amplitude on the likelihood of detecting the error, in which larger ERNs would appear when error corrections are more likely. Thus, the relationship between the amplitude of the ERN and parameters of error correction behavior could depend on the precise quantitative relationship between the detectability of an error (which itself could depend on the particular stimuli and movements involved) and the relative difficulty of correcting one type of error when another type is occurring. These and other task- and context-dependent factors have been mostly ignored by computational models and verbal theories.

STRATEGIC ADJUSTMENTS

Detection of an error can also lead to strategic adjustments—changes in strategy, attentional focus, response bias, or other parameters that will decrease the likelihood of future errors. The distinction between within-trial error corrections, reviewed above, and such strategic adjustments is similar to the distinction drawn by Donchin and coworkers (1978) between tactical and strategic information processing: the subject not only has the tactical goal to be correct on a single trial, but also strategic goals: to achieve an overall level of accuracy and response time, to please the experimenter, to obtain course credit or payment, and so on. Post-error slowing is generally assumed to be a strategic adjustment in which a subject slows down after making an error that happened when the subject responded too quickly (Rabbitt, 1981). If so, the size of the error signal could be related to the amount of slowing undertaken afterward. Gehring et al. (1993) used a single-trial measure of ERN amplitude and showed greater ERN amplitudes associated with increased post-error slowing. Other studies reported similar findings (e.g., Debener et al., 2005; Ladouceur et al., 2007; Rodríguez-Fornells et al., 2002). Some, however, failed to find a significant relationship (Dudschig & Jentzsch, 2009; Gehring & Fencsik, 2001; Hajcak et al., 2003b; see Figure 10.14).

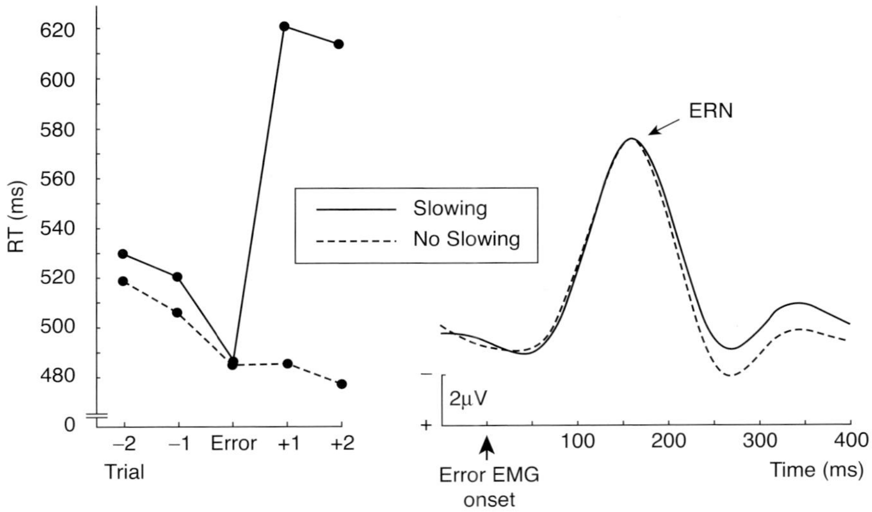

Fig. 10.14. Event-related potential waveforms from error trials that demonstrated post-error slowing (solid lines) or an absence of post-error slowing (dashed lines). The authors found no relationship between ERN amplitude and the degree of post-error slowing. Error trials in the two averages were paired according to reaction time to eliminate any confound between error reaction time and post-error slowing (as shown by the corresponding reaction time data on the left). From Gehring and Fencsik (2001), Figure 8, reprinted with permission from the *Journal of Neuroscience*.

A reason for some of the inconsistency may be that many processes can intervene between the ERN elicited by an error and the trial that follows that error, and that slowing depends in part on those processes that follow the ERN. Pe is one example of such a post-ERN process (Overbeek et al., 2005), but there are other post-ERN processes: Marco-Pallarés et al. (2008) found an oscillatory potential in the beta frequency range that correlated with the theta power underlying the ERN and (like the theta power itself) predicted post-error slowing.

Several dissociations between ERN amplitude and post-error slowing have appeared in the literature. Ullsperger and Szymanowsky (2004) found that emphasizing accuracy increased both the amplitude of the ERN and the degree of post-error slowing. Nieuwenhuis and coworkers (2001) compared errors of which subjects were aware with those that they did not detect. Perceived errors were associated with post-error slowing but unperceived errors were not, despite equivalent ERN amplitudes. Drug effects provide similar dissociations: for example, neither yohimbine (Riba et al. 2005a), alprazolam (Riba et al., 2005b), nor oxazepam (Johannes et al., 2001a) affected post-error slowing, although the former drug increased the size of the ERN and the latter two reduced it.

Various between-group associations and dissociations involving the ERN and post-error slowing have been reported. Alain and colleagues (2002) found both a reduced ERN and a reduction in post-error slowing in individuals with schizophrenia, whereas reduced ERNs with no changes in post-error slowing have been observed in schizophrenia (Mathalon et al., 2002) and Huntington's disease (Beste et al., 2008). Studies have shown that individuals scoring high on a worry questionnaire (Hajcak et al., 2003a), individuals with obsessive-compulsive symptoms (Hajcak & Simons, 2002), and individuals with major depression (Chiu & Deldin, 2007) showed increased ERNs relative to controls but the same level of post-error slowing. Gehring and Knight (2000) found that individuals with PFC lesions and age-matched (older) adults showed reduced ERNs, but equivalent post-error slowing, relative to young subjects. Swick and Turken's (2002) patient with ACC damage and a reduced ERN showed normal post-error slowing (complementing other reports of normal post-trial adjustments in individuals with ACC damage; Fellows & Farah, 2005; Modirrousta & Fellows, 2008). Band and Kok (2000), however, found that older subjects had a reduced ERN and an increase in post-error slowing (see also Falkenstein et al., 2000).

Several problems limit the utility of studies of post-error slowing in testing the putative link between the ERN and strategic adjustments. First, analysis of post-error effects is susceptible to confounds between previous-trial accuracy and previous-trial response latency. If error and correct trials show systematic differences in response time, then so-called post-error effects could be attributable to previous-trial response time instead of previous-trial accuracy. Few ERN studies have matched post-error and post-correct trials on foregoing reaction time; for exceptions, see Gehring and Fencsik (2001), Hajcak and Simons (2002), and Hajcak et al. (2003b).

A more insidious problem in studying post-error slowing is identifying the function of slowing. Most of the research has simply assumed that post-error slowing reflects a strategic increase in control (e.g., Botvinick et al., 2001). Suprisingly, this assumption has remained relatively untested until recently. Some evidence supports it: Hajcak and Simons (2008) found that errors that followed errors were associated with reduced post-error slowing, implicating insufficient strategic control as a cause of the double errors. Nevertheless, post-error slowing might instead occur because the same problem that caused the error in the first place (such as a lapse of attention) persists until the subsequent trial. Alternatively, subjects may divert attention to process the error, interfering with primary task processing. Supporting this explanation is evidence showing that post-error slowing is enhanced at short response–stimulus intervals (Dudschig & Jentzsch, 2009; Jentzsch & Dudschig, 2009). The infrequency of the error relative to other events may cause automatic attentional capture that is unrelated to the status of the event as an error (Notebaert et al., 2009), or the shift of attention may occur because of a capacity-limited error-monitoring process that can contribute to strategic control (Dudschig & Jentzsch, 2009; Jentzsch & Dudschig, 2009). Such alternative accounts are supported by reports of decreased accuracy on trials following errors relative to those following correct trials (Hajcak et al., 2003b) and of groups (such as the individuals with depression in Compton et al., 2008) that show enhanced post-error slowing but decreased accuracy following errors relative to controls. Future research would do well to take these alternative explanations for post-error slowing more seriously. In particular, it should no longer be assumed that post-error slowing is a straightforward index of cognitive control.

ERROR DETECTION

The issue of whether the ERN is related to conscious detection of the error is distinct from the issue of whether it is related to error correction: a correct response can follow an error simply because of continued processing of the stimulus and thus does not necessarily indicate that the error was detected (Gehring et al., 1995; Ullsperger & von Cramon, 2006a). Consequently, definitive evidence of error detection is possible only if the subject makes a response signaling that an error has occurred (Rabbitt, 1968). Findings relating the ERN to such error detection responses are inconsistent. Two studies using antisaccade tasks have reported that ERN activity is as large for errors that the subject fails to detect as it is for those that the subject can report (Endrass et al., 2007; Nieuwenhuis et al., 2001). In contrast, Wessel and colleagues (in press) recently reported two antisaccade experiments in which the ERN was larger in response to perceived errors than to unperceived errors. To explain the discrepancy, the authors pointed out that the ERN was numerically larger for perceived errors in the Endrass et al. study (implicating an issue with statistical power), and that task requirements in the Niewenhuis et al. study may have introduced a response bias toward not signaling an error.

Some studies of manual responses have shown larger ERNs when subjects signal that an error has occurred than when they are uncertain or fail to report the error (Scheffers & Coles, 2000; Ullsperger & von Cramon, 2006a), supporting a link between the ERN and awareness of the error, but other studies suggest the link might be tenuous. Steinhauser and Yeung (2010) reported that the ERN was larger for detected errors than for undetected errors, but (somewhat paradoxically) that the ERN was not affected by an incentive manipulation that changed the subjects' criterion for signalling that an error had occurred. Because of the latter finding, the authors suggested that the ERN was not directly involved in error detection, but instead that the difference between detected and undetected errors might have reflected different reaction times (and response conflict) associated with the two types of errors. Maier and coworkers (2008) reported a flanker study in which subjects detected errors on incongruent trials more often when the response did not match the flanker letter (nonflanker error) than when the response and the flanker letter matched (flanker error). Paradoxically, the latter condition showed the largest ERN. This result is difficult to interpret, however, because the two conditions were not matched: the a priori probability of an error being a nonflanker error was twice that of its being a flanker error, presenting a potential confound in the perceived likelihood of the two error types.

Evaluation

How the ERN relates to measures of compensatory behavior and conscious error detection is far from clear. At least some of this ambiguity may result from the difficulty of controlling for all possible confounds, as in the example of post-error slowing discussed above. The lack of robust findings relating the ERN to post-error behavior would seem to be evidence against the RL-ERN and conflict-monitoring theories, because both theories argue that some relationship should hold between the amount of ERN activity and post-error strategic adjustments (Botvinick et al., 2001; Holroyd & Coles, 2002). However, in both cases, there is no reason why one could not devise models that employ the same computations in the service of immediate error correction or inhibition rather than strategic control. Thus, some central ideas of these models—that the ERN represents a signal sensitive to the reward properties of the response outcome, or that it reflects conflict between responses—could potentially be instantiated in different computational models that make different predictions. This situation points to the need for greater effort in developing and evaluating competing computational models and alternative architectures. The ambiguity in these studies also points to a need for a better understanding of the underlying processes that result in post-error slowing and error detection responses.

Tests of the Theories

In this section, we focus on studies that seem to be more relevant to some subset of the theories than to the others. In some cases, it is apparent that the various theoretical perspectives on the ERN have led to different research agendas: some phenomena that are important to one theory are relatively unimportant to or discounted by others.

Error Detection and Conflict Monitoring

ERROR-CORRECT MISMATCH

Each of the major theories of the ERN links the ERN to a computation representing the difference between information corresponding to the correct response and information corresponding to the error. According to error-detection accounts, the ERN reflects a comparison between information corresponding to the actual (erroneous) and the intended (correct) response. The RL-ERN model links the ERN to learned values of stimulus–response conjunctions (Holroyd et al., 2005); the ERN reflects the comparison of the values of correct and incorrect stimulus–response conjunctions. Although the conflict-monitoring account is sometimes positioned as incompatible with one based on a mismatch-based error detector, the conflict computation is nevertheless sensitive to the difference between the response representation associated with an error and that derived from continued processing of the stimulus (see Holroyd et al., 2005, and Yeung et al., 2004b, for detailed comparisons of the mismatch theory and conflict-monitoring theory).

Because each of these models relates the ERN to error-correct mismatch in the sense we have described here, each one predicts (or could predict) that the amplitude of the ERN should be sensitive to the similarity of the representations involved in the comparison or conflict computation. Nevertheless, surprisingly, the bulk of the empirical studies and computational models in ERN research have limited themselves to two-choice tasks that permit little exploration of the role of representational mismatch. Here, we discuss the few exceptions to this tendency (for additional discussion of this issue, see Falkenstein et al., 2000).

A small number of studies have examined the effects of response similarity on the ERN. According to hierarchical models of response selection, the choice of an effector begins with relatively coarse decisions (e.g., right hand vs. left hand) and progresses to increasingly specific decisions (e.g., right index finger vs. right middle finger; Bernstein et al., 1995). The later in selection two responses diverge, the more similar they are. Using this logic, Bernstein et al. (1995) and Falkenstein and colleagues (1996) found that ERN amplitude increased as error and correct representations grew more dissimilar, consistent with a mismatch account. In contrast, Gehring and Fencsik (2001) reported greater ERN amplitudes for errors with representations similar to the correct response. Neither type of result, however, is definitive: the studies assume a certain kind of response representation (based on side or effector), and it is possible that similarity might be based on other movement parameters (such as velocity, force, etc.). Alternative computational models that incorporate different conceptions of response similarity would be helpful in testing the response mismatch hypothesis.

Another important unresolved issue is the role of stimulus similarity. Mismatch could involve the stimulus associated with (or predicted by) the error response and the stimulus that actually occurs (Schmidt & Gordon, 1977). Although Bernstein et al. (1995) found that stimulus similarity did not affect ERN amplitude significantly, the direction of the effect was consistent, with larger ERNs being associated with dissimilar stimuli than with similar stimuli. That study did not present waveforms, and it involved numerous other conditions, suggesting that additional study is warranted. Elton et al. (2004) also reported that stimulus mismatch did not affect the ERN in an experiment using auditory stimuli separated by different degrees of pitch (Figure 10.15A). However, their data actually showed a significant interaction with electrode site, such that at Cz the ERN amplitude was greater for errors with large stimulus deviations relative to small-deviation errors. Perhaps the most surprising study examining stimulus mismatch is that of Yeung et al. (2007b), who showed that simply increasing the brightness of the stimulus increased the amplitude of the ERN (Figure 10.15B). Orr and coworkers (2008; see Figure 10.15C) showed that ERN amplitude was greater when stimuli could be distinguished on the basis of simple visual features than when the stimuli shared common features. Similarly, Vocat et al. (2008) found that false-alarm errors in a go/no-go task caused by a salient stimulus dimension elicited a larger ERN than those caused by a less salient stimulus dimension. Although the idea of mismatch is generally associated with the error-detection theory, results of this kind are consistent with all extant theories of the ERN (as Yeung et al., 2007b, note): whether the ERN reflects error detection, conflict detection, or the motivational or emotional significance of error commission, stronger representations of the correct response should produce larger ERNs. Note that studies of mismatch must be careful that the representational mismatch manipulation does not produce stimuli that are so difficult to discriminate as to make subjects uncertain about which stimulus actually occurred (Pailing & Segalowitz, 2004a).

The congruency manipulation in conflict tasks provides another way to manipulate the level of mismatch: when the irrelevant and relevant dimensions of a stimulus both correspond to the correct response (as in a compatible flanker stimulus such as *HHHHH*), there is more information to serve as the basis of error detection or to cause response conflict than when only the relevant dimension signals the correct response (as in *SSHSS*). Here, the results seem to depend on the task. Scheffers and Coles (2000) found that congruent stimuli in a flanker task were associated with larger ERNs than were incongruent stimuli (but see Bartholow et al., 2005). More recently, Forster and Pavone (2008) found that congruent stimuli were associated with larger ERNs than incongruent stimuli in a task with tactile target stimuli and spatially incompatible visual distractors. Christ and colleagues (2000) found no difference in the ERNs elicited by congruent and incongruent stimuli in both a Simon and a spatial-Stroop task. There are a number of differences in the paradigms, methods, and subject performance that could account for these inconsistencies. In particular, it is important to note that although these studies all involve conflict tasks, they differ in the manner in which the overlap between stimulus and response dimensions engenders conflict (see Kornblum et al., 1990). That is, the studies differ in which parts of which representations are in conflict. It appears that to resolve some of these contradictory findings, the ERN literature would benefit from further study of how the effects of congruency on the ERN may depend on the specific aspects of the stimulus or response representations that are incongruent.

In sum, evidence for roles of both response and stimulus mismatch in ERN generation is limited and mixed. Studies supporting a role of response similarity in mismatch are contradictory and have not considered thoroughly the stimulus and response representations involved in the computations that result in an overt response. Evidence against a role of stimulus similarity is not convincing. The studies reviewed above employed a wide range of tasks and manipulations of response and stimulus similarity, complicating direct comparisons among studies and a synthesis of results. Based on the evidence, some role for stimulus representations in computing mismatch is plausible and worthy of further experimentation, although experiments must be careful to control for the effects of error probability and to dissociate the role of stimulus similarity in the mismatch computation from its role in activating the competing responses.

LEVEL OF RESPONSE CONFLICT

Conflict theory asserts that greater amounts of response conflict should be associated with larger ERNs, and several studies have tested this claim. Some investigators have attempted to measure response conflict directly, based on the assumption that error corrections will compete with the errors that they follow. Carbonnell and Falkenstein (2006), for example, measured the force of the error and correct responses, reasoning that, all things being equal, a more forceful error response will compete more with the subsequent correct response than will a less forceful error. Their results showed no difference in the size of the ERN on the two types of trials. In a similar fashion, Masaki and coworkers (2007) used EMG measures to show that conflict was greater in a difficult task condition than in an

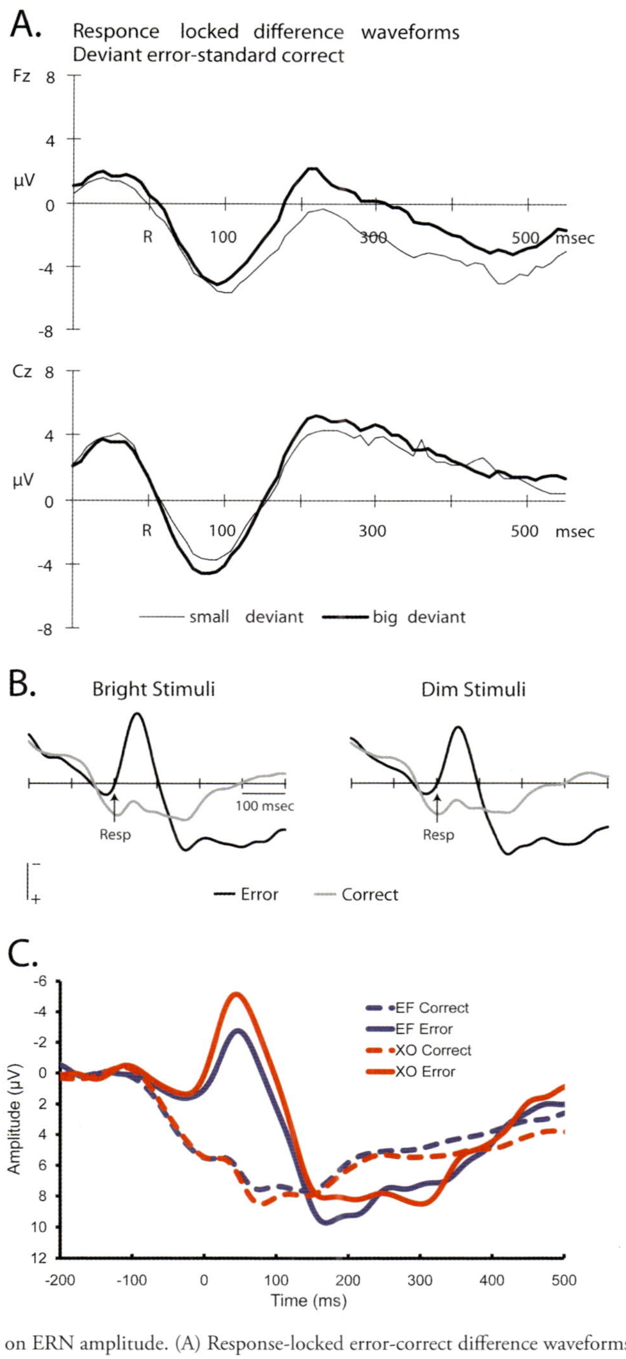

Fig. 10.15. Mismatch effects on ERN amplitude. (A) Response-locked error-correct difference waveforms elicited in a go/no-go task where the go stimulus was a standard auditory tone and the no-go stimuli were small or large auditory deviants. Errors were classified as no-go commission errors. There was no main effect of deviance on ERN amplitude, but a deviance by electrode (Fz vs. Cz) interaction revealed a significantly greater ERN at Cz for large deviants compared to small deviants. The opposite effect was reported at Fz, although the baseline and time course suggest that the Fz effect may have resulted from the Pe. From Elton et al. (2004), Figure 4, reprinted with permission of John Wiley & Sons, Inc. (B) Grand average response-locked ERP waveforms for correct and error trials in a flanker task with bright and dim (i.e., high- and low-luminance) stimuli. The authors found that ERN amplitude was reduced for dim stimuli. Modified from Yeung et al. (2007b), Figure 2, reprinted with permission from the authors. (C) Grand average response-locked ERP waveforms for correct and error trials in a variant of the flanker task examining the role of stimulus similarity on the ERN. Stimuli were composed of perceptually similar letters (EF trials; e.g., *EEEEE* and *FFEFF*) and perceptually dissimilar letters (XO trials; e.g., *XXXXX* and *OOXOO*). The ERN was larger on error trials in the perceptually dissimilar condition than in the perceptually similar condition. From Orr et al. (2008). Reprinted with permission from the authors.

easy one, yet the ERN amplitude was the same in the two conditions. Here, however—as we saw in the discussion of ERN latency above—the complicated logic of inhibitory processes makes the conclusions less compelling. If the error response on one trial is less forceful than the error response on another trial, it could be the case that the less forceful error was less forceful precisely because there was a great deal of response competition, and that the more forceful error occurred because there was little competition between the error and the correct response.

Studies such as this are on firmer ground in testing the conflict-monitoring model if they are based on simulations that establish the amount of response conflict in the conditions where ERNs are compared. For example, Burle et al. (2008) used conflict-model simulations to establish that—according to the computational model—there is less conflict when an error precedes an error correction by a long interval than when the two responses occur close together in time. Contrary to the predictions of the model, however, the ERN was larger when the interval between the error and the correction was long than when it was short (see Figure 10.13). Nevertheless, even this study might not be as damaging to the notion of conflict monitoring as it appears. To understand why this is the case, one must keep in mind that there is a distinction between the concept of conflict monitoring, which could be implemented computationally in a variety of ways, and the particular computational models of Botvinick et al. (2001) and Yeung et al. (2004b). The Burle et al. (2008) study showed that the predictions of that particular computational model did not hold. However, in alternative implementations of conflict monitoring, response conflict might not be evident in overt measures such as muscle activation or overt movement. Rather, the response conflict might be limited to covert activity, at the level of the primary motor cortex, or possibly even earlier in processing. Peripheral inhibitory processes can intervene between the cortex and overt behavior (De Jong et al., 1990; Ohtsuki, 1981), such that a high degree of conflict at the level of motor cortex might not result in overlapping movements. Such processes are not represented in the models of Botvinick et al. (2001) and Yeung et al. (2004b). Thus, it would appear worthwhile to explore alternative ways that conflict monitoring could be implemented computationally, with more detailed representations of the motor system, before drawing firm conclusions about how the size of the ERN should or should not vary with observable indications that conflict is present in the motor system.

N200/N450

According to the conflict model, ERN activity will occur on correct trials in conditions of high response conflict (Yeung et al., 2004b), appearing as an N200 preceding the correct response. Some studies, however, have reported dissociations of the N200 and ERN and argued against the similarity of these components. The ACC lesion patient of Swick and Turken (2002) showed a reduced ERN but a normal N200. Similarly, administering alcohol reduced the size of the ERN but did not affect the N200 (Ridderinkhof et al., 2002). Here, computational modeling has proven useful in showing that such dissociations need not imply that the components are distinct (Yeung & Cohen, 2006). In the case of alcohol, for example, a combined deficit in perceptual and attentional processing will reduce the size of the ERN but will not affect the amplitude of the N200 even if both represent the same conflict signal (Yeung & Cohen, 2006; Yeung et al., 2007b).

THE CRN

Observing the CRN sparked a challenge to the idea that the ERN represents part of an error-detection system: if an ERN can occur on a correct trial, then how could it represent the detection of an error (Vidal et al., 2000)? Recall that the CRN is an ERN-like potential whose peak follows the correct response by the same latency as the ERN peak follows the error response. As Vidal et al. point out, incorrect muscle activity cannot be the sole cause of the CRN on correct trials, because if there is some incorrect EMG activity, that activity will typically occur prior to the correct response, not after it. Possible explanations for the CRN include that it is an artifact of a stimulus-related N200 appearing in the response-locked waveforms, or that it is the result of correct responses that the subject's brain somehow mislabels as errors (Coles et al., 2001; Vidal et al., 2000). Supporting the latter interpretation are cases where an ERN-like potential accompanies slow, correct responses—responses that involve the correct effector but exceed a specific experimenter-imposed response deadline (Heldmann et al., 2008; Luu et al., 2000b). If subjects were to evaluate responses according to an internal deadline or some other subjective criterion, an ERN could occur (appearing as a CRN), despite the objective correctness of the response. This slow-trial CRN would grow larger as responses get later, making it consistent with a mismatch- or comparator-based error detector, where the standard that defines an error is determined in part by the subject's internal criteria. It is less clear how the conflict-monitoring theory would predict a slow-trial CRN. Not all instances of the CRN are easily explained by implicit response deadlines, however: Vidal and colleagues (2003) reported one study without an explicit deadline in which the CRNs did not differ on fast and slow correct trials. They argued that the CRN is therefore not likely to reflect error detection. They suggest that the ERN and CRN are the same component, which may represent a comparison process that precedes error detection per se, or else an emotional response to the error, rather than error detection itself.

If the ERN is a reflection of conflict monitoring, one can see that CRNs occurring in conditions of high response conflict could reflect the detection of conflict at the time of a correct response. Consistent with such an idea, Bartholow and colleagues (2005) found that the CRN in a flanker task was affected by the presence of incongruent flanker stimuli on correct trials, and that this effect was influenced by the probability of incongruent flankers: conditions where congruent stimuli were frequent—and the flanker (conflict) effect on RT was largest—showed the largest CRN on incongruent-flanker trials. Response conflict could also explain the CRNs observed when individuals respond deceptively in a recognition task (Johnson et al., 2004). However, the conflict monitoring model predicts ERN-like activity *before* the correct response in conditions of high conflict (Yeung et al., 2004b); thus, the CRN, if it is indeed an ERN on correct trials, would argue against the conflict-monitoring model. Nevertheless, the existence of CRNs could be consistent with some alternative implementation of conflict theory or alternative parameterization of the current models. To our knowledge, there have been no attempts to see whether a model based on conflict monitoring could produce a CRN, although early conceptions of conflict monitoring seem to allow for this possibility (Cohen et al., 2000). As for the other theories, it is also plausible that ideas of affective distress could accommodate the existence of the CRN. It is less clear how the RL-ERN theory will treat the CRN. In the past, this theory explicitly concerned the difference between error and correct trials, and thus appeared to say nothing about the CRN (Holroyd & Coles, 2002). More recent work, however, has argued for the existence of positive-polarity activity occurring on correct trials (Holroyd et al., 2008), suggesting that future RL-ERN modeling work may have to specify why some correct trials yield a CRN and some yield a positivity. One clue may lie in the role of uncertainty, as some evidence suggests that the CRN may occur when subjects are more uncertain of the accuracy of their responses (Pailing & Segalowitz, 2004a).

One criticism of the ERN literature in general, which is particularly important in the case of the CRN, is the lack of a systematic attempt to compare the ERN and CRN to other motor potentials, especially those potentials recorded in voluntary movement tasks. For example, classic studies of movement-related potentials recorded during self-paced finger-flexion responses showed an N+50 component with a latency and a scalp distribution very similar to those of the ERN, although with a slightly larger amplitude contralateral to the response (Shibasaki et al., 1980a, 1980b). It is possible that the CRN is a distinct movement-related potential like the N+50 whose appearance depends on particular task or response parameters. With several other components occurring within the 100 ms just prior to and just after a voluntary response, determining whether the ERN and CRN are fundamentally the same will require a larger comprehensive effort to identify and compare all movement-related potentials in that time range. Such an effort would be a welcome development in the ERN literature.

The existence of the CRN and the possibility that there are other negative-going potentials similar to the ERN at the time of a movement highlight the difficulty in determining how best to define and measure the ERN: defining it as the difference between error and correct trials ignores the possibility that a CRN exists, yet defining it as a negative-going peak in the unsubtracted error- or correct-trial waveform risks misidentifying some other movement-related potential (such as the N+50) as the ERN. It is probably best to adopt a research strategy that incorporates both kinds of analysis (see, e.g., Tops et al., 2006, who refer to the difference between correct and error trials as the Δ*ERN*).

PARSIMONY

One argument that has been put forth as favoring conflict theory is that the conflict model is more parsimonious than the error-detection model because the error-detection model requires a homunculus that can determine which response is correct (Botvinick et al., 2004; Carter et al., 1998; Yeung et al., 2004b). This argument has several weaknesses. First, claims of parsimony are impossible to support without a unified theory of cognition

(Newell, 1990) that describes a complete functional architecture of the brain, specifying how the elements in the model interact with other cognitive, perceptual, and motor systems. Simplicity in one part of such a theory (embodied in the model) might come about by making some part external to the model more complex. For example, because the conflict model does not represent the occurrence of an error, the model must be amended to achieve an explicit (conscious or declarative) representation that an error occurred. Yeung et al. (2004b) showed that adding a threshold to the conflict model can yield a signal that an error has occurred. Nevertheless, such a signal is achieved by sacrificing the simplicity that the conflict model claims as a virtue. In addition, the conflict-based error detector creates other issues to be resolved: is a homunculus necessary to set the threshold? Can a threshold-based error detector apply to more complicated movements that involve multiple effectors? Can the output of a threshold-based system yield a computation of error-correct similarity such as the one thought to underlie the mismatch computation in the error-detection model? Moreover, the work of Holroyd et al. (2005) showed that—contrary to the claims of the conflict theorists—an error-detection model need not invoke a homunculus. Perhaps even more important, claims of parsimony are not particularly relevant: in the philosophy of science, the notion that the simpler theory is more likely to be true is problematic (Hempel, 1966), one reason being that reality isn't necessarily simple. Or, as Gordon Logan once bluntly put it, "Parsimony is overrated" (G. Logan, personal communication).

EVALUATION

The debate between the conflict- and error-detection camps is far from over. All attempts thus far to measure conflict directly have disconfirmed, rather than confirmed, the predictions of the conflict-monitoring model. Nevertheless, difficulties with measuring conflict, and the possibility of amending the conflict model, suggest that these results are not conclusive. Despite the debate between the conflict monitoring theorists and others in the ERN community, some work has suggested that some of the findings explained by the conflict-monitoring theory can also be accommodated by a model with explicit error detection (Holroyd et al., 2005). With the exception of the CRN, which might disconfirm the timing of correct-trial ERNs predicted by the conflict-monitoring model, most findings could be accommodated within either framework. Better modeling work would attempt to compare error detection and conflict monitoring on an equal footing, within the same architecture, to come up with predictions that are unique to one or the other theory and would disconfirm the other (at least within that specific architecture). And for any model, alternative parameterizations should be explored that would make clearer the range of phenomena the model can predict (Roberts & Pashler, 2000).

The Reinforcement Learning Theory of the ERN

THE FRN

For the most part, the RL-ERN theory has been tested using the FRN rather than the classic ERN. The FRN has spawned its own active and growing literature (see Nieuwenhuis et al., 2004a). Some theorists propose that the FRN reflects another manifestation of the temporal-difference reinforcement-learning signal that causes the ERN (Holroyd & Coles, 2002). If so, studies of the FRN can provide a whole range of constraints on ERN theories. For example, the RL-ERN model of Holroyd and Coles (2002) predicts that early in learning, before subjects have learned the mapping between stimuli and responses, the error signal will tend to be elicited by feedback stimuli, but as learning progresses—once subjects know the mapping—the information necessary to produce the error signal will be available at the time of the error response. Thus, with practice, the FRN elicited by error feedback will grow smaller and the ERN elicited by an error response will grow larger. Several studies have confirmed that the ERN elicited by erroneous responses grows larger as subjects learn the stimulus–response mappings (Holroyd & Coles, 2002; Morris et al., 2008; Nieuwenhuis et al., 2002). The FRN results have been less consistent, with the predicted decrease absent in one study (Holroyd & Coles, 2002) and not significant in another (Nieuwenhuis et al., 2002). A third study showed a reversal in which, after learning, correct feedback elicited the FRN and error feedback did not (Morris et al., 2008). One key to understanding the discrepant results may lie in the findings of Eppinger and coworkers (2008), who did find the predicted learning-related decrease in the error-correct FRN difference waveform but also found that the learning-related effects were actually caused by the positivity elicited by correct feedback, not by the FRN.

Several studies have tested the claim that the FRN represents the detection of unexpected, unfavorable outcomes (Holroyd & Coles, 2002). The evidence is mixed. While several studies have shown that the more unexpected the negative feedback, the greater the amplitude of the FRN (Bellebaum & Daum, 2008; Hajcak et al., 2007; Holroyd et al., 2003; Potts et al., 2006b), some studies have shown that the FRN also responds to unexpected yet favorable outcomes (Donkers & van Boxtel, 2005; Donkers et al., 2005; Oliveira et al., 2007), and Hajcak and colleagues have shown that the FRN is not always sensitive to probability manipulations (Hajcak et al., 2005a, 2007). Rather, the sensitivity of the FRN to probability might be limited to situations in which there is a learnable contingency between responses and outcomes (Holroyd et al., 2009).

Until recently, the consensus in the literature has been that the FRN varies with the outcome value in a binary fashion, reflecting the evaluation of outcomes as good or bad but not capturing the gradations in outcome value that would seemingly be predicted by the RL-ERN theory. When subjects can gain money, lose money, or break even (gain or loss of zero), the FRN does not distinguish losses from neutral outcomes (Holroyd et al., 2004a, 2006). Moreover, when losses that differ in monetary value are possible, FRN amplitude is equivalent for large and small losses (Hajcak et al., 2006; Nieuwenhuis et al., 2004b; Yeung & Sanfey, 2004). Although a binary response seems inconsistent with the sensitivity to value implied by the RL-ERN model (Holroyd & Coles, 2002; Holroyd et al., 2005). Holroyd et al. (2006) suggested that cognitive categorization processes could be added to the RL-ERN model to account for binary ERN responses. More recent evidence, however, complicates this picture. Goyer et al. (2008) observed FRN activity that was modulated by the magnitude of the reward, such that larger losses elicited greater FRN activity. Nevertheless, the time course and scalp distribution of the magnitude effect differed from those of the outcome valence effect, suggesting that the magnitude effect most likely represents a different component. Similarly, Bellebaum et al. (2010) reported FRN amplitudes on loss trials that grew larger with the magnitude of an unobtained reward, although the positivity preceding the FRN may have contributed to the peak-to-peak measure underlying their effect.

Whereas most of the initial work on the FRN examined the role of negative events like losses in eliciting the component, a more recent trend is a focus on how gains can affect the FRN. In the literature on dopamine, an unexpected positive reward causes a phasic increase in dopamine that is involved in reinforcement learning (Schultz, 2002). The RL-ERN theory contends that a positive dopaminergic response should result in a positivity observable at the scalp, just as a negative response results in a negativity. Studies have therefore begun to focus on whether positive outcomes can affect the FRN waveform. Potts and colleagues (2006b) proposed that the P2a elicited by unpredicted reward feedback represents this positive dopaminergic response. Similarly, Holroyd and coworkers (2008) argue that the FRN is the same ERP component as the stimulus-locked N200 and that the error–correct difference in feedback-locked waveforms results from the summation of the N200 with a gain-related positivity—the feedback correct-related positivity—that occurs when there is correct feedback. This interpretation suggests that the null effects of reward expectation on loss waveforms may not be inconsistent with RL-ERN theory: expectation effects may simply be more evident in gain waveforms. Consistent with this account, Eppinger et al. (2008) found that effects due both to learning and to feedback validity were seen in a positivity—not FRN or ERN. Also consistent with this notion, Santesso et al. (2009) showed that pramiprexole, a dopamine agonist that reduces phasic dopamine levels, blocked a gain-related reduction in the FRN that appeared to occur in controls (although there was no control condition to verify that reduction). A special status for gains is also evident in the finding that reward expectation modulated ERPs and EEG frequency spectra related to gains but not losses in a gambling task (Cohen et al., 2007) and that increasing the frequency of large gain stimuli caused smaller gains to elicit an FRN, but changing the frequency of large losses did not have an effect on small-loss FRNs (Nittono et al., 2008).

There is still work to be done in resolving some inconsistencies among these studies. Some studies argue that unexpected gains will elicit the same effect as unexpected losses, namely, a greater negativity than the corresponding expected condition (e.g., Oliveira et al., 2007), whereas others argue that unexpected gains will cause a positivity rather than a negativity (Holroyd et al., 2008; Potts et al., 2006b). The truth might even be a mixture of these possibilities: single-unit recordings in the ACC have shown some neurons sensitive to rewards, some sensitive to omitted rewards, and others sensitive to both (Ito et al., 2003). Moreover, whether there is a purely feedback-related negativity at all is a new point of contention, with one position being that

FRN effects occur because of an overlapping correct-related positivity (Holroyd et al., 2008). Some recent evidence based on a spatiotemporal principal components analysis (PCA) may support the existence of a reward-related positivity: in data from a gambling task, the only component emerging from the PCA that matched the latency of the FRN was a positivity on gain trials, although the raw data showed the typical FRN on loss trials and no discernible peak on gain trials (Foti et al., 2011). Still, it would be reassuring to see some experimental condition that could actually produce a positive-going deflection at the latency of the FRN. One report showed a so-called reward positivity whose latency was earlier than the FRN, making an account of the FRN based on component overlap untenable (Holroyd et al., 2011). In learning and gambling tasks, the gain trial waveform often shows no peak—positive or negative—whose latency matches that of the FRN (see, e.g., Figure 10.7). To explain those cases as reward-related positivity effects, one would have to assume that there is always a negativity at the same latency that overlaps and precisely cancels out the positivity on those gain trials. Similar logic would also have to apply to the response-locked ERN.

Interestingly, as in the ERN studies reviewed earlier, little work has attempted to theorize about, model, or experimentally manipulate the stimulus representations that give rise to the feedback signal. Indeed, many studies confound the outcome valence with the stimulus representations that signal a gain or loss (e.g., +++ vs.−−−; Holroyd et al., 2004a). A complete theory of the FRN would have to specify the representation used in the underlying computation. In a recent study, we demonstrated a case where the FRN was elicited only when loss feedback could be distinguished from gain feedback on the basis of a distinctive visual feature, not when the discrimination was based on a conjunction of features (Liu & Gehring, 2009). In another study, we asked whether the perceptual similarity of the feedback stimuli affects the FRN (Liu, 2008). We manipulated the similarity of the gain and loss feedback stimuli (*E* vs. *F* in one case and *S* vs. *T* in another) and added irrelevant flanking stimuli. Our results showed that the FRN was larger when the feedback stimuli were dissimilar than when they were similar, and that the irrelevant flanking letters elicited a FRN-like negativity, suggesting that the FRN was modulated by the perceptual properties of feedback stimuli (Liu, 2008; see Figure 10.16). Other studies have found the FRN to be influenced by perceptual factors (Jia et al., 2007). Of course, a difficulty such studies face is that manipulating the stimulus representations that give rise to the FRN could affect other N2-like potentials that overlap the FRN, and not the FRN itself, if the FRN is indeed distinct

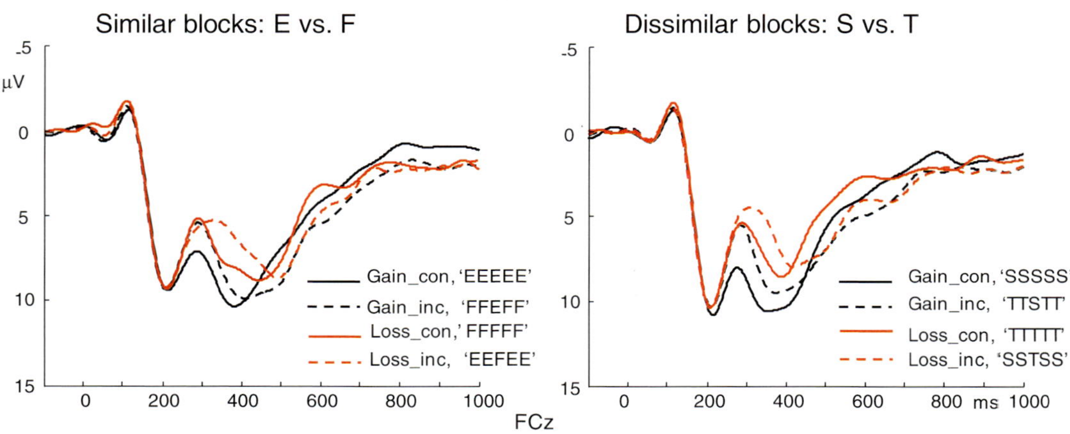

Fig. 10.16. Grand average ERP waveforms elicited by feedback stimuli indicating gains or losses in a gambling task. Feedback stimuli were composed of a central letter that indicated the subject's gain or loss, surrounded by irrelevant flanking letters that were either the same as or different from the central letter. Letters in the feedback stimuli were perceptually similar (*E* vs. *F*, left) or perceptually dissimilar (*S* vs. *T*, right). The feedback related negativity (FRN) was larger when the gain and loss feedback stimuli were perceptually dissimilar (*S* vs. *T*) than when they were similar (*E* vs. *F*). In addition, the FRN was affected by the presence of irrelevant flanking letters, such that a gain stimulus surrounded by irrelevant loss flanker stimuli elicited an FRN (i.e., the *FFEFF* and *TTSTT* waveforms). From Liu (2008). Reprinted with permission from the author.

from the other potentials. A challenge to the FRN literature is to disentangle the multiple N2-like potentials that occur after a feedback stimulus.

EXTENSION OF RL-ERN THEORY TO THE CLASSIC ERN

To extend the RL-ERN model to speeded response tasks, Holroyd et al. (2005) combined the reinforcement learning model with a model of speeded task performance similar to the conflict-monitoring model. The model is fairly complex, but the key insight is that states of the system are assigned values based on past performance outcomes. One layer of units within the system categorizes stimuli and activates responses, not unlike the mechanism in the conflict-monitoring model. Figure 10.12 shows the two additional layers responsible for the ERN. One layer represents the current stimulus and the current response. Critical to the processing in this layer are conjunction units, which are activated by a conjunction of stimulus and response units, such as the *H-left* (HL) unit that is activated when *H* is the stimulus and *left* is the current response. The second layer assigns values to different states within the state layer: if the instructions to the subject are to respond *H* with the left hand, then the *S-left* (SL) state would be assigned a negative value and the *H-left* (HL) state would receive a positive value. Finally, the temporal difference unit compares the current value layer with that predicted based on past reinforcements. If the value is negative, an error is signaled and adjustments are made to the motor control system.

EVALUATION

If the FRN reflects the same process as the ERN, the fact that it occurs well after the response is particularly damaging to a conflict-monitoring account of the ERN, because by the time feedback occurs, any response conflict has dissipated (although it is conceivable that conflict between the expected and actual feedback might be accommodated within a modified conflict model; van Veen et al., 2004). Nevertheless, there is some evidence suggesting that the two components are not identical. First, evidence reviewed below suggests that the ERN, but not the FRN, can be modulated by the value or motivational significance of the eliciting event. Also, although dissociations must be interpreted cautiously (e.g., Yeung & Cohen, 2006), ERN/FRN dissociations have been reported in OCD (Gründler et al., 2009), psychopathy and externalizing psychopathology (Bernat et al., 2011; Borries et al., 2010; Hall et al., 2007), trait anxiety (Gu et al., 2010; Hajcak et al., 2003a), and aging (Eppinger et al., 2008). Moreover, the classic ERN has a symmetrical midline scalp distribution, with its maximal amplitude at the scalp site FCz. The FRN, in contrast, typically has a scalp distribution that is more anterior and, in some cases, lateralized to the right (Gehring & Willoughby, 2004; Müller et al., 2005; Nieuwenhuis et al., 2004b; Potts et al., 2011), possibly because the FRN is generated by two sources, one in the posterior cingulate cortex and one in the rostral anterior cingulate or medial prefrontal areas (Müller et al., 2005; Nieuwenhuis et al., 2005a; van Veen et al., 2004). Consistent with a two-source model, Potts and colleagues (2011) reported a spatiotemporal principal components analysis supporting distinct frontal and central contributions to the FRN, with the latter possibly reflecting an ACC generator common to both the ERN and FRN.

Several other considerations suggest that it would be premature to conclude that the midbrain dopamine system is the sole or primary determinant of the ERN. First, as we review in the section on neurotransmitters below, the evidence linking the ERN to dopamine is mixed. Also, Jocham and Ullsperger (2009) and Frank and colleagues (2005, 2007) have suggested that the link between the ERN and dopamine may actually be opposite to that proposed by Holroyd and Coles (2002): The ERN could be caused by some nondopaminergic signal, with the ERN in turn causing a dopaminergic response in the basal ganglia (see Figure 10.17). Jocham and Ullsperger (2009) note that the midbrain dopaminergic system lacks the speed to generate the ERN as quickly as the Holroyd and Coles model would require, making it more plausible that some other neurotransmitter gives rise to the ERN before dopaminergic responses occur. Contradicting this assertion, however, is a report of intracranial field potentials in an individual with OCD showing error-related activity in the nucleus accumbens 40 ms prior to the ERN recorded at the scalp (Münte et al., 2008). The small sample size and likely presence of overlapping potentials point to the need for further study of this potentially important phenomenon.

ERN as an Affective Response

The proposition that the ERN reflects an affective response to error commission has motivated a search for the personality and emotional correlates of the

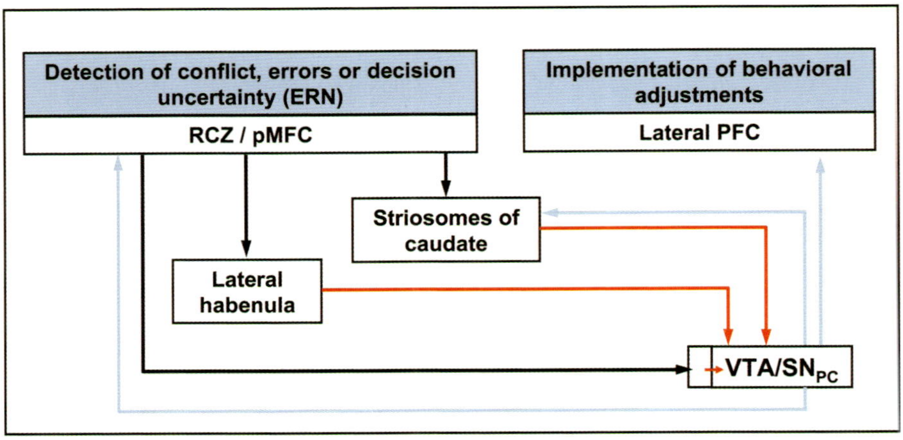

Fig. 10.17. A new model of the neuropharmacology of the ERN, showing three possible pathways through which an error signal from the ACC could inhibit midbrain dopamine neurons. One pathway projects directly from the rostral cingulate zone/posterior medial frontal cortex (RCZ/pMFC) to the ventral tegmental area and substantia nigra pars compacta (VTS/SNpc), which acts through GABAergic interneurons to inhibit dopamine neurons. Also, pathways through the lateral habenula and the striosomes of the caudate nucleus could inhibit midbrain dopamine neurons through GABAergic projections. The blue boxes show the corresponding cognitive processes; black arrows represent excitatory connections; red arrows represent inhibitory connections; blue arrows represent mesocortical and mesostriatal dopaminergic projections. From Jocham and Ullsperger (2009), Figure 2, reprinted with permission from Elsevier.

ERN. A variety of studies have reported links between emotional variables and ERN amplitude—often without providing an accompanying theory at the level of detail offered by the reinforcement-learning and conflict-monitoring models. It is important to note that, broadly speaking, affective and cognitive theories of the ERN are not mutually exclusive: affective states and traits could modulate the cognitive processes that underlie the ERN, or vice versa.

PATIENT STUDIES

The past decade has seen growing interest in the effects of neurological and psychiatric conditions on the ERN (for reviews, see Jocham & Ullsperger, 2009; Olvet & Hajcak, 2008; Ullsperger, 2006). Theories of obsessive-compulsive disorder (OCD) have attributed the exaggerated concerns and repetitive behaviors that characterize OCD to a hyperactive error signal (Pitman, 1987; Schwartz, 1997). Gehring and colleagues (2000) suggested that the ERN might represent such an error signal. They confirmed that individuals with OCD showed an exaggerated ERN relative to controls, and furthermore, that ERN amplitude was correlated with symptom severity (see Figure 10.18A). A number of studies have replicated and elaborated on this finding, both in individuals with OCD (Endrass et al., 2008; Johannes et al., 2001b; Ruchsow et al., 2005a; for an exception, see Nieuwenhuis et al., 2005b) and in individuals with subclinical symptoms (Hajcak & Simons, 2002). This effect appears to be unrelated to medication status when symptom severity is controlled (Hajcak et al., 2008; Stern et al., 2010). One might expect individuals with OCD to classify correct responses as errors, but some studies find elevated CRNs in OCD (e.g., Endrass et al., 2008; Hajcak & Simons, 2002) whereas others do not (e.g., Gehring et al., 2000; Hajcak et al., 2008; Stern et al. 2010). Interestingly, Hajcak and coworkers (2008) found that children with OCD show the exaggerated ERN both before and after treatment with cognitive-behavioral therapy, despite an improvement in symptoms. For this reason, they suggested that the ERN might serve as a useful endophenotype for OCD (Figure 10.18B).

Studies of the ERN in depression are somewhat inconsistent, with some studies reporting no difference between patients and controls (Ruchsow et al., 2004, 2006), others reporting an enhancement in patients (Chiu & Deldin, 2007; Holmes & Pizzagalli, 2008), and another finding a reduction, but only in depressed individuals who also showed psychomotor retardation (Schrijvers et al., 2008). Interpreting these discrepancies is difficult because of differences among the studies in whether patients were in remission or currently depressed, were taking medication, and had a comorbid anxiety disorder, as well as in the severity of their depression. Comorbid

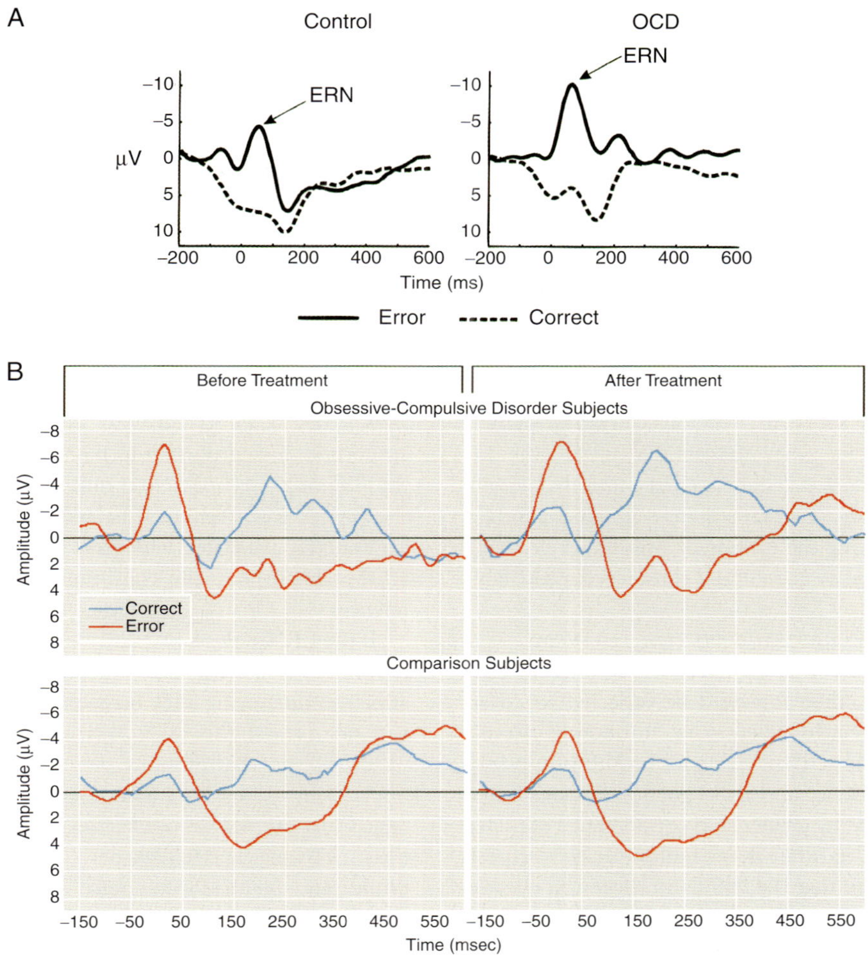

Fig. 10.18. (A) Error-related negativity waveforms from patients with OCD and healthy controls, showing a larger ERN in individuals with OCD. From Gehring et al. (2000), Figure 1, reprinted with permission of John Wiley & Sons, Inc. (B) Error-related negativity waveforms from children with OCD and controls, showing that the exaggerated ERN in OCD does not change with successful cognitive-behavioral treatment. From Hajcak et al. (2008), Figure 1, reprinted with permission from American Psychiatric Publishing, Inc.

anxiety, for example, was not excluded in either of the two studies reporting an enhancement. Schrijvers and colleagues (2009) suggested that ERN enhancement may occur in mild to moderate depression, in which anhedonia, apathy, and psychomotor slowing are less severe and the ERN-enhancing effects of comorbid affective distress are more apparent.

Studies using questionnaire measures are fairly consistent in showing elevated ERN amplitudes associated with negative affect and anxiety. Negative affect, as measured by self-report, has tended to be associated with increased ERN amplitudes (Hajcak et al., 2004; Luu et al., 2000a; but see Compton et al., 2008), although the effect may be limited to the period early in practice, before the individual with negative affect disengages from the task (Luu et al., 2000a). Questionnaire measures of trait anxiety (reflected in a high level of worry) and state anxiety are also associated with an increase in the size of the ERN (Hajcak et al., 2003a; Vocat et al., 2008, respectively). A few studies have examined the relationship between the ERN and the behavioral inhibition and activation systems (BIS/BAS) postulated by Gray (e.g., Gray, 1987). The BIS is associated with sensitivity to punishment, and greater BIS activation is associated with anxiety disorders. Boksem and colleagues (2006b) and Amodio and colleagues (2008) reported an enhancement of the ERN associated with high levels of behavioral inhibition. Cavanagh and Allen (2008), however, failed to find a global ERN difference between high- and low-BIS adults, although high-BIS individuals showed a relationship between flanker-task ERN amplitude and cortisol reactivity in a subsequent condition involving social evaluative threat. In a longitudinal study, McDermott and coworkers (2009) found that children who showed elevated behavioral inhibition in early childhood (4–7 years old) exhibited enhanced ERN activity as adolescents (15 years old). There was a tendency among a subset of children with high behavioral-inhibition scores to show a relationship between ERN amplitude and the risk of an anxiety diagnosis.

Belief systems that may affect anxiety may also affect the ERN. Inzlicht et al. (2009) reported that individuals with greater religious conviction showed reduced ERNs. They suggested that religion provides a buffer that reduces anxiety, and they argued that political conservatism may provide a similar function (citing the finding of Amodio et al., 2007, that conservative individuals show a reduced ERN).

AUTONOMIC NERVOUS SYSTEM ACTIVITY

If the ERN reflects an affective response to error commission, then it would be expected to covary with error-related changes in autonomic nervous system (ANS) activity. Evidence for this proposition is mixed. Hajcak and colleagues found that errors were followed by greater skin conductance responses (SCRs) and heart rate deceleration relative to correct responses, but that the ERN amplitude did not correlate with either of these measures (Hajcak et al., 2004; see also van Boxtel et al., 2005). Individuals high in trait-negative affect show larger error-related SCRs, as well as larger-amplitude ERNs (Hajcak et al., 2004). One positive finding was that of Hajcak and Foti (2008), who found that startle blinks following errors were greater than those following correct responses and that the ERN predicted the degree of blink potentiation. Also, Dywan and colleagues (2008) reported that respiratory sinus arrhythmia accounted for a relationship between a measure of the ERN and a measure of sadness.

AFFECTIVE AND MOTIVATIONAL VARIABLES

Other studies relating the ERN to affect have manipulated the motivational significance of the error response using monetary rewards and other incentives. Hajcak et al. (2005) manipulated the value of a correct response and found that errors that failed to earn a high-value (100-point) outcome elicited larger ERNs than those associated with a lower value (5 points). In a second experiment, Hajcak et al. (2005) showed that the size of the ERN was increased in the presence of a male experimenter who was seated next to the subject, evaluating the subject's performance. A similar finding was reported in children (Kim et al., 2005). Pailing and Segalowitz (2004b), however, reported that monetary incentive effects on the ERN were limited to subjects scoring high on a neuroticism scale. Ganushchak and Schiller (2008) compared blocks in which trials were associated with a monetary incentive to blocks without the incentive and reported that the incentive blocks yielded larger and later ERNs. Boksem and colleagues (2006a) reported a study of fatigue and the ERN in which a motivational incentive late in task performance caused an increase in response accuracy in some subjects that was associated with an increase in ERN amplitude. It is important to point out that, despite the differing incentives, the motivational effects in the Hajcak et al. and Ganushchak and Schiller studies were not associated with (and therefore were not attributable to) between-condition performance differences.

If the ERN reflects primarily an affective process, it would make sense that inducing short-duration affective states using affective pictures would modulate the size of the ERN. Larson and coworkers (2006) found that pleasant pictures superimposed on flanker stimuli increased the size of the ERN relative to neutral or unpleasant pictures. In contrast, Wiswede and colleagues (2009) found that unpleasant pictures presented 700 ms prior to flanker stimuli enhanced the ERN relative to the neutral or pleasant pictures. As Wiswede et al. note, the different demands on attention in the two tasks could account for the contradictory findings. Indeed, it is difficult to come up with a solid prediction about how such pictures will affect the ERN without a theory that specifies the processing evoked by the pictures in a particular task and how that process affects the computation reflected by the ERN. A gruesome autopsy photograph might make an error seem not so bad by comparison, or it could induce a bad mood that makes the error seem worse than it otherwise would. Either possibility could be compatible with a theory in which the ERN reflects an affective response to the error.

EVALUATION

The data reviewed here suggest that the ERN is sometimes related to variables that reflect or influence affective or motivational processing and is sometimes associated with changes in ANS activity. Nevertheless, such findings could be consistent with almost any theory of the ERN, because most computations that evaluate conflict or error processing could give rise to an affective response, and because affective and motivational manipulations could influence attention (Yeung, 2004). Thus, although the notion that the ERN is an affective response has had substantial heuristic value and has yielded a number of important findings, it suffers from the same problem as the other ERN theories, namely, that there are multiple alternative theories that can predict the same results and that theorists have thus far failed to carry out truly competitive tests of well-specified alternatives. Computational models that specify how the affective processes differ from the other cognitive and attentional processes would help in this endeavor.

Neurotransmitters

The reinforcement-learning and affective-response perspectives on the ERN have motivated studies of the neurotransmitters involved in the generation of the ERN. Specifically, the RL-ERN theory argues that the ERN is produced by a disinhibition of pyramidal neurons in the ACC following a phasic decrease in the activity of dopaminergic neurons in the basal ganglia (Holroyd & Coles, 2002). This claim has motivated several studies of drug effects on the ERN (for an excellent review of this topic, see Jocham & Ullsperger, 2009). Similarly, researchers have examined the ERN in the context of psychiatric disorders and genetic differences tied to changes in dopamine function. Nevertheless, a fair amount of evidence points to influences from other neurotransmitters.

Dopamine

Several drug studies report results consistent with a link between the ERN and dopamine. Haloperidol, a dopamine antagonist, reduces the amplitude of the ERN (de Bruijn et al., 2006b; Zirnheld et al., 2004), as does the atypical antipsychotic olanzapine (which also blocks serotonin and histamine; de Bruijn et al., 2006b). Amphetamine, which blocks dopamine uptake and promotes its release, enhances the ERN (de Bruijn et al., 2004). Similarly, caffeine, an indirect dopamine agonist, also causes increased ERN amplitude (Tieges et al., 2004). Nevertheless, there are questions about the specificity of both caffeine and amphetamine to the dopaminergic system (see Jocham & Ullsperger, 2009).

Because Parkinson's disease and Huntington's disease patients have low dopamine levels in the basal ganglia, RL-ERN theory predicts an attenuated ERN in those disorders. Several studies have confirmed this prediction (Beste et al., 2008; Falkenstein et al., 2001a; Ito & Kitagawa, 2006; Stemmer et al., 2007). Another found no difference in ERN amplitude between patients and controls (Holroyd et al., 2002). Although the balance of that evidence seems consistent with the RL-ERN theory, the results are complicated by performance differences between patients and healthy controls: in most studies showing a difference, patients were either slower (Beste et al., 2008), less accurate (Stemmer et al., 2007), or both (Falkenstein et al., 2001a; Ito & Kitagawa, 2006). In two studies the Parkinson's disease was relatively mild, so patient and control performance was roughly equal. One study failed to find a difference between groups (Holroyd et al., 2002) and another study, with a larger sample size, showed a reduction in the ERN but not in the CRN, effects that were equivalent in patients on and off medication (Willemssen et al., 2008). As Jocham and Ullsperger (2009) noted, the latter finding could suggest that the ERN is not sensitive to the acute administration of a dopamine agonist, although it is also possible that the 12-hr withdrawal period was simply not long enough to show effects.

Schizophrenia, also thought to involve dopamine dysfunction, is associated with ERN amplitudes that are reduced (Alain et al., 2002; Bates et al., 2002, 2004; Ford, 1999; Kim et al., 2006; Kopp & Rist, 1999; Mathalon et al., 2002; Morris et al., 2006) but can recover somewhat when patients are treated with antipsychotics (Bates et al., 2004). Some of these studies also reported increases of the CRN (Alain et al., 2002; Mathalon et al., 2002) and no effect on the Pe (Alain et al., 2002; Bates et al., 2004; Mathalon et al., 2002; Morris et al., 2006). It does not appear that performance differences alone can account for the ERN reductions (Bates et al., 2004), although the extent to which chronic medication may influence the results is unclear (see, e.g., Morris et al., 2008).

Genetic polymorphisms that affect neurotransmitter systems provide an additional way to assess dopamine contributions to the ERN. For example, the val/met polymorphism of the catechol-*O*-methyltransferase (COMT) gene is associated with levels of dopamine in the frontal cortex. Frank and colleagues (2007) found that the ERN did not differ between met/met individuals (with higher prefrontal dopamine levels) and val/met or val/val individuals. Interestingly, the polymorphism did affect the late Pe. Krämer and coworkers (2007) examined the COMT gene as well as the dopamine D4 receptor gene (DRD4) and found that individuals homozygous for the T allele of the DRD4 gene (associated with receptor responsiveness to dopamine) showed a larger ERN than individuals homozygous for the C allele. The COMT gene results were less clear, as there was a marginal effect such that the ERN was larger in val/val individuals than in met/met individuals, but only in a stop-signal task. Comparison of the Frank et al. and Krämer et al. studies shows that predictions in such studies are not always straightforward: according to Frank et al., the lower COMT in met/met individuals will cause higher levels of tonic dopamine and should thus result in a larger ERN than in val/val individuals. Krämer et al., however, predicted a smaller ERN in the met/met individuals, noting that lower COMT in those individuals should indeed be associated with greater levels of tonic dopamine but *lower* levels of *phasic* dopamine. (The Holroyd and Coles, 2002, model concerns phasic changes in dopamine.)

Other Neurotransmitters

Several studies have examined drug effects on other neurotransmitter systems. Alcohol reduces the size of the ERN (Easdon et al., 2005; Ridderinkhof et al., 2002), although the effect might be attributable to a degradation in stimulus processing (Yeung & Cohen, 2006; Yeung et al., 2007b) rather than to a direct role for enhancement of GABA receptors on generating the ERN. The benzodiazepines alprazolam (Riba et al., 2005b) and oxazepam (Johannes et al., 2001b) also reduce the ERN, perhaps consistent with a more direct role for GABA. Serotonin involvement is less clear: de Bruijn and colleagues (2006) found that the selective serotonin reuptake inhibitor (SSRI) paroxetine did not affect the ERN. Nevertheless, SSRI effects in the treatment of depression and OCD take several weeks, so that definitive conclusions regarding SSRI effects would seem to require longer-term administration of the drug. Gene studies seem to indicate some serotonin involvement: a study of individuals possessing one or two short alleles of the serotonin transporter gene 5-HTTLPR (thought to be associated with enhanced serotonin levels) showed those individuals to have a larger ERN than individuals with two long alleles (Fallgatter et al., 2004; Figure 10.19A). The finding was replicated in a group of children, although the ERN measure in that study could have been influenced by overlapping ERP components (Althaus et al., 2009). Beste et al. (2010a) reported that another polymorphism related to serotonin, the functional 5-HT1A C(-1019)G polymorphism, shows a relationship to ERN amplitude consistent with that reported by Fallgatter and colleagues, namely, a greater ERN being associated with increased serotonin: the CC genotype group showed a larger ERN than the CG and GG groups. Interestingly, the CRN did not differ between groups (see Figure 10.19B). Nevertheless, caution is warranted: a recent study failed to find a relationship between 5-HTTLPR and the ERN (Olvet et al., 2010).

Evaluation

The Holroyd and Coles (2002) theory has focused the spotlight on dopamine as an important contributor to the ERN, and some of the evidence is consistent with their model. Yet the evidence above suggests that there is room for neurotransmitter models that consider alternative roles for dopamine, such as the Jocham and Ullsperger (2009) model, as well as models incorporating other neurotransmitters. Jocham and Ullsperger (2009), for example, noted that acetylcholine has to date not been studied. Indeed, Sarter and coworkers (2006) pointed out that the functions of acetylcholine in effortful attention suggest that there is good reason to think that this neurotransmitter might be important for generating the ERN.

Development and Individual Differences

Initial studies of children suggested that clear ERNs that were distinguishable from CRNs were not evident until middle to late adolescence (Davies et al., 2004; Ladouceur et al., 2004, 2007). Nevertheless, Friedman et al. (2009) showed clear ERNs in a group of 10-year-old children. In addition, some children in that age range (10–11 years old) with anxiety disorders (Ladouceur et al., 2006) or greater socialization (Santesso et al., 2005) show a clearer ERN than age-matched controls. More recent evidence suggests that the ERN may be evident in children as young as 8 (McDermott & Fox, 2009) or even 5 to 7 years old (Torpey et al., 2009). A report of an ERN in 4-year olds (Brooker et al., 2011) is difficult to evaluate because it did not present an average waveform that isolated that age group. An ERN-like stimulus-locked potential has even been observed in 7-month-old infants observing impossible events (Berger et al., 2006). Although studies seem to indicate a trend for increasing ERN amplitude during adolescence (Davies et al., 2004; Ladouceur et al., 2004, 2007; Santesso & Segalowitz, 2008), there may be an interaction with task difficulty such that the developmental trend during adolescence is only seen for complex tasks (Hogan et al., 2005; but see Santesso & Segalowitz, 2008). The latter finding suggests that task differences might

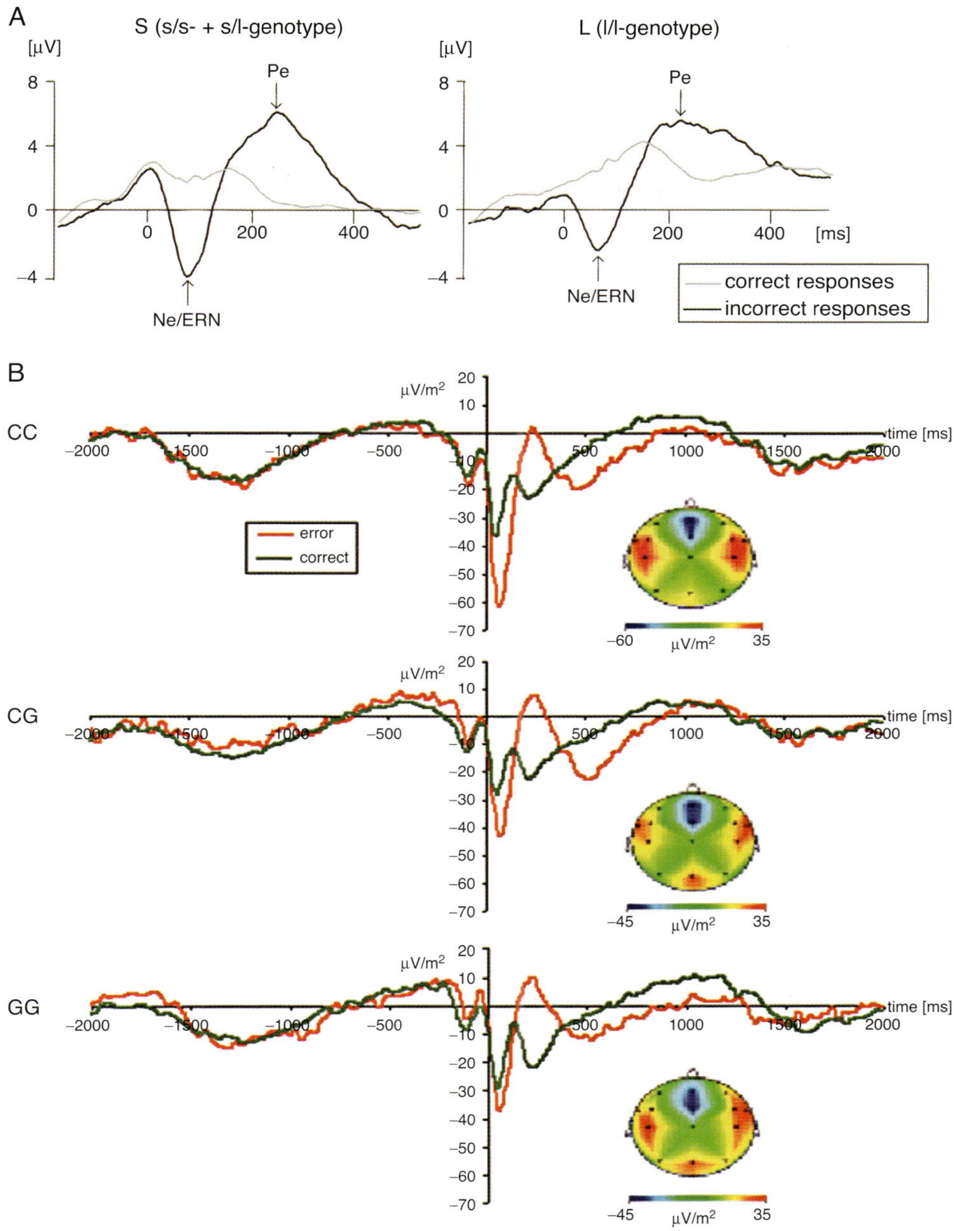

Fig. 10.19. Effects of serotonergic genotypes on the ERN. (A) Error-related negativity waveforms from subjects with short (left) or long (right) allelic variations of a serotonin transporter gene (5-HTTLPR). Individuals with one or two copies of the short allele showed a larger ERN than individuals homozygous for the long allele, suggesting that serotonin influences the generation of the ERN. From Fallgatter et al. (2004), Figure 1, reprinted with permission from Macmillan Publishers Ltd. (B) Response-locked ERPs at electrode Fz for correct (green) and error (red) trials. Separate plots show waveforms for different 5-HT1A genotype groups: top, CC; middle, CG; bottom, GG. This polymorphism influences serotonergic neurotransmission. From Beste et al. (2010), Figure 1, reprinted with permission of John Wiley & Sons, Inc.

explain some of the discrepant findings concerning younger children.

At the other end of the lifespan, numerous studies have reported a reduction in the amplitude of the ERN in older adults relative to younger controls (e.g., Band & Kok, 2000; Falkenstein et al., 2001b; Gehring & Knight, 2000; Nieuwenhuis et al., 2002), as well as an increase in its latency in some tasks (Falkenstein et al., 2001b). Two studies, however, in which older adults responded as accurately as younger adults failed to find a significant age-related ERN reduction (Eppinger et al., 2008; Friedman et al., 2009). Additional ambiguities remain. Pietschmann and colleagues (2008), for example, reported a study in which ERN amplitudes did not differ in older and younger adults, but in older adults there was no error effect—the CRN was just as large as the ERN. A CRN enhancement was also observed by Eppinger et al. (2008). One key to disambiguating these findings may lie in the effects of cardiorespiratory fitness and physical activity on the ERN in aging adults and the fact that different subject samples might differ in this important variable. Interestingly, however, one study examining this issue found that both younger and older physically active adults showed smaller ERNs than their less active counterparts, which is paradoxical if physical activity is thought to mitigate the effects of aging (Themanson et al., 2006). Nevertheless, studies of aging face the same challenge as studies of Parkinson's disease: the slowing of older adults presents a confound that cannot be corrected simply by speeding the responses of the slower group because of the confound introduced by equating the two groups' performance. In addition, the results of Yeung et al. (2007b) discussed above suggest that future studies should consider whether altered stimulus processing in older adults could be a source of age-related ERN reductions.

We have already reviewed some individual-difference studies related to negative affect and anxiety. Another category of individual differences with obvious relevance to the ERN is that involving measures of impulse control: if the ERN is larger in conditions where accuracy is emphasized over speed, then one might predict that the ERN would be smaller in individuals who are impulsive or have difficulty controlling their behavior. Findings have been inconsistent. Pailing and coworkers (2002) derived a simple measure of impulsivity based on each subject's error trial RT. As predicted, subjects showing the most impulsive behavior showed the smallest ERNs (see also Ruchsow et al., 2005b). It is not necessarily the case, however, that fast responding reflects an impulsive personality; thus, more definitive links to trait-level impulsivity come from studies using questionnaires and other diagnostic means of assessing the impulsive personality. Stahl and Gibbons (2007) used a questionnaire measure of impulsivity to categorize participants in a stop-signal task. They reported less ERN activity in the more impulsive subjects, although their analyses focused on an ERN-like stimulus-locked component. Potts et al. (2006) reported a similar reduction in high-impulsive individuals, although the effect was restricted to a condition where errors were penalized (as opposed to a condition where correct responses were rewarded, where no such difference was observed). At least two studies, however, have failed to find a relationship between the ERN and questionnaire measures of impulsivity (Santesso & Segalowitz, 2009; Vocat et al., 2008).

Studies of clinical impulse-control disorders are also inconsistent. There are reports of children with attention deficity hyperactivity disorder (ADHD) showing reduced (Groen et al., 2008; Liotti et al., 2005), equivalent (Jonkman et al., 2007; Wiersema et al., 2005), or enhanced (Burgio-Murphy et al., 2007) ERNs relative to controls. Wiersema and colleagues (2009) found no significant difference between adults with ADHD and controls, although their waveforms show a larger ERN in the controls. Across studies there are task differences, between-group performance differences, and ERN baseline/measurement issues that could contribute to the inconsistent findings (see Shiels & Hawk, 2010, for a review).

Among the personality disorders characterized by difficulty with impulse control is psychopathy. In the first study of this disorder, Dikman and Allen (2000) found a reduction in the ERN in conditions where errors were penalized for individuals scoring low on a socialization questionnaire thought to reflect proneness to psychopathy. Three studies of violent offenders with psychopathy offer mixed results: two reported a significant ERN reduction in violent offenders with psychopathy (Borries et al., 2010; Munro et al., 2007), although in one of these studies the effect was seen in a task involving face stimuli but not in a flanker task (Munro et al., 2007). Another study using a flanker task failed to find a significant reduction, although the reported ERN amplitude was 2 μV smaller in the offenders (Brazil et al., 2009). Psychopathy in these studies was assessed using the unidimensional Psychopathy Checklist-Revised (Hare, 1991), whereas a large body of evidence suggests that the disorder actually comprises at least two factors, one related to externalizing and the other involving trait fearlessness (Patrick & Bernat, 2009). Externalizing is a personality construct thought by some to be common to a number of different impulse-control disorders, including substance abuse, antisocial behavior, and, in children, conduct disorder. Hall et al. (2007) found that individuals with high scores on an externalizing scale showed a reduced ERN relative to low-externalizing individuals. As in the studies of ADHD, the numerous differences in subject characteristics, tasks, and ERN measurement methods make it difficult to draw definitive conclusions. If anything, the pattern seems consistent with at least some reduction associated with the externalizing component of psychopathy.

Other individual difference studies are generally consistent with a link between a reduced ERN and impulsive or risky behavior. De Bruijn and coworkers (2006) found a reduced ERN in borderline personality disorder, which is also characterized by impulse-control problems. Pailing and Segalowitz (2004b) found that individuals low in conscientiousness showed a reduced ERN response to monetary incentives. Santesso and Segalowitz (2009) found that individuals scoring high on a measure of risk propensity (comprising risk taking and sensation seeking) showed smaller ERNs. Frank and colleagues (2005, 2007) derived a measure of reward-based learning biases from a reinforcement-learning task, finding that subjects who learned to seek positive outcomes showed a smaller ERN than subjects who learned to avoid negative outcomes.

Evaluation

One criticism of these studies (and the clinical studies reviewed earlier) is that most of them focus on a single personality or clinical construct, without regard for other correlated personality variables or psychopathology. Studies rarely attempt empirically to rule out alternative hypotheses. Yet, it is clear that drawing inferences based on a single personality or psychopathology questionnaire is perilous. In clinical research, questionnaires can be checked against other clinical diagnostic assessments, and here the facts are sobering: the Obsessive-Compulsive Inventory-Revised used to measure obsessive-compulsive tendencies has a sensitivity of 65.6% and a specificity of 63.4% when a cut-off score of 21 is used to decide whether an individual has OCD (Foa et al., 2002). Still, according to Bayes' theorem, if 2% of the population has OCD, an individual scoring over 21 on the OCI-R has less than a 4% chance of actually having OCD.

It is therefore much more informative when studies, such as those of Santesso and Segalowitz (2009), Vocat et al. (2008), and Inzlicht et al. (2009), examine multiple personality constructs in the same individuals. Especially promising are studies that include measures both of personality traits and of psychiatric symptoms (e.g., Chang et al., 2010). In the future, it would seem more useful to have more large-scale studies that simultaneously assess numerous personality and psychopathology constructs rather than continuing to carry out small-scale, single-construct studies. But even research incorporating numerous control assessments leaves open the possibility that some construct that was not assessed is the one really driving the result. Cardiorespiratory fitness, for example, can be associated with changes in ERN amplitude (Themanson & Hillman, 2006; Themanson et al., 2006); in many studies, subject groups could have differed in their fitness levels. Task engagement is an even more troublesome variable. Tops and coworkers (2006) offer a simple explanation for a somewhat paradoxical finding—that higher agreeableness, a positive trait, and higher behavioral shame proneness, a negative one, are both associated with larger ERNs. According to Tops et al., subjects scoring high on either measure tend to be more engaged in the task than their low-scoring counterparts (recall also the finding of Luu et al., 2000a). Thus, individual differences might in many cases be explained by differences in task engagement: two groups of subjects whose brains have exactly the same ability to generate the ERN can show ERN amplitude differences simply because one group of subjects is less involved in the task. Similar undesirable confounds can arise if subjects differ in their representation of or sensitivity to the social evaluation inherent in an experiment (Cavanagh & Allen, 2008; Hajcak et al., 2005b).

A final important challenge for individual difference research is that it must seek a better understanding of the manner in which the outcome of a study can depend on the particular task used to elicit the ERN and the particular techniques used to measure it (Hogan et al., 2005; Munro et al., 2007). Studies of personality constructs and psychiatric disorders would have limited generalizability if simply changing the subjects' task altered the findings.

New Directions

The preceding sections have sketched some of the major findings and controversies in ERN research. In this section, we outline some of the issues that are emerging as important and will, in our opinion, shape the coming years of ERN research.

Beyond Choice Reaction Time

With the exception of the FRN studies, ERN research has focused almost exclusively on speeded reaction time (RT) tasks, such as choice RT and go/no-go RT tasks, and most of these have measured only button-press closure, neglecting richer measures such as response force, movement velocity, and EMG. A few recent studies have turned to more realistic aimed-movement and force-production tasks, beginning a likely trend in which the human movement control literature can influence ERN research. Such a trend may help to resolve the extent to which the ERN is involved in immediate error correction versus strategic control.

The complex motor tasks employed thus far include force production (de Bruijn et al., 2003), aimed movement (Anguera et al., 2009; Krigolson & Holroyd, 2007a), pointing (Vocat et al., 2011), and manual tracking tasks (Krigolson & Holroyd, 2006, 2007b). Negative-going potentials have been reported in such tasks when errors occur in the choice of force (de Bruijn et al., 2003) or when external perturbations cause an error in tracking or aimed movements (see Figure 10.20; Anguera et al., 2009; Krigolson & Holroyd, 2006, 2007a, 2007b). Some of these studies have concluded that the ERN is related to high-level errors, where the subject selects an action that violates a task goal, rather than low-level errors, where the subject chooses the correct action but the movement deviates from the planned trajectory (Krigolson & Holroyd, 2006, 2007a, 2007b). Another study has come to the opposite conclusion, however, based on evidence that ERN-like activity is sensitive to the extent of the low-level error (Anguera et al., 2009; see also Vocat et al., 2011). Nevertheless, there is little direct evidence that the slower potentials presented as evidence for these conclusions are indeed the ERN. Alternative possibilities are that the slow potentials represent contingent negative variations (CNVs; see Chapter 8, this volume) preceding anticipated target perturbations and movement-monitoring potentials (MMPs) associated with the ongoing movement (de Bruijn et al., 2003). Additional work is needed to determine more conclusively what components are elicited in these complex motor control tasks. An additional innovation involves measuring ERN-like potentials at the time of the response when an alternative in a gambling task is chosen, where the ERN-like potential appears to be larger for riskier choices (Yu & Zhou, 2009), an effect that is more pronounced in low-impulsive subjects (Martin & Potts, 2009). In these and any other extensions of the traditional paradigm, bridging studies are needed that elicit both the classic ERN in a simple motor task and the negativity in the alternative task, comparing the two components in the same subjects and, ideally, finding intermediate conditions in which the classic ERN transitions to the more atypical response.

The Bridge to the Brain

Another type of bridge that will play a critical role in unraveling the mystery of the ERN is the one that can be established between studies of scalp ERPs in humans and those of single-cell and LFP recordings in nonhuman primates. Thus far, neurophysiological studies of the ACC and other structures have yielded findings that are consistent with each of the major theories of the ERN reviewed here (e.g., Emeric et al., 2008; Ito et al., 2003), but there are also discrepancies (the most glaring being the lack of ACC conflict-related activity; Emeric et al., 2008; Ito et al., 2003). It is still not clear, however, how the various intracranial potentials propagate and summate to produce the scalp-recorded ERN. Links between the monkey studies and those in humans will be strengthened by further studies of saccade countermanding in humans that compare directly to those in monkeys, and by further studies of manual response tasks in monkeys that are more similar to those in humans. Moreover, the scalp-recorded ERN probably reflects the activity of more than one intracranial generator, and even individual LFP recordings can reflect several superposed sources; thus, recordings of current source density across a broad swath of the medial frontal cortex will be especially informative (Emeric et al., 2008).

ERN in the Social World

OBSERVED ERRORS

The idea that the ERN represents performance monitoring leads naturally to the question, does someone show an ERN when he or she sees someone else make an error? Several recent studies have reported evidence that direct experience and vicarious observation of negative events rely on common neural substrates. Both error commission and error observation elicit negative-going ERP components over medial-frontal scalp sites (Bates et al., 2005; Miltner et al., 2004; van Schie et al., 2004). Both components can be explained by a source in ACC (Miltner et al., 2004; van Schie et al., 2004), and both are accompanied by an increase in evoked theta power relative to correct responses (Bates et al., 2005). The so-called observation ERN is elicited by error observation in a choice response task (Miltner et al., 2004), a flanker task (van Schie et al., 2004), and a go/no-go task (Bates et al., 2005), as well as during observation of a human performer (Bates et al., 2005; Miltner et al., 2004)

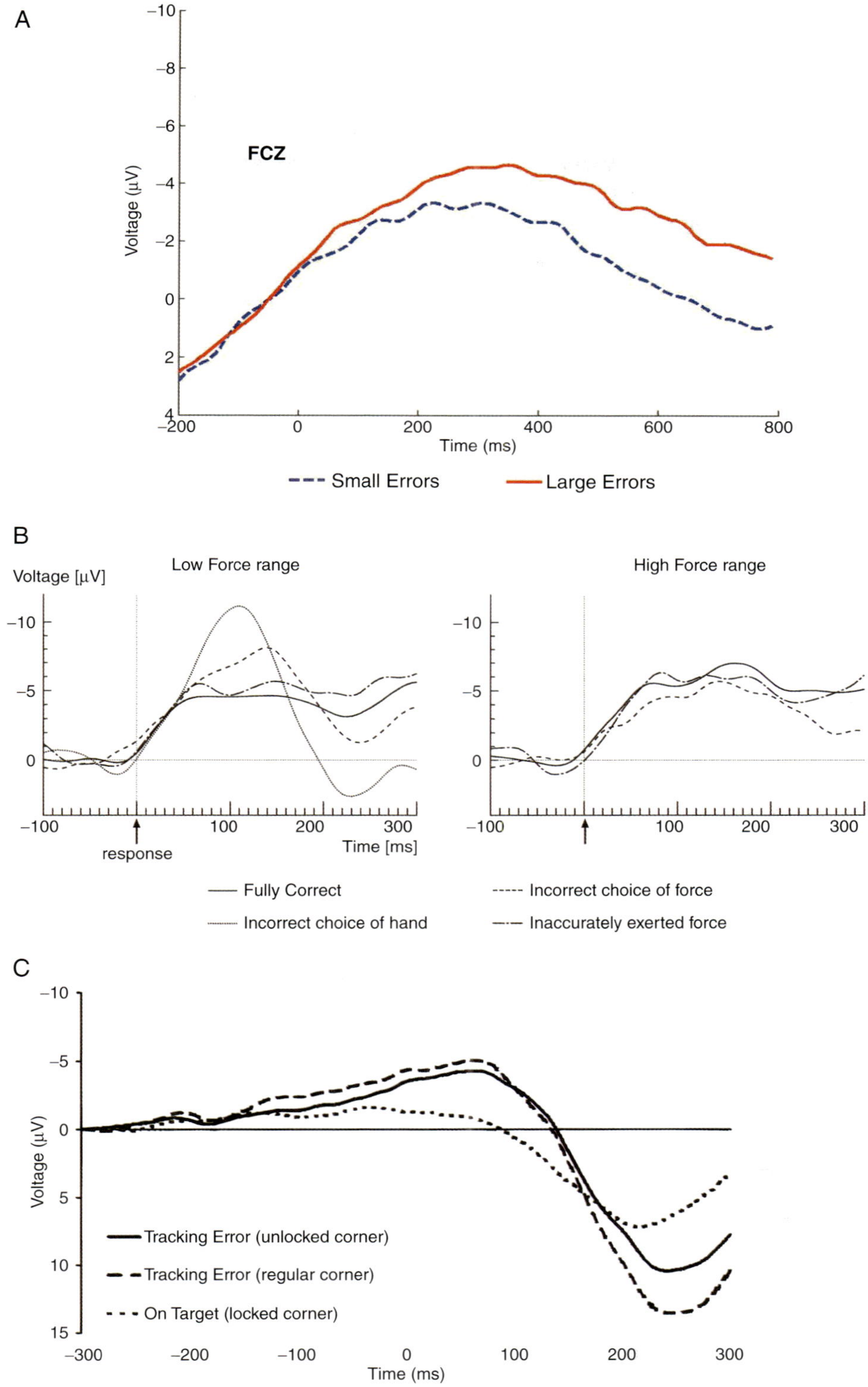

Fig. 10.20. Error-related negativity–like potentials elicited in motor control tasks, showing prolonged error-related activity that bears some similarity to the ERN. (A) Error-related negativity–like activity in a visuomotor adaptation task where participants adapted manual aiming movements to a 30° rotation of the visual feedback display. Movement errors were classified as high (red dashed) and low (blue solid) errors according to the distance of the initial endpoint of the movement from the target. From Anguera et al. (2009).

(Continued)

or "virtual feedback" regarding another person's performance (Miltner et al., 2004). An important consideration in these studies is that observed errors tend to be infrequent events, raising the possibility that the observer's ERN is an N200. Supporting this contention, de Bruijn et al. (2007) found that no ERN was elicited in an observer when the observed errors were as likely as correct responses (but see Bates et al., 2005).

Analogous results have been reported for the FRN. Both the direct experience and the observation of losses in a gambling task elicit an FRN-like wave (Fukushima & Hiraki, 2006; Yu & Zhou, 2006). However, the consequences of the observed individual's outcomes for the observer appear to be important: Itagaki and Katayama (2008) found that the FRN in an observer is larger when the observed individual loses than when that person gains—if the two individuals are cooperating. When the two individuals are competing, however, the gain of the observed individual will cause the larger ERN. Effects of cooperation versus competition on the observation FRN may be modulated by the gender and personality of the observer (Fukushima & Hiraki, 2006).

SOCIAL NEUROSCIENCE

The extension of cognitive neuroscience and psychophysiological methods to social, affective, and cultural neuroscience (e.g., Cacioppo et al., 2004) has led to ERN research being extended to the social world (e.g., Amodio et al., 2004, 2006; Inzlicht & Gutsell, 2007). For example, one issue of interest in contemporary social neuroscience concerns implicit attitudes—specifically, implicit race biases—and how individuals might use cognitive control when they are faced with social situations that activate the biases (e.g., Gehring et al., 2003; Richeson et al., 2003). Amodio and colleagues (2004, 2006) found an enhanced ERN in a task where individuals had to determine whether a masked stimulus was a gun or a tool. Preceding the stimulus was a prime stimulus consisting of a black or white face. Basing the study on the conflict model (Botvinick et al., 2001; Yeung et al., 2004b), Amodio and colleagues inferred that a tool stimulus following a black face would be analogous to an incongruent Stroop or flanker trial, involving response conflict between the racially biased, prepotent "gun" response associated with a black face and the correct "tool" response. Errors on such trials were indeed associated with a larger ERN than other trial types. It is not clear, however, that the prediction was derived correctly from the conflict model. The error on a "tool" trial is the "gun" response, which should not conflict with the prepotent "gun" response suggested by the black face.

Social and cultural neurosciences are exciting areas in which the ERN might prove a useful measure. Of course, such studies must include control conditions to show that the effects of a social-level construct on the ERN are not simply explained by the effects of some lower-level variable that is confounded with the construct of interest (such as the possibility that the ERN could be affected by the relative luminance of guns and tools or of black faces and white faces rather than their significance with regard to race biases (Yeung et al., 2007b). As with the individual difference studies discussed above, studies that compare groups of people must take care to eliminate alternative hypotheses by measuring all individual difference variables that might account for observed group differences, rather than presenting groups defined by their responses on a single questionnaire or by their membership in a single social category.

Genetics and the ERN

One of the more exciting areas of recent progress is in studies that relate the ERN to genetic polymorphisms involved in neurotransmitter and neurotrophin function. The studies of Beste et al. (2010a), Fallgatter et al. (2004), Frank et al. (2007), and Kramer et al. (2007) reviewed above show the potential for this technique to shed light on the neural basis of the ERN. It is clear that many more genes are worth investigating. One that seems promising is the glutamate transporter gene SLC1A1, which has been implicated in OCD (Arnold et al., 2006; Dickel et al., 2006; Stewart et al., 2007) and

Fig. 10.20. (Continued)
Reprinted with permission from the American Physiological Society. (B) Grand average response-locked waveforms elicited by a force production task. Errors were classified by errors of hand, errors of force selection, and force exertion. From de Bruijn et al (2003), Figure 3, reprinted with permission of John Wiley & Sons, Inc. (C) Grand average response-locked waveforms from a motor tracking task on trials in which no tracking error occurred, in which a tracking error occurred for a relatively easy condition ("regular corner"), and in which tracking errors occurred during a relatively difficult condition ("unlocked corner"). Subjects had to maintain a cursor between two moving barriers; zero indicates the time of barrier contact on an error trial and a matched time on correct trials. Modified from Krigolson and Holroyd (2006a), Figure 1, reprinted with permission from Elsevier.

would thus be of interest in ERN studies. Evidence from a twin study for heritability of ERN amplitude provides additional impetus to characterize the genetics underlying the ERN (Anokhin et al., 2008), although heritability of an ERP component could reflect not only neurotransmitter function, but also head shape and brain morphology.

Of course, new methods bring with them new challenges. Sample size is a methodological challenge: early findings of gene–phenotype relationships based on small samples often overestimate true effect sizes compared with later studies based on large samples (Green et al., 2008). Some of the discrepancies noted earlier among genetic association studies and pharmacological studies might result from sample sizes that are too small. Conceptual challenges also abound. Beste et al. (2010b), for example, found a rather striking difference between one group that carried the Met allele of the brain-derived neurotrophic factor (BDNF) Val66Met polymorphism and another that did not. Still, because BDNF controls neurogenesis and cortical morphology (see Beste et al., 2010b), it is not clear to what extent ERP differences could be caused by differences in structural factors such as cortical morphology or synaptic density. Nor is it clear what it would mean functionally if those structural differences did play a role.

Key Issues in Designing and Interpreting ERN Experiments

Despite all the exciting advances we have just reviewed, the ERN waters have been muddied by a number of problems with the quality of the experimentation and data. Here we review these problems and show some examples. Note that we do so not to criticize any individual investigator; therefore, we removed identifying information from the examples we have provided to the extent possible (and we've also included ourselves in the criticism).

Experimental Design and Signal-to-Noise Ratio

Research on the ERN requires a lot of data. In order to achieve an adequate signal-to-noise ratio in the error-trial average, the task design must provide for a relatively large number of trials. Several realities make this difficult. Once a task is learned, errors tend to occur infrequently. An additional complication is that, as reviewed earlier, the bulk of the evidence suggests that a low error rate is necessary to elicit a large ERN, further increasing the amount of data that must be collected. Finally, most within-subject experimental designs require the comparison of at least two experimental conditions, thus doubling the number of trials necessary for the analysis of interest. Unfortunately, investigators have too often allowed the practicalities of experimentation to override the need for clean waveforms, settling for a small number of error trials. Consequently, there are numerous papers where the conclusions are questionable because of the amount of noise in the waveforms.

There is a very simple heuristic that readers can use to evaluate the quality of data contributing to an ERP waveform. A figure comparing two waveforms should theoretically show no difference between the waveforms during the baseline interval (just prior to the stimulus in a stimulus-locked average, or in the corresponding period in a response-locked average, several hundred milliseconds prior to the response). The amount of noise in that interval thus gives a sense of the amount of noise in the portion of the waveform being analyzed. In short, if visual inspection of the ERPs suggests that altering the baseline used for computing a base-to-peak or amplitude measure would alter the outcome of the study, one should be skeptical of the conclusions raised in the study. Figure 10.21 shows several examples where a noisy baseline epoch leads to just this sort of skepticism. Note that in many cases, authors fail to show enough baseline data to draw firm conclusions one way or the other.

Component Overlap

Some of the difficulties with ERN research stem from issues of component overlap. The ERN is a negative-going peak that overlaps at least one major positivity (the P3 or Pe) and most likely other negativities, and investigators must take care to ensure that differences in the ERN waveform are not a result of these overlapping components. Although component overlap is a concern with any ERP component, the situation with the ERN is especially difficult, because the process of response locking does not eliminate stimulus-related potentials occurring at around the same time as the ERN. Despite the ubiquity of this issue in ERP research and the sophisticated analysis techniques that can be employed to deal with it (e.g., Bernat et al., 2005; Woldorff, 1993; Zhang, 1998), numerous examples exist in the literature where overlapping components, rather than the ERN itself, could be responsible for the reported experimental effects (see Figure 10.21). As with noisy data, the baseline interval and other neutral epochs in the waveform

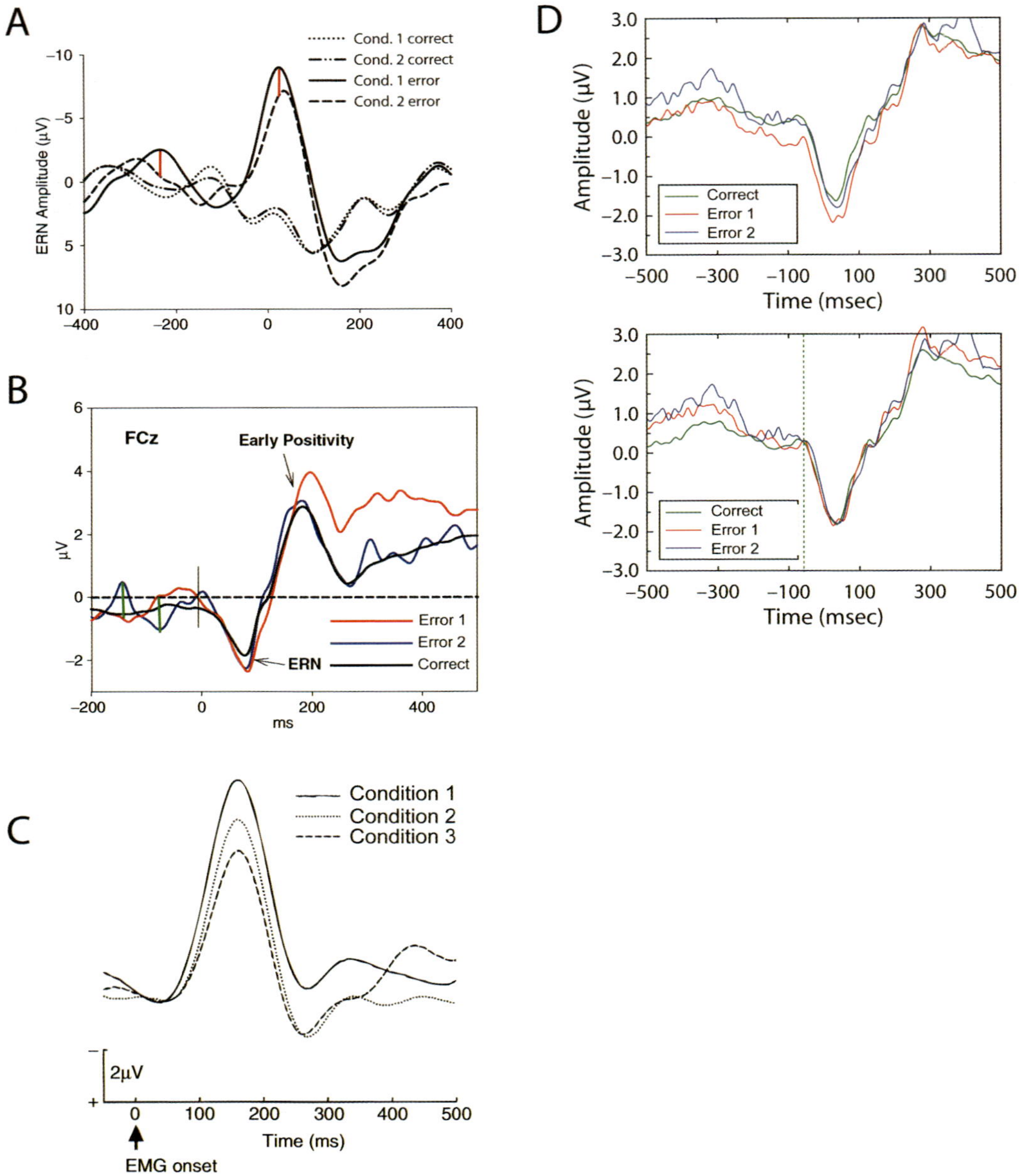

Fig. 10.21. Waveforms illustrating common problems in the ERN literature.

(A) The authors report a difference between the two error waveforms. The amplitude was computed as a negative peak from 50 to 150 ms following the response relative to a –400 to –50 ms preresponse baseline. The vertical red bars show how aligning the waveforms at a different point prior to the response would eliminate the reported effect.

(B) The authors report that there is no difference between the ERNs in the red and blue conditions. Nevertheless, the large, noisy fluctuations prior to the response show that taking the baseline at other plausible points (vertical green bars) would produce other effects, making the red larger than the blue in one case and the blue larger than the red in the other.

(C) The authors report ERN amplitude differences among the three waveforms. The differences in the baseline period and the presence of a difference that continues beyond the offset of the ERN suggest that overlapping components might contribute to the effect. Evaluating these data would be easier if the authors had plotted a longer preresponse interval.

(D) The authors reported that the three conditions differed, with Error-2 > Error-1 > Correct. The waveforms in the upper panel are shown as reported, using a preresponse baseline of –800 to –700 ms, which was not shown in the figure. The ERN was computed as the difference between the first negative peak following the response and the immediately preceding positive peak. Although peak-to-peak measures can help to remove overlapping low-frequency activity, the authors failed to report the time window in which the positive peak was detected. Alignment of the waveforms at the positive preresponse peak shows the amplitude of the ERN relative

(Continued)

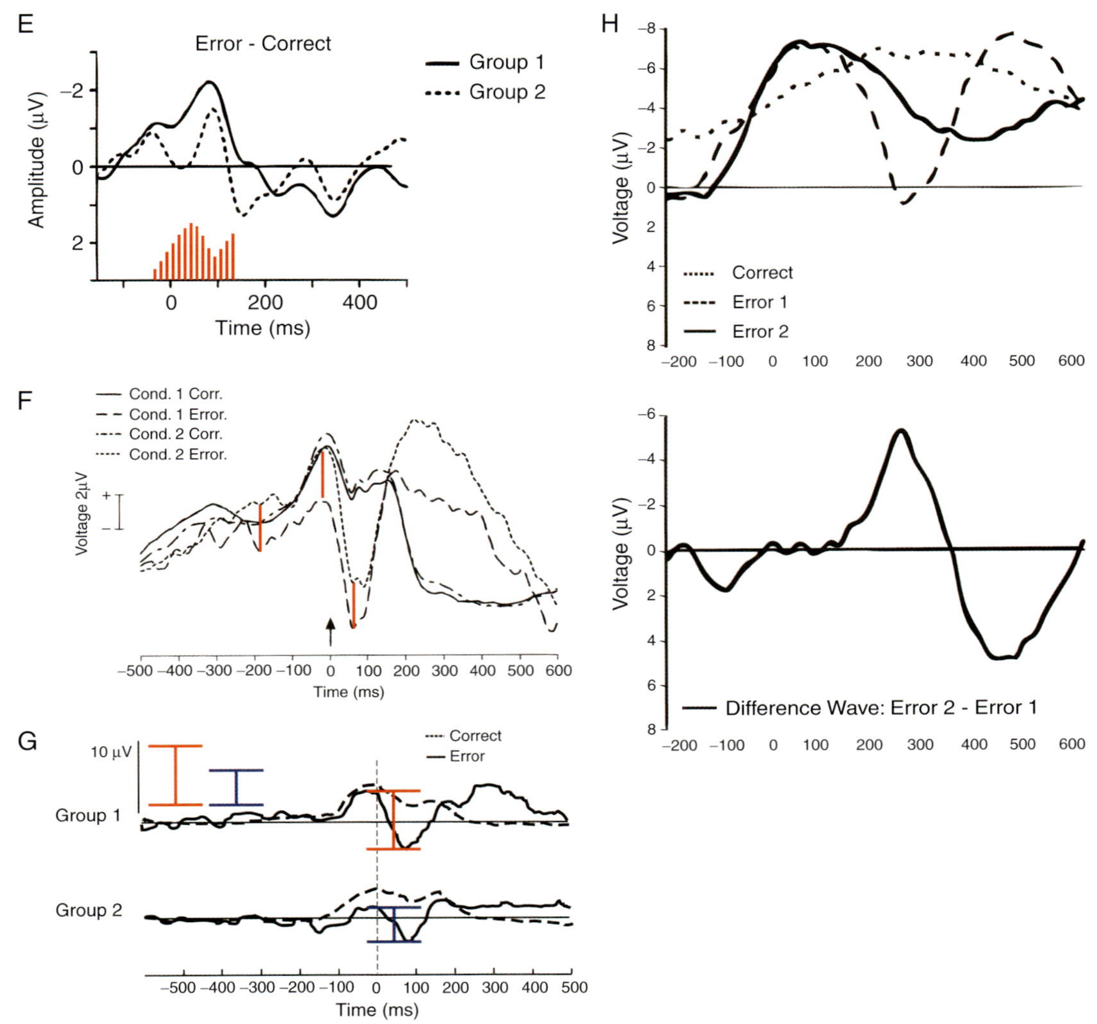

Fig. 10.21. (Continued)
to that peak to be approximately equal in the three conditions, suggesting that the authors' peak-to-peak measure was inappropriately influenced by the much earlier positivity. In addition, the amount of noise is a concern, as there are deflections in the Error-2 waveform prior to the response that exceed the putative postresponse amplitude difference in the ERN.
(E) The authors presented error-trial–correct-trial difference waveforms for two groups of subjects. Waveforms were baseline corrected to the average amplitude between 400 and 50 ms before the response. The entire baseline period is not shown. The height of the red vertical lines shows the difference between the two waveforms at each corresponding point in time. Note that the time course of the difference is certainly consistent with an ERN, but the onset and peak of the difference do not correspond to the onsets or peaks that visual inspection would assign to the ERN. Fortunately, individual averages for correct and error trials were also presented, but those waveforms present similar difficulties, including latency shifts in the underlying CRN and ERN peaks, making it difficult to interpret the effect shown here.
(F) The authors reported that Condition 1 Error shows a larger ERN than Condition 2 Error. The ERN was computed using a mean amplitude measure relative to a baseline interval (which was not shown) from 1000 to 750 ms prior to the response. The vertical red bars (of equal height) show that the difference between ERN peaks following the response is also present prior the response and that alternative baselines or peak-to-peak measures would eliminate the effect.
(G) The authors reported no difference between Group 1 and Group 2 in peak ERN amplitude. However, it appears that if they were to measure the ERN as the peak-to-peak difference (i.e., the difference in amplitude for the first positive peak preceding the response and the first negative peak following the response), there would be a difference between the two groups. The red bar denotes the peak-to-peak difference for Group 1, and the blue bar denotes the peak-to-peak difference for Group 2.
(H) The authors present the negative-going peak in the difference waveform (bottom panel) as evidence that there is an ERN in the Error-2 condition. The difference waveform was created by subtracting the Error-1 (dashed) waveform from the Error-2 (solid) waveform (shown in the top panel). Note that the large positive peak in the Error-1 condition appears to be responsible for the effect

(*Continued*)

(e.g., just prior to the response) can give the reader a sense of the likely contribution of component overlap. Judicious use of difference waveforms can help to increase one's confidence that observed effects show the time course and scalp distribution that an ERN effect should show (although difference waves should never be the sole basis of an analysis).

Baseline and Measurement Issues

Excessive noise and overlapping components both complicate the choice of how to quantify the amplitude or latency of the ERN—a choice that in itself can determine the outcome of an experiment. For example, most ERN researchers have had the experience of seeing a published waveform plot and wondering whether the conclusions of the study would have been eliminated or reversed simply by an alternative choice of baseline. Often the baseline interval is not even shown in the plot. Some investigators use a peak-to-peak measure of the ERN, and others filter the waveforms using a high-pass filter designed to eliminate low-frequency activity. Such precautions can help to reduce the contribution of low-frequency activity, but they are not infallible. Digital filters can introduce distortions (Yeung et al., 2004a, 2007a). In particular, while high-pass filters have been used to isolate the ERN from the slower overlapping Pe, they create an effect that is opposite in polarity and earlier in time from the activity in the unfiltered waveform. This artifactual activity can appear during the baseline epoch. Base-to-peak and peak-to-peak measures will thus still be influenced by the overlapping potential, and artificial oscillations will also appear (see Luck, 2005, chap. 5). Even without the use of high-pass filters, peak-to-peak measures are not as foolproof as they might seem for isolating the ERN from overlapping potentials. Peak-to-peak measures are sensitive to whatever is happening prior to the ERN peak, including slow potentials that could cause the peak-picking algorithm to choose the first point of the peak-picking window rather than a true positive peak. Peak measures are also especially susceptible to noise and the effects of component overlap (see Luck, 2005), both of which are concerns in ERN studies. Finally, while time-frequency analysis based on Morlet wavelets or time-frequency distributions might seem promising as a remedy for the problems of digital filters, the techniques are still new and their limitations are not yet well understood.

Difference Waveforms, Scatterplots, and Bar Graphs

When publishing a paper, it is tempting (especially when publishing for a non-ERP audience) to publish easily comprehended bar graphs, scatterplots, and difference waveforms (consisting of the ERP of one condition subtracted from that of another). Such figures simplify presentation of the data, but to a seasoned ERP investigator they are virtually useless unless supporting waveform data are presented with them. A key problem is that such figures do not permit the reader to determine the likelihood that a particular ERN effect could result from overlapping components, baseline problems, or noisy data. The problem is not solved in the numerous papers that show the ERN in an overall error-versus-correct-waveform comparison but then go on to bury the paper's most important finding in a scatterplot or bar graph. Supporting waveform data should always be included: average waveforms should accompany bar graphs or scatterplots, and the two single-condition waveforms used to create a difference waveform should be presented. Indeed, the publication guidelines of the Society for Psychophysiological Research mandate that these waveforms be presented (Picton et al., 2000). Inferences supported by a scatterplot can be strengthened by presenting average waveforms derived from a median split on the variable of interest (see, e.g., Inzlicht et al., 2009). In the case of the ERN, defining the component as the difference between error

Fig. 10.21. (Continued)
seen in the difference waveform. To accept the authors' alternative conclusion—that the negativity in the difference waveform is due to a negativity in the Error-2 condition—one must accept that the Error-2 negativity happens to coincide precisely in latency, frequency, and amplitude with a positivity identical to that seen in the Error-1 condition such that the two components sum together to produce the smooth Error-2 condition waveform. Although such a scenario is not impossible, it seems prudent to at least consider an alternative explanation, namely, that the peak in the difference waveform occurs solely because of the positive peak in the Error-1 condition.
Individual panels reprinted by permission. Copyright by Elsevier, John Wiley & Sons Inc., the Psychonomic Society, and Springer. Note that these waveforms are presented only as examples of common problems and are not meant as a criticism of the work of particular investigators; most other ERN investigators have published waveforms that could have contributed to this figure. Also, the most problematic figures are not included here, because those papers did not present relevant waveforms.

for performance monitoring theories of anterior cingulate cortex. *Brain and Cognition, 56*(2), 267–276.

Vidal, F., Burle, B., Bonnet, M., Grapperon, J., & Hasbroucq, T. (2003). Error negativity on correct trials: A reexamination of available data. *Biological Psychology, 64*(3), 265–282.

Vidal, F., Hasbroucq, T., Grapperon, J., & Bonnet, M. (2000). Is the "error negativity" specific to errors? *Biological Psychology, 51*(2–3), 109–128.

Vocat, R., Pourtois, G., & Vuilleumier, P. (2008). Unavoidable errors: A spatio-temporal analysis of time-course and neural sources of evoked potentials associated with error processing in a speeded task. *Neuropsychologia, 46*, 2545–2555.

Vocat, R., Pourtois, G., & Vuilleumier, P. (2011). Parametric modulation of error-related ERP components by the magnitude of visuo-motor mismatch. *Neuropsychologia, 49*, 360–367.

Weissman, D. H., Roberts, K. C., Visscher, K. M., & Woldorff, M. G. (2006). The neural bases of momentary lapses in attention. *Nature Neuroscience, 9*(7), 971–978.

Wessel, J. R., Danielmeier, C., & Ullsperger, M. (in press). Error awareness revisited: Accumulation of multimodal evidence from central and autonomic nervous systems. *Journal of Cognitive Neuroscience.*

West, R. (2003). Neural correlates of cognitive control and conflict detection in the stroop and digit-location tasks. *Neuropsychologia, 41*(8), 1122–1135.

Wiersema, J. R., van der Meere, J. J., & Roeyers, H. (2005). ERP correlates of impaired error monitoring in children with ADHD. *Journal of Neural Transmission, 112*(10), 1417–1430.

Wiersema, J. R., van der Meere, J. J., & Roeyers, H. (2009). ERP correlates of error monitoring in adult ADHD. *Journal of Neural Transmission, 116*(3), 371–379.

Willemssen, R., Müller, T., Schwarz, M., Hohnsbein, J., & Falkenstein, M. (2008). Error processing in patients with Parkinson's disease: The influence of medication state. *Journal of Neural Transmission, 115*, 461–468.

Wiswede, D., Munte, T. F., Goschke, T., & Russeler, J. (2009). Modulation of the error-related negativity by induction of short-term negative affect. *Neuropsychologia, 47*(1), 83–90.

Woldorff, M. G. (1993). Distortion of ERP averages due to overlap from temporally adjacent ERP— analysis and correction. *Psychophysiology, 30*(1), 98–119.

Wolpert, D. M., & Flanagan, J. R. (2001). Motor prediction. *Current Biology, 11*(18), R729–R732.

Yeung, N. (2004). Relating cognitive and affective theories of the error-related negativity. In M. Ullsperger & M. Falkenstein (Eds.), *Errors, conflicts, and the brain: Current opinions on performance monitoring* (pp. 63–70). Leipzig: Max Planck Institute of Cognitive Neuroscience.

Yeung, N., Bogacz, R., Holroyd, C. B., & Cohen, J. D. (2004a). Detection of synchronized oscillations in the electroencephalogram: An evaluation of methods. *Psychophysiology, 41*(6), 822–832.

Yeung, N., Bogacz, R., Holroyd, C. B., Nieuwenhuis, S., & Cohen, J. D. (2007a). Theta phase resetting and the error-related negativity. *Psychophysiology, 44*(1), 39–49.

Yeung, N., Botvinick, M. M., & Cohen, J. D. (2004b). The neural basis of error detection: Conflict monitoring and the error-related negativity. *Psychological Review, 111*(4), 931–959.

Yeung, N., & Cohen, J. D. (2006). The impact of cognitive deficits on conflict monitoring—predictable dissociations between the error-related negativity and N2. *Psychological Science, 17*(2), 164–171.

Yeung, N., Ralph, J., & Nieuwenhuis, S. (2007b). Drink alcohol and dim the fights: The impact of cognitive deficits on medial frontal cortex function. *Cognitive, Affective & Behavioral Neuroscience, 7*(4), 347–355.

Yeung, N., & Sanfey, A. G. (2004). Independent coding of reward magnitude and valence in the human brain. *Journal of Neuroscience, 24*(28), 6258–6264.

Yin, G., Zhang, J., Tian, Y., & Yao, D. (2009). A multi-component decomposition algorithm for event-related potentials. *Journal of Neuroscience Methods, 178*, 219–227.

Yordanova, J., Falkenstein, M., Hohnsbein, J., & Kolev, V. (2004). Parallel systems of error processing in the brain. *Neuroimage, 22*(2), 590–602.

Yu, R. J., & Zhou, X. L. (2006). Brain potentials associated with outcome expectation and outcome evaluation. *Neuroreport, 17*(15), 1649–1653.

Yu, R. J., & Zhou, X. L. (2009). To bet or not to bet? The error negativity or error-related negativity associated with risk-taking choices. *Journal of Cognitive Neuroscience, 21*(4), 684–696.

Zhang, J. (1998). Decomposing stimulus and response component waveforms in ERP. *Journal of Neuroscience Methods, 80*(1), 49–63.

Zirnheld, P. J., Carroll, C. A., Kieffaber, P. D., O'Donnell, B. F., Shekhar, A., & Hetrick, W. P. (2004). Haloperidol impairs learning and error-related negativity in humans. *Journal of Cognitive Neuroscience, 16*(6), 1098–1112.

PART 3

ERP Components in Specific Cognitive Domains

CHAPTER

11 ERP Components and Selective Attention

Steven J. Luck *and* Emily S. Kappenman

Abstract

This chapter describes how event-related potential (ERP) components have been used to answer questions about attentional processing. In particular, it discusses how attention modulates the flow of sensory processing in relatively simple tasks and how it operates at postperceptual levels in more complex dual-task paradigms. The chapter focuses primarily on the variety of attention called *selective attention*, the processes by which the brain selects some sources of inputs for enhanced processing. The first section describes how ERPs first became used in the study of attention, highlighting the unique ability of ERPs to answer questions that had puzzled attention researchers for decades. The second section describes major ERP attention studies in the auditory and visual modalities, respectively. The chapter concludes with a discussion of the operation of attention in postperceptual systems, such as working memory encoding and response selection.

Keywords: event-related potentials, auditory attention, visual attention, object-based attention, feature-based attention, attentional blink, psychological refractory period

The term *attention* has many meanings, but it typically refers to a set of processes that control the flow of information through the nervous system (see Luck & Gold, 2008; Luck & Vecera, 2002). Attention is not a cognitive "module" through which information flows, but is instead a collection of processes that operate by modulating the activity of other systems (e.g., perceptual systems, memory systems, response systems). Almost all cognitive systems are influenced by attention, and those that are not influenced by attention are considered interesting boundary cases by attention researchers. Consequently, attention influences almost all event-related potential (ERP) components, and those that are not influenced by attention are considered interesting boundary cases by ERP-oriented attention researchers. Thus, a summary of ERP components that are modulated by attention could easily turn into a summary of all ERP components.

The goal of this chapter is not simply to catalog the effects of attention on ERP components; instead, we will focus on those ERP components whose modulation by attention has provided insight into the nature of cognitive processing. In other words, we will describe how ERP components have been utilized to answer questions about attentional processing that are difficult to answer with behavioral data. In particular, this chapter will discuss how attention modulates the flow of sensory processing in relatively simple tasks and how it operates at postperceptual levels in more complex dual-task paradigms. The operation of attention in more complex perceptual tasks, such as visual search, is discussed in Chapter 12 of this volume.

This chapter focuses primarily on the variety of attention called *selective attention*, the processes by which the brain selects some sources of inputs for enhanced processing. Other varieties of attention, such as vigilance and executive control, have been studied much less extensively with ERPs (with the exception of error-related negativity studies, which are reviewed in Chapter 10 of this volume).

This chapter is divided into two main sections. In the first section, we describe how ERPs first became used in the study of attention, highlighting the unique ability of ERPs to answer questions that had puzzled attention researchers for decades. We also place special emphasis on experimental design issues that were first raised in studies of auditory attention but apply generally to the study of attention in all modalities. Although the focus of this chapter is on attention, we hope that this first section will prove useful to researchers interested in applying ERPs to answer questions in other cognitive domains. In the second half of the chapter, we describe major ERP attention studies in the auditory and visual modalities, respectively. The chapter concludes with a discussion of the operation of attention in postperceptual systems, such as working memory encoding and response selection.

ERP Approaches to the Study of Selective Attention

The Locus of Selection and the Cocktail Party Problem

Event-related potentials were first used to study mechanisms of attention in the late 1960s and early 1970s, a period when cognitive psychologists were vigorously debating the question of whether attention operates at an early stage or a late stage (the classic *locus-of-selection* question). Some researchers argued that the sensory systems are often overloaded by a multiplicity of inputs, leading to impaired sensory representations of the environment, and that attentional mechanisms are used to limit processing to a subset of the inputs so that those inputs will be accurately perceived (e.g., Broadbent, 1958; Treisman, 1964; Treisman & Geffen, 1967). This is called the *early selection* hypothesis. Other researchers argued that perceptual systems could process multiple inputs in parallel with no interference, whereas memory, decision, and response systems were highly limited; these researchers therefore believed that attention is used solely to modulate these postperceptual processes (e.g., Deutsch & Deutsch, 1963; Moray, 1959). This is called the *late selection* hypothesis. Although clever experimentalists designed numerous behavioral experiments that answered many questions about the nature of attentional selection, behavioral reaction time and accuracy measures were unable to discriminate definitively between early and late selection mechanisms, because behavior reflects the combined influence of early and late stages.

Unlike behavioral reaction time and accuracy measures, which (usually) cannot distinguish among the slowing of different component processes, the temporal resolution of ERPs can provide a millisecond-by-millisecond representation of processing as it unfolds over time. Therefore, by applying ERPs to the study of attentional selection, researchers have been able to directly measure the stages of processing that are influenced by attention. Specifically, if the early selection hypothesis is correct, then an attended stimulus should elicit larger sensory ERP components than an ignored stimulus; if the late selection hypothesis is correct, then attended and ignored stimuli should elicit equivalent sensory ERP components and differ only in terms of later, postperceptual components.

Early ERP research on this issue focused primarily on the auditory modality, using an experimental paradigm that was designed to be analogous to the *cocktail party* problem. This is the problem of processing the sensory information arising from one source (e.g., the person with whom you are having a conversation at a cocktail party) in the face of potentially interfering information from other concurrent sources (e.g., the other simultaneous conversations at the party). The relevant and irrelevant input sources are likely to differ in spatial location, pitch, or both, so it was hypothesized that listeners would use these cues to select information for further processing. That is, location and pitch information would be used to select the relevant stimuli, and more subtle features of the selected stimulus could then be perceived. For example, the location and pitch of a particular voice could be used to select that source of stimulation for further processing that would identify the words being spoken by that voice. Because this research was originally developed in the 1950s, when engineering concepts from telecommunications were being imported into the nascent field of cognitive psychology, attention researchers use the term *channel* to describe a source of information in the environment. Thus, an observer might attend to one channel (a particular voice arising from a particular location) and ignore other channels (other voices at other locations, other nonvocal sounds, etc.).

The experimental analog of this situation, as developed by Hillyard and his colleagues (Hansen & Hillyard, 1980; Hillyard et al., 1973; Schwent et al., 1976a, 1976b; Woods & Hillyard, 1978), is shown in Figure 11.1. A sequence of brief tones

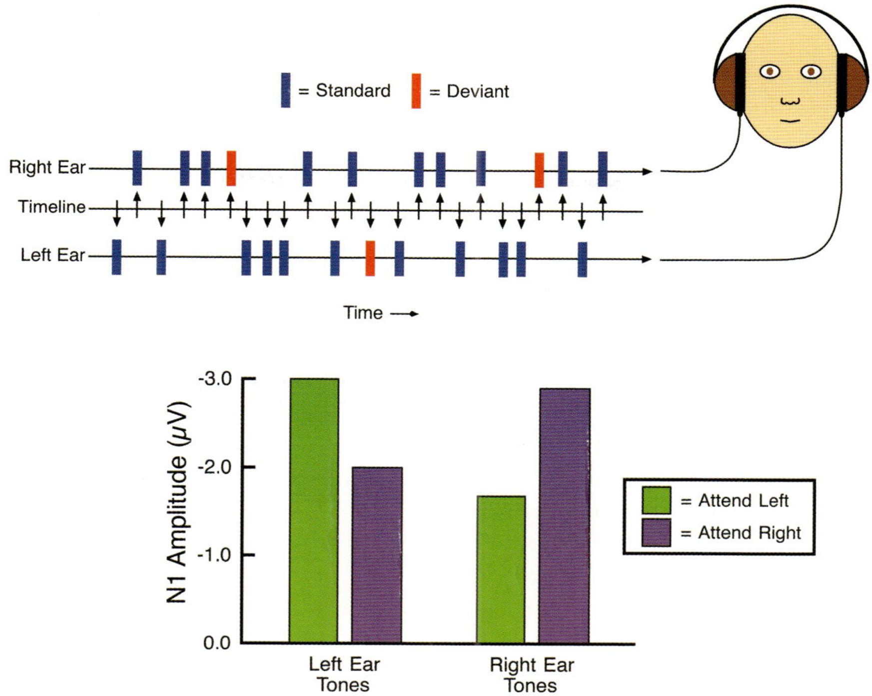

Fig. 11.1. Experimental paradigm and results of Hillyard et al. (1973).

is presented, with tones of one pitch presented to the left ear and tones of another pitch presented to the right ear (thus defining two input channels). Subjects are instructed to attend to the stimuli presented in one of these channels at the beginning of a block of trials and to press a button if they detect a slight deviation in this channel (e.g., a slightly softer tone presented in the attended pitch/ear combination). They are further instructed to ignore the other channel and not to press the button if a deviant tone is detected in that ear/pitch combination. The goal of this design is to determine the stage at which the ERP waveform for an attended stimulus differs from the waveform for an unattended stimulus. An early difference would imply that attention operates at an early stage of processing, before perception is complete, whereas a late difference would imply that attention operates only at postperceptual stages. Furthermore, it is possible to ask whether stimulus deviance is detected by the brain only in the attended channel.

As shown in Figure 11.1, the N1 wave (ca. 150 ms) is typically found to be larger for a given stimulus when it is presented in the attended ear than when it is presented in the unattended ear (this is true for both the standard and deviant stimuli). Because this effect began at a relatively early time, during which the brain is presumably engaged in perceptual processing, this pattern of results supports early selection theories of attention. More precisely, this pattern of results has led to the view that attention operates within both the auditory and visual modalities as a *sensory gain control*—like the volume control on a sound system or the brightness control on a video monitor—that serves to boost the effective intensity of the attended stimuli and reduce the effective intensity of the ignored stimuli. Moreover, this gain control can be set prior to the presentation of a stimulus, influencing the gain of the initial feedforward pass of information through sensory cortex (see the review by Hillyard et al., 1998).

The Design of Attention Experiments

More details about the results obtained with this paradigm will be provided in later sections. First, however, we would like to consider some important aspects of this experimental design—which we call the *Hillyard sustained attention paradigm*—because it has been used as the template for an enormous number of ERP attention experiments in the auditory, visual, and somatosensory modalities. There are two main issues addressed by this design. First, it is important to create a task environment that encourages subjects to focus attention appropriately. Second, it is important to rule out alternative explanations for differences in the ERP waveforms elicited by attended and unattended stimuli. It is important to consider these issues very carefully when designing an attention experiment or when evaluating attention experiments in the literature.

ENCOURAGING A STRONG FOCUS OF ATTENTION

Because the Hillyard sustained attention paradigm was originally developed to determine whether attention operates at an early stage of processing, experiments using variants of this paradigm are typically designed to both facilitate and encourage the focusing of attention within perceptual processing systems. In particular, the stimulus differences between the attended and unattended channels are large, making it easy to rapidly determine the channel to which a given stimulus belongs. In particular, if attention operates as a preset gate or gain control, then it is necessary to use channels that correspond to different pools of neurons at early stages in the sensory processing pathway. Otherwise, it would not be possible to increase the gain for the neurons corresponding to the attended channel prior to stimulus onset. In the case of the auditory system, tones that differ greatly in pitch stimulate different populations of neurons beginning in the cochlea, and of course, different pools of neurons code the two ears beginning at the cochlea as well. Thus, presenting stimuli at widely different pitches in separate ears optimizes the opportunity to observe early attention effects. If, in contrast, the attended and ignored stimuli are not coded by separate pools of neurons until a late stage of processing, then attention cannot act as a preset gain control at an early stage of processing. Indeed, the effects of attention at early points in the ERP waveform are abolished when the differences between the attended and ignored channels are made smaller. For example, Hansen and Hillyard (1983) used an attended channel defined by a particular pitch and spatial location, and they found that the N1 wave was nearly identical for the attended stimuli and for unattended stimuli that were similar in either pitch or spatial location to the attended pitch and location. Similar results were obtained in the visual modality by Hillyard and Münte (1984), who found that the earliest effects of spatial attention were eliminated when the attended and unattended locations were close together.

Whereas the differences between the attended and unattended channels are large in this paradigm, the differences between the standard and deviant stimuli within a channel are small. This increases the likelihood that subjects will be motivated to focus attention strongly on the attended channel. If the standard–deviant difference is large, focused attention may not be necessary for the brain to rapidly discriminate between the standard and deviant stimuli. Indeed, if the standard–deviant difference is large enough, the deviant stimulus may define a distinct sensory channel to which attention can be focused, and the standard stimuli may not fall within this channel, allowing them to be ignored along with the unattended channel. Thus, when the goal is to compare standard stimuli in the attended and ignored channels, it is important to ensure that the standard and deviant stimuli within a channel differ in a manner than cannot easily be discriminated.

It is also important to present the stimuli at a rapid rate (typically, two to four stimuli per second) if the goal is to test the effects of attention on sensory processing. A rapid rate is important because it is difficult to focus attention unless the brain faces an overload of stimulation (Lavie, 1995). If the stimuli are presented slowly, then the brain has sufficient time to fully process each stimulus, and there would be no reason to selectively perceive the stimuli in only the attended channel. Indeed, auditory attention has effects at an earlier point in the ERP waveform under higher than lower stimulation rates (Hansen & Hillyard, 1980; Schwent et al., 1976a; Woldorff & Hillyard, 1991).

If these conditions are not met by a given experiment, then the absence of effects at an early point in the ERP waveform (e.g., in the N1 latency range) cannot be used as evidence against early selection models of attention in general, but instead indicates the conditions under which early selection occurs (see Vogel et al., 2005, for an extensive discussion of how the locus of selection may vary according to task demands).

AVOIDING CONFOUNDS AND ALTERNATIVE EXPLANATIONS

Näätänen (1967) wrote an influential review of early ERP experiments on attention, pointing out a number of shortcomings of many previous experimental designs that permitted alternative explanations of their results (see also Näätänen, 1975, 1992), problems that the Hillyard sustained attention paradigm was designed to rule out. For example, the most obvious way to manipulate attention would be to have subjects perform an active discrimination task on a stream of stimuli in one condition (e.g., pressing one of two buttons, depending on the identity of the stimuli) and do some completely different task in another condition (e.g., reading a book). However, the subject's overall state of arousal might differ between these conditions, and this could influence sensory processing in a nonselective manner, yielding larger ERP responses to all stimuli in the active discrimination condition. Thus, to separate the effects of selective attention from the effects of global arousal, it is important to ensure that attentional manipulations influence the relative sensitivity to attended and unattended channels rather than the overall sensitivity of the system to all stimuli. In addition, the ERPs elicited by the stimuli in the active discrimination condition would be contaminated by motor potentials that would be absent in the other-task condition. It is also important to ensure that the experiment does not confound differences in the ERPs elicited by different physical stimuli with differences in the ERPs that arise from the effects of attention. That is, one cannot compare the ERP elicited by one stimulus that is being attended (e.g., a 1000 Hz target tone) with the ERP elicited by a different stimulus that is being ignored (e.g., a 500 Hz nontarget tone), because it will be impossible to determine whether any differences in the ERPs are a result of attention or a result of the physical stimulus properties.

The Hillyard sustained attention paradigm addresses these potential problems in the following manner. Rather than comparing an active discrimination condition with a condition in which subjects perform a completely different task, this paradigm involves comparing two different active discrimination conditions that are closely equated for task difficulty. In the original version of this task (Hillyard et al., 1973), subjects attended to 1500 Hz tone pips in the left ear in one condition and attended to 800 Hz tone pips in the right ear in the other condition; in both conditions, the task was to press a button when a slightly higher pitch was detected in the attended ear. Thus, a virtually identical task was performed in both conditions, which nearly perfectly equated overall arousal. In addition, the N1 wave for left-ear tones was larger when attention was direct to the left ear than when attention was directed to the right ear, and the N1 wave for right-ear tones was larger when attention was directed to the right ear than when attention was directed to the left ear. This rules out an explanation in terms of overall arousal. If, for example, attending to the left ear had been more arousing than attending to the right ear, then this should have led to a larger N1 wave when subjects attended to the left ear for both left-ear and right-ear tones, but attending to the left ear increased N1 amplitude only for left-ear tones.

It is not always easy to equate task difficulty in this manner. For example, imagine that a researcher wanted to examine the effects of attending to speech-like stimuli versus attending to pure tones. A single stream containing phonemes and pure tones could be presented, with subjects being asked to make a /ba/–/pa/ discrimination on the phonemes in one condition and a high-low pitch discrimination on the tones in another condition. It would be nontrivial to equate the difficulty of these two tasks, and subjects might therefore be in a different state of arousal in the two tasks. However, arousal-based explanations can be ruled out as long as the observed attention effects show a complementary pattern in which the ERP difference between the attended and ignored stimuli are the same for the two tasks. For example, if a larger N1 wave is observed for the phonemes in the attend-phonemes condition than in the attend-tones condition and a larger N1 wave is observed for the tones in the attend-tones condition than in the attend-phonemes condition, the results could not be explained by a greater level of overall arousal in one of these two conditions.

The Hillyard sustained attention paradigm also rules out a more subtle type of arousal-based explanation. Imagine that stimuli are presented every 300–700 ms in the left ear in one sequence, and that stimuli are also presented every 300–700 ms in the right ear in a completely independent sequence. Because the two sequences are independent, there are no constraints on the delay between a stimulus in one ear and a stimulus in the other ear. However, stimuli are always separated by at least 300 ms in a given ear. If subjects are asked to detect targets in one ear and ignore the other ear, they could potentially phasically increase their arousal level during the period in which a stimulus is expected in the attended ear and then decrease their arousal level for the 300 ms period during which no stimulus is expected in the attended ear. Because stimuli may be presented in the unattended ear during this 300 ms period, the subject could potentially be in a decreased state of global arousal when stimuli are presented in the unattended ear during this period. Thus, any differences in ERP activity between attended-ear and unattended-ear stimuli could be explained by transient changes in global responsiveness rather than by differences in responsiveness between the attended and unattended ears.

This problem was solved by Hillyard et al. (1973, Experiment 2) by constraining the timing of subsequent stimuli irrespective of the ear to which the stimuli were presented. That is, a single sequence of stimulus times was created (see the timeline in Figure 11.1), and each stimulus in this timeline was randomly assigned to the left ear or the right ear. With this constraint, subjects cannot phasically alter their arousal levels in a manner that is different for the attended-ear and unattended-ear stimuli. Hillyard et al. found the same pattern of N1 attention effects in this condition as in a condition in which the stimulus sequences in the two ears were independent, demonstrating that this pattern is not a result of phasic changes in arousal. However, almost all subsequent studies have used a single timeline (as in Figure 11.1) to ensure that the results are not confounded by phasic changes in arousal. An exception to this is a recent set of studies using steady-state visual evoked potentials (SSVEPs; see Di Russo et al., 2003b; Morgan et al., 1996; Müller & Hillyard, 2000; Müller et al., 1998), in which the rate of stimulation is too fast to permit phasic changes in arousal.

Because the stimuli in the Hillyard sustained attention paradigm are typically presented at a rapid rate, the ERP elicited by stimulus N-1 will overlap with the ERP elicited by stimulus N, and this will distort the observed waveform for stimulus N. However, the use of a single timeline guarantees that the stimuli preceding an attended stimulus will be identical (on average) to the stimuli preceding an unattended stimulus. Consequently, the overlapping activity will ordinarily be the same for the attended and unattended stimuli. However, subtle differences in overlap may remain that can confound the comparison between the ERPs elicited by attended and unattended stimuli (for a detailed discussion, see Woldorff, 1993). Moreover, if any nonrandom stimulus sequence constraints are imposed (e.g., no more than two targets in a row), substantial differences in overlapping activity may confound the ERPs. Thus, one must be careful to consider the effects of overlap in this paradigm.

This paradigm is also designed to rule out contamination from motor-related potentials as an explanation for the ERP difference between the attended and ignored stimuli. If subjects make a response to the attended stimuli and not to the unattended stimuli, then premotor ERP components will be present in the ERP waveform for the attended stimuli but not for the unattended stimuli. These potentials (e.g., the *Bereitschaftspotential* and the lateralized readiness potential) may begin hundreds of milliseconds before a response, and within 200 ms of stimulus onset (see Chapters 8 and 9, this volume), and so they can contaminate many ERP attention effects. To minimize this contamination, the Hillyard sustained attention paradigm uses frequent standard stimuli and rare deviant stimuli within both the attended and unattended channels and requires a motor response only for the deviant stimuli within the attended channel. The ERPs elicited by the standard stimuli in the attended and unattended channels can then be compared with minimal contamination from motor-related ERP activity. Because the standard and deviant stimuli are quite similar, they fall into the same sensory channel and are therefore subject to the same attention effects, at least at early stages of processing. Indeed, in most cases, the effects of attention at early points in the ERP waveform are the same for the standards and the deviants. It should be noted, however, that subthreshold motor activity may be elicited by the standards in the attended channel, and so limiting the analyses to the standard stimuli does not completely rule out the possibility of contamination from motor-related ERPs.

SUMMARY OF EXPERIMENTAL DESIGN CONSIDERATIONS

The application of ERPs to the study of selective attention has proved useful in distinguishing among early and late theories of selection mechanisms. However, it is important to note that although ERPs provide a wealth of information regarding the timing of specific cognitive processes, the ERP waveform is highly sensitive to physical and perceptual differences among stimuli (see Luck, 2005, for a full discussion of this issue). Therefore, merely collecting ERPs in conjunction with behavioral measures of attentional selection is not enough to disambiguate alternative explanations of results. Rather, in addition to the many experimental tricks necessary in behavioral experiments of selective attention, such as ensuring a strong focus of attention, ERP attention experiments require additional steps to equate the physical stimulus sequences. The Hillyard sustained attention paradigm—which was originally designed for use with auditory attention experiments but is easily modified for other modalities—provides an embodiment of these experiment design principles. In the following section, we will see how the use of this paradigm and its many variants has provided a wealth of information about the operation of selective attention in both the auditory and visual modalities.

Auditory Attention

Effects of Attention on Auditory Sensory ERP Components

As discussed at the beginning of this chapter, the early selection hypothesis predicts that sensory ERP components will be enhanced for stimuli presented in the attended channel compared to stimuli presented in the ignored channel. In addition, this effect should be the same for the standard and deviant stimuli, because attention is assumed by this hypothesis to operate before perception is complete (i.e., before the brain has determined whether a given stimulus is the target). Some initial studies suggested that the brainstem auditory evoked responses that are evoked within the first 10 ms following the onset of a stimulus (see Chapter 4, this volume) could be enhanced for attended stimuli (Lukas, 1980, 1981). However, a series of studies by Woldorff, Hackley, and Hillyard (Hackley et al., 1990; Woldorff, 1989; Woldorff et al., 1987) provided convincing evidence that these components are not influenced by attention, even when attention is highly focused. Thus, attention appears to have no influence on the very early transmission of auditory information from the cochlea through the brainstem auditory processing nuclei. However, under these highly focused conditions, attention did influence the ERP waveform in the 20–50 ms latency range, the time period of the midlatency responses. This is shown in Figure 11.2, which presents the responses to attended and unattended stimuli on three different time scales, one showing the auditory brainstem responses (0–14 ms), one showing the midlatency responses (0–70 ms), and one showing the longer latency responses (0–500 ms). The first effects of attention are visible in the midlatency responses. As described in Chapter 4 of

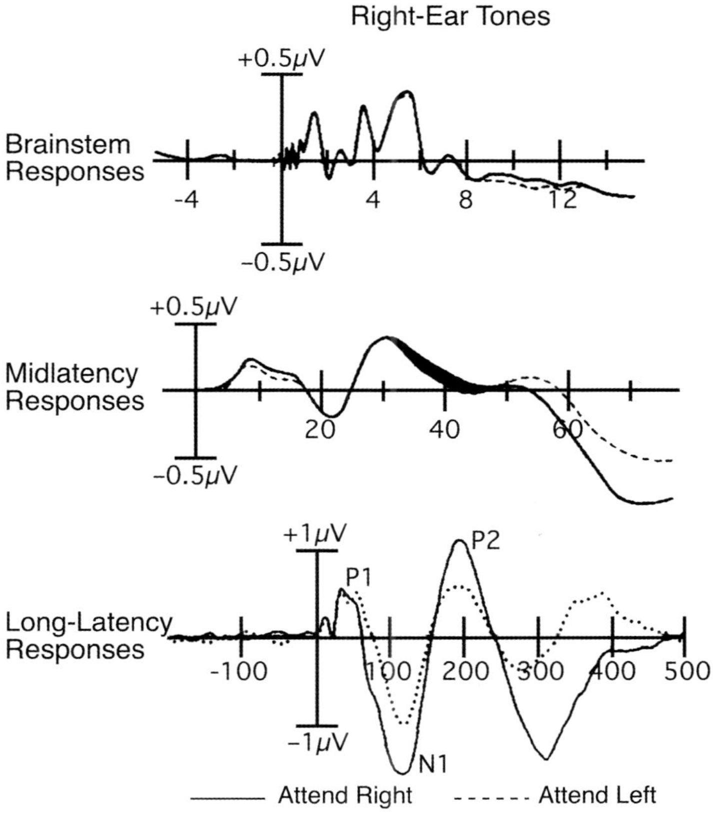

Fig. 11.2. Event-related potentials elicited by tones presented to the right ear when the right ear was attended versus when the left ear was attended from the studies of Woldorff et al. (1987) and Woldorff and Hillyard (1991). The same data are shown on three different time scales with different filter settings to focus on the brainstem responses (top), the midlatency responses (middle), and the long latency responses (bottom). The shaded region in the midlatency data shows the earliest effect of attention. Adapted with permission from Woldorff, M., Hansen, J. C., & Hillyard, S. A. (1987). Evidence for effects of selective attention to the midlatency range of the human auditory event related potential. In R. Johnson, J. W. Rohrbaugh & R. Parasuraman (Eds.), Current Trends in Event-Related Brain Potential Research (pp. 146–154). London: Elsevier Copyright (1987) and Woldorff and Hillyard (1991).

this volume, the auditory midlatency responses are thought to arise primarily from auditory cortex, with a possible contribution from upper brainstem and thalamic nuclei. Thus, auditory attention does appear to influence auditory sensory activity by the time it reaches cortex, consistent with the early selection hypothesis.

It is technically demanding to measure the effects of auditory attention on ERP activity within the first 50 ms following stimulus onset, because these early ERP components are small and may be influenced by attention only when attention is very highly focused. The effect of auditory attention that was observed first, and most commonly, is therefore a later effect in the latency range of the N1 wave (ca. 100 ms; see Chapter 4 of this volume for more details about the N1 wave). As discussed at the beginning of this chapter and illustrated in Figure 11.1, Hillyard et al. (1973) reported that the amplitude of the N1 wave was greater for a stimulus presented in a given ear/pitch combination when attention was directed to that ear/pitch combination compared to when attention was directed to the other ear/pitch combination (see the waveforms from a similar study in Figure 11.2). Because the N1 wave is a sensory response, Hillyard et al. concluded that attention operates during perception, consistent with the early selection hypothesis.

In addition, the N1 effect was the same for both the standard and deviant stimuli, indicating that it represents selective processing of the attended channel before the brain has determined whether the stimulus is a standard or a deviant stimulus. If attention operates only after stimulus identification is complete, as proposed by late selection theories, then only target stimuli should benefit from attention, and the attended-channel standards should elicit the same ERP response as the unattended-channel standards. Instead, the initial effects of attention did not distinguish between target and nontarget stimuli, supporting early selection theories. Moreover, later components, such as the P3 wave, were observed for the deviants but not for the standards in the attended channel, indicating that these components reflect processes that follow the classification of a stimulus as a standard or a deviant. Moreover, deviant stimuli in the unattended channel do not elicit a P3 wave, providing evidence for a suppression of perceptual processing within the unattended channel (but see the following section on the mismatch negativity for a more nuanced view).

Although Hillyard et al. (1973) clearly demonstrated that attention could have an impact in the N1 latency range, Näätänen (1975) proposed that this effect was not actually a modulation of the *exogenous* (stimulus-evoked) N1 wave. He noted that the effect of attention lasted considerably longer than the N1 wave (once appropriate filter settings were used), and from this observation he concluded that the effect of attention consisted of the addition of an *endogenous* (internally triggered) component, which he called the *processing negativity*. That is, he concluded that this endogenous negativity overlapped with the N1 wave for attended stimuli, artificially creating the appearance of a larger N1 component for the attended stimuli. As discussed in Chapter 1 of this volume, arguments based on differences in waveshape are weak, because the apparent duration (and scalp distribution) of a given observed peak depends a great deal on the nature of the overlapping components (see also Luck, 2005, chap. 2). For example, a long and broad underlying N1 component may appear to have a sharp and early peak if it is followed by a large and early P2 wave, but an attention-related modulation of the N1 component would have the long and broad time course of the underlying N1 component rather than showing a sharp and early peak (assuming that it is not cut off by a P2 attention effect). Thus, the finding that the effect of attention extends over a broader time range than the observed N1 peak does not necessarily indicate that the effect of attention reflects the addition of a new component.

Hansen and Hillyard (1980) replicated the finding of a sustained difference between the ERPs elicited by attended versus ignored stimuli, and they termed this difference the *Nd* wave (for *negative difference* wave). However, they also provided evidence that the early portion of the attention effect did indeed consist of a modulation of the amplitude of the exogenous N1 wave. This was supported by a magnetoencephalographic study showing that the N1 wave and the early portion of the attention effect both arise from the same brain region (Woldorff et al., 1993). However, the late part of the attention effect does appear to reflect a different neural process, as originally proposed by Näätänen (1975).

The term *processing negativity*, introduced by Näätänen (1975) to describe the difference in ERPs to attended versus unattended stimuli in the Hillyard sustained attention paradigm, is often used to refer to a more negative ERP for stimuli that contain attended features, and is used for both auditory and visual stimuli. Positive-going effects of this nature can also be observed (processing positivities; see, e.g., Anllo-Vento & Hillyard, 1996; Anllo-Vento et al., 1998).

The polarity of an ERP effect depends on several factors, including the location and orientation of the generator source and the choice of reference site (see Luck, 2005, chap.1), so not all negative-going effects necessarily reflect the same kind of neurocognitive process, and positive-going effects do not necessarily reflect fundamentally different kinds of processes than negative-going effects. Thus, the terms *processing negativity* and *processing positivity* should be taken as gross descriptions of experimental effects rather than as referring to distinct ERP components or distinct neural processes.

Effects of Attention on the Mismatch Negativity

Although research by Hillyard and his collaborators favored the early selection hypothesis, evidence from the mismatch negativity favored the late selection hypothesis. As described in detail in Chapter 6 of this volume, the mismatch negativity is a negative-going deflection peaking approximately 200 ms poststimulus that is observed when the eliciting stimulus differs from a set of preceding stimuli. In the classic oddball paradigm, for example, infrequent tones of one pitch might be interspersed with frequent tones of another pitch, and the infrequent tones will elicit a mismatch negativity (Näätänen et al., 1982). If subjects are actively discriminating the tones, the infrequent stimuli will also elicit an N2b component and a P3 component (see Chapter 7, this volume), but the mismatch negativity can be observed even if the subjects are ignoring the stimuli and engaging in a different task. The mismatch negativity can be observed even when the frequent and infrequent stimuli differ in a fairly subtle manner (e.g., the syllables "ba" versus "ga"). Thus, the mismatch negativity appears to reflect an automatic discrimination between stimulus categories, which is consistent with the notion that a stimulus can receive extensive sensory processing even when attention is not focused on it, as proposed by the late selection hypothesis.

How can we explain the finding of no effect of attention on the mismatch negativity at 200 ms when attention appears to influence sensory responses as early as 20–50 ms? There are at least two explanations (which are not mutually exclusive). First, attention must be very highly focused to observe a modulation of the earliest sensory responses, and Woldorff and his colleagues (Woldorff et al., 1991, 1998) found that the mismatch negativity is indeed suppressed for mismatching stimuli in one ear/pitch combination when attention is highly focused on a different ear/pitch combination. This is an example of the broad principle that attention operates at an early stage only when there is a high level of perceptual competition, which requires attention to be highly focused (Lavie, 1995; Luck & Hillyard, 1999; Vogel et al., 2005).

A second explanation is that the mismatch negativity is generally elicited by mismatches in relatively simple features; semantic deviance, for example, does not generate a mismatch. Thus, the presence of a mismatch negativity does not mean that the unattended stimuli were fully identified, but only that they were identified to a crude level. In other words, the effect of attention on the early components consists of only a partial suppression of the unattended stimuli, and the partial information from the unattended channel may be sufficient to detect simple mismatches. Nonetheless, the fact that the mismatch negativity can be observed even when subjects are distracted by reading a book does indicate that a reasonable level of processing can be carried out for auditory stimuli even when attention is not actively focused on these stimuli.

Attention and the Discrimination of Auditory Conjunctions

Unlike laboratory settings, the natural auditory environment is often quite complex, with multiple sources contributing a wide array of sounds that can overlap in features such as frequency, pitch, and location. Therefore, attending to a particular source in a natural auditory scene generally requires the processing of multiple simultaneously active features of acoustic input to isolate the source that has the appropriately matching feature dimensions. In other words, we are often looking not for specific features, but rather for conjunctions of features. Furthermore, different features are processed in separate regions of auditory cortex, begging the question of how the different features get organized into one auditory percept.

Researchers have debated whether the identification of an object on the basis of a conjunction of features occurs as a multistage process that involves sequentially assessing the presence or absence of each relevant feature, or whether all features can be assessed in a parallel fashion. For example, listeners may be able to first judge a source of sound on the basis of whether it matches the sound of interest on a dimension that is easily assessed before moving on to the analysis of a more difficult feature. Alternatively, the analysis of both simple and difficult feature dimensions may occur in parallel. Furthermore, it is possible that the features are processed in parallel, but the processing of a difficult feature may be terminated before it is complete once the brain determines that the more easily discriminable feature dimension does not contain the desired value. Alternatively, all features could be examined exhaustively before a decision is made about an object as a whole.

The use of ERPs has helped to resolve some of these issues regarding the processing of conjunctions of features. In the first of these studies, Hansen and Hillyard (1983) tested the predictions of parallel and serial models of auditory feature analysis using tones that varied orthogonally among dimensions of pitch, location, and duration. The attended channel was defined by a specified location and pitch, the values of which were varied across blocks of trials; targets were defined by stimulus duration. The pitch and location of a given tone could independently match or mismatch the relevant feature value, resulting in four classes of stimuli: (1) pitch match, location match; (2) pitch match, location mismatch; (3) pitch mismatch, location match; and (4) pitch mismatch, location mismatch. The authors used the negative difference wave (Nd) to assess whether each stimulus was analyzed exhaustively to determine whether it matched the specified pitch and location dimensions, or alternatively, whether a stimulus that mismatched on one dimension was disregarded from further processing on the other stimulus attribute. The predictions are illustrated in Figure 11.3. If the features are processed in parallel, as illustrated in Figure 11.3A, the difference between the match and mismatch Nd waves for one feature should be the same regardless of whether the stimulus matches or mismatches the other stimulus dimension. In other words, the difference between the Nd wave for stimuli that match and stimuli that mismatch the location feature should be the same regardless of whether the pitch matches or mismatches the channel of interest. This would demonstrate an independence of the analysis of the features. Alternatively, if easily discriminated features are processed in a contingent fashion, the amplitude of the Nd wave should be similar for the three classes of stimuli that mismatch on either one or both feature dimensions, whereas the Nd wave for stimuli matching both features should be larger (see Figure 11.3B).

Hansen and Hillyard (1983) further tested the viability of the contingent processing theory by independently varying the ease of discriminability of the two feature dimensions in separate groups of

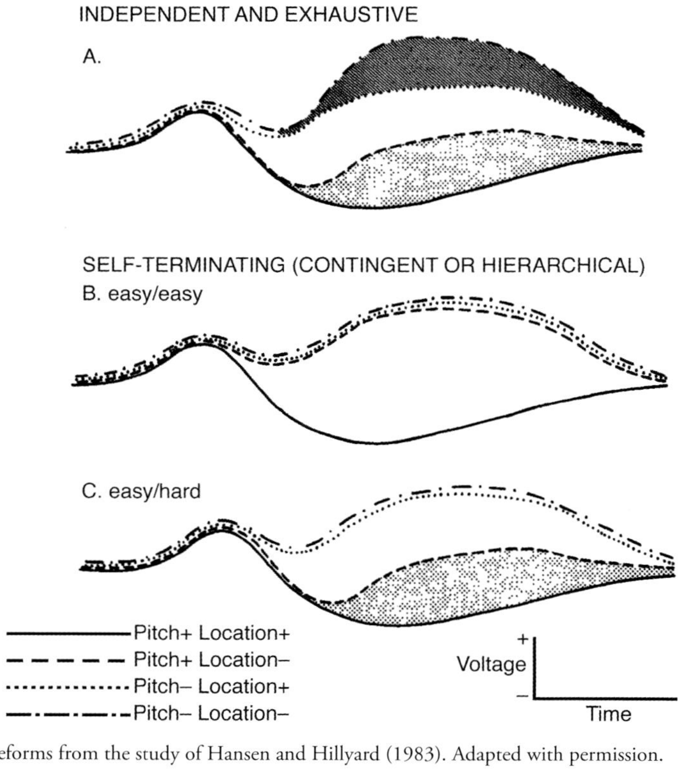

Fig. 11.3. Predicted waveforms from the study of Hansen and Hillyard (1983). Adapted with permission.

subjects. Specifically, both easy and hard discriminations of pitch and location were employed, such that one group of subjects received both easy discriminations, a second group received easy pitch and difficult location discriminations, and a third group received easy location and difficult pitch discriminations. For the group with both easy discriminations, the contingent processing predictions are as described earlier and depicted in Figure 11.3B, in which the mismatch of either of the easily discriminated features is similar to the Nd wave for stimuli that mismatch on both features. However, as depicted in Figure 11.3C, the contingent model predictions differ for conjunctions of features if one feature is easily discriminated and one feature is more difficult to discriminate. In this case, the contingent processing model predicts that the amplitude of the Nd wave will initially reflect whether the more easily discriminated feature matches the channel of interest, with the later part of the Nd wave reflecting whether the harder-to-discriminate feature matches the channel of interest. For example, if pitch is the more easily discriminated feature, the initial part of the Nd wave will be similar for pitch match stimuli, regardless of whether the location matches the channel of interest. However, once the more difficult-to-discriminate feature of location is analyzed, the later part of the Nd wave will reflect whether the location attribute matches or mismatches the channel of interest. This provides a further test of the predictions of the parallel and hierarchical models of conjunction feature analysis.

The results from all three conditions supported a contingent processing theory, in which the processing of one dimension is dependent on the other dimension. For example, in the group of subjects who encountered both easy pitch and easy location discriminations, tones that did not match one of the dimensions elicited an ERP that was similar to the ERP for tones that mismatched on both dimensions. Furthermore, for both of the groups with one easy and one hard discrimination, the amplitude in the later portion of the Nd wave was dependent on the negativity elicited by the easily discriminated attribute. Specifically, the difficult-to-discriminate attribute seemed to be analyzed only if the stimulus could not be rejected solely on the basis of the easily discriminated feature. The authors interpreted the pattern of results as being incompatible with independent processing of channel attributes and compatible with either (1) parallel self-terminating models or (2) serial processing models in which the easier feature is processed first.

Using a similar design, Woods and colleagues (1994) further examined the ERP signature of conjunction formation by assessing whether the ERP to stimuli containing multiple features of interest would merely be the sum of ERPs containing each of those features individually, or whether an additional conjunction-specific ERP signature would be present. Furthermore, the authors posited that the timing of the ERP components to simple features and conjunction-specific ERP signatures would provide insight into whether conjunctions were formed before or after the individual features were analyzed.

The authors examined multiple Nd waves that reflected either the ERP activity associated with the presence of one individual feature or the ERP activity associated with the conjunction of both features by using a subtraction procedure, in which the ERP to stimuli containing none of the features of interest was subtracted from the ERP to either individual feature only or conjunction stimulus ERPs. The authors further examined whether conjunction-specific ERP activity was elicited by subtracting both of the individual feature Nd waves from the conjunction Nd wave to assess the timing of the formation of auditory conjunctions. The results indicated that initially the individual features were processed separately and in parallel, with the conjunction ERP activity in the early portion of the waveform equal to the sum of the individual feature waveforms. However, as processing continued, an additional negativity appeared for conjunctions beginning at around 120 ms poststimulus. The authors interpreted the relatively early timing of this conjunction-specific activity as evidence that auditory conjunctions were formed before the individual features were fully analyzed.

Effects of Attention on Visual ERP Components

Overt and Covert Attention in Visual Perception

Although the general problem of perceiving one source of information in the presence of competing sources arises in the visual modality just as in the auditory and somatosensory modalities, the natures of the visual input and visual receptors are very different, and this leads to some differences in the way that attention operates. In the auditory modality, stimuli can travel around occluders and around the head, and the receptors are exquisitely tuned to temporal information but are organized into only two physically separable spatial channels (i.e., the left and right ears). In contrast, detection of a visual stimulus usually depends on an uninterrupted straight line between the stimulus and the receptors, and the receptors are spread out over millions of individual spatial locations but have relatively poor temporal resolution. Moreover, the eyes move continually to align the region of highest resolution (the fovea) with objects of interest in the environment, whereas ear movements (typically resulting from head movements) play only a modest role in human auditory perception. Most eye movements are sudden *saccades*, separated by periods of fixation, and vision is suppressed during the saccades. Thus, the input to the visual system consists of a series of brief snapshots that contain a precise and explicit representation of the spatial organization of the world. Consequently, an individual who is faced with many simultaneous objects can simply look at one of the objects to give that object preferential access to the fovea, which is the visual system's most fundamental processing resource. These shifts of gaze are called *overt attention*.

This does not mean, however, that the ability to make eye movements to an object of interest obviates the need for other mechanisms for focusing processing resources on objects that are not being foveated (for a detailed discussion, see Luck, 2009). First, because eye movements are relatively slow, it may be useful to facilitate the processing of objects at a potentially relevant spatial location prior to making an overt shift of gaze, which is called *covert attention*. These shifts of covert attention may serve several different functional roles, including determining whether an extrafoveal object is actually worth looking at and helping to precisely localize the object before the eyes move to it (Luck, 2009). Indeed, every saccade appears to be preceded by a shift of covert spatial attention (Deubel & Schneider, 1996; Hoffman & Subramaniam, 1995). Second, relevant visual information is often defined by nonspatial features (e.g., the color and shape of a pencil when one wishes to write something), and it would be incredibly inefficient to search for an object by moving the eyes randomly from location to location until that object is found. Thus, some mechanism is necessary to highlight objects that contain task-relevant features and are therefore good targets for future fixations. Third, once gaze is directed to the center of an object, some mechanism is necessary to allow perceptual processing mechanisms to expand across the extent of the object without also taking in other nearby objects. For example, after your eyes have been directed to one of the words on this page, some mechanism is necessary to allow the entire word to be processed without interference from other words on the lines above and below the fixated word.

Thus, in addition to overt shifts of gaze, we can easily identify at least three other mechanisms of attention that are important to the visual system: (1) covert spatial attention, which precedes overt shifts of gaze; (2) feature-based attention, which highlights objects containing relevant features; and (3) object-based attention, which allows processing to be determined by the shape of the attended object. We will now turn to each of these three varieties of visual attention. However, we would like to first note that the experiments and issues examined here are largely limited to those that have been explored in paradigms in which the attended and ignored stimuli are presented at different points in time (e.g., variants of the Hillyard sustained attention paradigm and the Posner cueing paradigm). In natural vision, attended and ignored objects are usually visible at the same time. The additional issues that arise in this situation will be discussed in Chapter 12 of this volume, which focuses on how various attentional mechanisms are coordinated in the perception of complex multiple-element stimulus arrays.

Covert Attention to Visual Locations

The Hillyard sustained attention paradigm has been used extensively to study the allocation of attention to visual locations. Eason and colleagues (1969) published an early study that used this paradigm, except that they did not use a single timeline for the attended and unattended locations (making it possible that the results were a consequence of phasic changes in arousal rather than spatially selective processing). Van Voorhis and Hillyard (1977) published the first study to implement a fully controlled ERP attention paradigm in the visual modality; this was followed by a large number of studies that yielded similar results (for reviews, see Hillyard et al., 1998; Mangun, 1995).

A prototypical example of the visuospatial variant of this paradigm is shown in Figure 11.4. Subjects fixate a central location (and it is absolutely crucial that they maintain fixation to avoid confounding changes in processing due to covert attention with changes in processing due to differences in the positions of the attended and unattended stimuli on the retina). At the beginning of each trial block, they are instructed to attend either to the left visual field (LVF) or to the right visual field (RVF) for the duration of that block. A sequence of stimuli

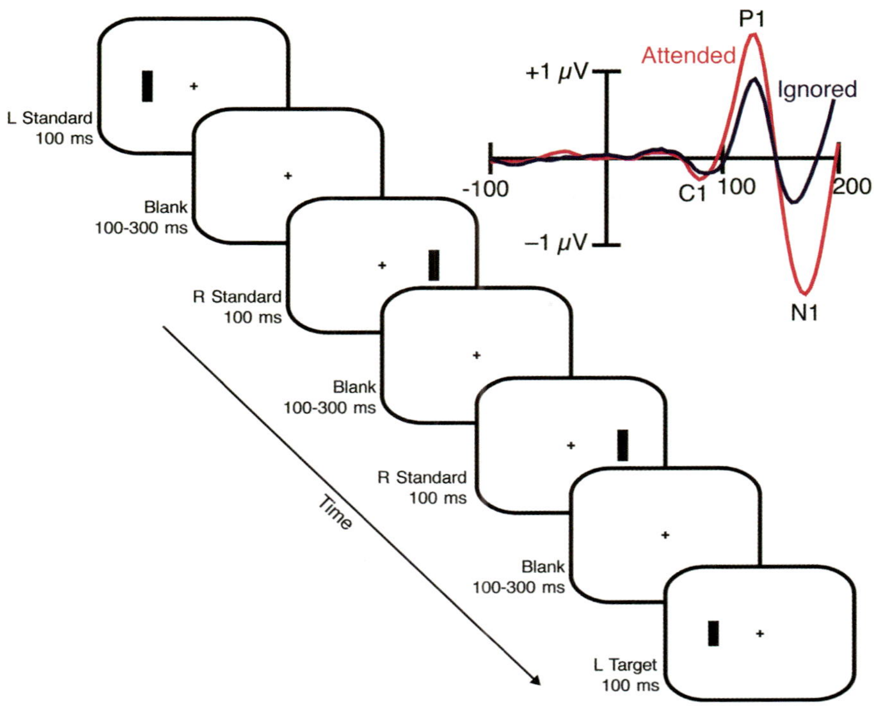

Fig. 11.4. Typical visual version of the Hillyard sustained attention paradigm, along with data from Gomez Gonzales et al. (1994) showing the results from a variant of this paradigm.

is then presented at a rapid rate (two to four stimuli per second), with any given item being equally likely to appear in the LVF or RVF (i.e., a single timeline is used, as illustrated in Figure 11.1). The subjects monitor the attended location for occasional deviant stimuli, pressing a button when they detect these stimuli (and not pressing for deviant stimuli presented in the unattended location). Many different types of stimuli have been used as the standard and deviant stimuli. For example, the standards and targets might be bars of slightly different sizes, different categories of words, faces with different expressions, or just about anything one could imagine. The main constraint is that, as in the auditory version of the paradigm, the target–standard distinction must be considerably more difficult to discriminate than the differences between the two locations, increasing the probability that (1) both the standards and deviants fall within the same perceptual channel and (2) the task is sufficiently difficult to motivate subjects to focus attention strongly on the to-be-attended location.

VISUOSPATIAL ATTENTION AND THE C1 WAVE

The typical pattern of results is shown in Figure 11.4 (data from the study of Gomez Gonzales et al., 1994). The first component that can be seen in these waveforms is the C1 wave, which arises from area V1 (Clark et al., 1994; Di Russo et al., 2002; Jeffreys & Axford, 1972). Because of the mapping of the visual field onto area V1 within the folds of the calcarine fissure, stimuli in the upper and lower visual fields project onto areas of V1 with opposite orientations with respect to the scalp, and this leads to a surface-negative voltage for upper field stimuli and a surface-positive voltage for lower field stimuli (for a detailed discussion, see Clark et al., 1994). When the stimuli are presented in the lower field or on the horizontal midline, the C1 is difficult to discern from the temporally overlapping P1 wave, because both are surface-positive voltages and therefore merge together. Consequently, a distinct C1 wave is not observed in most experiments.

When upper-field stimuli are used so that a distinct C1 wave can be observed, the C1 wave is not typically modulated by visuospatial attention (Clark & Hillyard, 1996; Di Russo et al., 2003a; Gomez Gonzales et al., 1994; Mangun et al., 1993). This corresponds with a study of single-unit responses in macaque monkeys that used a variant of the Hillyard sustained attention paradigm and found consistent attention effects in area V4, occasional attention effects in area V2, and no clear attention effects in area V1 (Luck et al., 1997). However, other single-unit studies have found attention effects in area V1 (McAdams & Reid, 2005; Roelfsema et al., 1998) and even in the lateral geniculate nucleus of the thalamus (McAlonan et al., 2008). Neuroimaging studies with human subjects have also found effects of spatial attention on the blood oxygen-level dependent (BOLD) response in area V1 (Gandhi et al., 1999; Somers et al., 1999).

To address these discrepancies, Martinez et al. (1999) used the same subjects and the same attention task with both functional magnetic resonance imaging (fMRI) and ERP recordings. They found no effect of attention on the C1 wave, as in prior ERP studies, and a significant effect of attention on V1 BOLD activity, as in prior fMRI studies. One likely explanation for the pattern of results within this study and across the previous studies is that attention does not modulate the first wave of feedforward sensory activity in area V1 (as reflected by a lack of C1 attention effects); however, later feedback signals sent to area V1 from other areas are modulated by attention (as reflected by the V1 BOLD attention effects). Indeed, source localization results in the Martinez et al. study suggested that attention did have an effect on ERP activity arising from area V1 later in the ERP waveform (see also Di Russo et al., 2003a). Moreover, some of the single-unit attention effects in area V1 clearly arise well after the initial wave of feedforward activation (Roelfsema et al., 1998).

However, it is possible that attention can modulate feedforward sensory activity in area V1 under appropriate task conditions, just as attention modulates auditory ERP responses in the 20–50 ms latency range only under conditions of highly focused attention, as discussed earlier in this chapter. One study has found an effect of visuospatial attention on C1 amplitude (Kelly et al., 2008), suggesting that attention can sometimes influence feedforward sensory activity in area V1, but it remains to be seen whether this result can be replicated.

BASIC P1 AND N1 ATTENTION EFFECTS

The earliest widely replicated ERP effect of visuospatial attention is a larger P1 wave for stimuli presented at the attended location than for stimuli presented at the unattended location, which is usually accompanied by a larger N1 wave for the stimuli at the attended location (for reviews, see Hillyard et al., 1998; Mangun, 1995). These P1 and N1 effects, which are illustrated in Figure 11.4, have been seen both in the Hillyard sustained attention paradigm (e.g., Hillyard & Münte, 1984; Mangun & Hillyard, 1987, 1988) and in the Posner cueing paradigm (e.g., Eimer, 1994a, 1994b; Hopfinger & Mangun, 1998; Luck et al., 1994; Mangun & Hillyard, 1991).

In the Posner cuing paradigm, each trial contains a cue and a target, and the cue indicates the likely location of the target for that trial and therefore induces the subject to attend to that location. On *valid* trials, the target appears at the cued location (which is presumably also the attended location). On *invalid* trials, the target appears at an uncued location. Some experiments also include *neutral* trials, on which the cue does not provide any information about the likely target location. In contrast to the Hillyard sustained attention paradigm, the Posner cuing paradigm requires subjects to respond to the target no matter where it appears. The advantage of this is that behavioral measures of attention can also be obtained. Specifically, the speed and/or accuracy of the target detection response are typically best on valid trials, worst on invalid trials, and intermediate on neutral trials (although the specific pattern depends on a variety of factors). The main disadvantage of the Posner cueing paradigm is that much more time is required to obtain a given number of trials in the ERP waveforms, especially for invalid trials. That is, only a fraction of the targets are presented at the unattended location (typically 10%–20%), and the need for a cue, a target, and a response on each trial leads to a relatively large amount of time between targets (typically 2–3 s). This contrasts with the Hillyard sustained attention paradigm, in which half of the stimuli are presented at the unattended location and two or three stimuli are presented per second. It is also possible that motor-related activity will contaminate the ERPs in the Posner cuing paradigm, although we now know enough about motor-related ERPs to distinguish between motor and sensory ERP responses in most cases (see Chapters 8 and 9, this volume). Thus, the choice between these paradigms typically reflects a balance between the desire to maximize the number of trials per waveform and the ability to relate the results to the extensive behavioral literature.[1]

In both paradigms, stimuli presented at the attended location typically elicit larger P1 and N1 waves than stimuli presented at the unattended location (i.e., larger for valid targets than for invalid targets in the Posner cuing paradigm). The P1 wave and the P1 attention effect both typically begin 70–100 ms after stimulus onset, with a scalp distribution that is maximal over the lateral occipital lobe, contralateral to the location of the eliciting stimulus.

An ipsilateral P1 wave is also observed, onsetting 10–20 ms after the contralateral P1, and it is also typically larger for attended stimuli than for unattended stimuli. The presence of an ipsilateral response suggests that the P1 wave is generated at a relatively late stage of the visual pathway, such as the lateral occipital complex, where receptive fields are very large and include both visual fields. It is unlikely that the P1 wave arises from an earlier stage, such as V4, where the receptive fields extend no more than 1° beyond the vertical meridian. However, V4 cells have large *silent suppressive surrounds* that extend far into the ipsilateral field (Desimone et al., 1985; Schein & Desimone, 1990), and there is some possibility that these inhibitory surrounds contribute to the P1 wave.

Heinze et al. (1994) used a variant of the Hillyard sustained attention paradigm with both ERPs and positron emission tomography (PET), and they found that spatial attention led to both an enlarged P1 wave and an increase in blood flow in ventrolateral occipital cortex. Source localization methods indicated that the location of the PET effect was consistent with the scalp distribution of the P1 effect. A similar approach was taken by Di Russo and colleagues (2003a), who used fMRI instead of PET and presented stimuli in upper and lower locations in the LVF and RVF. Subjects monitored one of these four locations for target stimuli embedded in a stream of standard stimuli (as in the typical Hillyard sustained attention paradigm). The waveforms and scalp distributions for upper-right stimuli are shown in Figure 11.5. The C1 wave—peaking 90 ms after stimulus onset with a maximum voltage over the occipital pole (electrode POz)—was not modulated by whether attention was directed toward or away from the location of the stimulus, replicating the studies described in the previous section. The P1 wave peaked at approximately 130 ms over contralateral occipito-temporal cortex (electrode PO7) and at approximately 150 ms over the ipsilateral hemisphere (electrode PO8), and it was larger over both hemispheres when attention was directed toward the location of the stimulus. A posterior N1 wave showed the same pattern but peaked approximately 50 ms later over each hemisphere. An additional N1 wave was also observed at anterior midline sites with a peak latency approximately halfway between the P1 and posterior N1 peak latencies (approximately 160 ms). This is the most common pattern of effects observed in visuospatial attention paradigms: a C1 wave over the occipital pole that is unaffected by the direction of attention; P1 and posterior N1 waves over lateral occipitotemporal cortex that are delayed over the ipsilateral hemisphere and influenced by the direction of attention; and a somewhat earlier and more bilateral N1 wave that is observed over fronto-central sites and is influenced by the direction of attention.

When an experimental variable such as the direction of attention is shown to increase the voltage in the time range of a given ERP component, this does not necessarily imply that the experimental effect reflects a simple increase in the amplitude of the underlying component (see Luck, 2005, chap. 2 and Chapter 1, this volume). As discussed previously, for example, Näätänen and his colleagues (Näätänen & Michie, 1979; Näätänen et al., 1978) argued that the auditory N1 attention effect consisted of the addition of an endogenous negativity rather than a modulation of the exogenous N1 wave (i.e., the addition of a process for attended stimuli that was not present for unattended stimuli rather than a difference in the amplitude of a process that was present for both the attended and unattended stimuli). Indeed, subsequent research has shown that this proposal was correct for at least part of the N1 attention effect (Hansen & Hillyard, 1980). In the visual modality, the evidence to date indicates that the P1, posterior N1, and anterior N1 attention effects do reflect modulations of the exogenous components (i.e., modulations of the stimulus-evoked neural activity rather than the addition of extra processes that are absent for unattended stimuli).

The general approach to this issue has been to compare the time course and scalp distributions of the results for the attended stimuli, the unattended stimuli, and the difference between the attended and unattended stimuli. The assumption is that, if attention operates by simply increasing or decreasing the amplitude of the exogenous components, then the timing and scalp distribution of the attended waveforms will be the same as those of the unattended waveforms, with only a change in amplitude. Moreover, the timing and scalp distribution of the attended-minus-unattended difference waveform should be the same as those of the unattended waveform. However, this is an oversimplification, because it is possible that some components will change by more or less than other components, and some might not change at all. Under these conditions, it is more difficult to compare the time courses and scalp distributions directly.

Consider, for example, the scalp distributions shown in Figure 11.5B. Because the C1 wave overlaps in time and space with the P1 wave but C1

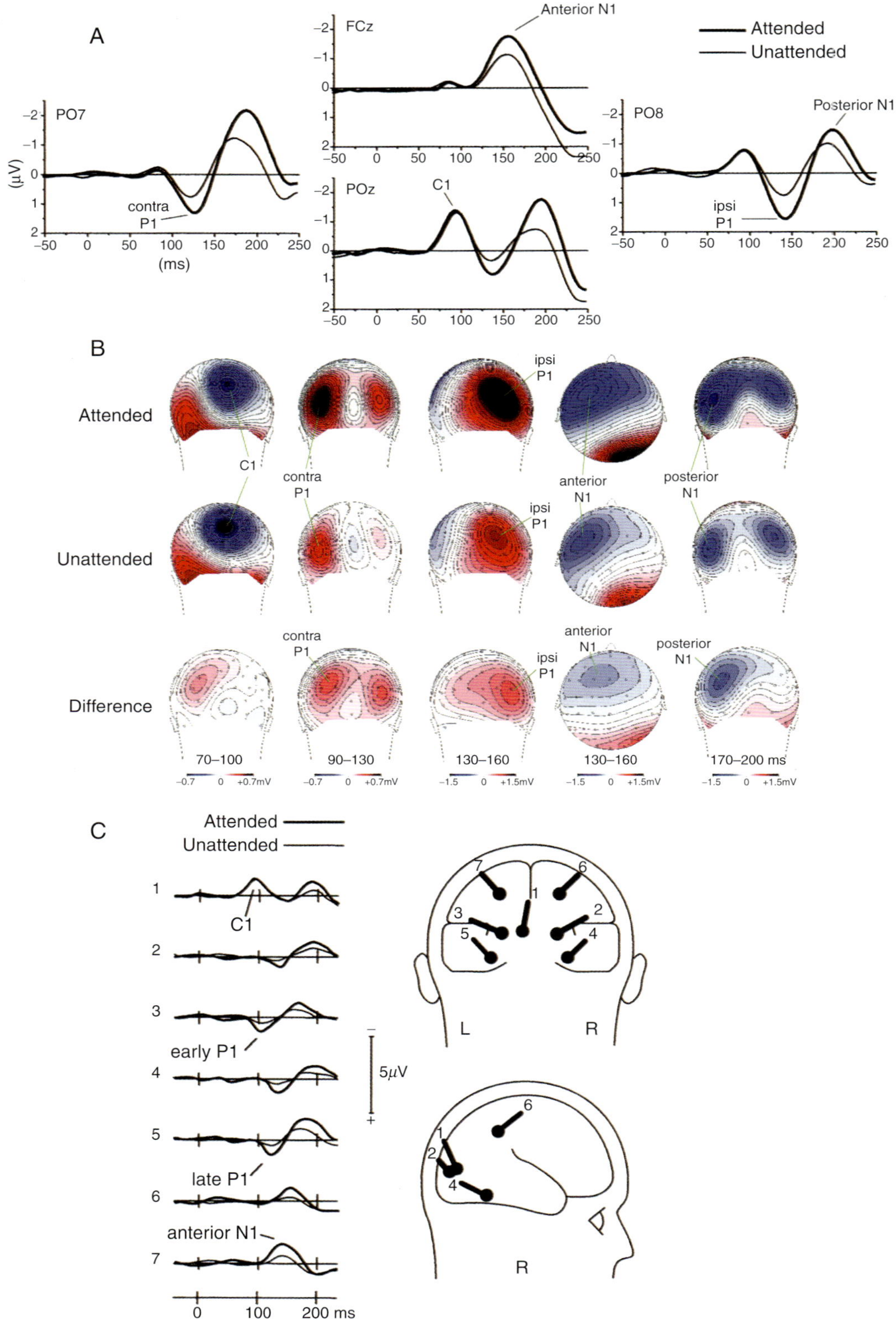

Fig. 11.5. Data from upper-right stimuli in the study of Di Russo et al. (2003a). (A) Grand average waveforms from selected scalp sites. (B) Scalp distribution of the mean voltage over several time ranges when the stimuli were attended, when the stimuli were unattended, and the difference between attended and unattended. (C) Estimates of the locations and time courses of the internal generator sources. Adapted with permission from Di Russo et al. (2003a).

amplitude is not modulated by attention, the difference between the scalp distributions of the attended and unattended waveforms will reflect only the P1 wave, whereas the scalp distribution for the unattended waveform will reflect both the C1 and P1 waves. Consequently, these scalp distributions look very different from each other from 70 to 100 ms, and they also look somewhat different from 90 to 130 ms. Thus, a simple comparison of scalp distributions or time courses is not enough to determine whether an attention effect consists of modulation of an exogenous component or the addition of an endogenous component.

To provide a more detailed analysis, Di Russo et al. (2003a) used inverse source localization methods to create a model of the neural generator sites of the ERP components, using fMRI results from the same paradigm to constrain the localization procedure. The results are summarized in Figure 11.5C, which shows the locations of the estimated sources and the activation time course for each estimated source. Seven dipoles were used to fit the data. The C1 wave was represented by a single dipole near striate cortex (dipole 1). The other components were represented by mirror symmetrical dipole pairs in the contralateral and ipsilateral hemispheres, each of which was activated 10–20 ms earlier in the contralateral hemisphere than in the ipsilateral hemisphere. The P1 was represented by two dipole pairs, one pair in the middle occipital gyrus (dipoles 2 and 3) and the other pair more ventrally in the fusiform gyrus (dipoles 4 and 5). The posterior N1 was also represented by these dipoles; that is, these dipoles showed a positive magnitude during the P1 latency range and a negative magnitude during the N1 latency range.[2] This should not be taken as evidence that the P1 and N1 are generated in precisely the same visual areas; the spatial resolution of the ERP technique may simply be too low to distinguish between the P1 and N1 generator sites. Finally, the anterior N1 wave was represented by a pair of dipoles in the parietal lobe, near the intraparietal sulcus, that pointed anteriorly. Note that the magnitude of the activity was greater for attended stimuli than for unattended stimuli for each dipole. However, this effect was not evident during the initial period of activation for the C1 wave; instead, attention appeared to influence only later activity coming from this source, consistent with the hypothesis that attention influences only feedback signals in area V1.

The model shown in Figure 11.5C was created on the basis of the data from the attended and unattended waveforms, which are dominated by stimulus-driven exogenous activity. A second model was created on the basis of the attended-minus-unattended difference wave, which reflects only the changes in activity caused by attention. The locations and time courses of this second model were very similar to those of the original model. If attention led to the addition of endogenous sources with distinctly different locations or orientations, then the second model should have been substantially different from the first model. Thus, the similarity of the two models is consistent with the hypothesis that attention simply modulates the amplitudes of the exogenous ERP components, as would be expected if attention operates as a gain control in extrastriate visual areas.

NATURE OF THE P1 AND N1 ATTENTION EFFECTS

The P1 and N1 waves are obligatory, exogenous sensory responses; this is, these waves are present for visual stimuli irrespective of top-down factors such as intentions, goals, tasks, and so on. These waves also vary in amplitude and latency according to low-level physical characteristics of the stimuli, such as luminance (Johannes et al., 1995). However, these voltage deflections represent the sum of several different underlying components, not all of which may be obligatory, exogenous sensory responses. The Di Russo et al. (2003) study, for example, found evidence for two separate attention-sensitive P1 generators and three separate attention-sensitive N1 generators, and similar conclusions were reached by Clark and Hillyard (1996). Given that human visual cortex contains more than 20 distinct areas, it is likely that even more generators contribute to the recorded voltage in the P1 and N1 latency ranges but cannot be resolved with the ERP technique. Thus, although the overall P1 and N1 voltage deflections are exogenous, it is not possible to conclude that all P1 and N1 subcomponents are exogenous. This should be kept in mind when thinking about the conclusions that have been drawn about the nature of the P1 and N1 attention effects.

The simple interpretation of the P1 and N1 attention effects is that they reflect a top-down modulation of the initial feedforward wave of sensory activity passing through extrastriate areas of visual cortex (for a detailed discussion, see Hillyard et al., 1998). The finding that the estimated source waveforms for the P1 and N1 sources are larger for attended stimuli than for unattended stimuli (as in Figure 11.5C) is consistent with this interpretation (although with all the caveats that accompany ERP source localization models). The P1 and N1 attention effects are also the same for target and nontarget stimuli, which is also consistent with a simple modulation of feedforward sensory activity. That is, if the target were discriminated prior to the operation of attention, then one would expect the target and nontarget to be differentially modulated by attention. The current evidence is consistent with a simple feedforward gain control explanation of the P1 effect, but the posterior N1 effect may not reflect a simple modulation of sensory gain. It is possible to obtain P1 attention effects without N1 attention effects and vice versa (e.g., see Mangun & Hillyard, 1991; Van Voorhis & Hillyard, 1977; Vogel & Luck, 2000), and this indicates that they do not reflect the operation of a simple gain control mechanism that operates prior to the time of the P1 wave in a single feedforward processing stream.

Luck and his colleagues (Hillyard et al., 1998; Luck, 1995, 1998a; Luck et al., 1994; Vogel & Luck, 2000) have proposed that the pattern of P1 and N1 dissociations across studies can be explained by the proposal that (1) the P1 attention effect reflects a suppression of feedforward sensory processing at unattended locations to reduce interference in the processing of information at the attended location, and (2) the N1 attention effect reflects the operation of a limited-capacity discrimination process directed to the information presented at the attended location. This proposal was initially spurred by the spatial cuing study illustrated in Figure 11.6. Each trial began with either a single cue arrow pointing to one of four possible target locations or a set of four arrows pointing to all four locations. When a single location was cued, the target appeared at that location on 75% of trials (*valid* trials) and at one of the other locations on 25% of trials (*invalid* trials). When all four locations were cued, the target was equally likely to appear at any of the locations (*neutral* trials). The P1 wave was reduced in amplitude on invalid trials compared to neutral trials, but it was not enlarged on valid trials compared to neutral trials. That is, compared to the neutral baseline condition, focusing attention on the cued location led to a suppressed P1 amplitude when the target appeared at one of the other locations, but it did not lead to an enhanced P1 amplitude when the target appeared at the cued location. In contrast, the N1 wave was enlarged on valid trials compared to neutral trials, but it was not reduced on invalid trials compared to neutral trials. That is, compared to the neutral baseline condition, focusing attention on the cued location led to an enhanced N1 amplitude when the target appeared at the cued location, but it

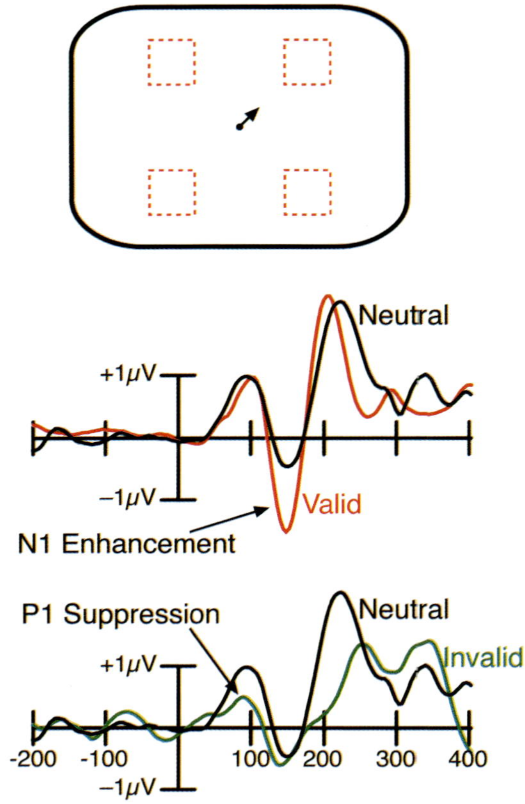

Fig. 11.6. Stimuli and grand average ERP waveforms from the study of Luck et al. (1994). Both rows of ERP waveforms show the ERPs elicited by the target on neutral trials, when attention was directed to all four locations. These neutral waveforms are overlaid by the waveforms elicited by the target when it appeared at the cued location (valid, top row of waveforms) and at the uncued location (invalid, bottom row of waveforms).

did not lead to a suppressed N1 amplitude when the target appeared at one of the other locations. This same pattern of P1 suppression at unattended locations and N1 enhancement at attended locations has also been observed in the context of a visual search paradigm, in which a very different type of neutral trial was used (Luck & Hillyard, 1995).

Other P1 and N1 dissociations support these hypotheses about the P1 and N1 attention effects, although somewhat less directly. First, the P1 effect is eliminated in visual search tasks when the task is changed from conjunction discrimination to feature detection (Luck & Hillyard, 1995). It has been proposed that feature values can be discriminated without filtering the features of other objects (Luck & Ford, 1998; Treisman, 1988), and the lack of a P1 effect under these conditions supports the hypothesis that this effect reflects a suppression of potentially interfering features from unattended locations. Second, the N1 effect was eliminated in a cuing experiment that required a simple target detection response, without a discrimination of target identity, but it returned when the task was changed to require a discrimination of target identity (Mangun & Hillyard, 1991). This fits with the hypothesis that the N1 effect reflects the operation of a discrimination mechanism. Third, the N1 effect but not the P1 effect is eliminated when items are presented in rapid succession at the attended location (Heinze et al., 1990; Luck et al., 1990), consistent with the hypothesis that the N1 effect reflects a mechanism that is highly limited in capacity and cannot operate repeatedly in a short period of time.

The hypothesis that the N1 wave is related to the operation of a limited-capacity discrimination mechanism has also been addressed outside the context of spatial attention experiments. In an early set of studies, Ritter and his colleagues (1982, 1983, 1988) compared the ERPs elicited by stimuli in two tasks: (1) a simple detection task in which subjects simply made a speeded response when they detected any stimulus and (2) a choice discrimination task in which they made different responses, depending on which of two stimulus forms was presented. They found that the voltage in the N1 latency range was greater in the discrimination task than in the detection task, and they concluded that this reflects the operation of a pattern recognition process. Vogel and Luck (2000) replicated this pattern of results and showed that the N1 was also larger for a color discrimination task than for a simple detection task. They also ruled out several confounding factors, such as differences in overlapping response-related ERP activity and differences in arousal. The scalp distribution of this effect was comparable to that of the posterior N1 attention effect in spatial attention experiments, and they concluded that both the N1 attention effect and the N1 discrimination effect reflect the operation of a general-purpose visual discrimination mechanism. This mechanism cannot be divided among multiple locations; it operates only when attention is allocated to a single spatial location. In the Posner cuing paradigm, therefore, this mechanism is present on valid trials (because attention is focused on the location of the target), but it is not present on neutral trials (because attention is not focused on a single location) or on invalid trials (because attention is focused on the wrong location). However, it is present in the Posner paradigm only when the task involves discrimination and not simple detection (as observed by Mangun & Hillyard, 1991).

Attention to Nonspatial Visual Features

Some theories of visual attention treat location differently from other features (e.g., Moore & Egeth, 1998; Nissen, 1985; Treisman, 1988; Wolfe, 1994), reflecting the fine-grained spatiotopic mapping that is present from the very first stages of the visual processing pathway. Other theories, however, do not accord any special status to location (e.g., Bundesen, 1990). Until recently, ERP studies appeared to provide strong support for "space-is-special" theories, showing that selection based on spatial location (*spatial attention*) operates at an earlier stage than selection based on nonspatial features (*featural attention*). In this section, we will review the typical pattern of featural attention effects, describe evidence that featural attention effects are eliminated when stimuli are presented in unattended locations, and then present new evidence indicating that featural attention can operate at an early stage and independently of spatial location under certain conditions.

TYPICAL FEATURAL ATTENTION EFFECTS

The Hillyard sustained attention paradigm can be easily modified to examine the effects of attending to specific orientations, colors, spatial frequencies, and other features. As an example, we will consider a study of color-based attention in which subjects viewed a random sequence of red and blue checkerboard patterns (Anllo-Vento et al., 1998). Each checkerboard was presented for 100 ms, followed by a blank interstimulus interval of 50–350 ms. The red and blue colors were presented at a standard brightness for 90% of the checkerboards and at a slightly dimmer brightness for the remaining 10%. Subjects were instructed to attend either to red or to blue at the beginning of a block of stimuli, pressing a button when they detected the slightly dimmer deviant stimuli in this color and ignoring the standard and deviant stimuli in the other color. Each subject attended to red for half of the trial blocks and to blue for the other half. This made it possible to examine the ERPs elicited by a particular color when that color was attended versus when that color was ignored, thus ruling out any differences in the ERPs due to color per se. This is exactly analogous to the visuospatial version of the Hillyard sustained attention paradigm, but using colors instead of locations to define the attended and unattended channels.

The results of this study were largely similar to those of many other feature-based attention studies (Harter & Guido, 1980; Harter & Previc, 1978; Hillyard & Münte, 1984; Kenemans et al., 1993; Previc & Harter, 1982; Wijers et al., 1989).

Specifically, when the ERPs were compared for standard stimuli presented in the attended and unattended colors, the ERP waveforms elicited by the attended color contained a broad negative-going response peaking at around 225 ms at posterior scalp sites and a positive-going response peaking at around 200 ms at anterior scalp sites. These effects are shown as attended-minus-unattended difference waves in Figure 11.7A. The broad posterior effect is typically termed the *selection negativity*, and the more temporally discrete anterior effect is called either the *anterior P2 attention effect* or the *anterior selection positivity*. The same pattern was observed for the deviant stimuli, except that the attended-color deviant stimuli (i.e., the targets) also elicited a large, broadly distributed P3 response (see Figure 11.7B). Similar effects are observed when attention is directed on the basis of other nonspatial dimensions, such as orientation, form, or direction of motion, although the scalp distributions may vary across dimensions (Anllo-Vento & Hillyard, 1996). The selection negativity is thought to reflect the continued processing of the stimuli containing the attended feature value, but little effort has been devoted to understanding the anterior selection positivity. Related effects can also be observed in visual search experiments, in which the ERPs are compared for arrays that do or do not contain target features (see Chapter 12, this volume).

Featural attention does not typically produce the same early P1 and N1 effects as spatial attention,[3] and some studies have directly examined the relationship between spatial and featural attention (e.g., Anllo-Vento & Hillyard, 1996; Hillyard & Münte, 1984). In the study of Hillyard and Münte (1984), for example, subjects attended to stimuli of a particular color at a particular location and were instructed to detect occasional size deviants among the stimuli of the attended color–location combination, ignoring size deviants that occurred in the unattended color and/or location. Thus, the attended channel was defined by a combination of color and location. Hillyard and Münte observed the typical P1 and N1 effects when comparing attended- and unattended-location stimuli, and these effects were the same whether the stimuli were presented in the attended color or the ignored color. They also observed the typical selection negativity and selection positivity effects when comparing attended- and unattended-color stimuli, but only for stimuli presented at the attended location. The ERPs did not differ between the attended and unattended colors for stimuli presented at the unattended location.

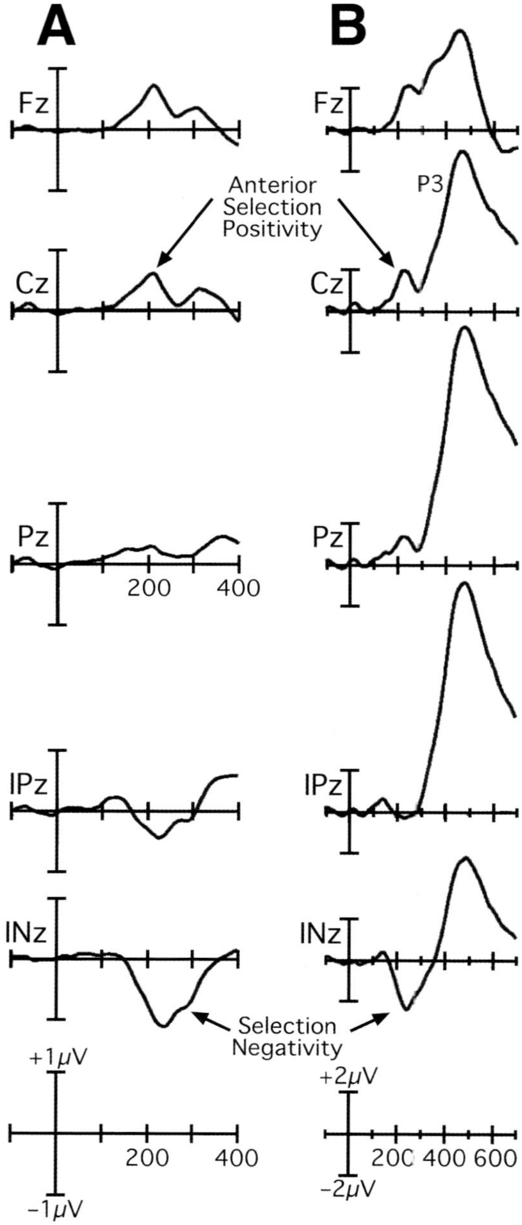

Fig. 11.7. Grand average difference waves (attended color minus unattended color) for standards (A) and deviants (B) in the study of Anllo-Vento et al. (1998).

From this result, they concluded that featural attention is not only later than spatial attention but is also contingent on spatial attention. That is, feature-based attention mechanisms are applied only to stimuli that pass through the initial spatial filter stage.

This proposal fits well with space-is-special theories of attention based on behavioral data (Moore & Egeth, 1998; Nissen, 1985; Treisman, 1988; Wolfe, 1994), and it makes a fundamental claim about the nature of visual attention. However, recent research suggests that featural attention can, under certain conditions, operate as early as spatial attention and independently of spatial attention. In particular, featural attention can influence the P1 wave under conditions that maximize competition between the attended and unattended feature values (Valdes-Sosa et al., 1998; Zhang & Luck, 2009).

According to the biased competition theory of attention (Desimone & Duncan, 1995), attention operates primarily to resolve competition between concurrent sources of stimuli. Both spatial attention effects and featural attention effects are larger when attended and unattended stimuli simultaneously compete for access to perceptual processing resources (Luck et al., 1997; Saenz et al., 2003). In addition, single-unit and fMRI experiments using simultaneously visible attended and ignored feature values have found featural attention effects in extrastriate visual cortex (Saenz et al., 2002; Treue & Maunsell, 1999). However, ERP studies of featural attention using the Hillyard sustained attention paradigm have presented the attended and ignored feature values sequentially rather than simultaneously, creating minimal competition. This may account for the lack of early featural attention effects in these studies. However, two ERP studies have used experimental designs in which attended and ignored colors were simultaneously visible, creating strong competition between the two colors, and both have reported color-based P1 attention effects (Valdes-Sosa et al., 1998; Zhang & Luck, 2009).

In the study of Valdes-Sosa et al. (1998), subjects viewed a cluster of red dots and a cluster of green dots that were spatially intermixed (see Figure 11.8). The red dots rotated in one direction and the green dots simultaneously rotated in the opposite direction, giving rise to the perception of two transparent overlapping surfaces. Subjects attended either to the red dots or to the green dots. Because the red and green dots were randomly distributed across the same region of space, the two colors were in strong competition with each other. The motion was present for an entire trial, and occasional deviation in the motion of either the red dots or the green dots was used to elicit an ERP. When the deviation occurred in the attended color, a larger P1 was elicited than when the deviation occurred in the unattended color. This was the first clear evidence that featural attention could influence P1 amplitude under conditions of simultaneous competition.

A subsequent study demonstrated that color-based attention could influence P1 amplitude even when the stimuli were presented in an unattended spatial location (Zhang & Luck, 2009). As in the Valdes-Sosa study, subjects were presented with interdigitated sets of red and green dots (see Figure 11.9). Each dot lasted for only a few video frames and was then replaced by a dot at a different location. This continued for an entire trial, leading to the perception of two sets of scintillating dots within the same region. Subjects attended either to red or to green dots and were instructed to look for occasional dimming of the dots in the attended color. The stream of interdigitated red and green dots was presented on one side of fixation, and unattended "probe" stimuli were flashed occasionally at an unattended location on the other side of fixation. These probe stimuli were sets of all-red or all-green dots, and they were used to assess the processing of the attended and ignored colors outside of the attended region. The P1 wave elicited by these probes was larger when the color of the probe matched the attended color than when it matched the unattended color. Thus, not only can featural attention operate as early as spatial attention, it can also operate at unattended spatial locations. To demonstrate that these effects were a consequence of the competition between the attended and ignored feature values,

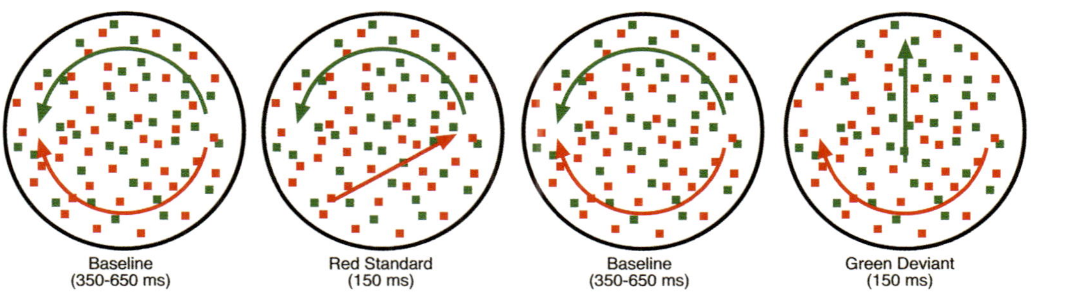

Fig. 11.8. Example sequence of stimuli from the study of Valdes-Sosa et al. (1998). The arrows show the direction of motion of the dots of the corresponding color and were not visible.

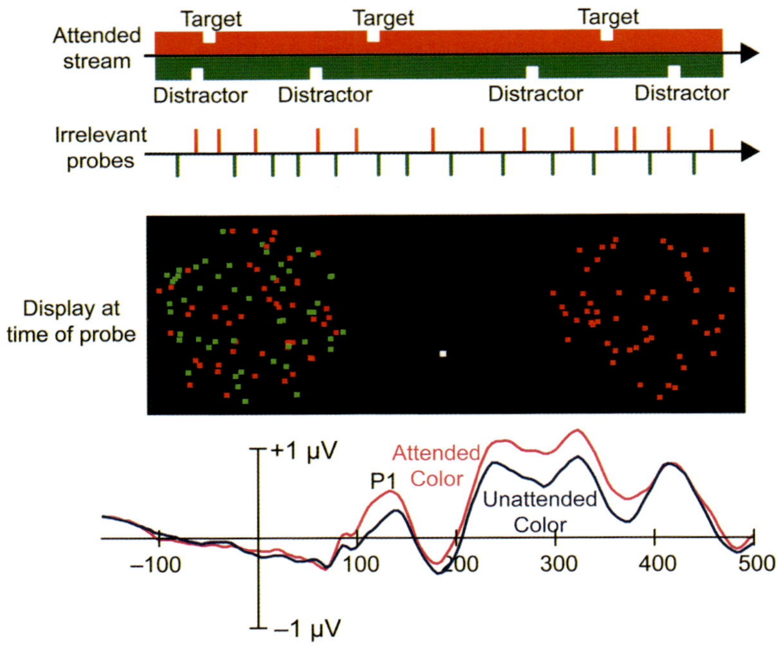

Fig. 11.9. Stimuli and grand average ERP waveforms from the study of Zhang and Luck (2009). Adapted with permission from Zhang and Luck (2009).

a follow-up experiment was conducted in which the attended stimulus stream consisted of a sequence of all-red dots and all-green dots rather than a stream of simultaneously interdigitated red and green dots. Under these conditions, the probes did not elicit a larger P1 when the probe color matched the attended color. Thus, simultaneous competition appears to be the key factor needed to observe early location-independent effects of featural attention.

It is not yet known whether these color-based P1 attention effects reflect a modulation of the same underlying ERP component as the previously observed location-based P1 attention effects. The timing and scalp distributions are at least grossly similar, but it is difficult to demonstrate that ERP effects observed in separate experiments actually arise from the same component (see Chapter 1, this volume). Thus, it is not yet possible to conclude that spatial and featural attention operate in exactly the same way. In addition, simultaneous competition does not appear to be necessary to observe location-based P1 attention effects, so location may still have a special status in attention.

Object-Based Visual Attention

In naturally occurring visually guided behavior, humans rarely focus attention on a peripheral location in anticipation of stimuli that might subsequently appear in that location. Instead, attention is usually directed to objects. Consider the example shown in Figure 11.10, in which attention might be directed to the apple. Identifying the apple requires separating it from the background, including the leaf that partially occludes the top of the apple. Thus, it is ultimately the object we wish to select rather than a region of space. The behavioral attention literature has extensively examined the issue of *object-based* attention, demonstrating that the allocation of attention is influenced by the nature of the objects in the visual input (see especially Driver & Baylis, 1989;

Fig. 11.10. Example of the need for object-based attentional selection in the perception of natural images.

Duncan, 1984; Egly et al., 1994; Kahneman et al., 1992; Vecera, 1994). As discussed by Vecera and Farah (1994), object-based attention could work in two ways. First, attention could operate on high-level object representations that have been abstracted away from specific spatial locations, such as the geon structural description representations proposed by Biederman (1987). In the example shown in Figure 11.10, this would involve forming a location-independent representation of the apple. Second, mechanisms of perceptual organization could link the micro elements of an object together into a spatial region that reflects the boundaries of the object, and attention could spread throughout this region. In the example shown in Figure 11.10, this would involve the spread of attention throughout the region defined by the apple.

Although object-based attention has been a topic of extensive research in the behavioral literature, it has been largely ignored by ERP researchers. Two studies have recorded ERPs in variants of the object-based attention paradigm of Egly et al. (1994). Figure 11.11 illustrates the task used for one of these studies (He et al., 2004). The display contains either two vertical bars or two horizontal bars. One end of one of the bars is cued, and a subsequent target is usually presented at the cued location (valid trials). In invalid trials, the target can be presented within the same rectangle as the cue (as shown in Figure 11.11) or at an equally distant location within the other rectangle. Reaction times (RTs) in this paradigm are fastest when the target appears at the cued location, slowest when the target appears at an uncued location within the uncued rectangle, and intermediate when the target appears an the uncued location within the cued rectangle. The faster RTs for the uncued location within the cued rectangle compared to the equidistant uncued location within the other rectangle are generally taken as evidence that attention "spreads" throughout the cued object. He et al. (2004) found that the N1 wave elicited by the target showed an analogous pattern: N1 amplitude was greatest for targets presented at the cued location, smallest for targets presented at the uncued location within the uncued rectangle, and intermediate for targets presented at the uncued location within the cued rectangle. The P1 wave, however, showed a pure spatial attention effect. That is, it was larger for targets presented at the cued location than for targets presented at the uncued locations, and it did not differ between the uncued locations in the cued and uncued rectangles. This same pattern was replicated in a study that included both ERPs and fMRI (Martinez et al., 2006), and the fMRI data suggested that the object-based and space-based components portions of the N1 attention effect both came from the same region of lateral occipital cortex. This is consistent with the proposal that object-based attention reflects a spatial spreading of attention within an object's boundaries.

Postperceptual Attention Effects

The studies described thus far have focused on the role of attention in the perception of stimuli. However, selective attention may also operate on postperceptual processes[4] (for a detailed discussion, see Luck & Vecera, 2002). Generally speaking, attention appears

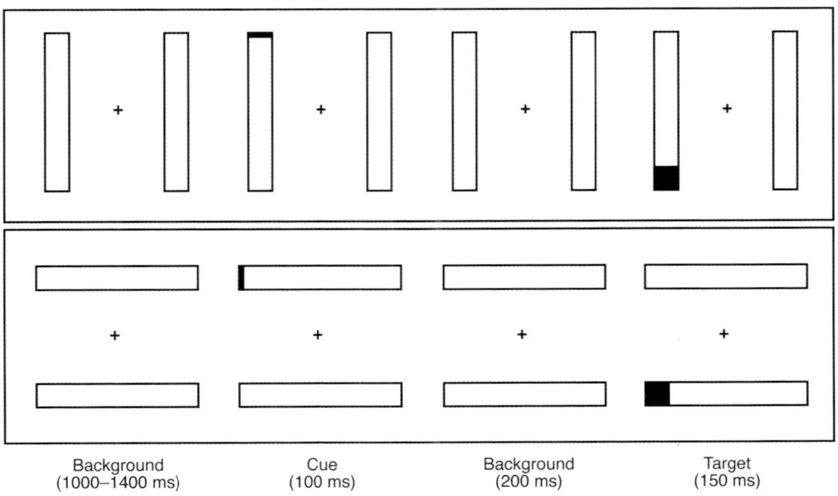

Fig. 11.11. Example stimuli from the study of He et al. (2004).

to operate within a given system when that system is overloaded (Lavie, 1995; Lavie & Cox, 1997; Lavie et al., 2004; Luck & Hillyard, 1999; Luck & Vecera, 2002). When multiple competing sensory inputs occur simultaneously or in rapid sequence, perceptual systems are overloaded and attention operates at an early stage to limit perceptual processing to a subset of the information. Under these conditions, attention influences the early sensory ERP components. However, it is possible to create situations in which the stimuli do not overload perceptual systems but overload memory encoding or response selection systems. In these tasks, all items are perceived but only a subset of the items are stored in memory or lead to behavioral responses, and this is accompanied by changes in ERP components that reflect memory- and response-related processes. The proposal that attention can operate at both early and late stages, depending on the nature of the stimuli and task, is called the *flexible selection* hypothesis (Vogel et al., 2005). In the following sections, we will discuss tasks in which attention modulates the N400 component, the P3 or P300 component, and the lateralized readiness potential (LRP). We will begin with studies that focus directly on the attention sensitivity of these components and then discuss studies that have used these components as a tool to answer broader questions about attention.

Attention and the N400 Component

The N400 component is elicited by stimuli that mismatch a previously established semantic context (see Chapter 15, this volume). Consider, for example, a sentence that begins "John put on his glasses and started to read a . . ." If this sentence ends with "book," then this final word will match the semantic context and generate very little N400 activity. If, however, this sentence ends with "mattress," then this final word will mismatch the context and generate a large N400. Similar effects can be obtained with simple word pairs. For example, the second word in the pair will elicit a small N400 for "read . . . book" but will elicit a large N400 for "read . . . mattress." This difference between an initial word or phrase and a subsequent target word is termed the *N400 priming effect.*

Three things are logically necessary for a subject to exhibit an N400 priming effect. First, the subject must understand and retain the information about the prime word or phrase. Second, the subject must identify the target word fully, at least to the point of lexical access. Third, the subject must actually compare the target word with the prime. All three of these steps can be influenced by attention.

The most basic way to manipulate attention in the context of words is to vary whether the task requires semantic processing of the words. Several studies have used such task manipulations to ask whether word recognition is automatic. For example, Kutas and Hillyard (1989) presented subjects with pairs of words followed by a single letter and asked the subjects to report whether that letter had been present in either of the words. The N400 was larger when the second word in each pair was semantically unrelated to the first word compared to when the two words were related, even though the meanings of the words were irrelevant to the task. Similarly, Connolly et al. (1990). presented spoken sentences and tested conditions in which the subjects were required (1) to simply listen to the sentences, (2) to detect whether a specific speech sound was in the final word of each sentence, or (3) to decide whether the final word of each sentence belonged to a specific semantic category. An equivalently large N400 was elicited by the final word of the sentences in all three conditions, despite the differences in the semantic requirements of the tasks.

Results of this nature might be taken as evidence that word identification is automatic. However, the perceptual load was so low in these experiments that subjects may have had sufficient cognitive resources to identify the meanings of the words while simultaneously performing the instructed task. And given that the instructed tasks were quite boring, the subjects may have engaged in semantic processing simply to reduce their level of boredom. To demonstrate that a process is automatic, one must go beyond showing that it occurs when it is not required; one must also show that it occurs when subjects are strongly motivated to avoid it (see Jonides, 1981). Heil et al. (2004) took a step in this direction by using a more difficult task to direct attention away from the meanings of the words. Specifically, they presented word pairs and asked subjects to detect whether the first word (the prime) contained a specific target letter and whether the second word (the target) was a real word or a pseudoword. The letter-search task for the prime word was designed to motivate subjects to ignore the meaning of this word and instead focus on its orthography. The N400 elicited by the target word was found to be larger when it was semantically unrelated to the prime than when it was semantically related. From this, Heil et al. (2004) concluded that subjects automatically extracted the meaning of the prime word. However, the N400 difference between related and unrelated trials was small compared to the N400 difference typically observed in studies in which subjects actively attend to the word meanings (and the Heil et al. study did not contain a condition in which the word meanings were relevant). Thus, it is entirely possible that the letter-search task was not sufficiently demanding to completely eliminate semantic processing of the prime. Indeed, as we will describe below, other attentional manipulations have eliminated the N400 priming effect, suggesting that semantic analysis is not fully automatic. Thus, it is important not to use the lack of an attention effect for a given ERP component to draw conclusions about the automaticity of that component unless a very strong manipulation of attention has been used.

Attention and the P3 Component

The P3 component is typically elicited by infrequent task-relevant events (see Chapter 7, this volume). There is no universally accepted theory of the P3 wave, but the most common view—proposed originally by Donchin (1981)—is that the P3 wave reflects an updating of *context* information. Most researchers assume that *context* is equivalent to *working memory*, but Donchin (1981) never made this claim. Nevertheless, researchers outside the ERP domain often treat *context* and *working memory* as equivalent concepts (e.g., Cohen et al., 1990), and the P3 wave has been used frequently as an index of working memory updating (e.g., Vogel et al., 1998).

Regardless of what process the P3 represents, there is clear evidence that P3 amplitude can be influenced by that amount of attention allocated to a stimulus. This is most clearly observed in dual-task experiments in which subjects are instructed to vary the relative allocation of attention between the two tasks. For example, Isreal et al. (1980b) asked subjects to perform an auditory oddball task, in which they mentally counted infrequent target tones, simultaneously with an air traffic control task in which they tracked either four or eight moving objects on a video screen. An increase in the number of objects being tracked increases the resources required for this task, which should in turn reduce the resources available for the auditory oddball task. Isreal et al. found that P3 amplitude for the infrequent target tones was reduced when the subjects tracked eight objects compared to four objects in the visual tracking task, consistent with the hypothesis that the amplitude of the P3 wave elicited by a given stimulus depends on the amount of resources available to process that stimulus (see also Isreal et al., 1980a; Kramer et al., 1983; Mangun & Hillyard, 1990; but see Kok, 2001, for caveats on the use of P3 as a measure of processing capacity).

The Attentional Blink Paradigm

The P3 and N400 components have been very useful in understanding the operation of attention in the *attentional blink* paradigm, in which subjects must discriminate two targets that appear in close succession. A typical attentional blink task is shown in Figure 11.12. Each trial consists of a sequence of letters presented very rapidly at fixation (one letter every 100 ms); two of the letters are red and must be reported at the end of the trial. The lag between the first target (*T1*) and the second target (*T2*) varies across trials (e.g., *lag 2* means that T2 was the second item after T1). If T1 and T2 are presented close together in time, attention may still be directed to T1 when T2 appears, making it difficult for the subject to report T2. This is called the *attentional blink* because it is as if T1 triggers a brief blink of attention, causing T2 to be missed. In most cases, this blink is triggered slowly enough that T2 is reported correctly when it is the item immediately following T1 (i.e., at lag 1). Thus, T2 accuracy is usually high at lag 1, drops to its lowest point at lag 3, and recovers by approximately lag 6 or 7 (see Figure 11.12). Many attentional blink studies also include a T2-only control condition, in which subjects ignore T1 and only report T2. In this condition, performance is largely independent of the T1–T2 lag.

The study that coined the term *attentional blink* assumed that subjects essentially failed to see T2 during the attentional blink period (Raymond et al., 1992). However, the visual system can discriminate complex scenes within 100–150 ms (Potter, 1976; Thorpe et al., 1996), so it should be able to discriminate relatively simple, highly discriminable, highly familiar stimuli such as letters even more rapidly. The attentional blink paradigm does not merely require perception of the two targets; the targets must also be stored in working memory so that they are not overwritten by the subsequent stimuli and can be reported at the end of the trial. It is therefore possible that T2 is identified but cannot be transferred to working memory when attention is occupied by T1. Indeed, Giesbrecht and Di Lollo (1998) showed that the perceptual representation of T2 is overwritten by the next item in the stimulus sequence during the attentional blink.

Event-related potentials have been used determine whether observers fail to report T2 correctly during the attentional blink period because they fail to see it or because they fail to encode it in working

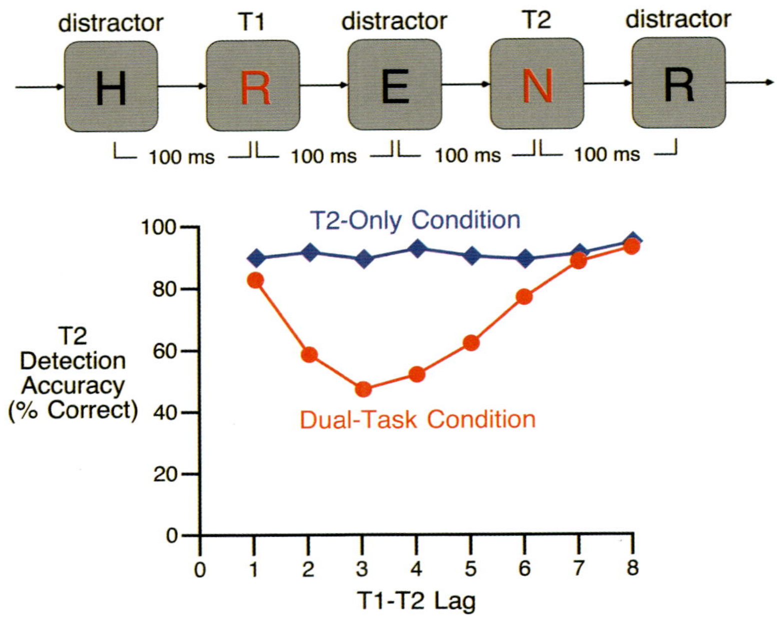

Fig. 11.12. Typical stimuli and results from the attentional blink paradigm.

memory (Luck et al., 1996; Vogel et al., 1998). The first experiment in this study examined the P1 and N1 components as measures of sensory processing, finding no reduction in amplitude for stimuli during the attentional blink period. This result indicated that these stages of sensory processing were not influenced by the attentional blink, but it is always possible that attention influenced some sensory process that did not have an ERP signature.

To definitively show that the attentional blink represents a postperceptual effect of attention, it was necessary to provide evidence that T2 was fully identified during the attentional blink period. However, there is no sensory ERP component that clearly reflects the *endpoint* of perception (if, indeed, there is such a thing). To solve this problem, Luck et al. (1996) employed a "trick of the trade," in which an ERP component is used as an index of the processes that necessarily precede it rather than the processes that directly generate the component. For example, as discussed in Chapter 9 of this volume, the subtraction procedure used to isolate the LRP guarantees that the voltage cannot exceed zero until the brain has determined which response is appropriate for the present stimulus. Thus, even though the LRP does not itself directly reflect the process of determining which response is appropriate, its latency can be used to measure the time required to make this determination. In the case of the attentional blink, the N400 component can be used in a similar manner to determine whether a word has been fully identified, even though the N400 component itself does not directly reflect word identification processes.

The logic behind this use of the N400 is as follows. When a given word is semantically unrelated to a previously established semantic context, that word will elicit a larger N400 than if the word is related to the context (e.g., in the context of *coffee*, the word *cream* is related and the word *cloud* is not). To generate a larger response for an unrelated word than for a related word, the brain must have already extracted the identity of the word (i.e., lexical access must have occurred). If the brain had not identified the word, how could it compare it to the semantic context and determine whether it matched or mismatched? Thus, the presence of a larger N400 for an unrelated word than for a related word implies that the brain must have identified the words. Thus, the presence of a larger N400 for unrelated words than for related words during the attentional blink would indicate that these words had been identified to a high level.

To implement this idea in an attentional blink experiment, Luck et al. (1996) preceded each trial with a 1000 ms *context word*, which the subjects simply stored in memory for that trial (see Figure 11.13A). After a 1000 ms delay, a stimulus stream was presented at a rate of 12 stimuli per second.

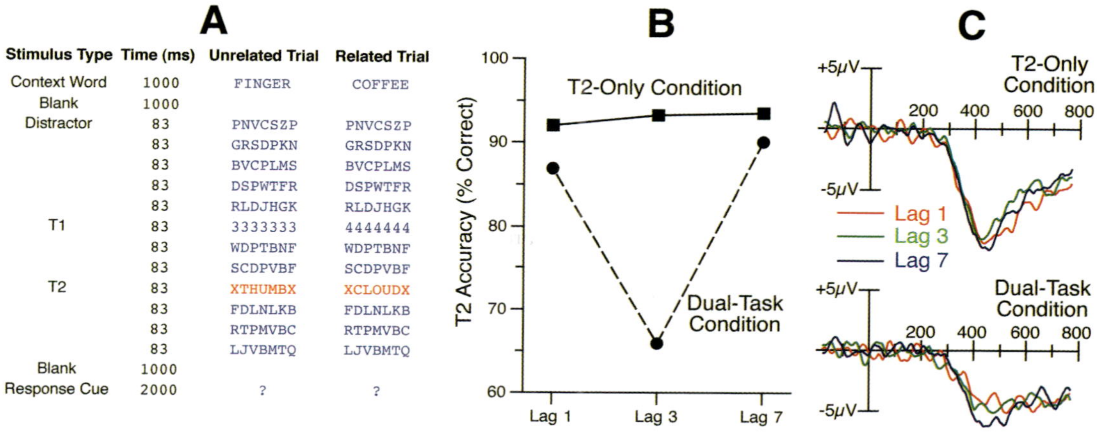

Fig. 11.13. Stimuli (A), behavioral results (B), and grand average ERP waveforms from the attentional blink study of Luck et al. (1996). The ERP waveforms are difference waves, formed by subtracting the ERP waveform elicited by T2 on related trials from the ERP waveform elicited by T2 on unrelated trials.

Most of the items in this stream were seven-character consonant strings, which served as distractors. T1 was a digit, repeated seven times to make a seven-character string. T2 was a word that was either semantically related or semantically unrelated to the context word for that trial (and was flanked by *X*s to make a seven-character string). All of the strings were drawn in blue, except that T2 was drawn in red to ensure that there would be no ambiguity about which word was T2 and to ensure good time locking in the ERPs. At the end of each trial, the subjects were instructed to make one button-press response to indicate whether T1 was an odd or even number and then to make a second button-press response to indicate whether T2 was semantically related or unrelated to the context word for that trial. This was the *dual-task* condition of the experiment. As is common, the experiment also included a *T2-only* condition, in which subjects ignored T1 and made only the T2 response on each trial. In both conditions, T2 was either the first item following T1 (*lag 1*), the third item following T1 (*lag 3*), or the seventh item following T1 (*lag 7*). Given the number of trials that must be averaged together to obtain clean ERP waveforms, it was not possible to test every lag, but lags 1, 3, and 7 provide the most important information because the attentional blink typically consists of a drop in performance between lag 1 and lag 3 with a recovery by lag 7.

In the T2-only condition, subjects were expected to have no trouble identifying T2 at any lag, leading to high levels of behavioral accuracy at all lags. Thus, the N400 was expected to be much larger when T2 mismatched the context word than when it matched the context word in this condition, irrespective of lag. In the dual-task condition, behavioral performance for T2 was expected to drop at lag 3, indicating that subjects could not report whether T2 was related or unrelated to the context word at this lag (which is the typical attentional blink pattern). If this decrement in behavioral performance occurs because subjects cannot identify T2 during the attentional blink period, then the N400 should be eliminated for T2 during the attentional blink. If, however, subjects can identify T2 during the N400 but fail to store it in working memory, then the brain would still be able to compare T2 with the context word and generate a larger N400 when T2 was unrelated than when it was related.

The behavioral results are shown in Figure 11.13B. As is typical in attentional blink experiments, T2 accuracy was high at all lags in the T2-only condition, but it dropped substantially at lag 3 in the dual-task condition. The ERP results are shown in Figure 11.13C. These ERPs are difference waves formed by subtracting the waveforms when T2 was semantically related to the context word from the waveforms when T2 was semantically unrelated to the context word. This subtraction was important for two reasons. First, it isolated the brains differential response to unrelated versus related words, eliminating any ERP activity related to other aspects of the processing of the stimuli. Second, it subtracted away the overlapping ERPs from the other stimuli in the stream. The resulting difference wave was somewhat larger in the T2-only condition than in the dual-task condition, which presumably reflects the general difficulty of coordinating two tasks in the dual-task condition. The key result is that the difference wave was not reduced in amplitude at lag 3 in the dual-task condition. That is, even though subjects were quite inaccurate in overtly reporting whether T2 was related or unrelated to the context word at lag 3 of this condition, their brains clearly made this discrimination well enough to generate a larger N400 for unrelated words than for related words. Thus, T2 is identified during the attentional blink period and the word identity is compared with the previously established semantic context, generating an N400 if the word does not match this context. However, this information either fails to reach awareness or is immediately forgotten.

It is conceivable that T2 was discriminated poorly during the attentional blink period, but that N400 amplitude is simply insensitive to the quality of the discrimination. To rule out this possibility, Vogel et al. (1998) ran a control experiment with the T2-only task in which varying levels of visual noise were added to the T2 word. It is well known that adding this sort of noise impairs the perceptual discriminability of visual stimuli. Indeed, discrimination accuracy decreased systematically as the noise level increased. The amplitude of the N400 difference wave also decreased systematically as the noise level increased, demonstrating that N400 amplitude is indeed sensitive to reductions in perceptual quality. Thus, the lack of a reduction in N400 amplitude during the attentional blink period in the main experiment provides strong evidence that the attentional blink does not reflect a reduction in perceptual discriminability (but see Giesbrecht et al., 2007, for conditions under which perception can be influenced by the attentional blink).

A leading hypothesis is that the attentional blink reflects a failure to store T2 in working memory, such that the representation of T2 is overwritten by the subsequent stimulus (Giesbrecht & Di Lollo, 1998). To test this hypothesis, Vogel et al. (1998) ran an experiment in which T2 was either a frequently occurring stimulus category or a rare stimulus category. The P3 wave was isolated with a rare-minus-frequent difference wave. The P3 difference wave was found to be completely eliminated during the attentional blink period. If the P3 wave reflects working memory encoding, as discussed above, then this finding provides support for the hypothesis that working memory encoding is disrupted during the attentional blink (for additional evidence from the P3 wave, see Vogel & Luck, 2002).

The Psychological Refractory Period Paradigm

The psychological refractory period (PRP) paradigm is similar to the attentional blink paradigm, except that immediate responses are made to the two targets and RT rather than accuracy is the main dependent variable. In a typical PRP experiment, subjects are presented with two targets (again called T1 and T2) on each trial (but without any additional distractor stimuli). The time between the onset of T1 and the onset of T2 is varied (the *stimulus onset asynchrony* or *SOA*). When the SOA between T1 and T2 is long, subjects are finished processing T1 by the time T2 is presented, and the RT for T2 is short. When the SOA between T1 and T2 is short, however, T1 is still being processed when T2 is presented, and the RT for T2 is prolonged. A leading theory proposes that the prolonged RTs at short

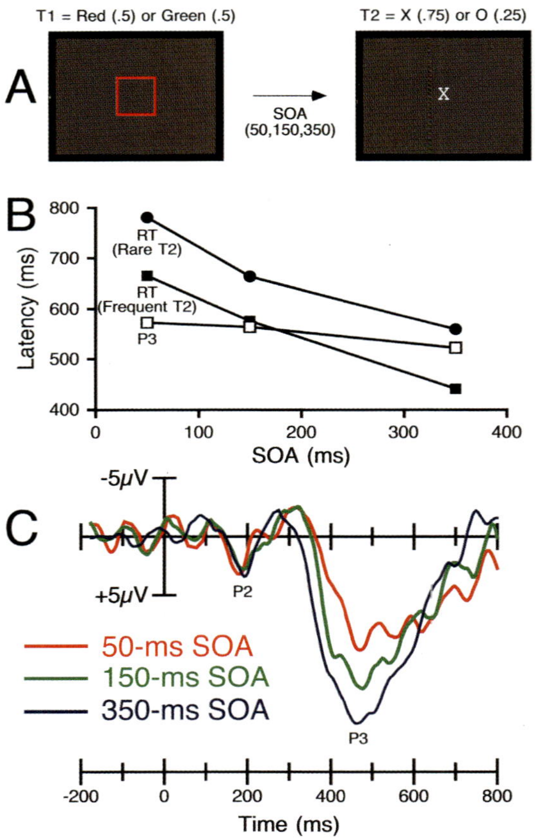

Fig. 11.14. Stimuli (A), behavioral results (B), and grand average ERP waveforms from the psychological refractory period study of Luck (1998). The ERP waveforms are difference waves, formed by subtracting the ERP waveform elicited by the frequent T2 stimulus from the ERP waveform elicited by the rare T2 stimulus.

SOAs are primarily a result of a bottleneck in the *response selection* process (i.e., the process of determining which response is appropriate once the stimulus has been identified). That is, the early stages of T2 processing can be carried out while subjects are still processing T1, but the response selection process for T2 cannot begin until the response selection process for T1 has finished (see the review by Pashler, 1994). Processing at other stages may also compete for limited processing resources, but this theory proposes that response selection is the only *single-channel* process (i.e., the only process that must be postponed for one stimulus if it is busy for another stimulus).

Event-related potentials have been used to test this proposal, examining the P3 wave (Arnell et al., 2004; Dell'Acqua et al., 2005; Luck, 1998b) and the LRP (Osman & Moore, 1993). The P3 wave is a sensitive index of stimulus evaluation time (see Chapter 7, this volume), and Pashler's (1994) bottleneck model would predict that stimulus evaluation is not postponed in the PRP paradigm. To isolate the P3 wave for T2 from other T2-related components and the overlapping T1 ERP waveform, studies have used rare and frequent T2 stimuli. In the study of Luck (1998), for example, T1 was either a red or a green square (with equal probabilities) and T2 was either an *X* or an *O* (with one being 75% probable and the other being 25% probable; see Figure 11.14A). The RT for T2 dropped steeply as the SOA between T1 and T2 increased, as is typical in PRP experiments, and this effect was the same for the rare and frequent T2 alternatives (see Figure 11.14B). The P3 wave was isolated by means of difference waves in which the ERP response to the frequent T2 was subtracted from the ERP response to the rare T2 (see Figure 11.14B). When the SOA became short, P3 amplitude was somewhat reduced and P3 latency was somewhat prolonged. However, the slowing of P3 latency was much less pronounced than the slowing of RT. These findings are consistent with the proposal that the increase in RT at short SOAs is not primarily caused by a slowing of the processes involved in perceiving and categorizing T2 (although there may be some competition for limited resources in these processes).

To provide evidence that response selection is the major locus of slowing in this paradigm, Osman and Moore (1993) examined the LRP, which reflects the difference in voltage between electrodes contralateral and ipsilateral to the hand of response (see Chapter 9, this volume). The LRP cannot be generated until the subject has determined which hand is appropriate for the given stimulus; that is, it cannot be generated until the response selection process has occurred. Osman and Moore found that the LRP was substantially delayed at short SOAs and was also reduced in amplitude. Importantly, the LRP peaked approximately 100 ms prior to the response at all SOAs, indicating that all of the slowing in RT could be explained by a slowing of the LRP. Together, the results of this study and of the Luck (1998) study provide strong evidence that the PRP is caused by a slowing of response selection (indexed by the LRP), with only modest interference with perception and categorization (indexed by the P3).

Notes

1 It should be noted that a very large number of trials must be averaged together to obtain a sufficient signal-to-noise ratio to observe reliable effects of attention on the early visual components, which are often well under 1 µV. In most cases, at least 200 trials must be averaged together in each waveform for each subject, and studies that require high signal-to-noise ratios (e.g., localization studies) may need over 1000 trials per waveform (see, e.g., Di Russo et al., 2003a).

2 Positive and negative for these dipoles reflect the polarity observed at the scalp for the corresponding ERP components and do not indicate anything about excitatory versus inhibitory activity.

3 A posterior selection positivity is sometimes observed beginning at around 100 ms in color-based attention experiments (see, e.g., the IPz site in Figure 11.7), and this effect partially overlaps with the P1 wave (Anllo-Vento et al., 1998; Zhang & Luck, 2009). However, the timing and scalp distribution of this effect indicate that it is not a modulation of the P1 wave.

4 The term *perception* is sometimes used to refer to a conscious experience of sensory inputs and is sometimes used simply to denote the late stages of sensory processing. For example, there is a large area of research on the topic of *perception without awareness*, which presupposes that perception could potentially be dissociated from conscious experience. Here we use the term *perception* in this latter sense, corresponding to the process of linking a sensory input with categories based on experience (e.g., determining that a particular sensory input is a horse rather than a cow, a fork, a pebble, etc.). Postperceptual processes, therefore, are those that operate on representations of sensory inputs that have already been categorized in this manner.

References

Anllo-Vento, L., & Hillyard, S. A. (1996). Selective attention to the color and direction of moving stimuli: Electrophysiological correlates of hierarchical feature selection. *Perception & Psychophysics*, *58*, 191–206.

Anllo-Vento, L., Luck, S. J., & Hillyard, S. A. (1998). Spatio-temporal dynamics of attention to color: Evidence from human electrophysiology. *Human Brain Mapping*, *6*, 216–238.

Arnell, K. M., Helion, A. M., Jurdelbrink, J. A., & Pasieka, B. (2004). Dissociating sources of dual-task interference using

human electrophysiology. *Psychonomic Bulletin & Review, 11*, 77–83.
Biederman, I. (1987). Recognition by components: A theory of human image understanding. *Psychological Review, 94*, 115–147.
Broadbent, D. E. (1958). *Perception and communication.* New York: Pergamon.
Bundesen, C. (1990). A theory of visual attention. *Psychological Review, 97*, 523–547.
Clark, V. P., Fan, S., & Hillyard, S. A. (1994). Identification of early visually evoked potential generators by retinotopic and topographic analyses. *Human Brain Mapping, 2*, 170–187.
Clark, V. P., & Hillyard, S. A. (1996). Spatial selective attention affects early extrastriate but not striate components of the visual evoked potential. *Journal of Cognitive Neuroscience, 8*, 387–402.
Cohen, J. D., Dunbar, K., & McClelland, J. L. (1990). On the control of automatic processes: A parallel distributed processing account of the Stroop effect. *Psychological Review, 97*, 332–361.
Connolly, J. F., Stewart, S. H., & Phillips, N. A. (1990). The effects of processing requirements on neurophysiological responses to spoken sentences. *Brain and Language, 39*, 302–318.
Dell'Acqua, R., Jolicoeur, P., Vespignani, F., & Toffanin, P. (2005). Central processing overlap modulates P3 latency. *Experimental Brain Research, 165*, 54–68.
Desimone, R., & Duncan, J. (1995). Neural mechanisms of selective visual attention. *Annual Review of Neuroscience, 18*, 193–222.
Desimone, R., Schein, S. J., Moran, J., & Ungerleider, L. G. (1985). Contour, color and shape analysis beyond the striate cortex. *Vision Research, 25*, 441–452.
Deubel, H., & Schneider, W. X. (1996). Saccade target selection and object recognition: Evidence for a common attentional mechanism. *Vision Research, 36*, 1827–1837.
Deutsch, J. A., & Deutsch, D. (1963). Attention: Some theoretical considerations. *Psychological Review, 70*, 80–90.
Di Russo, F., Martinez, A., & Hillyard, S. A. (2003a). Source analysis of event-related cortical activity during visuo-spatial attention. *Cerebral Cortex, 13*, 486–499.
Di Russo, F., Martinez, A., Sereno, M. I., Pitzalis, S., & Hillyard, S. A. (2002). Cortical sources of the early components of the visual evoked potential. *Human Brain Mapping, 15*, 95–111.
Di Russo, F., Teder-Sälejärvi, W. A., & Hillyard, S. A. (2003b). Steady-state VEP and attentional visual processing. In A. Zani & A. M. Proverbio (Eds.), *The cognitive electrophysiology of mind and brain* (pp. 259–274). San Diego, CA: Academic Press.
Donchin, E. (1981). Surprise! . . . Surprise? *Psychophysiology, 18*, 493–513.
Driver, J., & Baylis, G. C. (1989). Movement and visual attention: The spotlight metaphor breaks down. *Journal of Experimental Psychology: Human Perception and Performance, 15*, 448–456.
Duncan, J. (1984). Selective attention and the organization of visual information. *Journal of Experimental Psychology: General, 113*, 501–517.
Eason, R., Harter, M., & White, C. (1969). Effects of attention and arousal on visually evoked cortical potentials and reaction time in man. *Physiology and Behavior, 4*, 283–289.
Egly, R., Driver, J., & Rafal, R. D. (1994). Shifting visual attention between objects and locations: Evidence from normal and parietal lesion subjects. *Journal of Experimental Psychology: General, 123*, 161–177.
Eimer, M. (1994a). An ERP study on visual spatial priming with peripheral onsets. *Psychophysiology, 31*, 154–163.
Eimer, M. (1994b). "Sensory gating" as a mechanism for visuospatial orienting: Electrophysiological evidence from trial-by-trial cuing experiments. *Perception & Psychophysics, 55*, 667–675.
Gandhi, S. P., Heeger, D. J., & Boynton, G. M. (1999). Spatial attention affects brain activity in human primary visual cortex. *Proceedings of the National Academy of Sciences USA, 96*, 3314–3319.
Giesbrecht, B., Sy, J. L., & Elliott, J. C. (2007). Electrophysiological evidence for both perceptual and postperceptual seleciton during the attentional blink. *Journal of Cognitive Neuroscience, 19*, 2005–2018.
Giesbrecht, B. L., & Di Lollo, V. (1998). Beyond the attentional blink: Visual masking by object substitution. *Journal of Experimental Psychology: Human Perception & Performance, 24*, 1454–1466.
Gomez Gonzales, C. M., Clark, V. P., Fan, S., Luck, S. J., & Hillyard, S. A. (1994). Sources of attention-sensitive visual event-related potentials. *Brain Topography, 7*, 41–51.
Hackley, S. A., Woldorff, M., & Hillyard, S. A. (1990). Cross-modal selective attention effects on retinal, myogenic, brainstem and cerebral evoked potentials. *Psychophysiology, 27*, 195–208.
Hansen, J. C., & Hillyard, S. A. (1980). Endogenous brain potentials associated with selective auditory attention. *Electroencephalography and Clinical Neurophysiology, 49*, 277–290.
Hansen, J. C., & Hillyard, S. A. (1983). Selective attention to multidimensional auditory stimuli in man. *Journal of Experimental Psychology: Human Perception & Performance, 9*, 1–19.
Harter, M. R., & Guido, W. (1980). Attention to pattern orientation: Negative cortical potentials, reaction time, and the selection process. *Electroencephalography and Clinical Neurophysiology, 49*, 461–475.
Harter, M. R., & Previc, F. H. (1978). Size-specific information channels and selective attention: Visual evoked potential and behavioral measures. *Electroencephalography and Clinical Neurophysiology, 45*, 628–640.
He, X., Fan, S., Zhou, K., & Chen, L. (2004). Cue validity and object-based attention. *Journal of Cognitive Neuroscience, 16*, 1085–1097.
Heil, M., Roke, B., & Pecchinenda, A. (2004). Automatic semantic activation is no myth. *Psychological Science, 15*, 852–857.
Heinze, H. J., Luck, S. J., Mangun, G. R., & Hillyard, S. A. (1990). Visual event-related potentials index focused attention within bilateral stimulus arrays. I. Evidence for early selection. *Electroencephalography and Clinical Neurophysiology, 75*, 511–527.
Heinze, H. J., Mangun, G. R., Burchert, W., Hinrichs, H., Scholz, M., Münte, T. F., et al. (1994). Combined spatial and temporal imaging of brain activity during visual selective attention in humans. *Nature, 372*, 543–546.
Hillyard, S. A., Hink, R. F., Schwent, V. L., & Picton, T. W. (1973). Electrical signs of selective attention in the human brain. *Science, 182*, 177–179.
Hillyard, S. A., & Münte, T. F. (1984). Selective attention to color and location: An analysis with event-related brain potentials. *Perception & Psychophysics, 36*, 185–198.

Hillyard, S. A., Vogel, E. K., & Luck, S. J. (1998). Sensory gain control (amplification) as a mechanism of selective attention: Electrophysiological and neuroimaging evidence. *Philosophical Transactions of the Royal Society: Biological Sciences, 353*, 1257–1270.

Hoffman, J. E., & Subramaniam, B. (1995). The role of visual attention in saccadic eye movements. *Perception & Psychophysics, 57*, 787–795.

Hopfinger, J. B., & Mangun, G. R. (1998). Reflective attention modulates processing of visual stimuli in human extrastriate cortex. *Psychological Science, 9*, 441–447.

Isreal, J. B., Chesney, G. L., Wickens, C. D., & Donchin, E. (1980a). P300 and tracking difficulty: Evidence for multiple resources in dual-task performance. *Psychophysiology, 17*, 259–273.

Isreal, J. B., Wickens, C. D., Chesney, G. L., & Donchin, E. (1980b). The event-related brain potential as an index of display-monitoring workload. *Human Factors, 22*, 211–224.

Jeffreys, D. A., & Axford, J. G. (1972). Source locations of pattern-specific components of human visual evoked potentials. I: Components of striate cortical origin. *Experimental Brain Research, 16*, 1–21.

Johannes, S., Knalmann, U., Heinze, H. J., & Mangun, G. R. (1995). Luminance and spatial attention effects on early visual processing. *Cognitive Brain Research, 2*, 189–205.

Jonides, J. (1981). Voluntary versus automatic control over the mind's eye's movement. In J. B. Long & A. D. Baddeley (Eds.), *Attention and performance IX* (pp. 187–203). Hillsdale, NJ: Erlbaum.

Kahneman, D., Treisman, A., & Gibbs, B. J. (1992). The reviewing of object files: Object-specific integration of information. *Cognitive Psychology, 24*, 175–219.

Kelly, S. P., Gomez-Ramirez, M., & Foxe, J. J. (2008). Spatial attention modulates initial afferent activity in human primary visual cortex. *Cerebral Cortex, 18*, 2629–2636.

Kenemans, J. L., Kok, A., & Smulders, F. T. Y. (1993). Event-related potentials to conjunctions of spatial frequency and orientation as a function of stimulus parameters and response requirements. *Electroencephalography and Clinical Neurophysiology, 88*, 51–63.

Kok, A. (2001). On the utility of P3 amplitude as a measure of processing capacity. *Psychophysiology, 38*, 557–577.

Kramer, A. F., Wickens, C. D., & Donchin, E. (1983). An analysis of the processing requirements of a complex perceptual-motor task. *Human Factors, 25*, 597–621.

Kutas, M., & Hillyard, S. A. (1989). An electrophysiological probe of incidental semantic association. *Journal of Cognitive Neuroscience, 1*, 38–49.

Lavie, N. (1995). Perceptual load as a necessary condition for selective attention. *Journal of Experimental Psychology: Human Perception and Performance, 21*, 451–468.

Lavie, N., & Cox, S. (1997). On the efficiency of visual selective attention: Efficient visual search leads to inefficient distractor rejection. *Psychological Science, 8*, 395–398.

Lavie, N., Hirst, A., de Fockert, J. W., & Viding, E. (2004). Load theory of selective attention and cognitive control. *Journal of Experimental Psychology: General, 133*, 339–354.

Luck, S. J. (1995). Multiple mechanisms of visual-spatial attention: Recent evidence from human electrophysiology. *Behavioural Brain Research, 71*, 113–123.

Luck, S. J. (1998a). Neurophysiology of selective attention. In H. Pashler (Ed.), Attention (pp. 257–295). East Sussex, UK: Psychology Press.

Luck, S. J. (1998b). Sources of dual-task interference: Evidence from human electrophysiology. *Psychological Science, 9*, 223–227.

Luck, S. J. (2005). *An introduction to the event-related potential technique*. Cambridge, MA: MIT Press.

Luck, S. J. (2009). The spatiotemporal dynamics of visual-spatial attention. In F. Aboitiz & D. Cosmelli (Eds.), *From attention to goal-directed behavior: Neurodynamical, methodological, and clinical trends* (pp. 51–66). Berlin: Springer-Verlag.

Luck, S. J., Chelazzi, L., Hillyard, S. A., & Desimone, R. (1997). Neural mechanisms of spatial selective attention in areas V1, V2, and V4 of macaque visual cortex. *Journal of Neurophysiology, 77*, 24–42.

Luck, S. J., & Ford, M. A. (1998). On the role of selective attention in visual perception. *Proceedings of the National Academy of Sciences USA, 95*, 825–830.

Luck, S. J., & Gold, J. M. (2008). The construct of attention in schizophrenia. *Biological Psychiatry, 64*, 34–39.

Luck, S. J., Heinze, H. J., Mangun, G. R., & Hillyard, S. A. (1990). Visual event-related potentials index focused attention within bilateral stimulus arrays. II. Functional dissociation of P1 and N1 components. *Electroencephalography and Clinical Neurophysiology, 75*, 528–542.

Luck, S. J., & Hillyard, S. A. (1995). The role of attention in feature detection and conjunction discrimination: An electrophysiological analysis. *International Journal of Neuroscience, 80*, 281–297.

Luck, S. J., & Hillyard, S. A. (1999). The operation of selective attention at multiple stages of processing: Evidence from human and monkey electrophysiology. In M. S. Gazzaniga (Ed.), *The new cognitive neurosciences* (pp. 687–700). Cambridge, MA: MIT Press.

Luck, S. J., Hillyard, S. A., Mouloua, M., Woldorff, M. G., Clark, V. P., & Hawkins, H. L. (1994). Effects of spatial cuing on luminance detectability: Psychophysical and electrophysiological evidence for early selection. *Journal of Experimental Psychology: Human Perception and Performance, 20*, 887–904.

Luck, S. J., & Vecera, S. P. (2002). Attention. In S. Yantis (Ed.), *Stevens' handbook of experimental psychology: Vol. 1. Sensation and perception* (3rd ed., pp. 235–286). New York: Wiley.

Luck, S. J., Vogel, E. K., & Shapiro, K. L. (1996). Word meanings can be accessed but not reported during the attentional blink. *Nature, 382*, 616–618.

Lukas, J. H. (1980). Human attention: The olivo-cochlear bundle may function as a peripheral filter. *Psychophysiology, 17*, 444–452.

Lukas, J. H. (1981). The role of efferent inhibition in human auditory attention: An examination of the auditory brainstem potentials. *International Journal of Neuroscience, 12*, 137–145.

Mangun, G. R. (1995). Neural mechanisms of visual selective attention. *Psychophysiology, 32*, 4–18.

Mangun, G. R., & Hillyard, S. A. (1987). The spatial allocation of visual attention as indexed by event-related brain potentials. *Human Factors, 29*, 195–211.

Mangun, G. R., & Hillyard, S. A. (1988). Spatial gradients of visual attention: Behavioral and electrophysiological evidence. *Electroencephalography and Clinical Neurophysiology, 70*, 417–428.

Mangun, G. R., & Hillyard, S. A. (1990). Allocation of visual attention to spatial location: Event-related brain potentials and detection performance. *Perception and Psychophysics, 47*, 532–550.

Mangun, G. R., & Hillyard, S. A. (1991). Modulations of sensory-evoked brain potentials indicate changes in perceptual processing during visual-spatial priming. *Journal of Experimental Psychology: Human Perception and Performance, 17,* 1057–1074.

Mangun, G. R., Hillyard, S. A., & Luck, S. J. (1993). Electrocortical substrates of visual selective attention. In D. Meyer & S. Kornblum (Eds.), *Attention and performance XIV* (pp. 219–243). Cambridge, MA: MIT Press.

Martinez, A., Anllo-Vento, L., Sereno, M. I., Frank, L. R., Buxton, R. B., Dubowitz, D. J., et al. (1999). Involvement of striate and extrastriate visual cortical areas in spatial attention. *Nature Neuroscience, 2,* 364–369.

Martinez, A., Teder-Salejarvi, W., Vazquez, M., Molholm, S., Foxe, J. J., Javitt, D. C., et al. (2006). Objects are highlighted by spatial attention. *Journal of Cognitive Neuroscience, 18,* 298–310.

McAdams, C. J., & Reid, R. C. (2005). Attention modulates the responses of simple cells in monkey primary visual cortex. *Journal of Neuroscience, 25,* 11023–11033.

McAlonan, K., Cavanaugh, J., & Wurtz, R. H. (2008). Guarding the gateway to cortex with attention in visual thalamus. *Nature, 456,* 391–394.

Moore, C. M., & Egeth, H. (1998). How does feature-based attention affect visual processing? *Journal of Experimental Psychology: Human Perception & Performance, 24,* 1296–1310.

Moray, N. (1959). Attention in dichotic listening: Affective cues and the influence of instructions. *Quarterly Journal of Experimental Psychology, 11,* 56–60.

Morgan, S. T., Hansen, J. C., & Hillyard, S. A. (1996). Selective attention to stimulus locaitonmodulates the steady-state visual evoked potential. *Proceedings of the National Academy of Sciences USA, 93,* 4770–4774.

Müller, M. M., & Hillyard, S. A. (2000). Concurrent recordings of steady-state and transient event-related potentials as indices of visual-spatial attention. *Clinical Neurophysiology, 111,* 1544–1552.

Müller, M. M., Teder-Sälejärvi, W., & Hillyard, S. A. (1998). The time course of cortical facilitation during cued shifts of spatial attention. *Nature Neuroscience, 1,* 631–634.

Näätänen, R. (1967). Selective attention and evoked potentials. *Annals of the Finnish Academy of Science, 151,* 1–226.

Näätänen, R. (1975). Selective attention and evoked potentials in humans—A critical review. *Biological Psychology, 2,* 237–307.

Näätänen, R. (1992). *Attention and brain function.* Hillsdale, NJ: Erlbaum.

Näätänen, R., Gaillard, A. W. K., & Mantysalo, S. (1978). Early selective-attention effect on evoked potential reinterpreted. *Acta Psychologica,* 42, 313–329.

Näätänen, R., & Michie, P. T. (1979). Early selective attention effects on the evoked potential: A critical review and reinterpretation. *Biological Psychology, 8,* 81–136.

Näätänen, R., Simpson, M., & Loveless, N. E. (1982). Stimulus deviance and evoked potentials. *Biological Psychology, 14,* 53–98.

Nissen, M. J. (1985). Accessing features and objects: Is location special? In M. I. Posner & O. S. M. Marin (Eds.), *Attention and performance XI* (pp. 205–219). Hillsdale, NJ: Erlbaum.

Osman, A., & Moore, C. M. (1993). The locus of dual-task interference: Psychological refractory effects on movement-related brain potentials. *Journal of Experimental Psychology: Human Perception and Performance, 19,* 1292–1312.

Pashler, H. (1994). Dual-task interference in simple tasks: Data and theory. *Psychological Bulletin, 116,* 220–244.

Potter, M. C. (1976). Short-term conceptual memory for pictures. *Journal of Experimental Psychology: Human Learning and Memory, 2,* 509–522.

Previc, F. H., & Harter, M. F. (1982). Electrophysiological and behavioral indicants of selective attention to multifeature gratings. *Perception and Psychophysics, 32,* 465–472.

Raymond, J. E., Shapiro, K. L., & Arnell, K. M. (1992). Temporary suppression of visual processing in an RSVP task: An attentional blink? *Journal of Experimental Psychology: Human Perception and Performance, 18,* 849–860.

Ritter, W., Simson, R., & Vaughan, H. G. (1983). Event-related potential correlates of two stages of information processing in physical and semantic discrimination tasks. *Psychophysiology, 20,* 168–179.

Ritter, W., Simson, R., & Vaughan, H. G. (1988). Effects of the amount of stimulus information processed on negative event-related potentials. *Electroencephalography and Clinical Neurophysiology, 69,* 244–258.

Ritter, W., Simson, R., Vaughan, H. G., & Macht, M. (1982). Manipulation of event-related potential manifestations of information processing stages. *Science, 218,* 909–911.

Roelfsema, P. R., Lamme, V. A. F., & Spekreijse, H. (1998). Object-based attention in the primary visual cortex of the macaque monkey. *Nature, 395,* 376–381.

Saenz, M., Buracas, G. T., & Boynton, G. M. (2002). Global effects of feature-based attention in human visual cortex. *Nature Neuroscience, 5,* 631–632.

Saenz, M., Buracas, G. T., & Boynton, G. M. (2003). Global feature-based attention for motion and color. *Vision Research, 43,* 629–637.

Schein, S. J., & Desimone, R. (1990). Spectral properties of V4 neurons in the macaque. *Journal of Neuroscience, 10,* 3369–3389.

Schwent, V. L., Hillyard, S. A., & Galambos, R. (1976a). Selective attention and the auditory vertex potential. I: Effects of stimulus delivery rate. *Electroencephalography and Clinical Neurophysiology,* 40, 604–614.

Schwent, V. L., Hillyard, S. A., & Galambos, R. (1976b). Selective attention and the auditory vertex potential. II: Effects of signal intensity and masking noise. *Electroencephalography and Clinical Neurophysiology, 40,* 615–622.

Somers, D. C., Dale, A. M., Seiffert, A. E., & Tootell, R. B. H. (1999). Functional MRI reveals spatially specific attentional modulation in human primary visual cortex. *Proceedings of the National Academy of Sciences USA, 96,* 1663–1668.

Thorpe, S., Fize, D., & Marlot, C. (1996). Speed of processing in the human visual system. *Nature, 381,* 520–522.

Treisman, A. (1988). Features and objects: The fourteenth Bartlett memorial lecture. *Quarterly Journal of Experimental Psychology, 40,* 201–237.

Treisman, A. M. (1964). Selective attention in man. *British Medical Bulletin, 20,* 12–16.

Treisman, A. M., & Geffen, G. (1967). Selective attention: Perception or response?. *Quarterly Journal of Experimental Psychology, 19,* 1–17.

Treue, S., & Maunsell, J. H. (1999). Effects of attention on the processing of motion in macaque middle temporal and medial superior temporal visual cortical areas. *Journal of Neuroscience, 19,* 7591–7602.

Valdes-Sosa, M., Bobes, M. A., Rodriguez, V., & Pinilla, T. (1998). Switching attention without shifting the spotlight: Object-based attentional modulation of brain potentials. *Journal of Cognitive Neuroscience, 10,* 137–151.

Van Voorhis, S. T., & Hillyard, S. A. (1977). Visual evoked potentials and selective attention to points in space. *Perception & Psychophysics, 22,* 54–62.

Vecera, S. P. (1994). Grouped locations and object-based attention: Comment on Egly, Driver, and Rafal (1994). *Journal of Experimental Psychology: General, 123,* 316–320.

Vogel, E. K., & Luck, S. J. (2000). The visual N1 component as an index of a discrimination process. *Psychophysiology, 37,* 190–123.

Vogel, E. K., & Luck, S. J. (2002). Delayed working memory consolidation during the attentional blink. *Psychonomic Bulletin & Review, 9,* 739–743.

Vogel, E. K., Luck, S. J., & Shapiro, K. L. (1998). Electrophysiological evidence for a postperceptual locus of suppression during the attentional blink. *Journal of Experimental Psychology: Human Perception and Performance, 24,* 1656–1674.

Vogel, E. K., Woodman, G. F., & Luck, S. J. (2005). Pushing around the locus of selection: Evidence for the flexible-selection hypothesis. *Journal of Cognitive Neuroscience, 17,* 1907–1922.

Wijers, A. A., Mulder, G., Okita, T., Mulder, L. J. M., & Scheffers, M. K. (1989). Attention to color: An analysis of selection, controlled search, and motor activation, using event-related potentials. *Psychophysiology, 26,* 89–109.

Woldorff, M. (1989). *Auditory selective attention in humans: Analysis of mechanisms using event-related brain potentials.* Unpublished thesis, University of California, San Diego.

Woldorff, M. (1993). Distortion of ERP averages due to overlap from temporally adjacent ERPs: Analysis and correction. *Psychophysiology, 30,* 98–119.

Woldorff, M., Hansen, J. C., & Hillyard, S. A. (1987). Evidence for effects of selective attention to the midlatency range of the human auditory event related potential. In R. Johnson, J. W. Rohrbaugh, & R. Parasuraman (Eds.), *Current trends in event-related brain potential research* (pp. 146–154). London: Elsevier.

Woldorff, M., & Hillyard, S. A. (1991). Modulation of early auditory processing during selective listening to rapidly presented tones. *Electroencephalography and Clinical Neurophysiology, 79,* 170–191.

Woldorff, M. G., Gallen, C. C., Hampson, S. A., Hillyard, S. A., Pantev, C., Sobel, D., et al. (1993). Modulation of early sensory processing in human auditory cortex during auditory selective attention. *Proceedings of the National Academy of Sciences USA, 90,* 8722–8726.

Woldorff, M. G., Hackley, S. A., & Hillyard, S. A. (1991). The effects of channel-selective attention on the mismatch negativity wave elicited by deviant tones. *Psychophysiology, 28,* 30–42.

Woldorff, M. G., Hillyard, S. A., Gallen, C. C., Hampson, S. A., & Bloom, F. E. (1998). Magnetoencephalographic recordings demonstrate attentional modulation of mismatch-related neural activity in human auditory cortex. *Psychophysiology, 35,* 283–292.

Wolfe, J. M. (1994). Guided search 2.0: A revised model of visual search. *Psychonomic Bulletin & Review, 1,* 202–238.

Woods, D. L., Alho, K., & Algazi, A. (1994). Stages of auditory feature conjunciton: An event-related brain potential study. *Journal of Experimental Psychology: Human Perception and Performance, 20,* 81–94.

Woods, D. L., & Hillyard, S. A. (1978). Attention at the cocktail party: Brainstem evoked responses reveal no peripheral gating. In D. A. Otto (Ed.), *Multidisciplinary perspectives in event-related brain potential Research* (pp. 230–233). Washington, DC: U.S. Government Printing Office.

Zhang, W., & Luck, S. J. (2009). Feature-based attention modulates feedforward visual processing. *Nature Neuroscience, 12,* 24–25.

CHAPTER

12

Electrophysiological Correlates of the Focusing of Attention within Complex Visual Scenes: N2pc and Related ERP Components

Steven J. Luck

Abstract

As discussed in Chapter 11 of this volume, attention plays a role in the perception of simple stimuli, serving as a gain control that enhances the speed or accuracy of feedforward visual processing. However, additional attentional mechanisms are needed to support the iterative, reentrant processes and perception-action loops that are involved in the perception of complex, multiple-element stimulus arrays, and the event-related potential (ERP) correlates of these additional attentional mechanisms will be the focus of the present chapter. The first half of the chapter begins by discussing the computational problems that must be solved by these mechanisms and the sequence of steps involved in processing complex stimulus arrays, which include storing a target template in working memory, detecting relevant features, focusing attention on a peripheral location, shifting overt attention to the attended location, and then expanding or contracting attention around the object at this location. The ERP components corresponding to each of these steps are then described. The second half of the chapter provides a closer look at N2pc and related contralateral components that reflect the focusing of attention onto a peripheral location.

Keywords: event-related potential components, visual attention, P1, N1, N2pc, distractor positivity (P_D), sustained posterior contralateral negativity (SPCN)

The General Role of Attention in Visual Perception

The ultimate goal of research on visual perception is to understand how people perceive, remember, and act upon the objects in natural visual environments, such as the scene shown in Figure 12.1. Understanding these high-level aspects of vision requires that we understand the operation of almost the entire brain, because visually guided behavior depends on perceptual systems, memory systems, motor systems, motivational systems, and executive control systems. The perception of complex scenes also involves solving computational problems that are not present in laboratory experiments in which objects are presented one at a time on a blank background (see Feldman, 1985; Luck et al., 1997b). Whereas simple, isolated stimuli may be identified on the basis of a single feedforward sweep of neural activity, the perception of complex, multiple-element arrays involves iterative, reentrant processes and perception-action loops (see Di Lollo et al., 2000; Klyubin et al., 2004). As discussed in Chapter 11 of this volume, attention plays a role in the perception of simple stimuli, serving as a gain control that enhances the speed or accuracy of feedforward visual processing. However, additional attentional mechanisms are needed to support the iterative, reentrant processes and perception-action loops that are involved in the perception of complex, multiple-element stimulus arrays, and the event-related potential (ERP) correlates of these additional attentional mechanisms will be the focus of the present chapter. We will begin by discussing the computational problems that must be solved by these mechanisms and the sequence of steps involved in processing complex stimulus arrays.

Fig. 12.1. Example of a natural scene in which objects are distributed across the input, often overlapping each other. Note that the size of the retinal image created by a given object varies depending on both the size of the object and its distance from the observer.

A key factor limiting the perception of complex stimulus arrays is that high-resolution information about the visual environment is available to the brain only from the fovea, the central area of the retina (approximately 1 mm in diameter, which represents approximately 2° of visual angle). Receptive fields arising from the fovea are small, giving us the ability to resolve small details of objects presented at the center of gaze, whereas receptive fields from extrafoveal regions are large, making it impossible to resolve small details in objects presented in the periphery. In addition, approximately 50% of neurons in visual cortex are devoted to the tiny foveal region. Because of this limitation, saccadic eye movements are used to orient the direction of gaze so that an object of interest falls into the fovea. These eye movements dramatically improve the perception of the foveated object, and they can be considered a mechanism of attention. Indeed, foveating an object is arguably the visual system's most important and most potent mechanism of attention. Selecting an object by foveating it is often called *overt attention*, whereas attending to an object that is not foveated is often called *covert attention*.

Although foveating an object is a powerful means of focusing attention on it, it is also important to be able to focus processing resources on objects that are not being foveated (for a detailed discussion, see Luck, 2009). Relevant visual information is often defined by nonspatial features (e.g., the color and shape of a pencil when one wishes to write something), and it would be incredibly inefficient to search for a desired object by moving the eyes randomly until an object containing the relevant features appears in the fovea. Consequently, some sort of selective processing of relevant features is often necessary prior to an eye movement. Moreover, even if gaze is directed to the center of an object, some mechanism is necessary to allow perceptual processing mechanisms to expand across the extent of the object, allowing the entire object to be selected without also selecting surrounding objects. For example, after your eyes have been directed to one of the words on this page, some mechanism is necessary to allow the entire word to be processed without interference from other words on the lines above and below the fixated word. Moreover, the size of this attended region must be adjusted to reflect the size of the fixated object and the proximity of surrounding objects. If you bring this text closer to your eyes, for example, the size of the word on your retinas increases and the region of attention must expand. Thus, although shifts of gaze to an object are a powerful mechanism of attention (overt attention), it is also important to be able to allocate processing resources independently of eye position (covert attention).

The operation of overt and covert attention in natural vision typically follows a sequence of steps. First, a goal must be activated to guide the allocation of attention. For example, the shape and color of a pen might be stored as a *search template* in working memory to guide the process of searching for that pen (see., e.g., Duncan & Humphreys, 1989; Woodman et al., 2007). Second, sensitivity is increased for features specified by the search template so that objects containing those features will have priority for further processing (see, e.g., Hopf et al., 2004; Wolfe, 1994; Wolfe et al., 1989). Third, the presence of an object containing these features triggers a shift of covert spatial attention to the object's location, facilitating the perception of this object (Carrasco et al., 2000) and causing it to be stored in working memory (Hollingworth & Luck, 2009). Fourth if the now-attended peripheral object still appears to match the search template, this may trigger a shift of overt gaze to that location (Peterson et al., 2004). Fifth, once the eyes are centered on the object, attention will expand or contract around the object, depending on its size and the proximity of distracting objects (Hopf et al., 2006; LaBerge et al., 1991).

The remainder of this chapter will discuss the ERP components involved with each of these five steps of visual processing. Most of these components are described in detail in other chapters in this volume and therefore will be discussed only briefly. The N2pc component—which reflects the focusing of attention on a potential target in a visual search array—is not covered elsewhere in this volume and will therefore be discussed in detail in the second half of this chapter, along with other related components (including the distractor positivity and the sustained posterior contralateral negativity).

Step 1: Storing the Target Template in Working Memory (NSW and CDA)

Chapter 13 of this volume discusses the ERP correlates of working memory storage in detail, so here we will just note that sustained ERP activity is observed while an object is being maintained in working memory. This takes the form of a negative slow wave (NSW) over prefrontal cortex (Ruchkin et al., 2003) and contralateral delay activity (CDA) over temporal-parietal cortex (Vogel & Machizawa, 2004). Analogous single-unit activity has been observed in monkeys when they are preparing to search for a specific target (Chelazzi et al., 1993, 1998). These working memory representations may be used to guide attention to relevant objects when the to-be-attended object varies frequently, but working memory representations do not appear to be used to guide attention in typical visual search experiments in which the target remains the same for several minutes of testing (Woodman et al., 2001, 2007). Under these conditions, attention may be guided by long-term memory representations and by automatic priming from previous trials (Maljkovic & Nakayama, 1993; Meeter & Olivers, 2006). In either case, these representations provide the control signals that guide attention toward relevant features.

Step 2: Detecting Relevant Features with Feature-Based Attention (Bilateral P2, N2, and P3 Components)

Once the current goal has been activated in the form of a target template, feature-based attention mechanisms then operate across the visual field to highlight objects containing these features. Several ERP components related to feature-based attention were discussed extensively in Chapter 11. In the experiments described in that chapter, the attended and ignored stimuli were presented at different times, making it easy to isolate ERP activity related to the attended and ignored stimuli. In the natural environment, however, an observer is faced with a simultaneous array of relevant and irrelevant stimuli that are spatially intermixed. Consider, for example, the scene shown in Figure 12.1. If an observer is interested in eating a strawberry, this general goal does not specify the location to be attended. Instead, a *search template* is activated (i.e., a representation of the color and shape of a strawberry), and this is used to guide attention to locations that contain relevant features and are therefore likely to be the desired object. The visual search paradigm has been used extensively to simulate this sort of situation in the laboratory. In this paradigm, arrays of objects are presented, and the observer presses one of two buttons to indicate whether a predefined target object is present or absent in each array. Alternatively, each array might contain one of several different possible targets, and the observer must indicate which target is present in each array (see the reviews by Treisman, 1986; Wolfe, 1994). This section will discuss ERP components that are observed as subjects search for targets defined by specific features or combinations of features in the visual search paradigm. Some of these components are the same components observed when objects are presented one at a time, but others are found only when relevant and irrelevant objects are presented simultaneously.

When the target in a visual search task is defined by the presence of a simple feature value (e.g., the color red), and the distractor items are homogeneous along that feature dimension (e.g., all are blue), the target will "pop out" from the distractors. That is, subjects will be able to detect the target quickly, irrespective of the number of items in the display. Luck and Hillyard (1994a, 1994b) conducted several ERP experiments with popout stimuli of this nature.

As an example, consider the experiment shown in Figure 12.2A (Experiment 1 from Luck & Hillyard, 1994a). Each stimulus was either a homogeneous array of eight small blue vertical bars or an array of seven of these bars plus one "popout" bar that differed in color (green), orientation (horizontal), or size (large). Homogeneous arrays were presented on 50% of trials, and each of the three types of popout arrays was presented on 16.7% of trials. One of the three popout types was designated the target at the beginning of each trial block, and the subjects pressed one button when the target was present and another button when it was absent (the same nontarget response was made for both nontarget popouts and homogeneous arrays). Thus, different features were task-relevant in different trial blocks. In addition, a given popout type could be either the target or a nontarget, making it possible to assess the automatic allocation of attention to nontarget popout objects. That is, it is possible to ask whether a popout item automatically

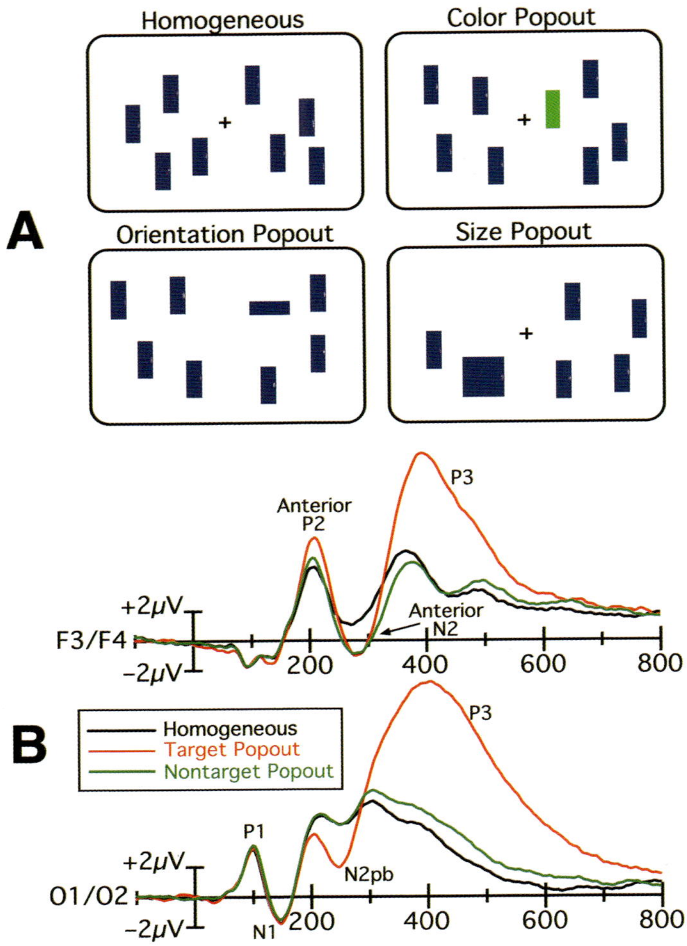

Fig. 12.2. Stimuli (A) and grand average ERP waveforms (B) from the study of Luck and Hillyard (1994a, Experiment 1). The locations of the items within a stimulus array varied randomly from trial to trial. Each stimulus sequence consisted of 50% homogeneous arrays and 16.7% of each of the three types of popout arrays. One popout type was designated the target at the beginning of each trial block; subjects were instructed to press one of two buttons for each array to indicate whether the target popout was present or absent. At frontal sites, both target and nontarget popouts elicited an enlarged N2 wave relative to homogeneous arrays, but only targets elicited enhanced P2 and P3 waves. At occipital sites, enlarged N2 and P3 waves were observed only for targets (the small difference in P3 amplitude between nontarget popouts and homogeneous arrays did not reach statistical significance).

attracts attention whether or not it matches the target template.

Figure 12.2B shows the ERPs recorded at frontal and occipital electrode sites for target popout arrays, nontarget popout arrays, and homogeneous arrays. Several attention-related ERP components can be identified in this experiment: an anterior P2 effect; an anterior N2 effect; a posterior N2 effect (N2pb); a prototypical P3b wave with a centroparietal maximum; and an occipital P3 wave.

THE ANTERIOR P2 COMPONENT

At the frontal sites shown in Figure 12.2B, the P2 wave was larger for target popouts than for either nontarget popouts or homogeneous arrays. This appears to be the same as the P2 effect observed in the one-object-at-a-time feature-based attention studies described in Chapter 11 of this volume. However, the waveforms shown in Figure 12.2B demonstrate that this effect can also be observed when only one object in an array contains the relevant feature and all of the other objects contain irrelevant features; that is, the presence of the relevant feature is the important factor, irrespective of the presence or absence of irrelevant features. Comparisons across experiments indicate that this P2 effect is approximately the same size whether one or all of the items in the array contain the relevant feature value, suggesting that it reflects the detection of a specific feature rather than the degree of match between the overall stimulus array and the relevant feature value.

Additional experiments demonstrated that this P2 effect is observed only when the target occurs infrequently and that it is also present for nontargets that are highly similar to the target (Luck & Hillyard, 1994a, 1994b). Very little is known about the functional significance of this anterior P2 effect. The fact that it is sensitive to the probability of the target stimulus makes it similar to the P3b component (see Chapter 7, this volume). However, the P2 probability effect is observed only when the target can be distinguished from nontargets on the basis of a fairly simple feature, such as color or letter shape, whereas the P3b probability effect is observed whenever the subject can correctly categorize the stimuli as targets and nontargets.

It should be noted that, although we are calling this a *P2 effect*, we cannot be certain that it reflects a change in the amplitude of a unitary P2 component that is simply larger for targets than for nontargets. Instead, this effect may reflect a neural generator source that is present for targets and absent for nontargets, overlapping in time with a different P2 wave that is present equally for targets and nontargets. The term *P2 effect* is intended to be descriptive, indicating that the effect occurs in approximately the same latency range as the P2 wave and has approximately the same scalp distribution. Indeed, it would be surprising if only one neural generator source were active from 150–250 ms at anterior scalp sites (see Chapters 1 and 3, this volume, for additional discussion of the problem of multiple overlapping components). The terminology used to describe attention-related effects throughout the rest of this chapter is intended to be superficially descriptive in this same manner. It is important, however, to avoid the common mistake of assuming that an effect in the time range of a given component reflects a change in the amplitude of that component.

THE ANTERIOR N2 COMPONENT

In the experiment shown in Figure 12.2, the P2 wave was followed by an N2 wave at anterior scalp sites that was larger for both target and nontarget popouts relative to the homogeneous arrays. The presence of this effect for nontarget as well as target popouts might suggest that it reflects an automatic detection of popout stimuli. However, when subjects were asked to discriminate the color of the whole stimulus array rather than the color of a single popout item, orientation popouts no longer generated an enlarged anterior N2 wave (Luck & Hillyard, 1994a, Experiment 3). Thus, the anterior N2 effect is observed only when the subjects are actively searching for an item that differs from the rest of the array (analogous behavioral results can be found in the studies of Bacon & Egeth, 1994; Pashler, 1988).

A similar anterior N2 effect has been observed in the flanker paradigm (Gehring et al., 1992) and in the Stroop paradigm (West & Alain, 1999, 2000; see Folstein & Van Petten, 2008, for an excellent review of these and related N2 effects). In these studies, a larger anterior N2 wave is observed for stimuli that contain conflicting features (e.g., incompatible trials). Conflict is produced in the flanker paradigm by surrounding the target with flanking distractor items that indicate a different response; conflict is produced in the Stroop paradigm by using ink of one color to write the name of a different color. The larger anterior N2 response on these conflict trials has been proposed to reflect the same neural system that generates the error-related negativity (see Chapter 10, this volume, and Yeung et al., 2004). It remains to be seen whether the anterior N2 effect observed for nontarget popout features in visual search is actually the same component as the anterior N2 effect observed on conflict trials in the flanker and Stroop paradigms.

THE POSTERIOR N2 COMPONENT

At posterior electrode sites, the N2 wave is enlarged for target popouts compared to nontarget popouts and homogeneous arrays. This is the same as the pattern observed for the anterior P2 component, which has a similar onset time at anterior sites. In addition, both of these effects become larger as the probability of the target decreases, and both are larger for color popouts than for orientation or size popouts (Luck & Hillyard, 1994a). It is therefore possible that they reflect the positive and negative sides of a single generator source.

The posterior N2 component can be divided into two subcomponents, one that is larger over the hemisphere contralateral to the side of the target item and one that is equivalent over the contralateral and ipsilateral hemispheres. The contralateral subcomponent is called *N2pc* (N2-posterior-contralateral) and the bilateral subcomponent is called *N2pb* (N2-posterior-bilateral). N2pb amplitude varies with target probability and is larger for color popouts, whereas N2pc amplitude is independent of probability and feature dimension (Luck & Hillyard, 1994a). Thus, only the N2pb component is similar to the anterior P2 component. We will consider the N2pc component in more detail in the second half of this chapter.

The specific neurocognitive process reflected by the anterior P2 and N2pb components is not clear. N2pb may be the same as the component labeled N2 in studies using the oddball paradigm (labeled N2c in the classification scheme proposed by Pritchard et al., 1991). In these studies, task-relevant oddballs will elicit an enlarged N2 wave relative to the standard stimuli, followed by the prototypical P3b component. This pattern is observed for both auditory and visual oddballs, but the scalp distribution is more anterior for auditory stimuli than for visual stimuli (Simson et al., 1977). Renault et al. (1982) proposed that this component reflects the process of categorizing a stimulus, because the duration of the effect depends on the difficulty of the categorization. However, this manipulation also increased the onset time of the P3 wave, and because the N2 wave was cut off by the larger and opposite-polarity P3 wave, it is not clear that the duration of the underlying N2 wave was actually longer for the more difficult categorization task.

THE P3 COMPONENT

In the study shown in Figure 12.2, an enlarged P3 wave was observed for target popouts relative to nontarget popouts and homogeneous arrays across the scalp (see Figure 12.2B). However, this P3 effect can be divided into two subcomponents. First, the prototypical P3b component (see Chapter 7, this volume) was present across the scalp with a centroparietal maximum. This component was probability sensitive and was not present in a follow-up experiment in which target and nontarget stimuli were equiprobable (Luck & Hillyard, 1994a, Experiment 2). Second, an occipitally maximal P3 effect was also present for target popouts, but it was present for popout targets even when targets and nontargets were equiprobable. The functional significance of this occipital P3 component is unknown.

THE P1 WAVE AND SENSORY CONFOUNDS

The paradigm shown in Figure 12.2A contains a small but important sensory confound, because the popout arrays are physically different from the homogeneous arrays. This confound has two consequences. First, sensory responses may be different to a popout item than to the surrounding items. Indeed, neurons in area V1 produce a larger response when the receptive field contains a popout item rather than a nonpopout item, probably because lateral inhibition is stronger between items that share the same feature value (Knierim & Van Essen, 1992). Second, because most of the items in each array consist of a particular feature value (e.g., small, blue, vertical in the example shown in Figure 12.2A), the visual system will become adapted to these features after several stimulus arrays have been presented, reducing the magnitude of the visual response to these values. Each singleton feature is presented relatively infrequently, especially at a given location in the display, and the visual system will therefore not be as adapted to these feature values, leading to a larger response. For example, blue-selective neurons with receptive fields in the upper-right quadrant of the display will be stimulated on every trial in the task shown in Figure 12.2A, leading to a reduced response when blue is presented in this location. In contrast, green-selective neurons with receptive fields in the upper-right quadrant will be stimulated rarely, leading to a large sensory response when green is presented in this location. Consequently, the overall ERP elicited by a stimulus array should contain a larger sensory response when a popout item is present, and this should be especially evident over the hemisphere containing the neurons that are selective for the popout feature (i.e., the hemisphere contralateral to the popout item).

Because of these factors, any differences between the ERPs elicited by popout arrays and homogeneous arrays could reflect low-level differences in sensory processing rather than higher-level attention effects. Fortunately, all of the ERP effects described so far have been shown to disappear under specific task conditions, demonstrating that they are not simple sensory effects. However, Luck and Hillyard (1994a) also found that the P1 wave was slightly but consistently larger over the hemisphere contralateral to the popout item than over the ipsilateral hemisphere, and this effect was present across all task manipulations, suggesting that it was a sensory effect rather than an attentional effect (note that this effect cannot be seen in Figure 12.2B, which is collapsed across hemispheres). To test whether this P1 effect might reflect adaptation of the neurons coding for the nonpopout items, Luck and Hillyard (1994a) conducted an experiment in which the stimulus arrays consisted of either a green popout item among blue distractor items or a blue popout item among green distractor items (randomly intermixed from trial to trial). This manipulation equated the level of adaptation of the blue- and green-selective neurons, and it eliminated the P1 effect, consistent with the hypothesis that the P1 effect was a consequence of sensory adaptation. Moreover, a sequential analysis demonstrated that the P1 was larger over the contralateral hemisphere than over the ipsilateral hemisphere for a given popout color when the preceding array consisted mostly of the other color (e.g., one green item and seven blue items on both trial *N* and trial *N* – 1), but this effect was absent when the preceding array consisted mostly of the other color (e.g., one green item and seven blue items on trial *N* and one blue item and seven green items on trial *N* – 1). Thus, this P1 effect does not appear to reflect the automatic allocation of attention to the popout item but instead reflects color-specific sensory adaptation.

This finding demonstrates the importance of using stimuli and experimental designs that control for bottom-up sensory effects (and especially adaptation effects, which can be quite strong). Even though the ERP consequences of the sensory confound in the experiments of Luck and Hillyard (1994a) were limited to the P1 latency range, sensory activity persists for hundreds of milliseconds in many cases, so sensory confounds could produce small effects well into the P2 and N2 latency ranges. It is often possible to design an experiment in a manner that eliminates sensory confounds of this nature (as described in the next section). When this is not possible, it is important to conduct control experiments showing that the effects can be eliminated by changes to the task (see, e.g., Hickey et al., 2009; Sawaki & Luck, submitted).

Step 3: Focusing Covert Attention on a Peripheral Location (N2pc, SPCN, Pd, and Probe-Elicited Sensory Responses)

Now that we have discussed how the target template is established and how it can be used to highlight objects containing relevant features, we will consider how attention is shifted to an object containing the relevant features. The ERP component most often used to study this focusing of attention in the context of multiple-object stimulus arrays is the N2pc component. In this section, we will briefly describe N2pc and related components, focusing on how they reflect the process of enhancing target processing and filtering distractors. A more detailed look at these components will be provided in the second half of this chapter.

A typical N2pc paradigm is shown in Figure 12.3 (from the study of Luck et al., 2006). Each stimulus array contains one red square, one green square, and a large number of black distractor squares. The locations of the individual squares vary at random from trial to trial, with the constraint that the two colored items are always on opposite sides of the display. At the beginning of each 5-min block of trials, the observer is instructed to attend either to red or to green. For each array, the observer must press one of two buttons to indicate whether the object drawn in the attended color contains a gap on the top or a gap on the bottom. To avoid spatial compatibility conflicts that arise when a subject must make a left-hand response for a right-side target or vice versa (known as the *Simon effect*), it is usually preferable to have subjects use two different fingers of the same hand to make the two different responses.

In the example shown in Figure 12.3, the subject would attend to an object in the left visual field (LVF) when red is the attended color and would attend to an object in the right visual field (RVF) when green is the attended color. This design therefore eliminates the sensory confounds described in the previous section.

The N2pc component consists of a greater negativity when the attended item is contralateral to the recording electrode than when the attended item is ipsilateral. It typically occurs during the time range of the N2 wave (200–300 ms) and is observed at posterior scalp sites over visual cortex, with a maximum voltage near the PO7 and PO8 electrodes. In Figure 12.3, the N2pc can be seen as a more negative voltage for RVF targets than for LVF targets over the left hemisphere and as a more negative voltage for LVF targets than for RVF targets over the right hemisphere. To minimize overall differences between LVF and RVF targets and between the left and right hemispheres, it is useful to create a contralateral waveform (average of RVF for the left hemisphere and LVF for the right hemisphere) and an ipsilateral waveform (average of LVF for the left hemisphere and RVF for the right hemisphere). The difference between these contralateral and ipsilateral waveforms is used to isolate the N2pc component from other overlapping ERP components. This is exactly analogous to the derivation of the lateralized readiness potential (LRP; see Chapter 9, this volume). Just as the LRP subtraction procedure guarantees that the brain must have begun to determine which response is associated with the stimulus when the LRP deviates from zero, the N2pc subtraction procedure guarantees that the brain must have begun to determine the location of the target when the N2pc deviates from zero. These two components therefore allow very strong conclusions to be drawn about the timing of neural events (for a detailed discussion, see Luck, 2005, chap. 2).

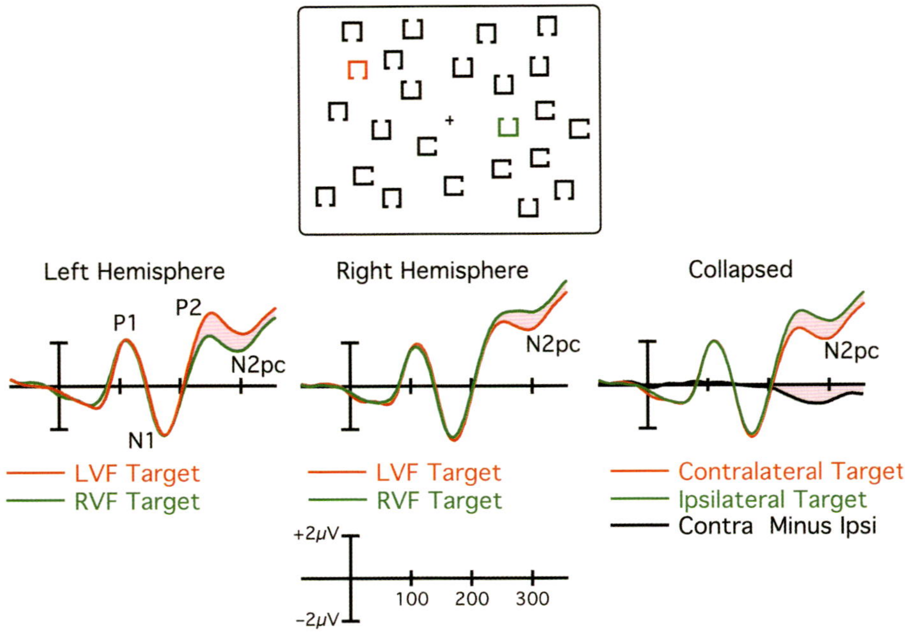

Fig. 12.3. Typical N2pc paradigm and grand average ERP waveforms from posterior occipitotemporal electrode sites (from the study of Luck et al., 2006). To avoid any possibility of physical stimulus confounds, each stimulus array contained a distinctly colored item on each side, and one of these two colors was designated as the target color in each trial block. Thus, the same stimulus array could be used to induce subjects to focus on either the LVF or RVF, depending on which color was defined as the target. The locations of the items varied at random from trial to trial, except that the two popout colors were always on opposite sides. In addition, a minimum interitem distance was imposed to avoid overlap among items. The side containing a particular popout color varied unpredictably from trial to trial so that subjects could not anticipate the target location prior to stimulus onset. The subject was required to press one of two buttons to indicate whether the gap in the target item was at the top or the bottom of the square. The voltage in the N2 latency range over the left hemisphere was more negative when the target was in the RVF than when it was in the LVF, and the voltage over the right hemisphere was more negative when the target was in the LVF than when it was in the RVF. These waveforms were combined into an ipsilateral waveform (left hemisphere/left target averaged with right hemisphere/right target) and a contralateral waveform (left hemisphere/right target averaged with right hemisphere/left target). The N2pc is defined as the difference between these contralateral and ipsilateral waveforms (shown as the shaded region), which was made explicit by constructing a contralateral-minus-ipsilateral difference wave.

As will be discussed in detail later in this chapter, the N2pc component appears to reflect the focusing of attention on a potential target item and the filtering of the surrounding distractor items. Thus, one can use the N2pc component to track the time course of the focusing of attention (see, e.g., Kiss et al., 2009; Lorenzo-Lopez et al., 2008; Woodman & Luck, 1999, 2003). For example, if the N2pc onsets at 180 ms, one can conclude that attention has been focused on the target by 180 ms (although note that the onset time in an averaged waveform reflects the trials with the earliest onset times, not the average onset time; see Luck, 2005, pp. 56–57). Of course, attention may have shifted at an earlier time within other brain systems that do not produce an ERP signature on the scalp. The N2pc component appears to be generated at intermediate and late stages of the ventral visual processing pathway, and single-unit studies suggest that shifts of attention in visual search are typically initiated in prefrontal cortex, the frontal eye fields, and posterior parietal cortex, followed at a later time by focusing within the ventral stream (Buschman & Miller, 2007). Indeed, a simultaneous ERP/single-unit study in monkeys found that selective processing occurred in the frontal eye fields prior to N2pc onset (Cohen et al., 2009). Thus, the N2pc component does not reflect the earliest time at which attention becomes focused on a potential target item, although it may reflect the earliest time at which attention becomes focused within the ventral stream.

Does the focusing of attention on a potential target item in visual search cause an increase in sensory transmission at the location of the target? This question was addressed in the paradigm shown in Figure 12.4A (from Luck et al., 1993). Subjects performed a visual search task in which each array contained a red *T* on one side and a green *T* on the

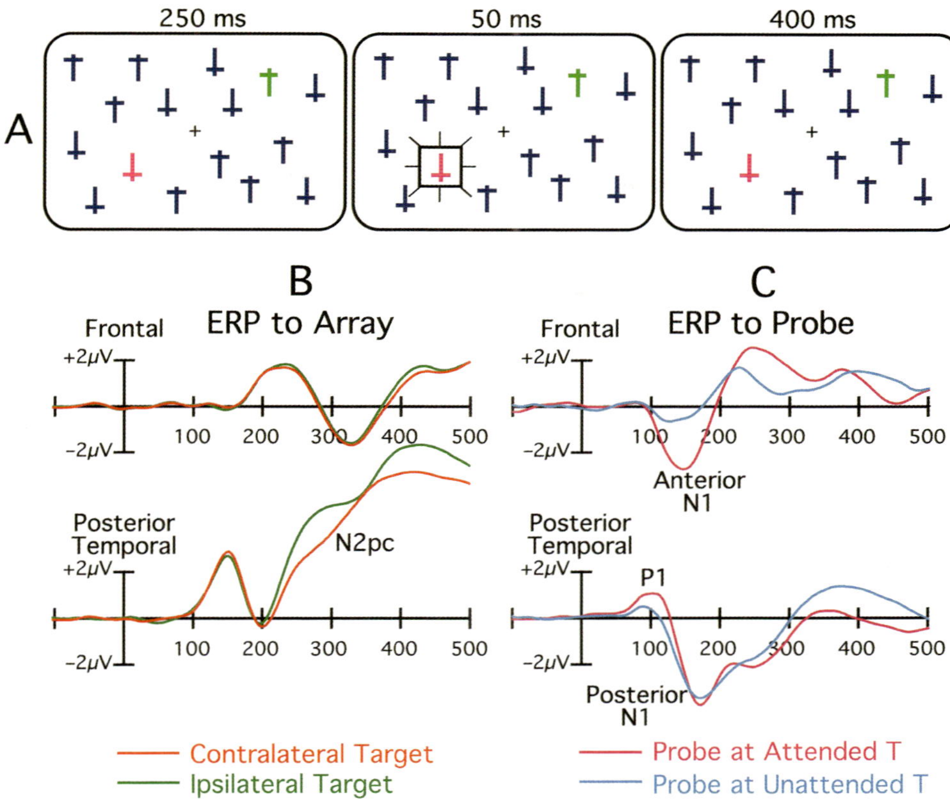

Fig. 12.4. (A) Stimuli from the study of Luck et al. (1993). Subjects attended to the red item in some trial blocks and the green item in others, and they pressed one of two buttons for each array to indicate whether the attended item was an upright or inverted *T*. When a probe stimulus appeared, it could appear around either the red item or the green item. (B) Grand average ERPs elicited by the visual search arrays on probe-absent trials, divided according to whether the attended item was contralateral or ipsilateral to the recording electrode. Note that the N2pc component can be seen at the posterior temporal electrode sites as the difference between the contralateral and ipsilateral waveforms from 200 to 325 ms poststimulus. (C) Event-related potentials elicited by the probe. The ERP elicited by the probe was overlapped by the ERP elicited by the visual search array. To isolate the activity elicited by the probe, the experiment included trials on which no probe was presented; the probe-elicited ERP was then isolated by subtracting probe-present trials from probe-absent trials. When isolated in this manner, the probe-elicited ERP waveform contained a larger P1 at posterior temporal sites and a larger N1 at frontal sites when the probe appeared at the location of the attended item than when it appeared in the opposite side of the array from the attended item.

other side (the sides were randomized from trial to trial). In some trial blocks subjects attended to the red *T*, and in other blocks they attended to the green *T*. The task was to press one of two buttons to indicate whether the attended *T* was upright or inverted. As shown in Figure 12.4b, an N2pc component was observed contralateral to the attended *T* beginning approximately 200 ms after the onset of the array.

On a subset of trials, a bright white "probe" square was presented around either the red *T* or the green *T*, onsetting 250 ms after the onset of the visual search array. The purpose of the probe square was to assess the processing of information at the probed location at the time of the probe. If sensory transmission is enhanced at the location of the attended *T*, then the sensory processing of the probe should be enhanced when it is presented at the location of the attended *T* compared to when it is presented at the location of the unattended *T*. Indeed, as shown in Figure 12.4C, the P1 and N1 waves elicited by the probe stimulus were larger when the probe was presented at the location of the attended *T* compared to when the probe was presented at the location of the unattended *T*.

Follow-up experiments showed that these P1 and N1 effects are very much like those observed in spatial cuing studies. When neutral trials are used to provide a baseline in cuing studies, the P1 is found to be suppressed on invalid trials compared to neutral trials, with no enhancement on valid trials, whereas the N1 is found to be enlarged on valid trials compared to neutral trials, with no suppression on invalid trials (see Luck et al., 1994, and Chapter 11, this volume). A similar pattern is found in visual search tasks when neutral trials are created by presenting arrays that do not contain the attended color (Luck & Hillyard, 1995). That is, the probe-elicited P1 is suppressed when the probe is presented at a location on the opposite side of the display from the attended item compared to when the attended item is absent, with no enhancement at the location of the attended item, whereas the N1 is enhanced when the probe is presented at the location of the attended item but is not suppressed (relative to the neutral trials) when the probe is presented on the opposite side of the display from the attended item. Moreover, the probe P1 effect is eliminated when the task simply involves detecting the attended color rather than discriminating the form of the object presented in this color. Thus, focusing attention on the location of an object in a visual search array leads to the same pattern of changes in sensory processing that occurs when attention is explicitly cued to a location in space.

Step 4: Shifting Overt Attention to the Attended Peripheral Object (Presaccadic Positivity and Spike Potential)

In natural vision, an individual will almost always fixate an important object that is being discriminated because this brings the object into the high-resolution foveal region of the retina. Thus, shifts of covert attention to the periphery do not usually serve the purpose of directly discriminating the identity of the attended object, but instead facilitate shifts of gaze (overt attention) to the object (for a detailed discussion, see Luck, 2009). For example, shifts of covert attention can be used to determine whether a peripheral object is likely to be a target and therefore worth fixating, and they may allow the location of the object to be determined more precisely so that a more accurate eye movement to this object can be programmed. Indeed, the N2pc component is larger when an object must be localized or fixated than when it must merely be detected (Hyun et al., 2009a;Luck et al., 1997b).

Only a few studies have examined the ERPs associated with saccadic eye movements, and these studies have identified several consistent components. First, there is a slow frontal negativity far in advance of the eye movement, which appears to be analogous to the contingent negative variation (see Chapter 8, this volume) and reflects preparation for the saccade-eliciting stimulus. This is followed by a parietally maximal component called the *presaccadic positivity* that onsets approximately 150 ms prior to the eye movement. Richards (2003) proposed that this component reflects the planning of the saccade. A sharp *spike potential* is then observed beginning approximately 20 ms prior to the saccade and peaking at saccade onset; this potential is negative at the frontal pole and positive at posterior scalp sites, and it is thought to reflect the motor activity associated with executing the saccade. Although behavioral studies suggest that shifts of gaze are necessarily preceded by shifts of covert attention (Deubel & Schneider, 1996; Hoffman & Subramaniam, 1995), no ERP component has been identified that reflects this specific shift of covert attention. This may reflect the fact that the eye movement studies have generally used tasks involving a single highly salient target stimulus, with no competition from simultaneous distractors. As described below, the N2pc component appears to depend on the presence of this kind of competition.

Step 5: Expanding or Contracting Attention Around the Fixated Object

Once the eyes have been directed to the desired object, attention presumably plays an important role in contracting or expanding around this object to minimize distraction from other surrounding objects. In the image shown in Figure 12.1, for example, an observer may wish to see the details of the sliced orange near the right edge of the image. Once the eyes have fixated this orange, attention may need to filter out the surrounding fruits and nuts to prevent the features of those other objects from interfering with the perception of the orange. The retinal size of an object will depend on both the object's actual size and its distance, and so the size of the attentional window surrounding the object will typically be adjusted on the basis of the incoming sensory information rather than being set in advance.

Although substantial behavioral research has examined variations in the size of the attended region in the periphery (e.g., Eriksen & St. James, 1986; Eriksen & Yeh, 1985), relatively little behavioral research—and almost no ERP research—has examined the expansion and contraction of attention around the fixated object. This is a ripe area for future research.

A Closer Look at N2pc and Related Components

Now that we have considered the ERP components associated with various processes involved with the perception of multiple-element scenes, we will take a closer look at the N2pc component and other lateralized components that reflect the focusing of attention on individual objects.

Discovery of the N2pc Component

The N2pc component was first observed in the two tasks shown in Figure 12.5 (Luck & Hillyard, 1990). In the *parallel search task,* the target was an arrow and the distractors were triangles. In this task, the horizontal line that forms the shaft of the arrow pops out from the display, and observers can detect the target rapidly and effortlessly without scanning through the array. In the *serial search task,* the triangle was the target and the arrows were distractors. In this task, the target does not contain a distinctive feature that is absent from the distractors, and observers must perform a slow and effortful search through the array to find the target (for an extensive discussion of this pattern of results, see Treisman & Gormican, 1988; Treisman & Souther, 1985).

In both of these conditions, the N2pc component can be seen as a more negative voltage when the target is contralateral to the electrode site relative to when it is ipsilateral. When the target popped out, this component was present from approximately 200–300 ms. When the target required a serial search, however, the N2pc was visible from 200 ms through the end of the recording epoch. This broad N2pc likely reflects the fact that the amount of time required to find the target is highly variable under serial search conditions, and the onset time of the N2pc component presumably varied widely from trial to trial in this condition, yielding a broad component when the data were averaged across trials (for similar results, see Wolber & Wascher, 2003). That is, even if the N2pc has a short duration on each individual trial, it will appear to have a long duration when trials with different N2pc onset times are averaged together. For this reason, most N2pc studies use targets that can be found in a short and consistent amount of time, as in the experiments shown in Figures 12.2 and 12.3.

Defining the N2pc Component as a Contralateral-Minus-Ipsilateral Difference

As discussed earlier in this chapter, the N2pc component is defined as the difference in amplitude between the contralateral and ipsilateral waveforms. Indeed, researchers sometimes construct contralateral-minus-ipsilateral difference waves to visualize the N2pc without distortions from overlapping components. This is shown in the right panel of Figure 12.3. Contralateral-minus-ipsilateral difference waves are particularly useful when one wishes to compare the time course of the N2pc

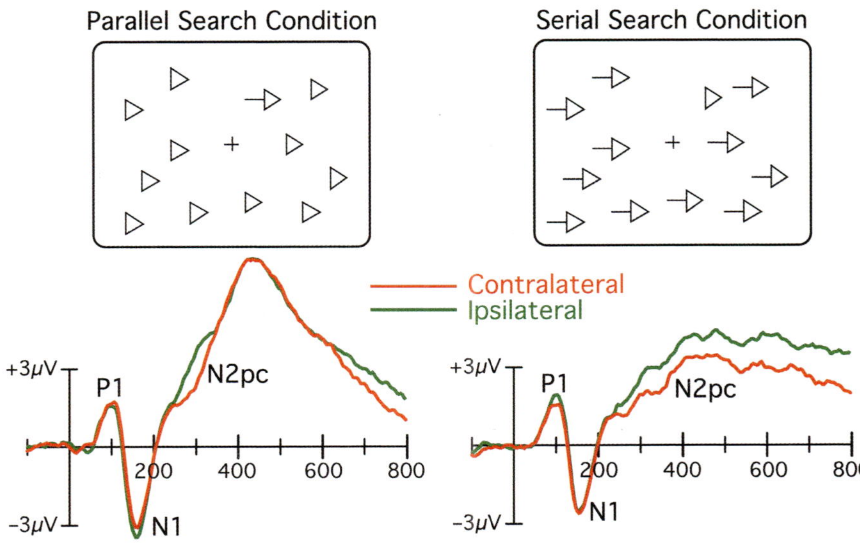

Fig. 12.5. Stimuli and grand average ERP waveforms from the study in which the N2pc component was first observed (Luck & Hillyard, 1990). In the parallel search condition, the arrow was the target and the triangles were the distractors; in the serial search condition, the triangle was the target and the arrows were the distractors. Subjects pressed one of two buttons for each array to indicate whether the target was present ($p = .5$) or absent ($p = .5$). The ERPs shown here were recorded from posterior temporal electrode sites (T5 and T6) and were averaged across arrays containing 4, 8, or 12 items. An N2pc component can be observed in both conditions as a difference between the contralateral and ipsilateral waveforms beginning at approximately 225 ms poststimulus, but the N2pc lasted much longer in the serial search condition than in the parallel search condition.

across different conditions. For example, Figure 12.6 shows the data from the experiment shown in Figure 12.3, but separated as a function of whether subjects attended to the red popout or the green popout. The red popouts were more salient than the green popouts in this experiment, which led to an earlier N2pc onset latency when the red item was attended than when the green item was attended. Although it is possible to see this when the contralateral and ipsilateral waveforms are overlaid (Figures 12.6A and 12.6B), it is much easier to see this latency difference when the contralateral-minus-ipsilateral difference waves for the red and green targets are overlaid (Figure 12.6C). Moreover, it would be very difficult to accurately measure the onset latency of the N2pc component without first constructing contralateral-minus-ipsilateral difference waves. That is, to measure the time at which the contralateral and ipsilateral waveforms diverged from each other, one cannot legitimately measure the onset latencies in the contralateral and ipsilateral waveforms separately. Note, however, that mean amplitude over a given time range does not need to be measured from the difference waves; because mean amplitude is a linear measure, the same results will be obtained whether you first make a difference wave and then measure mean amplitude or whether you first measure the mean amplitude from the contralateral and ipsilateral waveforms and then compute the difference between these measures (for a detailed discussion of linear versus nonlinear ERP measures, see Luck, 2005, pp. 232–235).

Because the N2pc component is defined as the difference between the contralateral and ipsilateral

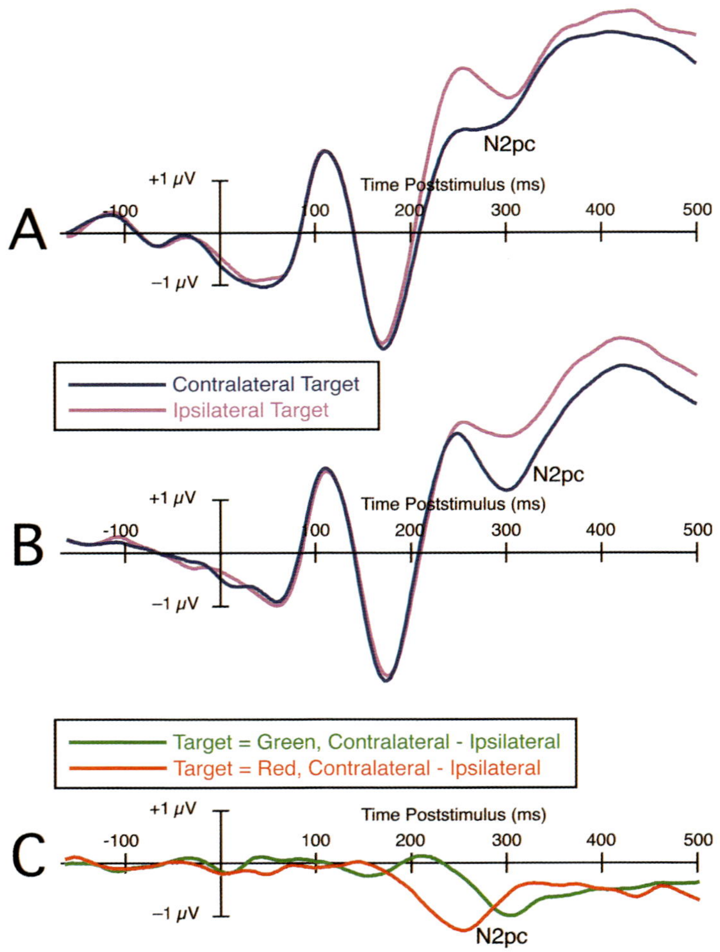

Fig. 12.6. Grand average ERP waveforms from a study using the stimuli shown in Figure 12.3 (Luck et al., 2006), showing the contralateral and ipsilateral waveforms for red targets (A) and green targets (B), along with the contralateral-minus-ipsilateral difference waveforms for the red and green targets (C)). This study examined the N2pc component in schizophrenic patients and healthy control subjects, and the data from the control subjects are shown here. Note that the red color was more salient than the green color, leading to faster reaction times and earlier N2pc onset and peak latencies.

waveforms, overlapping activity from other components does not distort the N2pc component. For example, an experimental manipulation that increases the amplitude of the P3 wave in one condition relative to another condition will influence the amplitude measured in the P2 and N2 waveforms, but it will not influence the difference in amplitude between the contralateral and ipsilateral electrode sites (assuming that the P3 wave is not lateralized with respect to the position of the attended item, which it is not). This is enormously important, because overlap among different components often makes ERP experiments difficult to interpret (for more discussion, see Chapter 1, this volume). As discussed earlier in this chapter, the LRP is also isolated by means of a contralateral-minus-ipsilateral difference wave, and both N2pc and LRP are therefore easier to isolate than most other ERP components.

However, it is worth considering in more detail the logic behind using this sort of difference to define and measure an ERP component. If a component is generated exclusively in the contralateral hemisphere, then the amplitude of the contralateral-minus-ipsilateral difference will be linearly proportional to the magnitude of the underlying generator source. This is true even if some voltage is conducted to the ipsilateral hemisphere. Imagine, for example, that X is the magnitude of the underlying generator source, 10% of X is the voltage measured over the contralateral hemisphere, and 3% of X is the voltage measured over the ipsilateral hemisphere. In this case, the amplitude of the contralateral-minus-ipsilateral difference wave will be 7% of X (because 10% – 3% = 7%). If we manipulate some factor that causes the magnitude of the generator to double (making it now $2X$), these percentages will still be true (because voltages propagate linearly through a conductor), and the new amplitude of the difference wave will be 7% of $2X$. That is, the voltage measured from the difference wave will double if the magnitude of the underlying generator source doubles.

What if the component is generated by sources in both hemispheres but is simply stronger in the contralateral hemisphere than in the ipsilateral hemisphere? If a given experimental effect produces a proportional change in the magnitudes of the contralateral and ipsilateral generator sources, then the amplitude measured from the contralateral-minus-ipsilateral difference wave will again be proportional to the change in the magnitudes of the internal generator sources. (The algebra is analogous to that described in the preceding paragraph.) Thus, measuring the amplitude of the difference between the contralateral and ipsilateral sites is a very reasonable way to assess the magnitude of the underlying generator sources.

However, there are at least three limitations on this approach. First, the difference between contralateral and ipsilateral activity will necessarily approach zero for electrode sites near the midline. Thus, one must be careful when using contralateral-minus-ipsilateral differences to assess the scalp distribution of the N2pc or LRP. Second, although this difference-based approach eliminates the contributions of most overlapping components, it will not eliminate overlap from other components that are lateralized with respect to the target (for the N2pc) or to the response (for the LRP). As discussed later in this chapter, several distinct components have now been identified that are lateralized with respect to the target, and these may distort measurements of N2pc amplitude or latency. Finally, any factor that changes the proportional difference between generators in the contralateral and ipsilateral hemispheres will be difficult to assess by measuring the difference between the contralateral and ipsilateral electrode sites. For example, a manipulation that caused an increase in the amplitude of the N2pc in the ipsilateral hemisphere but not in the contralateral hemisphere would lead to a decrease in the contralateral-minus-ipsilateral difference. Similarly, a manipulation that caused an increase in the activation of the incorrect response hand in an LRP study would lead to a decrease in the apparent LRP amplitude. Thus, although measuring N2pc and LRP amplitude as the difference in amplitude between the contralateral and ipsilateral waveforms has many advantages, one must think carefully when using this approach.

The N2pc component is typically superimposed on a large positive voltage produced by the P2 wave, and the waveform for both contralateral and ipsilateral targets is therefore on the positive side of the baseline in the N2pc latency range (see Figures 12.3–12.5). How, then, do we know that the N2pc is actually a negative-going component over the contralateral hemisphere rather than a positive-going component over the ipsilateral hemisphere? Defining N2pc amplitude as contralateral-minus-ipsilateral implicitly assumes that the N2pc is larger over the contralateral hemisphere, because it subtracts out the (supposedly) lesser amplitude recorded over the ipsilateral hemisphere. As we shall see later in this chapter, the N2pc may actually consist of the combination of a contralateral negativity and an ipsilateral positivity. However, the contralateral negativity appears to be the dominant subcomponent under most conditions.

In addition, if an experiment includes trials on which no target is present, the N2pc appears to consist primarily of an increased negativity over the contralateral hemisphere in the target waveforms compared to the no-target waveforms. For example, Figure 12.7A shows that the target-elicited N2pc consists mostly of an increased negativity for contralateral popouts rather than an increased positivity for ipsilateral popouts relative to arrays in which no popout was present (homogeneous arrays). Moreover, when the waveforms elicited by nontarget popouts are subtracted from the waveforms elicited by target popouts, the overlapping P2 component is largely subtracted away. In these difference waves, the N2pc clearly consists of a negativity contralateral to the target, with little or no voltage observed over the contralateral hemisphere (see Figures 12.14A and 12.15A, which will be discussed in detail later in this chapter). Together, these findings suggest that the N2pc component consists primarily of a negative potential over the hemisphere contralateral to the target.

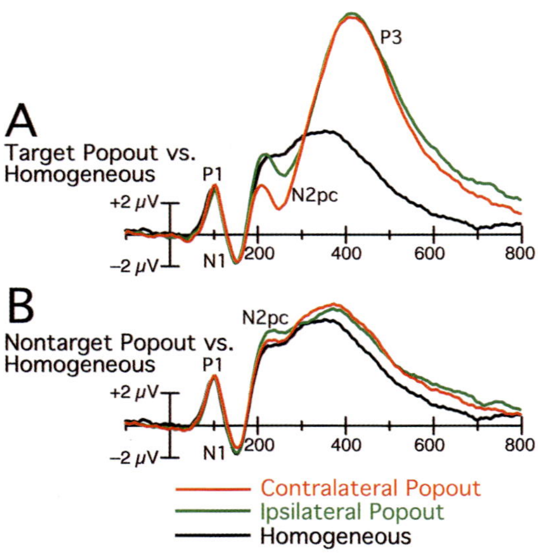

Fig. 12.7. Grand average ERP waveforms from the experiment shown in Figure 12.2A (Luck & Hillyard, 1994a, Experiment 1). Separate waveforms are shown for target (A) and nontarget (B) popouts at posterior temporal electrodes contralateral and ipsilateral to the side of the popout, and these are overlaid with the waveforms elicited by the homogeneous arrays (averaged over the left and right hemispheres because no popout was present in these arrays). A large N2pc was present for target popouts, but only a small N2pc was present for nontarget popouts (and this was confined to the color popouts, which were more salient than the orientation and size popouts).

When the waveforms from the left and right hemispheres are collapsed into contralateral and ipsilateral waveforms, we lose the ability to see any differences between the left and right hemispheres. It is possible, instead, to overlay the waveforms from the left and right hemispheres for targets on a particular side and see the differences between the two hemispheres. However, any overlapping components that differ in overall amplitude across the two hemispheres, irrespective of target location, will influence the difference in voltage between the left and right hemispheres, making this comparison difficult to interpret. For example, in the waveforms shown in Figure 12.3, the voltage at 300 ms is more positive over the right hemisphere than over the left hemisphere, irrespective of the position of the target. This difference between hemispheres probably reflects other components that overlap with N2pc.

A better approach is to overlay the waveforms elicited by LVF and RVF targets for a given hemisphere and ask whether the difference between these waveforms is bigger for the left hemisphere or the right hemisphere. In Figure 12.3, for example, the difference between LVF and RVF targets is somewhat larger over the left hemisphere than over the right hemisphere. This might suggest that attention operated more strongly in the left hemisphere than in the right hemisphere in that experiment. However, any differences in neural activity between LVF and RVF targets that are present in both hemispheres will distort this comparison (see Figure 9.1 in Chapter 9, this volume, for a discussion of the same set of issues in the context of the LRP). For example, if LVF targets elicited a larger P2 wave than RVF targets over both hemispheres in the experiment shown in Figure 12.3, this would increase the contralateral-minus-ipsilateral difference over the left hemisphere and decrease this difference over the right hemisphere. Thus, differences across the left and right hemispheres in the degree of lateralization cannot be unambiguously interpreted as differences in N2pc amplitude between the hemispheres.

As a result, most researchers collapse the two hemispheres into contralateral and ipsilateral waveforms to isolate the N2pc, and separate waveforms for the left and right hemispheres are not usually shown in publications. An interesting exception is that Eimer (1996) reported that the N2pc was present over both hemispheres for color and form targets but was present only over the left hemisphere for word targets (see also Dell'Acqua et al., 2007). This may reflect an important difference between the hemispheres in the role of attention for linguistic stimuli. However, it is possible that the apparent hemispheric asymmetry in the N2pc for words reflects an overall difference in some other component between word targets presented in the left and right hemispheres rather than a true asymmetry in the N2pc component. Other methods (e.g., current source density transformations) would be necessary to distinguish between these alternatives.

Sensitivity of N2pc to Basic Parameters

The N2pc component is typically observed for target stimuli but is absent or considerably smaller for nontarget stimuli. For example, Figure 12.7 shows the contralateral and ipsilateral waveforms for target and nontarget popouts in the experiment shown in Figure 12.2A. A substantial difference was observed between the contralateral and ipsilateral waveforms for the target popout stimuli, but this difference was much smaller for the nontarget popout stimuli. In fact, the orientation and size popouts produced no difference at all between contralateral and ipsilateral electrodes when they were nontargets. The small difference shown in Figure 12.7 was entirely due to the color popouts, which were highly salient. However, as we shall see in the next section, nontarget stimuli will elicit an N2pc if they are so similar to the target that focused attention is needed to determine whether the stimulus is a target or nontarget.

Unlike the anterior P2, posterior bilateral N2, and P3b components shown in Figure 12.2, the N2pc component is largely insensitive to the probability of the eliciting stimulus. An example is shown in Figure 12.8, which used the same stimuli shown in Figure 12.2 but varied the probability of the target popout stimulus (17% in one condition and 50% in another). The P3b component was substantially larger when the target was 17% probable than when it was 50% probable, whereas the N2pc component was equally large in these two conditions (see especially the contralateral-minus-ipsilateral difference waves).

The N2pc can be elicited by many types of lateralized visual targets. It has been observed in response to targets defined by color, orientation, size, motion, various types of shape, letter identity, word identity, Kanizsa figures, biological motion, facial expression, and direction of eye gaze (see, e.g., Conci et al., 2006; Doi et al., 2009; Eimer, 1996; Fenker et al., 2009; Girelli & Luck, 1997; Hirai & Hiraki, 2006; Holmes et al., 2009; Kiss et al., 2008; Luck & Hillyard, 1990, 1994a; Woodman & Luck, 1999). It has even been observed when subjects search stimuli stored in visual working memory rather than visible stimuli (Busch et al., 2009; Dell'Acqua et al., 2010), and it has also been observed when the target is defined by a difference between a stimulus array and a visual working memory representation of a previously presented array (Eimer & Mazza, 2005; Hyun et al., 2009b; Schankin & Wascher, 2008). No obvious differences in scalp distribution have been observed for these different target types, although it is possible that subtle differences exist.

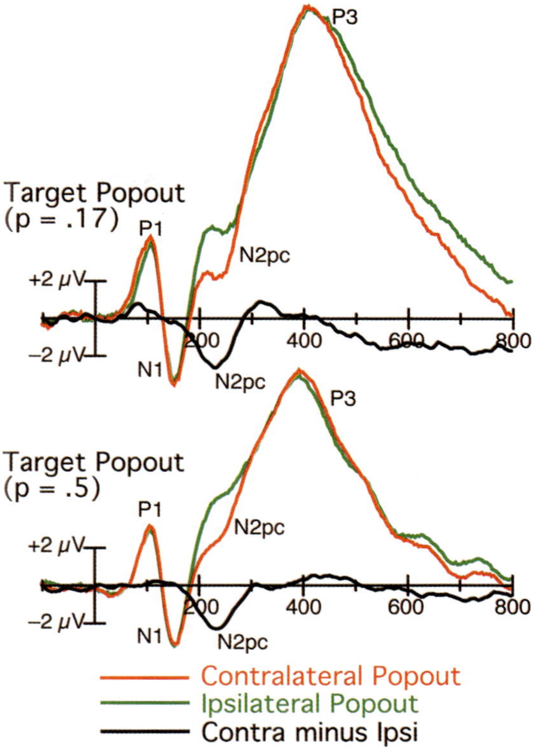

Fig. 12.8. Grand average ERP waveforms recorded at posterior temporal electrode sites in Experiment 2 of Luck and Hillyard (1994a). The stimuli and task were the same as in the experiment shown in Figures 12.2 and 12.7, except that the probability of the target array was either low ($p = .17$) or high ($p = .50$). Separate waveforms are shown for targets that were contralateral and ipsilateral to the recording electrode, and the difference between the contralateral and ipsilateral waveforms is also shown. The P3 wave was much larger when the target was rare than when it was frequent, but N2pc amplitude (i.e., the difference between the contralateral and ipsilateral waveforms) was unaffected by target probability.

Given that the N2pc component—defined as the contralateral-minus-ipsilateral difference—cannot begin until the brain has determined the location of the target stimulus, N2pc onset would be expected to depend on the salience of the target. Evidence for this has been observed in several experiments in which different types of popout stimuli were compared and both N2pc onset latency and reaction time were earlier for some popout stimuli than others (e.g., Girelli & Luck, 1997; Luck & Hillyard, 1994a); Luck et al., 2006). An example of this was described earlier in the context of Figure 12.6, which shows that the N2pc was earlier for the more salient red targets than for the less salient green targets. Both target discrimination reaction time and N2pc onset latency were 40–50 ms faster when the red item was the target than when the green item was the target.

Sequence of Lateralized Components

Figure 12.8 shows a commonly observed pattern in which the contralateral waveform is more negative than the ipsilateral waveform initially (from approximately 175 to 300 ms), then becomes more positive for approximately 100 ms, and then becomes more negative again for an extended period of time (this is more evident for the infrequent condition than for the frequent condition). The middle phase does not always cross over to become positive; instead, the initial contralateral negativity may fade to a near-zero value during this period and then become large again in the late period (see, e.g., Figure 12.4B, the parallel search waveforms in Figure 12.5, and Figure 12.7A). However, sometimes the waveform remains negative from the onset of the N2pc until the end of the averaging epoch (see, e.g., Figure 12.6). This middle positive phase has not been extensively studied. One study found that the middle phase was more laterally distributed than the initial phase and was enlarged when the distance between the target and a salient distractor was decreased (Hilimire et al., 2009), suggesting that it reflects a process that is triggered when short-range competition must be resolved. In a somewhat different paradigm, Luck et al. (1997b) found that the initial negative phase rather than the middle positive phase was increased when a distractor was placed next to the target. However, it is possible that the initial phase continued into the time range of the middle phase and the increased amplitude of the negative first phase canceled out the increased positivity of the second phase. This sort of overlap problem may make it difficult to determine the psychological correlates of the second phase.

Jolicoeur and his colleagues have called the late negative phase the *sustained posterior contralateral negativity* (SPCN), and they have proposed that it reflects continued processing of the attended item. For example, SPCN amplitude is increased when the difficulty of the discrimination is increased (Mazza et al., 2007; Prime & Jolicoeur, 2009). Jolicoeur and his colleagues have further proposed that it may be the same as the contralateral delay activity (CDA) observed during the delay period of visual working memory experiments (see Chapter 13, this volume). Perron et al. (2009) found that the N2pc and the SPCN/CDA have similar scalp distributions and that both are larger for lower-field than for upper-field stimuli, consistent with a shared neural generator source. However, McCullough and colleagues (2007) found that the SPCN/CDA had a more parietal focus than the N2pc, suggesting different generator sources for these two components. A magnetoencephalographic study also found that the SPCN/CDA was localized to posterior parietal cortex, which differs from the predominantly occipitotemporal source of the N2pc (Robitaille et al., 2009). Neuroimaging studies have shown that both parietal and occipitotemporal areas exhibit sustained activity during working memory maintenance (Xu & Chun, 2006), and it is therefore possible that the voltage over the broad SPCN/CDA period reflects a combination of activity in the N2pc generator and in other areas.

The N2pc, which is largest over lateral occipitotemporal electrode sites (e.g., PO7 and PO8), is sometimes accompanied by a component with the same time course over central electrode sites (e.g., C3 and C4). This more central component has been labeled *N2cc* (N2-central-contralateral) by Praamstra and his colleagues (Oostenveld et al., 2003; Praamstra, 2006, 2007; Praamstra & Oostenveld, 2003; Praamstra & Plat, 2001). N2pc and N2cc have overlapping scalp distributions, but it is possible to measure them separately by transforming voltage into current source density, which has a more narrowly focused scalp distribution. As discussed by Praamstra (2006), several sources of evidence indicate that N2cc is not merely volume-conducted voltage from the N2pc generator, but instead reflects a separate source (probably in motor or premotor cortex). N2cc is particularly prominent under conditions in which subjects are instructed to make either a left-hand or right-hand response, depending on the identity of the target presented on a given trial (e.g., left for an upright red *T* and right for an inverted red *T*, as in Figure 12.4), leading to the potential for Simon interference when the target appears on one side of the display but its identity indicates that the opposite hand should be used for the response. Praamstra (2006) has suggested that the appearance of a target on one side of the display leads to an automatic preparation of the ipsilateral hand, and that the N2cc reflects the suppression of this preparation so that the response can be made on the basis of the target's identity rather than its position.

Neural Generators of the N2pc Component

Evidence for the neural generator of the N2pc component has come from studies using combined ERP and ERMF (event-related magnetic field) recordings and from studies establishing homologies between the N2pc and single-unit attention effects observed in monkeys. The first ERP/ERMF study of the N2pc used a paradigm much like that shown in Figure 12.3, except that the items were rectangles and the subjects were required to indicate whether the rectangle of the attended color was horizontal or vertical. Figures 12.9A and 12.9B show the observed topography of the electrical N2pc and its magnetic analog, the M2pc, as isolated by means of LVF-target minus RVF-target difference waves (it is not appropriate to perform source localization on contralateral-minus-ipsilateral difference waves because this would artificially cause the voltage to approach zero for electrodes near the midline). The ERP difference waves were positive over the left posterior scalp and negative over the right posterior scalp. The negative voltage over the right hemisphere reflects the negative voltage of the N2pc for the LVF target, and the positive voltage over the left hemisphere reflects the subtraction of the negative voltage for the RVF target. The ERMF maps do not show opposite patterns over the left and right hemispheres, but this is exactly what would be expected from a LVF-minus-RVF ERMF difference wave. A dipole in the right hemisphere pointing downward would produce both a negative voltage over the right hemisphere (corresponding to the N2pc elicited by a LVF target) and a magnetic field that leaves the head medially (denoted by red) and returns into the head laterally (denoted by blue). A dipole in the left hemisphere pointing the opposite direction (because the RVF-target waveforms are subtracted to construct this difference wave) would produce both a positive voltage over the left hemisphere and a magnetic field that leaves the head medially (denoted by red) and returns into the head laterally (denoted by blue). Thus, even though the left-hemisphere and right-hemisphere dipoles point in opposite directions, producing opposite polarity voltages, they produce mirror-image magnetic fields.

Structural magnetic resonance imaging (MRI) data were obtained for each subject, and the minimum norm approach was used to estimate the distribution of current over the cortical surface corresponding to the left-minus-right target difference waves. This source localization approach was applied to the data from each subject individually (with 2400 trials per subject to obtain an adequate signal-to-noise ratio), and it was also applied to the grand average data (using the MRI data from a single subject to constrain the localization). The localization results from the grand average are shown in Figure 12.9C (similar, albeit noisier, results were obtained for the individual subjects). The estimated current distribution was focused over lateral occipitotemporal cortex, consistent with a set of generators in intermediate and/or high levels of the ventral visual processing pathway.

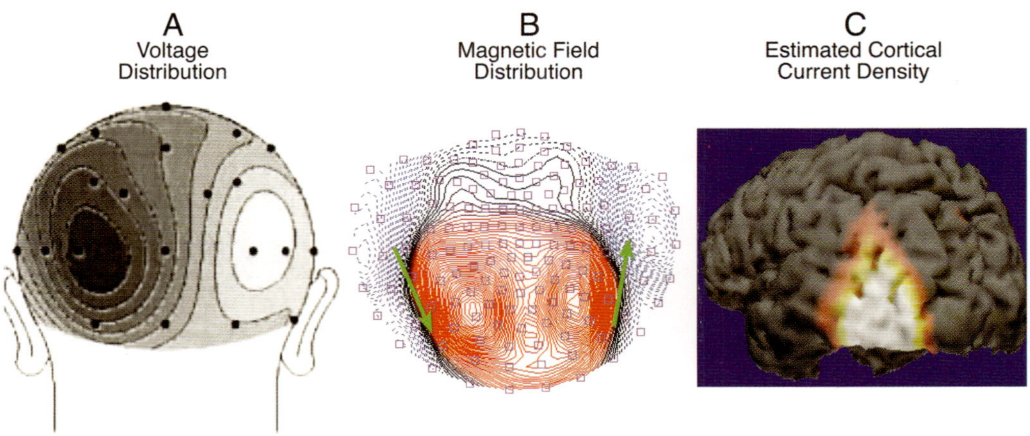

Fig. 12.9. Event-related potential and ERMF data from the study of Hopf et al. (2000). (A) Voltage distribution of the N2pc waveform, defined as the difference between trials with a target in the left and right visual fields. (B) Magnetic field distribution for the magnetic analog of the N2pc component, again defined as the difference between trials with a target in the LVF and RVF. (C) Estimated distribution of current density on the cortical surface for the combined electrical and magnetic data.

This study also found evidence for a parietal source during the initial portion of the N2pc (180–200 ms), which could reflect an attentional control signal that is then implemented in lateral occipitotemporal cortex during the main time period of the N2pc. However, follow-up studies have not replicated this parietal source (Hopf et al., 2004, 2006). A transcranial magnetic stimulation (TMS) study found that a TMS pulse over the right posterior parietal cortex 100 ms after stimulus onset, which presumably caused a temporary disruption of processing in that area, led to a delay in the onset of the N2pc (Fuggetta et al., 2006). This could reflect a disruption of the early parietal subcomponent, but it more likely reflects a disruption of parietal control circuitry that must be engaged for the N2pc to be elicited within ventral visual areas.

More precise localization of the N2pc was provided by a study that used a combination of ERPs, ERMFs, structural MRI, and functional MRI (fMRI; Hopf et al., 2006). This study tested the hypothesis that the locus of selection within the visual system depends on the locus of competition between the attended and unattended items. Recordings from monkeys have indicated that attention has much stronger effects on single-unit activity when both a target and a distractor are simultaneously present within the receptive field of the neuron being recorded (Luck et al., 1997a; Moran & Desimone, 1985; Reynolds et al., 1999), presumably because a neuron's firing rate becomes an ambiguous index of stimulus features when multiple objects are present inside the neuron's receptive field (Luck et al., 1997b). Because receptive field sizes are small at early stages of the ventral processing stream and become very large at late stages, neurons at late stages will have multiple objects inside the receptive field whenever the display contains multiple objects, whereas neurons at early stages will have multiple objects inside the receptive field only when the objects are very close to each other. Hopf et al. (2006) found that the estimated generator of the N2pc component also varied according to the spatial scale of the display. As illustrated in Figure 12.10, the N2pc was generated primarily in a late anterior region of the ventral stream when the display contained large-scale competition, whereas a more posterior intermediate region also contributed to the N2pc when the display contained small-scale competition. These estimated source locations were confirmed by fMRI data, which showed the same pattern and made it possible to identify the specific posterior and anterior regions (area V4 and the lateral occipital complex, respectively).

Converging evidence for sources in area V4 and the lateral occipital complex comes from homologies between the N2pc and monkey single-unit attention effects observed in visual search tasks in area V4 and in inferotemporal cortex (Chelazzi et al., 1993, 1998, 2001). As shown in Figure 12.11, neurons in these areas initially responded in an attention-independent manner. However, beginning

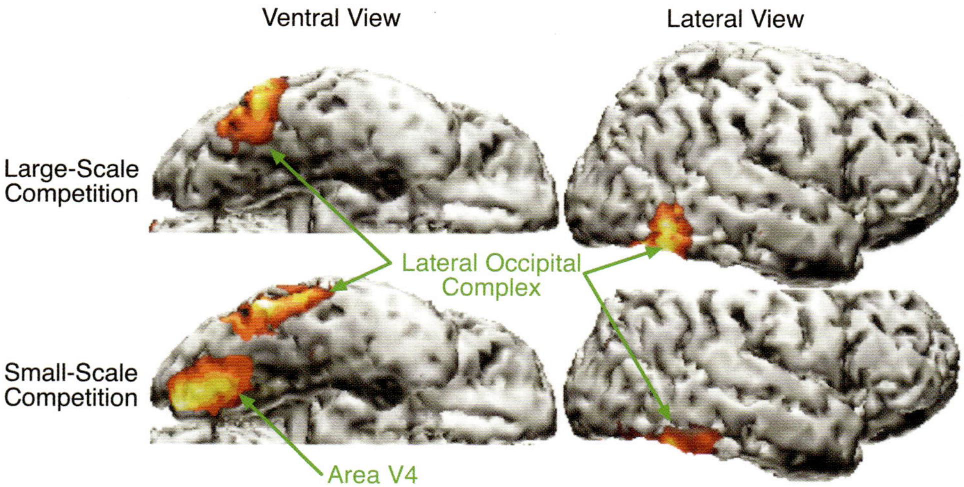

Fig. 12.10. Estimated distribution of current density on the cortical surface for the N2pc/M2pc component under conditions of large- or small-scale competition between the target and the distractors (Hopf et al., 2006). Under conditions of large-scale competition, the N2pc/M2pc was confined to the lateral occipital complex (the likely human homologue of macaque inferotemporal cortex). Under conditions of small-scale competition, the N2pc/M2pc was also present in area V4. The identity of these areas was verified using fMRI-based visual field mapping.

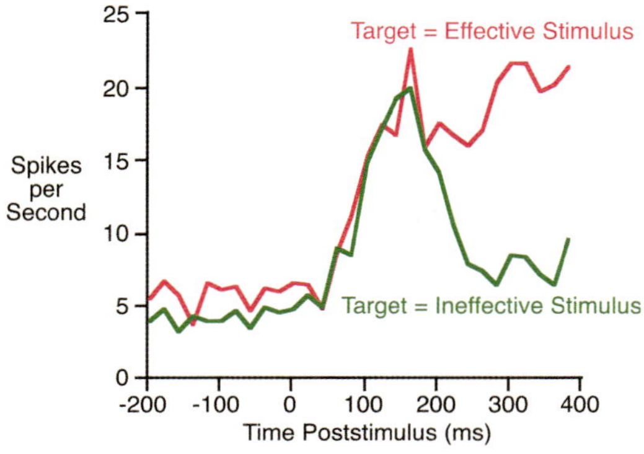

Fig. 12.11. Average firing rate of a population of inferotemporal neurons in response to a visual search array that contained both a target and a distractor inside the receptive fields of the neurons (Chelazzi et al., 1993). The neurons initially fired at the same rate independently of whether the target matched the stimulus preferences of a given neuron (i.e., whether the target, when presented alone, was an effective or ineffective stimulus for eliciting a response from the neuron). Beginning at 175 ms poststimulus, however, the firing rate became suppressed if the target did not match the neuron's stimulus preferences.

approximately 175 ms after stimulus onset, the neurons began firing differentially, depending on whether the target stimulus or a simultaneous distractor stimulus matched the neuron's stimulus preferences. The onset latency of this effect is similar to the onset latency of the human N2pc component. Moreover, several experimental manipulations have been shown to have analogous effects on the monkey single-unit attention effect and the N2pc component (Luck et al., 1997b). First, the single-unit effects are larger when the target is a complex image rather than a simple color patch, and the N2pc is also larger for complex multifeature targets than for simple single-feature targets. Second, the single-unit effects are larger when both the target and distractor items fall within the receptive field of the neuron being recorded, and the N2pc is larger when a distractor is located close to the target. Third, the single-unit effects for simple color targets are larger when the monkey is required to make an eye movement to the target rather than to make a simple manual detection response, and the N2pc component is also larger when an eye movement is required. These analogous effects of experimental manipulations across species and recording techniques are consistent with the hypothesis that the N2pc component arises from area V4 and the lateral occipital complex (which is the human homologue of monkey inferotemporal cortex).

N2pc in Special Populations

Most N2pc studies have examined the effects of within-subjects manipulations in healthy young adults. However, a few have examined the N2pc across different populations. N2pc amplitude is reduced and N2pc onset and peak latency are increased in older individuals (Lorenzo-Lopez et al., 2008), which is important to keep in mind when groups being compared may differ in age along with some other factor. N2pc onset and peak latency are also delayed in patients with hepatic encephalopathy, a neurodegenerative disorder that is a consequence of liver failure (Schiff et al., 2006). In contrast, no change in N2pc amplitude or latency has been observed in patients with schizophrenia (Luck et al., 2006) or Parkinson's disease (Praamstra & Plat, 2001) or in athletes with a history of multiple concussions (De Beaumont et al., 2007).

What Cognitive Process Is Indexed by N2pc?

The N2pc component is clearly related to attention, because the presence of an N2pc component for a given item depends on the task relevance of that item. However, *attention* is a broad umbrella term that includes many different types of processes (for reviews, see Luck & Gold, 2008; Luck & Vecera, 2002). Thus, it is important to provide a more specific description of the process reflected by N2pc.

THE FILTERING HYPOTHESIS

Luck and Hillyard (1994b) proposed that the N2pc component reflects a filtering process that is used to suppress the processing of distractor items surrounding a given object to reduce interference in the identification of the object of interest. This hypothesis was inspired by single-unit studies in monkeys showing that the effects of attention are much greater when a given neuron's receptive field contains both a to-be-discriminated target object and distractor objects (Luck et al., 1997a; Moran & Desimone, 1985; Reynolds et al., 1999; Treue & Maunsell, 1996). When multiple objects are present inside a neuron's receptive field, the neuron's firing rate no longer reflects the features present in a single object, which can lead to ambiguities in neural coding. If, for example, a red triangle and a blue circle are both present in the receptive field of a red-selective neuron, the neuron will fire at a high rate to indicate the presence of the color red, but it will not be clear whether the color red is present in the triangle or in the circle. Thus, attention may be used to focus a given neuron on the features of a single object, inhibiting inputs to the neuron from other objects (for an extensive discussion, see Luck et al., 1997b). As described in the influential *biased competition model of attention* (Desimone & Duncan, 1995), this may be implemented via inhibitory connections between objects: When top-down attentional control signals give a task-relevant object a competitive advantage over the distractor objects, the task-relevant object can inhibit the distractor objects more effectively than the distractor objects can inhibit the task-relevant object. Over a period of tens of milliseconds, these inhibitory interactions lead to a progressive increase in the competitive advantage of the task-relevant object and further inhibition of the distractors. Luck and Hillyard (1994b) proposed that the N2pc component reflects the inhibitory component of this overall scenario.

To test a hypothesis about the relationship between an ERP component and a specific psychological or physiological process, it is necessary to make predictions about how the component will vary as the hypothesized process varies across a range of experimental conditions. However, we rarely know enough about the hypothesized process to make strong predictions about how the ERP component should vary over experimental conditions. Indeed, the reason we wish to establish the link between an ERP component and a psychological or physiological process is that we do not understand the process very well and wish to use ERPs to study it. This leads to a chicken-and-egg problem: we cannot establish the relationship between the ERP component and the internal process unless we know enough about the process to manipulate it and test the effects on the ERP component; however, we need to know the relationship between the ERP component and the internal process to get to the point where we know enough about the process to manipulate it.

Studies of the N2pc component face this problem, because little is known about the filtering process that N2pc is hypothesized to reflect. To address this problem, Luck and Hillyard (1994b) performed a series of experiments in which they used extremely simple and obvious manipulations that must almost certainly influence filtering, even if we do not yet know the details of how the filtering process works. For example, there can be no filtering if there are no distractors to filter, so one experiment tested whether the N2pc was eliminated when the distractors were eliminated. However, even these simple manipulations have been the subject of dispute among researchers, and the relationship between the N2pc component and filtering remains controversial. The remainder of this chapter will discuss the evidence in favor of the filtering hypothesis, the evidence against this hypothesis, and a variant of the filtering hypothesis that can explain all of the existing evidence.

EVIDENCE SUPPORTING THE FILTERING HYPOTHESIS

The first experiment reported by Luck and Hillyard (1994b) tested the more general hypothesis that the N2pc component is related to attention. More specifically, this experiment asked whether the N2pc is observed when a given item must be scrutinized by attention for the observer to determine whether it is the target, but not when the item can be rejected preattentively on the basis of salient features. As illustrated in Figure 12.12, each array could contain a color popout item or an orientation popout item, and each popout item could be either large or small. At the beginning of each trial block, one of the four possible popout stimuli (large color popout, small color popout, large orientation popout, or small orientation popout) was designated the target for that block. The observers were instructed to press one button if the target was present and another button if it was absent.

The color and orientation differences were highly salient, but the size differences were subtle. No matter which popout was the target for a given trial block, one of the nontarget popouts was difficult to discriminate from the target (*difficult nontargets*) and two of the nontarget popouts were easy to discriminate from the target (*easy nontargets*). For example, when the small orientation popout was the target, the large orientation popout was the difficult nontarget and the small and large color popouts were the easy nontargets. If the N2pc component reflects the allocation of attention to a given item for the purpose of determining whether it is the target, then it should be present for both targets and difficult nontargets (because it is difficult to know whether these items are targets without focusing attention on them), whereas it should be absent for easy nontargets (because they can be rejected as nontargets without focusing spatial attention on them). This is exactly the pattern of results that was obtained: the N2pc was nearly identical for targets and difficult nontargets, but no significant N2pc was observed for easy nontargets.

A second experiment tested the filtering hypothesis by changing the task so that filtering would be unnecessary or even counterproductive. As illustrated in Figure 12.13, the nontargets were either

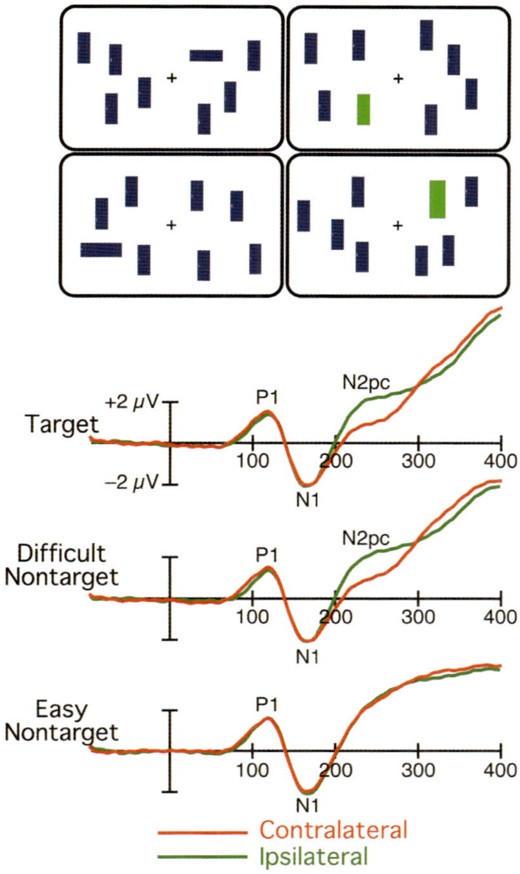

Fig. 12.12. Stimuli and grand average ERP waveforms recorded at posterior temporal electrode sites in Experiment 1 of Luck and Hillyard (1994b). In this experiment, the stimulus arrays could be homogeneous (p = .20) or could contain one of the four types of popout stimuli shown here (p = .20 for each type of popout array). One of the four popout stimuli was designated the target at the beginning of each trial block, and subjects pressed one of two buttons for each stimulus array to indicate whether the target was present or absent. For any given target, one of the nontarget popouts was difficult to discriminate from the target and the other two were easy to discriminate from the target. Both the target popout and the difficult nontarget popout elicited a large N2pc component, but no N2pc was observed for the easy nontarget popouts.

homogeneous arrays of vertical bars or homogeneous arrays of horizontal bars, and the targets were either arrays containing a vertical bar among horizontal bars or a horizontal bar among vertical bars. These different array types were randomly intermixed, so the subjects did not know in advance whether they were looking for a horizontal bar among vertical distractors or a vertical bar among horizontal distractors. Thus, the task required subjects to determine if one bar differed from the surrounding bars rather than identifying the popout bar's orientation, and filtering the bars surrounding the popout item could actually make it difficult to determine if the popout item differed from the surrounding items. Moreover, behavioral research has shown that tasks like this encourage subjects to adopt a *singleton detection mode*, in which they try to detect feature discontinuities rather than trying to identify specific target features (Bacon & Egeth, 1994). Thus, filtering should be minimized in this task, and little or no N2pc activity should be observed. Indeed, the observed contralateral-minus-ipsilateral difference was less than 0.2 μV and was not statistically significant (see Figure 12.13, bottom).

A third experiment tested the filtering hypothesis by manipulating the presence or absence of the distractor items. In the multiple-object condition (Figure 12.14A), each array contained a color popout, an orientation popout, a size popout, or no popout, and one of the three popout types was the target for a given block of trials. Subjects pressed one button if the target popout was present and

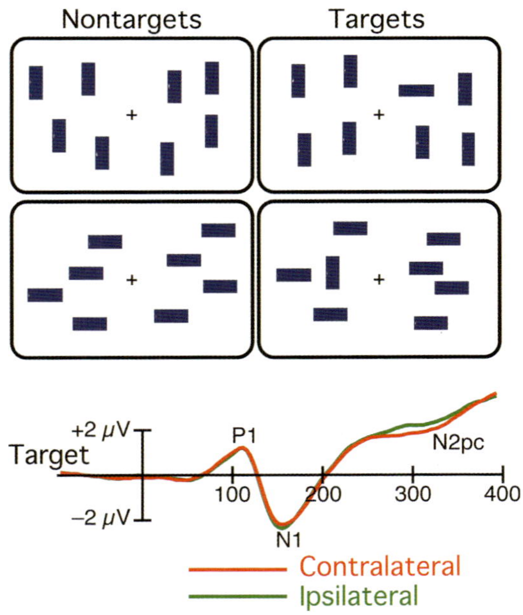

Fig. 12.13. Stimuli and grand average ERP waveforms recorded at posterior temporal electrode sites in Experiment 2 of Luck and Hillyard (1994b). Each of the four types of stimulus arrays shown here was equiprobable, and subjects pressed one of two buttons for each stimulus array to indicate whether a popout item was present or absent. Because the target was defined by the difference between the popout item and the surrounding items, there should have been no motivation to filter the items surrounding the popout. No significant difference between the contralateral and ipsilateral waveforms was observed in the N2pc latency range.

a different button if it was absent (i.e., if the array contained one of the other two popout features or no popout). Because this task required the subject to identify a specific target item that was surrounded by potentially distracting items, an N2pc component was expected for the target arrays. And because the nontarget popouts could be rejected on the basis of a simple feature, little or no N2pc was expected for the nontarget popout arrays (just as for the easy nontargets in the experiment shown in Figure 12.12). As predicted, a large N2pc was observed for the target popouts in this condition, but little or no N2pc was observed for the nontarget popouts. To isolate target-specific activity and subtract away all other components, the waveforms for the nontarget popouts were subtracted from the waveforms for the target popouts. This yielded a negative voltage from 200 to 300 ms in the contralateral waveform but little or no activity in the ipsilateral waveform (see Figure 12.14A, bottom). As mentioned earlier in this chapter, this is one piece of evidence supporting the claim that the N2pc reflects a negative voltage contralateral to the target rather than a positive voltage ipsilateral to the target.

In the single-object condition of this experiment (Figure 12.14B), the distractors were simply removed but the task remained the same. That is, one of the three items that served as a popout in the multiple-object condition was designated the target at the beginning of each trial block, and the other three

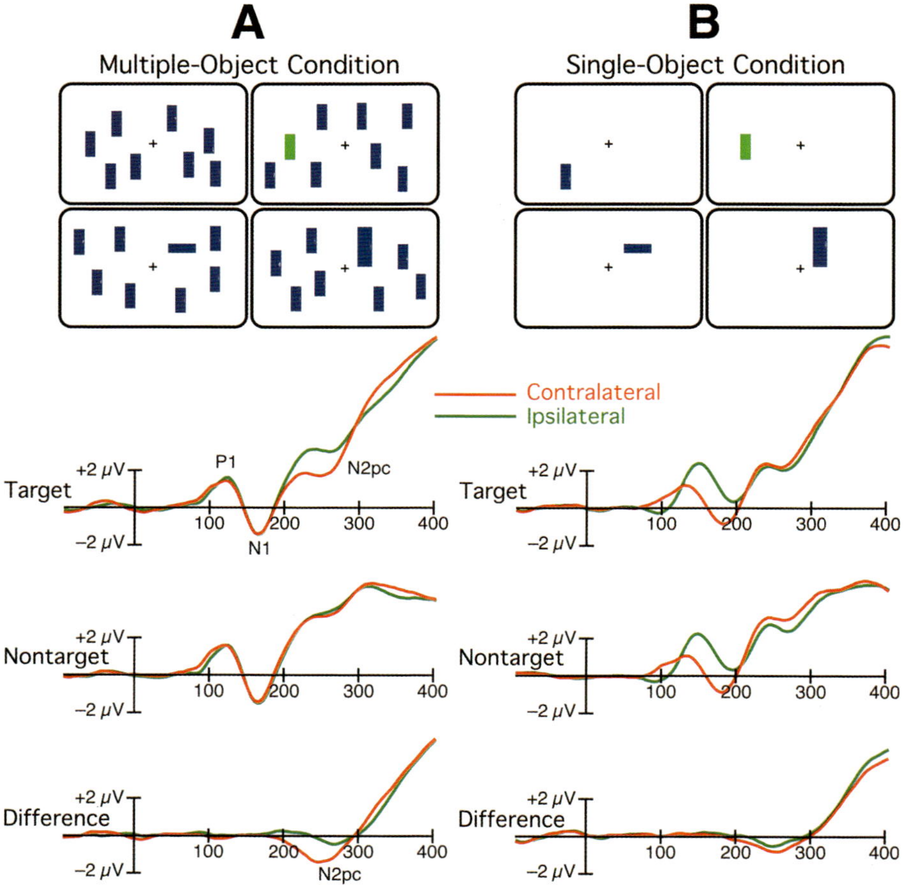

Fig. 12.14. Stimuli and grand average ERP waveforms recorded at posterior temporal electrode sites in the multiple-object condition (A) and the single-object condition (B) of Experiment 3 in Luck and Hillyard (1994b). In both conditions, each of the four types of stimulus arrays was equiprobable. The green, horizontal, or large bar was designated the target at the beginning of each trial block, and subjects pressed one of two buttons for each stimulus to indicate whether the target was present or absent. The small blue vertical bars were never targets. An N2pc component was clearly visible for the target stimuli but not for the nontarget stimuli in the multiple-object condition. Lateralized sensory responses made it difficult to determine whether an N2pc was present in the single-object condition. However, when the nontarget waveforms were subtracted from the target waveforms, eliminating any pure sensory activity, a large difference between contralateral and ipsilateral was present for the multiple-object condition but not for the single-object condition.

stimuli served as nontargets for that block. If the N2pc reflects filtering of distractors, then it should be eliminated in this condition. Indeed, little or no difference in voltage was observed between the contralateral and ipsilateral waveforms for either the targets or the nontargets. However, lateralized activity was observed in the P1 and N1 latency ranges, as would be expected given that the overall stimulus energy was highly lateralized in this condition. It is possible that a stimulus-evoked contralateral positivity would ordinarily be present in the N2 latency range for these stimuli and that this hypothetical positivity masked the negativity of the N2pc. To address this possibility, target-minus-nontarget difference waves were constructed to subtract away any sensory activity, making it possible to see any remaining target-specific activity (Figure 12.14B, bottom). These difference waves showed very little difference between the contralateral and ipsilateral sites, providing further evidence for the lack of an N2pc component in the single-object condition. In contrast, a substantial contralateral negativity was present in the corresponding subtraction from the multiple-object condition (Figure 12.14A, bottom). Thus, eliminating the distractors eliminated the N2pc component, consistent with the hypothesis that N2pc reflects filtering of the distractors.

An alternative explanation for these results is that both targets and nontargets elicited an N2pc in the single-item condition, because they consisted of lateralized stimuli with sudden onsets, which strongly capture attention under many conditions (see the review by Yantis, 1996). If the N2pc was present for both targets and nontargets, the target-minus-nontarget difference wave may have subtracted away the N2pc along with sensory-evoked activity. Thus, this experiment does not by itself provide definitive evidence for the filtering hypothesis.

The final experiment in this study was designed to rule out this alternative explanation and to provide a further test of the filtering hypothesis. The *popout condition* of this experiment was just like the multiple-object condition of the previous experiment, except that the entire array was shifted to the LVF or to the RVF (see Figure 12.15a). If a lateralized stimulus automatically attracts attention and elicits an N2pc, regardless of whether it contains any task-relevant features, then no N2pc should be visible in target-minus-nontarget difference waves in this condition. However, the voltage was more negative at contralateral than at ipsilateral scalp sites when the nontarget waveform was subtracted from the target waveform (Figure 12.15A, bottom), demonstrating that the N2pc is specific to arrays containing

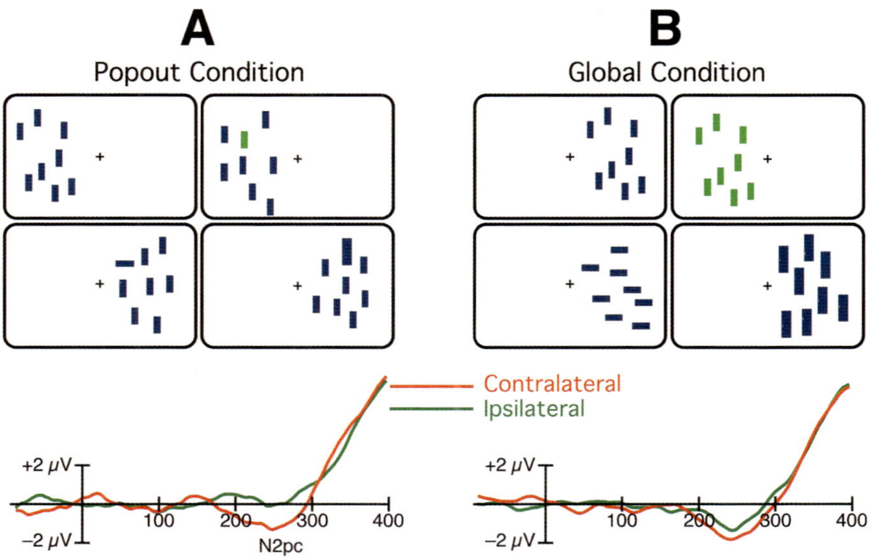

Fig. 12.15. Stimuli and grand average ERP difference waveforms recorded at posterior temporal electrode sites in the popout condition (A) and the global condition (B) of Experiment 4 in Luck and Hillyard (1994b). In both conditions, each of the four types of stimulus arrays was equiprobable. One of the popout stimuli was designated the target at the beginning of each trial block, and subjects pressed one of two buttons for each stimulus to indicate whether the target was present or absent. Because the stimulus arrays were entirely lateralized, the early sensory responses were lateralized, making it difficult to isolate the N2pc waveform by comparing contralateral and ipsilateral waveforms for the targets. To isolate the N2pc, contralateral-minus-ipsilateral waveforms were constructed (shown here). In the N2pc latency range, the contralateral response was substantially larger than the ipsilateral response in the popout condition but not in the global condition.

relevant features even when the lateralized appearance of the array might be expected to capture attention irrespective of its features. Thus, the lack of a differential N2pc component for targets relative to nontargets in the single-object condition of the previous experiment cannot easily be explained by an automatic capture of attention by both target and lateralized stimuli.

The experiment shown in Figure 12.15 also included a *global condition*, in which all of the items in each array were identical to each other (Figure 12.15B). As in the popout condition, the items were lateralized to either the LVF or the RVF. However, because the items were all identical to each other in the global condition, there was no need to filter out any of the items in the array. The target-minus-nontarget difference waveforms exhibited little or no difference between the contralateral and ipsilateral sites, providing additional evidence that the N2pc component is observed only under conditions in which filtering is necessary.

If the N2pc component reflects filtering, then it should be larger when the number of to-be-filtered distractors is increased. Two subsequent studies have provided evidence consistent with this prediction, demonstrating that the N2pc is larger when the number of distractors is increased from 1 to 3 (Luck et al., 1997b) or from 3 to 19 (Mazza et al., 2009b).

EVIDENCE AGAINST THE FILTERING HYPOTHESIS

Several studies have provided data challenging the filtering hypothesis. First, Eimer (1996) found that an N2pc component could be observed in a task in which each display contained only a target in one hemifield and a single distractor in the opposite hemifield, with no distractors near the target. This is not particularly strong evidence against the filtering hypothesis, because receptive fields at late stages of the ventral visual pathway are very large and bilateral (Gross & Mishkin, 1977). Consequently, both of the stimuli in the Eimer (1996) experiment would have appeared within the receptive field of a typical neuron in inferotemporal cortex, thus leading to the need to filter the distractor. Moreover, although a small N2pc can be observed with a single item in each hemifield, N2pc amplitude is increased if the number of distractors is increased (Luck et al., 1997b; Mazza et al., 2009b), consistent with the filtering hypothesis.

With an isolated target on one side of the display and an isolated distractor on the other side, might we expect to see the N2pc as a negativity contralateral to the distractor rather than contralateral to the target? Although inferotemporal receptive fields are typically bilateral, the response is typically larger for contralateral stimuli than for ipsilateral stimuli. If the purpose of filtering is to reduce interference with the representation of the target, then the filtering might be observed primarily in the hemisphere that most strongly codes the target. Thus, the filtering hypothesis does not predict a reversal of the N2pc when a single distractor is present in the hemifield opposite to the target.

A second challenge to the filtering hypothesis comes from studies by Mazza and colleagues (Mazza et al., 2009a, 2009b), who followed up on the experiment shown in Figure 12.13, in which the target could be either a vertical bar among horizontal distractors or a horizontal bar among vertical distractors. Because the target was defined by being surrounded by items of a different orientation, no filtering of the surrounding items was necessary and no N2pc was observed in this prior experiment. As illustrated in Figure 12.16, Mazza et al. (2009b) used an analogous design but with color rather than orientation as the popout feature. In the *constant condition*, the popout item was always red among green distractors (or the reverse for half of the subjects). In the *variable condition*, the popout item could either be red among green distractors or green among red distractors (randomly intermixed from trial to trial). Subjects were required to press one of two buttons on each trial to report whether the popout item contained a missing corner on the left side or the right side. A robust N2pc was observed in both the constant and variable conditions. This is inconsistent with the experiment shown in Figure 12.13, which was analogous to the variable condition of the Mazza et al. study but yielded no significant N2pc. This inconsistency could be explained by the fact that subjects in the Mazza et al. experiment were required to discriminate the shape of the popout item. That is, once the color had been used to locate the target item, it was presumably necessary to filter out surrounding distractors to accurately perceive the shape of this item.

A follow-up study examined this possibility by comparing *detection* and *discrimination* variants of the task (Mazza et al., 2009a). In the discrimination version, subjects were again required to report the position of the missing corner on the popout item. In the detection version, subjects were required to report only whether a popout item was present. The waveforms indicated that an N2pc component was present in all conditions, including the variable

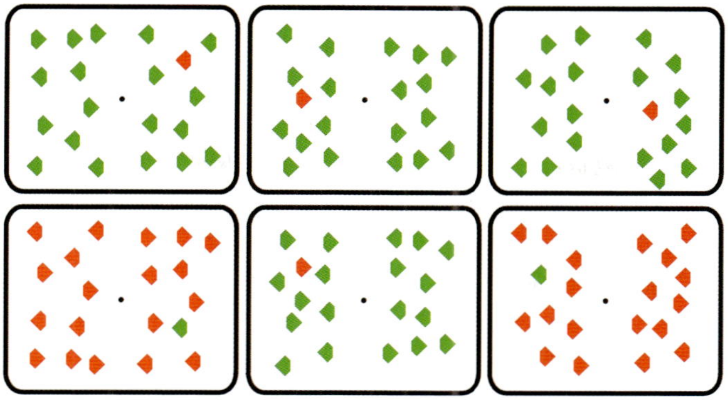

Fig. 12.16. Examples of the stimuli in the studies of Mazza and colleagues (Mazza et al., 2009a, 2009b). In the *constant* condition, the target was always red and the distractors were always green. In the *variable* condition, the target could be red among green distractors or green among red distractors.

condition with the detection task. Unfortunately, the data were not presented in a way that made it possible to determine if the N2pc was smaller for the variable condition than for the constant condition when subjects performed the detection task. Nonetheless, the presence of a substantial N2pc for the variable condition when subjects performed the detection task is a challenge to the filtering hypothesis.

Why might Mazza et al. (2009a) have found an N2pc for color popouts in the variable condition shown in Figure 12.16 even though Luck and Hillyard (1994b) did not find an N2pc for orientation popouts in the analogous experiment shown in Figure 12.13? One possibility is that the color popouts used by Mazza et al. (2009a) were substantially more salient than the orientation popouts used by Luck and Hillyard (1994b). That is, the color popouts may have triggered an automatic shift of attention, yielding an N2pc even though filtering was not needed for the task. Two previous findings make this explanation plausible. First, very salient singletons have been found to elicit an N2pc even when they are nontarget stimuli, including color singletons in the experiment shown in Figure 12.7 and motion singletons in the study of Girelli and Luck (1997). Second, a dual-task experiment demonstrated that an N2pc may be elicited by a popout target in low-load situations in which the subject has nothing better to do with attention, but the N2pc will be eliminated for the same target when a secondary task is used to minimize the availability of free attentional resources. Consequently, the presence of an N2pc for a highly salient popout does not indicate that the process reflected by the N2pc is necessary for performing the popout task.

Mazza et al. (2009a) presented two additional experiments that challenged the filtering hypothesis. Experiment 2 of this study varied the proximity of the distractors to the target. On *near* trials, two distractors were presented within approximately 1.25° of the target; on *far* trials, all distractors were at least 2.5° from the target. No significant increase in N2pc amplitude was observed for the near trials (although the later SPCN appeared to be increased). However, the distance between the target and distractors in the far condition may have been close enough that multiple items were still present inside of individual V4 receptive fields. Specifically, Motter (2009) has shown that almost all V4 receptive fields are at least 2.5° in diameter at the eccentricities used in the Mazza et al. (2009a) study, and the 2.5° spacing in the far condition was therefore insufficient to ensure that no distractors were in the same V4 receptive fields as the target. Receptive fields are even larger within the lateral occipital complex, and multiple distractors would have been present within the receptive fields of these neurons in all conditions of this experiment. Thus, equivalent filtering may have been necessary in both the near and far trials. To provide a valid test of the filtering hypothesis, it would be necessary to test a larger range of target-distractor distances.

In Experiment 3 of Mazza et al. (2009a), a *heterogeneous* condition was included, in which two different distractor colors were present in each array. Distractor heterogeneity is well known to increase the difficulty of visual search (see, e.g., Duncan & Humphreys, 1989), but the N2pc component was not found to be enlarged in this condition. However, distractor heterogeneity is thought to influence the difficulty of finding the target among the distractors, not the level of interference produced by the distractors once attention has been directed to the target. Indeed, Mazza et al. (2009a) found that N2pc onset latency was substantially increased in the heterogeneous condition. Thus, this experiment did not really test the filtering hypothesis.

Although it may be possible to explain away the findings of Mazza et al. (2009a), these explanations run the risk of making the filtering hypothesis unfalsifiable. The main difficulty in assessing the evidence for and against the filtering hypothesis is that we do not have an independent means of assessing whether the hypothesized filtering process was present or absent under a given set of experimental conditions. As discussed earlier in this chapter, this is a general problem in linking ERP components with specific cognitive or neural processes: unless we can make very strong predictions about when a particular process will be present or absent, it is difficult to provide very strong tests of the hypothesis that an ERP component reflects this process. Unfortunately, we are not yet at the point where we can know with absolute certainty when filtering will be present or absent, and this makes it difficult to conduct decisive tests of the N2pc filtering hypothesis.

FILTERING AND THE N2PC AND P_D COMPONENTS

Another challenge to the filtering hypothesis was provided by Hickey et al. (2009), who used the paradigm shown in Figure 12.17 to separately assess target processing and distractor suppression. Each stimulus array contained two objects, one located on the vertical midline and the other located to the left or right of fixation. One of the objects was a square or a diamond, and the other was a short or long horizontal line. In Experiment 1 of this study, the subjects' task was to press one of two buttons to indicate whether a square or a diamond was present; the horizontal line was always a distractor. Because one of the two objects was on the vertical midline, any ERP activity corresponding to this line was neither ipsilateral nor contralateral to a given electrode site, and any lateralized ERP activity must reflect the processing of the lateral stimulus (plus any interactions between the midline and lateral stimuli). Thus, this design made it possible to isolate lateralized processing of the target (when the target was the lateralized stimulus and the distractor was on the midline) and lateralized processing of the distractor (when the distractor was the lateralized stimulus and the target was on the midline). Hickey et al. reasoned that filtering should be seen in the contralateral-versus-ipsilateral comparisons when the distractor was lateralized, whereas target processing should be seen in these comparisons when the target was lateralized.

When the target was lateralized, the presence of lateralized sensory components made it difficult to discern whether an N2pc was present or absent (Figure 12.17A, bottom), just as in the single-item condition of the experiment shown in Figure 12.14. To minimize this sort of lateralized sensory response when the distractor item was lateralized, the distractor was presented at the same luminance as the background (the target, in contrast, was substantially brighter than the background). This substantially reduces P1 and N1 amplitude, which are primarily sensitive to information represented by the magnocellular pathway (see, e.g., Butler et al., 2007). As shown in Figure 12.17B (middle), the P1 and N1 waves were not lateralized when the distractor was lateralized, and the only lateralized effect was a greater positivity at contralateral relative to ipsilateral sites. Thus, distractors appear to elicit a contralateral positivity rather than a contralateral negativity in the N2 latency range. Hickey et al. (2009) called this effect P_D (distractor positivity), and they speculated that it reflects an active suppression process that is directed to distractors. Experiment 2 of this study demonstrated that the P_D was eliminated when the task was changed so that subjects simply had to detect the presence of a square or a diamond rather than discriminating its form. Hickey et al. (2009) speculated that the elimination of the P_D occurred because distractor suppression is unnecessary for a simple detection task.

Additional evidence that the P_D component reflects suppression was obtained in two recent studies examining the capture of attention by salient color popouts (Eimer & Kiss, 2008; Sawaki & Luck, 2010). When subjects were strongly motivated to restrict attention to form-defined targets, salient irrelevant color popouts did not yield an N2pc component, but instead yielded a P_D component. Other studies suggest that salient popouts are actively suppressed under these conditions, so the finding of a P_D component is consistent with the hypothesis that P_D reflects a suppressive mechanism.

When both the target and the distractors are lateralized, as in a typical N2pc paradigm, the N2pc and P_D components will presumably combine together. For example, if a visual search display consists of a target in one visual field and a single distractor in the opposite visual field, then the difference in amplitude between the electrodes contralateral

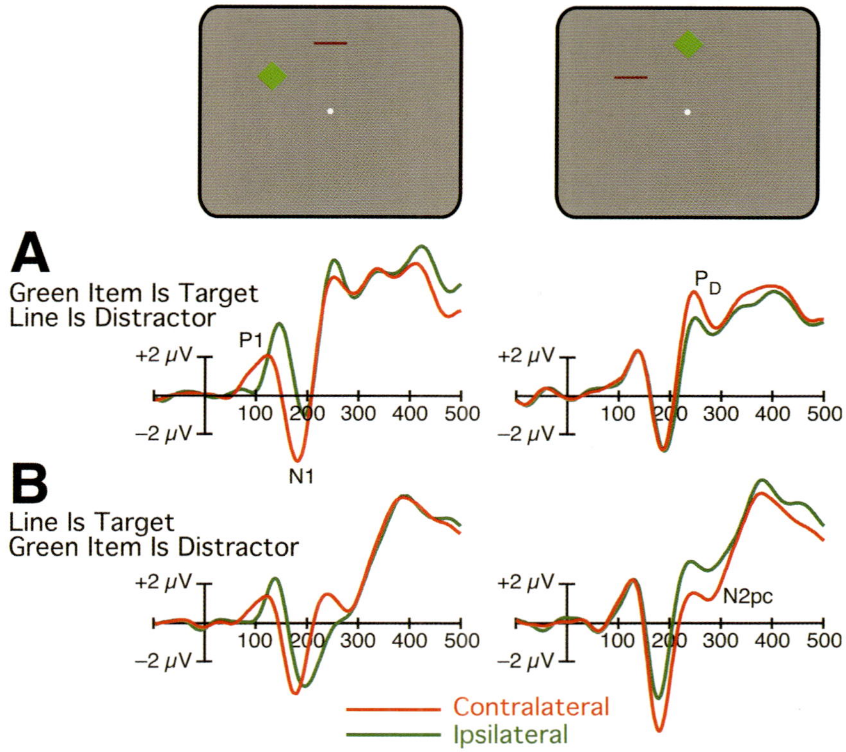

Fig. 12.17. Example stimuli and grand average ERP waveforms from the study of Hickey et al. (2009). Each array contained a bright green square or diamond along with a short or long red line that was isoluminant with the background. When the green stimulus was lateralized, lateralized sensory responses were observed (left column). When the red stimulus was lateralized, no lateralized sensory activity was observed in the first 200 ms, making it possible to attribute any later lateralized activity to top-down processing. When the green item was the target (A), the ERP was more positive contralateral to the red distractor item than ipsilateral to this item. This is the distractor positivity (P_D). When the red item was the target (B), a contralateral negativity (N2pc) was elicited by the red item. Waveforms courtesy of John McDonald.

versus ipsilateral to the target will be composed of both the N2pc elicited by the target and the P_D elicited by the distractor. Because the P_D is positive contralateral to the distractor, it will be negative contralateral to the target in this situation, and the N2pc and P_D will sum together. This will create a larger negativity contralateral to the target than would be present if only the target-related N2pc activity were present. However, if multiple identical distractors are present on both sides of the display, then the P_D will presumably be elicited by the distractors on both sides and will largely cancel out when the N2pc is isolated by means of a contralateral-minus-ipsilateral difference wave. If the display contains multiple identical distractors, a target that pops out on one side, and a distinctive distractor that pops out on the other side, as in the experiments shown in Figures 12.3 and 12.4, the distinctive distractor may require more suppression than the other distractors. In this case, a substantial P_D may be present contralateral to the distinctive distractor and therefore ipsilateral to the target, and the P_D will again sum together with the target-elicited N2pc component.

In addition to identifying the P_D component, Hickey et al. (2009) tested whether the N2pc component reflects target processing or distractor suppression. In Experiment 4 of this study, the short or long line was the target and the square or diamond was the distractor. The line was again isoluminant with the background, minimizing any lateralized P1 and N1 activity. When the line target was presented laterally, an N2pc component was observed (Figure 12.17). Thus, a lateralized line elicits a contralateral positivity (P_D) when it is a distractor but a contralateral negativity (N2pc) when it is a target.

Because the N2pc component was observed contralateral to the target when the distractor was on the vertical midline, Hickey et al. (2009) concluded that N2pc reflects a process that enhances the cortical representation of the target rather than a process that filters distractors. However, this conclusion was based on the assumption that suppression of a distractor presented on the midline will not lead to lateralized neural activity. This is not necessarily true. As was discussed in the context of Eimer's (1996) finding that the N2pc was present contralateral to the target when the only distractor was on the opposite side of the display, a distractor relatively far away from the target may still be present within the receptive field of a neuron that codes the target stimulus, and the filtering may therefore be carried out primarily in the hemisphere contralateral to the target. To rigorously test the filtering hypothesis, it would be necessary to present the isoluminant line stimulus at a lateral location without a distractor on the vertical midline. If the N2pc was still present in the complete absence of a distractor, then this would conclusively falsify the filtering hypothesis.

AN OVERALL APPRAISAL OF THE FILTERING HYPOTHESIS

As this discussion of the filtering hypothesis illustrates, it is quite difficult to link an ERP component with a specific cognitive or neural process. Unless we have an independent means of determining when a given process is present or absent, it is difficult to provide definitive tests of the relationship between an ERP component and that process. We do not have an independent means of determining when filtering is present or absent, and we do not have a sufficiently well established theory of filtering to make strong predictions about when filtering should be present or absent. Indeed, this is why it would be extremely valuable to have an ERP component that has been definitively linked to filtering.

What, then, can we conclude at this time about the relationship between the N2pc component and filtering? The most direct experiments have addressed the effects of the presence or absence of distractors on the N2pc component, and these experiments have found that (1) the N2pc is present when distractors are present (Luck & Hillyard, 1994a, 1994b); (2) the N2pc is reduced or eliminated when the distractors are removed (Experiments 3 and 4 in Luck & Hillyard, 1994b); and (3) the N2pc becomes larger when the number of distractors is increased (Experiments 1 and 2 in Luck et al., 1997b; Experiment 1 in Mazza et al., 2009b). These findings support the hypothesis that the N2pc component reflects filtering of distractors. The main evidence against this hypothesis is that the N2pc is observed contralateral to the target, even when the only distractor is on the midline or in the opposite visual field (Eimer, 1996; Hickey et al., 2009).

Although it is possible to explain away some of these results, a somewhat broader hypothesis could accommodate all of the existing results. Specifically, the N2pc may reflect a process that is applied to the representation of the target stimulus, but only under conditions of competition from simultaneous distractors. Indeed, Luck and Hillyard (1994a) noted that "the filtering process may be accomplished by activating the neural representation of the object undergoing identification rather than by inhibiting the distractor items, as long as this selection process is assumed to be necessary only when competing distractor items are present" (p. 1010). However, the term *filtering* tends to imply an operation that is applied to the distractors, and it may be more appropriate to say that the N2pc component reflects a process that resolves competition between the attended item and the distractor items. We call this the *competition resolution* hypothesis of the N2pc component. This hypothesis is explicitly agnostic about whether the competition resolution hypothesis acts directly on the representation of the target, directly on the representations of the distractors, or on interneurons that mediate the competition between the target and distractor representations. However, it proposes that the computational function of the process is to enhance the representation of the target by minimizing interference from simultaneous distractor items. This hypothesis appears to be fully compatible with all existing evidence.

LINKS TO THEORIES OF ATTENTION

The competition resolution hypothesis is closely related to the *biased competition model* of Desimone and Duncan (1995), which specifies that the main role of attention is to resolve competition between neural representations. It is also closely related to Treisman's *feature integration theory* (Treisman, 1988, 1999; Treisman & Gelade, 1980), which proposes that attention is needed to prevent features from different objects from combining together. Luck et al. (1997b) proposed an *ambiguity resolution theory*, which combines elements of these two theories. The ambiguity resolution theory proposes that the output of a neuron becomes ambiguous when multiple objects are present in the neuron's receptive field, because it is not clear which of the objects contains the feature that is represented by the neuron. As noted previously, if a red triangle and a blue circle are both present in the receptive field of a red-selective neuron, the neuron will fire at a high rate to indicate the presence of the color red, but it will not be clear whether the triangle or the circle is the red item. Attention can resolve this ambiguity by limiting processing to a single object at any time; when attention is focused on a single object, all of the remaining neural activity reflects the features present in the attended object. This is the type of attentional mechanism that the N2pc component appears to reflect.

A key prediction of feature integration theory and the ambiguity resolution theory is that this variety of attention should be unnecessary when the target is defined by the presence of a single feature (although other types of attentional mechanisms will obviously be necessary to implement the rules of the task). If, for example, the observer's task is to detect the color red, then the output of a red-selective neuron will be sufficient to perform this task even if non-red items are present along with a red item in a given neuron's receptive field. Consistent with this prediction, the N2pc component is larger for tasks that require subjects to combine multiple features than for tasks that require the mere detection of a simple feature (Luck & Ford, 1998; Luck et al., 1997b). However, a substantial N2pc is often present for simple feature-detection tasks (see, e.g., Figures 12.5, 12.14A, and 12.15A).

How can the presence of a substantial N2pc for simple feature targets be reconciled with the proposal that the variety of attention reflected by the N2pc component is not necessary for the detection of simple features? The answer to this question provides an important principle that is often overlooked in studies that attempt to link ERP components with neural or cognitive processes. Specifically, a given neural or cognitive process may be active in a laboratory experiment even though it is not strictly necessary for task performance. Most ERP experiments are long and boring, and they do not typically push subjects to the limits of their abilities. Consequently, a given process may occur not because it is necessary for task performance, but because there is no disincentive for using this process and because this process may ordinarily be necessary in natural task situations. In the case of the process reflected by the N2pc component, it may not be necessary for the detection of simple features, but most feature detection tasks are extremely simple and provide no disincentive for allocating attention to the target. Subjects may focus attention on the target because they have nothing else to do or because most real-world situations require the focusing of attention onto relevant objects.

To test this hypothesis, Luck and Ford (1998) performed a dual-task experiment in which subjects performed a visual search task simultaneously with a demanding task that required determining whether a degraded letter presented at fixation was a consonant or a vowel. In one set of conditions, the visual search task required subjects to merely report the presence or absence of a specific color in an array of colored squares. When this task was performed alone, an N2pc component was observed contralateral to the target color. However, when this task was performed simultaneously with the demanding central task, the N2pc component was eliminated. Subjects were still highly accurate at performing the visual search task, indicating that the process reflected by the N2pc component is not necessary for the detection of simple features. In a second set of conditions, the visual search task required the subject to report a conjunction of color and orientation. When this task was performed alone, a large N2pc component was observed. When it was performed simultaneously with the central task, the N2pc component was delayed rather than being eliminated. Thus, subjects could not perform a conjunction task without the process reflected by the N2pc component.

These results provide further support for the hypothesis that the N2pc component reflects a competition resolution process. They also provide a cautionary note about experiments that attempt to link an ERP component with a neural or cognitive process. That is, an ERP component that reflects a given neural or cognitive process may be observed under conditions in which that process is not necessary. Thus, even when a good theory of a process already exists, caution is needed in using the theory to predict the absence of the component under conditions in which the theory specifies that the process is not necessary.

References

Bacon, W. F., & Egeth, H. E. (1994). Overriding stimulus-driven attentional capture. *Perception & Psychophysics, 55*, 485–496.

Busch, N. A., Frund, I., & Herrmann, C. S. (2009). Electrophysiological evidence for different types of change detection and change blindness. *Journal of Cognitive Neuroscience, 22*, 1852–1869.

Buschman, T. J., & Miller, E. K. (2007). Top-down versus bottom-up control of attention in the prefrontal and posterior parietal cortices. *Science, 315*, 1860–1862.

Butler, P. D., Martinez, A., Foxe, J. J., Kim, D., Zemon, V., Silipo, G., et al. (2007). Subcortical visual dysfunction in schizophrenia drives secondary cortical impairments. *Brain, 130*, 417–430.

Carrasco, M., Penpeci-Talgar, C., & Eckstein, M. (2000). Spatial covert attention increases contrast sensitivity across the CSF: Support for signal enhancement. *Vision Research, 40*, 1203–1215.
Chelazzi, L., Duncan, J., Miller, E. K., & Desimone, R. (1998). Responses of neurons in inferior temporal cortex during memory-guided visual search. *Journal of Neurophysiology, 80*, 2918–2940.
Chelazzi, L., Miller, E. K., Duncan, J., & Desimone, R. (1993). A neural basis for visual search in inferior temporal cortex. *Nature, 363*, 345–347.
Chelazzi, L., Miller, E. K., Duncan, J., & Desimone, R. (2001). Responses of neurons in macaque area V4 during memory-guided visual search. *Cerebral Cortex, 11*, 761–772.
Cohen, J. Y., Heitz, R. P., Schall, J. D., & Woodman, G. F. (2009). On the origin of event-related potentials indexing covert attentional selection during visual search. *Journal of Neurophysiology, 101*, 1699–1704.
Conci, M., Gramann, K., Muller, H. J., & Elliott, M. A. (2006). Electrophysiological correlates of similarity-based interference during detection of visual forms. *Journal of Cognitive Neuroscience, 18*, 880–888.
De Beaumont, L., Brisson, B., Lassonde, M., & Jolicoeur, P. (2007). Long-term electrophysiological changes in athletes with a history of multiple concussions. *Brain Injury, 21*, 631–644.
Dell'Acqua, R., Pesciarelli, F., Jolicoeur, P., Eimer, M., & Peressotti, F. (2007). The interdependence of spatial attention and lexical access as revealed by early asymmetries in occipito-parietal ERP activity. *Psychophysiology, 44*, 436–443.
Dell'Acqua, R., Sessa, P., Toffanin, P., Luria, R., & Jolicoeur, P. (2010). Orienting attention to objects in visual short-term memory. *Neuropsychologia, 48*, 419–428.
Desimone, R., & Duncan, J. (1995). Neural mechanisms of selective visual attention. *Annual Review of Neuroscience, 18*, 193–222.
Deubel, H., & Schneider, W. X. (1996). Saccade target selection and object recognition: Evidence for a common attentional mechanism. *Vision Research, 36*, 1827–1837.
Di Lollo, V., Enns, J. T., & Rensink, R. A. (2000). Competition for consciousness among visual events: The psychophysics of reentrant visual processes. *Journal of Experimental Psychology: General, 129*, 481–507.
Doi, H., Ueda, K., & Shinohara, K. (2009). Neural correlates of the stare-in-the-crowd effect. *Neuropsychologia, 47*, 1053–1060.
Duncan, J., & Humphreys, G. (1989). Visual search and stimulus similarity. *Psychological Review, 96*, 433–458.
Eimer, M. (1996). The N2pc component as an indicator of attentional selectivity. *Electroencephalography and Clinical Neurophysiology, 99*, 225–234.
Eimer, M., & Kiss, M. (2008). Involuntary attentional capture is determined by task set: Evidence from event-related brain potentials. *Journal of Cognitive Neuroscience, 208*, 1423–1433.
Eimer, M., & Mazza, V. (2005). Electrophysiological correlates of change detection. *Psychophysiology, 42*, 328–342.
Eriksen, C. W., & St. James, J. D. (1986). Visual attention within and around the field of focal attention: A zoom lens model. *Perception & Psychophysics, 40*, 225–240.
Eriksen, C. W., & Yeh, Y. Y. (1985). Allocation of attention in the visual field. *Journal of Experimental Psychology: Human Perception and Performance, 11*, 583–597.
Feldman, J. (1985). Connectionist models and parallelism in high level vision. *Computer Vision, Graphics, and Image Processing, 31*, 178–200.
Fenker, D. B., Heipertz, D., Boehler, C. N., Schoenfeld, M. A., Noesselt, T., Heinze, H. J., et al. (2010). Mandatory processing of irrelevant fearful face features in visual search. *Journal of Cognitive Neuroscience, 22*, 2926–2938.
Folstein, J. R., & Van Petten, C. (2008). Influence of cognitive control and mismatch on the N2 component of the ERP: A review. *Psychophysiology, 45*, 152–170.
Fuggetta, G., Pavone, E. F., Walsh, V., Kiss, M., & Eimer, M. (2006). Cortico-cortical interactions in spatial attention: A combined ERP/TMS study. *Journal of Neurophysiology, 95*, 3277–3280.
Gehring, W. J., Gratton, G., Coles, M., & Donchin, E. (1992). Probability effects on stimulus evaluation and response processes. *Journal of Experimental Psychology: Human Perception and Performance, 18*, 198–216.
Girelli, M., & Luck, S. J. (1997). Are the same attentional mechanisms used to detect visual search targets defined by color, orientation, and motion? *Journal of Cognitive Neuroscience, 9*, 238–253.
Gross, C. G., & Mishkin, M. (1977). The neural basis of stimulus equivalence across retinal translation. In S. Harnad, R. W. Doty, L. Goldstein, J. Jaynes, & G. Krauthamer (Eds.), *Lateralization in the nervous system* (pp. 109–122). New York: Academic Press.
Hickey, C., Di Lollo, V., & McDonald, J. J. (2009). Electrophysiological indices of target and distractor processing in visual search. *Journal of Cognitive Neuroscience, 21*, 760–775.
Hilimire, M. R., Mounts, J. R., Parks, N. A., & Corballis, P. M. (2009). Competitive interaction degrades target selection: An ERP study. *Psychophysiology, 46*, 1080–1089.
Hirai, M., & Hiraki, K. (2006). Visual search for biological motion: An event-related potential study. *Neuroscience Letters, 403*, 299–304.
Hoffman, J. E., & Subramaniam, B. (1995). The role of visual attention in saccadic eye movements. *Perception & Psychophysics, 57*, 787–795.
Hollingworth, A., & Luck, S. J. (2009). The role of visual working memory (VWM) in the control of gaze during visual search. *Attention, Perception, & Psychophysics, 71*, 936–949.
Holmes, A., Bradley, B. P., Kragh Nielsen, M., & Mogg, K. (2009). Attentional selectivity for emotional faces: Evidence from human electrophysiology. *Psychophysiology, 46*, 62–68.
Hopf, J.-M., Boelmans, K., Schoenfeld, A. M., Luck, S. J., & Heinze, H.-J. (2004). Attention to features precedes attention to locations in visual search: Evidence from electromagnetic brain responses in humans. *Journal of Neuroscience, 24*, 1822–1832.
Hopf, J.-M., Luck, S. J., Boelmans, K., Schoenfeld, M. A., Boehler, N., Rieger, J., et al. (2006). The neural site of attention matches the spatial scale of perception. *Journal of Neuroscience, 26*, 3532–3540.
Hopf, J.-M., Luck, S. J., Girelli, M., Hagner, T., Mangun, G. R., Scheich, H., et al. (2000). Neural sources of focused attention in visual search. *Cerebral Cortex, 10*, 1233–1241.
Hyun, J.-S., Woodman, G. F., & Luck, S. J. (2009a). The role of attention in the binding of surface features to locations. *Visual Cognition, 17*, 10–24.
Hyun, J.-S., Woodman, G. F., Vogel, E. K., Hollingworth, A., & Luck, S. J. (2009b). The comparison of visual working

memory representations with perceptual inputs. *Journal of Experimental Psychology: Human Perception and Performance, 35*, 1140–1160.

Kiss, M., Driver, J., & Eimer, M. (2009). Reward priority of visual target singletons modulates event-related potential signatures of attentional selection. *Psychological Science, 20*, 245–251.

Kiss, M., Van Velzen, J., & Eimer, M. (2008). The N2pc component and its links to attention shifts and spatially selective visual processing. *Psychophysiology, 45*, 240–249.

Klyubin, A. S., Polani, D., & Nehaniv, C. L. (2004). Organization of the information flow in the perception-action loop of evolved agent. In R. S. Zebulum, D. Gwaltney, G. Hornby, D. Keymeulen, J. Lohn, & A. Stoica (Eds.), *Proceedings of 2004 NASA/DoD Conference on Evolvable Hardware* (pp. 177–180). Los Alamitos, CA: IEEE Computer Society.

Knierim, J. J., & Van Essen, D. C. (1992). Neuronal responses to static texture patterns in area V1 of the alert macaque monkey. *Journal of Neurophysiology, 67*, 961–980.

LaBerge, D., Brown, V., Carter, M., & Bash, D. (1991). Reducing the effects of adjacent distractors by narrowing attention. *Journal of Experimental Psychology: Human Perception and Performance, 17*, 65–76.

Lorenzo-Lopez, L., Amenedo, E., & Cadaveira, F. (2008). Feature processing during visual search in normal aging: Electrophysiological evidence. *Neurobiology of Aging, 29*, 1101–1110.

Luck, S. J. (2005). *An introduction to the event-related potential technique*. Cambridge, MA: MIT Press.

Luck, S. J. (2009). The spatiotemporal dynamics of visual-spatial attention. In F. Aboitiz & D. Cosmelli (Eds.), *From attention to goal-directed behavior: Neurodynamical, methodological, and clinical trends* (pp. 51–66). Berlin: Springer-Verlag.

Luck, S. J., Chelazzi, L., Hillyard, S. A., & Desimone, R. (1997a). Neural mechanisms of spatial selective attention in areas V1, V2, and V4 of macaque visual cortex. *Journal of Neurophysiology, 77*, 24–42.

Luck, S. J., Fan, S., & Hillyard, S. A. (1993). Attention-related modulation of sensory-evoked brain activity in a visual search task. *Journal of Cognitive Neuroscience, 5*, 188–195.

Luck, S. J., & Ford, M. A. (1998). On the role of selective attention in visual perception. *Proceedings of the National Academy of Sciences USA, 95*, 825–830.

Luck, S. J., Fuller, R. L., Braun, E. L., Robinson, B., Summerfelt, A., & Gold, J. M. (2006). The speed of visual attention in schizophrenia: Electrophysiological and behavioral evidence. *Schizophrenia Research, 85*, 174–195.

Luck, S. J., Girelli, M., McDermott, M. T., & Ford, M. A. (1997b). Bridging the gap between monkey neurophysiology and human perception: An ambiguity resolution theory of visual selective attention. *Cognitive Psychology, 33*, 64–87.

Luck, S. J., & Gold, J. M. (2008). The construct of attention in schizophrenia. *Biological Psychiatry, 64*, 34–39.

Luck, S. J., & Hillyard, S. A. (1990). Electrophysiological evidence for parallel and serial processing during visual search. *Perception & Psychophysics, 48*, 603–617.

Luck, S. J., & Hillyard, S. A. (1994a). Electrophysiological correlates of feature analysis during visual search. *Psychophysiology, 31*, 291–308.

Luck, S. J., & Hillyard, S. A. (1994b). Spatial filtering during visual search: Evidence from human electrophysiology. *Journal of Experimental Psychology: Human Perception and Performance, 20*, 1000–1014.

Luck, S. J., & Hillyard, S. A. (1995). The role of attention in feature detection and conjunction discrimination: An electrophysiological analysis. *International Journal of Neuroscience, 80*, 281–297.

Luck, S. J., Hillyard, S. A., Mouloua, M., Woldorff, M. G., Clark, V. P., & Hawkins, H. L. (1994). Effects of spatial cuing on luminance detectability: Psychophysical and electrophysiological evidence for early selection. *Journal of Experimental Psychology: Human Perception and Performance, 20*, 887–904.

Luck, S. J., & Vecera, S. P. (2002). Attention. In S. Yantis (Ed.), *Stevens' handbook of experimental psychology: Vol. 1: Sensation and perception* (3rd ed., pp. 235–286). New York: Wiley.

Maljkovic, V., & Nakayama, K. (1993). Priming of pop-out: An example of implicit short-term memory. *Society for Neuroscience Abstracts*, 19, 439.

Mazza, V., Turatto, M., & Caramazza, A. (2009a). An electrophysiological assessment of distractor suppression in visual search tasks. *Psychophysiology, 46*, 771–775.

Mazza, V., Turatto, M., & Caramazza, A. (2009b). Attention selection, distractor suppression and N2pc. *Cortex, 45*, 879–890.

Mazza, V., Turatto, M., Umiltá, C., & Eimer, M. (2007). Attentional selection and identification of visual objects are reflected by distinct electrophysiological responses. *Experimental Brain Research, 181*, 531–536.

McCollough, A. W., Machizawa, M. G., & Vogel, E. K. (2007). Electrophysiological measures of maintaining representations in visual working memory. *Cortex, 43*, 77–94.

Meeter, M., & Olivers, C. N. L. (2006). Intertrial priming stemming from ambiguity: A new account of priming in visual search. *Visual Cognition, 13*, 1–28.

Moran, J., & Desimone, R. (1985). Selective attention gates visual processing in the extrastriate cortex. *Science, 229*, 782–784.

Motter, B. C. (2009). Central V4 receptive fields are scaled by the V1 cortical magnification and correspond to a constant-sized sampling of the V1 surface. *Journal of Neuroscience, 29*, 5749–5757.

Oostenveld, R., Stegeman, D. F., Praamstra, P., & van Oosterom, A. (2003). Brain symmetry and topographic analysis of lateralized event-related potentials. *Clinical Neurophysiology, 114*, 1194–1202.

Pashler, H. (1988). Cross-dimensional interaction and texture segregation. *Perception and Psychophysiology, 43*, 307–318.

Perron, R., Lefebvre, C., Robitaille, N., Brisson, B., Gosselin, F., Arguin, M., et al. (2009). Attentional and anatomical considerations for the representation of simple stimuli in visual short-term memory: Evidence from human electrophysiology. *Psychological Research, 73*, 222–232.

Peterson, M. S., Kramer, A. F., & Irwin, D. E. (2004). Covert shifts of attention precede involuntary eye movements. *Perception & Psychophysics, 66*, 398–405.

Praamstra, P. (2006). Prior information of stimulus location: Effects on ERP measures of visual selection and response selection. *Brain Research, 1072*, 153–160.

Praamstra, P. (2007). Do's and don'ts with lateralized event-related brain potentials. *Journal of Experimental Psychology: Human Perception and Performance, 33*, 497–502.

Praamstra, P., & Oostenveld, R. (2003). Attention and movement-related motor cortex activation: A high-density EEG study of spatial stimulus–response compatibility. *Cognitive Brain Research, 16*, 309–322.

Praamstra, P., & Plat, F. M. (2001). Failed suppression of direct visuomotor activation in Parkinson's disease. *Journal of Cognitive Neuroscience, 13,* 31–43.

Prime, D. J., & Jolicoeur, P. (2010). Mental rotation requires visual ahort-term memory: Evidence from human electric cortical activity. *Journal of Cognitive Neuroscience, 22,* 2437–2446.

Pritchard, W. S., Shappell, S. A., & Brandt, M. E. (1991). Psychophysiology of N200/N400: A review and classification scheme. In J. R. Jennings & P. K. Ackles (Eds.), *Advances in psychophysiology* (pp. 43–106). London: Jessica Kingsley.

Renault, B., Ragot, R., Lesevre, N., & Remond, A. (1982). Onset and offset of brain events as indices of mental chronometry. *Science, 215,* 1413–1415.

Reynolds, J. H., Chelazzi, L., & Desimone, R. (1999). Competitive mechanisms subserve attention in macaque areas V2 and V4. *Journal of Neuroscience, 19,* 1736–1753.

Richards, J. E. (2003). Cortical sources of event-related potentials in the prosaccade and antisaccade task. *Psychophysiology, 40,* 878–894.

Robitaille, N., Grimault, S., Jolicoeur, P. (2009). Bilateral parietal and contralateral responses during maintenance of unilaterally-encoded objects in visual short-term memory: Evidence from magnetoencephalography. *Psychophysiology, 46,* 1090–1099.

Ruchkin, D. S., Grafman, J., Cameron, K., Berndt, R. S. (2003). Working memory retention systems: A state of activated long-term memory. *Behavioral and Brain Sciences, 26,* 709–728.

Sawaki, R., Luck, S. J. (2010). Capture versus suppression of attention by salient singletons: Electrophysiological evidence for an automatic attend-to-me signal. *Attention, Perception, & Psychophysics, 72,* 1455–1470.

Schankin, A., Wascher, E. (2008). Unvoluntary attentional capture in change blindness. *Psychophysiology, 45,* 742–750.

Schiff, S., Mapelli, D., Vallesi, A., Orsato, R., Gatta, A., Umilta, C., et al. (2006). Top-down and bottom-up processes in the extrastriate cortex of cirrhotic patients: An ERP study. *Clinical Neurophysiology, 117,* 1728–1736.

Simson, R., Vaughan, H. G., Ritter, W. (1977). The scalp topography of potentials in auditory and visual discrimination tasks. *Electroencephalography and Clinical Neurophysiology, 42,* 528–535.

Treisman, A. (1986). Features and objects in visual processing. *Scientific American, 255,* 114B–125B.

Treisman, A. (1988). Features and objects: The fourteenth Bartlett memorial lecture. *Quarterly Journal of Experimental Psychology, 40,* 201–237.

Treisman, A. (1999). Solutions to the binding problem: Progress through controversy and convergence. *Neuron, 24,* 105–110.

Treisman, A. M., Gelade, G. (1980). A feature-integration theory of attention. *Cognitive Psychology, 12,* 97–136.

Treisman, A., Gormican, S. (1988). Feature analysis in early vision: Evidence from search asymmetries. *Psychological Review, 95,* 15–48.

Treisman, A., Souther, J. (1985). Search asymmetry: A diagnostic for preattentive processing of separable features. *Journal of Experimental Psychology: General, 114,* 285–310.

Treue, S., Maunsell, J. H. R. (1996). Attentional modulation of visual motion processing in cortical areas MT and MST. *Nature, 382,* 539–541.

Vogel, E. K., Machizawa, M. G. (2004). Neural activity predicts individual differences in visual working memory capacity. *Nature, 428,* 748–751.

West, R., Alain, C. (1999). Event-related neural activity associated with the Stroop task. *Cognitive Brain Research, 8,* 157–164.

West, R., Alain, C. (2000). Effects of task context and fluctuations of attention on the neural activity supporting performance of the Stroop task. *Brain Research, 873,* 102–111.

Wolber, M., Wascher, E. (2003). Visual search strategies are indexed by event-related lateralizations of the EEG. *Biological Psychology, 63,* 79–100.

Wolfe, J. M. (1994). Guided search 2.0: A revised model of visual search. *Psychonomic Bulletin & Review, 1,* 202–238.

Wolfe, J. M., Cave, K. R., Franzel, S. L. (1989). Guided search: An alternative to the feature integration model for visual search. *Journal of Experimental Psychology: Human Perception and Performance, 15,* 419–433.

Woodman, G. F., Luck, S. J. (1999). Electrophysiological measurement of rapid shifts of attention during visual search. *Nature, 400,* 867–869.

Woodman, G. F., Luck, S. J. (2003). Serial deployment of attention during visual search. *Journal of Experimental Psychology: Human Perception and Performance, 29,* 121–138.

Woodman, G. F., Luck, S. J., Schall, J. D. (2007). The role of working memory representations in the control of attention. *Cerebral Cortex, 17,* i118–i124.

Woodman, G. F., Vogel, E. K., Luck, S. J. (2001). Visual search remains efficient when visual working memory is full. *Psychological Science, 12,* 219–224.

Xu, Y., Chun, M. M. (2006). Dissociable neural mechanisms supporting visual short-term memory for objects. *Nature, 440,* 91–95.

Yantis, S. (1996). Attentional capture in vision. In A. F. Kramer, M. G. H. Coles, & G. D. Logan (Eds.), *Converging operations in the study of selective attention* (pp. 45–76). Washington, DC: American Psychological Association.

Yeung, N., Cohen, J. D., & Botvinick, M. M. (2004). The neural basis of error detection: Conflict monitoring and the error-related negativity. *Psychological Review, 111,* 931–959.

CHAPTER

13

What ERPs Can Tell Us about Working Memory

Veronica B. Perez *and* Edward K. Vogel

Abstract

Our ability to perform a broad range of everyday cognitive tasks is thought to depend heavily upon the operation of the working memory system, which allows us to temporarily store information in the mind so that it may be manipulated or acted upon. Many cognitive processes that require information to be held in an online state are thought to utilize this memory system as a form of "mental workspace" to perform their requisite operations. This chapter discusses the definition of working memory, measuring visual working memory (VWM), event-related potential (ERP) studies of working memory, contralateral delay activity (CDA), and using the CDA to examine working memory.

Keywords: working memory, visual working memory, event-related potentials, contralateral delay activity

What Is Working Memory?

Our ability to perform a broad range of everyday cognitive tasks is thought to depend heavily upon the operation of the *working memory* system, which allows us to temporarily store information in the mind so that it may be manipulated or acted upon (e.g., Baddeley & Hitch, 1974; Cowan, 2001). Many cognitive processes that require information to be held in an online state are thought to utilize this memory system as a form of "mental workspace" to perform their requisite operations. A fundamental characteristic of working memory is that it is severely limited in its capacity to maintain information. In the visual working memory (VWM) domain, this capacity limit is thought to be approximately three or four simultaneous items (Cowan, 2001; Luck & Vogel, 1997; Sperling, 1960; Vogel et al., 2001). However, this capacity estimate actually varies substantially across individuals, ranging from as few as 1.5 objects up to about 6 (Vogel & Awh, 2008; Vogel & Machizawa, 2004). This intersubject variability in memory capacity appears to reflect an important and stable cognitive trait of the individual because it is strongly predictive of performance on a host of high-level aptitude measures. Individuals with high memory capacity tend to perform better on measures of fluid intelligence, abstract reasoning, and reading comprehension, and they tend to have higher grade point averages in school than individuals with low memory capacity (Cowan et al., 2006; Daneman & Carpenter, 1980; Engle et al., 1999; Kane, 2001; Kyllonen & Christal, 1990). Moreover, given the centrality of this memory system, it is not surprising that disruptions in working memory are often associated with pathological cognitive states, such as attention disorders, dementia, psychosis, and depression (Gold et al., 2003; Goldman-Rakic, 1999; Morris & Baddeley, 1988; Rinck & Becker, 2005; Sonuga-Barke et al., 2002).

Measuring Visual Working Memory

Over the years, researchers have developed a variety of tasks to measure working memory function. Many of them are variations of the match-to-sample task, in which the subject is initially shown a sample item and must then decide whether a later item matched the original one. One such task that we have used over the years to measure VWM capacity is the *change detection task* (Phillips, 1974). In this task, subjects are briefly shown an array of objects (e.g., colored squares) that they must attempt to remember. After a blank delay period of about 1 s, the objects reappear and subjects must report whether they are identical to the originals or if one of them is different. The beauty of this task is that it is extremely flexible, and simple manipulations of the task can be used to test a bevy of cognitive questions related to VWM and other related systems. For example, by varying the number of items in the memory array, one can measure VWM storage capacity (which averages about three or four items; Luck & Vogel, 1997); by varying the types of objects that need to be remembered, one can measure how information is represented in VWM (Alvarez & Cavanagh, 2004; Luck & Vogel, 1997); and by varying the magnitude of the changed item, one can measure the resolution of the information that is held in VWM (Awh et al., 2007). Thus, because of the elegance and flexibility of this task, change detection has become a staple measure of VWM.

Neurophysiological Measures of VWM

One important neural signature of VWM from single-unit studies in monkeys is referred to as *delay activity*, which is the increased and sustained firing rate of a neuron during the memory delay of a match-to-sample task (Fuster, 1973; Fuster & Alexander, 1971; Kubota & Niki, 1971). This delay activity appears to be necessary for accurate working memory performance because it is often reduced or absent on trials in which the incorrect response is made (Funahashi et al., 1989; Sakai et al., 2002). Although many cortical regions have cells that show delay activity, three primary areas have a large proportion of cells that show this property: the inferotemporal cortex, which often shows delay activity that is sensitive to the *identity* of the remembered stimulus (Chelazzi et al., 1998; Miller et al., 1993); the lateral intraparietal (LIP) cortex, which often shows *location*-specific delay activity (Chafee & Goldman-Rakic, 1998; Colby & Goldberg, 1999); and the prefrontal cortex, which often shows delay activity that is sensitive to the *rules* of the current task (Miller & Cohen, 2001; Wallis & Miller, 2003). The fact that the delay activity from a given area is often differentially sensitive to the attributes of the remembered material indicates that this activity is contributing to the specific representations held in memory rather than being the result of task-general processes such as task difficulty or fluctuations within the individual such as arousal or expectation.

Functional magnetic resonance imaging (fMRI) in humans has also been used to observe delay activity during working memory tasks, and the same three cortical areas appear to play substantial roles during online maintenance. Moreover, the sustained blood oxygen-level dependent (BOLD) responses in the inferior temporal cortex have been shown to be sensitive to the identity of the stimuli in memory (Druzgal & D'Esposito, 2001). Location-specific activity in the posterior parietal cortex has been shown to code for memory-driven maps of remembered locations (Sereno et al., 2001). Additionally, Pessoa et al. (2002) demonstrated that sustained BOLD activity in frontoparietal regions during a delayed response task was necessary for successful performance on a working memory task. In sum, this evidence demonstrates the crucial involvement of delay activity during the retention interval of VWM tasks.

Event-Related Potential Studies of Working Memory

Like single-unit recordings and fMRI, studies using event-related potentials (ERPs) have provided useful measures of delay activity in VWM tasks. Importantly, the high temporal resolution of ERPs allows for the isolation of activity during the retention period of memory tasks and can provide information about the timing of brain processes recruited in working memory. One ERP component that has been observed across many studies of working memory is the negative slow wave (NSW), which is a broadly distributed sustained negative wave that persists during the maintenance period of a memory task (Ruchkin et al., 1990). In one study testing visuospatial and verbal stimuli, the amplitude of this component was shown to increase as the memory load increased (Ruchkin et al., 1992). In the visuospatial task, subjects remembered the spatial arrangement of three, four, or five paired letters. The stimuli to be remembered in the phonological task were visually presented pronounceable nonwords consisting of either three, four, or five consonant-vowel syllables. The results (shown in Figure 13.1) showed a large negativity during the retention period of both the visuospatial and phonological memory tasks that scaled as a function of memory load. Importantly, they found that the NSW was maximal at parietotemporal sites for the

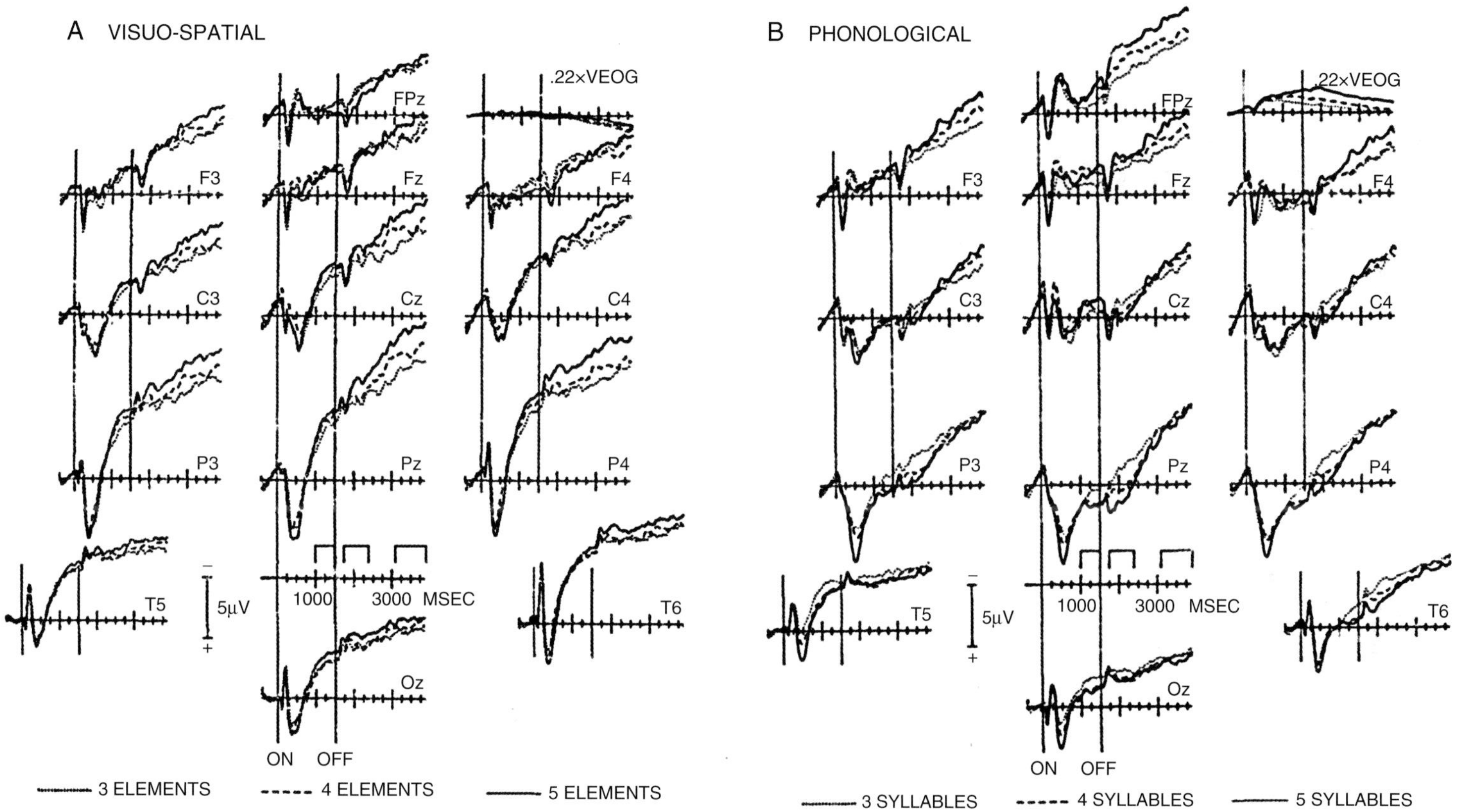

Fig. 13.1. (A) Example results of the NSW for a visuospatial working memory task in which subjects must remember pairs of letters at different spatial positions (adapted with permission from Ruchkin et al., 1992). (B) Example result of the NSW during the retention period for a phonological working memory task in which subjects remembered visually presented pronounceable nonwords of varying syllable lengths.

visuospatial material and maximal at frontal sites for the verbal material. This finding of scalp topography differences according to the type of information being held in VWM has been at least partially replicated, with a more posterior distribution for spatial memory tasks and a more frontal distribution for object memory tasks (Mecklinger & Muller, 1996). Further, Rosler et al. (1997) found that in trials where a larger NSW amplitude was observed during the retention period, there was a stronger probability of successfully remembering the information at test, which suggests that this activity is important for performance on the task.

Although it seems quite plausible that the NSW reflects the operation of VWM maintenance processes, a significant challenge for the validation of any neurophysiological measure of a cognitive process is to demonstrate that it is specific to that particular process. That is, performing any complex task enlists the engagement of a wide variety of processes, some of which are specifically relevant to memory (e.g., VWM maintenance), some are only partially relevant to memory (e.g., perceptual processing of stimuli), and some are task-general nonmnemonic processes such as arousal, effort, or the preparation of an upcoming response. Consequently, it is critical for any valid ERP measure of VWM maintenance to demonstrate that it is specifically related to VWM maintenance and not also to this set of nonmnemonic processes that are active during any task.

Contralateral Delay Activity

One useful approach to controlling for these types of task-general activity is referred to as the *contralateral control method* (Gratton, 1998). Essentially, the idea here is to exploit the fact that the visual system is primarily organized in a contralateral fashion. In these task designs, the subject fixates centrally and is presented with a bilateral display with equal amounts of stimuli in each hemifield. The subject is asked to remember or attend or make a decision about the stimuli in only one of these hemifields, and the activity of the process of interest can be isolated by examining the contralaterally specific activity with respect to the attended side of the display. That is, if subjects are asked to remember information in the right-hand hemifield of the display, we can choose to examine the neural response from the contralateral electrode sites in order to segregate task-relevant mnemonic information from task-general activity. The logic here is that most of the task-general activity (e.g., perceptual response, arousal, response preparation) will be equivalent for each hemisphere and that the primary differences between the hemispheres will be the result of the process of interest (see Chapters 9 and 12, this volume, for additional discussion in the context of the lateralized readiness potential and N2pc). Klaver et al. (1999) used this approach in a VWM task by presenting a bilateral array of two polygons (one in each hemifield) and cued subjects to remember one or both items. Beginning at about 250 ms after the onset of the memory array, they observed a sustained negative wave at posterior electrodes that was contralateral to the attended hemifield. This memory item–specific activity was isolated by subtracting the *ipsilateral activity* (which contained mostly task-general processes) from the *contralateral activity*. If subjects were asked to remember only one item—for example, in the right hemifield of the display—then task-general information measured by the ipsilateral electrode sites was subtracted from the contralateral information, yielding a large negativity throughout the retention period for the mnemonic information exclusively. For example, if both hemispheres had a voltage of –1 μV reflecting task-general processing but the left hemisphere had an additional –1 μV reflecting storage of the right-hemifield polygon, then the difference between the left and right hemispheres would reflect only the storage-related processing (see Chapter 12, this volume, for an extended discussion of the use of contralateral-minus-ipsilateral difference waves as a measure of lateralized processing).

However, if subjects were cued to remember both hemifields, then the negative waveforms reflecting the mnemonic information measured by each of the contralateral sites (i.e., left electrode sites for the right hemifield polygon and right electrode sites for the left hemifield polygon) would be equivalent because each hemifield housed exactly the same amount of mnemonic information. For example, if a particular subject yielded a voltage of –2 μV to the polygon in the right hemifield and the same negative voltage to the polygon in the left hemifield, then a difference wave between the two hemispheres would consist of a flat line at zero. Thus, when subjects must remember information in only one hemifield, this sustained contralateral component appears to be a good candidate for a highly specific measure of VWM maintenance. Indeed, this activity is likely a contralaterally distributed subcomponent of the NSW described above.

Using a similar contralateral control approach, our laboratory presented subjects with a bilateral display of colored squares and asked them to remember the items in a single hemifield over a 1 s retention interval (Vogel & Machizawa, 2004). After this interval, subjects were presented with a match or mismatch test array. Matched arrays were identical to the original array, and mismatched arrays contained one square that changed color on the cued hemifield (noncued hemifield stimuli were always held constant); subjects reported whether the two arrays were the same or different (see Figure 13.2). Similar to Klaver et al. (1999), we observed a large, sustained negative wave at posterior electrodes that was contralateral with respect to the side of the display that the subject was asked to remember on a given trial. This contralateral activity began approximately 275 ms following the onset of the memory array and persisted throughout the retention period until the test array appeared. Importantly, we found that the amplitude of this activity was significantly reduced when subjects made an incorrect response, which suggests that the wave reflects a process that is necessary for correct VWM performance. We refer to this wave as the *contralateral delay activity* (CDA) because of its apparent similarity to delay activity observed in monkey single-unit studies using delayed match-to-sample tasks.

In these tasks, it is critical that the subject fixates the center of the screen throughout the entire trial for several reasons. Most importantly, drifts in eye position toward the cued stimuli will reposition the subject's fixation on the stimuli themselves, thereby reducing, if not entirely eliminating, the contralaterally specific mnemonic information. Additionally, fixating the stimuli alters task performance since foveation typically yields much higher performance accuracy. Lastly, the eyes contain a strong fixed dipole, and a shift in eye position toward one side will lead to a sustained negative voltage over the contralateral hemisphere, which will artifactually increase the apparent amplitude of the CDA. Thus, covertly rather than overtly attending to the items in the cued hemifield is a fundamental requirement of studies that examine the CDA.

CDA Amplitude: Memory Load or Other Task Factors?

The most exciting attribute of this component is its acute sensitivity to the number of items that the subject remembers on a particular trial. Specifically, we found that CDA amplitude was smallest for a memory load of one item and rose monotonically as the memory load increased to two, three, and four items (see Figure 13.2). The fact that this component is sensitive to the present memory load suggests that it is a good candidate for an ERP-based measure of VWM and could possibly be used as an online measure of how much information is currently in mind. However, because several cognitive mechanisms in addition to VWM maintenance are likely involved in a task such as change detection, we see at least two potential alternative accounts of this amplitude increase. The first alternative is that as the number of memory items increases, the spatial scope of the memory array necessarily increases. Thus, the increase in amplitude may actually be the result of a broader spotlight of attention for larger arrays. We tested this in a later study by manipulating the spatial distance (close versus far) between the memory items and found that although CDA amplitude was not modulated by the distance between the items, it was again strongly modulated by the number of memory items (McCollough et al., 2007).

The second alternative account for the CDA amplitude increase is that as the memory load increases, the task becomes more difficult, and it is this general increase in effort or arousal that actually modulates the amplitude of the component. To test this, we also examined memory arrays that exceeded the known limits of VWM capacity (i.e., arrays of 6, 8, or 10 items; Vogel & Machizawa, 2004; McCollough et al., 2007). Our logic was that if the CDA was sensitive to the number of items that can be held in VWM, it should be constrained by the capacity limits of this system and reach a limit at approximately four items. Alternatively, if CDA amplitude was driven by the amount of effort required to perform the task, we would expect it to continue to increase as the difficulty of the task continues to increase—particularly for arrays of 6, 8, and 10 items. However, we found that CDA amplitude reached an asymptotic limit for arrays at around four items, showing no further increase for larger arrays. That is, despite continued increases in difficulty for supracapacity memory arrays, CDA amplitude did not increase, which indicates that task-general factors such as arousal and effort are insufficient to account for the rise in amplitude from one to four items.

CDA Amplitude: Sensitivity to Individual Differences in Memory Capacity

To further examine how sensitive this component is to memory capacity limitations, we tested whether the exact point at which the CDA reached a limit was different for each subject, depending upon his or her specific memory capacity. We reasoned that if

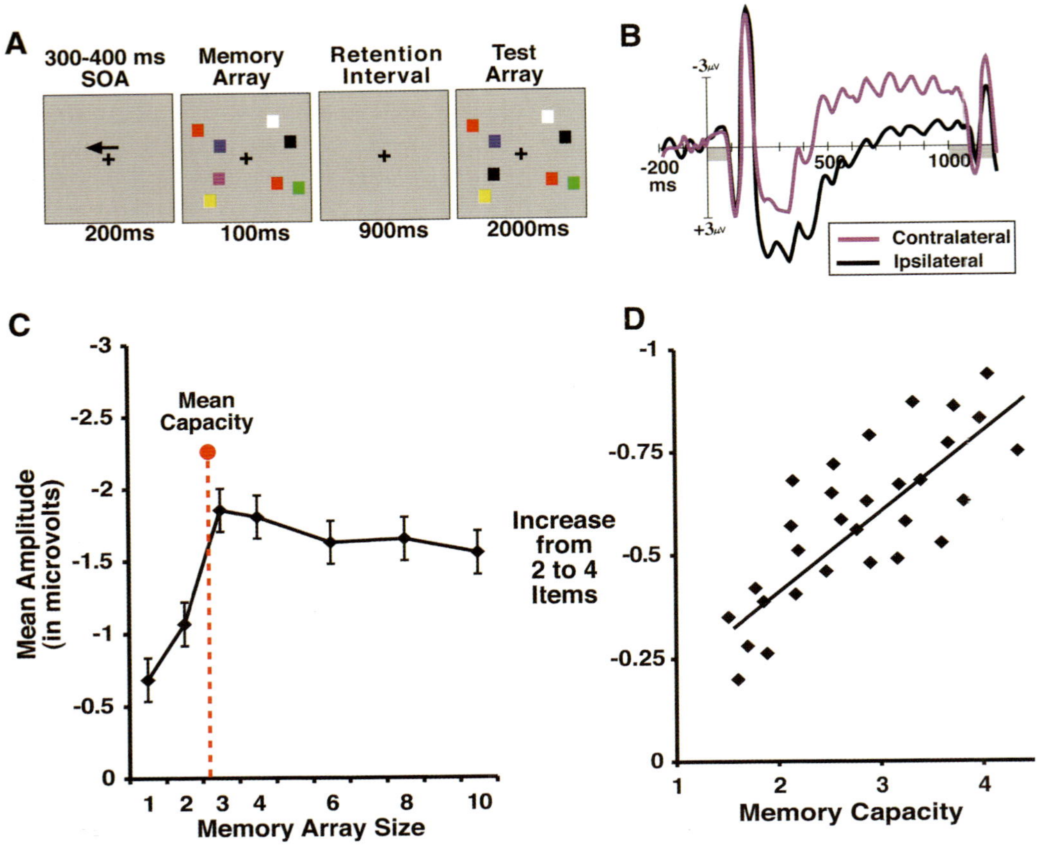

Fig. 13.2. (A) Example of the bilateral change detection paradigm for a "remember-left" trial. (B) Grand-averaged contralateral and ipsilateral waveforms time-locked to the memory array and averaged over posterior parietal, inferotemporal, and lateral occipital electrode sites. By convention, negative voltage is plotted upward. (C) Mean amplitude of CDA (300–900 ms) as a function of the number of items in the memory array. The dotted line represents the mean behavioral visual memory capacity for the sample. (D) Scatterplot of the rise in amplitude between two-item arrays and four-item arrays as a function of an individual's memory capacity ($r = .78$).

the CDA was a highly specific measure of the number of items that can be actively held in VWM, then the CDA for high-memory-capacity individuals, who can presumably hold more information in VWM, should reach a limit at larger array sizes than the CDA for low-memory-capacity individuals. To test this, we measured the rise in amplitude of the CDA from two items to four items and plotted this as a function of each subject's memory capacity, which we estimated from his or her behavioral performance. Indeed, there was a strong correlation ($r = .78$; see Figure 13.2) between an individual's memory capacity and the point at which the CDA reached asymptote (Vogel & Machizawa, 2004). Low-capacity individuals showed little rise in amplitude from two to four items, whereas high-capacity individuals showed large increases in amplitude. Thus, in addition to being highly sensitive to the number of items that are currently held in VWM, we found that the CDA is also very sensitive to individual differences in working memory ability.

CDA Amplitude: Sensitive to the Identity of Items in VWM?

While the *number of items* in memory appears to be a primary factor that modulates the amplitude of the CDA, it is still undetermined whether this component is sensitive to *what* information is currently being held in memory. One piece of evidence that has suggested that the CDA is at least somewhat sensitive to the identity of the information being held was recently provided in a study by Woodman and Vogel (2008). Here, we examined CDA amplitude for memory arrays in which subjects were presented colored, oriented rectangles and were asked to remember in separate blocks only the colors, only the orientations, or both the color and orientation of the items. Interestingly, we found that the overall amplitude of the CDA was significantly larger when the subjects held the orientations or both the color and orientation in memory than when they held only colors in memory, despite the fact that these three conditions used identical stimulus displays and only differed by the instructions given to the subject. Thus, one implication of this result is that the amplitude of the CDA appears to be determined by what *task-relevant information* is being held in memory rather than by what was simply present on the screen. Moreover, a critical aspect of this effect was that the increase in amplitude for orientations did not interact with the number of items the subject was remembering. That is, the increase in amplitude was simply an additive effect with set size: the amount of rise in amplitude from two to four items was equivalent for both color and orientation. These results suggest that at least some aspect of the identity of the remembered items is reflected in the CDA. If the CDA was completely insensitive to identity, we would expect no difference between conditions, particularly when the same physical stimuli (i.e., colored, oriented rectangles) are used in each condition. Precisely why orientation information generates larger CDA amplitudes is still a mystery. In subsequent unpublished studies in our laboratory, we observed that this increase in amplitude appears to occur for any stimulus with significant orientation information as part of its external contour (e.g., abstract line drawings). However, future work examining CDA amplitudes across a much broader range of stimulus types will be necessary to better characterize how much and what type of identity information is reflected in this component.

The Time Course of the CDA

Although most of the studies examining the CDA have focused on amplitude as the primary dependent measure, the time course of the component also appears to provide useful information regarding the operation of VWM. Of course, as with any sustained component, there are several temporal aspects of the wave that may each reflect different cognitive states. Here, we will discuss three aspects of CDA latency: onset, maximum, and duration. In terms of *onset*, the CDA generally begins at approximately 275 ms following the onset of the memory array. It is seen in the waveform as a large contralateral deflection that immediately follows the N2pc component (see Chapter 12, this volume). Interestingly, CDA onset is not at all affected by the number of items in the memory array, and this suggests to us that it may reflect the start of the process of storing the items in VWM. For example, Brisson and Jolicoeur (2007) observed that the onset of the CDA was delayed for targets that were presented during the psychological refractory period, suggesting that the encoding of items into VWM is delayed while attention is occupied with the first target.

While there is no clearly distinct peak, the CDA generally reaches maximum amplitude by approximately 450 ms following the presentation of the memory array. However, the latency at which it reaches maximum amplitude is strongly dependent upon the number of items that are to be stored in VWM. As can be seen in Figure 13.3, the latency to maximum amplitude increases linearly from one to four items, reaching an asymptotic limit for arrays exceeding capacity ($F(1,6) = 16.74; p < .01$). Several previous studies using psychophysical procedures have shown that the time required to consolidate items into VWM increases as the memory load increases up to capacity and that the slope of this increase ranges from 30 to 50 ms per object (Gegenfurtner & Sperling, 1993; Shibuya & Bundesen, 1988; Vogel et al., 2006). Consequently, the temporal sensitivity of the CDA suggests that it may reflect the point at which the memory items have reached a durable/stable state in VWM. Indeed, when we measure the slope of the CDA latency from one to four items, we find it to be 42 ms per object,

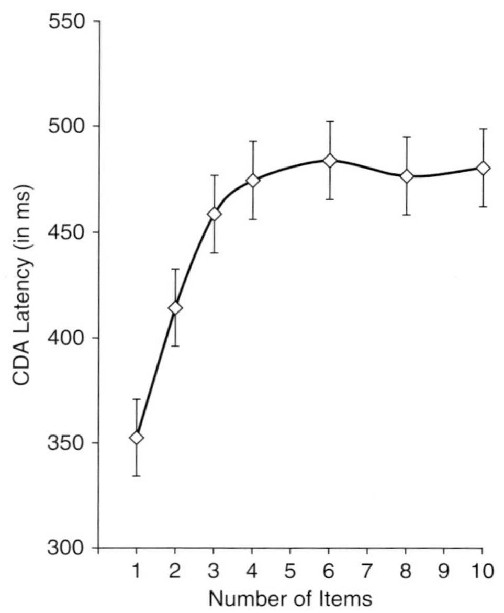

Fig. 13.3. Latency of CDA as a function of the number of items in the memory array. Fractional area latency was measured as the time point at which 75% of the CDA amplitude had been reached.

which is remarkably similar to the estimates of VWM consolidation time derived from psychophysical masking procedures.

How long does the CDA last? Generally speaking, the CDA persists throughout the retention period of the VWM task. In the vast majority of experiments we have used 900 ms retention intervals, but in a few experiments we have observed it out to about 4.5 s. Of course, there are several obvious reasons for using a shorter rather than a longer retention interval during ERP recordings: it is difficult for subjects to avoid making an eye blink during long trial periods, and long retention periods force a trade-off between very long experimental sessions and too few trials. However, one additional reason that we have used shorter retention periods is that after approximately 1000 ms, the CDA appears to decline in amplitude. Importantly, this apparent decline is not due to a decrease in the negative wave over contralateral electrodes but is the consequence of an increase in the amplitude of the ipsilateral electrodes (McCollough et al., 2007). That is, beginning at around 1000 ms, the contralateral-minus-ipsilateral difference wave appears to decline because the ipsilateral activity begins to rise. Precisely why this ipsilateral increase occurs is still a mystery and requires much further investigation. One plausible, but wholly unsubstantiated, explanation of this effect is that the memory representations initially begin as predominantly hemisphere-specific representations, but that after a period of time they become represented equally in both hemispheres. Perhaps, this could explain why many fMRI studies of working memory, which generally use retention intervals of several seconds, often do not find lateralized BOLD activations even when subjects remember items from a single hemifield (Ikkai & Curtis, 2008).

Scalp Topography and Possible Neural Sources of the CDA

The CDA is a broadly distributed wave that is primarily centered over the posterior electrodes (see Figure 13.4). The maximum amplitude of the component is often observed over the posterior parietal electrodes under the 10/20 system, including electrode sites P3/4, PO3/4, P7/8, O1/2, and PO7/8. While the CDA is typically measured by collapsing the left and right hemispheric responses for right- and left-cued hemifields, respectively, in previous work we measured the negative wave separately for each hemisphere and found that there was no significant difference in amplitude between them (McCollough et al., 2007). Moreover, in that study, we found that the scalp distribution of the CDA was significantly more dorsal than the N2pc component (see Chapter 11, this volume), which is an earlier contralateral component with a more ventral focus.

Although we have not attempted a formal source localization of the CDA, one cortical region appears to be a potential candidate source of this wave: the intraparietal sulcus (IPS). While the dorsal, posterior scalp topography of the CDA is generally consistent with such a locus, the strongest evidence in favor of an IPS source comes from neuroimaging experiments of VWM. In particular, several recent fMRI studies examining the BOLD response during

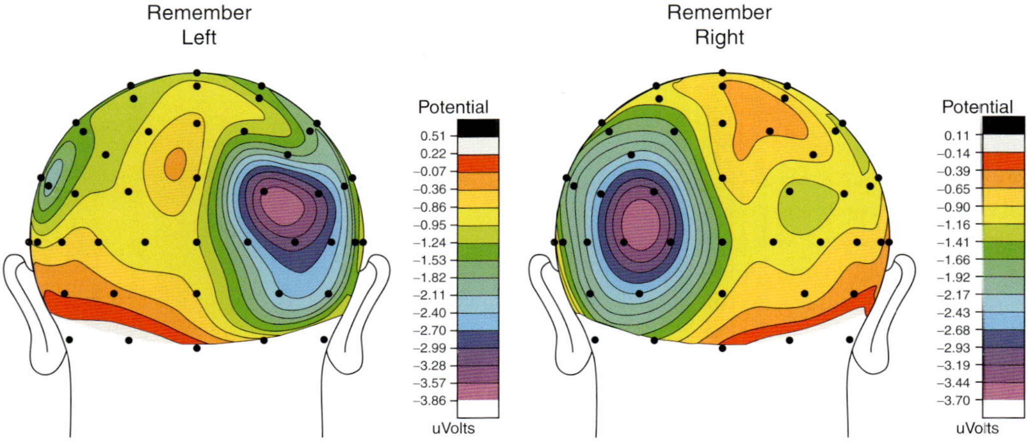

Fig. 13.4. Isocontour voltage maps fit to the spherical spline method of Perrin et al. (1989). The blue and purple regions represent increased delay activity in posterior scalp distributions responsive to remember-left and remember-right trials. Mean amplitude voltage was measured 300–900 ms after the onset of the sample array.

VWM tasks have found that the IPS is strongly modulated by the number of items currently being held in memory but reaches an asymptotic limit at approximately four items (Todd & Marois, 2004; Xu & Chun, 2006). Moreover, Todd and Marois (2005) found that this IPS activity was also sensitive to individual differences in memory capacity. Together, this pattern of functional properties for the IPS appears highly similar to that of the CDA and makes it a plausible candidate/contributor to the component. Indeed, Robitaille et al. (2009) have recorded the magnetic analog of the CDA and estimated a source in the parietal lobe. However, it seems fairly unlikely that such a large and sustained ERP component is generated by a single cortical source; it is more likely the result of several coordinated sources of which IPS may play a significant role.

Using the CDA to Examine Attentional Control Over Working Memory

The constructs of attention and working memory have historically been closely intertwined, and some theorists have even proposed that they are essentially the same mechanism (Cowan, 2001; Cowan et al., 2006). In recent work, we have begun to use the CDA as a tool to examine how attention controls the flow of information into VWM (Vogel et al., 2005a, 2005b). In particular, because the amplitude of the CDA is modulated by the number of items that are presently in VWM, we can use it as an online measure of how much information from a display was represented in memory. Moreover, because we have demonstrated that this component is acutely sensitive to individual variability in performance, it has the potential to help us begin to understand the nature of these individual differences.

One longtime question about individual differences in memory capacity is whether high-capacity subjects can maintain more representations in memory than low-capacity individuals, or whether this apparent difference is due to differences in the ability to control what is ultimately stored in VWM (Engle et al., 1999; Hasher & Zacks, 1988; Kane, 2001). That is, all individuals may be able to store roughly the same amount of information in VWM, but high-capacity individuals may simply be much better at storing only task-relevant information and filtering out task-irrelevant information. We tested this question by asking subjects to control which items from a display would be stored in VWM (Vogel et al., 2005a). In one experiment, subjects were asked to remember only the red items. On some trials, they were shown only red items (either two or four red rectangles). On other trials, they were presented a mix of two red items and two blue items, which required them to selectively keep the blue items out of VWM. We measured the efficiency of keeping the blue items out of memory by examining the CDA amplitude in relation to the red items–only conditions. If a subject was perfectly efficient at keeping the blue items out, then we would expect that CDA amplitude would be equivalent to the amplitude when they were presented with only two red items. By contrast, if the subject was perfectly inefficient at keeping the blue items out, then CDA amplitude should be equivalent to the amplitutde when he or she remembered four red items. Our results showed that high-capacity subjects were highly efficient at keeping the blue items out of memory and that low-capacity subjects unnecessarily stored the blue items in memory (see Figure 13.5). Thus, the CDA proved to be a useful tool for testing specific cognitive questions regarding how attention and working memory interact and gave us some insight into the nature of individual differences in VWM capacity. Moreover, this study helped establish one more important detail regarding the CDA: the amplitude is determined not only by how many items are present in the display, but also by an interaction between the number of task-relevant items present and how efficiently the subject can select those items.

Using the CDA to Examine the Role of VWM during Multiple Object Tracking

One of the more dramatic demonstrations of divided attention is the multiple object tracking (MOT) task, in which a subject must attempt to attentionally track the positions of multiple targets as they move among several identical distractors over the course of several seconds (Pylyshyn & Storm, 1988; Scholl et al., 2001). This task requires attention to dynamically update the positions of the moving targets and keep these representations segregated from the distractors. Several theorists have proposed that these target representations are held and updated in VWM (Cavanagh & Alvarez, 2005; Oksama & Hyönä, 2004). One piece of evidence in support of this proposal is that subjects can typically track up to a maximum of four targets, which is suspiciously similar to the capacity limit of VWM. Recently, we used the CDA as a means of further testing the role of VWM in MOT. In particular, because the CDA appears to be extremely sensitive to the number of items currently maintained in VWM, we reasoned that it should be present while

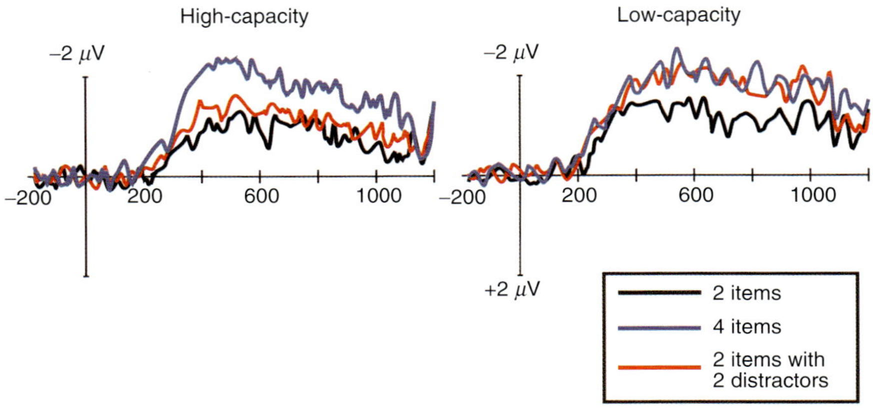

Fig. 13.5. Grand averaged difference waveforms (ipsilateral delay activity subtracted from contralateral delay activity) divided across high- and low-memory-capacity subjects from Vogel et al (2005). While high-capacity individuals showed a distractor-present amplitude that was equivalent to two items only, the low-capacity individuals showed a distractor-present amplitude that was equivalent to the four items only.

the subject performs an MOT task and should be modulated by the current number of tracked targets. Indeed, in a bilateral variation of a typical MOT task, we observed a large and sustained CDA that was strongly modulated by the number of targets that were being tracked on a given trial (Drew & Vogel, 2009). Moreover, we found that the rise in CDA amplitude between tracking one and tracking three targets was strongly predictive of an individual's specific tracking capacity ($r = .72$). While further research will be necessary to determine the degree of cognitive and neural overlap between the mechanisms facilitating the performance of MOT and VWM tasks, these results suggest that the same capacity-limited representational system may underlie both of them.

Future Directions and Unresolved Issues

We are optimistic about the future utility of the CDA as a tool to examine how VWM operates, as well as how various other cognitive mechanisms interact with this limited online memory system. Thus far, the property of the CDA that appears to have the most potential to help answer cognitive questions about capacity limits and representations is its sensitivity to the number of objects currently maintained in VWM. Of course, there are several remaining unanswered questions about this component that will need to be addressed before its full impact can be appreciated. Throughout this chapter, we have noted a few of these questions. For instance, what identity information about remembered objects is reflected in the CDA? We know it is modulated by the presence of orientation information, but we have no idea why this is the case. What is necessary to understand this property is a careful examination of CDA amplitudes across a very broad range of stimulus classes. If it turns out to be reliably sensitive to stimulus identities held in memory, it could open a whole new set of questions about the nature of memory representations that could be addressed with this ERP component.

A second unanswered question about the CDA is its neural origin. While recent fMRI studies have helped to implicate the IPS as a plausible source, and while recordings of the magnetic analog of the CDA have suggested a source in posterior parietal cortex, more precise methods could indicate whether there are multiple cortical sources contributing to this component, which would in turn help us determine the network of cortical activity that underlies this capacity-limited system. Of course, we understand that traditional source localization approaches can only go so far in establishing the neural sources of ERP components (Luck, 1999). Perhaps more traction could be gained on this issue by adopting the approach recently developed by Woodman et al. (2007), who measured surface ERPs while simultaneously recording single-unit responses in awake, behaving monkeys.

One critical but presently unexplored aspect of the CDA concerns the oscillations at various frequencies that underlie this averaged ERP component (see, e.g., Chapter 2, this volume). A recent study by van Dijk et al. (2008) has shown that the CDA at least partially reflects a difference in alpha activity between the contralateral and ipsilateral hemispheres. This may be related to computational work suggesting that synchronous firing across populations of cells may be the mechanism for maintaining object representations in VWM (Lisman & Idiart, 1995; Tallon-Baudry et al., 2001; Vogel et al., 2001). Indeed, it is even possible that VWM capacity limits are the consequence of a limited ability to keep the oscillations for a given object separate and asynchronous from the oscillations representing other objects in memory (Raffone & Wolters, 2001). Consequently, understanding the oscillations that give rise to the CDA will likely be critical to answering one of the most fundamental questions about working memory: why is its capacity limited in the first place?

References

Alvarez, G. A., & Cavanagh, P. (2004). The capacity of visual short-term memory is set both by visual information load and by number of objects. *Psychological Science, 15*(2), 106–111.

Awh, E., Barton, B., & Vogel, E. K. (2007). Visual working memory represents a fixed number of items regardless of complexity. *Psychological Science, 18*(7), 622–628.

Baddeley, A. D., & Hitch, G. J. (1974). Working memory. In G. H. Bower (Ed.), *The psychology of learning and motivation, Vol. VIII* (pp. 47–90). New York: Academic Press.

Brisson, B., & Jolicoeur, P. (2007). A psychological refractory period in access to visual short-term memory and the deployment of visual-spatial attention: Multitasking processing deficits revealed by event-related potentials. *Psychophysiology, 44*(2), 323–333.

Cavanagh, P., & Alvarez, G. A. (2005). Tracking multiple targets with multifocal attention. *Trends in Cognitive Sciences, 9*(7), 349–354.

Chafee, M., & Goldman-Rakic, P. (1998). Matching patterns of activity in primate prefrontal area 8a and parietal area 7ip neurons during a spatial working memory task. *Journal of Neurophysiology, 79*, 2919–2940.

Chelazzi, L., Duncan, J., Miller, E. K., & Desimone, R. (1998). Responses of neurons in inferior temporal cortex during memory-guided visual search. *Journal of Neurophysiology, 80*, 2918–2940.

Colby, C. L., & Goldberg, M. E. (1999). Space and attention in parietal cortex. *Annual Review of Neuroscience, 22*, 319–349.

Cowan, N. (2001). The magical number 4 in short-term memory: A reconsideration of mental storage capacity. *Behavioral and Brain Sciences, 24*, 87–185.

Cowan, N., Fristoe, N. M., Elliott, E. M., Brunner, R. P., & Saults, J. S. (2006). Scope of attention, control of attention, and intelligence in children and adults. *Memory & Cognition, 34*(8), 1754–1768.

Daneman, M., & Carpenter, P. A. (1980). Individual differences in working memory and reading. *Journal of Verbal Learning and Verbal Behavior, 19*, 450–466.

Drew, T., & Vogel, E. K. (2009). Neural measures of individual differences in selecting and tracking multiple moving objects. *Journal of Neuroscience, 28, 4183–4191.*

Druzgal, T. J., & D'Esposito, M. D. (2001). Activity in fusiform face area modulated as a function of working memory load. *Cognitive Brain Research, 10*, 355–364.

Engle, R. W., Kane, M. J., & Tuholski, S. W. (1999). Individual differences in working memory capacity and what they tell us about controlled attention, general fluid intelligence, and functions of the prefrontal cortex. In A. Miyake & P. Shah (Eds.), *Models of working memory: Mechanisms of active maintenance and executive control* (pp. 102–134). New York: Cambridge University Press.

Funahashi, S., Bruce, C. J., & Goldman-Rakic, P. S. (1989). Mnemonic coding of visual space in the monkey's dorsolateral prefrontal cortex. *Journal of Neurophysiology, 61*, 331–349.

Fuster, J. M. (1973). Unit activity in prefrontal cortex during delayed-response performance: Neuronal correlates of transient memory. *Journal of Neurophysiology, 36*, 61–78.

Fuster, J. M., & Alexander, G. E. (1971). Neuron activity related to short-term memory. *Science, 173*, 652–654.

Gegenfurtner, K. R., & Sperling, G. (1993). Information transfer in iconic memory experiments. *Journal of Experimental Psychology: Human Perception and Performance, 19*(4), 845–866.

Gold, J. M., Wilk, C., McMahon, R., & Luck, S. J. (2003). Working memory for visual features and conjunctions in schizophrenia. *Journal of Abnormal Psychology, 112*(1), 61–71.

Goldman-Rakic, P. S. (1999). The physiological approach: Functional architecture of working memory and disordered cognition in schizophrenia. *Biological Psychiatry, 46*(5), 650–661.

Gratton, G. (1998). The contralateral organization of visual memory: A theoretical concept and a research tool. *Psychophysiology, 35*, 638–647.

Hasher, L., & Zacks, R. (1988). Working memory, comprehension, and aging: A review and a new view In G. H. Bower (Ed.), *Psychology of Learning and Motivation* (pp. 193–225). San Diego, CA: Academic Press.

Ikkai, A., & Curtis, C. (2008). Cortical activity time locked to the shift and maintenance of spatial attention. *Cerebral Cortex, 18*(6):1384–1394.

Kane, M. J., Bleckly, M. K., Conway, A. R. A., & Engle, R. W. (2001). A controlled-attention view of working memory capacity. *Journal of Experimental Psychology: General, 130*(2), 169–183.

Klaver, P., Talsma, D., Wijers, A. A., Heinze, H. J., & Mulder, G. (1999). An event-related brain potential correlate of visual short-term memory. *Neuroreport, 10*, 2001–2005.

Kubota, K., & Niki, H. (1971). Prefrontal cortical unit activity and delayed alternation performance in monkeys. *Journal of Neurophysiology, 34*, 337–347.

Kyllonen, P. C., & Christal, R. E. (1990). Reasoning ability is (little more than) working memory capacity. *Intelligence, 14*, 398–433.

Lisman, J. E., & Idiart, M. A. P. (1995). Storage of 7+/-2 short-term memories in oscillatory subcycles. *Science, 267*, 1512–1515.

Luck, S. J. (1999). Direct and indirect integration of event-related potentials, functional magnetic resonance images, and single-unit recordings. *Human Brain Mapping, 8*, 15–20.

Luck, S. J., & Vogel, E. K. (1997). The capacity of visual working memory for features and conjunctions. *Nature, 390*, 279–281.

McCollough, A. W., Machizawa, M. G., & Vogel, E. K. (2007). Electrophysiological measures of maintaining representations in visual working memory. *Cortex, 43*(1), 77–94.

Mecklinger, A., & Muller, N. (1996). Dissociations in the processing of "What" and "Where" information in working memory: An event-related potential analysis. *Journal of Cognitive Neuroscience, 8*(5), 453–473.

Miller, E. K., & Cohen, J. D. (2001). An integrative theory of prefrontal cortex function. *Annual Review of Neuroscience, 24*, 167–202.

Miller, E. K., Li, L., & Desimone, R. (1993). Activity of neurons in anterior inferior temporal cortex during a short-term memory task. *Journal of Neuroscience, 13*, 1460–1478.

Morris, R. G., & Baddeley, A. D. (1988). Primary and working memory functioning in Alzheimer-type dementia. *Journal of Clinical and Experimental Neuropsychology, 10*(2), 279–296.

Oksama, L., & Hyönä, J. (2004). Is multiple object tracking carried out automatically by an early vision mechanism independent of higher-order cognition? An individual difference approach. *Visual Cognition, 11*(5), 631–671.

Perrin, F., Pernier, J., Bertrand, O., & Echallier, J. F. (1989). Spherical splines for scalp potential and current density mapping. *Electroencephalography and Clinical Neurophysiology, 72*: 184–187.

Pessoa, L., Gutierrez, E., Bandettini, P., & Ungerleider, L. G. (2002). Neural correlates of visual working memory: FMRI amplitude predicts task performance. *Neuron, 35*, 975–987.

Phillips, W. A. (1974). On the distinction between sensory storage and short-term visual memory. *Perception & Psychophysics, 16*, 283–290.

Pylyshyn, Z. W., & Storm, R. W. (1988). Tracking multiple independent targets: Evidence for a parallel tracking mechanism. *Spatial Vision, 3*(3), 179–197.

Raffone, A., & Wolters, G. (2001). A cortical mechanism for binding in visual working memory. *Journal of Cognitive Neuroscience, 13*, 766–785.

Rinck, M., & Becker, E. S. (2005). A comparison of attentional biases and memory biases in women with social phobia and major depression. *Journal of Abnormal Psychology, 114*(1), 62–74.

Robitaille, N., Grimault, S., & Jolicoeur, P. (2009). Bilateral parietal and contralateral responses during the maintenance of unilaterally-encoded objects in visual short-term memory: Evidence from magnetoencephalography. *Psychophysiology, 46*, 1090–1099.

Rosler, F., Heil, M., & Roder, B. (1997). Slow negative brain potentials as reflections of specific modular resources of cognition. *Biological Psychology, 45*, 109–141.

Ruchkin, D., Johnson, R., Canoune, H., & Ritter, W. (1990). Short-term memory storage and retention: An event-related brain potential study. *Electroencephalography and Clinical Neurophysiology, 76*, 419–439.

Ruchkin, D., Johnson, R., Grafman, J., Canoune, H., & Ritter, W. (1992). Distinctions and similarities among working memory processes: An event-related potential study. *Cognitive Brain Research, 1*, 53–66.

Sakai, K., Rowe, J. B., & Passingham, R. E. (2002). Active maintenance in prefrontal area 46 creates distractor-resistant memory. *Nature Neuroscience, 5*, 479–484.

Scholl, B. J., Pylyshyn, Z. W., & Feldman, J. (2001). What is a visual object? Evidence from target merging in multiple object tracking. *Cognition, 80*(1–2), 159–177.

Sereno, M. I., Pitzalis, S., & Martinez, A. (2001). Mapping of contralateral space in retinotopic coordinates by a parietal cortical area in humans. *Science, 294*, 1350–1354.

Shibuya, H., & Bundesen, C. (1988). Visual selection from multielement displays: Measuring and modeling effects of exposure duration. *Journal of Experimental Psychology: Human Perception and Performance, 14*(4), 591–600.

Sonuga-Barke, E. J., Dalen, L., Daley, D., & Remington, B. (2002). Are planning, working memory, and inhibition associated with individual differences in preschool ADHD symptoms? *Developmental Neuropsychology, 21*(3), 255–272.

Sperling, G. (1960). The information available in brief visual presentations. *Psychological Monographs, 74* (Whole No. 498).

Tallon-Baudry, C., Bertrand, O., & Fischer, C. (2001). Oscillatory synchrony between human extrastriate areas during visual short-term memory maintenance. *Journal of Neuroscience, 21*, RC177.

Todd, J. J., & Marois, R. (2004). Capacity limit of visual short-term memory in human posterior parietal cortex. *Nature, 428*, 751–754.

Todd, J. J., & Marois, R. (2005). Posterior parietal cortex activity predicts individual differences in visual short-term memory capacity. *Cognitive, Affective & Behavioral Neuroscience, 5*(2), 144–155.

van Dijk, H., van der Werf, J., Mazaheri, A., Medendorp, W. P., & Jensen, O. (2008). Modulations in oscillatory activity with amplitude asymmetry can produce cognitively relevant event-related responses. *Proceedings of the National Academy of Sciences USA, 28*, 1816–1823.

Vogel, E. K., & Awh, E. (2008). How to exploit diversity for scientific gain: Using individual differences to constrain cognitive theory. *Current Directions in Psychological Science, 17*, 171–176.

Vogel, E. K., & Machizawa, M. G. (2004). Neural activity predicts individual differences in visual working memory capacity. *Nature, 428*, 748–751.

Vogel, E. K., McCollough, A. W., & Machizawa, M. G. (2005a). Neural measures reveal individual differences in controlling access to working memory. *Nature, 438*(7067), 500–503.

Vogel, E. K., Woodman, G. F., & Luck, S. J. (2001). Storage of features, conjunctions, and objects in visual working memory. *Journal of Experimental Psychology: Human Perception and Performance, 27*, 92–114.

Vogel, E. K., Woodman, G. F., & Luck, S. J. (2005b). Pushing around the locus of selection: Evidence for the flexible-selection hypothesis. *Journal of Cognitive Neuroscience, 17*(12), 1907–1922.

Vogel, E. K., Woodman, G. F., & Luck, S. J. (2006). The time course of consolidation in visual working memory. *Journal of Experimental Psychology: Human Perception and Performance, 32*(6), 1436–1451.

Wallis, J. D., & Miller, E. K. (2003). From rule to response: Neuronal processes in the premotor and prefrontal cortex. *Journal of Neurophysiology, 90*, 1790–1806.

Woodman, G., Min-Suk, K., Rossi, A. F., & Schall, J. D. (2007). Nonhuman primate event-related potentials indexing covert shirts of attention. *Proceedings of the National Academy of Sciences USA, 104*(38), 15111–15116.

Woodman, G. F., & Vogel, E. K. (2008). Selective storage and maintenance of an object's features in visual working memory. *Psychonomic Bulletin and Review, 15*(1), 223–229.

Xu, Y. D., & Chun, M. M. (2006). Dissociable neural mechanisms supporting visual short-term memory for objects. *Nature, 440*(7080), 91–95.

CHAPTER

14

Electrophysiological Correlates of Episodic Memory Processes

Edward L. Wilding *and* Charan Ranganath

Abstract

The ability to remember an event relies on a complex interplay of processes that we are only beginning to understand. Event-related potential (ERP) studies have proven to be powerful tools in the delineation of these processes, and have provided insights into the nature and identity of the operations that support the encoding and retrieval of different kinds of events. This chapter first considers methodological issues in the use of ERPs to study memory and suggests how the ERP method can be used optimally to address psychologically relevant questions. Next, it presents a selective overview of the contributions of ERP research to the study of human long-term memory, focusing specifically on episodic memory retrieval, which supports memory for personally experienced events.

Keywords: event-related potential, long-term memory, memory retrieval

The ability to remember an event relies on a complex interplay of processes that we are only beginning to understand. Event-related potential (ERP) studies have proven to be powerful tools in the delineation of these processes, and have provided insights into the nature and identity of the operations that support the encoding and retrieval of different kinds of events. In this chapter, we will first consider methodological issues in the use of ERPs to study memory and suggest how the ERP method can be used optimally to address psychologically relevant questions. Next, we will present a selective overview of the contributions of ERP research to the study of human long-term memory, focusing specifically on episodic memory retrieval, which supports memory for personally experienced events.

Why ERPs Are Useful in the Study of Memory and How to Use Them

The basics of ERP methodology have been covered in detail elsewhere (Luck, 2005). Here, we emphasize only a few key points regarding why ERPs can and should be used to study memory processes, as well as some points that suggest caution when relying on interpretation of ERP components and suggestions on how to design memory experiments that take advantage of the strengths of the ERP method.

Why Use ERPs to Study Memory?

With the widespread availability of high-spatial-resolution neuroimaging techniques such as functional magnetic resonance imaging (fMRI), why should the modern cognitive neuroscientist use ERPs to study memory? As described earlier in this book, ERPs essentially reflect averaged recordings of (primarily cortical) field potentials that are volume conducted to the scalp. More specifically, ERPs most likely index the summated activity of large numbers of postsynaptic potentials, which might be inhibitory or excitatory. By contrast, fMRI measures changes in image intensity that are related to changes in local magnetic susceptibility over time, which in turn are influenced by changes in local blood oxygenation. Although local blood oxygenation changes are clearly correlated with changes in synaptic activity, the chain of events that links the two is complex (Buxton, 2002; Buxton et al., 1998; Heeger & Ress, 2002; Logothetis, 2002). As a consequence, there are significant differences between the kinds of information that fMRI and ERPs can provide. Because fMRI measures blood oxygenation, interpretation of fMRI data can be complicated by changes in the vasculature that may be separate from underlying neural changes. This places limitations on the use of fMRI if the intention is to compare groups of participants that might have vascular differences (D'Esposito, 2003) or to study participants who have suffered from a significant vascular event such as a stroke. ERPs, as a direct measure of neural activity, are less sensitive to such effects, so this method is a good source of converging evidence in the examination of differences between groups or in the investigation of plasticity in patients who have focal brain lesions.

Perhaps a more important advantage of the ERP method is the high temporal resolution that it offers. Again, as a direct measure of neural activity, ERPs can in principle be used to place an upper bound on the temporal onset of a particular process and to index distinct computations that are separated by only a few hundred milliseconds. In contrast, the temporal resolution of fMRI is limited ultimately by the fact that the sluggish hemodynamic response acts as a low-pass temporal filter of neural activity.

Real-time temporal resolution is critical if the goal is to provide a dynamic characterization of the neural and functional basis of human cognitive abilities. Like all higher-level cognitive functions, memory processing at encoding and retrieval involves a large number of distinct computations with overlapping time courses. These computations are probably implemented as synchronous patterns of activity across distributed cortical networks, so the ability to measure these in real time provides the starting point for separating the contributions of different networks. Furthermore, the relative timing of ERP effects can provide critical insights into the functional significance of any activity differences that are observed (for examples in the domain of memory, see Duarte et al., 2004; Duzel et al., 1999; Ranganath & Paller, 1999; Tsivilis et al., 2001; Wilding & Rugg, 1996).

Experimental Design and Interpretation of ERP Correlates of Memory

An example of a typical memory experiment is depicted in Figure 14.1. In such a study, ERPs could be recorded while participants study a series of items (the encoding phase) and/or while they are tested on recently studied items (the retrieval phase). Analyses can be completed to examine ERPs in several ways, including whether or not the item was successfully recognized at test, the way that an item was encoded, the kinds of instructions given at test, and the kinds of experiences that accompany test judgments. In general, the principles of design for an ERP study of memory are very similar to those for a behavioral study, although a few issues deserve special consideration. One issue concerns presentation parameters for studies that use visual stimuli. In such experiments, it is important to remember that participants might move their eyes to scan stimuli, and these eye movements can result in artifacts that will contaminate scalp-recorded ERPs. Although artifact rejection and correction methods can mitigate the influence of eye movements, it is preferable to design experiments to minimize the occurrence of artifacts in the first place. Three ways to do this are to keep the duration of stimulus presentation short (<1 s), to use stimuli that are small as well as relatively simple, and to present stimuli at or near a fixation point (Luck, 2005). A second issue that merits consideration in experiment design is that the experiment should yield a sufficiently large number of trials for each condition of interest, because the signal-to-noise ratio increases as a function of the square root of the number of trials (Rugg, 1995b). For most memory studies, a reasonable plan is to have an average of at least 30 trials per participant in each condition of interest. One can examine ERPs consisting of fewer trials per participant and condition, and of course, if the average is 30, some participants will have far fewer trials in some conditions. An approach often taken is to employ a criterion for the minimum number of trials (e.g., 16) that a participant must have in a given condition in order for the data from that condition to be included in a group-level data set that is submitted to statistical analysis (for comments relating signal-to-noise ratios to the different kinds of measures that can be extracted from ERP waveforms, see Luck, 2005). Fulfilling these criteria is not always straightforward in memory experiments, particularly ones in which contrasts are between neural activity associated with remembered and forgotten items. The challenge is to have a level of memory performance that is sufficiently high so that one can be confident that participants can complete the task and remain engaged with it, while at the same time having sufficient trials contributing to the forgotten conditions of interest.

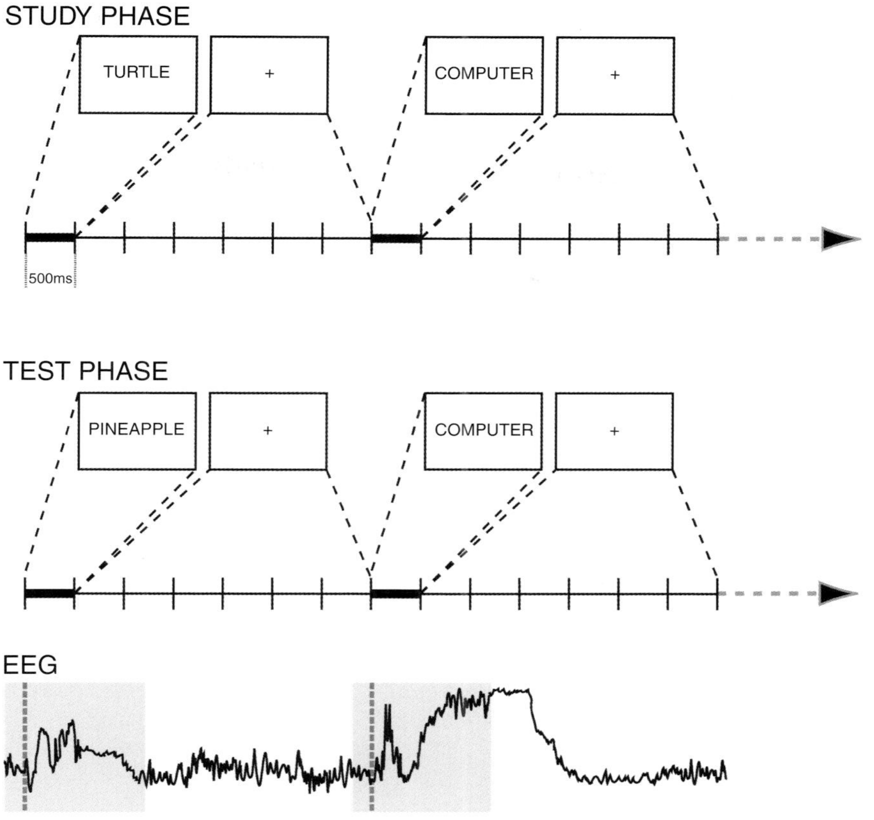

Fig. 14.1. A schematic depiction of a memory experiment. In this example, the participant is shown a series of words (each for 500 ms) during the study (encoding) phase. Usually, to ensure that participants attend to each item and process all items in a similar manner, they are asked to make a decision about each item. For instance, participants might be asked to indicate with a button press whether the concept or object denoted by the word is abstract or concrete in order to encourage them to think about the meaning of each word. Following presentation of the word, a fixation cross is shown in order to encourage participants to maintain central fixation, thereby minimizing the occurrence of artifacts related to eye movements. Next, a memory test is administered in which participants must indicate with a button press whether each item was or was not seen during the preceding study phase. In this case, an electroencephalogram (EEG; bottom) is recorded during the test phase and the time of onset of each test item is indicated with a dashed line. Event-related potentials might be constructed by isolating the segments of EEG occurring from 200 ms prior to stimulus onset to 1000 ms following stimulus onset (the averaging window is shown in a gray rectangle) and separately averaging these segments for trials with studied (e.g., "computer") and unstudied (e.g., "pineapple") items. Separate averages can then be formed according to whether responses at test were correct or incorrect.

Before proceeding, it is worth noting that, unlike some of the research domains covered in this book, the field of memory research has not been driven mainly by the need to understand the nature of specific ERP components, but rather by the use of experimental manipulations to identify electrophysiological correlates of the processes of interest. Typically, memory experiments are designed so that processes of interest will be engaged in some conditions but not others, or will be engaged to differing degrees across conditions. This is a departure from the approach taken in some other research fields, in which functional issues are addressed by assessing how ERP components (see Chapter 1, this volume) vary across conditions and/or populations. This is not to say, however, that ERP memory research has proceeded independently of knowledge about specific ERP components, as at least some of the ERP memory effects that have been the subject of considerable study have time courses and scalp distributions that overlap with those of well-characterized ERP components, including the P300 and the N400.

When such commonalities exist, they serve as important prompts for the researcher to consider in deciding whether an experiment contrast designed to isolate a given memory process can in fact be explained by appealing to knowledge about the conditions under which a particular ERP component might be modulated. This exercise may well then become the starting point for employing orthogonal manipulations that are implemented to enable separate assessments of changes that can be ascribed confidently to either the memory effect of interest or the ERP component with which the effect of interest bears temporal and spatial similarities (for relevant examples and discussion concerning the potential contributions of the P300 in retrieval tasks, see Friedman, 1990; Herron et al., 2003; Neville et al., 1986; Rugg & Nagy, 1989; Smith & Guster, 1993; Spencer et al., 2000).

Findings from ERP Studies of Memory Encoding and Retrieval Processes

It is a sign of the maturity of the ERP memory field that comprehensive reviews of several aspects of ERP memory research on memory encoding and retrieval are already available (Curran et al., 2006b; Donaldson et al., 2003; Friedman & Johnson, 2000; Mecklinger, 2000; Rugg et al., 2002; Wilding & Sharpe, 2003). In light of these reviews, the focus here is on emphasizing the broad utility of the ERP approach to characterizing human memory retrieval by focusing on relatively new developments, although necessarily this will involve some recapitulation of key findings. In addition, in keeping with the majority of the published literature, this review is concerned primarily with findings in studies where the participants were young healthy adults. This emphasis is not intended to give the impression, however, that ERP studies with other populations (in particular, older adults and patients with selective brain lesions) have not yielded influential findings that are relevant to questions about the functional neuroanatomy supporting human memory, as well as questions about the sensitivity of ERPs to certain memory processes. For reviews of applications of the ERP approach to studies of memory in older adults, see Friedman (2003), and for patient work, see Rugg (1992, 1995a).

Event-related potential studies of memory encoding are discussed first below, and complementary accounts can be found in the following papers: Otten et al. (2006), Otten and Rugg (2001a, 2001b), and Wilding and Sharpe (2003). The following section on ERP studies of memory encoding is followed by a review of studies of retrieval. In keeping with the greater relative emphasis on retrieval rather than on encoding in the published literature, this section comprises the bulk of this review. This section is also restricted for the most part to memory studies in which *direct* retrieval tasks were employed—that is, tasks in which participants are directed to make responses that are guided by their memories for stimuli encountered in prior learning episodes. For a review of *repetition effects* obtained in indirect memory tasks (where participants receive no information about the potential relevance of a learning episode to task performance), see Rugg and Doyle (1994).

ERP Studies of Memory Encoding Processes

We cannot remember all of our experiences, and processes that occur during encoding can critically influence whether an event will be remembered subsequently. In typical ERP studies of memory encoding, brain activity is recorded while a series of items is presented, and the participant is usually given an orienting task to constrain how each item will be processed (e.g., "Decide if this word corresponds to something that is living"). At some later point, the participant will be tested on the studied items, and then the activity elicited during the encoding of each item will be averaged according to whether or not it was remembered on the subsequent test (see Figure 14.1). Differences between the neural activities associated with studied stimuli that are subsequently recognized and those that are forgotten have been termed *Dm* (*difference due to memory*) *effects* (Paller, 1990; Paller et al., 1987a, 1987b) and more recently *subsequent memory effects* (e.g., Wagner & Paller, 2002). The key assumption is that differences between the neural activities elicited by study items that subsequently attract either correct or incorrect judgments are correlates of processes that are involved in successful memory formation.

What kinds of encoding processes might ERPs index? An early clue was the finding that subsequent memory effects are more robust if participants were asked to *recall* the studied items than if they were simply asked to *recognize* them (Johnson, 1995). In the simplest recall tasks, people are asked to remember as many items from a study list as possible. Variants of this task involve the use of partial cues at test. For example, people might see the first three letters of words (stem-cued recall) or a few letters of words selected randomly (fragment-cued recall). In standard recognition memory tasks, by contrast, people see re-presentations of studied items interspersed with unstudied items—the task being to distinguish between the two.

Why are ERP subsequent memory effects more robust for recall than for recognition memory tasks (Johnson, 1995; Paller, 1990; Paller et al., 1987a)? This may reflect the fact that the processes indexed by subsequent memory effects support performance on recall tasks to a greater degree than they do on recognition memory tasks. Alternatively, these differences according to the type of retrieval task may be a consequence of task demands. In forced-choice recognition tasks, a proportion of items will attract correct old judgments when little or no veridical information supported those judgments. A consequence of this would be dilution of the size of any subsequent memory effects, and this would not happen to the same degree for recall tasks. Support for this second account comes from recent studies in which people have been asked to indicate their confidence in their recognition memory judgments. This allows for analyses of subsequent memory effects that do not include activity associated with some low-confidence responses. Insofar as these responses are those that are less likely to be associated with neural activity that would differentiate between subsequent correct and incorrect judgments, their removal from a subsequent memory contrast should permit a cleaner separation between activity at encoding that supports later remembering or forgetting. In keeping with this view, robust subsequent memory effects have been observed when selective contrasts of this kind have been employed (Otten & Rugg, 2001a; Sommer et al., 1997).

It is also important that the scalp distributions of ERP subsequent memory effects differ according to whether the studied material is verbal or visual (Sommer et al., 1991, 1995, 1997) and whether words or nonwords are employed (Otten et al., 2007), and that, when study and test materials are held constant, the effects also vary according to the kinds of encoding operations that are required at study (Otten & Rugg, 2001a). These findings emphasize that, while there may be brain regions critical for remembering per se, successful remembering also depends upon activity in different neural networks according to specific stimulus properties as well as the cognitive operations that are engaged at the time of the initial processing of items. This observation is in one sense unsurprising, but presumably the level of activity in every one of the brain regions that are engaged during the initial processing of a stimulus is not a critical determinant of subsequent memorability. An important goal in ongoing research is to delineate those regions that play an influential role in remembering for particular stimulus types and cognitive operations and to separate that activity from two other kinds: that which plays no role in successful encoding and that which is influential across content and stimulus types.

In two recent influential studies, Otten and colleagues (2006, 2010) have shown that neural activity occurring *before* an item is presented in a study phase also predicts whether that item will be remembered subsequently. While the specific antecedents for this new kind of subsequent memory effect have not been delineated fully, it seems unlikely that these prestimulus differences reduce to an explanation in terms of levels of attention or resource allocation (Otten et al., 2006). In light of this, Otten et al. (2010) have suggested that the effects reflect differences in the degree of engagement of processes that enable the processing of semantic information associated with study items. Irrespective of the accuracy of this account, however, these findings highlight the utility and influence of real-time measures of neural activity in studies of memory encoding, as the temporal resolution necessary to distinguish between activity that occurs only a few hundred milliseconds before or after an event of interest is not available via hemodynamic imaging measures.

Finally, in other ERP subsequent memory studies, the effects have been partitioned according to whether, at the time of test, people were able to recover context information about a study encounter or had sufficient information only to make an accurate judgment about whether or not an item had been encountered previously. In these studies, the question of interest is the relationship between the subsequent memory effects when context information can and cannot be recovered. The results are mixed. In some studies, the critical subsequent memory effects were either identical or differed in magnitude only (Friedman & Trott, 2000; Guo et al., 2005, 2006; Smith, 1993). In others, however, distinct subsequent memory effects were associated with these two kinds of test outcomes (Duarte et al., 2004; Mangels et al., 2001; for what may be a comparable result, see Yovel & Paller, 2004). Taken as a whole, therefore, these findings are equivocal with respect to the question of whether ERPs provide evidence that these two kinds of subjective experience are a consequence of qualitatively different neural activity at the time of encoding, and thereby suggest the operation of distinct processes that support retrieval that either is or is not accompanied by recovery of contextual information. For reasons that will become apparent in the following section, in which ERP studies of retrieval are discussed, this is an important issue. For this reason, there is a strong incentive to pursue further work in which the focus is on the sensitivity of ERPs to encoding operations that predict when contextual information associated with study episodes will be available at the time of retrieval.

ERP Old-New Effects and Recognition Memory Processes

The overwhelming majority of ERP studies of memory have been designed to investigate the processes that support recognition memory. At present, there is significant controversy regarding how these processes should be modeled. Some researchers have argued that recognition is essentially a signal detection process—that is, when one encounters an item (e.g., a face), recognition of that item is based on a general assessment of the strength of the memory elicited by that item (e.g., Donaldson, 1996; Wixted, 2007). In these models, it is assumed that both previously encountered and novel stimuli will elicit some kind of memory *strength* signal, and on average, memory strength will be higher for previously encountered (studied) items. An alternative class of models assumes that recognition memory may be based on two processes, now commonly termed *recollection* and *familiarity* (Jacoby, 1998; Jacoby & Dallas, 1981; Mandler, 1980, 1991; Yonelinas, 2002). Recollection entails recovery of qualitative information from a prior episode. Familiarity provides only a quantitative basis for making judgments of prior occurrence.

In one widely cited *dual-process* model, it is assumed that familiarity resembles a signal detection process but that recollection resembles a threshold process, such that some items will elicit retrieval of associated contextual information, whereas others will fail to do so (Yonelinas, 1999, 2002). Although the body of behavioral evidence is in general more consistent with dual-process than with single-process models, many proponents of pure signal detection models have argued that these claims are overstated due to limitations in the methods that have been used to assess recollection and familiarity (e.g., Rotello et al., 2005; Squire et al., 2007; Wais et al., 2008; see also Yonelinas & Parks, 2007).

Findings from several ERP studies of recognition memory are germane to these issues. The relevant findings come from analyses of *old-new effects* (Rugg, 1994, 1995b). These are differences between the neural activity that is elicited by old (previously studied) and new (previously unstudied) items that attract correct task judgments. These effects are thus distinguishable from *repetition effects* (e.g., Friedman, 1990; Friedman et al., 1992), where the basis for separation is solely the history of prior exposure in a task.

The first report of an old-new ERP effect was provided by Sanquist and colleagues (1980). The task they employed bears similarities with those used in many of the studies described in the following sections and comprised study as well as test phases (see Figure 14.1 for an example). Sanquist et al. described a posterior midline maximum modulation that was more positive-going for old than for new test items that attracted correct old-new judgments on a recognition memory task. This old-new effect was larger for correct old judgments than for incorrect judgments to old and new items (misses and false alarms, respectively), thereby supporting the view that the effect indexes processes that support veridical memory judgments, and not simply (1) the belief that an item was encountered or (2) whether or not the item had been encountered at study (for other clear examples, see Bridson et al., 2006; Rugg et al., 1998; Wilding & Rugg, 1996).

Sanquist et al. (1980) recorded data from midline sites only, and in subsequent similar studies where montages including lateral sites were included, the effect identified by Sanquist et al. has commonly been found to be largest at left-parietal locations. For this reason, the effect has been termed the *left-parietal ERP old-new effect* (Rugg, 1994). The effect is typically evident from 500 to 800 ms after the presentation of test stimuli (see Figure 14.2), and it is preceded by a repetition effect that has a similar scalp distribution but is distinguishable from the left-parietal old-new effect because it does not predict the accuracy of test judgments (Bridson et al., 2006; Rugg et al., 1998).

It is generally agreed that the left-parietal ERP old-new effect is a correlate of recollection. Indeed, an indicator of the broad level of agreement regarding this account can be garnered from the growing number of studies in which the presence or absence of changes in this old-new effect has been employed in order to assess the degree to which recollection was engaged in a given task or condition, either in combination with converging behavioral data (e.g., Norman et al., 2008) or in the absence of direct behavioral support (Bergstrom et al., 2009a, 2009b; Czernochowski et al., 2005; Dzulkifli & Wilding, 2005; Dzulkifli et al., 2006; Herron & Rugg, 2003b; for a prescient prediction, see Paller et al., 1995).

The body of findings that supports the link between this ERP old-new effect and recollection has been reviewed in detail elsewhere (Curran et al., 2006b;

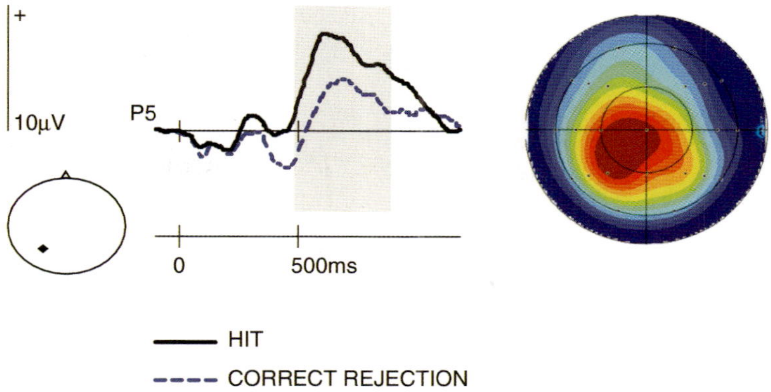

Fig. 14.2. The left-parietal ERP old-new effect comprises a greater relative positivity for old (black line) than for new (blue line) test items that attract correct memory judgments. The ERPs are shown here for a left-hemisphere parietal electrode location (P5) where the effect is commonly largest, and the scalp map on the right side depicts the typical distribution of the effect. The map is computed on the basis of the differences in mean voltage between the ERPs elicited by old and new items over the time period (500–800 ms) that is captured within the shaded rectangle.

Friedman & Johnson, 2000; Rugg & Yonelinas, 2003). Briefly, when people are asked to make forced-choice context judgments—for example, concerning the modality in which test items were presented at study—the magnitude of the left-parietal ERP old-new effect is correlated with the number of accurate context judgments that are made (Wilding, 2000; Wilding & Rugg, 1996; Wilding et al., 1995). These findings support the link between the left-parietal ERP old-new effect and recollection because recollection is the process that is assumed to support accurate context judgments. In a related vein, the magnitude of the left-parietal ERP old-new effect is correlated with participants' subjective reports of the amount of information retrieved (Vilberg & Rugg, 2009; Vilberg et al., 2006), and the effect is larger when participants claim that they can remember a prior event as opposed to having only a sense of familiarity (Duarte et al., 2004; Duzel et al., 1997; Smith, 1993). Furthermore, the size of the left-parietal ERP old-new effect is attenuated (relative to controls) in participants who, as determined by behavioral measures, have impairments in recollection as a result of brain damage, atrophy, or pharmacological interventions (Curran et al., 2006a; Duzel et al., 2001b; Mecklinger et al., 1998; Potter et al., 1992; Rugg et al., 1991; Smith & Halgren, 1989; Tendolkar et al., 1999).

Support for the view that two distinct processes can support recognition memory stems from the fact that ERP studies of recognition memory have revealed a second old-new effect that is temporally, topographically, and functionally distinct from the left-parietal old-new effect (Curran et al., 2006b; Rugg et al., 1998). This effect (see Figure 14.3) is typically evident between 300 and 500 ms after the presentation of a test stimulus and is largest at frontal midline scalp sites. Rugg et al. (1998) were the first to suggest tentatively that this *midfrontal old-new effect* may be an index of familiarity (although see Rugg & Allan, 2000), and support for this account has been inferred from findings in a number of subsequent studies.

In one set of studies, the critical experimental manipulation has been the presentation of two kinds of new (unstudied) test items. These differ in their degree of similarity to the studied items. For example, in the study reported by Curran (2000), study words were nouns, an equal number in their singular or plural forms. Some new test items (*similar lures*) comprised plurality-reversed forms of study words, with the remainder *(dissimilar lures)* being words for which a singular/plural counterpart had not been studied. Correct responses to studied words (plurality constant across study and test) elicited both midfrontal and left-parietal old-new effects, but incorrect (old) responses to similar lures elicited only a midfrontal effect. It has been argued that incorrect old responses (false alarms) to similar lures should be based upon familiarity (Hintzman & Curran, 1994), so the latter finding supports the link between the midfrontal old-new effect and familiarity. In addition, the fact that two distinct old-new effects were associated with correct responses to studied words is consistent with the view that two separable processes can contribute to recognition memory judgments. These claims are further substantiated by the fact that qualitatively similar findings have been reported in a number of

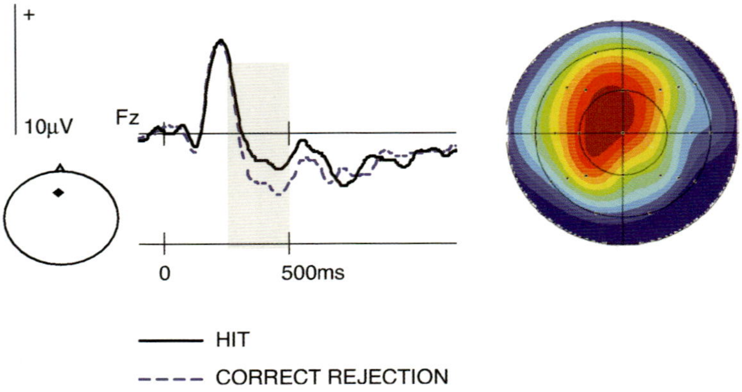

Fig. 14.3. The midfrontal ERP old-new effect comprises a greater relative positivity for old (black line) than for new (blue line) test items that attract correct memory judgments. The ERPs are shown here for a midline frontal electrode location (Fz) where the effect is commonly largest, and the scalp map on the right depicts the typical distribution of the effect. The map is computed on the basis of the differences in mean voltage between the ERPs elicited by old and new items over the time period (300–500 ms) that is captured within the shaded rectangle.

other studies using different old-lure manipulations (Curran & Cleary, 2003; Curran et al., 2002; Nessler et al., 2001; Penney et al., 2001).

These data points all comprise the same single dissociation between two ERP effects. It is therefore important that the reverse single dissociation—changes in the midfrontal old-new effect alongside nonsignificant changes in the parietal effect—has also been described (Azimian-Faridani & Wilding, 2004; Greve et al., 2007; Rhodes & Donaldson, 2007, 2008). Of greater import are within-experiment demonstrations of double dissociations between the effects of interest; the most compelling example of this kind was provided by Woodruff and colleagues (2006; also see Jager et al., 2006). Participants studied words and in a later test made a five-way response. They were asked to make a "remember" response to words for which contextual information about the previous encounter came to mind. For words not attracting a remember response, participants were asked to make a confidence judgment on a 4-point scale. The left-parietal old-new effect was reliable only for items attracting remember responses, while the midfrontal effect increased in magnitude across the transition from confident new responses to confident old responses (see Figure14.4; for other demonstrations that the midfrontal old-new effect indexes familiarity in a graded manner, see Azimian-Faridani & Wilding, 2004, 2006). These findings are consistent with dual-process models that characterize recollection as a threshold-like process and familiarity as a graded index of memory strength (for review, see Yonelinas, 2002).

Although the above evidence suggests a compelling correlation between the midfrontal old-new effect and familiarity, some have argued that this effect is a neural correlate of conceptual priming (Paller et al., 2007; Voss & Paller, 2006). *Priming* is a facilitation of processing by virtue of prior exposure (Schacter, 1987), with *conceptual priming* referring to facilitation of semantic processing (Gabrieli, 1998). As noted in Chapter 15 of this volume, N400 amplitude is modulated by semantic or conceptual priming, and the midfrontal old-new effect occurs in a time window similar to that of the N400. In keeping with our earlier comments, this component correspondence is sufficient to encourage consideration of the midfrontal old-new effect in terms of the antecedents of the N400. The midfrontal old-new effect does, however, have a somewhat more anterior maximum than the classical N400 in most circumstances (Kutas & Hillyard, 1984; Kutas & Van Petten, 1988). At the functional level, however, a conceptual priming account can accommodate some of the memory data described above if one assumes that repetition of meaningful studied stimuli at test results in conceptual priming, which in turn is associated with the midfrontal old-new effect (Olichney et al., 2000; Yovel & Paller, 2004). This account can, moreover, accommodate a larger subset of the available data if it is assumed either that conceptual priming can support familiarity-based recognition judgments (Yonelinas, 2002) or (albeit in a somewhat less theoretically obvious way) if it is assumed that, despite the independence of familiarity and priming, some of the same processes that influence familiarity also influence conceptual priming.

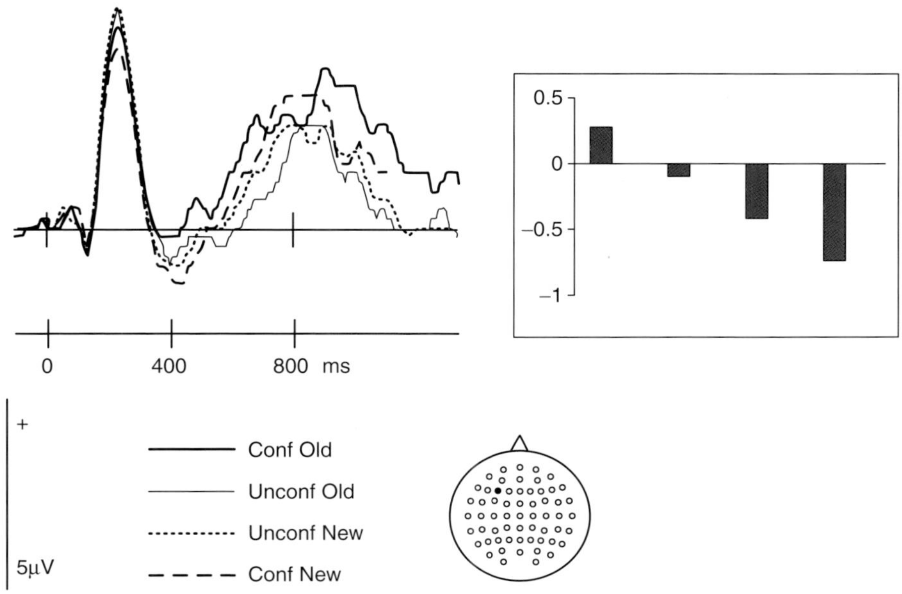

Fig. 14.4. The midfrontal ERP old-new effect varies according to response confidence. The waveforms on the left are for a left-frontal electrode location (F3). The bar graph on the right shows mean amplitude measures for the 300–500 ms poststimulus time period associated with (in order from left to right) confident old, unconfident old, unconfident new, and confident new responses. Figure reproduced with permission from Woodruff et al. (2006).

If the midfrontal old-new effect is a neural correlate of conceptual priming, then in principle, the effect should not be apparent during recognition of stimuli that have little semantic information associated with them, such as faces, fractals, or other novel visual stimuli. Consistent with this idea, in some studies a reliable midfrontal old-new effect has not been observed for unfamiliar faces (Mackenzie & Donaldson, 2007; Yovel & Paller, 2004). In other studies, however, reliable midfrontal old-new effects for faces have been obtained (Curran & Hancock, 2007; Nessler et al., 2005; Yick & Wilding, 2008; for a commentary, see Donaldson & Curran, 2007), and midfrontal old-new effects have also been reported for "meaningless" figures (Curran et al., 2002; Groh-Bordin et al., 2006, 2007). In defense of a conceptual priming account, however, Voss and Paller (2007) showed that participants can find subjective meaning even in ostensibly meaningless stimuli. Furthermore, they showed that the magnitude of the midfrontal old-new effect was larger for items people find meaningful than for those they find meaningless. This argument has itself been countered, based on the observation that the pattern of behavioral data reported by Voss and Paller is most readily interpreted as indicating that "meaningful" squiggles are associated with greater familiarity than are subjectively meaningless ones (Grove & Wilding, 2009). If this account is correct, then the data reported by Voss and Paller cannot adjudicate between a familiarity and a conceptual priming interpretation of the midfrontal old-new effect.

One set of data points that does adjudicate clearly between a conceptual priming account and a familiarity account comprises demonstrations that the magnitude of the midfrontal old-new effect changes according to manipulations of perceptual features of test stimuli. For example, the effect is in some cases larger when perceptual features are held constant across study and test phases of memory tasks in comparison to when those features change (Ecker et al., 2007a, 2007b; Groh-Bordin et al., 2005, 2006; Schloerscheidt & Rugg, 2004). It is difficult to explain such effects in terms of conceptual priming because, by definition, perceptual manipulations should have little effect on conceptual priming. These data, therefore, suggest that the midfrontal old-new effect is sensitive to variables other than conceptual overlap between study and test, and argue for a less restrictive functional account of the effect (for a recent energetic discussion, see Lucas et al., 2010; Stenberg et al., 2009, 2010).

In summary, the weight of evidence suggests that a familiarity account of the midfrontal ERP old-new effect is the strongest candidate at present.

Hence, alongside the evidence for the strong links between the left-parietal ERP old-new effect and recollection, the foregoing review emphasizes that the midfrontal and left-parietal old-new effects behave in a way that is compatible with the two distinct processes that are outlined in contemporary dual-process accounts of recognition memory (Curran et al., 2006b; Mecklinger, 2000; Rugg & Curran, 2007; Rugg & Yonelinas, 2003; Wilding & Sharpe, 2003).

Several questions remain, however, about the precise functional significance and neural bases of these two old-new effects. The left-parietal focus of the later effect is consistent with the view that it indexes activity in the underlying cortex, and this account is bolstered by findings from fMRI studies showing that activity in lateral parietal cortex is sensitive to recollection (Simons & Mayes, 2008; Wagner et al., 2005). According to one view, lateral parietal cortex may be involved in the maintenance and binding of recovered information during episodic retrieval (Vilberg & Rugg, 2008). This account can accommodate many of the existing ERP data points, as well as findings showing that the magnitude of the effect varies according to the amount of episodic information that is retrieved (Vilberg et al., 2006; Wilding, 2000).

For the midfrontal old-new effect, the scalp distribution of the effect suggests a generator in prefrontal cortex (PFC). While caution is necessary when inferring neural sources on the basis of scalp-recorded ERPs (Luck, 2005; Rugg, 1995b), it is also noteworthy that the time course of the effect overlaps with that shown by familiarity—sensitive cells in PFC single-cell and multicell recording studies (Xiang & Brown, 1998, 2004). These correspondences, in combination with some additional ERP data points (Tsivilis et al., 2001), have prompted the proposal that the effect is a downstream index of familiarity signals that are generated in the medial-temporal lobe (Rugg & Curran, 2007). In developing more refined functional accounts of this old-new effect than are available currently, it will also be important to accommodate findings that the conditions under which the effect occurs, and perhaps the time course of the effect, are subject to specific task requirements (Ecker et al., 2007c; Grove & Wilding, 2009; Tsivilis et al., 2001).

In addition, even if one is circumspect about the evidence linking the midfrontal old-new effect to familiarity, there is complementary evidence for dissociations between electrophysiological correlates of recollection and familiarity. For instance, a very early-onsetting (100–200 ms) old-new effect with a fronto-polar scalp topography has been reported (Tsivilis et al., 2001) for complex visual scenes. Duarte et al. (2004) reported a very similar effect and also showed that this effect was functionally distinct from a left-parietal ERP old-new effect.

Finally, it is also worth noting that, although the left-parietal ERP old-new effect is regarded as a content- and material-independent index of recollection (Allan et al., 2000), content-specific (Johnson et al., 2008) and material-specific (Yick & Wilding, 2008) indices of recollection have been described in recent studies. These findings are consistent with the view that recovery of information from episodic memory involves recapitulation of neural activity in the brain regions that were engaged during encoding (Damasio, 1989a, 1989b; Mesulam, 1990, 1998; Rubin & Greenberg, 1998). Whether material- and/or content-specific signatures of familiarity can be obtained, and indeed whether the midfrontal old-new effect is in fact a generic index of familiarity, are likely to be questions to which increasing attention will be paid in coming years.

ERPs and Memory Control and Monitoring Processes

In the previous section, the focus was on ERP modulations of retrieval processes that presumably arise because of interactions of some kind between internal representations of retrieval cues and memory traces (Schacter et al., 1978; Semon, 1921; Tulving, 1983). Of course, memory processes of this kind do not occur in a vacuum, and goal-directed control processes play a critical part in determining the outcome of the retrieval process and ultimately the memory judgment that is made. For example, the form that an internal representation of a retrieval cue takes might be influenced by the kinds of information that a person might want to retrieve. Similarly, the way in which (or the kind of) information that is subjected to monitoring and evaluation following retrieval can also vary with task demands. These examples encapsulate the widely accepted view that memory control processes are engaged before, during, *and* after interactions between retrieval cues and memory traces (Burgess & Shallice, 1996; Schacter et al., 1998), and ERP studies have provided insights into the nature of these processes.

There is considerable evidence that the integrity of the PFC is critical for cognitive control operations, including control over memory retrieval (for relevant reviews, see Ranganath, 2004; Ranganath &

Knight, 2003). For example, damage to the PFC can result in selective episodic memory impairments that are more marked on tasks requiring recovery of contextual details than on those that simply require judgments of whether an event had or had not occurred previously (Janowsky et al., 1989; Milner, 1971; Schacter et al., 1984; Shimamura & Squire, 1987; Squire, 1982). Damage to the PFC can also lead to confabulation, or more commonly to an increased likelihood of memory distortions (Burgess & Shallice, 1996). In keeping with such observations, it is not surprising that ERP correlates of memory control processes have scalp distributions that are consistent with generators located in the PFC. Below, we focus first on ERP indices of processes that operate after or in parallel with interactions between retrieval cues and memory traces (*postretrieval processes*). This class of operations is assumed to operate on the products of retrieval; hence, the ERP modulations that index them have (1) onset times no earlier than those of the left-parietal and midfrontal ERP old-new effects described above and (2) time courses that extend beyond the points at which the midfrontal and left-parietal effects terminate. After a review of this class of processes, the focus moves to those that are initiated before cue-trace interactions (*preretrieval processes*). These can be thought of as preparatory retrieval processes that influence the processes that are initiated when a retrieval cue is encountered (Rugg & Wilding, 2000; Rugg et al., 2002; Wilding & Herron, 2006).

POSTRETRIEVAL PROCESSING OPERATIONS

A number of studies contain reports of frontally distributed ERP old-new effects that have time courses extending well beyond those of the midfrontal and left-parietal ERP old-new effects described in the preceding sections. These effects often, but not always, have a right-lateralized scalp distribution. It is also important that these effects have been more commonly observed during tasks requiring recovery of contextual information, compared to tasks in which only recognition memory judgments were required (Johansson et al., 2002; Johnson et al., 1996, 1997; Senkfor & Van Petten, 1998; Van Petten et al., 2000). In one of the first studies in which a right-lateralized late frontal old-new effect was reported, Wilding and Rugg (1996) acquired ERPs during a task in which people made yes-no item recognition judgments to visually presented words and then made a two-alternative forced-choice judgment indicating whether old words had been spoken by a male or a female voice at study. The ERP old-new effects were contrasted for old words for which the source (voice) judgment was either correct or incorrect. A right-lateralized, frontally distributed, positive-going ERP old-new effect was observed. This effect began at around 800 ms poststimulus and lasted until the end of the 1400 ms recording epoch (see Figure 14.5). The effect was larger for items attracting correct rather than incorrect source judgments, and on the basis of this outcome, as well as the relatively protracted time

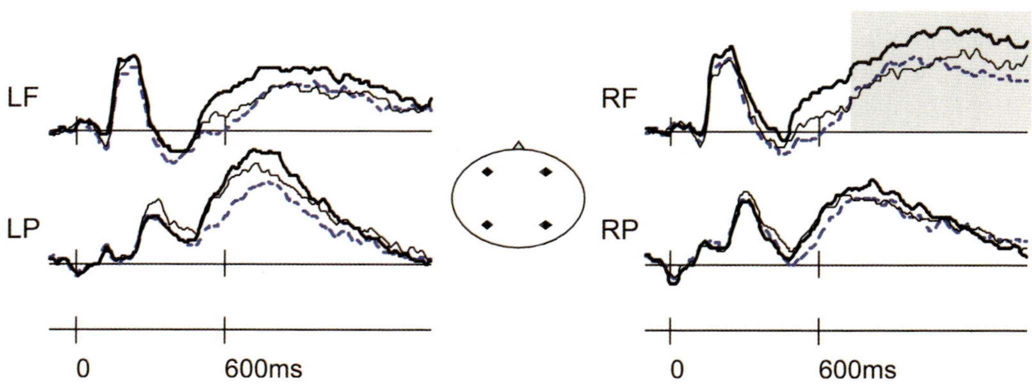

Fig. 14.5. The right-frontal ERP old-new effect. The shaded area illustrates a frontally distributed right-lateralized effect that, in this experiment, takes the form of greater relative positivity for test items that attracted correct old as well as correct context (voice) judgments (thick black line) relative to two other critical classes of ERPs: new items that attracted correct judgments (blue dotted line) and old items that attracted correct old judgments but incorrect voice judgments (thin black line). The ERPs are shown here for four electrode locations located over the left and right anterior as well as the posterior scalp. This is an adapted representation of the data presented by Wilding and Rugg (1996).

course of the effect, Wilding and Rugg (1996) proposed that it indexed postretrieval processes important for forming an integrated representation of a prior episode.

Subsequent research, however, has cast considerable doubt on this account. For instance, the right-frontal old-new effect is not always evident in circumstances where correct source judgments have been made (Swick et al., 2006; Wilding & Rugg, 1997), nor does the effect always predict the accuracy of source judgments (Senkfor & Van Petten, 1998; Vallesi & Shallice, 2006; Van Petten et al., 2000). Moreover, it has been shown recently that the effect is present in tasks where there is no episodic retrieval requirement. In two experiments, Hayama and colleagues (2008) contrasted the right-frontal old-new effects that were obtained in tasks requiring either source retrieval or semantic retrieval. Equivalent right-frontal old-new effects were obtained in both cases.

On the basis of these results, Hayama et al. (2008) proposed that the effect is likely to index generic monitoring operations that are engaged when assessing recovered information in a goal-directed way irrespective of the memory system or processes from which the information became available. This is probably the most parsimonious account available at present, but even this hypothesis has difficulty accounting for the inconsistent findings when ERPs are sorted according to the accuracy of source judgments. Another disparity stems from findings showing that right-frontal old-new effects are evident in some instances when only recognition memory judgments are required, but not in others (Allan & Rugg, 1997; Rugg et al., 2000), despite the use of ostensibly similar paradigms and similar levels of response accuracy.

In what ways might these issues be resolved? One possibility is that at least some of the disparities occur because not entirely the same frontal old-new effects were obtained in different studies. This possibility is supported by inspection of the data in published studies: there are marked differences in the degrees to which late-frontal old-new effects are lateralized to the right hemisphere (cf. Kuo & Van Petten, 2006, 2008; Tendolkar & Rugg, 1998; Trott et al., 1997, 1999; Wilding, 1999). There are to our knowledge, however, only two direct demonstrations of functionally and electrophysiologically separable late-frontal old-new effects (Cruse & Wilding, 2009; Woodruff et al., 2006). This is surprising, given that the findings from numerous hemodynamic brain imaging studies, as well as from studies of patients with focal prefrontal lesions, attest to the functional heterogeneity of the left and right prefrontal cortices (Nolde et al., 1998; Ranganath, 2004; Ranganath & Knight, 2004; Rugg & Henson, 2003; Stuss et al., 1994).

This observation can, of course, be countered by the argument that ERPs are sensitive to only a proportion of the total neural activity that is elicited in response to a given stimulus in a given context (Coles & Rugg, 1995; Coles et al., 1990; Hillyard & Kutas, 1983), but the findings of Woodruff et al. (2006) as well as those of Cruse and Wilding (2009) indicate that, at least under some circumstances, it is possible to distinguish between different late-frontal old-new effects that are likely to index control operations mediated by the PFC. There remain at present good grounds for assuming that the use of appropriate behavioral manipulations, perhaps in combination with increased spatial sampling, will lead to identification of a class of distinct late-occurring frontally distributed old-new effects, and following from this, an increased understanding of the nature and time courses of PFC contributions to memory retrieval processing operations.

PRERETRIEVAL PROCESSING OPERATIONS

Although cognitive theories have proposed a role for memory control processes that are engaged prior to retrieval (e.g., Burgess & Shallice, 1996; Schacter et al., 1998; Tulving, 1983; Wheeler et al., 1997), only recently have these processes been subject to empirical investigation, and this new strand of research has been driven by findings from ERP studies. In particular, the high temporal resolution of ERPs makes it possible to disentangle neural activity that is elicited immediately prior to a retrieval cue from activity that is triggered by the cue itself, something that is not straightforward using hemodynamic measures (Rugg & Wilding, 2000; for relevant fMRI work, see Simons et al., 2005a, 2005b).

In one important early ERP study, Duzel and colleagues (1999; also see Duzel et al., 2001a) acquired DC potentials while participants completed episodic (old-new item recognition judgments) or semantic retrieval (animacy judgments) tasks. A cue signaled which task to complete and was followed by presentation of four words one at a time. This study was novel in that the focus was on neural activity elicited by the cues that were used to signal which task was to be completed. The ERP data (see Figure 14.6) revealed a sustained positive shift in potentials following the episodic retrieval

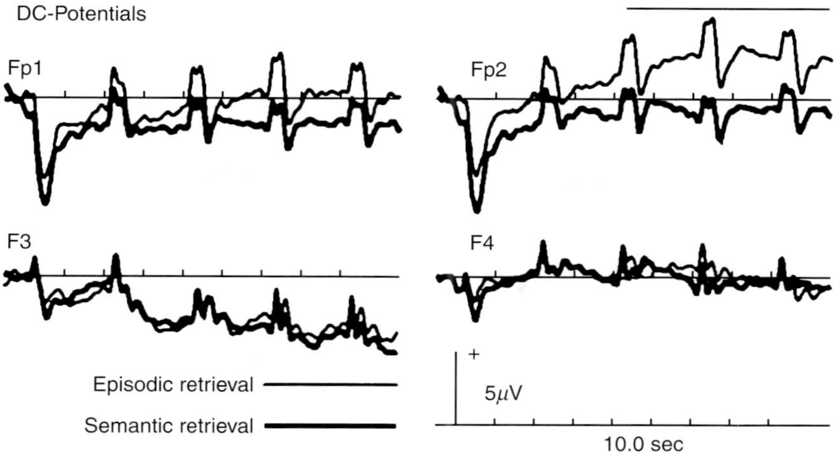

Fig. 14.6. Low-frequency changes in neural activity initiated in response to two different cues. The first (thin black line) signaled that participants should make episodic (old-new recognition memory) judgments to four successive test items. The second (thick black line) signaled that semantic (animate/inanimate) judgments should be made. The sustained greater relative positivity initiated by the episodic retrieval cue can be seen clearly at the right-hemisphere frontopolar site (FP2). FP1 is the homologous left hemisphere scalp site for FP2, while F4 and F3 are sites situated slightly posterior to the FP sites at right and left hemisphere locations, respectively. Figure reproduced with permission from Duzel, E., Cabeza, R., Picton, T. W., Yonelinas, A. P., Scheich, H., Heinze, H. J., Tulving, E., 1999. Task-related and item-related brain processes ofmemory retrieval. Proceedings of the National Academy of Sciences 96, 1794–1799. Copyright (1999) National Academy of Sciences, U.S.A.

task cue relative to the semantic retrieval task cue, the magnitude of which was largest over right-frontal scalp sites. The timing of the positive-going effect was such that it appeared soon after the onset of the cue and was sustained during the presentation of the four words for which the same task judgment was required. The authors interpreted the results as reflecting a process that was engaged during the episodic retrieval task rather than one engaged during the semantic task. More specifically, the right-frontal electrical shift was interpreted as an index of the development and/or maintenance of a preparatory retrieval set that people enter into when one attempts to recover information about a past episode, termed *episodic retrieval mode* (Duzel et al., 1999; Rugg & Wilding, 2000; Tulving, 1983).

In subsequent ERP investigations of retrieval mode, the effects of interest were obtained by using tasks in which people were cued trial-by-trial to make different kinds of task judgments. For instance, Morcom and Rugg (2002) required participants to switch between the episodic and semantic retrieval tasks used by Duzel et al. (1999), with a cue preceding each word that required a task judgment by a little less than 2 s. The principal contrasts of interest in these studies are between the ERPs that are elicited by the cues that signal preparation to retrieve either episodic or semantic information and the ERPs that precede the items to which the memory judgments are required.

Morcom and Rugg (2002) reported an ERP modulation that was more positive-going for cues signaling preparation for episodic rather than semantic retrieval. As in the study of Duzel et al. (1999), this effect was largest at right-frontal scalp locations. Critically, the effect was evident on trials on which the cue had been the same on the preceding trial (stay trials) but not when the preceding cue had been different (switch trials). This finding was interpreted as evidence that adopting a task-relevant set (retrieval mode in this instance) is resource demanding and takes time to accomplish. Herron and Wilding (2004, 2006a) replicated this finding and additionally showed that only activity associated with episodic cues changed over successive trials (Herron & Wilding, 2006a). This finding supports the claim that differences between the ERPs elicited by these two kinds of cues are correlates of episodic rather than semantic retrieval processing.

In keeping with the functional account offered by Duzel et al. (1999), these effects were characterized as indices of retrieval mode (Tulving, 1983; Wheeler et al., 1997). This claim is supported strongly by the fact that this index was observed for markedly different kinds of episodic retrieval tasks (in particular, see Herron & Wilding, 2004), since according to Tulving's initial account (Tulving, 1983), retrieval mode will be adopted whenever any form of episodic retrieval is required.

RETRIEVAL ORIENTATIONS

The ERP evidence for an episodic retrieval mode suggests that there may be one or more cognitive operations that are generally engaged when one is attempting to recover information about a past episode. In real-life situations, however, it is often the case that one is interested in recovering specific kinds of episodic information. For instance, if you are attempting to remember whether you turned off the stove before leaving the house or simply *thought* about turning it off, it would be useful to search specifically for vivid visual or tactile information that would indicate that you actually turned it off (Johnson & Raye, 1981; Johnson et al., 1993). Consistent with such experiences, ERP studies have revealed evidence that participants can adopt a *retrieval orientation* (Rugg & Wilding, 2000) that constrains the processing of retrieval cues based on the specific demands of episodic retrieval tasks. Like retrieval mode, retrieval orientations are cognitive sets that can be maintained while episodic retrieval is required and that can be initiated before a memory judgment is to be made (Rugg & Wilding, 2000). Orientations differ from mode in that they are cognitive operations engaged according to the specific demands of an episodic retrieval task, such as the kind of episodic content that might be required to make a task-relevant judgment.

Event-related potential correlates of retrieval orientation have been investigated in paradigms similar to those in which putative correlates of retrieval mode have been identified. The key difference, however, is the use of cues that signal preparation for different kinds of episodic memory retrieval, rather than enabling contrasts between preparation for episodic retrieval in comparison to a nonepisodic (typically semantic) retrieval condition. In the first demonstration of ERP indices of this class of preretrieval processes, Herron and Wilding (2004) showed that ERPs elicited by preparatory cues were more positive-going at left-frontal and left-central scalp sites when preparing to retrieve conceptual rather than perceptual information about items that had been encountered in a prior study phase (see Figure 14.7). The scalp distribution of these differences dissociates this orientation effect from the right-lateralized effects that have been linked to retrieval mode (Herron & Wilding, 2004).

This distinction between retrieval mode and retrieval orientations motivates questions about how these two classes of retrieval process confer benefits on memory retrieval and how they differ. The evidence to date suggests that adopting retrieval mode is beneficial for the accuracy of memory judgments, while orientations influence primarily the efficiency with which recovered information is processed (Herron & Wilding, 2006a, 2006b; Johnson & Rugg, 2006; Wilding & Herron, 2006). These data points are really, however, no more than a starting point for initiating directed studies that will delineate the ways in which retrieval mode and retrieval orientations operate at distinct retrieval processing stages in order to facilitate memory judgments. For example, benefits in the accuracy and/or the time course of memory judgments can come about for many reasons. One is that preretrieval processing influences cue-specification processes that will influence the outcomes of cue–trace interactions (Anderson, 2003; Anderson & Bjork, 1994). Another (not mutually exclusive) possibility is that the point of impact is further downstream, where outcomes of cue–trace interactions are evaluated in the service of task goals (Rugg & Wilding, 2000).

Additional insights into how preparing to retrieve influences memory decisions have also come from a related set of studies in which the focus remains on preretrieval processing, but where the inferences depend upon the outcomes of contrasts between neural activities that are triggered by the presentation of retrieval cues, rather than on analyses of neural activities that precede these cues. More specifically, the focus in this approach is on neural activities that are elicited by unstudied (new) test items under different task conditions (Rugg & Wilding, 2000). In the majority of studies to date in which this approach has been taken, the critical task conditions have had different episodic retrieval requirements (although see Hornberger et al., 2006), and any differences obtained are thus conceived of as being effects that are a consequence of adopting different retrieval orientations.

To index optimally these retrieval-cue specific indices of retrieval orientations, it has been argued that it is necessary to restrict comparisons to ERPs elicited by unstudied items, because ERP differences associated with studied items taken from two different memory tasks might index changes in retrieval orientation and/or processes related to successful recovery of different kinds of information (Donaldson et al., 2003; Rugg & Wilding, 2000; Rugg et al., 2002). It may be the case that comparable contrasts involving old items will under some circumstances index some of the same processes

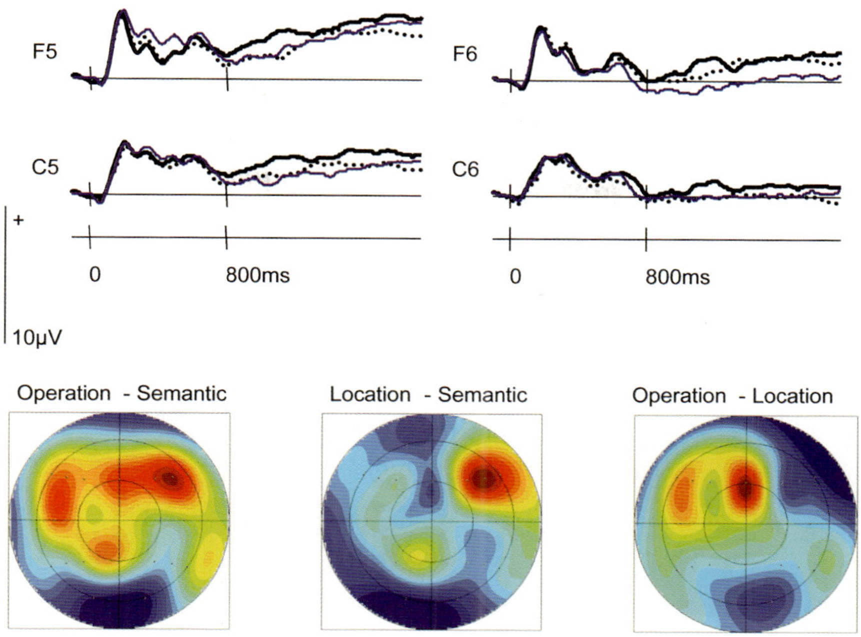

Fig. 14.7. Event-related potential waveforms showing the neural activity elicited by preparatory cues that signaled participants should prepare to make either semantic memory judgments (blue line), judgments about study location (dotted black line), or judgments about the cognitive operations engaged at study (continuous black line). The scalp maps show the distributions of the neural activity that differentiates among these three kinds of preparatory neural activity. The focal right-frontal positivity that is shared by both episodic preparatory conditions relative to the semantic condition can be seen clearly in the left-hand and middle scalp maps. The midline and left-sided activity differentiating between the two conditions associated with preparation for different kinds of episodic retrieval can be seen in the right-hand scalp map. This is an adapted representation of the data shown in Herron and Wilding (2004).

(e.g., Ranganath & Paller, 1999), but in addition to concerns about contamination from indices of successful retrieval, it is also possible that the time course over which processes related to retrieval attempts are made will vary for new and old items. For example, if these kinds of processes are terminated at around the time at which successful retrieval occurs, then they may be evident over a shorter period in contrasts between classes of old items than in contrasts restricted to new test items (Dzulkifli et al., 2004; Rugg et al., 2002).

The first ERP evidence that different retrieval task demands influenced the ERPs elicited by new test items was reported by Johnson and colleagues (1997). Waveforms were not shown in this short report, but in keeping with the theoretical framework outlined above, the authors proposed that the differences they observed reflected processes that were engaged in order to interrogate task-specific elements of memory traces. Further evidence of the sensitivity of ERPs to retrieval processing operations of this kind was provided by Ranganath and Paller (1999, 2000) and by Wilding (1999), who, using very different memory retrieval tasks, observed differences between ERPs elicited by classes of new items according to the kinds of information that were task-relevant.

Ranganath and Paller (1999) required participants to complete two different retrieval tasks (see Figure 14.8) assessing memory for previously studied line drawings of objects. In both test conditions, participants were shown studied objects, new objects, and studied objects for which the aspect ratio had been changed. In one of the two test conditions, the *general* test, participants were asked to make old-new recognition memory judgments, responding "old" to all studied items, irrespective of changes in aspect ratio. In the other test condition, the *specific* test, participants were told to make "old" responses only to objects presented in the same aspect ratio at study and test, and to treat each object with a changed aspect ratio as a new item. As shown in Figure 14.8, ERPs elicited by both old and new test items were more positive-going in the specific condition than in the general condition, particularly at left-frontal scalp locations. The fact that the ERP modulation was apparent for new items was interesting, because participants were instructed to treat new items in the

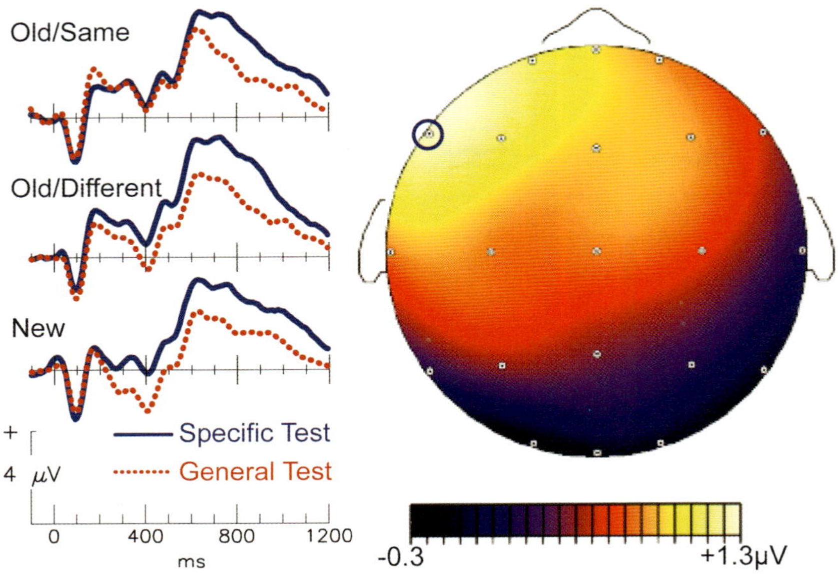

Fig. 14.8. Left-frontal ERPs are sensitive to retrieval orientation. In this experiment, participants were shown studied objects ("Old/Same"), studied objects for which the aspect ratio had been changed ("Old/Different"), and objects that were not previously studied ("New"). In the "Specific Test" condition (solid blue ERP trace), participants were told to pay careful attention to the aspect ratio of each object and determine whether the test item precisely matched the study item, whereas in the "General Test" (dotted brown trace) condition, participants were instructed to disregard aspect ratio changes and simply determine whether the item had been studied previously. Event-related potentials, shown for a left-frontal (F7) site, were more positive-going for specific test trials. As shown in the topographic map at the right, the effect was largest over left-frontal scalp sites. This is an adapted representation of the data first reported by Ranganath and Paller (1999).

same way in both test conditions, and accuracy and reaction times for new items did not differ across the two test conditions. The fact that the frontal ERP modulation was seen for new and old items, in combination with the early onset of the effect (~200 ms poststimulus), suggested that it indexed processes engaged prior to item recognition. More specifically, the data suggested that, across the two test conditions, participants changed the way in which they evaluated each test item, supporting the idea that participants adopt *retrieval orientations* (Rugg & Wilding, 2000) that constrain the processing of retrieval cues to optimize recovery of task-relevant information.

Wilding (1999) also observed reliable differences between ERPs elicited by new test items. In that experiment, study items were spoken words, an equal number spoken by a male and a female voice. Participants were cued trial-by-trial to complete one of two encoding tasks for each word. Encoding task was manipulated orthogonally to speaker gender. All stimuli were presented visually at test, and in separate blocks an equal number of old and new items were shown. In one block, the three-way test response requirement was to distinguish between new items and old items spoken in either the male or the female voice at study. In the other block, the three-way distinction was between new items and items encountered in the two different encoding tasks.

The ERPs elicited by new items in the voice retrieval condition were more positive-going than those in the encoding retrieval condition at left-frontal sites, but the reverse was true at right-frontal sites. It is tempting to link the left-frontal element of this difference across tasks to the comparable modulation described by Ranganath and Paller (1999, 2000), since in each case the greater relative positivity occurred in the task in which there was a heavier emphasis on recovery and assessment of perceptual content. There is, however, no direct evidence in the data set reported by Wilding (1999) that the polarity reversals at right- and left-hemisphere frontal sites index functionally distinct processes; as a result, it is equally plausible to offer a unitary account for the left- and right-hemisphere differences. One means of doing this is to surmise that the pattern of differences indexes a distinct set of processes that are engaged differentially across tasks where the retrieval emphasis is on perceptual or conceptual information.

Irrespective of the actual correspondence between the effects reported in these two studies, however, the frontal distributions of the effects in both cases implicate the PFC in retrieval processing operations that are not restricted to those that are engaged when episodic information is recovered successfully. The differences between the distributions of the effects across the two studies also indicate that separable processes of this kind are engaged according to the specific processing demands that are imposed at the time of retrieval.

Differences between ERPs elicited by new items have also been analyzed in tasks in which frequent switches between retrieval demands have been required (Herron & Wilding, 2006b; Johnson & Rugg, 2006; Werkle-Bergner et al., 2005; Wilding & Nobre, 2001). In each of these cases, participants were cued to complete one of two different episodic retrieval tasks. Herron and Wilding (2006b) as well as Johnson and Rugg (2006) contrasted ERPs elicited by new items on switch and stay trials. There were reliable differences only in the latter case. These findings therefore suggest a conclusion similar to that drawn above for the absence of differences between ERPs elicited by different preparatory retrieval cues on switch trials: namely, that a single exposure to a task cue is not sufficient to engage the operations that the cue signals. The correspondences between these sets of findings for ERPs elicited by preparatory retrieval cues and those elicited by new test items are consistent with the assumption that item-specific cue-processing operations are contingent on the successful adoption and maintenance of a relevant retrieval orientation, as described previously.

In general, it is reasonable to assume that the ERP modulations elicited by new test items described above reflect strategic changes in retrieval processing according to task demands. It is also important to note, however, that the scalp distributions of these ERP modulations have varied considerably across the studies of interest. This heterogeneity is consistent with the view that the specific processes that are engaged depend upon individual task demands; in line with this view, functional accounts of these task-specific effects have varied accordingly (e.g., Dzulkifli & Wilding, 2005; Dzulkifli et al., 2006; Herron & Rugg, 2003a; Ranganath & Paller, 1999, 2000; Werkle-Bergner et al., 2005).

That said, there are some commonalities across studies that allow some general (albeit tentative) conclusions to be made about the nature of the selective retrieval cue processing operations that are indexed by ERPs. First, a common finding across several studies is that ERP correlates of retrieval orientation have an extended duration and typically do not show marked changes in their scalp distributions with time (e.g., Dzulkifli & Wilding, 2005; Dzulkifli et al., 2006; Johnson & Rugg, 2006; Ranganath & Paller, 1999). This finding is consistent with the possibility that temporally extended ERP indices of task-specific cue processing reflect the maintenance of internal representations of retrieval cues in the service of recovery of task-relevant information and that these processes are terminated when a decision to respond "new" is made (Dzulkifli et al., 2004; Johnson & Rugg, 2006; Rugg et al., 2002). This account is consistent with the finding that between-task ERP differences elicited by unstudied items can emerge as early as 200 ms poststimulus (e.g., Hornberger et al., 2006; Werkle-Bergner et al., 2005), as well as the fact that ERP correlates of retrieval orientation are attenuated markedly when the task demands do not include an explicit requirement to recover information encoded in a prior episode (Hornberger et al., 2004). The latter finding rules out the alternative interpretation that effects ascribed to the consequences of having adopted a task-relevant orientation in fact reflect task differences that do not bear on explicit retrieval demands.

An important related finding is that these ERP indices of task-specific retrieval processing also predict response accuracy. Bridger and colleagues (2009) contrasted the differences between new test item ERPs in distinct retrieval tasks and plotted the magnitudes of these differences against task performance. The magnitudes of these differences—presumably indicative of an increased reliance on task-specific retrieval processing—were correlated with response accuracy. These findings provide evidence for the first time that the strategic processes indexed by differences between ERPs elicited by new test items do in fact influence the accuracy with which memory judgments are made. A goal for future work is identification of the specific processing stages that are involved in accomplishing this (Rugg & Wilding, 2000).

In summary, the work described above substantiates the view that people adopt a generic retrieval set—retrieval mode—that confers advantages on subsequent processing operations and ultimately on the accuracy of memory judgments (Tulving, 1983, 1993). The research has also indicated that the concept of retrieval mode is not in itself sufficient to capture the range of preparatory retrieval-related activity that is adopted when people prepare to retrieve. Support for the concept of retrieval orientations comes from the finding that neural activity differs according to the kinds of episodic memory judgments that people are preparing to make (Rugg & Wilding, 2000). Support for the view that adopting a task-relevant orientation influences the processes that are subsequently engaged in order to facilitate retrieval (and/or the processing of retrieved information) comes from contrasts between classes of ERPs that are elicited by new items in tasks having different retrieval demands (Johnson et al., 1997; Ranganath & Paller, 1999). In combination with the earlier description of processes that are indexed by ERP old-new effects, these findings emphasize the range of contributions that ERP research has made to our understanding of the cascade of processes that are engaged when memory judgments are required.

The foregoing accounts and data points highlight the usefulness of ERPs in the study of complex retrieval processing operations. The processes described above would be difficult to characterize fully by employing behavioral measures alone. In addition, hemodynamic imaging measures lack the necessary temporal precision to distinguish between processes that either immediately precede or follow a retrieval cue. These observations therefore emphasize the unique contributions that real-time measures of cognitive and neural processing can make to dynamic accounts of human memory retrieval operations.

Concluding Remarks

The past 20 years have seen tremendous growth in the field of ERP research on human memory, and this chapter provides indications of the directions in which the research is moving currently. The ERP research that is relevant to debates about single- versus dual-process models of recognition memory emphasizes the currency that the measure has for addressing ongoing controversies in human memory research. The ERP research that is concerned with functional accounts of preparatory retrieval processing and item-specific retrieval operations illustrates the ways in which ERPs have contributed to multi-stage models of human memory retrieval and the potential they have for continuing to do so. It is not an understatement to claim that, in the case of these latter research strands, ERP research has been influential in setting the agenda for subsequent behavioral and hemodynamic imaging studies. The use of careful experimental manipulations that isolate the processes of interest, when coupled with designs that exploit the temporal resolution of the ERP method, provides good grounds for believing that this leading role in memory research will continue in years to come.

References

Allan, K., Robb, W. G. K., & Rugg, M. D. (2000). The effect of encoding manipulations on neural correlates of episodic retrieval. *Neuropsychologia, 38*, 1188–1205.

Allan, K., & Rugg, M. D. (1997). An event-related potential study of explicit memory on tests of cued recall and recognition. *Neuropsychologia, 35*, 387–397.

Anderson, M. C. (2003). Rethinking interference theory: Executive control and the mechanisms of forgetting. *Journal of Memory and Language, 49*, 415–445.

Anderson, M. C., & Bjork, R. A. (1994). Mechanisms of inhibition in long term memory. In D. Dagenbach & T. H. Carr (Eds.), *Inhibitory processes in attention, memory, and language* (pp. 265–325). San Diego, CA: Academic Press.

Azimian-Faridani, N., & Wilding, E. L. (2004). An event-related potential study of the revelation effect. *Psychonomic Bulletin and Review, 11*, 926–931.

Azimian-Faridani, N., & Wilding, E. L. (2006). The influence of criterion shifts on electrophysiological correlates of recognition memory. *Journal of Cognitive Neuroscience, 18*, 1075–1086.

Bergsrtom, Z. M., de Fockert, J.W., & Richardson-Klavehn, A. (2009a). ERP and behavioural evidence for direct suppression of unwanted memories. *Neuroimage, 48*, 726–737.

Bergsrtom, Z. M., de Fockert, J.W., & Richardson-Klavehn, A. (2009b). Event-related potential evidence that automatic recollection can be voluntarily avoided. *Journal of Cognitive Neuroscience*, 21, 1280–1301.

Bridger, E. K., Herron, J. E., Elward, R. L., & Wilding, E. L. (2009). Neural correlates of individual differences in strategic retrieval processing. *Journal of Experimental Psychology: Learning, Memory and Cognition, 35*, 1175–1186.

Bridson, N. C., Fraser, C. S., Herron, J. E., & Wilding, E. L. (2006). Electrophysiological correlates of familiarity in recognition memory and exclusion tasks. *Brain Research, 1114*, 149–160.

Burgess, P. W., & Shallice, T. (1996). Confabulation and the control of recollection. *Memory, 4*, 359–411.

Buxton, R. B. (2002). *Introduction to functional magnetic resonance imaging: Principles and techniques*. Cambridge: Cambridge University Press.

Buxton, R. B., Wong, E. C., & Frank, L. R. (1998). Dynamics of blood flow and oxygenation changes during brain activation: The balloon model. *Magnetic Resonance in Medicine, 39*, 855–864.

Coles, M. G. H., Gratton, G., & Fabiani, M (1990). Event-related brain potentials. In J. T. Cacciopo & L. G. Tassinary (Eds.), *Principles of psychophysiology: Physical, social and inferential elements* (pp. 413–455). Cambridge: Cambridge University Press.

Coles, M. G. H., & Rugg, M. (1995). Event-related potentials: An introduction. In M. Rugg & M. G. H. Coles (Eds.), *Electrophysiology of mind* (pp. 1–23). Oxford: Oxford University Press.

Cruse, D. C., & Wilding, E. L. (2009). Prefrontal cortex contributions to episodic retrieval monitoring and evaluation. *Neuropsychologia, 47*, 2779–2789.

Curran, T. (2000). Brain potentials of recollection and familiarity. *Cognitive Brain Research, 28*, 923–938.
Curran, T., & Cleary, A. M. (2003). Using ERPs to dissociate familiarity from recollection in picture recognition. *Cognitive Brain Research, 15*, 191–205.
Curran, T., DeBuse, C., Woroch, B., & Hirshman, E. (2006a). Combined pharmacological and electrophysiological dissociation of familiarity and recollection. *Journal of Neuroscience, 26*, 1979–1985.
Curran, T., & Hancock, J. (2007). The FN400 indexes familiarity-based recognition of faces. *Neuroimage, 36*, 464–471.
Curran, T., Tanaka, J. W., & Weiskopf, D. M. (2002). An electrophysiological comparison of visual categorization and recognition memory. *Cognitive, Affective, & Behavioral Neuroscience, 2*, 1–18.
Curran, T., Tepe, K. L., & Piatt, C. (2006b). ERP explorations of dual processes in recognition memory. In H. D. Zimmer, A. Mecklinger, & U. Lindenberger (Eds.), *Binding in human memory: A neurocognitive approach* (pp. 139–179). Oxford: Oxford University Press.
Czernochowski, D., Mecklinger, A., Johansson, M., & Brinkmann, M. (2005). Age-related differences in familiarity and recollection: ERP evidence from a recognition memory study in children and young adults. *Cognitive, Affective, & Behavioral Neuroscience, 5*, 417–433.
D'Esposito, M. (2003). Alterations in the BOLD fMRI signal with aging and disease: A challenge for neuroimaging. *Nature Reviews Neuroscience, 4*, 863–872.
Damasio, A. R. (1989a). The brain binds entities and events by multiregional activation from convergence zones. *Neural Computation, 1*, 123–132.
Damasio, A. R. (1989b). Time-locked multiregional retroactivation: A systems-level proposal for the neural substrates of recall and recognition. *Cognition, 33*, 25–62.
Donaldson, D. I., & Curran, T. (2007). Potential (ERP) studies of recognition memory for faces. *Neuroimage, 36*, 488–489.
Donaldson, D. I., Wilding, E. L., & Allan, K. (2003). Fractionating retrieval from episodic memory using event-related potentials. In A. E. Parker, E. L. Wilding, & T. J. Bussey (Eds.), *The cognitive neuroscience of memory: Episodic encoding and retrieval* (pp. 39–58). New York: Psychology Press.
Donaldson, W. (1996). The role of decision processes in remembering and knowing. *Cognitive Brain Research, 24*, 523–533.
Duarte, A., Ranganath, C., Winward, L., Hayward, D., & Knight, R. T. (2004). Dissociable neural correlates for familiarity and recollection during the encoding and retrieval of pictures. *Cognitive Brain Research, 18*, 255–272.
Duzel, E., Cabeza, R., Picton, T. W., Yonelinas, A. P., Scheich, H., Heinze, H. J., & Tulving, E. (1999). Task-related and item-related brain processes of memory retrieval. *Proceedings of the National Academy of Sciences USA, 96*, 1794–1799.
Duzel, E., Picton, T. W., Cabeza, R., Yonelinas, A. P., Scheich, H., Heinze, H. J., & Tulving, E. (2001a). Comparing electrophysiological and hemodynamic measures of neural activation during memory retrieval. *Human Brain Mapping, 13*, 104–123.
Duzel, E., Vargha-Khadem, F., Heinze, H. J., & Mishkin, M. (2001b). Brain activity evidence for recognition without recollection after early hippocampal damage. *Proceedings of the National Academy of Sciences USA, 98*, 8101–8106.
Duzel, E., Yonelinas, A. P., Mangun, G. R., Heinze, H. J., & Tulving, E. (1997). Event-related brain potential correlates of two states of conscious awareness in memory. *Proceedings of the National Academy of Sciences USA, 94*, 5973–5978.
Dzulkifli, M. A., Herron, J. E., & Wilding, E. L. (2006). Memory retrieval processing: Neural indices of processes supporting selective episodic retrieval. *Neuropsychologia, 44*, 1120–1130.
Dzulkifli, M. A., Sharpe, H. L., & Wilding, E. L. (2004). Separating item-related electrophysiological indices of retrieval effort and retrieval orientation. *Brain and Cognition, 55*, 433–443.
Dzulkifli, M. A., & Wilding, E. L. (2005). Electrophysiological indices of strategic episodic retrieval processing. *Neuropsychologia, 43*, 1152–1162.
Ecker, U. K. H., Zimmer, H. D., & Groh-Bordin, C. (2007a). Color and context: An ERP study on intrinsic and extrinsic feature binding in episodic memory. *Cognitive Brain Research, 35*, 1483–1501.
Ecker, U. K. H., Zimmer, H. D., & Groh-Bordin, C. (2007b). The influence of object and background colour manipulations on the electrophysiological indices of recognition memory. *Brain Research, 1185*, 221–230.
Ecker, U. K. H., Zimmer, H. D., Groh-Bordin, C., & Mecklinger, A. (2007c). Context effects on familiarity are familiarity effects of context—An electrophysiological study. *International Journal of Psychophysiology, 64*, 146–156.
Friedman, D. (1990). ERPs during continuous recognition memory for words. *Biological Psychology, 30*, 61–87.
Friedman, D. (2003). Cognition and aging: A highly selective overview of event-related potential (ERP) data. *Journal of Clinical and Experimental Neurophysiology, 25*, 702–720.
Friedman, D., Hamberger, M., Stern, Y., & Marder, K. (1992). Event-related potentials during repetition priming in Alzheimer's patients and young and older controls. *Journal of Clinical and Experimental Neurophysiology, 14*, 448–462.
Friedman, D., & Johnson, R. (2000). Event-related potential (ERP) studies of memory encoding and retrieval: A selective review. *Microscopy Research and Techniques, 51*, 6–28.
Friedman, D., & Trott, C. (2000). An event-related potential study of encoding in young and older adults. *Neuropsychologia, 38*, 542–557.
Gabrieli, J. D. (1998). Cognitive neuroscience of human memory. *Annual Review of Psychology, 49*, 87–115.
Greve, A., van Rossum, M. C., & Donaldson, D. I. (2007). Investigating the functional interaction between semantic and episodic memory: Convergent behavioral and electrophysiological evidence for the role of familiarity. *Neuroimage, 34*, 801–814.
Groh-Bordin, C., Busch, N. A., Herrmann, C. S., & Zimmer, H. D. (2007). Event-related potential repetition effects at encoding predict memory performance at test. *Neuroreport, 18*, 1905–1909.
Groh-Bordin, C., Zimmer, H. D., & Ecker, U. K. H. (2006). Has the butcher on the bus dyed his hair? When color changes modulate ERP correlates of familiarity and recollection. *Neuroimage, 32*, 1879–1890.
Groh-Bordin, C., Zimmer, H. D., & Mecklinger, A. (2005). Feature binding in perceptual priming and in episodic object recognition: Evidence from event-related brain potentials. *Cognitive Brain Research, 24*, 556–567.
Grove, K. L., & Wilding, E. L. (2009). Retrieval processes supporting judgments of recency. *Journal of Cognitive Neuroscience, 21*, 461–473.
Guo, C., Duan, L., Li, W., & Paller, K. A. (2006). Distinguishing source memory and item memory: Brain potentials at encoding and retrieval. *Brain Research, 1118*, 142–154.

Guo, C., Voss, J. L., & Paller, K. A. (2005). Electrophysiological correlates of forming memories for faces, names, and face–name associations. *Cognitive Brain Research, 22*, 153–164.

Hayama, H. R., Johnson, J. D., & Rugg, M. D. (2008). The relationship between the right frontal old-new ERP effect and post-retrieval monitoring: Specific or non-specific? *Neuropsychologia, 46*, 1211–1223.

Heeger, D. J., & Ress, D. (2002). What does fMRI tell us about neural activity? *Nature Reviews Neuroscience, 3*, 142–151.

Herron, J. E., Quayle, A. H., & Rugg, M. D. (2003). Probability effects on event-related potential correlates of recognition memory. *Cognitive Brain Research, 16*, 66–73.

Herron, J. E., & Rugg, M. D. (2003a). Retrieval orientation and the control of recollection. *Journal of Cognitive Neuroscience, 15*, 843–854.

Herron, J. E., & Rugg, M. D. (2003b). Strategic influences on recollection in the exclusion task: Electrophysiological evidence. *Psychonomic Bulletin and Review, 10*, 703–710.

Herron, J. E., & Wilding, E. L. (2004). An electrophysiological dissociation of retrieval orientation and retrieval mode. *Neuroimage*, 22, 1554–1562.

Herron, J. E., & Wilding, E. L. (2006a). Brain and behavioral indices of retrieval mode. *Neuroimage, 32*, 863–870.

Herron, J. E., & Wilding, E. L. (2006b). Neural correlates of control processes engaged before and during recovery of information from episodic memory. *Neuroimage, 30*, 634–644.

Hillyard, S. A., & Kutas, M. (1983). Electrophysiology of cognitive processing. *Annual Review of Psychology, 34*, 33–61.

Hintzman, D. L., & Curran, T. (1994). Retrieval dynamics of recognition memory and frequency judgments: Evidence for separate processes of familiarity and recall. *Journal of Memory and Language, 33*, 1–18.

Hornberger, M., Morcom, A. M., & Rugg, M. D. (2004). Neural correlates of retrieval orientation: Effects of study-test similarity. *Journal of Cognitive Neuroscience, 16*, 1196–1210.

Hornberger, M., Rugg, M. D., & Henson, R. N. A. (2006). ERP correlates of retrieval orientation: Direct versus indirect memory tasks. *Brain Research, 107*, 124–136.

Jacoby, L. L. (1998). Invariance in automatic influences of memory: Toward a user's guide for the process-dissociation procedure. *Journal of Experimental Psychology: Learning, Memory and Cognition, 24*, 3–26.

Jacoby, L. L., & Dallas, M. (1981). On the relationship between autobiographical memory and perceptual learning. *Journal of Experimental Psychology: General, 3*, 306–340.

Jager, T., Mecklinger, A., & Kipp, K. H. (2006). Intra- and inter-item associations doubly dissociate the electrophysiological correlates of familiarity and recollection. *Neuron, 52*, 535–545.

Janowsky, J. S., Shimamura, A. P., & Squire, L. R. (1989). Source memory impairment in patients with frontal lobe lesions. *Neuropsychologia, 27*, 1043–1056.

Johansson, M., Stenberg, G., Lindberg, M., & Rosen, I. (2002). Memory for perceived and imagined pictures—an event-related potential study. *Neuropsychologia, 40*, 986–1002.

Johnson, J. D., Minton, B. R., & Rugg, M. D. (2008). Content dependence of the electrophysiological correlates of recollection. *Neuroimage, 39*, 406–416.

Johnson, J. D., & Rugg, M. D. (2006). Electrophysiological correlates of retrieval processing: Effects of consistent versus inconsistent retrieval demands. *Journal of Cognitive Neuroscience, 18*, 1531–1544.

Johnson, M. K., Hashtroudi, S., & Lindsay, D. S. (1993). A source monitoring framework. *Psychological Bulletin, 114*, 3–28.

Johnson, M. K., Kounios, J., & Nolde, S. F. (1997). Electrophysiological brain activity and memory source monitoring. *Neuroreport, 8*, 1317–1320.

Johnson, M. K., & Raye, C. L. (1981). Reality monitoring. *Psychology Review, 88*, 67–85.

Johnson, R. (1995). Event-related potential insights into the neurobiology of memory systems. In F. Boller & J.Grafman (Eds.), *Handbook of neuropsychology* (pp. 135–163). Amsterdam: Elsevier.

Kuo, T. Y., & Van Petten, C. (2006). Prefrontal engagement during source memory retrieval depends upon the prior encoding context. *Journal of Cognitive Neuroscience, 18*, 1133–1146.

Kuo, T. Y., & Van Petten, C. (2008). Perceptual difficulty in source memory encoding and retrieval: Prefrontal versus parietal electrical brain activity. *Neuropsychologia, 46*, 2243–2257.

Kutas, M., & Hillyard, S. A. (1984). Brain potentials during reading reflect word expectancy and semantic association. *Nature, 307*, 161–163.

Kutas, M., & Van Petten, C. (1988). ERP studies of language. In P. K. Ackles, J. R. Jennings & M. G. H. Coles (Eds.), *Advances in psychophysiology* (pp. 139–188). Greenwich, CT: JAI Press.

Logothetis, N. K. (2002). The neural basis of the BOLD fMRI signal. *Philosophical Transactions of the Royal Society of London B, 357*, 1003–1037.

Lucas, H. D., Voss, J. L., & Paller, K. A. (2010). Familiarity or conceptual priming? Good question! Comment on Stenberg, Hellman, Johansson and Rosen. *Journal of Cognitive Neuroscience, 22*, 615–617.

Luck, S. J. (2005). *An introduction to the event-related potential technique*. Cambridge, MA: MIT Press.

Mackenzie, G., & Donaldson, D. I. (2007). Dissociating recollection from familiarity: Electrophysiological evidence that familiarity for faces is associated with a posterior old-new effect. *Neuromage, 36*, 454–463.

Mandler, G. (1980). Recognising: The judgment of previous occurrence. *Psychology Review, 87*, 252–271.

Mandler, G. (1991). Your face looks familiar but I can't remember your name: A review of dual process theory. In W. E. Hockley & S. Lewandowsky (Eds.), *Relating theory and data: Essays on human memory in honor of Bennet B. Murdock* (pp. 207–225) Hillsdale, NJ: Erlbaum.

Mangels, J. A., Picton, T. W., & Craik, F. I. M. (2001). Attention and successful episodic encoding: An event-related potential study. *Cognitive Brain Research, 11*, 77–95.

Mecklinger, A. (2000). Interfacing mind and brain: A neurocognitive model of recognition memory. *Psychophysiology, 37*, 565–582.

Mecklinger, A., von Cramon, D. Y., & Matthes-von Cramon, G. (1998). Event-related potential evidence for a specific recognition memory deficit in adult survivors of cerebral hypoxia. *Brain, 121*, 1919–1935.

Mesulam, M. M. (1990). Large-scale neurocognitive networks and distributed processing for attention, language and memory. *Annals of Neurology, 28*, 597–613.

Mesulam, M. M. (1998). From sensation to cognition. *Brain, 121*, 1013–1052.

Milner, B. (1971). Interhemispheric differences in the location of psychological processes in man. *British Medical Bulletin, 27*, 272–277.

Morcom, A. M., & Rugg, M. D. (2002). Getting ready to remember: The neural correlates of task set during recognition memory. *Neuroreport, 13,* 149–152.

Nessler, D., Mecklinger, A., & Penney, T. B. (2001). Event related brain potentials and illusory memories: The effects of differential encoding. *Cognitive Brain Research, 10,* 283–301.

Nessler, D., Mecklinger, A., & Penney, T. B. (2005). Perceptual fluency, semantic familiarity and recognition-related familiarity: An electrophysiological exploration. *Cognitive Brain Research, 22,* 265–288.

Neville, H. J., Kutas, M., Chesney, G., & Schmidt, A. L. (1986). Event-related brain potentials during initial encoding and recognition memory of congruous and incongruous words. *Journal of Memory and Language, 25,* 75–92.

Nolde, S. F., Johnson, M. K., & Raye, C. L. (1998). The role of prefrontal cortex during tests of episodic memory. *Trends in Cognitive Sciences, 2,* 399–406.

Norman, K. A., Tepe, K., Nyhus, E., & Curran, T. (2008). Event-related potential correlates of interference effects on recognition memory. *Psychonomic Bulletin and Review, 15,* 36–43.

Olichney, J. M., Petten, C. V., Paller, K. A., Salmon, D. P., Iragui, V. J., & Kutas, M. (2000). Word repetition in amnesia: Electrophysiological measures of impaired and spared memory. *Brain, 123,* 1948–1963.

Otten, L. J., Quayle, A. H., Akram, S., Ditewig, T. A., & Rugg, M. D. (2006). Brain activity before an event predicts later recollection. *Nature Neuroscience, 9,* 489–491.

Otten, L. J., Quayle, A., & Puvaneswaran, B. (2010). Neural activity precedes memory formation of auditory and visual events. *Journal of Cognitive Neuroscience, 22,* 1212–1223.

Otten, L. J., & Rugg, M. D. (2001a). Electrophysiological correlates of memory encoding are task-dependent. *Cognitive Brain Research, 12,* 11–18.

Otten, L. J., & Rugg, M. D. (2001b). Task-dependency of the neural correlates of episodic encoding as measured by fMRI. *Cerebral Cortex, 11,* 1150–1160.

Otten, L. J., Sveen, J., & Quayle, A. H. (2007). Distinct patterns of neural activity during memory formation of nonwords versus words. *Journal of Cognitive Neuroscience, 19,* 1776–1789.

Paller, K. A. (1990). Recall and stem-completion have different electrophysiological correlates and are modified differentially by directed forgetting. *Journal of Experimental Psychology: Learning, Memory and Cognition, 16,* 1021–1032.

Paller, K. A., Kutas, M., & Mayes, A. R. (1987a). Neural correlates of encoding in an incidental learning paradigm. *Electroencephalography and Clinical Neurophysiology, 67,* 360–371.

Paller, K. A., Kutas, M., & McIsaac, H. K. (1995). Monitoring conscious recollection via the electrical activity of the brain. *Psychological Science, 6,* 107–111.

Paller, K. A., Kutas, M., Shimamura, A. P., & Squire, L. R. (1987b). Brain responses to concrete and abstract words reflect processes that correlate with later performance on a test of stem-completion priming. *Electroencephalography and Clinical Neurophysiology (Supplement), 40,* 360–365.

Paller, K. A., Voss, J. L., & Boehm, S. G. (2007). Validating neural correlates of familiarity. *Trends in Cognitive Sciences, 11,* 243–250.

Penney, T. B., Mecklinger, A., & Nessler, D. (2001). Repetition related ERP effects in a visual object target detection task. *Cognitive Brain Research, 10,* 239–250.

Potter, D. D., Pickles, C. D., Roberts, R. C., & Rugg, M. D. (1992). The effects of scopolamine on event-related potentials in a continuous recognition memory task. *Psychophysiology, 29,* 29–37.

Ranganath, C. (2004). The 3-D prefrontal cortex: Hemispheric asymmetries in prefrontal activity and their relation to memory retrieval processes. *Journal of Cognitive Neuroscience, 16,* 903–907.

Ranganath, C., & Knight, R. T. (2003). Prefrontal cortex and episodic memory: Integrating findings from neuropsychology and event-related functional neuroimaging. In A. E. Parker, E. L. Wilding, & T. J. Bussey (Eds.), *Memory encoding and retrieval: A cognitive neuroscience perspective* (pp. 83–99). New York: Psychology Press.

Ranganath, C., & Paller, K. A. (1999). Frontal brain potentials during recognition are modulated by requirements to retrieve perceptual details. *Neuron, 22,* 605–613.

Ranganath, C., & Paller, K. A. (2000). Neural correlates of memory retrieval and evaluation. *Cognitive Brain Research, 9,* 209–222.

Rhodes, S. M., & Donaldson, D. I. (2007). Electrophysiological evidence for the influence of unitization on the processes engaged during episodic retrieval: Enhancing familiarity-based remembering. *Neuropsychologia, 45,* 412–424.

Rhodes, S. M., & Donaldson, D. I. (2008). Electrophysiological evidence for the effect of interactive imagery on episodic memory: Encouraging familiarity for non-unitized stimuli during associative recognition. *Neuroimage, 39,* 873–884.

Rotello, C. M., Macmillan, N. A., Reeder, J. A., & Wong, M. (2005). The "remember" response: Subject to bias, graded, and not a process-pure indicator of recollection. *Psychonomic Bulletin and Review, 12,* 865–873.

Rubin, D. C., & Greenberg, D. L. (1998). Visual memory-deficit amnesia: A distinct amnesic presentation and etiology. *Proceedings of the National Academy of Sciences USA, 95,* 5413–5416.

Rugg, M. D. (1992). Event-related potentials in clinical neuropsychology. In J. R. Crawford, W. A. McKinlay, & D. M. Parker (Eds.), *A handbook of neuropsychological assessment* (pp. 393–411). Hillsdale, NJ: Erlbaum.

Rugg, M. D. (1994). Event-related potential studies of human memory. In M. S. Gazzaniga (Ed.), *The cognitive neurosciences* (pp. 789–801). Cambridge, MA: MIT Press.

Rugg, M. D. (1995a). Cognitive event-related potentials: Intracranial and lesion studies. In J. C. Baron & J. Grafman (Eds.), *Handbook of neuropsychology* (Vol. 10, pp. 380–420). Amsterdam: Elsevier.

Rugg, M. D. (1995b). ERP Studies of memory. In M. D. Rugg & M. G. H. Coles (Eds.), *Electrophysiology of mind: Event-related brain potentials and cognition* (pp. 132–170). Oxford: Oxford University Press.

Rugg, M. D., & Allan, K. (2000). Event-related potential studies of memory. In E. Tulving & F. I. M. Craik (Eds.), *The Oxford handbook of memory* (pp. 521–537). Oxford: Oxford University Press.

Rugg, M. D., Allan, K., & Birch, C. S. (2000). Electrophysiological evidence for the modulation of retrieval orientation by depth of study processing. *Journal of Cognitive Neuroscience, 12,* 664–678.

Rugg, M. D., & Curran, T. (2007). Event-related potentials and recognition memory. *Trends in Cognitive Sciences, 11,* 251–257.

Rugg, M. D., & Doyle, M. C. (1994). Event-related potentials and stimulus repetition in direct and indirect tests of memory.

In H. J. Heinze, T. F. Munte, & G. R. Mangun (Eds.), *Cognitive electrophysiology* (pp. 124–148). Boston: Birkhauser.

Rugg, M. D., & Henson, R. N. A. (2003). Episodic memory retrieval: An (event-related) functional neuroimaging perspective. In A. E. Parker, E. L. Wilding, & T. J. Bussey (Eds.), *The cognitive neuroscience of memory: Episodic encoding and retrieval* (pp. 5–38). New York: Psychology Press.

Rugg, M. D., Herron, J., & Morcom, A. (2002). Electrophysiological studies of retrieval processing. In L. R. Squire & D. L. Schacter (Eds.), *Neuropsychology of memory* (pp. 154–165). New York: Guilford Publications.

Rugg, M. D., Mark, R. E., Walla, P., Schloerscheidt, A., Birch, C. S., & Allan, K. (1998). Dissociation of the neural correlates of implicit and explicit memory. *Nature*, 392, 595–598.

Rugg, M. D., & Nagy, M. E. (1989). Event-related potentials and recognition memory for words. *Electroencephalography and Clinical Neurophysiology, 72*, 395–406.

Rugg, M. D., Roberts, R. C., Potter, D. D., Pickles, C. D., & Nagy, M. E. (1991). Event-related potentials related to recognition memory: Effects of unilateral temporal lobectomy and temporal lobe epilepsy. *Brain, 114*, 2313–2332.

Rugg, M. D., & Wilding, E. L. (2000). Retrieval processing and episodic memory. *Trends in Cognitive Sciences, 4*, 108–115.

Rugg, M. D., & Yonelinas, A. P. (2003). Human recognition memory: A cognitive neuroscience perspective. *Trends in Cognitive Sciences, 7*, 313–319.

Sanquist, T. F., Rohrbaugh, J. W., Syndulko, K., & Lindsley, D. B. (1980). Electrocortical signs of levels of processing: Perceptual analysis and recognition memory. *Psychophysiology, 17*, 568–576.

Schacter, D. L. (1987). Implicit memory: History and current status. *Journal of Experimental Psychology: Learning, Memory, and Cognition, 13*, 501–518.

Schacter, D. L., Eich, J. E., & Tulving, E. (1978). Richard Semon's theory of memory. *Journal of Verbal Learning and Verbal Behavior, 17*, 721–743.

Schacter, D. L., Harbluk, J. L., & McLachlan, D. R. (1984). Retrieval without recollection: An experimental analysis of source amnesia. *Journal of Verbal Learning and Verbal Behavior, 23*, 593–611.

Schacter, D. L., Norman, K. A., & Koutstaal, W. (1998). The cognitive neuroscience of constructive memory. *Annual Review of Psychology, 49*, 289–318.

Schloerscheidt, A. M., & Rugg, M. D. (2004). The impact of change in stimulus format on the electrophysiological indices of recognition. *Neuropsychologia, 42*, 451–466.

Semon, R. S. (1921). *The mneme*. London: George Allen and Unwin.

Senkfor, A. J., & Van Petten, C. (1998). Who said what: An event-related potential investigation of source and item memory. *Journal of Experimental Psychology: Learning, Memory and Cognition, 24*, 1005–1025.

Shimamura, A. P., & Squire, L. R. (1987). A neuropsychological study of fact memory and source amnesia. *Journal of Experimental Psychology: Learning, Memory, and Cognition, 13*, 464–473.

Simons, J. S., Gilbert, S. J., Owen, A. M., Fletcher, P. C., & Burgess, P. W. (2005a). Distinct roles for lateral and medial anterior prefrontal cortex in contextual recollection. *Journal of Neurophysiology, 94*, 813–820.

Simons, J. S., & Mayes, A. R. (2008). What is the parietal lobe contribution to human memory? *Neuropsychologia, 46*, 1739–1742.

Simons, J. S., Owen, A., Fletcher, P. C., & Burgess, P. W. (2005b). Anterior prefrontal cortex and the recollection of contextual information. *Neuropsychologia, 43*, 1774–1783.

Smith, M. E. (1993). Neurophysiological manifestations of recollective experience during recognition memory judgements. *Journal of Cognitive Neuroscience, 5*, 1–13.

Smith, M. E., & Guster, K. (1993). Decomposition of recognition memory event-related potentials yields target, repetition, and retrieval effects. *Electroencephalography and Clinical Neurophysiology, 86*, 335–343.

Smith, M. E., & Halgren, E. (1989). Dissociation of recognition memory components following temporal lobe lesions. *Journal of Experimental Psychology: Learning, Memory and Cognition, 15*, 50–60.

Sommer, W., Heinz, A., Leuthold, H., Matt, J., & Schweinberger, S. R. (1995). Metamemory, distinctiveness, and event-related potentials in recognition memory for faces. *Memory and Cognition, 23*, 1–11.

Sommer, W., Komoss, E., & Schweinberger, S. R. (1997). Differential localisation of brain systems subserving memory for names and faces in normal subjects with event-related potentials. *Electroencephalography and Clinical Neurophysiology, 102*, 192–199.

Sommer, W., Schweinberger, S. R., & Matt, J. (1991). Human brain potential correlates of face encoding into memory. *Electroencephalography and Clinical Neurophysiology, 79*, 457–463.

Spencer, K., Abad, E. V., & Donchin, E. (2000). On the search for the neurophysiological manifestation of recollective experience. *Psychophysiology, 37*, 494–506.

Squire, L. R. (1982). Comparisons between forms of amnesia: Some deficits are unique to Korsakoff's syndrome. *Journal of Experimental Psychology: Learning, Memory, and Cognition, 8*, 560–571.

Squire, L. R., Wixted, J. T., & Clark, R. E. (2007). Recognition memory and the medial temporal lobe: A new perspective. *Nature Reviews Neuroscience, 8*, 872–883.

Stenberg, G., Johansson, M., Hellman, J., & Rosen, I. (2009). Familiarity or conceptual priming: Event-related potentials in name recognition. *Journal of Cognitive Neuroscience, 21*, 447–460.

Stenberg, G., Johansson, M., Hellman, J., & Rosen, I. (2010). "Do you see yonder cloud?" On priming concepts, a new test and a familiar outcome. Reply to Lucas et al.: "Familiarity or conceptual priming? Good question! Comment on Stenberg, Hellman, Johansson and Rosen." *Journal of Cognitive Neuroscience, 4*, 618–620.

Stuss, D. T., Eskes, G. A., & Foster, J. K. (1994). Experimental neuropsychological studies of frontal lobe functions. In F. Boller & J. Grafman (Eds.), *Handbook of neuropsychology* (Vol. 9, pp. 149–183). New York: Elsevier.

Swick, D., Senkfor, A. J., & Van Petten, C. (2006). Source memory retrieval is affected by aging and prefrontal lesions: Behavioral and ERP evidence. *Brain Research, 1107*, 161–176.

Tendolkar, I., & Rugg, M. D. (1998). Electrophysiological dissociation of recency and recognition memory. *Neuropsychologia, 36*, 477–490.

Tendolkar, I., Schoenfeld, A., Golz, G., Fernandez, G., Kuhl, K., Ferszt, R., & Heinze, H. (1999). Neural correlates of recognition memory with and without recollection in patients with Alzheimer's disease and healthy controls. *Neuroscience Letters, 263*, 45–48.

Trott, C. T., Friedman, D., Ritter, W., & Fabiani, M. (1997). Item and source memory: Differential age effects revealed by event-related potentials. *Neuroreport, 8*, 3373–3378.

Trott, C. T., Friedman, D., Ritter, W., Fabiani, M., & Snodgrass, J. G. (1999). Episodic priming and memory for temporal source: Event-related potentials reveal age-related differences in prefrontal functioning. *Psychology and Aging, 14*, 390–413.

Tsivilis, D., Otten, L., & Rugg, M. D. (2001). Context effects on the neural correlates of recognition memory: An electrophysiological study. *Neuron, 31*, 497–505.

Tulving, E. (1983). *Elements of episodic memory*. Oxford: Oxford University Press.

Tulving, E. (1993). What is episodic memory? *Current Perspectives in Psychological Science, 2*, 67–70.

Vallesi, A., & Shallice, T. (2006). Prefrontal involvement in source memory: An electrophysiological investigation of accounts concerning confidence and accuracy. *Brain Research, 1124*, 111–125.

Van Petten, C., Senkfor, A. J., & Newberg, W. M. (2000). Memory for drawings in locations: Spatial source memory and event-related potentials. *Psychophysiology, 4*, 551–564.

Vilberg, K. L., Moosavi, R. F., & Rugg, M. D. (2006). The relationship between electrophysiological correlates of recollection and the amount of information retrieved. *Brain Research, 1122*, 161–170.

Vilberg, K. L., & Rugg, M. D. (2008). Memory retrieval and the parietal cortex: A review of evidence from a dual-process perspective. *Neuropsychologia, 46*, 1787–1799.

Vilberg, K. L., & Rugg, M. D. (2009). Functional significance of retrieval-related activity in lateral parietal cortex: Evidence from fMRI and ERPs. *Human Brain Mapping, 30*, 1490–1501.

Voss, J. L., & Paller, K. A. (2006). Fluent conceptual processing and explicit memory for faces are electrophysiologically distinct. *Journal of Neuroscience, 26*, 926–933.

Voss, J. L., & Paller, K. A. (2007). Neural correlates of conceptual implicit memory and their contamination of putative neural correlates of explicit memory. *Learning and Memory, 14*, 259–267.

Wagner, A. D., & Paller, K. A. (2002). Observing the transformation of experience into memory. *Trends in Cognitive Sciences, 6*, 93–102.

Wagner, A. D., Shannon, B. J., Kahn, I., & Buckner, R. L. (2005). Parietal lobe contributions to episodic memory retrieval. *Trends in Cognitive Sciences, 9*, 445–453.

Wais, P. E., Mickes, L., & Wixted, J. T. (2008). Remember/know judgments probe degrees of recollection. *Journal of Cognitive Neuroscience, 20*, 400–405.

Werkle-Bergner, M., Mecklinger, A., Kray, J., Meyer, P., & Duzel, E. (2005). The control of memory-retrieval: Insights from event-related potentials. *Cognitive Brain Research, 24*, 599–614.

Wheeler, M. A., Stuss, D., & Tulving, E. (1997). Toward a theory of episodic memory: The frontal lobes and autonoetic consciousness. *Psychological Bulletin, 121*, 331–354.

Wilding, E. L. (1999). Separating retrieval strategies from retrieval success: An event-related potential study of source memory. *Neuropsychologia, 37*, 441–454.

Wilding, E. L. (2000). In what way does the parietal ERP old-new effect index recollection? *International Journal of Psychophysiology, 35*, 81–87.

Wilding, E. L., Doyle, M. C., & Rugg, M. D. (1995). An event-related potential study of recognition memory with and without retrieval of context. *Neuropsychologia, 33*, 743–767.

Wilding, E. L., & Herron, J. E. (2006). Electrophysiological measures of episodic memory control and memory retrieval. *Electroencephalography and Clinical Neuroscience, 37*, 315–321.

Wilding, E. L., & Nobre, A. C. (2001). Task-switching and memory-retrieval processing: Electrophysiological evidence. *Neuroreport, 12*, 3613–3617.

Wilding, E. L., & Rugg, M. D. (1996). An event-related potential study of recognition memory with and without retrieval of source. *Brain, 119*, 889–905.

Wilding, E. L., & Rugg, M. D. (1997). Event-related potentials and the recognition memory exclusion task. *Neuropsychologia, 35*, 119–128.

Wilding, E. L., & Sharpe, H. (2003). Episodic memory encoding and retrieval: Recent insights from event-related potentials. In A. Zani & A. Mado-Proverbio (Eds.), *The cognitive electrophysiology of mind and brain* (pp. 169–196). San Diego, CA: Academic Press.

Wixted, J. T. (2007). Dual-process theory and signal-detection theory of recognition memory. *Psychology Review, 114*, 152–176.

Woodruff, C. C., Hayama, H. R., & Rugg, M. D. (2006). Electrophysiological dissociation of the neural correlates of recollection and familiarity. *Brain Research, 1100*, 125–135.

Xiang, J. Z., & Brown, M. W. (1998). Differential neuronal encoding of novelty, familiarity and recency in regions of the anterior temporal lobe. *Neuropharmacology, 37*, 657–676.

Xiang, J. Z., & Brown, M. W. (2004). Neuronal responses related to long-term recognition memory processes in the prefrontal cortex. *Neuron, 42*, 817–829.

Yick, Y. Y., & Wilding, E. L. (2008). Material-specific neural correlates of memory retrieval. *Neuroreport, 19*, 1463–1467.

Yonelinas, A. P. (1999). The contribution of recollection and familiarity to recognition and source-memory judgements: A formal dual-process model and an analysis of receiver operating characteristics. *Journal of Experimental Psychology: Learning, Memory and Cognition, 25*, 1415–1434.

Yonelinas, A. P. (2002). The nature of recollection and familiarity: A review of the 30 years of research. *Journal of Memory and Language, 46*, 441–517.

Yonelinas, A. P., & Parks, C. M. (2007). Receiver operating characteristics in recognition memory: A review. *Psychological Bulletin, 133*, 800–832.

Yovel, G., & Paller, K. A. (2004). The neural basis of the butcher-on-the-bus phenomenon: When a face seems familiar but is not remembered. *Neuroimage, 21*, 789–800.

CHAPTER

15 Language-Related ERP Components

Tamara Y. Swaab, Kerry Ledoux, C. Christine Camblin, *and* Megan A. Boudewyn

Abstract

Understanding the processes that permit us to extract meaning from spoken or written linguistic input requires elucidating how, when, and where in the brain sentences and stories, syllables and words are analyzed. Because human language is a cognitive function that is not readily investigated using neuroscience approaches in animal models, this task presents special challenges. In this chapter, we describe how event-related potentials (ERPs) have contributed to the understanding of language processes as they unfold in real-time. We will provide an overview of the many ERPs that have been used in language research, and will discuss the main models of what these ERPs reflect in terms of linguistic and neural processes. In addition, using examples from the literature, we will illustrate how ERPs can be used to study language comprehension, and will also outline methodological issues that are specific to using ERPs in language research.

Keywords: lexical processing, sentence processing, discourse processing, non-literal language, event-related potentials, N400, P600, Nref

Language is a central part of our everyday life. It enables us to accomplish a seemingly infinite number of uniquely human tasks: We talk to many people every day and answer hundreds of our children's questions, we tell each other jokes and recite poems to our loved ones, we negotiate deals and treaties, we listen to news reporters, read books, and (slowly) write chapters; and there are, of course, numerous other examples of our daily language use. Language is a defining feature of who we are as human beings. But understanding and producing language is extremely complex, and is subserved by many processes and many areas in the brain. This complexity arises in part as a function of the productivity and flexibility of language. We constantly produce novel utterances and find new ways to use existing words and phrases. In addition, words and sentences can assume diverse meanings, depending on a multitude of factors such as the context, the speaker, the listener, or the world knowledge relevant to the utterance. A popular bumper sticker from the San Francisco Bay Area illustrates this point. It reads: "It's great to be alive in Colma." To anyone unfamiliar with this Californian town, the statement appears to be extolling the virtues of Colma; however, residents will tell you that this statement actually refers to the large number of cemeteries to be found in this otherwise small town and quite literally means that it's great to be *alive* in Colma. This anecdote demonstrates the flexibility of meaning in language and how language comprehension is frequently not straightforward.

In order to understand the processes by which we extract meaning from spoken or written linguistic input, we must unravel how, when, and where our brain interprets sentences and stories, syllables and words. This is a difficult task, in part because human language is the only cognitive function that cannot easily be studied with animal models or traditional neuroscience methods. In this chapter, we aim to elucidate how language-related event-related potentials (ERPs) have contributed to our understanding of language processes as they unfold in real time. Specifically, we hope to accomplish the following: (1) provide an overview of the different ERP components that have been used in language research; (2) discuss the main theories of what these components may reflect; and (3) illustrate with examples from the literature how ERPs can be used to study language comprehension.[1] Whenever possible, we will include information on the neuronal generators of language-related ERPs. However, because all language ERPs occur (relatively) late and are long-latency components, it is very likely that multiple brain areas contribute to their generation, and (not surprisingly) little is known about possible generators of most of the language-related ERPs. Finally, we will discuss some methodological issues that are specific to using ERPs in language research.

The N400

The best-studied language-related ERP component is the N400. The N400 is a negative shift in the ERP waveform that is larger over centro-parietal than over anterior regions of the scalp (Kutas & Hillyard, 1983). In young adults, the N400 usually reaches its maximum amplitude between 380 and 440 ms after stimulus onset. However, this may be delayed in elderly persons (e.g., Gunter et al., 1992) and aphasic patients (Swaab et al., 1997). In the domain of language processing, the N400 is observed when words, sentences, or discourses are presented as written text (e.g., Kutas & Hillyard, 1980), as naturally produced connected speech (e.g., Holcomb & Neville, 1990, 1991), and with sign language (Kutas et al., 1987). When words are presented visually, the N400 effect typically onsets around 200 ms after the stimulus and lasts for about 300 ms. In the auditory modality, the N400 may begin as early as 100 ms after stimulus onset and last for 400 ms.

In a typical ERP N400 study, participants are presented with language stimuli, and the N400 effect is measured to critical words for which semantic information is manipulated. To illustrate this, let us consider the following example from a study by Sitnikova and colleagues (2002). They presented sentences in four conditions, as in the following example:

1. Diving was forbidden from the bridge because the river had rocks in it.
2. Diving was forbidden from the bridge because the river had cracks in it.
3. The guests played bridge because the river had rocks in it.
4. The guests played bridge because the river had cracks in it.

There were 40 stimuli in each condition, and the sentences of each quadruplet were presented in different lists to avoid repetition (word repetition leads to facilitated processing). Every content word in these conditions will elicit an N400. However, as the language comprehender proceeds through the sentence and contextual information accumulates, it becomes easier to predict or integrate the next word of the sentence, and this leads to a reduction in the amplitude of the N400. Thus, in normal sentence contexts, content words presented later in the sentence will show smaller amplitude N400s than words presented earlier in the sentence (e.g., van Petten & Kutas, 1991). In this example, however, Sitnikova and colleagues (2002) aimed to determine whether or not schizophrenic patients were able to select contextually appropriate meanings of ambiguous words that have one form representation but at least two unrelated meanings (e.g., *bridge*, meaning either the structure spanning the river or the card game). The critical word (*river* in the example) was always related to the most frequent meaning of the ambiguous word. The preceding context supported either the more frequent interpretation of the ambiguous word (Conditions 1 and 2) or the less frequent interpretation (Conditions 3 and 4). After the critical words, a control word was included that was congruent in meaning (or not) with the preceding context but was not related to any of the meanings of the ambiguous word (*rocks* vs. *cracks* in the example). Normal, neurologically unimpaired control subjects showed a reduced N400 to *river* when it was consistent with the meaning of the ambiguous word biased by the preceding context (1) relative to when it was inconsistent (3). A comparison of the control words in Conditions 1 and 2 revealed a reduced N400 when the control word was consistent with the meaning of the preceding context (1, *rocks*) relative to when it was not (2, *cracks*). The schizophrenic patients also showed this latter effect of semantic congruity, but in contrast to the normal control subjects, they did not show an N400 difference between the context-consistent critical word (*river* in Condition 1) and the context-inconsistent critical word (*river* in Condition 3). This indicates that schizophrenic patients have some deficit in using contextual information to resolve lexical ambiguities, even though they can still use the context to detect semantic anomalies.

What this study clearly demonstrates, in terms of the beauty of using ERPs in language research, is that one can measure ERPs to any and all words in the sentence without interrupting the language comprehender with a task. Tasks can be (and often have been) included *after* the language stimulus is presented. For example, one can ask subjects to make a "true" or "false" decision about a statement based on the content of the experimental sentence, or ask them to make a "good" or "bad" judgment about the sentences they have just read (or heard). However, quite a few ERP language researchers have argued that these types of tasks are no longer necessary, because N400 effects can be observed without the inclusion of a potentially interfering behavioral task (e.g., van Berkum, 2004). The ability to present stimuli without any task other than to listen or read can be essential in studies with patient populations, because they may not understand the behavioral task (e.g., aphasic patients: Swaab et al, 1997, 1998).

However, there are challenges too. One of the challenges in any study that uses ERPs is that blinking, eye movements, and other movements need to be minimized because they induce artifacts in the electroencephalography (EEG) signal. To accomplish this in ERP studies of reading, words are not presented all at once (as in this text), but instead one word at a time at the center of a computer screen, usually at a rate of one word every 500 ms with a 200 ms blank interval between words. Further, a fixation cross typically replaces words at the same central location on the screen so that subjects can fixate their eyes, usually 1000 ms before the onset of the first word and 2000 ms after the presentation of the last word of the critical stimulus. To further avoid blinks during presentation of the experimental stimuli, subjects are typically made aware of when it would be a good time to blink. This can be done, for example, by changing the color of the fixation cross to signal a "rest" period, or during a task following the experimental materials, provided that the task (e.g., involving true/false questions) does not contain a critical manipulation. Often, subjects can control when they want to proceed to the next experimental stimulus by pressing a button on a button box. This gives them a chance to blink and move their eyes for a longer period of time, if necessary. In addition, participants are usually given a break after a 5–10 min block of an experiment. In general, experimental blocks are preceded by a practice block, both to familiarize the subjects with the language task and to train them to avoid moving their eyes or blinking during presentation of the experimental stimuli. One big hurdle in ERP studies of language is that it is usually not possible to generate more than 40 stimuli in a condition; this, of course, makes it even more important not to lose many trials to subject-generated artifacts (see also the methods section at the end of this chapter).

In the auditory modality, language stimuli have typically been presented as naturally produced connected speech. The onset of critical words in the speech stream is located with a speech-editing program, and this measured latency is used to send out a trigger or event code; this marks the onset of the critical word to which the ERPs can be time-locked. Since spoken words vary in duration, the onset of the critical words typically varies across experimental stimuli in language experiments. This can be avoided by using identical speech input up to the time of the critical manipulation, which is accomplished by splicing the critical words in all conditions (for individual item sets) onto the same spoken stem (e.g., "[She drank her coffee with cream and]^{Stem} sugar$^{\text{Condition 1}}$/dog$^{\text{Condition 2}}$). The previous example would involve three audio files, would create two experimental items in different conditions, and, critically, would allow participants to hear identical spoken input in each condition up to the presentation of the critical word. Because the incoming speech signal is continuous, ERP components to the different words in the sentences overlap, resulting in the absence of clear N1 and P2 components in the averaged ERP; yet, (fortunately) a robust N400 effect can be measured in studies of comprehension of spoken language.

The Discovery of the N400

The existence of the N400 was first reported by Marta Kutas and Steven Hillyard in a seminal study published in *Science* in 1980. At the time, no language-related ERPs had been discovered and electrophysiological methods were used almost exclusively to study other perceptual and cognitive processes, such as attention and memory. Marta Kutas, who had performed studies with Emanuel Donchin using the P300, wondered whether or not this component would also be sensitive to "oddballs" of language. Kutas and Hillyard (1980) asked subjects to read sentences such as "He spread the warm bread with socks." These sentences were presented one word at a time at fixation, as is now typical for ERP studies of reading (see the previous section). In this same study, subjects also read normal sentences (e.g., "It was my first day at work") and sentences that ended with a word that was normal in meaning,

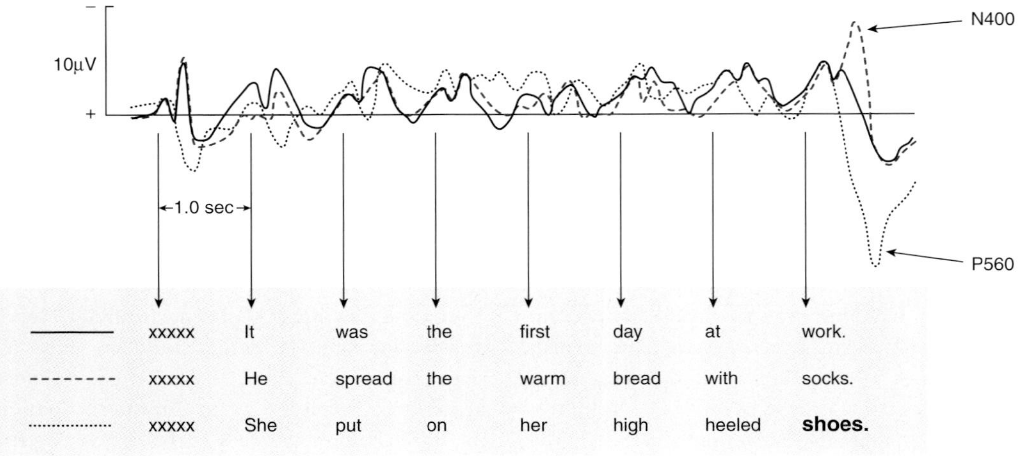

Fig. 15.1. The discovery of the N400 (Kutas and Hillyard, 1980). Subjects were presented with sentences one word at a time in three conditions. Arrows show when each word was presented. In this and all other figures, the vertical axis shows the amplitude in microvolts and the horizontal axis shows the time in milliseconds. Negative polarity is depicted upward. Event-related potentials were compared to the final words in three conditions. In the congruent condition (solid line), the last word of the sentence was semantically appropriate given the context. In the anomalous condition (dashed line), the sentence-final word was semantically inappropriate given the context, and this elicited the N400. The dotted line shows that a sentence-final word that was semantically congruent but physically deviant (a different font size) elicited a large positive shift (P650) but no N400. Redrawn with permission from Kutas and Hillyard (1980), Figure 1.

given the context, but anomalous because of a change in the physical appearance of the critical word (e.g., "She put on her high-heeled SHOES"). The physical oddball indeed resulted in a positive deflection in the ERP waveform. The semantic anomaly, however, elicited a negative ERP that peaked at around 400 ms and was maximal over centro-parietal electrode sites; this component was labeled the N400 (see Figure 15.1).

The discovery of the N400 has led to a flurry of ERP studies of word, sentence, and discourse comprehension (see Figures 15.2 and 15.3). Many of the early studies were devoted to identifying the processing nature of the N400. These studies showed that the N400 is modality independent; that is, N400 effects are observed to words whether they are written, spoken, or signed[2] (e.g., Bentin et al., 1985; Holcomb & Neville, 1990; Kutas & Hillyard, 1980; Kutas et al., 1987; McCallum et al., 1984). In addition, many studies have shown that the N400 is not only sensitive to semantic violations, but is also found when words are semantically appropriate but less expected in the context—for example *wasp* in "She was stung by a wasp," where *bee* would be the most expected completion (Kutas & Hillyard, 1984; Kutas et al., 1984). Further studies have shown that the N400 is not restricted to manipulation of meaning in sentences, but is also found for manipulations of discourse contexts, on the one hand (e.g., van Berkum et al., 1999), and semantic or repetition priming manipulations, on the other hand, where only one word serves as the context (e.g., Bentin et al., 1985). Van Berkum and colleagues (1999) manipulated whether or not the final word of the last sentence in a short passage was consistent in meaning with the preceding discourse context (e.g., "He ate a juicy steak" preceded by a discourse context that had introduced a vegetarian versus a discourse context that had introduced a person who loves to eat meat). They found a reduced N400 to critical (final) words that matched versus those that did not match the meaning of the global discourse contexts, even when these words were semantically appropriate in the local sentence context. In studies of semantic priming, the amplitude of the N400 is reduced to words that are associatively or semantically related to the preceding context word relative to when they are not (e.g., *doctor–nurse* vs. *table–nurse*; e.g., Bentin et al., 1985, Brown & Hagoort, 1993; Chwilla et al., 1998, 2000; Holcomb, 1993).

Other studies have shown that the N400 is sensitive to lexical properties of words. For example, real words (e.g., *plant*) elicit smaller N400s than pseudowords (*orthographically legal*, pronounceable nonwords, e.g., *plunt*), but random letter strings do not produce an N400 component (e.g., *ntlpu*).

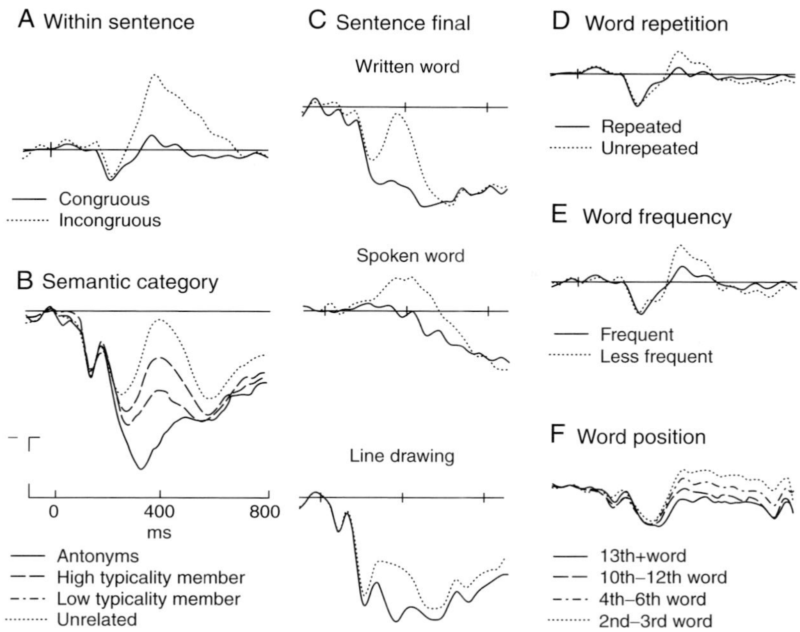

Fig. 15.2. N400 findings for different experimental manipulations. Solid lines in this figure always reflect the semantically easier processing condition. Negative amplitude is depicted upward. Semantically congruent words in sentences elicit a reduced N400 relative to semantically incongruent words; this is found when the violation occurs in midsentence (A), but also at the end of a sentence for written and spoken words and even when the final word is replaced with a line drawing that is consistent or not with the preceding sentence context (C), although this latter ERP effect has a more anterior distribution than the canonical N400. The amplitude of the N400 is reduced to words that occur at a later position in semantically congruent sentences (F) and also to words that are preceded by one context word that is related in meaning (semantic priming, B) or by an identical word (repetition priming, D). Words that are used frequently in our language elicit a reduced-amplitude N400 relative to less frequently used words (E). Reprinted with permission from Kutas and Federmeier (2000), Trends in Cognitive Sciences (Cell Press).

Frequently used (*high-frequency*) words show smaller amplitude N400s than do *infrequently used* (*low-frequency*) words (e.g., Barber et al., 2004), but this effect is modulated by the context, such that words later in the sentence no longer show lexical frequency effects (Van Petten & Kutas, 1991). Additionally, words with a small *orthographic neighborhood* (i.e., words that can be formed by changing one letter of another existing word, such as *fun* and *fan*) show reduced N400s relative to words with large orthographic neighborhoods (e.g., Holcomb et al., 2002; for a review, see Grainger & Holcomb, 2009).[3]

Some studies that have manipulated the concreteness or imageability of words have observed an increased negative shift to high-imageable/concrete words (e.g., *"banana"*) than to low-imageable/abstract words (e.g., *"justice"*). This effect occurs in the same latency range of the N400, but the topographic distribution of this effect of imageability or concreteness is anterior, instead of the typical centro-parietal scalp distribution of the canonical N400 (Holcomb et al., 1999; Kounios & Holcomb, 1994; Swaab et al., 2002; West & Holcomb, 2000, 2002). Different topographic distributions of scalp-recorded ERPs may indicate that (partially) non-overlapping neural sources have contributed to their generation and that they are not actually the same component (see Figure 15.4).

The effect of imageability or concreteness is not modulated by semantic relatedness in a semantic priming paradigm (Swaab et al., 2002), but larger effects of concreteness are observed for words in anomalous compared to congruent contexts (Holcomb et al., 1999). These results are consistent with the idea that verbal- and image-based representations may be stored separately, but that effects of context can be greater for highly imageable or concrete words in an image-based memory system. Finally, N400-like components have been observed in nonlinguistic meaningful contexts such as line drawings (Ganis et al., 1996), stories that are formed by a series of cartoon-like pictures (West & Holcomb, 2002), and short movies (Sitnikova et al., 2003). These studies have found a negative-polarity ERP in the same time window as the canonical

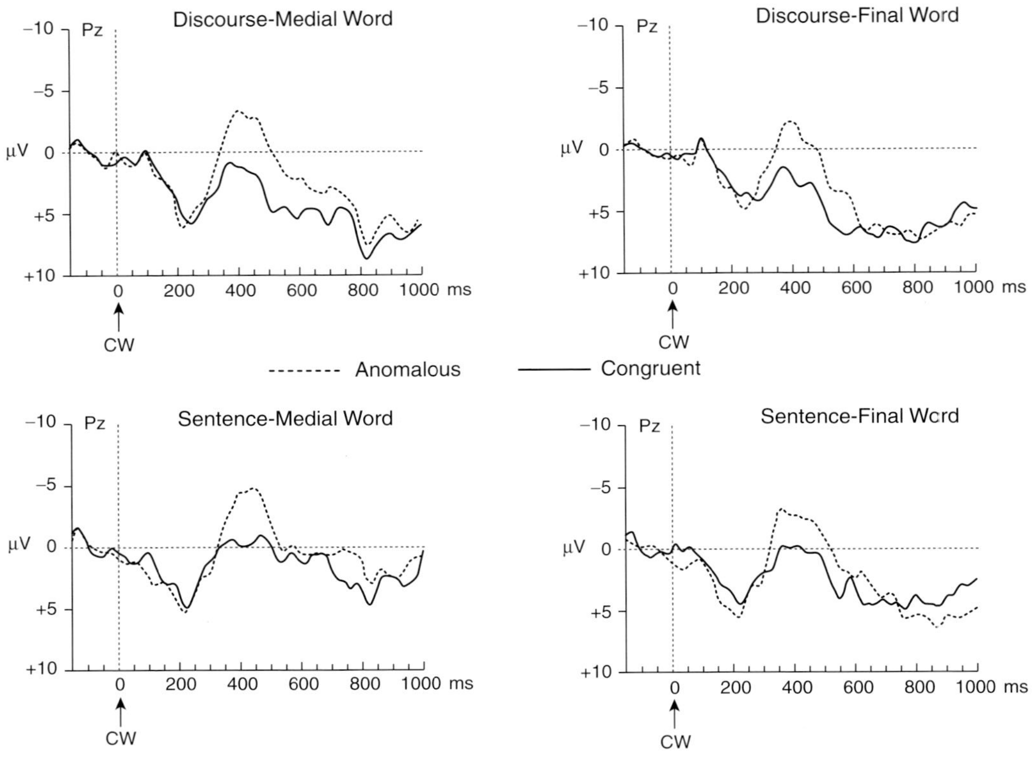

Fig. 15.3. A comparison of semantic violations in discourse and sentence contexts. The top part of this figure shows a reduced N400 (solid line) to critical words (CW) that violate the meaning of the discourse context, even though these words are appropriate in the local sentence context (e.g., "Yesterday he ate a big juicy steak," when the previous discourse has been about a vegetarian). This finding is observed for both discourse-final and discourse-medial critical words. The discourse N400 effect does not differ from the effect of semantic anomalies in single-sentence contexts (bottom of the figure). Redrawn with permission from Figure 5 in van Berkum, Brown & Hagoort, Journal of Cognitive Neuroscience, 1999 (MIT Press).

N400 but with a more anterior distribution (but see Willems et al., 2008). If the differences in the topographic distribution of the canonical N400 and the ERP elicited in other meaningful contexts indeed indicate that these ERPs are generated by (partially) nonoverlapping neuronal sources in the brain, then this would be relevant to the question of whether the semantic processing system that enables language comprehension is the same as or different from the system that must exist to process meaning in nonlinguistic contexts. As can be seen in the bottom part of Figure 15.4, functional magnetic resonance imaging (fMRI) findings from the same paradigm showed separable areas of activation for the effects of priming and imageability, which is further indirect evidence that the ERP effect of imageability is not generated by completely overlapping neuronal sources, as is the canonical N400 (Giesbrecht et al., 2004).

In the following sections, we will discuss in more detail some of the major findings with the N400 in different language contexts for reading comprehension and comprehension of spoken language.

N400 and Lexical Context

As discussed above, N400 effects are found even when the preceding context consists of a list of words that do not form sentences or discourse. These N400 effects have been observed in studies of semantic and repetition priming (e.g., Bentin & Peled, 1990; Bentin et al., 1985; Boddy, 1986; Holcomb & Neville, 1990; Joyce et al., 1999; Kutas & Hillyard, 1989; Rugg et al., 1993; Swaab et al., 2002; see Figure 15.2).

In semantic priming studies, a reduced N400 is obtained to target words that are semantically and/or associatively related to their preceding prime (relative to unrelated target words: e.g., *doctor–nurse* vs. *table–nurse*). The N400 priming effect has been observed in a range of tasks, including semantic judgment (e.g., is this word related in meaning to the preceding word?) and lexical decision (is this

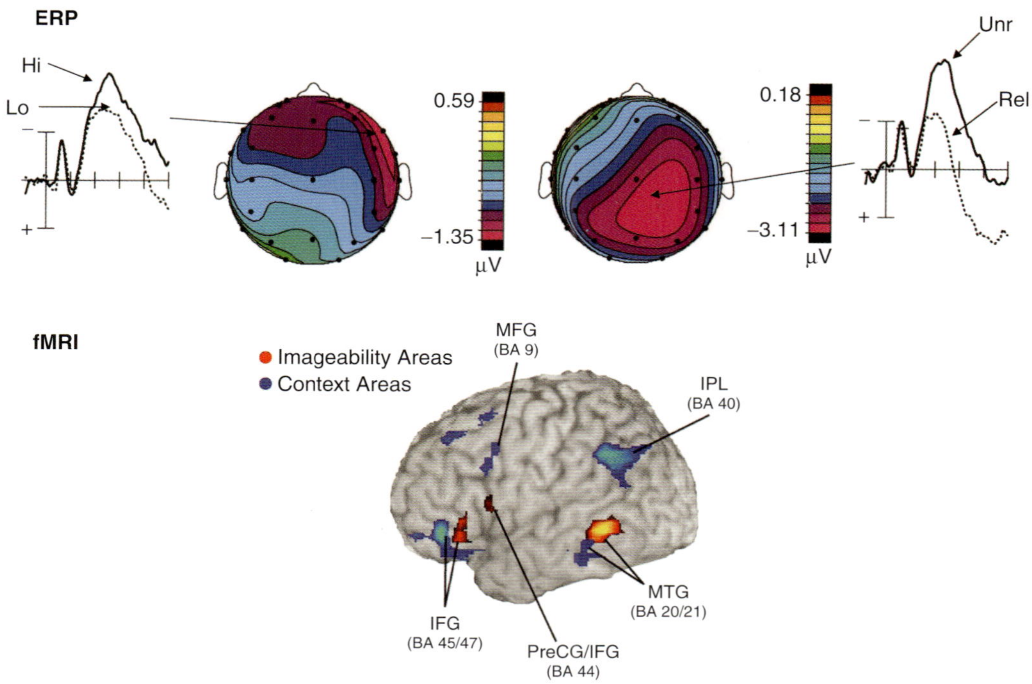

Fig. 15.4. Event-related potential and functional magnetic resonance imaging findings for effects of imageability and priming (see text for explanation of these terms). The top left part of the figure shows that high-imageable words (Hi, solid line) elicit a larger negative shift than do low-imageable words (Lo, dotted line. This effect has a more anterior distribution than the canonical N400 effect of priming shown on the right top part of this figure. The canonical N400 has a parietal distribution, and a reduced N400 to related words (Rel, dotted line) is found relative to unrelated words (Unr, solid line). The distribution of the effects of imageability and priming is evident from the topographic maps that show the distribution of voltage of the ERP effects over the head. The pink color displays the largest effects. The bottom part of this figure shows the fMRI effects of imageability and priming from the same paradigm. Effects of priming were observed in the middle temporal gyrus (MTG), inferior parietal lobe (IPL), and inferior and middle frontal gyrus (IFG and MFG). Effects of imageability were found in the middle temporal gyrus (MTG) and inferior frontal gyrus (IFG). Note that the effects of imageability never overlap with the effects of priming. Redrawn from Figures 1, 2 and 3 in Swaab, Baynes and Knight, Cognitive Brain Research, 2002 (Elsevier), and Figure 4 in Giesbrecht, Camblin and Swaab, Cerebral Cortex, 2004 (Oxford University Press).

stimulus a word or not?), but also in no-task situations, where participants were asked to just listen to or read pairs of words for comprehension.

Lexical repetition also leads to a reduced N400 (see Figure 15.2,D; Bentin & Peled, 1990; Joyce et al., 1999; Rugg et al., 1993), and this effect is greatest at shorter lags (e.g., zero to six intervening items; Rugg & Nagy, 1989). The N400 repetition effect has been observed across a variety of tasks, including lexical decision (Bentin & Peled, 1990; Karayanidis et al., 1991; Rugg et al., 1988), semantic classification (Hamberger & Friedman, 1992; Rugg et al., 1988), and recognition memory tasks (Bentin & Peled, 1990). The effect is restricted to repeated items that are semantically meaningful; geometric line drawings, for example, do not show a reduced negativity in the N400 time range as a result of repetition (Rugg et al., 1995; Van Petten & Senkfor, 1996).

Reductions in the amplitude of the N400 to semantically related and repeated words have been obtained in the visual and auditory modalities, as well as in studies that have used cross-modal presentation (see Figure 15.2,B; Domalski et al., 1991; Holcomb et al., 2005; Joyce et al., 1999; Rugg et al., 1993). Sometimes, the onset of the N400 priming effect occurs earlier in the auditory modality (Holcomb & Neville, 1990), which is consistent with findings that the unique identification of words occurs well before the whole speech signal is heard (Grosjean, 1980).

A large number of ERP semantic priming studies have been devoted to the processing nature of the N400 effect (e.g., Bentin et al., 1985; Brown & Hagoort, 1993; Chwilla et al., 2000; Friederici, 1995; Holcomb, 1993; Holcomb & Neville, 1990; Kutas & Hillyard, 1989). In order to better understand the nature of the N400 effects that have been obtained, we will first briefly review one dominant behavioral account of lexical semantic priming that was proposed by Neely (1991), who postulated three different priming mechanisms. The first of these mechanisms is automatic spread of activation within a lexical-semantic network (Collins & Loftus, 1975; Neely, 1977). That is, activation spreads from the semantic node associated with the prime to the semantic node associated with the target, thereby reducing the processing time of the target upon its presentation. This spread of activation is assumed to be an automatic process that cannot be influenced by any subject strategies. The other two priming mechanisms were postulated because different patterns of results had been obtained as a function of the task that was used in the priming paradigm (e.g., lexical decision vs. naming). The first of these additional mechanisms is expectancy-induced priming: subjects use the meaning of the primes to generate an expectancy set of possible target words (Becker, 1980, 1985; Posner & Snyder, 1975). Expectancy-induced priming reflects controlled processing, and is capacity consuming, relatively slow, and presumably under the subjects' strategic control. The other additional priming mechanism has been labeled *semantic matching* (Neely & Keefe, 1989) or *postlexical meaning integration* (De Groot, 1985). Semantic matching is not unlike the integration process that occurs in the more common processing of sentences or discourse (Brown & Hagoort, 1993; Chwilla et al., 1998, de Groot, 1985), and may be automatic in nature since it does not require conscious awareness of the prime (e.g., Bodner & Masson, 2003). Several studies have shown that the N400 priming effect is greater when the task requires participants to pay attention to the meaning of the words. Additionally, no N400 priming effects are found for words presented in an unattended visual location (McCarthy & Nobre, 1993) or in the unattended ear in a dichotic listening task (Bentin, Kutas & Hillyard, 1995). But expectancy-induced processing is not required to elicit an N400. Chwilla and colleagues (1998) prevented the contribution of expectancy-induced priming in a study that utilized a backward priming paradigm. *Backward priming* refers to the paradoxical finding of facilitation of the processing of a target word when there is only an association from the target to the prime, in the absence of an association from the prime to the target. Consider, for example, the prime–target pair *baby–stork*. In this case, *baby* has a forward association to words such as *mother* and *infant*, but no such forward association exists between *baby* (prime) and *stork* (target). Thus, neither forward spread of activation in a lexical network nor the generation of an expectancy set for the target word can explain the finding of facilitated processing of the target word in this case. Koriat (1981) was the first to observe backward priming and suggested that these effects might be attributed to spread of activation in a lexical network in a backward direction. At longer intervals between prime and targets, backward priming may also result from a postlexical relatedness-checking strategy. That is, given enough time, participants in a priming study may realize that there is a backward meaning relation between target and prime after the target word has been presented (Seidenberg et al., 1984). In order to prevent the use of this strategy, Chwilla and colleagues (1998) made the interstimulus interval between prime and target 0 ms, theoretically too short for a postlexical relatedness-checking strategy. Therefore, the curious finding of facilitated processing of a target word that only has a backward meaning relation to the prime can then be explained only in terms of a backward association. Under these circumstances, clear N400 effects of backward priming were obtained (e.g., the N400 was reduced to *stork* when preceded by *baby*). Findings of this study are therefore consistent with the idea that N400 priming effects can occur in the absence of controlled processes, such as participants reflecting on the relation of *stork* back to *baby*. N400 effects have also been found when participants are engaged in a shallow processing task that is orthogonal to the relationship between words (e.g., letter search; Kutas & Hillyard, 1989). Furthermore, Luck et al. (1996) demonstrated that awareness of the target word is not required, since they found preserved N400 effects in the attentional blink even when the second of two targets presented in rapid serial visual presentation (RSVP) could not be accurately reported. Other studies have shown effects of semantic priming when perception of the prime was masked (e.g., Deacon et al., 2000; for a discussion of masked N400 repetition effects, see Holcomb & Grainger, 2007). It appears, then, that N400 effects can be obtained under automatic processing conditions. However, at this point, there is no conclusive evidence from semantic priming studies to suggest that the N400 is exclusively modulated by automatic integration and not by spread of activation in a semantic network.

N400 and Sentence Context[4]

In the previous section, we presented an overview of some of the many studies that have found N400 priming effects to semantically associated and repeated words. These studies have provided us with insights about the organization and processing of lexical representations when presented outside of a structured linguistic context. The sentence, as the smallest complete unit of language with a cohesive structure, provides a valuable springboard to study language processing in a meaningful and structured environment. This section will focus on N400 effects that have been found in ERP studies of the processing of the meaning of words embedded in sentence contexts.

ERP STUDIES OF SENTENCE READING

A seminal question in studies of sentence comprehension (that have not focused on syntactic aspects of processing) concerns the point at which lexical processing is influenced by the meaning of the larger sentence context. As was discussed earlier, much of the initial work on the functional nature of the N400 was performed with sentences, and in the 1980s it was shown that the amplitude of the N400 varied as a function of cloze probability and sentential constraint (Kutas et al., 1984). Cloze probability measures are obtained by asking participants to finish sentences from which the last word is omitted, such as "I drink my coffee with sugar and _____." The cloze probability is measured as the percentage of people that finish the sentence with a specific word. In this example, *cream* might have a cloze probability of 60% and *milk* might have a cloze probability of 30%. In this case, *cream* is considered the best completion of the sentence. Sentential constraint is determined by the range of possible continuations of a particular sentence context. In this example, the sentence can be finished in a meaningful way only with a very limited number of words and is therefore highly constraining. Other sentences, like "Yesterday she went to the _______" are not very constraining and can be completed with many possible words (e.g., *store, theatre, movies, ski slopes, school, dentist, doctor, hospital*). Strongly constraining sentences will necessarily always end with a word that also has a very high cloze probability. But both high- and low-constraint sentences can be completed with a low-cloze word, that is, a word that is appropriate given the context but does not provide the best completion. In this case, one can distinguish the effects of constraint separately from the effects of cloze probability; that is, one can study the processing of unexpected words in sentences of varying constraint. This is important in identifying the scope and nature of the influence of sentence contexts on the processing of upcoming words in the sentence. In other words, is there a difference in the processing of unexpected (low-cloze) words in high- versus low-constraint sentences? Some models assume that lexical processing is impervious to contextual constraint and would therefore predict no difference in processing as a function of contextual constraint. Models that assume the immediate influence of context on lexical processing would predict a larger impairment in the processing of unexpected words in highly rather than in minimally constraining sentences. Behavioral studies have used lexical decision and naming methods to investigate the influence of contextual constraint on lexical processing of upcoming words. These studies have shown that the strength of the sentential constraint and the degree of semantic relation between the best and the actual completion both influence processing (e.g., Schwanenflugel & LaCount, 1988; Schwanenflugel & Shoben, 1985). Schwanenflugel and colleagues found evidence of faster processing only for words that were related in meaning to the best completion for low-constraint sentences, such as "She cleaned the dirt from her sandals," where *shoes* is the best completion, but not for high-constraint sentence, such as "On a hot summer's day, many people go to the lake," where *beach* is the best completion. On the basis of these results, it had been assumed that strongly constraining sentences result in a narrow scope of activation of lexical candidates, possibly because these contexts activate a larger set of featural restrictions for upcoming words in the sentence (e.g., Schwanenflugel & LaCount, 1988; Schwanenflugel & Shoben, 1985).

However, N400 ERP studies have produced contradictory results to the aforementioned behavioral studies: words that are unexpected but related in meaning to the best sentence completion are actually processed more easily in high- than in low-constraint sentences (Federmeier & Kutas, 1999; Kutas et al., 1984). Federmeier and Kutas (1999) have suggested that this paradox could be resolved if one assumes that high contextual constraint can lead to both benefits and impairments in processing of upcoming words if context exerts its influence at different stages of processing. That is, initially high sentential constraint can facilitate processing of words that are unexpected but related in meaning to the best completions because the context has activated a set of semantic features that matches the meaning of both words. The impairments in processing that have been observed in the behavioral literature may then indicate a later stage of contextual influence that does not affect the amplitude of the N400.

Federmeier and colleagues (2007) have examined this idea by testing whether the contextual effects of sentence constraint and cloze probability jointly affect the N400 or have independent influences on processing. They manipulated sentence constraint and cloze probability as in the following example:

Strong Constraint:

The children went outside to *play* (high cloze—expected)/*look* (low cloze—unexpected)

Weak Constraint:

Joy was too frightened to *move* (medium cloze—expected)/*look* (low cloze—unexpected).

For the unexpected completions, the critical words were the same in both constraint conditions (*look* in the example) and formed plausible but unexpected completions with very low cloze probabilities (3.1%). For the expected completions, contextual constraint and cloze probability were necessarily confounded (high-constraint sentences, by definition, will yield less possible responses than low-constraint sentences) such that high-constraint sentences ended with critical words of very high cloze probability (85.3%) and low-constraint sentences ended with critical words of medium cloze probability (26.9%). Federmeier et al. (2007) replicated previous findings of the effects of cloze probability on the N400 such that congruent sentence-final words with high cloze probability resulted in reduced amplitude N400s relative to words that were congruent but with medium cloze probability (*play* and *move* in the example, respectively). Interestingly, there was no modulation of the amplitude of the N400 to the low-cloze words as a function of sentential constraint. Instead, low-cloze words in highly constraining sentences elicited a positive shift with a frontal distribution that occurred after the N400 (see Figure 15.5). This result has important theoretical implications. In contrast to the behavioral results discussed earlier, the ERP results show that both cloze probability and sentence constraint affect processing, but that the effects of contextual constraint are actually delayed relative to the effects of cloze probability.

Other studies of sentence comprehension have tested if and when effects of sentence congruence influence lexical priming effects. These studies also allow us to ask whether N400 congruency effects in sentence contexts can be distinguished from N400 priming effects. Van Petten (1993) has shown that both lexical association and sentence congruence modulated the N400 to critical words that were preceded by an associatively related or unrelated word that was embedded in meaningful sentences or in

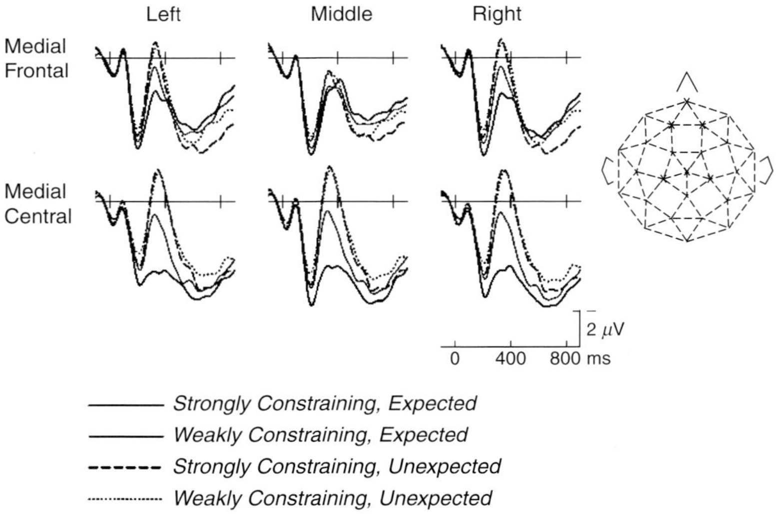

Fig. 15.5 Event-related potential findings for critical words in different sentences of different constraint and cloze probability (see text for explanation). The *x* marks on the schematic of the positions of the electrodes on the head indicate the sites from which the ERP results are displayed. The results show that contextual constraint had no immediate effect on the processing of unexpected words in the sentence; the N400 amplitude between the strongly constraining and weakly constraining unexpected words was not different. The effects of constraint occur later in the ERP waveform; a larger positive shift with a medial frontal distribution is elicited to unexpected words in highly constraining sentences. Redrawn with permission from Federmeier, Wlotko, Ochoa-Dewald & Kutas, Brain Research, 2007 (Elsevier).

so-called syntactic prose (i.e., meaningless sentences with a preserved structure; e.g., Marslen-Wilson & Tyler, 1980), as in the following example:

a Congruent/Associated
When the *moon* is full, it is hard to see many *stars* or the Milky Way.

b Congruent/Unassociated
When the *insurance* investigators found that he'd been drinking, they *refused* to pay the claim.

c Syntactic Prose/Associated
When the *moon* is rusted, it is available to buy many *stars* or the Santa Ana.

d Syntactic Prose/Unassociated
When the *insurance* supplies explained that he'd been complaining, they *refused* to speak the keys.

Modulation of the N400 was found as a function of sentence congruency and lexical association, with the biggest effect in the congruent/associated condition (Condition a). However, the effects of sentence congruency were subject to great individual variability. This finding was followed up in a subsequent study that examined whether effects of sentential congruency and lexical association varied as a function of individual differences in reading span (Van Petten et al., 1997; see Daneman & Carpenter, 1980, for a discussion of reading span). Reading span presumably measures working memory capacity in relation to sentence processing, and many studies have observed correlations between reading span and sentence-processing ability (but see Caplan & Waters, 1999, for a critique on the reading span measure). Van Petten and colleagues (1997) presented participants with the same stimuli as in their study discussed above (Van Petten, 1993), but this time, participants were divided into low-, medium-, and high-span groups according to their scores on the reading span task. The sentences were presented one word at a time at a rapid serial presentation rate of one word every 300 ms, which is much faster then the average rate of RSVP that has been used in ERP experiments in general (~500 ms) but closer to the average fixation time to words under natural reading conditions (e.g., Camblin et al., 2007b). As can be seen in Figure 15.6, medium- and high-span readers showed the same N400 results as in the Van Petten (1993) study. However, low-span readers did not show N400 effects of sentence congruence in the absence of lexical association (Condition b). The authors concluded that lexical association effects are not modulated as a function of reading span,[5]

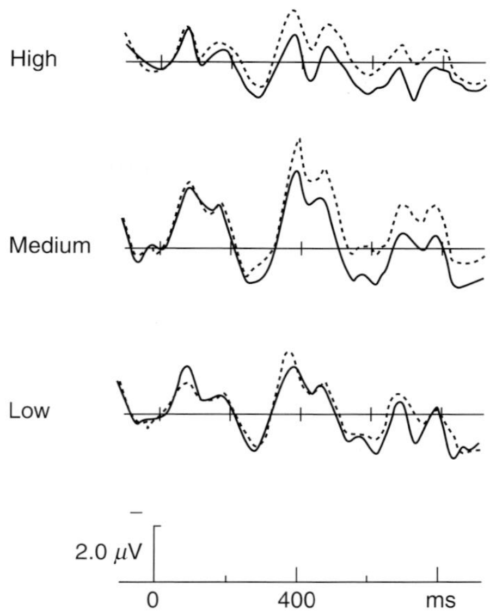

Fig. 15.6 In contrast to medium- and high-span subjects, low-span subjects do not show an N400 effect of overall sentence congruence in the absence of lexical association (see text for explanation). Redrawn with permission from Van Petten, Weckerly, McIsaac & Kutas, Psychological Science, 1997 (APS).

but that low-span readers benefit from the effects of sentence congruence only when lexical associations can aid in building sentence-level context representations.

It is prudent at this point to raise two methodological issues. One is not ERP-specific but is relevant to the present study: namely, the use of a probe recognition task. Gordon et al. (2000) have shown that the probe word recognition task induces readers to read sentences more like incoherent lists, which may enhance effects of lexical association and decrease effects of sentence coherence. The other, more ERP-specific issue is the necessity to use RSVP paradigms with ERPs to avoid eye movements that will contaminate the ERPs. As mentioned before, van Petten et al. (1997) used an RSVP rate of one word every 300 ms to more closely approximate normal reading rates. However, since words are presented one at a time, this fast rate may impose additional demands on the reader that may influence overall sentence integration (Camblin et al., 2007b). This issue will be discussed in more detail later in the chapter ("Using ERPs to Study Language: Methodological Issues").

Interestingly, Coulson, and colleagues (2005) observed hemispheric asymmetries for the ERP effects of lexical association and sentence congruence. In this study, critical words were presented in

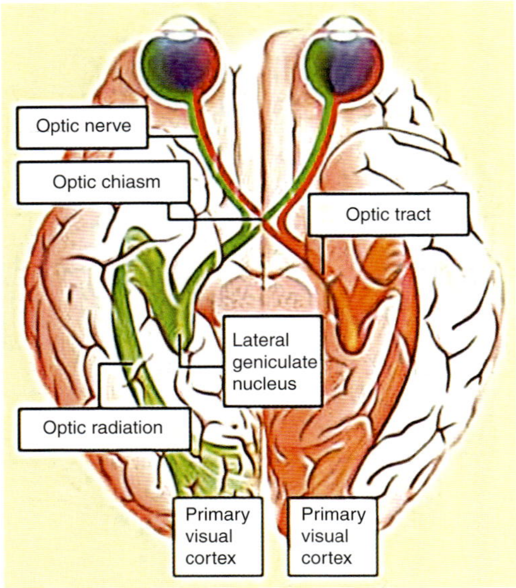

Fig. 15.7 The VHF method has been used to study the contribution of the LH and RH to language processing. The VHF studies present stimuli laterally to either the right visual field (RVF, in red) or the left visual field (LVF, in green). The RVF input hits the left side of the retinas of both eyes and is processed first by the LH, and the LVF input hits the right side of the retinas of both eyes and is processed first by the RH. This is due to the organization of the visual system, where the optic nerves from both eyes cross at the optic chiasm and are sent to the primary visual cortices of the LH and RH of the brain. Reprinted with permission from Bruno Dubuc, Canadian Institutes of Health Research: Institute of Neurosciences, Mental Health and Addiction, www.thebrain.mcgill.ca

the left visual field/right hemisphere (LVF/RH) or in the right visual field/left hemisphere (RVF/LH) by making use of the visual half field (VHF) method (see Figure 15.7).

This method takes advantage of the neuroanatomical organization of the visual system, in which the visual input from the left VHF is first projected (via the optic chiasm) to the RH and vice versa. This has the consequence that any stimulus presented to the RVF is initially processed by the LH and any stimulus presented in the LVF is initially processed by the RH, after which the information is transferred via the corpus callosum to the other hemisphere. Briefly presenting a stimulus to a specific visual field can allow researchers to draw inferences about the nature of processing in the contralateral cerebral hemisphere. In VHF experiments, it is crucial that participants keep their eyes focused centrally to make sure that the stimulus is actually presented to one of the VHFs (*and* to obtain ERPs that are not contaminated by eye movements). The following procedures can be implemented to make sure that eye movements have not contaminated the results: (1) monitor eye movements throughout the experimental session in order to ascertain that all subjects maintain fixation, (2) conduct an eye-calibration procedure for each subject, and (3) inspect VHF ERP data for a larger N1 component on electrode sites contralateral to the visual field of presentation; the presence of a larger N1 indicates that the visual stimulus was indeed first processed by the contralateral hemisphere.

To study the contributions of the LH and RH to effects of sentence congruence and lexical association, Coulson et al. (2005) utilized a counterbalanced design in which sentence congruence and lexical association were manipulated such that the last word of the sentence was either congruent or not with the whole-sentence context and was associated in meaning or not with an immediately preceding word, as in the following example (critical words are underlined):

a Congruent/Associated
The Italian cook always added too much olive <u>oil</u>.
b Congruent/Unassociated
They were hard to walk in, but she loved her olive <u>shoes</u>.
c Incongruent/Associated
During the test, Ellen leaned over and borrowed my spare <u>tire</u>.
d Incongruent/Unassociated
They were truly stuck, since she didn't have a spare <u>pencil</u>.

Coulson et al. (2005) showed N400 effects of sentence congruence for both hemifields that were identical in size and onset. This finding suggests that the RH is in fact sensitive to sentence-level semantic constraints, which confirms earlier behavioral findings of Faust and colleagues (Faust, 1998; Faust & Gernsbacher, 1996; Faust et al., 1993, 1995). Effects of lexical priming, on the other hand, varied as a function of hemifield of presentation: whereas in the LVF/RH, effects of lexical priming were observed for both congruent and incongruent sentence endings, in the RVH/LH effects of lexical priming were observed only for sentences that ended with an incongruent word. Thus, these results suggest that the two hemispheres are differentially sensitive to effects of lexical priming in sentence contexts. The LH "uses" lexical associations only when no overall integration of the sentential context is possible, whereas the RH uses lexical associations regardless of the overall congruence of the sentence contexts.

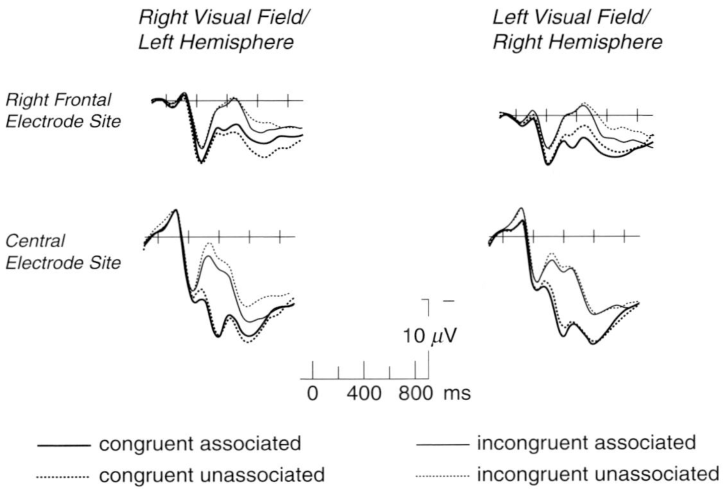

Fig. 15.8 Only LVF presentations result in effects of association in both congruent- and incongruent-context sentences. This effect was not observed on the N400, but rather on a frontally distributed positive shift—more positive to associated than unassociated words in both congruent and incongruent conditions. The RVF presentation resulted in an N400 effect of association for the incongruent condition only (bottom left quadrant). Redrawn from Figures 4 and 5 in Coulson, Federmeier, van Petten and Kutas, Journal of Experimental Psychology: Learning, Memory and Cognition, 2005 (APA).

In addition, the effects of lexical priming only showed up as a canonical N400 effect for the incongruent/associated critical words presented to the RVF/LH. For the LVF/RH, a frontally distributed ERP effect of lexical association was observed (see Figure 15.8). Coulson et al. (2005) state that this effect needs further study, but it does open the intriguing possibility that under certain circumstances, effects of lexical priming may have a distinct electrophysiological signature from the classical N400 effect of sentence congruence.

In sum, ERP studies that have compared effects of lexical priming and overall congruence in sentence comprehension have found that effects of lexical association can be modulated as a function of overall sentence congruence. Relative to the Van Petten (1993) study, the Coulson et al. (2005) study shows less robust effects of lexical association, which may be explained in terms of differential cloze probability. In the Coulson et al. (2005) study, the cloze probability of the sentence-final words was higher than the cloze probability of the critical words in sentence-intermediate positions in the Van Petten (1993) study, and this may indicate that effects of lexical association contribute to processing only in sentence contexts that are not highly constraining or incongruent. In addition, Van Petten et al. (1997) have shown that low-span readers rely more on lexical associations than do high-span readers, and Coulson et al. (2005) have shown that lexical association contributes to RH sentence processing regardless of the overall congruence of the sentence. In general, the effects of lexical association and sentence congruence both appear to affect the canonical N400 ERP, except when the critical associate is presented in the LVF/RH; in this case, a right frontal effect of association is obtained.

ERP STUDIES OF SPOKEN SENTENCE COMPREHENSION

Event-related potential studies of spoken sentence comprehension have been conducted much less frequently than ERP studies of reading. In part this may be due to the fact that it is relatively time-consuming to identify the onset of the critical stimulus in the continuous speech signal, which is, of course, necessary in order to accurately time-lock the ERPs. Also, because the speech signal is continuous, ERPs from previous input will overlap with the ERP of interest, and early P1, N1, and P2 components cannot be discerned in the ERPs to critical stimuli embedded in continuous speech.[6] Nevertheless, several studies have used the N400 to investigate the influence of word or sentential context on the process of spoken word recognition (e.g., Connolly & Phillips, 1994; Diaz & Swaab, 2007; Friederici et al., 2004; Friedrich & Kotz, 2007; Friedrich et al., 2004; Hagoort & Brown, 2000; O'Rourke & Holcomb, 2002; Praamstra et al., 1994; Radeau et al., 1998; van den Brink & Hagoort, 2004; van den Brink et al., 2001, 2006; Van Petten et al., 1999). These studies have confirmed behavioral findings of a very rapid influence of context on spoken word processing. But it has been a matter of debate whether or not the electrophysiological manifestation of these effects was found on the N400 (Van Petten et al., 1999) or instead on a separable component that would be specifically sensitive to phonological mismatch (Connolly & Phillips, 1994) or lexical selection (van den Brink & Hagoort, 2004) during spoken word recognition. To address this issue, Diaz and Swaab (2007) directly compared the processing of words in lists to the processing of words in meaningful sentences. Phonological and semantic congruence was manipulated in both contexts. To assess the effects of phonological mismatch on ERPs per se, participants were asked to listen to a series of eight words in which the final word was either phonologically congruent or incongruent with respect to the onset of the preceding words but none of the words were semantically or associatively related to each other. The second list condition was included to elicit a canonical N400 effect, and here the eighth word of the list was either semantically congruent or incongruent with respect to the semantic category of the preceding words, but there was no alliterative overlap in word onsets. In addition, participants listened to four types of sentences in which the semantic and phonological congruence of the terminal word was manipulated (as in Connolly & Phillips, 1994). These manipulations are illustrated in the following examples (critical final words are underlined):

Lists

ALLITERATIVE CONDITION

Congruent: chat, champ, chaff, chant, challis, chad, chap, chapter

Incongruent: chat, champ, chaff, chant, challis, chad, chap, address.

CATEGORY CONDITION

Congruent: giraffe, sheep, bear, wolf, rabbit, lamb, elephant, dog

Incongruent: giraffe, sheep, bear, wolf, rabbit, lamb, elephant, desk.

Sentences

High Congruent: He mailed the letter without a stamp.

Phonologically Congruent: He mailed the letter without a stance.

Low Congruent: He mailed the letter without a thought.

Incongruent: He mailed the letter without a roof.

In lists of words, the lexical-phonological processes could not be influenced by a meaningful context, and separable effects of phonological and semantic information were predicted on the processing of the critical target words in the lists. The results indeed confirmed this (see Figure 15.9): an early, topographically distinct electrophysiological manifestation of phonological incongruence was observed that preceded the N400 effects of semantic incongruence. In contrast, no such separation was obtained in the meaningful sentence contexts: there was only evidence for an early N400 to semantic incongruency, but no separable effect of phonological incongruence was obtained.

This indicates that semantic information of the sentence context very rapidly influences lexical processing in meaningful contexts. Together, these results provide evidence that the moment in time at which context starts to exert its influence on lexical processing depends on how constraining the previous context is, with lists of words at one end of the spectrum and highly constraining sentence contexts at the other end.

N400 and Discourse Contexts

Language comprehension critically depends on successfully integrating the meaning of individual words into the meaning of larger discourse contexts. Extracting meaning from the discourse requires rapid processing of several different sources of linguistic information, including phonological, syntactic, and semantic aspects of the language input. In addition, word and sentence meaning critically depend on the context of the utterance. For example, even though the word *steak* makes perfect sense in the isolated sentence context "Yesterday my uncle ate a big juicy steak," it is not sensible when the global discourse context has just established that my uncle is a vegetarian: "A few years ago, my uncle decided that he should become a vegetarian. He had not found it very difficult to stay away from meat, because there are so many delicious other foods. Yesterday he ate a big juicy steak."

Previous research has shown that such contextual information influences the identification of word meanings in sentences and discourse; that is, listeners and readers employ the meaning of the preceding words and sentences in determining the meaning of the current word. Thus, language comprehenders do not represent the surface (literal word-by-word) information of a context, but rely on the integrated

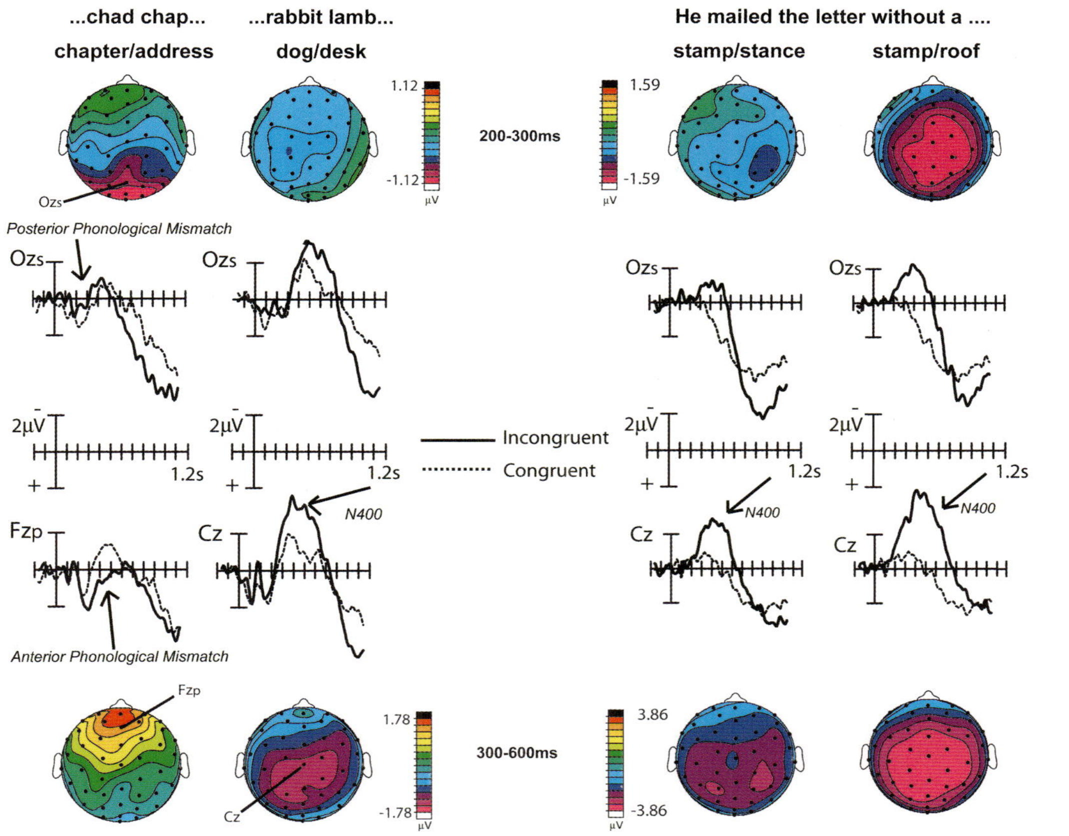

Fig. 15.9 Event-related potential effects and topographic distribution of these effects in 200–300 ms (top) and 300–600 ms (bottom) epochs. On the left side are displayed the effects of violation of phonological or semantic congruence in lists of words. Effects of violation of phonological congruence were found for occipital sites (more negative to violation of expectancy) and frontal sites (more positive to violation of expectancy). A canonical N400 effect was obtained when a violation of congruence of a semantic category occurred. For the sentences (right side of the figure), both semantically anomalous words that share the speech onset with the word that is the best completion for the sentence and those that do not result in a canonical N400 effect. Redrawn from Figures 2 and 3 in Diaz & Swaab, Brain Research, 2007 (Elsevier).

meaning of the overall representation (Bransford & Johnson, 1972). However, the time course of the activation and integration of information from the discourse context during lexical processing is a subject of much debate. Some theories of language comprehension assume that sentences (or constituents) are initially processed independently and that discourse information only becomes available at later stages of processing (see Chomsky, 1975; Fodor et al.,1974; Forster, 1989; Katz, 1972; Searle, 1979; Sperber & Wilson, 1986). Other models predict a briefer delay of the influence of discourse information on real-time comprehension, such that discourse can influence processing after each word, but not until syntax and sentence-level meaning have been integrated (Fodor et al., 1996); Frazier, 1999). In contrast, constraint-based models argue for the simultaneous activation of phonological, syntactic, and semantic information during lexical processing. Even though no explicit predictions have been made with respect to discourse information, it would be in the spirit of these latter models to assume that this simultaneously activated information can be used immediately to comprehend the preceding utterance (Jackendoff, 2002, 2007; MacDonald et al., 1994; Marslen-Wilson & Tyler, 1980; Tanenhaus & Trueswell, 1995). In other words, all sources of contextual information are integrated during lexical-semantic processing, not thereafter.

Several ERP studies have used the exquisite temporal resolution of ERPs in general and of the N400 in particular to investigate whether the wider discourse context can immediately influence lexical semantic processing in the local sentence context, or alternatively, whether there is a delay in the influence of discourse context on lexical semantic processing in the local sentence (e.g., Camblin et al., 2007a; Federmeier & Kutas, 1999; Nieuwland & van Berkum, 2006; Nieuwland et al., 2007; St. George et al., 1994; van Berkum, 2004; van Berkum et al., 1999, 2003, 2007, 2008; for a review, see van Berkum, 2009).

St. George et al. (1994) were the first to show that the N400 is sensitive to global discourse-level effects on the processing of the meaning of single words. Subjects read ambiguous paragraphs from a behavioral study by Bransford and Johnson (1972, p. 722), in which everyday activities were described that did not make sense unless a disambiguating title was provided, as in the following example:

> "The procedure is actually quite simple. First you arrange things into different groups depending on their makeup. Of course, one pile may be sufficient depending on how much there is to do. If you have to go somewhere due to lack of facilities that is the next step, otherwise you are pretty well set. It is important not to overdo any particular endeavor. That is, it is better to do too few things at once than too many. In the shorter run this may not seem important, but complications from doing too many can easily arise. A mistake can be expensive as well. The manipulation of the appropriate mechanism should be self-explanatory, and we need not dwell on it here. At first the whole procedure will seem complicated. Soon, however, it will become just another facet of life. It is difficult to foresee any end to the necessity of this task in the immediate future, but then one can never tell."

Even though the individual sentences of this paragraph make sense, it is very difficult to make sense of the whole passage. However if we give you the title "A Procedure for Washing Clothes," then everything starts to fall into place (although admittedly the passage remains awkward).

Event-related potentials were obtained to all content words in paragraphs, as in the example above, and the N400 was significantly reduced to these words when presented in the titled relative to the untitled condition. This suggests that providing a title facilitated the generation of a discourse model, which in turn facilitated the integration of the meaning of single words into a representation of the overall context.

Federmeier and Kutas (1999) used the N400 to investigate the lexical-semantic processing of words within and across category boundaries when a preceding context strongly favored one specific lexical candidate from a semantic category ("They wanted to make the hotel look more like a tropical resort. So along the driveway they planted rows of *palms/pines/tulips*"). They found that the amplitude of the N400 varied as a function of the match of the critical word with the semantic category biased by the overall context, with a reduction of the N400 seen to the most expected final word given the discourse context (*palms*). Importantly, the amplitude of the N400 to the discourse-unexpected words varied as a function of the semantic relationship of these words with the most expected word, such that a smaller N400 was found to *pines* (which is a close semantic associate of *palms*) than to *tulips* (which has a more distant semantic relationship). These findings not only illustrate discourse context effects on processing but also, according to the authors, indicate that long-term memory organization influences language processing.

We will now turn to some N400 studies showing that a cohesive, supportive discourse can delay or even override local effects of lexical-semantic processing. Nieuwland and van Berkum (2006) presented participants with short, cartoon-like passages featuring animacy violations. Animacy violations produce an N400 effect, presumably because human-exclusive actions or emotions, such as talking to a therapist or singing a love song, are ascribed to inanimate objects, such as a yacht or a peanut, and this creates a kind of semantic anomaly. The authors constructed 60 six-sentence cartoon-like stories that featured a cartoon-style, animacy-violating character, such as a peanut singing about his girlfriend. The experimental manipulation involved the penultimate sentence of each story, which contained either an animate (yet context-appropriate) or inanimate (yet context-inappropriate) description of the main character (e.g., the peanut). In the animate, context-appropriate condition, the peanut would be described as being in love, for example, while the inanimate, context-inappropriate condition would describe the peanut as salted. All inanimate, context-inappropriate words were selected so as to be *canonical*, or common descriptors of that particular object ("The peanut was *salted*"). It is important to note that such common descriptors would ordinarily be expected to reduce the N400 amplitude, particularly relative to animacy-violating descriptors (*in love*). In contrast, as can be seen in Figure 15.10, Nieuwland and van Berkum (2006) found a reduction of the N400 amplitude to the noncanonical but context-appropriate words (*in love*). This suggests that a discourse context can

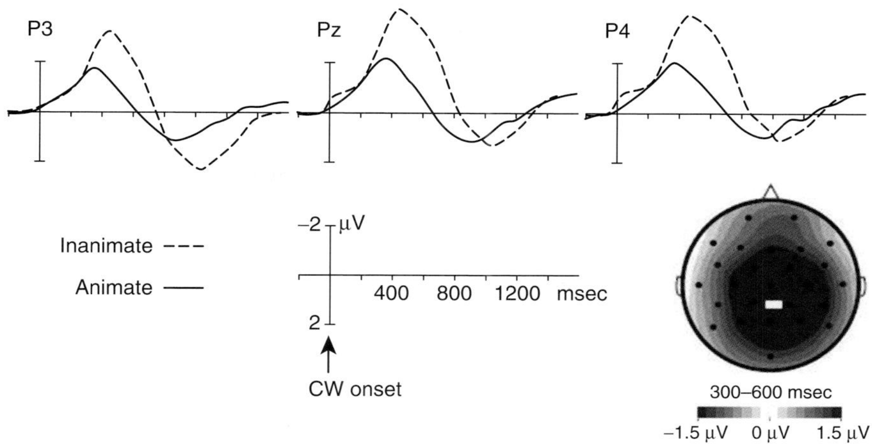

Fig. 15.10 A reduced N400 is found to words that violate animacy (dotted line; e.g., *the peanut fell in love*) relative to those that do not (solid line; e.g., *the peanut was salted*). This paradoxical result is obtained when the violation of animacy is context appropriate in the discourse (e.g., in a story of a peanut falling in love). CW onset indicates the moment in time when the critical word to which the ERP was measured was presented. Reprinted with permission from Nieuwland & van Berkum, Journal of Cognitive Neuroscience, 2006 (MIT press).

actually override the effects of both animacy violations and plausibility.

Camblin and colleagues (2007a) have provided other N400 evidence of the rapid influence of discourse context on lexical processing. They varied semantic association and discourse congruity in multisentence passages (e.g., "The movie was applauded by adults and *children/toddlers*," preceded by a context referring either to a Disney film or to a Holocaust documentary). Participants were asked to read these passages for comprehension. Discourse congruity was found to have an earlier effect on lexical processing than association, as evidenced by an earlier onset of the N400 effect for the discourse manipulation. Recently, this time course effect was replicated when the same passages were presented as naturally produced speech and a very clear delay in the N400 effect to lexical associations was observed relative to the N400 effect of discourse congruence effect, which onset much earlier (Boudewyn et al., in press; see Figure 15.11).[7]

Other studies have shown that N400 effects of repetition priming are not immune to discourse-level effects either (e.g., Camblin et al., 2007b; Gordon et al., 2004; Johns et al., under review; Ledoux et al., 2006, 2007; Swaab et al., 2004). For example, studies of both visual and auditory modalities have shown that classic N400 effects of repetition priming with repeated-name coreference (when two instances of a name refer to the same person) were found only when the sentence structure was conducive to this type of coreference. Compare, for example, the following sentences:

(1) At the office Daniel moved the desk because Daniel needed room for the filing cabinet.

(2) At the office Daniel and Amanda moved the desk because Daniel needed room for the filing cabinet.

The repeated name "Daniel" (the anaphor) in Sentence (1) is awkward (the pronoun *he* would be preferred in this case), whereas in Sentence (2) the use of a repeated name is a perfectly acceptable vehicle for coreference. When the sentence structure is not conducive to repeated name coreference (as in "At the office Daniel moved the desk because Daniel . . . ,") the repetition priming benefit is eliminated. In the behavioral literature, Gordon and colleagues have labeled this effect the *repeated name penalty* and have observed that this type of penalty occurs when a repeated name refers to an antecedent in discourse focus (e.g., as in Sentence 1 above; see, e.g., Gordon & Hendrick, 1997; Gordon et al., 1999). Gordon and Hendrick (1997) proposed that the repeated name penalty results from "disjoint reference": when the second *Daniel* is encountered, it is initially processed as a new entity in the discourse and additional processing is required to determine that the anaphor and the antecedent *Daniel* refer to the same person. Ledoux and colleagues (2007) used ERPs to examine why this penalty might occur. In other words, why is it so awkward to repeat

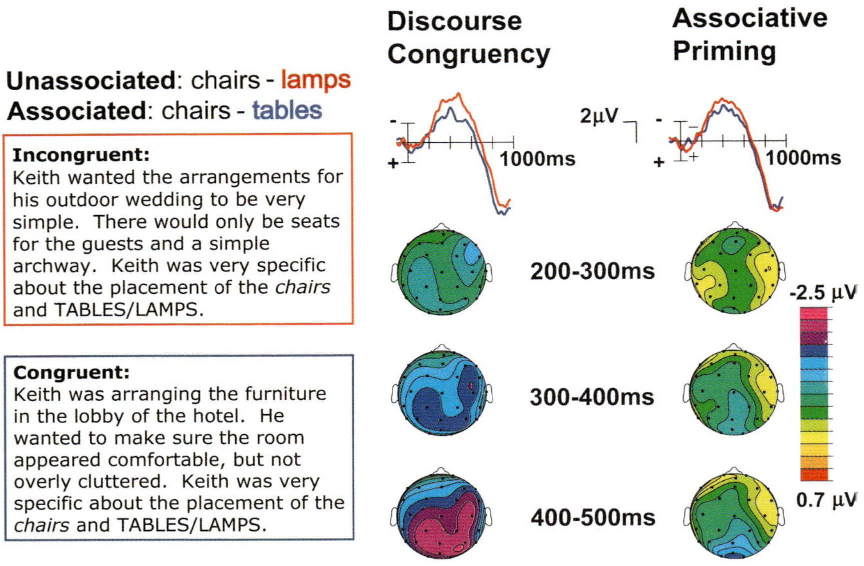

Fig. 15.11 In this study by Boudewyn and colleagues (in press), participants heard stories that ended with a word that was either congruent or not with the preceding discourse and was associated or not with a word immediately preceding the final word of the last sentence (see examples on the left side of the figure). The ERP waveforms and topographic maps of the effects of discourse congruence and associative priming are displayed. Red lines show ERPs to discourse-incongruent and unrelated words, respectively, and blue lines show ERPs to discourse-congruent and related words, respectively. The N400 effects of lexical association are delayed relative to the effects of discourse context in spoken language comprehension; whereas significant effects of discourse congruence were obtained in all three epochs shown for the topographic maps, the effects of association were not obtained until after 400 ms.

a name when the antecedent is in discourse focus? If the repeated name penalty would be reflected in a modulation of the amplitude of the N400, then this would indicate processing difficulty of a semantic nature, possibly because the anaphor is more difficult to integrate in the context or because initially the repeated name does not activate any retrieval cues. Ledoux et al. presented sentences as in the example above. In addition, they added a control condition in which the anaphors were replaced with new names to examine the interaction of lexical repetition and discourse prominence. This was done to test the prediction that coreference to a prominent antecedent causes a repeated name to be processed as if it were a new name, and this is exactly what they found (see Figure 15.12).

Discourse context can thus be seen to override the more purely lexical-level benefits of both semantic association and repetition (for review, see Ledoux et al., 2006).

Another essential aspect of discourse processing is to establish the referents of the currently expressed entity. This entity will often be a noun phrase. An essential function of noun phrases such as *the morning star* is that they refer to a particular discourse entity, the referent (i.e., Venus; even though Venus is actually a planet, as pointed out by Alex, the the son of the first author of this chapter, who was 7 at the time). Referents can be entities in the actual world, entities in some possible world, or even entities that don't exist. Discourse contexts are often needed to determine the intended referent of an expression because multiple expressions can have the same referent (e.g., *morning star* and *evening star* both refer to the planet Venus) and the same noun phrase can have more than one referent (*my son Alex* can refer to various people). As discussed above, *co*referential processing is specifically used to establish whether or not two linguistic expressions refer to the same semantic entity (e.g., "*Tamara Swaab's son Alex* knows that Venus is not a star because *he* likes to read about the universe," where *he* corefers to *Tamara Swaab's son*). Van Berkum and colleagues have performed ERP studies of referential processing in discourse contexts, and have consistently found a negative shift over frontal electrode sites for expressions that might be linked to more than one referent in the preceding discourse (ambiguous referents; Nieuwland et al., 2007; for reviews, see van Berkum, 2009; Van Berkum et al., 2007). They have labeled this ERP the *Nref* (see Figure 15.13).

Taken together, the N400 studies discussed in this section indicate the rapid and sometimes dominating effect of discourse representations on the

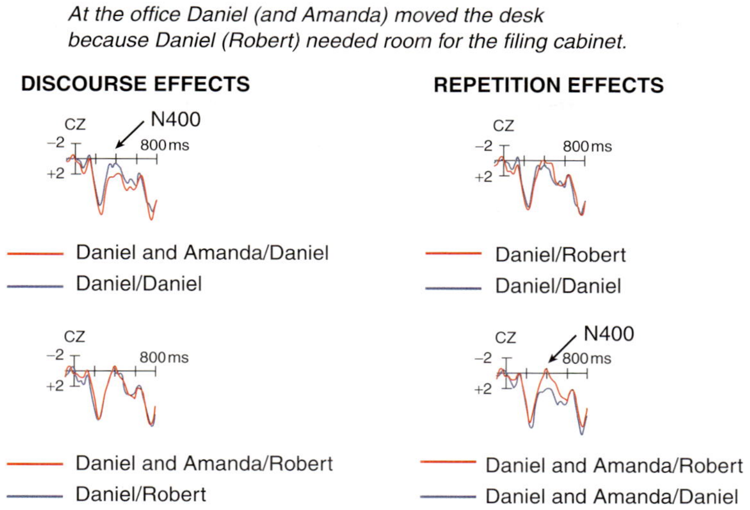

Fig. 15.12 Effects of lexical repetition (right side) and discourse prominence (left side) during reading of sentences. Event-related potentials are shown for a central site. A repeated name penalty is found for repeated names with antecedents in discourse prominence (*Daniel/Daniel*, blue line, vs. *Daniel* and *Amanda/Daniel*, red line, top left quadrant). Effects of discourse prominence are not obtained when a new name is entered in the discourse (*Robert*, bottom left quadrant). Effects of lexical repetition are only obtained for repeated names with antecedents that are not prominent in the discourse (*Daniel* and *Amanda/Daniel*, blue line, vs. *Daniel* and *Amanda/Robert*, red line, bottom right quadrant) but not for repeated names with prominent antecedents (top right quadrant). Data from Ledoux et al. (2007).

processing of incoming words. Overall, the results are more consistent with interactive models of processing, where contextual information can have an immediate impact on language comprehension and processing.[8]

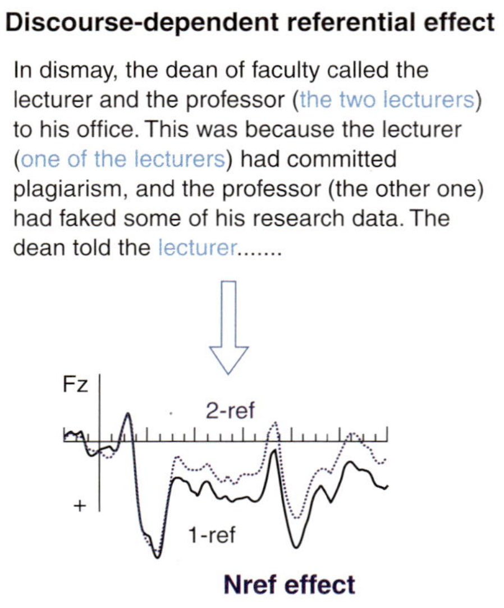

Fig. 15.13 The Nref is obtained to critical words that can refer to more than one antecedent (blue line) relative to those with unambiguous single antecedents (black line). Adapted with permission from van Berkum (2009).

N400 and Nonliteral Language

Language input is richly ambiguous and requires processing that goes well beyond literal and straightforward interpretations. For example, the following sentence illustrates that many words have more than one meaning, some of which are more literal than others: "I still miss my wife, but I have improved my aim" (Coulson & Williams, 2005). One of the first studies using ERPs to examine nonliteral language processing was done by Pynte and colleagues (1996). In their design, they tried to tease apart the effect of metaphor familiarity and the presence of supporting context. In the absence of greater context, short, familiar metaphors (e.g., "Those fighters are lions") produced larger N400s to the terminal, metaphorical word than N400s elicited by the same terminal word in a nonmetaphorical sentence (e.g., "Those animals are lions"). There was also a trend whereby unfamiliar metaphors (e.g., "Those apprentices are lions") produced larger N400s than familiar metaphors, perhaps reflecting the association between the nouns in the familiar metaphor. However, this result was flipped when the unfamiliar metaphor was paired with a supportive context (e.g., "They are not cowardly: Those apprentices are lions") and the familiar metaphor was paired with an unsupportive context (e.g., "They are not naive: Those fighters are lions"): In this case, the unfamiliar metaphor elicited N400s of smaller amplitudes. This indicates that the context provided in support of a metaphor frame may be more influential in processing than more basic properties of the metaphor like familiarity and association.

In a later study, Coulson and Van Petten (2002) examined the processing of metaphors by comparing them not only to straightforward literal controls but also to literal mappings. For example, consider the word *syrup* in the following sentences:

Literal control: I read that one of Canada's major exports is maple syrup.
Literal mapping: In the movie *Psycho*, the blood was really cherry syrup.
Metaphor: He didn't understand the words, but her voice was sweet syrup.

While the word "*syrup*" is not being used metaphorically in the literal mapping condition, it does elicit some of the same processing that is required to understand metaphors according to the conceptual blending hypothesis: mappings need to be produced between disparate domains, and integration across those domain backgrounds needs to take place (Fauconnier & Turner, 1998). In the example above, the qualities of blood and cherry syrup both need to be activated and mappings need to be made between the two regarding their similarities specific to that given context. Coulson and Van Petten (2002) found a graded N400 effect, such that the largest N400s were produced to metaphors, the smallest to literal controls, and those of intermediate amplitude were found in response to literal mapping. Taken together, the findings of Coulson and Van Petten (2002) and Pynte et al. (1996), show that while processing metaphors does appear more effortful than processing sentences that are transparent literally, this difficulty can be reduced by providing a supporting context and may not be entirely unique to nonliteral language comprehension.

Many studies of nonliteral language comprehension have focused on whether or not its representation and processing are different from those of literal language. Some studies indicate a special role for the RH in the representation and processing of nonliteral language. A few ERP studies have been conducted with the VHF technique to investigate whether or not the RH is involved in the processing of words that have both a literal and a nonliteral sense and in the processing of jokes (Coulson & Lovett, 2004; Coulson & Severens, 2007; Coulson & Van Petten, 2007; Coulson & Williams, 2005; Coulson & Wu, 2005). Kacinik and colleagues (2008) used ERPs and the VHF technique to study processing and representation of polysemy in language. Polysemous words such as *bright* have one form representation but multiple senses that can be related to the literal meaning (e.g., *bright light*) or to the metaphoric meaning (e.g., *bright student*). The RH has been proposed to be preferentially involved in comprehending subordinate figurative meanings (Anaki et al., 1998; Beeman, 1998; Brownell, 2000; Jung-Beeman, 2005). However, a series of behavioral VHF and central ERP experiments (Kacinik & Chiarello, 2007; Kacinik et al., in preparation) have repeatedly failed to show differences for the integration of literal and figurative meanings into ambiguous contexts. Kacinik and colleagues (2008) measured ERPs to lateralized sentence-final words related to the literal or figurative sense of polysemous words in ambiguous contexts (e.g., "The girl did not approach the **slimy** frog/clerk"). Participants were asked to read these sentences for comprehension and to answer a true/false comprehension question that followed the presentation of each of the stimuli. No significant differences between the integration of literal and figurative meanings were found in either visual field with respect to both N400 and late positive effects. The more imageable literal endings, however, did show a bigger anterior imageability effect in the LVF/RH than in the RVF/LH, supporting prior indications that brain activity differences in understanding literal and figurative meanings mainly reflect differences in imageability rather than in literalness or figurativeness per se. Semantically incongruent endings in the LVF/RH also resulted in a larger N400 than for the RVF/LH, providing further evidence of RH involvement in sentence comprehension and sensitivity to message-level meaning.

Coulson and Van Petten (2007) also did not find VHF ERP evidence of an RH advantage for figurative language comprehension. Participants read sentences that ended with either highly predictable (high-cloze) words or appropriate but low-cloze literal or low-cloze metaphorical words that were presented in either the left or right VHF. A significant effect of sentence type was found, such that both low-cloze literal and low-cloze metaphorical sentence endings elicited relatively larger N400 amplitudes than high-cloze endings, with the metaphorical endings eliciting the largest N400 amplitudes of all (see Figure 15.14). This implies that metaphorical processing is more difficult than its literal counterpart; however, the pattern of differences

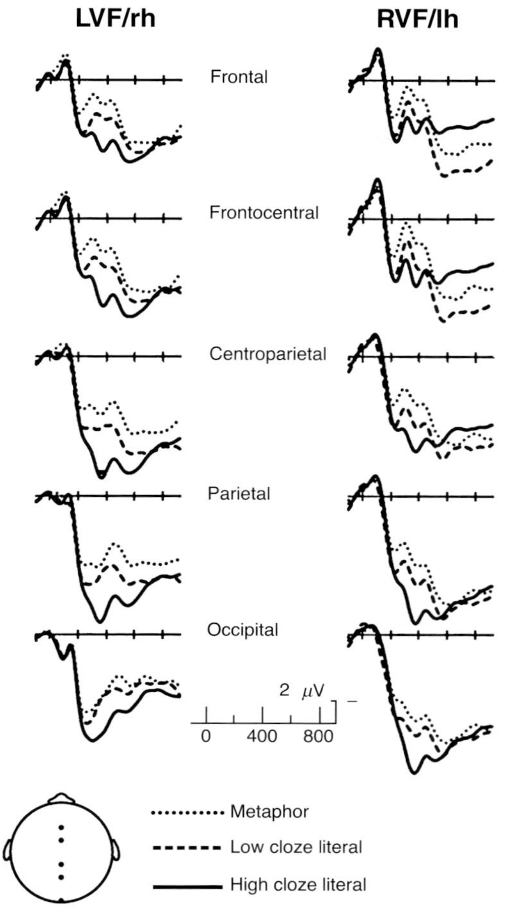

Fig. 15.14 Event-related potentials to metaphorical (dotted line), low-cloze literal (dashed line), and high-cloze literal sentence-final words. The largest-amplitude N400 was obtained for the metaphorical sentence final words for both RVF and LVF presentations. Reprinted with permission from Coulson & Van Petten, Brain Research, 2007 (Elsevier).

in N400 amplitude among the conditions was found for both left visual hemifield (LVHF) and right visual hemifield (RVHF) presentation, suggesting that the RH does not have a special role in metaphorical processing.

A series of studies by Coulson and colleagues previously found, however, that the RH may be essential for joke comprehension (Coulson & Lovett, 2004; Coulson & Severens, 2007; Coulson & Williams, 2005; Coulson & Wu, 2005). It has been shown that compared to an unfunny sentence ending, joke endings to sentences (e.g., "Statistics indicate that Americans spend 80 million a year on games of chance, mostly *weddings/dice*") evoke relatively larger N400 waveforms (Coulson & Kutas, 2001). Coulson and Williams (2005) went on to show that when presented using the VHF technique, "one-liners" compared to nonjoke sentences such as "I still miss my wife, but I am improving my *aim/ego*" replicate this effect when presented to the RVF/LH but a null effect when presented to the LVF/RH. This suggests that the RH has no more difficulty integrating a joke ending into the sentence than a nonjoke ending (Coulson & Williams, 2005).

The Processing Nature of the N400

Taken together, the findings discussed in the previous section clearly show that the N400 is modulated by semantic aspects of the input, and specifically, that the amplitude of the N400 is reduced to words that can be easily related to the meaning of the overall semantic context. But the exact processing nature of the N400 is still a matter of debate. Broadly, two accounts of the N400 have emerged. Kutas and colleagues (2006, p. 669) have proposed that "Overall, the extant data suggest that N400 amplitude is a general index of the ease or difficulty of retrieving stored conceptual knowledge associated with a word (or other meaningful stimuli), which is dependent on both the stored representation itself, and the retrieval cues provided by the preceding context." Thus, according to this account, the amplitude of the N400 is modulated by the ease of semantic retrieval and by the top-down contextual influence on this process. That is, when reading a sentence such as "He spread the bread with butter/cream," the word *cream* elicits an N400 because the sentence context has been used to predict and preactivate lexical-semantic features of the expected word *butter*, and *cream* does not match all of these semantic features (and is therefore more difficult to retrieve).

Hagoort (2005), on the other hand, has proposed that the N400 is a reflection of a semantic integration or unification process, such that words that can be easily integrated into the preceding conceptual context generate a reduced N400. Hence, this latter account assumes that the process of lexical-semantic integration or semantic unification is the only driving force behind the N400 and that the ease of semantic retrieval per se is *not* reflected by the N400 (see also Friederici, 2002, for a comparable account of the N400). Thus, in the sentence "He spread the bread with butter/cream" the N400 is reduced to "*butter*" because it is more easily integrated with the higher-order meaning representation of the preceding sentence context than is "*cream*".

Some of the findings that we have discussed appear more difficult to reconcile with the integration account of the N400. First, there is the finding that the N400 is sensitive to lexical factors such as word frequency and orthographic neighborhood, which have little to do with combinatorial semantics. Also, findings showing the modulation of the N400 as a function of preactivation of semantic features of words in sentence or discourse contexts are more easily explained in terms of facilitated retrieval than in terms of ease of integration.

Thus, these empirical findings with the N400 appear more consistent with the retrieval view of Kutas and colleagues (2000, 2006; see also Lau et al., 2008). Recently, van Berkum (2009 has proposed an extension of the retrieval model of the N400, labeled the *multiple cause intensified etrieval model* (MIR), to take into account some of the more recent findings with the N400 (see notes 7 and 8). His model assumes that "The amplitude of the word elicited N400 reflects the computational resources used in retrieving the relatively invariant 'coded' meaning(s) stored in semantic long-term memory for, and made available by, the word at hand" (van Berkum, 2009 p 12). As in the Kutas et al. (2006) model, van Berkum (2009) assumes that retrieval is facilitated (i.e., requires fewer computational resources) when the meaning of a word is consistent with contextually preactivated semantic features. But he also assumes that the N400 is not only dependent on the semantic context per se, but is also modulated as a function of emotional connotation, linguistic focus, or preword hesitation, factors that can all lead to the retrieval of a richer set of semantic features that requires increased computational resources. Van Berkum (2009) also broadens the array of factors that may generate contextual expectations, including nonlinguistic ones (e.g., a mental representation of the sensory context, a mental model of the situation being discussed, and some metalinguistic representation of the discourse). The amplitude of the N400 is modulated (i.e., reduced) when these contextual factors facilitate retrieval of the currently processed word.

Hagoort and colleagues (2009) have suggested that these different accounts of the processing nature of the N400 might be reconciled if the LH and RH make different contributions to the creation of a meaning representation of the overall context, as proposed by Federmeier and colleagues (Federmeier, 2007; Federmeier & Kutas, 1999b; Kutas & Federmeier, 2000). They propose that the LH is involved in predictive and the RH in integrative semantic processing. In other words, the language-dominant LH generates contextually consistent semantic predictions that will facilitate retrieval and reduce the amplitude of the N400 (i.e., if the prediction is met). The RH, on the other hand, activates semantic information on the basis of the input and incrementally integrates the semantic information of the current input with that of previously activated semantic information. If this information matches, the integration is facilitated and a reduction of the N400 ensues.

Possible Neuronal Generators of the N400

Because scalp-recorded ERPs cannot directly be related to their generating source(s) as a result of the inverse problem (i.e., different internal source configurations can provide identical external electromagnetic fields), knowledge of the possible brain areas that contribute to the N400 has been gathered from studies using methods with better spatial resolution (magnetoencephalography [MEG] and fMRI), studies using intracranial recordings from presurgical epileptic patients, and studies in patients with lesions in verified locations. Many of these studies have shown evidence that the left (and to a lesser extent possibly the right) temporal lobe is a likely contributor to the scalp-recorded N400 (e.g., Nobre & McCarthy, 1995; Nobre et al., 1994; for reviews, see Halgren, 2002; Lau et al., 2008; Van Petten & Luka, 2006). These cortices have long been considered important in the representation and retrieval of semantic information (e.g., Beeman & Chiarello, 1998; Bright et al., 2007; Damasio et al., 1996; Hagoort et al., 1996; Martin, 2007). But evidence from fMRI and MEG studies also suggests a generator in the left inferior frontal cortex (Hagoort et al., 2004; Halgren et al., 2002; see Figure 15.15 for a depiction of these possible neural sources of the N400).

Recent studies have clearly implicated a function of the left inferior frontal cortex in semantic processing as well (e.g., Giesbrecht et al., 2004; Gold & Buckner, 2002; Hagoort et al., 1996; Polcrack et al., 1999; Swaab et al., 1997, 1998; Thompson-Schill et al., 1997, 1998, 1999; Wagner et al., 2000, 2001). Even though the exact nature of this contribution is still a matter of debate, the different proposals converge on the idea that this area is not sensitive to semantic retrieval per se, but instead may be involved in context-sensitive response selection (Kerns et al., 2004), lexical selection or competition among semantic features (Thompson-Schill et al., 1997), semantic unification (Hagoort et al., 2005), or processes of response selection for semantic information (Gabrieli et al., 1998).

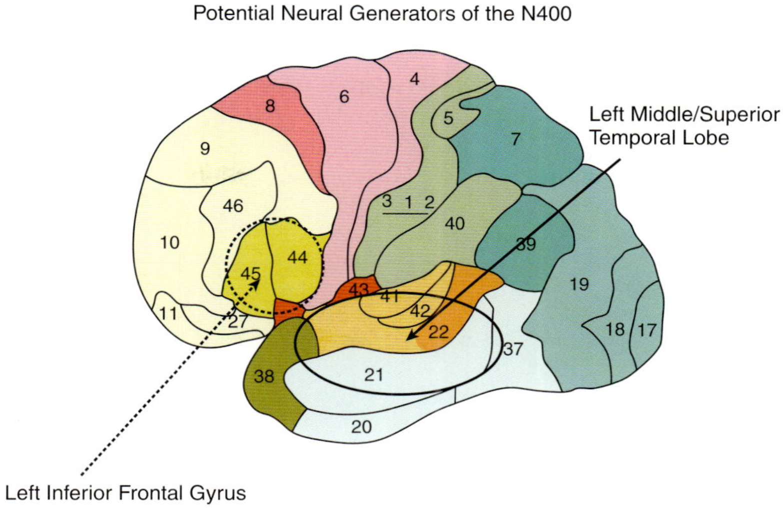

Fig. 15.15 Plausible N400 generators may be localized in the (left) middle/superior temporal lobes. Some findings suggest a generator in the left inferior frontal gyrus.

The P600

In the previous section, we described the N400 ERP component associated with semantic processing. Following its discovery, researchers sought to identify other ERP components that might be similarly associated with other aspects of language processing. Perhaps foremost among the candidate processes were those associated with building or extracting a syntactic structure, a fundamental and essential aspect of language production and comprehension. And, before long, evidence emerged that seemed to implicate one ERP component as a reflection of such processing: the P600.

The P600 (Osterhout & Holcomb, 1992; also sometimes called the *syntactic positive shift*, or SPS; Hagoort et al., 1993) is a slow late positive shift in the ERP waveform. It typically onsets around 500 ms after the onset of a stimulus (although earlier positive shifts have also been observed; Mecklinger et al., 1995) and lasts for several hundred milliseconds; its peak amplitude is generally observed at around 600 ms (if at all; the component often appears as more of a shift without a clear peak). The P600 is generally maximal over posterior electrode sites (but sometimes a more anterior distribution has been observed; Friederici et al., 2002; Hagoort et al., 1999; Kaan & Swaab, 2003b) and is generally widespread, without distinct laterality. The P600 has been observed in response to both written and auditory stimuli (Hagoort & Brown, 2000).

Initially, the observation of the modulation of the P600 component in response to syntactic manipulations seemed a perfect complement to the N400 as a marker of semantic processing. However, the functional interpretation of the P600 component is probably not quite as straightforward as was once believed. Below, we briefly review some of the situations in which modulation of the P600 is observed in an attempt to come to some tentative conclusions about what this component might be able to tell us about language processing in the brain.

The P600 and Syntactic Anomaly

An initial set of studies examined the ERP response to syntactic anomalies, that is, sentences that contained some kind of violation of syntactic principles. In an early study of this type, Osterhout and Holcomb (1992) had participants read sentences like the following:

a The broker hoped to sell the stock.
b The broker persuaded to sell the stock.

Sentence (a) conforms to the phrase-structure principles of the verb "*hope*" (which, as an intransitive verb, easily accepts the clausal complement). Sentence b, on the other hand, violates the phrase-structure principles of the verb "*persuade*"; as a transitive verb, "*persuade*" requires a noun phrase that can act as a direct object, and can only accept the clausal complement as part of a reduced relative

Fig. 15.16. A P600 effect of a syntactic violation (red line) relative to a syntactically legal continuation of a sentence (blue line). Data from Hagoort, Brown & Groothusen, Language and Cognitive Processes, (1993).

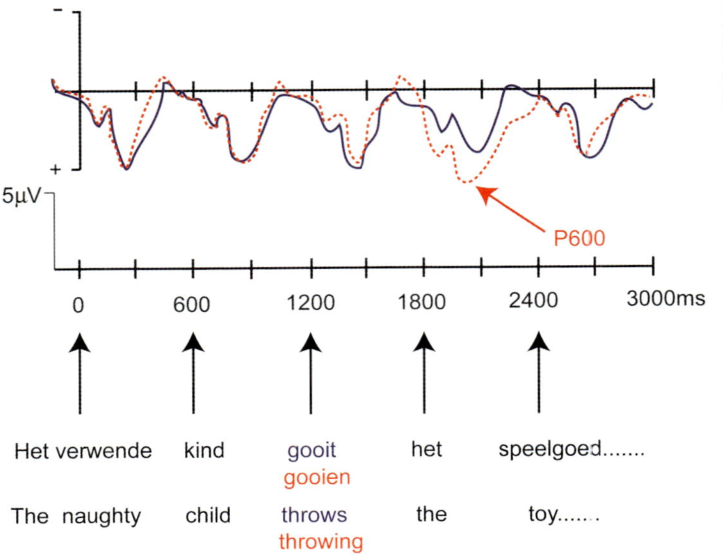

clause (as in "The broker persuaded to sell the stock was the one who got rich"), which Sentence b also does not allow. Osterhout and Holcomb (1992; see also Osterhout & Holcomb, 1993) showed that sentences with such violations of phrase structure elicited a larger P600 relative to similar sentences that did not contain violations. At around the same time, Hagoort and colleagues (1993) showed a similar effect in response to violations of subject–verb agreement in Dutch (as in "Het verwende kind *gooien het speelgood op de grond"/"The naughty child *throwing the toy on the floor") when compared to structurally well-formed sentences (see Figure 15.16).

Many studies have shown P600 effects to a broad range of syntactic anomalies, including the aforementioned phrase-structure (see also Friederici et al., 1996; Neville et al., 1991) and number agreement violations, as well as other types of agreement violations, including gender and case marking (Coulson et al., 1998; Friederici et al., 1993; Osterhout, 1997; Osterhout & Mobley, 1995), verb tense violations (Osterhout & Nicol, 1999), subcategorization violations (Ainsworth-Darnell et al., 1998; Osterhout et al., 1994), and violations of subjacency (McKinnon & Osterhout, 1996; Neville et al., 1991). Interestingly, some studies have shown P600 effects in response to syntactic violations even in sentences that are otherwise meaningless (e.g., "The boiled watering-can smokes/*smoke the telephone in the cat"; Hagoort & Brown, 1994; but see also Münte et al., 1997), reinforcing the conclusion that it is something about linguistic structure (as separate from meaning) that is reflected in this component. P600-type effects are not necessarily restricted to domains of language processing. In fact, they have been observed in response to violations of several different types of structure, including those seen in music (Besson & Macar, 1987; Janata, 1995; Patel et al., 1998), mathematical rules (Núñez-Peña & Honrubia-Serrano, 2004), and abstract sequences (Lelekov et al., 2000). These findings suggest that the P600 may index processes of structure building quite generally, of which syntactic processing is one example.

The P600 and Syntactic Ambiguity

Additionally, P600 effects are not limited to cases of outright structural violation. Several studies have demonstrated that the P600 component is also sensitive to syntactic complexity within sentences that do not contain any structural violation. That is, sentences that are grammatically well formed, but syntactically more difficult or less preferred, can also elicit a larger-amplitude P600 relative to sentences that are easier to parse or in some other way more preferred. P600 effects of this type have been seen in response to sentences that contain some temporary syntactic ambiguity, in which at least two alternative syntactic parses can be entertained at some point during sentence processing (Kaan & Swaab, 2003a; Mecklinger et al., 1995; Osterhout, 1997; Osterhout & Holcomb, 1992, 1993;

Osterhout et al., 1994). Consider the following example (from Kaan & Swaab, 2003a):

a The man is painting the house but the garage is already finished.

b The man is painting the house and the garage is already finished.

Both (a) and (b) are grammatical sentences; however, (b) contains a temporary syntactic ambiguity that is absent in a. Until the verb in the second clause ("*is*") is encountered in (b), two plausible and formally permissible structures can be generated from the sentence fragment (one in which "*and*" combines "*the house*" and "*the garage*" into a conjoined noun phrase, both of which are being painted, and one in which "*the garage*" is the head noun of a new clause, as is ultimately forced upon encountering "*is*"). Behavioral research has demonstrated that most readers prefer the first structure, the one that must be abandoned upon encountering "*is*". The ERP results mirror the behavioral results: at the verb in the second clause, the amplitude of the P600 is larger in (b) than in (a), suggesting that readers have to abandon the preferred structural interpretation at this point in favor of the less preferred one (Kaan & Swaab, 2003a).

Other research has shown that it is not necessary for a sentence to include this kind of "garden path" (in which one is first led down a specific syntactic path before recognizing the need to change directions toward another) to elicit a larger P600; sentences that are unambiguous, but syntactically more difficult or sophisticated, can lead to increases in the amplitude of the P600 as well (Kaan & Swaab, 2003b, Kaan et al., 2000; see Figure 15.17).

Based on results such as those presented above, several prominent interpretations have been offered concerning the functional significance of the P600. Some of these models describe a role for the P600 in processes of syntactic analysis and reanalysis or repair, as needed. Osterhout and colleagues (1994) suggested that the P600 reflects the cost of reprocessing that is necessary when an initial parse is disconfirmed. Kaan and colleagues (2000; see also Fiebach et al., 2002) proposed that the P600 reflects the difficulty of syntactic integration, a process that

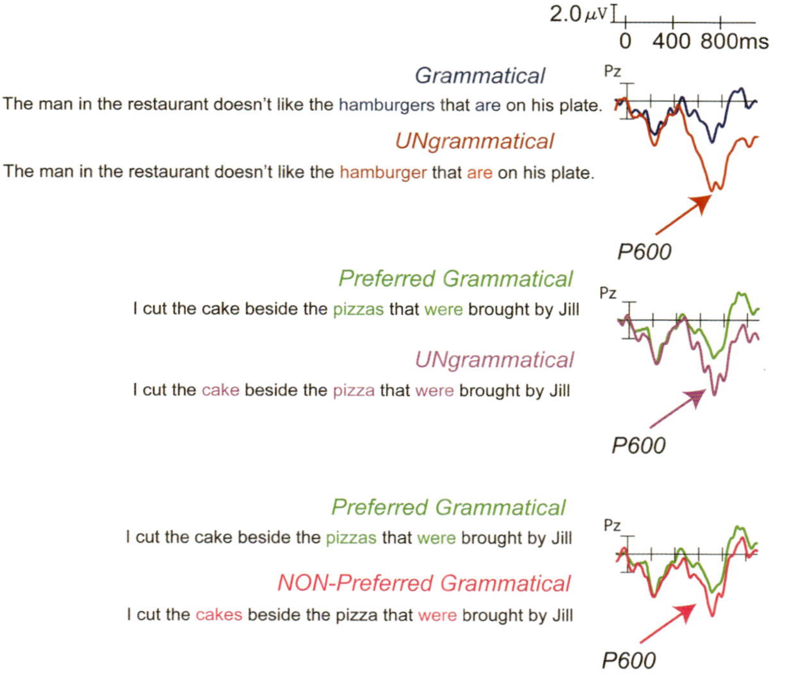

Fig. 15.17 P600 effects are also found to grammatical but less preferred syntactic continuations of sentences. In the top panel, the typical P600 effect to ungrammatical continuations is shown; when a verb and a noun do not agree in number (i.e., ". . . the hamburger that *are* . . ."), a greater P600 is found to the verb (red line) compared to a grammatical continuation (blue line). In the bottom panel, preferred and non-preferred grammatical continuations are compared; i.e., it is easier to attach the verb (were) to the most recent noun phrase (pizzas), than to a noun phrase earlier in the sentence (cakes), even though this latter continuation is also grammatical. A greater P600 is found to the non-preferred continuation (pink line), than to the preferred continuation (green line). For comparison, the middle panel shows a greater P600 when the verb cannot be attached to any of the preceding noun phrases (UNgrammatical, purple line), relative to the preferred grammatical condition (green, line). (Data from Kaan & Swaab, 2003b).

is made easier when the current syntactic structure is predictable (and thus becomes readily activated). Friederici (1995, 2002; see also Friederici & Kotz, 2003; Friederici & Weissenborn, 2007) has proposed a three-stage model of language comprehension in which the P600 corresponds to the final stage of syntactic reanalysis that may arise when information from the initial two stages (early phrase-structure building and semantic/verb-argument information activation) cannot be readily reconciled. Hagoort (2003; see also Hagoort, 2005) has proposed a unification model of syntactic processing in which the P600 indexes the amount of time required to unify syntactic frames into one phrasal configuration. This unification takes longer (and thus the amplitude of the P600 is larger) when, for instance, syntactic ambiguity (temporarily) introduces more than one possible syntactic configuration and competition among the alternatives results.

Finally, it may well be that P600 effects are not of a piece, but instead comprise a family of effects that may reflect separable underlying functional processes. Hagoort and colleagues (1999) suggested that the topographic distribution of the P600 might differ, depending on the type of demand placed on the parser. Specifically, they suggested that P600 effects tend to be more frontally distributed in cases in which syntactic preferences are not met, but tend to be more posterior in cases of outright syntactic violations.

The P600 and Semantic-Thematic Integration

Until quite recently, despite the lack of agreement on the exact functional nature of the P600 component, most researchers would have at least felt comfortable with a characterization of this component as reflecting some aspect of syntactic processing, as compared with the N400 and its role as an index of semantic processing. However, even this rather general depiction of the P600 has been called into question by a recent series of studies that report P600-type effects to stimuli that contain seeming semantic violations and thus might otherwise have been reasonably expected to elicit N400 effects.

One example of such a study comes from Kuperberg et al. (2003b), who presented participants with sentences like the following:

a For breakfast the *boys* would only eat toast and jam.

b For breakfast the *eggs* would only eat toast and jam.

c For breakfast the *boys* would only bury toast and jam.

All three sentences are syntactically well formed. Sentence (a) is also semantically well formed. Sentences (b) and (c) both contain semantic/pragmatic violations. In (b), the incongruity results from an animacy violation of thematic roles: eggs are inanimate and thus cannot fill the Agent thematic role demanded by the verb "*eat*". (An inanimate entity like "*eggs*" is better suited to the Theme thematic role of "*eat*".) In "c", the incongruity arises from a pragmatic violation: boys can indeed bury things (even toast and jam), but are not expected to do so at breakfast time. As expected, sentences like (c) show an increased N400 at the critical verb ("*eat*" vs. "*bury*") relative to the semantically well-formed controls (a). A surprising pattern of results was found, however, to the critical verb in sentences like (b): instead of an N400 to the critical verb in these sentences, a large P600 was elicited when compared to the controls (Kuperberg et al., 2003b; see Figure 15.18).

This effect cannot be explained just in terms of component overlap with the N400, since it seems unlikely that *eat* in *the eggs would eat* would elicit

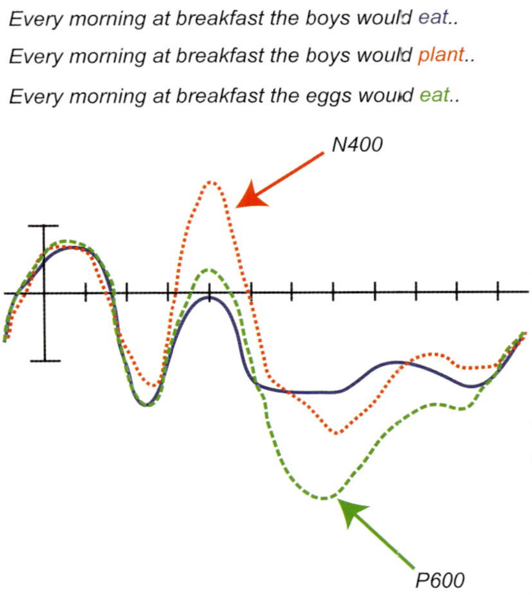

Fig. 15.18 Event-related potentials to verbs that form a normal (blue solid line), an implausible but possible (red dotted line), and a thematically violated continuation of the sentence (dashed green line). Adapted from Kuperberg, Holcomb, Sitnikova, Greve & Caplan, 2003, Cognitive Brain Research (Elsevier).

a reduction in the N400 relative to the control condition *the boys would eat.*

Several subsequent studies have also demonstrated P600 effects to what would generally be characterized as semantic violations (Hoeks et al., 2004; Kim & Osterhout, 2005; Kolk et al., 2003; Kuperberg et al., 2006, 2007; Nakano & Swaab, 2005; van Herten et al., 2005). For example, Kim and Osterhout (2005) found a P600 effect to sentences like "The hearty meal was devouring the kids" when compared to semantically well-formed passive and active control sentences. However, they demonstrated that this effect depended, in part, upon the "semantic attraction" of a given thematic role assignment, given a particular verb. So, for a verb like "*devour*", for which "*meal*" is an appropriate Theme but not an appropriate Agent, there is a semantically attractive alternative thematic role assignment available to readers. On the other hand, for sentences such as "The dusty tabletops were devouring with gusto", in which "*tabletops*" is not appropriate for any thematic role associated with "*devour*", semantic attraction is low. And indeed, Kim and Osterhout found, for sentences with low semantic attraction, an N400 effect instead of a P600 effect. However, Kuperberg and colleagues (2006) demonstrated a similar (if not larger) P600 effect to sentences in which reassignment of thematic roles did not lead to repair (as in "To make good documentaries cameras must interview . . ."), suggesting that semantic attraction or fit cannot be the only underlying cause of this effect. Instead, violations of animacy in semantic–thematic relationships may play an important role in eliciting the P600 effect.

Results such as these throw into doubt the traditional interpretation of the P600 component as an index of syntactic analysis and repair and, more broadly, raise serious and interesting questions about the relationship between semantic and syntactic processes in the brain.

The P600 and Syntactic Priming

Recently, we used the responsiveness of the P600 component to aspects of syntactic processing to examine the nature of syntactic priming (Ledoux et al., 2007; see also Tooley et al., 2009). *Syntactic priming* is the facilitation of sentence processing that occurs when a sentence has the same syntactic form as a preceding sentence. Behavioral studies have shown syntactic priming effects repeatedly in studies of language production. However, such effects have been less consistently demonstrated in studies of language comprehension. When priming effects *have* been demonstrated in comprehension, the effects seemed to depend on the repetition of lexical items (especially verbs) across sentences. In these cases, it was difficult to disentangle the contribution of lexical repetition priming effects and true syntactic priming effects using behavioral measures alone. On the other hand, ERPs seemed well suited to this task, given that lexical repetition priming has been shown to influence the N400 component and syntactic processing to influence the P600 component.

We thus designed an ERP experiment to attempt to dissociate these two types of effects. We presented participants with sentences such as the following (from Ledoux et al., 2007):

> *Main-clause (MC) prime*:
> The speaker proposed the solution to the group at the space program.
> *Reduced-relative (RR) prime*:
> The speaker proposed by the group would work perfectly for the program.
> *Target (always RR)*:
> The manager proposed by the directors was a bitter old man.

For each stimulus set, participants read one of two types of prime sentences: MC and RR forms. The MC interpretation is preferred, and behaviorally, readers have temporary difficulty parsing the RR version. We expected to see similar evidence of this difficulty electrophysiologically, and that is what we found: a larger P600 to the disambiguating region (following the verb) for RR prime sentences relative to MC prime sentences. After each prime sentence, readers were presented with another sentence that contained the same main verb presented in the prime sentence. This target sentence was always of the RR form, regardless of the type of prime sentence that had preceded it. So, half of the participants saw an MC prime followed by an RR target, and half of them saw an RR prime followed by an RR target. We looked at the ERPs to see if the response to the same target sentence differed purely as a function of the type of prime sentence that had been read before it. We found evidence that this was the case: the P600 was reduced for the disambiguating region following the verb in the RR target sentences that had been preceded by a prime sentence with a similar syntactic construction (RR) relative to when the same target sentences had been preceded by a prime sentence with a different syntactic construction (MC). We took this reduction in the amplitude of the P600 to be evidence of syntactic priming. We were also able to demonstrate that this syntactic priming effect was separate from the priming effect that arose from lexical repetition. We found a reduction in the amplitude of the N400 component to the second presentation of the verb preceding the disambiguating region (in the target sentences) relative to its first presentation (in the prime sentences). It seems, then, that the context in which a sentence is presented (being in close proximity to sentences of similar construction) can influence the brain's response during syntactic parsing, and that this response can be dissociated from the processing benefit conferred by lexical repetition (Ledoux et al., 2007).

The Processing Nature of the P600

Even though many experiments have shown that the P600 is sensitive to syntactic aspects of the linguistic input, the finding of a P600 to violations of thematic constraints (as in "The eggs would eat toast with jam at breakfast") calls into question whether or not the P600 is uniquely evoked by syntactic processing. Several new hypotheses about the functional interpretation of the P600 have been offered in light of these recent results (for reviews, see Bornkessel-Schlesewsky & Schlesewsky, 2008; Kolk & Chwilla, 2007; Kuperberg, 2007; Stroud & Phillips, 2009). One proposal is that the P600 effect in these experiments arises as a result of strong semantic-thematic attraction (Kim & Osterhout, 2005) or fit (Kuperberg et al., 2006) in sentences in which a plausible meaning can be derived if thematic roles are reassigned (see also Kemmerer et al., 2007, for another explanation based on the temporary syntactic reanalysis of grammatical-semantic violations). More recently, Kuperberg (2007) has proposed a model of language comprehension in which two processing streams act in parallel. One, the semantic memory–based stream, computes semantic features and relationships among sentence components and is primarily reflected in the N400 component. The other, the combinatorial stream, is sensitive to a multitude of linguistic constraints, including constraints of morphosyntax and of semantic–thematic relationships (including animacy). When the two streams provide contradictory output (i.e., when the semantic interpretation output by the first stream contradicts morphosyntactic or semantic-thematic information in the sentence), continued analysis must be undertaken to resolve the inconsistency, and it is this extended analysis that is reflected in the P600 component. A rather different proposal was presented by van Herten and colleagues (2006; see also Kolk & Chwilla, 2007), who suggested that the P600 might instead reflect the engagement of executive or cognitive control processes in the service of error monitoring and reprocessing in order to resolve response uncertainty during language processing (see also Vissers et al., 2006, 2007, 2008). This last proposal is most damaging to the idea that the P600 is sensitive to syntax, because it suggests that the P600 is not even language-specific (which will be discussed further in the next section).

While P600 effects of this type are still open to interpretation and further study, they do suggest that the interaction between semantic and syntactic processes in the brain may be more dynamic than was previously supposed. This conclusion is further supported by other studies that have reported an influence of lexical/semantic and discourse factors on syntactic parsing processes (Brown et al., 2000; Gunter et al., 2000; Osterhout et al., 1994; van Berkum et al., 1999, 2003; Weckerly & Kutas, 1999; see also Bornkessel & Schlesewky, 2006). What is clear at this point is the importance of further studies of factors, such as those described above, that seem to be at the interface of semantic and syntactic processing. This will be necessary for a more complete understanding of the P600 component in language processing.

Is the P600 Distinct from the P300? Task Sensitivity and Possible Neural Generators

Very soon after the discovery of the P600 in the early 1990s (Hagoort et al., 1993; Osterhout & Holcomb, 1992), a debate emerged in the literature on whether or not the P600 is in fact just another manifestation of the P3b, a member of the P300 family (see Chapter 7, this volume). This would imply that the P600 may be related to cognitive processing that is not specific to the building of hierarchical structure (Coulson et al., 1998; Gunter et al., 1997; but see Osterhout & Hagoort, 1999). Coulson and colleagues (1998) published a study that suggested that the P600 is a member of the P300 family. Specifically, they argued that manipulations known to modulate the P3b, such as salience of the stimulus, probability of occurrence, and task relevance, also modulate the P600. Further, they found no significant differences in the topographic distribution of the P3b and the P600, which also challenges the idea that these ERPs are distinct. If the P600 and the P3b are not distinct, then this would further call into question the assumption of a syntax-sensitive ERP, although it would not dispute the fact that this positive-deflecting ERP is sensitive to manipulations of syntax as well. Next, we will discuss the results of some studies that suggest that the P3b and the P600 may in fact *not* be the same ERP component.

In 1996, Osterhout and colleagues (Osterhout et al., 1996) conducted a study on whether or not the P600 and the P3b could be differentiated on the basis of their sensitivity to syntactic manipulations of subject–verb agreement and nonsyntactic manipulations of saliency, probability of occurrence, and task relevance. Participants read sentences such as the following:

a *Nonanomalous Control*: The doctors believe the patient will recover.
b *Agreement Violation*: The doctors *believes the patient will recover.
c *Letter Case Violation*: The doctors BELIEVE the patient will recover.
d *Double Anomaly*: The doctors *BELIEVES the patients will recover.

Event-related potentials were measured to the critical words that were violations (the verbs in the examples) relative to the control condition (a). It was expected that the comparison of (a) and (b) would show a P600 and the comparison of (a) and (c) would show a P3b. Importantly, only the P600 should be sensitive to the syntactic violation and only the P3b should be sensitive to the manipulations of probability of occurrence and task relevance. Further, in the Double Anomaly condition (d), additive effects of P3b and P600 would also indicate that these ERPs are separable. When probability was manipulated (20% vs. 60% violations), a modulation of the P3b was obtained in the Letter Case Violation condition (c). In contrast, the amplitude of the P600 was not affected by the probability manipulation (see Figure 15.19).

The latency, amplitude, morphology, and topographic distribution of the P3b and P600 effects differed as well; the P3b effect was larger in amplitude than the P600 effect and was maximal over right posterior electrode sites in a 400–800 ms time window; the P600 was maximal in a later 500–900 ms time window and was more evenly distributed over posterior sites of both hemispheres. When task relevance was manipulated (sentence acceptability task vs. reading task), P3bs were obtained to Letter Case violations for both tasks, but the effect was greatly increased in the sentence acceptability judgment task. P600s were found to syntactic violations in both task conditions as well and the P600 was also larger in the task-relevant condition, but the modulation of the P600 was much less robust than that found for the P3b in the Letter Case violation. Finally, when participants were presented with a Double Anomaly (condition d), a large-amplitude broad positivity was obtained that was larger than the P3b to the Letter Case manipulation alone and

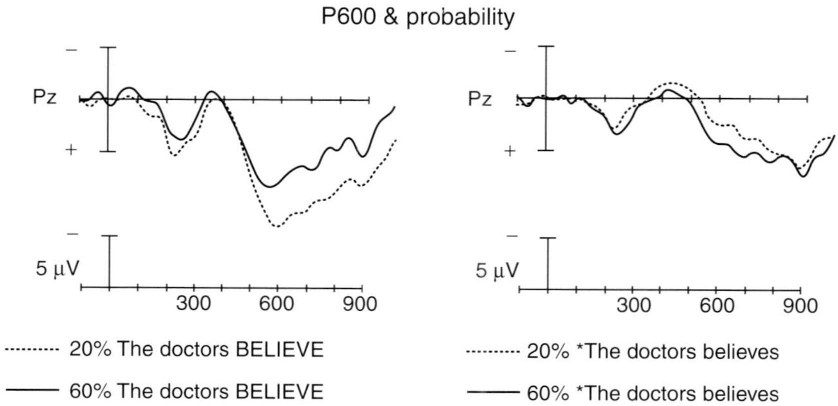

Fig. 15.19. Event-related potentials to critical words that were syntactically correct but physically deviant relative to the preceding context (uppercase letters) in 20% (dotted line) and 60% (solid line) of the experimental materials. A larger positive shift or P3b is obtained in the 20% condition. The P600, which is obtained to the syntactic violation of subject–verb agreement (**The doctors believes . . .*) is not sensitive to this probability manipulation (right side). Adapted from Osterhout, McKinnon, Bersick & Corey, Journal of Cognitive Neuroscience, 1996 (MIT press).

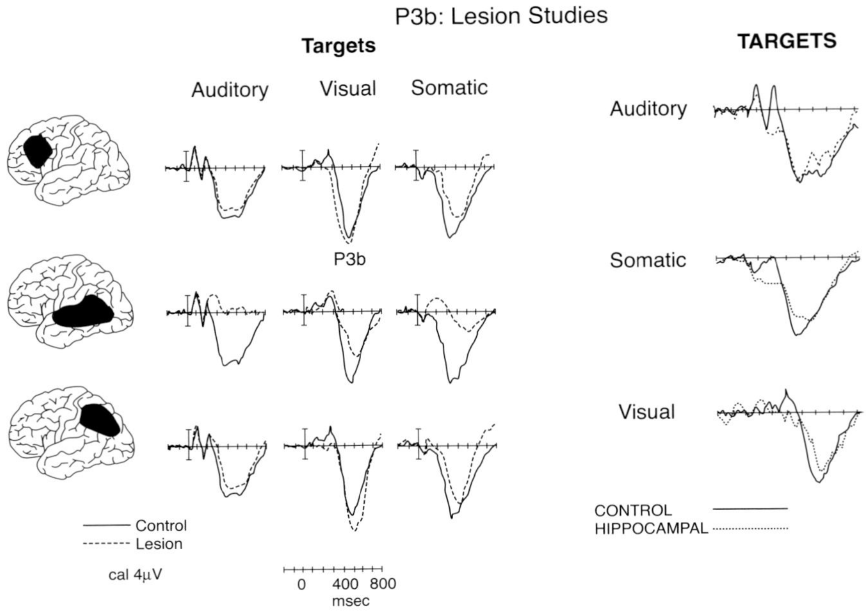

Fig. 15.20 P3b effects in a standard oddball paradigm (see Chapter 7, this volume) for visual, auditory, or somatic target stimuli. Results are compared between normal, neurologically unimpaired subjects (control, solid line), and patients (dotted line) with focal lesions in the left frontal cortex (top left), left temporal-parietal junction (middle left), left parietal cortex (bottom left), and hippocampus (top, middle, and bottom right). Only patients with temporal parietal lesions show marked attenuations of the P3b response. Adapted with permission from Knight and Scabini (1998).

larger than the P600 to the syntactic violation, indicating that this effect was additive. One could argue that this study in fact showed that the P600 is sensitive to task relevance. However, in a study that compared syntactic violations to syntactic preference, Kaan and Swaab (2003a) found that the amplitude of the P600 does not vary as a function of the task when syntactic preference is manipulated.

Other dissociations with regard to P600 and P3b have been observed in a study by Hagoort and colleagues (2003), who found that aphasic patients with a syntactic deficit did not show a P600 to syntactic violations of word order but did show a P3b to target stimuli in a standard oddball paradigm (see Chapter 7, this volume). Further evidence that the P600 and the P3b may be distinct ERP components comes from studies in patients with localized brain lesions that are indicative of separable neuronal sources. The work of Knight and colleagues has shown that the P3b is significantly attenuated in patients with temporal-parietal lesions (Knight et al., 1988, 1989). Frisch and colleagues, on the other hand, have shown the presence of a P3b and the absence of a P600 in patients with lesions in the basal ganglia (Frisch et al., 2003). The results of these latter two studies are shown in Figures 15.20 and 15.21.

Less direct but nevertheless suggestive evidence comes from fMRI studies that show that largely nonoverlapping brain regions become activated in oddball experiments and syntactic experiments. Oddball fMRI studies have shown activations in brainstem, temporal lobes, and medial frontal lobes (e.g., Calhoun et al., 2006; McCarthy et al., 1997). Functional MRI studies of syntax have observed activations in anterior regions of the superior temporal gyrus (STG; Friederici et al., 2003; Meyer et al., 2000) and posterior portions of the inferior frontal gyrus (Broca's area; e.g., Caplan et al., 2008; Friederici et al., 2003; Kuperberg et al., 2003; but see January et al., 2009). Furthermore, recent studies that performed repeated transcranial magnetic stimulation of Broca's area show performance improvements during syntactic processing (Sakai et al., 2002) and processing of artificial grammar (Uddén et al., 2008).

Finally, it has been shown that the P600 and the P3b have different oscillatory signatures; whereas an increase in P600 amplitude is correlated with a decrease in alpha and beta bands (Davidson &

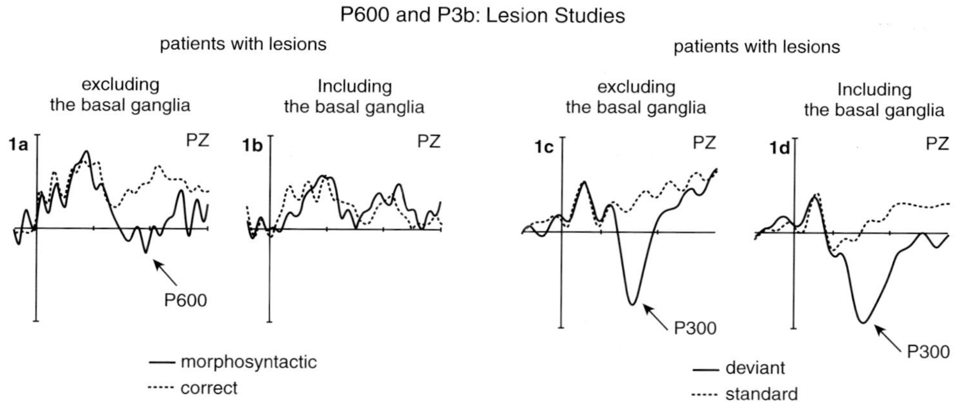

Fig. 15.21 Comparison of P600 and P3b in patients with and without lesions that include the basal ganglia. Patients with lesions that include the basal ganglia do not show a P600 to morphosyntactic violations (solid line) when compared to correct continuations (dotted line) of sentences (left panel), but they do show a P3b response (right panel), with a larger P300 to deviant (solid line) than to standard (dotted line) target stimuli. Reprinted with permission from Frisch, Kotz, von Cramon & Friederici, Clinical Neurophysiology (2003).

Indefrey, 2007), a larger amplitude of the P3b is associated with a tighter synchronization in the gamma band and a reduction of power in the gamma band (Ford et al., 2008).

In sum, the extant evidence generally supports a separation between P3b and P600 ERPs. Studies have shown that the P600 is not sensitive to probability (Osterhout et al., 1996) and that manipulation of syntactic preference results in a P600 that is not sensitive to task relevance (Kaan & Swaab, 2003b). In addition, separable neural sources may be involved in the generation of the P3b and P600.

The E/LAN

In addition to the modulations of the P600 component described in the previous section, syntactic anomalies have also been shown to elicit earlier negative shifts in the ERP waveform (Friederici et al., 1993; Münte et al., 1993; Neville et al., 1991; see Figure 15.22).

These shifts appear over anterior electrodes and in some cases (though not all) have been shown to be lateralized to the left side of the head. For this reason, this class of ERP component is generally referred to as a *left anterior negativity* (LAN). (In most cases, when not left-lateralized, the anterior negativity is bilaterally distributed.) Researchers have reported LAN effects at varying latencies following a stimulus. Early LAN (ELAN) effects have been observed as early as 100–300 ms poststimulus onset. These early effects have been distinguished from later LAN effects in the same latency window as the N400 (300–500 ms) but have been differentiated from that component by its anterior distribution (and by a different set of eliciting conditions). Whether the ELAN and LAN are truly two different components indexing functionally distinct language processes, or whether they reflect a single process that varies in onset, is a matter of great debate.

The ELAN and LAN effects have been observed to word category violations, that is, when the parser anticipates that an upcoming word will be of a particular grammatical category (noun, verb, etc.) but is presented with a word that violates that expectation, as in "The young apprentice went to see the new **designing* in the museum," where a verb (designing) is in a noun position (design) (Friederici et al., 1996; Hagoort et al., 2003; Hahne & Friederici, 1999; Münte et al., 1993). In German, the latency of the anterior negativity has been shown to depend in part on the location of the violation in the critical word: the onset was earlier when the violation was part of the prefix of the critical word and was later when it occurred in the suffix (Friederici et al., 1993, 1996; Hahne & Friederici, 1999, 2002). It is under conditions of word category violation that the ELAN has been most reliably elicited. The later LAN effect has been elicited under a broader range of conditions, especially those involving violations of morphosyntax (number, case, gender, and tense violations; Deutsch & Bentin, 2001; Gunter et al., 1997, 2000; Osterhout & Mobley, 1995; Penke et al., 1997).

A LAN has also been consistently observed to fully grammatical sentences that contain long-distance dependency constructions, such as filler-gap sentences (Felser et al., 2003; Kluender & Kutas, 1993). In such

Fig. 15.22 The LAN and the P600 response to syntactic violations (dotted line). Reprinted with permission from Friederici et al. (2004).

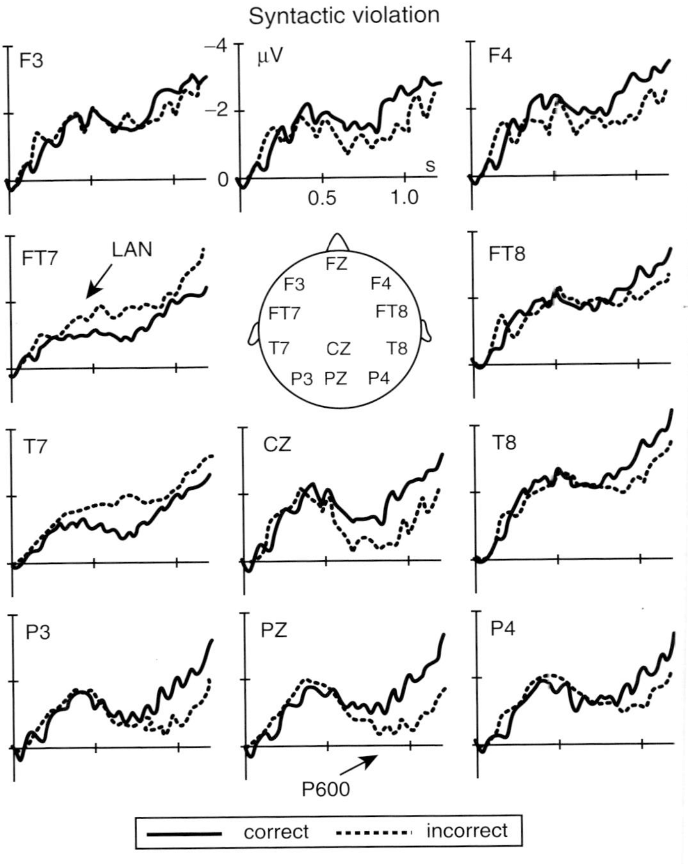

sentences, one sentence component (the filler) has been omitted or displaced from its original position, leaving behind a trace (the gap) that must be filled with the missing component. An example of gap filling comes from WH-constructions, such as "Which movie did John like?" In this case, the filler "*movie*" has been moved from its position as the object of the verb "*like*", and has created a gap after the verb that must be detected and filled with the displaced filler. A LAN has been observed at the position of the gap (Fiebach et al., 2001, 2002; Kluender & Kutas, 1993). Another example comes from verb gapping, in which a verb is omitted in a sentence that contains two conjoined clauses, as in "Mary ate the hamburger, and Susie the salad". The detection of the gap in sentences such as these has also been shown to elicit an E/LAN effect (Kaan et al., 2004).

The Functional Nature of the E/LAN

The functional significance of the E/LAN is the subject of much debate. Because the effect is most consistently elicited only under conditions of structural violation, it seems to index some aspect of difficulty during syntactic processing. (Also, because E/LAN effects are seen primarily to violations, they are usually accompanied by a P600 effect, but the converse is not always true: P600 effects have been observed in the absence of E/LAN effects.) Hagoort (2003), in his unification model, suggested that anterior negativities result when a syntactic element cannot be bound to any other element in the current parse (either because of a violation of word category or because of a violation of morphosyntactic principles). Friederici (1995, 2002) and Friederici and Weissenborn (2007) propose a functional distinction between the ELAN and the LAN. According to this model, the ELAN indexes difficulty during an initial stage (Phase 1) of phrase-structure building during which word category information is identified. The LAN, on the other hand, indexes difficulty during a stage (Phase 2) in which morphosyntactic information is processed and integrated. (Semantic information is also integrated during Phase 2, a process reflected in this model by the amplitude of the N400.) As mentioned previously, Friederici proposed that the amplitude of the P600 reflects processing during a final stage (Phase 3), in which reanalysis and repair are engaged as needed to fully integrate the syntactic and semantic outputs of Phases 1 and 2.

An alternative proposal suggests that anterior negativities may more broadly index the working memory operations involved in the processing of verbal material (Fiebach, et al., 2001; King & Kutas, 1995; Kluender & Kutas, 1993). Some have suggested that the LAN may be a "time slice" of the slow negative waves that have been observed with working memory manipulations. Anterior negativities similar to the LAN have been observed under other conditions that could be thought to tax verbal working memory. Verbal working memory would certainly be recruited during the processing of filler-gap constructions and other long-distance dependencies, since sentence constituents must be maintained and manipulated over the course of the sentence. It is more difficult to reconcile this view of the LAN as an index of increased working memory load with its response to morphosyntactic violations, where the form of a word does not obey the structure of the sentence, such as in violations of tense, gender, number, and case (e.g., "The young apprentice go to the museum"). Another possibility is that there are multiple distinct processes occurring under different conditions that produce many ERP components that are difficult to differentiate due to the relatively limited spatial resolution of the technique. Future research may help to answer some of these questions.

Using ERPs to Study Language: Methodological Issues

Anyone who studies language processing is aware of the creativity, time, and effort it takes to construct the stimuli for an experiment. With pretesting and norming, stimulus construction can sometimes take up to 1 year and very often requires at least 3 months. In ERP research, this problem is compounded by the need to have enough stimuli in each condition to achieve the appropriate signal-to-noise ratio (see Luck, 2005). Fortunately, because many of the language-related ERP effects have now been well documented in the literature, a workable rule of thumb is to have a minimum of 25 trials in each condition after artifact rejection. Given the rate at which artifacts (primarily due to blinks and eye movements) occur, even when instructions are given to minimize them, this generally means that it is wise to start with at least 40 trials in each condition. However, this is still a sizable number. Unlike studies of visual spatial attention, for example, where the same stimulus is often repeated many times, repetition of the same language stimulus is often detrimental (unless the repetition serves a language function, as for example in coreferential processing with repeated names, e.g., Swaab Camblin & Gordon, 2004, or when repetition priming is studied). In fact, repetition can lead to changes in subject strategies and diminishing alertness. In addition, lexical repetition per se has substantial effects on the amplitude of the N400 (e.g., Besson & Kutas, 1993). As in behavioral studies of language, in order to avoid repetition of stimuli, a Latin square design is frequently applied, such that a given experimental item appears in each condition across different lists that are counterbalanced across participants. To the extent possible, it is advisable to keep the critical word (to which the ERPs will be measured) the same across the different versions of the experimental item. When critical words are different across conditions, they should be carefully matched on relevant lexical properties known to influence processing/ERPs (such as word length, lexical frequency, part of speech, and concreteness and age of acquisition). Useful databases characterizing such lexical features can be found online (e.g., at http://www.psych.rl.ac.uk/MRC_Psych_Db.html; http://elexicon.wustl.edu/). Other factors (such as cloze probability, plausibility, and task-induced processing requirements) may also influence the amplitude of language-related ERP components and should be carefully controlled unless they are the specific object of study.

While the same lexical items can be used for the comparison across conditions, other problems may arise when the critical word in the sentence is not in the same position across conditions. As in behavioral studies, ERP effects of the sentential position of the critical word have been demonstrated, such that words presented at later positions in the sentence are more easily integrated, resulting in a general reduction in the amplitude of the N400 to these words toward the relative end of a sentence. In addition, to avoid sentence wrap-up effects involving the integration of the overall meaning of the sentence, it is best to avoid presenting the critical words in the sentence-final position.

Furthermore, ERP baseline issues may occur when the words preceding and following the critical word are different across experimental items, as in the following example:

Anomalous: They admired my of sketch the landscape.
Control: They admired my sketch of the landscape.

The critical word *of* in this example introduces a word-order violation in the Anomalous condition and follows a normal word order in the Control condition. However, to see the effect of the word order violation on the critical word *of*, the prestimulus baseline is affected by a difference in word class across conditions: in the Control condition the open-class word *sketch* will elicit an N400 that is not found to the closed-class word *my* in the Anomalous condition. If a typical presentation rate of one word every 500 ms is used, then this will affect a 100 to 200 ms prestimulus baseline of the critical word. Ideally then, the same words should directly precede (and follow) the critical word.

RSVP Requirements

Because ERPs are vulnerable to artifacts from eye movements, ERP reading experiments typically present participants with sentence or discourse stimuli one word at a time, usually at a rate of 500 ms (300 ms per word, with an interstimulus interval of 200 ms). The reading conditions during an ERP experiment thus differ from natural reading conditions in at least two ways. First, under normal reading conditions, not every word in the sentence or discourse is read with the same speed, and some words are skipped altogether. Second, the average reading speed is about 200 ms per word, which is much shorter than the RSVP rate typically used. Third, under normal reading conditions, participants can and often will return to words earlier in the sentence when they are confronted with ambiguity or another difficulty later in the sentence, or when they temporarily "zone out" during reading and find that they are not consciously aware of the content of a passage that they have just read. This option is not available to ERP participants when they are forced to read one word at the time.

To more closely mimic natural reading speed, reading studies with ERPs have been performed at faster RSVP rates of 200–250 ms (e.g., Camblin et al., 2007b; van Petten et al., 1997). Under these circumstances, short-latency ERP components such as N1 and P2 do not clearly appear in the ERP waveform because of the overlap of ERPs from previous words. However, even at these fast rates, it has been possible to observe distinct N400 effects (van Petten et al., 1997). Interestingly, in at least one study, the fast rate of presentation led to changes in the typical pattern of results (Camblin et al., 2007b), presumably because readers lacked control over their reading input, which they would still have in natural reading conditions. This may place additional demands on the reader that interfere with the normal reading process. Recently, Van Berkum and colleagues introduced the variable serial visual presentation (VSVP) technique in concert with ERPs, in which the presentation duration of each word depends on its length (for details, see Otten & van Berkum., 2007). This procedure would seem to better approximate more natural reading conditions when compared with a fixed fast RSVP of 200 ms per word.

Ditman and colleagues (2007) found that the self-paced reading paradigm can be used while recording ERPs, allowing for comparison of ERP results within participants. In self-paced reading paradigms, subjects are asked to press a button each time they want to advance to the next word of a sentence. The latency of these button presses is assumed to correlate with processing difficulty, such that subjects will take longer to press the button for the next word when they have more difficulty processing the current word. Ditman et al. showed typical effects of pragmatic and syntactic violations that had been established in previous ERP studies and self-paced reading studies with the same paradigms. Importantly, motor artifacts from motor preparation and the button press itself did not appear to adversely affect the EEG recording.

Other studies have combined behavioral methods with ERPs to investigate language processing during reading. As discussed previously, Gordon and colleagues have studied coreferential processing during reading with identical or very similar stimuli using ERPs with both visual and auditory presentations and eye-tracking methods (Camblin et al., 2007; Gordon et al., 2004; Swaab et al., 2004). Two linguistic expressions are said to be coreferential if they refer to the same semantic entity, as in "Emily asked for a definition of coreferential processing because she wants to avoid confusion for the readers of this chapter," where *Emily* and *she* refer to the same person (and the observant reader probably also realizes that this request was made by one of the editors of this book). Interestingly, typical effects of coreferential difficulty were observed with ERP studies that used standard RSVP rates of 500 ms (Ledoux et al., 2007; Swaab et al., 2004) in eye-tracking studies when participants read at their own speed (Ledoux et al., 2007) and also in studies with naturally connected speech (Camblin et al., 2007). However, when faster RSVP rates were used, Camblin and colleagues observed effects of lexical repetition but not of coreferential difficulty. This may indicate that fast RSVP rates will prompt the reader to process words in a sentence more like words in lists, which will preserve lexical effects of repetition but interfere with higher-order processing that requires integration of context, as for example when readers are establishing coreferential relations in a text.

We hope we have shown that ERPs provide a very useful tool in the study of language comprehension. Even though much work lies ahead of us in unraveling the mysteries of meaning in language, when all is said and done, "Language is to the mind more than light is to the eye" (Gibson, *The Miracle Worker*, p. 25).

Notes

1 A review of the psycholinguistics models of language processing is outside the scope of this chapter (see Traxler & Gernsbacher, 2006). Also, we cannot review all relevant research involving language-related ERPs: we will not separately review research with language-related ERPs in brain damaged and/or psychiatric populations (e.g., Bonte & Blomert, 2004; Ditman & Kuperberg, 2007; Swaab et al., 1997,1998; Wassenaar & Hagoort, 2007; Wassenaar et al., 2004; for reviews, see Kuperberg & Caplan, 2003; Kuperberg et al., 2010; Münte et al., 2000; Swaab, 1998). Event-related potential studies of language production will not be reviewed here either. These studies are rare, in large part because speaking induces large motor artifacts in the EEG. However, clever designs have been used to identify the relative timing in the access of different types of information during word production (Levelt, 1999) in the incredibly fast processes that lead from thought to speech production (Schmitt et al., 2001; van Turennout et al., 1998). Recently, ERPs have also been used to test the theories of gesture and language (e.g., Holle & Gunter, 2007; Özyürek et al., 2007), embodied language (e.g., Chwilla et al., 2007), and bilingualism (e.g., Elston-Güttler et al., 2005; Proverbio et al., 2004).

2 The topographic distribution of the N400 varies to some extent for written and spoken language. In the visual modality the N400 is maximal over centro-parietal sites over the right hemisphere, whereas for spoken language the N400 is more equally distributed over centro-parietal sites of the left and right hemispheres.

In addition, the onset latency of the N400 in written language is around 200 ms, whereas in natural spoken language the N400 may start to diverge as early as 50 ms after the onset of a critical word because of coarticulatory information from the preceding speech.

3 Recent studies have shown separable effects of early (pre)-lexical reading processes on the N250 ERP component (e.g., Grainger and Holcomb, 2006) and a P250 during early speech recognition (e.g., Friedrich and Kotz, 2007).

4 Some researchers have also looked at slow wave responses elicited to multiple words in sentences instead of to each individual word. In general, these studies have found that more negative slow waves are observed to sentences that are more difficult to process (relative to easier sentences). For examples, see King and Kutas (1995) and Munte et al. (1998).

5 Interestingly, the effects of both lexical association and sentence congruence varied as a function of the number of intervening words between prime and target. N400 priming effects were *not* obtained for lexical associates that were separated by an average of 4.8 intervening words but were present when 1 or no words intervened, whereas the reverse was true for sentence congruence effects. This is consistent with findings that effects of lexical association are short-lived (e.g., Chwilla et al., 2000) but that sentence context effects build up over time.

6 Studies of speech perception have used the sensitivity of the mismatch negativity (MMN; see Chapter 6, this volume) to deviations in auditory input to successfully investigate acoustic and phonological aspects of speech (e.g., Kaan, 2008; Kaan et al., 2007; Näätänen et al., 1997; Phillips, 2001).

7 Other ERP studies have shown that the reader and listener may actually use discourse information to anticipate and predict the upcoming words in the sentence or story (e.g., Delong et al., 2005; Nieuwland & Van Berkum, 2006a; Otten & Van Berkum, 2008; Otten et al., 2007; Van Berkum et al., 2005; Wicha et al., 2004).

8 Work of Hagoort and colleagues has also shown that world knowledge is immediately integrated during normal language comprehension (Hagoort et al., 2004; Hald et al., 2007). In addition, nonlinguistic (pragmatic) information, such as whether or not the voice of the speaker matches the message (e.g., a male talking about being pregnant; van Berkum et al., 2008) and even whether or not the attitude and moral values of the comprehender clash with the message (Van Berkum et al., 2009), also immediately influence the amplitude of the N400. This further illustrates that real-time comprehension takes immediate advantage of many sources of contextual information.

References

Ainsworth-Darnell, K., Shulman, H. G., & Boland, J.E. (1998). Dissociating brain responses to syntactic and semantic anomalies: Evidence from event-related potentials. *Journal of Memory and Language*, *38*, 112–130.

Anaki, D., Faust, M., & Kravetz, S. (1998). Cerebral hemispheric asymmetries in processing lexical metaphors. *Neuropsychologia*, *36*(4), 353–362.

Barber, H., Vergara, M., & Carreiras, M. (2004). Syllable-frequency effects in visual word recognition: Evidence from ERPs. *Neuroreport*, *15*(3), 545–548.

Becker, C. A. (1980). Semantic context effects in visual word recognition: An analysis of semantic strategies. *Memory & Cognition*, *8*(6), 493–512.

Becker, C.A. (1985). What do we really know about semantic context effects during reading?. In Besner, D., Waller, T.G., & MacKinnon, G.E. (Eds.), *Reading research: advances in theory and practice* (pp. 125–166). New York: Academic Press.

Beeman, M., & Chiarello, C. (Eds.). (1998). *Right hemisphere language comprehension: Perspectives from cognitive neuroscience*. Mahwah, NJ: Erlbaum.

Bentin, S., McCarthy, G., & Wood, C. C. (1985). Event-related potentials, lexical decision, and semantic priming. *Electroencephalography and Clinical Neurophysiology*, *60*(4), 343–355.

Bentin, S., & Peled, B. (1990). The contribution of task-related factors to ERP repetition effects at short and long lags. *Memory & Cognition, 18*(4), 359–366.

Bentin, S., Kutas, M. & Hillyard, S.A. (1995). Semantic processing and memory for attended and unattended words in dichotic listening: behavioral and electrophysiological evidence. *Journal of Experimental Psychology: Human Perception and Performance, 21*(1), 54–67.

Besson, M., & Kutas, M. (1993). The many facets of repetition: A cued-recall and event-related potential analysis of repeating words in same versus different sentence contexts. *Journal of Experimental Psychology: Learning, Memory, and Cognition, 19*(5), 1115–1133.

Besson, M., & Macar, F. (1987). An event-related potential analysis of incongruity in music and other non-linguistic contexts. *Psychophysiology, 24,* 14–25.

Boddy, J. (1986). Event-related potentials in chronometric analysis of primed word recognition with different stimulus onset asynchronies. *Psychophysiology, 23,* 232–245.

Bodner, G. E., & Masson, M. E. J. (2003). Beyond spreading activation: An influence of relatedness proportion on masked semantic priming. *Psychonomic Bulletin and Review, 10*(3), 645–652.

Bonte, M. L., & Blomert, L. (2004). Developmental dyslexia: ERP correlates of anomalous phonological processing during spoken word recognition. *Cognitive Brain Research, 21*(3), 360–376.

Bornkessel, I., & Schlesewsky, M. (2006). The extended argument dependency model: A neurocognitive approach to sentence comprehension across languages. *Psychological Review, 113*(4), 787–821.

Bornkessel-Schlesewsky, I., & Schlesewsky, M. (2008). An alternative perspective on "semantic P600" effects in language comprehension. *Brain Research Reviews, 59*(1), 55–73.

Boudewyn, M., Gordon, P., Long, D., Polse, L., & Swaab, T. Y. (in press). (i). Does discourse congruence influence spoken language comprehension before lexical association? Evidence from event-related potentials. *Language and Cognitive Processes.*

Bransford, J. D., & Johnson, M. K. (1972). Contextual prerequisites for understanding: Some investigations of comprehension and recall. *Journal of Verbal Learning and Verbal Behavior, 11,* 717–726.

Bright, P., Moss, H. E., Longe, O., Stamatakis, E. A., & Tyler, L. K. (2007). Conceptual structure modulates anteromedial temporal involvement in processing verbally presented object properties. *Cerebral Cortex, 17*(5), 1066–1073.

Brown, C., & Hagoort, P. (1993). The processing nature of the N400: Evidence from masked priming. *Journal of Cognitive Neuroscience, 5,* 34–44.

Brown, C. M., van Berkum, J. J. A., & Hagoort, P. (2000). Discourse before gender: An event-related brain potential study on the interplay of semantic and syntactic information during spoken language understanding. *Journal of Psycholinguistic Research, 29,* 53–68.

Calhoun, V. D., Adali, T., Pearlson, G. D., & Kiehl, K. A. (2006). Neuronal chronometry of target detection: Fusion of hemodynamic and event-related potential data. *Neuroimage, 30,* 544–553.

Camblin, C. C., Gordon, P. C., & Swaab, T. Y. (2007a). The interplay of discourse congruence and lexical association during sentence processing: Evidence from ERPs and eye-tracking. *Journal of Memory and Language, 56,* 103–128.

Camblin, C. C., Ledoux, K., Boudewyn, M., Gordon, P. C., & Swaab, T. Y. (2007b). Processing new and repeated names: Effects of coreference on repetition priming with speech and fast RSVP. *Brain Research, 1146,* 172–184.

Caplan, D., Chen, E., & Waters, G. (2008). Task-dependent and task-independent neurovascular responses to syntactic processing. *Cortex, 44*(3), 257–275.

Caplan, D., & Waters, G. S. (1999). Verbal working memory and sentence comprehension. *Behavioral and Brain Sciences, 22*(1), 77–126.

Chomsky, N. (1975). Reflections on language, *Pantheon.*

Chwilla, D. J., Hagoort, P., & Brown, C. M. (1998). The mechanism underlying backward priming in a lexical decision task: Spreading activation versus semantic matching. *Quarterly Journal of Experimental Psychology, 51A*(3), 531–560.

Chwilla, D. J., Kolk, H. H. J., & Mulder, G. (2000). Mediated priming in the lexical decision task: Evidence from event-related potentials and reaction time. *Journal of Memory and Language, 42*(3), 314–341.

Chwilla, D. J., Kolk, H. H. J., & Vissers, C. T. W. M. (2007). Immediate integration of novel meanings: N400 support for an embodied view of language comprehension. *Brain Research, 1183,* 109–123.

Collins, A. M., & Loftus, E. F. (1975). A spreading-activation theory of semantic processing. *Psychological Review, 82,* 407–428.

Connolly, J. F., & Phillips, N. A. (1994). Event-related potential components reflect phonological and semantic processing of the terminal word of spoken sentences. *Journal of Cognitive Neuroscience, 6,* 256–266.

Coulson, S., Federmeier, K. D., Van Petten, C., & Kutas, M. (2005). Right hemisphere sensitivity to word- and sentence-level context: Evidence from event-related brain potentials. *Journal of Experimental Psychology: Learning, Memory, and Cognition, 31*(1), 129–147.

Coulson, S., King, J., & Kutas, M. (1998). Expect the unexpected: Event-related brain responses to morphosyntactic violations. *Language and Cognitive Processes, 13,* 21–58.

Coulson, S., & Kutas, M. (2001). Getting it: Human event-related brain response to jokes in good and poor comprehenders. *Neuroscience Letters, 316,* 71–74.

Coulson, S., & Lovett, C. (2004). Handedness, hemispheric asymmetries, and joke comprehension. *Cognitive Brain Research, 19,* 275–288.

Coulson, S., & Severens, E. (2007). Hemispheric asymmetry and pun comprehension: When cowboys have sore calves. *Brain and Language, 100,* 172–187.

Coulson, S., & Van Petten, C. (2002). Conceptual integration and metaphor: An event-related potential study. *Memory and Cognition, 30,* 958–968.

Coulson, S., & Van Petten, C. (2007). A special role for the right hemisphere in metaphor comprehension? ERP evidence from hemifield presentation. *Brain Research, 1146,* 128–145.

Coulson, S., & Williams, R. F. (2005). Hemishperic asymmetries and joke comprehension. *Neuropsychologia, 43,* 128–141.

Damasio, H., Grabowski, T. J., Tranel, D., Hichwa, R. D., et al (1996). A neural basis for lexical retrieval. *Nature, 380*(6574), 499–505.

Daneman, M., & Carpenter, P. A. (1980). Individual differences in working memory and reading. *Journal of Verbal Learning and Verbal Behavior, 19,* 450–466.

Davidson, D. J., & Indefrey, P. (2007). An inverse relation between event-related and time-frequency violation responses in sentence processing. *Brain Research, 1158*, 81–92.

Deacon, D., Hewitt, S., Yang, C., & Nagata, M. (2000). Event-related potential indices of semantic priming using masked and unmasked words: Evidence that the N400 does not reflect a post-lexical process. *Cognitive Brain Research, 9*(2), 137–146.

De Groot, A. M. (1985). Word-context effects in word naming and lexical decision. *Quarterly Journal of Experimental Psychology A: Human Experimental Psychology, 37*, 281–297.

DeLong, K.A., Urbach, T.P., & Kutas, M. (2005). Probabilistic word pre-activation during language comprehension inferred from electrical brain activity. *Nature Neuroscience, 8*(8), 1117–1121.

Deutsch, A. & Bentin, S. (2001). Syntactic and semantic factors in processing gender agreement in Hebrew: evidence from ERPs and eye movements. *Journal of Memory and Language, 45*, 200–224.

Diaz, M. T., & Swaab, T. Y. (2007). Electrophysiological differentiation of phonological and semantic integration in word and sentence contexts. *Brain Research, 1146*, 85–100.

Ditman, T., Holcomb, P. J., & Kuperberg, G. R. (2007). An investigation of concurrent ERP and self-paced reading methodologies. *Psychophysiology, 44*, 927–935.

Domalski, P., Smith, M. E., & Halgren, E. (1991). Cross-modal repetition effects on the N4. *Psychological Science, 2*, 173–178.

Elston-Güttler, K. E., Paulmann, S., & Kotz, S. A. (2005). Who's in control? Proficiency and L1 influence on L2 processing. *Journal of Cognitive Neuroscience, 17*(10), 1593–1610.

Fauconnier, G., & Turner, M. (1998). Conceptual integration networks. *Cognitive Science, 22*, 133–187.

Faust, M. (1998). Obtaining evidence of language comprehension from sentence priming. In M. Beeman & C. Chiarello (Eds.), *Right hemisphere language comprehension: Perspectives from cognitive neuroscience* (pp. 161–186). Hillsdale, NJ: Erlbaum.

Faust, M., Babkoff, H., & Kravets, S. (1995). Linguistic processes in the two cerebral hemispheres: Implications for modularity vs. interactionism. *Journal of Clinical and Experimental Neuropsychology, 17*, 171–192.

Faust, M., & Gernsbacher, M. (1996). Cerebral mechanisms for suppression of inappropriate information during sentence comprehension. *Brain and Language, 53*, 234–259.

Faust, M., Kravetz, S., & Babkoff, H. (1993). Hemisphericity and top-down processing of language. *Brain and Language, 53*, 234–259.

Federmeier, K. D. (2007). Thinking ahead: The role and roots of prediction in language comprehension. *Psychophysiology, 44*, 491–505.

Federmeier, K. D., & Kutas, M. (1999). A rose by any other name: Long-term memory structure and sentence processing. *Journal of Memory and Language, 41*, 469–495.

Federmeier, K. D., Wlotko, E., De Ochoa-Dewald, E., & Kutas, M. (2007). Multiple effects of sentential constraint on word processing. *Brain Research, 1146*, 75–84.

Felser, C., Clahsen, H., & Munte, T. F. (2003). Storage and integration in the processing of filler-gap dependencies: An ERP study of topicalization and wh-movement in German. *Brain and Language, 87*, 345–354.

Fiebach, C. J., Schlesewsky, M., & Friederici, A. D. (2001). Syntactic working memory and the establishment of filler-gap dependencies: insights from ERPs and fMRI. *Journal of Psycholinguistic Research, 30*(3), 321–338.

Fiebach, C. J., Schlesewsky, M., & Friederici, A. D. (2002). Separating syntactic memory costs and syntactic integration costs during parsing: The processing of German WH-questions. *Journal of Memory and Language, 47*, 250–272.

Fodor, J.A., Bever, T.G., & Garrett, M. (1974). *The psychology of language: an introduction to psycholinguistics and generative grammar*. New York: McGraw-Hill.

Fodor, J. D., Ni, W., Crain, S., & Shankweiler, D. (1996). Tasks and timing in the perception of linguistic anomaly. *Journal of Psycholinguistic Research, 25*, 25–57.

Ford, J. M., Roach, B. J., Hoffman, R. S., & Mathalona, D. H. (2008). The dependence of P300 amplitude on gamma synchrony breaks down in schizophrenia. *Brain Research, 1235*, 133–142.

Forster, K. I. (1989). Basic issues in lexical processing. In W. Marslen Wilson (Ed.), *Lexical representation and process* (pp. 75–107), Cambridge MA: MIT Press.

Frazier, L. (1999). *On sentence interpretation*. The Hague: Kluwer.

Friederici, A. D. (1995). The time course of syntactic activation during language processing: A model based on neuropsychological and neurophysiological data. *Brain and Language, 50*, 259–281.

Friederici, A. D. (2002). Towards a neural basis of auditory sentence processing. *Trends in Cognitive Sciences, 6*(2), 78–84.

Friederici, A. D., Gunter, T. C., Hahne, A., & Mauth, K. (2004). The relative timing of syntactic and semantic processes in sentence comprehension. *Neuroreport, 15*, 165–169.

Friederici, A. D., Hahne, A., & Mecklinger, A. (1996). Temporal structure of syntactic parsing: Early and late event-related brain potential effects. *Journal of Experimental Psychology: Learning, Memory and Cognition, 22*, 1219–1248.

Friederici, A. D., Hahne, A., & Saddy, D. (2002). Distinct neurophysiological patterns reflecting aspects of syntactic complexity and syntactic repair. *Journal of Psycholinguistic Research, 31*, 45–63.

Friederici, A. D., & Kotz, S. A. (2003). The brain basis of syntactic processes: Functional imaging and lesion studies. *Neuroimage, 20*, 8–17.

Friederici, A. D., Pfeifer, E., & Hahne, A. (1993). Event-related brain potentials during natural speech processing: Effects of semantic, morphological and syntactic violations. *Cognitive Brain Research, 1*, 183–192.

Friederici, A. D., Rüschemeyer, S., Hahne, A., & Fiebach, C. J. (2003). The role of left inferior frontal and superior temporal cortex in sentence comprehension: Localizing syntactic and semantic processes. *Cerebral Cortex, 13*(2), 170–177.

Friederici, A. D., & Weissenborn, J. (2007). Mapping sentence form onto meaning: The syntax–semantic interface. *Brain Research, 1146*, 50–58.

Friedrich, C. K., & Kotz, S. A. (2007). Event-related potential evidence of form and meaning coding during online speech recognition. *Journal of Cognitive Neuroscience, 19*(4), 594–604.

Friedrich, C. K., Kotz, S. A., Friederici, A. D., & Gunter, T. C. (2004). ERPs reflect lexical identification in word fragment priming. *Journal of Cognitive Neuroscience, 16*, 541–552.

Frisch, S., Kotz, S. A., von Cramon, D. Y., & Friederici, A. D. (2003). Why the P600 is not just a P300: The role of the basal ganglia. *Clinical Neurophysiology, 114*(2), 336–340.

Gabrieli, J.D.E., Poldrack, R.A., & Desmond, J.E. (1998). The role of the left prefrontal cortex in language and memory. *Proceedings of the National Academy of Sciences, 95*(3), 906–913.

Ganis, G., Kutas, M., & Sereno, M. I. (1996). The search for "common sense": An electrophysiological study of the comprehension of words and pictures in reading. *Journal of Cognitive Neuroscience, 8*(2), 89–106.

Gibson, W. (1957). *The Miracle Worker*. New York: Simon & Schuster.

Giesbrecht, B., Camblin, C. C., & Swaab, T. Y. (2004). Separable effects of semantic priming and imageability on word processing in human cortex. *Cerebral Cortex, 14*, 521–529.

Gold, B. T., & Buckner, R. L. (2002). Common prefrontal regions co-activate with dissociable posterior regions during controlled semantic and phonological tasks. *Neuron, 35*, 803–812.

Gordon, P. C., Camblin, C. C., & Swaab, T. Y. (2004). On-line measures of coreferential processing. In M. Carreiras & C. Clifton (Eds.), *The on-line study of sentence comprehension: Eyetracking, ERP and beyond* (pp. 103–128). New York: Psychology Press.

Gordon, P. C., & Hendrick, R. (1997). Intuitive knowledge of linguistic coreference. *Cognition, 62*, 325–370.

Gordon, P. C., Hendrick, R., & Ledoux K. (2000). Language comprehension and probe-list memory. *Journal of Experimental Psychology: Learning, Memory, and Cognition, 26*(3), 766–775.

Gordon, P. C., Hendrick, R., Ledoux, K., & Yang, C. L. (1999). Processing of reference and the structure of language: An analysis of complex noun phrases. *Language and Cognitive Processes, 14*, 353–379.

Grainger, J., & Holcomb, P. J. (2009). Watching the word go by: On the time-course of component processes in visual word recognition. *Language and Linguistic Compass, 3*(1), 128–156.

Grainger, J., Kiyonaga, K., & Holcomb, P. (2006). The time course of orthographic and phonological code activation. *Psychological Science, 17*(12), 1021–1026.

Grosjean, F. (1980). Spoken word recognition processes and the gating paradigm. *Perception & Psychophysics, 28*, 267–283.

Gunter, T. C., Friederici, A. D., & Schriefers, H. (2000). Syntactic gender and semantic expectancy: ERPs reveal early autonomy and late interaction. *Journal of Cognitive Neuroscience, 12*, 556–568.

Gunter, T. C., Stowe, L. A., & Mulder, G. (1997). When syntax meets semantics. *Psychophysiology, 34*, 660–676.

Hagoort, P. (2003). How the brain solves the binding problem for language: A neurocomputational model of syntactic processing. *Neuroimage, 20*, S18–S29.

Hagoort, P. (2005). On Broca, brain, and binding: A new framework. *Trends in Cognitive Sciences, 9*, 416–423.

Hagoort, P., Baggio, G., & Willems, R.M. (2009). Semantic unification. In: M. S. Gazzaniga (Ed.), *The new cognitive neurosciences*. (pp. 1–49). Cambridge, MA: MIT Press.

Hagoort, P., & Brown, C. M. (1994). Brain responses to lexical ambiguity resolution and parsing. In C. Clifton, Jr., L. Frazier, & K. Rayner (Eds.), *Perspectives on sentence processing* (pp. 45–80). Hillsdale, NJ: Erlbaum.

Hagoort, P., & Brown, C. M. (2000). ERP effects of listening to speech compared to reading: The P600/SPS to syntactic violations in spoken sentences and rapid serial visual presentation. *Neuropsychologia, 38*, 1531–1549.

Hagoort, P., Brown, C., & Groothusen, J. (1993). The syntactic positive shift (SPS) as an ERP measure of syntactic processing. *Language and Cognitive Processes, 8*, 439–483.

Hagoort, P., Brown, C., & Osterhout, L. (1999). The neurocognition of syntactic processing. In C. Brown & P. Hagoort (Eds.), *Neurocognition of language* (pp. 273–316). Oxford: Oxford University Press.

Hagoort, P., Brown, C. M., & Swaab, T. Y. (1996). Lexical-semantic event-related potential effects in patients with left hemisphere lesions and aphasia, and patients with right hemisphere lesions without aphasia. *Brain, 119*, 627–649.

Hagoort, P., Hald, L., Bastiaansen, M., & Petersson, K. M. (2004). Integration of word meaning and world knowledge in language comprehension. *Science, 304*, 438–441.

Hagoort, P., Wassenaar, M., & Brown, C. (2003). Real-time semantic compensation in patients with agrammatic comprehension: Electrophysiological evidence for multiple-route plasticity. *Proceedings of the National Academy of Sciences USA, 100*(7), 4340–4345.

Hahne, A. & Friederici, A.D. (1999). Electrophysiological evidence for two steps in syntactic analysis: early automatic and late controlled processes. *Journal of Cognitive Neuroscience, 11*(2), 194–205.

Hahne, A. & Friederici, A.D. (2003). Differential task effects on semantic and syntactic processes as revealed by ERPs. *Cognitive Brain Research, 13*(3), 339–356.

Hald, L. A., Steenbeek-Planting, E. G., & Hagoort, P. (2007). The interaction of discourse context and world knowledge in online sentence comprehension. Evidence from the N400. *Brain Research, 1146*, 210–218.

Halgren, E., Boujon, C., Clarke, J., Wang, C. & Chauvel, P. (2002). Rapid distributed fronto-parieto-occipital processing stages during working memory in humans. *Cerebral Cortex, 12*, 710–728.

Hamberger, M. & Friedman, D. (1992). Event-related potential correlates of repetition priming and stimulus classification in young, middle-aged, and older adults. *The Journal of Geruntology, 47*(6), 395–405.

Hoeks, J. C. J., Stowe, L. A., & Doedens, G. (2004). Seeing words in context: The interaction of lexical and sentence level information during reading. *Cognitive Brain Research, 19*, 59–73.

Holcomb, P. J. (1993). Semantic priming and stimulus degradation: Implications for the role of the N400 in language processing. *Psychophysiology, 30*(1), 47–61.

Holcomb, P. J., Anderson, J., & Grainger, J. (2005). An electrophysiological study of cross-modal repetition priming. *Psychophysiology, 42*, 493–507.

Holcomb, P. J, & Grainger, J. (2007). Exploring the temporal dynamics of visual word recognition in the masked repetition priming paradigm using event-related potentials. *Brain Research, 1180*, 39–58.

Holcomb, P. J., Grainger, J., & O'Rourke, T. (2002). An electrophysiological study of the effects of orthographic neighborhood size on printed word perception. *Journal of Cognitive Neuroscience, 14*(6), 938–950.

Holcomb, P. J., Kounios, J., Anderson, J. E., & West, W. C. (1999). Dual-coding, context availability, and concreteness effects n sentence comprehension: An electrophysiological investigation. *Journal of Experimental Psychology: Learning, Memory, and Cognition, 25*, 721–742.

Holcomb, P. J., & Neville, H. J. (1990). Auditory and visual semantic priming in lexical decision: A comparison using

event-related brain potentials. *Language and Cognitive Processes, 5*(4), 281–312.
Holcomb, P. J., & Neville, H. J. (1991). Natural speech processing: an analysis using event-related potentials. *Psychobiology, 19*(4), 286–300.
Jackendoff, R. (2002). *Foundations of language: brain, meaning, grammar, evolution*. Oxford: University Press.
Jackendoff, R. (2007). A parallel architecture perspective on language processing. *Brain Research, 1146*, 2–22.
Janata, P. (1995). ERP measures assay the degree of expectancy violation of harmonic contexts in music. *Journal of Cognitive Neuroscience, 7*, 153–164.
January, D., Trueswell, J. C., & Thompson-Schill, S. L. (2009). Co-localization of Stroop and syntactic ambiguity resolution in Broca's area: Implications for the neural basis of sentence processing. *Journal of Cognitive Neuroscience. 21*, 2434–2444.
Johns, C., Gordon, P. C., Long, D., & Swaab, T. Y. (under review). An ERP investigation of referential accessibility.
Joyce, C. A., Paller, K. A., Schwartz, T. J., & Kutas, M. (1999). An electrophysiological analysis of modality-specific aspects of word repetition. *Psychophysiology, 36*, 655–665.
Jung-Beeman, M. (2005). Bilateral brain processes for comprehending natural language. *Trends in Cognitive Sciences, 9*, 712–718.
Kaan, E. (2007). Event-related potentials and language processing: A brief introduction. *Language and Linguistics Compass, 1*(6), 571–591.
Kaan, E., Harris, A., Gibson, E., & Holcomb, P. (2000). The P600 as an index of syntactic integration difficulty. *Language & Cognitive Processes, 15*, 159–201.
Kaan, E., & Swaab, T. Y. (2003a). Electrophysiological evidence for serial sentence processing: A comparison between non-preferred and ungrammatical continuations. *Cognitive Brain Research, 17*, 621–635.
Kaan, E., & Swaab, T. Y. (2003b). Repair, revision, and complexity in syntactic analysis: An electrophysiological differentiation. *Journal of Cognitive Neuroscience, 15*, 98–110.
Kaan, E., Wijnen, F., & Swaab, T. Y. (2004). Gapping: electrophysioglogical evidence for immediate processing of "missing" verbs in sentence comprehension. *Brain and Language, 74*, 1–25.
Kaan, E., Wayland, R., Bao, M., & Barkley, C.M. (2007). Effects of native language and training on lexical tone perception: An ERP study. *Brain Research, 1148*, 113–122.
Kacinik, N. A., Aragon, J., & Swaab, T. Y. (2008). An ERP study of cerebral asymmetries for processing literal and figurative meanings. *Journal of Cognitive Neuroscience, Supplement*,
Kacinik, N. A., & Chiarello, C. (2007). Understanding metaphors: Is the right hemisphere uniquely involved? *Brain and Language, 100*, 188–207.
Kacinik, N. A., Polse, L. R., Baynes, K., & Swaab, T. Y. (in preparation). Figurative meanings are not more difficult to integrate into sentence contexts than literal meanings: Evidence from ERPs.
Karayanidis, F., Andrews, S., Ward, P. B., & McConaghy, N. (1991). Effects of inter-item lag on word repetition: an event-related potential study. *Psychophysiology, 28*(3), 307–318.
Katz, J.J. . . . (1972). *Semantic theory*. New York: Harper and Row.
Kemmerer, D., Weber-Fox, C., Price, K., Zdanczyk, C., & Way, H. (2007). Big brown dog or brown big dog? An electrophysiological study of semantic constraints on prenominal adjective order. *Brain and Language, 100*, 238–256.
Kerns, J. G., Cohen, J. D., Stenger, V. A., & Carter, C. S. (2004). Prefrontal cortex guides context-appropriate responding during language production. *Neuron, 43*(2), 283–291.
Kim, A., & Osterhout, L. (2005). The independence of combinatory semantic processing: Evidence from event-related potentials. *Journal of Memory and Language, 52*, 205–225.
King, J. W., & Kutas, M. (1995). Who did what and when? Using word- and clause-level ERPs to monitor working memory usage in reading. *Journal of Cognitive Neuroscience, 7*, 376–395.
Kluender, R. & Kutas, M. (1993). Bridging the gap: evidence from ERPs on the processing of unbounded dependencies. *Journal of Cognitive Neuroscience, 5*(2), 196–214.
Knight, R. T., & Scabini, D. (1998). Anatomic bases of event-related potentials and their relationship to novelty detection in humans. *Journal of Clinical Neurophysiology, 15*, 3–13.
Knight, R. T., Scabini, D., & Clayworth, C. C. (1988). The effects of lesions of superior temporal gyrus and inferior parietal lobe on temporal and vertex components of the human AEP. *Electroencephalography and Clinicial Neurophysiology, 70*(6), 499–509.
Knight, R. T., Scabini, D., Woods, D. L., & Clayworth, C. C. (1989). Contributions of temporal-parietal junction to the human auditory P3. *Brain Research, 502*(1), 109–116.
Kolk, H., & Chwilla, D. (2007). Late positivities in unusual situations. *Brain and Language, 100*, 257–261.
Kolk, H. H. J., Chwilla, D. J., van Herten, M., & Oor, P. J. W. (2003). Structure and limited capacity in verbal working memory: A study with event-related potentials. *Brain and Language, 85*, 1–36.
Koriat, A. (1981). Semantic facilitation in lexical decision as a function of prime-target association. *Memory and Cognition, 9*(6), 587–598.
Kounios, J., & Holcomb, P. J. (1994). Concreteness effects in semantic processing: ERP evidence supporting dual-coding theory. *Journal of Experimental Psychology: Learning, Memory, and Cognition, 20*, 804–823.
Kuperberg, G. R. (2007). Neural mechanisms of language comprehension: Challenges to syntax. *Brain Research, 1146*, 23–49.
Kuperberg, G. R., & Caplan, D. (2003). Language dysfunction in schizophrenia. In R. B. Schiffer, S. M. Rao, & B. S. Fogel (Eds.), *Neuropsychiatry* (2nd ed., pp. 444–466). Philadelphia: Lippincott, Williams and Wilkins.
Kuperberg, G. R., Caplan, D., Sitnikova, T., Eddy, M., & Holcomb, P. J. (2006). Neural correlates of processing syntactic, semantic, and thematic relationships in sentences. *Language and Cognitive Processes, 21*, 489–530.
Kuperberg, G. R., Holcomb, P. J., Sitnikova, T., Greve, D., Dale, A., & Caplan, D. (2003a). Distinct patterns of neural modulation during the processing of conceptual and syntactic anomalies. *Journal of Cognitive Neuroscience, 15*(2), 272–293.
Kuperberg, G. R., Kreher, D. A., & Ditman, T. (2010). What can event-related potentials tell us about language, and perhaps even thought, in schizophrenia? *International Journal of Psychophysiology, "Language and Psychophysiology" Special Issue. 75*(2), 66–76.
Kuperberg, G. R., Sitnikova, T., Caplan, D., & Holcomb, P. J. (2003b). Electrophysiological distinctions in processing conceptual relationships within simple sentences. *Cognitive Brain Research, 17*, 117–129.

Kutas, M., & Hillyard, S. A. (1980). Reading senseless sentences: Brain potentials reflect semantic incongruity. *Science, 207,* 203–205.

Kutas, M., & Hillyard, S. A. (1983). Event-related brain potentials to grammatical errors and semantic anomalies. *Memory & Cognition, 11*(5), 539–550.

Kutas, M., & Hillyard, S. A. (1984). Brain potentials reflect word expectancy and semantic association during reading. *Nature, 307,* 161–163.

Kutas, M., & Hillyard, S. A. (1989). An electrophysiological probe of incidental semantic association. *Journal of Cognitive Neuroscience, 1*(1), 38–49.

Kutas, M., & Federmeier, K.D. (2000). Electrophysiology reveals semantic memory use in language comprehension. *Trends in Cognitive Sciences, 4*(12), 463–470.

Kutas, M., Neville, H. J., & Holcomb, P. J. (1987). A preliminary comparison of the N400 response to semantic anomalies during reading, listening and signing. *Electroencephalography and Clinical Neurophysiology, Supplement. 39,* 325–330.

Kutas, M., van Petten, C. K., & Kluender, R. (2006). Psycholinguistics electrified II (1994–2005). In M. J. Traxler & M. A. Gernsbacher. (Eds.), *Handbook of psycholinguistics* (2nd ed., pp. 1–67). New York: Elsevier Press.

Lau, E. F., Phillips, C., & Poeppel, D. (2008). A cortical network for semantics: (de)constructing the N400. *Nature Reviews Neuroscience, 9,* 921–933.

Ledoux, K., Camblin, C. C., Swaab, T. Y., & Gordon, P. C. (2006). Reading words in discourse: The modulation of lexical priming effects by message-level context. *Behavioral and Cognitive Neuroscience Reviews, 5,* 107–127.

Ledoux, K., Traxler, M. J., & Swaab, T. Y. (2007). Syntactic priming in comprehension: Evidence from event-related potentials. *Psychological Science, 18,* 135–143.

Lelekov, T., Franck, N., Dominey, P. F., & Georgieff, N. (2000). Dissociable ERP profiles for processing rules vs. instances in a cognitive sequencing task. *Neuroreport, 11,* 2145–2149.

Levelt, W.J.M. (1999). Models of word production. *Trends in Cognitive Sciences, 3*(6), 223–232.

Luck, S. J. (2005). *An introduction to the event-related potential technique.* Cambridge, MA: MIT Press.

Luck, S. J., Vogel, E. K., & Shapiro, K. L. (1996). Word meanings can be accessed but not reported during the attentional blink. *Nature, 383,* 616–618.

MacDonald, M. C., Pearlmutter, N. J., & Seidenberg, M. S. (1994). The lexical nature of syntactic ambiguity resolution. *Psychological Review, 101,* 676–703.

Marslen-Wilson, W., & Tyler, L. K. (1980). The temporal structure of spoken language understanding. *Cognition, 8,* 1–71.

Martin, A. (2007). The representation of object concepts in the brain. *Annual Review of Psychology, 58,* 25–45.

McCallum, W. C., Farmer, S. F., & Pocock, P. V. (1984). The effects of physical and semantic incongruities of auditory event-related potentials. *Electroencephalography and Clinical Neurophysiology: Evoked Potentials, 59,* 477–488.

McCarthy, G. & Nobre, A.C. (1993). Modulation of semantic processing by spatial selective attention. *Electroencephalography and Clinical Neurophysiology, 88,* 210–219.

McCarthy, G., Luby, M., Gore, J., & Goldman-Rakic, P. (1997). Infrequent events transiently activate human prefrontal and parietal cortex as measured by functional MRI. *Journal of Neurophysiology, 77*(3), 1630–1634.

McKinnon, R., & Osterhout, L. (1996). Constraints on movement phenomena in sentence processing: Evidence from event-related brain potentials. *Language and Cognitive Processes, 11,* 495–523.

Mecklinger, A., Schriefers, H., Steinhauer, K., & Friederici, A. D. (1995). Processing relative clauses varying on syntactic and semantic dimensions: An analysis with event-related potentials. *Memory and Cognition, 23,* 477–494.

Meyer, M., Friederici, A. D., & von Cramon, D. Y. (2000). Neurocognition of auditory sentence comprehension: Event related fMRI reveals sensitivity to syntactic violations and task demands. *Cognitive Brain Research, 9*(1), 19–33.

Münte, T. F., Heinze, H., & Mangun, G. R. (1993). Dissociation of brain activity related to syntactic and semantic aspects of language. *Journal of Cognitive Neuroscience, 5*(3), 335–344.

Münte, T. F., Matzke, M., & Johannes, S. (1997) Brain activity associated with syntactic incongruities in words and pseudowords. *Journal of Cognitive Neuroscience, 9,* 318–329.

Münte, T. F., Schiltz, K., & Kutas, Marta (1998). When temporal terms belie conceptual order. *Nature, 395*(5697), 71–73.

Münte, T. F., Urbach, T. P., Düzel, E., & Kutas, M. (2000). Event-related brain potentials in the study of human cognition and neuropsychology. In F. Boller, J. Grafman, & G. Rizzolatti (Eds.), *Handbook of neuropsychology* (Vol. 1, 2nd ed., pp. 139–235). Amsterdam: Elsevier Science.

Näätänen, R., Lehtokoski, A., Lennes, M., Cheour, M., Huotilainen, M., Iivonen, A., Vainio, M., Alku, P., Iimoniemi, R., Luuk, A., Allik, J., Sinkkonen, J., & Alho, K. (1997). Language-specific phoneme representations revealed by electric and magnetic brain responses. *Nature, 385,* 432–434.

Neely, J. H. (1977). Semantic priming and retrieval from lexical memory: Roles of inhibitionless spreading activation and limited-capacity attention. *Journal of Experimental Psychology, 106,* 226–254.

Neely, J.H. (1991). Semantic priming effects in visual word recognition: a selective review of current findings and theories. In Besner, D. & Humphreys, G. W. (Eds.), *Basic Processes in Reading: Visual Word Recognition* (pp. 264–353). Hillsdale, NJ: Erlbaum.

Neely, J. H., Keefe, D. E., & Ross, K. L. (1989). Semantic priming in the lexical decision task: Roles of prospective prime-generated expectancies and retrospective semantic matching. *Journal of Experimental Psychology: Learning, Memory, and Cognition, 15,* 1003–1019.

Neville, H. J., Nicol, J. L., Barss, A., Forster, K. I., & Garrett, M. F. (1991). Syntactically based sentence processing classes: Evidence from event-related brain potentials. *Journal of Cognitive Neuroscience, 3,* 151–165.

Nieuwland, M. S., Otten, M., & van Berkum, J. A. (2007). Who are you talking about? Tracking discourse-level referential processing with event-related brain potentials. *Journal of Cognitive Neuroscience, 19*(2), 228–236.

Nieuwland, M. S., & Van Berkum, J. A. (2006). When peanuts fall in love: N400 evidence for the power of discourse. *Journal of Cognitive Neuroscience, 18,* 1098–1111.

Nobre, A. C., & McCarthy, G. (1994). Language-related ERPs: Scalp distributions and modulation by word type and semantic priming. *Journal of Cognitive Neuroscience, 6*(3), 233–255.

Nobre, A. C., & McCarthy, M. (1995). Language-related field potentials in the anterior-medial temporal lobe: Effects of word type and semantic priming. *Journal of Neuroscience, 15,* 1090–1099.

Núñez-Peña, M. I., & Honrubia-Serrano, M. L. (2004). P600 related to rule violation in an arithmetic task. *Cognitive Brain Research, 18*, 130–141.

O'Rourke, T. B., & Holcomb, P. J. (2002). Electrophysiological evidence for the efficiency of spoken word processing. *Biological Psychology, 60*(2–3), 121–150.

Osterhout, L. (1997). On the brain response to syntactic anomalies: Manipulations of word position and word class reveal individual differences. *Brain and Language, 59*, 494–522.

Osterhout, L., & Hagoort, P. (1999). A superficial resemblance does not necessarily mean you are part of the family: Counterarguments to Coulson, King, and Kutas (1998) in the P600/SPS-P300 debate. *Language and Cognitive Processes, 14*, 1–14.

Osterhout, L., & Holcomb, P. J. (1992). Event-related potentials elicited by syntactic anomaly. *Journal of Memory and Language, 31*, 785–806.

Osterhout, L., & Holcomb, P. J. (1993). Event-related potentials and syntactic anomaly: Evidence of anomaly detection during the perception of continuous speech. *Language and Cognitive Processes, 8*, 413–437.

Osterhout, L., Holcomb, P. J., & Swinney, D. A. (1994). Brain potentials elicited by garden-path sentences: Evidence of the application of verb information during parsing. *Journal of Experimental Psychology: Learning, Memory, and Cognition, 20*, 786–803.

Osterhout, L., McKinnon, R., Bersick, M., & Corey, V. (1996). On the language specificity of the brain response to syntactic anomalies: Is the syntactic positive shift a member of the P300 family? *Journal of Cognitive Neuroscience, 8*, 507–526.

Osterhout, L., & Mobley, L. A. (1995). Event-related brain potentials elicited by failure to agree. *Journal of Memory and Language, 34*, 739–773.

Osterhout, L., & Nicol, J. (1999). On the distinctiveness, independence, and time course of the brain responses to syntactic and semantic anomalies. *Language and Cognitive Processes, 14*, 283–317.

Otten, M. & van Berkum, J.J.A. (2007). What makes a discourse constraining? Comparing the effects of discourse message and scenario fit on the discourse-dependent N400 effect. *Brain Research, 1153*, 166–177.

Otten, M. & van Berkum, J.J.A. (2008). Discourse-based word anticipation during language processing: prediction or priming? *Discourse Processes, 45*(6), 464–496.

Özyürek, A., Willems, R. M., Kita, S., & Hagoort, P. (2007). On-line integration of semantic information from speech and gesture: Insights from event-related brain potentials. *Journal of Cognitive Neuroscience, 19*(4), 605–616.

Patel, A. D., Gibson, E., Ratner, J., Besson, M., & Holcomb, P. J. (1998). Processing syntactic relations in language and music: An event-related potential study. *Journal of Cognitive Neuroscience, 10*, 717–733.

Penke, M., Weyerts, H., Gross, M., Zander, E., Munte, T. F., & Clahsen, H. (1997). How the brain processes complex words: An event-related potential study of German verb inflections. *Cognitive Brain Research, 6*, 37–52.

Phillips, C. (2001). Levels of representation in the electrophysiology of speech perception. *Cognitive Science, 25*, 711–731.

Poldrack, R. A., Wagner, A. D., Prull, M. W., Desmond, J. E., Glover, G. H., & Gabrieli J. D. E. (1999). Functional specialization for semantic and phonological processing in the left inferior prefrontal cortex. *Neuroimage, 10*, 15–35.

Posner, M.I. & Snyder, C.R.R. (1975). Attention and cognitive control. In Solso, R.L. (Ed.), *Information processing and cognition* (pp. 55–85). Hillsdale, NJ: Erlbaum.

Praamstra, P., Meyer, A. S., & Levelt, W. J. M. (1994). Neurophysiological manifestations of phonological processing: Latency variations of a negative ERP component time-locked to phonological mismatch. *Journal of Cognitive Neuroscience, 6*, 204–209.

Proverbio, A. M., Leoni, G., & Zani, A. (2004). Language switching mechanisms in simultaneous interpreters: An ERP study. *Neuropsychologia, 42*(12), 1636–1656.

Pynte, J., Besson, M., Robichon, F.-H., & Poli, J. (1996). The time-course of metaphor comprehension: An event-related potential study. *Brain and Language, 55*, 293–316.

Radeau, M., Besson, M., Fonteneau, E., & Castro, S. L. (1998). Semantic, repetition and rime priming between spoken words: Behavioral and electrophysiological evidence. *Biological Psychology, 48*(2), 183–204.

Rugg, M. D., Cowan, C. P., Nagy, M. E., & Milner, A. D. (1988). Event related potentials from close head injury patients in an auditory "oddball" task: Evidence of dysfunction in stimulus categorization. *Journal of Neurology, Neurosurgery and Psychiatry, 51*, 691–698.

Rugg. M. D., Doyle, M. C., & Melan, C. (1993). An event-related potential study of the effects of within- and across-modality word repetition. *Language and Cognitive Processes, 8*, 357–377.

Rugg, M. D., & Nagy, M. E. (1989). Event-related potentials and recognition memory for words. *Electroencephalography and Clinical Neurophysiology, 72*(5), 395–406.

Rugg, M. D., Soardi, M., & Doyle, M. C. (1995). Modulation of event-related potentials by the repetition of drawings of novel objects. *Cognitive Brain Research, 3*, 17–24.

Sakai, K. L., Noguchi, Y., Takeuchi, T., & Watanabe, E., (2002). Selective priming of syntactic processing by event-related transcranial magnetic stimulation of Broca's area. *Neuron, 35*, 1177–1182.

Schlesewsky, M., & Bornkessel, I. (2006). Context-sensitive neural responses to conflict resolution: Electrophysiological evidence from subject–object ambiguities in language comprehension. *Brain Research, 1098*(1), 139–152.

Schmitt, B. M., Schiltz, K., Zaake, W., Kutas, M., & Münte, T. F. (2001). An electrophysiological analysis of the time course of conceptual and syntactic encoding during tacit picture naming. *Journal of Cognitive Neuroscience, 13*(4), 510–522.

Schwanenflugel, P. J., & LaCount, K. L., (1988). Semantic relatedness and the scope of facilitation for upcoming words in sentences. *Journal of Experimental Psychology, Learning, Memory, and Cognition, 14*, 344–354.

Schwanenflugel, P. J., & Shoben, E. J. (1985). The influence of sentence constraint on the scope of facilitation for upcoming words. *Journal of Memory and Language, 24*, 232–252.

Searle, J. R. (1979). *Expression and meaning: Studies in the theory of speech acts*. London: Cambridge University Press.

Seidenberg, M. S., Waters, G. S., Sanders, M., & Langer, P. (1984). Pre- and postlexical loci of contextual effects on word recognition. *Memory and Cognition, 12*(4), 315–328.

Sitnikova, T., Kuperberg, G., & Holcomb, P. J. (2003). Semantic integration in videos of real-world events: An electrophysiological investigation. *Psychophysiology, 40*(1), 160–164.

Sitnikova, T., Salisbury, D. F., Kuperberg, G. R., & Holcomb, P. J. (2002). Electrophysiological insights into language processing in schizophrenia. *Psychophysiology, 39*, 851–860.

Sperber, D., & Wilson, D. (1986). *Relevance: Communication and cognition*. Oxford: Blackwell.

St. George, M., Mannes, S., & Hoffmann, J. E. (1994). Global semantic expectancy and language comprehension. *Journal of Cognitive Neuroscience, 6*, 70–83.

Stroud, C. & Phillips, C. (2009). The structural and semantic selectivity of the "thematic" P600 in sentence comprehension. *Poster presented at twenty-second annual CUNY conference, Davis, California.*

Swaab, T. Y. (1998). Event-related potentials in cognitive neuropsychology: Methodological considerations and an example from studies of aphasia. *Behavior, Research Methods, Instruments and Computers, 30*, 157–170.

Swaab, T. Y., Baynes, K., & Knight, R. T. (2002). Separable effects of priming and imageability on word processing: An ERP study. *Cognitive Brain Research, 15*, 99–103.

Swaab, T. Y., Brown, C., & Hagoort, P. (1997). Spoken sentence comprehension in aphasia: Event-related potential evidence for a lexical integration deficit. *Journal of Cognitive Neuroscience, 9*, 39–66.

Swaab, T. Y., Brown, C., & Hagoort, P. (1998). Understanding ambiguous words in sentence contexts: Electrophysiological evidence for delayed contextual selection in Broca's aphasia. *Neuropsychologia, 36*, 737–761.

Swaab, T. Y., Camblin, C. C., & Gordon, P. C. (2004). Reversed lexical repetition effects in language processing. *Journal of Cognitive Neuroscience, 16*, 715–726.

Tanenhaus, M. K. & Trueswell, J. C. (1995). Sentence comprehension. In J. L. Miller & P. D. Eimas (Eds.), *Speech, language, and communication: Handbook of perception and cognition* (2nd ed., pp. 217–262). San Diego, CA: Academic Press.

Thompson-Schill, S. L., D'Esposito, M., Aguirre, G. K., & Farah, M. J. (1997). Role of left inferior prefrontal cortex in retrieval of semantic knowledge: A reevaluation. *Proceedings of the National Academy of Sciences USA, 94*, 14792–14797.

Thompson-Schill, S. L., D'Esposito, M., & Kan, I. P. (1999). Effects of repetition and competition on activity in left prefrontal cortex during word generation. *Neuron, 23*, 513–522.

Thompson-Schill, S. L., Swick, D., Farah, M. J., D'Esposito, M., Kan, I. P., & Knight, R. T. (1998). Verb generation in patients with focal frontal lesions: A neuropsychological test of neuroimaging findings. *Proceedings of the National Academy of Sciences USA, 9*, 15855–15860.

Tooley, K., Traxler, M., & Swaab, T. Y. (2009). Electrophysiological evidence of syntactic priming in sentence comprehension. *Journal of Experimental Psychology: Language, Memory and Cognition, 23*(5), 609–645.

Traxler, M. J., & Gernsbacher, M. A. (2006). *Handbook of psycholinguistics* (2nd ed.). Oxford: Blackwell.

Uddén, J., Folia, V., Forkstam, C., Ingvar, M., Fernandez, G., Overeem, S., van Elswijk, G., Hagoort, P., & Petersson, K. M. (2008). The inferior frontal cortex in artificial syntax processing: An rTMS study. *Brain Research, 1224*, 69–78.

van Berkum, J. J. A. (2004). Sentence comprehension in a wider discourse: Can we use ERPs to keep track of things? In M. Carreiras & C. Clifton, Jr. (Eds.), *The on-line study of sentence comprehension: Eyetracking, ERPs and beyond* (pp. 229–270). New York: Psychology Press.

van Berkum, J. J. A. (2009). The neuropragmatics of "simple" utterance comprehension: An ERP review. In U. Sauerland & K. Yatsushiro (Eds.), *Semantic and pragmatics: From experiment to theory*. 276–316. Basingstoke, UK: Palgrave Macmillan.

van Berkum, J. J. A., Brown, C. M., & Hagoort, P. (1999). Early referential context effects in sentence processing: Evidence from event-related brain potentials. *Journal of Memory and Language, 41*, 147–182.

van Berkum, J. J. A., Brown, C. M., Hagoort, P., & Zwitserlood, P. (2003). Event-related brain potentials reflect discourse-referential ambiguity in spoken language comprehension. *Psychophysiology, 40*, 235–248.

van Berkum, J. J. A., Brown, C. M., Zwitserlood, P., Kooijman, V. & Hagoort, P. (2005). Anticpating upcoming words in discourse: evidence from ERPs and reading times. *Journal of Experimental Psychology: Learning, Memory, and Cognition, 31*(3), 443–467.

Van Berkum, J. J. A., Holleman, B., Nieuwland, M. S., Otten, M., & Murre, J. (2009). Right or wrong? The brain's fast response to morally objectionable statements. *Psychological Science. 20*(9), 1092–1099.

van Berkum, J. J. A., Koornneef, A. W., Otten, M., & Nieuwland, M. S. (2007). Establishing reference in language comprehension: An electrophysiological perspective. *Brain Research, 1146*, 158–171.

van Berkum, J. J. A., van den Brink, D., Tesink, C. M. J. Y., Kos, M., & Hagoort, P. (2008). The neural integration of speaker and message. *Journal of Cognitive Neuroscience, 20*(4), 80–591.

van den Brink, D., Brown, C. M., & Hagoort, P. (2001). Electrophysiological evidence for early contextual influences during spoken-word recognition: N200 versus N400 effects. *Journal of Cognitive Neuroscience, 13*, 967–985.

van den Brink, D., Brown, C. M., & Hagoort, P. (2006). The cascaded nature of lexical selection and integration in auditory sentence processing. *Journal of Experimental Psychology: Learning, Memory, and Cognition, 32*, 364–372.

van den Brink, D., & Hagoort, P. (2004). The influence of semantic and syntactic context constraints on lexical selection and integration in spoken-word comprehension as revealed by ERPs. *Journal of Cognitive Neuroscience, 16*, 1068–1084.

van Herten, M., Kolk, H. H. J., & Chwilla, D. J. (2005). An ERP study of P600 effects elicited by semantic anomalies. *Cognitive Brain Research, 22*, 241–255.

Van Petten, C. (1993). A comparison of lexical and sentence-level context effects in event related potentials. *Language and Cognitive Processes, 8*, 485–531.

Van Petten, C., Coulson, S., Weckerly, J., Federmeier, K. D., Folstein, J., & Kutas, M. (1999). Lexical association and higher-level semantic context: An ERP study. *Journal of Cognitive Neuroscience Supplement, 46.*

Van Petten, C., & Kutas, M. (1991). Influences of semantic and syntactic context on open- and closed-class words. *Memory and Cognition, 19*, 95–112.

Van Petten, C., & Luka, B. J. (2006). Neural localization of semantic context effects in electromagnetic and hemodynamic studies. *Brain and Language, 97*, 279–293.

Van Petten, C., & Senkfor, A. J. (1996). Memory for words and novel visual patterns: Repetition, recognition, and encoding effects in the event-related brain potential. *Psychophysiology, 33*, 491–506.

Van Petten, C., Weckerly, C., McIsaac, H. K., & Kutas, M. (1997). Working memory capacity dissociates lexical and sentential context effects. *Psychological Science, 8*, 238–242.

Van Turennout, M., Hagoort, P. & Brown, C.M. (1998). Brain activity during speaking: from syntax to phonology in 40 milliseconds. *Science, 280*, 572–574.

Vissers, C. Th. W. M., Chwilla, D. J., & Kolk, H. H. J. (2006). Monitoring in language perception: The effect of misspellings of words in highly constrained sentences. *Brain Research, 1106*(1), 150–163.

Vissers, C. Th. W. M., Chwilla, D. J., & Kolk, H. H. J. (2007). The interplay of heuristics and parsing routines in sentence comprehension: Evidence from ERPs and reaction times. *Biological Psychology, 75*(1), 8–18.

Vissers, C. Th. W. M., Kolk, H. H. J., van de Meerendonk, N., & Chwilla, D. J. (2008). Monitoring in language perception: Evidence from ERPs in a picture-sentence matching task. *Neuropsychologia, 46*(4), 967–982.

Wagner, A. D., Koustaal, W., Maril, A., Schacter, D. L., & Buckner, R. L. (2000). Task-specific repetition priming in left inferior prefrontal cortex. *Cerebral Cortex, 10*, 1176–1184.

Wagner, A. D., Paré-Blagoev, E. J., Clark, J., & Poldrack, R. A. (2001). Recovering meaning: Left prefrontal cortex guides controlled semantic retrieval. *Neuron, 31*. 329–338.

Wassenaar, M., Brown, C. M., & Hagoort, P., (2004). ERP effects of subject–verb agreement violations in patients with Broca's aphasia. *Journal of Cognitive Neuroscience, 16*(4), 553–576.

Wassenaar, M., & Hagoort, P. (2007). Thematic role assignment in patients with Broca's aphasia: Sentence–picture matching electrified. *Neuropsychologia, 45*(4), 716–740.

Weckerly, J., & Kutas, M. (1999). An electrophysiological analysis of animacy effects in the processing of objective relative sentences. *Psychophysiology, 36*, 559–570.

West, W. C., & Holcomb, P. J. (2000). Imaginal, semantic, and surface-level processing of concrete and abstract words: An electrophysiological investigation. *Journal of Cognitive Neuroscience, 12*, 1024–1037.

West, W. C., & Holcomb, P. J. (2002). Event-related potentials during discourse-level semantic integration of complex pictures. *Cognitive Brain Research, 13*(3), 363–375.

Wicha, N.Y.Y., Moreno, E.M., & Kutas, M. (2004). Anticipating words and their gender: an event-related potential study of semantic integration, gender expectancy, and gender agreement in Spanish sentence reading. *Journal of Cognitive Neuroscience, 16*(7), 1272–1288.

Willems, R. M., Özyürek, A., & Hagoort, P. (2008). Similar neural correlates for the integration of words and pictures at the sentence level: Evidence from ERPs and fMRI. *Journal of Cognitive Neuroscience, 20*(7), 1235–1249.

Wu, Y. C., & Coulson, S. (2005). Meaningful gestures: Electrophysiological indices of iconic gesture comprehension. *Psychophysiology, 42*, 654–667.

CHAPTER

16 ERPs and the Study of Emotion

Greg Hajcak, Anna Weinberg, Annmarie MacNamara, *and* Dan Foti

Abstract

Interest in the neuroscience of emotion has increased dramatically over the course of the last two decades. The rapid growth and popularity, however, have come with a definitional imbroglio, as there seem to be as many conceptualizations of emotion as there are emotion researchers. This chapter begins by presenting an increasingly common conceptualization of emotion and emphasizes key distinctions used in emotion research. Next, multiple event-related potential (ERP) components sensitive to emotional content and the time course of emotional processing are highlighted. Then, how that time course can be clarified through data reduction techniques is examined, with examples provided. Subsequently, methodological issues are outlined, with the key decisions about ERP elicitation and measurement specified. Event-related potential findings related to clinical, developmental, and aging applications in the psychology of emotion are summarized. Finally, speculation on the future of emotion research using ERPs is proffered in terms of key questions to be answered.

Keywords: event-related potentials, emotion, emotion research, ERP components

Introduction

Interest in the neuroscience of emotion has increased dramatically over the course of the last two decades. The rapid growth and popularity, however, have come with a definitional imbroglio, as there seem to be as many conceptualizations of emotion as there are emotion researchers. And yet, there is growing consensus about critical issues in the field. This chapter begins by presenting an increasingly common conceptualization of emotion and emphasizes key distinctions used in emotion research. Next, multiple event-related potential (ERP) components sensitive to emotional content and the time course of emotional processing are highlighted. Then, how that time course can be clarified through data reduction techniques is examined, with examples provided. Subsequently, methodological issues are outlined, with the key decisions about ERP elicitation and measurement specified. Next, ERP findings related to clinical, developmental, and aging applications in the psychology of emotion are summarized. Finally, speculation on the future of emotion research using ERPs is proffered in terms of key questions to be answered.

Emotion: An Overview

On May 11, 1997, a computer named Deep Blue was completing a six-game chess match with Garry Kasparov. Despite enormous pressure with the world watching, Deep Blue was never anxious about losing, and never experienced appreciation for or frustration with Kasparov's strategy. Deep Blue took no joy in its landmark victory against the world champion. Though most of us would agree that computers like Deep Blue lack emotion, and that emotions are integral to human experience, there is less agreement about what emotions *are*.

Consider someone alone walking through a dark urban alley. The individual's attention may be captured by a sound in the night and then by the form of a person approaching. Physiological changes unfold as the brain processes the potential threat. The pupils dilate to improve the detection of faint signals, the heart pounds to deliver more blood to large muscle groups, and breathing becomes shallow and rapid as oxygen consumption increases. These changes herald readiness for protection, a proneness to scream, and preparation to run or fight if necessary. Only after the stranger passes do the physiological mechanisms begin to normalize. The individual later might describe the subjective experience as terrifying and recall thinking, "This is bad; I'm in trouble."

This example captures the *function* of fear—an emotion that prepares the organism to act defensively in the face of danger. Although the subjective experience of fear may be its most *salient* feature, the fact that fear motivates animals to seek safety is likely the most important consequence. Hence, emotions are thought to have evolved to serve both intrapersonal and interpersonal functions: fear aids in attaining safety, and facial expressions of fear alert others to potential danger in the environment (Darwin, 1872; Davidson, 1993).

The form of emotion follows from its function. As fear mobilizes an organism for action, emotion can be measured across multiple response systems (e.g., pupil diameter, changes in heart rate, self-report, and an organism's actual movement). Emotions can therefore be characterized in terms of relatively short-lasting physiological, experiential, and behavioral responses to motivationally salient internal and external stimuli (e.g., Gross & Thompson, 2007). Indeed, emotions might be best conceptualized in terms of the dynamic interplay between specific stimuli and the changes they elicit in the individual (Bradley & Lang, 2000; Lang et al., 1997). From this perspective, it is possible to distinguish among emotion, affect, and mood. *Affect* is a superordinate category that encompasses any valenced (i.e., pleasant or unpleasant) state (e.g., thirsty), whereas *mood* is an affective response that fluctuates more slowly, lasts longer, and is less influenced by particular eliciting stimuli than emotion (e.g., Rottenberg & Gross, 2003; Scherer, 1984).

It is important to note that there is little agreement about how many emotions exist or whether it is even reasonable to conceptualize emotions as discrete. Although some have focused on distinct emotions such as anger, fear, and happiness (Ekman, 1992; Izard et al., 1972; Plutchik, 1980), others have suggested that emotions might be better classified into superordinate *dimensions* of affective experience (e.g., Christie & Friedman, 2004; Ekman, 1992; Russell, 1980). The perspective in this chapter assumes that emotion is rooted in motivational states and that motivation is governed by two primary parameters: direction (i.e., movement toward or away from a stimulus) and intensity (i.e., the strength, speed, or vigor of that movement; Bradley, 2000; Davidson et al., 1990). Whether or not discrete emotions can be fully understood in terms of these more basic dimensions is unclear, but no ERP component has been found that reflects a specific emotion, and variation in the timing and amplitude of stimulus-elicited ERPs appears to relate to broad dimensions of emotion and motivation. This perspective provides a parsimonious account of ERP data in emotion research and facilitates the formulation of more specific questions about emotion using ERPs.

If emotions are defined by the interaction between motivationally salient stimuli and organisms, studying emotions in the laboratory presents a particular difficulty. The very best manipulations of emotion (e.g., extreme fear or pleasure) cannot be performed because of ethical concerns, despite the scientific interest in engaging the neural circuits that respond to imperatives for survival or reproduction. The technical challenges are also important, as a single-trial ERP typically contains more noise than signal, so that the same manipulation must be repeated many times and the neuroelectric measure averaged over many stimulus presentations to isolate the resulting signal. A common approach is to assess the transient and stimulus-bound neural response to multiple presentations of pictures. Humans are visual creatures, so images may engage the motivational circuits that become activated in the real world. Consistent with this possibility, pictures elicit robust changes in both peripheral and central nervous system activity, which implies that emotional pictures engage motivational systems (Bradley et al., 2001; Lang et al., 1993, 1997). Pictures are not perfect stimuli, however, and it is unlikely that any picture can induce the same degree of fear as an assailant in a dark alley. Moreover, eliciting certain emotions (e.g., anger) may require specific participant and picture selection (Harmon-Jones et al., 2006). These limitations are not fatal, as many studies suggest that picture-viewing paradigms are a viable scientific method for studying emotion—one that is ideally suited for ERPs.

Emotional face displays and emotional words also modulate ERPs (Bernat et al., 2001; Herbert et al., 2006; Kanske & Kotz, 2007). Although these topics will be briefly addressed, the major focus of the present review concerns ERP measures elicited by complex visual images from the International Affective Picture System (IAPS; Lang et al., 2005). The IAPS is a freely available set of thousands of images that have been used in most published ERP studies on emotion (http://csea.phhp.ufl.edu/media/). These stimuli depict a wide range of human experience and vary on arousal level and valence. Pleasant pictures include children playing, cute animals, delicious foods, and nudes; neutral pictures include household objects and everyday landscapes/scenes; unpleasant pictures depict human and animal threat, dirty toilets, and scenes of injury and death. The images are well selected such that, compared to emotional facial expressions, IAPS images elicit relatively more intense emotions (e.g., Britton et al., 2006; Dolcos & Cabeza, 2002).

The IAPS images have been rated by young adults on two scales that indicate valence and arousal (e.g., Lang, 1994). *Valence* refers to the pleasantness or unpleasantness evoked by a picture; "happy" is pleasant, "sad" is unpleasant. For instance, both a sunset and an erotic image tend to be rated as pleasant; a graveyard and an attack scene are both rated as unpleasant. Regardless of whether something is pleasant or unpleasant, *arousal* refers to the intensity of emotion, which can vary from relaxed, calm, dull, or sleepy to stimulated, jittery, or wide awake. Both pleasant and unpleasant stimuli can be either low or high on the arousal dimension in principle; for example, a sunset and a graveyard are both rated low in arousal; erotica and attack scenes are rated as more arousing.

Arousal is not completely independent of valence: pictures rated as either very pleasant *or* very unpleasant also are rated as eliciting a more intense emotional response (Lang et al., 1999). Relative to neutral valence, both increases and decreases in valence ratings are accompanied by increased ratings of arousal; stimuli rated low in arousal tend to be rated more neutrally with respect to valence. Similar patterns have been reported for both words and sounds, which suggests that a biphasic structure may generalize across stimulus modalities (Bradley & Lang, 1999). This pattern of self-report data is consistent with the existence of two fundamental motivational systems that support emotional responses: pleasant ratings reflect *appetitive* activation, whereas unpleasant ratings reflect *defensive* activation (Bradley & Lang, 2000; Bradley et al., 2001; Codispoti et al., 2001; Lang et al., 1997). Ratings of arousal, then, reflect the degree to which appetitive or defensive systems are activated by a stimulus.

This point is crucial for emotion research, since stimuli can differ on valence, arousal, or both of these dimensions. If a neural structure like the amygdala responds more to pictures of guns than to pictures of forks, one might be tempted to conclude that the amygdala is uniquely sensitive to aversive stimuli. To draw this conclusion, one would also need to demonstrate that the amygdala does not respond to very pleasant stimuli as well. Many neural measures appear to respond more to *both* pleasant and unpleasant visual stimuli relative to neutral stimuli (Zald, 2003). This chapter therefore uses the term *emotional* to refer to both unpleasant and pleasant stimuli that are greater than neutral in arousal. This approach implies that a measure distinguishing emotional from neutral pictures would be increased (or decreased) for both unpleasant *and* pleasant pictures compared to neutral pictures and explicitly notes in other instances when a measure differentiates pleasant from unpleasant stimuli.

From the perspective that emotional responding is rooted in an organism's basic motivational systems, it is reasonable to assume that attention might be preferentially allocated to stimuli that are motivationally salient (Mack & Rock, 1998; Most et al., 2007; Neisser, 1979). Stimuli that most efficiently capture attention do tend to be the ones critical to survival, such as those pertaining to reproduction or danger (e.g., Bradley et al., 2001; Lang et al., 1997). The preferential allocation of attention to emotional stimuli has been described as *motivated attention*, such that the engagement of motivational systems increases attention to stimuli, facilitating memory, perception, elaborated processing, and preparation for action (Bradley, 2000; Lang et al., 1997; Sabatinelli et al., 2005). In this way, emotion and attention may be inextricably related. Indeed, the majority of extant data indicate that emotional stimuli are processed preferentially and demand increased mobilization of resources: both pleasant and unpleasant stimuli are viewed for longer periods than neutral pictures (Bradley & Lang, 2000; Bradley et al., 2001; Calvo & Lang, 2004; Lang et al., 1993, 1997, 1998; Nummenmaa et al., 2006)—even when participants are instructed to look only at neutral images (Nummenmaa et al., 2006). Compared to neutral images, both appetitive and aversive emotional stimuli more effectively capture and hold attention (Armony & Dolan, 2002; Mogg et al., 1997; Öhman, 1992; Schupp et al., 2007a) and are subject to increased processing even when unattended

(Anderson & Phelps, 2001; Esteves et al., 1994). Emotional stimuli, regardless of valence, are also more likely to be recalled than nonemotional stimuli (Buchanan & Adolphs, 2002; Hamann, 2001; Hamann et al., 1997; Phelp et al., 1997).

Emotion research has moved away from the study of purely subjective measures; even though *feelings* seem like the sine qua non of emotion, this method has a number of methodological shortcomings. Participants are typically asked to report on their emotional experience following the presentation of a stimulus, so that self-report measures tend to be retrospective and likely reflect a subjective averaging of the emotional experience. Kahneman (1999) has described the tendency of people to report affective states according to a *peak-end* rule insofar as the most intense and recent experiences tend to determine the nature of retrospective affective reports. Moreover, implicit in the use of self-report methodology is the assumption that emotions are accessible to consciousness or that emotion is equivalent to feeling (Davidson, 2003). Self-report may be relatively sensitive to perceived or actual demand characteristics, and its measurement assumes that individuals can report on their feelings accurately and reliably. Although feelings may have a role in adaptive functioning (Bechara et al., 2000; Damasio, 1996), there is evidence that the generation of emotion takes place unconsciously (Berridge, 2003; Monahan et al., 2000; Ohman, 2005), with even the suggestion that the study of feelings is unlikely to advance the scientific understanding of emotion (LeDoux, 1998). In sum, characterizing emotion requires complementing self-report measures with more objective indices of emotion that do not rely on introspection.

Several psychophysiological measures are sensitive to the perception of emotional stimuli, including heart rate, skin conductance, facial muscle activity, pupil diameter, and the hemodynamic response measured using functional neuroimaging (Bradley, 2000; Bradley et al., 2001; Lang et al., 1993; Sabatinelli et al., 2007). All of these measures have relatively similar and comparatively poor temporal resolution—which is potentially problematic if emotional stimuli alter multiple early information processing stages. Event-related potentials directly index neural activity with a time scale on the order of milliseconds and are ideal for tracking the temporal dynamics of neural response to emotional stimuli.

The World of Emotion and ERPs

Because of their excellent temporal resolution, ERP techniques may offer unique insights into *affective chronometry*, or the unfolding of an emotional response over time (Davidson, 1998, 2008; Rothbart & Derryberry, 1981). Researchers have used ERPs to ask: what is the earliest point in information processing that is sensitive to emotional stimuli? Many ERP components are sensitive to both positively and negatively valenced stimuli compared to neutral stimuli (Cacioppo et al., 1994; Cuthbert et al., 2000; Foti et al., 2009; Keil et al., 2002; Schupp et al., 2004a, 2006a), and enhanced electrocortical responding in the context of emotional material has been reported both in anticipation of (Poli et al., 2007), as well as following, the presentation of emotional stimuli (Hajcak & Olvet, 2008). In what follows, we review the literature on emotional modulation of a wide range of ERP components as well as how these data have been interpreted.

Contingent Negative Variation/ Stimulus-Preceding Negativity

Enhanced neural activity in anticipation of emotional stimuli has been observed using paradigms in which a neutral stimulus cues the presentation of a subsequent and potentially emotional stimulus. If the interval between the first and second stimuli exceeds 2 s (Poli et al., 2007), a pair of negative-going slow waves time-locked to the cue known as the *contingent negative variation* (CNV; Brunia & Damen, 1988; Loveless & Sanford, 1974a, 1974b; Walter et al., 1964) are observed. The earlier slow wave is associated with processing the cuing stimulus and appears maximal at fronto-central sites (Loveless & Sanford, 1974a, 1974b). When the second stimulus requires a motor response, a later centro-parietal negativity is observed and is associated with response preparation (Rohrbaugh & Gaillard, 1983) and expectancy (Loveless & Sanford, 1974a, 1974b). This later wave has also been observed when no motor response is required, provided that the cued stimuli were sufficiently motivationally relevant (e.g., emotional IAPS images or electrical shock)—and in these cases, the late CNV is referred to as the *nonmotor CNV* or *stimulus-preceding negativity* (SPN; Brunia, 1988). The SPN is larger when participants believe that an upcoming stimulus is either pleasant or unpleasant compared to neutral (Amrhein et al., 2005; Howard et al., 1992; Klorman & Ryan, 1979; Poli et al., 2007; Simons et al., 1978; Takeuchi et al., 2005). Further, the magnitude of the SPN is largest in anticipation of more arousing affective stimuli and those stimuli pertaining more directly to biological imperatives (e.g., stimuli related to mutilation/threat and erotica; Poli et al., 2007; Takeuchi et al., 2005), leading many to suggest that the component might be a valuable index of the role of arousal in affective anticipation.

Early Poststimulus Components

Early visual ERP components that appear sensitive to emotional content include the P1, N1, and P2, which peak between 100 and 200 ms following stimulus onset (Batty & Taylor, 2003; Carretié et al., 2004, 2007; Foti et al., 2009; Keil et al., 2002). Emotional images appear to impact the magnitude of the P1, which generally appears largest at lateral occipital sites and peaks between 100 and 130 ms following picture onset—although the P1 has been scored as early as 80 ms after stimulus onset (Mueller et al., 2008; Muhlberger et al., 2009; Olofsson & Polich, 2007). The P1 literature has been mixed. An enhanced P1 to emotional images has been reported at both occipital (Carretié et al., 2004; Delplanque et al., 2004; Holmes et al., 2008; Mueller et al., 2008; Muhlberger et al., 2009) and frontal sites (Carretié et al., 2007). Although some research demonstrates an enhanced P1 to emotional images (Delplanque et al., 2004; Hot et al., 2006; Smith et al., 2003), other studies suggest P1 reduction to emotional images (Rigoulot et al., 2008). Variation in low-level sensory features between different stimuli may account for some of these inconsistencies. Modulation of the P1 by emotional stimuli seems more reliably elicited when faces (Holmes et al., 2008; Mueller et al., 2008; Muhlberger et al., 2009) rather than more complex IAPS images (Foti et al., 2009; Olofsson & Polich, 2007) are used as stimuli (cf., Delplanque et al., 2004; Hot et al., 2006). There is evidence that task differences may interact with emotional factors to determine the magnitude of the P1: studies using modified oddball or categorization tasks report a larger P1 for emotional stimuli (Carretié et al., 2004; Delplanque et al., 2004; Rigoulot et al., 2008), whereas studies employing passive viewing paradigms have not found an effect of emotion on the P1 (Foti et al., 2009; Weinberg & Hajcak, 2010). An example of the P1 is presented in Figure 16.1, which depicts ERP responses from 44 individuals while they passively viewed emotional faces. The P1 is observed as a positive deflection that peaks at approximately 120 ms at electrode Oz, and it is enhanced for fearful compared to neutral faces.

Within the time range of the P1 is a centroparietal negative deflection in the ERP waveform referred to as the *N1* that peaks at around 130 ms after stimulus onset (Foti et al., 2009; Keil et al., 2001). The N1 is sensitive to the emotional content of visual stimuli, is larger for both pleasant and unpleasant compared to neutral images (Begleiter et al., 1979; Carretié et al., 2007; Foti et al., 2009; Keil et al., 2001; Weinberg & Hajcak, 2010, 2011) and has been interpreted as reflecting increased early visual processing of emotional content. Figure 16.2 presents ERP data recorded from 64 participants during a passive viewing task in which IAPS stimuli were presented in valence-specific blocks. Event-related potentials averaged from two midline central recording sites (i.e., Cz and CPz) are presented (right), as well as the scalp distributions in the time range of the N1 for pleasant and unpleasant (relative to neutral) pictures (top, left). As illustrated in Figure 16.2, emotional (both pleasant and unpleasant) compared to neutral stimuli elicit a larger N1. There is also some evidence that the N1 is resistant to habituation specifically for highly arousing unpleasant pictures compared to both pleasant and neutral ones (Carretié et al., 2003), though this effect has not always been replicated with larger stimulus samples (Codispoti et al., 2007; Olofsson & Polich, 2007).

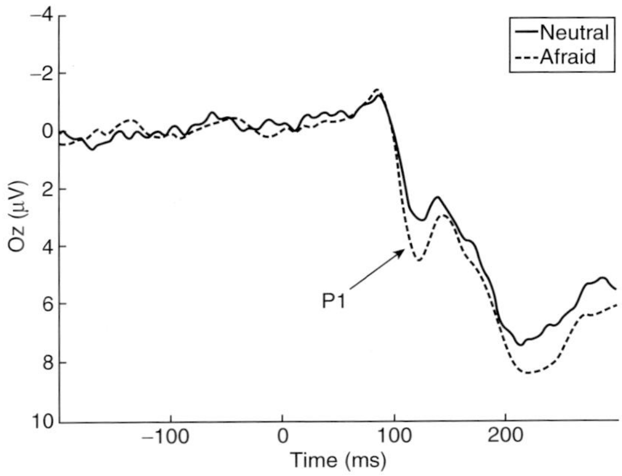

Fig. 16.1. Event-related potentials in response to neutral and afraid faces at channel Oz recorded from 44 individuals during a passive viewing task. Data from Foti et al. (2010).

Following the N1 is the P2, which peaks approximately 180 ms after stimulus onset (Carretié et al., 2004) and appears maximal and most distinct from the preceding and overlapping P1/N1 waves at anterior and central sites (Carretié et al., 2001, 2004; Luck & Hillyard, 1994). In nonaffective research, the magnitude of the P2 is enhanced for target stimuli, particularly when targets are infrequent (Luck & Hillyard, 1994), suggesting that the P2 indexes postperceptual selective attention.

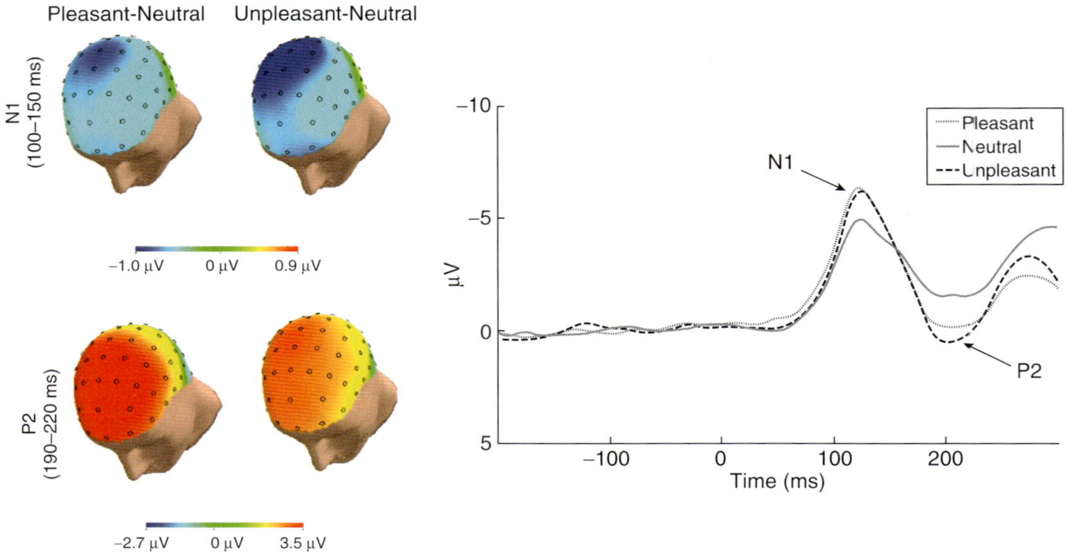

Fig. 16.2. Event-related potentials recorded from 64 participants during a passive viewing task in which IAPS stimuli were presented in valence-specific blocks. The waveforms (right) represent the ERPs from two central recording sites (i.e., average of Cz and CPz). Also presented are scalp distributions (left) of pleasant and unpleasant (relative to neutral) pictures in the time window of the N1 (i.e., 100–150 ms; top) and the P2 (i.e., 190–220 ms; bottom). Data from Weinberg and Hajcak (2010).

In keeping with the notion that emotional stimuli are inherently motivationally relevant, early studies examining ERP responses to line drawings conditioned to have distinct affective connotations (pleasant, neutral, or unpleasant) indicate that the magnitude of the P2 is sensitive to affective evaluations (Begleiter et al., 1979). There is also evidence that the P2 is enhanced for emotional words (Kanske & Kotz, 2007; Kissler et al., 2006; Schapkin et al., 2000), facial expressions (Eimer et al., 2003), and pictures (Carretié et al., 2001, 2004; Delplanque et al., 2004; Olofsson & Polich, 2007). In Figure 16.2 (right), the P2 is evident as a positive peak immediately following the N1. Scalp distributions for pleasant and unpleasant compared to neutral stimuli are presented in Figure 16.2 (left, bottom).

N170/Vertex Positive Potential

Although faces elicit a *relatively* weak emotional response (Britton et al., 2006), they are important signals of social and emotional information, and a number of studies have examined the impact of emotion on face-sensitive ERP components (Bentin et al., 1996, 2007; Jeffreys, 1989; Rossion & Jacques, 2008; Watanabe et al., 2003). The *N170/vertex positive potential* (VPP) discriminates faces from other objects: this component peaks between 140 and 180 ms and is observed as either a negative-going potential at bilateral occipitotemporal sites (the N170) when an average of all electrodes is used as a reference or a centrally distributed, positive-going potential (VPP) when an average of two mastoid electrodes is used as a reference (see Chapter 5, this volume; Joyce & Rossion, 2005). In addition to being sensitive to faces, the N170/VPP appears larger for emotional than neutral facial expressions (Batty & Taylor, 2003; Blau et al., 2007; Gur et al., 2002; Righart & de Gelder, 2008a; Willis et al., 2009); however, some studies have failed to find an effect of emotional expression on the N170/VPP (e.g., Eimer & Holmes, 2007; Eimer et al., 2003; Holmes et al., 2005). It is possible that the N170/VPP is more enhanced to fearful and angry expressions compared to both pleasant and neutral expressions (Batty & Taylor, 2003; Caharel et al., 2005; Righart & de Gelder, 2006, 2008b; Stekelenburg & Gelder, 2004; Williams et al., 2006). Recent research suggests that happy, but not sad, facial expressions elicit an enhanced VPP in adults. Figure 16.3 displays data recorded from a passive viewing task in which 46 healthy volunteers viewed happy, sad, neutral, angry, and fearful facial expressions from a standardized set of facial expressions (Tottenham et al., 2009). In this experiment, 130 faces were presented in random order, and each face was viewed twice. Stimulus-locked ERPs at Cz are shown for each expression type (right). The scalp distributions in the time window of the VPP (150–200 ms) for each emotional expression compared to neutral are also presented in Figure 16.3 (left). As suggested by

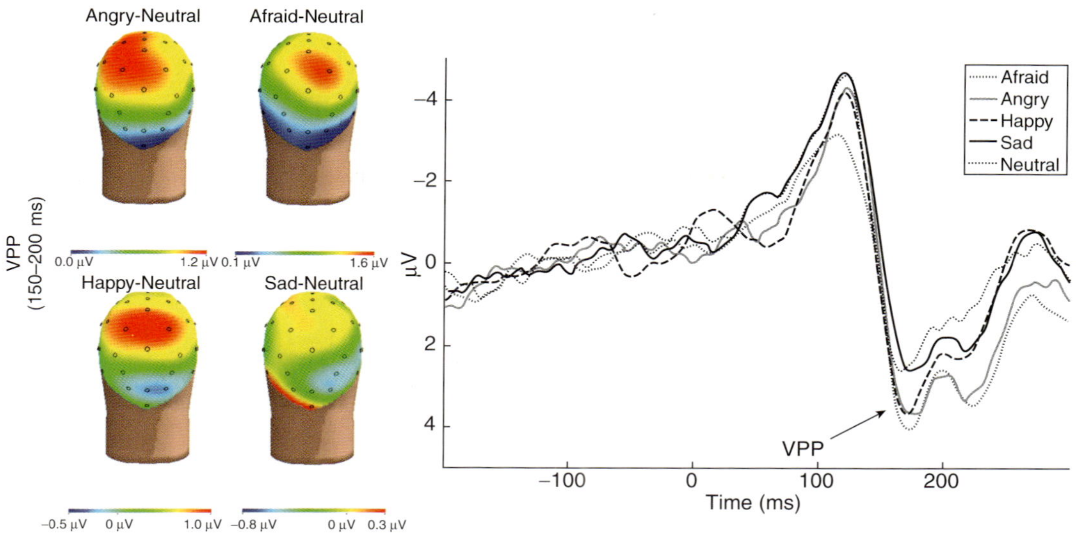

Fig. 16.3. Event-related potentials recorded from 46 participants during a passive viewing task in which 130 happy, sad, neutral, angry, and fearful facial expressions were presented in random order. Stimulus-locked ERPs at Cz are shown for each expression type (right). Also shown (left) are scalp distributions in the time window of the VPP (150–200 ms) for each emotional expression minus neutral. A portion of these data are from Foti et al. (2010).

Figure 16.3, happy, fearful, and angry faces elicited a larger VPP than neutral ones; sad faces and neutral faces did not differ from one another.

N2/Early Posterior Negativity

The N2 manifests as a central negativity (Carretié et al., 2004) peaking approximately 250 ms after stimulus presentation (though some studies have reported a more parietal distribution; e.g., Olofsson & Polich, 2007) and appears to index selective attention to specific stimulus features (e.g., color, shape, form; Codispoti et al., 2006b). Emotional stimuli have also been shown to influence the magnitude of the N2 (Olofsson & Polich, 2007; Palomba et al., 1997), though there are mixed reports as to whether this effect is equal for both pleasant and unpleasant stimuli (e.g., Carretié et al., 2004). There is evidence for lateralization of this effect, such that greater emotional modulation of the N2 has been observed over the right hemisphere than the left (Junghöfer et al., 2001; Palomba et al., 1997; Schupp et al., 2003, 2006b).

Also in the time range of the N2 is the *early posterior negativity* (EPN), which is generally observed as a *relative* negativity following emotional content between 200 and 300 ms at occipital sites. Functionally, the EPN appears sensitive to perceptual aspects of stimuli, including emotional content, that relate to increased selective attention (Bradley et al., 2007; Schupp et al., 2006a); the EPN has been associated with increased visual processing of emotional compared to neutral stimuli (e.g., De Cesarei & Codispoti, 2006; Foti et al., 2009; Junghöfer et al., 2001; Schupp et al., 2003, 2006a; Weinberg & Hajcak, in press). At short stimulus durations (i.e., three stimuli per second), the EPN may present as an absolute negativity (Herbert et al., 2008; Junghöfer et al., 2001), while at longer stimulus presentations, it presents instead as a reduction in a positivity for emotional compared to neutral stimuli (De Cesarei & Codispoti, 2006; Foti et al., 2009; Schupp et al., 2003; Weinberg & Hajcak, in press). Emotional modulation of the EPN appears fairly robust, as it has been observed across multiple tasks (e.g., passive viewing, speeded response tasks), stimulus types (e.g., IAPS, emotional adjectives, hand gestures; Herbert et al., 2008; Junghöfer et al., 2001), and stimulus durations (ranging from 120 to 1500 ms). Figure 16.4 presents data recorded from 64 participants in a passive viewing task in which IAPS images were presented for 1500 ms in valence-specific blocks; on the left are topographic maps depicting voltage differences (in microvolts) for erotic minus neutral images (top) and mutilation minus neutral images (bottom) in the time range of the EPN (200–280 ms following picture onset). Also shown (right) are stimulus-locked ERPs averaged over electrode sites Iz, P9, and P10 for pleasant, neutral, and unpleasant pictures. As suggested in Figure 16.4, the EPN presents as a

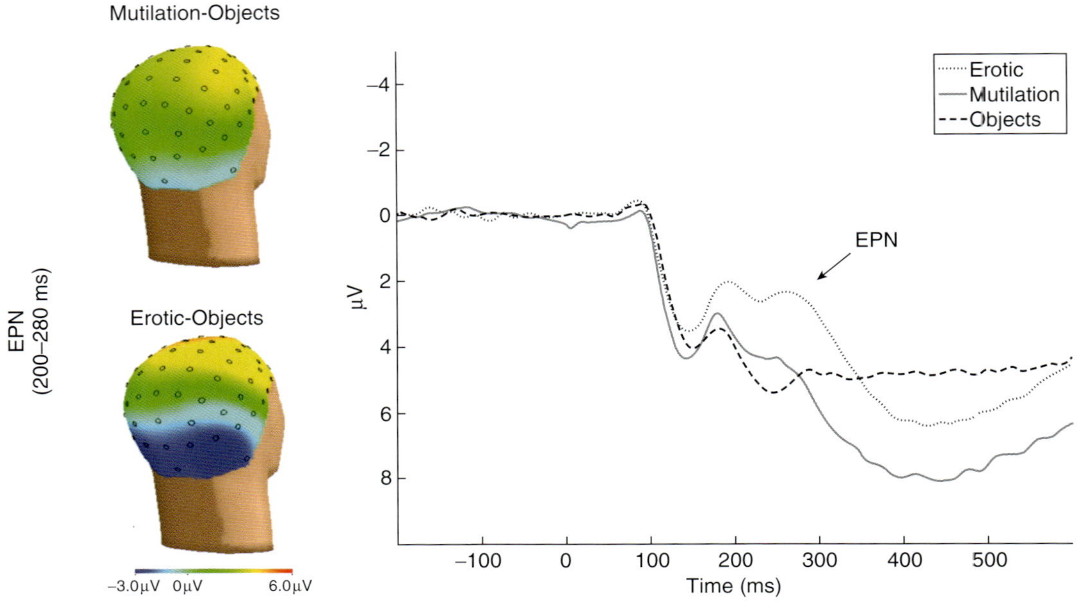

Fig. 16.4. Stimulus-locked ERPs averaged at Iz, P9, and P10 recorded from 64 participants in a passive viewing task in which IAPS images were presented in valence-specific blocks (right); to the left are topographic maps depicting voltage differences (in microvolts) for mutilation minus neutral images (top) and erotic minus neutral images (bottom) in the time range of the EPN (200–280 ms following picture onset). Data from Weinberg and Hajcak (2010).

relative negativity (at this longer picture presentation, i.e., 1500 ms) for both pleasant and unpleasant compared to neutral stimuli; however, there is accumulating evidence that the EPN may be uniquely sensitive to pleasant stimulus content compared to both neutral and unpleasant images (Schupp et al., 2004b, 2006a; Weinberg & Hajcak, 2010). Whether the EPN is observed, however, may depend heavily on the reference selected (this is discussed further below in the section on referencing).

P300

Following the EPN is a series of overlapping positivities that begins with the P300 (e.g., Foti & Hajcak, 2008; Foti et al., 2009; Hajcak et al., 2010b; MacNamara & Hajcak, 2009; MacNamara et al., 2009; Weinberg & Hajcak, in press). Because the P300 (also P3, or P3b) is among the most thoroughly researched ERP components (Luck, 2005; Polich, 2007), and because evidence from nonaffective research may shed light on affective processes indexed by these components, we provide a brief outline of the P300 (see Chapter 7, this volume, for a much more extensive review). First reported in 1965 (Sutton et al., 1965), the P300 is a midline, parietally maximal positive deflection that typically peaks between 300 and 500 ms after stimulus onset and appears sensitive to the motivational significance of stimuli. In the majority of P300 research, a variation on oddball designs is used, such that an infrequently presented *target* stimulus is either covertly or overtly identified in a stream of *standard* stimuli (Polich, 2007). In a basic example, participants might see *Y*s and *X*s on 80% and 20% of trials, respectively, and would be instructed to count the relatively infrequent *X*s (Donchin & Heffley, 1978; Pritchard, 1981). The P300 is larger for infrequent target stimuli, and its amplitude is further increased when target frequency is decreased (Duncan-Johnson & Donchin, 1977; Squires et al., 1976). There is evidence that the P300 is larger for targets even when targets and standards are equated for probability, suggesting that task relevance itself is sufficient to potentiate the P300 (Duncan-Johnson & Donchin, 1977). However, this enhancement appears to depend heavily on attention: if attention is divided between two tasks, or if stimuli in the oddball task are not attended to, targets do not elicit a P300 (Duncan-Johnson & Donchin, 1977; Hillyard et al., 1973).

Task relevance is not, however, the only factor that influences the P300. In studies dating back almost 50 years, emotional compared to neutral images also elicit an increased positivity 300–500 ms following picture presentation, which is similar to the P300 observed for explicitly designated targets

(Johnston et al., 1986; Lifshitz, 1966; Radilova, 1982). In one of the earliest investigations using emotional stimuli and ERP waveforms, images of neutral scenes, female nudes, and medical photographs of ulcerated legs were presented both in and out of focus (Lifshitz, 1966). Sustained differentiation between emotional (both pleasant and unpleasant) and neutral stimuli was evident from about 200 to 1000 ms following picture presentation across four subjects.

Emotional modulation of positivities in the time range of the P300 has since been observed repeatedly for both pleasant and unpleasant stimuli (Johnston et al., 1986; Palomba et al., 1997; Radilova, 1982), suggesting that processes indexed by the P300 may be linked broadly to motivation. Similar results have been reported for emotional adjectives (Naumann et al., 1992), faces (Allison et al., 1999; Cacioppo et al., 1993; Schupp et al., 2004c), and even lines implicitly conditioned to have emotional meaning (Begleiter et al., 1979). Thus, in addition to the influence of top-down imperatives of task demands, intrinsic motivational properties of stimuli may further modulate the P300.

Although it is difficult to demonstrate that two ERPs are "the same," the P300 elicited in traditional (i.e., nonaffective) oddball tasks has similar timing and scalp topography as variation in the P300 by emotional stimuli—and both effects may reflect the allocation of attentional resources based on motivational salience. As described above, the designation of target stimuli in an oddball task is arbitrary; there is nothing inherently motivationally salient about *X*s compared to *Y*s. And yet, as we have discussed, there is ample evidence that certain types of stimuli—namely, emotional stimuli—*are* inherently motivationally salient and may capture attention automatically. Moreover, apart from topography and timing, the nonaffective and affective P300 appear to respond similarly to manipulations of attention; neither unattended targets nor unattended emotional images elicit an enhanced P300 (Duncan-Johnson & Donchin, 1977; Hillyard et al., 1973; MacNamara & Hajcak, 2009). In the language of the P300, emotional stimuli might be considered *natural targets.*

Late Positive Potential

Although most of the early studies reporting increased positivities following the presentation of emotional stimuli described effects in the time window of the P300, increased positivities extending beyond this time range have been observed (Cacioppo et al., 1993; Cuthbert et al., 2000; Foti et al., 2009; Johnston et al., 1986; Keil et al., 2002; Lang et al., 1997; Palomba et al., 1997). More recent work in emotion has referred to the *late positive potential* (LPP), commonly identified as a midline centroparietal ERP that becomes evident by 300 ms following stimulus onset, and that is larger following the presentation of both pleasant and unpleasant compared to neutral pictures, words, and faces (Cacioppo et al., 1993; Cuthbert et al., 2000; Foti & Hajcak, 2008; Hajcak et al., 2006, 2007; Hajcak & Nieuwenhuis, 2006; Hajcak & Olvet, 2008; MacNamara & Hajcak, 2009, 2010; Moser et al., 2006; Schupp et al., 2000, 2003, 2004b; Weinberg & Hajcak, 2010). The LPP may also shift spatially over the course of affective processing, progressing from a parietal distribution to a more centrally maximal distribution, and may reflect separable components (Foti et al., 2009; MacNamara et al., 2009). Figure 16.5 presents ERP averages collapsed across five centroparietal sites (Pz, CPz, Cz, CP1, and CP2) from a recent study in which 64 participants passively viewed 135 pleasant, unpleasant, and neutral pictures (45 from each category). Each picture was presented twice for 1500 ms within a valence-specific block. The LPP is evident here as a sustained positive deflection in the stimulus-locked ERP following the presentation of pleasant and unpleasant compared to neutral images (right). The scalp distribution of pleasant minus neutral images and unpleasant minus neutral images is also depicted in Figure 16.5 (left) from 400 to 1000 ms (top) and from 1000 to 1500 ms (bottom) following picture onset.

Although many studies have demonstrated emotional modulation of the LPP lasting for the full duration of stimulus presentation (e.g., 300–1500 ms; Cuthbert et al., 2000; Foti & Hajcak, 2008; Hajcak et al., 2007; Hajcak & Nieuwenhuis, 2006; Hajcak & Olvet, 2008; Junghöfer et al., 2001; Lang et al., 1997; Weinberg & Hajcak, 2010, 2011), there is evidence that an increased LPP persists even into the period following emotional picture offset (Hajcak & Olvet, 2008; Hajcak et al., 2010b; MacNamara & Hajcak, 2010).

A positive association between the magnitude of the LPP and subjective arousal ratings of emotional stimuli (Cuthbert et al., 2000; Schupp et al., 2004a; Weinberg & Hajcak, 2010), as well as the motivational salience of stimulus categories, has been reported. The LPP increases more in amplitude to stimuli most directly relevant to biological

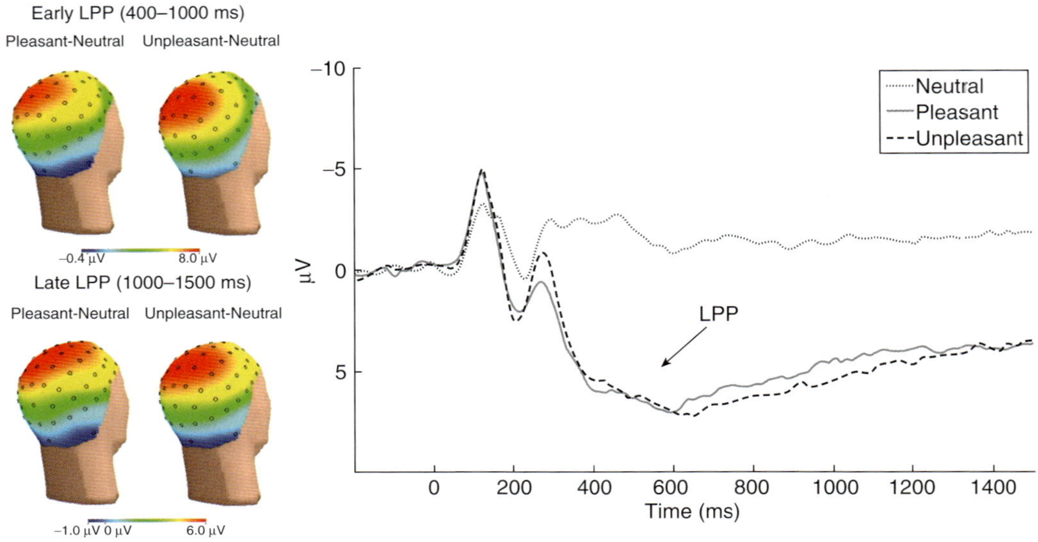

Fig. 16.5. Event-related potential averages over five centroparietal sites (Pz, CPz, Cz, CP1, and CP2) recorded in a recent study in which 64 participants passively viewed 135 pleasant, unpleasant, and neutral pictures (45 from each category). Each picture was presented twice for 1500 ms within a valence-specific block. The LPP is evident here as a sustained positive deflection in the stimulus-locked ERP following the presentation of pleasant and unpleasant compared to neutral images (right). The scalp distribution of pleasant minus neutral images and unpleasant minus neutral images is also depicted (left), from 400 to 1000 ms (top) and from 1000 to 1500 ms (bottom) following picture onset. Data from Weinberg and Hajcak (2010).

imperatives (e.g., threat, mutilation, and erotic images; Briggs & Martin, 2009; Schupp et al., 2004a; Weinberg & Hajcak, 2010). Further, emotional modulation of the LPP is remarkably stable; whereas multiple peripheral and central measures sensitive to emotional stimuli habituate over repeated presentations, including skin conductance, heart rate, facial muscle activity, and amygdala activation measured using functional magnetic resonance imaging (fMRI; Codispoti & De Cesarei, 2007; Codispoti et al., 2006a, 2007; Liberzon et al., 2003), the emotional modulation of the LPP does not appear to habituate (Codispoti et al., 2006a, 2007; Olofsson & Polich, 2007). Combined, this research suggests that the LPP is a valuable tool in measuring sustained attention to emotional content.

Differentiating ERPs in Emotion Research

In light of consistent evidence that emotional stimuli enhance the amplitude of components ranging from the P1 to the LPP, obtaining a parsimonious account of emotional modulation of ERPs is challenging. For example, although the onset of the EPN typically precedes that of the LPP, the boundary between these two components is often not clear. The EPN is typically described as a temporo-occipital negativity, but this only characterizes half of the actual electrical dipole. In fact, in the time range of the EPN, the difference between emotional and neutral pictures is more accurately described as a dipole that is negative at temporo-occipital sites and positive at central sites; indeed, some studies score the EPN at both locations (Flaisch et al., 2008a; Schupp et al., 2004b, 2006b). This raises the question of when the sustained positivity at central sites ceases to be the EPN and becomes the LPP, and where to draw the boundary. Other issues are whether and how the P300 and LPP differ from one another.

Factor analytic approaches can be applied to address the issue of component overlap and to better differentiate emotional modulation of ERP components. Specifically, temporal principal components analysis (PCA) is a method that extracts linear combinations of time points that capture unique portions of variance in the ERP, thereby providing a data-driven technique for separating neural responses that overlap in time (for a related technique, see Chapter 3, this volume, on independent component analysis). Temporal PCA can be used to identify a weighted sum of time points that best represent a component, independent of other overlapping sources of variance in the waveform. Principal components analysis has been successfully applied to the analysis of ERP data (Donchin & Heffley, 1978), and a number of studies of emotion have used temporal PCA to quantify a range of components, including P1, N1, P2, P3a, P3b, and LPP (Carretié et al., 2003; Codispoti et al., 2007; Delplanque et al., 2004, 2005, 2006; Kayser et al., 2000; Smith et al., 2003; Vanderploeg et al., 1987).

Similarly, when applied to electrode sites rather than time points, PCA can be used to reduce the spatial dimensions of ERP data and separate components with overlapping scalp distributions. Temporal and spatial PCA complement one another and can be used in conjunction as part of a two-step PCA. Several ERP studies of emotion have employed a two-step PCA (e.g., a temporal PCA is performed first to identify components based on variation in timing; this is followed by a spatial PCA to identify components based on variation in scalp distribution), and this approach has been effective for exploring the underlying structure of ERP modulation by emotional stimuli, yielding several consistent results (Carretié et al., 2004; Foti et al., 2009; Hot et al., 2006; MacNamara et al., 2009; Rigoulot et al., 2008; Weinberg & Hajcak, in press).

Two early (i.e., <300 ms) effects emerge: between 100 and 150 ms, emotional pictures are associated with a dipole that is positive at frontal sites and negative at parietal/occipital sites. This is followed from 150 to 300 ms by a separate dipole that is positive at frontal/central sites and negative at temporal/occipital sites. Independent of these early effects, several late (i.e., >300 ms) relative positivities for emotional pictures also emerge. From approximately 300 to 600 ms, there is a positivity at parietal sites, followed by additional positivities at occipital, parietal, and central sites that peak approximately 800–1000 ms after stimulus presentation. Finally, there is a positivity at frontal/central sites that peaks at approximately 1600 ms. Together, these PCA results indicate that the N1, EPN, and LPP ought to be considered unique indices of emotional processing, and that the LPP may be more accurately described as the sum of several overlapping positivities at posterior and superior sites rather than a single sustained positivity.

Temporal-spatial PCA has shed light on the *when* and *where* of ERP responses to emotional pictures; however, a crucial further question is *what* psychological process this emotional modulation reflects. That is, given that the ERP can be reliably decomposed into a set of unique effects, are these effects *functionally* dissociable? Using PCA, we found that the LPP factor, but not the temporally earlier EPN and P300 components, is uniquely associated with behavioral interference on a speeded response task (Weinberg & Hajcak, in press). In particular, we examined whether the LPP elicited by emotional and neutral pictures predicted reaction times to subsequently presented target stimuli: In both within- and between-subjects analyses, longer reaction times were characterized by an increased LPP; the EPN and P300 factors, on the other hand, did not predict reaction time. Unlike earlier components, the LPP has also been uniquely linked to memory encoding and storage (Azizian & Polich, 2007; Dolcos & Cabeza, 2002; Palomba et al., 1997). These findings are consistent with research suggesting that early and late components may index distinct facets of emotional processing (Azizian & Polich, 2007; Bradley et al., 2007; Olofsson & Polich, 2007; Olofsson et al., 2008).

There is also emerging evidence that picture complexity may differentially impact early versus late components, which has important implications for stimulus selection in affective investigations (Bradley et al., 2007). Early (150–250 ms) components appear more sensitive to the perceptual organization of IAPS images (e.g., simple compositions such as a single gun in the foreground with a white background, compared to complex compositions such as a car accident) than to emotional content, while the LPP is sensitive to *both* picture complexity and emotionality. However, research using emotional faces—in which emotional and neutral faces are presumably fairly well matched on complexity—has also demonstrated affective modulation of both early and late ERP components (Holmes et al., 2008; Mueller et al., 2008), suggesting that previously observed effects on early components cannot be attributed solely to differences in picture complexity between emotional and neutral stimuli. Along the same lines, studies have examined how being paired with an aversive unconditioned stimulus impacts the ERP response to a conditioned stimulus (CS+). These studies find both early (i.e., N1/P2) and later (i.e., P300/LPP) effects on the ERP as a result of conditioning, which cannot be attributed to physical properties of the CS+ (Pizzagalli et al., 2003; Skrandies & Jedynak, 2000).

Some have suggested that early components reflect obligatory attentional capture (Foti et al., 2009; Olofsson et al., 2008), while the later processing evident in the LPP may be associated with more flexible, sustained, and elaborative processes (e.g., Azizian & Polich, 2007; Codispoti et al., 2007; Foti et al., 2009; Hajcak & Olvet, 2008; Olofsson & Polich, 2007; Olofsson et al., 2008; Weinberg & Hajcak, 2010, in press). Consistent with this possibility, early components (<300 ms) appear to index relatively gross discrimination between affective and nonaffective stimuli (Weinberg & Hajcak, 2010). On the other hand, the LPP appears to differentiate among more specific picture content within the broad categories of pleasant, neutral, and unpleasant (Briggs & Martin, 2009; Schupp et al., 2004a; Weinberg & Hajcak, 2010). For example, Figure 16.6 presents stimulus-locked ERPs averaged over five centroparietal sites (Pz, CPz, Cz, CP1, and CP2) for nine specific picture types within the broad semantic categories of unpleasant (top), pleasant (middle), and neutral (bottom). Scalp topographies in two time windows (400–1000 ms and 1000–1500 ms) representing the difference between eight specific picture categories and neutral images of objects are presented in Figure 16.7. The most motivationally salient images elicit the largest LPPs even within the broad unpleasant, neutral, and pleasant picture categories.

There is also evidence that earlier versus later portions of the LPP differentially relate to various within- and between-group comparisons. Erotica are characterized by an unusually large and early portion of the LPP (i.e., 400–1000 ms), whereas the later portion of the LPP elicited by erotica is more similar to that elicited by other pleasant picture categories (i.e., Figures 16.6 and 16.7). Although both pleasant and unpleasant pictures modulate the early LPP, only modulation of the later portion of the LPP predicts better memory for pictures (Koenig & Mecklinger, 2008). Because of its sustained nature, later portions of the LPP can uniquely be manipulated *during* picture viewing (e.g., Hajcak et al., 2010b; Holmes et al., 2008; Weinberg & Hajcak, 2011). For instance, online manipulations that direct attention to more or less arousing portions of aversive stimuli impact the amplitude of the later portion of the LPP (Dunning & Hajcak, 2009; Hajcak et al., 2009).

The later portion of the LPP may also uniquely distinguish various clinical groups. For instance, we found that cocaine-addicted individuals were characterized by an increased early LPP to cocaine-related stimuli (Dunning et al., 2011); in addition, only those individuals who tested positive for cocaine on the day of testing were characterized by a reduced later portion of the LPP to all emotional stimuli, including cocaine stimuli. A recent report from Horan and colleagues (2010) found that individuals with schizophrenia had intact emotional modulation of the P300 but reduced later activity in the LPP to emotional pictures.

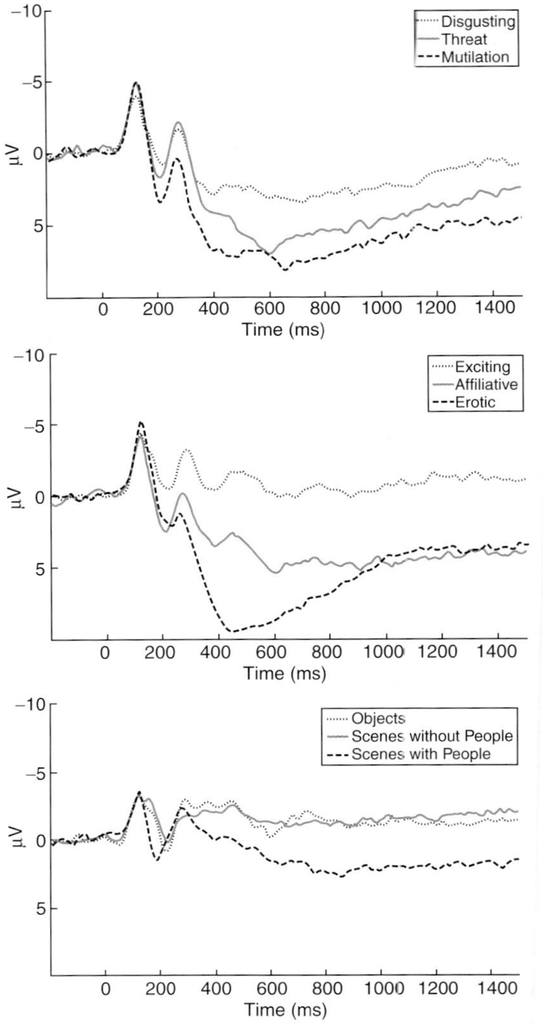

Fig. 16.6. Stimulus-locked ERPs averaged over five centroparietal sites (Pz, CPz, Cz, CP1, and CP2) for nine specific picture types within the broad semantic categories of unpleasant (top), pleasant (middle), and neutral (bottom). Data recorded from 64 subjects engaged in a passive affective viewing task in which images were presented in valence-specific blocks. Data from Weinberg and Hajcak (2010).

Combined, evidence suggests that affective factors exert a potent influence on ERPs from very early on—in some cases, prior even to the presentation of emotional stimuli—and that this effect persists beyond the presentation of emotional stimuli. A number of electrocortical indices of emotional processing are useful in tracking processing stages in emotion perception, and may yield novel information about neural activity involved in stimulus discrimination, categorization, and encoding.

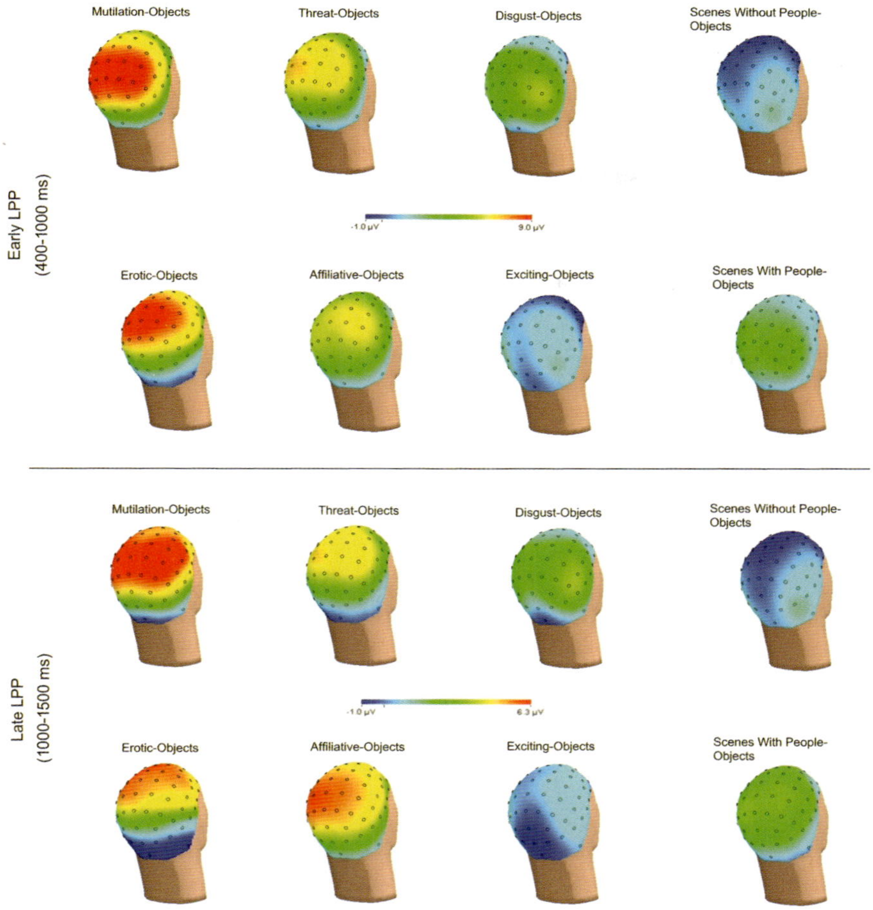

Fig. 16.7. Scalp topographies in two time windows (400–1000 ms, top; 1000–1500 ms, bottom) representing the difference between eight specific picture categories and neutral images of objects. Data recorded from 64 subjects engaged in a passive affective viewing task in which images were presented in valence-specific blocks. Data from Weinberg and Hajcak (2010).

Methodology and Measurement

Numerous hurdles arise when applying ERPs to the study of emotion, particularly when attempting to integrate findings across studies. In this section, we consider three critical methodological issues that strongly influence the interpretation of ERP data: filter settings, reference scheme, and component scoring strategy. These issues are by no means unique to the study of emotion, but they are particularly relevant in light of the wide range of ERP components that are sensitive to emotional content. There is no correct solution to these methodological issues; rather, we suggest that researchers consider how the use of different approaches influences observed effects and conclusions drawn.

Filter Settings

Grand average ERP waveforms represent summed stimulus- or response-locked activity across a spectrum of frequencies, and filters are typically applied as part of data processing to reduce the influence of unwanted activity. For example, a low-pass filter can be used to eliminate high-frequency oscillations resulting from electrical noise, and a high-pass filter can be used to eliminate slow drift resulting from skin potentials. The high-pass filter choice is particularly important when the LPP is a dependent measure of interest. Relative to components such as the N1 or the EPN, the LPP represents slower frequency activity that unfolds over the course of hundreds of milliseconds or more. As such, the application of a high-pass filter designed to remove slow wave activity from the data can drastically alter the observed LPP. Figure 16.8 depicts the LPP in response to neutral, pleasant, and unpleasant IAPS high-pass filtered with a Butterworth filter at three different half-power filter cutoffs. As seen in Figure 16.8 (top), the application of a 0.1 Hz filter yields a sustained

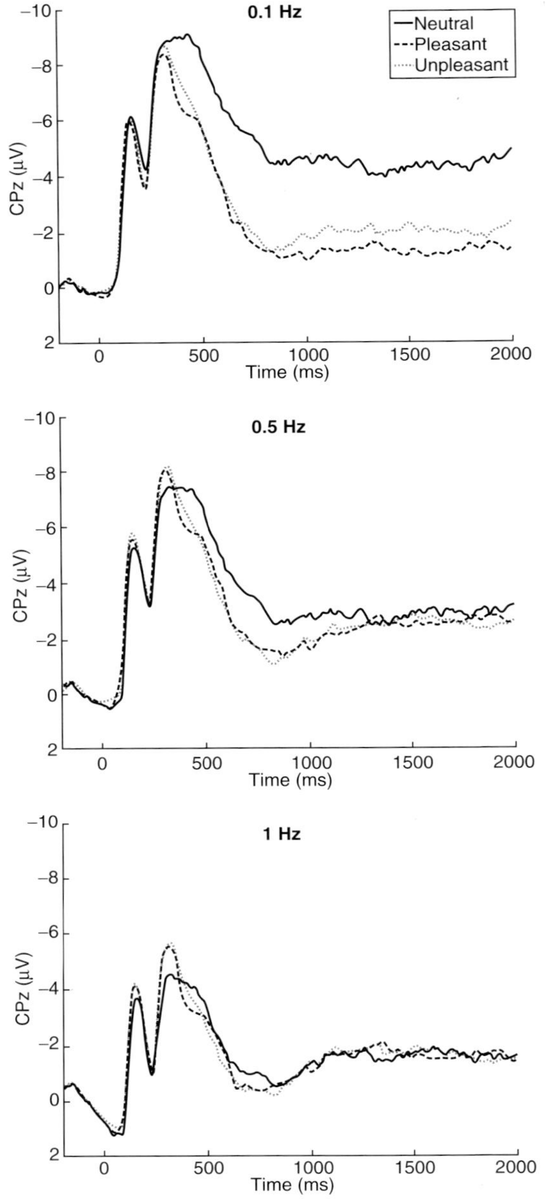

Fig. 16.8. Event-related potentials in response to neutral, pleasant, and unpleasant IAPS pictures at channel CPz recorded from 83 individuals during a passive viewing task. The three graphs depict the same data after a 0.1 Hz (top), 0.5 Hz (middle), or 1 Hz (bottom) high-pass filter was applied. Data from Foti et al. (2009).

LPP that is approximately 4 μV larger for emotional compared to neutral stimuli. When a 0.5 Hz filter is applied, the magnitude of the LPP becomes significantly attenuated, and little affective modulation is apparent after 1500 ms. When a 1 Hz filter is applied to the same data, the LPP is entirely removed from the waveform. Because each filter setting could lead to very different interpretations of the timing and magnitude of the LPP, researchers should take care when processing ERP data to ensure that the effect of interest is not being inadvertently removed through filtering. As a point of reference, we typically employ a 0.1 Hz high-pass filter (Dunning & Hajcak, 2009; Foti & Hajcak, 2008; Foti et al., 2009; Hajcak & Olvet, 2008; Hajcak et al., 2007, 2009; MacNamara & Hajcak, 2009, 2010; MacNamara et al., 2009), although a more conservative cutoff of 0.01 Hz is frequently used as well and may be particularly useful when effects beyond 1000 ms are of interest (Bradley et al., 2007; De Cesarei & Codispoti, 2006; Flaisch et al., 2008b; Hajcak & Dennis, 2009; Olofsson & Polich, 2007; Pastor et al., 2008; Schupp et al., 2004c, 2007b). It should be noted that the optimal cutoff will depend both on the particular type of filter employed and on whether the cutoff is the half-power or half-amplitude value. The recommended values above were determined using a Butterworth filter with a half-power cutoff and may need to be converted for use with other filters or filter settings.

Reference Selection

By definition, voltage is the potential for current to flow between two points. In ERP studies, voltage is the potential between an electrode of interest and a chosen reference, and it depends equally on neural activity at both sites. The choice of reference in ERP studies can drastically change the pattern of observed effects. One prominent example of this is the influence of the reference scheme on the EPN and the LPP. To illustrate the effect of the reference scheme on the EPN and LPP, Figure 16.9 presents the spatial distribution of the difference between pleasant and neutral IAPS pictures recorded from 83 participants; the data were rereferenced offline to either an averaged mastoid reference or an average electrode reference (i.e., average of all 64 electrode sites). The affective modulation of the EPN at temporo-occipital sites is prominent when an average electrode reference is used, and the EPN is nearly absent when a mastoid reference is used. Given that the EPN is maximal at scalp electrodes near the mastoids when using an average reference, this pattern is expected: rereferencing to a mastoid scheme effectively subtracts a large portion of the EPN from the data. Indeed, studies measuring the EPN most often employ an average electrode reference (e.g., Bradley et al., 2007; Codispoti et al., 2007; De Cesarei & Codispoti, 2006; Flaisch et al., 2008a, 2008b; Foti et al., 2009; Herbert et al., 2008;

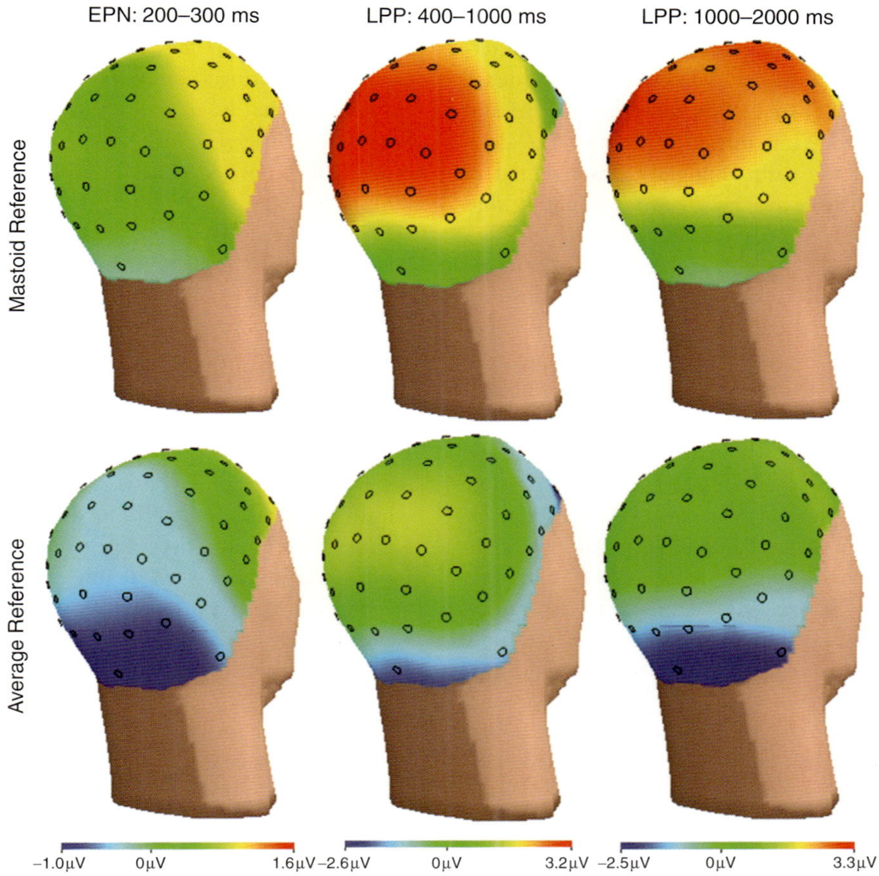

Fig. 16.9. Scalp distributions of the voltage difference between pleasant and neutral pictures, recorded during a passive viewing paradigm among 83 participants using a mastoid reference (top) and an average reference (bottom) in the time range of the EPN (left), early LPP (middle), and late LPP (right). Data from Foti et al. (2009).

Herrmann et al., 2007; Kissler et al., 2009; Muhlberger et al., 2009; Pastor et al., 2008; Peyk et al., 2009; Schacht & Sommer, 2009; Schupp et al., 2004b, 2004c, 2006b; Werheid et al., 2007; Wieser et al., 2006a, 2006b), although a mastoid reference has also been used in some studies that quantify the EPN (Grossmann et al., 2007; Scott et al., 2009).

Although the emotional modulation of the LPP at centroparietal sites is observable under both referencing schemes in Figure 16.9, the effect of emotion is more prominent when a mastoid reference is used. The LPP is apparent across a wide range of posterior and superior sites, and employing an average electrode reference will result in subtracting a portion of this activity from the data. It is also worth noting that there is a frontward shift in the scalp topography of the LPP over time in mastoid-referenced data. The LPP is maximal at centroparietal sites from 300 to 1000 ms and is more broadly distributed across all superior sites from 1000 to 2000 ms. This change in scalp distribution is largely lost when an average electrode reference is used. Within the literature, the LPP is most frequently considered under a mastoid reference scheme (Anokhin et al., 2006; Carretié et al., 2006; Codispoti et al., 2006a; Crites et al., 1995; Cuthbert et al., 2000; Dunning & Hajcak, 2009; Hajcak et al., 2007, 2009; Hajcak & Nieuwenhuis, 2006; Hajcak & Olvet, 2008; Krompinger et al., 2008; MacNamara & Hajcak, 2009; Moser et al., 2006; Schupp et al., 2000, 2004a), although an average electrode reference is also used, particularly in studies examining both the LPP and the EPN (Bradley et al., 2007; Codispoti et al., 2007; De Cesarei & Codispoti, 2006; Flaisch et al., 2008b; Foti et al., 2009; Hajcak & Dennis, 2009; Herbert et al., 2008; Keil et al., 2002; Pastor et al., 2008; Sabatinelli et al., 2007; Schupp et al., 2004c, 2007b).

In both average and mastoid reference schemes, the EPN and LPP are evident; the choice of reference scheme, however, will strongly influence whether the EPN or LPP is more prominent and can lead to differing interpretations of the data. For example, the scalp topographies presented in Figure 16.9 for the average electrode reference suggest notable emotional modulation of the EPN, a small LPP in the early window, and almost no effects in the later window of the LPP. In contrast, the mastoid-referenced data indicate a small EPN and a robust LPP in both the early and late windows. The two referencing schemes suggest different patterns of ERP activity between pleasant and neutral pictures over time. Finally, it is worth noting that specific reference schemes may not be feasible in all cases. In particular, an average electrode reference may only be appropriate when using a high-density electroencephalography (EEG) recording system with at least 64 scalp electrodes (Luck, 2005). In cases where fewer electrodes are used, an alternative reference such as mastoids or the nose tip (e.g., Herrmann et al., 2008) may be more appropriate. No referencing scheme is more correct than any other; the best reference choice will depend on the particulars of the data, and it is advisable to plot the data using multiple reference schemes to visualize the impact of this decision.

Component Scoring

Another key methodological decision is how to quantify ERP effects of interest. The most straightforward scoring method is to calculate the mean activity level or the peak deflection within a time window at one or more electrode sites. For example, the EPN has been scored as the average activity from approximately 200 to 300 ms at occipital and lateral parietal sites such as P7/8 and O1/2 (Holmes et al., 2008; Muhlberger et al., 2009). Unlike the EPN, however, the LPP is sustained and becomes broadly distributed across posterior and superior sites over time, thereby presenting a challenge in choosing a particular time window and location for scoring purposes. One method for dealing with this is to choose comprehensive sets of electrode sites and windows that can then be entered as their own within-subjects factors in subsequent statistical analysis. For example, Cuthbert and colleagues (2000) conducted a study in which emotional images were presented for 6 s each. They reduced the ERP over this time interval by creating averages of activity within 1 s blocks (i.e., 1–2 s, 2–3 s, etc.) and then checked for an interaction between picture type and time window in predicting LPP amplitude. A similar approach has been used in the spatial domain by creating a comprehensive set of electrode clusters that can be used to examine whether LPP modulation interacts with scalp region (Codispoti et al., 2007; Foti & Hajcak, 2008; Hajcak et al., 2007).

Although standard time-window approaches are relatively easy to implement and successfully reduce the number of time points to be analyzed, this comes at the cost of temporal precision. Guthrie and Buchwald (1991) proposed a technique in which the difference between the ERP waveform on two trials types could be compared at each time point. To control for familywise error, significance testing is corrected by using the autocorrelation within the dataset to determine the number of consecutive time points necessary to indicate a meaningful difference between the waveforms. This method can be used to identify when in the ERP there is a significant difference between neutral and emotional stimuli. For example, Hajcak and colleagues (2009) applied this method to analyze the LPP in a paradigm in which, on each trial, participants first passively viewed an image for 3 s and then received an instruction to focus their attention on either a more or less arousing portion of that picture while they viewed the image for an additional 3 s (i.e., 6 s total). The LPP was significantly increased for unpleasant pictures across a window from 160 to 3000 ms during the passive viewing phase. Further, on trials where participants were asked to focus on a less arousing portion of the picture, the LPP was significantly decreased across a window of 620–3000 ms following the instruction. In this way, shifts in LPP amplitude were analyzed while preserving the temporal resolution of the recorded ERP waveform. Other methods for analyzing the time course of ERPs could similarly be applied to examine when the LPP diverges between emotional and neutral pictures. Specifically, the jackknifing method that has previously been utilized to estimate the onset of the lateralized readiness potential could be utilized to examine timing differences in emotion research (Miller et al., 1998).

Principal Component Analysis

Many of the ERP effects linked to emotion can overlap in space and time. While standard component scoring techniques are limited in their ability to separate overlapping responses, temporal-spatial PCA has emerged as a powerful tool for understanding and quantifying the structure of ERP modulation by emotional stimuli. Two-step PCA has received increased attention over the past several years, with recent publications providing specific guidelines for optimally applying temporal-spatial PCA to ERP data (Dien, 2010b; Dien & Frishkoff, 2005; Dien et al., 2005, 2007) and the release of an open-source Matlab program that was developed for this purpose (Dien, 2010a).

A visual example of this ERP reduction is presented in Figure 16.10, which depicts ERP data from the same 83 participants described in the referencing example above. As seen in the top portion of the figure, the original ERP waveform is characterized by an enhanced N1 for unpleasant pictures at CPz, followed by an enhanced LPP from approximately 400 to 2000 ms. Principal components analysis was performed on the data as follows:

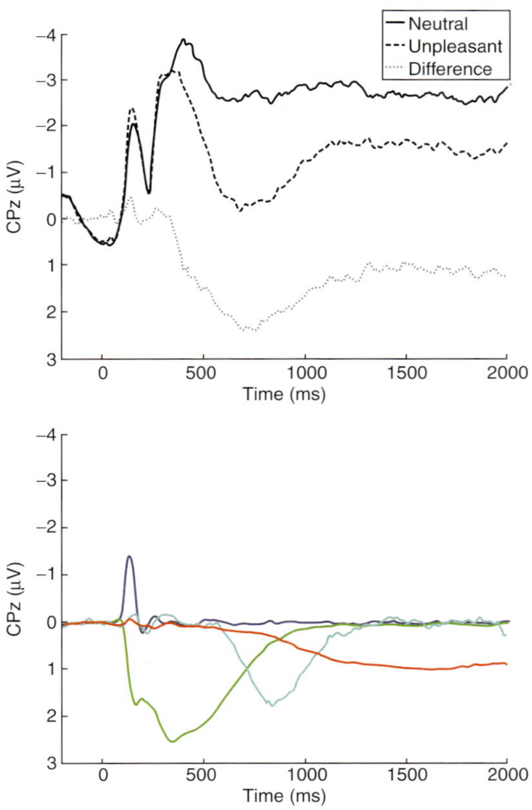

Fig. 16.10. Top: ERPs in response to neutral and unpleasant pictures at channel CPz, recorded during a passive viewing paradigm. Bottom: microvolt-scaled loadings of four temporal-spatial PCA factors representing the difference between unpleasant and neutral pictures at CPz. The dark blue line represents the N1. The green, light blue, and red lines represent three overlapping positivities that comprise the LPP. Data from Foti et al. (2009).

a temporal PCA was conducted first, using all time points as variables and considering all participants, electrode sites, and stimuli as observations. This yielded 1126 temporal factors (i.e., 1 for each time point), and visual inspection of the resulting Scree plot indicated that 12 ought to be retained for Varimax rotation. Following this step, a separate spatial PCA was then performed on each of the 12 temporal factors, this time using all electrode sites as variables and all participants and stimuli as observations. This yielded 64 spatial factors (i.e., 1 for each electrode), and visual inspection of this Scree plot indicated that 4 factors ought to be retained for Promax rotation. In total, this yielded 48 PCA factors (4 spatial factors for each of 12 temporal factors) that could then be subjected to statistical analysis by examining the associated scores for each participant and stimulus type. The bottom portion of Figure 16.10 depicts the temporal loadings at CPz for four factors that significantly differentiated unpleasant and neutral pictures, with the overlaid waveforms representing the difference between unpleasant and neutral pictures. The PCA waveforms have been rescaled to microvolts, thereby providing a sense of the relative contribution of each factor to the original data plotted immediately above. As seen in Figure 16.10, an early negativity corresponding to the N1 (dark blue waveform) can be separated from three later positivities that comprise the LPP (green, light blue, and red waveforms). Further, the timing of these later positivities reveals that the LPP in this dataset can be divided into three time windows: 300–600, 600–1200, and 1200–2000 ms.

Like the Guthrie and Buchwald (1991) technique, PCA offers a data-driven method for identifying variation in ERP data over time. An additional advantage of PCA is that ERPs are decomposed across both time points and electrodes. This yields a relatively small set of unique factors that can be further analyzed. Temporal-spatial PCA is a powerful tool for separating overlapping responses and exploring the underlying nature of a dataset, and it has proven useful for providing a parsimonious account of emotional modulation of ERPs during the first 2 s of stimulus presentation. PCA is not infallible; in the absence of prior knowledge about an ERP component's morphology, choosing a PCA factor to analyze is no less arbitrary than choosing a time window or electrode based on visual inspection of the grand average waveform. Because PCA identifies combinations of electrodes and time points that account for the maximum amount of variance, there is an inherent risk of finding statistically significant differences between trial types on particular factors that are not obviously interpretable. This risk is best mitigated by selecting PCA factors that are grounded in results of prior studies for statistical analysis (Dien et al., 2005).

Ultimately, well-chosen PCA factors, time windows, and electrode sites ought to converge on the same results. Indeed, for the example presented in Figure 16.10, the mean level of activity at a pooling of centroparietal electrodes (Cz, CPz, CP1/2, and Pz) from 300 to 600 ms was highly correlated with scores on the corresponding temporal-spatial factor ($r = .81$). Scoring that portion of the LPP with either the time window average or the PCA factor, therefore, would yield similar results. Scoring the LPP by averaging across electrode sites and time points can provide a perfectly reasonable estimate of the underlying component magnitude. As such, different component scoring strategies should be mutually informative: PCA results can guide the selection of relevant electrodes and/or time windows and vice versa.

Utilizing ERP Studies of Emotion to Inform Psychological Theory

The goal of many ERP studies is *not* simply to ask when emotional stimuli differ from neutral. Rather, established ERP components can be used as dependent measures to draw conclusions about broader psychological processes. In what follows, we consider how many of the basic findings outlined above have been recruited to ask questions about attentional biases, emotion–cognition interactions, individual differences and developmental processes. For example, although there is ample evidence that multiple ERPs are sensitive to the *arousal* dimension of emotional stimuli, some have argued that valence might also play a critical role (see the section "Negativity Bias" below). In addition, the dynamic and reciprocal interactions between emotional stimuli and other cognitive processes have been studied using ERPs, as we discuss below in the section "Emotion–Cognition Interactions and ERPs." Although enhanced attention to motivationally relevant stimuli can be adaptive, we discuss instances in which the processing of emotional stimuli is exaggerated or impaired in the section "Individual Differences/Psychopathology." Finally, the vast majority of research discussed above was conducted with college and adult populations—and yet there may be important differences in the way emotional stimuli are processed across the lifespan, from infancy to old age (see the section "Development and Aging").

Negativity Bias

Although both appetitive and aversive stimuli may capture attention and receive increased processing resources, prioritizing threatening above appetitive stimuli may have been beneficial for survival from an evolutionary perspective. The notion that individuals may preferentially process unpleasant compared to pleasant stimuli is referred to as a *negativity bias* (Cacioppo et al., 1999; Ito et al., 1998). A critical issue regarding the negativity bias is the ability to match, or equate, the *potency* of negative and positive stimuli (Taylor, 1991). A negativity *bias* ought to reflect increased processing of aversive compared to appetitive stimuli when all else is equal (e.g., probability of occurrence, intensity of pleasant and unpleasant stimuli). Studies have often equated pleasant and unpleasant stimuli on the arousal dimension that captures the intensity of appetitive or defensive activation, and the aim has been to demonstrate a negativity bias for aversive stimuli matched to appetitive stimuli on arousal ratings (Ito et al., 1998).

Several ERP studies have reported that aversive compared to appetitive stimuli elicit greater ERP indices of neural processing. For example, a larger P1 (Smith et al., 2003), P2 (Carretié et al., 2001; Huang & Luo, 2006), and P3a (Delplanque et al., 2006) in response to aversive compared to appetitive stimuli have been reported. The majority of negativity bias studies, however, have focused on the LPP as a measure of elaborative processing of emotional stimuli. Many studies have found larger-amplitude LPPs to unpleasant compared to pleasant pictures (Foti et al., 2009; Hajcak & Olvet, 2008; Huang & Luo, 2006; Ito & Cacioppo, 2000; Ito et al., 1998; Schupp et al., 2000).

There may be significant variation in neural response across picture content. For instance, erotic and mutilation pictures appear to elicit larger LPPs than other images in their respective pleasant and unpleasant categories (Briggs & Martin, 2009; Schupp et al., 2004a, 2004b; Weinberg & Hajcak, 2010). By contrast, exciting images of sports and roller coasters are encompassed in the broad pleasant category and appear to elicit LPPs that do not differ from those elicited by neutral pictures (though the two subtypes were not directly compared; Schupp et al., 2004a). Importantly, even when exciting IAPS images are explicitly matched to erotic stimuli in terms of arousal ratings, erotic images continue to elicit larger LPPs than exciting stimuli (Briggs & Martin, 2009; Weinberg & Hajcak, 2010). Thus, certain picture subtypes—such as erotic stimuli and pictures of mutilation—may contain information that is more biologically relevant than pictures of sports, babies, or happy families and may therefore elicit larger LPPs (Briggs & Martin, 2009; Weinberg & Hajcak, 2010). These data also suggest that arousal ratings may relate imperfectly to motivational salience and that both are important for understanding variation in the LPP.

This issue is especially germane in studies of the negativity bias, which have primarily utilized unpleasant stimuli of biologically relevant content such as mutilation and threat, whereas pleasant pictures have contained images of roller coasters and sports as well as pictures of food and of cute animals, which may be less biologically relevant (Briggs & Martin, 2009; Franken et al., 2008; Weinberg & Hajcak, 2010). By contrast, when weak positive exemplars (i.e., exciting images) have been excluded from analyses, no evidence of a negativity bias has been found (Franken et al., 2008; Weinberg & Hajcak, 2010). Moreover, when more salient positive stimuli have been used, some studies have suggested that pleasant pictures may elicit *larger* ERPs than unpleasant stimuli (the LPP: Briggs & Martin, 2009; the EPN: Schupp et al., 2004b, 2006b). Overall, then, whether a negativity bias can be indexed with ERPs, and the LPP in particular, likely depends heavily on picture selection.

Emotion–Cognition Interactions and ERPs

Emotional stimuli seem to capture attention in a relatively obligatory or bottom-up fashion. Emotional stimuli can also influence attention indirectly by interacting with top-down cognitive control processes (Taylor & Fragopanagos, 2005). Event-related potentials have been used to examine the ways in which the emotional content of pictures interacts with a number of other factors, including task relevance and manipulations of stimulus meaning. The majority of these studies have focused on mid- to late-latency components such as the EPN, P300, and LPP.

Task Relevance, Attention, and Emotional Stimuli

Oddball and related paradigms have traditionally been used in ERP research to manipulate top-down stimulus salience. For example, stimuli that are denoted as targets reliably elicit larger P300s than nontarget stimuli. To determine whether emotional salience might interact with target status to impact the EPN and parietal positivities occurring approximately 400–600 ms following stimulus onset, Schupp and colleagues (2007b) modified the traditional target paradigm by designating affective stimulus categories (pleasant, unpleasant, or neutral pictures) as targets during each of three separate runs. A target effect and an effect of emotion were observed on the EPN such that both target and emotional pictures were associated with larger (i.e., more negative) amplitudes. As expected, all targets also elicited an increased positivity in the time range of the P300/early portion of the LPP. Emotional stimuli that were targets elicited the largest amplitudes, suggesting that top-down and bottom-up manipulations of stimulus salience both potentiate neural metrics of attention at later processing stages. Similar work by Ferrari and colleagues (2008) required participants to indicate whether pictures contained animals/no animals or people/no people in separate runs. Pictures that were both target stimuli (i.e., animals or people) and emotional elicited the largest LPPs, indicating an additive effect of emotion and target status. Together, these studies suggest that stimulus importance—whether cognitively or affectively denoted—operates to increase occipital and parietal ERPs from about 200 to 600 ms after stimulus onset.

Attending to affective versus nonaffective stimulus dimensions has also been shown to modulate parietal positivities in much the same way as bottom-up manipulations of emotion. For example, when Hajcak and colleagues instructed participants to categorize emotional IAPS as pleasant or unpleasant (i.e., along an affective dimension), they elicited larger LPPs compared to when the same pictures were categorized according to the number of people they contained (i.e., along a nonaffective dimension: Hajcak et al., 2006; see also Krolak-Salmon et al., 2001). Thus, manipulating the top-down focus on emotional compared to nonemotional factors seems to increase the LPP in much the same way as picture content itself.

Manipulations of *spatial* attention also modulate the LPP to emotional pictures. For example, the LPP can be increased or decreased by instructing participants to direct attention to more or less arousing parts of pictures (Dunning & Hajcak, 2009; Hajcak et al., 2009). The LPP is enhanced for aversive IAPS presented in attended locations, but not when the same stimuli are presented in unattended spatial locations (MacNamara & Hajcak, 2009, 2010). These results are in line with work by Eimer and colleagues (2003) and Holmes and colleagues (2003), who found that emotional faces elicited increased positivities, but only when presented in spatially attended locations. Together, these results suggest that the LPP is gated by top-down modulations of spatial attention.

Another question is whether the concurrent performance of a nonaffective task can reduce or attenuate affective modulation of ERPs. Schupp and colleagues found that when a concurrent task is relatively easy, no interference with emotional processing of pictures is observed, as indexed by the EPN (Schupp et al., 2007a; see also Schupp et al., 2003). Hajcak and colleagues (2007) similarly found that simple arithmetic was not sufficient to interfere with emotional modulation of the LPP to affective pictures. However, tasks that are more perceptually or cognitively demanding have been shown to interfere with the processing of emotional stimuli, as indexed by reduced affective modulation of the EPN (Schupp et al., 2007a). Future studies might similarly examine whether performing a difficult concurrent task is sufficient to alter the LPP.

Stimulus Meaning/Reappraisal

People have the capacity to alter the nature and trajectory of their response to emotional stimuli. For instance, *cognitive reappraisal* is an emotion regulation technique in which individuals alter their response to emotional events by changing the way they think (e.g., Gross & Thompson, 2007). During a scary movie, for example, an individual can engage in cognitive reappraisal by reminding himself that the events onscreen are not real and that he is simply watching a film, or he can tell himself that the film will end on a positive note and that the protagonist will likely be saved. Regardless of how one accomplishes cognitive reappraisal, the aim is to reduce emotional intensity by *reinterpreting meaning*.

To determine whether the effects of reappraisal are evident in the ERP response to emotional stimuli, Hajcak and Nieuwenhuis (2006) asked participants to willfully reinterpret emotional pictures or to simply attend to pictures the way they normally would. Both subjective ratings of emotional response and the LPP were reduced following reappraisal instructions, suggesting that the voluntary modulation of the affective value of emotional stimuli was sufficient to alter the amplitude of the LPP. Similar findings have also been reported using more open-ended emotion regulation techniques (Krompinger et al., 2008; Moser et al., 2006).

Because instructions in emotion regulation studies are somewhat unspecific, these results may be attributable to changes in stimulus meaning or other processes (e.g., thinking of something else). To determine whether meaning change itself is capable of modulating neural and subjective response to emotional stimuli, Foti and Hajcak (2008) and MacNamara and colleagues (2009) provided participants with either negative (e.g., "This plane was the target of a terrorist bomb") or neutral (e.g., "These people are boarding an early morning flight") descriptions of unpleasant or neutral pictures. Unpleasant and negatively described pictures elicited larger LPPs and were rated as more emotionally arousing and more unpleasant than neutral and neutrally described pictures, respectively. Manipulations of meaning can even evoke lasting change: when participants listened to negative or neutral descriptions of unpleasant and neutral pictures and then viewed the same pictures 30 min later—this time *without* the preceding descriptions—the LPP and subjective picture ratings were reduced for those pictures that had been preceded by neutral compared to negative descriptions (MacNamara et al., 2011). Thus, meaning change is capable of modulating both subjective responses and neural responses (measured via the LPP) to emotional and neutral pictures.

Memory

Emotional material is remembered better than nonemotional material (Bradley et al., 1992; Christianson, 1992), and studies using ERPs have begun to investigate neural correlates of the encoding and retrieval of emotional compared to neutral stimuli. A common finding in the ERP memory literature is the *old-new* effect (see Chapter 14, this volume). When participants are presented with previously seen and novel stimuli during a recognition paradigm, previously seen (i.e., old) stimuli elicit larger midfrontal and parietal positivities than correctly categorized new stimuli, beginning approximately 250 ms after stimulus onset. The old-new effect is believed to index the successful recognition of stimuli (Rugg & Curran, 2007; Rugg & Nagy, 1989).

Unpleasant compared to neutral pictures (Johansson et al., 2004) and emotional compared to neutral words (Dietrich et al., 2001) elicit an enhanced old-new effect, suggesting that processes underlying recognition may be facilitated for affective stimuli. However, extreme arousal may compromise memory. Schaefer and colleagues found that compared to moderately arousing affective stimuli, stimuli both high and low in arousal elicited decreased old-new ERPs (Schaefer et al., 2009). Koenig and Mecklinger (2008) also found that only moderately arousing positive stimuli elicited increased slow wave activity during encoding and enhanced memory during a subsequent recognition test.

The tendency for stimuli that are correctly remembered compared to those that are later forgotten to elicit more positive ERPs during *encoding* is referred to as the *subsequent memory effect* (Paller et al., 1987). Dolcos and Cabeza (2002) found that during encoding, correctly remembered emotional compared to neutral pictures elicited larger centroparietal positivities from 400 to 600 ms following stimulus onset, suggesting that the increased processing elicited by these stimuli may have facilitated later recall.

Individual Differences/Psychopathology

Event-related potentials have been used to index abnormal emotional processing in a variety of psychopathologies including anxiety, depression, and schizophrenia. Relative to functional neuroimaging, the measurement of ERPs is potentially better tolerated in patient samples and will likely remain less expensive for decades to come. Recent evidence suggests that substantial variation in the emotional modulation of at least some ERP components (e.g., the P300) is heritable (Anokhin et al., 2010), lending support to the notion that ERP indices of emotional processing might serve as biomarkers of risk for certain forms of psychopathology. To the extent that the time course of emotional processing is important for understanding psychopathology, ERPs may be an ideal measure for translational work in affective neuroscience.

Anxiety

The majority of ERP work on anxiety has used briefly presented stimuli (on the order of 150–600 ms) in the context of behavioral tasks that require a response (e.g., cue-target paradigms). Consistent with behavioral work indicating an attentional bias toward threatening stimuli among anxious individuals (Bar-Haim et al., 2007), threatening pictures have been found to elicit a larger P1 (Li et al., 2005), N2pc (Fox et al., 2008), P2 (Bar-Haim et al., 2005), and LPP (MacNamara & Hajcak, 2009, 2010; see also Holmes et al., 2009) among high- compared to low-anxious participants. In several of these studies, increased ERPs to threatening stimuli among anxious participants were observed despite the absence of between-group behavioral effects as a function of emotion (Bar-Haim et al., 2005; Holmes et al., 2009; MacNamara & Hajcak, 2009). Event-related potentials may provide a more sensitive index of attentional biases toward threatening stimuli than behavioral measures.

Specific forms of anxiety and fear, including phobias, have also been examined using ERPs. Phobic individuals have faster reaction times than controls when asked to search for their feared stimuli among distracters (Flykt & Caldara, 2006; Öhman et al., 2001), and ERP studies support the notion that phobic individuals attend preferentially to their feared stimuli. In comparison to control subjects, phobic individuals exhibit larger EPNs (Michalowski et al., 2009; van Strien et al., 2009), P300s (Schienle et al., 2008), and LPPs (Flykt & Caldara, 2006; Michalowski et al., 2009; Schienle et al., 2008) to their feared stimuli.

Event-related potential work further suggests that attentional biases in phobias primarily impact later processing of feared stimuli. Phobic patients showed increased P1s to all stimuli, suggesting greater overall vigilance compared to controls, although later components (i.e., the LPP) differentiated phobic and control participants in terms of response to feared stimuli in particular (Michalowski et al., 2009). In another study, P1s and N1s did not differ for phobic participants and controls, although the LPP to feared stimuli was larger for phobic compared to control participants (Flykt & Caldara, 2006). In addition, ERPs have been used to examine the neural processing of feared stimuli in spider-phobic individuals following treatment using cognitive-behavioral therapy (CBT). Phobic individuals exhibited larger P300s and LPPs to their feared stimuli in comparison to nonphobic individuals prior to treatment; after treatment, however, the LPP to feared stimuli was *enhanced* for treated phobics in comparison to phobic participants who did not receive treatment (Leutgeb et al., 2009). Though seemingly counterintuitive, CBT treatment for specific phobias involves reducing avoidance. Thus, it is possible that treated individuals subsequently engaged in less avoidance of their feared stimuli—and less avoidance would likely be characterized by an increased LPP.

Some ERP studies suggest that anxiety is associated with the increased processing of emotional stimuli in general, not just threatening stimuli (e.g., Sass et al., 2009). Several studies have demonstrated that socially anxious individuals exhibit larger P1 amplitudes to both neutral and emotional faces (Kolassa et al., 2007, 2009). Thus, even innocuous social stimuli may elicit increased processing in social anxiety, and this increased processing can be evident quite early after stimulus onset. Moreover, greater P1 amplitudes for all face types were evident despite the fact that socially anxious participants rated angry faces as more arousing than controls (Kolassa et al., 2009). Event-related potentials may reveal processing biases of social stimuli that are not apparent in self-report ratings.

As suggested earlier, event-related potentials may be most useful in shedding light on the *time course* of attentional biases in anxiety. The timing of attentional biases can be difficult to index with behavioral or self-report measures alone. Initial work suggests that some forms of anxiety (e.g., social anxiety and generalized anxiety disorder, GAD) may be associated with increased processing of threatening stimuli soon after stimulus presentation, followed by strategic avoidance of threatening stimuli at later processing stages (Holmes et al., 2008; Mercado et al., 2009). This vigilance-avoidance model would predict increased early ERPs and attenuated later ERPs in response to emotional stimuli as a function of individual differences in anxiety. Consistent with this possibility, larger EPNs for angry and fearful faces have been found among high compared to low socially anxious individuals; however, the LPP was characterized by emotional enhancement for low, but not high, socially anxious individuals (Muhlberger et al., 2009). Weinberg and Hajcak (2011) also examined the early and late processing of emotional and neutral stimuli using a passive viewing paradigm in individuals with GAD and controls. Individuals with GAD had larger N1s yet smaller LPPs for unpleasant compared to neutral stimuli. Moreover, larger N1s predicted smaller LPPs, but only in the GAD group. Together, these studies suggest that certain kinds of anxiety may be characterized by early hypervigilance for threat and later avoidance of emotionally arousing stimuli.

Depression

Depression is associated with decreased P300s in auditory oddball tasks, suggesting the presence of an overall attentional deficit (see Chapter 20, this volume). Two studies employed an affective oddball paradigm among depressed participants. Using neutral faces as standards and both happy and fearful faces as targets, Cavanagh and Geisler (2006) found that compared to controls, depressed individuals had smaller P300s to happy targets. Using negative and neutral words, however, Ilardi and colleagues (2007) found that compared to controls, depressed participants had increased P300s to negative oddball stimuli. Thus, results from the two affective oddball paradigms in depression are mixed: one study suggests reduced processing of pleasant stimuli, whereas the other indicates increased processing of unpleasant stimuli in depression.

Indeed, there is debate regarding whether depression is better characterized by abnormal processing of appetitive stimuli specifically or emotional stimuli more generally. The positive attenuation hypothesis asserts that depression involves decreased reactivity to pleasant stimuli (Clark & Watson, 1991). By this account, the processing of negative stimuli should be intact or even increased in depression. The emotion context insensitivity (ECI) hypothesis, by contrast, suggests that depression may involve decreased responsivity to both pleasant *and* unpleasant stimuli (Rottenberg et al., 2005).

Studies that utilize passive viewing tasks are generally consistent with the ECI hypothesis and have reported reduced emotional modulation of ERPs such as the LPP in depression (Foti et al., 2010; Kayser et al., 2000; Williams et al., 2007). In one study, both individuals with major depressive disorder and healthy controls had increased VPPs in response to angry and fearful compared to neutral faces; however, only healthy controls were characterized by an increased LPP to angry and fearful faces (Foti et al., 2010). Consistent results indicating reduced differentiation between emotional and neutral stimuli in depression in the context of memory tasks have been reported (Deldin et al., 2001, 2009; Deveney & Deldin, 2004). Overall, these results suggest that depressed participants may exhibit blunted emotional processing particularly at later stages of processing.

Schizophrenia

Schizophrenia is associated with a variety of negative symptoms such as anhedonia, avolition, and asociality. Impairments in the processing and discrimination of facial emotions also characterize schizophrenia (e.g., Mandal et al., 1998; Morrison et al., 1988) and have been associated with increased symptom severity (Kohler et al., 2000). Whether deficits in the processing of facial stimuli in schizophrenia stem from general cognitive deficits that affect the early perceptual processing of faces, or whether they represent more specific emotion-related deficits, has been a recent research focus explored using ERPs.

Several studies have found that facial stimuli elicit decreased N170 amplitudes among schizophrenic patients (Johnston et al., 2005; Lee et al., 2010; Turetsky et al., 2007), indicating some abnormality in discriminating facial stimuli. Other research, however, suggests difficulties in later, higher-order processes such as the discernment and processing of facial *expression*: An and colleagues (2003) found reduced P300s to negative faces among schizophrenic compared to control participants. Work examining *both* the N170 and the P300, however, suggests that group differences in higher-order processing of faces (i.e., the P300) may be attributable to earlier differences in the N170 (i.e., a "flow-on" effect; Johnston et al., 2005; Turetsky et al., 2007). By using IAPS pictures instead of emotional faces, other work has suggested deficits in the sustained processing of emotional stimuli in schizophrenic individuals. Specifically, schizophrenics exhibited intact emotional modulation of earlier components (P1, P2, and P3) but reduced emotional modulation of the LPP (Horan et al., 2010). Although the available evidence is mixed, examining emotional modulation of ERP components may be a promising strategy for delineating processes that are both compromised and intact among individuals with schizophrenia—particularly in regards to specific types of stimuli (i.e., facial stimuli compared to non-facial stimuli).

Development and Aging

Because ERPs are relatively noninvasive and because they do not rely on introspection, they may provide an ideal means of examining emotional response across the lifespan. Studies have used ERPs to investigate: infants' ability to discriminate emotional expressions; changes in emotional processing throughout childhood and adolescence; emotion regulation abilities in children; and emotional processing in old age.

Infants and Children

Studies in infancy have yielded mixed results regarding the discrimination of facial expressions of emotion. Several studies have investigated the negative central (Nc) ERP component, a negative-going amplitude occurring approximately 400 ms after stimulus onset that is maximal at frontal and central sites and is believed to index attention and orienting in infants (Courchesne, 1977; Richards, 2003). Nelson and DeHaan (1996) reported that among 7-month-olds, the Nc distinguished between happy and fearful faces but not between fearful and angry faces. By contrast, Kobiella and colleagues (2008) found that angry compared to fearful faces elicited larger Ncs in 7-month-olds. Stahl and colleagues (2010) reported that among 6-month-olds, neutral compared to angry faces elicited larger Nc amplitudes. Thus, more work is needed in this area to determine whether and how specific facial expressions differentiate electrocortical activity in infants.

Event-related potentials have also been used to examine developmental changes in the processing of emotional stimuli throughout childhood and into adolescence. For example, P1 latency has been shown to vary by facial expression for younger (i.e., 4- to 7-year-olds) but not older children (i.e., 8- to 15-year-olds); N170 amplitudes may begin to be modulated by emotion beginning in early adolescence (14- to 15-year-olds), suggesting that the processing of emotional stimuli changes throughout childhood and into adolescence (Batty & Taylor, 2006). In addition, ERP amplitudes evoked by emotional stimuli may be related to emotional regulation abilities in children: Dennis and colleagues (2009) found that larger Nc and P1 amplitudes to fearful and sad faces in 5- to 9-year-olds were associated with better attentional control and fewer maternal reports of emotion dysregulation.

Most studies of emotion in children have examined the processing of facial expressions (e.g., fearful versus angry). Using age-appropriate IAPS pictures, Hajcak and Dennis (2009) found that the LPP but not the EPN was larger for emotional compared to neutral pictures in children 5–8 years old. These data further suggest that important developmental changes in emotional processing from childhood through adolescence might be better understood using ERPs.

Top-down modulations of emotional salience also appear to affect the LPP to emotional and neutral pictures in children: Dennis and Hajcak (2009) found that negatively compared to neutrally described IAPS pictures elicited larger LPPs in 5- to 10-year-olds, though this effect was evidenced at later latencies than in adults. Moreover, greater top-down modulation of the LPP in this study was associated with fewer symptoms of anxiety and depression; larger LPPs, on the other hand, were associated with more mood symptoms and poorer parental reports of emotional regulation abilities (Dennis & Hajcak, 2009), suggesting that increases in the elaborated processing of emotional stimuli may signal decreased emotional well-being in children of this age.

Older Adults

A variety of studies have suggested that older adults may experience relatively less negative, and possibly more positive, emotion than younger adults (see Mather & Carstensen, 2005). For example, in comparison to younger adults, older adults show a positivity bias when recalling and recognizing emotional pictures (Charles et al., 2003); they spend less time looking at negative faces (Rösler et al., 2005) and spend more time considering the positive compared to negative aspects of a decision (Mather et al., 2005). Work using ERPs has been consistent with the notion that older age is associated with changes in the way emotional stimuli are processed. Wood and Kisley (2006) found that pleasant and unpleasant compared to neutral IAPS images elicited smaller LPPs in older (56–81 years) compared to younger (19–22 years) adults; in addition, older adults were uniquely characterized by the absence of a negativity bias. There were no differences in earlier components (the P2), suggesting that age-related differences may be specific to later-latency components reflecting sustained attention to emotional stimuli. A follow-up study by Kisley and colleagues (2007) examined the LPP to emotional and neutral pictures in adults from 18 to 81 years of age and found that decreases in the negativity bias with advancing age could be attributed to the reduced processing of negative stimuli. Along similar lines, Langeslag and van Strien (2009) found evidence of a negativity bias in younger but not older adults, as indexed by the LPP and memory performance. In sum, these studies support the notion that sustained attention to negative stimuli declines with advancing age.

Conclusions and Future Directions

Many ERP components are sensitive to emotional compared to neutral visual stimuli, and these ERPs directly reflect underlying neural variation in response to emotional stimuli. We do not believe that *any* neural activity *is* an emotion (or is "emotional"). A racing heart *can* be part of a coordinated emotional response, but not always. Rather, emotion describes the activation of an organism when confronted by motivationally salient stimuli. Attention, perception and memory are some of the processes that are increased in emotion—and variation in these processes can be measured across a range of ERPs.

Some researchers present emotional stimuli in blocks (e.g., all unpleasant pictures together), whereas others employ randomized presentation or modified oddball paradigms. It is not clear whether or how such task-related differences impact ERPs to emotional stimuli. Though we have focused on the temporal resolution of ERPs to highlight the *when* of emotional processing, less is known about the *how* and *where*. Based on recent theorizing on the P300 (Nieuwenhuis et al., 2005), we suggested that variation in the LPP may also reflect neuromodulatory activity of the locus coeruleus norepinephrine system (Hajcak et al., 2010b). This possibility might be substantiated by measuring ERPs in combination with a pharmacological challenge, (e.g., Kerestes et al., 2009; Labuschagne et al., 2010). In terms of neural generators of the LPP, Keil and colleagues (2002) estimated the neural source of the LPP in occipital and posterior parietal cortex. Consistent with these data, another study found that variation in the LPP related to neural activity in occipital, parietal, and inferotemporal regions of the brain using fMRI (Sabatinelli et al., 2007). Although the findings of these studies are generally consistent with the notion that the LPP indexes activity in parietal and occipital regions implicated in visual attention, understanding the circuitry that supports emotional modulation of ERPs requires more research that combines ERP and other neuroimaging techniques.

Along similar lines, though we know that the LPP can be altered by contextual, attentional, and other top-down factors, the neural systems that modulate the LPP are less clear. In a small clinical sample, we found that stimulating specific regions of the prefrontal cortex altered the amplitude of the LPP: the amplitude of the LPP was reduced while the prefrontal cortex was stimulated; this effect was unique to stimulation of the dorsolateral prefrontal cortex (DLPFC) and was not observed when the frontopolar prefrontal cortex was stimulated (Hajcak et al., 2010a). Although these results must be interpreted with care, they suggest that DLPFC activation may reduce the impact of emotional stimuli on ERPs.

For the foreseeable future, it is the temporal resolution of ERPs that is particularly appealing for research in emotion. It will be important for future studies to carefully parse the functional significance of ERP variation at multiple time points and to relate ERPs to other measures of emotional processing. This strategy will permit ERPs to inform research on emotion in an increasingly specific manner.

Acknowledgment

We thank Eddie Harmon-Jones, Andreas Keil, K Luan Phan, John Polich, and Robert F. Simons for perspicacious editorial suggestions.

References

Allison, T., Puce, A., Spencer, D., & McCarthy, G. (1999). Electrophysiological studies of human face perception. I: Potentials generated in occipitotemporal cortex by face and non-face stimuli. *Cerebral Cortex, 9*(5), 415–430.

Amrhein, C., Pauli, P., Dengler, W., & Wiedemann, G. (2005). Covariation bias and its physiological correlates in panic disorder patients. *Journal of Anxiety Disorders, 19*(2), 177–191.

An, S., Lee, S., Lee, C., Cho, H., Lee, P., Lee, C., et al. (2003). Reduced P3 amplitudes by negative facial emotional photographs in schizophrenia. *Schizophrenia Research, 64*(2–3), 125–135.

Anderson, A., & Phelps, E. (2001). Lesions of the human amygdala impair enhanced perception of emotionally salient events. *Nature, 411*(6835), 305–308.

Anokhin, A., Golosheykin, S., & Heath, A. (2010). Heritability of individual differences in cortical processing of facial affect. *Behavior Genetics, 40*(2), 178–185.

Anokhin, A., Golosheykin, S., Sirevaag, E., Kristjansson, S., Rohrbaugh, J., & Heath, A. (2006). Rapid discrimination of visual scene content in the human brain. *Brain Research, 1093*(1), 167–177.

Armony, J., & Dolan, R. (2002). Modulation of spatial attention by fear-conditioned stimuli: An event-related fMRI study. *Neuropsychologia, 40*(7), 817–826.

Azizian, A., & Polich, J. (2007). Evidence for attentional gradient in the serial position memory curve from event-related potentials. *Journal of Cognitive Neuroscience, 19*(12), 2071–2081.

Bar-Haim, Y., Lamy, D., & Glickman, S. (2005). Attentional bias in anxiety: A behavioral and ERP study. *Brain and Cognition, 59*(1), 11–22.

Bar-Haim, Y., Lamy, D., Pergamin, L., Bakermans-Kranenburg, M., & van IJzendoorn, M. (2007). Threat-related attentional bias in anxious and nonanxious individuals: A meta-analytic study. *Psychological Bulletin, 133*(1), 1–24.

Batty, M., & Taylor, M. (2003). Early processing of the six basic facial emotional expressions. *Cognitive Brain Research, 17*(3), 613–620.

Batty, M., & Taylor, M. (2006). The development of emotional face processing during childhood. *Developmental Science, 9*(2), 207–220.

Bechara, A., Damasio, H., & Damasio, A. R. (2000). Emotion, decision making and the orbitofrontal cortex. *Cerebral Cortex, 10*(3), 295–307.

Begleiter, H., Porjesz, B., & Garozzo, R. (1979). Visual evoked potentials and affective ratings of semantic stimuli. *Evoked Brain Potentials and Behavior*, 127–141.

Bentin, S., Allison, T., Puce, A., Perez, E., & McCarthy, G. (1996). Electrophysiological studies of face perception in human. *Journal of Cognitive Neuroscience, 8*(6), 551–565.

Bentin, S., Taylor, M., Rousselet, G., Itier, R., Caldara, R., Schyns, P., et al. (2007). Controlling interstimulus perceptual variance does not abolish N170 face sensitivity. *Nature Neuroscience, 10*(7), 801–802.

Bernat, E., Bunce, S., & Shevrin, H. (2001). Event-related brain potentials differentiate positive and negative mood adjectives during both supraliminal and subliminal visual processing. *International Journal of Psychophysiology, 42*(1), 11–34.

Berridge, K. C. (2003). Pleasures of the brain. *Brain Cogn, 52*(1), 106–128.

Blau, V., Maurer, U., Tottenham, N., & McCandliss, B. (2007). The face-specific N170 component is modulated by emotional facial expression. *Behavioral and Brain Functions, 3*(7),

Bradley, M. (2000). Emotion and motivation. In J. Cacioppo, L. Tassinary, & G. Berntson (Eds.), *Handbook of psychophysiology* (2nd ed., pp. 602–642). Cambridge: Cambridge University Press.

Bradley, M., Codispoti, M., Cuthbert, B., & Lang, P. (2001). Emotion and motivation I: Defensive and appetitive reactions in picture processing. *Emotion, 1*(3), 276–298.

Bradley, M., Greenwald, M., Petry, M., & Lang, P. (1992). Remembering pictures: Pleasure and arousal in memory. *Journal of Experimental Psychology: Learning, Memory, and Cognition, 18*(2), 379–390.

Bradley, M., Hamby, S., Löw, A., & Lang, P. (2007). Brain potentials in perception: Picture complexity and emotional arousal. *Psychophysiology, 44*(3), 364–373.

Bradley, M., & Lang, P. (1999). *International affective digitized sounds (IADS): Stimuli, instruction manual and affective ratings.* Technical Report B-2. Gainesville, FL: Center for Research in Psychophysiology, University of Florida.

Bradley, M., & Lang, P. (2000). Measuring emotion: Behavior, feeling, and physiology. *Cognitive Neuroscience of Emotion, 25*, 49–59.

Briggs, K., & Martin, F. (2009). Affective picture processing and motivational relevance: Arousal and valence effects on ERPs in an oddball task. *International Journal of Psychophysiology, 72*(3), 299–306.

Britton, J., Taylor, S., Sudheimer, K., & Liberzon, I. (2006). Facial expressions and complex IAPS pictures: Common and differential networks. *Neuroimage, 31*(2), 906–919.

Brunia, C. (1988). Movement and stimulus preceding negativity. *Biological Psychology, 26*(1–3), 165–178.

Brunia, C., & Damen, E. (1988). Distribution of slow brain potentials related to motor preparation and stimulus anticipation in a time estimation task. *Electroencephalography and Clinical Neurophysiology, 69*(3), 234–243.

Buchanan, T., & Adolphs, R. (2002). The role of the human amygdala in emotional modulation of long-term declarative memory. In S. Moore & M. Oaksford (Eds.), *Emotional cognition: From brain to behaviour* (pp. 9–34). London: Benjamins.

Cacioppo, J., Crites, S., Jr., Berntson, G., & Coles, M. (1993). If attitudes affect how stimuli are processed, should they not affect the event-related brain potential? *Psychological Science, 4*(2), 108–112.

Cacioppo, J., Crites, S., Gardner, W., & Berntson, G. (1994). Bioelectrical echoes from evaluative categorizations. I: A late positive brain potential that varies as a function of trait negativity and extremity. *Journal of Personality and Social Psychology, 67*(1), 115–125.

Cacioppo, J., Gardner, W., & Berntson, G. (1999). The affect system has parallel and integrative processing components: Form follows function. *Journal of Personality and Social Psychology, 76*(5), 839–855.

Caharel, S., Courtay, N., Bernard, C., Lalonde, R., & Rebai, M. (2005). Familiarity and emotional expression influence an early stage of face processing: An electrophysiological study. *Brain and Cognition, 59*(1), 96–100.

Calvo, M., & Lang, P. (2004). Gaze patterns when looking at emotional pictures: Motivationally biased attention. *Motivation and Emotion, 28*(3), 221–243.

Carretié, L., Hinojosa, J., Albert, J., & Mercado, F. (2006). Neural response to sustained affective visual stimulation using an indirect task. *Experimental Brain Research, 174*(4), 630–637.

Carretié, L., Hinojosa, J., López-Martín, S., & Tapia, M. (2007). An electrophysiological study on the interaction between emotional content and spatial frequency of visual stimuli. *Neuropsychologia, 45*(6), 1187–1195.

Carretié, L., Hinojosa, J., Martín-Loeches, M., Mercado, F., & Tapia, M. (2004). Automatic attention to emotional stimuli: Neural correlates. *Human Brain Mapping, 22*(4), 290–299.

Carretié, L., Hinojosa, J., & Mercado, F. (2003). Cerebral patterns of attentional habituation to emotional visual stimuli. *Psychophysiology, 40*(3), 381–388.

Carretié, L., Mercado, F., Tapia, M., & Hinojosa, J. (2001). Emotion, attention, and the "negativity bias," studied through event-related potentials. *International Journal of Psychophysiology, 41*(1), 75–85.

Cavanagh, J., & Geisler, M. (2006). Mood effects on the ERP processing of emotional intensity in faces: A P3 investigation with depressed students. *International Journal of Psychophysiology, 60*(1), 27–33.

Charles, S., Mather, M., & Carstensen, L. (2003). Aging and emotional memory: The forgettable nature of negative images for older adults. *Journal of Experimental Psychology: General, 132*(2), 310–324.

Christianson, S. (1992). *The handbook of emotion and memory: Research and theory.* Hillsdale, NJ: Erlbaum.

Christie, I., & Friedman, B. (2004). Autonomic specificity of discrete emotion and dimensions of affective space: A multivariate approach. *International Journal of Psychophysiology, 51*(2), 143–153.

Clark, L., & Watson, D. (1991). Tripartite model of anxiety and depression: Psychometric evidence and taxonomic implications. *Journal of Abnormal Psychology, 100*(3), 316–336.

Codispoti, M., Bradley, M., & Lang, P. (2001). Affective reactions to briefly presented pictures. *Psychophysiology, 38*(3), 474–478.

Codispoti, M., & De Cesarei, A. (2007). Arousal and attention: Picture size and emotional reactions. *Psychophysiology, 44*(5), 680–686.

Codispoti, M., Ferrari, V., & Bradley, M. (2006a). Repetitive picture processing: Autonomic and cortical correlates. *Brain Research, 1068*(1), 213–220.

Codispoti, M., Ferrari, V., & Bradley, M. (2007). Repetition and event-related potentials: Distinguishing early and late processes in affective picture perception. *Journal of Cognitive Neuroscience, 19*(4), 577–586.

Codispoti, M., Ferrari, V., Junghöfer, M., & Schupp, H. (2006b). The categorization of natural scenes: Brain attention networks revealed by dense sensor ERPs. *Neuroimage, 32*(2), 583–591.

Courchesne, E. (1977). Event-related brain potentials: Comparison between children and adults. *Science, 197*(4303), 589–592.

Crites, S., Jr., Cacioppo, J., Gardner, W., & Berntson, G. (1995). Bioelectrical echoes from evaluative categorization: II. A late positive brain potential that varies as a function of attitude registration rather than attitude report. *Journal of Personality and Social Psychology, 68*(6), 997–1013.

Cuthbert, B., Schupp, H., Bradley, M., Birbaumer, N., & Lang, P. (2000). Brain potentials in affective picture processing: Covariation with autonomic arousal and affective report. *Biological Psychology, 52*(2), 95–111.

Damasio, A. R. (1996). The somatic marker hypothesis and the possible functions of the prefrontal cortex. *Philosophical Transactions of the Royal Society of London B: Biological Science, 351*(1346), 1413–1420.

Darwin, C. (1872). *The expression of the emotions in man and animals; with an introduction, afterword, and commentaries by Paul Ekman.* New York: Oxford University Press.

Davidson, R. (1993). Parsing affective space: Perspectives from neuropsychology and psychophysiology. *Neuropsychology, 7*(4), 464–475.

Davidson, R. (1998). Affective style and affective disorders: Perspectives from affective neuroscience. *Cognition & Emotion, 12*(3), 307–330.

Davidson, R. (2003). Seven sins in the study of emotion: Correctives from affective neuroscience. *Brain and Cognition, 52*(1), 129–132.

Davidson, R. (2008). Asymmetric brain function, affective style, and psychopathology: The role of early experience and plasticity. *Development and Psychopathology, 6*(4), 741–758.

Davidson, R., Ekman, P., Saron, C., Senulis, J., & Friesen, W. (1990). Approach-withdrawal and cerebral asymmetry: Emotional expression and brain physiology I. *Journal of Personality and Social Psychology, 58*(2), 330–341.

De Cesarei, A., & Codispoti, M. (2006). When does size not matter? Effects of stimulus size on affective modulation. *Psychophysiology, 43*(2), 207–215.

Deldin, P., Keller, J., Gergen, J., & Miller, G. (2001). Cognitive bias and emotion in neuropsychological models of depression. *Cognition & Emotion, 15*(6), 787–802.

Deldin, P., Naidu, S., Shestyuk, A., & Casas, B. (2009). Neurophysiological indices of free recall memory biases in major depression: The impact of stimulus arousal and valence. *Cognition & Emotion, 23*(5), 1002–1020.

Delplanque, S., Lavoie, M., Hot, P., Silvert, L., & Sequeira, H. (2004). Modulation of cognitive processing by emotional valence studied through event-related potentials in humans. *Neuroscience Letters, 356*(1), 1–4.

Delplanque, S., Silvert, L., Hot, P., Rigoulot, S., & Sequeira, H. (2006). Arousal and valence effects on event-related P3a and P3b during emotional categorization. *International Journal of Psychophysiology, 60*(3), 315–322.

Delplanque, S., Silvert, L., Hot, P., & Sequeira, H. (2005). Event-related P3a and P3b in response to unpredictable emotional stimuli. *Biological Psychology, 68*(2), 107–120.

Dennis, T., & Hajcak, G. (2009). The late positive potential: A neurophysiological marker for emotion regulation in children. *Journal of Child Psychology and Psychiatry, 50*(11), 1373–1383.

Dennis, T., Malone, M., & Chen, C. (2009). Emotional face processing and emotion regulation in children: An ERP study. *Developmental Neuropsychology, 34*(1), 85–102.

Deveney, C., & Deldin, P. (2004). Memory of faces: A slow wave ERP study of major depression. *Emotion, 4*, 295–304.

Dien, J. (2010a). The ERP PCA toolkit: An open source program for advanced statistical analysis of event-related potential data. *Journal of Neuroscience Methods, 187*(1), 138–145.

Dien, J. (2010b). Evaluating two-step PCA of ERP data with Geomin, Infomax, Oblimin, Promax, and Varimax rotations. *Psychophysiology, 47*(1), 170–183.

Dien, J., Beal, D., & Berg, P. (2005). Optimizing principal components analysis of event-related potentials: Matrix type, factor loading weighting, extraction, and rotations. *Clinical Neurophysiology, 116*(8), 1808–1825.

Dien, J., & Frishkoff, G. (2005). Principal components analysis of event-related potential datasets. In T. C. Handy (Ed.),

Event-related potentials: A methods handbook (pp. 189–208). Cambridge, MA: MIT Press.

Dien, J., Khoe, W., & Mangun, G. (2007). Evaluation of PCA and ICA of simulated ERPs: Promax vs. Infomax rotations. *Human Brain Mapping, 28*(8), 742–763.

Dietrich, D., Waller, C., Johannes, S., Wieringa, B., Emrich, H., & Münte, T. (2001). Differential effects of emotional content on event-related potentials in word recognition memory. *Neuropsychobiology, 43*(2), 96–101.

Dolcos, F., & Cabeza, R. (2002). Event-related potentials of emotional memory: Encoding pleasant, unpleasant, and neutral pictures. *Cognitive, Affective, and Behavioral Neuroscience, 2*(3), 252–263.

Donchin, E., & Heffley, E. (1978). Multivariate analysis of event-related potential data: A tutorial review. In D. Otto (Ed.), *Multidisciplinary perspectives in event-related brain potential research* (pp. 555–572). Washington, DC: Environmental Protection Agency.

Duncan-Johnson, C., & Donchin, E. (1977). On quantifying surprise: The variation of even related potentials of schizophrenics, patients with Parkinson's disease and controls. *Psychophysiology, 14*(5), 456–467.

Dunning, J., & Hajcak, G. (2009). See no evil: Directing visual attention within unpleasant images modulates the electrocortical response. *Psychophysiology, 46*(1), 28–33.

Dunning, J., Parvaz, M., Hajcak, G., Maloney, T., Alia-Klein, N., Woicik, P., et al. (2011). Motivated attention to cocaine and emotional cues in abstinent and current cocaine users: An ERP study. *European Journal of Neuroscience, 33*(9), 1716–1723.

Eimer, M., & Holmes, A. (2007). Event-related brain potential correlates of emotional face processing. *Neuropsychologia, 45*(1), 15–31.

Eimer, M., Holmes, A., & McGlone, F. (2003). The role of spatial attention in the processing of facial expression: An ERP study of rapid brain responses to six basic emotions. *Cognitive, Affective, and Behavioral Neuroscience, 3*(2), 97–110.

Ekman, P. (1992). An argument for basic emotions. *Cognition & Emotion, 6*(3), 169–200.

Esteves, F., Parra, C., Dimberg, U., & Ohman, A. (1994). Nonconscious associative learning: Pavlovian conditioning of skin conductance responses to masked fear-relevant facial stimuli. *Psychophysiology, 31*(4), 375–385.

Ferrari, V., Codispoti, M., Cardinale, R., & Bradley, M. (2008). Directed and motivated attention during processing of natural scenes. *Journal of Cognitive Neuroscience, 20*(10), 1753–1761.

Flaisch, T., Junghöfer, M., Bradley, M., Schupp, H., & Lang, P. (2008a). Rapid picture processing: Affective primes and targets. *Psychophysiology, 45*(1), 1–10.

Flaisch, T., Stockburger, J., & Schupp, H. (2008b). Affective prime and target picture processing: An ERP analysis of early and late interference effects. *Brain Topography, 20*(4), 183–191.

Flykt, A., & Caldara, R. (2006). Tracking fear in snake and spider fearful participants during visual search: A multi-response domain study. *Cognition & Emotion, 20*(8), 1075–1091.

Foti, D., & Hajcak, G. (2008). Deconstructing reappraisal: Descriptions preceding arousing pictures modulate the subsequent neural response. *Journal of Cognitive Neuroscience, 20*(6), 977–988.

Foti, D., Hajcak, G., & Dien, J. (2009). Differentiating neural responses to emotional pictures: Evidence from temporal-spatial PCA. *Psychophysiology, 46*(3), 521–530.

Foti, D., Olvet, D., Klein, D., & Hajcak, G. (2010). Reduced electrocortical response to threatening faces in major depressive disorder. *Depression and Anxiety, 27*(9), 813–820.

Fox, E., Derakshan, N., & Shoker, L. (2008). Trait anxiety modulates the electrophysiological indices of rapid spatial orienting towards angry faces. *Neuroreport, 19*(3), 259–263.

Franken, I., Muris, P., Nijs, I., & van Strien, J. (2008). Processing of pleasant information can be as fast and strong as unpleasant information: Implications for the negativity bias. *Netherlands Journal of Psychology, 64*(4), 168–176.

Gross, J., & Thompson, R. (2007). Emotion regulation: Conceptual foundations. In J.J. Gross (Ed.), *Handbook of emotion regulation* (pp. 3–24). New York, NY: Guilford.

Grossmann, T., Striano, T., & Friederici, A. (2007). Developmental changes in infants' processing of happy and angry facial expressions: a neurobehavioral study. *Brain Cognition, 64*(1), 30–41.

Gur, R., Sara, R., Hagendoorn, M., Marom, O., Hughett, P., Macy, L., et al. (2002). A method for obtaining 3-dimensional facial expressions and its standardization for use in neurocognitive studies. *Journal of Neuroscience Methods, 115*(2), 137–143.

Guthrie, D., & Buchwald, J. (1991). Significance testing of difference potentials. *Psychophysiology, 28*(2), 240–244.

Hajcak, G., Anderson, B., Arana, A., Borckardt, J., Takacs, I., George, M., et al. (2010a). Dorsolateral prefrontal cortex stimulation modulates electrocortical measures of visual attention: Evidence from direct bilateral epidural cortical stimulation in treatment-resistant mood disorder. *Neuroscience, 170*(1), 281–288.

Hajcak, G., & Dennis, T. (2009). Brain potentials during affective picture processing in children. *Biological Psychology, 80*(3), 333–338.

Hajcak, G., Dunning, J., & Foti, D. (2007). Neural response to emotional pictures is unaffected by concurrent task difficulty: An event-related potential study. *Behavioral Neuroscience, 121*(6), 1156–1162.

Hajcak, G., Dunning, J., & Foti, D. (2009). Motivated and controlled attention to emotion: Time-course of the late positive potential. *Clinical Neurophysiology, 120*(3), 505–510.

Hajcak, G., MacNamara, A., & Olvet, D. (2010b). Event-related potentials, emotion, and emotion regulation: An integrative review. *Developmental Neuropsychology, 35*(2), 129–155.

Hajcak, G., Moser, J., & Simons, R. (2006). Attending to affect: Appraisal strategies modulate the electrocortical response to arousing pictures. *Emotion, 6*(3), 517–522.

Hajcak, G., & Nieuwenhuis, S. (2006). Reappraisal modulates the electrocortical response to unpleasant pictures. *Cognitive, Affective and Behavioral Neuroscience, 6*(4), 291–297.

Hajcak, G., & Olvet, D. (2008). The persistence of attention to emotion: Brain potentials during and after picture presentation. *Emotion, 8*(2), 250–255.

Hamann, S. (2001). Cognitive and neural mechanisms of emotional memory. *Trends in Cognitive Sciences, 5*(9), 394–400.

Hamann, S., Cahill, L., McGaugh, J., & Squire, L. (1997). Intact enhancement of declarative memory for emotional material in amnesia. *Learning & Memory, 4*(3), 301–309.

Harmon-Jones, E., Lueck, L., Fearn, M., & Harmon-Jones, C. (2006). The effect of personal relevance and approach-related action expectation on relative left frontal cortical activity. *Psychological Science, 17*(5), 434–440.

Herbert, C., Junghöfer, M., & Kissler, J. (2008). Event related potentials to emotional adjectives during reading. *Psychophysiology, 45*(3), 487–498.

Herbert, C., Kissler, J., Junghöfer, M., Peyk, P., & Rockstroh, B. (2006). Processing of emotional adjectives: Evidence from startle EMG and ERPs. *Psychophysiology, 43*(2), 197–206.

Herrmann, M., Huter, T., Muller, F., Muhlberger, A., Pauli, P., Reif, A., et al. (2007). Additive effects of serotonin transporter and tryptophan hydroxylase-2 gene variation on emotional processing. *Cerebral Cortex, 17*(5), 1160–1163.

Herrmann, M., Huter, T., Plichta, M., Ehlis, A., Alpers, G., Muhlberger, A., et al. (2008). Enhancement of activity of the primary visual cortex during processing of emotional stimuli as measured with event-related functional near-infrared spectroscopy and event-related potentials. *Human Brain Mapping, 29*(1), 28–35.

Hillyard, S., Hink, R., Schwent, V., & Picton, T. (1973). Electrical signs of selective attention in the human brain. *Science, 182*(4108), 177–180.

Holmes, A., Nielsen, M., & Green, S. (2008). Effects of anxiety on the processing of fearful and happy faces: An event-related potential study. *Biological Psychology, 77*(2), 159–173.

Holmes, A., Nielsen, M., Tipper, S., & Green, S. (2009). An electrophysiological investigation into the automaticity of emotional face processing in high versus low trait anxious individuals. *Cognitive, Affective and Behavioral Neuroscience, 9*(3), 323–334.

Holmes, A., Vuilleumier, P., & Eimer, M. (2003). The processing of emotional facial expression is gated by spatial attention: Evidence from event-related brain potentials. *Cognitive Brain Research, 16*(2), 174–184.

Holmes, A., Winston, J., & Eimer, M. (2005). The role of spatial frequency information for ERP components sensitive to faces and emotional facial expression. *Cognitive Brain Research, 25*(2), 508–520.

Horan, W., Wynn, J., Kring, A., Simons, R., & Green, M. (2010). Electrophysiological correlates of emotional responding in schizophrenia. *Journal of Abnormal Psychology, 119*(1), 18–30.

Hot, P., Saito, Y., Mandai, O., Kobayashi, T., & Sequeira, H. (2006). An ERP investigation of emotional processing in European and Japanese individuals. *Brain Research, 1122*(1), 171–178.

Howard, R., Longmore, F., & Mason, P. (1992). Contingent negative variation as an indicator of sexual object preference: Revisited. *International Journal of Psychophysiology, 13*(2), 185–188.

Huang, Y., & Luo, Y. (2006). Temporal course of emotional negativity bias: An ERP study. *Neuroscience Letters, 398*(1–2), 91–96.

Ilardi, S., Atchley, R., Enloe, A., Kwasny, K., & Garratt, G. (2007). Disentangling attentional biases and attentional deficits in depression: An event-related potential P300 analysis. *Cognitive Therapy and Research, 31*(2), 175–187.

Ito, T., & Cacioppo, J. (2000). Electrophysiological evidence of implicit and explicit categorization processes. *Journal of Experimental Social Psychology, 36*(6), 660–676.

Ito, T., Larsen, J., Smith, N., & Cacioppo, J. (1998). Negative information weighs more heavily on the brain: The negativity bias in evaluative categorizations. *Journal of Personality and Social Psychology, 75*(4), 887–900.

Izard, C., Bartlett, E., & Marshall, A. (1972). *Patterns of emotions: A new analysis of anxiety and depression.* New York: Academic Press.

Jeffreys, D. (1989). A face-responsive potential recorded from the human scalp. *Experimental Brain Research, 78*(1), 193–202.

Johansson, M., Mecklinger, A., & Treese, A. (2004). Recognition memory for emotional and neutral faces: An event-related potential study. *Journal of Cognitive Neuroscience, 16*(10), 1840–1853.

Johnston, P., Stojanov, W., Devir, H., & Schall, U. (2005). Functional MRI of facial emotion recognition deficits in schizophrenia and their electrophysiological correlates. *European Journal of Neuroscience, 22*(5), 1221–1232.

Johnston, V., Miller, D., & Burleson, M. (1986). Multiple P3s to emotional stimuli and their theoretical significance. *Psychophysiology, 23*(6), 684–694.

Joyce, C., & Rossion, B. (2005). The face-sensitive N170 and VPP components manifest the same brain processes: The effect of reference electrode site. *Clinical Neurophysiology, 116*(11), 2613–2631.

Junghöfer, M., Bradley, M., Elbert, T., & Lang, P. (2001). Fleeting images: A new look at early emotion discrimination. *Psychophysiology, 38*(2), 175–178.

Kahneman, D. (1999). Objective happiness. In D. Kahneman, E. Diener, & N. Schwarz (Eds.), *Well-being: The foundations of hedonic psychology* (pp. 3–25). New York: Russell Sage Foundation.

Kanske, P., & Kotz, S. (2007). Concreteness in emotional words: ERP evidence from a hemifield study. *Brain Research, 1148*, 138–148.

Kayser, J., Bruder, G., Tenke, C., Stewart, J., & Quitkin, F. (2000). Event-related potentials (ERPs) to hemifield presentations of emotional stimuli: Differences between depressed patients and healthy adults in P3 amplitude and asymmetry. *International Journal of Psychophysiology, 36*(3), 211–236.

Keil, A., Bradley, M., Hauk, O., Rockstroh, B., Elbert, T., & Lang, P. (2002). Large-scale neural correlates of affective picture processing. *Psychophysiology, 39*(5), 641–649.

Keil, A., Müller, M., Gruber, T., Wienbruch, C., Stolarova, M., & Elbert, T. (2001). Effects of emotional arousal in the cerebral hemispheres: A study of oscillatory brain activity and event-related potentials. *Clinical Neurophysiology, 112*(11), 2057–2068.

Kerestes, R., Labuschagne, I., Croft, R., O'Neill, B., Bhagwagar, Z., Phan, K., et al. (2009). Evidence for modulation of facial emotional processing bias during emotional expression decoding by serotonergic and noradrenergic antidepressants: An event-related potential (ERP) study. *Psychopharmacology, 202*(4), 621–634.

Kisley, M., Wood, S., & Burrows, C. (2007). Looking at the sunny side of life. *Psychological Science, 18*(9), 838–843.

Kissler, J., Assadollahi, R., & Herbert, C. (2006). Emotional and semantic networks in visual word processing: Insights from ERP studies. *Progress in Brain Research, 156*, 147–183.

Kissler, J., Herbert, C., Winkler, I., & Junghöfer, M. (2009). Emotion and attention in visual word processing: An ERP study. *Biological Psychology, 80*(1), 75–83.

Klorman, R., & Ryan, R. (1979). Heart rate, contingent negative variation, and evoked potentials during anticipation of affective stimulation. *Psychophysiology, 17*(6), 513–523.

Kobiella, A., Grossmann, T., Reid, V., & Striano, T. (2008). The discrimination of angry and fearful facial expressions in 7-month-old infants: An event-related potential study. *Cognition & Emotion, 22*(1), 134–146.

Koenig, S., & Mecklinger, A. (2008). Electrophysiological correlates of encoding and retrieving emotional events. *Emotion, 8*(2), 162–173.

Kohler, C., Bilker, W., Hagendoorn, M., Gur, R., & Gur, R. (2000). Emotion recognition deficit in schizophrenia: Association with symptomatology and cognition. *Biological Psychiatry, 48*(2), 127–136.

Kolassa, I., Kolassa, S., Bergmann, S., Lauche, R., Dilger, S., Miltner, W., et al. (2009). Interpretive bias in social phobia: An ERP study with morphed emotional schematic faces. *Cognition & Emotion, 23*(1), 69–95.

Kolassa, I., Kolassa, S., Musial, F., & Miltner, W. (2007). Event-related potentials to schematic faces in social phobia. *Cognition & Emotion, 21*(8), 1721–1744.

Krolak-Salmon, P., Fischer, C., Vighetto, A., & Mauguiere, F. (2001). Processing of facial emotional expression: Spatio-temporal data as assessed by scalp event related potentials. *European Journal of Neuroscience, 13*(5), 987–994.

Krompinger, J., Moser, J., & Simons, R. (2008). Modulations of the electrophysiological response to pleasant stimuli by cognitive reappraisal. *Emotion, 8*(1), 132–137.

Labuschagne, I., Croft, R., Phan, K., & Nathan, P. (2010). Augmenting serotonin neurotransmission with citalopram modulates emotional expression decoding but not structural encoding of moderate intensity sad facial emotional stimuli: An event-related potential (ERP) investigation. *Journal of Psychopharmacology, 24*(8), 1153–1164.

Lang, P. (1994). The motivational organization of emotion: Affect–reflex connections. In S. Van Goozen, N. E. Van de Poll, & J.A. Sergeant (Eds.), *Emotions: Essays on emotion theory* (pp. 61–93). Hillsdale, NJ: Erlbaum.

Lang, P., Bradley, M., & Cuthbert, B. (1997). Motivated attention: Affect, activation, and action. In P. J. Lang, R.F. Simons, & M.T. Balaban (Eds.), *Attention and orienting: Sensory and motivational processes*, (pp. 97–135). Hillsdale, NJ: Erlbaum.

Lang, P., Bradley, M., & Cuthbert, B. (1998). Emotion, motivation, and anxiety: Brain mechanisms and psychophysiology. *Biological Psychiatry, 44*(12), 1248–1263.

Lang, P., Bradley, M., & Cuthbert, B. (1999). *International Affective Picture System (IAPS): Instruction manual and affective ratings*. Gainesville, FL: Center for Research in Psychophysiology, University of Florida.

Lang, P., Bradley, M., & Cuthbert, B. (2005). *International Affective Picture System (IAPS): Affective ratings of pictures and instruction manual.* Gainesville FL: Center for Research in Psychophysiology, University of Florida.

Lang, P., Greenwald, M., Bradley, M., & Hamm, A. (1993). Looking at pictures: Affective, facial, visceral, and behavioral reactions. *Psychophysiology, 30*(3), 261–273.

Langeslag, S., & van Strien, J. (2009). Aging and emotional memory: The co-occurrence of neurophysiological and behavioral positivity effects. *Emotion, 9*(3), 369–377.

LeDoux, J. (1998). *The emotional brain: The mysterious underpinnings of emotional life*. New York: Touchstone Books.

Lee, S., Kim, E., Kim, S., & Bae, S. (2010). Event-related potential patterns and gender effects underlying facial affect processing in schizophrenia patients. *Neuroscience Research, 67*(2), 172–180.

Leutgeb, V., Schäfer, A., & Schienle, A. (2009). An event-related potential study on exposure therapy for patients suffering from spider phobia. *Biological Psychology, 82*(3), 293–300.

Li, X., Li, X., & Luo, Y. (2005). Anxiety and attentional bias for threat: An event-related potential study. *Neuroreport, 16*(13), 1501–1505.

Lifshitz, K. (1966). The averaged evoked cortical response to complex visual stimuli. *Psychophysiology, 3*(1), 55–68.

Loveless, N., & Sanford, A. (1974a). Effects of age on the contingent negative variation and preparatory set in a reaction-time task. *Journal of Gerontology, 29*(1), 52–63.

Loveless, N., & Sanford, A. (1974b). Slow potential correlates of preparatory set. *Biological Psychology, 1*(4), 303–314.

Luck, S. (2005). *An introduction to the event-related potential technique* Cambridge, MA: MIT Press.

Luck, S., & Hillyard, S. (1994). Electrophysiological correlates of feature analysis during visual search. *Psychophysiology, 31*(3), 291–308.

Mack, A., & Rock, I. (1998). Inattentional blindness: Perception without attention. In R. D. Wright (Ed.), *Visual Attention*, (pp. 55–76). Oxford: Oxford University Press.

MacNamara, A., Foti, D., & Hajcak, G. (2009). Tell me about it: Neural activity elicited by emotional stimuli and preceding descriptions. *Emotion, 9*(4), 531–543.

MacNamara, A., & Hajcak, G. (2009). Anxiety and spatial attention moderate the electrocortical response to aversive pictures. *Neuropsychologia, 47*(13), 2975–2980.

MacNamara, A., & Hajcak, G. (2010). Distinct electrocortical and behavioral evidence for increased attention to threat in generalized anxiety disorder. *Depression and Anxiety, 27*(3), 234–243.

MacNamara, A., Ochsner, K., & Hajcak, G. (2011). Previously reappraised: The lasting effect of description type on picture-elicited electrocortical activity. *Social Cognitive and Affective Neuroscience, 6*(3), 348–358.

Mandal, M., Pandey, R., & Prasad, A. (1998). Facial expressions of emotions and schizophrenia: A review. *Schizophrenia Bulletin, 24*(3), 399–412.

Mather, M., & Carstensen, L. (2005). Aging and motivated cognition: The positivity effect in attention and memory. *Trends in Cognitive Sciences, 9*(10), 496–502.

Mather, M., Knight, M., & McCaffrey, M. (2005). The allure of the alignable: Younger and older adults' false memories of choice features. *Journal of Experimental Psychology: General, 134*(1), 38–51.

Mercado, F., Carretié, L., Hinojosa, J., & Peñacoba, C. (2009). Two successive phases in the threat-related attentional response of anxious subjects: Neural correlates. *Depression and Anxiety, 26*(12), 1141–1150.

Michalowski, J., Melzig, C., Weike, A., Stockburger, J., Schupp, H., & Hamm, A. (2009). Brain dynamics in spider-phobic individuals exposed to phobia-relevant and other emotional stimuli. *Emotion, 9*(3), 306–315.

Miller, J., Patterson, T., & Ulrich, R. (1998). Jackknife-based method for measuring LRP onset latency differences. *Psychophysiology, 35*(1), 99–115.

Mogg, K., Bradley, B., De Bono, J., & Painter, M. (1997). Time course of attentional bias for threat information in non-clinical anxiety. *Behaviour Research and Therapy, 35*(4), 297–303.

Monahan, J. L., Murphy, S. T., & Zajonc, R. B. (2000). Subliminal mere exposure: Specific, general, and diffuse effects. *Psychological Science, 11*(6), 462–466.

Morrison, R., Bellack, A., & Mueser, K. (1988). Deficits in facial-affect recognition and schizophrenia. *Schizophrenia Bulletin, 14*(1), 67–83.

Moser, J., Hajcak, G., Bukay, E., & Simons, R. (2006). Intentional modulation of emotional responding to unpleasant pictures: An ERP study. *Psychophysiology, 43*(3), 292–296.

Most, S., Smith, S., Cooter, A., Levy, B., & Zald, D. (2007). The naked truth: Positive, arousing distractors impair rapid target perception. *Cognition & Emotion, 21*(5), 964–981.

Mueller, E., Hofmann, S., Santesso, D., Meuret, A., Bitran, S., & Pizzagalli, D. (2008). Electrophysiological evidence of attentional biases in social anxiety disorder. *Psychological Medicine, 39*(7), 1141–1152.

Muhlberger, A., Wieser, M., Herrmann, M., Weyers, P., Troger, C., & Pauli, P. (2009). Early cortical processing of natural and artificial emotional faces differs between lower and higher socially anxious persons. *Journal of Neural Transmission, 116*(6), 735–746.

Naumann, E., Bartussek, D., Diedrich, O., & Laufer, M. (1992). Assessing cognitive and affective information processing functions of the brain by means of the late positive complex of the event-related potential. *Journal of Psychophysiology, 6,* 285–285.

Neisser, U. (1979). The control of information pickup in selective looking. In A.D. Pick (Ed.), *Perception and its development: A tribute to Eleanor J. Gibson* (pp. 201–219). Hillsdale, NJ: Erlbaum.

Nelson, C., & De Haan, M. (1996). Neural correlates of infants' visual responsiveness to facial expressions of emotion. *Developmental Psychobiology, 29*(7), 577–595.

Nieuwenhuis, S., Aston-Jones, G., & Cohen, J. (2005). Decision making, the P3, and the locus coeruleus-norepinephrine system. *Psychological Bulletin, 131*(4), 510–532.

Nummenmaa, L., Hyönä, J., & Calvo, M. (2006). Eye movement assessment of selective attentional capture by emotional pictures. *Emotion, 6*(2), 257–268.

Öhman, A. (1992). Orienting and attention: Preferred preattentive processing of potentially phobic stimuli. In B.A. Campbell, H. Haynes, & R. Richardson (Eds.), *Attention and information processing in infants and adults: Perspectives from human and animal research* (pp. 263–295). Hillsdale, NJ: Erlbaum.

Ohman, A. (2005). The role of the amygdala in human fear: Automatic detection of threat. *Psychoneuroendocrinology, 30*(10), 953–958.

Öhman, A., Flykt, A., & Esteves, F. (2001). Emotion drives attention: Detecting the snake in the grass. *Journal of Experimental Psychology: General, 130*(3), 466–478.

Olofsson, J., Nordin, S., Sequeira, H., & Polich, J. (2008). Affective picture processing: An integrative review of ERP findings. *Biological Psychology, 77*(3), 247–265.

Olofsson, J., & Polich, J. (2007). Affective visual event-related potentials: Arousal, repetition, and time-on-task. *Biological Psychology, 75*(1), 101–108.

Paller, K., Kutas, M., & Mayes, A. (1987). Neural correlates of encoding in an incidental learning paradigm. *Electroencephalography and Clinical Neurophysiology, 67*(4), 360–371.

Palomba, D., Angrilli, A., & Mini, A. (1997). Visual evoked potentials, heart rate responses and memory to emotional pictorial stimuli. *International Journal of Psychophysiology, 27*(1), 55–67.

Pastor, M., Bradley, M., Low, A., Versace, F., Molto, J., & Lang, P. (2008). Affective picture perception: Emotion, context, and the late positive potential. *Brain Research, 1189,* 145–151.

Peyk, P., Schupp, H., Keil, A., Elbert, T., & Junghöfer, M. (2009). Parallel processing of affective visual stimuli. *Psychophysiology, 46*(1), 200–208.

Phan, K., Liberzon, I., Welsh, R., Britton, J., & Taylor, S. (2003). Habituation of rostral anterior cingulate cortex to repeated emotionally salient pictures. *Neuropsychopharmacology, 28*(7), 1344–1350.

Phelps, E., LaBar, K., & Spencer, D. (1997). Memory for emotional words following unilateral temporal lobectomy. *Brain and Cognition, 35*(1), 85–109.

Pizzagalli, D., Greischar, L., & Davidson, R. (2003). Spatiotemporal dynamics of brain mechanisms in aversive classical conditioning: High-density event-related potential and brain electrical tomography analyses. *Neuropsychologia, 41*(2), 184–194.

Plutchik, R. (1980). *Emotion: A psychoevolutionary synthesis.* New York: Harper Collins.

Poli, S., Sarlo, M., Bortoletto, M., Buodo, G., & Palomba, D. (2007). Stimulus-preceding negativity and heart rate changes in anticipation of affective pictures. *International Journal of Psychophysiology, 65*(1), 32–39.

Polich, J. (2007). Updating P300: An integrative theory of P3a and P3b. *Clinical Neurophysiology, 118*(10), 2128–2148.

Pritchard, W. (1981). Psychophysiology of P300. *Psychological Bulletin, 89*(3), 506–540.

Radilova, J. (1982). The late positive component of visual evoked response sensitive to emotional factors. *Activitas Nervosa Superior, Supplement 3*(Pt 2), 334–337.

Richards, J. (2003). Attention affects the recognition of briefly presented visual stimuli in infants: An ERP study. *Developmental Science, 6*(3), 312.

Righart, R., & de Gelder, B. (2006). Context influences early perceptual analysis of faces—an electrophysiological study. *Cerebral Cortex, 16*(9), 1249–1257.

Righart, R., & de Gelder, B. (2008a). Rapid influence of emotional scenes on encoding of facial expressions: An ERP study. *Social Cognitive and Affective Neuroscience, 3*(3), 270–278.

Righart, R., & de Gelder, B. (2008b). Recognition of facial expressions is influenced by emotional scene gist. *Cognitive, Affective and Behavioral Neuroscience, 8*(3), 264–272.

Rigoulot, S., Delplanque, S., Despretz, P., Defoort-Dhellemmes, S., Honoré, J., & Sequeira, H. (2008). Peripherally presented emotional scenes: A spatiotemporal analysis of early ERP responses. *Brain Topography, 20*(4), 216–223.

Rohrbaugh, J., & Gaillard, A. (1983). Sensory and motor aspects of the contingent negative variation. In A.W.K. Gaillaird & W. Ritter (Eds.), *Tutorials in event related potential research: Endogenous components* (pp. 269–310). Amsterdam: North-Holland.

Rösler, A., Ulrich, C., Billino, J., Sterzer, P., Weidauer, S., Bernhardt, T., et al. (2005). Effects of arousing emotional scenes on the distribution of visuospatial attention: Changes with aging and early subcortical vascular dementia. *Journal of the Neurological Sciences, 229,* 109–116.

Rossion, B., & Jacques, C. (2008). Does physical interstimulus variance account for early electrophysiological face sensitive responses in the human brain? Ten lessons on the N170. *Neuroimage, 39*(4), 1959–1979.

Rothbart, M., & Derryberry, D. (1981). *Development of individual differences in temperament.* Hillsdale, NJ: Erlbaum.

Rottenberg, J., & Gross, J. (2003). When emotion goes wrong: Realizing the promise of affective science. *Clinical Psychology Science and Practice, 10*(2), 227–232.

Rottenberg, J., Gross, J., & Gotlib, I. (2005). Emotion context insensitivity in major depressive disorder. *Journal of Abnormal Psychology, 114*(4), 627–639.

Rugg, M., & Curran, T. (2007). Event-related potentials and recognition memory. *Trends in Cognitive Sciences, 11*(6), 251–257.

Rugg, M., & Nagy, M. (1989). Event-related potentials and recognition memory for words. *Electroencephalography and Clinical Neurophysiology, 72*(5), 395–406.

Russell, J. (1980). A circumplex model of affect. *Journal of Personality and Social Psychology, 39*(6), 1161–1178.

Sabatinelli, D., Bradley, M., Fitzsimmons, J., & Lang, P. (2005). Parallel amygdala and inferotemporal activation reflect emotional intensity and fear relevance. *Neuroimage, 24*(4), 1265–1270.

Sabatinelli, D., Lang, P., Keil, A., & Bradley, M. (2007). Emotional perception: Correlation of functional MRI and event-related potentials. *Cerebral Cortex, 17*(5), 1085–1091.

Sass, S., Heller, W., Stewart, J., Silton, R., Edgar, J., Fisher, J., et al. (2009). Time course of attentional bias in anxiety: Emotion and gender specificity. *Psychophysiology, 47*(2), 247–259.

Schacht, A., & Sommer, W. (2009). Emotions in word and face processing: Early and late cortical responses. *Brain Cognition, 69*(3), 538–550.

Schaefer, A., Fletcher, K., Pottage, C., Alexander, K., & Brown, C. (2009). The effects of emotional intensity on ERP correlates of recognition memory. *Neuroreport, 20*(3), 319–324.

Schapkin, S., Gusev, A., & Kuhl, J. (2000). Categorization of unilaterally presented emotional words: an ERP analysis. *Acta Neurobiologiae Experimentalis, 60*(1), 17–28.

Scherer, K. (1984). On the nature and function of emotion: A component process approach. In K.R. Scherer & P.E. Ekman (Eds.), *Approaches to emotion* (pp. 293–317). Hillsdale, NJ: Erlbaum.

Schienle, A., Schäfer, A., & Naumann, E. (2008). Event-related brain potentials of spider phobics to disorder-relevant, generally disgust-and fear-inducing pictures. *Journal of Psychophysiology, 22*(1), 5–13.

Schupp, H., Cuthbert, B., Bradley, M., Cacioppo, J., Ito, T., & Lang, P. (2000). Affective picture processing: The late positive potential is modulated by motivational relevance. *Psychophysiology, 37*(2), 257–261.

Schupp, H., Cuthbert, B., Bradley, M., Hillman, C., Hamm, A., & Lang, P. (2004a). Brain processes in emotional perception: Motivated attention. *Cognition & Emotion, 18*(5), 593–611.

Schupp, H., Flaisch, T., Stockburger, J., & Junghöfer, M. (2006a). Emotion and attention: Event-related brain potential studies. In S. Anders, G. Ende, M. Junghöfer, J. Kissler, & D. Wildgruber (Eds.), *Progress in brain research: Understanding emotions* (Vol. 156, pp. 31–51). Amsterdam: Elsevier.

Schupp, H., Junghöfer, M., Weike, A., & Hamm, A. (2003). Attention and emotion: An ERP analysis of facilitated emotional stimulus processing. *Neuroreport, 14*(8), 1107–1110.

Schupp, H., Junghöfer, M., Weike, A., & Hamm, A. (2004b). The selective processing of briefly presented affective pictures: An ERP analysis. *Psychophysiology, 41*(3), 441–449.

Schupp, H., Ohman, A., Junghöfer, M., Weike, A., Stockburger, J., & Hamm, A. (2004c). The facilitated processing of threatening faces: An ERP analysis. *Emotion, 4*(2), 189–200.

Schupp, H., Stockburger, J., Bublatzky, F., Junghöfer, M., Weike, A., & Hamm, A. (2007a). Explicit attention interferes with selective emotion processing in human extrastriate cortex. *BMC Neuroscience, 8*(1), 16.

Schupp, H., Stockburger, J., Codispoti, M., Junghöfer, M., Weike, A., & Hamm, A. (2006b). Stimulus novelty and emotion perception: the near absence of habituation in the visual cortex. *Neuroreport, 17*(4), 365–369.

Schupp, H., Stockburger, J., Codispoti, M., Junghöfer, M., Weike, A., & Hamm, A. (2007b). Selective visual attention to emotion. *Journal of Neuroscience, 27*(5), 1082–1089.

Scott, G., O'Donnell, P., Leuthold, H., & Sereno, S. (2009). Early emotion word processing: Evidence from event-related potentials. *Biological Psychology, 80*(1), 95–104.

Simons, R., Öhman, A., & Lang, P. (1978). Anticipation and response set: Cortical, cardiac, and electrodermal correlates. *Psychophysiology, 16*(3), 222–233.

Skrandies, W., & Jedynak, A. (2000). Associative learning in humans—conditioning of sensory-evoked brain activity. *Behavioural Brain Research, 107*(1–2), 1–8.

Smith, N., Cacioppo, J., Larsen, J., & Chartrand, T. (2003). May I have your attention, please: Electrocortical responses to positive and negative stimuli. *Neuropsychologia, 41*(2), 171–183.

Squires, K., Wickens, C., Squires, N., & Donchin, E. (1976). The effect of stimulus sequence on the waveform of the cortical event-related potential. *Science, 193*(4258), 1142–1146.

Stahl, D., Parise, E., Hoehl, S., & Striano, T. (2010). Eye contact and emotional face processing in 6 month-old infants: Advanced statistical methods applied to event-related potentials. *Brain and Development, 32*(4), 305–317.

Stekelenburg, J., & Gelder, B. (2004). The neural correlates of perceiving human bodies: An ERP study on the body-inversion effect. *Neuroreport, 15*(5), 777–780.

Sutton, S., Braren, M., Zubin, J., & John, E. (1965). Evoked-potential correlates of stimulus uncertainty. *Science, 150*(3700), 1187–1188.

Takeuchi, S., Mochizuki, Y., Masaki, H., Takasawa, N., & Yamazaki, K. (2005). Stimulus preceding negativity represents arousal induced by affective picture. *International Congress Series, 1278*, 385–388.

Taylor, J., & Fragopanagos, N. (2005). The interaction of attention and emotion. *Neural Networks, 18*(4), 353–369.

Taylor, S. (1991). Asymmetrical effects of positive and negative events: The mobilization-minimization hypothesis. *Psychological Bulletin, 110*(1), 67–85.

Tottenham, N., Tanaka, J., Leon, A., McCarry, T., Nurse, M., Hare, T., et al. (2009). The NimStim set of facial expressions: Judgments from untrained research participants. *Psychiatry Research, 168*(3), 242–249.

Turetsky, B., Kohler, C., Indersmitten, T., Bhati, M., Charbonnier, D., & Gur, R. (2007). Facial emotion recognition in schizophrenia: When and why does it go awry? *Schizophrenia Research, 94*(1–3), 253–263.

van Strien, J., Franken, I., & Huijding, J. (2009). Phobic spider fear is associated with enhanced attentional capture by spider pictures: A rapid serial presentation event-related potential study. *Neuroreport, 20*(4), 445–449.

Vanderploeg, R., Brown, W., & Marsh, J. (1987). Judgments of emotion in words and faces: ERP correlates. *International Journal of Psychophysiology, 5*(3), 193–205.

Walter, W., Cooper, R., Aldridge, V., McCallum, W., & Winter, A. (1964). Contingent negative variation: An electric sign of sensorimotor association and expectancy in the human brain. *Nature, 203*(168), 380–384.

Watanabe, S., Kakigi, R., & Puce, A. (2003). The spatiotemporal dynamics of the face inversion effect: A magneto- and electro-encephalographic study. *Neuroscience, 116*(3), 879–895.

Weinberg, A., & Hajcak, G. (2010). Beyond Good and Evil: The time-course of neural activity elicited by specific picture subtypes. *Emotion, 10*(6), 767–782.

Weinberg, A., & Hajcak, G. (2011). Electrocortical evidence for vigilance-avoidance in Generalized Anxiety Disorder. *Psychophysiology, 48*(6), 842–851.

Weinberg, A., & Hajcak, G. (in press). The Late Positive Potential predicts subsequent interference with target processing. *Journal of Cognitive Neuroscience, 23.*

Werheid, K., Schacht, A., & Sommer, W. (2007). Facial attractiveness modulates early and late event-related brain potentials. *Biological Psychology, 76*(1–2), 100–108.

Wieser, M., Mühlberger, A., Alpers, G., Macht, M., Ellgring, H., & Pauli, P. (2006a). Emotion processing in Parkinson's disease: Dissociation between early neuronal processing and explicit ratings. *Clinical Neurophysiology, 117*(1), 94–102.

Wieser, M., Mühlberger, A., Kenntner-Mabiala, R., & Pauli, P. (2006b). Is emotion processing affected by advancing age? An event-related brain potential study. *Brain Research, 1096*(1), 138–147.

Williams, L., Kemp, A., Felmingham, K., Liddell, B., Palmer, D., & Bryant, R. (2007). Neural biases to covert and overt signals of fear: Dissociation by trait anxiety and depression. *Journal of Cognitive Neuroscience, 19*(10), 1595–1608.

Williams, L., Palmer, D., Liddell, B., Song, L., & Gordon, E. (2006). The "when" and "where" of perceiving signals of threat versus non-threat. *Neuroimage, 31*(1), 458–467.

Willis, M., Palermo, R., Burke, D., Atkinson, C., & McArthur, G. (2009). Switching associations between facial identity and emotional expression: A behavioural and ERP study. *NeuroImazge, 50*(1), 329–339.

Wood, S., & Kisley, M. (2006). The negativity bias is eliminated in older adults: Age-related reduction in event-related brain potentials associated with evaluative categorization. *Psychology and Aging, 21*(4), 815–820.

Zald, D. (2003). The human amygdala and the emotional evaluation of sensory stimuli. *Brain Research Reviews, 41*(1), 88–123.

PART 4

ERPs in Special Populations

CHAPTER

17 Event-Related Potentials and Development

Donna Coch *and* Margaret M. Gullick

Abstract

Brain development is a complex and dynamic process dependent on the interaction of multiple internal and external factors. Currently, there are very few neuroimaging tools that are amenable to measuring brain development in typically developing infants and children. Of the tools available, the recording of event-related potentials (ERPs) is perhaps uniquely well suited to the challenges of conducting research with developing populations. This chapter reviews research on brain development relevant to developmental ERP researchers, some of the issues involved in identifying ERP components in developing populations, and some of the literature related to the development of specific ERP components.

Keywords: brain development, event-related potentials, ERP component, child development

Introduction

Brain development is a complex and dynamic process dependent on the interaction of multiple internal and external factors. In many ways, behavioral development is no less complex. Thus, charting the course of human development at multiple levels of analysis by making brain–behavior connections is a demanding and formidable task in terms of both theory (e.g., Fischer & Rose, 1994; Johnson, 2001) and measurement (e.g., Segalowitz & Schmidt, 2008). Currently, there are very few neuroimaging tools that are amenable to measuring brain development in typically developing infants and children (e.g., see Casey & de Haan, 2002). Of the tools available, the recording of event-related potentials (ERPs) is perhaps uniquely well suited to the challenges of conducting research with developing populations. Here, we selectively review research on brain development relevant to developmental ERP researchers, some of the issues involved in identifying ERP components in developing populations, and some of the literature related to the development of specific ERP components.

Brain Development and ERPs

Event-related potentials are a measure of the synchronous activation of neurons that are aligned so as to generate electrical fields that can be recorded at the scalp and likely primarily reflect postsynaptic activity (e.g., see Coles & Rugg, 1995; Luck, 2005; Nunez & Srinivasan, 2006, for reviews). As such, ERPs recorded at the scalp are dependent on brain development. A growing number of studies have indicated that the human brain continues to develop at least through the third decade of life, and that significant developmental changes in synaptic density, myelination, and other maturational processes are characteristic of early development (see Nelson & Luciana, 2001, for a review). The amplitude and latency of ERPs may index developmental neural changes involving "localization of function as the brain becomes tuned to the experienced world (related to synaptic pruning) and a speeding up of transmission as pathways become efficient (related to myelination)" (Picton & Taylor, 2007, p. 249).

Based on studies of postmortem specimens, Yakovlev and Lecours (1967) reported that there are myelogenetic cycles of development such that neurons within different brain regions become myelinated at different times; not surprisingly, neurons in primary sensory and motor areas are myelinated before neurons in higher-level association areas. More recent studies have confirmed this pattern of a predictable topographical and chronological sequence of myelination extending into adulthood, with significant development occurring prenatally and within the first 2 years of infancy (e.g., Benes, 1998; Sampaio & Truwit, 2001). Studies have also linked myelination and maturation of white matter to behavioral skills and cognitive development (e.g., Luna et al., 2001; Nagy et al., 2004; Paus et al., 1999; Pujol et al., 2006). Such changes in myelination and synaptic efficiency are thought to underlie developmental changes in the latency of ERP components (de Haan, 2007a; Eggermont, 1988; Picton & Taylor, 2007).

Due to methodological limitations, comparatively little is known about synaptogenesis in the human brain. Documenting changes in synaptic density is important to developmental ERP research because, although the exact neurophysiological mechanism is unknown (e.g., Mitzdorf, 1994), changes in the amplitude of ERP components over time are thought to be due to such changes in synaptic density (Courchesne, 1990; de Haan, 2007a; Vaughan & Kurtzberg, 1992). In a series of reports, Huttenlocher and colleagues have documented different developmental patterns of synaptogenesis in different neural regions (e.g., Huttenlocher, 1979, 1994; Huttenlocher & Dabholkar, 1997; Huttenlocher et al., 1982). For example, synaptogenesis begins prenatally in both auditory cortex (Heschl's gyrus) and prefrontal cortex (middle frontal gyrus), but synaptic density increases more rapidly in auditory cortex, with maximum density reached at about 3 postnatal months, compared to 15 months for prefrontal cortex. In both regions, a phase of net synapse elimination follows, which comes to a close in auditory cortex by about age 12 years but extends into midadolescence in prefrontal cortex (Huttenlocher & Dabholkar, 1997). Others have confirmed a nonlinear pattern of spine and synapse development, following a general pattern across the cortex of early overproduction, decrease, and plateau (e.g., Bourgeois, 2001; Michel & Garey, 1984). Although the exact mechanistic relationship is unknown, the amplitude of an ERP component likely "reflects a combination of synaptogenesis and synchronization" (Picton & Taylor, 2007, p. 253). Within the first 2 years of life, "reduced synaptic efficiency (perhaps related to early overproduction) results in greater slow wave activity rather than peaked activity, the latter being more typical of adult ERPs . . . [but] the characteristics of the peaked adult waveform typically begin to emerge when children reach 4 years of age (perhaps related to more efficient pruning), and continue to develop well into adolescence" (DeBoer et al., 2007, p. 9).

Broader measures such as brain weight have also revealed both remarkable growth during infancy and a lengthy course of development, with the human brain quadrupling in weight during the first 3 years but not reaching maximum weight until about age 19 (e.g., Dekaban, 1959; Dekaban & Sadowsky, 1978). In general, human brain development appears to be characterized by both progressive and regressive patterns of growth at multiple levels of analysis, from the synapse to gross anatomy (e.g., see Webb et al., 2001, for a review). A number of magnetic resonance imaging (MRI) studies conducted with children and adolescents have reported regionally specific, nonlinear, and extended structural development of the human brain (e.g., Caviness et al., 1996; Courchesne et al., 2000; De Bellis et al., 2001; Giedd et al., 1999; Giedd et al., 1996a, 1996b; Holland et al., 1986; Lenroot & Giedd, 2007; Matsuzawa et al., 2001; Pfefferbaum et al., 1994; Reiss et al., 1996; Sowell et al., 2003). Cortically, prefrontal and posterior temporal areas in particular appear to have lengthy developmental time courses (e.g., Gogtay et al., 2004; Sowell et al., 2001). There is also evidence for both hierarchical (i.e., lower-level sensory before higher-level cognitive) and concurrent brain maturation (Guillery, 2005). Similar patterns have been observed functionally in terms of regional cerebral blood flow (rCBF) and glucose metabolism measures (e.g., Chiron et al., 1992; Chugani, 1994; Chugani et al., 1993; Van Bogaert et al., 1998). Interestingly, despite such evidence for marked brain development throughout infancy, childhood, and adolescence, the blood oxygenation level dependent (BOLD) effect upon which many developmental functional magnetic resonance imaging (fMRI) studies rely appears to be similar in children and adults (e.g., Kang et al., 2003; Richter & Richter, 2003).

ERP Measures of Development

Given the considerable evidence for synaptic and neuronal change throughout infancy, childhood, and adolescence, it would be expected that neural activation patterns as reflected in ERPs would differ across development. Indeed, "it is clear that ERP components change in complex ways during development, so that rules for identifying components in adults cannot be applied with confidence in infants, children, and preadolescents. . . . Identifying components at different ages brings the added difficulty that, although the sequence of developmental changes in a particular component may be comparable for all normal children, the exact rate of change may vary from child to child. Moreover, knowledge of the stage of development of one component may not accurately predict the developmental stage of another" (Courchesne, 1990, p. 237). That is, variability and heterochronicity are often hallmarks of ERPs recorded from developing populations.

Determining whether a given ERP component recorded from a child and a given ERP component recorded from an adult are indeed the same component is a difficult task. Equivalency may be defined in a number of ways; for example, the components may be produced by the same neural generator (although that dipole may have a different orientation within the developing brain compared to the mature brain) or the components may index similar task-related processing (although the child may perform the task differently than the adult, reflecting activity of different neural generators with age). Generally, a "series of converging operations (relationship to RT, to chronological age, to stimulus probability, etc.) is necessary to determine the equivalence of components recorded at different ages" (Kurtzberg et al., 1984b, p. 315). In this sort of functional approach, child and adult participants are typically tested under the same conditions and components that show similar task-related modulations across age are identified (e.g., de Haan, 2007a).

Traditional characterization of components in terms of amplitude, distribution, latency, and responsiveness to experimental manipulations (e.g., Courchesne, 1983) is necessary but not always sufficient to identify ERP components in developing populations. In part, this is because "almost nothing is known about how physiologic activity in the developing brain propagates to the scalp surface, and thus we do not know what the relation is between activity *in* the brain vs. *at* the scalp. [In addition], the field has only begun to define ERP components of interest . . . thus much discontinuity remains between age groups and across areas of investigation" (DeBoer et al., 2007, p. 7). There are a number of extant reviews of developmental ERP research, some of which expertly present the technical and practical issues involved in recording ERPs from infants and children but none of which offers a definite solution to the problem of identifying ERP components across ages (e.g., Bashore, 1990; Courchesne, 1979; DeBoer et al., 2005, 2007; de Haan, 2008; Fox et al., 2007; Gavin & Davies, 2008; Klorman et al., 1978; Nelson & Monk, 2001; Picton & Taylor, 2007; Steinschneider et al., 1992; Taylor & Baldeweg, 2002; Thierry, 2005; Trainor, 2008; Vaughan & Kurtzberg, 1992). Here, we discuss briefly response variability, differences in spatial distribution, and contrasts between age effects and performance effects both as important considerations in addressing the issue of identifying ERP components in children and as challenges in developmental ERP research more generally.

Response Variability

De Haan (2007a) noted that variability in the response of the neural generators of a given component could affect the amplitude and latency of that component; developmentally, "an increase in the consistency of the brain's response with age . . . could lead to a decrease in latency variability across trials, which could in turn contribute to shortening peak latencies and growing peak amplitudes" (pp. 308–309). Inconsistent timing or strength of the neural firing patterns underlying the ERP response would lead to a lower signal-to-noise ratio (SNR) and more smear in children's waveforms. However, Segalowitz and Davies (2004, p. 129) concluded that a lower SNR was not a general property of children's developing brains based on their finding that some components recorded from children in a go/no-go paradigm appeared adult-like, while others did not. Given the possibility of this type of response variability, it has been recommended that the age ranges of participants be relatively circumscribed in developmental ERP studies: infant participants should be no more than 1 to 2 months apart in age, children should be no more than 1 to 2 years apart in age, and adolescents should be no more than 2 to 3 years apart in age (Picton et al., 2000). Even stricter participant inclusion criteria have been suggested more recently: within 10 days in infants, 1 month in children, and 1 year in adolescents (DeBoer et al., 2007).

The refractory rates of neural generators in infants and children may play a role in this type of response variability. Developmentally, little is known about the refractoriness of even basic visual and auditory processing systems in the human brain (see Chapter 4, this volume), but there is some ERP evidence suggesting that refractory effects are similar in children and adults, although not entirely adult-like in young children (e.g., Coch et al., 2005d; Cohn et al., 1985). This means that adjustments may need to be made to rates of stimulus presentation (interstimulus intervals [ISIs]) for infants and children compared to adults, in addition to factoring in children's slower behavioral reaction times (intertrial intervals; e.g., Picton & Taylor, 2007). Longer reaction times in children are also a consideration when recording components time-locked to responses such as the error-related negativity (ERN; e.g., Segalowitz & Davies, 2004).

Another source of response variability in developmental ERP research involves the relatively high rate of trial rejection. Due to eye movements, blinks, other motion artifacts (e.g., wiggling, giggling), noncompliance, lapses in attention, and limited recording time with breaks interspersed, fewer usable trials in data collected from infants and children result in noisier individual and grand averages and less reliable analyses. Both within-participant differences in completion of one condition compared to another (or recording of more usable trials in one condition compared to another) and between-participant differences must be considered as sources of response variability; indeed, the latter "have been associated with both amplitude and latency differences in certain components" (DeBoer et al., 2007, p. 25). Because the total number of trials completed may in part reflect developmental differences of experimental interest, it has been stated that "age differences in the total number of trials obtained could confound age differences in the ERP components of interest and should therefore be considered as a potential covariate in statistical analyses" (Scerif et al., 2006, p. 364).

In their go/no-go study with 7- to 17-year-olds, Segalowitz and colleagues reported that the developmental heterogeneity of the ERP components recorded was greater than the variability in behavioral measures of speed of processing (Segalowitz & Davies, 2004, p. 128). Moreover, they noted that individual differences can result in a large range of ERP responses within an age group, but that researchers "must also be careful that what we are seeing as heterogeneous results in children is not just a function of greater day-to-day variability. Consistent responses that vary over time are not a result of lower SNR but of true lability. The appropriate control for this is careful attention to motivational factors and task instructions" (Segalowitz & Davies, 2004, p. 130). Others have also commented on the vast individual differences in developmental ERP waveforms, particularly when recording from infants. One approach to address this issue is to test infants multiple times and calculate test-retest reliability in order to determine if the differences are real; another is to conduct longitudinal studies to track the sequence of development of ERP components in order to have baseline, normative data (e.g., Trainor, 2008, pp. 85–86).

Spatial Distribution

Developmental differences in component amplitude are due not only to synaptic density, neuronal alignment, and synchronization of activation but also to differences in the tissues through which the electrical signal propagates; the ERP of an infant or a young child "may be larger than that of an older child simply because the generator is relatively closer to the recording electrodes" (Picton & Taylor, 2007, p. 253). Studies directly comparing scalp and epidural recordings have shown that the skull markedly attenuates the amplitude of electrical potentials (Domino et al., 1964). Thus, smaller cerebral volumes and thinner skulls (e.g., Roche, 1953) in younger children may result in larger-amplitude components (e.g., McCulloch, 2007); indeed, although not in developmental studies, skull thickness has been shown to influence P300 amplitude (Pfefferbaum & Rosenblum, 1987) and to influence electroencephalographic (EEG)/ERP recording and localization (Chauveau et al., 2004). Such observations have led to specific recommendations regarding recording sites: "In the adult, the relatively broad scalp distribution of most ERP components, due to thicker and less conductive tissue between the generator and scalp electrodes, renders precise positioning of electrodes less critical. By contrast, in infants major differences in ERP waveshape may be observed over a 2 cm distance . . . a small cluster of recording electrodes is desirable in order to optimize the recording of the desired ERP . . . between 10 and 16 scalp electrodes are required to sample adequately the entire scalp distribution" (Kurtzberg et al., 1984b, p. 303). While topographic information can help to differentiate components, many studies with infants and children have used only a few recording sites, often limited to the midline (e.g., McIsaac & Polich, 1992).

Another way that cortical maturation may influence developmental differences in component distribution is that brain development may alter the orientation of the generating dipoles. Segalowitz et al. note that "[i]f brain growth in the child alters the orientation of these dipoles by changing the cortical folding, then the scalp ERP pattern would be considerably different. However, if the orientation angle of the dipole producing the ERN, for example, is different in children but still functional, then we should pick up the negativity at a different spot. For this, we need to examine the waveforms more carefully around the entire scalp, something which has not yet been done" (Segalowitz & Davies, 2004, p. 129). This point is, of course, related to recording across the entire scalp and carefully analyzing data from all active electrodes. It also suggests that measuring components where they are maximal, as recommended (Picton et al., 2000), may involve measuring components from different sites for children and adults (Taylor & Baldeweg, 2002). Consideration of differential distribution due to dipole orientation is also particularly important in terms of interpreting the absence of a significant ERP response developmentally. Generally, given the SNR of developmental ERP data, issues of localization are difficult to interpret without a solid background in comparative neurophysiology and strong theoretical and empirical foundations (Hood, 2001), which are often lacking in developmental research.

Developmental localization differences in ERP componentry may also reflect functional differences between age groups. Theoretically, the brains of infants and young children are less functionally specialized than the brains of adults, given their relative lack of experience with the world, such that differences in spatial distribution of components may reflect recruitment of larger regions of cortex for the same task in younger or less experienced compared to older or more experienced participants (e.g., de Haan, 2007a; Johnson, 2001). For example, in 6-month-old infants, the ERP response to auditory stimuli is of equal amplitude over the auditory and visual cortices, but between 6 and 36 months of age, the amplitude of the ERP response to auditory stimuli decreases at occipital sites while remaining relatively stable at temporal sites, suggesting development of cerebral specialization (Neville, 1995). This same sort of pattern was observed in slightly older children (20- to 42-month-olds) in a study in which auditory ERPs were recorded to known open-class (nouns, verbs, adjectives) and closed-class (prepositions, articles) words: at the earliest stages of language development, open- and closed-class words elicited similar ERPs; at 28 to 30 months of age, when the children were speaking in short phrases, the two classes of words elicited different patterns of brain activity, but the adult pattern of a left hemisphere asymmetry for closed-class words was not yet apparent; by 3 years of age, when most of the children spoke in sentences and used closed-class words appropriately, an adult-like left hemisphere asymmetry to closed-class words was observed (Mills et al., 1993; Neville & Mills, 1997).

Age and Performance Effects

Although one of the strengths of the ERP technique for use with infants is that an overt response is not required in order to elicit an ERP, having a "behavioral anchor" in studies with children and adolescents can provide another important source of information for interpretation of ERP components across ages (Kurtzberg et al., 1984b, p. 313). While behavioral measures that can be recorded during the ERP paradigm and directly correlated with the electrophysiological measure are probably most informative (DeBoer et al., 2007, p. 11), standardized assessments of behaviors with a known developmental time course (e.g., a standardized test of phonological processing) related to the ERP task (e.g., a rhyming paradigm) can also be useful; it is also possible to employ both of these types of behavioral measures in developmental studies (e.g., Coch et al., 2005a). Such behavioral measures can provide converging and constraining evidence that grounds interpretation of the observed ERP effects, although the relation between brain and behavior is associational rather than causal (Hood, 2001). Moreover, the behavioral performance measures may be more useful for grouping participants than chronological age. For example, in an auditory nonword rhyming study with 6- to 8-year-olds, a median split based on scores on a standardized test of phonological awareness created two groups of participants that differed by phonological awareness but not by age; in brain–behavior analyses, the onset of the ERP rhyming effect was found to be 80 ms later in the group with poorer phonological awareness than in the group with better phonological awareness (Coch et al., 2005a).

This distinction between age and performance effects is not often made in the ERP literature (but see, e.g., Scerif et al., 2006), even with adult participants. In a recent example of research measuring performance effects, Landi and Perfetti (2007) found that N400 amplitude in a word priming task with college students varied as a function of comprehension skills, such that poorer comprehenders showed smaller N400s. Developmentally, in their study with 28- to 30-month-olds, Mills and colleagues reported that children with larger vocabularies showed more focal distribution of ERP differences to known and unknown words than children with smaller vocabularies, regardless of age (Mills et al., 1993; Neville & Mills, 1997). Teasing apart the effects of chronological age and experience is crucial to charting the developmental course of specific skills and the neural systems underlying those skills. Children may use the same cognitive and neural processes as adults to complete a given task, or children may use different cognitive strategies and different neural systems than adults, or different neural systems may underlie seemingly similar behavioral performances in children and adults. Event-related potentials, in combination with other evidence such as error rates, response times, and response variability, can help to distinguish among these possibilities; while electrophysiological data cannot reveal exactly which cognitive or perceptual strategies children, compared to adults, are using for a specific task, they can show that brain activation patterns are different in children and adults in response to the same task (Segalowitz & Davies, 2004, p. 130). Differences in componentry may reflect maturational differences or differences in cognitive abilities with development; to control for effects of cognitive ability, equating task demands across groups and providing consistent instructions and practice trials with feedback are recommended in developmental ERP research (Kurtzberg et al., 1984b, p. 315).

In the context of developmental fMRI research, Casey (e.g., Casey, 2002) has outlined three options for distinguishing age from performance differences that may also be useful to consider for developmental ERP research. The first option is to create paradigms and tasks with parametrically differentiated levels of difficulty a priori, allowing for comparison of children and adults not only at the same level but also at different levels equated for behavioral performance. The second option is to correlate age and behavioral performance separately with measures of brain activity; such correlations would require a significant amount of variability within each of the measures of interest, and consideration of likely correlation between age and behavioral performance on many behavioral tasks. The third option is to group individuals based on their performance post hoc, thus allowing for comparison of "different age groups with similar behavioral performance or the same age groups with different performance . . . [however, this] approach is valuable only when the different age groups have overlapping distributions in response latency and accuracy" (p. 1409).

As with many developmental neuroimaging methods, basic brain development, response variability, shifting spatial distribution of activation, and age and performance effects are all crucial issues to consider when conducting developmental ERP research. These are just a few of the "added elements of developmental change [that] create a new level of complexity . . . in contrast to the psychophysiological study of adult populations" (Fox et al., 2007, p. 454). In the next section, we review selectively some of the developmental literature related to specific ERP components, summarizing some of the known changes in componentry over time.

Developmental Studies of ERP Components

Early Auditory Components

The perception and processing of sound elicits a complex electrophysiological response that develops over time. As reviewed by Trainor (2008), the auditory brainstem response (ABR), a series of seven peaks, occurs during the first 10 ms of the auditory ERP; these peaks represent activity in subcortical regions along the auditory pathway from the cochlear nucleus to the thalamus. The so-called middle latency responses during the subsequent 50 ms represent initial activities in auditory cortex, and the "late" auditory components, consisting of, in adults, a P1 (peaking at about 50 ms), N1 (peaking at about 100 ms), and P2 (peaking at about 180 ms), follow (see Chapter 4, this volume). The ABR can be recorded from newborns and is often used as a screening tool for detecting hearing difficulties in infants and children (e.g., Stapells, 1989). The components of the ABR appear to be mature by about age 3, developing during infancy but showing little change from ages 4 through 17 (e.g., Allison et al., 1984; Klein, 1984; Ponton et al., 1993; Salamy et al., 1982; Stapells, 1989). While the ABR can be recorded in infants, Trainor (2008, p. 75) states that it is "very difficult, if not impossible," to measure the midlatency auditory components in infants due to their small amplitude and inconsistency (cf. Galambos, 1982). Below, we focus on the exogenous "late" components P1, N1, P2, and N2 for which there are complex longitudinal and cross-sectional developmental data and evidence of change well into adolescence (e.g., see Oades et al., 1997; Ponton et al., 2000, 2002; Sussman et al., 2008; Wunderlich & Cone-Wesson, 2006), consistent with the lengthy time course of anatomical maturation of the human auditory cortex (e.g., Huttenlocher & Dabholkar, 1997; Moore & Guan, 2001). Although paradigms, stimuli, and parameters (e.g., volume, duration, temporal frequency) vary widely across auditory ERP studies, the typical study with infants and children

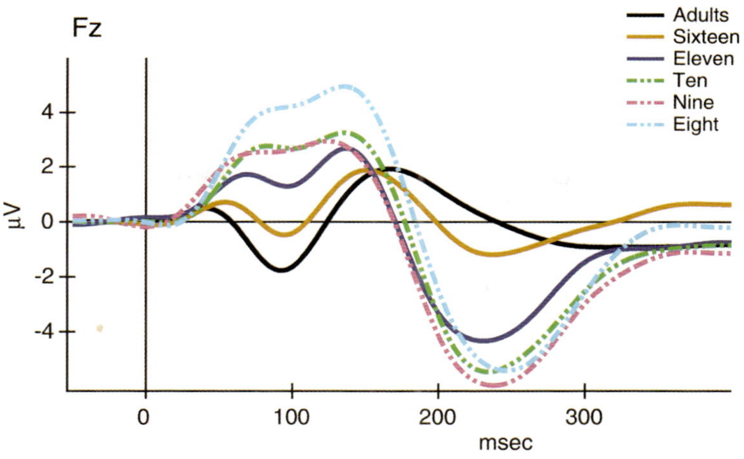

Fig. 17.1. Major maturational changes in the auditory ERP waveform are evident across six age groups from 8 years of age to adulthood. A decrease in P1 latency and an increase in P2 latency as the N1 emerges from a broad positive peak in childhood are illustrated. An N2 dominates the child waveforms and is still present in adolescents but absent in adults. These data clearly indicate that the auditory ERP response continues to develop into adolescence. Unattended pure tone standard stimuli (50 ms duration) were presented binaurally through earphones with an 800 ms stimulus onset asynchrony (SOA) at 75 dB SPL. Recordings from Fz are shown. Note that negative is plotted down. Reprinted with permission from Sussman, E., Steinschneider, M., Gumenyuk, V., Grushko, J., & Lawson, K. (2008). The maturation of human evoked brain potentials to sounds presented at different stimulus rates. *Hearing Research, 236*, 61–79 (modified Figure 2, p. 66).

involves passive presentation of pure tones, speech sounds, or clicks often accompanied by visual stimuli (e.g., bubbles, screensavers, puppet shows) not time-locked to auditory presentations in order to maintain engagement and keep participants from wiggling; auditory studies can also be conducted with sleeping infants, reducing motor artifacts. For a more detailed review of the dramatic developmental changes in the ERP waveform elicited by auditory stimuli during infancy, refer to Trainor (2008, p. 87); also see Figure 17.1 for an illustrated summary of the effects discussed below.

P1

During infancy, the typical auditory ERP waveform is characterized by a broad positive peak (P2) followed by a broad negative wave (N2; e.g., see Wunderlich & Cone-Wesson, 2006, for a review). A more adult-like, earlier P1-N1 pattern of components that increase in amplitude across infancy has been reported (e.g., Ohlrich et al., 1978; Weitzman & Graziani, 1968), but these components are less frequently evoked in young infants (Wunderlich & Cone-Wesson, 2006, p. 215).

The typical adult P1-N1-P2 sequence is also not observed in children and does not appear to emerge fully until adolescence (e.g., Albrecht et al., 2000; Gilley et al., 2005; Jing & Benasich, 2006; Korpilahti & Lang, 1994; Ponton et al., 2000; Sussman et al., 2008). The increasing amplitude of the P1 and N1 and the decreasing amplitude of the N2 appear to drive this developmental change in morphology of the auditory ERP waveform (Johnstone et al., 1996; Wunderlich & Cone-Wesson, 2006). The P1 can be identified in 3- and 5-year-olds (e.g., Gilley et al., 2005; Ponton et al., 2000; Sussman et al., 2008), and its amplitude increases from early to late childhood, perhaps with marked growth at around age 10, then decreases to adulthood (Ceponiene et al., 2002b; Ponton et al., 2000). In contrast, the latency of the P1 shortens across childhood to adulthood (e.g., Ceponiene et al., 2002b; Eggermont, 1988; Kraus et al., 1993; McArthur & Bishop, 2002; Oades et al., 1997; Ponton et al., 2000; Shahin et al., 2004; Sharma et al., 1997; Sussman et al., 2008). For example, in one study, P1 latency decreased in amplitude from 85 ms in 8-year-olds to 73 ms in 11-year-olds, to 66 ms in adolescents, to 40 ms in adults (Sussman et al., 2008). The decreases in amplitude and latency of the P1 observed in early adolescence may be a result of the simultaneous emergence of the N1 (Ceponiene et al., 2002b; Sussman et al., 2008; Trainor, 2008). P1 is maximal at the vertex in adults, and the peak of the P1 moves posteriorly from a frontal maximum with increasing age throughout adolescence (e.g., Oades et al., 1997; Sussman et al., 2008). Interestingly, in a study comparing the response to piano, violin, and pure tones in 4- and 5-year-olds learning to be musicians via the Suzuki method and their age-matched nonmusician controls, P1 was larger in the music students for all tones, suggesting some degree of experiential plasticity in this component (Shahin et al., 2004).

N1

Trainor (2008) summarizes that the N1 component in adults is comprised of the N100 or N1b, which is generated in the supratemporal plane of the auditory cortex and recorded maximally at central sites; the N1a (peaking at about 75 ms), which is recorded maximally at temporal and frontal-pole sites; and the N1c (peaking at about 130 ms), which is recorded maximally at temporal sites (see also Näätänen & Picton, 1987; Chapter 4, this volume). However, studies with infants usually report on a general "N1" component. The N1 is typically small or absent in newborns, although it has been reported that N1 latency to clicks does not change with age from 2 weeks to 3 years (Barnet et al., 1975; Davis & Onishi, 1969; Little et al., 1999; Ohlrich et al., 1978; Weitzman & Graziani, 1968). In another study with newborns, toddlers, 4- to 6-year-olds, and adults, the latency of the N1 reportedly did not change during infancy and early childhood (Wunderlich et al., 2006). Trainor (2008, p. 88) notes that although a broad positive wave characterizes infant auditory ERPs (see above and below), a number of studies have reported a "negative trough" appearing by 4 months that separates the broad positivity into two peaks (e.g., Dehaene-Lambertz, 2000; Novak et al., 1989).

The central N1 (N1b) is only observed in young children when the auditory stimulus presentation rate is slow (ISI greater than about 1 s): N1b has been reported in 3- to 9-year-olds in paradigms with long ISIs (e.g., Bruneau et al., 1997; Ceponiene et al., 1998; Gilley et al., 2005; Karhu et al., 1997; Pang & Taylor, 2000; Takeshita et al., 2002). At the faster stimulus presentation rates typically used in studies with adults, the N1b is not apparent as a separate component until early adolescence (e.g., Albrecht et al., 2000; Ponton et al., 2000; Sharma et al., 1997; Sussman et al., 2008). However, if slow N2 activity is filtered, the N1 can be identified in 9-year-olds with stimuli presented at short ISIs, but not in 4-year-olds (Ceponiene et al., 2002b). There is also evidence from magnetoencephalography studies indicating that the N1b in children is sensitive to the rate of stimulus presentation, probably due to refractoriness (see also Coch et al., 2005d, for ERP evidence; Paetau et al., 1995). The latency of the N1b decreases with increasing age, while N1b amplitude increases until about age 10–12 years, then decreases (Bender et al., 2006; Borg et al., 1988; Gomes et al., 2001; Goodin et al., 1978; Kraus et al., 1993; McArthur & Bishop, 2002; Oades et al., 1996; Pang & Taylor, 2000; Ponton et al., 2000; Sharma et al., 1997; Tonnquist-Uhlén et al., 1995). By the late teens, the parietal distribution of the N1b has shifted anteriorly to a frontocentral distribution similar to that of adults (Bender et al., 2006; Borg et al., 1988; Goodin et al., 1978; Oades et al., 1997; Pang & Taylor, 2000; Tonnquist-Uhlén et al., 1995). A smaller N1b peak amplitude is associated with poorer behavioral performance on auditory frequency discrimination and masking tasks in both children and adults (McArthur & Bishop, 2002).

At slow stimulus presentation rates of 1 s or longer, both an N1b and a lateral N1 can be identified in 6-year-olds (Gomes et al., 2001; Sharma et al., 1997). The lateral N1 is later and larger in 9-year-olds than in adults (Karhu et al., 1997), and its amplitude and latency decrease over developmental time from ages 6 to 12 to adulthood (Gomes et al., 2001; Pang & Taylor, 2000). The lateral N1 may be either the temporal N1a or N1c. In a study with 3- to 16-year-olds and adults, the N1a recorded from the left hemisphere was mature by age 3, while the N1a recorded from the right hemisphere was not adult-like until age 7 or 8 (Pang & Taylor, 2000). Temporal N1a and N1c can be elicited by click stimuli in 5- to 20-year-olds, but no N1b has been observed for children between 5 and 8 years of age in this paradigm (Tonnquist-Uhlén et al., 2003). In another study with children aged 4 to 8 years old, sounds similarly elicited an N1c (peaking at about 170 ms) with a midtemporal distribution instead of an N1 (peaking at about 100 ms, N1b) with a frontocentral distribution, as seen in adults (Bruneau et al., 1997). The N1c also matures earlier over the left hemisphere than over the right hemisphere and matures earlier to speech sounds than to tones (Pang & Taylor, 2000). N1c amplitude decreases with age from 6 to 18 years, while N1c latency increases slightly over the same time period (Bender et al., 2006). Thus, these three subcomponents of the adult N1 appear to have different developmental time courses, with the temporal N1a and N1c maturing earlier than the N1b.

Functionally, the N1 is perhaps most closely associated with attentional processing. In a classic ERP auditory attention study with adults, Hillyard and colleagues dichotically presented similar series of tone pips and instructed participants to attend only to one stream; comparison of the ERPs to the same tones when attended versus unattended showed an enhanced N1 to attended tones (Hillyard et al., 1973; see also Chapter 11, this volume). A similar N1 auditory attention effect has been reported in adults for syllables and environmental sounds (e.g., Hink et al., 1977, 1978; Woods et al., 1984). In a child-friendly modification of this task, children and adults listened to one of two concurrently presented narratives as it periodically switched between a speaker at their right and a speaker at their left (as the other narrative played from the opposite speaker) while a visual image reminded them of which side to attend to and ERPs were recorded to probe stimuli embedded in the attended and unattended narratives (Coch et al., 2005b; Sanders et al., 2006). While adults showed the typical N1 auditory attention effect, 3- to 5- and 6- to 8-year-old children showed a broad positivity during the first 300 ms after probe stimulus onset; probes embedded in the attended narrative elicited a larger positivity than the same probes embedded in the unattended narrative (Coch et al., 2005b; Sanders et al., 2006). Thus, despite the marked morphological differences between the child and adult waveforms elicited by the auditory probe stimuli, selective auditory attention enhanced processing within 100 ms of probe stimulus presentation in all age groups, suggesting that "the neural mechanisms by which selective attention affects auditory processing are remarkably adult-like" by the age of 3 in typically developing children (Sanders et al., 2006, p. 2126).

P2

A number of authors have reported that the newborn and young infant auditory ERP waveform is characterized by a large, broad positivity lasting from roughly 100 to 500 ms over frontocentral and temporal regions (e.g., Ceponiene et al., 2002a; Kraus et al., 1993; Kurtzberg et al., 1984a; Novak et al., 1989); this positivity is generally referred to as the *P2* (Wunderlich & Cone-Wesson, 2006). Over the first months of life, a small negativity between 100 and 400 ms recorded at temporal sites becomes more positive, resulting in 3-month-olds showing widespread positivity across frontal, central, and temporal regions in response to auditory stimuli (Trainor, 2008, p. 87). During early infancy, the amplitude of the P2 to tones or clicks increases linearly from 5 to 17 weeks of age with no shift in latency (Little et al., 1999). In most 6- to 7-month-old infants, the latency of the P2 is modulated by deviance (infrequent silent gaps in stimuli elicit shorter-latency P2s; Trainor et al., 2001). Typically, the P2 has a peak latency between 200 and 250 ms in infants (Wunderlich & Cone-Wesson, 2006).

While it has been claimed that the P2 matures early, appearing adult-like by 2 or 3 years of age (Crowley & Colrain, 2004), there is some evidence for lengthy maturation of aspects of the P2 response. Indeed, many investigators have reported that the amplitude of the P2 increases from infancy until late childhood (e.g., Kraus et al., 1993; Oades et al., 1997; Ponton et al., 2000; Wunderlich et al., 2006). While some have reported that the latency of the P2 decreases markedly at lateral sites with increasing age across childhood and adolescence (Oades et al., 1997), others have reported no change in P2 latency (peak at about 140–210 ms) from age 5 to 20 (Ponton et al., 2000; Wunderlich & Cone-Wesson, 2006). This effect may be task-specific: in an oddball task, P2 peak latency decreased with age from 6 to 15 years old (Goodin et al., 1978); in a contingent negative variation-type task with 6- to 18-year-olds, P2 peak latency (average 199 ms) did not change with age (Bender et al., 2006), while with rapid presentation of tones, P2 peak latency was similar in 9- to 16-year-olds (stable in latency across childhood, with an average 140 ms peak latency) but longer in adults than in children (Sussman et al., 2008). P2 latency may increase and amplitude decreases with the emergence of the N1 during late childhood and early adolescence (Sussman et al., 2008). While 9-year-olds and adults both show a P2, the P2 in children is distributed posteriorly to the vertex, while the P2 in adults is widely distributed across the scalp (Ceponiene et al., 2002b). An anterior shift from a predominantly parietal to a more centrally distributed P2 with age has been reported in a number of studies (e.g., Goodin et al., 1978; Oades et al., 1997; Ponton et al., 2000; Sussman et al., 2008). In the study with 4- and 5-year-old Suzuki musicians and their nonmusician controls, the P2 was larger for the musicians only for tones from the instrument practiced (piano or violin; Shahin et al., 2004). Indeed, the P2 appears enhanced in both adult and child musicians compared to matched nonmusicians and can be enhanced in adult nonmusicians with auditory training (Trainor et al., 2003).

N2

The N2 may not be present at birth (Kushnerenko et al., 2001), but the auditory ERP waveform becomes increasingly complex from 5 to 17 weeks of age, and N2 amplitude and latency change differentially for click and tone stimuli during this time (Little et al., 1999). In 14- to 18-week-old infants, the amplitude of the N2 does not habituate with repeated presentation of pure tones or broadband stimuli (Weber, 1972). The peak latency of the N2 in infants is usually about 300–550 ms (Wunderlich & Cone-Wesson, 2006); by 5 months of age, an N2 peaking at around 350 ms is evoked by both tone and click stimuli (e.g., Barnet et al., 1975; Jing & Benasich, 2006; Little et al., 1999; Thomas & Lykins, 1995). The amplitude of the N2 appears to increase from 3 to 6 months of age (Pivik et al., 2007). In premature infants, a vertex N2 peaking at 500–700 ms has been reported at 23 to 29 weeks (Weitzman & Graziani, 1968). As noted above, by 6 months of age in typically developing infants, earlier auditory components (P1/N1) are present in addition to the later N2 but are not yet adult-like (Trainor, 2008).

The N2 or N250 is considered a "classic characteristic" of the child auditory ERP waveform (Sussman et al., 2008, p. 78). The connection between the childhood N250 and the N2 observed in infancy is unclear, as few developmental data are available. The N2/N250 becomes increasingly prominent in the ERP waveform during early childhood (3–6 years) and thereafter dominates auditory ERPs until early adolescence; it then decreases in amplitude across adolescence (e.g., Ceponiene et al., 1998, 2002b; Enoki et al., 1993; Pang & Taylor, 2000; Ponton et al., 2000; Sussman et al., 2008). For example, Enoki and colleagues (1993) reported no N2 in any of the 4-year-olds tested but an N2 in half of the 5- and 6-year-olds, and Takeshita and colleagues (2002) reported an N250 in the majority of their 6- to 14-year-old participants but in only 4 of 10 adult participants. In 9-year-olds but not adults, a vertex N2 is enhanced for stimuli toward the end of a repeated stimulus train (Karhu et al., 1997). The latency of the N2 may decrease from age 4 to adolescence (Fuchigami et al., 1993; Goodin et al., 1978; Oades et al., 1997; Sussman et al., 2008) or increase with age from 5 to 20 (Ponton et al., 2000). According to some researchers, the frontocentral scalp distribution of the N2 remains stable from age 4 to adulthood, suggesting that the childhood N2 corresponds to the adult N2 (Ceponiene et al., 2002b, p. 880). Others have reported that with rapid stimulus presentation, the N2 is apparent in the waveforms of 8- to 16-year-olds but absent in adult waveforms (Sussman et al., 2008), while with slow stimulus presentation (8 s ISI) the N2 is not present in 8- to 12-year-olds (Gomes et al., 1999). Alternately, Ceponiene and colleagues (1998) have reported that the N2 in children is not sensitive to stimulus rate but is sensitive to stimulus complexity (being larger to complex tones than to simple tones; Ceponiene et al., 2001).

A number of researchers have investigated the development of these early auditory components in populations with or at risk for language and reading disorders (e.g., for a review, see Leppänen & Lyytinen, 1997). For example, in 3- to 8-year-old children with specific language impairment (SLI), there was no evidence of modulatory selective attention effects on early sensorineural processing in the child-friendly auditory attention paradigm described above due to reduced amplification of the neural response to probes in the attended stream; that is, the early broad positivity was not enhanced by attention in children with SLI, as it was in typically developing children of the same age (Stevens et al., 2006). However, intensive computerized training does enhance the effect of attention on the early positivity in children with SLI, specifically by increasing the amplitude of the response to attended stimuli (Stevens et al., 2008). Others have also reported differential auditory processing in terms of ERPs in young children with SLI (e.g., McArthur & Bishop, 2004, 2005; Neville et al., 1993) and in children at familial risk for dyslexia (e.g., Hämäläinen et al., 2007; van Herten et al., 2008).

Across these studies with both typically and atypically developing populations, it is clear that the auditory ERP waveform continues to develop well into adolescence (for a discussion of how these changes reflect the maturation of the auditory system, see Trainor, 2008). It is also clear that both age and ISI need to be considered carefully in ERP studies of auditory development, as both affect ERP componentry (e.g., Sussman et al., 2008). Finally, caution must be used in summarizing across studies that have used different auditory stimuli (e.g., clicks, tones of different frequencies, speech) and different recording electrode montages, "as the components are differentially visible and can appear to have different maturational curves, depending on where the responses are recorded" (Ponton et al., 2000; Taylor & Baldeweg, 2002, p. 325). Each of these factors likely contributes to the considerable variability in morphology observed in auditory ERP waveforms developmentally (e.g., Wunderlich & Cone-Wesson, 2006).

Early Visual Components

Typically, the visual evoked potential (VEP) is elicited experimentally in infants and children by simple visual stimuli and can be used as an index of the development of specific aspects of the visual system (for reviews, see Sokol, 1982; Taylor & McCulloch, 1992). With young participants, VEPs are often evoked by passive viewing of light flashes or flickering black-and-white checkerboards, sometimes with an extra incentive (such as an experimenter dangling an attractive toy in front of the monitor) to attend to the stimuli. As in adults, VEPs in infants and children typically include an N1, P1, N170 (specific to face processing), N2, and P2; an N3 is observed only in infants. Exogenously generated visual waveforms can be recorded in premature infants as early as 24 weeks gestational age (Pike et al., 1999; Taylor & McCulloch, 1992; Taylor et al., 1987), but there is great morphological variability in infants' VEPs (Benavente et al., 2005). Hartmann (1995, p. 218) suggested a "2 stage linear model" of VEP development in which rapid change characterizes the first 15 postnatal weeks, followed by more gradual development (see also McCulloch et al., 1993, 1999; Spekreijse, 1983). Infants' VEP to simple light flashes may be similar to adults' by 4 months of age (e.g., Dustman & Beck, 1969; McCulloch, 2007; Taylor & McCulloch, 1992), while infants' VEP to checkerboards and more complex visual stimuli has a longer developmental time course (e.g., Crognale, 2002; Crognale et al., 1997; Hartmann, 1995; Pompe et al., 2006; Taylor & McCulloch, 1992); sharper peaks and more complex waveforms are present by 4 years of age for checkerboards with larger checks (Moskowitz & Sokol, 1983) but VEP development continues through late childhood and into adolescence for more complex stimuli (e.g., Allison et al., 1983; Balachandran et al., 2004; Brecelj, 2003; Brecelj et al., 2002; Courchesne, 1990; Dustman & Beck, 1969; Spekreijse, 1978).

Visual evoked potentials have been used to investigate not only the general development of the visual system but also, more specifically, the development of binocularity, motion, color, and orientation processing in both research and clinical settings (e.g., Braddick & Atkinson, 1983; Braddick et al., 2005; Brecelj, 2003; Coch et al., 2005c; Crognale, 2002; Crognale et al., 1998; Fulton et al., 2006; Gordon & McCulloch, 1999; Hartmann, 1995; Madrid & Crognale, 2000; Pompe et al., 2006; Sloper & Collins, 1998). For example, the VEP to colored stimuli has a prolonged developmental time course compared to the VEP to flashes or black-and-white stimuli, and peaked responses to green and red (medium and long wavelength) stimuli are observed earlier (at 4 weeks) than responses to blue (short wavelength, at 6–8 weeks) stimuli (e.g., Crognale, 2002; Crognale et al., 1998). These early, immature positive-negative waveforms to chromatic stimuli (e.g., Boon et al., 2007; Madrid & Crognale, 2000; Pompe et al., 2006) eventually develop into a more mature triphasic complex by 6 years of age but continue developing through adolescence (e.g., Crognale, 2002; Madrid & Crognale, 2000; Pompe et al., 2006). A number of group differences in VEP development have also been investigated; for example, preterm and full-term infants of the same chronological age show different VEPs (e.g., Sokol & Jones, 1979; Taylor et al., 1987; Tsuneishi & Casaer, 2000) and female infants and children tend to have shorter-latency and higher-amplitude VEPs than males (e.g., Benavente et al., 2005; Brecelj, 2003; Malcolm et al., 2002; Pryds et al., 1989). In addition, connections between visual electrophysiology and visual behavior may differ developmentally: psychophysical thresholds and VEP measures may be more discrepant in 5- to 12-year-old children than in adults, at least for color vision (Boon et al., 2007). Below, we review selectively developmental findings related to some of the primary visual components.

N1

The N1 visual evoked peak in the ERP waveform appears as early as 24 weeks gestational age (Taylor & McCulloch, 1992; Taylor et al., 1987). The N1, sometimes called an *N300* because it peaks just before 300 ms in infants, can be observed consistently by 30–31 weeks gestational age (e.g., Taylor et al., 1987; Tsuneishi & Casaer, 1997). Some report that the N1 consists of an earlier N1a and a later N1b that dominates the waveform by term (e.g., Kraemer et al., 1999; Tsuneishi & Casaer, 1997, 2000; Tsuneishi et al., 1995), while others do not find a bifid peak in infants as consistently (e.g., Taylor et al., 1987). While both N1a and N1b peak latencies decrease with increasing age during early infancy, N1a latency decreases more rapidly and in a stepwise manner, likely reflecting myelination of the optic radiations (Tsuneishi & Casaer, 1997; Tsuneishi et al., 1995). The peak latency of the N1 is about 234–298 ms in premature infants from 25 to 30 weeks gestational age (Pryds et al., 1989) and decreases with age from 157 ms at 33–34 weeks postconception

(Pike et al., 1999) to 85–95 ms at 8 weeks after term (Moskowitz & Sokol, 1983), 73 ms by 12 weeks (Benavente et al., 2005), and 61 ms by 8 years of age (Voll et al., 1982), then continues to shorten through adolescence (Batty & Taylor, 2002). The latency of the N1 also decreases with increasing stimulus intensity in 8- to 15-year-olds (Carrillo-de-la-Peña et al., 1999). Some have reported that N1 amplitude is relatively stable across infancy and childhood, with a decrease in adolescence that may be due to the disappearance of a positive slow wave underlying P1 (Batty & Taylor, 2002, p. 488), while others have reported an increase in N1 amplitude from childhood to adolescence (Carrillo-de-la-Peña et al., 1999). An increase in N1 amplitude with selective attention is not apparent until at least age 12 (Shibasaki & Miyazaki, 1992). Courchesne (1990) reported no visual N1 in infants and children but a positivity within the 85–130 ms epoch instead.

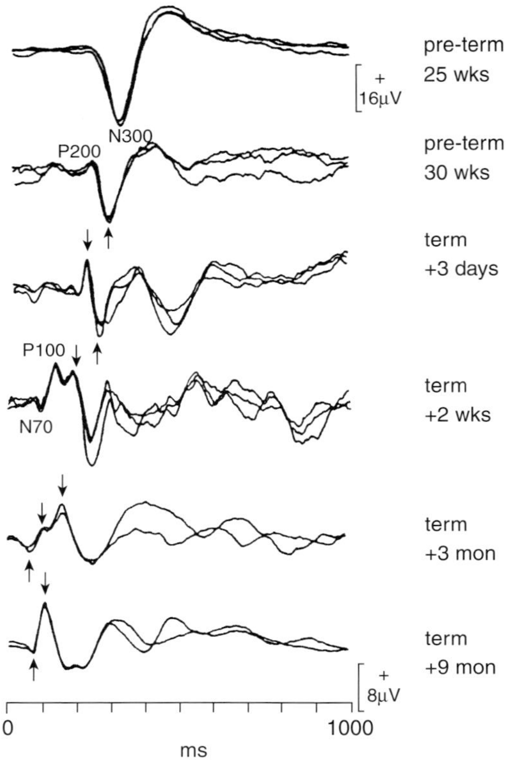

Fig. 17.2. Maturational changes in luminance VEPs from preterm to 9 months. In the preterm period (top two traces), only the N300 is evident and then the P200 emerges. By term (third trace), the P200 is more prominent. Postterm, the P100 emerges (fourth trace from the top). The latencies of both the P100 and P200 decrease rapidly over the next several months and begin to merge (fifth trace from the top). By 9 months postterm, usually only a single, early positive component (P100) characterizes the waveform. The N70 (N1) is clear from the early postterm period on. Recordings from Oz are shown. Note that negative is plotted down. Reprinted with permission from Taylor, M. J., & McCulloch, D. L. (1992). Visual evoked potentials in infants and children. *Journal of Clinical Neurophysiology, 9*(3), 357–372 (Figure 1, p. 360).

P1

A P1 component has been identified as early as 32 weeks postconception, although it may not be present consistently until 36 weeks (Ferriss et al., 1967; Harding et al., 1989; Kraemer et al., 1999; Pike et al., 1999; Roy et al., 2004; Tsuneishi et al., 1995). The P1 to flash stimuli observed in childhood and adolescence appears to stem from the P200, which bifurcates between 0 and 4 weeks after birth to form both a P100 and a P200; these two waves then merge to form another P1 by 6 months (Taylor & McCulloch, 1992, p. 360; see Figure 17.2). The P1 to checkerboard stimuli can be elicited by very large checks in newborns (Kos-Pietro et al., 1997) but is often not observed until after 2 months of age for smaller checks (Moskowitz & Sokol, 1983; Spekreijse, 1978; see Figure 17.3). The amplitude of the P1 increases over the first 3 months of postnatal life, perhaps until about 20 months, then decreases with age across childhood (Crognale et al., 1997; Kos-Pietro et al., 1997; Pompe et al., 2006; Shaw & Cant, 1981); however, the amplitude of the P1 evoked by achromatic pattern-reversal stimuli increases gradually until about 10 years of age (Madrid & Crognale, 2000). The peak latency of the P1 decreases from 250–350 ms in newborns (McCulloch et al., 1999) to 190 ms by 2 months and to 100 ms by 3 or 6 months (Aso et al., 1988; Crognale et al., 1997, 1998; Moskowitz & Sokol, 1983; Pike et al., 1999; Spekreijse, 1978); for checkerboard stimuli with larger checks, the P1 after 1 year has an adult-like latency of 100 ms (Crognale et al., 1997; Kos-Pietro et al., 1997; Malcolm et al., 2002; McCulloch & Skarf, 1991), while for more complex stimuli such as smaller checks or black-and-white patterns, P1 latency has been shown to shorten over time until at least age 5, and perhaps until 10 years of age or even to adulthood (Allison et al., 1983; Batty & Taylor, 2002; Brecelj, 2003; Brecelj et al., 2002; Pompe et al., 2006; Sloper & Collins, 1998; Sokol & Jones, 1979; Sokol et al., 1981; Spekreijse, 1978; Voll et al., 1982; Zemon et al., 1995). For example, in one study comparing processing of larger and smaller checks, P1 latency to larger checks was adult-like by 16 to 20 weeks, but P1 latency to smaller checks continued to develop until 5 or 6 years of age (Sokol & Jones, 1979). Recent research has shown that the

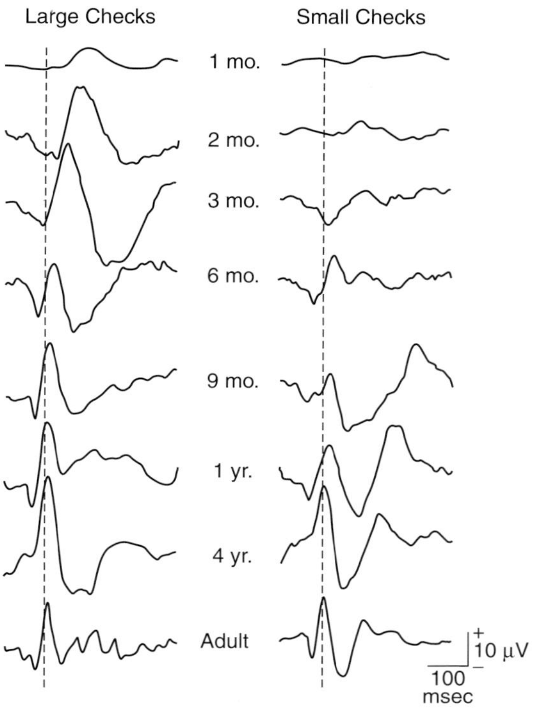

Fig. 17.3. Visual evoked potential waveforms obtained from individual subjects ranging from 1 month to 4 years of age in response to large (60 arcmin) and small (15 arcmin) checks. Visual evoked potentials from an adult (bottom trace) are shown for comparison. The adult response shows an N1 (85–95 ms) followed by a P1 (100–120 ms) for both large and small checks; for small checks, an N2 (140–160 ms) and a P2 (200 ms) are also observed. The traces from infants and children show that the latency of all components of the pattern reversal VEP decreases with increasing age. At 1 month, the VEP to large checks is characterized by a P1, with no measurable response to small checks. At 2 months, the N1 and N2 are elicited by large checks and a P1 is evident to small checks. At 3 months for large checks and at 9 months for small checks, a late positive component has emerged; between 9 months and 4 years, the latency of this component decreases dramatically. By 4 years of age, the VEP is quite similar to that evoked in the adult. The checks reversed at a rate of 1.88 alternations per second. Recordings from an electrode approximately 1 cm above the inion on the midline are shown. Note that negative is plotted down. Reprinted with permission from Moskowitz, A., & Sokol, S. (1983). Developmental changes in the human visual system as reflected by the latency of the pattern reversal VEP. *Electroencephalography and Clinical Neurophysiology*, *56*, 1–15 (Figure 1, p. 3).

latency of the P1 in 1- to 4-month-old infants is correlated with age and diffusion tensor imaging (DTI) measures of the optic radiations, suggesting that P1 latency is related to developmental myelination (Dubois et al., 2008). A model of P1 development in early infancy, based on normative data from studies using pattern reversal stimuli, has been developed (McCulloch et al., 1999).

N170

The N170, specifically sensitive to eye and face stimuli in adults, has become one of the most studied early visual components. Developmentally, the N170 observed in children may be somewhat more generalized than that observed in adults (de Haan et al., 2007). The infant N290 may be the progenitor of the N170: it appears to have response properties similar to those of the N170 and decreases in latency with age, peaking at 190 ms by 12 months of age (de Haan et al., 2007). A more adult-like N170 appears at around 4 years of age and continues to decrease in amplitude until age 7 (de Haan et al., 2007; Taylor et al., 1999, 2001). A number of articles and chapters review the development of the N170 in detail (e.g., de Haan, 2001; de Haan et al., 2002, 2003, 2007; Heisz et al., 2006).

N2

Some report that the N2 is present to flash stimuli by 27 weeks postconception (Pike et al., 1999) or by birth (Ferriss, 1970), while others report that it is not elicited until 2 months of age (Barnet et al., 1980). Checkerboard stimuli, even with large checks, do not elicit an N2 until 2 months of age (Moskowitz & Sokol, 1983). N2 mean peak latency decreases gradually throughout infancy and childhood, beginning at around 300 ms at 27 weeks postconception and decreasing to 250 ms by term (Pike et al., 1999), 248 ms by 0–5 days after birth, 208 ms by 26–30 days old, 150 ms by 51–60 days old, and 148 ms by 81–90 days old (Ferriss, 1970). This developmental change is slow, as no latency differences have been found in age cohorts from 5 to 16 years old (Batty & Taylor, 2002; Horst et al., 1982) for either flashes or picture stimuli, although each child group had longer-latency N2s than a comparison group of adults (Batty & Taylor, 2002). The amplitude of the N2 may increase (DeFrance et al., 1997) or decrease (Batty & Taylor, 2002; Satterfield et al., 1984) with age. In motion processing paradigms, the N2 is dominant ipsilateral to stimulation in children at least through age 16, a distributional pattern not observed in adults (Hollants-Gilhuijs et al., 2000).

P2

The P2 component of the VEP appears to have a particularly short developmental time course. A P2 to light flashes has been observed at 30–35 weeks gestational age and may develop from a bifurcation in the N3 at around 35 weeks postconception (e.g., Ellingson et al., 1973; McCulloch, 2007);

it is elicited consistently by 36 weeks gestational age (Taylor et al., 1987). The P2 is the most reliably evoked visual component in newborns (e.g., Benavente et al., 2005; Ellingson et al., 1973); given this, Ellingson et al. (1973, p. 120) suggested using the P2 as a "marker" component from which to identify existing components when not all components typical in the adult VEP are observed in the infant VEP. The latency of the P2 evoked by pattern reversal stimuli decreases from about 205 ms in 32-week gestational age infants to 197 ms at 1 to 2 months of age to 111 ms at 7 to 8 months of age and then remains stable at about 107 to 109 ms from 1 to 19 years of age (Aso et al., 1988); simple flashes also elicit a P2 with latency of between 184 and 219 ms in newborns (Ellingson et al., 1973; Ferriss, 1970). In infants 55 to 107 days old, the amplitude of the P2 is sensitive to the contour density of checkerboard stimuli (as represented by an inverted-U-shaped function the peak of which shifts with age (Karmel et al., 1974)).

N3

The N3 peaks at about 470–600 ms in its initial appearance at 27 weeks gestational age, then bifurcates between 30 and 35 weeks (Pike et al., 1999). By 36 weeks gestational age it includes several other components, including the P2 (Ellingson et al., 1973; McCulloch, 2007). While observable at term with a peak latency of 256 ms (Benavente et al., 2005), the infant N3 disappears sometime between 17 days and 15 weeks postterm (Kraemer et al., 1999; McCulloch, 2007; McCulloch et al., 1993) and does not appear to have a correlate in the adult VEP.

Contingent Negative Variation

A number of developmental studies of the contingent negative variation (CNV; see Chapter 8, this volume) have employed paradigms similar to that used in the original report of the CNV, involving dependence on a predictable relationship between a conditional and an imperative stimulus (Walter et al., 1964). Due to the response requirements of the typical CNV-eliciting paradigm, there are few reliable reports of a CNV in children under about the age of 5 years (cf. Gullickson, 1973; Prevec et al., 1984). The response requirement, in combination with the limited language skills, attention spans, and difficulty keeping still of most young children, are issues that seem to preclude the use of traditional CNV paradigms with very young children (Otto & Reiter, 1984, p. 377).

One of the earliest reports of a CNV in children involved a cross-modal paradigm with a visual conditional stimulus (flash) and an auditory imperative stimulus (tone). In this study with children and adolescents from ages 6 to 18, Cohen (1973) reported that the CNV reached adult amplitude (about $-20\ \mu V$) by age 15, although a smaller CNV (about $-11\ \mu V$) was present even in the 6-year-olds. In addition, the distribution of the CNV (earliest peak amplitude) shifted from more parietal in the youngest children to more frontocentral in the older participants (Cohen, 1973). In another cross-modal paradigm using auditory words as the conditional stimuli and a flash as the imperative stimulus, no CNV was elicited in any child younger than age 5 and a reliable CNV was not observed until age 9 or 10 (Lenzi et al., 1978). Sustained attention and motivation to cooperate in an essentially uninteresting and repetitive task seem to be key issues in elicitation of the CNV in young children; as one author claimed, "a CNV can be elicited more and more easily if the child is given firm social support, reassurance, and encouragement during the experiment" (Walter, 1966, p. 17).

Indeed, in a study designed in part to investigate such issues, Prevec and colleagues (1984) reported that of 23 children aged 5 to 7, only 10 generated a "clearly recognizable CNV" when tested in the traditional paradigm. When the paradigm was modified to prolong the duration of the conditional stimulus and employ pictures from a well-known story as the imperative stimuli, nine of the children randomly chosen for retesting showed a marked CNV (with an average amplitude of $-10\ \mu V$). When the paradigm was further modified such that the auditory conditional stimulus extended nearly to the presentation of the imperative stimulus and a slower repetition rate was used (and "if the measurement was carried out smoothly"), eighteen 3- to 5-year-olds showed evidence of a CNV (-5 to $-15\ \mu V$), although no behavioral response was required; in this same design, the CNV reportedly was observed in most children at age 3½ and in all children tested at age 4 (Prevec et al., 1984, p. 121). Thus, the authors concluded that the basic problem of successful CNV recording with children is to attract and maintain their attention (Prevec et al., 1984, p. 114).

More recent studies of CNV development have employed variants of the Posner cuing paradigm and go/no-go tasks. In a study with 6- to 9-year-olds (average age 7.2 years) and adults using a classic Posner paradigm, only the adults showed a CNV, while the children showed instead an N2-P3 complex during the cue-to-target period (Perchet & Garcia-Larrea, 2005). However, given that the spatial cue (conditional stimulus) and the visual target (imperative stimulus) were less than 1 s apart, the "CNV" observed in adults may have been more a motor readiness potential (Perchet & Garcia-Larrea, 2005). In a larger study with 57 children ages 7 to 17, Segalowitz and Davies (2004) modified the standard CNV paradigm into a simple go/no-go task, using green (go) and red (no-go) circles as conditional stimuli and an image of a race car as the imperative stimulus. They found that age was positively correlated with CNV amplitude measured at Cz (within the 400–2000 ms time window) for go trials but not for no-go trials. However, they reported a qualitative difference in the ERPs related to age: go trials elicited a larger CNV than no-go trials in adults and older children, while the youngest children tended to show a reversal of this pattern such that go trials elicited a prolonged positivity (although "some children despite their youth [did] produce a negativity on CNV go trials, similar to those of adults"; Segalowitz & Davies, 2004, p. 123). The authors speculated that this pattern might have been influenced by a later and more prolonged P3 in the younger children. Interestingly, CNV amplitude was correlated with behavioral performance on a number of executive function tasks (e.g., two-back, word recall, and Stroop; Segalowitz & Davies, 2004). Overall, the authors concluded that younger children are able to sustain attention but may use different neural mechanisms to do so than adults, and that frontal lobe maturation as indexed by the CNV continues at least through adolescence (Segalowitz & Davies, 2004, p. 123).

Findings from other studies that have considered the early and late aspects of the CNV have confirmed a lengthy developmental time course for this component. There appears to be some agreement that the CNV consists of an early, anteriorly distributed component that reflects orienting and a late, more centrally distributed component that reflects (in part) response preparation (e.g., Gaillard, 1977; Loveless & Sanford, 1974, 1975; Rosahl & Knight, 1995). In an early report of a standard CNV paradigm with a tone and a light used with 10-, 14-, and 19-year-olds, 10-year-olds did not show the early CNV but did show the late CNV, such that an adult-like CNV was not apparent until early adolescence (Klorman, 1975). A more recent report using a cued go/no-go task with 9-year-olds and adults showed a similar pattern: a smaller early CNV (850–1250 ms) in children at frontocentral sites but a mature late CNV in children (Jonkman et al., 2003). Behaviorally, children also performed more poorly than adults on the task; the authors note that "because inhibition processes are thought to reside in the prefrontal cortex, and maturation of frontal cortex continues into puberty, this is not a surprising finding" (Jonkman et al., 2003, p. 759). However, in another continuous performance task study conducted by Jonkman and colleagues with groups of 6- and 7-year-olds, 9- and 10-year-olds, and 19- to 23-year-olds, younger children showed a reduced late CNV: as measured across four midline leads, the youngest age group had significantly smaller late CNVs than both the older children and the adults, while the older children (9–10) showed late CNVs of similar amplitude to adults' (Jonkman, 2006). The authors reported that late CNV amplitude increased linearly with age (Jonkman, 2006). Finally, in a study employing visual conditional and imperative stimuli and conducted with two groups of 12-year-olds, "bright" children and their peers, the early CNV (600–940 ms) was not different between the child groups but was smaller in the 12-year-olds than in an adult control group; the amplitude of the late CNV (940–2300 ms) was comparable in the children and adults (Segalowitz et al., 1992a). The smaller early CNV may have been caused by an overlapping late P3 in the children (cf. Jonkman et al., 2003).

From the earliest developmental reports on the CNV, there has been an interest in comparing the CNV in typically and atypically developing populations; indeed, Cohen (1973) reported that, although a CNV was present by age 6 in all of the typically developing children whom he tested, a reduced or absent CNV characterized the ERP waveforms of 60% of the learning-disabled (LD) children whom he tested. Since then, adolescent poor readers have been shown to have a smaller or absent parietal CNV compared to good readers (Chayo-Dichy et al., 1990) and frontal CNV amplitude has been shown to correlate with reading skill for poor readers but not for good readers at age 15 (Segalowitz et al., 1992b).

A number of developmental CNV studies have also been conducted with populations with attention deficit/hyperactivity disorder (AD/HD). There have been reports of either a smaller CNV in children (9- to 12-year-olds) with AD/HD or poor concentration compared to age-matched controls (e.g., Aydin et al., 1987; Banaschewski et al., 2008; Grünewald-Zuberbier et al., 1978; Sartory et al., 2002) or a larger CNV in children with AD/HD compared to age-matched controls (e.g., Newton & Oglesby, 1994) or no difference in CNV amplitude between children and adolescents with AD/HD and controls (e.g., Strandburg et al., 1996). The early CNV also has a less anterior distribution in 8- to 15-year-olds with AD/HD compared to age-matched controls (Dumais-Huber & Rothenberger, 1992). In a training study, Heinrich et al. (2004) reported that 7- to 13-year-olds with AD/HD showed larger CNVs in a cued continuous performance task after learning to self-control their slow cortical potentials, as well as reduced symptom severity on a standard rating scale. Overall, while there are relatively few studies of the CNV in children and relatively little is known about the typical developmental course of the CNV as an index of response preparation (e.g., Segalowitz & Davies, 2004), such information is "crucial for the understanding of underlying factors and processes involved in developmental disorders of response inhibition and preparation" (Jonkman et al., 2003, p. 759).

Error-Related Negativity

The error-related negativity (ERN or NE) is a response-locked frontocentral component that peaks at about 50–100 ms following an incorrect response (error); it appears to be generated by the anterior cingulate cortex (ACC) and reflects conflict or error monitoring processes (e.g., see Falkenstein et al., 2000; Yeung et al., 2004; see also Chapter 10, this volume). Recently, there has been growing interest in tracking the developmental time course of the ERN, particularly as an index of action monitoring in adolescents.

Using traditional flanker tasks, a number of investigators have reported that ERN amplitude on error trials increases with increasing age across childhood and adolescence. The ERN may be small or absent in young children (age 7 to 12) and smaller in early adolescence than in late adolescence (e.g., Davies et al., 2004b; Ladouceur et al., 2004, 2007; Santesso & Segalowitz, 2008; Santesso et al., 2006a; Segalowitz & Davies, 2004). The ERN amplitude on error trials may not become adult-like until the late teens; a nonlinear quadratic growth pattern with an initial reduction in amplitude in late childhood and a subsequent rise throughout adolescence reportedly characterizes ERN development (Davies et al., 2004a, 2004b; Segalowitz & Davies, 2004). In one study using a visual flanker task with 7- to 17-year-olds and adults (19 to 25 years old), the correlation between ERN amplitude and age was –.47 (Segalowitz & Davies, 2004).

Using go/no-go tasks, the ERN has been reported in 7- to 9-year-olds in conditions of working alone and working under the observation of a friend (Kim et al., 2005). More recently, Kim et al. have reported no significant difference overall between ERN peak amplitude in children (ages 7–11) and young adults (age 21–25) in a go/no-go task, despite a trend for increasing ERN amplitude with age (Kim et al., 2007). Other reports indicate a "nearly absent" ERN in 7- to 8-year-olds in comparison to 13- to 14-year-olds and adults in a go/no-go task, but no difference in ERN amplitude between the adolescents and adults, perhaps due to the relative ease of the task (Wiersema et al., 2007). With the number of errors partialed, the correlation between age and ERN amplitude in this study was –.39 (Wiersema et al., 2007). In another study directly comparing less and more complex task conditions, smaller ERN amplitudes in adolescents than in adults were reported only for more difficult tasks, consistent with the behavioral finding that adolescents corrected fewer errors in incompatible trials (Hogan et al., 2005). However, Santesso and Segalowitz (2008) reported a smaller ERN in both visual flanker and go/no-go tasks in males ages 15 to 16 compared to males ages 18 to 20. The ERN amplitude may also increase across task learning, as shown in a probabilistic learning study with 10- to 12-year-olds (Groen et al., 2007).

A handful of developmental ERN studies have been conducted with special populations. In 10-year-olds, standardized test scores indicating poor socialization were associated with smaller ERNs (Santesso et al., 2005), while parent-reported obsessive-compulsive behaviors were associated with larger ERNs (Santesso et al., 2006b); similarly, 8- to 14-year-olds diagnosed with anxiety showed a larger ERN than age-matched controls (Ladouceur et al., 2006), and 8- to 17-year-olds seeking treatment for obsessive-compulsive disorder (OCD) symptoms showed a larger ERN than controls, even after receiving treatment (Hajcak et al., 2008). This pattern is consistent with studies indicating hyperactivity of the ACC in adults with OCD (e.g., Fitzgerald et al., 2005; Gehring et al., 2000). Error-related negativity studies of children with AD/HD have reported conflicting results, some showing no differences between children ages 7 and 13 with AD/HD and controls (Jonkman et al., 2007; Wiersema et al., 2005), some showing a larger ERN in children with AD/HD (Burgio-Murphy et al., 2007), and some showing a smaller ERN in children with AD/HD (Liotti et al., 2005; van Meel et al., 2007); it is likely that task differences affected these differential findings (e.g., see Jonkman et al., 2007, for a discussion).

P300

The P300 in adults reflects working memory processes such as context updating or attentional resource allocation (e.g., Polich, 2007; see also Chapter 7, this volume), and the P300 in children is thought to reflect similar processes (e.g., Curry & Polich, 1992; Friedman et al., 1989; Gill & Polich, 2002; Kilpelainen et al., 1999). Most studies focusing on the P300 in children have used an oddball detection task in which participants are required to respond to a target stimulus and withhold responses to background, nontarget stimuli (standards). In more complex designs, a third type of stimulus, a rare or unique nontarget distractor, is also used.

In early infancy, rare target events (10%–20% probability) presented in passive viewing or listening oddball paradigms do not elicit a typical P300 consistently; instead, a complex of a negative wave (Nc, see below) followed by a positive wave (either Pc or PSW) may be elicited. The Pc, a "broader positive peak," and the PSW, a "sustained positive slow wave," occur in the same temporal window and have the same distribution; thus, the distinction is unclear (de Haan, 2007b, p. 115). For infants ages 3 to 6 months, Pc amplitude does not vary with target probability (from 10% to 50%; Ackles & Cook, 1998) and, in 6-month-olds, it increases with increased target exposure (Nikkel & Karrer, 1994). Nevertheless, this positive slow wave is hypothesized to be the precursor of the adult P300, and the PSW/Pc appears to develop into the P300 across infancy (Nelson et al., 1998). Although some have reported that the PSW may not be present before 2 months of age (Karrer & Monti, 1995), others have reported a late positive wave to rare target tones in infants aged 0–3 months (Tokioka et al., 1987), and clicks and tones have been shown to elicit a PSW that increases in amplitude and decreases in latency from 5 to 17 weeks of age (Little et al., 1999). In the auditory modality, the P300 appears to emerge at about 7 months of age: In a study with 6- to 10-month-old infants (average age 7.4 months), McIsaac and Polich (1992) reported that infants showed a centroparietal P300 similar to the adult P300, but with smaller amplitude and longer latency (peaking at about 620 ms, versus 300 ms in adults). In the visual modality, unique pictures have been shown to elicit positive deflections in infants 7 weeks through 12 months of age (Schulman-Galambos & Galambos, 1978), and target faces have been shown to elicit a Pc peaking at 1300 ms by 7 months of age (Courchesne et al., 1981). Also using face stimuli, Nelson and colleagues (Nelson & Collins, 1991, 1992; Nelson & deRegnier, 1992; Nelson & Salapatek, 1986) found that 4-month-olds did not distinguish between familiarized face stimuli presented as infrequent targets and frequent standards, while 6-month-olds showed a PSW beginning at around 700–800 ms to infrequent targets, but 8-month-olds did not; at 12 months, a PSW was again in evidence to the infrequent familiar stimuli. Thus, both "the response to the familiar stimulus and the influence of frequency on familiarity change across the first year" (Webb, 2007, p. 161).

Theoretically, the immature PSW form of the P300 observed in young infants may be evoked by partially encoded stimuli: infants may not fully encode rare stimulus presentations and their memory traces may decay more, or more quickly, between these infrequent stimulus presentations compared to adults' (Nelson & Collins, 1991, 1992; Nelson & Monk, 2001, p. 131). By 7 or 8 months of age, infants are "able to ignore how often a stimulus is presented, and respond instead to the stimulus on the basis of whether it has been seen before" (Nelson & Collins, 1992, p. 119), thus invoking the memory updating processes indexed by the P300. An alternative theory is that the P300 reflects the ability to categorize events in an abstract, high-level manner, while the PSW indexes the limited ability of young infants to categorize "on the basis of evident, concrete properties" (Courchesne et al., 1981, p. 809). However, such higher-level categorization may be predicated on complete encoding and a strong memory trace (Nelson & Collins, 1992).

Across childhood and adolescence, P300 latencies have been reported to shorten with increasing age in both the auditory and visual modalities in oddball paradigms (e.g., Courchesne, 1977, 1978; Goodin et al., 1978; Johnstone et al., 1996; Ladish & Polich, 1989; Pearce et al., 1989; Polich et al., 1985, 1990). Similar decreases in P300 latency across childhood have been reported to go targets in go/no-go paradigms (Davis et al., 2003; Dimoska et al., 2007; Johnstone et al., 2005, 2007; Jonkman, 2006; Lewis et al., 2006). In the auditory modality, latency of the P300 has been reported to both decrease sharply, by more than 50 ms, in early adolescence (Johnson, 1989, p. 140; Oades et al., 1997) and decrease rapidly until around age 16 or 18 (Enoki et al., 1990; Fuchigami et al., 1993; Martin et al., 1988). Others have also reported a decrease in

P300 latency from ages 6 to 14 in both passive (–22.1 ms/year) and active (–15.8 ms/year) auditory oddball tasks (Zenker & Barajas, 1999). Interestingly, the decrease in P300 latency with age is correlated with memory and digit span increases from ages 5 to 14 and beyond (Howard & Polich, 1985; Polich et al., 1983); similarly, in a go/no-go task, age-related decreases in P300 latency to no-go stimuli from ages 7 to 12 (–42 ms/year) were correlated with reduced reaction times (to go stimuli) across this age range (Johnstone et al., 2007).

Also in the auditory modality, using an oddball paradigm with participants ages 5 to 19, Ladish and Polich (1989, p. 221) reported that the "mechanism which controls P3 amplitude, at least as derived from auditory stimuli, is fundamentally the same from 5 years of age throughout adulthood," and it is simply the speed of P300 production that changes over development. Other reports have also indicated that the amplitude and parietal distribution of the P300 elicited by auditory oddballs do not change significantly across childhood and adolescence (e.g., Goodin et al., 1978; Johnson, 1989; Pearce et al., 1989) and that P300 amplitude is modulated similarly by target frequency in children and adults (Curry & Polich, 1992; Ladish & Polich, 1989; Polich et al., 1990), although the effects of stimulus sequence on P300 amplitude may be stronger in children (Kilpelainen et al., 1999). However, in some oddball and go/no-go paradigms (e.g., Dimoska et al., 2007; Johnstone et al., 2005; Oades et al., 1997), the amplitude of a more frontocentral P300 has been reported to increase with age from early childhood to adulthood.

In the visual modality, the latency of the P300 elicited in oddball tasks has been shown to decline steadily throughout childhood and into young adulthood (Courchesne, 1990; Johnson, 1989; Mullis et al., 1985; Taylor & Eals, 1996; van Baal et al., 1998), decreasing from a peak latency of 800–900 ms at age 3 to approximately 400 ms in late adolescence (Courchesne, 1990, p. 229). While some have reported adult-like visual P300 latencies at age 11 or 12 (Batty & Taylor, 2002), others have reported decreases in visual P300 latency until the late teens or early 20s (Mullis et al., 1985; Taylor & Eals, 1996); more marked differences may be observed in late childhood (9–11) than across adolescence (12–19; Taylor & Smith, 1995). P3 latency in a visual oddball paradigm with children followed longitudinally from age 5 to age 7 appears to be influenced strongly by genetic effects (van Baal et al., 1998, p. 39), although similar twin studies with adults have not found a genetic influence on P3 latency (e.g., O'Connor et al., 1994). Visual P300 amplitude may continue to mature throughout childhood and adolescence and into the early 30s, and an adult-like, anteriorly distributed P300 to visual oddballs is not observed until age 17 or 18 (e.g., Berman et al., 1990; Johnson, 1989; Mullis et al., 1985; Stauder et al., 1999). The declines in P3 amplitude and latency with age across childhood and adolescence may vary by task (e.g., verbal or nonverbal; Taylor & Smith, 1995). While adults show anterior P300s to both rare targets and novel nontargets in an oddball paradigm (Courchesne, 1977, 1978), children and adolescents (to age 17) show a parietal PSW to novel nontargets (Courchesne, 1977, 1978; Mullis et al., 1985; Thomas & Nelson, 1996). In go/no-go paradigms, the developmental time course of a no-go P3 is unclear: Jonkman (2006) reported no no-go P3 in 6- and 7-year-olds, a small no-go P3 in 9- and 10-year-olds, and a large no-go P3 in young adults, while Davis et al. (2003) and Lewis et al. (2006) observed a large no-go P3 already in 5- and 6-year-olds.

While much of the adult P300 literature makes a clear distinction between the P3a and P3b components (e.g., Polich, 2007; see also Chapter 7, this volume), relatively little of the developmental P300 literature has explicitly addressed these subcomponents (but see, e.g., Polich et al., 1983, 1985). The P3a is usually elicited by novel or rare items, is more anteriorly distributed, and is thought to reflect attention shift, while the P3b is evoked by task-relevant target stimuli and is more posteriorly distributed (Tokioka et al., 1987); thus, the oddball paradigms and other simple target detection tasks discussed above likely elicited a P3b. Some authors further distinguish between an early P3a (eP3a) and a late P3a (lP3a), suggesting that the former indexes an automatic distraction detector, while the latter represents the actual attention switch (Ceponiene et al., 2004). While these various subcomponents of the P300 have not been studied extensively across age groups, the P3a and P3b appear to have different maturational time courses. A P3a has been identified in children as young as 2 years old (Kushnerenko et al., 2002) and is clearly in evidence in the waveforms of 6- and 7-year-olds (Stige et al., 2007; Wetzel & Schröger, 2007a, 2007b), but with a larger amplitude and a more frontocentral distribution in younger children that become smaller and more central over time (Wetzel & Schröger, 2007b). While larger deviance elicits larger P3as from 6 years old to adulthood, the P3a is elicited by deviant pitch changes in young children (but not adults) under "ignore" conditions

(Wetzel et al., 2006). The eP3a and lP3a are also identifiable in middle childhood, although the lP3a has a more posterior distribution than the eP3a in children, while the opposite is true in adults (Ceponiene et al., 2004, p. 137; Gumenuyk et al. 2004; Wetzel & Schröger, 2007b). In the auditory modality, the eP3a appears to be mature by age 12 or 13 (Fuchigami et al., 1995; Segalowitz & Davies, 2004), while the lP3a may have a longer developmental time course (Ceponiene et al., 2004).

The P3b has been identified as such in subjects as young as 5 to 9 years old (e.g., Batty & Taylor, 2002; Gill & Polich, 2002; Gomes et al., 2007; Jonkman et al., 2003; Polich et al., 1985; Segalowitz & Davies, 2004; Stige et al., 2007). In younger children (6–8) compared to older children (10–12) and adults, both irrelevant and relevant cues may elicit a P3b (Perchet & Garcia-Larrea, 2005; Wetzel & Schröger, 2007a). While the morphology of the P3b does not seem to change over developmental time, the latency and amplitude of the P3b tend to decrease with age to the late teens and the scalp distribution may shift from a parietal to a centroparietal maximum (Friedman, 1991; Friedman et al., 1997; Fuchigami et al., 1995; Johnson, 1989). Thus, studies that have reported on the P3b component specifically are consistent with the general "P300" studies reviewed above in suggesting a developmental time course for this component that extends into adolescence (Courchesne, 1977, 1978; Johnson, 1989; Mullis et al., 1985; Taylor & Eals, 1996; Thomas & Nelson, 1996).

Nc

The Nc, or negative central wave, is often associated with the P300 because it is also elicited in children participating in oddball paradigms. Present at birth and considered the earliest-developing endogenous component (Kurtzberg et al., 1986), the Nc is modality independent (Karrer & Ackles, 1987) and is maximal at frontocentral sites. Like the P300, the Nc is affected by stimulus familiarity and probability; however, in infants it tends to be larger for familiar, infrequent events (e.g., de Haan et al., 2003; de Haan & Nelson, 1997; Karrer & Ackles, 1987), while the P300 is largest to unfamiliar, infrequent events (e.g., Polich, 2007).

Mostly measured in habituation and oddball face processing studies, the latency of the Nc steadily decreases across early development, peaking at around 1000–1200 ms in newborns (Nelson, 1997, p. 101), 800 ms in 1-month-olds (Karrer & Monti, 1995), 700 ms in 4- to 7-month-olds (Courchesne et al., 1981), 550–600 ms or 400–800 ms by 6 months (Ackles & Cook, 1998; Karrer & Monti, 1995; Nelson & Salapatek, 1986; Thomas & Nelson, 1996), and 400 ms by 1 year of age (e.g., Nelson & deRegnier, 1992). The peak amplitude of the Nc also decreases over developmental time (Ceponiene et al., 2004; Courchesne, 1978; de Haan, 2007b; Gumenuyk et al., 2004; Webb et al., 2005), but the functional significance of this pattern is unclear; in part, this may be due to the finding that individual differences in the number of trials included in an infant's average ERP waveform are associated with differences in the amplitude of the Nc (Snyder et al., 2002). Webb and colleagues (2005) reported that a mother's face stimulus elicited a larger Nc than a female stranger's face stimulus in 4-month-olds but that this pattern reversed between 6 and 8 months. In a similar study with 6-month-olds, Swingler and colleagues (2007) found that Nc amplitude was related to individual differences in proximity- and interaction-seeking behaviors: infants who showed more such behaviors had larger Nc waves to a stranger's face stimulus than to a mother's face stimulus. In a study with toddlers, participants between the ages of 18 and 24 months showed a greater Nc to a mother's face stimulus than to a stranger's face stimulus, while participants aged 45 to 54 months showed the reverse pattern; Nc amplitude to the two stimulus types was similar for children between 24 and 45 months (Carver et al., 2003). By 6 to 8 years of age, children show a larger Nc to infrequent novel faces than to frequent or infrequent familiar faces (Thomas & Nelson, 1996). There has also been at least one report of different topographies for the Nc elicited by familiar versus novel stimuli in infants, suggesting involvement of different neural circuits in processing these stimulus types (Snyder et al., 2002).

Overall, the exact processes underlying the Nc remain unclear (de Haan, 2007b, p. 103). Nc has been described as an index of either nonobligatory selective attention (Nikkel & Karrer, 1994), or the perception of relevant stimuli (Courchesne et al., 1981), or an automatic orienting response (Vaughan & Kurtzberg, 1992). Although not common in the literature, some have even argued that the Nc is not an evoked wave itself, but emerges as a ridge between two large positive components (Tokioka et al., 1987, p. 381). Courchesne (1978) proposed that Nc development was complete by age 7, and several authors have noted that Nc is not typically elicited in adults in oddball paradigms and may

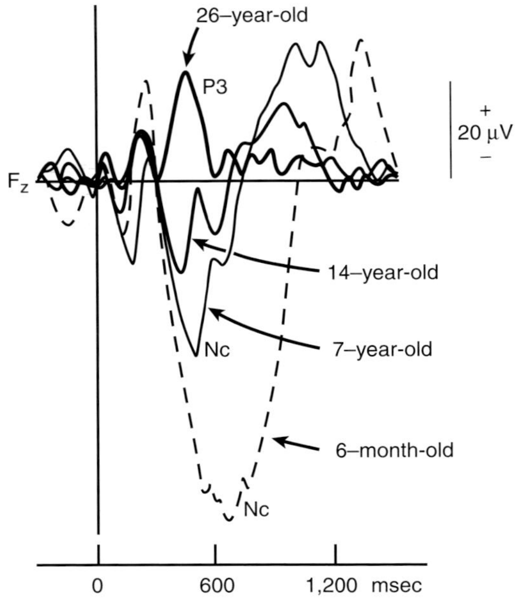

Fig. 17.4. Event-related potentials recorded to infrequently presented novel stimuli in a 6-month-old and in a 7-, 14-, and 26-year-old. Nc is prominent in the waves elicited from the infant and children (with decreasing amplitude and latency with increasing age), while the P3 is the most prominent component in the adult. Recordings from Fz are shown. Note that negative is plotted down. Reprinted with permission from Courchesne, E. (1979). From infancy to adulthood: The neurophysiological correlates of cognition. In J. E. Desmedt (Ed.), *Progress in clinical neurophysiology: Vol. 6. Cognitive components in cerebral event-related potentials and selective attention* (pp. 224–242). Basel: Karger (Figure 3, p. 229).

disappear between the ages of 10 and 14 or sometime in early adolescence (e.g., Courchesne, 1978; Courchesne et al., 1981; Oades et al., 1997; Thomas & Nelson, 1996; see Figure 17.4). Courchesne (1978, p. 474) further proposed that the Nc was an immature form of an N2, and that the disappearance of the Nc in early adolescence corresponded with the appearance of an N2 in the 14- to 17-year-old subjects in his oddball paradigm. Another theory, based on Eimer's (2000) findings with the N400 to faces ("N400f") and its role in identity recognition, is that the Nc eventually develops into this N400 (de Haan et al., 2003).

Most theories about the functional significance of the Nc focus on recognition memory or attention. For example, de Haan and Nelson (1997, 1999) concluded that the Nc was an index of recognition memory based on their finding that Nc was larger to both familiar faces and toys in 6-month-olds compared to unfamiliar faces and toys (although the Nc to faces was more right-lateralized than the Nc to objects), with identical Nc responses to two unfamiliar faces or toys (de Haan & Nelson, 1997, 1999). In another view, Nc represents the allocation of attention, with infrequent and familiar stimuli garnering more resources (e.g., Ackles & Cook, 1998, 2007). Kobiella and colleagues (2007) reported a larger Nc to angry than to fearful faces in 7-month-olds and interpreted the Nc as an orienting response, arguing that angry faces were more "arousing" and required greater allocation of attention. Reynolds and Richards (2005, p. 612) similarly hypothesized that the Nc reflects stimulus saliency and allocation of attention "regardless of novelty versus familiarity or frequency." In an oddball paradigm with 6-month-olds, Ackles and Cook (2007) reported an Nc of similar amplitude to rare oddball faces, a new face on every trial, or the same face on every trial, suggesting that Nc amplitude is not always modulated by familiarity/recognition memory and that attentional resources may have been allocated similarly across these three conditions. Other work also suggests that Nc indexes selective attention in infants (e.g., Nelson & Collins, 1991, 1992; Nelson et al., 1998). Bridging the two most common views of the Nc, it has been proposed that this component may "reflect the influence of recognition on attention" (Snyder et al., 2002, p. 486).

The Nc has been further specified in recent research; for example, Karrer and colleagues (1998) measured both an Nc (400–800 ms) and an Nc2 (800–1200 ms) in infant participants in their oddball paradigm. Similarly, some groups have identified the negative wave preceding the P3 as Nc (Ackles & Cook, 1998; Carver et al., 2003; de Haan et al., 2003; Karrer & Monti, 1995; Schulman-Galambos & Galambos, 1978), while others have identified a negative wave following the P3 as Nc (e.g., Gumenuyk et al., 2004); a double-peaked Nc has also been observed in children ages 9 to 13 (Ceponiene et al., 2004). The first of the two negative waves, "Nc1," precedes the P300, and is thought to reflect processing of "salient stimuli" or an allocation of attention (Ceponiene et al., 2004; de Haan, 2007b, pp. 109–110), as it is elicited by rare target events in oddball paradigms, is larger in "attend" versus "ignore" conditions (Richards, 2003), and decreases in amplitude with repeated experience with stimuli (Nikkel & Karrer, 1994). The second of the two negative waves, "Nc2," begins at around 500 ms, peaks between 800 and 1100 ms, and may be a "reorienting" wave (Ceponiene et al., 2004) similar to a "reorienting negativity" that reflects return of attention to the task and matures by age 11 (Gumenuyk et al., 2004). Both the Nc1 and the Nc2 are maximal at frontal sites, but source localization suggests that the Nc1 has a prefrontal generator, while the Nc2 originates from the frontal pole (Reynolds & Richards, 2005). This differentiation of the Nc into subcomponents is relatively recent, and further research is needed to investigate the functional significance of these subcomponents and their developmental time courses.

N400

In a seminal study conducted in 1980, it was shown that semantically anomalous words presented in sentences (e.g., *He spread the warm bread with socks.*) elicited a negative-going ERP component peaking at about 400 ms compared to canonical completions (e.g., *He spread the warm bread with butter*; Kutas & Hillyard, 1980). Over a decade later, a similar N400 effect was demonstrated in children and adolescents, in both the auditory and visual modalities, in a study with participants ages 7 to 26 years old (Holcomb et al., 1992). However, a number of developmental trends were noted in this study. First, in the youngest children (ages 7–12), a marked N400 was evident to both best completions and anomalous completions (although significantly larger to anomalous completions), while only anomalous completions elicited an N400 in older participants. Second, across modalities, the younger participants showed larger effects over parietal sites, while the older participants showed larger effects at anterior sites; moreover, a left-hemisphere asymmetry appeared to increase with age and was not observed until age 13. Third, the peak latency of the N400 decreased until age 19 and then increased. This pattern of findings indicates that the N400 elicited in spoken and written language contexts may have a long developmental time course, with some aspects of N400 development extending into late adolescence.

Other N400 sentence studies with children have reported similar findings. For example, in an auditory sentence study with children ages 5 to 11 (mean age 8) and adults, Juottonen and colleagues (1996) found a larger N400 effect in children than in adults and an N400 to both congruous and incongruous terminal words in children, but primarily to incongruous terminal words in adults. Friederici and Hahne (2001) also reported a widely distributed N400 in children ages 6 through 9 to semantic errors in auditory sentences, as well as a broadly distributed N400 decreasing in latency and becoming more localized with increasing age in a study with 6- to 13-year-olds (Hahne et al., 2004). More recently, Atchley et al. (2006) reported an N400 to semantic anomalies in spoken sentences in children ages 8 to 13 that peaked later and had a more anterior distribution than the N400 in an adult comparison group, and a larger N400 effect in the children. In younger children (36- and 48-month-olds), semantic anomalies in spoken sentences elicit a negative slow wave reminiscent of the Nc with peaks at 400, 600, and 800 ms (Silva-Pereyra et al., 2005); a long-lasting (300–1200 ms) N400 to semantic anomalies in spoken sentences maximal at centroparietal sites has also been reported in 19- and 24-month-olds (Friedrich & Friederici, 2005c).

Given the time intensiveness and difficulty of such sentence processing tasks for young children, particularly in terms of sustained attention, a number of researchers have investigated the N400 to words in simpler contexts such as lexical decision tasks or semantic categorization tasks. For example, in a study with four recording sites and verbal responses, first, third, and sixth graders showed a semantic priming effect (reduction of N400 amplitude for primed targets) for written word pairs consisting of category name primes and high-frequency exemplar targets compared to nonexemplar targets (Gonzalez-Garrido et al., 1997). In a word/pseudoword lexical decision task with 9- to 11-year-olds, children showed a larger N400 to pseudowords that was later than that observed in adults (Fonseca et al., 2006). In a semantic categorization task (press a button to indicate an animal name), 10- and 11-year-olds showed a marked N400 to real words, pseudowords, unpronounceable letter strings, and strings of false font characters, although the amplitude of the N400 distinguished legal from illegal strings (Coch et al., 2002b). In a study with a similar design, beginning readers 6 and 7 years old showed a similar pattern, but a median split of participants based on scores on standardized reading tests revealed that better readers had markedly larger N400s (Coch & Holcomb, 2003). Finally, in a primed lexical decision task with 9- to 15-year-olds who had been labeled as having failure to thrive as infants and their controls, only the controls showed an ERP priming effect in the N400 time window, which the authors interpreted as an indicator of less automatized word processing in the children who had been labeled as having failure to thrive (Dykman et al., 2000).

Investigators have also explored N400 development to spoken word stimuli in infants and toddlers. In a paradigm in which a picture is presented followed by an auditory word that either matches or mismatches the picture, 14-month-olds showed an N400-like incongruity effect at parietal sites (Friedrich & Friederici, 2005a). At 19 months, a similar incongruity effect was apparent, but toddlers who scored lower on comprehension tests showed a smaller incongruity effect much later than adults (Friedrich & Friederici, 2004). In contrast, high comprehenders showed an effect starting in the same time window as adults and stronger over the right hemisphere, as in adults, although more anteriorly distributed than in adults (Friedrich & Friederici, 2004; see also Torkildsen et al., 2006). In a similar study comparing 12-month-olds, 19-month-olds, and adults, participants were required to look at pictures presented serially and listen to pseudowords, nonwords, or real words (congruous or incongruous with the picture; Friedrich & Friederici, 2005b). In the 19-month-olds and adults, only incongruous words and pseudowords elicited an N400, suggesting that the 19-month-olds treated pseudowords but not nonwords as potential word candidates; all types of word stimuli elicited an N400 in the 12-month-olds (Friedrich & Friederici, 2005b). Finally, in a longitudinal study, the occurrence of an N400 elicited by words and pseudowords presented in picture-auditory stimulus pairs at 19 months was associated with language skills at 30 months: children with later age-level expressive language skills already showed an N400 at 19 months, but children with poor language skills at 30 months did not show an N400 at 19 months (Friedrich & Friederici, 2006).

In studies investigating the processing of known, unknown, and backward auditory words in infants, Mills and colleagues (1993) have reported that known words, compared to unknown words, elicit a bilateral temporoparietal N350 in 13- to 17-month-olds and a left hemisphere N350 in 20-month-olds, while backward words do not elicit an N350 (see Figure 17.5). The authors interpreted these findings as indicating that specialized neural word processing is already established by 20 months; as mentioned above, infants in a high language production group showed smaller, earlier, and more focal ERP components than infants in a low language production group, suggesting that even in

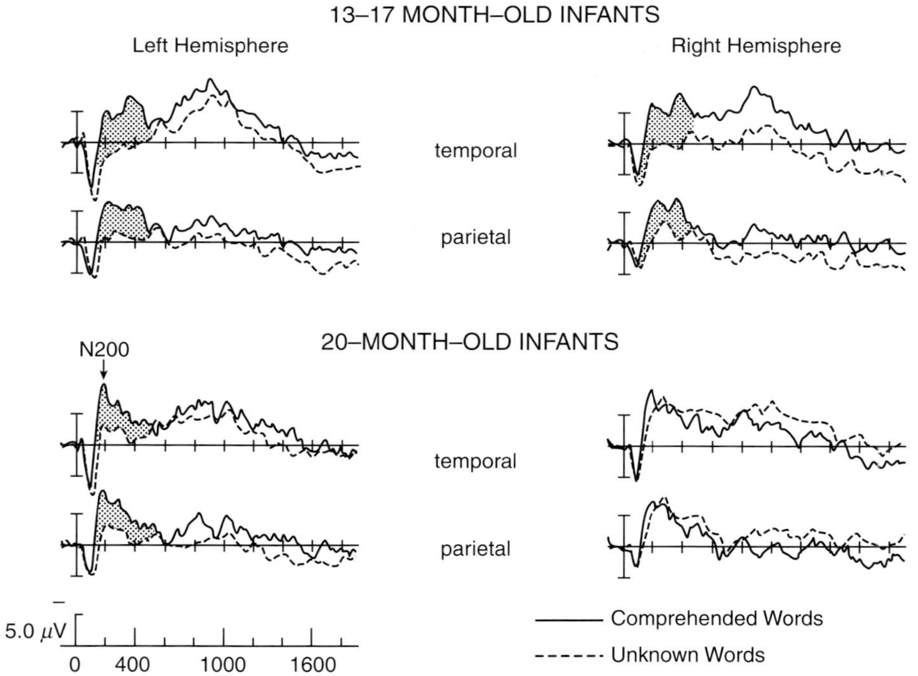

Fig. 17.5. Event-related potentials elicited by auditory comprehended (solid line) and unknown (dashed line) words in 13- to 17-month-olds (top) and 20-month-olds (bottom). At 13–17 months, comprehended words elicited a larger N350 than unknown words at left and right temporoparietal sites. By 20 months, that pattern was limited to temporal and parietal regions of the left hemisphere, suggesting increases in functional specialization of neural systems mediating language comprehension from 13 to 20 months of age. Note that negative is plotted up. Reprinted with permission from Neville, H. J., & Mills, D. L. (1997). Epigenesis of language. *Mental Retardation and Developmental Disabilities Research Reviews, 3*, 282–292 (Figure 5, p. 287).

children of the same age, language abilities are positively correlated with neural specialization for word processing in terms of the N350 (Mills et al., 1993, 1994). Mills and colleagues have also reported a larger N200–400/500 to known words in 17- to 22-month-olds in other paradigms, including bilingual infants (e.g., Conboy & Mills, 2006; Mills et al., 2004, 2005); for a recent review of this work, refer to Mills and Sheehan (2007). In contrast, Molfese and colleagues have reported larger negative peaks in the 180–340 ms and 580–700 ms time windows for unknown words than for known words in 16-month-old infants (Molfese, 1990) and a greater positivity in the 520–600 ms window for labels that did not match objects in 14-month-old infants in a training study (Molfese et al., 1990).

The N400 component has also been investigated in populations diagnosed with or at risk for reading disorders. For example, in 20- to 24-month-olds, in both a cross-modal picture-word priming paradigm and an unimodal auditory word pair priming paradigm, children at risk for dyslexia due to a family history showed a later and more circumscribed N400-like priming effect (Torkildsen et al., 2007). However, by ages 7 to 10 years, the N400 priming effect for spoken word and nonword pairs appeared comparable in a group of children with dyslexia and controls (Bonte & Blomert, 2004b); in 9- to 12-year-olds in an auditory sentence task, semantic anomalies elicited similar N400s in children with dyslexia and controls (Sabisch et al., 2006); and there were no differences in the N400 in a word categorization task with 10-year-old children with dyslexia and controls (Silva-Pereyra et al., 2003).

Finally, within the N400 family of components, an N350/N400 has been shown to be sensitive to manipulations of rhyme and may reflect phonological processing related to rhyme decisions (e.g., Rugg, 1984a, 1984b). In studies in the visual (Grossi et al., 2001) and auditory (Coch et al., 2002a) modalities with participants from age 7 to adulthood, simple prime-target word pairs requiring a rhyme judgment have been shown to elicit an ERP rhyming effect such that targets that rhyme with primes elicit a smaller N350/N400 than targets that do not rhyme with primes. This effect appears to be essentially adult-like by the age of 7 in terms of amplitude, latency, and distribution (Coch et al., 2002a; Grossi et al., 2001; see Figure 17.6). In a similar pseudoword rhyming paradigm with 6- to 8-year-olds, mentioned above, the onset of the rhyming effect was shown to correlate positively with a standardized behavioral measure of phonological awareness (Coch et al., 2005a). Others have reported a similar phonological priming effect on the N400 for alliterating targets paired with spoken pseudoword primes in 5- to 8-year-olds and adults (Bonte & Blomert, 2004a). In adults and 9- and 10-year-olds, the orthographic and phonological incongruity between written primes and targets in a rhyming

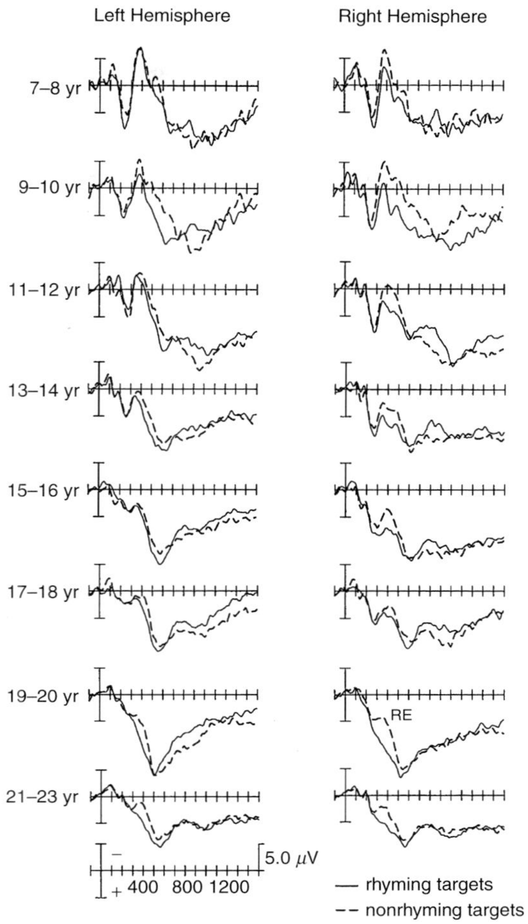

Fig. 17.6. The typical N350/N400 ERP rhyming effect (labeled RE), such that nonrhyming targets (e.g., *chair-moose*) elicit more negative-going waveforms than rhyming targets (e.g., *juice-moose*), is apparent for all age groups, from 7- and 8-year-olds to adults, at left and right parietal sites. The distribution and onset timing of the rhyming effect did not vary with age. Visual words with a duration of 300 ms were presented in pairs with an 1167 ms SOA. Participants were asked to decide if the words in each pair rhymed and entered their responses with a button press (only correct responses were averaged); the two youngest age groups (7–8 and 9–10) were less accurate than the older age groups. Note that negative is plotted up. Reprinted with permission from Grossi, G., Coch, D., Coffey-Corina, S., Holcomb, P. J., & Neville, H. J. (2001). Phonological processing in visual rhyming: A developmental ERP study. *Journal of Cognitive Neuroscience, 13*(5), 610–625 (Figure 4D, p. 616).

paradigm affects the rhyming effect for both age groups (Weber-Fox et al., 2003). Finally, some authors have reported that the rhyming effect is similar in children with dyslexia and controls (e.g., Ackerman et al., 1994), while others have reported differences (e.g., Lovrich et al., 1996; McPherson et al., 1998).

Conclusion

Event-related potentials provide a unique perspective on human development. From basic perception to higher cognitive functions, the ERP method represents another level of analysis for developmentalists, one that may provide both converging and constraining evidence with respect to other neuroimaging measures (both structural and functional) and behavioral measures of human development. The ERP findings reviewed above characterize a complex pattern of development over time both within and across systems and provide evidence for brain–behavior relations. Researchers designing developmental ERP studies have "the opportunity to study not only changes in response systems over age, but also the processes underlying this change. Not only must change in the response system be measured but it must be understood as a function of the physical and psychological maturation" of the child (Fox et al., 2007, p. 454). Such evidence from multiple levels of analysis informs a fuller understanding of child development and is consistent with a recent emphasis on the "interrelated and interdependent" nature of development (e.g., Ansari & Coch, 2006; Diamond, 2007).

Well-controlled, carefully designed longitudinal studies tracking the development of ERP components throughout infancy, childhood, and adolescence are generally lacking but necessary; in particular, likely due to compliance issues, very little is known about the electrophysiological response within the age range from 1 to 3 years. Such longitudinal studies, or combined longitudinal and cross-sectional studies, based on knowledge of behavioral data and developmental theory, would not only provide more thorough developmental ERP data but also provide evidence that could refine understanding of the functional significance of ERP components in all age groups (e.g., Jonkman, 2006). Others have also noted recently that key future directions in developmental ERP research include grounding ERP data in behavioral data, designing new paradigms that are child-friendly, and designing studies to explore the development of ERP components specifically (DeBoer et al., 2007). Matching infant and child components with adult components is just one avenue to charting development with ERPs, and perhaps it need not be a focus of all developmental ERP research; investigating multiple developmental pathways in the service of understanding the dynamics of development, rather than an outcome, may also be a fruitful approach (e.g., Karmiloff-Smith, 1997). A further area of study is exploration of the influence of slow-wave activity on ERP components; for example, Dimoska and colleagues (2007, p. 109) have claimed recently that "underlying slow-wave activity accounts for a large number of developmental effects in the traditionally quantified ERP components, but may also obscure effects occurring in residual activity."

In addition to charting systematically the typical course of development of ERP components and the processes that they index, ERPs are a promising measure of atypical development. For example, ERPs can be used to determine how specific prenatal teratogens or experiences might alter neural functioning and lead to atypical neural and behavioral development (e.g., Nelson, 2007; Pivik et al., 2007), not only bridging the gap between brain and behavior but also potentially revealing specific causes for disorders (Monk et al., 2001, p. 230). Importantly, ERPs may be used as diagnostic tools earlier in development than is possible with behavioral testing (Trainor, 2008, p. 86). Indeed, Otto (1984, p. 321) noted a number of possible uses of ERPs in atypical development over two decades ago, including diagnostic value, indexing therapeutic efficiency, culture-free assessment, and early detection, but few developmental researchers have used ERPs in these ways. Extant studies that have taken such approaches, beyond clinical screening for basic auditory and visual deficits (e.g., Stapells, 1989; Taylor & McCulloch, 1992), such as those measuring early auditory language processing in typically developing children and children at family risk for dyslexia in connection with later language and reading difficulties (e.g., Leppänen et al., 1999; Lyytinen et al., 2005; D. L. Molfese et al., 2007; V. J. Molfese et al., 2008), have illustrated the exciting prospect of using ERPs predictively in development. In addition, ERPs are beginning to be used as an index of therapeutic and educational efficiency, and, moreover, to pinpoint the locus of therapeutic and educational effects and the mechanism by which specific remediation methods work (e.g., see Goswami, 2005; Stevens et al., 2008). However, in order for ERPs to be a truly useful diagnostic tool, methods to analyze and measure ERP waveforms on an individual rather than a group level will need to be developed (e.g., Trainor, 2008); this is not an issue specific to developmental ERP research.

As the field of child development becomes increasingly integrated and interdisciplinary (e.g., Diamond, 2007), developmental ERP researchers have a bridging role to play: making connections across brain and behavior, providing evidence not only about the development of ERP components themselves but also about the neurocognitive and perceptual processes that those components index and how development of those processes may be influenced by genes and environment. For both typically and atypically developing populations, there are many questions and many avenues of research that have yet to be explored with ERPs; the future is filled with possibilities for developmental ERP investigations.

Acknowledgment

Writing of this chapter was supported in part by Grant No. R03HD053362 from the National Institute of Child Health and Human Development to DC.

References

Ackerman, P. T., Dykman, R. A., & Oglesby, D. M. (1994). Visual event-related potentials of dyslexic children to rhyming and nonrhyming stimuli. *Journal of Clinical and Experimental Neuropsychology, 16*(1), 138–154.

Ackles, P. K., & Cook, K. G. (1998). Stimulus probability and event-related potentials of the brain in 6-month-old human infants: A parametric study. *International Journal of Psychophysiology, 29*, 115–143.

Ackles, P. K., & Cook, K. G. (2007). Attention or memory? Effects of familiarity and novelty on the Nc component of event-related brain potentials in six-month-old-infants. *International Journal of Neuroscience, 117*, 837–867.

Albrecht, R., Suchodoletz, W. v., & Uwer, R. (2000). The development of auditory evoked dipole source activity from childhood to adulthood. *Clinical Neurophysiology, 111*, 2268–2276.

Allison, T., Hume, A. L., Wood, C. C., & Goff, W. R. (1984). Developmental and aging changes in somatosensory, auditory and visual evoked potentials. *Electroencephalography and Clinical Neurophysiology, 58*, 14–24.

Allison, T., Wood, C. C., & Goff, W. R. (1983). Brain stem auditory, pattern-reversal visual, and short-latency somatosensory evoked potentials: Latencies in relation to age, sex, and brain and body size. *Electroencephalography and Clinical Neurophysiology, 55*, 619–636.

Ansari, D., & Coch, D. (2006). Bridges over troubled waters: Education and cognitive neuroscience. *Trends in Cognitive Sciences, 10*(4), 146–151.

Aso, K., Watanabe, K., Negoro, T., Takaetsi, E., Furune, S., Takahashi, I., et al. (1988). Developmental changes of pattern reversal visual evoked potentials. *Brain Development, 10*, 154–159.

Atchley, R. A., Rice, M. L., Betz, S. K., Kwasny, K. M., Sereno, J. A., & Jongman, A. (2006). A comparison of semantic and syntactic event related potentials generated by children and adults. *Brain and Language, 99*, 236–246.

Aydin, C., Idiman, F., & Idiman, E. (1987). Contingent negative variation in normal children and in children with attention deficit disorder. *Advances in Biological Psychiatry, 16*, 178–190.

Balachandran, C., Klistorner, A., & Billson, F. (2004). Multifocal VEP in children: Its maturation and clinical application. *British Journal of Ophthalmology, 88*, 226–232.

Banaschewski, T., Yordanova, J., Kolev, V., Heinrich, H., Albrecht, B., & Rothenberger, A. (2008). Stimulus context and motor preparation in attention-deficit/hyperactivity disorder. *Biological Psychology, 77*, 53–62.

Barnet, A. B., Friedman, S. L., Weiss, I. P., Ohlrich, E. S., Shanks, B., & Lodge, A. (1980). VEP development in infancy and early childhood: A longitudinal study. *Electroencephalography and Clinical Neurophysiology, 49*, 476–489.

Barnet, A. B., Ohlrich, E. S., Weiss, I. P., & Shanks, B. (1975). Auditory evoked potentials during sleep in normal children from ten days to three years of age. *Electroencephalography and Clinical Neurophysiology, 39*, 29–41.

Bashore, T. R. (1990). Age-related changes in mental processing revealed by analyses of event-related brain potentials. In J. W. Rohrbaugh, R. Parasuraman, & R. Johnson (Eds.), *Event-related brain potentials: Basic issues and applications* (pp. 242–275). New York: Oxford University Press.

Batty, M., & Taylor, M. J. (2002). Visual categorization during childhood: An ERP study. *Psychophysiology, 39*, 482–490.

Benavente, I., Tamargo, P., Tajada, N., Yuste, V., & Oliván, M. J. (2005). Flash visually evoked potentials in the newborn and their maturation during the first six months of life. *Documenta Opthalmologica, 110*, 255–263.

Bender, S., Oelkers-Ax, R., Resch, F., & Weisbrod, M. (2006). Frontal lobe involvement in the processing of meaningful auditory stimuli develops during childhood and adolescence. *Neuroimage, 33*, 759–773.

Benes, F. M. (1998). Human brain growth spans decades. *American Journal of Psychiatry, 155*(11), 1489.

Berman, S., Friedman, D., & Cramer, M. (1990). A developmental study of event-related potentials during explicit and implicit memory. *International Journal of Psychophysiology, 10*, 191–197.

Bonte, M. L., & Blomert, L. (2004a). Developmental changes in ERP correlates of spoken word recognition during early school years: A phonological priming study. *Clinical Neurophysiology, 115*, 409–423.

Bonte, M. L., & Blomert, L. (2004b). Developmental dyslexia: ERP correlates of anomalous phonological processing during spoken word recognition. *Cognitive Brain Research, 21*, 360–376.

Boon, M. Y., Suttle, C. M., & Dain, S. J. (2007). Transient VEP and psychophysical chromatic contrast thresholds in children and adults. *Vision Research, 47*, 2124–2133.

Borg, E., Spens, K. E., & Tonnquist, I. (1988). Auditory brain map, effects of age. *Scandinavian Audiology Supplementum, 30*, 161–164.

Bourgeois, J.-P. (2001). Synaptogenesis in the neocortex of the newborn: The ultimate frontier for individuation? In C. A. Nelson & M. Luciana (Eds.), *Handbook of developmental cognitive neuroscience* (pp. 23–34). Cambridge, MA: MIT Press.

Braddick, O. J., & Atkinson, J. (1983). Some recent findings on the development of human binocularity: A review. *Behavioural Brain Research, 10*, 141–150.
Braddick, O. J., Birtles, D., Wattam-Bell, J., & Atkinson, J. (2005). Motion- and orientation-specific cortical responses in infancy. *Vision Research, 45*, 3169–3179.
Brecelj, J. (2003). From immature to mature pattern ERG and VEP. *Documenta Opthalmologica, 107*, 215–224.
Brecelj, J., Strucl, M., Zidar, I., & Tekavcic-Pompe, M. (2002). Pattern ERG and VEP maturation in schoolchildren. *Clinical Neurophysiology, 113*, 1764–1770.
Bruneau, N., Roux, S., Guérin, P., Barthélémy, C., & Lelord, G. (1997). Temporal prominence of auditory evoked potentials (N1 wave) in 4–8-year-old children. *Psychophysiology, 34*, 32–38.
Burgio-Murphy, A., Klorman, R., Shaywitz, S. E., Fletcher, J. M., Marchione, K. E., Holahan, J. M., et al. (2007). Error-related event-related potentials in children with attention-deficit hyperactivity disorder, oppositional defiant disorder, reading disorder, and math disorder. *Biological Psychology, 75*, 75–86.
Carrillo-de-la-Peña, M., Holguín, S. R., Corral, M., & Cadaveira, F. (1999). The effects of stimulus intensity and age on visual-evoked potentials (VEPs) in normal children. *Psychophysiology, 36*, 693–698.
Carver, L. J., Dawson, G., Panagiotides, H., Meltzoff, A. N., McPartland, J., Gray, J., et al. (2003). Age-related differences in neural correlates of face recognition during the toddler and preschool years. *Developmental Psychobiology, 42*, 148–159.
Casey, B. J. (2002). Windows into the human brain. *Science, 296*, 1408–1409.
Casey, B. J., & de Haan, M. (2002). Introduction: New methods in developmental science. *Developmental Science, 5*(3), 265–267.
Caviness, V. S., Kennedy, D. N., Richelme, C., Rademacher, J., & Filipek, P. A. (1996). The human brain age 7–11 years: A volumetric analysis based on magnetic resonance images. *Cerebral Cortex, 6*, 726–736.
Ceponiene, R., Cheour, M., & Näätänen, R. (1998). Interstimulus interval and auditory event-related potentials in children: Evidence for multiple generators. *Electroencephalography and Clinical Neurophysiology, 108*, 345–354.
Ceponiene, R., Kushnerenko, E., Fellman, V., Renlund, M., Suominen, K., & Näätänen, R. (2002a). Event-related potential features indexing auditory discrimination in newborns. *Cognitive Brain Research, 13*, 101–113.
Ceponiene, R., Lepisto, T., Soininen, M., Aronen, E., Alku, P., & Näätänen, R. (2004). Event-related potentials associated with sound discrimination versus novelty detection in children. *Psychophysiology, 41*, 130–141.
Ceponiene, R., Rinne, T., & Näätänen, R. (2002b). Maturation of cortical sound processing as indexed by event-related potentials. *Clinical Neurophysiology, 113*, 870–882.
Ceponiene, R., Shestakova, A., Balan, P., Alku, P., Yiaguchi, K., & Näätänen, R. (2001). Children's auditory event-related potentials index sound complexity and "speechness." *International Journal of Neuroscience, 109*, 245–260.
Chauveau, N., Franceries, X., Doyon, B., Rigaud, B., Morucci, J. P., & Celsis, P. (2004). Effects of skull thickness, anisotropy, and inhomogeneity on forward EEG/ERP computations using a spherical three-dimensional resistor mesh model. *Human Brain Mapping, 21*, 86–97.
Chayo-Dichy, R., Ostrosky-Solis, F., Meneses, S., Harmony, T., & Guevara, M. A. (1990). The late event related potential CNV and PINV in normal and dyslexic subjects. *International Journal of Neuroscience, 54*, 347–357.
Chiron, C., Raynaud, C., Mazière, B., Zilbovicius, M., Laflamme, L., Masure, M.-C., et al. (1992) Changes in regional cerebral blood flow during brain maturation in children and adolescents. *Journal of Nuclear Medicine, 33*(5), 696–703.
Chugani, H. T. (1994). Development of regional brain glucose metabolism in relation to behavior and plasticity. In G. Dawson & K. W. Fischer (Eds.), *Human behavior and the developing brain* (pp. 153–175). New York: Guilford Press.
Chugani, H. T., Phelps, M. E., & Mazziotta, J. C. (1993). Positron emission tomography study of human brain functional development. In M. H. Johnson (Ed.), *Brain development and cognition* (pp. 125–143). Cambridge, MA: Blackwell.
Coch, D., Grossi, G., Coffey-Corina, S., Holcomb, P. J., & Neville, H. J. (2002a). A developmental investigation of ERP auditory rhyming effects. *Developmental Science, 5*(4), 467–489.
Coch, D., Grossi, G., Skendzel, W., & Neville, H. (2005a). ERP nonword rhyming effects in children and adults. *Journal of Cognitive Neuroscience, 17*(1), 168–182.
Coch, D., & Holcomb, P. J. (2003). The N400 in beginning readers. *Developmental Psychobiology, 43*(2), 146–166.
Coch, D., Maron, L., Wolf, M., & Holcomb, P. J. (2002b). Word- and picture-processing in children: An event-related potential study. *Developmental Neuropsychology, 22*(1), 373–406.
Coch, D., Sanders, L. D., & Neville, H. J. (2005b). An ERP study of selective auditory attention in children and adults. *Journal of Cognitive Neuroscience, 17*(4), 605–622.
Coch, D., Skendzel, W., Grossi, G., & Neville, H. J. (2005c). Motion and color processing in school-age children and adults: An ERP study. *Developmental Science, 8*(4), 372–386.
Coch, D., Skendzel, W., & Neville, H. J. (2005d). Auditory and visual refractory period effects in children and adults: An ERP study. *Clinical Neurophysiology, 116*, 2184–2203.
Cohen, J. (1973). The CNV in children with special reference to learning disabilities. *Electroencephalography and Clinical Neurophysiology*, (*Suppl 33*), 151–154.
Cohn, N. B., Dustman, R. E., & Shearer, D. E. (1985). The effect of age, sex and interstimulus interval on augmenting and reducing occipital VEPs. *Electroencephalography and Clinical Neurophysiology, 62*, 177–183.
Coles, M. G. H., & Rugg, M. D. (1995). Event-related brain potentials: An introduction. In M. D. Rugg & M. G. H. Coles (Eds.), *Electrophysiology of mind: Event-related brain potentials and cognition* (pp. 1–26). New York: Oxford University Press.
Conboy, B. T., & Mills, D. L. (2006). Two languages, one developing brain: Event-related potentials to words in bilingual toddlers. *Developmental Science, 9*(1), F1–F12.
Courchesne, E. (1977). Event-related brain potentials: A comparison between children and adults. *Science, 197*, 589–592
Courchesne, E. (1978). Neurophysiological correlates of cognitive development: Changes in long-latency event-related potentials from childhood to adulthood. *Electroencephalography and Clinical Neurophysiology, 45*, 468–482.

Courchesne, E. (1979). From infancy to adulthood: The neurophysiological correlates of cognition. In J. E. Desmedt (Ed.), *Progress in clinical neurophysiology: Vol. 6. Cognitive components in cerebral event-related potentials and selective attention* (pp. 224–242). Basel: Karger.

Courchesne, E. (1983). Cognitive components of the event-related brain potential: Changes associated with development. In A. W. K. Gaillard & W. Ritter (Eds.), *Tutorials in ERP research: Endogenous components* (pp. 329–344). New York: North Holland.

Courchesne, E. (1990). Chronology of postnatal human brain development: Event-related potentials, positron emission tomography, myelinogenesis, and synaptogenesis studies. In J. R. Rohrbaugh, R. Parasuraman, & R. Johnson (Eds.), *Event-related brain potentials: Basic issues and applications* (pp. 210–241). New York: Oxford University Press.

Courchesne, E., Chisum, H. J., Townsend, J., Cowles, A., Covington, J., Egaas, B., et al. (2000). Normal brain development and aging: Quantitative analysis at in vivo MR imaging in healthy volunteers. *Radiology, 216*, 672–682.

Courchesne, E., Ganz, L., & Norcia, A. M. (1981). Event-related brain potentials to human faces in infants. *Child Development, 52*, 804–811.

Crognale, M. A. (2002). Development, maturation, and aging of chromatic visual pathways: VEP results. *Journal of Vision, 2*, 438–450.

Crognale, M. A., Kelly, J. P., Chang, S. K., Weiss, A. H., & Teller, D. Y. (1997). Development of pattern visual evoked potentials: Longitudinal measurements in human infants. *Optometry and Vision Science, 74*(10), 808–815.

Crognale, M. A., Kelly, J. P., Weiss, A. H., & Teller, D. Y. (1998). Development of the spatio-chromatic visual evoked potential (VEP): A longitudinal study. *Vision Research, 38*, 3283–3292.

Crowley, K. E., & Colrain, I. M. (2004). A review of the evidence for P2 being an independent component process: Age, sleep and modality. *Clinical Neurophysiology, 115*, 732–744.

Curry, J. G., & Polich, J. (1992). P300, global probability and stimulus sequence effects in children. *Developmental Neuropsychology, 8*, 185–202.

Davies, P. L., Segalowitz, S. J., & Gavin, W. J. (2004a). Development of error-monitoring event-related potentials in adolescents. *Annals of the New York Academy of Sciences USA, 1021*, 324–328.

Davies, P. L., Segalowitz, S. J., & Gavin, W. J. (2004b). Development of response-monitoring ERPs in 7- to 25-year-olds. *Developmental Neuropsychology, 25*(3), 355–376.

Davis, E. P., Bruce, J., Snyder, K., & Nelson, C. A. (2003). The x-trials: Neural correlates of an inhibitory control task in children and adults. *Journal of Cognitive Neuroscience, 15*(3), 432–443.

Davis, H., & Onishi, S. (1969). Maturation of auditory evoked potentials. *International Audiology, 8*, 24–33.

De Bellis, M. D., Keshavan, M. S., Beers, S. R., Hall, J., Frustaci, K., Masalehdan, A., et al. (2001). Sex differences in brain maturation during childhood and adolescence. *Cerebral Cortex, 11*, 552–557.

de Haan, M. (2001). The neurophysiology of face processing during infancy and childhood. In C. A. Nelson & M. Luciana (Eds.), *Handbook of developmental cognitive neuroscience* (pp. 381–414). Cambridge, MA: MIT Press.

de Haan, M. (2007a). Current and future directions in infant electrophysiology. In M. de Haan (Ed.), *Infant EEG and event-related potentials* (pp. 305–316). New York: Psychology Press.

de Haan, M. (2007b). Visual attention and recognition memory in infancy. In M. de Haan (Ed.), *Infant EEG and event-related potentials* (pp. 101–143). New York: Psychology Press.

de Haan, M. (2008). Event-related potential (ERP) measures in visual development research. In L. A. Schmidt & S. J. Segalowitz (Eds.), *Developmental psychophysiology: Theory, systems, and methods* (pp. 103–126). New York: Cambridge University Press.

de Haan, M., Johnson, M. H., & Halit, H. (2003). Development of face-sensitive event-related potentials during infancy: A review. *International Journal of Psychophysiology, 51*, 45–58.

de Haan, M., Johnson, M. H., & Halit, H. (2007). Development of face-sensitive event-related potentials during infancy. In M. de Haan (Ed.), *Infant EEG and event-related potentials* (pp. 77–100). New York: Psychology Press.

de Haan, M., & Nelson, C. A. (1997). Recognition of the mother's face by six-month-old infants: A neurobehavioral study. *Child Development, 68*(4), 187–210.

de Haan, M., & Nelson, C. A. (1999). Brain activity differentiates face and object processing in 6-month-old infants. *Developmental Psychology, 35*(4), 1113–1121.

de Haan, M., Pascalis, O., & Johnson, M. H. (2002). Specialization of neural mechanisms underlying face recognition in human infants. *Journal of Cognitive Neuroscience, 14*(2), 199–209.

DeBoer, T., Scott, L. S., & Nelson, C. A. (2005). ERPs in developmental populations. In T. C. Handy (Ed.), *Event-related potentials: A methods handbook* (pp. 263–298). Cambridge, MA: MIT Press.

DeBoer, T., Scott, L. S., & Nelson, C. A. (2007). Methods for acquiring and analyzing infant event-related potentials. In M. de Haan (Ed.), *Infant EEG and event-related potentials* (pp. 5–37). New York: Psychology Press.

DeFrance, J. F., Sands, S., Schweitzer, F. C., Ginsberg, L., & Sharma, J. C. (1997). Age-related changes in cognitive ERPs of attenuation. *Brain Topography, 9*(4), 283–293.

Dehaene-Lambertz, G. (2000). Cerebral specialization for speech and non-speech stimuli in infants. *Journal of Cognitive Neuroscience, 12*(3), 449–460.

Dekaban, A. S. (1959). Developmental anatomy and physiology of the central nervous system from birth to two years of age. In A. S. Dekaban (Ed.), *Neurology of infants* (pp. 1–38). Baltimore: Williams & Wilkins.

Dekaban, A. S., & Sadowsky, D. (1978). Changes in brain weights during the span of human life: Relation of brain weights to body heights and body weights. *Annals of Neurology, 4*, 345–356.

Diamond, A. (2007). Interrelated and interdependent. *Developmental Science, 10*(1), 152–158.

Dimoska, A., Johnstone, S. A., Chiswick, D., Barry, R. J., & Clarke, A. R. (2007). A developmental investigation of stop-signal inhibition. *Journal of Psychophysiology, 21*(2), 109–126.

Domino, E. F., Matsuoka, S., Waltz, J., & Cooper, I. (1964). Simultaneous recording of scalp and epidural somatosensory-evoked responses in man. *Science, 145*(3637), 1199–1200.

Dubois, J., Dehaene-Lambertz, G., Soarès, C., Cointepas, Y., Le Bihan, D., & Hertz-Pannier, L. (2008). Microstructural correlates of infant functional development: Example of the visual pathways. *Journal of Neuroscience, 28*(8), 1943–1948.

Dumais-Huber, C., & Rothenberger, A. (1992). Psychophysiological correlates of orienting, anticipation and contingency changes in children with psychiatric disorders. *Journal of Psychophysiology, 6*, 225–239.

Dustman, R. E., & Beck, E. C. (1969). The effects of maturation and aging on the wave form of visually evoked potentials. *Electroencephalography and Clinical Neurophysiology, 26*, 2–11.

Dykman, R. A., Ackerman, P., Loizou, P. C., & Casey, P. H. (2000). An event-related potential study of older children with an early history of failure to thrive. *Developmental Neuropsychology, 18*(2), 187–212.

Eggermont, J. J. (1988). On the rate of maturation of sensory evoked potentials. *Electroencephalography and Clinical Neurophysiology, 70*, 293–305.

Eimer, M. (2000). Event-related brain potentials distinguish processing stages involved in face perception and recognition. *Clinical Neurophysiology, 111*, 694–705.

Ellingson, R. J., Lathrop, G. H., Danahy, T., & Nelson, B. (1973). Variability of visual evoked potentials in human infants and adults. *Electroencephalography and Clinical Neurophysiology, 34*, 113–124.

Enoki, H., Sanada, S., & Ohtahara, S. (1990). P300 in children: Developmental change and its clinical application. *Electroencephalography and Clinical Neurophysiology, 76*, 104.

Enoki, H., Sanada, S., Yoshinaga, H., Oka, E., & Ohtahara, S. (1993). The effects of age on the N200 component of the auditory event-related potentials. *Cognitive Brain Research, 1*, 161–167.

Falkenstein, M., Hoorman, J., Christ, S., & Hohnsbein, J. (2000). ERP components on reaction errors and their functional significance: A tutorial. *Biological Psychology, 51*, 87–107.

Ferriss, G. S. (1970). Evoked responses and their clinical application. *American Journal of EEG Technology, 10*(1), 17–26.

Ferriss, G. S., Davis, G. D., Dorsen, M. M., & Hackett, E. R. (1967). Changes in latency and form of the photically induced average evoked response in human infants. *Electroencephalography and Clinical Neurophysiology, 22*, 305–312.

Fischer, K. W., & Rose, S. P. (1994). Dynamic development of coordination of components in brain and behavior: A framework for theory and research. In G. Dawson & K. W. Fischer (Eds.), *Human behavior and the developing brain* (pp. 3–66). New York: Guilford Press.

Fitzgerald, K. D., Welsh, R. C., Gehring, W. J., Abelson, J. L., Himle, J. A., Liberzon, I., & Taylor, S. F. (2005). Error-related hyperactivity of the anterior cingulate cortex in obsessive-compulsive disorder. *Biological Psychiatry, 57*, 287–294.

Fonseca, L. C., Tedrus, G. M. A. S., & Gilbert, M. A. P. (2006). Event related potentials during the visual discrimination of words and pseudowords by children. *Arquivos de neuro-psiquiatria, 64*(3-A), 553–558.

Fox, N. A., Schmidt, L. A., Henderson, H. A., & Marshall, P. J. (2007). Developmental psychophysiology: Conceptual and methodological issues. In J. T. Cacioppo, L. G. Tassinary, & G. Berntson (Eds.), *Handbook of psychophysiology* (3rd ed., pp. 453–481). Cambridge: Cambridge University Press.

Friederici, A. D., & Hahne, A. (2001). Development patterns of brain activity reflecting semantic and syntactic processes. In J. Weissenborn & B. Höhle (Eds.), *Approaches to bootstrapping: Phonological, lexical, syntactic and neurophysiological aspects of early language acquisition* (Vol. 24, pp. 231–246). Philadelphia: John Benjamins.

Friedman, D. (1991). The endogenous scalp-recorded brain potentials and their relationship to cognitive development. In J. R. Jennings & M. G. H. Coles (Eds.), *Handbook of cognitive psychophysiology: Central and autonomic nervous system approaches* (pp. 621–656). New York: Wiley.

Friedman, D., Kazmerski, V., & Fabiani, M. (1997). An overview of age-related changes in the scalp distribution of P3b. *Electroencephalography and Clinical Neurophysiology, 104*, 498–513.

Friedman, D., Putnam, L., & Sutton, S. (1989). Cognitive brain potentials in children, young adults, and senior citizens: Homologous components and changes in scalp distribution. *Developmental Neuropsychology, 5*(1), 33–60.

Friedrich, M., & Friederici, A. D. (2004). N400-like semantic incongruity effect in 19-month-olds: Processing known words in picture contexts. *Journal of Cognitive Neuroscience, 16*(8), 1465–1477.

Friedrich, M., & Friederici, A. D. (2005a). Lexical priming and semantic integration reflected in the event-related potential of 14-month-olds. *Neuroreport, 16*(6), 653–656.

Friedrich, M., & Friederici, A. D. (2005b). Phonotactic knowledge and lexical-semantic processing in one-year-olds: Brain responses to words and nonsense words in picture contexts. *Journal of Cognitive Neuroscience, 17*(11), 1785–1802.

Friedrich, M., & Friederici, A. D. (2005c). Semantic sentence processing reflected in the event-related potentials of one- and two-year-old children. *Neuroreport, 16*(16), 1801–1804.

Friedrich, M., & Friederici, A. D. (2006). Early N400 development and later language acquisition. *Psychophysiology, 43*(1), 1–12.

Fuchigami, T., Okubo, O., Ejiri, K., Fujita, Y., Kohira, R., Noguchi, Y., et al. (1995). Developmental changes in P300 wave elicited during two different experimental conditions. *Pediatric Neurology, 13*, 25–28.

Fuchigami, T., Okubo, O., Fujita, Y., Okuni, M., Noguchi, Y., & Yamada, T. (1993). Auditory event-related potentials and reaction time in children: Evaluation of cognitive development. *Developmental Medicine and Child Neurology, 35*, 230–237.

Fulton, A. B., Brecelj, J., Lorenz, B., Moskowitz, A., Thompson, D., & Westall, C. A. (2006). Pediatric clinical visual electrophysiology: A survey of actual practice. ISCEV Committee for Pediatric Clinical Electrophysiology Guidelines. *Documenta Opthalmologica, 113*, 193–204.

Gaillard, A. W. (1977). The late CNV wave: Preparation versus expectancy. *Psychophysiology, 14*(6), 563–568.

Galambos, R. (1982). Maturation of auditory evoked potentials. In G. A. Chiarenza & D. Papakostopoulos (Eds.), *Clinical application of cerebral evoked potentials in pediatric medicine* (pp. 323–343). Amsterdam: Excerpta Medica.

Gavin, W. J., & Davies, P. L. (2008). Obtaining reliable psychophysiological data with child participants: Methodological considerations. In L. A. Schmidt & S. J. Segalowitz (Eds.), *Developmental psychophysiology: Theory, systems and methods* (pp. 424–447). Cambridge: Cambridge University Press.

Gehring, W. J., Himle, J., & Nisenson, L. G. (2000). Action-monitoring dysfunction in obsessive-compulsive disorder. *Psychological Science, 11*(1), 1–6.

Giedd, J. N., Blumenthal, J., Jeffries, N. O., Castellanos, F. X., Liu, H., Zijdenbos, A., et al. (1999). Brain development during childhood and adolescence: A longitudinal MRI study. *Nature Neuroscience*, *2*(10), 861–863.

Giedd, J. N., Snell, J. W., Lange, N., Rajapaske, J. C., Casey, B. J., Kozuch, P. L., et al. (1996a). Quantitative magnetic resonance imaging of human brain development: Ages 4–18. *Cerebral Cortex*, *6*, 551–560.

Giedd, J. N., Vaituzis, C., Hamburger, S. D., Lange, N., Rajapakse, J. C., Kaysen, D., et al. (1996b). Quantitative MRI of the temporal lobe, amygdala, and hippocampus in normal human development: Ages 4–18 years. *Journal of Comparative Neurology*, *366*, 223–230.

Gill, O., & Polich, J. (2002). P300 stimulus sequence effects in children and adults. *Perceptual and Motor Skills*, *94*, 509–520.

Gilley, P. M., Sharma, A., Dorman, M. F., & Martin, K. (2005). Developmental changes in refractoriness of the cortical auditory evoked potential. *Clinical Neurophysiology*, *116*, 648–657.

Gogtay, N., Giedd, J. N., Lusk, L., Hayashi, K. M., Greenstein, D., Vaituzis, A. C., et al. (2004). Dynamic mapping of human cortical development during childhood through early adulthood. *Proceedings of the National Academy of Sciences USA*, *101*(21), 8174–8179.

Gomes, H., Duff, M., Barnhardt, J., Barrett, S., & Ritter, W. (2007). Development of auditory selective attention: Event-related measures of channel selection and target detection. *Psychophysiology*, *44*, 711–727.

Gomes, H., Dunn, M., Ritter, W., Kurtzberg, D., Brattson, A., Kruezer, J. A., et al. (2001). Spatiotemporal maturation of the central and lateral N1 components to tones. *Developmental Brain Research*, *129*, 147–155.

Gomes, H., Sussman, E., Ritter, W., Kurtzberg, D., Cowan, N., & Vaughan, H. G. (1999). Electrophysiological evidence of developmental changes in the duration of auditory sensory memory. *Developmental Psychology*, *35*(5), 294–302.

Gonzalez-Garrido, A. A., Oropeza de Alba, J. L., Castañeda, R. R., Castañeda, J. M. R., Avalos, M. C. P., & Sosa, M. V. (1997). Event related brain potentials to semantically incongruent words in children of different ages. *Archives of Medical Research*, *28*(1), 109–113.

Goodin, D. S., Squires, K. C., Henderson, B. H., & Starr, A. (1978). Age-related variations in evoked potentials to auditory stimuli in normal human subjects. *Electroencephalography and Clinical Neurophysiology*, *44*, 447–458.

Gordon, G. E., & McCulloch, D. L. (1999). A VEP investigation of parallel visual pathway development in primary school age children. *Documenta Opthalmologica*, *99*, 1–10.

Goswami, U. (2005). The use of event related potentials in the study of early cognitive development. *Infant and Child Development*, *14*, 95–98.

Groen, Y., Wijers, A. A., Mulder, L. J. M., Minderaa, R. B., & Althaus, M. (2007). Physiological correlates of learning by performance feedback in children: A study of EEG event-related potentials and evoked heart rate. *Biological Psychology*, *76*, 174–1887.

Grossi, G., Coch, D., Coffey-Corina, S., Holcomb, P. J., & Neville, H. J. (2001). Phonological processing in visual rhyming: A developmental ERP study. *Journal of Cognitive Neuroscience*, *13*(5), 610–625.

Grünewald-Zuberbier, E., Grünewald, G., Rasche, A., & Netz, J. (1978). Contingent negative variation and alpha attenuation responses in children with different abilities to concentrate. *Electroencephalography and Clinical Neurophysiology*, *44*, 37–47.

Guillery, G. W. (2005). Is postnatal neocortical maturation hierarchical? *Trends in Cognitive Sciences*, *28*(10), 513–517.

Gullickson, G. R. (1973). CNV and behavioral attention to a glide-tone warning of interesting non-moving or kaleidoscopic visual or auditory patterns in 2- and 3-year-old children. *Electroencephalography and Clinical Neurophysiology*, (*Suppl 33*), 145–150.

Gumenuyk, V., Korzyukov, O., Alho, K., Escera, C., & Näätänen, R. (2004). Effects of auditory distraction on electrophysiological brain activity and performance in children aged 8–13 years. *Psychophysiology*, *41*, 30–36.

Hahne, A., Eckstein, K., & Friederici, A. D. (2004). Brain signatures of syntactic and semantic processes during children's language development. *Journal of Cognitive Neuroscience*, *16*(7), 1302–1318.

Hajcak, G., Franklin, M. E., Foa, E. B., & Simons, R. F. (2008). Increased error-related brain activity in pediatric obsessive-compulsive disorder before and after treatment. *American Journal of Psychiatry*, *165*(1), 116–123.

Hämäläinen, J. A., Leppänen, P. H. T., Guttorm, T. K., & Lyytinen, H. (2007). N1 and P2 components of auditory event-related potentials in children with and without reading disabilities. *Clinical Neurophysiology*, *118*, 2263–2275.

Harding, G. F. A., Grose, J. H., Wilson, A. Y., & Bissenden, J. G. (1989). The pattern reversal VEP in short-gestation infants. *Electroencephalography and Clinical Neurophysiology*, *74*, 76–80.

Hartmann, E. E. (1995). Infant visual development: An overview of studies using visual evoked potential measures from Harter to the present. *International Journal of Neuroscience*, *80*, 203–235.

Heinrich, H., Gevensleben, H., Freisleder, F. J., Moll, G. H., & Rothenberger, A. (2004). Training of slow cortical potentials in attention-deficit/hyperactivity disorder: Evidence of positive behavioral and neurophsyiological effects. *Biological Psychiatry*, *55*, 772–775.

Heisz, J. J., Watter, S., & Shedden, J. M. (2006). Automatic face identity encoding at the N170. *Vision Research*, *46*(28), 4604–4614.

Hillyard, S. A., Hink, R. F., Schwent, V. L., & Picton, T. W. (1973). Electrical signs of selective attention in the human brain. *Science*, *182*, 177–180.

Hink, R. F., Hillyard, S. A., & Benson, P. J. (1978). Event-related brain potentials and selective auditory attention to acoustic and phonetic cues. *Biological Psychology*, *6*, 1–16.

Hink, R. F., van Voorhis, S. T., Hillyard, S. A., & Smith, T. S. (1977). The division of attention and the human auditory evoked potential. *Neuropsychologia*, *15*, 597–605.

Hogan, A. M., Vargha-Khadem, F., Kirkham, F. J., & Baldeweg, T. (2005). Maturation of action monitoring from adolescence to adulthood: An ERP study. *Developmental Science*, *8*(6), 525–534.

Holcomb, P. J., Coffey, S. A., & Neville, H. J. (1992). Visual and auditory sentence processing: A developmental analysis using event-related brain potentials. *Developmental Neuropsychology*, *8*, 203–241.

Holland, B. A., Haas, D., Norman, D., Brant-Zawadzki, M., & Newton, T. H. (1986). MRI of normal brain maturation. *American Journal of Neuroradiology*, *7*, 201–208.

Hollants-Gilhuijs, M. A. M., De Munck, J. C., Kubova, Z., van Royen, E., & Spekreijse, H. (2000). The development of

hemispheric asymmetry in human motion VEPs. *Vision Research, 40*, 1–11.
Hood, B. M. (2001). Combined electrophysiological and behavioral measurement in developmental cognitive neuroscience: Some cautionary notes. *Infancy, 2*(2), 213–217.
Horst, R. L., Thatcher, R. W., Lester, M. L., & McAlester, R. (1982). Differential effects of age on sensory evoked potentials in children. In A. Rothenberger (Ed.), *Event-related potentials in children: Basic concepts and clinical applications* (pp. 71–76). New York: Elsevier.
Howard, L., & Polich, J. (1985). P300 latency and memory span development. *Developmental Psychology, 21*, 283–289.
Huttenlocher, P. R. (1979). Synaptic density in human frontal cortex—developmental changes and effects of aging. *Brain Research, 163*, 195–205.
Huttenlocher, P. R. (1994). Synaptogenesis in human cerebral cortex. In G. Dawson & K. W. Fischer (Eds.), *Human behavior and the developing brain* (pp. 137–152). New York: Guilford Press.
Huttenlocher, P. R., & Dabholkar, A. S. (1997). Regional differences in synaptogenesis in human cerebral cortex. *Journal of Comparative Neurology, 387*, 167–178.
Huttenlocher, P. R., de Courten, C., Garey, L. J., & van der Loos, H. (1982). Synaptic development in human cerebral cortex. *International Journal of Neurology, 16–17*, 144–154.
Jing, H., & Benasich, A. A. (2006). Brain responses to tonal changes in the first two years of life. *Brain & Development, 28*, 247–256.
Johnson, M. H. (2001). Functional brain development in humans. *Nature Reviews Neuroscience, 2*, 475–483.
Johnson, R. (1989). Developmental evidence for modality-dependent P300 generators: A normative study. *Psychophysiology, 26*, 651–667.
Johnstone, S. J., Barry, R. J., Anderson, J. W., & Coyle, S. F. (1996). Age-related changes in child and adolescent event-related potential component morphology, amplitude, and latency to standard and target stimuli in an auditory oddball task. *International Journal of Psychophysiology, 24*, 223–238.
Johnstone, S. J., Dimoska, A., Smith, J. L., Barry, R. J., Pleffer, C. B., Chiswick, D., et al. (2007). The development of stop-signal and Go/Nogo response inhibition in children aged 7–12 years: Performance and event-related potential indices. *International Journal of Psychophysiology, 63*, 25–38.
Johnstone, S. J., Pleffer, C. B., Barry, R. J., Clarke, A. R., & Smith, J. L. (2005). Development of inhibitory processing during the Go/Nogo task: A behavioral and event-related potential study of children and adults. *Journal of Psychophysiology, 19*(1), 11–23.
Jonkman, L. M. (2006). The development of preparation, conflict monitoring and inhibition from early childhood to young adulthood: A Go/Nogo ERP study. *Brain Research, 1097*, 181–193.
Jonkman, L. M., Lansbergen, M., & Stauder, J. E. A. (2003). Developmental differences in behavioral and event-related brain responses associated with response preparation in a go/nogo task. *Psychophysiology, 40*, 752–761.
Jonkman, L. M., van Melis, J. J. M., Kemner, C., & Markus, C. R. (2007). Methylphenidate improves deficient error evaluation in children with ADHD: An event-related brain potential study. *Biological Psychology, 76*, 217–229.
Juottonen, K., Revonsuo, A., & Lang, H. (1996). Dissimilar age influences on two ERP waveforms (LPC and N400) reflecting semantic context effects. *Cognitive Brain Research, 4*, 99–107.
Kang, H. C., Burgund, E. D., Lugar, H. M., Petersen, S. E., & Schlagger, B. L. (2003). Comparison of functional activation foci in children and adults using common stereotactic space. *Neuroimage, 19*, 16–28.
Karhu, J., Herrgård, E., Pääkkönen, A., Luoma, L., Airaksinen, E., & Partanen, J. (1997). Dual cerebral processing of elementary auditory input in children. *Neuroreport, 8*, 1327–1330.
Karmel, B. Z., Hoffman, R. F., & Fegy, M. J. (1974). Processing of contour information by human infants evidenced by pattern-dependent evoked potentials. *Child Development, 45*(1), 39–48.
Karmiloff-Smith, A. (1997). Crucial differences between developmental cognitive neuroscience and adult neuropsychology. *Developmental Neuropsychology, 13*(4), 513–524.
Karrer, J. H., Karrer, R., Bloom, D., Chaney, L., & Davis, R. (1998). Event-related potentials during an extended visual recognition memory task depict delayed development of cerebral inhibitory processes among 6-month-old infants with Down syndrome. *International Journal of Psychophysiology, 29*(2), 167–200.
Karrer, R., & Ackles, P. K. (1987). Visual event-related potentials of infants during a modified oddball procedure. In R. Johnson, J. Rohrbaugh, & R. Parasuraman (Eds.), *Current trends in event-related potential research* (pp. 603–608). New York: Elsevier Science.
Karrer, R., & Monti, L. A. (1995). Event-related potentials of 4–7-week-old infants in a visual recognition memory task. *Electroencephalography and Clinical Neurophysiology, 94*, 414–424.
Kilpelainen, R., Koistinen, A., Kononen, M., Herrgard, E., Partanen, J., & Karhu, J. (1999). P300 sequence effects differ between children and adults for auditory stimuli. *Psychophysiology, 36*, 343–350.
Kim, E. Y., Iwaki, N., Imashioya, H., Uno, H., & Fujita, T. (2007). Error-related negativity in a visual go/no-go task: Children vs. adults. *Developmental Neuropsychology, 31*(2), 181–191.
Kim, E. Y., Iwaki, N., Uno, H., & Fujita, T. (2005). Error-related negativity in children: Effect of an observer. *Developmental Neuropsychology, 28*(3), 871–883.
Klein, A. J. (1984). Frequency and age-dependent auditory evoked potential thresholds in infants. *Hearing Research, 16*, 291–297.
Klorman, R. (1975). Contingent negative variation and cardiac deceleration in a long preparatory interval: A developmental study. *Psychophysiology, 12*(6), 609–617.
Klorman, R., Thompson, L. W., & Ellingson, R. J. (1978). Event-related brain potentials across the life span. In E. Callaway, P. Tueting, & S. Koslow (Eds.), *Event-related brain potentials in man* (pp. 511–583). New York: Academic Press.
Kobiella, A., Grossmann, T., Reid, V. M., & Striano, T. (2007). The discrimination of angry and fearful face expressions in 7-month-old infants: An event-related potential study. *Cognition & Emotion, 22*(1), 134–146.
Korpilahti, P., & Lang, H. A. (1994). Auditory ERP components and mismatch negativity in dysphasic children. *Electroencephalography and Clinical Neurophysiology, 91*, 256–264.
Kos-Pietro, S., Towle, V. L., Cakmur, R., & Spire, J.-P. (1997). Maturation of human visual evoked potentials: 27 weeks conceptional age to 2 years. *Neuropediatrics, 28*, 318–323.

Kraemer, M., Abrahamsson, M., & Sjöström, A. (1999). The neonatal development of the light flash visual evoked potential. *Documenta Opthalmologica, 99*, 21–39.

Kraus, N., McGee, T., Carrell, T., Sharma, A., Micco, A., & Nicol, T. (1993). Speech-evoked cortical potentials in children. *Journal of the American Academy of Audiology, 4*, 238–248.

Kurtzberg, D., Hilpert, P. J., Kreuzer, J. A., & Vaughan, H. G. (1984a). Differential maturation of cortical auditory evoked potentials to speech sounds in normal fullterm and very low-birthweight infants. *Developmental Medicine and Child Neurology, 26*, 466–475.

Kurtzberg, D., Vaughan, H. G., Courchesne, E., Friedman, D., Harter, M. R., & Putnam, L. (1984b). Developmental aspects of event-related potentials. In R. Karrer, J. Cohen, & P. Tueting (Eds.), *Annals of the New York Academy of Sciences: Vol. 425. Brain and information: Event related potentials* (pp. 300–318). New York: New York Academy of Sciences.

Kurtzberg, D., Vaughan, H. G., & Novak, G. P. (1986). Discriminative brain responses to speech sounds in the newborn high risk infant. In V. Gallai (Ed.), *Maturation of the CNS and evoked potentials* (pp. 253–259). Amsterdam: Elsevier.

Kushnerenko, E., Ceponiene, R., Balan, P., Fellman, V., & Näätänen, R. (2002). Maturation of the auditory change detection response in infants: A longitudinal ERP study. *Brain Imaging, 13*(15), 1843–1848.

Kushnerenko, E., Ceponiene, R., Fellman, V., Huotilainen, M., & Winkler, I. (2001). Event-related potential correlates of sound duration: Similar pattern from birth to adulthood. *Neuroreport, 12*, 3777–3781.

Kutas, M., & Hillyard, S. A. (1980). Reading senseless sentences: Brain potentials reflect semantic incongruity. *Science, 207*, 203–204.

Ladish, C., & Polich, J. (1989). P300 and probability in children. *Journal of Experimental Child Psychology, 48*, 212–223.

Ladouceur, C. D., Dahl, R. E., Birmaher, B., Axelson, D. A., & Ryan, N. D. (2006). Increased error-related negativity (ERN) in childhood anxiety disorders: ERP and source localization. *Journal of Child Psychology and Psychiatry, 47*(10), 1073–1082.

Ladouceur, C. D., Dahl, R. E., & Carter, C. S. (2004). ERP correlates of action monitoring in adolescence. *Annals of the New York Academy of Sciences USA, 1021*, 329–336.

Ladouceur, C. D., Dahl, R. E., & Carter, C. S. (2007). Development of action monitoring through adolescence into adulthood: ERP and source localization. *Developmental Science, 10*(6), 874–891.

Landi, N., & Perfetti, C. A. (2007). An electrophysiological investigation of semantic and phonological processing in skilled and less-skilled comprehenders. *Brain and Language, 102*, 30–45.

Lenroot, R. K., & Giedd, J. N. (2007). The structural development of the human brain as measured longitudinally with magnetic resonance imaging. In D. Coch, K. W. Fischer, & G. Dawson (Eds.), *Human behavior, learning, and the developing brain: Typical development* (pp. 50–73). New York: Guilford Press.

Lenzi, P., Conti, A., & Arpini, A. (1978). Late evoked responses to acoustic stimuli in relation to age. *Audiology, 17*, 152–158.

Leppänen, P. H. T., & Lyytinen, H. (1997). Auditory event-related potentials in the study of developmental language-related disorders. *Audiology and Neuro-otology, 2*, 308–340.

Leppänen, P. H. T., Pihko, E., Eklund, K. M., & Lyytinen, H. (1999). Cortical responses of infants with and without a genetic risk for dyslexia: II. Group effects. *Neuroreport, 10*(5), 969–973.

Lewis, M. D., Lamm, C., Segalowitz, S. J., Stieben, J., & Zelazo, P. D. (2006). Neurophysiological correlates of emotion regulation in children and adolescents. *Journal of Cognitive Neuroscience, 18*(3), 430–443.

Liotti, M., Pliszka, S. R., Perez, R., Kothmann, D., & Woldorff, M. G. (2005). Abnormal brain activity related to performance monitoring and error detection in children with ADHD. *Cortex, 41*, 377–388.

Little, V. M., Thomas, D. G., & Letterman, M. R. (1999). Single-trial analyses of developmental trends in infant auditory event-related potentials. *Developmental Neuropsychology, 16*(3), 455–478.

Loveless, N. E., & Sanford, A. J. (1974). Slow potential correlates of preparatory set. *Biological Psychology, 1*, 303–314.

Loveless, N. E., & Sanford, A. J. (1975). The impact of warning signal intensity on reaction time and components of the contingent negative variation. *Biological Psychology, 2*, 217–226.

Lovrich, D., Cheng, J. C., & Velting, D. M. (1996). Late cognitive brain potentials, phonological and semantic classification of spoken words, and reading ability in children. *Journal of Clinical and Experimental Neuropsychology, 18*(2), 161–177.

Luck, S. J. (2005). An introduction to the event-related potential technique. Cambridge, MA: MIT Press.

Luna, B., Thulborn, K. R., Munoz, D. P., Merriam, E. P., Garver, K. E., Minshew, N. J., et al. (2001). Maturation of widely distributed brain function subserves cognitive development. *Neuroimage, 13*(5), 786–793.

Lyytinen, H., Guttorm, T. K., Huttunen, T., Hämäläinen, J., Leppänen, P. H. T., & Vesterinen, M. (2005). Psychophysiology of developmental dyslexia: A review of findings including studies of children at risk for dyslexia. *Journal of Neurolinguistics, 18*, 167–195.

Madrid, M., & Crognale, M. A. (2000). Long-term maturation of visual pathways. *Visual Neuroscience, 17*, 831–837.

Malcolm, C. A., McCulloch, D. L., & Shepherd, A. J. (2002). Pattern-reversal visual evoked potentials in infants: Gender differences during early visual maturation. *Developmental Medicine and Child Neurology, 44*, 345–351.

Martin, L., Barajas, J. J., Fernandez, R., & Torres, E. (1988). Auditory event-related potentials in well-characterized groups of children. *Electroencephalography and Clinical Neurophysiology, 71*, 375–381.

Matsuzawa, J., Matsui, M., Konishi, T., Noguchi, K., Gur, R. C., Bilker, W., et al. (2001). Age-related volumetric changes of brain gray and white matter in healthy infants and children. *Cerebral Cortex, 11*, 335–342.

McArthur, G. M., & Bishop, D. V. M. (2002). Event-related potentials reflect individual differences in age-invariant auditory skills. *Neuroreport, 13*, 1079–1082.

McArthur, G. M., & Bishop, D. V. M. (2004). Which people with Specific Language Impairment have auditory processing deficits? *Cognitive Neuropsychology, 21*(1), 79–94.

McArthur, G. M., & Bishop, D. V. M. (2005). Speech and non-speech processing in people with Specific Language Impairment: A behavioural and electrophysiological study. *Brain and Language, 94*, 260–273.

McCulloch, D. L. (2007). Visual evoked potentials in infants. In M. de Haan (Ed.), *Infant EEG and event-related potentials* (pp. 39–76). New York: Psychology Press.

McCulloch, D. L., Orbach, H., & Skarf, B. (1999). Maturation of the pattern-reversal VEP in human infants: A theoretical framework. *Vision Research, 29*, 3673–3680.

McCulloch, D. L., & Skarf, B. (1991). Development of the human visual system: Monocular and binocular pattern VEP latency. *Investigative Ophthalmology and Visual Science, 32*(8), 2372–2381.

McCulloch, D. L., Skarf, B., & Taylor, M. J. (1993). A two-stage linear model for maturation of visual evoked potential (VEP) latency in infants. *Investigative Ophthalmology and Visual Science, 34*, 1353.

McIsaac, H., & Polich, J. (1992). Comparison of infant and adult P300 from auditory stimuli. *Journal of Experimental Child Psychology, 53*, 115–128.

McPherson, W. B., Ackerman, P. T., Holcomb, P. J., & Dykman, R. A. (1998). Event-related brain potentials elicited during phonological processing differentiate subgroups of reading disabled adolescents. *Brain and Language, 62*, 163–185.

Michel, A. E., & Garey, L. J. (1984). The development of dendritic spines in the human visual cortex. *Human Neurobiology, 3*, 223–227.

Mills, D. L., Coffey-Corina, S. A., & Neville, H. J. (1993). Language acquisition and cerebral specialization in 20-month-old infants. *Journal of Cognitive Neuroscience, 5*, 317–334.

Mills, D. L., Coffey-Corina, S. A., & Neville, H. J. (1994). Variability in cerebral organization during primary language acquisition. In K. W. Fischer & G. Dawson (Eds.), *Human behavior and the developing brain* (pp. 427–455). New York: Guilford Press.

Mills, D. L., Plunkett, K., Prat, C., & Schafer, G. (2005). Watching the infant brain learn words: Effects of vocabulary size and experience. *Cognitive Development, 20*, 19–31.

Mills, D. L., Prat, C., Zangl, R., Stager, C. L., Neville, H. J., & Werker, J. F. (2004). Language experience and the organization of brain activity to phonetically similar words: ERP evidence from 14- and 20-month-olds. *Journal of Cognitive Neuroscience, 16*(8), 1452–1464.

Mills, D. L., & Sheehan, E. A. (2007). Experience and developmental changes in the organization of language-relevant brain activity. In D. Coch, K. W. Fischer, & G. Dawson (Eds.), *Human behavior, learning, and the developing brain: Typical development* (pp. 183–218). New York: Guilford Press.

Mitzdorf, U. (1994). Properties of cortical generators of event-related potentials. *Pharmacopsychiatry, 27*, 49–51.

Molfese, D. L. (1990). Auditory evoked responses recorded from 16-month-old human infants to words they did and did not know. *Brain and Language, 38*, 345–363.

Molfese, D. L., Molfese, V. J., & Molfese, P. J. (2007). Relation between early measures of brain responses to language stimuli and childhood performance on language and language-related tasks. In D. Coch, G. Dawson, & K. W. Fischer (Eds.), *Human behavior, learning, and the developing brain: Atypical development* (pp. 191–211). New York: Guilford Press.

Molfese, D. L., Morse, P. A., & Peters, C. J. (1990). Auditory evoked responses to names for different objects: Cross-modal processing as a basis for infant language acquisition. *Developmental Psychology, 26*(5), 780–795.

Molfese, V. J., Molfese, D. L., Beswick, J. L., Jacobi-Vessels, J., Molfese, P. J., Molnar, A. E., et al. (2008). Use of event-related potentials to identify language and reading skills. *Topics in Language Disorders, 28*(1), 28–45.

Monk, C. S., Webb, S. J., & Nelson, C. A. (2001). Prenatal neurobiological development: Molecular mechanisms and anatomical change. *Developmental Neuropsychology, 19*(2), 211–236.

Moore, J. K., & Guan, Y.-L. (2001). Cytoarchitectural and axonal maturation in human auditory cortex. *Journal of the Association for Research in Otolaryngology, 2*, 297–311.

Moskowitz, A., & Sokol, S. (1983). Developmental changes in the human visual system as reflected by the latency of the pattern reversal VEP. *Electroencephalography and Clinical Neurophysiology, 56*, 1–15.

Mullis, R. J., Holcomb, P. J., Diner, B. C., & Dykman, R. A. (1985). The effects of aging on the P3 component of the visual event-related potential. *Electroencephalography and Clinical Neurophysiology, 62*, 141–149.

Näätänen, R., & Picton, T. (1987). The N1 wave of the human electric and magnetic response to sound: A review and an analysis of the component structure. *Psychophysiology, 24*(4), 375–425.

Nagy, Z., Westerberg, H., & Klingberg, T. (2004). Maturation of white matter is associated with the development of cognitive functions during childhood. *Journal of Cognitive Neuroscience, 16*(7), 1227–1233.

Nelson, C. A. (1997). Electrophysiological correlates of memory development in the first year of life. In H. Reese & M. Franzen (Eds.), *Biological and neuropsychological mechanisms: Life-span developmental psychology* (pp. 95–131). Mahwah, NJ: Erlbaum.

Nelson, C. A. (2007). A developmental cognitive neuroscience approach to the study of atypical development: A model system involving infants of diabetic mothers. In D. Coch, G. Dawson, & K. W. Fischer (Eds.), *Human behavior, learning, and the developing brain: Atypical development* (pp. 1–27). New York: Guilford Press.

Nelson, C. A., & Collins, P. F. (1991). Event-related potentials and looking-time analysis of infants' responses to familiar and novel events: Implications for visual recognition memory. *Developmental Psychology, 27*(1), 50–58.

Nelson, C. A., & Collins, P. F. (1992). Neural and behavioral correlates of visual recognition memory in 4- and 8-month-old infants. *Brain and Cognition, 19*, 105–121.

Nelson, C. A., & deRegnier, R.-A. (1992). Neural correlates of attention and memory in the first year of life. *Developmental Neuropsychology, 8*(2&3), 119–134.

Nelson, C. A., & Luciana, M. (Eds.). (2001). *Handbook of developmental cognitive neuroscience*. Cambridge MA: MIT Press.

Nelson, C. A., & Monk, C. S. (2001). The use of event-related potentials in the study of cognitive development. In C. A. Nelson & M. Luciana (Eds.), *Handbook of development cognitive neuroscience* (pp. 125–136). Cambridge, MA: MIT Press.

Nelson, C. A., & Salapatek, P. (1986). Electrophysiological correlates of infant recognition memory. *Child Development, 57*, 1483–1497.

Nelson, C. A., Thomas, K. M., de Haan, M., & Wewerka, S. S. (1998). Delayed recognition memory in infants and adults as revealed by event-related potentials. *International Journal of Psychophysiology, 29*, 145–165.

Neville, H. J. (1995). Developmental specificity in neurocognitive development in humans. In M. S. Gazzaniga (Ed.), *The cognitive neurosciences* (pp. 219–231). Cambridge, MA: MIT Press.

Neville, H. J., Coffey, S. A., Holcomb, P. J., & Tallal, P. (1993). The neurobiology of sensory and language processing in language-impaired children. *Journal of Cognitive Neuroscience, 5*, 235–253.

Neville, H. J., & Mills, D. L. (1997). Epigenesis of language. *Mental Retardation and Developmental Disabilities Research Reviews, 3*, 282–292.

Newton, J. E. O., & Oglesby, D. M. (1994). Visual slow brain potentials in children with attention deficit disorder. *Integrative Physiological and Behavioral Science, 29*(1), 39–55.

Nikkel, L., & Karrer, R. (1994). Differential effects of experience on the ERP and behavior of 6-month-old infants: Trends during repeated stimulus presentations. *Developmental Neuropsychology, 10*(1), 1–11.

Novak, G. P., Kurtzberg, D., Kreuzer, J., & Vaughan, H. G. (1989). Cortical responses to speech sounds and their formants in normal infants: Maturational sequence and spatiotemporal analysis. *Electroencephalography and Clinical Neurophysiology, 73*, 295–305.

Nunez, P. L., & Srinivasan, R. (2006). *Electric fields of the brain: The neurophysics of EEG* (2nd ed.). New York: Oxford University Press.

Oades, R. D., Dittman-Balcar, A., Schepker, R., Eggers, C., & Zerbin, D. (1996). Auditory event-related potentials (ERPs) and mismatch negativity (MMN) in healthy children and those with attention-deficit or tourette/tic symptoms. *Biological Psychology, 43*, 163–185.

Oades, R. D., Dittmann-Balcar, A., & Zerbin, D. (1997). Development and topography of auditory event-related potentials (ERPs): Mismatch and processing negativity in individuals 8–22 years of age. *Psychophysiology, 34*, 677–693.

O'Connor, S., Morzorati, S., Christian, J. C., & Li, T.-K. (1994). Heritable features of the auditory oddball event-related potential: Peaks, latencies, morphology and topography. *Electroencephalography and Clinical Neurophysiology, 92*, 115–125.

Ohlrich, E. S., Barnet, A. B., Weiss, I. P., & Shanks, B. L. (1978). Auditory evoked potential development in early childhood: A longitudinal study. *Electroencephalography and Clinical Neurophysiology, 44*, 411–423.

Otto, D., Karrer, R., Halliday, R. A., Horst, R. L., Klorman, R., Squires, N., et al. (1984). Developmental aspects of event-related potentials: Aberrant development. In R. Karrer, J. Cohen, & P. Tueting (Eds.), *Annals of the New York Academy of Sciences: Vol. 425. Brain and information: Event related potentials* (pp. 319–337). New York: New York Academy of Sciences.

Otto, D., & Reiter, L. (1984). Developmental changes in slow cortical potentials of young children with elevated body lead burden: Neurophysiological considerations. In R. Karrer, J. Cohen, & P. Tueting (Eds.), *Annals of the New York Academy of Sciences: Vol. 425. Brain and information: Event-related potentials* (pp. 377–383). New York: New York Academy of Sciences.

Paetau, R., Ahonen, A., Salonen, O., & Sams, M. (1995). Auditory evoked magnetic fields to tones and pseudowords in healthy children and adults. *Journal of Clinical Neurophysiology, 12*(2), 177–185.

Pang, E. W., & Taylor, M. J. (2000). Tracking the development of the N1 from age 3 to adulthood: An examination of speech and non-speech stimuli. *Clinical Neurophysiology, 111*, 388–397.

Paus, T., Zijdenbos, A., Worsley, K., Collins, D. L., Blumenthal, J., Giedd, J. N., et al. (1999). Structural maturation of neural pathways in children and adolescents: In vivo study. *Science, 283*, 1908–1911.

Pearce, J. W., Crowell, D. H., Tokioka, A., & Pacheco, G. P. (1989). Childhood developmental changes in the auditory P300. *Journal of Child Neurology, 4*, 100–106.

Perchet, C., & Garcia-Larrea, L. (2005). Learning to react: Anticipatory mechanisms in children and adults during a visuospatial attention task. *Clinical Neurophysiology, 116*, 1906–1917.

Pfefferbaum, A., Mathalon, D. H., Sullivan, E. V., Rawles, J. M., Zipursky, R. B., & Lim, K. O. (1994). A quantitative magnetic resonance imaging study of changes in brain morphology from infancy to late adulthood. *Archives of Neurology, 51*, 874–887.

Pfefferbaum, A., & Rosenblum, M. (1987). Skull thickness influences P3 amplitude. *Psychopharmacology Bulletin, 23*(3), 493–496.

Picton, T. W., Bentin, S., Berg, P., Donchin, E., Hillyard, S. A., Johnson, R., et al. (2000). Guidelines for using human event-related potentials to study cognition: Recording standards and publication criteria. *Psychophysiology, 37*, 127–152.

Picton, T. W., & Taylor, M. J. (2007). Electrophysiological evaluation of human brain development. *Developmental Neuropsychology, 31*(3), 249–278.

Pike, A. A., Marlow, N., & Reber, C. (1999). Maturation of the flash visual evoked potential in preterm infants. *Early Human Development, 54*, 215–222.

Pivik, R. T., Dykman, R. A., Jing, H., Gilchrist, J. M., & Badger, T. M. (2007). The influence of infant diet on early developmental changes in processing human voice speech stimuli: Variations in breast and milk formula-fed infants at 3 and 6 months after birth. *Developmental Neuropsychology, 31*(3), 279–335.

Polich, J. (2007). Updating P300: An integrative theory of P3a and P3b. *Clinical Neurophysiology, 118*, 2128–2148.

Polich, J., Howard, L., & Starr, A. (1983). P300 latency correlates with digit span. *Psychophysiology, 20*(6), 665–669.

Polich, J., Howard, L., & Starr, A. (1985). Effects of age on the P300 component of the event-related potential from auditory stimuli: Peak definition, variation, and measurement. *Journal of Gerontology, 40*(6), 721–726.

Polich, J., Ladish, C., & Burns, T. (1990). Normal variation of P300 in children: Age, memory span, and head size. *International Journal of Psychophysiology, 9*, 237–248.

Pompe, M. T., Kranjc, B. S., & Brecelj, J. (2006). Visual evoked potentials to red-green stimulation in schoolchildren. *Visual Neuroscience, 23*, 447–451.

Ponton, C. W., Eggermont, J. J., Coupland, S. G., & Winkelaar, R. (1993). The relation between head size and auditory brain-stem response interpeak latency maturation. *Journal of the Acoustical Society of America, 94*(4), 2149–2158.

Ponton, C. W., Eggermont, J. J., Khosla, D., Kwong, B., & Don, M. (2002). Maturation of human central auditory system activity: Separating auditory evoked potentials by dipole source modeling. *Clinical Neurophysiology, 113*, 407–420.

Ponton, C. W., Eggermont, J. J., Kwong, B., & Don, M. (2000). Maturation of human central auditory system activity: Evidence from multi-channel evoked potentials. *Clinical Neurophysiology, 111*, 22–236.

Prevec, T. S., Ribaric, K., & Butinar, D. (1984). Contingent negative variation audiometry in children. *Audiology, 23*, 114–126.

Pryds, O., Trojaborg, W., Carlsen, J., & Jensen, J. (1989). Determinants of visual evoked potentials in preterm infants. *Early Human Development, 19*, 117–125.

Pujol, J., Soriano-Mas, C., Ortiz, H., Sebastián-Gallés, N., Losilla, J. M., & Deus, J. (2006). Myelination of language-related areas in the developing brain. *Neurology, 66*, 339–343.

Reiss, A. L., Abrams, M. T., Singer, H. S., Ross, J. L., & Denckla, M. B. (1996). Brain development, gender and IQ in children: A volumetric imaging study. *Brain, 119*, 1763–1774.

Reynolds, G. D., & Richards, J. E. (2005). Familiarization, attention, and recognition memory in infancy: An event-related potential and cortical source localization study. *Developmental Psychology, 41*(4), 598–615.

Richards, J. E. (2003). Attention affects the recognition of briefly presented visual stimuli in infants: An ERP study. *Developmental Science, 6*(3), 312–328.

Richter, W., & Richter, M. (2003). The shape of the fMRI BOLD response in children and adults changes systematically with age. *Neuroimage, 20*, 1122–1131.

Roche, A. F. (1953). Increase in cranial thickness during growth. *Human Biology, 25*(2), 81–92.

Rosahl, S. K., & Knight, R. T. (1995). Role of prefrontal cortex in generation of the contingent negative variation. *Cerebral Cortex, 5*(2), 123–134.

Roy, M. S., Gosselin, J., Hanna, N., Orquin, J., & Chemtob, S. (2004). Influence of the state of alertness of the pattern visual evoked potentials (PVERP) in very young infant. *Brain Development, 26*, 197–202.

Rugg, M. D. (1984a). Event-related potentials and the phonological processing of words and non-words. *Neuropsychologia, 22*, 435–443.

Rugg, M. D. (1984b). Event-related potentials in phonological matching tasks. *Brain and Language, 23*, 225–240.

Sabisch, B., Hahne, A., Glass, E., von Suchodoletz, W., & Friederici, A. D. (2006). Auditory language comprehension in children with developmental dyslexia: Evidence from event-related brain potentials. *Journal of Cognitive Neuroscience, 18*(10), 1676–1695.

Salamy, A., Mendleson, T., & Tooley, W. H. (1982). Developmental profiles for the brainstem auditory evoked potential. *Early Human Development, 6*, 331–339.

Sampaio, R. C., & Truwit, C. L. (2001). Myelination in the developing human brain. In C. A. Nelson & M. Luciana (Eds.), *Handbook of developmental cognitive neuroscience* (pp. 35–44). Cambridge, MA: MIT Press.

Sanders, L. D., Stevens, C., Coch, D., & Neville, H. J. (2006). Selective auditory attention in 3- to 5-year-old children: An event-related potential study. *Neuropsychologia, 44*, 2126–2138.

Santesso, D. L., & Segalowitz, S. J. (2008). Developmental differences in error-related ERPs in middle- to late-adolescent males. *Developmental Psychology, 44*(1), 205–217.

Santesso, D. L., Segalowitz, S. J., & Schmidt, L. A. (2005). ERP correlates of error monitoring in 10-year olds are related to socialization. *Biological Psychology, 70*, 79–87.

Santesso, D. L., Segalowitz, S. J., & Schmidt, L. A. (2006a). Error-related electrocortical responses in 10-year-old children and young adults. *Developmental Science, 9*(5), 473–481.

Santesso, D. L., Segalowitz, S. J., & Schmidt, L. A. (2006b). Error-related electrocortical responses are enhanced in children with obsessive-compulsive behaviors. *Developmental Neuropsychology, 29*(3), 431–445.

Sartory, G., Heine, A., Müller, B. W., & Elvermann-Hallner, A. (2002). Event- and motor-related potentials during the continuous performance task in attention-deficit/hyperactivity disorder. *Journal of Psychophysiology, 16*, 97–106.

Satterfield, J. H., Schell, A. M., Backs, R. W., & Hidaka, K. C. (1984). A cross-sectional and longitudinal study of age effects of electrophysiological measures in hyperactive and normal children. *Biological Psychiatry, 19*(7), 973–990.

Scerif, G., Kotsoni, E., & Casey, B. J. (2006). Functional neuroimaging of early cognitive development. In R. Cabeza & A. Kingstone (Eds.), *Handbook of functional neuroimaging of cognition* (2nd ed., pp. 351–378). Cambridge, MA: MIT Press.

Schulman-Galambos, C., & Galambos, R. (1978). Cortical responses from adults and infants to complex visual stimuli. *Electroencephalography and Clinical Neurophysiology, 45*, 425–435.

Segalowitz, S. J., & Davies, P. L. (2004). Charting the maturation of the frontal lobe: An electrophysiological strategy. *Brain and Cognition, 55*, 116–133.

Segalowitz, S. J., & Schmidt, L. A. (2008). Capturing the dynamic endophenotype: A developmental psychophysiological manifesto. In L. A. Schmidt & S. J. Segalowitz (Eds.), *Developmental psychophysiology: Theory, systems, and methods* (pp. 1–12). New York: Cambridge University Press.

Segalowitz, S. J., Unsal, A., & Dywan, J. (1992a). Cleverness and wisdom in 12-year-olds: Electrophysiological evidence for late maturation of the frontal lobe. *Developmental Neuropsychology, 8*, 279–298.

Segalowitz, S. J., Wagner, W. J., & Menna, R. (1992b). Lateral versus frontal ERP predictors of reading skill. *Brain and Cognition, 20*, 85–103.

Shahin, A., Roberts, L. E., & Trainor, L. J. (2004). Enhancement of auditory cortical development by musical experience in children. *Neuroreport, 15*, 1917–1921.

Sharma, A., Kraus, N., McGee, T. J., & Nicol, T. G. (1997). Developmental changes in P1 and N1 central auditory responses elicited by consonant-vowel syllables. *Electroencephalography and Clinical Neurophysiology, 104*, 540–545.

Shaw, N. A., & Cant, B. R. (1981). Age-dependent changes in the amplitude of the pattern visual evoked potential. *Electroencephalography and Clinical Neurophysiology, 51*, 671–673.

Shibasaki, H., & Miyazaki, M. (1992). Event-related potential studies in infants and children. *Journal of Clinical Neurophysiology, 9*(3), 408–418.

Silva-Pereyra, J., Rivera-Gaxiola, M., Fernández, T., Díaz-Comas, L., Harmony, T., Fernández-Bouzas, A., et al. (2003). Are poor readers semantically challenged? An event-related brain potential assessment. *International Journal of Psychophysiology, 49*, 187–199.

Silva-Pereyra, J., Rivera-Gaxiola, M., & Kuhl, P. K. (2005). An event-related brain potential study of sentence comprehension in preschoolers: Semantic and morphosyntactic processing. *Cognitive Brain Research, 23*, 247–258.

Sloper, J. J., & Collins, A. D. (1998). Reduction in binocular enhancement of the visual-evoked potential during development accompanies increasing stereoacuity. *Journal of Pediatric Ophthalmology and Strabismus, 35*, 154–158.

Snyder, K., Webb, S. J., & Nelson, C. A. (2002). Theoretical and methodological implications of variability in infant brain response during a recognition memory paradigm. *Infant Behavior and Development, 25*, 466–494.

Sokol, S. (1982). Infant visual development: Evoked potential estimates. *Annals of the New York Academy of Sciences, 388*, 514–525.

Sokol, S., & Jones, K. (1979). Implicit time of pattern evoked potentials in infants: An index of maturation of spatial vision. *Vision Research, 19*, 747–755.

Sokol, S., Moskowitz, A., & Towle, V. L. (1981). Age-related changes in the latency of the visual evoked potential: Influences of check size. *Electroencephalography and Clinical Neurophysiology, 51*, 559–562.

Sowell, E. R., Peterson, B. S., Thompson, P. M., Welcome, S. E., Henkenius, A. L., & Toga, A. W. (2003). Mapping cortical change across the human life span. *Nature Neuroscience, 6*(3), 309–315.

Sowell, E. R., Thompson, P. M., Tessner, K. D., & Toga, A. W. (2001). Mapping continued brain growth and gray matter density reduction in dorsal frontal cortex: Inverse relationships during postadolescent brain maturation. *Journal of Neuroscience, 21*(22), 8819–8829.

Spekreijse, H. (1978). Maturation of contrast EPs and development of visual resolution. *Archives Italiennes de Biologie, 116*(3–4), 358–369.

Spekreijse, H. (1983). Comparison of acuity tests and pattern evoked potential criteria: Two mechanisms underlay acuity maturation in man. *Behavioural Brain Research, 10*, 107–117.

Stapells, D. R. (1989). Auditory brainstem response assessment of infants and children. *Seminars in Hearing, 10*(3), 229–251.

Stauder, J. E. A., Molenaar, P. C. M., & van der Molen, M. W. (1999). Brain activity and cognitive transition during childhood: A longitudinal event-related brain potential study. *Child Neuropsychology, 5*(1), 41–59.

Steinschneider, M., Kurtzberg, D., & Vaughan, H. G. (1992). Event-related potentials in developmental neuropsychology. In I. Rapin & S. J. Segalowitz (Eds.), *Handbook of neuropsychology: Vol. 6. Child neuropsychology* (pp. 239–299). New York: Elsevier Science.

Stevens, C., Fanning, J., Coch, D., Sanders, L., & Neville, H. (2008). Neural mechanisms of selective attention are enhanced by computer training. *Brain Research, 1205*, 55–69.

Stevens, C., Sanders, L., & Neville, H. (2006). Neurophysiological evidence for selective auditory attention deficits in children with specific language impairment. *Brain Research, 1111*, 143–152.

Stige, S., Fjell, A. M., Smith, L., Lingren, M., & Walhovd, K. B. (2007). The development of visual P3a and P3b. *Developmental Neuropsychology, 31*(1), 563–584.

Strandburg, R. J., Marsh, J. T., Brown, W. S., Asarnow, R. F., Higa, J., Harper, R., et al. (1996). Continuous-processing-related event-related potentials in children with Attention Deficit Hyperactivity Disorder. *Biological Psychiatry, 40*, 964–980.

Sussman, E., Steinschneider, M., Gumenyuk, V., Grushko, J., & Lawson, K. (2008). The maturation of human evoked brain potentials to sounds presented at different stimulus rates. *Hearing Research, 236*, 61–79.

Swingler, M. M., Sweet, M. A., & Carver, L. J. (2007). Relations between mother–child interactions and the neural correlates of face processing in 6-month-olds. *Infancy, 11*(1), 63–86.

Takeshita, K., Nagamine, T., Thuy, D. H. D., Satow, T., Matsuhashi, M., Yamamoto, J., et al. (2002). Maturational change of parallel auditory processing in school-aged children revealed by simultaneous recording of magnetic and electric cortical responses. *Clinical Neurophysiology, 113*, 1470–1484.

Taylor, M. J., & Baldeweg, T. (2002). Application of EEG, ERP and intracranial recordings to the investigation of cognitive functions in children. *Developmental Science, 5*(3), 318–334.

Taylor, M. J., & Eals, M. (1996). An event-related potential study of development using visual semantic tasks. *Journal of Psychophysiology, 10*, 125–139.

Taylor, M. J., Edmonds, G. E., McCarthy, G., & Allison, T. (2001). Eyes first! Eye processing develops before face processing in children. *Neuroreport, 12*(8), 1671–1676.

Taylor, M. J., McCarthy, G., Saliba, E., & Degiovanni, E. (1999). ERP evidence of developmental changes in processing of faces. *Clinical Neurophysiology, 110*, 910–915.

Taylor, M. J., & McCulloch, D. L. (1992). Visual evoked potentials in infants and children. *Journal of Clinical Neurophysiology, 9*(3), 357–372.

Taylor, M. J., Menzies, R., MacMillan, L. J., & Whyte, H. E. (1987). VEPs in normal full-term and premature neonates: Longitudinal versus cross-sectional data. *Electroencephalography and Clinical Neurophysiology, 68*, 20–27.

Taylor, M. J., & Smith, M. L. (1995). Age-related ERP changes in verbal and nonverbal memory tasks. *Journal of Psychophysiology, 9*, 283–297.

Thierry, G. (2005). The use of event-related potentials in the study of early cognitive development. *Infant and Child Development, 14*, 85–94.

Thomas, D. G., & Lykins, M. S. (1995). Event-related potential measures of 24-hour retention in 5-month-old infants. *Developmental Psychology, 31*(6), 946–957.

Thomas, K. M., & Nelson, C. A. (1996). Age-related changes in the electrophysiological response to visual stimulus novelty: A topographical approach. *Electroencephalography and Clinical Neurophysiology, 98*, 294–308.

Tokioka, A., Crowell, D. H., & Pearce, J. W. (1987). Electrophysiological investigation of cognition in infants. In D. M. Topping, D. Crowell, & V. N. Kobayashi (Eds.), *Thinking across cultures: The Third International Conference on Thinking* (pp. 369–385). Hillsdale, NJ: Erlbaum.

Tonnquist-Uhlén, I., Borg, E., & Spens, K. E. (1995). Topography of auditory evoked long-latency potentials in normal children, with particular reference to the N1 component. *Electroencephalography and Clinical Neurophysiology, 95*, 34–41.

Tonnquist-Uhlén, I., Ponton, C. W., Eggermont, J. J., Kwong, B., & Don, M. (2003). Maturation of human central auditory system activity: The T-complex. *Clinical Neurophysiology, 114*, 685–701.

Torkildsen, J. v. K., Sannerud, T., Syversen, G., Thormodsen, R., Simonsen, H. G., Moen, I., et al. (2006). Semantic organization of basic-level words in 20-month-olds: An ERP study. *Journal of Neurolinguistics, 19*, 431–454.

Torkildsen, J. v. K., Syversen, G., Simonsen, H. G., Moen, I., & Lindgren, M. (2007). Brain responses to lexical-semantic priming in children at-risk for dyslexia. *Brain and Language, 102*, 243–261.

Trainor, L. J. (2008). Event-related potential (ERP) measures in auditory development research. In L. A. Schmidt & S. J. Segalowitz (Eds.), *Developmental psychophysiology: Theory, systems, and methods* (pp. 69–102). New York: Cambridge University Press.

Trainor, L. J., Samuel, S. S., Desjardins, R. N., & Sonnadara, R. R. (2001). Measuring temporal resolution in infants using mismatch negativity. *Neuroreport, 12*, 2443–2448.

Trainor, L. J., Shahin, A., & Roberts, L. E. (2003). Effects of musical training on the auditory cortex in children. *Annals of the New York Academy of Sciences, 999*, 506–513.

Tsuneishi, S., & Casaer, P. (1997). Stepwise decrease in VEP latencies and the process of myelination in the human visual pathway. *Brain and Development, 19*, 547–551.

Tsuneishi, S., & Casaer, P. (2000). Effects of preterm extrauterine visual experience on the development of the human visual system: A flash VEP study. *Developmental Medicine and Child Neurology, 42*, 663–668.

Tsuneishi, S., Casaer, P., Fock, J. M., & Hirano, S. (1995). Establishment of normal values for flash visual-evoked potentials (VEPs) in preterm infants—a longitudinal study with special reference to 2 components in the N1 wave. *Electroencephalography and Clinical Neurophysiology, 96*(4), 291–299.

van Baal, G. C. M., de Geus, E. J. C., & Boomsma, D. I. (1998). Longitudinal study of genetic influences on ERP-P3 during childhood. *Developmental Neuropsychology, 14*(1), 19–45.

Van Bogaert, P., Wikler, D., Damhaut, P., Szliwowski, H. B., & Goldman, S. (1998). Regional changes in glucose metabolism during brain development from the age of 6 years. *Neuroimage, 8*, 62–68.

van Herten, M., Pasman, J., van Leeuwen, T. H., Been, P. H., van der Leij, A., Zwarts, F., et al. (2008). Differences in AERP responses and atypical hemispheric specialization in 17-month-old children at risk of dyslexia. *Brain Research, 1201*, 100–105.

van Meel, C. S., Heslenfeld, D. J., Oosterlaan, J., & Sergeant, J. A. (2007). Adaptive control deficits in attention-deficit/hyperactivity disorder (ADHD): The role of error processing. *Psychiatry Research, 151*, 211–220.

Vaughan, H. G., & Kurtzberg, D. (1992). Electrophysiologic indices of human brain maturation and cognitive development. In M. R. Gunnar & C. A. Nelson (Eds.), *The Minnesota symposia on child psychology: Vol. 24. Developmental behavioral neuroscience* (pp. 1–36). Hillsdale, NJ: Erlbaum.

Voll, R., Rieger, H., Allehoff, W. H., Eisert, H. G., & Schmidt, M. H. (1982). VEP in eight-year-old children: Normative data and relation to cerebral dysfunction. In A. Rothenberger (Ed.), *Event-related potentials in children: Basic concepts and clinical application* (pp. 67–70). New York: Elsevier.

Walter, W. G. (1966). Electrophysiological contributions to psychiatric therapy. *Current Psychiatric Therapies, 6*, 13–25.

Walter, W. G., Cooper, R., Aldridge, V. J., McCallum, W. C., & Winter, A. L. (1964). Contingent negative variation: An electric sign of sensorimotor association and expectancy in human brain. *Nature, 203*(4943), 380–384.

Webb, S. J. (2007). Recognition memory: Brain–behavior relations from 0 to 3. In D. Coch, K. W. Fischer, & G. Dawson (Eds.), *Human behavior, learning, and the developing brain: Typical development* (pp. 138–182). New York: Guilford Press.

Webb, S. J., Long, J. D., & Nelson, C. A. (2005). A longitudinal investigation of visual event-related potentials in the first year of life. *Developmental Science, 8*(6), 605–616.

Webb, S. J., Monk, C. S., & Nelson, C. A. (2001). Mechanisms of postnatal neurobiological development: Implications for human development. *Developmental Neuropsychology, 19*(2), 147–171.

Weber, B. A. (1972). Short-term habituation of the averaged electroencephalic response in infants. *Journal of Speech and Hearing Research, 15*, 753–756.

Weber-Fox, C., Spencer, R., Cuadrado, E., & Smith, A. (2003). Development of neural processes mediating rhyme judgments: Phonological and orthographic interactions. *Developmental Psychobiology, 43*, 128–145.

Weitzman, E. D., & Graziani, L. J. (1968). Maturation and topography of the auditory evoked response of the prematurely born infant. *Developmental Psychobiology, 1*(2), 79–89.

Wetzel, N., & Schröger, E. (2007a). Cognitive control of involuntary attention and distraction in children and adolescents. *Brain Research, 1155*, 134–146.

Wetzel, N., & Schröger, E. (2007b). Modulation of involuntary attention by the duration of novel and pitch deviant sounds in children and adolescents. *Biological Psychology, 75*, 24–31.

Wetzel, N., Widmann, A., Berti, S., & Schröger, E. (2006). The development of involuntary and voluntary attention from childhood to adulthood: A combined behavioral and event-related potential study. *Clinical Neurophysiology, 117*, 2191–2203.

Wiersema, J. R., van der Meere, J. J., & Roeyers, H. (2005). ERP correlates of impaired error monitoring in children with ADHD. *Journal of Neural Transmission, 112*, 1417–1430.

Wiersema, J. R., van der Meere, J. J., & Roeyers, H. (2007). Developmental changes in error monitoring: An event-related potential study. *Neuropsychologia, 45*, 1649–1657.

Woods, D. L., Hillyard, S. A., & Hansen, J. C. (1984). Event-related brain potentials reveal similar attentional mechanisms during selective listening and shadowing. *Journal of Experimental Psychology: Human Perception and Performance, 10*(6), 761–777.

Wunderlich, J. L., & Cone-Wesson, B. K. (2006). Maturation of CAEP in infants and children: A review. *Hearing Research, 212*, 212–223.

Wunderlich, J. L., Cone-Wesson, B. K., & Shepherd, R. (2006). Maturation of the cortical auditory evoked potential in infants and young children. *Hearing Research, 212*, 185–202.

Yakovlev, P. I., & Lecours, A.-R. (1967). The myelogenetic cycles of regional maturation of the brain. In A. Minkowski (Ed.),

Regional development of the brain in early life (pp. 3–69). Oxford: Blackwell Scientific.
Yeung, N., Botvinick, M. M., & Cohen, J. D. (2004). The neural basis of error detection: Conflict monitoring and the error-related negativity. *Psychological Review, 111*(4), 931–959.
Zemon, V., Eisner, W., Gordon, J., Grose-Fifer, J., Tenedios, F., & Shoup, H. (1995). Contrast-dependent responses in the human visual system: Childhood through adulthood. *International Journal of Neuroscience, 80*, 181–201.
Zenker, F., & Barajas, J. J. (1999). Auditory P300 development from an active, passive, and single-tone paradigms. *International Journal of Psychophysiology, 33*, 99–111.

CHAPTER

18 The Components of Aging

David Friedman

Abstract

This chapter reviews the effects of aging on a variety of ERP components. These include the relatively early-latency P50, N1 and P2 potentials, which are thought to reflect primarily the processing of sensory aspects of experience. Although more data are required, study of these components suggests that older adults do not exhibit sensory deficits per se but may exhibit deficiencies in inhibitory processing at the relatively early stages of information processing. Other longer-latency components, such as the N2b, P3a (novelty P3) and P3b reflect higher-order cognitive functions and are modulated more by task characteristics than by the sensory properties of the events that elicit them. Age-related studies indicate prolonged latency of the N2b and P3b, consistent with the general slowing reported in the behavioral literature. The P3a, a neural sign of the orienting response thought to depend upon prefrontal cortex and its interconnections, does not habituate in older adults, suggesting that older adults continue to recruit prefrontal cortical mechanisms for events that should no longer capture their attention. Several ERP modulations are defined on the basis of a difference between electrical activity in one condition compared to that in another. These comprise the mismatch negativity, reflecting relatively automatic deviance detection, the parietal old/new or episodic memory (EM) effect, an index of recollective processing, the N400, a measure of semantic processing, the error-related negativity (ERN) and the medial-frontal negativity (MFN), the latter two reflecting executive aspects of cognition. Studies of the MMN indicate less sensitivity to deviance as we age. The parietal EM effect is sometimes smaller in older relative to young adults, suggesting an age-related reduction in the quality of information retrieved from episodic memory. The N400 literature indicates that, although the N400 shows age-related diminution, semantic processing is generally intact in older adults. Too little is known about the ERN and MFN to come to conclusions at this time. In general, this review indicates a mixed picture of spared and impaired cognitive functions as individuals age.

Keywords: cognitive aging; deviance detection; executive function; recollection; event-related potential (ERP); mismatch negativity; parietal EM effect; error-related negativity; medial-frontal negativity; P50; N1 (N100); P2 (P200); N400; N2b; P3a; P3b

Normal aging carries certain risks, not the least of which is change in those aspects of cognition, such as top-down, executive control, and mnemonic processing, that are critical for navigating everyday life and, therefore, successful aging. However, not all aspects of cognition are impaired as we age. The age-related pattern of spared and relatively compromised cognition in aging populations appears to be reflected in a pattern of spared and relatively compromised event-related potential (ERP) components. Unfortunately, the amount of research work devoted to higher-order cognitive processes as opposed to putatively sensory processes is disproportionately weighted toward the endogenous components (Friedman et al., 1997; Polich, 1996). Endogenous components are relatively insensitive to the stimulus's physical properties, but their amplitudes and latencies are intimately linked to the nature of the task in which those stimuli are embedded (Sutton et al., 1967). As a result, the current review of age-related change in ERP components will, of necessity, be slanted toward these components.

Within the endogenous domain, the greatest amount of research attention has been paid to the P300 family of components (see Chapter 7, this volume), primarily using the ubiquitous oddball paradigm (Donchin & Coles, 1988). Despite what could be considered an overemphasis on the oddball task, P300s have also been recorded in an extremely wide variety of other task paradigms, including working memory (McEvoy et al., 2001), episodic memory (Friedman, 2007; Friedman et al., 2007a), task switching (Friedman et al., 2007b; Themanson et al., 2006), repetition priming (Hamberger & Friedman, 1992; Rugg et al., 1997), semantic priming (Kutas & Iragui, 1998), and selective attention (Gaeta et al., 2003; Karayanidis et al., 1995).

Because a large number of studies on aging have involved the auditory oddball paradigm (Iragui et al., 1993) and to a lesser extent its visual counterpart (Beck et al., 1980), there are a number of investigations of the aging brain's response to the sensory information inherent in the standard, target, and novel stimuli typically presented during this task (manifested in the middle- and longer-latency components between 50 and 100 ms poststimulus), such as P1 or P50 and N1 or N100 (Reinvang et al., 2005). These relatively early components are thought to reflect the arrival of sensory information at midbrain and higher, modality-specific, cortical processing centers. Very little is known about the impact of aging on the very-early-latency auditory brainstem responses (ABRs) occurring within the first 10 ms following stimulus onset (see Chapter 4, this volume). These ABRs reflect the arrival of auditory information in the nuclei of the brainstem.

Several ERP modulations are defined on the basis of a difference between electrical activity in one condition compared to that in another, including the mismatch negativity (MMN; Näätänen et al., 2007), the negative difference (Nd) waveform in studies of selective attention (Hansen & Hillyard, 1980), the N400 in linguistic and semantic memory assessments (Chapters 6, 11, and 15, this volume), the lateralized readiness potential (LRP; Coles et al., 1988; Chapter 9, this volume), and the error-related (Falkenstein et al., 2001) and medial frontal (Gehring & Willoughby., 2002) negativities in studies of executive control (see Chapter 10, this volume). With the exception of the Nd (Gaeta et al., 2003; Karayanidis et al., 1995; Woods, 1992) and the LRP (e.g., Falkenstein et al., 2006; Zeef & Kok, 1993), for which only a small number of studies exist, age-related changes in these remaining components will be reviewed below.

I will first review aging effects on the components between 50 and 200 ms poststimulus recorded primarily in passive and active auditory oddball paradigms. Although similar studies have been conducted with comparable findings in the visual modality (see De Sanctis et al., 2008, for a review), a much greater number have used auditory stimuli. Four components have been assessed: P50, N1 (N100), P2 (P200), and N2 (N200). The MMN is part of the "N2 complex" (Näätänen et al., 2007), which includes the N2b.

Relatively Early-Latency Components

Because some of these ERP activities appear to be affected by attention, it is safest to measure them when attention is not an issue, as in the passive oddball paradigm, in which participants are instructed to ignore the stimuli. One can also measure the frequent standards within the active oddball paradigm, as these presumably do not recruit the same degree of attentional processing as targets (which typically require a reaction time, or RT, response) or novel stimuli; the latter, if sufficiently surprising, typically capture attention involuntarily (Friedman et al., 2001).

To illustrate the basic phenomena that will be reviewed in this section, Figure 18.1 (data modified from Gaeta et al., 2001a) depicts the grand-averaged young- (18–30 years old) and older-adult (65–85 years old) ERPs elicited by first-position standards occurring in a train of eight identical standards each separated by a 500 ms interstimulus-interval (ISI). Participants watched a silent movie while ignoring the trains of stimuli. Prominent P50 (also known as *P1*), N1, and P2 components are present in both the young and older adult grand mean data. The P50 appears larger in the older adult waveforms, whereas the N1 component is of similar magnitude in the two groups. The P2 is larger in the young adult data but only at the 8 s intertrial interval (ITI; i.e., the interval between the last tone in one train and the first tone in the next train). Furthermore, there is a marked effect of ITI on the N1 and P2 components, both of which are larger after an 8 s interval. By contrast, the P50 does not appear to be influenced by the amount of time between stimulus trains. The greater reduction in the N1 and P2 components following a 1 s compared to an 8 s ITI is

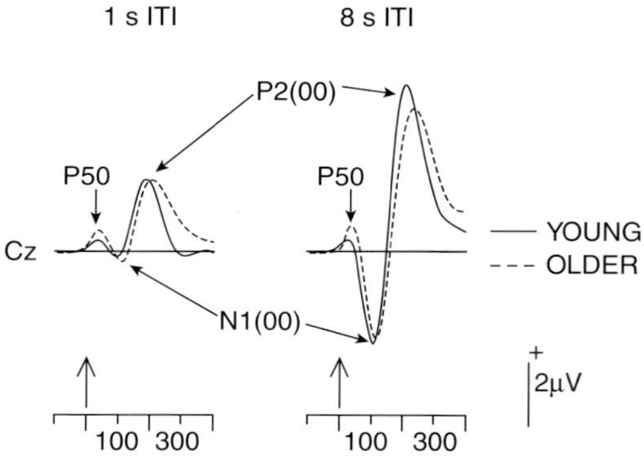

Fig. 18.1. Grand mean ERPs averaged across the 16 participants within each age group. The ERPs were elicited by the first tone in an eight-tone train (500 ms ISI) of frequently occurring standard stimuli in a passive paradigm in which participants watched a silent movie while ignoring the tones. The stimuli consisted of standard (50%) and deviant (50%) trains of auditory stimuli separated by either 1 s or 8 s ITIs. All tones in a standard train were standards. In deviant trains, the first tone was a deviant, whereas the remaining seven tones were standards. Arrows mark stimulus onset with time lines every 100 ms. Data modified from Gaeta et al. (2001a).

most likely due to the refractoriness or "fatigue" of the neural generators. This is explained a few paragraphs below in greater detail. The effects of the ITI appear to be similar in both age groups because both show larger N1 and P2 components at the longer ITI. Some of these observations are mirrored in the age-related studies reviewed below.

In reviewing the evidence, there appears to be a consensus that the P50 is larger in older than young adults when it has been measured directly and reported in the published reports (Amenedo & Diaz, 1998b; Bennett et al., 2004; Bertoli et al., 2005; Chao & Knight, 1997; Czigler et al., 1992; Fabiani et al., 2006; Smith et al., 1980; Snyder & Alain, 2005) or observed by visual inspection of the published waveforms by the current author (Bertoli & Probst, 2005; Gaeta et al., 1998; Golob et al., 2007; but see Bertoli et al., 2002). Interestingly, there does not appear to be any evidence that P50 is of longer latency in older relative to young adults

The enhanced P50 amplitudes observed in older adults might reflect a deficit in inhibitory function. For example, Knight and colleagues (1989) recorded P50 components to auditory clicks in patients with dorsolateral prefrontal lesions. P50 amplitudes were reliably larger in the patients compared to the age-matched controls. A highly similar result was reported by Alho and coworkers (1994). Knight and colleagues (Alho et al., 1994; Knight et al., 1989) interpreted this to mean that the prefrontal lesion patients had lost inhibitory control over thalamically mediated gating of inputs to sensory cortex. Thus, extrapolating this result to normally aging older adults might implicate a deficit in prefrontal inhibitory control as a cause of age-related enhancement of P50 amplitudes (see also Chao & Knight, 1997; see also the review of N1 amplitude below).

Similarly, based on P50 magnitude findings, there is some support for the possibility that probable Alzheimer's disease (PAD) patients may be characterized by deficits in inhibitory control. Further, there is limited evidence that some individuals within the aging population who show poor inhibitory control may eventually develop PAD. In healthy persons, the second of a two-click pair separated by a 500 ms interval elicits a smaller P50 than the first click (i.e., the P50 to the second click is suppressed), which has been labeled *sensory gating* (Freedman et al., 1987). Reduced suppression of P50 relative to normal controls was originally reported in schizophrenia patients (i.e., the response to the second click was less suppressed in the patients; Adler et al., 1985; Siegel et al., 1984; Chapter 19, this volume). The reduced suppression was interpreted as indexing a deficit in inhibition, or sensory gating (i.e., a failure to filter irrelevant input) and a proclivity to sensory overload, consistent with some schizophrenia symptomatology (Braff & Geyer, 1990; Freedman et al., 1987). Based on these findings, there is some very limited evidence that altered P50 suppression might be a marker of PAD (Ally et al., 2006; Cancelli et al., 2006; Jessen et al., 2001), because, as in schizophrenia patients, the suppression has been reported to be smaller in PAD patients than in age-matched controls. Applying the interpretation proffered for schizophrenia patients would suggest that PAD patients may also be characterized by sensory gating difficulties. However, the results of the studies of P50 suppression in PAD are difficult to evaluate because, in two of them, the waveforms from PAD patients and controls were not depicted, precluding a determination of the quality of the data and the measurement technique (Donchin et al., 1977). Other studies have indicated that the P50 component itself is larger in patients with mild cognitive impairment (MCI), a putative precursor stage to PAD, compared to age-matched, healthy controls. Hence, poor inhibitory control (i.e., enhanced P50 magnitudes relative to age-matched controls) may be a predictor of conversion to PAD (Golob et al., 2001, 2007; Irimajiri et al., 2005).

Like the P50, the N1 has also been reported to be larger in older than young adults (Alain & Woods, 1999; Amenedo & Diaz, 1999; Anderer et al., 1996; Chao & Knight, 1997; Gaeta et al., 2002; Karayanidis et al., 1995; Kisley et al., 2005; Snyder & Alain, 2005), although there is little evidence for age-related prolongation of N1 latency (Anderer et al., 1996; Goodin et al., 1978; Iragui et al., 1993). On the other hand, there are also data suggesting that young, relative to older, adults show larger amplitudes (Bennett et al., 2004; Bertoli & Probst, 2005; Cooper et al., 2006; Ford et al., 1995; Golob et al., 2001). Confusing the picture further, still other studies have reported equivalent amplitudes between young and older adults (Bertoli et al., 2002; Czigler et al., 1992; Ford et al., 1995; Friedman et al., 1993b; Gaeta et al., 1998, 2001b, 2003; Iragui et al., 1993; Pekkonen et al., 1996; Picton et al., 1984; Woods, 1992). One possible explanation for these disparate results is the quite different paradigms under which the N1 has been measured, including passive and active oddballs, startle-noise oddball, delayed match-to-sample, and selective and cued attention tasks. As noted, N1 is modulated by attention; for example, it is larger to targets (deviants) than standards and to attended compared to unattended standards during selective attention tasks (Hillyard et al., 1973; Näätänen & Picton, 1987; Chapter 11, this volume). To the extent that these attentional effects differ for young and older adults, some of these differences could be accounted for on this basis. However, as shown by Amenedo and Diaz (1999), older adults demonstrated larger N1 magnitudes whether the standards were attended or unattended. Hence, although the data are clearly limited, age-related differences in attentional function may not explain these discrepancies (see also Gaeta et al., 2003).

An alternative explanation might lie in age-related differences in the recovery cycle or refractory period of the N1. That is, if the time between successive auditory events is short (e.g., 500 ms), the N1 to the second in a series of tones will be markedly smaller than that to the first. As the interval between successive stimuli becomes longer, the N1 generators "recover" some of their amplitude, producing equivalent magnitude to the first tone at an interval between 6 and 10 s (Davis et al., 1966). There is some recent evidence that older adults do show different N1 recovery functions than their young adult counterparts (Fabiani et al., 2006), although the underlying mechanism may not be a simple age-related change in the refractory period of N1 generators (Sable et al., 2004). Fabiani et al. (2006; see also Gaeta et al., 2001a) presented trains of five auditory events (400 ms ISIs), with each train separated by either 1 or 5 s ITIs. Fabiani et al. (2006) reported that the recovery rate for N1 was similar for young and older adults (i.e., the N1 to the first standard in the train was larger for the 5 s compared to the 1 s ITI in both older and young adults; see also Figure 18.1). However, relative to the first tone of a train, older adults showed reliably less N1 suppression to the remaining tones compared to young adults, suggesting the possibility, as noted earlier, of age-related inefficiency in inhibitory control, or sensory gating (Chao & Knight, 1997; Hasher et al.,1991). Because N1 amplitude to the first tone in the 1 s ITI was equivalent in both age groups and was as large in older adults as in young adults in the 5 s ITI condition, the authors concluded that sensory memory was intact in older adults. On the other hand, evidence based on the MMN, suggests that sensory memory decays faster in older adults (Pekkonen, 2000), but this interpretation is open to question. I consider this issue further in the section on the MMN.

There is a dearth of studies of the N1 in PAD and MCI. Intriguingly, however, in a study in which MCI patients and controls listened passively to tones at two ISIs (2/s and 1.5/s), the two groups had equivalent N1s at the longer ISI, whereas MCI subjects showed greater N1 at the shorter ISI (Irimajiri et al., 2005). On the other hand, there is, to my knowledge, no evidence that this is also the case in PAD. In fact, one study (Pekkonen et al., 1994) found equivalent N1 magnitudes in PAD patients and controls at both 1 s and 3 s ISIs. In another study, Yamaguchi and colleagues (2000) reported reliably smaller N1 magnitudes in PAD patients relative to age-matched controls. Hence, these results must be interpreted cautiously. Nonetheless, they suggest that this phenomenon should be explored further in PAD patients, MCI patients, and controls, in the hope that reduced suppression may help identify older persons at greatest risk for the development of PAD.

One could use N1 amplitude (or, for that matter, any component's amplitude) as a "biomarker" for a given disease without knowledge of the underlying processes. However, it would arguably be more useful diagnostically if those processes were known. N1, like the earlier P50, is an "obligatory" component of the auditory ERP because it is elicited whether the stimuli are attended or ignored, although its magnitude can be affected by attention.

According to Näätänen and Picton's (1987) exhaustive review, there are potentially three sets of processes contributing to the "true N1 component" recorded at the scalp (i.e., those processes not reflecting aspects of the MMN and the Nd). Two of these are thought to reflect the activity of cerebral generators in and around primary and secondary auditory cortex. Hypotheses advanced by Näätänen and Picton (1987) suggest that one or both of these cortical generators could reflect the representation of sensory information and/or the formation of a sensory memory within the auditory cortex (see also Näätänen et al., 2005; Chapter 4, this volume). Based on the Fabiani et al. (2006) data detailed above, it does not appear that aging adversely affects the representation or formation of auditory sensory memories, but these mechanisms might be disrupted in aging individuals with PAD or in those with a diagnosis of MCI who progress to PAD. However, there are simply too few data to determine whether or not this is the case.

Relative to the P50 and the N1, age-related change in the P2 component has been less well studied, and only a handful of investigators have directly measured it. Czigler and colleagues (1992) used ISIs of 800, 2400, and 7200 ms in a passive oddball task. While P2 magnitude increased with ISI (as would be expected based on a refractory-period explanation, just as for the N1), this increment was much less dramatic for older adults. A similar finding was also reported in the Fabiani et al. (2006) study described earlier and is reminiscent of their N1 finding. Amenedo and Diaz (1999) reported larger P2 amplitudes in older compared to younger adults, again regardless of whether the standards were attended or unattended. Similar age-related enhancements have been described by Anderer et al. (1996), although one study found a larger P2 magnitude in young compared to older adults (Bertoli et al., 2002) and another (Iragui et al., 1993) reported no difference between young and older adults.

Early-latency components in the visual modality analogous to the P50 and N1 have also been reported to show age-related change. However, this literature is as mixed in its results as the data on age-related change in the auditory modality described above. For the sake of brevity, I refer the reader to a recent paper in which these studies are reviewed (De Sanctis et al., 2008). In this paper, De Sanctis and colleagues observed that the amplitude of the visual N1 (which was significantly larger in older adults) was much more variable in older than young adults. Whereas young adults showed a fairly tight distribution of amplitude values, older adults showed a bimodal distribution; 10 older adults with lower-amplitude N1s showed values similar to those of the young adults, whereas the remaining 9 produced the highest-magnitude N1s. This potentially important observation might explain some of the variability among studies in both the auditory and visual modalities.

To summarize, relative to the N1 and P2, the age-related enhancement of auditory P50 (P1) seems to be better established in the literature. The lack of suppression of the N1 observed by Fabiani et al. (2006) is intriguing and deserves follow-up. Both of these phenomena may be indicative of an age-related inhibitory deficit and could provide additional evidence for the frontal-lobe and inhibitory-control deficit hypotheses of cognitive aging (West, 1996). In addition, there is some, albeit limited, evidence that P50 enhancement may be a marker of risk status in some older adults who progress to PAD. Hence, this phenomenon may be worthy of pursuit in large samples of PAD patients, MCI patients, and age-matched controls.

The MMN

A reasonably large age-related literature has accumulated on the automatic deviance detection system, as reflected by the MMN (Pekkonen, 2000). Figure 18.2 presents data recorded during the ignore paradigm described earlier (Gaeta et al., 2001a). In the figure, the ERPs to first-position standards and deviants from the 1 s ITI condition are depicted. In addition, grand-averaged preliminary data from five PAD patients are shown. Two basic age-related phenomena are reflected in these data: the MMN is somewhat smaller and of longer latency in healthy older and PAD participants compared to young adults, and PAD patients appear to produce MMN magnitudes and latencies similar to those of healthy controls. Despite the overall age-related difference in MMN amplitude, its scalp distribution appears similar in all three groups. The current source density (CSD) maps are consistent with bilateral generators in and around auditory cortex, which is thought to be a primary contributor to the MMN recorded at the scalp (Giard et al., 1990; Näätänen, 1992; Näätänen et al., 2007).

As described by Näätänen in this volume (Chapter 6) and elsewhere (Näätänen et al., 2007), the relatively automatic, preattentive, MMN-based deviance detection system is recruited in a wide variety of situations, including simple changes in

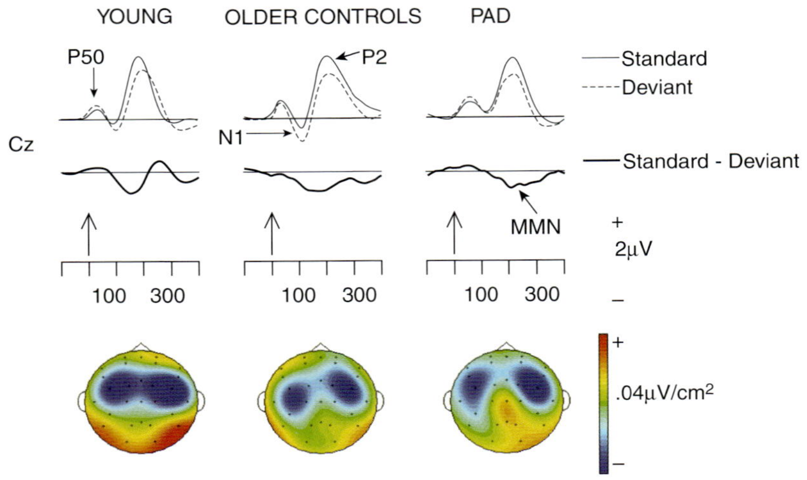

Fig. 18.2. Grand mean ERPs averaged across 16 young adults (left), 16 older adults (middle), and 5 PAD patients (right) during a passive paradigm (Gaeta et al., 2001a). The data were recorded at a midline frontal scalp location (Cz) while subjects watched a silent movie and heard eight-tone trains of stimuli (see the caption for Figure 18.1). The top row of waveforms depicts the superimposed ERPs elicited by the first standard and deviant in a train; the second row shows the standard minus deviant difference waveforms; the third row depicts the CSD scalp distributions of the difference waveforms in the second row.

pitch, intensity, duration, phonetic characteristics, and spatial location, as well as by more "abstract" changes in the acoustic background, such as whether two tones are rising or falling in pitch (Paavilainen et al., 1998). Moreover, on the basis of this latter finding, the MMN deviance detection system cannot reflect a simple sensory (echoic) memory of the acoustic past but rather appears to reflect a memory of the regularities or invariances in the acoustic stream (whether it is for pitch or the abstract nature of the regularity). Because of these properties, it has been natural for investigators to ask whether aging affects any aspects of the MMN deviance detection system.

In the early stages of research on the MMN's characteristics, it was thought that it reflected a short-lived sensory memory (Sams et al., 1993). On this basis, several of the early studies of aging compared short (e.g., 200 ms) and long (e.g., 4000 ms) stimulus onset asynchronies (SOAs) to determine whether the memory on which the MMN was based had a shorter duration in older than young adults. However, as noted above, the MMN is not thought to reflect a simple sensory memory. There are also methodological limitations in assessing the effects of SOA even in young adults (see Ritter et al., 2002, for a discussion). Therefore, these investigations are equivocal with respect to age-related changes in the duration of the memory on which the MMN is based at short and long SOAs using the typical oddball paradigm (Gaeta et al., 2001a; Ritter et al., 1998; Winkler et al., 2002). Hence, the review below only considers the results of experiments in which auditory stimuli were ignored and a short SOA was used (between 0.5 and 2 s) even if a long SOA was also included. When considered in this fashion, although there are exceptions, there appears to be a consensus that the MMN is reduced in older adults. This suggests the possibility that sensitivity to regularities in the acoustic environment decreases as we age.

In one of the first of these investigations, Czigler et al. (1992) used infrequent changes in pitch. They found that older adults showed smaller-amplitude MMNs than young adults. Similarly, Cooper et al. (2006) reported smaller MMNs in older compared to young adults whether deviance was defined by pitch or duration. Furthermore, two reports suggested that when the MMN was recorded to duration deviants in an unattended channel (i.e., the ear of input) during selective attention paradigms, it was smaller in older compared to young adults (Karayanidis et al., 1995; Woods, 1992; see also Gaeta et al., 2001b, for frequency deviants; but see Pekkonen et al., 1996) even at very short ISIs between 200 and 400 ms. Gaeta et al. (1998) also reported smaller MMNs in older relative to young adults whether elicited by small-frequency (50 Hz difference from the standard) or large-frequency (300 Hz difference) deviants or novel environmental sounds. In fact, older adults did not show a reliable MMN to the small-frequency deviants. Alain and Woods (1999) assessed MMN magnitudes to both frequency changes (small [122 Hz] and large [414 Hz]) and pattern deviants (tones of two different pitches alternated and interrupted infrequently by a repeat). They observed reduced MMNs to small- and large-frequency as well as pattern deviants. In another study from this same group, Alain and colleagues (2004) assessed the sensitivity of older adults' deviance detection system by varying systematically the gap between tone pips constituting a deviant event. Tones with gaps occurred infrequently, and continuous tones served as standards. The MMN amplitudes were smaller in older than young adults. Importantly, when MMNs were computed to near-threshold gap deviants, thereby matching the performance of young and older subjects, young but not older adults showed reliable MMNs. In a very similar paradigm, Bertoli et al. (2002) varied the gap duration of deviant stimuli between 6 and 24 ms. The MMNs were smaller in older compared to young participants. Moreover, similar to the results of Alain et al. (2004), it took a longer-duration gap (15 ms) to produce a reliable MMN in older compared to young (9 ms) adults.

The results of these investigations suggest that the preattentive deviance detection system of older adults is less sensitive than that of young adults. Employing rule-based auditory features to create invariances in the acoustic environment, the results of a study by Gaeta et al. (2002) support this hypothesis. Stimuli were either a frequent ascending tone pair or an infrequent descending tone pair. Tone pairs were presented under three conditions: (1) physical feature monaural (1 tone pair), (2) abstract feature monaural (10 tone pairs of different pitches), and (3) abstract feature binaural (10 tone pairs of different pitches; the first presented to the left ear and the second to the right). Relative to young adults, older adults showed smaller-magnitude MMNs, which were elicited under all three conditions for young adults but only in the monaural conditions for older adults. Thus, rule-based neural representations were created by both age groups under monaural conditions, but only by the young adults in the binaural condition. Reliable MMNs in the rule-based conditions were present despite the fact that behavioral discrimination (after the MMN recordings) fell to near chance levels for both age groups, suggesting an age-related decline in the efficacy of integrating multiple sources into a single auditory stream.

By contrast with the studies reviewed above, there are some reports of age-equivalent MMN amplitudes. For example, with pitch deviants and a 1 s ISI, Pekkonen and colleagues (1993) reported that MMN magnitude was similar in young and older adults, a finding also reported by Gunter and coworkers (1996) with a 1 s ISI. In a follow-up of their earlier study, Pekkonen et al. (1996) found age-equivalent MMN magnitudes to pitch deviants at .5 and 1.5 s ISIs. Using frequency deviants and a selective attention procedure similar to those mentioned earlier with 600 ms ISIs, Amenedo and Diaz (1998a) did not observe age-related differences in MMN magnitudes. However, these latter results are equivocal because there appeared to have been no effect of selective attention on the ERP waveforms, calling into question the sensitivity of the paradigm.

Using the design described above in the section on relatively early-latency components, Gaeta et al. (2001a; see Figure 18.2) showed that at a 1 s ITI, the vast majority of young and older subjects showed a robust MMN. However, at the 8 s ITI, only six young and five older adults showed robust MMNs. Although this might suggest that the memory on which the MMN was based had decayed at the 8 s ITI for both age groups, Gaeta et al. (2001a) argued that it was the nature of the perceptual grouping of the trains that was modulating the MMN magnitude. That is, when the SOA is constant between the standards within a train and the interval between standards and deviants is short (1 s ITI), all tones are treated as belonging to the same *perceptual group*. Therefore, a deviant is detected as a departure from invariance, resulting in a robust MMN. When the SOA between standards within a train is short and the interval between standards and deviants is long (8 s ITI), the perceptual group that comprises the invariance is the train of standards and the deviant lies outside the frame of temporal relevance for some subjects. Hence, presentation of a deviant is not perceived by the MMN system as a divergence from regularity and an MMN is not elicited for these subjects. A similar interpretation is possible to account for the data of Fabiani et al. (2006), who also reported age-invariant MMNs using a highly similar design with 1 and 5 s ITIs. Hence, although the findings of these two studies are ostensibly at odds with much of the data reviewed above, it is likely that the age-equivalent MMNs in these studies is peculiar to the types of stimuli used (trains rather than single events) and reflect the maintenance with age of perceptual grouping of acoustic stimuli.

With respect to abnormal aging, very few investigations exist. In the earliest of these, Pekkonen et al. (1994) reported that the MMN was of similar amplitude in PAD patients and controls at a 1 s ISI. This was confirmed in a subsequent study by the same group using the magnetoencephalographic analog of the electrical MMN (Pekkonen et al., 2001). Similarly, Gaeta and colleagues (1999) observed that MMN magnitudes were similar in PAD patients and age-matched controls.

In summary, it seems fairly clear that the system upon which the MMN relies is relatively less sensitive in older compared to young adults and, based on very limited evidence, is similar in PAD patients and controls, at least in the fairly simple paradigms used to date. What is less clear is whether there is an age- and/or PAD-related reduction in the duration of the memory upon which the MMN is based. This will certainly require further work.

The Novelty P3 (P3a)

Although the MMN reflects the detection of a change in the invariance of the acoustic environment, it does not reflect the capture of attention by that change. The involuntary capture of attention is reflected by the novelty P3 or P3a, which is elicited if the change in the background is sufficiently deviant (Friedman et al., 2001; see also Chapter 7, this volume). The novelty P3, therefore, reflects an aspect of the orienting response, a fundamental biological mechanism necessary for survival (Sokolov, 1990).

An example of the dissociation between the MMN and the novelty P3 is depicted in Figure 18.3, which illustrates the averaged ERPs elicited by standards and small (50 Hz) and large (i.e., novel environmental sound) deviants recorded while young and older adults watched a silent movie and ignored the background auditory events (Gaeta et al., 1998). While a MMN was elicited by both small and large deviants in young as well as older adults (although the MMN was not reliable in the latter), only the environmental-sound deviants elicited the novelty P3. Note also that the novelty P3 was reduced in older compared to young adults, an age-related phenomenon that has been replicated most often with auditory stimuli but has also been observed in the visual and somatosensory modalities (Czigler et al., 2006; Fabiani & Friedman, 1995; Friedman & Simpson, 1994; Friedman et al., 1993b, 1998; Gaeta et al., 1998, 2001b; Knight, 1987; Weisz & Czigler, 2006; Yamaguchi & Knight, 1991a; but see Daffner et al., 2006b).

Although the majority of studies have employed active oddball designs, in which the participant is asked to respond to targets via RT and withhold responding to standards (and novel stimuli), it is perhaps more ecologically valid to assess the ERP concomitants of the orienting response while participants ignore the background events. Using this method, one can obtain a truer assessment of how and to what extent deviant events involuntarily capture attention, because the participant is engaged in the primary task of reading, watching a silent movie, or performing a visual discrimination (Alain &

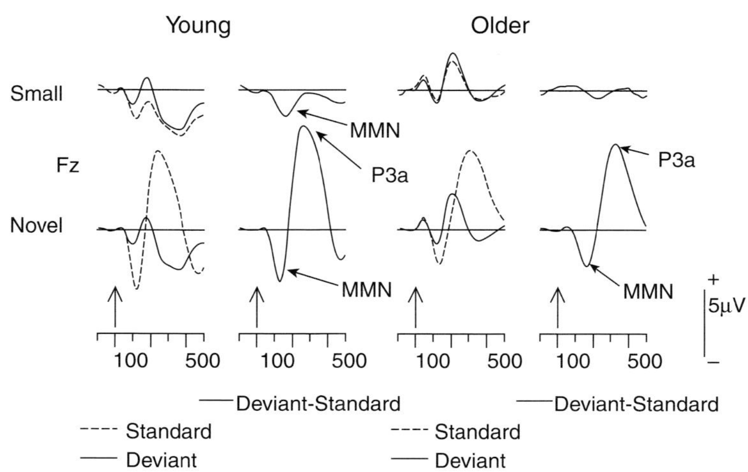

Fig. 18.3. Grand mean ERPs averaged across 26 young adults (left two columns) and 20 older adults (right two columns) during a passive oddball paradigm (Gaeta et al., 1998). The data were recorded at a midline frontal scalp location (Fz) while subjects watched a silent movie and heard low-probability, small (950 Hz), large (700 Hz), and environmental-sound, novel-deviant auditory events randomly intermixed with frequently occurring standards (1000 Hz). Only the data for the small and novel deviants are presented. Participants were instructed to ignore the auditory events. The left column of each group's waveforms shows the standard and deviant waveforms, while the right column illustrates the deviant minus standard difference waveforms.

Woods, 1999; Friedman et al., 1998; Gaeta et al., 1998; Yago et al., 2003; see Figure 18.3).

Like older adults in the Gaeta et al. (1998) study, PAD patients showed a novelty P3 only in response to the environmental-sound deviants that did not differ in amplitude from that of age-matched controls (Gaeta et al., 1999). A similar finding was reported by Yamaguchi et al. (2000), also in the auditory modality, although with an active paradigm. The results of these latter two studies contrast with those from an investigation by Daffner et al. (2001) using visual stimuli. They reported that the novelty P3a was reliably smaller in PAD patients than in age-matched controls. The methods used by Daffner et al. (2001) are quite different than those typically employed to assess either age- or dementia-related changes in the processing reflected by the novelty P3. For example, these investigators collect *looking times* (i.e., how long a participant spends viewing a given standard, target, or novel visual event), as well as RTs to predesignated targets. Participants are allowed to view the stimuli for as long as they deem necessary. Although an intriguing method for assessing the extent to which novel events are processed, this technique may change the oddball task significantly, such that brain systems other than those reflecting the involuntary capture of attention (novelty P3) may be recruited. It may be one or more of those very systems, such as controlled attention (effortful processing resources), that are dysfunctional in PAD (Parasuraman & Haxby, 1993), leading to the reduction in the P3 elicited by the novel objects.

A critical aspect of the orienting response is its habituation over time (Lynn, 1966). Accordingly, several investigators have shown that, like other ubiquitous markers of the orienting response such as the galvanic skin response, novelty P3 amplitude diminishes in young adults as more and more novel events are experienced or the same novel event is repeated (Czigler et al., 2006; Friedman & Simpson, 1994; Kazmerski & Friedman, 1995; Knight, 1984, 1996; Yamaguchi & Knight, 1991b; see the reviews by Friedman et al., 2001, and Ranganath & Rainer, 2003). By contrast, in older adults, the novelty P3 has been shown not to habituate (Czigler et al., 2006; Fabiani & Friedman, 1995; Friedman & Simpson, 1994; Kazmerski & Friedman, 1995; Weisz & Czigler, 2006), whether those events are attended or ignored (Friedman et al., 1998). Because the scalp-recorded novelty P3 receives contributions from prefrontal cortex (Daffner et al., 2000; Halgren et al., 1998a; Knight, 1984) and because patients with prefrontal damage do not show habituation to these types of stimuli (Knight, 1984; Woods & Knight, 1986), this type of finding has been interpreted by some to indicate support for the *frontal-lobe deficit* hypothesis of cognitive aging (Friedman et al., 1998; see also Buckner, 2004, and West, 1996; but see Greenwood, 2000, 2007).

This hypothesis has also been motivated by the well-replicated finding that older, relative to young, adults show more frontally oriented scalp distributions to both oddball targets and deviant, novel events (Fabiani & Friedman, 1995; Fjell & Walhovd, 2003; Friedman & Simpson, 1994; Friedman et al., 1993b; Kutas et al., 1994; Pfefferbaum et al., 1984a; Yamaguchi & Knight, 1991a; see Figure 18.4), suggesting that older adults may call on frontal lobe resources to a greater extent than young adults even for events that should have been well encoded and categorized. However, recent evidence suggests that some older adults may call on these resources to a greater extent than others (Daffner et al., 2006a; Fabiani et al., 1998; Riis et al., 2008). An open question is whether the change in scalp distribution (and, presumably, the underlying generators) reflects a compensatory mechanism in high-functioning older adults whose brains might have the capacity to produce this change. That is, relative to young adults, the topographic change in older adults ought to be associated with equivalent or near-equivalent performance that, without such compensatory activity, would have been lower. On the other hand, the change could represent an attempt on the part of low-functioning older adults performing poorly on the task to compensate for the deleterious effects of brain aging (Raz, 2000). Adjudicating between these alternatives is difficult and, in fact, examples of both interpretations have been advanced. For example, Fabiani and colleagues (1998) divided their older participants into good and poor frontal-lobe test performers. The poor performers showed target P3 scalp distributions that had frontal maxima (the more typical finding when averages are computed across all participants' data within the older age group), whereas the good performers showed parietal-maximal scalp topographies. In this case, the data suggested that the frontal scalp topography reflected less efficient processing. On the other hand, in a series of investigations, Daffner and coworkers, using the visual oddball paradigm described earlier, came to the opposite conclusion (Daffner et al., 2005, 2006a, 2006b; Riis et al., 2008): that more frontal P3 activity indicates greater compensatory recruitment of frontal resources and reflects

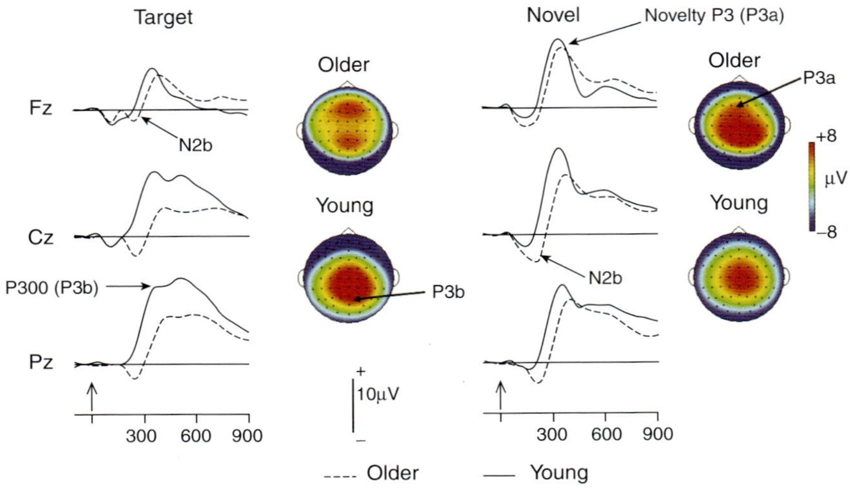

Fig. 18.4. Grand mean ERPs averaged across young (left; $N = 10$) and older (right; $N = 10$) adults to targets (solid lines) and novel environmental sounds (dashed lines) during novelty oddball blocks. Surface potential maps of the novelty P3 and target P3b appear to the right of each set of waveforms. The older adult maps appear on top of the young adult maps. The small dots represent the electrode locations. The maps were computed at the time point that displayed the largest amplitude. Note that, for both the novelty P3 and the target P3b, peak latency is longer for older than for young adults. Arrows mark stimulus onset with time lines every 300 ms. The data are from an unpublished study by Friedman.

successful cognitive aging. Notwithstanding the differences in design and stimulus materials between the Daffner experiments and those predominating in the literature (for discussion, see Fjell & Walhovd, 2005, and Daffner et al., 2005), the compensation hypothesis is currently highly controversial (Colcombe et al., 2005; Friedman, 2003; Greenwood, 2007; Zarahn et al., 2007). Determining whether a pattern of brain activity is compensatory, inefficient, detrimental, or unrelated to performance will require more precise definitions of what is meant by *compensation* (see Davis et al., 2007, and Stern, 2002, for examples), and under which conditions such compensation might or might not be expected.

To summarize, with the exception of the studies by Daffner and colleagues, the majority of investigations of age-related change in the novelty P3 have shown it to be reduced and its topography to be more frontally oriented in older compared to young adults. Consistent with these findings, some data suggest that only highly deviant events are likely to capture the older adult's attention. Moreover, unlike young adults, who show a reduction in amplitude after the first few presentations of unexpected novel events, older adults do not, suggesting that they repeatedly recruit prefrontal cortical mechanisms for stimulus events that should no longer involuntarily capture attention. Finally, the extent to which compensatory mechanisms are responsible for the recruitment of frontal resources in response to novel events by some older adults is currently equivocal and requires further experimentation and documentation.

N2b and P3b

By contrast with the MMN, which appears to reflect a relatively automatic, preattentive mechanism, the N2b (at approximately 200–300 ms poststimulus), like the P3b, is typically observed only when participants are focusing attention on the sequence of events in order to make a task-relevant decision (Ritter et al., 1979; Chapter 11, this volume). The N2b is also elicited in the passive oddball paradigm when highly deviant stimuli, such as environmental sounds, involuntarily capture attention (Kazmerski et al., 1997). The P3b component is arguably the best-studied of the ERP components. It is, therefore, not surprising that much more research effort has been expended in assessing age-related changes in its amplitude, latency, and scalp topography compared to the N2b.

To illustrate some of the effects of aging on the N2b and P3b components, Figure 18.4 depicts the grand mean ERPs elicited by targets and, for comparison, novel environmental sounds in young ($N = 10$) and older ($N = 10$) adults recorded during a version of the novelty oddball task (Friedman, unpublished data). The scalp distribution maps to the right of the waveforms depict, as noted earlier, one of the most consistent findings in the ERP/aging literature: relative to young adults, older adults' novel (P3a) and target (P3b) topographies are more frontally oriented. Another ubiquitous finding evident in the figure is the age-related prolongation of the latency of both novel and target P3 components (e.g., Brown et al., 1983; Fjell & Walhovd, 2003; Iragui et al., 1993; Picton et al., 1984). The most comprehensive assessment to date of age-related effects on P3b latency has been published by Polich (1996), and the reader is referred to that publication for more details. A third well-replicated finding observable in Figure 18.4 is the smaller-amplitude P3b component in older compared to young adults. This phenomenon has been reported in a wide variety of tasks in addition to the oddball paradigm (e.g., Ford et al., 1997; Friedman et al., 1997, 2007b; Gaeta et al., 2003; Iragui et al., 1993; McEvoy et al., 2001).

The N2b also shows age-related variation. For example, its latency increases with age in highly similar fashion to that demonstrated for the P3b (Anderer et al., 1996; Enoki et al., 1993; Goodin et al., 1978a; Iragui et al., 1993). This can be seen in Figure 18.4, where, relative to young adults, a prominent though prolonged-latency N2b component is evident in the target ERPs of the older adults. By contrast with latency, one of the difficulties in assessing the consistency of age-related change in N2b magnitude is that, although it is typically recorded in a wide range of paradigms, it has often not been measured directly. Nonetheless, when N2b amplitude has been calculated, there is a somewhat greater number of studies whose results show larger N2b magnitudes in older compared to young adults (Anderer et al., 1996; Czigler et al., 2006; Friedman et al., 1993b; Gaeta et al., 2003; Woods, 1992) than indicate the converse (Bertoli et al., 2005; Czigler & Balazs, 2005; Karayanidis et al., 1995).

With respect to pathological aging, the results are quite mixed. Although, as might be expected, PAD patients sometimes show prolonged-latency P3b components relative to age-matched controls (e.g., Bennys et al., 2007; Golob & Starr, 2000; O'Donnell et al., 1990; Patterson et al., 1988; Williams et al., 1991), other research groups have reported a failure to distinguish PAD participants from controls on the basis of this metric (Gordon et al., 1986; Kazmerski & Friedman, 1998; Kraiuhin et al., 1990; Pfefferbaum et al., 1984b). Whether the P3b latency prolongation has sufficient sensitivity and specificity for clinical diagnosis is, therefore, open to question (Gordon et al., 1986; Patterson et al., 1988). Similarly, P3b amplitudes are sometimes (Frodl et al., 2002; Goodin et al., 1978b; Saito et al., 2001), but not always (Golob & Starr, 2000; Kazmerski & Friedman, 1998) reported to be smaller in PAD patients compared to controls. Moreover, the scalp distribution of P3b does not appear to differ between PAD patients and controls, at least in the auditory modality (Ford et al., 1997; Kazmerski & Friedman, 1998). Hence, as for P3b latency, the utility of amplitude and topography for clinical diagnosis is equivocal. This conclusion is further supported by the results of a recent study by Golob et al. (2007), who reported that neither P3b latency prolongation nor amplitude reduction was able to predict which MCI patients converted to PAD (see Taylor & Olichney, 2007, for a review of ERPs in dementia).

In sum, the evidence to date is fairly conclusive that the mental operations indexed by N2b and P3b increase in latency as individuals age, consistent with the phenomenon of general slowing reported in the behavioral literature (Salthouse, 1991). Whether P3b latency slowing is exacerbated in pathological aging is not yet certain. The tenuousness of this finding in PAD may be due to the fairly simple oddball paradigms that have most often been used to assess P3b latency. N2b is thought to be the first brain event indicating that a conscious sensory discrimination has been made (Ritter et al., 1979). P3b, therefore, must reflect a subsequent stage of processing, although a consensus on its functional significance has yet to be reached (see below).

One of the major theoretical positions advanced to account for age-related changes in cognition postulates that even relatively early stages of processing are slowed (Salthouse, 1996). If this hypothesis is valid, then the early latency ERP components (e.g., P50, N100) should be of longer latency in older adults compared to young adult controls. The prolongation of these relatively early-latency components could engender a cascade of slowing that might be manifested in the age-related RT retardation that has been ubiquitously observed in a wide variety of cognitive paradigms (Salthouse, 1996) and prolonged P3b latencies, one of the most often replicated findings in the ERP aging literature (Polich, 1996). However, there is very little evidence for retardation in the early-latency components (Anderer et al., 1996; Goodin et al., 1978a; Iragui et al., 1993). By contrast, the ERP data suggest that the slowing is restricted to the later stages of information processing, as indexed by N2b and P3b. Hence, because of this fact, the retardation in N2b and P3b latencies in older adults cannot be due simply to information loss from slowed operations at earlier stages in the processing stream.

Like P3 latency, the evidence for an age-related reduction in P3b amplitude is compelling. By contrast, the evidence for further reduction by PAD and MCI is equivocal. However, the implications of the age-related reduction in P3b amplitude for understanding the underpinnings of cognitive aging phenomena are not clear at this time. Complicating the picture further is the age-related change in P3b scalp topography. This is somewhat problematic for interpretation of the age-related significance of the changes in cognition that underlie the P3b recorded at the scalp because it is well known that it receives contributions from a widespread intracranial network (Halgren et al., 1998b), any aspect of which could be altered with aging. Hence, these alterations could be functionally driven, reflecting, for example, an age-related change in the source regions that contribute to the P3b or, less interestingly, to age-related structural changes within the brain. For example, age-related shrinkage in brain volume, which is well documented (Raz et al., 2004), might alter the orientation of the brain generators, thereby modifying older adults' P3b scalp topography even though the processes reflected by the P3b do not change with age. However, for this account to be viable, the topographic change would most likely have to be highly similar across tasks (e.g., oddball, Sternberg short-term memory, repetition priming). Although this proposition has, to my knowledge, not been tested directly, there is some evidence that, although older adults typically produce more frontally oriented P3b distributions in a variety of cognitive paradigms, those topographies can be modified by task demands (Friedman et al., 1997). Then again, the Friedman et al. (1997) topographic comparisons were made between independent samples of subjects who had participated in similar, though not identical, experiments. Thus, the evidence from the Friedman et al. investigation is limited and needs to be bolstered by within-subject, task-related comparisons of P3b topography in older adults.

One influential theory of the functional significance of the P3b posits that it reflects updating when the subject's model of the environment requires revision (Donchin & Coles, 1988), a key aspect of working memory (Baddeley, 1992). Furthermore, a major hypothesis used to account for cognitive decline in aging is that working memory, which has been conceptualized as the amount of resources available to process information online, is reduced (Craik & Byrd, 1982; Park, 2000a). Therefore, it could be the case that older adults call upon this type of general-purpose, working memory/attentional resource much more often than young adults, thereby accounting for the ubiquity and prominence of the frontal aspect of the P3b distribution in a wide variety of tasks (Fabiani & Friedman, 1995; Ford et al., 1997; Friedman et al., 2007b; see Friedman et al., 1997, for discussion). However, as for the novelty P3, whether this topographic change reflects a compensatory modification of brain activity to counteract the deleterious effects of brain aging is unknown at this time and clearly requires further research effort (Friedman, 2003, 2007).

N400

The presentation of a phrase- or sentence-ending pictorial or verbal concept that is incongruent with the meaning of the preceding material produces a large-amplitude N400 effect in young adults (Kutas & Hillyard, 1980; Chapter 15, this volume). This incongruity effect, defined by subtracting the ERP to incongruous endings from that to congruous endings, has been assessed in a number of age-related investigations (see the reviews by Federmeier, 2007; King & Kutas, 1995; Kok, 2000). Generally, the N400 effect shows an amplitude reduction and a latency prolongation with increasing age (Cameli & Phillips, 2000; Gunter et al., 1992, 1996; Hamberger & Friedman, 1992; Hamberger et al., 1995; Harbin et al., 1984; Iragui et al., 1996; Kutas & Iragui, 1998; Phillips & Lesperance, 2003), with latency retardation similar to those observed for the N2b and P3b. Like the P3b latency findings, age-related retardation in N400 latency does not appear to be accounted for by delays in earlier, primary visual components (Gunter et al., 1992; Kutas & Iragui, 1998). The age-related N400 reduction appears to be generalizable across a variety of conditions, including, but not limited to, category and repetition priming (Hamberger & Friedman, 1992; Harbin et al., 1984; Kutas & Iragui, 1998), antonymic contexts (Kutas & Iragui, 1998), and visual as well as auditory sentence contexts (Cameli & Phillips, 2000; Federmeier et al., 2003; Hamberger et al., 1995; Woodward et al., 1993). Interestingly, when sentences were presented in the more natural context of connected speech, the age-related prolongation in N400 latency reported for visually presented completions was not observed (Federmeier et al., 2003).

Because semantic-memory impairments are often observed in Alzheimer's disease (Nebes, 1989), the semantic incongruity effect has also been studied in patients with a diagnosis of PAD. Although there are exceptions (Ford et al., 2001; Hamberger et al., 1995), the results suggest that, by and large, the N400 effect produced by PAD patients is reduced and its latency is prolonged compared to age-matched controls (Ford et al., 1996; Iragui et al., 1996; Ostrosky-Solis et al., 1998; Revonsuo et al., 1998; Schwartz et al., 1996; see the review by Taylor & Olichney, 2007). Moreover, some very limited evidence suggests that the reduction in the N400 effect may predict conversion from MCI to PAD (Olichney et al., 2002).

As reviewed briefly above, the majority of evidence indicates that the magnitude of the N400 effect is reduced in older adults and further reduced in PAD. In addition, N400 latency shows a similar age- and disease-related pattern. Nonetheless, as in young adults, the N400 in older adults, and to some extent in PAD patients, is inversely related to the degree to which a given context "primes" the eliciting word (or picture). Hence, the data could be interpreted to indicate that the semantic memory network is intact in older adults, consistent with a large behavioral literature suggesting that semantic knowledge is maintained and/or increases with age (Park et al., 2002; Salthouse, 1993). Therefore, N400 latency prolongation might be a consequence of an age-related increment in the size of the semantic network—it takes longer to search a larger network. The amplitude reductions in the N400 effect may also have a similar cause: in older relative to young adults, the preceding context could have primed a wider range of words related to the best completion. Alternatively, the possibility exists that older adults activate a wider range of terminal words due to age-related alteration in inhibitory processing (Phillips & Lesperance, 2003), consistent with the inhibitory-deficit hypothesis of cognitive aging (Hasher & Zacks, 1988). Age-related limitation in working memory/processing resources might also account for these data, since when these resources are taxed, older adults show subtle differences in the N400 effect (Federmeier, 2007; Federmeier et al., 2003).

Episodic Memory Components

Mnemonic function has proven to be an important aspect of research on cognitive aging due to the well-documented age-related deficits in episodic memory (EM; Light, 1991). Several ERP investigations of recognition memory aging are available (Friedman et al., 2007a). The ERP episodic-recognition memory effects are computed by subtracting the ERP to unstudied (i.e., new) items from those studied previously (i.e., old), yielding the old/new or EM effect (Friedman & Johnson, 2000; Rugg & Curran, 2007; Chapter 14, this volume). In studies of young adults, two of these EM effects, the medial-prefrontal and left-parietal, have been consistently recorded in the retrieval phases of recognition-memory paradigms. They have been associated with, respectively, familiarity- and recollection-based processes (Friedman & Johnson, 2000; Rugg & Curran, 2007; Rugg & Yonelinas, 2003), which have been hypothesized by dual-process theorists to underlie recognition-memory performance (Mandler, 1980; Yonelinas, 2002). However, the extent to which the medial-prefrontal effect reflects episodic, familiarity-driven processes as opposed to semantically driven conceptual processing is currently the subject of much debate (Paller et al., 2007). Figure 18.5 depicts these two EM effects in the grand-averaged ERP waveforms elicited by correctly rejected new and correctly recognized old items in young and older adults from a study by Nessler et al. (2007a). As can be observed, these EM effects overlap large-amplitude potentials such as the N400 and P3b. Nonetheless, they have been shown to be relatively independent of these components on the basis of differences in scalp topography (see Johnson, 1993, for a rationale and overview; see Friedman, 2000, for an application to aging). Note that the medial-prefrontal EM effect is of similar magnitude in young and older adults, whereas the left-parietal EM effect is reduced in older relative to young adults. In age-related studies that have measured directly the medial-prefrontal EM effect (Nessler et al., 2007a, 2008; Trott et al., 1999; Wegesin et al., 2002), its amplitude has been found to be age equivalent. By contrast, in some investigations, the left-parietal EM effect has been reported to be smaller in older relative to young adults (Friedman et al., 1993a; Nessler et al., 2007a, 2008; Rugg et al., 1997; Swick & Knight, 1997), primarily when participants are asked to make simple old/new recognition judgments. Under these relatively simple old/new retrieval conditions, older adults often do not show impaired recognition memory or their performance does not differ markedly from that of young adults (Craik & McDowd, 1987). Then again, when source- or contextual-memory judgments are required, age equivalence of the left-parietal EM effect has been observed (Mark & Rugg, 1998; Trott et al., 1999; Wegesin et al., 2002), despite the fact that older adults typically perform worse than young adults under these more taxing retrieval circumstances (Spencer &

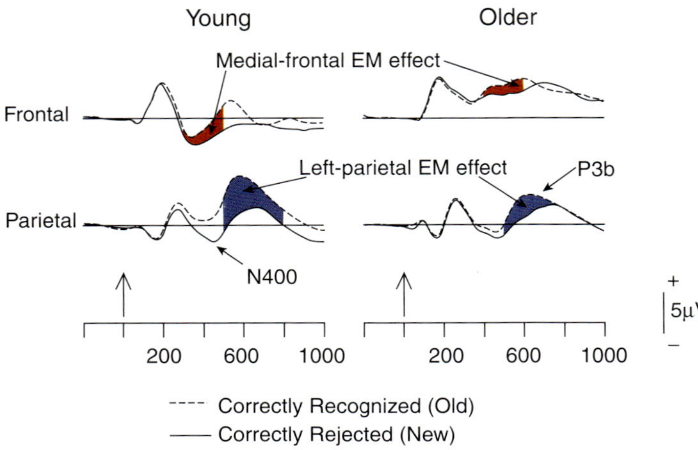

Fig. 18.5. Grand mean ERPs averaged across 16 participants in young (left) and older (right) adult groups during the second of two retrieval phases of a verbal recognition memory experiment (Nessler et al., 2007b). For each group, the ERPs elicited by correct rejections of new or unstudied items and correctly recognized old items are superimposed. Arrows mark stimulus onset with timelines every 200 ms. Red shading indicates the medial-prefrontal EM effect; blue shading indicates the later-onsetting left-parietal EM effect.

Raz, 1995). The findings for the medial-prefrontal EM effect have generally been interpreted as reflecting maintained familiarity-based processes in aging, consistent with current theories of memory aging (Yonelinas, 2002). The paradoxical finding of smaller left-parietal EM effects in the face of relatively preserved performance during old/new recognition memory might then indicate that, in these retrieval situations, older adults rely primarily on familiarity to make their recognition decisions (Nessler et al., 2008). This may be due to the possibility that older adults retrieve a smaller amount of contextual detail because their memory representations are qualitatively diminished as a result of inefficient encoding (Nessler, Johnson, Bersick, & Friedman, 2006) However, when necessary, as in source-memory paradigms, older adults can use recollection-based processes to support their retrieval performance, albeit to a lesser extent than their young adult counterparts. Hence, these data as a whole suggest a default retrieval strategy in which familiarity is the primary basis for recognition decisions in older individuals, perhaps due to the more effortful, resource-demanding nature of recollective processing (Jennings & Jacoby, 1997).

In some memory retrieval conditions, a few studies have reported the presence of electrical activity in older adult waveforms that could be interpreted as compensatory. For example, a left-frontal negative activity (~400–1000 ms poststimulus) has been reported only in older adult ERPs during the retrieval phases of recognition-memory investigations (Czernochowski et al., 2007; Li et al., 2004; Swick et al., 2006). Because this negativity has been observed most often during source-memory paradigms (but see Nessler et al., 2007a, for left-frontal negativity in a standard old/new recognition task), some authors have suggested that the negativity reflects alternate retrieval strategies necessitated, in older adults, by the greater demands on control processes required to retrieve contextual information from stored memory traces (Czernochowski et al., 2007; Li et al., 2004; Swick et al., 2006; Wegesin et al., 2002). However, in direct opposition to this argument, Duarte and colleagues (2006) reported what appears to be a similar negativity in a source-memory paradigm in the ERPs of low-performing older adults, but not in those of high-performing older or young adults. Hence, in the current state of knowledge, the functional significance of this activity is unclear.

To sum up briefly, to the extent that the medial-prefrontal and left-parietal EM effects reflect, respectively, familiarity- and recollection-based processes (see Rugg & Curran, 2007, and Paller et al., 2007, for extended discussions), the data indicate that both are relatively intact in older adults. Nonetheless, the predominant retrieval mode in older adults appears to be one based on familiarity. Whether, on the basis of ERP studies, older adults compensate to overcome these episodic recognition deficits is an open question; the data are simply too sparse to permit a definitive conclusion. Categorization of older (and young) adults as low and high performers on a variable of theoretical importance might aid in that endeavor, and investigators should consider this in future studies (Czernochowski et al., 2007; Duarte et al., 2006).

Error-Related and Medial Frontal Negativities

The error-related negativity (ERN; Falkenstein et al., 2001) occurs on trials in which an error has been made (see Chapter 10, this volume), whereas the medial-frontal negativity or (MFN; Gehring &

Willoughby., 2002; Nessler et al., 2007b) occurs on trials in which a correct response has been made. These two ERP modulations are best observed in response as compared to the typical stimulus-locked average (Figure 18.6). Both are thought to reflect cognitive-control functions (Botvinick et al., 2001, 2004; Ridderinkhof et al., 2004), although there is reasonably good evidence to indicate that they signal at least partially nonoverlapping behavioral processes with neural generators located in different regions of the anterior cingulate cortex (ACC; Friedman et al., 2007b; Johnson et al., 2004; Masaki et al., 2007; Nessler et al., 2007). The ACC monitors for and detects response conflict and plays a critical role in goal-directed action by signaling other prefrontal regions to adjust behavior in the presence of competing-response information (Botvinick et al., 2001).

As shown in Figure 18.6, the ERN has consistently been found to be smaller in older relative to young adults (Band & Kok, 2000; Falkenstein et al., 2001; Gehring & Knight, 2000; Mathalon et al., 2003; Nessler et al., 2007b; Themanson et al., 2006; West, 2004; but see Eppinger et al., 2008). Figure 18.6 depicts the response-locked ERPs recorded during a modified version of the Eriksen flanker task (Friedman et al., 2009). Note that, whereas the ERN on error relative to correct trials is larger in young relative to older adults, the MFN on high-response-conflict (see the Figure 18.6 caption), incompatible-response trials relative to lower-conflict, compatible-response trials is larger in older than young adults, suggesting an age-related dissociation in the processes reflected by the ERN and MFN (see also Nessler et al., 2007b). Studies showing that the amplitude of the MFN increases with the degree of conflicting response information suggest that the MFN is an indicator of the amount of response conflict detected (Botvinick et al., 2004; Hogan et al., 2005; Johnson et al., 2004; Kray et al., 2005; Nessler et al., 2007b; West, 2004). Although the ERN has also been interpreted as reflecting some aspects of conflict monitoring (e.g., an error is in conflict with the appropriate response), it likely reflects additional error-specific processes that appear to undergo age-related change (Nieuwenhuis et al., 2002; Taylor et al., 2007).

Unlike the ERN, there is a paucity of age-related studies of response conflict monitoring and detection as reflected by the MFN. In three investigations from this laboratory, including response-competition (Nessler et al., 2007), task switching (Friedman

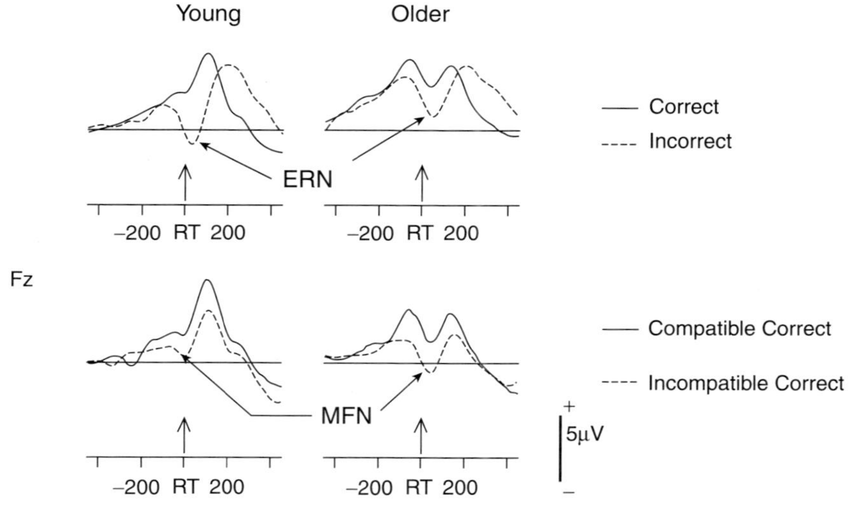

Fig. 18.6. Grand mean response-locked ERPs averaged across 16 participants in young (left) and older (right) adult groups during a version of the Eriksen flanker task (Eriksen & Eriksen, 1974). The central stimulus was an arrow pointing either to the right or to the left (the direction indicated the to-be-pressed response key). The flankers on either side of the target arrow were congruent or incongruent with the central arrow or were neutral. In the Compatible condition, participants pressed the left or right button in accord with the direction of the central arrow, whereas in the Incompatible condition, they responded in the direction opposite that indicated by the central arrow. In order to perform well in the Incompatible condition, it was necessary to implement cognitive control by monitoring, detecting, and reducing the ensuing response conflict through inhibition or suppression of the incorrect but prepotent response tendency. The top row of ERPs illustrates the brain's response on erroneous-compared to correct-response trials, yielding the ERN peaking at about 70 ms following mean RT. The bottom row of waveforms depicts the brain's response on correct incompatible- compared to correct compatible-response trials, resulting in the MFN. Arrows mark mean RT with timelines every 200 ms.

et al., 2007b), and Eriksen flanker paradigms (Friedman et al., 2009), MFN magnitude was similar in young and older adults at low to medium levels of cognitive demand (but see Kray et al., 2005). At high levels of cognitive demand (such as in incompatible-response trials following an error), older adults produced larger MFNs than younger adults (Nessler et al., 2007b), indicating that they detected a greater amount of response conflict than young adults. One reason for the greater MFN magnitudes in older adults under high-demand conditions might have been that attempts to resolve the heightened response conflict by upregulating cognitive control were relatively unsuccessful, thereby resulting in high residual levels of response conflict. This is attested to by the fact that in these high-demand situations, older adults performed more poorly than young adults.

In summary, although the ERN and MFN have the potential to shed light on different aspects of cognitive control in older adults, the scarcity of age-related data in which the two have been compared hinders interpretation at this time. Nonetheless, based on the extant ERN data, there do appear to be age-related changes in error-monitoring processes, although a recent study strongly questions this conclusion (Eppinger et al., 2008). The age-related MFN literature is scant, and more studies are called for. The limited data suggest that the processes involved in response-conflict monitoring and detection are relatively intact. Hence, the older adult difficulty is more likely attributable to a deficit in some aspects of top-down, executive-control processing (Friedman et al., 2009; Nessler et al., 2007b).

Overall Conclusions and Directions for Future Work

Several unifying theories have been proposed to account for the wide variety of cognitive aging phenomena that have been described in the literature (Craik & Salthouse, 2007; Park, 2000b). These include deficits in working memory/processing resources (Craik & Byrd, 1982), general slowing (Salthouse, 1996), deficits in inhibitory control (Hasher & Zacks, 1988), and losses in sensory acuity (Baltes & Lindenberger, 1997). Unfortunately, the study of ERP components has not always been directed explicitly at providing support for or against these theoretical positions. Nonetheless, some of the ERP findings presented above have the potential, at least preliminarily, to contribute evidence to these hypotheses. For example, the fairly consistent finding that P3b amplitude is reduced in a wide variety of paradigms in older adults suggests that age-related cognitive deficits may be due to reduced processing resources and/or alterations in working memory (Donchin & Coles, 1988; Kramer et al., 1986; but see Polich, 2007). On the other hand, age-related changes in primary components such as the P50 and N1 do not appear likely, at least currently, to account for a large percentage of the variance in cognitive change with aging. Age-related alterations in these early-latency components may fit better with the inhibitory-rather than the sensory-deficit theory of cognitive aging, but much more work is required before this will be known with any certainty. Nonetheless, the interaction between age-related sensory processing deficits and downstream cognitive processes, which receives support in the behavioral literature (Wingfield et al., 2005), has rarely been considered in ERP/aging research. This might prove to be a fruitful research area. The processing speed theory of cognitive aging does not receive much support from the extant data because age-related delays in early-latency components (such as P50 and N1) have not been reported or, if they have, those delays are much too small to account for the long delays in the latencies of the N2b, P3a, and P3b components.

Despite these hints at putative mechanisms, there are difficulties that need to be overcome before a clearer picture of the underlying causes of cognitive change with aging using ERP data can be painted. The greater variability observed in older adult data needs to be exploited, as some individuals fare better than others as they age, and ERP components may aid in understanding why. Although attempts in this direction have been made (Czernochowski et al., 2007; Duarte et al., 2006; Fabiani et al., 1998; Riis et al., 2008), more work is certainly necessary. The omnipresence of the frontal-maximum scalp distribution of the P3a and P3b in older adults presents some difficulty for interpretation of the age-related functional significance of these components. Future studies need to attempt to manipulate this topographic feature via experimental design to determine whether it can be modified and, if so, by which variables. Moreover, this feature of older adults' scalp distribution may aid in understanding individual differences in cognitive aging and what they imply about preserved and impaired cognition (Fabiani et al., 1998; Riis et al., 2008). Similarly, tantalizing hints at difficulties in inhibitory control based on relatively early-latency components such as the P50 and N1 need to be followed up, employing an individual differences approach. For example, because a decrease in inhibitory control has been hypothesized to underlie the age-related enhancement of the P50 and N1 components (Alain & Woods, 1999; Fabiani et al., 2006; Chao & Knight, 1997), perhaps older adults with enhanced amplitudes are those who also show deficits in other, higher-order cognitive tasks that require executive processes, such as inhibitory control, for good task performance (e.g., selective attention, task switching, flanker paradigms). More research effort directed at the validity of the compensation hypothesis, which is based almost exclusively on functional magnetic resonance imaging (fMRI) and positron emission tomography (PET) data, is necessary (Cabeza et al., 2002). Because of its high temporal precision, the ERP technique may contribute information not available from fMRI or PET.

Moreover, a great deal of age-related ERP data has been obtained using some variant of the oddball paradigm. It is difficult to assess the validity of the major cognitive-aging hypotheses using this task because levels of cognitive complexity are generally not built into the oddball design. Therefore, it assesses cognition only in a very general fashion. Hence, ERP paradigms that assess directly such constructs as attentional resources, inhibitory control, and working and episodic memory, separately as well as in combination, are required. For instance, ERP investigations of age-related change in episodic memory have not usually incorporated assessments of executive control, although deficits in cognitive control are clearly present as we age and may be a cause of episodic memory failure (Braver & West, 2007; Friedman, 2007). Hence, investigators should take greater advantage of the ERN and MFN in studies of episodic memory (Curran et al., 2007), as well as in other domains such as attentional and inhibitory control. Further research using the full armamentarium of ERP components in the variety of domains reviewed above will certainly provide a better understanding of the mechanisms underlying the preservation and disruption of neurocognitive functions as individuals age.

Acknowledgments

The research reported in this chapter was supported by NIA Grants AG005213 and AG009988 and by the New York State Department of Mental Hygiene. I thank my colleagues Drs. Helen Gaeta, Ray Johnson, Jr., Doreen Nessler, and Walter Ritter for many fruitful discussions concerning the data reported here. I also thank Mr. Charles Brown, Jr., for technical assistance and software development.

Address correspondence and reprint requests to: David Friedman, Ph.D., Cognitive Electrophysiology Laboratory, New York State Psychiatric Institute, 1051 Riverside Drive, Unit 6, New York City, NY 10032, USA. Phone: 212–543-5476; Fax: 212–543-6002; E-mail: mailto:df12@columbia.edu.

References

Adler, L. E., Waldo, M. C., & Freedman, R. (1985). Neurophysiologic studies of sensory gating in schizophrenia: Comparison of auditory and visual responses. *Biological Psychiatry, 20*, 1284–1296.

Alain, C., McDonald, K. L., Ostroff, J. M., & Schneider, B. (2004). Aging: A switch from automatic to controlled processing of sounds? *Psychology and Aging, 19*, 125–133.

Alain, C., & Woods, D. L. (1999). Age-related changes in processing auditory stimuli during visual attention: Evidence for deficits in inhibitory control and sensory memory. *Psychology and Aging, 14*(3), 507–519.

Alho, K., Woods, D. L., Algazi, A., Knight, R. T., & Näätänen, R. (1994). Lesions of frontal cortex diminish the auditory mismatch negativity. *Electroencephalography and Clinical Neurophysiolgy, 91*, 353–362.

Ally, B. A., Jones, G. E., Cole, J. A., & Budson, A. E. (2006). Sensory gating in patients with Alzheimer's disease and their biological children. *American Journal of Alzheimer's Disease and Other Dementias, 21*(6), 439–447.

Amenedo, E., & Diaz, F. (1998a). Automatic and effortful processes in auditory memory reflected by event-related potentials: Age-related findings. *Electroencephalography and Clinical Neurophysiology, 108*(4), 361–369.

Amenedo, E., & Diaz, F. (1998b). Effects of aging on middle-latency auditory evoked potentials: A cross-sectional study. *Biological Psychiatry, 43*(3), 210–219.

Amenedo, E., & Diaz, F. (1999). Ageing-related changes in the processing of attended and unattended standard stimuli. *Neuroreport, 10*(11), 2383–2388.

Anderer, P., Semlitsch, H. V., & Saletu, B. (1996). Multichannel auditory event-related brain potentials: Effects of normal aging on the scalp distribution of N1, P2, N2 and P300 latencies and amplitudes. *Electroencephalography and Clinical Neurophysiology, 99*(5), 458–472.

Baddeley, A. (1992). Working memory. *Science, 255*, 556–559.

Baltes, P. B., & Lindenberger, U. (1997). Emergence of a powerful connection between sensory and cognitive functions across the adult life span: A new window to the study of cognitive aging? *Psychology and Aging, 12*(1), 12–21.

Band, G. P., & Kok, A. (2000). Age effects on response monitoring in a mental-rotation task. *Biological Psychology, 51*(2–3), 201–221.

Beck, E. C., Swanson, C., & Dustman, R. E. (1980). Long latency components of the visually evoked potential in man: Effects of aging. *Experimental Aging Research, 6*(6), 523–545.

Bennett, I. J., Golob, E. J., & Starr, A. (2004). Age-related differences in auditory event-related potentials during a cued attention task. *Clinical Neurophysiology, 115*(11), 2602–2615.

Bennys, K., Portet, F., Touchon, J., & Rondouin, G. (2007). Diagnostic value of event-related evoked potentials N200 and P300 subcomponents in early diagnosis of Alzheimer's disease and mild cognitive impairment. *Journal of Clinical Neurophysiology, 24*(5), 405–412.

Bertoli, S., & Probst, R. (2005). Lack of standard N2 in elderly participants indicates inhibitory processing deficit. *Neuroreport, 16*(17), 1933–1937.

Bertoli, S., Smurzynski, J., & Probst, R. (2002). Temporal resolution in young and elderly subjects as measured by mismatch negativity and a psychoacoustic gap detection task. *Clinical Neurophysiology, 113*(3), 396–406.

Bertoli, S., Smurzynski, J., & Probst, R. (2005). Effects of age, age-related hearing loss, and contralateral cafeteria noise on the discrimination of small frequency changes: Psychoacoustic and electrophysiological measures. *Journal of the Association of Research on Otolaryngology, 6*(3), 207–222.

Botvinick, M. M., Braver, T. S., Barch, D. M., Carter, C. S., & Cohen, J. D. (2001). Conflict monitoring and cognitive control. *Psychological Review, 108*(3), 624–652.

Botvinick, M. M., Cohen, J. D., & Carter, C. S. (2004). Conflict monitoring and anterior cingulate cortex: An update. *Trends in Cognitive Sciences, 8*(12), 539–546.

Braff, D. L., & Geyer, M. A. (1990). Sensorimotor gating and schizophrenia: Human and animal model studies. *Archives of General Psychiatry, 47*, 181–188.

Braver, T. S., & West, R. (2007). Working memory, executive control and aging. In F. I. M. Craik & T. A. Salthouse (Eds.), *The handbook of cognition and aging* (Vol. 3, pp. 657). New York: Psychology Press.

Brown, W. S., Marsh, J. T., & Agenath, L. (1983). Exponential electrophysiological aging: P3 latency. *Electroencephalography and Clinical Neurophysiology, 55*, 277–285.

Buckner, R. L. (2004). Memory and executive function in aging and AD: Multiple factors that cause decline and reserve factors that compensate. *Neuron, 44*(1), 195–208.

Cabeza, R., Anderson, N. D., Locantore, J. K., & McIntosh, A. R. (2002). Aging gracefully: Compensatory brain activity in high-performing older adults. *Neuroimage, 17*(3), 1394–1402.

Cameli, L., & Phillips, N. A. (2000). Age-related differences in semantic priming: Evidence from event-related brain potentials. *Brain Cognition, 43*(1–3), 69–73.

Cancelli, I., Cadore, I. P., Merlino, G., Valentinis, L., Moratti, U., Bergonzi, P., et al. (2006). Sensory gating deficit assessed by P50/Pb middle latency event related potential in Alzheimer's disease. *Journal of Clinical Neurophysiology, 23*(5), 421–425.

Chao, L. L., & Knight, R. T. (1997). Prefrontal deficits in attention and inhibitory control with aging. *Cerebral Cortex, 7*, 63–69.

Colcombe, S. J., Kramer, A. F., Erickson, K. I., & Scalf, P. (2005). The implications of cortical recruitment and brain morphology for individual differences in inhibitory function in aging humans. *Psychology and Aging, 20*(3), 363–375.

Coles, M. G., Gratton, G., & Donchin, E. (1988). Detecting early communication: Using measures of movement-related potentials to illuminate human information processing. *Biological Psychology, 26*(1–3), 69–89.

Cooper, R. J., Todd, J., McGill, K., & Michie, P. T. (2006). Auditory sensory memory and the aging brain: A mismatch negativity study. *Neurobiology of Aging, 27*(5), 752–762.

Craik, F. I. M., & Byrd, M. (1982). Aging and cognitive deficits: The role of attentional resources. In F. I. M. Craik & S. Trehub (Eds.), *Aging and cognitive processes* (pp. 191–211). New York: Plenum Press.

Craik, F. I. M., & McDowd, J. M. (1987). Age differences in recall and recognition. *Journal of Experimental Psychology: Learning, Memory, and Cognition, 13*, 474–479.

Craik, F. I. M., & Salthouse, T. A. (2007). *The handbook of aging and cognition* (p. 657). New York: Psychology Press.

Curran, T., DeBuse, C., & Leynes, P. A. (2007). Conflict and criterion setting in recognition memory. *Journal of Experimental Psychology: Learning, Memory, and Cognition, 33*(1), 2–17.

Czernochowski, D., Fabiani, M., & Friedman, D. (2007). Use it or lose it? SES mitigates age-related decline in a recency/recognition task. *Neurobiology of Aging, 29*, 945–958.

Czigler, I., & Balazs, L. (2005). Age-related effects of novel visual stimuli in a letter-matching task: An event-related potential study. *Biological Psychology, 69*(2), 229–242.

Czigler, I., Csibra, G., & Csontos, A. (1992). Age and interstimulus interval effects on event-related potentials to frequent and infrequent auditory stimuli. *Biological Psychology, 33*, 195–206.

Czigler, I., Pato, L., Poszet, E., & Balazs, L. (2006). Age and novelty: Event-related potentials to visual stimuli within an auditory oddball—visual detection task. *International Journal of Psychophysiology, 62*(2), 290–299.

Daffner, K. R., Mesulam, M. M., Scinto, L. F. M., Acar, D., Calvo, V., Fausti, R., et al. (2000). The central role of the prefrontal cortex in directing attention to novel events. *Brain, 123*, 927–939.

Daffner, K. R., Rentz, D. M., Scinto, L. F., Faust, R., Budson, A. E., & Holcomb, P. J. (2001). Pathophysiology underlying diminished attention to novel events in patients with early AD. *Neurology, 56*(10), 1377–1383.

Daffner, K. R., Ryan, K. K., Williams, D. M., Budson, A. E., Rentz, D. M., Scinto, L. F., et al. (2005a). Age-related differences in novelty and target processing among cognitively high performing adults. *Neurobiology of Aging, 26*(9), 1283–1295.

Daffner, K. R., Ryan, K. K., Williams, D. M., Budson, A. E., Rentz, D. M., Scinto, L. F. M., et al. (2005b). Age-sensitivity of the P3 in cognitively high-performing adults: Unsettled issues. *Neurobiology of Aging*, 26, 1301–1304.

Daffner, K. R., Ryan, K. K., Williams, D. M., Budson, A. E., Rentz, D. M., Wolk, D. A., et al. (2006a). Age-related differences in attention to novelty among cognitively high performing adults. *Biological Psychology, 72*(1), 67–77.

Daffner, K. R., Ryan, K. K., Williams, D. M., Budson, A. E., Rentz, D. M., Wolk, D. A., et al. (2006b). Increased responsiveness to novelty is associated with successful cognitive aging. *Journal of Cognitive Neuroscience, 18*(10), 1759–1773.

Davis, H., Mast, T., Yoshie, N., & Zerlin, S. (1966). The slow response of the human cortex to auditory stimuli: Recovery process. *Electroencephalography and Clinical Neurophysiology, 21*(2), 105–113.

Davis, S. W., Dennis, N. A., Daselaar, S. M., Fleck, M. S., & Cabeza, R. (2007). Que PASA? The posterior anterior shift in aging. *Cerebral Cortex, 18*, 1201–1209.

De Sanctis, P., Katz, R., Wylie, G. R., Sehatpour, P., Alexopoulos, G. S., & Foxe, J. J. (2008). Enhanced and bilateralized visual sensory processing in the ventral stream may be a feature of normal aging. *Neurobiology of Aging, 29*, 1576–1586.

Donchin, E., Callaway, E., Cooper, R., Desmedt, J. E., Goff, W. R., Hillyard, S. A., et al. (1977). Publication criteria for studies of evoked potentials in man: Report of a committee. In J. E. Desmedt (Ed.), *Attention, voluntary contraction, and event-related cerebral potentials: Progress in clinical neurophysiology* (pp. 1–11). Basel: S. Karger.

Donchin, E., & Coles, M. G. H. (1988). Is the P300 component a manifestation of context updating? *Behavioral and Brain Sciences, 11*, 357–374.

Duarte, A., Ranganath, C., Trujillo, C., & Knight, R. T. (2006). Intact recollection memory in high-performing older adults: ERP and behavioral evidence. *Journal of Cognitive Neuroscience, 18*(1), 33–47.

Enoki, H., Sanada, S., Yoshinaga, H., Oka, E., & Ohtahara, S. (1993). The effects of age on the N200 component of the auditory event-related potentials. *Brain Research and Cognitive Brain Research, 1*(3), 161–167.

Eppinger, B., Kray, J., Mock, B., & Mecklinger, A. (2008). Better or worse than expected? Aging, learning, and the ERN. *Neuropsychologia, 46*, 521–539.

Eriksen, B. A., & Eriksen, C. W. (1974). Effects of noise letters upon the identification of a target letter in a non-search task. *Perception & Psychophysics, 16*, 143–149.

Fabiani, M., & Friedman, D. (1995). Changes in brain activity patterns in aging: The novelty oddball. *Psychophysiology, 32*(6), 579–594.

Fabiani, M., Friedman, D., & Cheng, J. C. (1998). Individual differences in P3 scalp distribution in older adults, and their relationship to frontal lobe function. *Psychophysiology, 35*(6), 698–708.

Fabiani, M., Low, K. A., Wee, E., Sable, J. J., & Gratton, G. (2006). Reduced suppression or labile memory? Mechanisms of inefficient filtering of irrelevant information in older adults. *Journal of Cognitive Neuroscience, 18*(4), 637–650.

Falkenstein, M., Hoormann, J., & Hohnsbein, J. (2001). Changes of error-related ERPs with age. *Experimental Brain Research, 138*(2), 258–262.

Falkenstein, M., Yordanova, J., & Kolev, V. (2006). Effects of aging on slowing of motor-response generation. *International Journal of Psychophysiology, 59*(1), 22–29.

Federmeier, K. D. (2007). Thinking ahead: The role and roots of prediction in language comprehension. *Psychophysiology, 44*(4), 491–505.

Federmeier, K. D., Van Petten, C., Schwartz, T. J., & Kutas, M. (2003). Sounds, words, sentences: Age-related changes across levels of language processing. *Psychology and Aging, 18*(4), 858–872.

Fjell, A. M., & Walhovd, K. B. (2003). On the topography of P3a and P3b across the adult lifespan—a factor-analytic study using orthogonal procrustes rotation. *Brain Topography, 15*(3), 153–164.

Fjell, A. M., & Walhovd, K. B. (2005). Age-sensitivity of P3 in high-functioning adults. *Neurobiology of Aging, 26*(9), 1297–1299; discussion 1301–1296.

Ford, J., Roth, W., Isaacks, B., White, P., Hood, S., & Pfefferbaum, A. (1995). Elderly men and women are less responsive to startling noises: N1, P3 and blink evidence. *Biological Psychology, 39*, 57–80.

Ford, J. M., Askari, N., Mathalon, D. H., Menon, V., Gabrieli, J. D., Tinklenberg, J. R., et al. (2001). Event-related brain potential evidence of spared knowledge in Alzheimer's disease. *Psychology and Aging, 16*(1), 161–176.

Ford, J. M., Roth, W. T., Isaacks, B. G., Tinklenberg, J. R., Yesavage, J., & Pfefferbaum, A. (1997). Automatic and effortful processing in aging and dementia: Event-related brain potentials. *Neurobiology of Aging, 18*(2), 169–180.

Ford, J. M., Woodward, S. H., Sullivan, E. V., Isaacks, B. G., Tinklenberg, J. R., Yesavage, J. A., et al. (1996). N400 evidence of abnormal responses to speech in Alzheimer's disease. *Electroencephalography and Clinical Neurophysiology, 99*, 235–246.

Freedman, R., Adler, L. E., Gerhardt, G. A., Maldo, M., Baker, N., Rose, G. M., et al. (1987). Neurobiological studies of sensory gating in schizophrenia. *Schizophrenia Bulletin, 13*, 669–667.

Friedman, D. (2000). Event-related brain potential investigations of memory and aging. *Biological Psychology, 54*(1–3), 175–206.

Friedman, D. (2003). Cognition and aging: A highly selective overview of event-related potential (ERP) data. *Journal of Clinical and Experimental Neuropsychology, 25*, 702–720.

Friedman, D. (2007). A neurocognitive overview of aging phenomena based on the event-related potential (ERP). In Y. Stern (Ed.), *Cognitive reserve* (pp. 285–303). Amsterdam: Swets.

Friedman, D., Berman, S., & Hamberger, M. (1993a). Recognition memory and ERPs: Age-related changes in young, middle-aged and elderly adults. *Journal of Psychophysiology, 7*, 181–201.

Friedman, D., Cycowicz, Y. M., & Gaeta, H. (2001). The novelty P3: An event-related brain potential (ERP) sign of the brain's evaluation of novelty. *Neuroscience and Biobehavioral Reviews, 25*(4), 355–373.

Friedman, D., & Johnson, R. (2000). Event-related potential (ERP) studies of memory encoding and retrieval: A selective review. *Microscopy Research and Technology, 51*(1), 6–28.

Friedman, D., Kazmerski, V. A., & Cycowicz, Y. M. (1998). Effects of aging on the novelty P3 during attend and ignore oddball tasks. *Psychophysiology, 35*(5), 508–520.

Friedman, D., Kazmerski, V., & Fabiani, M. (1997). An overview of age-related changes in the scalp distribution of P3b. *Electroencephalography and Clinical Neurophysiology, 104*(6), 498–513.

Friedman, D., Nessler, D., Cycowicz, Y. M., & Horton, C. (2009). Development of and change in cognitive control: A comparison of children, young and older adults. *Cognitive Affective and Behavioral Neuroscience, 9(1)*, 91–102.

Friedman, D., Nessler, D., & R. Johnson, J. (2007a). Memory encoding and retrieval in the aging brain. *Clinical Electroencephalography and Neuroscience, 38*(1), 2–7.

Friedman, D., Nessler, D., Johnson, R., Jr., Ritter, W., & Bersick, M. (2007b). Age-related changes in executive function: An event-related potential (ERP) investigation of task-switching. *Neuropsychology of Development and Cognition B, Aging, Neuropsychology and Cognition, 15 (1)*, 95–128.

Friedman, D., & Simpson, G. V. (1994). ERP amplitude and scalp distribution to target and novel events: effects of temporal order in young, middle-aged and older adults. *Brain Research Cognitive Brain Research, 2*(1), 49–63.

Friedman, D., Simpson, G., & Hamberger, M. (1993b). Age-related changes in scalp topography to novel and target stimuli. *Psychophysiology, 30*(4), 383–396.

Frodl, T., Hampel, H., Juckel, G., Burger, K., Padberg, F., Engel, R. R., et al. (2002). Value of event-related P300 subcomponents in the clinical diagnosis of mild cognitive impairment and Alzheimer's disease. *Psychophysiology, 39*(2), 175–181.

Gaeta, H., Friedman, D., & Ritter, W. (2003). Auditory selective attention in young and elderly adults: The selection of single versus conjoint features. *Psychophysiology, 40*, 389–406.

Gaeta, H., Friedman, D., Ritter, W., & Cheng, J. (1998). An event-related potential study of age-related changes in sensitivity to stimulus deviance. *Neurobiology of Aging, 19*(5), 447–459.

Gaeta, H., Friedman, D., Ritter, W., & Cheng, J. (1999). Changes in sensitivity to stimulus deviance in Alzheimer's disease: An ERP perspective. *Neuroreport, 10*(2), 281–287.

Gaeta, H., Friedman, D., Ritter, W., & Cheng, J. (2001a). The effect of perceptual grouping on the mismatch negativity. *Psychophysiology, 38*(2), 316–324.

Gaeta, H., Friedman, D., Ritter, W., & Cheng, J. (2001b). An event-related potential evaluation of involuntary attentional shifts in young and older adults. *Psychology and Aging, 16*(1), 55–68.

Gaeta, H., Friedman, D., Ritter, W., & Hunt, G. (2002). Age-related changes in neural trace generation of rule-based auditory features. *Neurobiology of Aging, 23*(3), 443–455.

Gehring, W. J., & Knight, R. T. (2000). Prefrontal–cingulate interactions in action monitoring. *Nature Neuroscience, 3*(5), 516–520.

Gehring, W. J., & Willoughby., A. R. (2002). The medial frontal cortex and the rapid processing of monetary gains and losses. *Science, 295*, 2279–2282.

Giard, M. H., Perrin, F., Pernier, J., & Bouchet, P. (1990). Brain generators implicated in the processing of auditory stimulus deviance: A topographic event-related potential study. *Psychophysiology, 27*, 627–640.

Golob, E. J., Irimajiri, R., & Starr, A. (2007). Auditory cortical activity in amnestic mild cognitive impairment: Relationship to subtype and conversion to dementia. *Brain, 130*(Pt 3), 740–752.

Golob, E. J., Miranda, G. G., Johnson, J. K., & Starr, A. (2001). Sensory cortical interactions in aging, mild cognitive impairment, and Alzheimer's disease. *Neurobiology of Aging, 22*(5), 755–763.

Golob, E. J., & Starr, A. (2000). Effects of stimulus sequence on event-related potentials and reaction time during target detection in Alzheimer's disease. *Clinical Neurophysiology, 111*(8), 1438–1449.

Goodin, D. C., Squires, K. C., Henderson, B. H., & Starr, A. (1978a). Age-related variations in evoked potentials to auditory stimuli in normal human subjects. *Electroencephalography and Clinical Neurophysiology, 44*, 447–458.

Goodin, D. S., Squires, K. C., & Starr, A. (1978b). Long latency event-related components of the auditory evoked potential in dementia. *Brain, 101*(4), 635–648.

Gordon, E., Kraiuhin, C., Stanfield, P., & Meares, R. (1986). The prediction of normal P3 latency and the diagnosis of dementia. *Neuropsychologia, 24*(6), 823–830.

Greenwood, P. M. (2000). The frontal aging hypothesis evaluated. *Journal of the International Neuropsychological Society, 6*(6), 705–726.

Greenwood, P. M. (2007). Functional plasticity in cognitive aging: Review and hypothesis. *Neuropsychology, 21*(6), 657–673.

Gunter, T. G., Jackson, J. L., & Mulder, G. (1992). An electrophysiological study of semantic processing in young and middle-aged academics. *Psychophysiology, 29*, 38–54.

Gunter, T. C., Jackson, J. L., & Mulder, G. (1996). Focussing on aging: An electrophysiological exploration of spatial and attentional processing during reading. *Biological Psychology, 43*(2), 103–145.

Halgren, E., Marinkovic, K., & Chauvel, P. (1998a). Generators of the late cognitive potentials in auditory and visual oddball tasks. *Electroencephalography and Clinical Neurophysiology, 106*, 156–164.

Halgren, E., Marinkovic, K., & Chauvel, P. (1998b). Generators of the late cognitive potentials in auditory and visual oddball tasks. *Electroencephalography and Clinical Neurophysiology, 106*(2), 156–164.

Hamberger, M., & Friedman, D. (1992). Event-related potential correlates of repetition priming and stimulus classification in young, middle-aged, and older adults. *Journal of Gerontology, 47*(6), 395–405.

Hamberger, M. J., Friedman, D., Ritter, W., & Rosen, J. (1995). Event-related potential and behavioral correlates of semantic processing in Alzheimer's patients and normal controls. *Brain and Language, 48*(1), 33–68.

Hansen, J. C., & Hillyard, S. A. (1980). Endogenous brain potentials associated with selective auditory attention. *Electroencephalography and Clinical Neurophysiology, 49*, 277–290.

Harbin, T. J., Marsh, G. R., & Harvey, M. T. (1984). Differences in the late components of the event-related potential due to age and to semantic and non-semantic tasks. *Electroencephalography and Clinical Neurophysiology, 59*, 489–496.

Hasher, L., Stoltzfus, E. R., Zacks, R. T., & Rypma, B. (1991). Age and inhibition. *Journal of Experimental Psychology: Learning Memory, and Cognition, 17*(1), 163–169.

Hasher, L., & Zacks, R. T. (1988). Working memory, comprehension, and aging: A review and a new view. In G. H. Bower (Ed.), *The psychology of learning and motivation* (pp. 193–225). Orlando, FL: Academic Press.

Hillyard, S. A., Hink, R. F., Schwent, V. L., & Picton, T. W. (1973). Electrical signs of selective attention in the human brain. *Science, 182*, 177–180.

Hogan, A. M., Vargha-Khadem, F., Kirkham, F. J., & Baldeweg, T. (2005). Maturation of action monitoring from adolescence to adulthood: An ERP study. *Developmental Science, 8*(6), 525–534.

Iragui, V., Kutas, M., & Salmon, D. P. (1996). Event-related brain potentials during semantic categorization in normal aging and senile dementia of the Alzheimer type. *Electroencephalography and Clinical Neurophysiology, 100*, 392–406.

Iragui, V. J., Kutas, M., Mitchiner, M. R., & Hillyard, S. A. (1993). Effects of aging on event-related brain potentials and reaction times in an auditory oddball task. *Psychophysiology, 30*(1), 10–22.

Irimajiri, R., Golob, E. J., & Starr, A. (2005). Auditory brainstem, middle- and long-latency evoked potentials in mild cognitive impairment. *Clinical Neurophysiology, 116*(8) 1918–1929.

Jennings, J. M., & Jacoby, L. L. (1997). An opposition procedure for detecting age-related deficits in recollection: Telling effects of repetition. *Psychology and Aging, 12*(2), 352–361.

Jessen, F., Kucharski, C., Fries, T., Papassotiropoulos, A., Hoenig, K., Maier, W., et al. (2001). Sensory gating deficit expressed by a disturbed suppression of the P50 event-related potential in patients with Alzheimer's disease. *American Journal of Psychiatry, 158*(8), 1319–1321.

Johnson, R., Jr. (1993). On the neural generators of the P300 component of the event-related potential. *Psychophysiology, 30*, 94–101.

Johnson, R., Jr., Barnhardt, J., & Zhu, J. (2004). The contribution of executive processes to deceptive responding. *Neuropsychologia, 42*(7), 878–901.

Karayanidis, F., Andrews, S., Ward, P. B., & Michie, P. T. (1995). ERP indices of auditory selective attention in aging and Parkinson's disease. *Psychophysiology, 32*(4), 335–350.

Kazmerski, V., & Friedman, D. (1995). Repetition of novel stimuli in an ERP oddball paradigm: Aging effects. *Journal of Psychophysiology, 9*, 238–311.

Kazmerski, V. A., & Friedman, D. (1998). The scalp topography of P3b in early Alzheimer's disease. *Journal of Psychophysiology, 12*, 127–143.

Kazmerski, V. A., Friedman, D., & Ritter, W. (1997). Mismatch negativity during attend and ignore conditions in Alzheimer's disease. *Biological Psychiatry, 42*(5), 382–402.

King, J. W., & Kutas, M. (1995). Do the waves begin to waver? ERP studies of language processing in the elderly. In P. A. Allen & T. R. Bashore (Eds.), *Age differences in word and language processing* (Vol. 110, pp. 314–344). Amsterdam: North Holland.

Kisley, M. A., Davalos, D. B., Engleman, L. L., Guinther, P. M., & Davis, H. P. (2005). Age-related change in neural processing of time-dependent stimulus features. *Brain Research Cognition and Brain Research, 25*(3), 913–925.

Knight, R. T. (1984). Decreased response to novel stimuli after prefrontal lesions in man. *Electroencephalography and Clinical Neurophysiology, 59*, 9–20.

Knight, R. T. (1987). Aging decreases auditory event-related potentials to unexpected stimuli in humans. *Neurobiology of Aging, 8*(2), 109–113.

Knight, R. T. (1996). Contribution of human hippocampal region to novelty detection. *Nature, 383*, 256–259.

Knight, R. T., Scabini, D., & Woods, D. L. (1989). Prefrontal cortex gating of auditory transmission in humans. *Brain Research, 504*, 338–342.

Kok, A. (2000). Age-related changes in involuntary and voluntary attention as reflected in components of the event-related potential (ERP). *Biological Psychology, 54*(1–3), 107–143.

Kraiuhin, C., Gordon, E., Coyle, S., Sara, G., Rennie, C., Howson, A., et al. (1990). Normal latency of P300 event-related potential in mild-to-moderate Alzheimer's disease and depression. *Biological Psychiatry, 28*, 372–386.

Kramer, A., Schneider, W., Fisk, A., & Donchin, E. (1986). The effects of practice and task structure on components of the event- related brain potential. *Psychophysiology, 23*, 33–46.

Kray, J., Eppinger, B., & Mecklinger, A. (2005). Age differences in attentional control: An event-related potential approach. *Psychophysiology, 42*(4), 407–416.

Kutas, M., & Hillyard, S. A. (1980). Reading between the lines: Event-related brain potentials during natural sentence processing. *Brain and Language, 11*, 354–373.

Kutas, M., & Iragui, V. (1998). The N400 in a semantic categorization task across 6 decades. *Electroencephalography and Clinical Neurophysiology, 108*(5), 456–471.

Kutas, M., Iragui, V., & Hillyard, S. A. (1994). Effects of aging on event-related brain potentials (ERPs) in a visual detection task. *Electroencephalography and Clinical Neurophysiology, 92*(2), 126–139.

Li, J., Morcom, A. M., & Rugg, M. D. (2004). The effects of age on the neural correlates of successful episodic retrieval: An ERP study. *Cognitive and Affective Behavioral Neuroscience, 4*(3), 279–293.

Light, L. L. (1991). Memory and aging: Four hypotheses in search of data. *Annual Review of Psychology, 42*, 333–376.

Lynn, R. (1966). *Attention, arousal and the orientation reaction* (Vol. 3). London: Pergamon.

Mandler, G. (1980). Recognizing: The judgement of previous occurrence. *Psychological Review, 87*, 252–271.

Mark, R. E., & Rugg, M. D. (1998). Age effects on brain activity associated with episodic memory retrieval: An electrophysiological study. *Brain, 121*, 861–873.

Masaki, H., Falkenstein, M., Sturmer, B., Pinkpank, T., & Sommer, W. (2007). Does the error negativity reflect response conflict strength? Evidence from a Simon task. *Psychophysiology, 44*(4), 579–585.

Mathalon, D. H., Bennett, A., Askari, N., Gray, M., Rosenbloom, M. J., & Ford, J. M. (2003). Response-monitoring dysfunction in aging and Alzheimer's disease: An event-related potential study. *Neurobiology of Aging, 24*, 675–685.

McEvoy, L. K., Pellouchoud, E., Smith, M. E., & Gevins, A. (2001). Neurophysiological signals of working memory in normal aging. *Cognitive Brain Research, 11*(3), 363–376.

Näätänen, R. (1992). *Attention and brain function*. Hillsdale, NJ: Erlbaum.

Näätänen, R., Jacobsen, T., & Winkler, I. (2005). Memory-based or afferent processes in mismatch negativity (MMN): A review of the evidence. *Psychophysiology, 42*(1), 25–32.

Näätänen, R., Paavilainen, P., Rinne, T., & Alho, K. (2007). The mismatch negativity (MMN) in basic research of central auditory processing: A review. *Clinical Neurophysiology, 118 (12)*, 2544–2590.

Näätänen, R., & Picton, T. (1987). The N1 wave of the human electric and magnetic response to sound: A review and an analysis of the component structure. *Psychophysiology, 24*(4), 375–425.

Nebes, R. D. (1989). Semantic memory in Alzheimer's disease. *Psychological Bulletin, 106*, 377–394.

Nessler, D., Friedman, D., Johnson, R., Jr., & Bersick, M. (2007a). Does repetition engender the same retrieval processes in young and older adults? *Neuroreport, 18*(17), 1837–1840.

Nessler, D., Friedman, D., Johnson, R., Jr., & Bersick, M. (2007b). ERPs suggest that age affects cognitive control but not response conflict detection. *Neurobiology of Aging, 28*(11), 1769–1782.

Nessler, D., Johnson, R., Jr., Bersick, M., & Friedman, D. (2006). On why the elderly have normal semantic retrieval but deficient episodic encoding: A study of left inferior frontal ERP activity. *Neuroimage, 30(1)*, 299–312.

Nessler, D., Johnson, R., Jr., Bersick, M., & Friedman, D. (2008). Age-related ERP differences at retrieval persist despite age-invariant performance and left-frontal negativity during encoding. *Neuroscience Letters, 432 (2)*, 151–156.

Nieuwenhuis, S., Ridderinkhof, K. R., Talsma, D., Coles, M. G., Holroyd, C. B., Kok, A., et al. (2002). A computational account of altered error processing in older age: Dopamine and the error-related negativity. *Cognitive and Affective Behavioral Neuroscience, 2*(1), 19–36.

O'Donnell, B. F., Friedman, S., Squires, N. K., & Maloon, A. (1990). Active and passive P3 latency in dementia: Relationship to psychometric, electroencephalographic, and computed tomographic measures. *Neuropsychiatry, Neuropsychology, and Behavioral Neurology, 3*(3), 164–179.

Olichney, J. M., Morris, S. K., Ochoa, C., Salmon, D. P., Thal, L. J., Kutas, M., et al. (2002). Abnormal verbal event related potentials in mild cognitive impairment and incipient

Alzheimer's disease. *Journal of Neurology, Neurosurgery and Psychiatry, 73*(4), 377–384.

Ostrosky-Solis, F., Castaneda, M., Perez, M., Castillo, G. A., & Bobes, M. A. (1998). Cognitive brain activity in Alzheimer's disease: Electrophysiological response during picture semantic categorization. *Journal of the International Neuropsychological Society, 4*, 415–425.

Paavilainen, P., Jaramillo, M., & Näätänen, R. (1998). Binaural information can converge in abstract memory traces. *Psychophysiology, 35*, 483–487.

Paller, K. A., Voss, J. L., & Boehm, S. G. (2007). Validating neural correlates of familiarity. *Trends in Cognitive Sciences, 11* (6), 243–250.

Parasuraman, R., & Haxby, J. V. (1993). Attention and brain function in Alzheimer's disease: A review. *Neuropsychology, 7*, 242–272.

Park, D. (2000a). The basic mechanisms accounting for age-related decline in cognitive function. In D. Park (Ed.), *Cognitive aging: A primer* (pp. 3–21). Philadelphia: Psychology Press.

Park, D. (2000b). *Cognitive aging: A primer.* Philadelphia: Psychology Press.

Park, D. C., Lautenschlager, G., Hedden, T., Davidson, N. S., Smith, A. D., & Smith, P. K. (2002). Models of visuospatial and verbal memory across the adult life span. *Psychology and Aging, 17*(2), 299–320.

Patterson, J. V., Michalewski, H. J., & Starr, A. (1988). Latency variability of the components of auditory event-related potentials to infrequent stimuli in aging, Alzheimer-type dementia, and depression. *Electroencephalography and Clinical Neurophysiology Evoked Potentials, 71*(6), 450–460.

Pekkonen, E. (2000). Mismatch negativity in aging and in Alzheimer's and Parkinson's diseases. *Audiology and Neuro-Otology, 5*(3–4), 216–224.

Pekkonen, E., Jaaskelainen, L. P., Erkinjuntti, T., Hietanen, M., Huotilainen, M., Ilmoniemi, R. J., et al. (2001). Preserved stimulus deviance detection in Alzheimer's disease. *Neuroreport, 12*(8), 1649–1652.

Pekkonen, E., Jousmaki, V., Kononen, M., Reinikainen, K., & Partanen, J. (1994). Auditory sensory memory impairment in Alzheimer's disease: An event-related potential study. *Neuroreport, 5*, 2537–2540.

Pekkonen, E., Jousmaki, V., Partanen, J., & Karhu, J. (1993). Mismatch negativity area and age-related auditory memory. *Electroencephalography and Clinical Neurophysiology, 87*, 321–325.

Pekkonen, E., Rinne, T., Reinikainen, K., Kujala, T., Alho, K., & Näätänen, R. (1996). Aging effects on auditory processing: An event-related potential study. *Experimental Aging Research, 22*, 171–184.

Pfefferbaum, A., Ford, J. M., Wenegrat, B. G., Roth, W. T., & Kopell, B. S. (1984a). Clinical application of the P3 component of event-related potentials. I. Normal aging. *Electroencephalography and Clinical Neurophysiology, 59*(2), 85–103.

Pfefferbaum, A., Wenegrat, B., Ford, J. M., Roth, W. T., & Kopell, B. S. (1984b). Clinical application of the P3 component of event-related potentials II. Dementia, depression and schizophrenia. *Electroencephalography and Clinical Neurophysiology, 59*, 104–124.

Phillips, N. A., & Lesperance, D. (2003). Breaking the waves: Age differences in electrical brain activity when reading text with distractors. *Psychology and Aging, 18*(1), 126–139.

Picton, T. W., Stuss, D. T., Champagne, S. C., & Nelson, R. F. (1984). The effects of age on human event-related potentials. *Psychophysiology, 21*, 312–325.

Polich, J. (1996). Meta-analysis of P300 normative aging studies. *Psychophysiology, 33*, 334–353.

Polich, J. (2007). Updating P300: An integrative theory of P3a and P3b. *Clinical Neurophysiology, 118*(10), 2128–2148.

Ranganath, C., & Rainer, G. (2003). Neural mechanisms for detecting and remembering novel events. *Nature Review Neuroscience, 4*(3), 193–202.

Raz, N. (2000). Aging of the brain and its impact on cognitive performance: Integration of structural and functional findings. In F. I. M. Craik & T. A. Salthouse (Eds.), *Handbook of aging and cognition* - (Vol. 2, pp. 1–90). Mahwah, NJ: Erlbaum.

Raz, N., Gunning-Dixon, F., Head, D., Rodrigue, K. M., Williamson, A., & Acker, J. D. (2004). Aging, sexual dimorphism, and hemispheric asymmetry of the cerebral cortex: Replicability of regional differences in volume. *Neurobiology of Aging, 25*(3), 377–396.

Reinvang, I., Espeseth, T., & Gjerstad, L. (2005). Cognitive ERPs are related to ApoE allelic variation in mildly cognitively impaired patients. *Neuroscience Letters, 382*(3), 346–351.

Revonsuo, A., Portin, R., Juottonen, K., & Rinne, J. O. (1998). Semantic processing of spoken words in Alzheimer's disease: An electrophysiological study. *Journal of Cognitive Neuroscience, 10*(3), 408–420.

Ridderinkhof, K. R., Ullsperger, M., Crone, E. A., & Nieuwenhuis, S. (2004). The role of the medial frontal cortex in cognitive control. *Science, 306*(5695), 443–447.

Riis, J. L., Chong, H., Ryan, K. K., Wolk, D. A., Rentz, D. M., Holcomb, P. J., et al. (2008). Compensatory neural activity distinguishes different patterns of normal cognitive aging. *Neuroimage, 39*(1), 441–454.

Ritter, W., Gomes, H., Cowan, N., Sussman, E., & H.G. Vaughan, J. (1998). Reactivation of a dormant representation of an auditory stimulus feature. *Journal of Cognitive Neuroscience, 10*, 605–614.

Ritter, W., Simson, R., Vaughan, H. G., & Friedman, D. (1979). A brain event related to the making of a sensory discrimination. *Science, 203*, 1358–1361.

Ritter, W., Sussman, E., Molholm, S., & Foxe, J. J. (2002). Memory reactivation or reinstatement and the mismatch negativity. *Psychophysiology, 39*(2), 158–165.

Rugg, M. D., & Curran, T. (2007). Event-related potentials and recognition memory. *Trends in Cognitive Sciences, 11*(6), 251–257.

Rugg, M. D., Mark, R. E., Gilchrist, J., & Roberts, R. C. (1997). ERP repetition effects in indirect and direct tasks: Effects of age and interitem lag. *Psychophysiology, 34*(5), 572–586.

Rugg, M. D., & Yonelinas, A. P. (2003). Human recognition memory: A cognitive neuroscience perspective. *Trends in Cognitive Sciences, 7*(7), 313–319.

Sable, J. J., Low, K. A., Maclin, E. L., Fabiani, M., & Gratton, G. (2004). Latent inhibition mediates N1 attenuation to repeating sounds. *Psychophysiology, 41*(4), 636–642.

Saito, H., Yamazaki, H., Matsuoka, H., Matsumoto, K., Numachi, Y., Yoshida, S., et al. (2001). Visual event-related potential in mild dementia of the Alzheimer's type. *Psychiatry and Clinical Neuroscience, 55*(4), 365–371.

Salthouse, T. A. (1991). *Theoretical perspectives on cognitive aging* Hillsdale, NJ: Erlbaum.

Salthouse, T. A. (1993). Speed and knowledge as determinants of adult age differences in verbal tasks. *Journal of Gerontology, 48*(1), P29–P36.

Salthouse, T. A. (1996). The processing-speed theory of adult age differences in cognition. *Psychological Review, 103*(3), 403–428.

Sams, M., Hari, R., Rif, J., & Knuutila, J. (1993). The human auditory sensory memory trace persists about 10 sec: Neuromagnetic evidence. *Journal of Cognitive Neuroscience, 5,* 363–370.

Schwartz, T. J., Kutas, M., Butters, N., Paulsen, J. S., & Salmon, D. P. (1996). Electrophysiological insights into the nature of the semantic deficit in Alzheimer's disease. *Neuropsychologia, 34,* 827–841.

Siegel, C., Waldo, M., Mizner, G., Adler, L. E., & Freedman, R. (1984). Deficits in sensory gating in schizophrenic patients and their relatives. *Archives of General Psychiatry, 41,* 607–612.

Smith, D. B., Michalewski, H. J., Brent, G. A., & Thompson, L. W. (1980). Auditory averaged evoked potentials and aging: Factors of stimulus, task and topography. *Biological Psychology, 11*(2), 135–151.

Snyder, J. S., & Alain, C. (2005). Age-related changes in neural activity associated with concurrent vowel segregation. *Brain Research and Cognitive Brain Research, 24*(3), 492–499.

Sokolov, E. N. (1990). The orienting response, and future directions of its development. *Pavlovian Journal of Biological Science, 25*(3), 142–150.

Spencer, W. D., & Raz, N. (1995). Differential effects of aging on memory for content and context: A meta-analysis. *Psychology and Aging, 10,* 527–539.

Stern, Y. (2002). What is cognitive reserve? Theory and research application of the reserve concept. *Journal of the International Neuropsychological Society, 8*(3), 448–460.

Sutton, S., Tueting, P., Zubin, J., & John, E. R. (1967). Information delivery and the sensory evoked potential. *Science, 155*(768), 1436–1439.

Swick, D., & Knight, R. T. (1997). Event-related potentials differentiate the effects of aging on word and nonword repetition in explicit and implicit memory tasks. *Journal of Experimental Psychology: Learning, Memory, and Cognition, 23*(1), 123–142.

Swick, D., Senkfor, A. J., & Van Petten, C. (2006). Source memory retrieval is affected by aging and prefrontal lesions: Behavioral and ERP evidence. *Brain Research, 1107 (1),* 161–176.

Taylor, J. R., & Olichney, J. M. (2007). From amnesia to dementia: ERP studies of memory and language. *Clinical Electroencephalography and Neuroscience, 38*(1), 8–17.

Taylor, S. F., Stern, E. R., & Gehring, W. J. (2007). Neural systems for error monitoring: Recent findings and theoretical perspectives. *Neuroscientist, 13*(2), 160–172.

Themanson, J. R., Hillman, C. H., & Curtin, J. J. (2006). Age and physical activity influences on action monitoring during task switching. *Neurobiology of Aging, 27*(9), 1335–1345.

Trott, C. T., Friedman, D., Ritter, W., Fabiani, M., & Snodgrass, J. G. (1999). Episodic priming and memory for temporal source: Event-related potentials reveal age-related differences in prefrontal functioning. *Psychology and Aging, 14*(3), 390–413.

Wegesin, D. J., Friedman, D., Varughese, N., & Stern, Y. (2002). Age-related changes in source memory retrieval: An ERP replication and extension. *Cognitive Brain Research, 13*(3), 323–338.

Weisz, J., & Czigler, I. (2006). Age and novelty: Event-related brain potentials and autonomic activity. *Psychophysiology, 43*(3), 261–271.

West, R. (2004). The effects of aging on controlled attention and conflict processing in the Stroop task. *Journal of Cognitive Neuroscience, 16*(1), 103–113.

West, R. L. (1996). An application of prefrontal cortex theory to cognitive aging. *Psychological Bulletin, 120,* 272–292.

Williams, P. A., Jones, G. H., Briscoe, M., & Thomas, R. (1991). P300 and reaction-time measures in senile dementia of the Alzheimer type. *British Journal of Psychiatry, 159,* 410–414.

Wingfield, A., Tun, P. A., & McCoy, S. L. (2005). Hearing loss in older adulthood: What it is and how it interacts with cognitive performance. *Current Directions in Psychological Science, 14*(3), 144–148.

Winkler, I., Korzyukov, O., Gumenyuk, V., Cowan, N., Linkenkaer-Hansen, K., Ilmoniemi, R. J., et al. (2002). Temporary and longer term retention of acoustic information. *Psychophysiology, 39*(4), 530–534.

Woods, D. L. (1992). Auditory selective attention in middle-aged and elderly subjects: An event-related brain potential study. *Electroencephalography and Clinical Neurophysiology, 84,* 456–468.

Woods, D. L., & Knight, R. T. (1986). Electrophysiologic evidence of increased distractibility after dorsolateral prefrontal lesions. *Neurology, 36*(2), 212–216.

Woodward, S. H., Ford, J. M., & Hammett, S. C. (1993). N4 to spoken sentences in young and older subjects. *Electroencephalography and Clinical Neurophysiology, 87,* 306–320.

Yago, E., Escera, C., Alho, K., Giard, M. H., & Serra-Grabulosa, J. M. (2003). Spatiotemporal dynamics of the auditory novelty-P3 event-related brain potential. *Brain Research and Cognitive Brain Research, 16*(3), 383–390.

Yamaguchi, S., & Knight, R. T. (1991a). Age effects on the P300 to novel somatosensory stimuli. *Electroencephalography and Clinical Neurophysiology, 78,* 297–301.

Yamaguchi, S., & Knight, R. T. (1991b). P300 generation by novel somatosensory stimuli. *Electroencephalography and Clinical Neurophysiology, 78*(1), 50–55.

Yamaguchi, S., Tsuchiya, H., Yamagata, S., Toyoda, G., & Kobayashi, S. (2000). Event-related brain potentials in response to novel sounds in dementia. *Clinical Neurophysiology, 111,* 195–203.

Yonelinas, A. P. (2002). The nature of recollection and familiarity: A review of 30 years of research. *Journal of Memory and Language, 46,* 441–517.

Zarahn, E., Rakitin, B., Abela, D., Flynn, J., & Stern, Y. (2007). Age-related changes in brain activation during a delayed item recognition task. *Neurobiology of Aging, 28 (5),* 784–798.

Zeef, E. J., & Kok, A. (1993). Age-related differences in the timing of stimulus and response processes during visual selective attention: Performance and psychophysiological analyses. *Psychophysiology, 30,* 138–151.

CHAPTER

19 Abnormalities of Event-Related Potential Components in Schizophrenia

Brian F. O'Donnell, Dean F. Salisbury, Margaret A. Niznikiewicz, Colleen A. Brenner, *and* Jenifer L. Vohs

Abstract

Schizophrenia is a disabling psychotic illness that has been associated with alterations in synaptic connectivity and neurotransmission. Since event-related potential (ERP) components are typically generated by the summation of postsynaptic potentials produced by neural populations, these measures are well suited to assess such pathophysiological alterations. This chapter reviews the utility of ERP components in the investigation of the cognitive and neural mechanisms affected by schizophrenia. It focuses on five components: mismatch negativity (MMN), P50 measures of sensory gating, N100 and P300 in the oddball discrimination paradigms, and the N400 component elicited during language processing. These components test key cognitive systems affected by schizophrenia: sensory memory (MMN), sensory processing and inhibition (P50, N1), selective attention and working memory (P300), and semantic processing (N400). These components are discussed with respect to the following issues: (1) cognitive and neural systems indexed by the component, (2) abnormalities in schizophrenia, (3) sensitivity and specificity to schizophrenia, (4) clinical correlates, and (5) relationship to genetic variation. ERP components are well validated biomarkers for schizophrenia which have significant promise in the characterization of genomic and epigenomic factors, pharmacological response in humans and animal models, and the developmental and cognitive expression of the illness.

Keywords: schizophrenia, event-related potentials, ERP components, cognitive mechanism, neural mechanism

Introduction

Schizophrenia is a psychotic disorder that affects about 0.8% of the global population (Saha et al., 2005). The initial episode of schizophrenia is associated with a marked decline in psychosocial function, coupled with symptoms such as hallucinations, delusions, anhedonia, avolition, and disordered thinking and behavior. The clinical course of schizophrenia is heterogeneous, but long-term outcome studies indicate that the majority of patients show persistent cognitive, occupational, and interpersonal dysfunction (Hafner & van der Heiden, 1999; Sullivan, 2005). While the onset of schizophrenia usually occurs in adolescence or early adulthood, accumulating evidence indicates that neurodevelopmental disturbances are evident much earlier in life (Lewis & Lieberman, 2000). Cognitive, motor, and psychosocial impairments are evident in childhood and adolescence in individuals who later develop schizophrenia (Davies et al., 1998; O'Donnell, 2007). Behavioral genetic studies indicate that genetic variation plays a major role in the disorder, with estimated heritability of over 60% (Sullivan, 2005; Tsuang, 2004). Risk factors, such as viral infection of the mother during pregnancy and prenatal and delivery complications, suggest that both prenatal and postnatal insults may increase vulnerability to the illness and interact with genetic risk (Tsuang, 2004).

The severe and often persistent disturbances of cognition and behavior characteristic of schizophrenia were suggestive of a major disturbance of brain function to early investigators. Consistent neuropathological findings, however, were elusive. Definitive evidence of neurobiological abnormalities in schizophrenia first emerged with the advent of noninvasive techniques to assess brain function (event-related potentials, positron emission tomography [PET], functional magnetic resonance imaging [fMRI]) and anatomy (computed tomography, structural MRI; McCarley et al., 1996; Ross et al., 2006).

Event-related potential (ERP) paradigms have been used to investigate schizophrenia for over three decades, and have characterized the effect of the disorder on the neural mechanisms associated with sensory, attentional, working memory, linguistic, and decision-making deficits (Nestor & O'Donnell, 1998; Turetsky et al., 2007). Event-related potentials have several advantages in the study of the pathophysiology of schizophrenia. They are direct measures of electrical current produced by neural activity and have high temporal resolution. Since models of the neurodevelopment and pathophysiology of schizophrenia emphasize disturbances of neurotransmission, connectivity, and synchronization (Lewis & Lieberman, 2000; Ross et al., 2006), ERPs are likely to be highly sensitive to the consequences of such abnormalities. Event-related potential components can provide well-validated probes for sensory, cognitive, and motor anomalies in the disorder. Typically, ERP component abnormalities have been identified using signal averaging techniques, but recent analytic techniques allow identification of trial-to-trial variation in component-related activity (Makeig et al., 2004). Finally, ERP measurements are relatively inexpensive and applicable to most subjects, making testing of large samples for functional genomic analysis feasible.

This chapter will review the use of ERP components to investigate the cognitive and neural mechanisms affected by schizophrenia. It will focus on five components: mismatch negativity (MMN), P50 measures of sensory gating, N100 and P300 in the oddball discrimination paradigms, and the N400 component elicited during language processing. These components test key cognitive systems affected by schizophrenia: sensory memory (MMN), sensory processing and inhibition (P50, N1), selective attention and working memory (P300), and semantic processing (N400). These components will be evaluated with respect to the following issues: (1) cognitive and neural systems indexed by the component, (2) abnormalities in schizophrenia, (3) sensitivity and specificity to schizophrenia, (4) clinical correlates, and (5) relationship to genetic variation.

Mismatch Negativity

Cognitive and Neural Mechanisms

Mismatch negativity (MMN; see Chapter 6, this volume) is a negative deflection in the ERP that peaks about 150 to 250 ms after the presentation of a deviant auditory stimulus interspersed among more frequent standard stimuli (Figure 19.1). For example, if a 1000 Hz tone is a standard stimulus that occurs on 90% of trials, a 2000 Hz tone that occurred on 10% of the trials would elicit an MMN. The MMN is usually measured by subtracting the ERP to the standard stimulus from the ERP to the deviant stimulus. The MMN appears in the absence of task demands or even when a subject is engaged in a task in another sensory modality. Thus, MMN is thought to be evoked preattentively and to index

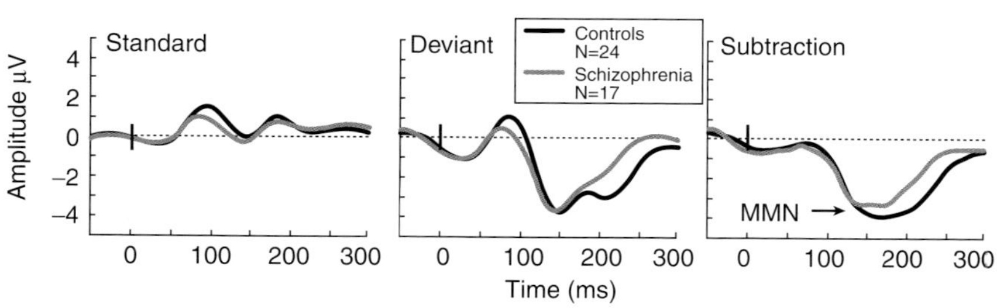

Fig. 19.1. Deviant tones elicit the MMN component, which can be visualized by subtraction of the standard from the deviant ERP. Left panel: Grand average ERP response to standard stimuli. P1 (~90 ms), N1 (~140 ms), and P2 (~180 ms) components are apparent. Middle panel: ERP response to deviant stimuli. The N1 and P2 are superimposed on a larger negative deflection, the mismatch negativity. Right panel: MMN is visualized by subtracting the ERP to standard stimuli from the ERP to deviant stimuli. Note that patients with schizophrenia have a smaller MMN than psychiatrically well controls subjects. The ERPs were recorded from Fz, the anterior midline electrode.

echoic memory comparison of a stimulus to a neural trace of previous stimuli that is little affected by top-down executive functions (Näätänen, 1990). These deviations include slight changes in the pitch, duration, or location of short tones (oddball stimuli). The MMN can also be elicited by phonemes (e.g., Aaltonen et al., 1987; see Näätänen, 2001, for review) or by deviations from learned patterns of sounds (see Kujala et al., 2007, for review), indicating that acquired knowledge may influence low-level perception.

Depth recordings in animals (Csépe, 1995; Javitt et al., 1994) and humans (Kropotov et al., 1995; Rosburg et al., 2005), as well as source analysis studies using magnetoencephalography (MEG) and electroencephalography, have localized the primary generators of the auditory MMN to superficial layers within primary auditory cortices or the secondary auditory cortices just anterior to primary auditory cortices (e.g., Alho, 1995; Frodl Bauch et al., 1997; Hari et al., 1992; Kasai et al., 1999a; Näätänen & Alho, 1995). Most of the scalp-recorded activity reflected in the MMN can be attributed to temporal lobe generators within the superior temporal plane, with relatively modest contributions from other cortical areas (e.g., a right frontal source invisible to MEG: Giard et al., 1990; Rinne et al., 2000; a possible parietal source: Kasai et al., 1999a; see Molholm et al., 2005, for fMRI localization of MMN sources). When inferring brain abnormalities based on scalp-recorded ERP defects, it is often difficult or impossible to localize affected regions when a component has multiple distributed sources. For example, since P300 likely has three or four (or more) bilateral sources, localization of the source or sources of a scalp-recorded abnormality may not be feasible. Because the MMN is generated by a small number of cortical generators, primarily in the superior temporal cortices, it may serve as a precise tool for establishing links between neurophysiology and underlying cortical pathophysiology in schizophrenia.

Generation of the MMN likely involves activity in local interneuron processing and is extremely sensitive to activity at the *N*-methyl-D-aspartate (NMDA) receptor. In monkeys, administration of NMDA antagonists leads to reduction of the MMN but no change to the initial sensory ERPs, like N1 (Javitt et al., 1996). In humans, administration of psychotomimetic drugs that act at the NMDA receptor leads to a reduction of the MMN (Umbricht et al., 2000). Recent models of schizophrenia have posited a role for dysfunction in synaptic plasticity and related activity within the NMDA receptor (e.g., Javitt et al., 2000b; Krystal et al., 2003). Thus, MMN may serve as an index of function within a neural circuit that depends heavily on a particular type of interneuronal communication and processing thought to be abnormal in schizophrenia.

MMN in Schizophrenia

This review will focus on the MMN elicited by deviance in the frequency (pitch) and duration of tone pips or phonemes, which have been used in the majority of studies of schizophrenia. Most studies also use deviant-minus-standard subtraction waveforms for measurement of the MMN, which detects an interaction between stimulus probability and diagnostic group. Reductions in schizophrenia of pitch and duration deviant MMN are generally highly replicable and are illustrated in Figure 19.1. Two groups were largely responsible for the interest in examination of MMN in schizophrenia. Shelley and colleagues (1991) in Australia examined early ERP indices of preattentive stimulus processing and were the first to report MMN deficits in schizophrenia. The deviant stimuli (10%) were slightly longer than the standard stimuli in half of the blocks of tones or slightly shorter for the remaining half. Javitt et al. (1993) in the United States replicated the reduced MMN in schizophrenia using infrequent pitch deviants. These groups have continued to probe cortical dysfunction in schizophrenia using MMN and to further delineate the underlying biological deficits giving rise to pitch and duration deviant MMN responses. Javitt et al. (1995) replicated the MMN reduction to pitch deviants (15%), and Javitt et al. (1998) showed that MMN amplitudes quickly asymptote in schizophrenia with increasing pitch deviance and with decreasing deviant probability, so that patient–control differences were exacerbated in conditions where controls generated larger MMN amplitudes. Shelley and colleagues (1999) similarly showed that decreasing probability did not augment the MMN in schizophrenia, as it did in controls, but that changes in the interstimulus interval (ISI) had little differential effect in these groups. These initial reports were soon replicated by a number of other researchers worldwide. Hirayasu et al. (1998b) replicated the reduction of MMN to frequency deviants, as did Alain et al. (1998). Kreitsmann-Andermahr et al. (1999) replicated the MMN reduction in schizophrenia using MEG. For the traditional small-frequency-deviance MMN, Javitt et al. (1993), Hirayasu et al. (1998b), and Kreitsmann-Andermahr et al. (1999) reported greater left hemisphere reductions in schizophrenia, similar to the left-greater than right-sided P300 deficit reported in schizophrenia (e.g., Salisbury et al., 1998a, 1998b), although hemispheric asymmetry in schizophrenia has not subsequently been a robust finding. Kasai et al. (1999b) showed reduced duration deviant MMN in schizophrenia regardless of the direction of attention in a dichotic task. Baldeweg et al. (2002) replicated MMN reductions in schizophrenia, as did Park et al. (2002). Baldeweg et al. suggested differential frontal and temporal component deficits, although source modeling was not performed. Park et al. (2002) used individual MRI and source modeling, and suggested reductions of left temporal gyrus sources in schizophrenia, along with reductions in a putative inferior parietal lobule source. Kasai et al. (2002) showed reductions in the MMN elicited by phonemes in schizophrenia. Thus, MMN to a wide variety of auditory features appears to be impaired in schizophrenia.

Studies of MMN deficits in schizophrenia have usually used paradigms in which the deviant tone differed either in pitch from the standard tone (e.g. 1000 vs. 2000 Hz) or in duration (e.g. 50 vs. 100 ms). In her influential review, Michie (2001) suggested that the brain defect in schizophrenia may be more sensitive to duration deviance than pitch deviance. Nearly all studies of pitch deviants find reductions in MMN, with some notable exceptions associated with methodological differences, as discussed by Michie (2001). Yet, the more simple detection of a pitch difference, apparent at the start of a tone, might be less computationally demanding than detection of a duration difference, and thus might be easier for even a compromised system to perform. Group differences on a measure can be quantified using an estimate of the effect size. *Effect size* (*d*) refers to the standardized mean difference between groups (Cohen, 1992) and has been applied to estimate the sensitivity of MMN to schizophrenia. A recent meta-analysis by Umbricht and Krljes (2005) suggests that studies using pitch deviants showed an overall effect size of 0.94, whereas studies using duration deviants showed an overall effect size of 1.24. Here we suggest that a crucial test for assessing any differential sensitivity for pitch or duration differences in schizophrenia needs to be conducted. The psychophysical matching of discriminability of the different deviants needs to be assessed so that the deviant stimuli can be matched cross-dimensionally. Until such a mapping is performed, it remains unclear whether the pitch deviance and duration deviance are, in fact, of equal perceptual size.

In an attempt to relate the impairments in pitch-deviant MMN to perception using psychophysical methods, Rabinowicz et al. (2000) showed that schizophrenia is characterized by increases in the difference limen, or just noticeable difference, between frequencies. Although not directly tested, they suggested that this auditory frequency imprecision was related to the reductions observed in MMN. Javitt et al. (2000a) showed that tone-matching performance and MMN amplitude were highly correlated. Todd et al. (2003) likewise used psychophysics to examine temporal discrimination and demonstrated higher discrimination thresholds in schizophrenia coupled with reduced-duration deviant MMN amplitudes. Kawakubo et al. (2006) demonstrated that reductions in the phoneme MMN were associated with poorer verbal memory performance in schizophrenia. Hence, the reductions in MMN appear to be related to auditory perceptual and cognitive dysfunctions.

Sensitivity and Specificity

Auditory MMN deficits are present in many different disorders (see Näätänen, 2003, and Chapter 6, this volume, for reviews). The presence of an intact MMN may predict eventual recovery from coma (e.g., Kane et al., 1996), and MMN to phonemes is sensitive to dyslexia (see Kujala & Näätänen, 2001, for review) and aphasia (e.g., Csépe et al., 2001). The MMN amplitude may be pathologically large in Post-Traumatic Stress Disorder (PTSD; Morgan & Grillon, 1999). Yet, within the major psychotic illnesses, MMN reductions may be specific to schizophrenia. Several studies have examined the MMN in schizophrenia contrasted with that in bipolar disorder. Catts et al. (1995) reported unaffected MMN amplitude in bipolar affective disorder. Umbricht et al. (2003) reported no reductions in bipolar disorder, including patients who were psychotic, or in depressed individuals. Hall et al. (2007a) examined well twins and twins concordant for and discordant for bipolar disorder and found no genetic link between bipolar disorder and MMN reduction. Salisbury et al. (2007) found no MMN reductions in bipolar patients early in the course of the disease. However, these bipolar patients had been acutely psychotic, and tended to have an overall MMN group means at the extreme limit of normal (Salisbury, unpublished data). The possibility remains that acutely psychotic bipolar disorder may be associated with small MMN reductions, but the extant reports suggest that MMN is normal in bipolar disorder. Thus, MMN may be selectively reduced in schizophrenia within the major psychotic and mood disorders, although it is also reduced in a variety of illnesses that impact cortical function.

Clinical Correlates

Mismatch negativity to pitch-deviant tones is not impaired at the first hospitalization for schizophrenia (Salisbury et al., 2002a). This was demonstrated independently by Umbricht et al. (2006), who also showed a relatively normal MMN in first hospitalized patients to duration deviants. Umbricht et al. did show that lower MMN amplitudes were associated with poorer premorbid functioning, so it is possible that patients who are more ill at initial presentation may have MMN impairments. Salisbury et al. (2007) showed that MMN was rapidly reduced in amplitude during the early course of schizophrenia. This finding was consistent with demonstrations of reduced-pitch MMN in patients with recent-onset schizophrenia (Javitt et al., 2000b; Umbricht et al., 2006). Todd et al. (2008) showed reductions in duration and intensity, but not pitch, MMN during the early course of schizophrenia. Pitch MMN showed a greater reduction with aging in schizophrenic patients than in control subjects. Because Salisbury et al. (2007) tested subjects longitudinally, the findings suggest an active period peri-onset of MMN reduction. The degree of reduction was correlated with cortical gray matter volume in the left superior temporal plane area containing Heschl's gyrus, which encompasses primary and portions of secondary auditory cortices and which shows marked reductions after the first hospitalization (Kasai et al 2003). Light and Braff (2005a, 2005b) showed that MMN deficits were stable over the later course of schizophrenia and were associated with current functional status: impaired MMN was associated with impaired everyday functioning. Thus, MMN seems to be relatively unaffected near the onset of schizophrenia, although it may be impaired before first hospitalization in the more poorly functioning individuals, becomes progressively impaired in the first few years after first hospitalization, and remains impaired later in the disease, with the more functionally impaired patients showing the smallest MMN amplitudes.

Recently, there has been great interest in MMN before first hospitalization, during the so-called prodrome phase. Typically, adolescent subjects showing transient psychotic-like symptoms are examined. Brockhaus-Dumke et al. (2005) failed to detect differences in MMN to pitch or duration deviants in prodromal subjects, although the overall mean was intermediate between control and schizophrenic MMNs. Whether there were subsets of prodromal subjects with abnormal MMNs who subsequently converted to schizophrenia was not assessed.

Genetics

Due to its reduction in schizophrenia, its relative specificity within psychotic disorders for schizophrenia, and its ease of measurement, MMN could have potential as an endophenotype for underlying genetic abnormalities in schizophrenia. Initial studies showed reductions of MMN in the relatives of schizophrenia patients (Michie et al., 2002), further strengthening the possibility for using MMN as an endophenotype. Jessen et al. (2001) reported significant reductions of MMN in relatives of schizophrenia patients, but the lack of MMN reductions in the schizophrenia patients in this study raised interpretive problems. Most studies of relatives of probands have failed to reveal MMN deficits (Bramon et al., 2004; Schreiber et al., 1992). When coupled with the demonstration that MMN was relatively normal in first hospitalized patients (Salisbury et al., 2002a), it became unlikely that MMN could serve as a stable endophenotype. Definitive studies of inheritance by Hall and colleagues (2006b) revealed that MMN was highly heritable and showed higher correlations in monozygotic twins than dizygotic twins and was associated with a different subset of genes than P50 gating and P300 (Hall et al., 2006a), but these studies showed only weak associations with schizophrenia (Hall et al., 2007b). Hall et al. suggested that the demonstration of normal MMN at first hospitalization (Salisbury et al., 2002a; Umbricht et al., 2006), relatively normal MMN in most family studies, a substantial environmental component, and a significant association with schizophrenia only when the highest schizophrenia heritability estimates were used made MMN a poor choice for an endophenotype. Likewise, Ahveninen et al. (2006) found normal MMN amplitudes in the dizygotic twins of probands who were discordant for schizophrenia. Hence, MMN does not appear to show abnormalities in most prodromal subjects, family members, or identical twins discordant for schizophrenia. The progressive MMN reduction observed after onset of the illness may instead indicate a pathophysiological process associated with the expression of susceptibility genes during the early stages of psychosis.

Summary

Mismatch negativity is reduced in schizophrenia to both pitch and duration deviance, indicative of a disturbance in sensory memory mechanisms. It is not clear whether the greater sensitivity of duration-deviant MMN to schizophrenia compared to pitch-deviant MMN reflects underlying cortical substrate differences due to processing requirements or psychophysical differences between the deviants. At the first hospitalization for schizophrenia, MMN is relatively normal, but it rapidly becomes abnormal during the course of disease. This rapid reduction may reflect cortical gray matter changes as a primary consequence of the disease and a putative synaptic plasticity defect related to NMDA receptors or a secondary effect of antipsychotic medications. Before the first hospitalization, MMN amplitude is likely related to the severity of symptoms, and may be reduced in prodromal subjects who later convert to schizophrenia, although definitive data are needed. Later in the disease MMN deficits are stable, are present even after medication withdrawal, and are correlated with the current level of everyday functioning. The MMN has a genetic component, but also a shared environmental component, and is only weakly associated with schizophrenia. Thus, it may not serve as a sensitive intermediate endophenotype, although it may reflect the activity of genes that do not become active until adolescence and young adulthood. During the early course of the disease, MMN is associated with changes in left temporal cortical gray matter and may provide an index of the efficacy of adjunctive drug therapies aimed at halting this peri-onset brain reduction.

P50 Sensory Gating

Cognitive and Neural Mechanisms

The P50 ERP is a midlatency cortical response associated with sensory stage auditory processing in humans (Jerger et al. 1992). When two identical clicks are presented in a paired-stimulus (S1, S2) procedure, midlatency ERPs evoked by the second (test stimulus) are normally attenuated relative to those produced by the first (conditioning stimulus; Fruhstorfer et al., 1970). P50 responses to S1 and S2 in healthy subjects are shown in Figure 19.2. In the control subjects, the amplitude of S2 is reduced compared to S1, indicative of an inhibitory or gating effect. This effect is apparent in control subjects for ISIs ranging from 75 to 500 ms (Nagamoto et al., 1989), and effect size diminishes with intervals longer than 1 s (Adler et al., 1982). P50 suppression has long been considered a test of the strength of inhibitory neuronal mechanisms for sensory input (Adler et al., 1982; Freedman et al., 1996). This inhibitory effect is often interpreted as an index of sensory gating, or the automatic suppression of responses to redundant stimuli. The magnitude of P50 amplitude suppression is often expressed as a ratio of the amplitude of the neural response to the second click (S2) divided by the amplitude of the response to the first click (S1). The gating ratio is usually expressed as the amplitude of $S2/S1^{*}100$. Low gating ratios are interpreted to indicate strong suppression or sensory gating, as seen in healthy participants who usually exhibit a reduction in P50 amplitude to S2 by 50% or more and a mean gating ratio of approximately 39% (Adler et al., 1982; Boutros et al., 2004; Clementz et al., 1998a; Patterson et al., 2008). In contrast, individuals with schizophrenia generally exhibit higher ratios, with a recent meta-analysis reporting a mean gating ratio of 80% (Patterson et al., 2008). This deficiency in inhibition or sensory gating has been interpreted to

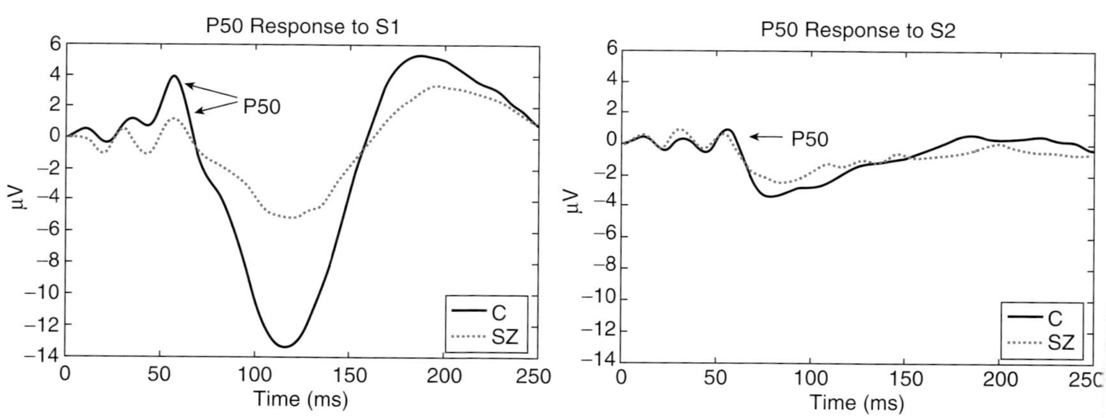

Fig. 19.2. Grand average ERPs showing P50 suppression or gating in subjects with schizophrenia and healthy comparison subjects. The left figure shows the ERPs of control subjects (solid line) and patients (dashed line) to the first click (S1). The control subjects show a P50 deflection at about 50 ms, while the patients show reduced P50 amplitude. The right figure shows P50 to S2. The S2 P50 in control subjects is reduced in amplitude compared to S1, while the control amplitude is unaffected.

indicate that patients with schizophrenia fail to filter or gate out redundant sensory stimuli, resulting in experiences of sensory flooding and distraction by irrelevant events.

The anatomical and neurophysiological systems subserving the P50 sensory gating effect have received extensive investigation in humans and animal models, and have implicated hippocampal, temporal, and frontal lobe involvement (Garcia-Rill et al., 2008; Grunwald et al., 2003; Jensen et al., 2008; Krause et al., 2003; Weisser et al., 2001). Attenuation of P50 amplitudes to repeated stimuli appears to involve habituation at prefrontal and temporoparietal cortices and, at a later stage, the hippocampus (Grunwald et al., 2003). In humans, Thoma et al. (2004) found that thinner auditory cortex as measured by three-dimensional (3-D) structural MRI was correlated with P50 gating deficits in patients with schizophrenia. Furthermore, Korzyukov et al. (2007) used intracranial electrode grids and source localization procedures to determine that interaction between the temporal and frontal lobes influenced P50 generation and sensory gating. Tregellas and colleagues (2007) investigated P50 sensory gating in an fMRI environment and found increased hippocampal, thalamic, and dorsolateral prefrontal cortex activity in patients with schizophrenia. They interpreted these results to indicate diminished inhibitory function within these regions as associated with schizophrenia sensory gating deficits. Sensory gating of the auditory ERP has also been studied in rat models of schizophrenia, usually by measurement of suppression of the N40 component in a dual click paradigm (Stevens et al., 1995; Swerdlow et al., 2006). Amphetamine (a dopamine agonist), which is commonly used to produce rat models of psychoses, reduces rat N40 at S1 and produces abnormal suppression ratios (Adler et al., 1986; Stevens et al., 1995; Swerdlow et al., 2006). Furthermore, administration of haloperidol, a dopamine antagonist that ameliorates psychotic symptoms in humans, also blocks the amphetamine-induced gating deficit. The nicotinic cholinergic system also appears to be important to normal sensory gating in both humans (Adler et al., 1992) and animals (Simosky et al., 2001). A computer simulation by Moxon and colleagues (2003) proposed that gating deficits found in schizophrenia may be the consequence of a twofold system failure: (1) dysfunction in the alpha-7 nicotinic acetylcholinergic reaction with GABA-B interneurons may cause abnormal suppression to S2, whereas (2) increased dopamine in hippocampal layer CA3 creates a lower amplitude of both S1 and S2 responses. In summary, the P50 sensory gating effect appears to be dependent on the interaction of superior temporal gyrus, hippocampal, and frontal circuits. Moreover, pharmacological manipulations of receptors thought to be involved in schizophrenia can induce or ameliorate sensory gating deficits.

While early studies suggested that the P50 response occurs during a preattentive stage of auditory processing, and thus was presumed to be unaffected by attention or alertness, later studies indicated that the effect can be affected by task demands and stimulus characteristics (Freedman et al., 1987; Waldo & Freedman, 1986). Other studies have found that psychological stress results in higher sensory gating ratios due to a decreased response to S1 (White & Yee, 1997; Yee & White, 2001; see also White & Yee, 2006). The P50 response to S2 also appears to be sensitive to task demands and manipulations of attention. Clementz et al. (2002) manipulated the expectancies of healthy participants by presenting single, paired, or triple auditory clicks in either a predictable or an unpredictable sequence. They found that the unpredictable stimulus configuration resulted in larger P50 and N100 amplitudes and greater low-frequency (1–20 Hz) power in response to the second and third stimuli, with no differences in response to the first stimulus. Similarly, Guterman and coworkers (1992) examined the effects of attentional manipulations on responses elicited in the dual-click procedure and found that sensory gating was diminished when subjects were required to attend to the second stimulus. Boutros et al. (1995) and Zouridakis et al. (1997) found that stimulus novelty or probability affected S2 amplitude. These findings indicate that the amplitude of midlatency evoked responses to the second (or repeated) stimuli can be influenced by the participant's expectancy and task manipulations of attention.

Abnormalities in Schizophrenia

The P50 sensory gating effect has been hypothesized to test the ability of the nervous system to filter out extraneous, repetitive information so that higher-order auditory processing is not overloaded by irrelevant information. This inhibitory deficit has been interpreted to reflect sensory gating deficits in schizophrenia, and may be related to subjective reports of sensory inundation and "flooding" by patients (Bunney et al., 1999; Freedman et al., 1987).

Recently, the inhibitory nature of the P50 sensory gating construct has come under scrutiny. Since the gating effect is conventionally measured as a ratio, either a diminished S1 or a larger S2 could result in an abnormal response. Several studies suggest that elevated gating ratios in schizophrenia may also be influenced by a smaller P50 response to the S1 stimulus (Clementz, 1998; Johannesen et al., 2005), with an S2 response comparable to that of control subjects. Greater latency variability of P50 may contribute to this decreased S1 response in schizophrenia (Jin et al., 1997). A smaller S1 P50 amplitude in schizophrenia would be consistent with other findings suggestive of auditory cortex abnormalities in schizophrenia, attentional deficits, or irregular neural timing (Boutros et al., 1995; Zouridakis et al., 1997).

Sensitivity and Specificity

The effect size for the sensory gating deficit is large, with a measure across studies of $d = 1.28$ (de Wilde et al., 2007). The magnitude of the effect, however, has been highly variable across studies. These differences may be related to methodological factors such as stimulus sound intensity and subject postural position, as well as subject variables such as medication and smoking (de Wilde et al., 2007). While the finding of sensory gating deficits in schizophrenia appears to be robust, these deficits are not specific to this diagnostic category. Olincy and Martin (2005) found that patients with schizoaffective disorder bipolar type, those with schizophrenia, and those with bipolar disorder and a history of psychosis had impaired P50 performance, while those with bipolar disorder and no history of psychosis were not impaired. This suggests that the P50 gating ratio may represent a physiological mechanism associated with psychosis and is not specific to one diagnostic category. Hall et al. (2007b) used a combined twin and family dataset to perform structural equation modeling of P50 performance and bipolar disorder. They found that bipolar disorder was significantly associated with a lower P50 suppression ratio and a lower difference score, and that shared genetic factors accounted for these associations. These findings support Olincy and Martin's theory that P50 is associated with shared psychosis susceptibility genes in schizophrenia. Outside of the schizophrenia spectrum, Boutros and colleagues (2006) reported decreased P50 sensory gating in 3-week abstinent cocaine users, and Marco and coworkers (2005) reported similar findings in those with alcohol dependence. Interestingly, within these populations, length of abstinence was positively correlated with ERP amplitudes, indicating possible recovery of neural network activity with abstinence. Finally, some studies have reported P50 gating deficits in children with autism and mental retardation (Orekhova et al., 2008), as well as in those with Huntington's disease (Uc et al., 2003). In summary, while P50 sensory gating is sensitive to schizophrenia, it is not specific to this disorder.

Clinical Correlates

Surprisingly, relatively few studies have investigated the relationship between reported perceptual inundation and P50 ERP suppression. Several recent studies using the Sensory Gating Inventory, which is a self-report measure of sensory disturbances, have reported associations between perceptual anomalies and P50 suppression in both healthy controls (Kisley et al., 2004) and schizophrenia patients (Johannesen et al., 2008). These studies provide construct validity for the use of P50 suppression as an objective measure of sensory modulation in schizophrenia.

Of further interest is the relationship between the ability of the nervous system to adaptively modulate sensory information and the characteristic symptoms of schizophrenia. Because more severe negative symptomatology has been associated with poorer premorbid functioning and poorer outcomes, several studies have investigated the relationship between P50 sensory gating and the severity of positive and negative symptoms. Some investigators have found that more severe negative symptoms are associated with more severe gating deficits in schizophrenia (Louchart-de la Chapelle et al., 2005; Ringel et al., 2004). However, other studies have failed to demonstrate a relationship between measures of symptom characterization in schizophrenia and P50 sensory gating (Adler et al., 1990; Boutros et al., 2004; Brockhaus-Dumke et al., 2008). These findings indicate that P50 sensory gating may reflect a more stable, biologically mediated trait than clinical symptoms, which vary over time.

Some of the inconsistencies in the P50 literature may also be due to medication effects. P50 gating may be differentially affected by different types of antipsychotic medication. Several studies suggest that novel antipsychotic medications, particularly clozapine, may ameliorate the P50 sensory gating deficit, while conventional antipsychotic medication does not (Adler et al., 2004; Light et al., 2000). Nicotine use can also reduce or eliminate the gating deficit in schizophrenia (e.g., Leonard et al., 1998).

Genetics

P50 suppression has been proposed as an endophenotype for schizophrenia that represents a stable biological marker that is closer to the genetic underpinnings of the disorder than the heterogeneous clinical manifestation. Several studies have shown sensory gating deficits in the first-degree biological relatives of those with schizophrenia (Clementz et al., 1998b; Siegel et al., 1984) and bipolar disorder with psychotic features (Schulze et al., 2007), while Young et al. (1996) established statistical heritability when measuring P50 gating in healthy monozygotic and dizygotic twin pairs. P50 suppression ratios have been linked to specific genetic variants of the 15 alpha-7 acetylcholine nicotinic receptor gene (*CHRNA7*) in healthy controls and in those with schizophrenia. Moreover, Martin et al. (2007) found that those with schizophrenia and schizoaffective disorder bipolar type with a variant allele also had abnormal P50 gating ratios, while only schizoaffective participants with a common allele had normal mean P50 ratios. The alpha-7 nicotinic agonist 3–2,4 dimethoxybenzylidene anabaseine (DMXBA) has recently been shown to enhance auditory sensory gating in both humans and animals and may prove to be a useful treatment for cognitive dysfunction in patients with schizophrenia (Martin et al., 2004). These findings strongly implicate the *CHRNA7* gene in the process of auditory sensory gating, although the exact allelic associations in the literature vary (Houy et al., 2004; Martin et al., 2007; Raux et al., 2002).

Summary

The P50 sensory gating deficit is one of the most robust neurobiological findings in schizophrenia. However, interpretation of this deficit is complicated by a variety of factors that may exacerbate or ameliorate the deficit. These include pharmacological treatment, nicotine use, methodological features, stress, and attention. Familial studies suggest that the deficit is moderately to highly heritable. The specific association of the deficit with the *CHRNA7* nicotinic receptor gene and its modulation by cholinergic drugs may provide guidance for pharmacological interventions and assessment of treatment effects.

ERP Components from Oddball Discrimination Tasks: P300

Cognitive and Neural Mechanisms

Oddball discrimination tasks have been extensively used to study ERP components in schizophrenia, particularly to evaluate the effects of the illness on attentional modulation of processing. The oddball discrimination task requires discrimination of infrequent target stimuli interspersed among frequent nontarget stimuli, with stimulus onset asynchronies in a range of about 500 to 2000 ms. For example, in the auditory modality, a subject might be required to respond to infrequent (10%) high-pitched tones interspersed amid more frequent low-pitched tones. Infrequent target stimuli are usually presented on 5% to 20% of trials. The frequent nontarget stimuli elicit prominent N1 and P2 components. The P300 component is only elicited by the infrequent target stimulus in this paradigm, as shown in Figure 19.3. P300 is exquisitely sensitive to a wide variety of experimental manipulations, including task relevance, attentional resources, discrimination difficulty, event probability, and ISI (Picton, 1992; Polich & Kok, 1995). The P300 component has been conceptualized to index such diverse processes

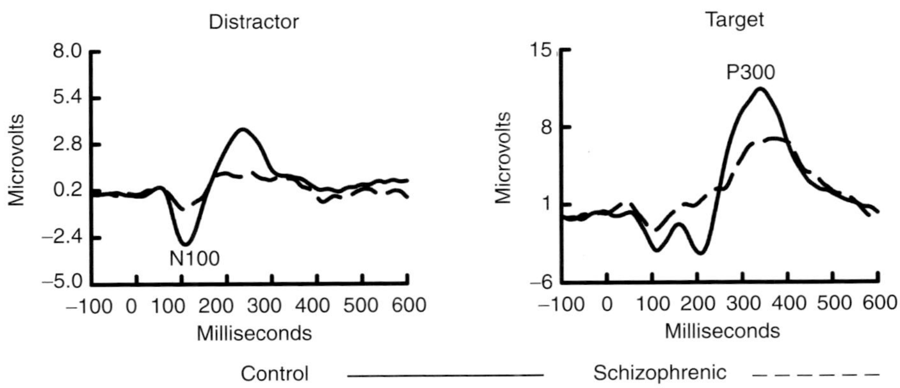

Fig. 19.3. Grand average ERPs elicited by frequent distractor tones and infrequent target tones in control subjects (solid line) and patients with schizophrenia (dashed line). Note the reduction of P300 amplitude to target stimuli as well as the reduction of N100 to frequent distractor stimuli in patients with schizophrenia.

as selective attention, expectancies, and contextual updating (Donchin & Coles, 1988; Johnson, 1986; Picton, 1992; Polich & Kok, 1995; see also Chapter 7, this volume). These components can be further decomposed on the basis of sensitivity to experimental manipulations and signal analytic techniques such as principal components analysis (PCA; see Chapter 1, this volume). However, because the majority of studies use latency and amplitude measures from averaged ERPs, we will focus on these approaches to component characterization. The neural generators of N1 and P3 appear to differ as a function of modality of stimulation, as described in subsequent sections.

Abnormalities in Schizophrenia

P300 has received much more attention in studies of schizophrenia compared to earlier components. Unless otherwise noted, P300 refers to the "P300b" elicited by task relevant stimuli rather than the "P300a" component to novel, but task-irrelevant, distractor stimuli. The auditory P300 to infrequent target stimuli is typically attenuated in amplitude in patients with schizophrenia (Ford, 1999; Jeon & Polich, 2003; Roth et al., 1980). P300 latency prolongation is a less consistent finding, reflecting a smaller effect size (Jeon & Polich, 2003). Attention and working memory operations are severely affected by schizophrenia, and it is likely that P300 amplitude reduction reflects these disturbances (Nestor & O'Donnell, 1998). An interpretative issue is whether P300 amplitude reduction in the averaged ERP is the result of reduced P300 amplitudes for single trials or normal amplitudes with greater between-trial variability in patients. Single-trial analysis indicates that patients with schizophrenia have smaller P300 amplitudes on individual trials, as well as higher variability in P300 latency across trials (Ford et al., 1994b). Evidence from EEG and MEG source analyses, intracranial recordings, and patients with focal lesions suggests that the auditory P300 component is generated by a distributed neural circuit, with critical structures including the posterior superior temporal gyrus and adjacent parietal-temporal junction and lateral frontal cortex (Soltani & Knight, 2000). Consistent with these findings, the left posterior superior gyrus volume in schizophrenia has been reported to be negatively correlated with auditory P300 amplitude in chronic patients (O'Donnell et al., 1999) and in first-episode patients with schizophrenia (McCarley et al., 2002).

Visual P300 amplitude has been reported to be reduced in schizophrenia by some (Brecher et al., 1987; Pfefferbaum et al., 1984; Sponheim et al., 2006) but not other investigators (Bruder et al., 1998; Doniger et al., 2002; Egan et al., 1994; Ford et al., 1994a, 1994b; Ohta et al., 1996; Pfefferbaum et al., 1989; Potts et al., 2002). Meta-analytic studies indicate a smaller effect size for visual compared to auditory P300 amplitude deficits (Jeon & Polich, 2003). This differential sensitivity may be due to the relationship of auditory P300 to superior temporal gyrus abnormalities, which are characteristic of schizophrenia. Human lesion, depth recording, and fMRI data suggest that the visual P300, like the auditory P300, is dependent on temporal-parietal activation. In addition, the visual P300 is associated with visual cortical circuits as well, suggestive of an interplay between posterior sensory regions and more anterior multimodal circuits (Soltani & Knight, 2000).

Sensitivity and Specificity

Meta-analytic studies indicate that P300 amplitude and latency abnormalities show moderate sensitivity to schizophrenia, with the auditory P300 more sensitive than the visual P300 and amplitude more sensitive than latency. In a meta-analytic study evaluating the sensitivity of P300 measures to schizophrenia, Jeon and Polich (2003) estimated a visual P300 effect size (d) of 0.39, compared to an auditory P300 effect size of 0.89. P300 latency prolongation was associated with an effect size of $d = 0.59$ and a visual effect size of $d = 0.49$. While an effect size of 0.80 is considered large, this degree of sensitivity still indicates an overlap of 52.6% between control and schizophrenic populations. Consequently, none of the P300 measures provides sufficient sensitivity for clinical diagnosis.

P300 deficits are not specific to schizophrenia; P300 amplitude reduction or latency prolongation has been reported in bipolar disorder, depression, and Alzheimer's disease, in individuals at high risk for alcoholism, and in old compared to young subjects (O'Donnell et al., 2004; Pfefferbaum et al., 1984; Picton, 1992; Polich & Kok, 1995; see also Chapter 7, this volume). Many, but not all, studies have shown a topographic asymmetry of the auditory P300 component in schizophrenia, with a larger reduction on left compared to right temporal electrode sites (see Jeon & Polich, 2001; O'Donnell et al., 1999; van der Stelt, 2004). This asymmetry is present at the first episode of the illness and differentiates schizophrenia from affective psychosis (McCarley et al., 2002; O'Donnell et al., 1999). In summary, while P300 amplitude reduction is not specific to schizophrenia, a greater left compared to right reduction in P300 topography may be specific to schizophrenia compared to affective psychoses.

Clinical Correlates

Most ERP studies of patients with schizophrenia test medicated patients. Since antipsychotic medication has pervasive effects on a variety of neurotransmission systems, such as the dopamine and serotonin systems, it is possible that ERP differences in schizophrenia may be secondary to pharmacological effects. However, auditory P300 amplitude reduction has been found in patients withdrawn from medication (Ford et al., 1994a; Hirayasu et al., 1998a) and in never-medicated patients (Hirayasu et al., 1998a), indicating that medication does not cause this abnormality. Moreover, in a well-designed placebo control study, medicated patients with schizophrenia did not show an effect of medication on P300 amplitude deficit compared to baseline or compared to patients receiving a placebo (Ford et al., 1994a).

Several studies have addressed the issue of whether P300 abnormalities change with age or stage of illness. Auditory P300 amplitude reduction (Brown et al., 2002; Hirayasu et al., 1998a) and topographic abnormalities are present at the first episode (Salisbury et al., 1998b; but not Hirayasu et al., 1998a). Short-term follow-up studies indicate that P300 amplitude reductions are stable over time (e.g., Turetsky et al., 1998). Cross-sectional studies, however, suggest that auditory P300 abnormalities may interact with illness duration or age. Several groups have found that the slope of auditory P300 latency on age is increased in males with chronic schizophrenia (Mathalon et al., 2000; O'Donnell et. al, 1995) and in first-episode male patients (Wang et al., 2003). Since P300 latency increases in neurodegenerative disorders, this finding is suggestive of a progressive pathophysiological process in male patients. However, since the increased slope is also found in first-episode patients, this effect is not dependent on treatment with medication or necessarily long illness duration. Mathalon et al. (2000), van der Stelt (2004), and Olichney et al. (1998) found that amplitude correlated negatively with illness duration and positively with age of onset. These findings suggest that age or illness duration may interact with P300 abnormalities in schizophrenia, but these cross-sectional studies do not provide unambiguous evidence of such effects. These results are consistent with recent evidence from longitudinal structural MRI studies of progressive changes after the first episode of schizophrenia (Shenton et al., 2001).

Genetics

Schizophrenia is associated with high heritability, although the genetic mechanisms are almost certainly polygenic (Tsuang, 2004). Twin and family studies indicate that P300 amplitude has a heritability of about 60% and that P300 latency has a heritability of about 51% (van Beijstervelt & van Baal, 2002). Siblings of patients with schizophrenia have often been reported to show auditory P300 amplitude reduction compared to healthy control subjects (Hall et al., 2007b; Turetsky et al., 2000; Winterer et al., 2003). Winterer et al. (2003) found that siblings showed increased P300 amplitude at frontal sites, while Turetsky et al. (2000) found that siblings had decreased P300 amplitude only at frontal sites. Other studies have failed to find differences between siblings and probands, but this may be due to lack of power or genetic heterogeneity among samples. In a study of twins concordant and discordant for schizophrenia and healthy control twins, the phenotypic correlation between P300 amplitude and schizophrenia was–.35, of which 75% was attributable to genetic effects (Hall et al., 2007b), indicating that this measure has potential as an endophenotypic marker.

The role of specific genetic variations on P300 amplitude in schizophrenia has recently come under scrutiny. P300 has amplitude has been associated with translocation at 1q42.2 of chromosome 1 in a family with a high prevalence of psychiatric illness including schizophrenia (Blackwood et al., 2001). This genetic abnormality affects the gene Disrupted in Schizophrenia, or *DISC1*, which increases the risk for schizophrenia. Catechol-*O*-methyltransferase (*COMT*) is a postsynaptic enzyme involved in dopamine catabolism that has also been associated with schizophrenia (Egan et al., 2001). A single *COMT* gene polymorphism includes replacement of a valine peptide sequence by a methionine sequence at position 158. The Met allele produces lower rates of enzymatic degradation of dopamine, and thus higher levels of dopaminergic activity, compared to the Val allele. The Met allele has been associated with lower frontal P3 amplitude, particularly in schizophrenia (Gallinat et al., 2007).

Summary

The P300 components elicited by auditory and visual oddball discrimination tasks have been extensively studied in schizophrenia. While both components are affected, the auditory P300 is more sensitive to schizophrenia than the visual P300. These abnormalities suggest pervasive impairments of selective attention and working memory in the disorder. For auditory P300, the effect appears larger at left temporal compared to right temporal recording sites, similar to asymmetries reported in MRI studies of the gray matter volume of the superior temporal gyrus. While P300 abnormalities are not specific to schizophrenia, the auditory P300 component appears to be sensitive to a genetic risk, and to be relatively insensitive to medication or changes in clinical state.

ERP Components from Oddball Discrimination Tasks: N1

Cognitive and Neural Mechanisms

While both nontarget and target stimuli in oddball discrimination tasks evoke an N1 component (see Figure 19.3), this component has received less systematic investigation compared with the P300 component described in the previous section. Nevertheless, there is accumulating evidence that the N1 component may be affected in schizophrenia and that this may be a more specific deficit than P300 abnormalities. The auditory N1 and visual N1 are negative-going ERP components relative to ear, mastoid, or nose references (see Chapter 4, this volume). The auditory N1 component has a peak latency of about 100 ms, while the latency of the visual N1 is usually about 150 ms. Both can be elicited by stimuli in the absence of explicit task demands but are also sensitive to selective attention.

Abnormalities in Schizophrenia

The auditory N1 component has been shown to be reduced in amplitude in patients with schizophrenia to standard as well as target stimuli (Brown et al., 2002; Ford et al., 1994a, 1994b; O'Donnell et al., 2004; Pfefferbaum et al., 1984; Roth et al., 1980). Auditory N1 amplitude reduction occurs even when attention is directed to a concurrent reading task, indicating that this deficit reflects alterations in preattentive sensory processing (O'Donnell et al., 1994). Electroencephalographic, MEG, fMRI, and human depth electrode recordings suggest that the auditory N1 is generated by auditory cortex within the superior temporal area, including Heschl's gyrus and the planum temporale (Picton et al., 1999; Yvert et al., 2005). This cortical region is integral to audition and, on the left side, to language processing. The reduction in N1 amplitude is consistent with clinical symptoms of auditory distortions, auditory hallucinations, and disorganized speech, as well as structural evidence of reduction of gray matter in the superior temporal gyrus, including the planum temporale (Henn & Braus, 1999; Shenton et al., 2001).

The visual N1 component, like the auditory N1 component, is elicited by stimuli regardless of task demands, although task demands do modulate amplitude and topography. In discrimination tasks, visual N1 amplitude or topography varies with stimulus features, including location, spatial frequency and orientation, and spatial attention (Mangun & Hillyard, 1988; O'Donnell et al., 1997). Visual N1 amplitude reduction in schizophrenia has been reported by some (Bruder et al, 1998; Ford et al., 1999; Ohta et al., 1996; Pfefferbaum et al., 1984; Potts et al., 2002) but not all investigators (Doninger et al., 2000; Egan et al., 1994; Sponheim et al., 2006; Vohs et al., 2008). Visual N1 amplitude reduction is therefore consistent with behavioral evidence of perceptual disturbances in schizophrenia that affect visual pathways (Butler & Javitt, 2005; O'Donnell et al., 2006), but it appears to be a less robust abnormality than auditory N1 amplitude reduction.

Sensitivity and Specificity: Clinical Correlates

Auditory N1 has been reported to be reduced in amplitude in first-episode studies of schizophrenia (Brown et al., 2002) and in unmedicated patients (Pfefferbaum et al, 1989), indicating that this abnormality is not secondary to the general effects of chronic psychiatric illness or treatment with antipsychotic medication. With respect to specificity, while several studies suggest that auditory N1 amplitude reduction may be more sensitive to schizophrenia than to other neuropsychiatric disorders such as bipolar disorder (Force et al., 2008; O'Donnell et al., 2004) and seizure disorders (Ford, 2001), other studies show reduced N1 amplitude in depression and Alzheimer's disease (Pfefferbaum et al., 1984).

Genetics

Force et al. (2008) have reported that the auditory N1 obtained in a dichotic listening task was reduced in amplitude in subjects with schizophrenia and in their first-degree relatives. In contrast, N1 was unaffected in subjects with bipolar disorder or their first-degree relatives. This finding suggests that auditory N1 may serve as a more specific endophenotype than P50 gating or P300 for schizophrenia. This finding should motivate the use of auditory N1 in genomic studies of schizophrenia and at-risk populations.

Summary

N1 is elicited by transient auditory and visual stimuli. Auditory N1 appears to be more sensitive to schizophrenia than visual N1, with auditory N1 amplitude reduction reported in both medicated and unmedicated patients. Auditory N1 may also be sensitive to the genetic risk for schizophrenia. Several studies suggest that auditory N1 amplitude reduction may differentiate schizophrenia and bipolar disorder.

N400

Cognitive and Neural Mechanisms

Eugene Bleuler's (1950) original descriptions of schizophrenia emphasized disorganized thinking and speech as hallmarks of the disorder. He attributed the disorganized thinking to a splitting of associational threads and loss of goal-oriented ideation. Similarly, contemporary accounts of the mechanism for disorganized thinking and discourse in schizophrenia frequently emphasize a disturbance in the organization of semantic memory or of processes that operate upon it (Chapman & Chapman, 1973). Semantic memory has been hypothesized to be organized into a network of semantic features and concepts in which words close in meaning, such as *pencil* and *pen*, are strongly associated with each other, or are closer in semantic space; and words that are distant in meaning, such as *pen* and *ocean*, are weakly associated. Associative activation is thought to occur quickly, within 500 ms after presentation of a word. Activation of one word or node in the network spreads to related items but is usually limited by decay or inhibition. Controlled processes may inhibit the spread of activation to contextually appropriate words. For example, the meaning of *pen* depends on whether a speaker is discussing enclosures for animals or a writing implement.

Two types of disturbance have accrued support from experimental studies: disturbances within the associational linkages within a semantic network (Kwapil et al., 1990; Spitzer, 1997) and a failure in contextual control of semantic search (Nestor & O'Donnell, 1998). The hypothesis of overactivation within semantic networks assumes that initial activation spreads too widely in the network or is dominated by strong associations, which results in loose associations and derailed thinking. Thus, in behavioral priming studies using short stimulus onset asynchronies (SOAs), faster reaction times (RTs) to related words in patients compared to control subjects are treated as evidence for overactivation in semantic networks (e.g., Kwapil et al., 1990; Maher et al., 1996; Moritz et al., 2003; Peled et al., 2005; Quelen et al., 2005; Spitzer et al., 1997). Experimental and computational evidence suggests that the disturbance in schizophrenia is influenced by altered associational connectivity rather than by network size (i.e., number of associates; Nestor et al., 2001). Controlled processes also appear to be affected, since priming paradigms using long SOAs that demand controlled processes demonstrate longer RTs to related words (hypo-priming) in patients (e.g., Barch et al., 1996; Blum & Freides, 1995; Chapin et al., 1989; Cohen et al., 1999; Passerieux et al., 1997; Titone et al., 2000; Vinogradov et al., 1992).

These semantic processing effects have also been explored with ERPs, in particular the N400 component. N400 is a negative-going deflection that peaks about 400 ms after stimulus onset. In general, N400 amplitude is inversely proportional to the likelihood that a word is congruent or consistent with the prior semantic context (e.g., Kutas & Hillyard, 1980; see also Chapter 15, this volume). N400 latency reflects the speed of linguistic operations associated with semantic search (Van Petten & Kutas, 1990). The specific operations indexed by N400 are influenced by the linguistic and temporal characteristics of a paradigm. Sentence paradigms are well suited to probe contextual effects on lexical processing, while semantic priming paradigms can test for associational modulation of processing as well. In studies of N400 in schizophrenia, the N400 component has generally been elicited either to words during sentence processing or to word pairs in semantic priming paradigms.

SENTENCE PROCESSING

A word that is nonsensical or incongruent with the preceding context of a sentence generates a larger N400 than a word that is congruent with the context of the sentence (see Chapter 15, this volume, for a detailed discussion). For example, N400 would be larger (more negative) if the terminal word in a sentence is nonsensical with respect to the preceding context (e.g., *She spread her bread with socks.*) relative to when it is sensible (*She spread her bread with butter.*) An example of N400s elicited by incongruent sentence endings is shown in Figure 19.4. In general, the amplitude of N400 is larger (more negative) with smaller cloze probabilities for the target word. This effect has been used in schizophrenia to examine the neurophysiological correlates of integration of a word with the sentence context. An important interpretive issue is whether N400 is

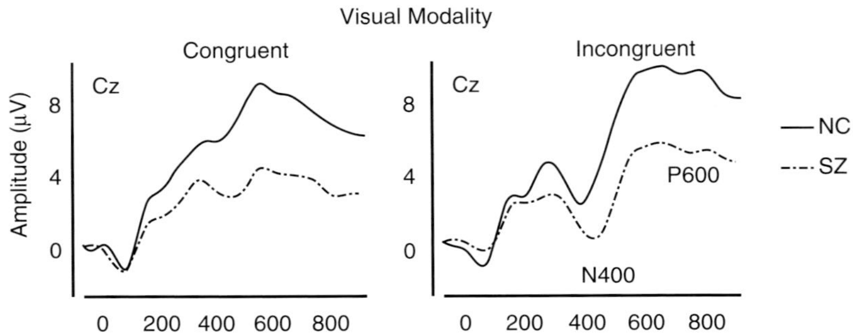

Fig. 19.4. Grand average ERPs to congruent and incongruent terminal words in sentences in subjects with schizophrenia (SZ) and 16 normal control (NC) subjects. Incongruent words elicit a prominent negative deflection about 400 ms after stimulus onset, the N400 component. Subjects with schizophrenia show a more negative and delayed N400 component compared to control subjects.

measured separately for congruent and incongruent ERPs or from the waveform generated by the difference between congruent and incongruent ERPs. Difference waveforms are often used to characterize the N400 effect, but they may be difficult to interpret if clinical groups differ from control groups with respect to the N400 elicited by the baseline congruent condition.

WORD PRIMING

N400 is also sensitive to the semantic relationship between two words presented in rapid succession. In word priming studies in healthy controls, a smaller N400 (less negative) is recorded to target words semantically related to a prime (e.g., *cat* and *dog*) and a larger N400 (more negative) is recorded to target words semantically unrelated to a prime (e.g., *cat* and *paper*; e.g., Holcomb & Neville, 1990). This N400 priming effect has been observed at both long and short SOAs, in the time interval between the onset of words (e.g., Hill & Weisbrod, 2005; Hill et al., 2002), and across visual and auditory modalities (Holcomb et al., 2005) and suggests that N400 is also sensitive to lexical processes. At short SOAs (500 ms or less), N400 probably reflects automatic processes of semantic activation, while at long SOAs (more than 500 ms), it is more influenced by processes of context utilization.

The N400 component may also be informative regarding the integrity of specific cortical regions responsible for semantic processing in schizophrenia. In human intracranial recordings, the N400 component has been associated with depth activity in the anterior medial temporal lobe, such as the hippocampus and the parahippocampal gyrus, as well as the superior temporal sulcus (Guillem et al., 1995; Halgren et al., 1994b). In addition to medial temporal and temporal cortices, posterior parietal regions (Guillem et al., 1999; Halgren et al., 1994b) as well as fusiform and lingual gyri (McCarthy et al., 1995; Nobre & McCarthy, 1995) have been implicated. Studies by Guillem et al. (1999) and Halgren et al. (1994a) point to posterior-ventral prefrontal cortex. Intracranial studies show the largest contribution from the left hemisphere, with relatively little activity observed in the homologous structures of the right hemisphere (Halgren, 2004; Van Petten & Luka, 2006). Most fMRI studies of similar semantic processing paradigms report both temporal and prefrontal areas of activation (Halgren, 2004; Van Petten & Luka, 2006).

N400 in Schizophrenia

SENTENCE PROCESSING

Both sentence and word priming studies have been used to examine semantic processing in schizophrenia. In studies of N400 elicited by congruent and incongruent words embedded in a sentence (usually in the terminal position), several findings have emerged. First, N400 is often delayed in latency, suggesting that semantic activation is more diffuse or that integration is slowed in schizophrenia (Nestor et al., 1997; Niznikiewicz et al., 1997). Early ERP components are not delayed in latency, indicating that this disturbance may be specific to semantic processes (Niznikiewicz et al., 1997). Second, the difference measure of the N400 amplitude effect is sometimes (Adams et al., 1993; Niznikiewicz et al., 1997) but not always (Andrews et al., 1993; Mitchell et al., 1991; Nestor et al., 1997; Ruchsow et al., 2003) reduced in schizophrenia. Third, N400 amplitude elicited by congruent words may be more negative in patients with schizophrenia than in control subjects, suggestive of a deficit in contextual integration that occurs even when discourse is fully coherent (Nestor et al., 1997; Niznikiewicz et al., 1997; see Figure 19.4). These disturbances in contextual integration occur even when patients and control subjects perform the task with comparable accuracy and in both the visual and auditory modalities, suggestive of a supramodal linguistic deficit (Niznikiewicz et al., 1997).

SEMANTIC PRIMING STUDIES

Since short SOAs bias processing toward automatic semantic activation and long SOAs are more influenced by controlled processes, N400 recorded during priming paradigms can test both mechanisms. As in RT studies, N400 studies suggest that both automatic and controlled processes are affected. In ERP studies, hyperpriming is supported by a reduced N400 at short SOAs in patients relative to controls (e.g., Condray et al., 1999, 2003; Kiang et al. 2008; Kreher et al., 2007; Mathalon et al., 2002; Niznikiewicz et al., 2002), indicating a greater influence of immediate associational connections. On the other hand, disturbed controlled processes are implicated in ERP studies using long SOAs and word-pair or sentence stimuli that find a larger amplitude N400 in schizophrenia patients, suggesting difficulties in integrating context (e.g., Adams et al., 1993; Ditman & Kuperberg, 2007; Grillon et al., 1991; Hokama et al., 2003; Kiang et al., 2008; Kostova et al., 2003; Koyama et al., 1991; Nestor et al., 1997; Niznikiewicz et al., 1997; Ruchsow et al., 2003; Salisbury et al., 2002b; Sitnikova et al., 2002). Additional variables that may influence these results are the length of illness (Maher et al., 1996) and the degree of thought disorder (Aloia et al., 1998; Gouzoulis-Mayfrank et al., 2003; Kostova et al., 2005). Differences among studies may be related to the severity of thought disorder, which can affect behavioral measures of semantic priming as well (Spitzer, 1997).

Sensitivity and Specificity

The sensitivity and specificity of N400 deficits to schizophrenia remain to be systematically evaluated. Aging and neurodegenerative disorders may also affect N400. N400 abnormalities have been reported in Alzheimer's disease (Olinchey et al., 2007). The N400 difference effect becomes smaller, slower, and more variable with age (Kutas & Iragui, 1998).

Clinical Correlates

Most studies of N400 use patients who are receiving antipsychotic medication, which could perturb linguistic and related cognitive processes or ameliorate disturbed functions. Hokama et al. (2003) used a semantic priming task with a 1000 ms SOA to evaluate semantic processing in patients with schizophrenia who had been unmedicated for at least 1 month. Patients with schizophrenia showed prolonged N400 latency and a smaller N400 effect, indicating that these language abnormalities are not secondary to antipsychotic medication. Controlled pharmacological trials will be required to determine whether medication lessens N400 deficits.

Genetics

Unlike P300 and MMN, there has been little investigation of genetic factors and N400 in schizophrenia. Kimble et al. (2000) found that N400 during sentence processing was unaffected in relatives of individuals with schizophrenia.

Summary

Studies using lexical priming and sentence processing paradigms suggest that N400 may be sensitive to disturbances in semantic processing in schizophrenia. Evidence across paradigms suggests that both automatic hyperactivation of associated words in semantic networks and disturbances in controlled search and integration are affected in schizophrenia. The influence of medication, genetic factors, and severity of thought disorder may contribute to differences in findings among studies.

Conclusions

Several decades of research have shown that a variety of ERP components are reliably affected by schizophrenia, and that these disturbances are due to effects of the illness rather than to medication, comorbid disorders, or general effects of chronic mental illness. These components provide a rich source of information regarding specific cognitive processes, neural systems, pathophysiological mechanisms, and genetic factors involved in schizophrenia.

Event-related potentials can provide information regarding perceptual or cognitive processes that are not associated with a behavioral response, and can provide a time marker for neurophysiological processes elicited by a stimulus or a covert event. Reduction of P50 amplitude to the first click in the sensory gating paradigm, and failure of inhibition to the second click, suggest a disruption in automatic sensory processes. Other ERP indices of sensory processing, such as the N1 component to visual and auditory stimuli, are also affected (Ford et al., 1999; Pfefferbaum et al., 1984). These findings are consistent with a large body of evidence of subjective sensory distortions in the auditory and visual modalities in schizophrenia (Bunney et al. 1999; Cutting & Dunne, 1986; Phillipson & Harris, 1985) and with objective evidence of impaired psychophysical thresholds for sensory features (Butler & Javitt, 2005; O'Donnell et al., 2006). The MMN amplitude reduction indicates that automatic comparison of sensory features is disrupted, since the MMN depends on comparison of the current stimulus features with a short-term representation derived from regularities in preceding auditory stimuli (Näätänen et al., 2003). This deficit is also evident within 100 to 200 ms of stimulus onset and, like the P50 deficit, can be detected in the absence of a behavioral response or other task demands. P300 and N400 disturbances, in contrast, are influenced by attention and other cognitive processes. P300 amplitude decrements and latency prolongation in a wide variety of paradigms and across modalities suggest a pervasive disturbance of cognitive and working memory mechanisms. N400 component abnormalities suggest that two basic mechanisms associated with semantic processing are affected by schizophrenia. First, semantic activation spreads too broadly within the associative connections that support semantic networks. Second, slower controlled processes that integrate a word with the current context appear to be impaired. Auditory processes appear to be more impaired than visual processes for the nonverbal discrimination tasks typically associated with the N1 and P300 components, while N400 disturbances appear to be supramodal. In summary, ERP findings suggest that the effects of schizophrenia are evident across stages and modalities of information processing rather than being confined to a specific aberrant mechanism.

Event-related potential abnormalities have implicated several brain regions, in particular the temporal-parietal cortex, hippocampus, and prefrontal cortex. The auditory P50, N1, and MMN all depend on circuits in the superior temporal cortex, as do the auditory and visual P300 components. P50 inhibition or gating is thought to involve medial temporal circuits as well as auditory cortex. Moreover, the auditory P300 deficit is more robust than the visual P300 deficit, again suggesting more severe disruption of the temporal lobes. The N400 component has also been associated with activity in the superior temporal cortex, hippocampus, and prefrontal regions. These ERP findings are congruent with structural MRI studies indicating that reductions of hippocampal and superior temporal gyrus volume are among the most consistent findings in schizophrenia (Shenton et al., 2001).

Twin and family studies suggest that genetic factors make a major contribution to the MMN, P50, and P300 components. In addition, first-degree relatives of patients with schizophrenia have been reported to show reduced P300 amplitude, reduced auditory N1, and a P50 gating deficit, suggesting that these components may have potential as endophenotypes for schizophrenia. An *endophenotype* has been defined as a measurable feature associated with an illness invisible to the unaided eye that is intermediate in the pathway between a genotype and the illness phenotype (Gottesman & Gould, 2003). Gottesman and Gould propose five criteria for an endophenotype: (1) association with the illness in the population, (2) heritability, (3) state independence, (4) endophenotype and illness cosegregation in families, and (5) presence of the endophenotype in nonaffected family members at higher rates than in the general population. The auditory P300 and P50 gating effects appear to robustly meet the criteria for association with the illness and heritability. Several reports of associations between specific genetic polymorphisms and P300 and P50 gating deficits have already been reported. The MMN, on the other hand, while heritable, may only occur after the onset of psychotic symptoms and has not been reliably associated with familial risk. There are insufficient data to determine whether N400 has potential as an endophenotype. The relationship of ERP abnormalities to genetic variation has just begun to be investigated. Because ERP measures can be inexpensively obtained from large numbers of subjects and the variables of interest quickly quantified from data, these measures hold particular promise in functional genomic studies of schizophrenia.

The relationship of ERP abnormalities to pathophysiological changes at the cellular level in schizophrenia is not well understood. Since transient ERP components primarily reflect the superposition of synchronous postsynaptic potentials generated by cortical tissue (Olejniczak, 2006), the reduction in amplitude observed for P50, MMN, and P300 likely indicates alterations in these synaptic structures. Reductions in synaptic connectivity, neural organization, and neurotransmission that have been associated with schizophrenia could therefore directly affect the amplitude of these components (Ross et al., 2006; Selemon & Goldman-Rakic, 1999). Such a relationship is also supported by correlations found between gray matter volume in the superior temporal gyrus and P300 amplitude

(McCarley et al., 2002; O'Donnell et al., 1999). Enhancement of N400 reported by some researchers, on the other hand, may indicate a failure to inhibit inappropriate cortical regions involved in semantic associative processes. While amplitude reduction may affect sensory as well as later-stage components, prolongation of latency is most consistently reported for later components that index attention, working memory, and semantic processing, such as P300 and N400. This suggests that the slowing of processing that is also evident in RT measures may be due to failures of integration across regions of the brain involved in these cognitive processes rather than a general slowing of neurotransmission.

The development of ERP deficits over the lifespan in schizophrenia is not well understood, particularly prior to the first hospital admission for psychotic symptoms. A common model of schizophrenia posits that a genotypic risk factor interacts with insults or stressors early in development, such as obstetric complications, viral infection, or hormonal anomalies in utero, to increase the risk for schizophrenia later in life. These early developmental anomalies may be evident long before a psychosis develops and may also be present in relatives as well. Consistent with an early neurodevelopmental model, longitudinal studies show that cognitive and social deficits are often present in childhood and early adolescence in individuals prone to schizophrenia, and these deficits may worsen over time (O'Donnell, 2007). While longitudinal studies of P50 gating, MMN, P300, or N400 from childhood to the onset of schizophrenia have not been conducted, first-episode and longitudinal studies after onset indicate that P300 and MMN differ in their relationship to the course of the illness. P300 amplitude is reduced at the first episode, while the MMN deficit may only occur well after the onset of psychotic symptoms. The MMN deficit may therefore be the marker of illness progression, paralleling MRI studies indicating that gray matter loss in some regions may worsen following the first psychotic episode, particularly in early-onset cases. (Rapoport et al., 2005). Prodromal and longitudinal studies with ERP measures could yield valuable insights into the neurobiological evolution of schizophrenia across the lifespan.

With respect to neurotransmission, dopaminergic dysregulation has long been suspected in schizophrenia, since the efficacy of typical antipsychotic medications depends on D2 receptor antagonism. However, these medications usually do not restore cognitive or psychosocial functioning to premorbid levels, and hypofunction of the NMDA receptor has also been implicated (Lewis & Lieberman, 2000; McCarley et al., 1996; Ross et al., 2006). Both dopaminergic and glutamatergic modulation affect ERPs. Dopaminergic depletion has been reported to reduce the P300 response in animals by some investigators, with the auditory P300 found to be more sensitive than the visual P300 (Soltani & Knight, 2000). Both P300 (Oranje et al., 2000) and MMN amplitude are sensitive to administration of NMDA antagonists in humans (Umbricht et al., 2000). P50 gating appears to be sensitive to psychopharmacological manipulations as well. Animal models of schizophrenia may be better suited than human studies for testing pharmacological manipulations on ERP components and for charting the relationship between ERP deficits and neurodevelopment.

Acknowledgments

We are grateful for support from NIMH RO1 MH62150 (BFO), NIMH R21 MH0917774 (BFO), the Indiana University Foundation (BFO), NARSAD (DFS), NIMH RO1 MH 58704 (DFS), NIMH RO1 40799 (DFS), NIMH RO1 MH 63360 (MAN), and NIMH R03 MH 078036 (MAN), the Research Training in Clinical Science Award, NIMH T32 MH17146 (JLV), and the Indiana CTSI Predoctoral Training Award: PHS (NCCR) Grant No: TL1RR02575 (JLV).

References

Aaltonen, O., Niemi, P., Nyrke, T., & Tuhkanen, M. (1987). Event-related brain potentials and the perception of a phonetic continuum. *Biological Psychology, 24*, 197–207.

Adams, J., Faux, S. F., Nestor, P. G., Shenton, M. E., Marcy, B., Smith, S., et al. (1993). ERP abnormalities during semantic processing in schizophrenia. *Schizophrenia Research, 10*, 247–257.

Adler, L. E., Gerhardt, G. A., Franks, R., Baker, N., Nagamoto, H., Drebing, C., et al. (1990). Sensory physiology and catecholamines in schizophrenia and mania. *Psychiatry Research, 31*, 297–309.

Adler, L. E., Hoffer, L. J., Griffith, J., Waldo, M. C., & Freedman, R. (1992). Normalization by nicotine of deficient auditory sensory gating in the relatives of schizophrenics. *Biological Psychiatry, 32*(7), 607–616.

Adler, L. E., Olincy, A., Cawthra, E., McRae, K., Harris, J., Nagamoto, H., et al. (2004). Varied effects of atypical neuroleptics on P50 auditory gating in schizophrenia patients. *American Journal of Psychiatry, 161*(10), 1822–1828.

Adler, L. E., Pachtman, E., Franks, R. D., Pecevich, M., Waldo, M. C., & Freedman, R. (1982). Neurophysiological evidence for a defect in neuronal mechanisms involved in sensory gating in schizophrenia. *Biological Psychiatry, 17*(6), 639–654.

Adler, L. E., Rose, G., & Freedman, R. (1986). Neurophysiological studies of sensory gating in rats: Effects of amphetamine,

phencyclidine, and haloperidol. *Biological Psychiatry, 21*(8–9), 787–798.
Ahveninen, J., Jääskeläinen, I. P., Osipova, D., Huttunen, M. O., Ilooniemi, R. J., Kaprio, J., et al., (2006). Inherited auditory-cortical dysfunction in twin pairs discordant for schizophrenia. *Biological Psychiatry, 60*, 612–620.
Alain, C., Hargrave, R., & Woods, D. L. (1998). Processing of auditory stimuli during visual attention in patients with schizophrenia. *Biological Psychiatry, 44*, 1151–1159.
Alho, K. (1995). Cerebral generators of mismatch negativity (MMN) and its magnetic counterpart (MMNm) elicited by sound changes. *Ear and Hearing, 16*, 38–51.
Aloia, M. S., Gourovitch, M. L., Missar, D., Pickar, D., Weinberger, D. R., & Goldberg, T. E. (1998). Cognitive substrates of thought disorder, II: Specifying a candidate cognitive mechanism. *American Journal of Psychiatry, 155*, 1677–1684.
Andrews, S., Shelley, A. M., Ward, P. B., Fox, A., Catts, S. V., & McConaghy, N. (1993). Event-related potential indices of semantic processing in schizophrenia. *Biological Psychiatry, 34*, 443–458.
Baldeweg, T., Klugman, A., Gruzelier, J. H., & Hirsch, S. R. (2002). Impairment in frontal but not temporal components of mismatch negativity in schizophrenia. *International Journal of Psychophysiology, 43*, 111–122.
Barch, D. M., Cohen, J. D., Serven-Schreiber, D., Steingard, S., Steinhauer, S. R., & van Kammen, D. P. (1996). Semantic priming in schizophrenia: An examination of spreading activation using word pronunciation and multiple SOAs. *Journal of Abnormal Psychology, 105*, 592–601.
Blackwood, D. H. R., Fordyce, A., Walker, M. T., St. Clair, D. M., Porteous, D. J., & Muir, W. J. (2001). Schizophrenia and affective disorders—cosegregation with a translocation at chromosome 1q42 that directly disrupts brain-expressed genes: Clinical and P300 findings in a family. *American Journal of Human Genetics, 69*, 428–433.
Bleuler, T. (1950). *Dementia praecox of the group of schizophrenias* (H. Zinkin, Trans.). New York: International Universities Press.
Blum, N. A., & Freides, D. (1995). Investigating thought disorder in schizophrenia with the lexical decision task. *Schizophrenia Research, 16*, 217–224.
Boutros, N. N., Gooding, D., Sundaresan, K., Burroughs, S., & Johanson, C. E. (2006). Cocaine-dependence and cocaine-induced paranoia and mid-latency auditory evoked responses and sensory gating. *Psychiatry Research, 145*(2–3), 147–154.
Boutros, N. N., Korzyukov, O., Jansen, B., Feingold, A., & Bell, M. (2004). Sensory gating deficits during the mid-latency phase of information processing in medicated schizophrenia patients. *Psychiatry Research, 126*, 203–215.
Boutros, N. N., Torello, M. W., Barker, B. A., Tueting, P. A., Wu, S.-C., & Nasrallah, H. A. (1995). The P50 evoked potential component and mismatch detection in normal volunteers: Implications for the study of sensory gating. *Psychiatry Research, 57*, 83–88.
Bramon, E., Croft, R. J., McDonald, C., Virdi, G. K., Gruzelier, J. G., Baldeweg, T., et al. (2004). Mismatch negativity in schizophrenia: A family study. *Schizophrenia Research, 67*, 1–10.
Brecher, M., Porjesz, B., & Begleiter, H. (1987). Late positive component amplitude in schizophrenics and alcoholics in two different paradigms. *Biological Psychiatry, 22*, 848–856.
Brockhaus-Dumke, A., Schultze-Lutter, F., Mueller, R., Tendolkar, I., Bechdolf, A., Pukrop, R., et al. (2008). Sensory gating in schizophrenia: P50 and N100 gating in antipsychotic-free subjects at risk, first-episode, and chronic patients. *Biological Psychiatry, 64*, 376–384.
Brockhaus-Dumke, A., Tendokar, I., Pukrop, R., Schultze-Lutter, F., Klosterkötter, J., & Ruhrmann, S. (2005). Impaired mismatch negativity in prodromal subjects and patients with schizophrenia. *Schizophrenia Research, 73*, 297–310.
Brown, K. J., Gonsalvez, C. J., Harris, A. W., Williams, L. M., & Gordon, E. (2002). Target and non-target ERP disturbances in first episode vs. chronic schizophrenia. *Clinical Neurophysiology, 113*, 1754–1763.
Bruder, G., Kayser, J., Tenke, C., Rabinowicz, E., Friedman, M., Amador, X., et al. (1998). The time course of visuospatial processing deficits in schizophrenia: An event-related brain potential study. *Journal of Abnormal Psychology, 107*, 399–411.
Bunney, W. E., Hetrick, W. P., Bunney, B. G., Patterson, J. V., Jin, Y., Potkin, S. G., et al. (1999). Structured interview for assessing perceptual anomalies (SIAPA). *Schizophrenia Bulletin, 25*, 577–592.
Butler, P. D., & Javitt, D. C. (2005). Early-stage visual processing deficits in schizophrenia. *Current Opinion in Psychiatry, 18*, 151–157.
Catts, S. V., Shelley, A. M., Ward, P. B., Liebert, B., McConaghy, N., Andrews, S., et al. (1995). Brain potential evidence for an auditory sensory memory deficit in schizophrenia. *American Journal of Psychiatry, 152*, 213–219.
Chapin, K., Vann, L. E., Lycaki, N., Josef, N., & Meyendorff, X. (1989). Investigation of the associative network in schizophrenia using the semantic priming paradigm. *Schizophrenia Research, 2*, 355–360.
Chapman, L. J., & Chapman, J. P. (1973). *Disordered thought in schizophrenia*. Englewood Cliffs, NJ: Prentice-Hall.
Clementz, B., Barber, S., & Dzau, J. (2002). Knowledge of stimulus repetition affects the magnitude and spatial distribution of low frequency event-related brain potentials. *Audiology and Neuro-Otology, 7*(5), 303–314.
Clementz, B. A. (1998). Psychophysiological measures of (dis)inhibition as liability indicators for schizophrenia. *Psychophysiology, 35*, 648–668.
Clementz, B. A., Geyer, M. A., & Braff, D. L. (1998a). Multiple site evaluation of P50 suppression among schizophrenia and normal comparison subjects. *Schizophrenia Research, 30*(1), 71–80.
Clementz, B. A., Geyer, M. A., & Braff, D. L. (1998b). Poor P50 suppression among schizophrenia patients and their first-degree biological relatives. *American Journal of Psychiatry, 155*(12), 1691–1694.
Cohen, J. (1992). A power primer. *Psychological Bulletin, 112*, 155–159.
Cohen, J. D., Barch, D. M., Carter, C., & Servan-Schreiber, D. (1999). Context-processing deficits in schizophrenia: Converging evidence from three theoretically motivated cognitive tasks. *Journal of Abnormal Psychology, 108*, 120–133.
Condray, R., Siegle, G. J., Cohen, J. D., van Kammen D. P., & Steinhauer, S. R. (2003). Automatic activation of the semantic network in schizophrenia: Evidence from event-related brain potentials. *Biological Psychiatry, 54*, 1134–1148.
Condray, R., Steinhauer, S. R., Cohen, J. D., van Kammen, D. P., & Kasparek, A. (1999). Modulation of language processing in schizophrenia: Effects of context and haloperidol on the event-related potential. *Biological Psychiatry, 45*, 1336–1355.

Csépe, V. (1995). On the origin and development of the mismatch negativity. *Ear and Hearing, 16*, 91–104.
Csépe, V., Osman-Sági, J., Molnár, M., & Gósy, M. (2001). Impaired speech perception in aphasic patients: Event-related potential and neuropsychological assessment. *Neuropsychologia, 39*, 1194–1208.
Cutting, J., & Dunne, F. (1986). The nature of abnormal perceptual experiences at the onset of schizophrenia. *Psychopathology, 19*, 347–352.
Davies, N., Russell, A., Jones, P., & Murray, R. M. (1998): Which characteristics of schizophrenia predate psychosis? *Journal of Psychiatric Research, 32*, 121–131.
de Wilde, O., Bour, L., Dingemans, P., Koelman, J., & Linszen, D. (2007). A meta-analysis of P50 studies in patients with schizophrenia and relatives: Differences in methodology between research groups. *Schizophrenia Research*, 97, 137–51.
Ditman, T., & Kuperberg, G. R. (2007). The time course of building discourse coherence in schizophrenia: An ERP investigation. *Psychophysiology, 44*, 991–1001.
Donchin, E., & Coles, M. G. H. (1988). Is the P300 component a manifestation of context updating? *Behavioral and Brain Sciences, 11*, 357–373.
Doniger, G.M., Foxe, J.J., Murray, M.M., Higgins, B.A., Javitt, D.C. (2002). Impaired visual object recognition and dorsal/ventral stream interaction in schizophrenia. *Archives of General Psychiatry. 59*, 1011–1020.
Egan, M. F., Duncan, C. C., Suddath, R. L., Kirsch, D. G., Mirsky, A. F., & Wyatt, R. J. (1994). Event-related potential abnormalities correlate with structural brain alternations and clinical features in patients with chronic schizophrenia. *Schizophrenia Research, 11*, 259–271.
Egan, M. F., Goldberg, T. E., Kolachana, B. S., Callicott, C. M., Mazzanti, R. E., Straub, B., et al. (2001). Effect of COMT Val108/158Met genotype on frontal lobe function and risk for schizophrenia. *Proceedings of the National Academy of Sciences USA, 98*, 6917–6922.
Force, R. B., Venables, N. C., & Sponheim, S. R (2008). An auditory processing abnormality specific to liability for schizophrenia. *Schizophrenia Research, 103*(1–3), 298–310.
Ford, J. M. (1999). Schizophrenia: The broken P300 and beyond. *Psychophysiology, 36*, 667–682.
Ford, J. M., Mathalon, D. H., Kalba, S., Marsh, L., & Pfefferbaum, A. (2001). N1 and P300 abnormalities in patients with schizophrenia, epilepsy, and epilepsy with schizophrenia-like features. *Biological Psychiatry, 49*, 848–860.
Ford, J. M., White, P. M., Csernansy, J. G., Faustman, W. O., Roth, W. T., & Pfefferbaum, A. (1994a). ERPs in schizophrenia: Effects of antipsychotic medication. *Biological Psychiatry, 36*, 153–170.
Ford, J. M., White, P. M., Lim, K. O., & Pfefferbaum, A. (1994b). Schizophrenics have fewer and smaller P300s: A single-trial analysis. *Biological Psychiatry, 35*, 96–103.
Freedman, R., Adler, L. E., Gerhardt, G. A., Waldo, M., Baker, N., Rose, G. M., et al. (1987). Neurobiological studies of sensory gating in schizophrenia. *Schizophrenia Bulletin, 13*(4), 669–678.
Freedman, R., Adler, L. E., Myles-Worsley, M., Nagamoto, H. T., Miller, C., Kisley, M., et al. (1996). Inhibitory gating of an evoked response to repeated auditory stimuli in schizophrenic and normal subjects: Human recordings, computer simulation, and an animal model. *Archives of General Psychiatry, 53*(12), 1114–1121.
Frodl-Bauch, T., Kathmann, N., Möller, H. J., & Hegerl, U. (1997). Dipole localization and test-retest reliability of frequency and duration mismatch negativity generator processes. *Brain Topography, 10*, 3–8.
Fruhstorfer, H., Soveri, P., & Jarvilehto, T. (1970). Short-term habituation of the auditory evoked response in man. *Electroencephalography and Clinical Neurophysiology, 28*(2), 153–161.
Gallinat, J., Bajbouj, M., Sander, T., Schlattmann, P., Xu, K., Ferro, E. F., et al. (2007). Association of the G1947A COMT (Val108/158Met) gene polymorphism with prefrontal P300 during information processing. *Biological Psychiatry, 54*, 40–48.
Garcia-Rill, E., Moran, K., Garcia, J., Findley, W. M., Walton, K., Strotman, B., et al. (2008). Magnetic sources of the M50 response are localized to frontal cortex. *Clinical Neurophysiology, 119*(2), 388–398.
Giard, M. H., Perrin, F., Pernier, J., & Bouchet, P. (1990). Brain generators implicated in the processing of auditory stimulus deviance: A topographic event-related potential study. *Psychophysiology, 27*, 627–640.
Gottesman II, & Gould TD. (2003). The endophenotype concept in psychiatry: etymology and strategic intentions. *American Journal of Psychiatry, 160*, 636–45.
Gouzoulis-Mayfrank, E., Voss, T., Morth, D., Thelen, B., Spitzer, M., & Meincke, U. (2003). Semantic hyperpriming in thought-disordered patients with schizophrenia: State or trait? A longitudinal investigation. *Schizophrenia Research, 65*, 65–73.
Grillon, C., Ameli, R., & Glazer, W. (1991). N400 and semantic categorization in schizophrenia. *Biological Psychiatry, 29*, 467–480.
Grunwald, T., Boutros, N. N., Pezer, N., von Oertzen, J., Fernandez, G., Schaller, C., et al. (2003). Neuronal substrates of sensory gating within the human brain. *Biological Psychiatry, 53*(6), 511–519.
Guillem, F., N'Kaoua, B., Rougier, A., & Claverie, B. (1995). Intracranial topography of event- related potentials (N400/P600) elicited during a continuous recognition memory task. *Psychophysiology, 32*, 382–392.
Guillem, F., Rougier, A., & Claverie, B. (1999). Short- and long-delay intracranial ERP repetition effects dissociate memory systems in the human brain. *Journal of Cognitive Neuroscience, 11*, 437–458.
Guterman, Y., Josiassen, R. C., & Bashore, T. R. J. (1992). Attentional influence on the P50 component of the auditory event-related brain potential. *International Journal of Psychophysiology, 12*, 197–209.
Hafner, H., & van der Heiden, W. (1999). The course of schizophrenia in the light of modern follow-up studies: The ABC and WHO studies. *European Archives of Psychiatry in Clinical Neuroscience, 249*, 14–26.
Halgren, E. (2004). How can intracranial recordings assist MEG source localization? *Neurology & Clinical Neurophysiology, 86*, 1–18.
Halgren, E., Baudena, P., Heit, G., Clarke, J. M., Marinkovic, K., & Clarke, M. (1994a). Spatio- temporal stages in face and word processing. 1. Depth-recorded potentials in the human occipital, temporal and parietal lobes. *Journal of Physiology (Paris), 88*, 1–50.
Halgren, E., Baudena, P., Heit, G., Clarke, J. M., Marinkovic, K., Chauvel, P., et al. (1994b). Spatio-temporal stages in face and word processing. 2. Depth-recorded potentials in the human

frontal and Rolandic cortices. *Journal of Physiology (Paris), 88*, 51–80.

Hall, M. H., Rijsdijk, F., Kalidini, S., Schulze, K., Kravariti, E., Kane, F., et al. (2007a). Genetic overlap between bipolar illness and event-related potentials. *Psychological Medicine, 37*, 667–678.

Hall, M. H., Rijsdijk, F., Picchioni, M., Schulze, K., Ettinger, U., Toulopoulou, T., et al. (2007b). Substantial shared genetic influences on schizophrenia and event-related potentials. *American Journal of Psychiatry, 164*, 804–812.

Hall, M. H., Schulze, K., Bramon, E., Murray, R. M., Sham, P., & Rijsdijk, F. (2006a). Genetic overlap between P300, P50, and duration mismatch negativity. *American Journal of Medical Genetics Part B (Neuropsychiatric Genetics), 141B*, 336–343.

Hall, M. H., Schulze, K., Rijsdijk, F., Picchioni, M., Ettinger, U., Bramon, E., et al., (2006b). Heritability and reliability of P300, P50 and duration mismatch negativity. *Behavioral Genetics, 36*, 845–857.

Hari, R., Rif, J., Tiihonen, J., & Sams, M. (1992). Neuromagnetic mismatch fields to single and paired tones. *Electroencephalography and Clinical Neurophysiology, 82*, 152–154.

Henn, F. A., & Braus, D. F. (1999). Structural neuroimaging in schizophrenia: An integrative view of neuromorphology. *European Archives of Psychiatry and Clinical Neuroscience, 249*, S48–S56.

Hill, H., Ott, F., & Weisbrod, M. (2005). SOA-dependent N400 and P300 semantic priming effects using pseudoword primes and a delayed lexical decision. *International Journal of Psychophysiology, 56*, 209–221.

Hill, H., Strube, M., Roesch-Ely, D., & Weisbrod, M. (2002). Automatic vs. controlled processes in semantic priming—differentiation by event-related potentials. *International Journal of Psychophysiology, 44*, 197–218.

Hirayasu, Y., Asato, N., Ohta, H., Hokama, H., Arakaki, H., & Ogura, C. (1998a). Abnormalities of auditory event-related potentials prior to treatment. *Biological Psychiatry, 43*, 244–253.

Hirayasu, Y., Potts, G. F., O'Donnell, B. F., Kwon, J. S., Arakaki, H., Akdag, J. S., et al, (1998b). Auditory mismatch negativity in schizophrenia: Topographic evaluation with a high-density recording montage. *American Journal of Psychiatry, 155*, 1281–1284.

Hokama, H., Hiramatsu, K., Wang, J., O'Donnell, B. F., & Ogura, C. (2003). N400 abnormalities in unmedicated patients with schizophrenia during a lexical decision task. *International Journal of Psychophysiology, 48*, 1–10.

Holcomb, P. J., & Neville, H. J. (1990). Auditory and visual semantic priming in lexical decision: A comparison using evoked potentials. *Language and Cognitive Processes, 5*, 281–312.

Holcomb, P. J., Reder, L., Misra, M., & Grainger, J. (2005). The effects of prime visibility on ERP measures of masked priming. *Cognitive Brain Research, 24*, 155–172.

Houy, E., Raux, G., Thibaut, F., Belmont, A., Demily, C., Allio, G., et al. (2004). The promoter-194 C polymorphism of the nicotinic alpha 7 receptor gene has a protective effect against the P50 sensory gating deficit. *Molecular Psychiatry, 9*(3), 320–322.

Javitt, D. C., Doneshka, P., Grochowski, S., & Ritter, W. (1995). Impaired mismatch negativity generation reflects widespread dysfunction of working memory in schizophrenia. *Archives of General Psychiatry, 52*, 550–558.

Javitt, D. C., Doneshka, P., Zylberman, I., Ritter, W., & Vaughan, H. G., Jr. (1993). Impairment of early cortical processing in schizophrenia: An event-related potential confirmation study. *Biological Psychiatry, 33*, 513–519.

Javitt, D. C., Grochowski, S., Shelley, A. M., & Ritter, W. (1998). Impaired mismatch negativity (MMN) generation in schizophrenia as a function of stimulus deviance, probability, and interstimulus/interdeviant interval. *Electroencephalography and Clinical Neurophysiology, 108*, 143–153.

Javitt, D. C., Shelley, A. M., & Ritter, W. (2000a). Associated deficits in mismatch negativity generation and tone matching in schizophrenia. *Clinical Neurophysiology, 111*, 1733–1737.

Javitt, D. C., Shelley, A. M., Silipo, G., & Lieberman, J. A. (2000b). Deficits in auditory and visual context-dependent processing in schizophrenia: Defining the pattern. *Archives of General Psychiatry, 57*, 1131–1137.

Javitt, D. C., Steinschneider, M., Schroeder, C. E., & Arezzo, J. C. (1996). Role of cortical *N*-methyl-D-aspartate receptors in auditory sensory memory and mismatch negativity generation: Implications for schizophrenia. *Proceedings of the National Academy of Sciences USA, 93*, 11962–11967.

Javitt, D. C., Steinschneider, M., Schroeder, C. E., Vaughan, H. G., Jr., & Arezzo, J. C. (1994). Detection of stimulus deviance within primate primary auditory cortex: Intracortical mechanisms of mismatch negativity (MMN) generation. *Brain Research, 667*, 192–200.

Jensen, K. S., Oranje, B., Wienberg, M., & Glenthoj, B. Y. (2008). The effects of increased serotonergic activity on human sensory gating and its neural generators. *Psychopharmacology (Berlin), 196*(4), 631–641.

Jeon, Y. W., & Polich, J. (2001). P300 asymmetry in schizophrenia: A meta-analysis. *Psychiatry Research, 104*, 61–74.

Jeon, Y. W., & Polich, J. (2003). Meta-analysis of P300 and schizophrenia: Patients, paradigms and practical implications. *Psychophysiology, 40*, 684–701.

Jerger, K., Biggins, C., & Fein, G. (1992). P50 suppression is not affected by attentional manipulations. *Biological Psychiatry, 31*, 365–377.

Jessen, F., Fries, T., Kucharski, C., Nishimura, T., Hoenig, K., Maier, W., et al. (2001). Amplitude reduction of the mismatch negativity in first-degree relatives of patients with schizophrenia. *Neuroscience Letters, 309*, 185–188.

Jin, Y., Potkin, S. G., Sandman, C. A., & Bunney, W. E. (1997). Electroencephalographic photic driving in patients with schizophrenia and depression. *Biological Psychiatry, 41*, 496–499.

Johannesen, J. K., Bodkins, M., O'Donnell, B. F., Shekhar, A., & Hetrick, W. P. (2008). Perceptual anomalies in schizophrenia co-occur with selective impairments in the gamma frequency component of midlatency auditory ERPs. *Journal of Abnormal Psychology, 117*(1), 106–118.

Johannesen, J. K., Kieffaber, P. D., O'Donnell, B. F., Shekhar, A., Evans, J. D., & Hetrick, W. P. (2005). Contributions of subtype and spectral frequency analyses to the study of P50 ERP amplitude and suppression in schizophrenia. *Schizophrenia Research, 78*, 269–284.

Johnson, R., Jr. (1986). A triarchic model of of P300 amplitude. *Psychophysiology, 23*, 367–384.

Kane, N. M., Curry, S. H., Rowlands, C. A., Manara, A. R., Lewis, T., Moss, T., et al. (1996). Event-related potentials—neuropsychological tools for predicting emergence and early outcome from traumatic coma. *Intensive Care Medicine, 22*, 39–46.

Kasai, K., Nakagome, K., Itoh, I., Koshida, I., Hata, A., Iwanami, A., et al. (1999a). Multiple generators in the auditory automatic discrimination process in humans. *Neuroreport, 10*, 2267–2271.

Kasai, K., Nakagome, K., Itoh, I., Koshida, I., Hata, A., Iwanami, A., et al. (2002). Impaired cortical network for preattentive detection of change in speech sounds in schizophrenia: A high-resolution event-related potential study. *American Journal of Psychiatry, 159*, 546–553.

Kasai, K., Okazawa, K., Nakagome, K., Hiramatsu, K., Hata, A., Fukuda, M., et al. (1999b). Mismatch negativity and N2b attenuation as an indicator for dysfunction of the preattentive and controlled processing for deviance detection in schizophrenia: A topographic event-related potential study. *Schizophrenia Research, 35*, 141–156.

Kasai, K., Shenton, M. E., Salisbury, D. F., Hirayasu, Y., Onitsuka, T., Spencer, M., et al., (2003). Progressive decrease of left Heschl's gyrus and planum temporale gray matter volume in first-episode schizophrenia: A longitudinal magnetic resonance imaging study. *Archives of General Psychiatry, 60*, 766–765.

Kawakubo, Y., Kasai, K., Kudo, N., Rogers, M. A., Nakagome, K., Itoh, K., et al. (2006). Phonetic mismatch negativity predicts verbal memory deficits in schizophrenia. *Neuroreport, 17*, 1043–1046.

Kiang, M., Kutas, M., Light, G. A., & Braff, D. L. (2008). An event-related brain potential study of direct and indirect semantic priming in schizophrenia. *American Journal of Psychiatry, 165*, 74–81.

Kimble, M., Lyons, M., O'Donnell, B., Nestor, P., Niznikiewicz, M., & Toomey, R. (2000). The effect of family status and schizotypy on electrophysiologic measures of attention and semantic processing. *Biological Psychiatry, 47*, 402–412.

Kisley, M. A., Noecker, T. L., & Guinther, P. M. (2004). Comparison of sensory gating to mismatch negativity and self-reported perceptual phenomena in healthy adults. *Psychophysiology, 41*(4), 604–612.

Korzyukov, O., Pflieger, M. E., Wagner, M., Bowyer, S. M., Rosburg, T., Sundaresan, K., et al. (2007). Generators of the intracranial P50 response in auditory sensory gating. *Neuroimage, 35*(2), 814–826.

Kostova, M., Passerieux, C., Laurent, J. P., & Hardy-Bayle, M. C. (2005). N400 anomalies in schizophrenia are correlated with the severity of formal thought disorder. *Schizophrenia Research, 78*, 285–291.

Kostova, M., Passerieux, C., Laurent, J. P., Saint-Georges, C., & Hardy-Bayle, M. C. (2003). Functional analysis of the deficit in semantic context processes in schizophrenic patients: An event-related potentials study. *Neurophysiologie Clinique/Clinical Neurophysiology, 33*, 11–22.

Koyama, S., Nageishi, Y., Shimokochi, M., Hokama, H., Myiazato, Y., Miyatani, M., & Ogura, C. (1991). The N400 component of event-related potentials in schizophrenic patients: A preliminary study. *Electroencephalography and Clinical Neurophysiology, 78*, 124–132.

Krause, M., Hoffmann, W. E., & Hajos, M. (2003). Auditory sensory gating in hippocampus and reticular thalamic neurons in anesthetized rats. *Biological Psychiatry, 53*(3), 244–253.

Kreher, D. A., Holcomb, P. J., Goff, D., & Kuperberg, G. R. (2008). Neural evidence for faster and further automatic spreading activation in schizophrenic thought disorder. *Schizophrenia Bulletin* 34, 473–82.

Kreitschmann-Andermahr, T., Rosburg, T., Meier, T., Volz, H.-P., Nowak, H., & Sauer, H. (1999). Impaired sensory processing in male patients with schizophrenia: A magnetoencephalographic study of auditory mismatch detection. *Schizophrenia Research, 35*, 121–129.

Kropotov, J. D., Näätänen, R., Sevostianov, A. V., Alho, K., Reinikainen, K., & Kropotova, O. V. (1995). Mismatch negativity to auditory stimulus change recorded directly from the human temporal cortex. *Psychophysiology, 32*, 418–422.

Krystal, J., D'Souza, D. C., Mathalon, D., Perry, E., Belger, A., & Hoffman, R. (2003). NMDA receptor antagonist effects, cortical glutamatergic function, and schizophrenia: Toward a paradigm shift in medication development. *Psychopharmacology, 169*, 215–233.

Kujala, T., & Näätänen, R. (2001). The mismatch negativity in evaluating central auditory dysfunction in dyslexia. *Neuroscience and Biobehavioral Review, 25*, 535–543.

Kujala, T., Tervaniemi, M., & Schröger, E. (2007). The mismatch negativity in cognitive and clinical neuroscience: Theoretical and methodological considerations. *Biological Psychology, 74*, 1–19.

Kutas, M., & Hillyard, S. A. (1980). Reading senseless sentences: Brain potentials reflect semantic incongruity. *Science, 207*(4427), 203–205.

Kutas, M., & Iragui, V. (1998). The N400 in a semantic categorization task across 6 decades. *Electroencephalography and Clinical Neurophysiology, 108*, 456–471.

Kwapil, T. R., Hegley, D. C., Chapman, L. J., & Chapman, J. P. (1990). Facilitation of word recognition by semantic priming in schizophrenia. *Journal of Abnormal Psychology, 99*, 215–221.

Laurent, A., Garcia-Larra, L., d'Amato, T., Bosson, J. L., Saoud, M., Marie-Cardine, M., et al. (1999). Auditory event-related potentials and clinical scores in unmedicated schizophrenic patients. *Psychiatry Research, 86*, 229–238.

Leonard, S., Gault, J., Adams, C., Breese, C. R., Rollins, Y., Adler, L. E., et al. (1998). Nicotinic receptors, smoking and schizophrenia. *Restorative neurology and neuroscience, 12*, 195–201.

Lewis, D., & Lieberman, J. (2000). Catching up on schizophrenia: Natural history and neurobiology. *Neuron, 28*, 325–334.

Light, G., & Braff, D. (2005a). Mismatch negativity deficits are associated with poor functioning in schizophrenia patients. *Archives of General Psychiatry, 62*, 127–136.

Light, G., & Braff, D. (2005b). Stability of mismatch negativity deficits and their relationship to functional impairments in chronic schizophrenia. *American Journal of Psychiatry, 162*, 1741–1743.

Light, G., Geyer, M. A., Clementz, B., Cadenhead, K., & Braff, D. L. (2000). Normal P50 suppression in schizophrenia patients treated with atypical antipsychotic medications. *American Journal of Psychiatry, 157*(5), 767–771.

Louchart-de la Chapelle, S., Levillain, D., Menard, J.-F., Van der Elst, A., Allio, G., Haouzir, S., et al. (2005). P50 inhibitory gating deficit is correlated with the negative symptomatology of schizophrenia. *Psychiatry Research, 136*, 27–34.

Maher, B. A., Manschreck, T. C., Redmond, D., & Beaudette, S. (1996). Length of illness and the gradient from positive to negative semantic priming in schizophrenic patients. *Schizophrenia Research, 22*, 127–132.

Makeig, S., Debener, S., Onton, J., & Delorme, A. (2004). Mining event-related brain dynamics. *Trends in Cognitive Sciences, 8*, 204–210.

Mangun, G. R., & Hillyard, S. A. (1988). Spatial gradients of visual attention: Behavioral and electrophysiological evidence. *Electroencephalography and Clinical Neurophysiology, 70*, 417–428.

Marco, J., Fuentemilla, L., & Grau, C. (2005). Auditory sensory gating deficit in abstinent chronic alcoholics. *Neuroscience Letters, 375*(3), 174–177.

Martin, L. F., Kem, W. R., & Freedman, R. (2004). Alpha-7 nicotinic receptor agonists: Potential new candidates for the treatment of schizophrenia. *Psychopharmacology (Berlin), 174*(1), 54–64.

Martin, L. F., Leonard, S., Hall, M. H., Tregellas, J. R., Freedman, R., & Olincy, A. (2007). Sensory gating and alpha-7 nicotinic receptor gene allelic variants in schizoaffective disorder, bipolar type. *American Journal of Medical Genetics B, Neuropsychiatric Genetics, 144B*(5), 611–614.

Mathalon, D. H., Ford, J. M., Rosenbloom, M., & Pfefferbaum, A. (2000). P300 reduction and prolongation with illness duration in schizophrenia. *Biological Psychiatry, 47*, 413–427.

McCarley, R. W., Hsiao, J. K., Freedman, R., et al. (1996). Neuroimaging and the cognitive neuroscience of schizophrenia. *Schizophrenia Bulletin, 22*, 703–725.

McCarley, R. W., Salisbury, D. F., Hirayasu, Y., Yurgelun-Todd, D. A., Tohen, M., Zarate, C., et al. (2002). Association between smaller left posterior superior temporal gyrus volume on magnetic resonance imaging and smaller left temporal P300 amplitude in first-episode schizophrenia. *Archives of General Psychiatry, 59*, 321–331.

McCarthy, G., Nobre, A. C., Bentin, S., & Spencer, D. D. Language-related field potentials in the anterior-medial temporal lobe: I. Intracranial distribution and neural generators. *Journal of Neuroscience, 15*, 1080–1089.

Michie, P. T. (2001). What has MMN revealed about the auditory system in schizophrenia? *International Journal of Psychophysiology, 42*, 177–194.

Michie, P. T., Innes-Brown, H., Todd, J., & Jablensky, A. V. (2002). Duration mismatch negativity in biological relatives of patients with schizophrenia spectrum disorders. *Biological Psychiatry, 52*, 749–758.

Mitchell, P. F., Andrews, S., Fox, A., Catts, P. B., & McConghy, N. (1991). Active and passive attention in schizophrenia: An ERP study of information processing in a linguistic task. *Journal of Abnormal Psychology, 32*, 101–124.

Molholm, S., Martinez, A., Ritter, W., Javitt, D. C., & Foxe, J. J. (2005). The neural circuitry of pre-attentive auditory change-detection: An fMRI study of pitch and duration mismatch negativity generators. *Cerebral Cortex, 15*, 545–551.

Morgan, C. A., III, & Grillon, C. (1999). Abnormal mismatch negativity in women with sexual assault-related posttraumatic stress disorder. *Biological Psychiatry, 45*, 827–832.

Moritz, S., Woodward, T. S., Kuppers, D., Lausen, A., & Schickel, M. (2003). Increased automatic spreading of activation in thought-disordered schizophrenic patients. *Schizophrenia Research, 59*, 181–186.

Moxon, K. A., Gerhardt, G. A., & Adler, L. E. (2003). Dopaminergic modulation of the P50 auditory-evoked potential in a computer model of the CA3 region of the hippocampus: Its relationship to sensory gating in schizophrenia. *Biological Cybernetics, 88*(4), 265–275.

Näätänen, R. (1990). The role of attention in auditory information processing as revealed by event-related potentials and other measures of cognitive function. *Behavioral Brain Science, 13*, 201–288.

Näätänen, R. (2001). The perception of speech sounds by the human brain as reflected by the mismatch negativity (MMN) and its magnetic counterpart (MMNm). *Psychophysiology, 38*, 1–21.

Näätänen, R. (2003). Mismatch negativity: Clinical research and possible applications. *International Journal of Psychophysiology, 48*, 179–188.

Näätänen, R., & Alho, K. (1995). Generators of electrical and magnetic mismatch response in humans. *Brain Topography, 7*, 315–320.

Nagamoto, H. T., Adler, L. E., Waldo, M. C., & Freedman, R. (1989). Sensory gating in schizophrenics and normal controls: Effects of changing stimulation interval. *Biological Psychiatry, 25*(5), 549–561.

Nestor, P. G., Han, S. D., Niznikiewicz, M., Salisbury, D., Spencer, K., Shenton, M. E., et al.(2001). Semantic disturbance in schizophrenia and its relationship to the cognitive neuroscience of attention. *Biological Psychology, 57*, 23–46.

Nestor, P. G., Kimble, M. W., O'Donnell, B. F., Smith, R. S., Niznikiewicz, M., Shenton, M. E., et al. (1997). Aberrant semantic activation in schizophrenia: A neurophysiological study. *American Journal of Psychiatry, 154*, 640–646.

Nestor, P. G., & O'Donnell, B. F. (1998). The mind adrift: Attention dysregulation in schizophrenia. In R. Parasuraman (Ed.), *The attentive brain* (pp. 527–546). Boston: MIT Press.

Niznikiewicz, M. A., O'Donnell, B. F., Nestor, P. G., Smith, L., Law, S., Karapelou, M., et al. (1997). ERP assessment of visual and auditory language processing in schizophrenia. *Journal of Abnormal Psychology, 106*, 85–94.

Niznikiewicz, M. A., Shenton, M. E., Voglmaier, M., Nestor, P. G., Dickey, C. C., Frumin, M., et al. (2002). Semantic dysfunction in women with schizotypal personality disorder. *American Journal of Psychiatry, 159*, 1767–1774.

Nobre, A. C., & McCarthy, G. (1995). Language-related field potentials in the anterior-medial temporal lobe: II. Effects of word type and semantic priming. *Journal of Neuroscience, 15*, 1090–1098.

O'Donnell, B. F. (2007). Cognitive impairment in schizophrenia: A life span perspective. *American Journal of Alzheimer's Disease and Other Dementias, 22*, 398–405.

O'Donnell, B. F., Bismark, A., Hetrick, W. P., Bodkins, M., Vohs, J. L., & Shekhar, A. (2006). Early stage vision in schizophrenia and SPD personality disorder. *Schizophrenia Research, 86*, 89–98.

O'Donnell, B. F., Faux, S. F., McCarley, R. W., Kimble, M. O., Salisbury, D. F., Nestor, P. G., et al. (1995). Increased rate of P300 latency prolongation with age in schizophrenia. *Archives of General Psychiatry, 52*, 544–549.

O'Donnell, B. F., Hokama, H., McCarley, R. W., Smith, R. S., Mondrow, E., Salisbury, D., et al. (1994). Auditory ERPs to non-target stimuli in schizophrenia: Relationship to probability, task-demands, and target ERPs. *International Journal of Psychophysiology, 17*, 219–231.

O'Donnell, B. F., McCarley, R. W., Potts, G. F., Salisbury, D. F., Nestor, P. G., Hirayasu, Y., et al. (1999). Identification of neural circuits underlying P300 abnormalities in schizophrenia. *Psychophysiology, 36*, 388–398.

O'Donnell, B. F., Swearer, J. M., Smith, L. T., Hokama, H., & McCarley, R. W. (1997). A topographic study of ERPs

elicited by visual feature discrimination. *Brain Topography, 10*, 133–143.
O'Donnell, B. F., Vohs, J. L., Hetrick, W. P., Carroll, C. A., & Shekhar, A. (2004). Auditory event-related potential abnormalities in bipolar disorder and schizophrenia. *International Journal of Psychophysiology, 53*, 45–55.
Ohta, H., O'Donnell, B. F., McCarley, R., Hokama, H., Karapelou, M. E., & Law, S. E. (1996). Visual event-related potential abnormalities during spatial localization tasks in schizophrenia. In C. Ogura, Y. Koga, & M. Shimokochi (Eds.), *Proceedings of the XIth International Conference on Event-Related Potentials of the Brain* (pp. 946–950). New York: Elsevier.
Olejniczak, P. (2006). Neurophysiologic basis of EEG. *Journal of Clinical Neurophysiology, 23*, 186–189.
Olichney, J. M., Iragui, V. J., Kutas, M., Nowacki, R., Morris, S., & Jeste, D. V. (1998). Relationship between auditory P300 amplitude and age of onset of schizophrenia in older patients. *Psychiatry Research, 79*, 241–254.
Olichney, J. M., Taylor, J. R., Gatherwright, J., Salmon, D. P., Bressler, A. J., Kutas, M., & Iragui-Madoz, V. J. (2008). Patients with MCI and N400 or P600 abnormalities are at very high risk for conversion to dementia. *Neurology, 70*, 1763–1770.
Olincy, A., & Martin, L. (2005). Diminished suppression of the P50 auditory evoked potential in bipolar disorder subjects with a history of psychosis. *American Journal of Psychiatry, 162*(1), 43–49.
Oranje, B., van Berckel, B. N., Kemner, C., van Ree, J. M., Kahn, R. S., & Verbaten, M. N. (2000). The effects of a subanaesthetic dose of ketamine on human selective attention. *Neuropsychopharmacologia, 22*, 293–302.
Orekhova, E. V., Stroganova, T. A., Prokofyev, A. O., Nygren, G., Gillberg, C., & Elam, M. (2008). Sensory gating in young children with autism: Relation to age, IQ, and EEG gamma oscillations. *Neuroscience Letters, 434*(2), 218–223.
Park, H. J., Kwon, J. S., Youn, T., Pae, J. S., Kim, J. J., Kim, M. S., et al. (2002). Statistical parametric mapping of LORETA using high density EEG and individual MRI: Application to mismatch negativities in schizophrenia. *Human Brain Mapping, 17*, 168–178.
Passerieaux, C., Segui, J., Besche, C., Chevalier, J. F., Widlöcher, D., & Hardy-Baylé, M. C. (1997). Heterogeneity in cognitive functioning of schizophrenic patients evaluated by a lexical decision task. *Psychological Medicine, 27*, 1295–1302.
Patterson, J. V., Hetrick, W. P., Boutros, N. N., Jin, Y., Sandman, C., Stern, H., et al. (2008). P50 sensory gating ratios in schizophrenics and controls: A review and data analysis. *Psychiatry Research, 158*(2), 226–247.
Peled, A., Netzer, I., & Modai, I. (2005). Rating of textual associations in organized and nonorganized sentences for the assessment of semantic networks in schizophrenia. *Comprehensive Psychiatry, 46*, 176–180.
Pfefferbaum, A., Ford, J. M., White, P. M., & Roth, W. T. (1989). P3 in schizophrenia is affected by stimulus modality, response requirements, medication status, and negative symptoms. *Archives of General Psychiatry, 46*, 1035–1044.
Pfefferbaum, A., Wenegrat, B. G., Ford, J. M., Roth, W. T., & Kopell, B. S. (1984). Clinical application of the P3 component of event-related potentials. II. Dementia, depression, and schizophrenia. *Electroencephalography and Clinical Neurophysiology, 59*, 104–124.
Phillipson, O. T., & Harris, J. P. (1985). Perceptual changes in schizophrenia: A questionnaire survey. *Psychological Medicine, 15*, 859–866.
Picton, T. W. (1992). The P300 wave of the human event-related potential. *Journal of Clinical Neurophysiology, 9*, 456–479.
Picton, T. W., Alain, C., Woods, D. L., John, M. S., Scherg, M., Valdes-Sosa, P., Bosch-Bayard, J., & Trujillo, N.J. (1999). Intracerebral sources of human auditory-evoked potentials. *Audiology and Neuro-Otology, 4*, 64–79.
Polich, J., & Kok, A. (1995). Cognitive and biological determinants of P300: An integrative review. *Biological Psychology, 41*, 103–146.
Potts, G. F., O'Donnell, B. F., Hiyasu, Y., & McCarley, R. W. (2002). Disruption of neural systems of visual attention in schizophrenia. *Archives of General Psychiatry, 59*, 418–424.
Quelen, F., Grainger, J., & Raymondet, P. (2005). An investigation of semantic priming in schizophrenia using a new priming paradigm. *Schizophrenia Research, 80*, 173–183.
Rabinowicz, E. F., Silipo, G., Goldman, R., & Javitt, D. C. (2000). Auditory sensory dysfunction in schizophrenia. *Archives of General Psychiatry, 57*, 1149–1155.
Rapoport, J. L., Addington, A. M., & Frangou, S. (2005). Psych MRC: The neurodevelopmental model of schizophrenia: Update 2005. *Molecular Psychiatry, 10*, 434–449.
Raux, G., Bonnet-Brilhault, F., Louchart, S., Houy, E., Gantier, R., Levillain, D., et al. (2002). The–2 bp deletion in exon 6 of the "alpha 7-like" nicotinic receptor subunit gene is a risk factor for the P50 sensory gating deficit. *Molecular Psychiatry, 7*(9), 1006–1011.
Ringel, T. M., Heidrich, A., Jacob, C. P., Pfuhlmann, B., Stoeber, G., & Fallgatter, A. J. (2004). Sensory gating deficit in a subtype of chronic schizophrenic patients. *Psychiatry Research, 125*(3), 237–245.
Rinne, T., Alho, K., Ilmoniemi, R. J., Virtanen, J., & Näätänen, R. (2000). Separate time behaviors of the temporal and frontal mismatch negativity sources. *Neuroimage, 12*, 14–19.
Rosburg, T., Trautner, P., Dietl, T., Korzyukov, O. A., Boutros, N. N., Schaller, C., et al. (2005). Subdural recordings of the mismatch negativity (MMN) in patients with focal epilepsy. *Brain, 128*, 819–828.
Ross, C. A., Margolis, R. L., Reading, S. A. J., Pletnikov, M., & Coyle, J. T. (2006). Neurobiology of schizophrenia. *Neuron, 52*, 139–153.
Roth, W. T., Pfefferbaum, A., Horvath, T. B., Berger, P. A., & Kopell, B. (1980). P3 reduction in auditory evoked potentials of schizophrenics. *Electroencephalography and Clinical Neurophysiology, 49*, 497–505.
Ruchsow, M., Trippel, N., Groen, G., Spitzer, M., & Kiefer, M. (2003). Semantic and syntactic processes during sentence comprehension in patients with schizophrenia: Evidence from event-related potentials. *Schizophrenia Research, 64*, 147–156.
Saha, S., Chant, D., Welham, J., & McGrath, J. (2005). A systematic review of the prevalence of schizophrenia. *PLoS Medicine*, 2, e141.
Salisbury, D. F., Kasai, K., Kuroki, N., Shenton, M. E., & McCarley, R. W. (2007). Progressive and interrelated functional and structural evidence of post-onset brain reduction in schizophrenia. *Archives of General Psychiatry, 64*, 521–529.
Salisbury, D. F., Shenton, M. E., Griggs, C. B., Bonner-Jackson, A., & McCarley, R. W. (2002a). Mismatch

negativity in chronic schizophrenia and first-episode schizophrenia. *Archives of General Psychiatry, 59*, 686–694.

Salisbury, D. F., Shenton, M. E., & McCarley, R. W. (1998a). P300 topography differs in schizophrenia and manic psychosis. *Biological Psychiatry, 45*, 98–106.

Salisbury, D. F., Shenton, M. E., Nestor, P. G., & McCarley, R. W. (2002b). Semantic bias, homograph comprehension, and event-related potentials in schizophrenia. *Clinical Neurophysiology, 113*, 383–395.

Salisbury, D. F., Shenton, M. E., Sherwood, A. R., Fischer, I. A., Yurgelun-Todd, D. A., Tohen, M., et al. (1998b). First-episode schizophrenic psychosis differs from first-episode affective psychosis and controls in P300 amplitude over left temporal lobe. *Archives of General Psychiatry, 55*, 173–180.

Schreiber, H., Stolz-Born, G., Kornhuber, H. H., & Born, H. (1992). Event-related potential correlates of impaired selective attention in children at high risk for schizophrenia. *Biological Psychiatry, 32*, 634–651.

Schulze, K. K., Hall, M. H., McDonald, C., Marshall, N., Walshe, M., Murray, R. M., et al. (2007). P50 auditory evoked potential suppression in bipolar disorder patients with psychotic features and their unaffected relatives. *Biological Psychiatry, 62*(2), 121–128.

Selemon, L. D., & Goldman-Rakic, P. S. (1999). The reduced neuropil hypothesis: A circuit based model of schizophrenia. *Biological Psychiatry, 45*, 17–25.

Shelley, A. M., Silipo, G., & Javitt, D. (1999). Diminished responsiveness of ERPs in schizophrenic subjects to changes in auditory stimulation parameters: Implications for theories of cortical dysfunction. *Schizophrenia Research, 37*, 65–79.

Shelley, A. M., Ward, P. B., Catts, S. V., Michie, P. T., Andrews, A., & McConaghy, N. (1991). Mismatch negativity: An index of preattentive processing deficit in schizophrenia. *Biological Psychiatry, 30*, 1059–1062.

Shenton, M. E., Dickey, C. C., Frumin, M., & McCarley, R. W. (2001). A review of MRI findings in schizophrenia. *Schizophrenia Research, 49*, 1–52.

Siegel, C., Waldo, M., Mizner, G., Adler, L. E., & Freedman, R. (1984). Deficits in sensory gating in schizophrenic patients and their relatives: Evidence obtained with auditory evoked responses. *Archives of General Psychiatry, 41*(6), 607–612.

Simosky, J. K., Stevens, K. E., Kem, W. R., & Freedman, R. (2001). Intragastric DMXB-A, an alpha7 nicotinic agonist, improves deficient sensory inhibition in DBA/2 mice. *Biological Psychiatry, 50*(7), 493–500.

Sitnikova, T., Salisbury, D. F., Kuperberg, G., & Holcomb, P. I. (2002). Electrophysiological insights into language processing in schizophrenia. *Psychophysiology, 39*, 851–860.

Soltani, M., & Knight, R. T. (2000). Neural origins of the P300. *Critical Reviews in Neurobiology, 14*, 199–224.

Spitzer, M. (1997). A cognitive neuroscience view of schizophrenic thought disorder. *Schizophrenia Bulletin, 23*, 29–50.

Sponheim, S. R., McGuire, K. A., & Stanwick, J. J. (2006). Neural anomalies during sustained attention in first-degree biological relatives of schizophrenia patients. *Biological Psychiatry, 60*, 242–252.

Stevens, K. E., Meltzer, J., & Rose, G. M. (1995). Nicotinic cholinergic normalization of amphetamine-induced loss of auditory gating in freely moving rats. *Psychopharmacology (Berlin), 119*(2), 163–170.

Sullivan, P. F. (2005). The genetics of schizophrenia. *PLoS Medicine, 2*, e212.

Swerdlow, N. R., Geyer, M. A., Shoemaker, J. M., Light, G. A., Braff, D. L., Stevens, K. E., et al. (2006). Convergence and divergence in the neurochemical regulation of prepulse inhibition of startle and N40 suppression in rats. *Neuropsychopharmacology, 31*(3), 506–515.

Thoma, R. J., Hanlon, F. M., Sanchez, N., Weisend, M. P., Huang, M., Jones, A., et al. (2004). Auditory sensory gating deficit and cortical thickness in schizophrenia. *Neurology & Clinical Neurophysiology, 62*, 1–7.

Titone, D., Levy, D. L., & Holzman, P. S. (2000). Contextual insensitivity in schizophrenic language processing: Evidence from lexical ambiguity. *Journal of Abnormal Psychology, 109*, 761–767.

Todd, J., Michie, P. T., & Jablensky, A. V. (2003). Association between reduced mismatch negativity (MMN) and raised temporal discrimination thresholds in schizophrenia. *Clinical Neurophysiology, 114*, 2061–2070.

Todd, J., Michie, P. T., Schall, U., Karayanidis, F., Yabe, H., & Näätänen, R. (2008). Deviant matters: Duration, frequency, and intensity deviants reveal different patterns of mismatch negativity reduction in early and late schizophrenia. *Biological Psychiatry, 63*, 58–64.

Tregellas, J. R., Shatti, S., Tanabe, J. L., Martin, L. F., Gibson, L., Wylie, K., et al. (2007). Gray matter volume differences and the effects of smoking on gray matter in schizophrenia. *Schizophrenia Research, 97*(1–3), 242–249.

Tsuang, M. (2004). Schizophrenia: Genes and environment. *Biological Psychiatry, 47*, 210–220.

Turetsky, B. I., Calkins, M. E., Light, G. A., Olincy, A., Radant, A. E., & Swerdlow, N. R. (2007). Neurophysiological endophenotypes of schizophrenia: The viability of selected candidate measures. *Schizophrenia Bulletin, 33*, 1–26.

Turetsky, B. I., Cannon, T. D., & Gur, R. E. (2000). P300 subcomponent abnormalities in schizophrenia: III. Deficits in unaffected siblings of schizophrenic probands. *Biological Psychiatry, 47*, 380–390.

Turetsky, B. I., Colbath, E. A., & Gur, R. E. (1998). P300 subcomponent abnormalities in schizophrenia: II. Longitudinal stability and relationship to symptom change. *Biological Psychiatry, 43*, 31–39.

Uc, E. Y., Skinner, R. D., Rodnitzky, R. L., & Garcia-Rill, E. (2003). The midlatency auditory evoked potential P50 is abnormal in Huntington's disease. *Journal of Neurological Science, 212*(1–2), 1–5.

Umbricht, D., Bates, J. A., Lieberman, J. A., Kane, J. M., & Javitt, D. C. (2006). Electrophysiological indices of automatic and controlled auditory information processing in first-episode, recent-onset and chronic schizophrenia. *Biological Psychiatry, 59*, 762–772.

Umbricht, D., Koller, R., Schmid, L., Skrabo, A., Grübel, C., Huber, T., et al. (2003). How specific are deficits in mismatch negativity generation to schizophrenia? *Biological Psychiatry, 53*, 1120–1131.

Umbricht, D., & Krljes, S. (2005). Mismatch negativity in schizophrenia: A meta-analysis. *Schizophrenia Research, 76*, 1–23.

Umbricht, D., Schmid, L., Koller, R., Vollenweider, F. X., Hell, D., & Javitt, D. (2000). Ketamine-induced deficits in auditory and visual context-dependent processing in healthy volunteers. Implications for models of cognitive deficits in schizophrenia. *Archives of General Psychiatry, 57*, 1139–1147.

van Beijstervelt, C. E. M., & van Baal, G. C. M. (2002). Twin and family studies of the human electroencephalogram: A review and a meta-analysis. *Biological Psychiatry, 61*, 111–138.

van der Stelt, O., Frye, J., Lieberman, J. A., & Belger, A. (2004). Impaired P3 generation reflects high-level and progressive neurocognitive dysfunction in schizophrenia. *Archives of General Psychiatry, 61*, 237–248.

Van Petten, C., & Kutas, M. (1990). Interactions between sentence context and word frequency in event-related brain potentials. *Memory and Cognition, 18*, 380–393.

Van Petten, C., & Luka, B. J. (2006). Neural localization of semantic context effects in electromagnetic and hemodynamic studies. *Brain and Language, 97*, 279–293.

Vinogradov, S., Ober, B. A., & Shenaut, G. K. (1992). Semantic priming of word pronunciation and lexical decision in schizophrenia. *Schizophrenia Research, 8*, 171–181.

Vohs, J. L., Hetrick, W. P., Bodkins, M., Bismark, A., Shekhar, A., & O'Donnell, B. F. (2008). Visual event-related potentials in schizotypal personality and schizophrenia. *Journal of Abnormal Psychology, 117*, 119–131.

Waldo, M., & Freedman, R. (1986). Gating of auditory evoked responses in normal college students. *Psychiatry Research, 19*, 233–239.

Wang, J., Hirayasu, Y., Hiramatsu, K., Hokama, H., Miyazato, H., & Ogura, C. (2003). Increased rate of P300 latency prolongation with age in drug-naive and first episode schizophrenia. *Clinical Neurophysiology, 114*, 2029–2035.

Weisser, R., Weisbrod, M., Roehrig, M., Rupp, A., Schroeder, J., & Scherg, M. (2001). Is frontal lobe involved in the generation of auditory evoked P50? *Neuroreport, 12*(15), 3303–3307.

White, P. M., & Yee, C. M. (1997). Effects of attentional and stressor manipulations on the P50 gating response. *Psychophysiology, 34*, 703–711.

White, P. M., & Yee, C. M. (2006). P50 sensitivity to physical and psychological state influences. *Psychophysiology, 43*(3), 320–328.

Winterer, G., Egan, M. F., Raedler, T., Sanchez, C., Jones, D. W., Coppola, R., et al. (2003). P300 and genetic risk for schizophrenia. *Archives of General Psychiatry, 60*, 1158–1167.

Yee, C. M., & White, P. M. (2001). Experimental modification of P50 suppression. *Psychophysiology, 38*, 531–539.

Young, D. A., Waldo, M., Rutledge, J. H., 3rd, & Freedman, R. (1996). Heritability of inhibitory gating of the P50 auditory-evoked potential in monozygotic and dizygotic twins. *Neuropsychobiology, 33*(3), 113–117.

Yvert, B., Fischer, C., Betrand, O., & Pernier, J. (2005). Localization of human supratemporal auditory areas from intracerebral auditory evoked potentials using distributed source models. *Neuroimage, 28*(1), 140–153.

Zouridakis, G., Boutros, N. N., & Jansen, B. H. (1997). A fuzzy clustering approach to study the auditory P50 component in schizophrenia. *Psychiatry Research, 69*(2–3), 169–181.

CHAPTER

20 Event-Related Brain Potentials in Depression: Clinical, Cognitive, and Neurophysiological Implications

Gerard E. Bruder, Jürgen Kayser, *and* Craig E. Tenke

Abstract

Individuals who have a depressive disorder commonly experience difficulties with concentration, attention, and other cognitive functions, such as memory and executive control. The recording of event-related brain potentials (ERPs) provides a noninvasive means for studying cognitive deficits in depressive disorders and their underlying neurophysiological mechanisms. This chapter reviews the findings of studies measuring ERPs in depressed patients during a variety of sensory, cognitive, and emotional tasks in order to contribute to a better understanding of the specific processes and neurophysiological mechanisms that are dysfunctional in depressive disorders. It highlights the clinical relevance of ERP findings in depressed patients by describing the relation of patients' ERPs to their clinical features, most notably severity of depressive symptoms, diagnostic subtype, and therapeutic response to treatments. From a methodological perspective, it presents new findings illustrating the power of combining current source density (CSD) and principal components analysis (PCA) techniques, which take better advantage of both the temporal resolution of ERPs and the spatial resolution of dense electrode arrays than traditional analysis methods of reference-dependent surface potentials.

Keywords: depression, depressive disorders, event-related potentials, cognitive deficits, current source density analysis, principal components analysis

Introduction

Individuals who have a depressive disorder commonly experience difficulties with concentration, attention, and other cognitive functions, such as memory and executive control (Austin et al., 2001; Porter et al., 2003). The recording of event-related brain potentials (ERPs) provides a noninvasive means for studying cognitive deficits in depressive disorders and their underlying neurophysiological mechanisms. The precise temporal resolution of ERPs can reveal unique information about the specific stage of processing that may lead to disruption of performance on cognitive tasks—for example, early sensory/attentional processing, as reflected in the N1 potential, or later cognitive evaluation, as reflected in the P3 potential. Moreover, ERPs can provide noninvasive biological markers for assessing treatment effects and, most promisingly, for determining who will benefit from a particular course of treatment.

By far the largest number of ERP studies of depression have focused on the cognitive P3 potential during target detection oddball tasks. We will review the findings of these studies and focus on recent studies that have examined P3 subcomponents, which provide new evidence concerning specific cognitive operations that may be disturbed in depression. After reviewing these findings, we will examine ERP findings in depressed patients obtained during more challenging cognitive paradigms, including more demanding auditory or visual discrimination tasks. We will also review studies that have recorded ERPs in depressed patients during recognition memory tasks, which provide information on ERP correlates of episodic memory. Surprisingly few studies have measured ERPs of depressed patients during processing of emotional stimuli; such data may have particular relevance to mood disorders and will therefore be reviewed. A number of recent studies in depressed patients have found abnormalities of negative brain potentials associated with monitoring of cognitive performance, such as error-related negativity (ERN). These studies, as well as others measuring the intensity-dependency of auditory N1-P2 potentials, will be highlighted because they suggest the potential value of these ERP measures for predicting the clinical response to antidepressants.

One aim of this review is therefore to bring together the findings of studies measuring ERPs in depressed patients during a variety of sensory, cognitive, and emotional tasks in order to contribute to a better understanding of the specific processes and neurophysiological mechanisms that are dysfunctional in depressive disorders. For instance, evidence of ERP abnormalities related to attentional or cognitive control processes is suggestive of deficits involving frontal or anterior cingulate cortex. Another aim is to highlight the clinical relevance of ERP findings in depressed patients by pointing to the relation of patients' ERPs to their clinical features, most notably severity of depressive symptoms, diagnostic subtype, and therapeutic response to treatments. From a more methodological perspective, we will present new findings illustrating the power of combining current source density (CSD) and principal components analysis (PCA) techniques, which take better advantage of both the temporal resolution of ERPs and the spatial resolution of dense electrode arrays than traditional analysis methods of reference-dependent surface potentials (Kayser & Tenke, 2006a, 2006b).

P3 in Auditory and Visual Oddball Tasks

The P3 or P300 potential provides physiological measures associated with attentional and working memory operations during cognitive task performance (see Polich, 2007; Chapter 7, this volume). It has typically been measured during oddball tasks, in which a subject responds to an infrequent target stimulus in a series of frequent nontarget standard stimuli. In the typical study, subjects hear a pseudorandom sequence of 90% low-pitched and 10% high-pitched tones, each presented for 50 ms at a rate of one per second, and the subject's task is to respond to the infrequent high-pitched tone (e.g., by pressing a button or silently counting). With all common EEG recording reference schemes (nose, linked mastoids, average reference), the classical P3 potential (P3b) is maximal over midline parietal scalp sites and has a peak latency ranging from 300 to 500 ms. Figure 20.1 illustrates the average waveforms for healthy adults at midline frontal (Fz), central (Cz), parietal (Pz), and occipital (Oz) electrode sites (nose reference) to infrequent targets (solid line) and frequent nontargets (dashed line) in an oddball task. The waveforms are typical of those seen in auditory oddball tasks, consisting of early N1 and P2 peaks to both targets and nontargets, followed by a negative peak and a late positive peak occurring about 200 ms (N2) and 350 ms (P3) relative to the onset of only the target stimuli. The P3b component has its maximum at Pz.

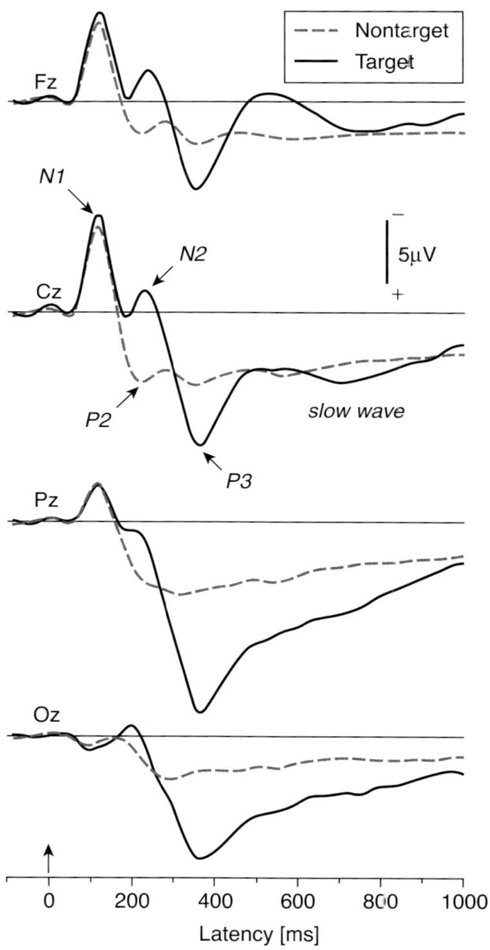

Fig. 20.1. Grand mean, nose-referenced ERP waveforms for 26 healthy adults comparing targets (solid lines) and nontargets (dashed lines) in an auditory oddball task at frontal (Fz), central (Cz), parietal (Pz), and occipital (Oz) midline electrode sites. Reprinted with permission from Kayser et al. (1998).

Most studies in depressed patients have used an auditory oddball task. Although specific procedures vary from study to study (e.g., frequency of target and nontarget tones, stimulus duration, interstimulus intervals, response mode), the use of the same basic task facilitates the comparison and summary of P3 findings across studies. However, despite the use of largely comparable oddball tasks, there have been conflicting findings as to whether depressed patients have reduced P3 amplitude. A review of early studies (Roth et al., 1986) using mostly oddball tasks found that only about half showed reduced P3 amplitude in depressed patients when compared to healthy controls. Table 20.1 summarizes the findings of more recent studies published over the last 20 years that compared P3 amplitudes of depressed patients and healthy controls in auditory oddball tasks. Sixty percent (12 of 20) of the comparisons listed in Table 20.1 found significantly smaller P3 amplitude in patients who had a major depressive disorder (MDD) compared to healthy controls (HCs). These studies had moderate to large effect sizes, which ranged widely, from 0.52 to 2.25 (Cohen's *d*). Among the studies that failed to find significant differences, there were often trends for depressed patients to have smaller P3 than controls, but with small effect sizes ranging from 0.11 to 0.52. The mean effect size of studies reported in Table 20.1 is 0.85 (SD = 0.75; median = 0.79), indicative of a moderate group difference. Thus, while there continue to be conflicting findings, the overall trend is for most studies using an auditory oddball paradigm to show at least some reduction of P3 amplitude in depressed patients.

The large difference in effect size across studies does, however, suggest that differences in the clinical characteristics of the patients in these studies may have played a role. Although differences in P3 amplitude among patients have generally not been found to be related to their overall severity of depression, there is evidence that some subtypes of depression show the greatest reductions of P3 amplitude. All three studies testing patients who had a major depression with melancholic features found reduced P3, with large effect sizes of 0.85, 0.98, and 2.25 (Ancy et al., 1996; Gangadhar et al., 1993; Urretavizcaya et al., 2003). Melancholic features include profound loss of interest or pleasure, lack of reactivity to usual pleasurable stimuli, and associated symptoms, such as early morning awakening, depression worse in the morning, psychomotor retardation, weight loss, and excessive guilt (American Psychiatric Association, 1994). Also, P3 has been found to be more reduced in patients who had a psychotic rather than a nonpsychotic depression (Karaaslan et al., 2003; Kaustio et al., 2002) and in patients who attempted suicide compared to those without a suicidal history (Hansenne et al., 1996). Smaller P3 amplitude was associated with higher scores on scales for assessing suicidal risk (Hansenne et al., 1996) and psychotic symptoms (Santosh et al., 1994). Greater P3 reduction in psychotic depression is consistent with evidence that cognitive deficits on neuropsychological tests are more severe in psychotic than nonpsychotic depression (Castaneda et al., 2008) and with the robust P3 reduction seen in schizophrenia (Jeon & Polich, 2003; Chapter 19, this volume).

The patients in all but two of the studies in Table 20.1 were unmedicated at the time of testing. Although one of these studies found no difference in P3 between medicated and unmedicated patients (Sara et al., 1994), studies have generally found P3 amplitude to increase or normalize following treatment with antidepressants or electroconvulsive therapy (ECT; Blackwood et al., 1987; Gangadhar et al., 1993; Nurminen et al., 2005; but see the negative findings of Vandoolaeghe et al., 1998). Following vagus nerve stimulation, treatment responders but not nonresponders showed an increase in P3 amplitude, but no control group was tested and so it is not known whether this treatment normalized P3 (Neuhaus et al., 2007). These findings indicate that reduced P3 in depressed patients during an auditory oddball task is at least partially state-dependent and may normalize with improvement of depression during successful treatment. This possibility is also supported by the finding that young women with a history of a major depressive episode, but no current depressive disorder, did not differ from normal controls in P3 amplitude during an auditory oddball task (Houston et al., 2004).

Fewer studies have measured P3 in depressed patients during visual oddball tasks, and as is the case for the auditory modality, there have been conflicting findings. Diner et al. (1985) conducted one of the first studies, in which 10 depressed patients and 10 controls were tested in a variant of a three-stimulus visual oddball task with an infrequent target (letter string *DTM*), a frequent standard (*RSC*), and infrequent nontarget three-letter words. P3 amplitude to targets was significantly smaller in depressed patients when compared to controls, and greater severity of depression was associated with smaller P3. In contrast, Bange and Bathien (1998) found no difference in P3 amplitude between

Table 20.1. Auditory oddball studies comparing depressed patients and healthy controls

Study	Sample[a]	EEG Montage	EEG Reference	P3 Amplitude	Effect Size[b]
Blackwood et al. (1987)	16 MDD (med-free), 59 HC	Cz	Left ear	MDD < HC	0.79
Muir et al. (1991)	46 MDD (35 med-free), 212 HC	Cz	Left ear	MDD < HC	0.52
Gangadhar et al. (1993)	17 MDD (med-free), 22 HC	Cz	Mastoids	MDD < HC	0.98
Sara et al. (1994)	14 MDD (med-free), 27 HC 13 MDD (medicated)	Fz, Cz, Pz	Linked ears	MDD = HC MDD = HC	0.18 0.31
Hansenne et al. (1996)	10 MDDwS (med-free), 20 HC 10 MDDwoS (med-free)	Cz	Left ear	MDDwS < HC MDDwoS = HC	1.72 –0.12
Ancy et al. (1996)	17 MDD (15 med-free), 15 HC	Cz	Mastoids	MDD < HC	0.85
Yanai et al. (1997)	16 MDD (med-free), 17 HC	Pz	Linked ears	MDD < HC	2.18
Wagner et al. (1997)	11 MDD (med-free), 10 HC	Fz, Cz	Right mastoid	MDD < HC	—
Bruder et al. (1998)	40 MDD/DYS (med-free), 22 HC	12 sites	Nose	MDD/DYS = HC	—
Vandoolacghe et al. (1998)	35 MDD (med-free), 11 HC	Cz	Mastoids	MDD = HC	0.52
Kaustio et al. (2002)	22 MDD/DYS (med-free), 22 HC	16 sites	Right mastoid	MDD/DYS = HC	—
Anderer et al. (2002)	60 MDD (med-free), 29 HC	19 sites	Average mastoid	MDD < HC	—
Röschke & Wagner (2003)	21 MDD (med-free), 21 HC	Cz, Pz	Right mastoid	MDD < HC	—
Urretavizcaya et al. (2003)	50 MDD (med-free), 31 HC	C_3, Cz, C_4	Linked ears	MDD < HC	2.25
Kaiser et al. (2003)	16 MDD (medicated), 16 HC	18 sites	Average	MDD = HC	0.11
Karaaslan et al. (2003)	16 MDDwP (med-free), 20 HC 20 MDDwoP (med-free)	Cz	Linked mastoid	MDDwP < HC MDDwoP = HC	1.43 0.08
Kawasaki et al. (2004)	22 MDD (med-free), 22 HC	16 sites	Linked ears	MDD < HC	0.90

[a]MDD = major depressive disorder, HC = healthy controls, DYS = dysthymic disorder, MDDwS = MDD with suicide attempt, MDDwoS = MDD without suicide attempt, MDDwP = MDD with psychotic features, MDDwoP = MDD without psychotic features.
[b]Cohen's d effect size.

patients having either unipolar major depressive disorder (n = 12) or bipolar depressive disorder (n = 11) and healthy controls (n = 20) in either single-stimulus or two-stimulus visual oddball tasks. They did, however, report that patients with a bipolar depressive disorder had significantly longer latency of the P3 peak compared to controls, and the depressed patients showed a reduction in P3 latency in remission. Although some studies have not found a difference in P3 latency between depressed patients and healthy controls in visual or auditory oddball tasks (Blackwood et al., 1987; Diner et al., 1985; Gangadhar et al., 1993), longer P3 latency in bipolar depressed patients, but *not* unipolar depressed patients, parallels the findings of Muir et al. (1991) for the auditory modality. This suggests that patients who typically display psychomotor retardation, such as those having bipolar or melancholic depression, may be most likely to show longer P3 latency suggestive of a slowing of cognitive processing. Schlegel et al. (1991) also found that longer P3 latency for depressed patients (n = 36) in an auditory task was correlated with their total score on the Bech-Rafaelsen Melancholia Scale and the four retardation items on this scale.

P3 in Cognitively Challenging Auditory and Visual Tasks

The conflicting findings for P3 amplitude in depressed patients may be due in part to the use of simple oddball tasks that are not cognitively challenging enough to elicit robust P3 reductions in patients having subtle cognitive deficits. We have argued that it would be more fruitful to measure P3 in depressed patients during cognitively demanding tasks (Bruder, 1992). Given evidence from neuropsychological and dichotic listening tests suggestive of right parietotemporal dysfunction in depression (Bruder et al., 1989; Heller et al., 1995), we reasoned that depressed patients might show greater P3 deficits in tasks that tap right hemispheric processing, such as those involving spatial or complex tonal processing. Event-related potentials were measured in 25 unmedicated depressed patients and 27 healthy controls during spatial and temporal discrimination tasks in the auditory modality (Bruder et al., 1991). The spatial task used a dichotic paradigm to manipulate the apparent location of a click, and the subject's task was to discriminate a difference in the location of standard and test stimuli. The temporal task required discriminating a difference in the duration of a standard click train and a test click train. A titration procedure was used to determine the difference between standard and test stimuli in each task that would yield 75% correct responses for each subject, and these threshold values were used during the ERP measurements. To evaluate differences between subtypes of depression, patients were divided into those having either a typical melancholic form of depression or an atypical depression. Patients meeting criteria for atypical depression showed symptoms that are in some respects the opposite of those seen for melancholia, that is, reactivity of mood with preserved pleasure capacity and one or more associated features—hypersomnia, overeating, rejection sensitivity, or bodily inertia. There was no difference among the patient subgroups and healthy controls in behavioral thresholds for discriminating stimuli in the spatial or temporal tasks and no difference in their P3 amplitudes. However, patients having a typical melancholic depression had considerably longer P3 latency in the spatial task when compared to patients having atypical depression and healthy controls. In contrast, there was no difference among groups in P3 latency during the temporal discrimination task, which indicates that the cognitive task in which the P3 is measured is an important factor. The melancholic subgroup showed evidence of a slowing of cognitive processing only in the spatial task that involves predominantly right hemisphere processing. Moreover, the longer P3 latency in melancholic than atypical depression supports findings from oddball tasks suggesting that longer P3 latency is most evident in specific diagnostic subtypes, that is, melancholic and bipolar depression.

Given evidence of right hemisphere dysfunction in depression, a subsequent study measured ERPs of 44 unmedicated depressed patients and 19 healthy controls during a complex tone test (Bruder et al., 1995). This is a cognitively demanding dichotic listening task that yields a left ear (right hemisphere) advantage in healthy adults for perceiving complex tones (Sidtis, 1981; Tenke et al., 1993). Depressed patients had significantly smaller P3 amplitude compared to controls and also failed to show either the behavioral left ear (right hemisphere) advantage or the hemispheric asymmetry of P3 seen for controls. The absence of any difference in early sensory potentials (e.g., N1) between depressed patients and controls supports the conclusion that the lack of a right hemisphere advantage for perceiving complex tones is related to a relatively late stage of cognitive processing reflected in the P3.

Arguing that binaural oddball tasks are too simple to consistently reveal cognitive dysfunction in depression, Tenke et al. (2008) developed a dichotic oddball task that increases the cognitive challenge. Event-related potentials of 38 unmedicated depressed patients and 26 healthy controls were measured in tonal and phonetic tasks with dichotic presentation of stimuli. Tonal nontargets were pairs of complex tones (corresponding to musical notes G and B above middle C) presented simultaneously to each ear (L/R) in an alternating series (G/B or B/G). A different target tone (note A) replaced one stimulus of the pair on 20% of the trials. Phonetic nontargets were pairs of syllables (/ba/, /da/) presented simultaneously to each ear (L/R) in an alternating series, and the target was a different syllable (/ta/) that replaced one syllable in the pair 20% of the trials. The subject's task was to respond to the target with a button press. Target detection was poorer in depressed patients than controls for both tones and syllables. Patients also showed reductions of CSD for parietal and temporal lobe sources corresponding to P3. While reduction of the parietal source was related to the patients' poorer performance, temporal lobe source reduction was not. Given the involvement of primary and secondary auditory cortex in tonal and phonetic processing (Zatorre et al., 1992), these findings support evidence of temporoparietal dysfunction in depression (e.g., Bruder et al., 1995; Deldin et al., 2000; Heller et al., 1995; Post et al., 1987). The P3 source reduction in depressed patients was not lateralized to one hemisphere, and the tonal and phonetic tasks did not yield consistent behavioral ear advantages in healthy adults. The above findings indicate that cognitively challenging dichotic listening tasks yield consistently smaller P3 amplitudes in depressed patients when compared to healthy controls.

Two studies measuring ERPs during cognitively demanding visual tasks agreed in showing that individuals at risk for later development of depressive disorders had reduced P3 amplitude. Houston et al. (2003) used a visuospatial oddball task that challenged attention and a complex cognitive skill (i.e., mental rotation). Young women with a history of a major depressive episode but no current depressive disorder (n = 29) had smaller P3 amplitude when compared to those with no history of depression (n = 101). Moreover, topographic maps of CSD measures corresponding to P3 indicated that the difference between the previously depressed and nondepressed groups was maximal over the right prefrontal region. Similarly, Zhang et al. (2007) measured ERPs of healthy adults with or without a family history of depression (n = 14 per group). The task was a visual go/no-go task in which large or small letters *H* and *O* were presented on a monitor and subjects were required to respond with the right hand to a large *H* or with the left hand to a large *O*, and no response was required for smaller letters. Subjects with a family history of depression, who are at increased risk for developing a depressive disorder, showed smaller P3 amplitudes over temporoparietal regions when compared to low-risk subjects. Low resolution electromagnetic tomography (LORETA) source localization methods pointed to decreased activation of the left middle temporal gyrus in the high-risk subjects. The authors suggested that a P3 decrement in visual tasks is a vulnerability marker for developing depression. This contrasts with the P3 findings for simple auditory oddball tasks, where subjects at high risk for depression did not differ from low-risk subjects in P3 amplitude (Houston et al., 2004) and where the P3 increase following remission of depression was suggestive of a more state-dependent effect. Thus, while P3 reductions in a simple auditory oddball task appear to reflect the patient's current clinical state, P3 reductions in more demanding visual tasks appear to reflect underlying vulnerability for a depressive disorder. It is interesting to note in this regard that we have found electroencephalographic (EEG) evidence of reduced right posterior activity in offspring at risk for depressive disorders (Bruder et al., 2007), which implicates cortical regions known to mediate visual attention and perception as possible vulnerability indicators for depression.

P3 Subcomponents

P3 is not a unitary phenomenon but consists of two or more subcomponents associated with different cognitive operations and neural generators (see Chapter 7, this volume). Although the focus of most studies in depressed patients has been on the parietal maximum P3b, this component is often preceded by a component with a more frontocentral topography, that is, P3a. This frontal aspect of P3 is prominent to novel distracter stimuli (e.g., environmental sounds) that are interspersed with target and standard stimuli in a three-stimulus oddball task (Polich & Criado, 2006; Simons et al., 2001; Spencer et al., 1999). The importance of differentiating between P3 subcomponents is that the novelty P3 or P3a is thought to reflect frontal attention or orienting mechanisms, whereas P3b reflects temporoparietal mechanisms associated with context updating and memory processing (Polich, 2007). Studies examining P3 subcomponents in depressed patients could therefore provide new information concerning the nature of their cognitive deficit and underlying neurophysiological mechanisms.

The first studies to examine P3a and P3b subcomponents in depressed patients used go/no-go reaction time tasks (Pierson et al., 1996) and divided patients into two subgroups to deal with the issue of clinical heterogeneity of depression. The authors referred to an initial study, in which they recorded ERPs during a simple forewarned reaction time task and reported that a subgroup of anxious-agitated-impulsive patients had *greater* amplitude of the frontal P3a when compared to a subgroup of patients having retarded-blunted affect. In their subsequent study, they used a complex forewarned choice reaction time task so as to measure P3 subcomponents in a more effortful and cognitively demanding task. Although they reported finding no difference in P3a amplitude among groups, peak-to-peak measures of N2b-P3a amplitudes were smaller in depressed patients than in controls, and retarded-blunted-affect patients had smaller N2b-P3a amplitudes than the anxious-agitated-impulsive patients, with the same tendency when compared to controls. Also, the anxious-agitated-impulsive subgroup had larger P3b amplitudes when compared to either the retarded-blunted-affect subgroup or controls. These findings supported the importance of differentiating between P3 subcomponents and also between depressed patients with different symptom features.

In a study measuring ERPs during tonal or phonetic two-stimulus oddball tasks (Bruder et al., 2002), we used PCA to identify and measure overlapping P3 subcomponents in patients having a depressive disorder alone ($n = 58$), an anxiety disorder alone ($n = 22$), comorbidity of these disorders ($n = 18$), and healthy controls ($n = 49$). An early P3 subcomponent (peak latency 315 ms) was *larger* in patients having an anxiety disorder alone (primarily social phobia or panic disorder) when compared to depressed patients or healthy controls. Depressed patients having a comorbid anxiety disorder tended to have a smaller early P3 than healthy controls, but those having a depressive disorder alone did not. The timing and frontocentral topography of this early P3 subcomponent resemble those seen for P3a. It should be noted, however, that our study used a nose recording reference, as opposed to linked ears in Pierson et al. (1996), which has implications for P3 morphology and topography. Nevertheless, these findings appear to agree with other evidence that P3a or novelty P3 is *heightened* in patients having an anxiety disorder. Thus, patients having a posttraumatic stress disorder were reported to have a larger novelty P3 at frontal sites when compared to normal controls (Kimble et al., 2000). We also found that a later positive subcomponent (peak latency 400 ms) with a parietal maximum typical of P3b did not differ between patients having a depressive disorder alone and controls, but it was larger in depressed patients having a comorbid anxiety disorder when compared to the other groups. The above findings suggest that patients having a depressive disorder, an anxiety disorder, or comorbidity of these disorders differ in the amplitude of P3 subcomponents.

A limitation of the above studies is that P3 subcomponents in depressed patients were measured in paradigms that are not ideal for measuring P3a or novelty P3. Bruder et al. (2009) measured ERPs of 20 unmedicated depressed patients and 20 healthy controls recorded from a 30-channel montage (nose reference) during a novelty oddball task (Friedman et al., 1993) with three stimuli: infrequent target tones ($p = .12$), frequent nontarget tones ($p = .76$), and infrequent novel stimuli (e.g., animal or environmental sounds; $p = .12$). Subjects responded as quickly as possible to target tones only. There was no difference between patients and controls in accuracy or reaction time. Figure 20.2 shows the grand average waveforms at midline electrode sites for patients and controls. The waveforms show the expected N1 and N2 potentials, which are most evident at vertex (Cz), and a novelty P3 that is also evident at this central site. The P3 to targets is largest at the midparietal site (Pz), which is typical of the P3b component. The greater P3 to novels than targets at Fz and Cz reflects the more frontocentral distribution of the novelty P3. As can be seen in Figure 20.2, both the novelty P3 and target P3 were reduced in depressed patients when compared to controls. Average ERP waveforms for each stimulus condition and for each subject were carefully inspected to select time windows that bracketed the peaks and optimized the measurement of mean integrated amplitude of the novelty P3 (220–375 ms) and target P3 (280–470 ms). Amplitude of the novelty P3 was significantly smaller in patients than in controls at frontal ($p < .05$), central ($p < .05$), and parietal ($p < .01$) sites (see the top portion of Figure 20.3). Patients also tended to have smaller P3 amplitude to targets at central ($p < .05$) and parietal ($p < .10$) sites (see the lower portion of Figure 20.3). The difference between patients and controls at the parietal site had a large effect size for the novelty P3 (1.0) and a smaller effect size for

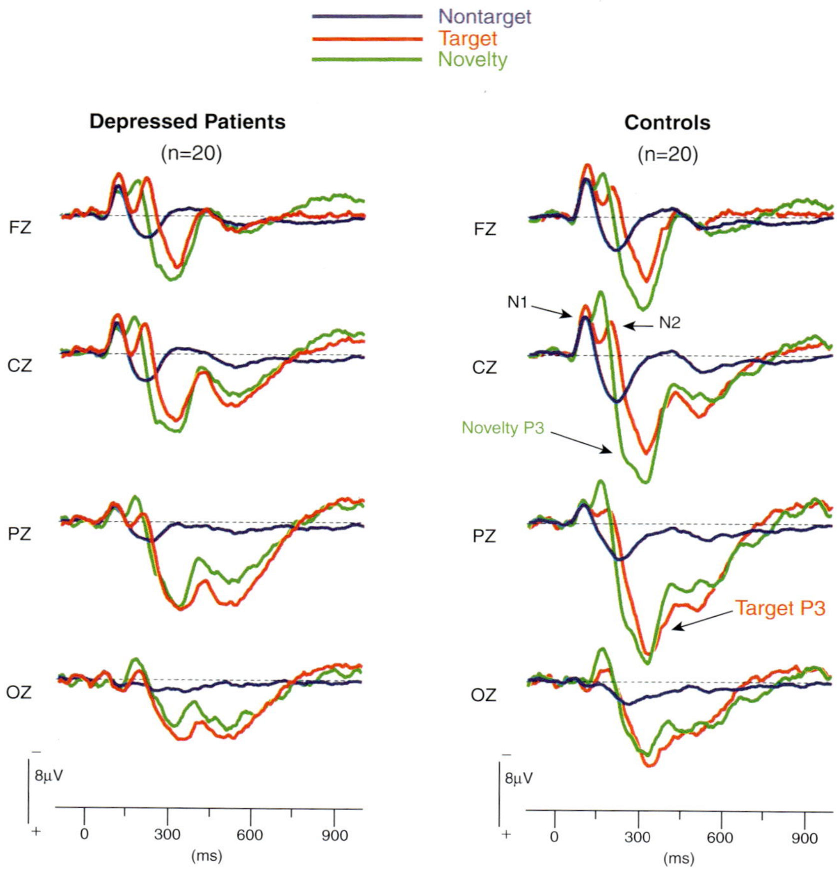

Fig. 20.2. Grand average, nose-referenced ERP waveforms for 20 depressed patients and 20 healthy controls to nontarget, target, and novel stimuli at midline sites (Fz, Cz, Pz, Oz).

the target P3 (0.61). There was no significant difference in the mean integrated amplitude between patients and controls in the N1 (70–145 ms) and N2 (150–240 ms) windows, which indicates that the reduced novelty P3 in patients was likely not due to an earlier deficit in detection of the deviant novel sounds.

The novelty P3 reduction in depressed patients is suggestive of a deficit in automatic shifting of attention (orienting) and evaluation of novel environmental sounds (Friedman et al., 2001; Polich, 2007). There are, however, two issues that needed further study. First, the novelty P3 component overlaps with the P3b component to targets, which leaves open the possible contribution of P3b to the group differences in mean amplitude in the novelty P3 window. The use of multivariate statistics, such as PCA, could aide in identifying and measuring these separate P3 subcomponents. Second, both neuroimaging and ERP studies have found evidence that prefrontal, anterior cingulate, and hippocampal regions are involved in novelty processing (Halgren et al., 1995; Knight et al., 1998; Polich, 2007), but the neural generators underlying the novelty P3 reductions in depressed patients remain unknown. An independent replication and extension of the above study was therefore performed, in which ERPs of a larger sample of depressed patients (n = 49) and healthy controls (n = 49) were recorded from 67 channels during the same novelty oddball task (Tenke et al., 2010). Most importantly, we used a combined CSD-PCA approach to help identify neural sources corresponding to P3 subcomponents (see Kayser & Tenke, 2006a, 2006b, for details).

In the initial step of this approach, all averaged ERP waveforms are transformed into reference-free CSD estimates using a spherical spline surface Laplacian algorithm suggestged by Perrin et al. (1989). The CSD is a mathematical transformation (second spatial derivative) that provides a representation of the direction, location, and intensity of current generators that underlie an ERP topography.

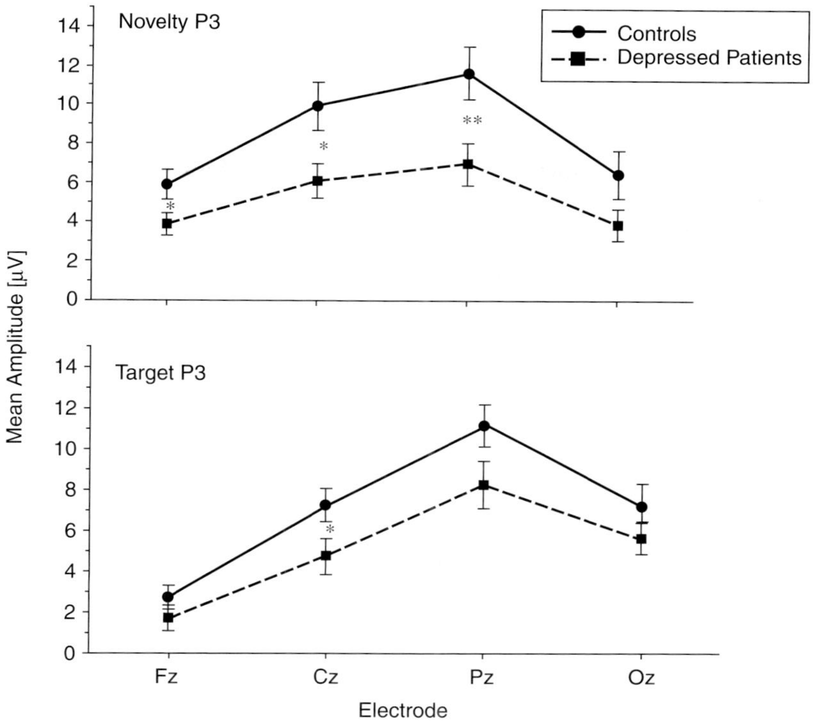

Fig. 20.3. Mean integrated amplitude (± SEM) of novelty P3 and target P3 in 20 depressed patients and 20 healthy controls at frontal (Fz), central (Cz), parietal (Pz), and occipital (Oz) midline sites. Significant simple group effects at each site are indicated by * < .05, ** < .01.

Current source density maps represent the magnitude of the radial (transcranial) current flow entering (sinks) and leaving (sources) the scalp (Nunez, 1981; Nunez, & Srinivasan, 2006). The CSD is a reference-free technique that provides topographies with more sharply localized peaks than those of scalp potentials and eliminates volume-conducted activity from distant regions (Tenke & Kayser, 2005). In the next step, the averaged CSD waveforms are submitted to an unrestricted temporal PCA, followed by Varimax rotation of covariance loadings (Kayser & Tenke, 2006a). This approach yields distinctive PCA components (factor loadings) and corresponding weighting coefficients (factor scores), which provide a concise, efficient simplification of the temporal and spatial distribution of neuronal generators (Kayser & Tenke, 2003, 2006c). Temporal PCA not only aids in determining the relevant statistically independent components within a data set, it also generates efficient measurements of these overlapping components. The combined CSD-PCA method overcomes two critical limitations of ERP research: (1) the dependence of ERP surface potentials on a reference location (e.g., linked mastoids or nose)[1] and (2) the definition and measurement of ERP components (e.g., peak or integrated amplitudes in specified time windows). The use of reference-free CSD measures sharpens topographies related to underlying neuronal generators, and PCA allows identification and quantification of statistically independent factors (sources and sinks) corresponding to ERP/CSD components. The CSD-PCA technique provides a conservative source localization method that avoids any biophysical assumptions, unlike other popular tools—for example, Brain Electrical Source Analysis (BESA) or LORETA.

Figure 20.4A shows the grand mean CSD waveforms for novel, target, and nontarget stimuli for depressed patients and controls. In addition to the expected N1 and N2 sinks to target stimuli at vertex (Cz), there was an early source to novel stimuli but *not* to targets. There was also a prominent source over midparietal sites (Pz), which corresponds to the late-positive, parietal-maximum P3b component. The extracted CSD-PCA factor loadings (orange inset in Figure 20.4A) and topographies separate the early vertex source unique to novels (241 ms peak latency of factor loadings) from parietotemporal P3 source activity common to targets and novels (343 ms loadings peak) and from

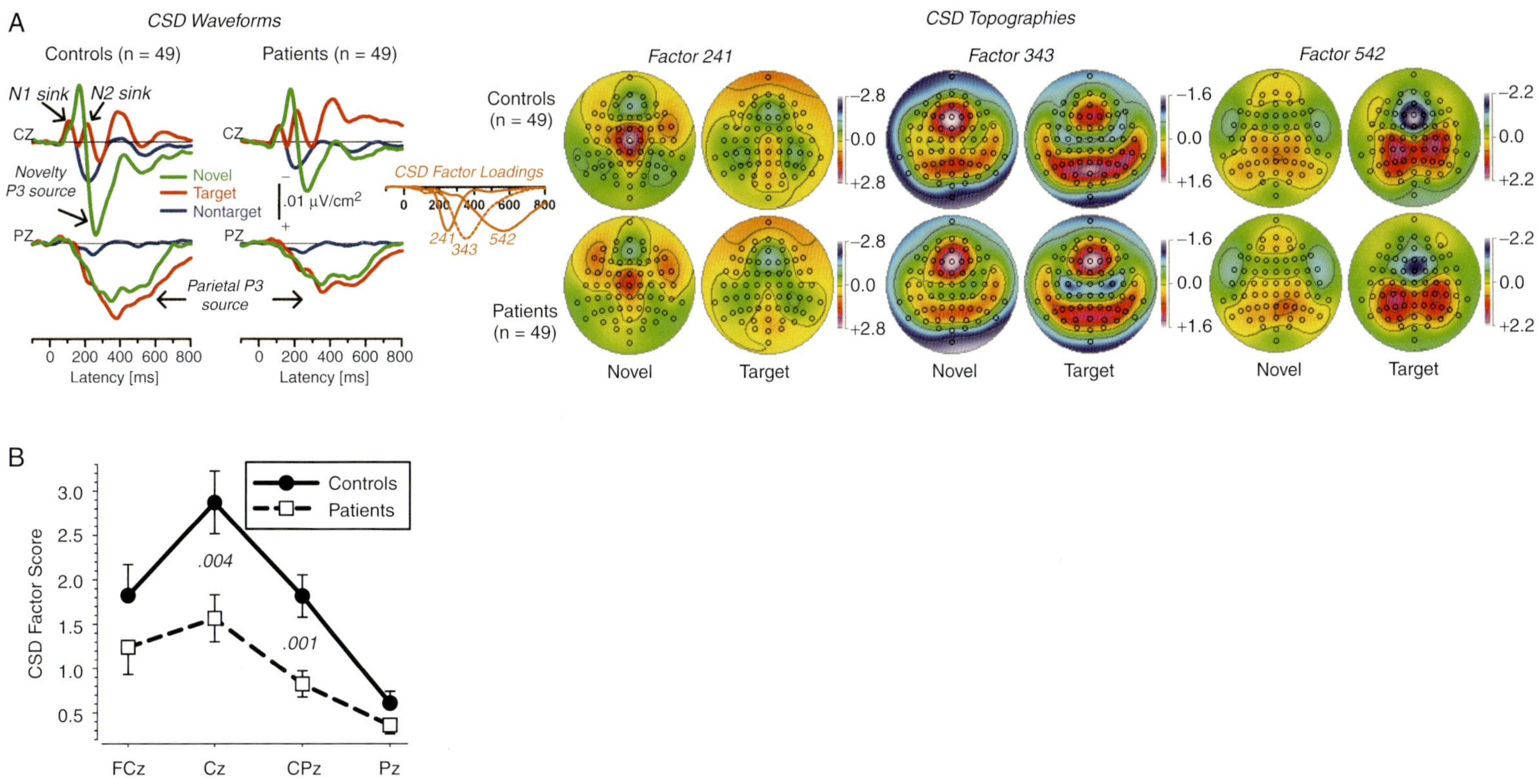

Fig. 20.4. (A) Grand mean CSD waveforms for novel, target, and nontarget stimuli illustrating the difference between depressed patients and controls in novelty P3 source at vertex (Cz) and the lack of a difference in source at Pz. The CSD-PCA factor loadings (orange inset) and topographies separate an early vertex source unique to novels (factor 241) from a subsequent temporoparietal P3 source common to novels and targets (343) and from a later target-specific centroparietal source (542). (B) Selected means (± SEM) of the novelty vertex source (241), which was significantly reduced in depressed patients at midcentral sites (Cz and CPz). Reprinted with permission from Tenke et al. (2010).

later target-specific centroparietal source activity (542 ms loadings peak). The novelty vertex source (factor 241) was markedly reduced in depressed patients when compared to controls ($p = .01$), with the largest group difference at the midline central site (Group by Electrode interaction, $p < .001$; see Figure 20.4B). Group differences were less evident for the later P3 sources to targets (factors 343 and 542). Thus, a vertex source that was *only* present to novel distracter stimuli, and had a shorter latency (241 ms) than the source corresponding to the parietal P3b, was markedly reduced in depressed patients when compared to healthy controls.

Our studies indicate that the novelty P3 is reduced in depressed patients. The findings using the CSD-PCA technique are remarkable for two reasons. First, the novelty vertex source that discriminated between patients and controls had a shorter peak latency, that is, 241 ms from onset of novel stimuli, than the sources corresponding to the parietal P3b component. This suggests that the novelty P3 reduction in depressed patients is indicative of a deficit in early shifting of attention (i.e., orienting) to novel distracter stimuli and not to later cognitive evaluation of these stimuli. Second, the novelty vertex source was localizable to the frontocentral region within and along the longitudinal fissure. Studies using other source localization techniques have localized generators of novelty P3 to the region of the anterior cingulate cortex (ACC), whereas P3b to target stimuli has prominent sources in the region of the temporal-parietal junction (Dien et al., 2003; Mecklinger & Ullsperger, 1995). Contributions of other cortical areas (including frontal gyrus, insula, and posterior cingulate) to these components are, however, also known (Kiehl et al., 2001). Despite convergent evidence for involvement of ACC in the novelty P3, anatomical and biophysical considerations demand caution when interpreting putative generators of midline ERPs. The radial orientation of an equivalent dipole within the longitudinal fissure that is typical of inverse solutions is not normal to the surface of the cingulate gyrus, but rather is tangential to the local alignment of cortical neurons. This paradoxical alignment requires additional assumptions before the generator can be considered to be physiologically plausible (Kayser & Tenke, 2006a; Kayser et al., 2007; Tenke & Kayser, 2005, 2008). Frontal cortex, including the anterior cingulate, is known to be of key importance for attention and has been found to be dysfunctional in depressed patients (Bremmer et al., 2004; Drevets et al., 1997; Siegle et al., 2004). Although this may point to the frontal cortex—and in particular the ACC—as being responsible for novelty P3 reductions in depression, studies indicate that the hippocampus and other cortical structures are also involved in the generation of the novelty P3 (Halgren et al., 1995; Kiehl et al., 2001; Knight, 1996); therefore, further research is needed to pinpoint the origins of this deficit in depressed patients.

ERPs during Processing of Emotional Words or Pictures

Studies of healthy adults have consistently found that emotionally arousing words or pictures elicit a late positive potential (i.e., beyond 300 ms) extending into a slow wave, and the amplitude of this potential is greater for negative or positive emotional stimuli than for neutral stimuli (Johnston et al., 1986; Kayser et al., 1997; Naumann et al., 1992; Palomba et al., 1997). Several studies of depressed patients have reported abnormalities of P3 to visually presented emotional stimuli. In one of the first studies, Blackburn et al. (1990) recorded ERPs from three midline sites referenced to the left ear and found that depressed patients ($n = 15$) had *smaller* P3 amplitude to negative than to neutral or positive words, whereas healthy controls ($n = 15$) had *larger* P3 amplitude to negative than to neutral or positive words. Inspection of their data also suggests that depressed patients had smaller P3 amplitude than controls to negative words, but not to neutral or positive words, but no statistics were presented to support the significance of group differences in P3 amplitude. All the patients in the Blackburn et al. study were taking antidepressant medications, and the impact of these medications on their findings is unknown. Using a 30-channel montage, Kayser et al. (2000) measured nose-referenced ERPs of 30 unmedicated depressed patients and 16 healthy controls during passive viewing of negative pictures of patients with dermatological diseases or neutral control pictures of these patients after surgical treatment. As shown in Figure 20.5A, depressed patients had significantly smaller amplitude of a late P3 potential (460 ms peak latency of a surface potential factor derived from unrestricted temporal PCA followed by Varimax rotation) when compared to controls. As in prior studies (Cacioppo et al., 1993, 1996; Kayser et al., 1997), healthy controls showed enhanced late P3 (P460) amplitude to negative compared to neutral stimuli, and this enhancement was greatest over the right parietal region (see Figure 20.5B). In contrast, depressed patients did not show this increase in late P3 to

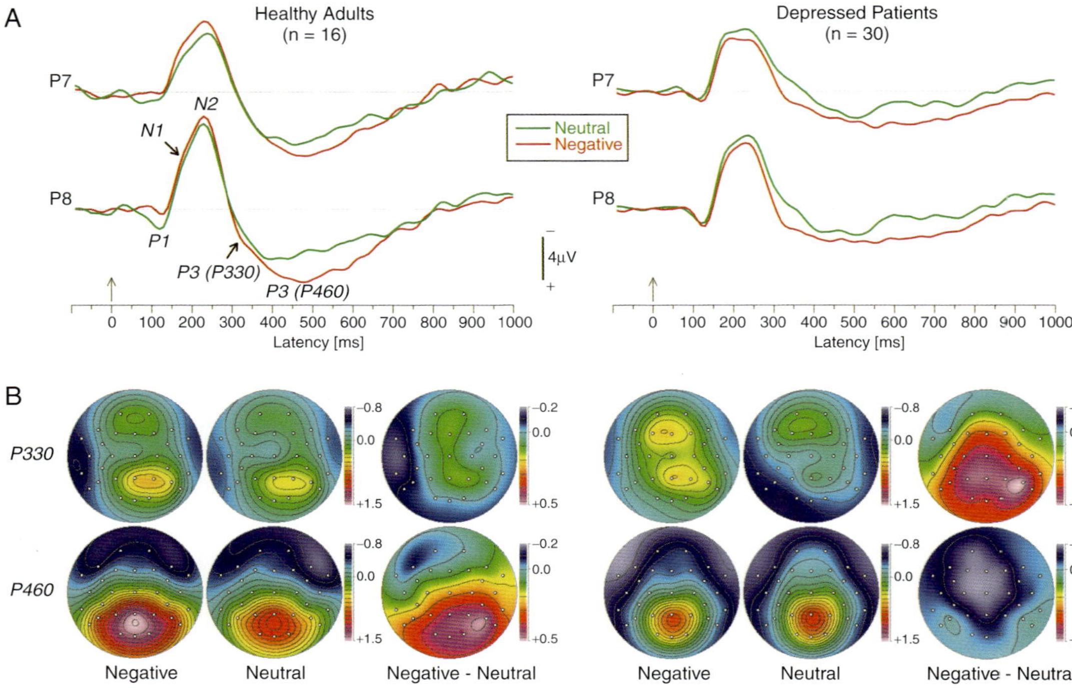

Fig. 20.5. (A) Grand mean, nose-referenced ERP waveforms at lateral-parietal (P7/8) sites comparing neutral and negative stimuli for healthy and depressed participants. Distinct ERP components (P1, N1, N2, early and late P3) are labeled for healthy adults at site P8. Data-driven ERP measures of P3 subcomponents were determined by means of PCA (factors P330 and P460; cf. Kayser et al., 2000). (B) Topographies of PCA factor scores for factor P330 (early P3 rising phase) and factor P460 (classic parietal P3) comparing negative and neutral stimuli and their differences (pooled across visual fields of lateralized presentations) for healthy and depressed participants. Unlike healthy participants, depressed patients showed no effect of emotional content for factor P460, but showed instead an emotional content effect for factor P330 with comparable right-posterior lateralization.

negative as compared to neutral stimuli over either hemisphere. Interestingly, the PCA-based ERP decomposition also revealed an early P3 subcomponent (330 ms peak latency of factor loadings) consisting of a right parietal and frontal positivity, which showed a right-lateralized, negative-larger-than-neutral emotion effect in patients, suggesting intact early classification but impaired late evaluation of affective significance in depression. There was also no difference between depressed patients and healthy controls in valence and arousal self-report measures to these stimuli, which further suggests preserved (cognitive) classification of emotional stimuli in depression.

While subjects in the Kayser et al. (2000) study passively viewed the emotional pictures to reduce the impact of cognitive processing resulting from specific task demands (e.g., target detection, matching paradigm), Deldin et al. (2000) recorded ERPs (linked mastoids) from nine sites to positive, neutral, and negative face and word stimuli during a recognition memory task. They found a lateralized abnormality of the N2 potential in depressed patients (n = 19) when compared to healthy controls (n = 15). N2 amplitude over the right parietal region was reduced in the depressed patients, and this reduction was most evident during the processing of pleasant faces. If one assumes that the recording location of referenced surface potentials reflects differential activation of the underlying cortical regions, the findings of both Kayser et al. (2000) and Deldin et al. (2000), involving different ERP components and methods, appear to be consistent with the hypothesis that depressed patients have impaired activation of right parietal regions during the processing of emotional stimuli (Heller, 1990, 1993). Of course, further study of the neural generators of these effects is needed before this conclusion can be drawn with confidence. Additional evidence of reduced P3 amplitude to emotional stimuli in depressed subjects was obtained by Cavanagh and Geisler (2006), but it was present only for midline electrode sites. They recorded ERPs (linked ears) from seven sites during a visual oddball task in which neutral faces served as standards and happy or fearful faces were targets.

Depressed subjects (n = 36) had reduced P3 amplitude to happy faces when compared to nondepressed controls (n = 18). In summary, the most consistent finding across studies was reduced late P3 (P3b) amplitude to emotional stimuli in depressed subjects, but the valence of stimuli to which this occurs and the laterality of this P3 deficit are less clear. However, only Kayser et al. (2000) used a sufficiently dense EEG montage to evaluate lateralized P3 activity over inferior temporal and parietal regions.

Given evidence that depressed patients have a negative bias for processing information during memory tasks, several studies have measured ERPs of depressed patients in memory tasks with stimuli of different valence to assess processing bias. Studies recording ERPs during recognition memory tasks have found that the influence of the emotional valence of stimuli on late positive or P3 potentials in healthy adults was less evident for depressed subjects (Deldin et al., 2001b; Dietrich et al., 2000a). Although the specific findings differ across studies using word or face stimuli, evidence of mood-congruent biases in depressed patients has been reported in studies measuring slow wave amplitudes during sustained processing of positive, neutral, or negative stimuli in working memory tasks (Deldin et al., 2001a; Deveney & Deldin, 2004; Shestyuk et al., 2005).

ERPs to Olfactory Stimuli

Given the overlapping cortical and limbic systems involved in olfaction, emotional processing, and depression (most notably, the amygdala and orbitofrontal cortex), the study of ERPs to olfactory stimuli may hold particular promise for elucidating the neurophysiological dysfunctions responsible for abnormalities of emotional reactivity in depressed patients. Odors appear to be powerful emotional stimuli with distinctive hedonic valence (Pause et al., 2003). Importantly, the emotional content of odors can be perceived with little cognitive mediation (Ehrlichman & Bastone, 1992), allowing a more direct assessment of emotional processing in depression. Although there is evidence of differences in emotional evaluation of odors between depressed patients and controls (Pause et al., 2000; Steiner et al., 1993), we know of only one study that used ERPs to study olfactory processing in depressed patients. Pause et al. (2003) measured olfactory ERPs at 30 scalp locations (linked ears reference) in 22 patients having an MDD and 22 healthy controls in a task requiring discrimination of pleasant (phenyl-ethyl alcohol = rose) and unpleasant (isobutyraldehyde = rotten butter) odors presented using a constant-flow olfactometer. Control tasks measured visual ERPs to colors or emotional pictures (Lang et al., 1999). Although patients performed as well as controls, they showed reduced amplitude of P2 and early P3 potentials at frontal sites. In contrast, only visual ERPs reflecting later cognitive processing (P3b and slow wave) were reduced in depressed patients to colors or emotional slides. The authors attributed the reduction of the olfactory P2 potential in depressed patients to a deficit in the ability to preattentively encode the pleasantness of odors. Reduced early P3 at frontal sites in depressed patients was thought to reflect a reduction in early cognitive evaluative processes. The authors further proposed that reduced olfactory P2 and P3 in depressed patients may be related to specific alterations in the amygdala and the orbitofrontal cortex, respectively. When 14 of the 22 patients were retested after successful antidepressant treatment, these patients no longer showed smaller olfactory ERPs. However, reduced sample power and the possibility of selective patient dropout or repeated testing may have contributed to these null findings. In addition to having significant problems regarding olfactory ERP component definition and measurement, this study did not control for medication and no information was given about the relation of the olfactory ERP deficits to the severity of depressive symptoms. Further studies recording ERPs to odors of positive and negative valence are needed to replicate and expand on the findings of Pause et al. (2003).

ERPs during Recognition Memory Tasks

A meta-analysis indicated that depression is associated with memory impairments for tests of both recall and recognition (Burt et al., 1995). This memory loss is not universal but appears to depend on patient characteristics, such as diagnostic subtype, severity of depression, and age (Purcell et al., 1997). Unmedicated outpatients having an MDD demonstrated a deficit in verbal episodic memory on the California Verbal Learning Test (Otto et al., 1994). Impaired verbal episodic memory in depressed patients may stem from left prefrontal and medial temporal deficits, in particular involving the hippocampus. Thus, Sapolsky (2000) reviewed evidence from volumetric magnetic resonance imaging (MRI) studies of patients having severe, repeated depressive episodes and found evidence of hippocampal atrophy, which was greater on the left side. These hippocampal deficits in depressed patients have been linked to explicit memory impairments (Sapolsky, 2000; Shah et al., 1998). However, studies have rarely measured the neurophysiological functioning of depressed patients while they were engaged in a memory task.

Event-related potential correlates of memory processes have been examined during a continuous word recognition memory task (Friedman, 1990). Subjects viewed a series of words, some of which were repeated after a number of intervening words, and their task was to decide whether each word was new (not previously presented) or old (previously presented). A robust, replicable finding in healthy adults has been a more positive-going potential for correctly recognized old than new words about 250 to 800 ms after word onset, referred to as the *old-new effect* (see Chapter 14, this volume). Intracranial recordings in and around medial temporal structures of epilepsy patients have shown similar old-new effects, suggesting generators in the hippocampus, parahippocampal gyrus, or amygdala (Elger et al., 1997; Smith et al., 1986). Moreover, patients with left anterior temporal lobectomy showed a dramatic reduction of the old-new effect for word recognition when compared to patients with right temporal lobectomy or controls (Johnson, 1995; Rugg et al., 1991; Smith & Halgren, 1989).

Given evidence of memory impairments and hippocampal deficits in depressed patients, one would predict that they will show a reduced old-new effect during a word recognition task. One study reported findings consistent with this prediction (Dietrich et al., 2000a). The authors measured ERPs of 11 unmedicated depressed patients and 11 healthy controls during a continuous word recognition memory task. The depressed patients were significantly poorer in recognizing the repeated (old) items and showed a smaller old-new effect when compared to the controls. Moreover, the reduction of the old-new effect for words persisted following clinical improvement of depression (Dietrich et al., 2000b). This study, however, had several methodological weaknesses, such as the use of a right mastoid reference (which is problematic for examining laterality effects), a small sample size, and a younger control group showing unusually large P3 amplitudes (group mean > 20 μV), which limit the impact of the findings.

We measured 31-channel ERPs from 37 right-handed, unmedicated depressed patients (21 men) and 40 right-handed, healthy controls (19 men) during continuous recognition memory tasks in which a series of words were presented in either the visual or auditory modality (Kayser et al., in press). Subjects indicated for each word whether it was new or old by pressing one of two buttons (for procedural details, see Kayser et al., 2007). Although all subjects had adequate above-chance performance (86.4% overall correct recognition of repeated words; *SD* = 12.7), depressed women showed poorer recognition memory than healthy women, but there was no group difference in men (Group by Gender interaction, $p < .05$). There was, however, no significant group difference in response latency for visual or auditory word presentations. For improved spatial and temporal characterization of the ERP old-new effects, the data were analyzed using our CSD-PCA technique (Kayser & Tenke, 2006a, 2006b; Kayser et al., 2007).

Both patients and controls showed the expected old-new effects, with greater late source activity (positivity) at posterior sites to correctly recognized old words for both auditory (Figure 20.6A) and visual (Figure 20.6B) modalities. This source activity, corresponding to the late P3b potential, was identified in separate PCAs for each modality by the auditory CSD factor 650 (peak latency of factor loadings in milliseconds) and the visual CSD factor 475 (peak latency in milliseconds; see the green factor loading waveforms in Figures 20.6A and 20.6B). Based on response-locked averages, these P3 source factors peaked about 170 ms and 140 ms, respectively, prior to response onset in either modality (cf. Kayser et al., 2007). As is evident for both groups in the CSD factor topographies, increased lateral-parietal P3 sources (warm colors) for old as compared to new auditory stimuli were accompanied by increased lateral-frontal sinks (right portion of Figure 20.6A), and similar old-new effects with midparietal and midfrontal P3 sources were found for visual stimuli (right portion of Figure 20.6B). These old-new effects were present for both groups, as indicated by the significant pairwise max (T^2) randomization tests (Maris, 2004) for each group and modality (last column in Figures 20.6A and 20.6B). However, there were notable topographic group differences in the visual P3 old-new effect, which was shifted toward occipital sites in depressed patients. A repeated measures analysis of variance (ANOVA) of P3 source factor scores from both modalities was computed at homologous left and right lateral-parietal sites (P3/4, P7/8, CP5/6), where the P3 source was prominent. This analysis revealed a significant Group main effect ($p < .001$) and interactions of Group by Hemisphere ($p = .001$)

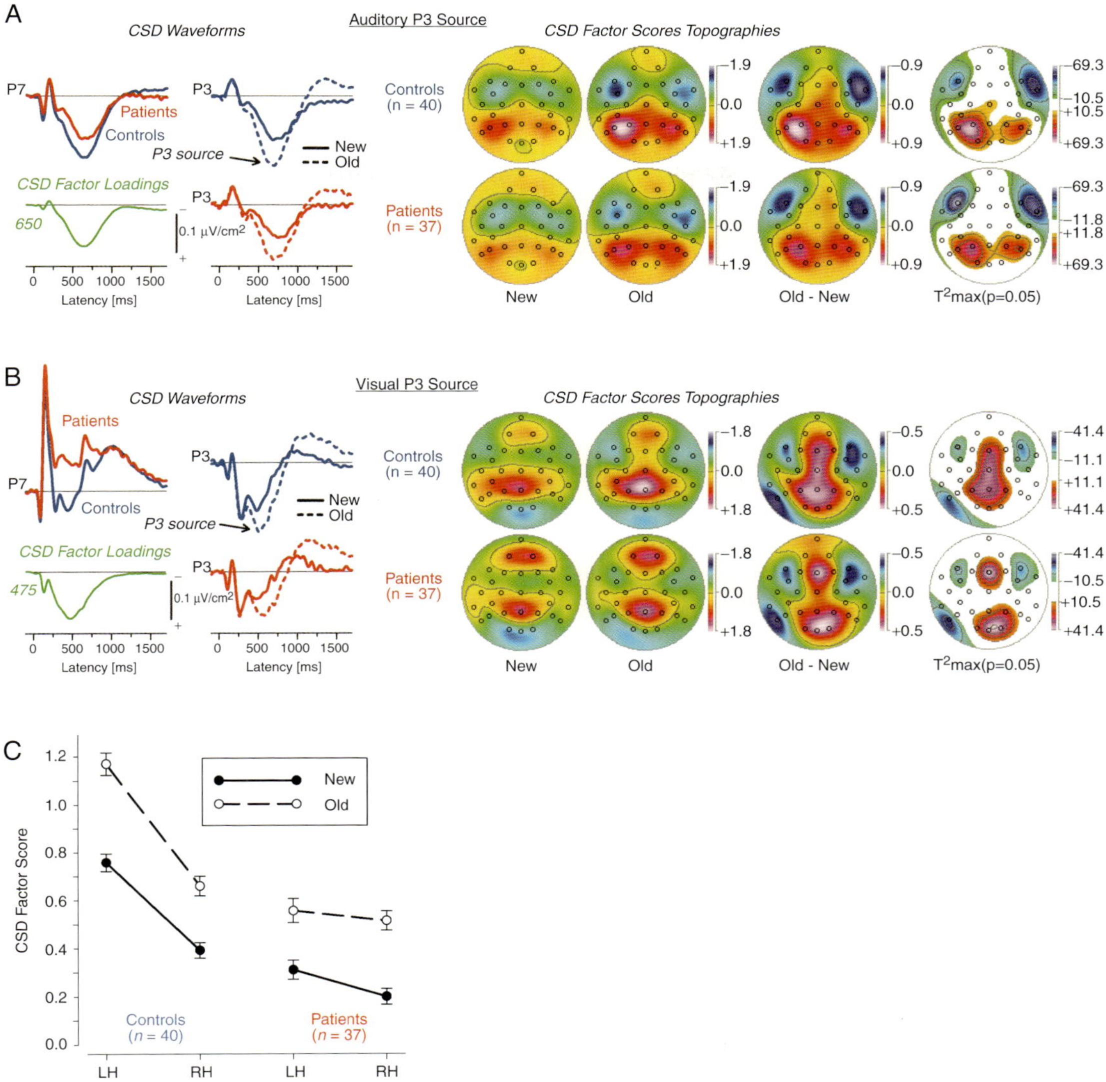

Fig. 20.6. P3 source activity during auditory (A) and visual (B) word recognition memory tasks in 37 depressed patients and 40 healthy controls. Grand mean CSD waveforms at selected left parietal sites illustrate group differences at site P7 and old-new effects for each group at site P3. Significant topographic old-new effects (max T^2 randomization tests) are shown for each modality and group for CSD factors corresponding to the modality-specific P3 source (cf. the CSD factor loadings in the green inset). (C) P3 source means (± SEM) for old and new items (across modality) at lateral parietal sites of the left (LH: P7, P3, CP5) and right (RH: P8, P4, CP6) hemispheres.

and Group by Hemisphere by Condition (new, old; $p < .01$). Healthy adults had overall greater P3 source activity at lateral-parietal sites when compared to depressed patients, particularly over the left hemisphere (Figure 20.6C). Although the condition main effects, indicative of the old-new effect, were highly significant for both groups ($p < .0001$), the old-new effect was larger over the left than the right hemisphere in controls ($p = .01$), but not in patients, and there was a significant simple Group by Condition interaction at the left ($p < .05$) but not the right hemisphere. An analogous ANOVA for the accompanying sink activity at lateral-frontal sites (FC5/6, F7/8, FT9/10) revealed only a marginally significant Group main effect ($p = .07$), but a significant Group by Gender interaction ($p < .05$), stemming from reduced sinks in depressed compared to healthy women, but no group difference in men.

In summary, although the findings show only small behavioral impairment of recognition memory for words in depressed women and none in depressed men, they indicate that the ERP correlate of conscious episodic memory retrieval is reduced in depressed patients over the left parietal region and that this reduction is largely independent of processing modality, which suggests a deficit in accessing semantic (i.e., lexicon) information during continuous word recognition. Event-related functional MRI (fMRI) studies of healthy adults have found that recognition of old words involves a left-lateralized network including the frontal, lateral parietal, and posterior cingulate and the precuneous (Henson et al., 2000). Also, given the evidence of left medial temporal lobe involvement in the old-new effect for words (Johnson, 1995; Rugg et al., 1991; Smith & Halgren, 1989), a distributed network including the hippocampus or other medial temporal lobe structures may also contribute to the reduced old-new effect for depressed patients. Further studies using ERP measures in conjunction with neuroimaging techniques are needed to further clarify the neural basis of the episodic memory deficit in depression.

N1 and Intensity Dependence of Auditory ERPs

Up to this point, we have focused on late cognitive potentials, but studies have also examined earlier negative brain potentials (N1 or N2) in depressed subjects. The N1 potential is known to reflect early sensory processing of stimuli and is also modulated by attention and arousal level (see Chapters 4 and 11, this volume). However, unlike the omnipresent mid-parietal P3b potential, the amplitude and topography of N1 and N2 change considerably with the recording reference, dependent on processing modality. For example, for a visual N1 peaking at approximately 140 ms, the nose-referenced ERP morphology will reveal a distinct inferior parietal negativity, which will reverse into a distinct positive deflection at midparietal sites when rereferencing these ERPs to linked mastoids, leaving only a substantially reduced negative deflection over lateral parietal regions (e.g., Kayser et al., 2003, 2007). In contrast, the auditory N1 peaking at about 100 ms will maintain a central maximum with most common reference schemes, because the direction and location of the known underlying generator within the primary auditory cortex will always result in a midcentral negativity unless a vertex reference is used. The reason is that the reference location, like all other electrodes included in the EEG montage, is an active site, and the differential activity (i.e., the potential difference or ERP) between any two recording sites will tend to be smaller with closer proximity or larger with increasing distance (e.g., cf. Luck, 2005, chap. 3).

There have been reports of reduced N1 amplitude in depressed subjects when compared to non-depressed controls (Burkhart & Thomas, 1993; El Massioui & Lesevre, 1988; Knott & Lapierre, 1987; Sandman et al., 1987). These four studies recorded ERPs primarily at central sites (linked-ears reference) in dichotic listening or tone-counting tasks, but one study used a visual reaction time (RT) task (Knott & Lapierre, 1987). Twice as many studies did not, however, find evidence of N1 reduction in depressed patients in auditory tasks (Blackwood et al., 1987; Bruder et al., 1995, 1998; El Massioui et al., 1996; Knott et al., 1991; Ogura et al., 1993; Sara et al., 1994; Tenke et al., 2008). All four studies that recorded ERPs mainly at midline sites (two linked ears, one left ear, and one nose reference) during binaural oddball tasks found no difference in N1 amplitude between depressed patients and controls (Blackwood et al., 1987; Bruder et al., 1998; Ogura et al., 1993; Sara et al., 1994). The remaining four studies that found no N1 reduction in depressed patients recorded ERPs during different dichotic listening tasks (two linked ears and two nose references). The lack of an N1 reduction in depressed patients was also evident in our findings for a novelty oddball task (see Figures 20.2 and 20.4) and for auditory and visual word recognition memory tasks (Figure 20.6). Although medication differences across studies is not an issue (all patients were off medication), the extent to which differences in clinical characteristics of patients could account for the conflicting findings is unclear.

Although our CSD-PCA study using the novelty oddball task did not focus on N1 sink activity (Tenke et al., 2010), a subsequent analysis of a subgroup of depressed patients who responded favorably to antidepressants showed reduced amplitude of an N1 sink (120 ms loadings peak) to novel sounds. The depressed patients were tested during a pretreatment session and subsequently were treated as part of ongoing clinical trials in which they received 8–12 weeks of monotherapy with escitalopram or another selective serotonin reuptake inhibitor (SSRI), the noradrenaline/dopamine reuptake inhibitor (NDRI) bupropion, or dual therapy with both SSRI and NDRI antidepressants. Following treatment, the Clinical Global Impression: Improvement (CGI-I) scale was used by an independent clinician to rate the treatment response of the patients. Responders (rated as being much or very much improved) showed reduced N1 sink activity (maximum anterior to the Sylvian fissure) compared to either nonresponders ($p < .05$) or healthy controls ($p = .01$), whereas no difference was found between nonresponders and controls (see the blue regions in Figure 20.7A). Although samples were small, it is interesting to note that responders to monotherapy had the smallest N1 and nonresponders to dual therapy had the largest N1 (see Figure 20.7B). The CSD-PCA topographies (Figure 20.7A) indicate that N1 sinks were coupled with sources posterior to the Sylvian fissure and are thereby consistent with tangentially oriented generators in or adjacent to primary auditory cortex. Given the high serotonergic innervation of primary auditory cortex (Campbell et al., 1987; Lewis et al., 1986), it is possible that reduced pretreatment N1 to novel distracter sounds may reflect a lower level of serotonin neuronal activity in responders. Interestingly, a study of the effects of tryptophan depletion on mismatch negativity (MMN) in healthy adults suggested that decreased serotonin may reduce involuntary attention shifting to task-irrelevant sounds (Ahveninen et al., 2002). Further study should therefore examine whether the reduced N1 sink activity in depressed patients who respond favorably to antidepressants may be associated with decreased automatic directing of attention to task-irrelevant novel sounds.

There is also evidence that the intensity dependence of early auditory ERPs (N1-P2) may be of value for identifying a subgroup of depressed patients with a serotonin deficit responsive to treatment with antidepressants that act on the serotonergic neural system. Increases in tone intensity from 60 to 100 dB are known to result in a linear increase in N1-P2 amplitude in healthy adults. Hegerl and Juckel (1993) reviewed findings from basic and clinical studies suggesting that the slope of the function relating tone intensity and N1-P2 amplitude provides a noninvasive indicator of central serotonergic activity. Juckel et al. (1999) found direct evidence of an inverse relationship between serotonergic neural activity in the dorsal raphé and intensity dependence of auditory ERPs recorded from primary auditory cortex in cats. Hegerl and Juckel (2001) suggest that serotonergic neurons modulate activity in primary auditory cortex by providing a stable tonic firing rate. A high firing rate of serotonergic neurons is associated with a weak intensity dependence, that is, only a small increase in N1/P2 amplitude with increasing tone intensity, whereas a low tonic firing rate is related to a strong intensity dependence, that is, a large increase in

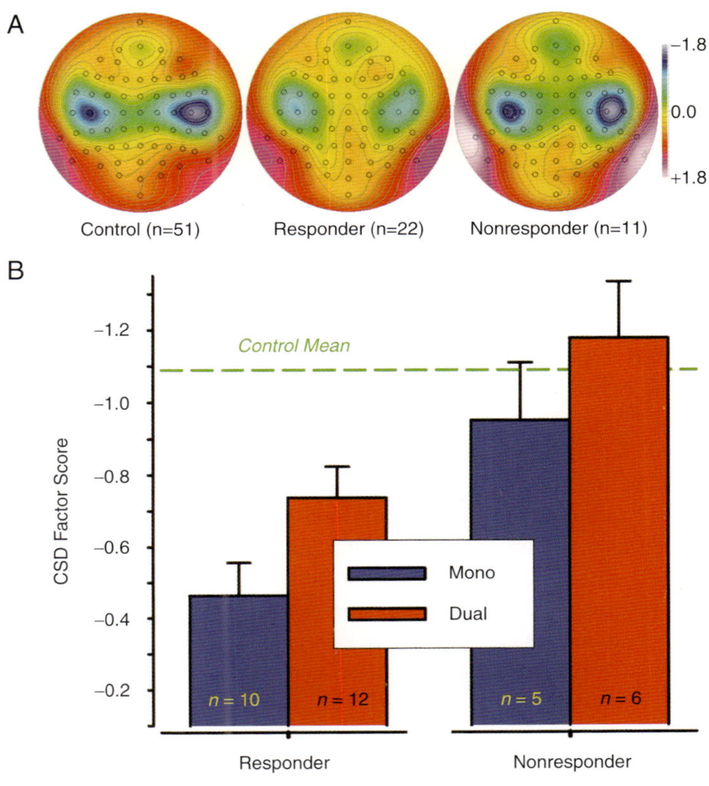

Fig. 20.7. (A) N1 sink (factor 120) topography for 51 healthy controls, 22 treatment responders, and 11 treatment nonresponders. (B) N1 sink means (± SEM) for subgroups of treatment responders and nonresponders to monotherapy or dual therapy.

N1/P2 amplitude. Depressed patients having low serotonergic activity, as evidenced by pronounced intensity dependence of N1-P2 potentials before treatment, responded better to an SSRI antidepressant compared to patients having evidence of high serotonergic activity (Gallinat et al., 2000; Hegerl & Juckel, 2001; Paige et al., 1994). Paige et al. (1995) also found that small samples of responders (n = 4) and nonresponders (n = 4) to the NDRI bupropion showed similar differences in intensity dependence, which could raise questions about the specificity of this finding to SSRI antidepressants. Three studies do, however, suggest that the relation of intensity dependence of auditory ERPs and clinical improvement differs for serotonergic and noradrenergic antidepressants. Linka et al. (2004) tested 16 inpatients having an MDD episode before receiving 3–4 weeks of treatment with the SSRI citalopram. Stronger intensity dependence of N1 was associated with a greater decrease in depression following treatment, which is in accord with earlier findings for SSRIs (Gallinat et al., 2000; Hegerl & Juckel, 2001; Paige et al., 1994). In their next study (Linka et al., 2005), 14 inpatients having a major depressive episode were tested before receiving the selective noradrenaline reuptake inhibitor (NARI) reboxetine. In contrast to findings for SSRIs, *smaller* intensity dependence of N1 was associated with greater improvement in depression following 3–4 weeks of treatment with an NARI antidepressant. Patients were not, however, randomly assigned to treatment, which weakens the comparison of findings for the SSRI and NARI antidepressants. More recently, Mulert et al. (2007) measured the intensity dependence in depressed patients who were randomly assigned to treatment with either the SSRI citalopram or the noradrenergic antidepressant reboxetine. Indices of intensity dependence were obtained using LORETA analyses to measure the tomographic current source distribution in primary auditory cortex for the latency window 60–240 ms following stimulus onset. The authors found a significant difference between citalopram responders (n = 7) and nonresponders (n = 4), with responders showing the expected stronger intensity dependence. In contrast, reboxetine responders (n = 3) and nonresponders (n = 6) did not show a significant difference in intensity dependence. These are encouraging findings, but given the small samples in these studies, further research is needed to investigate the specificity of intensity dependence as a predictor of SSRI treatment response.

If intensity dependence of auditory ERPs provides a marker of central serotonergic activity, the slope of this function would be expected to decrease following treatment with an SSRI. Gallinat et al. (2000) retested 19 depressed patients following 4 weeks of treatment with an SSRI and found no change in the intensity dependence function, which agrees with prior findings for two studies using SSRI or other antidepressants (Paige et al., 1994, 1995). In contrast, a double-blind, placebo-controlled study in healthy adults did find a decrease in the slope of the N1-P2 function during acute administration with a single dose of the SSRI citalopram (Nathan et al., 2006). However, acute depletion of serotonin in healthy adults after tryptophan administration did not affect intensity dependence (Debener et al., 2002; Dierks et al., 1999).

Studies of intensity dependence of auditory ERPs as predictors of the response to antidepressants are of particular interest because of the potential for clinical application, but they have suffered from a number of limitations. Sample sizes were generally small, most studies used open treatment with only a single antidepressant, and retest intervals when patients were on an SSRI were too short to expect significant enhancement of serotonin. Also, a variety of methods have been used to measure intensity dependence, including measures of scalp potentials, LORETA, and dipole source analysis of N1, P2, or N1-P2 difference waveforms. Interestingly, the reliabilities (temporal stability, internal consistency) of intensity-dependent ERP amplitude slope estimates can be substantially improved by using PCA-based as opposed to peak-based amplitude measures (Beauducel et al., 2000), which suggests possible avenues of improvement for predicting treatment response.

Nd, N2, and MMN

Studies have also used auditory ERPs to study selective attention in depressed subjects. Negativity in the region of N1 is known to be greater to attended than unattended stimuli, which has allowed the measurement of "attention-related" N1 and, more specifically, the negative difference (Nd) potential, that is, the difference in ERP to attended and ignored stimuli (see Chapter 11, this volume). Three studies have agreed in finding no difference in attention-related N1 or Nd in depressed subjects and healthy controls (Burkhart & Thomas, 1993; Knott et al., 1991; Massioui & Lesèvre, 1988). Thus, depression does *not* appear to involve a deficit in voluntarily directing attention to specific stimuli.

There is, however, less agreement concerning the N2 potential in depression, with some studies finding increased N2 amplitude in depressed or dysthymic subjects when compared to nondepressed controls (Bruder et al., 1998; Giese-Davis et al., 1993; Sandman et al., 1992) and others finding no difference (Blackwood et al., 1987; Kaiser et al., 2003) or reduced N2 in depressed subjects (Deldin et al., 2000; Massioui & Lesèvre, 1988; Massioui et al., 1996; Sandman et al., 1987). While this difference in N2 findings could stem in part from differences in clinical characteristics of patients, Sara et al. (1994) found no evidence of a relation between N2 amplitude and severity of depression. They did find that drug-free patients having a major depression showed greater N2 amplitude when compared to medicated depressed patients and healthy controls, which differs from the lack of medication effects on P3 in their study. However, the subjects in most studies were off medication, and this is therefore not likely to be a factor. Differences in the tasks used in the above studies may well have contributed to the different findings. Specifically, the three studies finding increased N2 amplitude in depressed subjects used auditory oddball (Bruder et al., 1998), tone discrimination (Giese-Davis et al., 1993), or tone-counting tasks (Sandman et al., 1992). In contrast, two studies finding decreased N2 amplitude used auditory selective attention tasks (Massioui & Lesèvre, 1988; Massioui et al., 1996) and one used a visual recognition memory task (Deldin et al., 2000). Moreover, the studies differed widely in the methods used to compute N2 amplitude. Some studies computed N2 amplitude based on difference waveforms (e.g., target minus nontargets) to reduce the influence of exogenous components (N1, P2), while others used baseline-to-peak or peak-to-peak measures that may have been more affected by these overlapping components. Moreover, N2 identification, and accordingly the experimenter's decision of how and where to measure it, are considerably affected by the choice of ERP recording reference.

As seen for P3, N2 is composed of two or more overlapping subcomponents (Näätänen & Gaillard, 1983). Mismatch negativity or N2a is associated with automatic detection of a mismatch between stimuli (see Chapter 6, this volume). This precedes and overlaps N2b, which is associated with categorization and controlled processing of target stimuli. Both are typically computed by obtaining difference waveforms, subtracting the waveforms for frequent from rare stimuli. The problem is that little attention has been directed to obtaining separate measures of MMN and N2b in depressed patients. In an auditory oddball task, Ogura et al. (1993) measured the mean integrated amplitude of N2 from difference waveforms (rare minus frequent stimuli; linked ears) in 36 unmedicated depressed patients and 36 healthy controls. To obtain estimates of N2 subcomponents, they measured the mean amplitude of N2a in the latency range of 120–165 ms and N2b in the latency range of 170–235 ms. The mean amplitudes were smaller in depressed patients in both the early and late windows. While the N2a estimate for rare stimuli was reduced in depressed patients compared to controls, negativity in the N2b latency range was *greater* to frequent stimuli in depressed patients. The authors concluded that the automatic processing of mismatch was reduced in depressed patients, whereas the later controlled processing of nontargets was more activated in these patients. Another possibility, not considered by the authors, is that N2 and P2 typically overlap in auditory oddball tasks, such that P2 is present for frequent nontarget tones but replaced (or overlapped) by N2 for infrequent target tones (cf. Kayser et al., 1998). In this case, the authors' findings for frequent stimuli could be interpreted as a reduced nontarget P2 in depressed patients. A critical limitation of this study, however, is that N2a was not obtained in a standard MMN paradigm, where subjects do not attend to tones and parameters are optimized for measuring MMN. Giese-Davis et al. (1993) used the paradigm of Sams et al. (1983) to provide separate measures of N2a (150–250 ms) and N2b (150–350 ms) using difference waveforms (linked mastoids). They found no difference between dysthymic subjects and controls in N2a in an ignore condition, but dysthymics had markedly greater N2b than controls. Umbricht et al. (2003) also found no difference in MMN (nose reference) between 22 depressed patients and 25 healthy controls in a standard paradigm.

Sumich et al. (2006) compared the amplitude of N2 (linked mastoids) to target tones in 70 subclinically depressed subjects (i.e., those scoring 2 or more on the Depression Anxiety and Stress Scale) and 70 subjects with no signs of depression. While these groups did not differ in overall level of N2, the nondepressed subjects showed greater N2 amplitude over the right than left central sites, but the subclinically depressed subjects did not show a hemispheric asymmetry of N2. Although in a different modality, this parallels the finding of reduced N2 amplitude over right parietal sites in depressed patients during the processing of pleasant faces (Deldin et al., 2000). These findings are also of interest given reports that depressed patients show reduced P3 amplitude over right temporoparietal sites (Kawasaki et al., 2004) or fail to show the right-greater-than-left P3 asymmetry seen in healthy controls for tonal stimuli (Bruder et al., 1998) or emotional pictures (Kayser et al., 2000). The above findings support the hypothesis that depression is associated with reduced activation of right temporoparietal regions during the processing of tonal or emotional stimuli.

N2 has also been measured in tasks designed to study conflict processing or response inhibition in depressed patients. In the visual modality, a negative potential, N270, was measured in 25 unmedicated depressed patients and 25 matched controls at frontal (F3/4) and parietal (P3/4) electrodes (linked ears reference) during an S1-S2 paradigm (Mao et al., 2005). Subjects indicated whether the S2 stimulus (colored dot) matched the S1 stimulus or was a mismatch. The N270 potential was elicited to S2 stimuli that differed from the S1 stimulus and was measured from its peak amplitude in the difference waveform (mismatch minus match conditions). Depressed patients had smaller N270 amplitude in the difference waveforms compared to controls at frontal and parietal electrode sites. Mao et al. interpreted the reduced N270 as evidence of impairment of a "conflict processing system" involving anterior cingulate and dorsal lateral prefrontal cortex. This system, which is active under mismatch or stimulus discrepancy conditions, is thought to involve the same brain processes as response conflict or error detection. In the auditory modality, Kaiser et al. (2003) measured 61-channel ERPs (average reference) of 16 medicated depressed patients and 16 healthy controls during a go/no-go task. The go task was a modification of an auditory oddball task, but the no-go task required inhibition of responses to rare tones. Depressed patients did not differ from controls in performance or ERPs during the go task, but they performed more poorly than controls in the no-go task. Also, the patients showed a reduction of inferior frontotemporal positivity in the N2 latency range (i.e., polarity-inverted N2) during the no-go task. The authors interpreted this as suggesting a deficit in response inhibition in depression, which is thought to involve a prefrontal executive control system.

Error-Related Negativity and Posterror Processing

Following errors in two-choice RT tasks, such as go/no-go or Eriksen flanker tasks, there is an increase in response-locked frontocentral negativity referred to as *error-related negativity* (ERN) or *error negativity* (N_e; see Chapter 10, this volume). This component peaks 50–150 ms following an incorrect response and is maximum over midline frontocentral sites. The ERN has been considered an electrophysiological index of a response-monitoring or conflict detection with likely generators in the region of the ACC (Dehaene et al., 1994; Ruchsow et al., 2002; van Veen & Carter, 2002; see Falkenstein et al., 2000, for a review). Given the substantial evidence for the role of the ACC in depression (Drevets, 2000), it is not surprising that studies have found abnormalities of ERN in depressed subjects. Chiu and Deldin (2007) measured ERN (linked mastoids) in 18 individuals having a current major depressive episode and 17 nondepressed controls during an arrow flanker task in which a target arrow was flanked by congruent, incongruent, or neutral distracters and subjects responded in the direction of the target arrow. Subjects were also given accuracy feedback under reward, punishment, or neutral conditions. The amplitude of the ERN was greatest at frontal and frontocentral sites, and the depressed group showed greater ERN amplitude than the controls, particularly in the punishment condition. More recently, Holmes and Pizzagalli (2008) measured the ERN (129-channel montage, average reference) of 20 unmedicated patients with MDD and 20 matched healthy controls during a Stroop task. The depressed patients had significantly larger ERN than the controls. Using LORETA analyses, the authors found that depressed patients, relative to controls, showed greater current density in rostral ACC and medial prefrontal cortex at the time of maximal ERN (80 ms following errors). Moreover, functional connectivity analyses revealed that activity in these regions was correlated with subsequent activity in left dorsolateral prefrontal cortex in healthy controls but not in depressed patients. This supported the authors' hypothesis that exaggerated error processing (i.e., increased ERN) in depressed patients is not followed by recruitment of prefrontal-based cognitive control.

There is evidence that enhanced ERN depends on the severity of depression or negative affect. First, in the study by Chiu and Deldin (2007), the magnitude of ERN in their neutral condition was larger in subjects with more severe self-ratings of depression. Second, Tucker et al. (2003) found evidence of larger feedback-related negativity in subjects having a major depression when compared to nondepressed controls; this difference was greatest in subjects with moderate depression but was less in those with more severe depression. Third, two studies measured ERN during an Eriksen flanker task or a go/no-go task in patients having a major depressive disorder in remission and found no difference between the patients and controls (Ruchsow et al., 2004, 2006). Moreover, remitted depressed patients in both studies showed *less* ERN than controls for error trials following another error.

Enhanced ERN is not specific to depression but is seen in children and adults having an obsessive-compulsive disorder (Gehring et al., 2000; Hajcak et al., 2008). Moreover, college students with high scores on scales measuring general negative affect have greater ERN amplitude than those with lower negative affect (Hajcak et al., 2004; Luu et al., 2000). Given that negative affect is evident in both depression and anxiety, these findings are consistent with the conclusion that enhanced ERN is present in both depressive and anxiety disorders. This raises the question of whether increased ERN is associated with depression per se or with the anxiety states that often accompany depressive disorders. No study has directly compared ERN in patients having a depressive disorder alone, an anxiety disorder alone, or comorbidity of these disorders. Importantly, two studies in elderly depressed patients suggest that elevated ERN is associated with a poor outcome of treatment with an SSRI antidepressant. In their initial study, Kalayam and Alexopoulos (2003) measured ERN (linked mastoids) in 22 elderly depressed patients (over 60 years old) during a Stroop interference task. They compared the ERN of 13 patients whose depression remitted during 6 weeks of treatment with citalopram and 9 unremitted patients. The unremitted patients had greater ERN than the remitted patients, with the greatest difference between groups at the left frontal site (F3). Greater ERN amplitude at this site was correlated with less change in depressive symptoms during treatment and with abnormal initiation/perseveration scores on the Mattis Dementia Rating Scale. The authors hypothesized that ACC dysfunction contributes to limited improvement in depression during treatment. In their next study, Alexopoulos et al. (2007) recorded ERPs (128-channel montage, average reference) of 12 elderly depressed patients in an emotional go/no-go task, citing neuroimaging evidence that this task activates the rostral ACC. The six patients who remained symptomatic after 8 weeks of treatment had larger ERNs at midline frontal and frontocentral sites when compared to the six patients who remitted. The nonremitters also had a smaller amplitude of error-related positivity 150–350 ms after an incorrect response. These findings are intriguing given evidence from neuroimaging and electrophysiological studies linking increased rostral ACC activity and clinical response to antidepressants (Mayberg et al., 1997; Pizzagalli et al., 2001). The studies do, however, have several limitations. The samples were extremely small, there was no placebo control group, and the lack of a normal control group makes it difficult to know whether the nonremitters had abnormally large ERNs or remitters abnormally small ERNs. Also, findings for geriatric depression may not generalize to younger depressed patients.

There is also the question of why *both* increased ERN and dysfunction of rostral ACC should be related to a poorer response to antidepressants. A possible explanation is provided by the EEG findings of Pizzagalli and his associates (2006). They measured the resting EEG before subjects performed an Eriksen flanker task. Subjects having high scores on the Beck Depression Inventory, unlike subjects with low scores, showed lower accuracy after incorrect than correct trials. Also, topographic analyses of resting EEG using LORETA indicated that depressed subjects had reduced pretask gamma band activity localized to the region of the rostral ACC. Also, higher pretask gamma was predictive of posterror adjustment in behavioral performance. The authors interpreted these findings as suggesting that depressed subjects have deficits in pretask tonic activity in rostral ACC and in making behavioral adjustments after errors. Although increased affective reactions to errors in depressed patients might be expected to be associated with greater ERN during task performance, no ERP data were reported in this study.

Pizzagalli et al. (2001) previously reported that greater resting EEG theta activity, localized by LORETA to the rostral region of the ACC, was predictive of a more favorable response to treatment with the antidepressant nortriptyline. They suggested that in treatment responders, rostral ACC hyperactivity prior to treatment may reflect increased sensitivity to affective conflict or the ability to monitor the outcome of actions and adjust behavior—for example, by making posterror behavioral adjustments. In treatment nonresponders, this adaptive action monitoring may be dysfunctional, leading to reduced posterror behavioral adjustments. Following suggestions that resting and task-related theta at anterior midline sites may reflect a common process (Tenke & Kayser, 2005; Tzur &

Berger, 2007; Wang et al., 2005), it may be predicted that treatment nonresponders, as compared to responders, will have smaller resting EEG theta activity and greater ERN associated with failure to adjust their behavior during task performance. Future studies measuring both resting EEG and ERN in depressed adults prior to treatment are needed to evaluate this hypothesis.

Further study should also examine the possible relation of ERN abnormalities in depression and anxiety disorders to neurotransmitter systems. In this regard, there is evidence that the serotonin transporter gene (5-HTTLRP) is associated with ERN amplitude in healthy subjects (Fallgatter et al., 2004). Subjects with one or two copies of the low-activity 5-HTTLRP short genotype showed greater ERN compared to those homozygous for the long allele. Individuals with the short allele are at increased risk for developing depression in response to stress, which raises the possibility that heightened ERN may be a risk marker for a form of depression that responds poorly to treatment with antidepressants.

Conclusions

Cognitive P3 Potential

The classical P3 potential with a parietal maximum has been extensively studied in depressed patients. Although there have been conflicting findings, the general trend is for depressed patients to show some reduction of P3 to target stimuli in an auditory oddball task (mean effect size = 0.85). Differences in P3 findings across studies could well stem from differences in the clinical features of the patients, with patients having melancholic or psychotic depressions having the largest P3 reductions. Reduced P3 amplitude in depressed patients is at least partially state-dependent, in that clinical improvement during treatment is accompanied by an increase in P3. Although there are fewer reports of abnormal P3 latency in depression, there is evidence that P3 latency is longer in patients having a bipolar or melancholic depression who typically show psychomotor retardation or cognitive slowing.

The moderate size of the P3 reduction in depression may also stem from the use of simple oddball tasks with relatively little cognitive demand. Studies measuring P3 in cognitively challenging auditory or visual tasks have more consistently found a P3 reduction in depressed patients. Also, depressed patients show an overall reduced P3 to visually presented affective stimuli, and they fail to show greater late P3 amplitude to negative as compared to neutral stimuli, which is seen in healthy adults. Overall, the P3 decrement in depressed patients suggests a deficit in temporoparietal regions involved in context updating, memory, and emotional processing, although frontal regions may also play a role. P3 reduction is not, however, specific to depression, but is also seen in other neuropsychiatric disorders that display cognitive deficits, such as schizophrenia, alcoholism, Alzheimer's disease, and Parkinson's disease (Jeon & Polich, 2003; Polich & Herbst, 2000; see also Chapters 18, 19, and 21, this volume).

One of the problems is that most studies of depressed subjects have not differentiated P3 subcomponents. The P3a or novelty P3 has a more frontocentral distribution than the classical P3b potential and may contribute to the P3 reduction in depressed patients. P3a or novelty P3 is reduced in depressed patients but is *increased* in patients having an anxiety disorder. This also underscores the importance of taking the patient's specific clinical features, and particularly comorbidity of depression and anxiety, into account. The reduced novelty P3 in depression suggests an orienting deficit or dysfunction of automatic switching of attention to task-irrelevant stimuli, which is thought to involve prefrontal, anterior cingulate, and hippocampal regions.

Old-New Effect during Recognition Memory Tasks

The increased late positive potential to correctly recognized words, that is, the old-new effect, is thought to be a neurophysiological correlate of conscious episodic memory retrieval. Studies of continuous word recognition indicate that this old-new effect is reduced in depressed patients. Given evidence of left medial temporal involvement in the old-new effect for words, the reduced old-new effect is consistent with neuroimaging findings of reduced hippocampal volumes in depressed patients. Although the old-new deficit appears to be greatest over the left parietal region, further study is needed to determine whether this is specific to recognition memory for words or occurs for nonverbal stimuli as well.

N1 and Nd Potentials

Most studies have not found a reduction of N1 amplitude in depressed patients. Also, studies have agreed in finding no difference between depressed subjects and controls in attention-related N1 or the Nd potential. These findings argue against any deficit in early sensory processing or voluntarily directing attention in depressed patients. A subgroup of depressed patients who responded favorably to antidepressants was found to have reduced auditory N1 to novel distracters, which may be related to the extensive serotonergic innervation of primary auditory cortex. There is evidence that the intensity dependence of early auditory potentials (N1-P2) provides an index of central serotonergic activity. Depressed patients with pronounced intensity dependence of N1-P2 prior to treatment had a better response to treatment with SSRI antidepressants when compared to patients with less intensity dependence. Small sample sizes and methodological weaknesses in studies of intensity dependence do, however, limit the strength of the conclusions from these studies. Also, further study is needed to examine the extent to which findings are specific to SSRI antidepressants or generalize to other classes of antidepressants with different mechanisms of action.

N2 and MMN Potentials

N2 amplitude in depressed or dysthymic subjects has been reported to be increased, decreased, or no different when compared to healthy controls. Differences in the tasks, clinical characteristics of patients, or medication status may have contributed to these different findings. Methodological difficulties in identifying and measuring N2, particularly when using an EEG montage with a limited number of recording channels, have also contributed to inconsistent findings. Tasks involving simple discrimination or counting of tones showed increased N2 in depressed patients, whereas those involving more complex decisions or response inhibition (e.g., selective attention or recognition memory tasks) were more likely to show decreased N2. Studies have reported that depressed patients have reduced amplitude of potentials in the N2 latency range during visual S1-S2 or auditory go/no-go tasks under stimulus mismatch or response conflict conditions. It was suggested that executive control systems, involving prefrontal cortex and ACC, may be responsible for these deficits. As is the case for P3 studies of depressed subjects, little attention has been directed to separating N2 subcomponents (i.e., MMN and N2b). Two studies that measured MMN under standard ignore conditions found no difference between dysthymic or depressed patients and controls, while two studies found evidence of enhanced N2b in dysthymic or depressed patients. Further study of MMN and N2b in depressed patients using conditions known to maximize these potentials, as well as techniques for separately measuring them, are needed to draw more definitive conclusions.

Performance-Monitoring Potentials

Healthy adults show an increase in frontocentral negativity following errors in two-choice RT tasks or following negative feedback. Studies have found that ERN is *heightened* not only in depressed subjects, but also children or adults having anxiety disorders and in college students with general negative affect. This condition is therefore not specific to depressive disorders and is likely to be particularly high in subjects having comorbidity of depression and anxiety, but this has yet to be studied. The clinical relevance stems from the finding that ERN is greater in elderly patients whose depression failed to remit following treatment with an SSRI antidepressant when compared to remitters. Given evidence that ERN is generated in medial frontal areas in or near the ACC, these findings are consistent with EEG and neuroimaging evidence that the rostral ACC is associated with a clinical response to antidepressants. However, the sample sizes in these studies were very small and the studies lacked healthy adult or placebo control groups. Heightened ERN has also been found in healthy adults with the low-activity serotonin transporter allele (5-HTTLRP short), suggesting that heightened ERN may be a risk marker for developing a form of depression that responds poorly to treatment.

Future Directions

Given the conflicting P3 findings for depressed patients during a standard two-stimulus auditory oddball task, continued use of this task in depressed patients would appear to be of limited value. On the other hand, ERPs continue to be a useful tool for studying the nature of cognitive deficits in depression and for providing information about their neurophysiological underpinnings. It is of value to measure ERPs during more challenging cognitive tasks that allow evaluation of hypotheses concerning the specific cognitive or neurophysiological deficits in depression. For instance, the measurement of N2 and P3 during go/no-go tasks can be used to test hypotheses about response inhibition or conflict monitoring in depression (Donkers & van Boxtel, 2004; Kaiser et al., 2003). Event-related negativity can also be measured during go/no-go or flanker tasks to test hypotheses concerning performance monitoring and ACC dysfunction in depression (Chiu & Deldin, 2007; Ruchsow et al., 2006).

Similarly, measuring the old-new effect during recognition memory tasks provides a means for evaluating hypotheses concerning conscious memory retrieval deficits in depression and their neurophysiolological correlates (Kayser et al., 2007).

Future studies of the N2 and P3 potentials in depression should also differentiate subcomponents that involve different cognitive operations and neuronal generators. The novelty oddball task can be of particular value for measuring P3a or novelty P3 in depression (Bruder et al., 2009; Tenke et al., 2010). It is also important to use techniques that are capable of providing separate measures of neuronal sources underlying these subcomponents—for example, combined CSD-PCA measures (Kayser & Tenke, 2006a, 2006b). In general, the use of denser EEG montages and appropriate methods to capture the temporal and spatial dynamics of ERPs and define ERP components would be advantageous (Kayser & Tenke, 2005).

Surprisingly few studies have used ERPs to study the processing of emotional information in depressed patients. Event-related potentials can be used to measure neurophysiological responses to emotional stimuli without a cognitive task in order to yield a purer measure of affective processing in depressed patients (Kayser et al., 1997, 2000). In this regard, one area that is ripe for exploration is the measurement of ERPs to olfactory stimuli (Pause et al., 2003). Measurement of ERPs to olfactory stimuli provides a window for studying neurophysiological correlates of emotional processing because of the direct projections to cortical and limbic structures that are known to be involved in emotional processing and depression, in particular the amygdala, orbitofrontal cortex, and medial temporal cortex. On the other hand, there is also growing interest in the interaction of cognitive control and emotional processing regions (Ochsner & Gross, 2005), which can be readily studied with ERPs—for example, using cognitive control or interference tasks with emotional distracters (Fales et al., 2008).

In future ERP studies of cognitive and affective processing, it will also be important to examine subtypes of depression, but this requires sufficiently large samples because small subgroups ($n < 10$) are of limited value. Comparison of ERPs in depressed versus control groups is a useful first step, but it fails to account adequately for the clinical and biological heterogeneity of depressive disorders. Studies comparing ERPs in subtypes with different symptom features are particularly important when studying P3 subcomponents (Bruder et al., 2002; Pierson et al., 1996). The influence of comorbidity with anxiety also needs further attention in these studies, as well as in those of performance monitoring (i.e., ERN) in depression. Another useful approach is to subtype patients on the basis of their clinical response to antidepressants with a specific mechanism of action, which would require even larger samples. Some of the most promising findings concern the relation between ERN and intensity dependence of auditory ERPs (N1-P2) to the outcome of treatment with antidepressants. Studies with larger samples comparing the value of these measures for predicting response to different classes of antidepressants (e.g., SSRI or NDRI antidepressants) are needed to confirm the specificity of their relation to SSRI antidepressants. Further study should also be directed to determining the neurotransmitter systems that may be related to ERP abnormalities in depressed patients. In addition to the relation of the serotonin system to intensity dependence of N1-P2 and ERN, there is evidence that P3 subcomponents are dependent on dopamine or norepinephrine systems (Polich, 2007; Turetsky & Fein, 2002). Pharmacological studies measuring ERPs before and during acute treatment with drugs that selectively act on specific neurotransmitter systems would be of value here.

Findings linking the serotonin transporter gene (5-HTTLPR) with both intensity dependence of N1-P2 (Strobel et al., 2003) and ERN (Fallgatter et al., 2004) also indicate the importance of additional study of the genetic correlates of ERP abnormalities in depressed patients. Lastly, both intensity dependence of N1-P2 (Hegerl & Juckel, 1993) and ERN amplitude in depressed subjects (Pizzagalli et al., 2006) have been hypothesized to be related to pretest tonic EEG activity in specific frequency bands. Studies should therefore examine how findings for antidepressant responders and nonresponders on these ERP measures depend on differences in resting EEG oscillations.

Acknowledgments

This work was supported by National Institute of Mental Health grants MH36295 (GEB) and MH58346 (JK). We thank Drs. Jonathan Stewart and Patrick McGrath and the staff of the Depression Evaluation Service at the New York State Psychiatric Institute, where the patients in our studies were recruited, diagnosed, and treated with antidepressants.

Note

1 No recording reference anywhere on the human body can be considered neutral or inactive (e.g., sternum, neck, mastoid, nose, ear lobe), and any site will be differentially affected by a given combination of neuronal generators through volume-conducted activity (see Kayser & Tenke, 2006a). The choice of the reference is, therefore, essential for identifying both spatial and temporal information in ERP recordings, as the reference will invariably affect the spatiotemporal activation of ERP generator patterns. Although some reference choices may enhance or reduce a particular generator topography, all reference schemes, including a montage-dependent average reference, are subject to the same reference problem. Using multiple reference schemes may help the recognition of distinct ERP components, but it will not solve the reference problem.

References

Ahveninen, J., Kahkonen, S., Pennanen, S., Liesivuori, J., Ilmoniemi, R. J., & Jaaskelainen, I. P. (2002). Tryptophan depletion effects on EEG and MEG responses suggest serotonergic modulation of auditory involuntary attention in humans. *Neuroimage, 16*, 1052–1061.

Alexopoulos, G. S., Murphy, C. F., Gunning-Dixon, F. M., Kalayam, B., Katz, R., Kanellopoulos, D., Etwaroo, G. R., Klimstra, S., & Foxe, J. J. (2007). Event-related potentials in an emotional go/no-go task and remission of geriatric depression. *Neuroreport*, 18, 217–221.

American Psychiatric Association (1994). *Diagnostic and statistical manual of mental disorders* (4th ed.). Washington, DC: Author.

Ancy, J., Gangadhar, B. N., & Janakiramaiah, N. (1996). "Normal" P300 amplitude predicts rapid response to ECT in melancholia. *Journal of Affective Disorders, 41*, 211–215.

Anderer, P., Saletu, B., Semlitsch, H. V., & Pascual-Marqui, R. D. (2002). Structural and energetic processes related to P300: LORETA findings in depression and effects of antidepressant drugs. *Methods and Findings in Experimental and Clinical Pharmacology, 24*, 85–91.

Austin, M. P., Mitchell, P., & Goodwin, G. M. (2001). Cognitive deficits in depression. *British Journal of Psychiatry, 178*, 200–206.

Bange, F., & Bathien, N. (1998). Visual cognitive dysfunction in depression: An event- related potential study. *Electroencephalography and Clinical Neurophysiology, 108*, 472–481.

Beauducel, A., Debener, S., Brocke, B., & Kayser, J. (2000). On the reliability of augmenting/reducing: Peak amplitudes and principal components analysis of auditory evoked potentials. *Journal of Psychophysiology, 14*, 226–240.

Blackburn, I. M., Roxborough, H. M., Muir, W. J., Glabus, M., & Blackwood, D. H. R. (1990). Perceptual and physiological dysfunction in depression. *Psychological Medicine, 20*, 95–103.

Blackwood, D. H. R., Whalley, L. J., Christie, J. E., Blackburn, I. M., St. Clair D. M., & McInnes, A. (1987). Changes in auditory P3 event-related potential in schizophrenia and depression. *British Journal of Psychiatry, 150*, 154–160.

Bremner, J. D., Vythilingam, M., Vermetten, E., Vaccarino, V., & Charney, D. S. (2004). Deficits in hippocampal and anterior cingulate functioning during verbal declarative memory encoding in midlife major depression. *American Journal of Psychiatry, 161*, 637–645.

Bruder, G.E. (1992). P300 findings for depressive and anxiety disorders. *Annuals of the New York Academy of Sciences, 658*, 205–222.

Bruder, G.E., Kayser, J., Tenke, C. E., Leite, P., Schneier, F. R., Stewart, J. W., & Quitkin, F. M. (2002). Cognitive ERPs in depressive and anxiety disorders during tonal and phonetic oddball tasks. *Clinical Electroencephalography, 33*, 119–124.

Bruder, G.E., Kroppmann, C.J., Kayser, J., Stewart, J.W., McGrath, P.J., & Tenke, C.E. (2009). Reduced brain responses to novel sounds in depression: P3 finding in a novelty oddball task. *Psychiatry Research, 170*, 218–223.

Bruder, G. E., Quitkin, F. M., Stewart, J. W., Martin, C., Voglmaier, M. M., & Harrison, W. M. (1989). Cerebral laterality and depression: Differences in perceptual asymmetry among diagnostic subtypes. *Journal of Abnormal Psychology, 98*, 177–186.

Bruder, G. E., Tenke, C. E., Stewart, J. W., Towey, J. P., Leite, P., Voglmaier, M., & Quitkin, F. M. (1995). Brain event-related potentials to complex tones in depressed patients: Relations to perceptual asymmetry and clinical features. *Psychophysiology, 32*, 373–381.

Bruder, G. E., Tenke, C. E., Towey, J. P., Leite, P., Fong, R., Stewart, J. E., McGrath, P. J., & Quitkin, F. M. (1998). Brain ERPs of depressed patients to complex tones in an oddball task: Relation of reduced P3 asymmetry to physical anhedonia. *Psychophysiology, 35*, 54–63.

Bruder, G. E., Tenke, C. E., Warner, V., & Weissman, M. M. (2007). Grandchildren at high and low risk for depression differ in EEG measures of regional brain asymmetry. *Biological Psychiatry, 62*, 1317–1323.

Bruder, G. E., Towey, J. P., Stewart, J. W., Friedman, D., Tenke, C., & Quitkin, F. M. (1991). Event-related potentials in depression: Influence of task, stimulus hemifield and clinical features on P3 latency. *Biological Psychiatry, 30*, 233–246.

Burkhart, M. A., & Thomas, D. G. (1993). Event-related potential measures of attention in moderately depressed subjects. *Electroencephalography and Clinical Neurophysiology, 88*, 42–50.

Burt, D. B., Zembar, M. J., & Niederehe, G. (1995). Depression and memory impairment: A meta-analysis of the association, its pattern, and specificity. *Psychological Bulletin, 117*, 285–305.

Cacioppo, J. T., Crites, S. L., Bernston, G. G., & Coles, M. G. (1993). If attitudes affect how stimuli are processed, should they not affect the event-related potential? *Psychological Science, 4*, 108–112.

Cacioppo, J. T., Crites, S. L., & Gardner, W. L. (1996). Attitudes to the right: Evaluative processing is associated with lateralized late positive event-related potentials. *Personality and Social Psychology Bulletin, 22*, 1205–1219.

Campbell, M. J., Lewis, D. A., Foote, S. L., & Morrison, J. H. (1987). Distribution of choline acetyltransferase-, serotonin-, dopamine-beta-hydroxylase-, tyrosine hydroxylase-immunoreactive fibers in monkey primary auditory cortex. *Journal of Comparative Neurology, 261*, 209–220.

Castaneda, A. E., Tuulio-Henriksson, A., Marttunen, M., Suvisaari, J., & Lönnqvist, J. (2008). A review on cognitive impairments in depressive and anxiety disorders with a focus on young adults. *Journal of Affective Disorders, 106*, 1–27.

Cavanagh, J., & Geisler, M. W. (2006). Mood effects on the ERP processing of emotional intensity in faces: A P3 investigation with depressed students. *International Journal of Psychophysiology, 60*, 27–33.

Chiu, P. H., & Deldin, P. J. (2007). Neural evidence for enhanced error detection in major depressive disorder. *American Journal of Psychiatry, 164*, 608–616.

Debener, S., Strobel, A., Kurschner, K., Kranczioch, C., Hebenstreit, J., Maercker, A., Beauducel, A., & Brocke, B. (2002). Is auditory evoked potential augmenting/reducing affected by acute tryptophan depletion? *Biological Psychology, 59*, 121–133.

Dehaene, S., Posner, M., & Tucker, D. (1994). Commentary: Localization of a neural system for error detection and compensation. *Psychological Medicine, 5*, 303–305.

Deldin, P. J., Deveney, C. M., Kim, A. S., Casas, B. R., & Best, J. L. (2001a). A slow wave investigation of working memory biases in mood disorders. *Journal of Abnormal Psychology, 110*, 267–281.

Deldin, P. J., Keller, J., Gergen, J. A., & Miller, G. A. (2000). Right-posterior face processing anomaly in depression. *Journal of Abnormal Psychology, 109*, 116–121.

Deldin, P. J., Keller, J., Gergen, J. A., & Miller, G. A. (2001b). Cognitive bias and emotion in neuropsychological models of depression. *Cognition and Emotion, 15*, 787–802.

Deveney, C. M., & Deldin, P. J. (2004). Memory of faces: A slow wave ERP study of major depression. *Emotion, 4*, 295–304.

Dien, J., Spencer, K. M., & Donchin., E. (2003). Localization of the event-related potential novelty response as defined by principal components analysis. *Cognitive Brain Research, 17*, 637–650.

Dierks, T., Barta, S., Demisch, L., Schmeck, K., Englert, E., Kewitz, A., Maurer, K., & Poustka, F. (1999). Intensity dependence of auditory evoked potentials (AEPs) as biological marker for cerebral serotonin levels: Effects of tryptophan depletion in healthy subjects. *Psychopharmacology, 146*, 101–107.

Dietrich, D. E., Emrich, H. M., Waller, C., Wieringa, B. M., Johannes, S., & Munte, T. F. (2000a). Emotion/cognition-coupling in word recognition memory of depressive patients: An event-related potential study. *Psychiatry Research, 96*, 15–29.

Dietrich, D. E., Kleinschmidt, A., Hauser, U., Schneider, U., Spannhuth, C. W., Kipp, K., Huber, T. J., Wieringa, B. M., Emrich, H. M., & Johannes, S. (2000b). Word recognition memory before and after successful treatment of depression. *Pharmacopsychiatry, 33*, 221–228.

Diner, B. C., Holcomb, P. J., & Dykman, R. A. (1985). P300 in major depressive disorder. *Psychiatry Research, 15*, 175–184.

Donkers, F. C. L., & van Boxtel, G. J. M. (2004). The N2 in go/no-go tasks reflects conflict monitoring, not response inhibition. *Brain and Cognition, 56*, 165–176.

Drevets, W. (2000). Neuroimaging studies of mood disorders. *Biological Psychiatry, 48*, 813–829.

Drevets, W. C., Price, J. L., Simpson, J. R., Jr., Todd, R. D., Reich, T., Vannier, M., & Raichle, M. E. (1997). Subgenual prefrontal cortex abnormalities in mood disorders. *Nature, 386*, 824–827.

Ehrlichman, H., & Bastone, L. (1992). Olfaction and emotion. In J. Serby & K. L. Chobor (Eds.), *Science of olfaction* (pp. 410–438). New York: Springer.

El Massioui, F., Everett, J., Martin, M. T., Jouvent, R., & Widlocher, D. (1996). Attention deficits in depression: An electrophysiological marker. *Neuroreport, 7*, 2483–2486.

El Massioui, F., & Lesèvre, N. (1988). Attention impairment and psychomotor retardation in depressed patients: An event-related potential study. *Electroencephalography and Clinical Neurophysiology, 70*, 46–55.

Elger, C. E., Grunwald, T., Lehnertz, K., Kutas, M., Helmstaedter, C., Brockhaus A, Van Roost, D., & Heinze, H. J. (1997). Human temporal lobe potentials in verbal learning and memory processes. *Neuropsychologia, 35*, 657–667.

Fales, C. L., Barch, D. M., Rundle, M. M., Mintun, M. A., Snyder, A. Z., Cohen, J. D., Mathews, J., & Sheline, Y. I. (2008). Altered emotional interference processing in affective and cognitive-control brain circuitry in major depression. *Biological Psychiatry, 63*, 377–384.

Falkenstein, M., Hoormann, J., Christ, S., & Hohnsbein, J. (2000). ERP components on reaction errors and their functional significance: A tutorial. *Biological Psychology, 51*, 87–107.

Fallgatter, A. J., Herrmann, M. J., Roemmler, J., Ehlis, A., Wagener, A., Heidrich, A., Ortega, G., Zeng, Y., & Lesch, K. (2004). Allelic variation of serotonin transporter function modulates the brain electrical response for error processing. *Neuropsychopharmacology, 29*, 1506–1511.

Friedman, D. (1990). ERPs during continuous recognition memory for words. *Biological Psychology, 30*, 61–87.

Friedman, D., Cycowicz, Y. M., & Gaeta, H. (2001). The novelty P3: An event-related brain potential (ERP) sign of the brain's evaluation of novelty. *Neuroscience and Biobehavioral Reviews, 25*, 355–373.

Friedman, D., Simpson, G., & Hamberger, M. (1993). Age-related changes in scalp topography to novel and target stimuli. *Psychophysiology, 30*, 383–396.

Gallinat, J., Bottlender, R., Juckel, G., Munke Puchner, A., Stotz, G., Kuss, H. J., Mavrogiorgou, P., & Hegerl, U. (2000). The loudness dependency of the auditory evoked N1/P2-component as a predictor of the acute SSRI response in depression. *Psychopharmacology, 148*, 404–411.

Gangadhar, B. N., Ancy, J., Janakiramaiah, N., & Umapathy, C. (1993). P300 amplitude in non-bipolar, melancholic depression. *Journal of Affective Disorders, 28*, 57–60.

Gehring, W. J., Himle, J., & Nisenson, L. G. (2000). Action-monitoring dysfunction in obsessive-compulsive disorder. *Psychological Science, 11*, 1–6.

Giese-Davis, J. E., Miller, G. A., & Knight, R. A. (1993). Memory template comparison processes in anhedonia and dysthymia. *Psychophysiology, 30*, 646–656.

Hajcak, G., Franklin, M. E., Foa, E. B., & Simons, R. F. (2008). Increased error-related brain activity in pediatric obsessive-compulsive disorder before and after treatment. *American Journal of Psychiatry, 165*, 116–123.

Hajcak, G., McDonald, N., & Simons, R. F. (2004). Error-related psychophysiology and negative affect. *Brain and Cognition, 56*, 189–197.

Halgren, E., Baudena, P., Clarke, J. M., Heit, G., Marinkovic, K., Devaux, B., Vignal, J. P., & Biraben, A. (1995). Intracerebral potentials to rare target and distractor auditory and visual stimuli. II. Medial, lateral and posterior temporal lobe. *Electroencephalography and Clinical Neurophysiology, 94*, 229–250.

Hansenne, M., Pitchot, W., Moreno, A. G., Zaldua, I. U., & Ansseau, M. (1996). Suicidal behavior in depressive disorder: An event-related potential study. *Biological Psychiatry, 40*, 116–122.

Hegerl, U., Gallinat, J., & Juckel, G. (2001). Event-related potentials: Do they reflect central serotonergic neurotransmission and do they predict clinical response to serotonin agonists? *Journal of Affective Disorders, 62*, 93–100.

Hegerl, U., & Juckel, G. (1993). Intensity dependence of auditory evoked potentials as an indicator of central serotonergic neurotransmission: A new hypothesis. *Biological Psychiatry, 33*, 173–187.

Heller, W. (1990). The neuropsychology of emotion: Developmental patterns and implications for psychopathology. In N. L. Stein, B. Leventhal, & T. Trabasso (Eds.), *Psychological and biological approaches to emotion*, (pp. 167–211). Hillside, NJ: Erlbaum.

Heller, W. (1993). Neuropsychological mechanisms of individual differences in emotion, personality and arousal. *Neuropsychology, 7*, 476–489.

Heller, W., Etienne, M. A., & Miller, G. A. (1995). Patterns of perceptual asymmetry in depression and anxiety: Implications for neuropsychological models of emotion and psychopathology. *Journal of Abnormal Psychology, 104*, 327–333.

Henson, R. N., Rugg, M. D., Shallice, T., & Dolan, R. J. (2000). Confidence in recognition memory for words: Dissociating right prefrontal roles in episodic retrieval. *Journal of Cognitive Neuroscience, 12*, 913–923.

Holmes, A. J., & Pizzagalli, D. A. (2008). Spatiotemporal dynamics of error processing dysfunctions in major depressive disorder. *Archives of General Psychiatry, 65*, 179–188.

Houston, R. J., Bauer, L. O., & Hesselbrock, V. M. (2003). Depression and familial risk for substance dependence: A P300 study of young women. *Psychiatry Research: Neuroimaging, 124*, 49–62.

Houston, R. J., Bauer, L. O., & Hesselbrock, V. M. (2004). P300 evidence of cognitive inflexibility in female adolescents at risk for recurrent depression. *Progress in Neuro-Psychopharmacology and Biological Psychiatry, 28*, 529–536.

Jeon, Y.W., & Polich, J. (2003). Meta-analysis of P300 and schizophrenia: Patients, paradigms, and practical implications. *Psychophysiology, 40*, 684–701.

Johnson, R., Jr. (1995). Event-related potential insights into the neurobiology of memory systems. In F. Boller & J. Grafman (Eds.), *Handbook of neuropsychology* (Vol. 10, pp. 135–164). Amsterdam: Elsevier.

Johnston, V. S., Miller, D. R., & Burleson, M. H. (1986). Multiple P3s to emotional stimuli and their theoretical significance. *Psychophysiology, 23*, 684–694.

Juckel, G., Hegerl, U., Molnar, M., Csepe, V., & Karmos, G. (1999). Auditory evoked potentials reflect serotonergic neuronal activity—a study in behaving cats administered drugs acting on 5-HT1A autoreceptors in the dorsal raphe nucleus. *Neuropsychopharmacology, 21*, 710–716.

Kaiser, S., Unger, J., Kiefer, M., Markela, J., Mundt, C., & Weisbrod, M. (2003). Executive control deficit in depression: Event-related potentials in a Go/Nogo task. *Psychiatry Research*, 122, 169–184.

Kalayam, B., & Alexopoulos, G. S. (2003). A preliminary study of left frontal region error negativity and symptom improvement in geriatric depression. *American Journal of Psychiatry, 160*, 2054–2056.

Karaaslan, F., Gonul, A. S., Oguz, A., Erdinc, E., & Esel, E. (2003). P300 changes in major depressive disorders with and without psychotic features. *Journal of Affective Disorders, 73*, 283–287.

Kaustio, O., Partanen, J., Valkonen-Korhonen, M., Viinamäki, H., & Lehtonen, J. (2002). Affective and psychotic symptoms relate to different types of P300 alteration in depressive disorder. *Journal of Affective Disorders, 71*, 43–50.

Kawasaki, T., Tanaka, S., Wang, J., Hokama, H., & Hiramatsu, K. (2004). Abnormalities of P300 cortical current density in unmedicated depressed patients revealed by LORETA analysis of event-related potentials. *Psychiatry and Clinical Neurosciences, 58*, 68–75.

Kayser, J., Bruder, G. E., Tenke, C. E., Stewart, J. W., & Quitkin, F. M. (2000). Event-related potentials (ERPs) to hemifield presentations of emotional stimuli: Differences between depressed patients and healthy adults in P3 amplitude and asymmetry. *International Journal of Psychophysiology, 36*, 211–236.

Kayser, J., Fong, R., Tenke, C. E., & Bruder, G. E. (2003). Event-related brain potentials during auditory and visual word recognition memory tasks. *Cognitive Brain Research, 16*, 11–25.

Kayser, J., & Tenke, C.E. (2003). Optimizing PCA methodology for ERP component identification and measurement: Theoretical rationale and empirical evaluation. *Clinical Neurophysiology, 114*, 2307–2325.

Kayser, J., & Tenke, C. (2005). Trusting in or breaking with convention: Towards a renaissance of principal components analysis in electrophysiology. *Clinical Neurophysiology, 116*(8), 1747–1753.

Kayser, J., & Tenke, C. E. (2006a). Principal components analysis of Laplacian waveforms as a generic method for identifying ERP generator patterns: I. Evaluation with auditory oddball tasks. *Clinical Neurophysiology, 117*, 348–368.

Kayser, J., & Tenke, C. E. (2006b). Principal components analysis of Laplacian waveforms as a generic method for identifying ERP generator patterns: II. Adequacy of low-density estimates. *Clinical Neurophysiology, 117*, 369–380.

Kayser, J., & Tenke, C. E. (2006c). Consensus on PCA for ERP data, and sensibility of unrestricted solutions. *Clinical Neurophysiology, 117*, 703–707.

Kayser, J., Tenke, C. E., & Bruder, G. E. (1998). Dissociation of brain ERP topographies for tonal and phonetic oddball tasks. *Psychophysiology, 35*, 576–590.

Kayser, J., Tenke, C. E., Gates, N. A., & Bruder, G. E. (2007). Reference-independent ERP old/new effects of auditory and visual word recognition memory: Joint extraction of stimulus- and response-locked neuronal generator patterns. *Psychophysiology, 44*, 949–967.

Kayser, J., Tenke, C. E., Gates, N. A., Stewart, J. W., McGrath, P. J., & Bruder, G.E. (in press). Event-related brain potentials (ERPs) during auditory and visual word recognition memory tasks in affective disorders.

Kayser, J., Tenke, C., Nordby, H., Hammerborg, D., Hugdahl, K., & Erdmann, G. (1997). Event-related potential (ERP) asymmetries to emotional stimuli in a visual half-field paradigm. *Psychophysiology, 34*, 414–426.

Kiehl, K. A., Laurens, K. R., Duty, T. L., Forster, B. B., & Liddle, P. F. (2001). Neural sources involved in auditory target detection and novelty processing: An event-related fMRI study. *Psychophysiology, 38*, 133–142.

Kimble, M., Kaloupek, D., Kaufman, M., & Deldin, P. (2000). Stimulus novelty differentially affects attentional allocation in PTSD. *Biological Psychiatry, 47*, 880–890.

Knight, R. (1996). Contribution of human hippocampal region to novelty detection. *Nature, 383*, 256–259.

Knight, R. T., & Scabiani, D. (1998). Anatomic bases of event-related potentials and their relationship to novelty detection in humans. *Journal of Clinical Neurophysiology, 15*, 3–13.

Knott, V., Lapierre, Y., Griffiths, L., de Lugt, D., & Bakish, D. (1991). Event-related potentials and selective attention in major depressive illness. *Journal of Affective Disorders, 23*, 43–48.

Knott, V. J., & Lapierre, Y. D. (1987). Electrophysiological and behavioral correlates of pychomotor responsivity in depression. *Biological Psychiatry, 22*, 313–324.

Lang, P. J., Bradley, M. M., & Cuthbert, B. N. (1999). *International Affective Picture System (IAPS): Technical manual and affective ratings* Gainesville: Center for Research in Psychophysiology, University of Florida.

Lewis, D. A., Campbell, M. J., Foote, S. L., & Morrison, J. H. (1986). The monoaminergic innervation of primate neocortex. *Human Neurobiology, 5*, 181–188.

Linka, T., Müller, B. W., Bender, S., & Sartory, G. (2004). The STET intensity dependence of the auditory evoked N1 component as a predictor of response to Citalopram treatment in patients with major depression. *Neuroscience Letters, 367*, 375–378.

Linka, T., Müller, B. W., Bender, S., Sartory, G., & Gastpar, M. (2005). The intensity dependence of auditory evoked ERP components predicts responsiveness to Reboxetine treatment in major depression. *Pharmacopsychiatry, 38*, 139–143.

Luck, S. J. (2005). *An introduction to the event-related potential technique*. Cambridge, MA: MIT Press.

Luu, P., Collins, P., & Tucker, D. M. (2000). Mood, personality, and self-monitoring: Negative affect and emotionality in relation to frontal lobe mechanisms of error monitoring. *Journal of Experimental Psychology: General, 129*, 43–60.

Mao, W., Wang, Y., & Wang, D. (2005). Cognitive impairment in major depressive disorder revealed by event-related potential N270. *Clinical Electroencephalography and Neuroscience, 36*, 9–14.

Maris, E. (2004). Randomization tests for ERP topographies and whole spatiotemporal data matrices. *Psychophysiology, 41*, 142–151.

Massioui, F., Everett, J., Martin, M. T., Jouvent, R., & Widlöcher, D. (1996). Attention deficits in depression: An electrophysiological marker. *Neuroreport, 7*, 2483–2486.

Massioui, F., & Lesèvre, N. (1988). Attention impairment and psychomotor retardation in depressed patients: An event-related potential study. *Electroencephalography and Clinical Neurophysiology*, 70, 46–55.

Mayberg, H. S., Brannan, S. K., Mahurin, R. K., Jerabek, P. A., Brickman, J. S., Tekell, J. L., Silva, J. A., McGinnis, S., Glass, T. G., Martin, C. C., & Fox, P. T. (1997). Cingulate function in depression: A potential predictor of treatment resonse. *Neuroreport, 8*, 1057–1061.

Mecklinger, A., & Ullsperger, P. (1995). The P300 to novel and target events: A spatiotemporal dipole model analysis. *Neuroreport, 7*, 241–245.

Muir, W. J., St. Clair, D. M., & Blackwood, D. H. R. (1991). Long-latency auditory event-related potentials in schizophrenia and in bipolar and unipolar affective disorder. *Psychological Medicine, 21*, 867–879.

Mulert, C., Juckel, G., Brunnmeier, M., Karch, S. Leicht, G., Mergl, R., Moller, H. J., Hegerl, U., & Pogarell, O. (2007). Prediction of treatment response in major depression: Integration of concepts. *Journal of Affective Disorders, 98*, 215–225.

Näätänen, R., & Gaillard, A. W. K. (1983). The orienting reflex and the N2 deflection of the event-related potential (ERP). In A. W. K. Gaillard & W. Ritter (Eds.), *Tutorials in event related potential research: Endogenous components* (pp. 119–141). Amsterdam: North-Holland.

Nathan, P. J., Segrave, R., Phan, K. L., O'Neill, B., & Croft, R. J. (2006). Direct evidence that acutely enhancing serotonin with the selective serotonin reuptake inhibitor citalopram modulates the loudness dependence of the auditory evoked potential (LDAEP) marker of central serotonin function. *Human Psychopharmacology, 21*, 47–52.

Naumann, E., Bartussek, D., Diedrich, O., & Laufer, M. (1992). Assessing cognitive and affective information processing functions of the brain by the means of the late positive complex of the event-related potential. *Journal of Psychophysiology, 6*, 285–298.

Neuhaus, A. H., Luborzewski, A., Rentzsch, J., Brakemeier, E. L., Opgen-Rhein, C., Gallinat, J., & Bajbouj, M. (2007). P300 is enhanced in responders to vagus nerve stimulation for treatment of major depressive disorder. *Journal of Affective Disorders, 100*, 123–128.

Nunez, P. (1981). *Electric fields of the brain* New York: Oxford University Press.

Nunez, P. L., & Srinivasan, R. (2006). *Electric fields of the brain: The neurophysics of EEG* New York: Oxford University Press.

Nurminen, M., Valkonen-Korhonen, M., Mervaala, E., Paakkonen, A., Partanen, J., Viinamaki, H., & Lehtonen, J. (2005). Enhanced attention-dependent auditory processing by electroconvulsive therapy in psychotic depression. *Journal of ECT, 21*, 19–24.

Ochsner, K., & Gross, J. J. (2005). The cognitive control of emotion. *Trends in Cognitive Sciences, 9*, 242–249.

Ogura, C., Nageishi,Y., Omura, F., Fukao, K., Ohta, H., Kishimoto, A., & Matsubayashi, M. (1993). N200 component of event-related potentials in depression. *Biological Psychiatry, 33*, 720–726.

Otto, M. W., Bruder, G. E., Fava, M., Delis, D. C., Quitkin, F. M., & Rosenbaum, J. F. (1994). Norms for depressed patients for the California Verbal Learning Test: Associations with depression severity and self-report of cognitive difficulties. *Archives of Clinical Neuropsychology, 9*, 81–88.

Paige, S. R., Fitzpatrick, D. F., Kline, J. P., Balogh, S. E., & Hendricks, S. E. (1994). Event-related potential amplitude/intensity slopes predict response to antidepressants. *Neuropsychobiology, 30*, 197–201.

Paige, S. R., Hendricks, S. E., Fitzpatrick, D. F., Balogh, S., & Burke, W. J. (1995). Amplitude/intensity functions of auditory event-related potentials predict responsiveness to bupropion in major depressive disorder. *Psychopharmacology Bulletin, 31*, 243–248.

Palomba, D., Angrilli, A., & Mini, A. (1997). Visual evoked potentials, heart rate responses and memory to emotional pictorial stimuli. *International Journal of Psychophysiology, 27*, 55–67.

Pause, B. M., Miranda, A., Nysterud, M., & Ferstl, R. (2000). Olfactory and emotional stimulus evaluation in patients with major depression. *Zeitschrift für Klinische Psychololgie und Psychotherapie, 29*, 16–23.

Pause, B. M., Raack, N., Sojka, B., Goder, R., Aldenhoff, J. B., & Ferstl, R. (2003). Convergent and divergent effects of odors and emotions in depression. *Psychophysiology, 40*, 209–225.

Perrin, F., Pernier, J., Bertrand, O., & Echallier, J. F. (1989). Spherical splines for scalp potential and current density mapping [Corrigenda EEG 02274, EEG Neurophysiol, 1990, 76, 565]. *Electroencephalography and Clinical Neurophysiology, 72*, 184–187.

Pierson, A., Ragot, R., Van Hooff, J., Partiot, A., Renault, B., & Jouvent, R. (1996). Heterogeneity of information-processing alterations according to dimensions of depression: An event-related potentials study. *Biological Psychiatry, 40*, 98–115.

Pizzagalli, D., Pascual-Marqui, R. D., Nitschke, J. B., Oakes, T. R., Larson, C. L., Abercrombie, H. C., Schaefer, S. M., Koger, J. V., Benca, R. M., & Davidson, R. J. (2001). Anterior cingulate activity as a predictor of degree of treatment response in major depression: Evidence from brain electrical tomography analysis. *American Journal of Psychiatry, 158*, 405–415.

Pizzagalli, D. A., Peccoralo, L. A., Davidson, R. J., & Cohen, J. D. (2006). Resting anterior cingulate activity and abnormal responses to errors in subjects with elevated depressive symptoms: A 128-channel EEG study. *Human Brain Mapping, 27*, 185–201.

Polich, J. (2007). Updating P300: An integrative theory of P3a and P3b. *Clinical Neurophysiology, 118*, 2128–2148.

Polich, J., & Criado, J. R. (2006). Neuropsychology and neuropharmacology of P3a and P3b. *International Journal of Psychophysiology, 60*, 172–185.

Polich, J., & Herbst, K.L. (2000). P300 as a clinical assay: Rationale, evaluation, and findings. *International Journal of Psychophysiology, 38*, 3–19.

Porter, R. J., Gallagher, P., Thompson, J. M., & Young, A. H. (2003). Neurocognitive impairment in drug-free patients with major depressive disorder. *British Journal of Psychiatry, 182*, 214–220.

Post, R. M., DeLisi, L. E., Holcomb, H. H., Uhde, T. W., Cohen, R., & Buchsbaum, M. S. (1987). Glucose utilization in the temporal cortex of affectively ill patients: Positron emission tomography. *Biological Psychiatry, 22*, 545–553.

Purcell, R., Maruff, P., Kyrios, M., & Pantelis, C. (1997). Neuropsychological function in young patients with unipolar major depression. *Psychological Medicine, 27*, 1277–1285.

Röschke, J., & Wagner, P. (2003). A confirmatory study on the mechanisms behind reduced P300 waves in depression. *Neuropsychopharmacology, 28*, S9–S12.

Roth, W. T., Duncan, C. C., Pfefferbaum, A., & Timsit-Berthier, M. (1986). Applications of cognitive ERPs in psychiatric patients. In W. C. McCallum, R. Zappoli, & F. Denoth (Eds.), *Cerebral psychophysiology: Studies in event-related potentials*, (Suppl. 38, pp. 419–438). Amsterdam: Elsevier.

Ruchsow, M., Grothe, J., Spitzer, M., & Keifer, M. (2002). Human anterior cingulate cortex is activated by negative feedback: Evidence from event-related potentials in a guessing task. *Neuroscience Letters, 325*, 203–206.

Ruchsow, M., Herrnberger, B., Beschoner, P., Grön, G., Spitzer, M., & Kiefer, M. (2006). Error processing in major depressive disorder: Evidence from event-related potentials. *Journal of Psychiatric Research, 40*, 37–46.

Ruchsow, M., Herrnberger, B., Wiesend, C., Gron, G., Spitzer, M., & Kiefer, M. (2004). The effect of erroneous responses on response monitoring in patients with major depressive disorder: A study with event-related potentials. *Psychophysiology, 41*, 833–840.

Rugg, M. D., Roberts, R. C., Potter, D. D., Pickles, C. D., & Nagy, M. E. (1991). Event-related potentials related to recognition memory. Effects of unilateral temporal lobectomy and temporal lobe epilepsy. *Brain, 114*(5), 2313–2332.

Sams, M., Alho, K., & Näätänen, R. (1983). Sequential effects on the ERP in discriminating two stimuli. *Biological Psychology, 17*, 41–58.

Sandman, C., Gerner, R., O'Halloran, J., & Isenhart, R. (1987). Event-related potentials and item recognition in depressed, schizophrenic and alcoholic patients. *International Journal of Pyschophysiology, 5*, 215–225.

Sandman, C. A., Vigor-Zierk, C. S., Isenhart, R., Wu, J., & Zetin, M. (1992). Cardiovascular phase relationships to the cortical event-related potential of schizophrenic, depressed and normal subjects. *Biological Psychiatry, 32*, 778–789.

Santosh, P. J., Malhotra, S., Raghunathan, M., & Mehra, Y. N. (1994). A study of P300 in melancholic depression—correlation with psychotic features. *Biological Psychiatry, 35*, 474–479.

Sapolsky, R. M. (2000). Glucocorticoids and hippocampal atrophy in neuropsychiatric disorders. *Archives of General Psychiatry, 57*, 925–935.

Sara, G., Gordon, E., Kraiuhin, C., Coyle, S., Howson, A., & Meares, R. (1994). The P300 ERP component: An index of cognitive dysfunction in depression? *Journal of Affective Disorders, 31*, 29–38.

Schlegel, S., Nieber, D., Herrmann, C., & Bakauski, E. (1991). Latencies of the P300 component of the auditory event-related potential in depression are related to the Bech-Rafaelsen Melancholia Scale but not to the Hamilton Rating Scale for Depression. *Acta Psychiatrica Scandinavica, 83*, 438–440.

Shah, P. J., Ebmeier, K. P., Glabus, M. F., & Goodwin, G. M. (1998). Cortical grey matter reductions associated with treatment-resistant chronic unipolar depression. Controlled magnetic resonance imaging study. *British Journal of Psychiatry, 172*, 527–532.

Shestyuk, A. Y., Deldin, P. J., Brand, J. E., & Deveney, C. M. (2005). Reduced sustained brain activity during processing of positive emotional stimuli in major depression. *Biological Psychiatry, 57*, 1089–1096.

Sidtis, J. J. (1981). The complex tone test: Implication for the assessment of auditory laterality effects. *Neuropsychologia, 19*, 103–112.

Siegle, G. J., Steinhauer, S. R., & Thase, M. E., (2004). Pupillary assessment and computational modeling of the Stroop task in depression. *International Journal of Psychophysiology, 52*, 63–76.

Simons, R., Graham, F., Miles, M., & Chen, X. (2001). On the relationship of P3a and the novelty-P3. *Biological Psychology, 56*, 207–218.

Smith, M. E., & Halgren, E. (1989). Dissociation of recognition memory components following temporal lobe lesions. *Journal of Experimental Psychology: Learning, Memory, and Cognition, 15*, 50–60.

Smith, M. E., Stapleton, J. M., & Halgren, E. (1986). Human medial temporal lobe potentials evoked in memory and language tasks. *Electroencephalography and Clinical Neurophysiology, 63*, 145–159.

Spencer, K., Dien, J., & Donchin, E. (1999). A componential analysis of the ERP elicited by novel events using a dense electrode array. *Psychophysiology, 36*, 409–414.

Steiner, J. E., Lidar-Lifschitz, D., & Perl, E. (1993). Taste and odor: Reactivity in depressive disorders, a multidisciplinary approach. *Perceptual and Motor Skills, 77,* 1331–1346.

Strobel, A., Debener, S., Schmidt, D., Hünnerkopf, R., Lesch, K. P., & Brocke, B. (2003). Allelic variation in serotonin transporter function associated with the intensity dependence of the auditory evoked potential. *American Journal of Medical Genetics, 118B,* 41–47.

Sumich, A. L., Kumari, V., Heasman, B. C., Gordon, E., & Brammer, M. (2006). Abnormal asymmetry of N200 and P300 event-related potentials in subclinical depression. *Journal of Affective Disorders, 92,* 171–183.

Tenke, C. E., Bruder, G. E., Towey, J. P., Leite, P., & Sidtis, J. J. (1993). Correspondence between brain ERP and behavioral asymmetries in a dichotic complex tone test. *Psychophysiology, 30,* 62–70.

Tenke, C. E., & Kayser, J. (2005). Reference-free quantification of EEG spectra: Combining current source density (CSD) and frequency principal components analysis (fPCA). *Clinical Neurophysiology, 116,* 2826–2846.

Tenke, C. E., & Kayser, J. (2008). ERP generators within the longitudinal fissure: Are putative inverses flawed? Program No. 872.28. 2008 Neuroscience Meeting Planner. Washington, DC: Society for Neuroscience. Available at http://psychophysiology.cpmc.columbia.edu/mmedia/sfn2008/sfn2008.pdf.

Tenke, C. E., Kayser, J., Shankman, S. A., Griggs, C. B., Leite, P., Stewart, J. W., & Bruder, G. E. (2008). Hemispatial PCA dissociates temporal from parietal ERP generator patterns: CSD components in healthy adults and depressed patients during a dichotic oddball task. *International Journal of Psychophysiology, 67,* 1–16.

Tenke, C. E., Kayser, J., Stewart, J. W., & Bruder, G. E. (2010). Novelty P3 reductions in depression: Characterization using principal components analysis (PCA) of current source density (CSD) waveforms. *Psychophysiology, 47,* 133–146.

Tucker, D., Luu, P., Frishkoff, F., Quiring, J., & Poulsen, C. (2003). Frontolimbic response to negative feedback in clinical depression. *Journal of Abnormal Psychology, 112,* 667–678.

Turetsky, B. I., & Fein, G. (2002). Alpha2-noradrenergic effects on ERP and behavioral indices of auditory information processing. *Psychophysiology, 39,* 147–157.

Tzur, G., & Berger, A. (2007). When things look wrong: Theta activity in rule violation. *Neuropsychologia, 45,* 3122–3126.

Umbricht, D., Koller, R., Schmid, L., Skrabo, A., Grübel, C., Huber, T., & Stassen, H. (2003). How specific are deficits in mismatch negativity generation to schizophrenia? *Biological Psychiatry, 53,* 1120–1131.

Urretavizcaya, M., Moreno, I., Benlloch, L., Cardoner, N., Serrallonga, J., Menchon, J. M., & Vallejo, J. (2003). Auditory event-related potentials in 50 melancholic patients: Increased N100, N200 and P300 latencies and diminished P300 amplitude. *Journal of Affective Disorders, 74,* 293–297.

Wagner, P., Röschke, J., Fell, J., & Frank, C. (1997). Differential pathophysiological mechanisms of reduced P300 amplitude in schizophrenia and depression: A single trial analysis. *Schizophrenia Research, 25,* 221–229.

Van Veen, V., & Carter, C. (2002). The timing of action-monitoring processes in the anterior cingulate cortex. *Journal of Cognitive Neuroscience, 14,* 593–602.

Vandoolaeghe, E., van Hunsel, F., Nuyten, D., & Maes, M. (1998). Auditory event related potentials in major depression: Prolonged P300 latency and increased P200 amplitude. *Journal of Affective Disorders, 48,* 105–113.

Wang, C., Ulbert, I., Schomer, D. L., Marinkovic, K., & Halgren, E. (2005). Responses of human anterior cingulate cortex microdomains to error detection, conflict monitoring, stimulus-response mapping, familiarity, and orienting. *Journal of Neuroscience, 25,* 604–613.

Yanai, I., Fujikawa, T., Osada, M., Yamawaki, S., & Touhouda, Y. (1997). Changes in auditory P300 in patients with major depression and silent cerebral infarction. *Journal of Affective Disorders, 46,* 263–271.

Zatorre, R. J., Evans, A. C., Meyer, E., & Gjedde, A. (1992). Lateralization of phonetic and pitch discrimination in speech processing. *Science, 256,* 846–849.

Zhang, Y., Hauser, U., Conty, C., Emrich, H. M., & Dietrich, D. E. (2007). Familial risk for depression and P3b component as a possible neurocognitive vulnerability marker. *Neuropsychobiology, 55,* 14–20.

CHAPTER

21

Alterations of ERP Components in Neurodegenerative Diseases

Rolf Verleger

Abstract

This chapter reviews event-related potential (ERP) studies in patients suffering from neurodegenerative diseases. Such studies have been conducted from two different points of view: using ERPs to learn something about the disease and using the disease to learn something about ERPs. This review focuses on the former aspect: the utility of ERPs in the clinic. Thus, ERP research in neurodegenerative diseases will be discussed from the perspective of the insights gained from ERPs (1) for diagnosis, (2) for delineating and understanding the consequences of the disease for cognition, and (3) for determining the prognosis about the course of the disease.

Keywords: event-related potential, neurodegenerative diseases, diagnosis, prognosis

This chapter reviews event-related potential (ERP) studies in patients suffering from neurodegenerative diseases. Such studies have been conducted from two different points of view: using ERPs to learn something about the disease and using the disease to learn something about ERPs. This review will focus on the former aspect: the utility of ERPs in the clinic. Thus, ERP research in neurodegenerative diseases will be discussed from the perspective of the insights gained from ERPs (1) for diagnosis, (2) for delineating and understanding the consequences of the disease for cognition, and (3) for determining the prognosis about the course of the disease.

A further important aim in research on neurodegenerative diseases is to develop effective therapies. To achieve this, mechanisms of neuronal degeneration have to be understood on a molecular level. Except for Parkinson's disease, this has so far not been achieved to such a degree that therapeutic treatment would significantly ameliorate the symptoms or even halt the progress of the disease. The molecular level is not readily accessible to ERPs.

The present chapter updates two previous reviews on this topic. The more recent one (Verleger, 2004) focused on ERP correlates of movement-related problems, whereas in the present review, ERP correlates of cognitive problems are also covered. This topic was also covered in an earlier review (Verleger, 2003), which additionally included neurological syndromes produced by causes other than neurodegeneration: infarction of blood vessels, inflammation, and epilepsy. The present chapter, completed in May 2008, updates the neurodegenerative part of the 2003 review.

Alzheimer's Disease

Alzheimer's disease (AD) is the most common cause of dementia in elderly people. Pathological markers are plaques and tangles within neuronal tissue, but these markers can so far be identified only by neuropathological examination (i.e., after the patient's death). Thus, standard criteria recommend the diagnosis of "probable" AD by diagnosing dementia and excluding other possible causes of dementia (vascular encephalopathy, lack of certain vitamins or hormones, and others).

The Diagnostic Problem

In clinical practice, several diagnostic problems may arise in which ERPs might be useful: (1) Does the person suffer from a dementing illness at all? (2) If so, is it due to AD? (3) Will the person become demented in the near future?

Problem 1 has been tackled in ERP research by measuring delays of P3 latency in the oddball task. Such delays have been found reliably since the study of Goodin et al. (1978; for reviews, see Polich, 1991; Polich & Herbst, 2000; for more recent studies, see, e.g., Frodl et al, 2002; Golob et al., 2007; Muscoso et al., 2006) when patients were more than mildly demented, but not always in those cases of beginning dementia where there was a real need for diagnostic information that would complement the results of neuropsychological testing (e.g., Daffner et al., 2001; Gordon et al., 1986; Polich et al., 1986; Verleger et al., 1992). Other approaches used to distinguish AD patients from healthy persons include measuring the novelty P3 induced by drawings of nonexistent objects (Daffner et al., 2001), using a combination of different ERP components (in a task requiring selective comparison of two out of four consecutive stimuli; Chapman et al., 2007), and predicting scores on neuropsychological tests of memory by structural equation of behavioral and ERP measures recorded in tasks measuring explicit or implicit memory (Hogan et al., 2006). Somewhat disappointingly, in the last approach, the overt behavior rather than the ERPs proved to be a direct predictor of memory-test scores.

With regard to problem 2, vascular dementia (Yamaguchi et al., 2000) and Huntington's disease (Goodin & Aminoff, 1986) could be distinguished rather well from Alzheimer-type dementia by diagnosis-specific abnormalities. However, the relevance of these findings for diagnosing AD is limited, because in practice there is no problem with diagnosing Huntington's disease (by genetic testing) and measuring the scarcity of the blood supply to the brain, leading to vascular dementia, so there is not much need for additional diagnostic help by ERPs.

Problem 3 has become a current topic by focusing on conditions that predispose to the disease.

One such condition may be defined from genetic analysis. Comparing healthy middle-aged persons without a genetic risk to ones who carried the apolipoprotein E ε4 allele and had both one parent and an additional relative affected by AD, Green and Levey (1999) obtained delayed N2 and P3 latencies in the persons at risk. Event-related potentials proved very sensitive, because response times were not delayed and the two groups did not differ in neuropsychological tests of memory (California Verbal Learning Test) and praxis (block design, digit symbol), both of which are very sensitive to AD. A similar result was obtained by Ally et al. (2006) in a less well-controlled study where risk was defined by having a parent suffering from AD without an additional check for the apolipoprotein E ε4 allele, no neuropsychological tests were reported except the mini-mental state test, and there were no overt responses to the oddball targets. Again, P3 latencies were delayed, and additionally P3 amplitudes reduced, in persons at risk.[1]

Another condition predisposing to the development of AD is defined in terms of behavior. *Mild cognitive impairment* (MCI) is the term used to describe elderly individuals who have some decline in their cognitive functions (preferably but not necessarily memory) while still preserving their ability to perform basic everyday activities (Winblad et al., 2004). These persons have a high risk of becoming demented. Golob et al. (2007) studied persons with MCI, patients with AD, and healthy persons in an auditory oddball task (see also Golob et al., 2002, for preliminary results). After 5 years, about half of the MCI group had become demented (MCI-AD). Two features of the ERP at study onset distinguished MCI-AD persons: their P3 latencies were delayed compared to those of the MCI patients who remained stable (MCI-MCI) and to those of the healthy persons, tending toward the markedly delayed latencies of AD patients, and their P50 amplitudes were larger than those of all other groups (healthy people, MCI-MCI patients, and AD patients). Again, ERPs proved to be a sensitive measure, because MCI-AD and MCI-MCI patients differed neither in their response times to targets nor in neuropsychological tests of memory and learning, psychomotor speed, word fluency, and praxis conducted at study onset, except for the Boston Naming Test. However, as the authors conceded, the sensitivity and specificity of even the better predictor, P50, was "well below the level necessary for clinical use as a predictor for future dementia" and "The specificity . . . relative to other types of dementia and other neurological disorders is unknown" (Golob et al., 2007, p. 750).

The P50 increase in the Golob et al. (2007) study, if replicable in other laboratories, is interesting because the U-shaped course from healthy status to AD (small P50 in healthy people and MCI-MCI, large P50 in MCI-AD, reduced P50 in AD) suggests that this increase reflects some process that precedes the overt onset of AD, either pathogenic or compensatory. P50 is an early component originating in auditory cortex. Alzheimer-typical plaques and tangles are usually not found there. Therefore, Golob et al. (2007) speculated that the P50 increase might reflect the lack of inhibition from the affected frontal cortex onto auditory cortex. However, in patients who actually had frontal lesions due to tumor or infarction, P50 was not found to be altered (Chao & Knight, 1998; Knight et al., 1980). Unfortunately, studies from other laboratories that compared patients with AD, patients with MCI, and healthy controls did not report results of this early component (Bennys et al., 2007; Frodl et al., 2002; Gironell et al, 2005; data of the Bennys et al. and Gironell et al. studies also appear to suffer from methodological problems).

Probing the Memory Deficit

In focusing on cortical reflections of AD patients' memory deficit, some studies focused on better diagnosing AD, while others aimed at delineating and understanding the consequences of the disease for cognition. Tools used in this research were mismatch negativity, priming positivity, N400, and the P3 measured in Sternberg's task.

Mismatch negativity (MMN), a measure of preconscious auditory memory (Kujala et al., 2007; see also Chapter 6, this volume), was of equal size in AD patients and healthy controls in Gaeta et al. (1999) and in the 1 s interstimulus interval (ISI) condition of Pekkonen et al. (1994), but it tended to be smaller than in healthy controls in the study of Kazmerski et al. (1997) and was smaller in the 3 s ISI condition of Pekkonen et al. (1994). Thus, the results are equivocal.

Priming positivity is evoked by attended items of a large set (words or pictures) being repeated, either immediately or (with a usually weaker effect) with some items inbetween. This positivity is thought to consist both of a decrease of N400, reflecting facilitated (post)lexical processing, and an increase of P600, reflecting some conscious recollection, though this latter component might be of less relevance in elderly people (Rugg et al., 1997). The priming positivity of AD patients did not differ reliably from that of age-matched healthy participants in the studies of Rugg et al. (1994), Kazmerski et al. (1995), and Kazmerski and Friedman (1997). Thus, these studies demonstrate preserved priming by earlier presentation in AD. However, when repetitions occurred infrequently and at very long intervals, priming positivity disappeared in AD patients, in contrast to their preserved short-term repetition priming and in contrast to the response of healthy persons (Schnyer et al., 1999).

Working memory is needed to judge whether the final word of a heard sentence fits the preceding context. This processing may be measured by the difference in N400 amplitude between nonfitting and fitting words (see Chapter 15, this volume). This difference was greatly reduced in AD patients in two studies (Ford et al., 1996; Revonsuo et al., 1998). Less marked effects were obtained when words were primed by only one preceding word (Schwartz et al., 1996) or picture (Ford et al., 2001) or when spoken sentences were simultaneously presented visually (Wolk et al., 2005), which supports the interpretation that the working memory deficit is the important factor.

In Sternberg's memory search task, a *memory set* of varying size (one to five letters) is presented, followed by single items that either were or were not members of the memory set, requiring an appropriate choice response. As could be expected, performance became worse in AD patients when the memory set was larger, also reflected in the disappearance of the P3 component evoked by single letters (De Toledo-Morrell et al., 1991), but this finding was not replicated, somewhat surprisingly (Swanwick et al., 1997). More data are needed to clarify this issue.

Using the rationale of repetition priming, Olichney et al. (2002, 2006) delineated different degrees of memory affect in AD and in MCI, its possible precursor. Participants had to state whether a visually presented word was or was not an exemplar of some category defined by the experimenter about 1 s before word presentation. Some category–word pairs were repeated across trials. The late positivity evoked by words fitting the category and the N400 evoked by nonfitting words were reduced in these repeating words, obviously reflecting some memory process (as confirmed by the correlation of late-positivity reduction with the results of neuropsychological memory tests). Patients with MCI who developed AD (after 2-year catamneses) lacked the late-positivity reduction when fitting repeated words, but they did not differ clearly from MCI-MCI patients and controls in the N400 reduction with repetition of nonfitting words (Olichney et al., 2002). Thus, they were intermediate between MCI-MCI and AD patients because AD patients lacked both the late-positivity reduction when fitting words were repeated and the N400 reduction when nonfitting words were repeated (Olichney et al., 2006). It is not entirely clear from the study of Olichney et al. (2002) whether the lacking late-positivity reduction allowed a better prediction than neuropsychological tests about who would become demented.

Conclusion

The contribution of ERP research to problems posed by AD has been limited. Most promising appear to be the findings by Golob et al. (2007) and Olichney et al. (2002) about the value of ERP components for predicting whether persons with MCI will become demented. Both findings need replication from other laboratories.

Parkinson's Disease

The main pathological mechanism in Parkinson's disease (PD) is degeneration of dopamine-producing neurons in the brainstem (substantia nigra), depriving the basal ganglia of the dopamine needed for their adequate functioning. The basal ganglia project in different ways via the thalamus to cortical areas, one important target area being the supplementary motor area (SMA). The cardinal symptoms of PD are stiffness, lack of movement, and tremor during rest. There is still a gap between what is known about the neuroanatomical basis of PD and what this means in terms of behavior. Patients with PD frequently suffer from a slight diffuse impairment of cognitive functions, reminiscent of frontal-lobe pathology. Event-related potential research might play a role in filling this gap, providing insights into the motor and nonmotor impairments of PD patients. A methodological obstacle for ERPs is that activity of the basal ganglia cannot be measured at the scalp, simply because of their distance to the scalp and also presumably because the basal ganglia constitute a network of mutually activating and inhibiting modules (e.g., Gilbert, 2001; Stocco et al., 2010), which further decreases the probability that their sum of activation will have an impact on scalp measurements. What can be measured by ERPs are cortical consequences of the basal ganglia deficit.

Movement-Related Activity

The problem of PD patients in initiating movements is most obvious in global movements of the whole body such as standing up and walking. However, methodological considerations (movement artifacts and immobile recording devices) have forced researchers to focus on finger movements, with only few exceptions (Vidailhet et al., 1993).

SELF-INITIATED MOVEMENTS

There is a slight but rather consistent deficit in PD patients' amplitude of the *Bereitschaftspotential* (BP; see Chapter 9, this volume), either at an early phase of the BP or generally throughout its time-course but not focused on the actual motor execution part (Dick et al., 1989; Jahanshahi et al., 1995; Touge et al., 1995; "cues absent" condition of Cunnington et al., 1995). In view of the severe impairment that PD may cause in movement initiation, this deficit is surprisingly small. It was found that BPs were even larger in PD patients than in healthy controls before two key presses performed in a quick, precise sequence by the left and right hands, with an interval of 40–60 ms counted as correct (Fattapposta et al., 2000). This deviant result might be due to the extraordinary effort made by the patients in order to do well. These larger BP amplitudes were reduced to normal values after administration of L-dopa medication (Fattapposta et al., 2002).

A difference pointed out by Cunnington et al. (1995) and occurring only after movement onset remained literally out of sight in other studies: at least in later stages of the disease (Cunnington et al., 1997), PD patients' BPs do not return to baseline as readily as those of healthy subjects, possibly indicating difficulties in the release of motor activation (see also Devos et al., 2002; Verleger et al., 1999a). Further, none of the BP studies used multichannel recordings to find subtle topographical differences between PD patients and healthy persons. Thus, there is a need for further studies on the BP before and after self-initiated movements.

MOVEMENT PREPARATION BEFORE IMPERATIVE SIGNALS

It might be argued (Rockstroh et al., 1989) that PD patients' deficit in movement initiation is underestimated in BP studies because self-initiated movements are actually those in which the deficit was successfully overcome. Perhaps for this reason, the bulk of movement-related research in PD has shifted from the BP to the contingent negative variation (CNV; see Chapter 8, this volume), that is, from self-initiated movements to movement preparation between some announcing signal and an imperative stimulus. Empirical reasons for this shift have not been convincing so far, because in three studies using both approaches, PD patients' reduction of amplitudes was not more distinct with CNV than with BP (Cunnington et al., 1995, though not using an announcing signal proper; Ikeda et al., 1997; Oishi et al., 1995).

The overall finding in these studies that measured CNV before the imperative stimulus was a reduction of PD patients' amplitudes (Cunnington et al., 1995, 1997, 1999; Gerschlager et al., 1999; Ikeda et al., 1997; Linden et al., 1990; Praamstra et al., 1996; Pulvermüller et al., 1996; Wascher et al., 1997; Wright et al., 1993), with the exception of the previously mentioned report by Oishi et al. (1995) and a study on PD patients with marked hemi-parkinsonism (i.e., patients who were much more affected on one side of the body than on the other; Cunnington et al., 2001). The amount and topography of CNV reduction depended on the task, both between and within studies (see Verleger, 2004, for details). This variability is hard to subsume under a general rule and might be related to the fact that CNV is a differing mixture of activations of response preparation, stimulus expectation, and effort (van Boxtel, 1994; Verleger et al., 2000), so each of these processes might be differentially affected in PD, depending on the task and on the patient's status. Of special interest is therefore the study of Mattox et al. (2006), where the stimulus-preceding negativity (SPN) was measured after participants predicted some yes/no outcome and were expecting feedback stimuli indicating gain or loss. The SPN preceding feedback is probably devoid of any component related to motor preparation. In blocks when the amount of money to be gained or lost was high, SPN was larger in healthy participants than in PD patients, because SPN decreased in PD patients from low to high reward blocks while tending to increase in healthy participants.

The reduction of patients' CNV amplitudes may be subject to change: in PD patients, the CNV became indistinguishable from that of normal participants when their subthalamic nuclei were stimulated by implanted electrodes (Gerschlager et al., 1999). In addition, PD patients' amplitudes could be boosted by insertion of a block in which, rather than performing actual movements, they imagined the feelings and sensations when performing the movements (Lim et al., 2006).

Studies that measured the contralateral preponderance of the CNV before the imperative lateralized readiness potential (LRP) signal (after warning signals that had already indicated which hand to use) found that contralateral preponderance was more extended from central sites to frontal sites in PD, both when comparing PD patients to healthy participants (Praamstra et al., 1996) and when comparing the use of the more severely affected hand to the use of the less affected one in hemi-parkinson patients (Cunnington et al., 2001).

Again, as with the BP, these studies converge to the conclusion that it is not movement execution above all that is impaired in PD. Rather, a number of processes involved in response preparation and anticipation in general, depending on the particular task, appear to be vulnerable to this dysfunction of the basal ganglia.

MOVEMENT PREPARATION AFTER AMBIGUOUS IMPERATIVE SIGNALS

Due to overlap of stimulus-related potentials, movement preparation after imperative signals is best investigated by means of the LRP (see Chapter 9, this volume). Two lines of research have used this measure.

In choice-response tasks without a warning signal, Low et al. (2002) found later onset of the LRP in PD patients than in the control group with reference to the imperative stimulus, but not a generally prolonged LRP with reference to response onset. They concluded that the patients' main problem was their impairment in the cognitive processes of mapping the stimuli to associated responses rather than in motor preparation and execution.

Praamstra et al. (1998) explained their above-mentioned 1996 finding of an enlarged area of contralateral activation in PD patients by assuming that these patients were more dependent on the lateral premotor system, making more use of visual stimulation for selecting movements. Indeed, when making imperative arrows ambiguous by surrounding them with irrelevant arrows, Praamstra et al. (1998, 1999) demonstrated that PD patients were more affected than healthy participants by those flanking stimuli: when these flankers suggested responses different from the imperative stimulus, LRPs went farther in the wrong direction in PD patients than in healthy participants. Similarly, the irrelevant lateral position of a relevant stimulus (the *Simon effect*) affected PD patients' motor systems more than those of healthy participants (Praamstra & Plat, 2001): lateral posterior activation contralateral to the relevant stimulus (N2pc) spread to a larger extent to (pre)motor sites in PD patients than in healthy participants. Only weak support for Praamstra's general thesis was provided in a study on subliminal priming (Seiss & Praamstra, 2004), where PD patients' LRPs were enhanced in one condition only, and only if scaled in proportion to the general size of the amplitudes. But findings from the flanker task and from the Simon effect converge on the conclusion that visuomotor transmission is enhanced in PD. This might be interpreted as learned compensatory behavior that is applied by the patients even in inappropriate situations or, as favored by Praamstra and Plat (2001), as pathological reduction of executive control over the motor system. We will come back to this issue below in the discussion of measures of executive control.

Note that Praamstra and colleagues assumed that the effect of enhanced LRP deflections with irrelevant stimulation originates from the premotor cortex rather than the motor cortex. In line with this conclusion, this effect was always obtained with stimuli that maximized spatial information, that is, laterally presented relevant stimuli in Praamstra and Plat (2001) and arrays of arrows in the flanker task (Praamstra et al., 1998, 1999). Thus, this effect is probably not on the LRP proper but on its premotor attentional component, denoted N2cc by Praamstra and Oostenveld (2003). Therefore, the use of arrows might be necessary to obtain this effect in the flanker task (see Wascher et al., 1999, for a direct comparison of LRP effects evoked by letters vs. arrows in this task).

Finally, although not a traditional ERP effect, I would like to mention the promising results obtained by cross-correlating activity of the motor cortex with electromyographic (EMG) firing patterns. Deficits in these correlation patterns could be pinpointed in PD—for example, by Salenius et al. (2002) and by Timmermann et al. (2003).

CONCLUSION

Taken together, the ERP findings on PD patients' movement control appear to be relevant for understanding the disease. Based on the large task-dependent variations of CNV and on the LRP effects mentioned directly above, one general conclusion from these studies might be that the impairment of movement is more intimately linked to the general cognitive syndrome of PD than is frequently realized.

Other Approaches

ODDBALL TASK

Beginning with Hansch et al. (1982), more than a dozen studies used the oddball task to investigate PD. The aims were

- to establish differences among PD patients related to neuropsychological performance, to the presence of dementia, or to medication status
- to investigate whether PD patients differ from healthy participants
- to investigate how PD patients with dementia differ from other demented patients

In retrospect, these goals do not appear to be of continuing relevance. This is not to say that they were irrelevant a priori. However, as with AD, interindividual variability proved to be too large to allow ERPs to be used for diagnosis. The main result, summarized by Ebmeier (1992), was that P2, N2, and P3 peak latencies tended to be delayed in PD, more so in more cognitively impaired patients. Note that this is not dissimilar to AD and, moreover, does not allow for the clinically relevant distinction between PD and multisystem atrophy (Pirtošek et al., 2001).

MEASURES OF EXECUTIVE CONTROL

Some diffuse cognitive impairment is common in PD, reminiscent of frontal lobe pathology. Event-related potentials offer the opportunity to measure signs of this pathology uncontaminated by PD patients' impairment in overt behavior.

Tsuchiya et al. (2000) interspersed occasional novel, unusual sounds among target and nontarget sounds in an oddball task. The orienting responses of PD patients to novel sounds were impaired, as reflected by reduced and delayed frontal P3 specifically to the novel stimuli. Since such responses to novel sounds are drastically reduced in patients with lesions of the frontal lobe (Knight, 1984), this is evidence for impaired frontal-lobe functioning.

A similar argument may be made for the no-go P3, which is often larger and more anteriorly distributed than the go-P3 in situations where go and no-go stimuli are equally probable (e.g., Verleger et al., 2006). Therefore, no-go P3 might reflect some aspect of the inhibitory function of the frontal lobe (Pfefferbaum et al., 1985). Indeed, the two studies that compared PD patients' go and no-go P3s found their no-go P3s to be specifically reduced (Bokura et al., 2005; Pulvermüller et al., 1996) and delayed (Bokura et al., 2005). However, the role of no-go P3 is still in debate. It might reflect release of motor activation rather than its inhibition (Verleger et al., 2006). Therefore, these results might reflect PD patients' problems with the dynamics of motor control, indicating difficulties in the release of motor activation similar to the prolonged duration of PD patients' BP after movement onset (Cunnington et al., 1995, 1997; Devos et al., 2002).

When oddball sequences are presented to both ears separately and participants are instructed to attend to sounds in one ear only, pressing a key to occasional targets, then the ERPs evoked by standard sounds are more negative for sounds in the attended than in the unattended ear (Nd: Hansen & Hillyard, 1980; viz. "Processing Negativity," Näätänen, 1990; see also Chapter 11, this volume). The Nd of PD patients did not differ from that of healthy subjects when ISIs were brief (0.5 s in Vieregge et al., 1994; 0.35 s in Karayanidis et al., 1995), but they differed markedly when ISIs were 1 s (Stam et al., 1993; Vieregge et al., 1994). Thus, it appears that PD patients have a deficit in maintaining the attentional trace of the standard sound whenever intervals between the sounds get too long.

Finally, error negativity (see Chapter 10, this volume) was found to be reduced in PD (Falkenstein et al., 2001; Ito & Kitagawa, 2006; Stemmer et al., 2007) except in the study of Holroyd et al. (2002). Reasons for the variation between studies are not clear, but if the deficit was obtained, it appeared quite specific to Ne, in view of the lack of difference between PD patients and healthy subjects for Pe (the positivity following the error negativity; Falkenstein et al., 2005) and for the N1, P2, N2, and P3 components (Stemmer et al., 2007). The possible deficit in error negativity in PD patients might be interpreted either as a deficit in the actual error-monitoring process or as a consequence of being overloaded in moderately complex tasks. Either alternative would reflect a deficit in executive control.

Above, it was noted that PD patients' increased LRP amplitudes in the flanker task might indicate either learned compensatory behavior or pathological reduction of executive control. In those studies, no mention was made of the N2 component evoked by stimuli with incompatible flankers, although there is evidence that N2 in this task is a measure of executive control (Folstein & van Petten, 2008; Kopp et al., 1996). Indeed, Verleger et al. (2010) found that N2 was reduced in healthy middle-aged carriers of heterozygous gene mutations that increase the susceptibility to develop PD (*Parkin* and PINK1), whereas, unlike in PD, their LRPs were reduced rather than increased and their key-press responses delayed compared to healthy age-matched persons. Verleger et al. concluded from this pattern that the mutation carriers already tended to have a PD-like deficit in the mechanisms of cognitive control but that they possibly compensated for the resulting hypersensitivity of the motor system to irrelevant stimulation by strategically reducing their response speed, thereby making the system less sensitive to this stimulation.

MEMORY

Mismatch negativity serves as a measure of preconscious auditory memory. Of particular interest in the context of the preceding paragraph, there is evidence for the involvement of a frontal-lobe component of MMN (Shalgi & Deouell, 2007). However, to my knowledge, the report of a reduced MMN in PD (Pekkonen et al., 1995) has not been replicated so far (cf. Pekkonen et al., 1998; Vieregge et al., 1994).

Tachibana and colleagues investigated repetition priming, reflected as a reduction of N400, where repetitions were either task-relevant (Minamoto et al., 2001) or not (Tachibana et al., 1999b). A clear differential effect of priming on PD patients' and control participants' N400 could not be established. However, even at first presentation, the PD patients had consistently smaller N400 amplitudes. This might reflect shallower processing of the presented words, reflecting the parkinsonian frontal-lesion-like impairment.

CONCLUSION

Taken together, of the ERP studies used to investigate PD beyond movement disorders, those reflecting impaired frontal-lobe function have provided the most consistent results. Pursuing this line of research further and making direct comparisons to patients with frontal lobe lesions will certainly shed new light on the parkinsonian syndrome of cognitive impairment.

Cerebellar Atrophy and Other Diseases of the Cerebellum

Cerebellar atrophy (CA) denotes a class of diseases, some certainly hereditary, others of still unclear etiology, that are characterized by impairments of movement precision and of balance related to shrinkage of cerebellar volume. Often, in the course of the disease, pathology extends from the cerebellum to neighboring structures (olivo-ponto-cerebellar atrophy, OPCA).

Questions can be asked about CA in analogy to the questions about PD. The cerebellum exerts its main influence on movement control only indirectly, by projecting to cortical areas (via the thalamus) rather than directly to the effectors. The motor cortex is not the only target area of cerebellar projections (Clower et al., 2005; Kelly & Strick, 2003; Schmahmann, 1996). Thus, the functions of the cerebellum are under debate; thus, as in PD, precisely describing the motor and nonmotor impairments of CA patients is still a challenge. Recording ERPs from the cerebellum is problematic because only part of the cerebellum makes contact with the skull, and this part is the very low posterior side of the skull, where the long back muscles are attached to the skull. In fact, none of the studies on CA patients to be referenced here recorded ERP differences directly ascribed to cerebellar activity. What were measured were cerebrocortical consequences of the cerebellar dysfunction.

For simplicity, studies of CA patients will be discussed together with studies of patients whose damage to the cerebellum was caused by other diseases, namely, infarctions of arteries supplying the cerebellum (the single-case studies reported by Ikeda et al., 1994, and Gerloff et al., 1996; most of the patients in the studies of Kitamura et al., 1999, and Restuccia et al., 2007; some of the patients in the studies of Shibasaki et al., 1978, and Daum et al., 1993) and tumors of the cerebellum that were resected (all patients in the study of Akshoomoff & Courchesne, 1994; two patients in the study of Restuccia et al., 2007; one patient each in the studies of Daum et al., 1993, and Kitamura et al., 1999).

Movement-Related Activity

Cerebellar atrophy patients' BPs preceding self-initiated finger movements have drastically reduced amplitudes (Gerloff et al., 1996; Ikeda et al., 1994; Shibasaki et al., 1978; Wessel et al., 1994), earlier onsets (i.e., patients need more time to prepare movements: Shibasaki et al., 1978; Wessel et al., 1994), and more diffuse topographies, which lack clear central-midline foci (Gerloff et al., 1996; Tarkka et al., 1993; Wessel et al., 1994). Lesions of the cerebellar dentate nucleus appear to be particularly harmful (Kitamura et al., 1999; Shibasaki et al., 1978, 1986). Thus, differences from healthy persons are more consistent and more marked in CA patients than in PD patients.

There is some inconsistency in the CNV in CA. Ikeda et al. (1994) presented a single patient whose BP was entirely absent, whereas the CNV was not (though amplitudes were not compared to normal values). In the same vein, CA patients' CNV amplitudes in the study of Daum et al. (1993) were not significantly smaller than those of healthy participants. However, the studies by Yamaguchi et al. (1998), Verleger et al. (1999b), and Trillenberg et al. (2004) demonstrated drastically reduced CNV amplitudes and lack of a clear central-midline focus in these patients. Thus, CA patients' CNV results do not differ qualitatively from their BP results. This amplitude reduction was independent of movement complexity in the study of Verleger et al. (1999b). That is, while being generally smaller in the patients than in healthy persons, the CNV became larger by the same amount in patients and in healthy persons when S1 indicated a difficult bimanual movement, although the CA patients often failed to perform these movements successfully. This led us to suggest that there is general input from the cerebellum to the motor cortex, deficient in the patients, and a specific increase in activity with fine movements generated by the motor cortex itself, equally present in the patients. The constant input by the cerebellum might involve general routines needed when expecting and preparing for events in time, abilities that are deficient in the patients (e.g., Trillenberg et al., 2004). From the patients' successful performance of simple movements, in spite of their severe deficit in slow potentials, Verleger et al. (1999b) concluded that the motor cortex does not necessarily need the cerebellar information: such information is, however, needed when movements have to be coordinated (Verleger et al., 1999b) or when the eliciting events vary in time (Trillenberg et al., 2004).

Requiring participants to exert force in a graded manner for several hundred milliseconds following the imperative signal, Verleger et al. (1999b) recorded a "movement-accompanying negativity," probably composed of efferent motor control and of somatosensory reafference. (See also work by Slobounov and colleagues on this component in healthy participants, e.g., Slobounov & Ray, 1998, or, more recently, by Kirsch et al., 2010.) Like the CNV, this component was drastically reduced in cerebellar patients.

To my knowledge, no study has reported LRPs in CA patients so far. A priori, the processing of conflicting information, as reflected by reversals of LRP, seems to be an interesting issue in understanding the deficits of CA patients. Coordination of agonist and antagonist activity is known to be deficient in these patients, and it would be interesting to know how this would generalize to finding the correct balance between conflicting information that activates both motor cortices.

Search for Consequences of Cerebellar Lesions on Cognition

Akshoomoff and Courchesne (1994) had participants respond to targets in the auditory or visual modality and, having detected the target, shift their attention to the other modality. Their cerebellar patients (tumor-resected 10-year-old children) had difficulty detecting targets when the preceding target in the other modality had been presented less than 2.5 s ago. This was also reflected by a lack of difference between the patients' P3 waves evoked by targets in the to-be-attended modality and in the to-be-ignored modality when the preceding target had been presented less than 2.5 s prior to the current target. In a different paradigm, Yamaguchi et al. (1998) did not find evidence for impaired capacity for fast shifts of spatial attention in the visual modality: In an S1-S2 task, they measured the contra-ipsilateral difference evoked by arrow cues or peripheral cues within the 0.8 s S1-S2 interval. Neither the parietal difference at about 300 ms ("*early directed-attention negativity*", EDAN) nor the frontocentral difference at about 400 ms ("*anterior directed-attention negativity*", ADAN) was reduced or delayed in CA patients. The diverging results between these two studies may be due to the different causes of cerebellar damage, and to the differences in participants' age, in paradigm, and in the type of attentional shift.

Tachibana et al. (1995) measured CA patients' performance on a visual oddball task (plant names were targets, animal names were the frequent nontargets) relative to a simple-response task. Patients' Na component, peaking at about 200 ms, was delayed (measured in the ERP difference between nontargets in the oddball and a simple-response condition), causing a delay of the ensuing N2 and P3 latencies in the target waveshapes. Following Ritter et al.'s (1983) interpretation of Na, Tachibana et al. (1995) suggest that the cerebellar lesions cause impairments in pattern classification, a skill that is obviously needed in reading Japanese *kana* symbols. However, since Na is obtained for any kind of visual discrimination (Vogel & Luck, 2000), the deficit might be more general than just the one in pattern classification. In a following study, Tachibana et al. (1999a) found that the patients who had delayed P3s in this task also had lower frontal blood perfusion, as measured by single photon emission computed tomography (SPECT).

Finally, Restuccia et al. (2007) obtained mismatch responses to electrical finger stimulation, consisting of a posterior negativity at about 150 ms. This somatosensory MMN was missing when the affected hand was stimulated. (All patients had unilateral lesions.)

In summary, ERP evidence of cognitive impairments in CA is interesting but scarce.

Other Degenerative Diseases

Other degenerative neurological diseases, fortunately relatively infrequent, include Huntington's disease, progressive supranuclear palsy, and amyotrophic lateral sclerosis.

Huntington's Disease

Huntington's disease (HD) is a hereditary degenerative disease of the brain focusing on parts of the basal ganglia (caudate nucleus and putamen), producing hyperkinesia, akinesia, and dementia.

Although excess movements are a core symptom of the disease, only a few studies have investigated movement-related potentials in HD. Johnson et al. (2001) found reduced amplitudes in movement preparation, very similar to the results obtained for PD in the same task studied by Cunnington et al. (1997). But recordings were made at Cz only, and there was no EOG recording for artifact control. The similarity to PD might seem surprising, because PD leads to hypokinesia, whereas HD leads to hyperkinesia, but these findings were confirmed in a recently published study in which the CNV was measured in a warned simple-response task (de Tommaso et al., 2007).

As in other degenerative disorders (AD and PD), P3 latencies in oddball tasks are delayed in HD (Filipović et al., 1990; Goodin & Aminoff, 1986; Rosenberg et al., 1985) but the finding that latencies of earlier components (N1, P2) were specifically delayed in HD (Goodin & Aminoff, 1986) was not replicated by Rosenberg et al. (1985) and Filipović et al. (1990). In a large study of 30 patients, of 40 of the patients' sons and daughters, and of 60 healthy controls, Hömberg et al. (1986) established that HD patients' P3 latencies (as well as latencies of the earlier components N1, P2, and N2) were delayed in an auditory oddball task, and that the P3 delay occurred not only in patients but also in their offspring who are at risk of developing HD, correlating with deficits in psychometric tests. Promising as this finding was for identifying persons who are at risk for developing HD, such identification has become reliably possible by genetic testing, which is much more precise than the necessarily variable ERP measures.

More recently, Münte et al. (1997) used visual search tasks and a continuous word recognition task to characterize more precisely HD patients' cognitive impairment by means of ERPs. Several remarkable results were obtained. First, in both tasks, the patients' P1 was markedly reduced and the following N1 was massively delayed. As Münte et al. note, this is in contrast to findings of earlier studies. (For example, Rosenberg et al., 1985, who also presented visual stimuli, did not obtain a delay of N1 in HD patients; a more recent study, by Antal et al., 2003, reported no difference in P1 and a reduction rather than a delay of N1, but recordings were not made from temporo-occipital sites.) This might mean that HD patients have a deficit specifically in perceiving the complex stimuli used in these tasks. (Below, compare the findings in PSP by Johnson, 1992.) Further, in both tasks, the patients' ERP waveshapes were not modulated by task demands in the P3 latency range: patients displayed neither a P3 in response to targets in the visual search task nor a priming-and-recognition positivity in the word recognition task. Although it can be argued that these missing modulations are a trivial consequence of the distorted input indicated by the P1-N1 abnormalities, the lack of priming positivity is remarkable because even AD patients were shown to have priming positivity (see above). Thus, ERP studies might well contribute to understanding the specific mechanisms of impairment in HD.

Note added in print: Just after completion of this review, a series of studies had been published by Beste and colleagues (Beste et al., 2006; 2007; 2008a; 2008b; 2008c; 2008d; 2009a, 2009b, 2010). These studies would require more careful discussion than can be inserted at this stage.

Progressive Supranuclear Palsy

Progressive supranuclear palsy (PSP) is characterized by palsy of vertical saccades, loss of voluntary facial movements, axial dystonia, gait disturbance, and dementia due to pathological alterations in several subcortical regions. As in other dementing diseases, P3 latency has been found to be delayed both in visual (Johnson et al., 1991; Pierrot-Deseilligny et al., 1989) and in auditory (Takeda et al., 1998) oddball tasks and P3 amplitude to be reduced (Johnson et al., 1991; Pirtošek et al., 2001; Wang et al., 2000). In addition, anterior visual P200 and auditory P2 were small and delayed in the patients (Johnson et al., 1991; Pirtošek et al., 2001), and target-evoked auditory N2 was delayed as well (Pirtošek et al., 2001). Johnson (1992) described the results of a comprehensive series of tasks, including the oddball and three other tasks: Sternberg's memory-scanning task, a final-word verification task (sentences had to be read, with the final word being either congruent or incongruent with the preceding context), and a mental-rotation task using pictures of rotated right and left hands as stimuli. P200 and P3 (as well as response times) were found to be delayed, which replicated the just-mentioned oddball results. Moreover, whereas the memory-scanning task resulted in principally similar patterns of patients and controls with increasing load, the other two tasks yielded unexpected results: in the final-word verification task, no differentiation between congruent and incongruent words was visible in the patients' ERPs in the N400 time range, and in the mental-rotation task the hand stimuli elicited very large occipital N1 components in the control group but not in the patients. The missing N400 effect is reminiscent of Minamoto et al.'s (2001) finding in PD, and the missing N1 enhancement, probably related to lack of a discrimination process (Vogel & Luck, 2000), is reminiscent of Münte et al.'s (1997) finding in HD. More generally, Johnson's (1992) study is a nice example of applying a kind of ERP test battery, probing several diverse but well-investigated ERP effects, from which a neurophysiological-cognitive profile can be built of the investigated group.

Amyotrophic Lateral Sclerosis

Amyotrophic lateral sclerosis (ALS) involves degeneration of neurons of the pyramidal tract and of the consecutive spinal neurons, progressively reducing the patients' ability to move. Questions open to research are (1) how the patients' motor cortex deals with this efferent blockade, (2) whether other cognitive functions are affected by the degenerative process, and (3) whether the patients' ERPs can be used to communicate with their environment when the patients are no longer able to move.

MOVEMENT-RELATED POTENTIALS

Two studies measured movement-related potentials in ALS patients. Westphal et al. (1998) recorded BPs before self-paced movements. The total group of ALS patients did not differ from healthy controls, but a subgroup of patients with increased spasticity had lower BP amplitudes. Hanagasi et al. (2002)

recorded the CNV before warned simple responses. The CNV amplitudes were larger in the patients than in healthy participants. In our own CNV study (unpublished), we did not find differences between patients and healthy controls. Hanagasi et al.'s (2002) finding might be due either to pathological hyperexcitability in ALS, as suggested by those authors, or to the patients' increased effort to overcome the efferent blockade. The latter interpretation would match the results obtained in a positron emission tomography (PET) study by Kew et al. (1993), who found enlargement of the ALS patients' hand motor area. The two studies by Westphal et al. (1998) and Hanagasi et al. (2002) and our unpublished data at least converge on the conclusion that movement-related potentials are not generally smaller in ALS patients. This is in strong contrast to patients with striatocapsular infarction (Verleger et al., 2003), who also have to deal with the problem of efferent blockade. The mechanisms causing this different pattern of results have to be further investigated.

EFFECT ON NONMOTORIC FUNCTIONS

There is accumulating evidence for abnormalities in relatively early sensory ERP components in ALS patients. Their auditory N100 is reduced (Raggi et al., 2008; Vieregge et al., 1999). This can be said even though Haganasi et al. (2002) did not find such a difference. However, their Figures 2 and 3 strongly suggests that even though patients' N100 amplitudes were not reduced at Cz, which is where the authors measured this component, patients' amplitudes were actually reduced at Fz. This topographic pattern is in full agreement with Raggi et al.'s (2008) results. Similarly, in the visual modality, ALS patients' temporo-occipital P100 components were found to be virtually absent in response to words or to visual-search displays (Münte et al., 1998a, 1999), whereas the following N1 was not found to differ. Such differences in early components were not obtained in two other studies (Gil et al., 1995: auditory; Paulus et al., 2002: auditory and visual), but this might be due to methodological issues inherent in these studies. For example, Paulus et al. did not report measurements of P1 and N1 from temporo-occipital electrodes.

Moreover, differences between ALS patients and healthy persons were reported for processing negativity (Nd) and MMN. Vieregge et al. (1999) presented oddball sequences to both ears separately, instructing participants to attend to sounds in one ear only. Nd was greatly reduced in ALS patients, both with slow and with fast presentation. Raggi et al. (2008) reported a MMN reduction in ALS patients in response to infrequent high-pitched sounds. However, the status of this finding is unclear, because this difference was significant at Pz only, rather than at Fz or the mastoids where MMN is usually measured, and because Hanagasi et al. (2002) did not obtain a difference in MMN between ALS patients and controls. Peculiarly, though entirely independent of any overt response, the Nd reduction in the study of Vieregge et al. (1999) correlated with the patients' motor disability but not with neuropsychological tests (in contrast to healthy subjects, whose Nd reduction did correlate with tests sensitive to frontal-lobe function).

Evidence is mixed for differences in the N2 and P3 components. Gil et al. (1995) and Paulus et al. (2002) reported delayed latencies of P3 and (Gil et al.) of N2, but these delays were not replicated by Hanagasi et al. (2002) and Vieregge et al. (1999; not reported in that publication). Hanagasi et al. (2002) and Raggi et al. (2008) did obtain group differences in P3, but in amplitudes rather than latencies: amplitudes both of the target-evoked P3 in an active oddball task (Hanagasi et al.) and of the novelty-evoked P3 (Hanagasi et al., Raggi et al., 2002) were smaller in ALS patients. Perhaps in line with these findings, Münte et al. (1998b) reported recognition positivity to be absent in ALS patients. Perhaps also in line with these results is the finding reported by Kotchoubey et al. (2003) in a totally locked-in ALS patient who produced a tonic and a phasic negativity in response to oddball targets rather than the typical P3 complex.

To summarize, there is surprising though still sparse evidence for abnormalities in relatively early sensory ERP components in ALS: auditory N1, visual P1, and auditory processing negativity. Whether this reflects some damage of afferent pathways in ALS or the possibility that efferent pathways must be intact in order to obtain normal sensory components is an open question. One might argue that ALS patients are permanently distracted in such tasks by their greater difficulties with the occasional responses, which is why even their potentials to standards might become affected, but it is doubtful whether this argument can account for the reported abnormalities of early components. For example, no responses whatsoever were required in Raggi et al.'s (2008) study.

COMMUNICATION OF LOCKED-IN PATIENTS

Severe ALS is a model case for the locked-in syndrome, where patients might use their ERPs to communicate with their environment. Use of the P3 component evoked by target stimuli for this purpose was suggested and demonstrated in healthy subjects by Farwell and Donchin (1988): participants might select letters to spell words and sentences by emitting enhanced P3s to arrays when these arrays contain the intended letter. Indeed, a locked-in patient (after ischemia of the basilar cerebral artery) studied by Onofrj et al. (1996) emitted N2-P3 complexes to target sounds and thus, in principle, would have been able to use this means of communication. In recent years, this line of applied research has been pursued systematically by Birbaumer's group in patients, mostly suffering from chronic ALS, whose remaining control over muscles (e.g., for moving the eyes) had become too unreliable to be used for communication (Birbaumer et al., 1999, 2000). In the *thought translation device*, patients selected letters by enhancing the positive voltage level measured from Cz 1.5 to 2 s after letter presentation. In doing so, patients became able to produce messages, though at a very slow speed (e.g., in Kübler et al., 1999: 12 s per letter at best, more than 3 min per letter at worst; similarly in Neshige et al., 2007), which might perhaps be improved by optimization of parameters (Krusienski et al., 2008).

Alternative ways of using the EEG to help locked-in patients communicate with their environment have used modulations of selected frequency spectra of the spontaneous EEG. These approaches (e.g., Pfurtscheller et al., 1993; Wolpaw et al., 1991) will not be discussed further here. Comprehensive reviews of all of these techniques have been published by Kübler et al. (2001) and Kübler and Kotchoubey (2007).

Concluding Remarks

In this review, I outlined strengths and drawbacks of using ERP components in research on neurological patients. Due to space limitations, this review was restricted to research on neurodegenerative diseases. With regard to the immediate utility of ERPs for diagnostic purposes, it is obvious that ERPs (in contrast to early, "hard-wired" components of evoked potentials) do not play a relevant role in the diagnosis of these syndromes. However, interesting proposals have been put forward for prognosis, most prominently a recent one by Golob et al. (2007) for predicting whether some patients will develop a full-blown dementia. As might be apparent, the major contribution of ERPs to clinical neurology lies in understanding the mechanisms of cognitive impairments of diseases, and ERPs have continued to contribute relevant knowledge to this area.

Note

1 Defining the genetic risk by having one parent or sibling affected by AD, Boutros et al. (1995) reported enhanced rather than diminished ERP amplitudes for persons at risk, in particular enhanced P300 for persons whose relatives had a diagnosis of definite AD (post mortem). However, artifacts were rejected by a pure amplitude criterion applied at Cz and Pz that was set too liberally (Verleger, 1993) and, possibly for that reason, the reported large P300 amplitudes had large variance, too.

References

Akshoomoff, N. A., & Courchesne, E. (1994). ERP evidence for a shifting attention deficit in patients with damage to the cerebellum. *Journal of Cognitive Neuroscience, 6*, 388–399.

Ally, B. A., Jones, G. E., Cole, J. A., & Budson, A. E. (2006). The P300 component in patients with Alzheimer's disease and their biological children. *Biological Psychology, 72*, 180–187.

Antal, A., Beniczky, S., Kincses, T. Z., Jakab, K., Benedek, G., & Vécsei, L. (2003). Perceptual categorization is impaired in Huntington's disease: An electrophysiological study. *Dementia and Geriatric Cognitive Disorders, 16*, 187–192.

Bennys, K., Portet, F., Touchon, J., & Rondouin, G. (2007). Diagnostic value of event-related evoked potentials N200 and P300 subcomponents in early diagnosis of Alzheimer's disease and mild cognitive impairment. *Journal of Clinical Neurophysiology, 24*, 405–412.

Beste, C., Konrad, C., Saft, C., Ukas, T., Andrich, J., Pfleiderer, B., Hausmann, M., & Falkenstein, M. (2009a). Alterations in voluntary movement execution in Huntington's disease are related to the dominant motor system: evidence from event-related potentials. *Experimental Neurology, 216*, 148–157.

Beste, C., Saft, C., Andrich, J., Gold, R., & Falkenstein, M. (2006). Error processing in Huntington's disease. *PLoS One, 1*, e86.

Beste, C., Saft, C., Andrich, J., Gold, R., & Falkenstein, M. (2008a). Stimulus-response compatibility in Huntington's disease: a cognitive-neurophysiological analysis. *Journal of Neurophysiology, 99*, 1213–1223.

Beste, C., Saft, C., Andrich, J., Gold, R., & Falkenstein, M. (2008b). Response inhibition in Huntington's disease-A study using ERPs and sLORETA. *Neuropsychologia, 46*, 1290–1297.

Beste, C., Saft, C., Güntürkün, O., & Falkenstein, M. (2008c). Increased cognitive functioning in symptomatic Huntington's disease as revealed by behavioral and event-related potential indices of auditory sensory memory and attention. *Journal of Neuroscience, 28*, 11695–11702.

Beste, C., Saft, C., Konrad, C., Andrich, J., Habbel, A., Schepers, I., Jansen, A., Pfleiderer, B., & Falkenstein, M. (2008d). Levels of error processing in Huntington's disease: a combined study using event-related potentials and voxel-based morphometry. *Human Brain Mapping, 29*, 121–130.

Beste, C., Saft, C., Yordanova, J., Andrich, J., Gold, R., Falkenstein, M., & Kolev, V. (2007). Functional compensation of pathology in cortico-subcortical interactions in preclinical Huntington's disease? *Neuropsychologia, 45*, 2922–2930.

Beste, C., Willemssen, R., Saft, C., & Falkenstein, M. (2009b). Error processing in normal aging and in basal ganglia disorders. *Neuroscience, 159*, 143–149.

Beste, C., Willemssen, R., Saft, C., & Falkenstein, M. (2010). Response inhibition subprocesses and dopaminergic pathways: basal ganglia disease effects. *Neuropsychologia, 48*, 366–373.

Birbaumer, N., Ghanayim, N., Hinterberger, T., Iversen, I., Kotchoubey, B., Kübler, A., Perelmouter, J., Taub, E., & Flor, H. (1999). A spelling device for the paralysed. *Nature, 398*, 297–298.

Birbaumer, N., Kübler, A., Ghanayim, N., Hinterberger, T., Perelmouter, J., Kaiser, J., Iversen, I., Kotchoubey, B., Neumann, N., & Flor, H. (2000). The Thought Translation Device (TTD) for completely paralyzed patients. *IEEE Transactions on Rehabilitation Engineering, 8*, 190–193.

Bokura, H., Yamaguchi, S., & Kobayashi, S. (2005). Event-related potentials for response inhibition in Parkinson's disease. *Neuropsychologia, 43*, 967–975.

Boutros, N., Torello, M. W., Burns, E. M., Wu, S.-S., & Nasrallah, H. A. (1995). Evoked potentials in subjects at risk for Alzheimer's disease. *Psychiatry Research, 57*, 57–63.

Chao, L. L., & Knight, R. T. (1998). Contribution of human prefrontal cortex to delay performance. *Journal of Cognitive Neuroscience, 10*, 167–177.

Chapman, R. M., Nowlis, G. H., McCrary, J. W., Chapman, J. A., Sandoval, T. C., Guillily, M. D., Gardner, M. N., & Reilly, L. A. (2007). Brain event-related potentials: Diagnosing early-stage Alzheimer's disease. *Neurobiology of Aging, 28*, 194–201.

Clower, D. M., Dum, R. P., & Strick, P. (2005). Basal ganglia and cerebellar inputs to "AIP." *Cerebral Cortex, 15*, 913–920.

Cunnington, R., Iansek, R., & Bradshaw, J. L. (1999). Movement-related potentials in Parkinson's disease: External cues and attentional strategies. *Movement Disorders, 14*, 63–68.

Cunnington, R., Iansek, R., Bradshaw, J. L., & Phillips, J. G. (1995). Movement-related potentials in Parkinson's disease: Presence and predictability of temporal and spatial cues. *Brain, 118*, 935–950.

Cunnington, R., Iansek, R., Johnson, K. A., & Bradshaw, J. L. (1997). Movement-related potentials in Parkinson's disease: Motor imagery and movement preparation. *Brain, 120*, 1339–1353.

Cunnington, R., Lalouschek, W., Dirnberger, G., Walla, P., Lindinger, G., Asenbaum, S., Brücke, T., & Lang, W. (2001). A medial to lateral shift in pre-movement cortical activity in hemi-Parkinson's disease. *Clinical Neurophysiology, 112*, 608–618.

Daffner, K. R., Rentz, D. M., Scinto, L. F. M., Faust, R., Budson, A. E., & Holcomb, P. J. (2001). Pathophysiology underlying diminished attention to novel events in patients with early AD. *Neurology, 56*, 1377–1383.

Daum, I., Schugens, M. M., Ackermann, H., Lutzenberger, W., Dichgans, J., & Birbaumer, N. (1993). Classical conditioning after cerebellar lesions in humans. *Behavioral Neuroscience, 107*, 748–756.

De Toledo-Morrell, L., Evers, S., Hoeppner, T. J., Morrell, F., Garron, D. C., & Fox, J. H. (1991). A "stress" test for memory dysfunction. Electrophysiologic manifestations of early Alzheimer's disease. *Archives of Neurology, 48*, 605–609.

De Tommaso, M., Difruscolo, O., Sciruicchio, V., Specchio, N., & Livrea, P. (2007). Abnormalities of the contingent negative variation in Huntington's disease: Correlations with clinical features. *Journal of the Neurological Sciences, 254*, 84–89.

Devos, D., Derambure, P., Bourriez, J. L., Cassim, D. F., Blond, S., Guieu, J. D., Destée, A., & Defebvre, L. (2002). Influence of internal globus pallidus stimulation on motor cortex activation pattern in Parkinson's disease. *Clinical Neurophysiology, 113*, 1110–1120.

Dick, J. P. R., Rothwell, J. C., Day, B. L., Cantello, R., Buruma, O., Gioux, M., Benecke, R., Berardelli, A., Thompson, P. D., & Marsden, C. D. (1989). The Bereitschaftspotential is abnormal in Parkinson's disease. *Brain, 112*, 233–244.

Ebmeier, K. P. (1992). A quantitative method for the assessment of overall effects from a number of similar electrophysiological studies: Description and application to event-related potentials in Parkinson's disease. *Electroencephalography and Clinical Neurophysiology, 84*, 440–446.

Falkenstein, M., Hielscher, H., Dziobek, I., Schwarzenau, P., Hoormann, J., Sundermann, B., & Hohnsbein, J. (2001). Action monitoring, error detection, and the basal ganglia: An ERP study. *Neuroreport, 12*, 157–161.

Falkenstein, M., Willemssen, R., Hohnsbein, J., & Hielscher, H. (2005). Error processing in Parkinson's disease: The error positivity (Pe). *Journal of Psychophysiology, 19*, 305–310.

Farwell, L. A., & Donchin, E. (1988). Talking off the top of your head: Toward a mental prosthesis utilizing event-related brain potentials. *Electroencephalography and Clinical Neurophysiology, 70*, 510–523.

Fattapposta, F., Pierelli, F., Traversa, G., My, F., Mostarda, M., D'Alessio, C., Soldati, G., Osborn, J., & Amabile, G. (2000). Preprogramming and control activity of bimanual self-paced motor task in Parkinson's disease. *Clinical Neurophysiology, 111*, 873–883.

Fattapposta, F., Pierelli, F., Traversa, G., My, F., Mostarda, M., Del Monte, S., Parisi, L., Serrao, M., Morocutti, A., & Amabile, G. (2002). L-dopa effects on preprogramming and control activity in a skilled motor act in Parkinson's disease. *Clinical Neurophysiology, 113*, 243–253.

Filipović, S., Kostić, V. S., Šternić, N., Marinković, Z., & Ocić, G. (1990). Auditory event-related potentials in different types of dementia. *European Neurology, 30*, 189–193.

Folstein, J. R., & van Petten, C. (2008). Influence of cognitive control and mismatch on the N2 component of the ERP: A review. *Psychophysiology, 45*, 152–170.

Ford, J. M., Askari, N., Mathalon, D. H., Menon, V., Gabrieli, J. D. E., Tinklenberg, J. R., & Yesavage, J. (2001). Event-related brain potential evidence of spared knowledge in Alzheimer's disease. *Psychology and Aging, 16*, 161–176.

Ford, J. M., Woodward, S. H., Sullivan, E. H., Isaacks, B. G., Tinklenberg, J. R., Yesavage, J. A., & Roth, W. T. (1996). N400 evidence of abnormal responses to speech in Alzheimer's disease. *Electroencephalography and Clinical Neurophysiology, 99*, 235–246.

Frodl, T., Hampel, H., Juckel, G., Bürger, K., Padberg, F., Engel, R. R., Möller, H.-J., & Hegerl, U. (2002). Value of event-related P300 subcomponents in the clinical diagnosis of mild cognitive impairment and Alzheimer's disease. *Psychophysiology, 39*, 175–181.

Gaeta, H., Friedman, D., Ritter, W., & Cheng, J. (1999). Changes in sensitivity to stimulus deviance in Alzheimer's disease: An ERP perspective. *Neuroreport, 10*, 281–287.

Gerloff, C., Altenmüller, E., & Dichgans, J. (1996). Disintegration and reorganization of cortical motor processing in two patients with cerebellar stroke. *Electroencephalography and Clinical Neurophysiology, 98*, 59–68.

Gerschlager, W., Alesch, F., Cunnington, R., Deecke, L., Dirnberger, G., Endl, W., Lindinger, G., & Lang, W. (1999). Bilateral subthalamic nucleus stimulation improves frontal cortex function in Parkinson's disease. An electrophysiological study of the contingent negative variation. *Brain, 122*, 2365–2373.

Gil, R., Neau, J. P., Dary-Auriol, M., Agbo, C., Tantot, A. M., & Ingrand, P. (1995). Event-related auditory evoked potentials and amyotrophic lateral sclerosis. *Archives of Neurology, 52*, 890–896.

Gilbert, P. F. C. (2001). An outline of brain function. *Cognitive Brain Research* 12, 61–74.

Gironell, A., García-Sánchez, C., Estévez-González, A., Boltes, A., & Kulisevsky, J. (2005). Usefulness of P300 in subjective memory complaints: A prospective study. *Journal of Clinical Neurophysiology, 22*, 279–284.

Golob, E. J., Irimajiri, R., & Starr, A. (2007). Auditory cortical activity in amnestic mild cognitive impairment: Relationship to subtype and conversion to dementia. *Brain, 130*, 740–752.

Golob, E. J., Johnson, J. K., & Starr, A. (2002). Auditory event-related potentials during target detection are abnormal in mild cognitive impairment. *Clinical Neurophysiology, 113*, 151–161.

Goodin, D. S., & Aminoff, M. J. (1986). Electrophysiological differences between subtypes of dementia. *Brain, 109*, 1103–1113.

Goodin, D. S., Squires, K. C., & Starr, A. (1978). Long latency event-related components of the auditory evoked potential in dementia. *Brain, 101*, 635–648.

Gordon, E., Kraiuhin, C., Harris, A., Meares, R., & Howson, A. (1986). The differential diagnosis of dementia using P300 latency. *Biological Psychiatry, 21*, 1123–1132.

Green, J., & Levey, A.J. (1999). Event-related potential changes in groups at increased risk for Alzheimer disease. *Archives of Neurology, 56*, 1398–1403.

Hanagasi, H. A., Gurvit, I. H., Ermutlu, N., Kaptanoglu, G., Karamursel, S., Idrisoglu, H. A., Emre, M., & Demiralp, T. (2002). Cognitive impairment in amyotrophic lateral sclerosis: Evidence from neuropsychological investigation and event-related potentials. *Cognitive Brain Research, 14*, 234–244.

Hansch, E. C., Syndulko, K., Cohen, S. N., Goldberg, Z. I., Potvin, A. R., & Tourtellotte, W. W. (1982). Cognition in Parkinson disease: An event-related potential perspective. *Annals of Neurology, 11*, 599–607.

Hansen, J. C., & Hillyard, S. A. (1980). Endogenous brain potentials associated with selective auditory attention. *Electroencephalography and Clinical Neurophysiology, 49*, 277–290.

Hogan, M. J., Carolan, L., Roche, R. A. P., Dockree, P. M., Kaiser, J., Bunting, B. P., Robertson, I. H., & Lawlor, B. A. (2006). Electrophysiological and information processing variability predicts memory decrements associated with normal age-related cognitive decline and Alzheimer's disease (AD). *Brain Research, 1119*, 215–226.

Holroyd, C. B., Praamstra, P., Plat, E., & Coles, M. G. H. (2002). Spared error-related potentials in mild to moderate Parkinson's disease. *Neuropsychologia, 40*, 2116–2124.

Hömberg, V., Hefter, H., Granseyer, G., Strauss, W., Lange, H., & Hennerici, M. (1986). Event-related potentials in patients with Huntington's disease and relatives at risk in relation to detailed psychometry. *Electroencephalography and Clinical Neurophysiology, 63*, 552–569.

Ikeda, A., Shibasaki, H., Kaji, R., Terada, K., Nagamine, T., Honda, M., & Kimura, J. (1997). Dissociation between contingent negative variation (CNV) and Bereitschaftspotential (BP) in patients with parkinsonism. *Electroencephalography and Clinical Neurophysiology, 102*, 142–151.

Ikeda, A., Shibasaki, H., Nagamine, T., Terada, K., Kaji, R., Fukuyama, H., & Kimura, J. (1994). Dissociation between contingent negative variation and Bereitschaftspotential in a patient with cerebellar efferent lesion. *Electroencephalography and Clinical Neurophysiology, 90*, 359–364.

Ito, J., & Kitagawa, J. (2006). Performance monitoring and error processing during a lexical decision task in patients with Parkinson's disease. *Journal of Geriatric Psychiatry and Neurology, 19*, 46–54.

Jahanshahi, M., Jenkins, I. H., Brown, R. G., Marsden, C. D., Passingham, R. E., & Brooks, D. J. (1995). Self-initiated versus externally triggered movements. I. An investigation using measurement of regional cerebral blood flow with PET and movement-related potentials in normal and Parkinson's disease subjects. *Brain, 118*, 913–933.

Johnson, K. A., Cunnington, R., Iansek, R., Bradshaw, J. L., Georgiou, N., & Chiu, E. (2001). Movement-related potentials in Huntington's disease: Movement preparation and execution. *Experimental Brain Research, 138*, 492–499.

Johnson, R., Jr. (1992). Event-related brain potentials. In I. Litvan & Y. Agid (Eds.), *Progressive supranuclear palsy: Clinical and research approache* (pp. 122–154). New York: Oxford University Press.

Johnson, R., Jr., Litvan, I., & Grafman, J. (1991). Progressive supranuclear palsy: Altered sensory processing leads to degraded cognition. *Neurology, 41*, 1257–1262.

Karayanidis, F., Andrews, S., Ward, P. B., & Michie, P. T. (1995). ERP indices of auditory selective attention in aging and Parkinson's disease. *Psychophysiology, 32*, 335–350.

Kazmerski, V. A., & Friedman, D. (1997). Effect of multiple presentation of words on event-related potential and reaction time repetition effects in Alzheimer's patients and young and older controls. *Neuropsychiatry, Neuropsychology, and Behavioral Neurology*, 10, 32–47.

Kazmerski, V. A., Friedman, D., & Hewitt, S. (1995). Event-related potential repetition effect in Alzheimer's patients: Multiple repetition priming with pictures. *Aging and Cognition, 2*, 169–191.

Kazmerski, V. A., Friedman, D., & Ritter, W. (1997). Mismatch negativity during attend and ignore conditions in Alzheimer's disease. *Biological Psychiatry, 42*, 382–402.

Kelly, R. M., & Strick, P. L. (2003). Cerebellar loops with motor cortex and prefrontal cortex of a nonhuman primate. *Journal of Neuroscience, 23*, 8432–8444.

Kew, J. J. M., Leigh, P. N., Playford, E. D., Passingham, R. E., Goldstein, L. H., Frackowiak, R. S. J., & Brooks, D. J. (1993). Cortical function in amyotrophic lateral sclerosis: A positron emission tomography study. *Brain, 116*, 655–680.

Kirsch, W., Hennighausen, E., & Rösler, F. (2010). ERP correlates of linear hand movements in a motor reproduction task. *Psychophysiology 47*, 486–500.

Kitamura, J.-I., Shibasaki, H., Terashi, A., & Tashima, K. (1999). Cortical potentials preceding voluntary finger movement in patients with focal cerebellar lesion. *Clinical Neurophysiology, 110*, 126–132.

Knight, R. T. (1984). Decreased response to novel stimuli after prefrontal lesions in man. *Electroencephalography and Clinical Neurophysiology, 59*, 9–20.

Knight, R. T., Hillyard, S. A., Woods, D. L., & Neville, H. J. (1980). The effects of frontal and temporal-parietal lesions on the auditory evoked potential in man. *Electroencephalography and Clinical Neurophysiology, 50,* 112–124.

Kopp, B., Rist, F., & Mattler, U. (1996). N200 in the flanker task as a neurobehavioral tool for investigating executive control. *Psychophysiology, 33,* 282–294.

Kotchoubey, B., Lang, S., Winter, S, & Birbaumer, N. (2003). Cognitive processing in completely paralyzed patients with amyotrophic lateral sclerosis. *European Journal of Neurology, 10,* 551–558.

Krusienski, D. J., Sellers, E. W., McFarland, D. J., Vaughan, T. M., & Wolpaw, J. R. (2008). Toward enhanced P300 speller performance. *Journal of Neuroscience Methods, 167,* 15–21.

Kübler, A., & Kotchoubey, B. (2007). Brain–computer interfaces in the continuum of consciousness. *Current Opinion in Neurology, 20,* 643–649.

Kübler, A., Kotchoubey, B., Hinterberger, T., Ghanayim, N., Perelmouter, J., Schauer, M., Fritsch, C., Taub, E., & Birbaumer, N. (1999). The thought translation device: A neurophysiological approach to communication in total motor paralysis. *Experimental Brain Research, 124,* 223–232.

Kübler, A., Kotchoubey, B., Kaiser, J., Wolpaw, J., & Birbaumer, N. (2001). Brain–computer communication: Unlock the locked-in. *Psychological Bulletin, 127,* 358–375.

Kujala, T., Tervaniemi, M., & Schröger, E. (2007). The mismatch negativity in cognitive and clinical neuroscience: Theoretical and methodological considerations. *Biological Psychology, 74,* 1–19.

Lim, V. K., Polych, M. A., Holländer, A., Byblow, W. D., Kirk, I. J., & Hamm, J. P. (2006). Kinesthetic but not visual imagery assists in normalizing the CNV in Parkinson's disease. *Clinical Neurophysiology, 117,* 2308–2314.

Linden, A., Bracke-Tolkmitt, R., Lutzenberger, W., Canavan, A. G. M., Scholz, E., Diener, H.-C., & Birbaumer, N. (1990). Slow cortical potentials in Parkinsonian patients during the course of an associative learning task. *Journal of Psychophysiology, 4,* 145–162.

Low, K. A., Miller, J., & Vierck, E. (2002). Response slowing in Parkinson's disease: A psychophysiological analysis of premotor and motor processes. *Brain, 125,* 1980–1994.

Mattox, S. T., Valle-Inclán, F., & Hackley, S. A. (2006). Psychophysiological evidence for impaired reward anticipation in Parkinson's disease. *Clinical Neurophysiology, 117,* 2144–2153.

Minamoto, H., Tachibana, H., Sugita, M., & Okita, T. (2001). Recognition memory in normal aging and Parkinson's disease: Behavioral and electrophysiological evidence. *Cognitive Brain Research, 11,* 23–32.

Münte, T. F., Ridao-Alonso, M. E., Preinfalk, J., Jung, A., Wieringa, B. M., Matzke, M., Dengler, R., & Johannes, S. (1997). An electrophysiological analysis of altered cognitive functions in Huntington disease. *Archives of Neurology, 54,* 1089–1098.

Münte, T. F., Tröger, M., Nusser, I., Wieringa, B. M., Johannes, S., Matzke, M., & Dengler, R. (1998a). Alterations of early components of the visual evoked potential in amyotrophic lateral sclerosis. *Journal of Neurology, 245,* 206–210.

Münte, T. F., Tröger, M., Nusser, I., Wieringa, B. M., Matzke, M., Johannes, S., & Dengler, R. (1998b). Recognition memory deficits in amyotrophic lateral sclerosis assessed with event-related brain potentials. *Acta Neurologica Scandinavica, 98,* 110–115.

Münte, T. F., Tröger, M. C., Nusser, I., Wieringa, B. M., Matzke, M., Johannes, S., & Dengler, R. (1999). Abnormalities of visual search behaviour in ALS patients detected with event-related brain potentials. *Amyotrophic Lateral Sclerosis and Other Motor Neuron Disorders, 1,* 21–27.

Muscoso, E. G., Costanzo, E., Daniele, O., Maugeri, D., Natale, E., & Caravaglios, G. (2006). Auditory event-related potentials in subcortical vascular cognitive impairment and in Alzheimer's disease. *Journal of Neural Transmission, 113,* 1779–1786.

Näätänen, R. (1990). The role of attention in auditory information processing as revealed by event-related potentials and other brain measures of cognitive function. *Behavioral and Brain Sciences, 13,* 201–288.

Neshige, R., Murayama, N., Igasaki, T., Tanoue, K., Kurokawa, H., & Asayama, S. (2007). Communication aid device utilizing event-related potentials for patients with severe motor impairment. *Brain Research, 1141,* 218–227.

Oishi, M., Mochizuki, Y., Du, C., & Takasu, T. (1995). Contingent negative variation and movement-related cortical potentials in parkinsonism. *Electroencephalography and Clinical Neurophysiology, 95,* 346–349.

Olichney, J. M., Iragui, V. J., Salmon, D. P., Riggins, B. R., Morris, S. K, & Kutas, M. (2006). Absent event-related potential (ERP) word repetition effect in mild Alzheimer's disease. *Clinical Neurophysiology, 117,* 1319–1330.

Olichney, J. M., Morris, S. K., Ochoa, C., Salmon, D. P., Thal, L. J., Kutas, M., & Iragui, V. J. (2002). Abnormal verbal event-related potentials in mild cognitive impairment and incipient Alzheimer's disease. *Journal of Neurology, Neurosurgery and Psychiatry, 73,* 377–384.

Onofrj, M., Melchionda, D., Thomas, A., & Fulgente, T. (1996). Reappearance of event-related P3 potential in locked-in syndrome. *Cognitive Brain Research,* 4, 95–97.

Paulus, K. S., Magnano, I., Piras, M. R., Solinas, M. A., Solinas, G., Sau, G. F., & Aiello, I. (2002). Visual and auditory event-related potentials in sporadic amyotrophic lateral sclerosis. *Clinical Neurophysiology, 113,* 853–861.

Pekkonen, E., Ahveninen, J., Virtanen, J., & Teräväinen, H. (1998). Parkinson's disease selectively impairs preattentive auditory processing: An MEG study. *Neuroreport, 9,* 2949–2952.

Pekkonen, E., Jousmäki, V., Kononen, M., Reinikainen, K., & Partanen, J. (1994). Auditory sensory memory impairment in Alzheimer's disease: An event-related potential study. *Neuroreport, 5,* 2537–2540.

Pekkonen, E., Jousmäki, V., Reinikainen, K., & Partanen, J. (1995). Automatic auditory discrimination is impaired in Parkinson's disease. *Electroencephalography and Clinical Neurophysiology, 95,* 47–52.

Pfefferbaum, A., Ford, J. M., Weller, B. J., & Kopell, B. S. (1985). ERPs to response production and inhibition. *Electroencephalography and Clinical Neurophysiology, 60,* 423–434.

Pfurtscheller, G., Flotzinger, D., & Kalcher, J. (1993). Brain–computer interface—a new communication device for handicapped persons. *Journal of Microcomputer Applications, 16,* 293–299.

Pierrot-Deseilligny, C., Turell, E., Penet, C., Lebrigand, D., Pillon, B., Chain, F., & Agid, Y. (1989). Increased wave P 300 latency in progressive supranuclear palsy. *Journal of Neurology, Neurosurgery, and Psychiatry,* 52, 656–658.

Pirtošek, Z., Jahanshahi, M., Barrett, G., & Lees, A. J. (2001). Attention and cognition in bradykinetic-rigid syndromes: An event-related potential study. *Annals of Neurology, 50,* 567–573.

Polich, J. (1991). P300 in the evaluation of aging and dementia. In C. H. M. Brunia, G. Mulder, & M. N. Verbaten (Eds.), *Event-related brain research, electroencephalography and clinical neurophysiology* (suppl. 42., pp. 304–323). Amsterdam: Elsevier.

Polich, J., Ehlers, C. L., Otis, S., Mandell, A. J., & Bloom, F. E. (1986). P300 latency reflects the degree of cognitive decline in dementing illness. *Electroencephalography and Clinical Neurophysiology, 63*, 138–144.

Polich, J., & Herbst, K. L. (2000). P300 as a clinical assay: Rationale, evaluation, and findings. *International Journal of Psychophysiology, 38*, 3–19.

Praamstra, P., Meyer, A. S., Cools, A. R., Horstink, M. W. I. M., & Stegeman, D. F. (1996). Movement preparation in Parkinson's disease: Time course and distribution of movement-related potentials in a movement precueing task. *Brain, 119*, 1689–1704.

Praamstra, P., & Oostenveld, R. (2003). Attention and movement-related motor cortex activation: A high-density EEG study of spatial stimulus-response compatibility. *Cognitive Brain Research, 16*, 309–322.

Praamstra, P., & Plat, F. M. (2001). Failed suppression of direct visuomotor activation in Parkinson's disease. *Journal of Cognitive Neuroscience, 13*, 31–43.

Praamstra, P., Plat, E. M., Meyer, A. S., & Horstink, M. W. I. M. (1999). Motor cortex activation in Parkinson's disease: Dissociation of electrocortical and peripheral measures of response generation. *Movement Disorders, 14*, 790–799.

Praamstra, P., Stegeman, D. F., Cools, A. R., & Horstink, M. W. I. M. (1998). Reliance on external cues for movement initiation in Parkinson's disease: Evidence from movement-related potentials. *Brain, 121*, 167–177.

Pulvermüller, F., Lutzenberger, W., Müller, V., Mohr, B., Dichgans, J., & Birbaumer, N. (1996). P3 and contingent negative variation in Parkinson's disease. *Electroencephalography and Clinical Neurophysiology, 98*, 456–467.

Raggi, A., Consonni, M., Iannaccone, S., Perani, D., Zamboni, M., Sferrazza, B., & Cappa, S. F. (2008). Auditory event-related potentials in non-demented patients with sporadic amyotrophic lateral sclerosis. *Clinical Neurophysiology, 119*, 342–350.

Restuccia, D., Della Marca, G., Valeriani, M., Leggio, M. G., & Molinari, M. (2007). Cerebellar damage impairs detection of somatosensory input changes: A somatosensory mismatch negativity study. *Brain, 130*, 276–287.

Revonsuo, A., Portin, R., Juottonen, K., & Rinne, J. O. (1998). Semantic processing of spoken words in Alzheimer's disease: An electrophysiological study. *Journal of Cognitive Neuroscience, 10*, 408–420.

Ritter, W., Simson, R., & Vaughan, H. G., Jr. (1983). Event-related potential correlates of two stages of information processing in physical and semantic discrimination tasks. *Psychophysiology, 20*, 168–179.

Rockstroh, B., Elbert, T., Canavan, A., Lutzenberger, W., & Birbaumer, N. (1989). *Slow cortical potentials and behaviour.* Baltimore, München, and Wien: Urban & Schwarzenberg.

Rosenberg, C., Nudleman, K., & Starr, A. (1985). Cognitive evoked potentials (P300) in early Huntington's disease. *Archives of Neurology, 42*, 984–987.

Rugg, M. D., Mark, R. E., Gilchrist, J., & Roberts, R. C. (1997). ERP repetition effects in indirect and direct tasks: Effects of age and interitem lag. *Psychophysiology, 34*, 572–586.

Rugg, M. D., Pearl, S., Walker, P., Roberts, R. C., & Holdstock, J. S. (1994). Word repetition effects on event-related potentials in healthy young and old subjects, and in patients with Alzheimer-type dementia. *Neuropsychologia, 32*, 381–398.

Salenius, S., Avikainen, S., Kaakkola, S., Hari, R., & Brown, P. (2002). Defective cortical drive to muscle in Parkinson's disease and its improvement with levodopa. *Brain, 125*, 491–500.

Schmahmann, J. D. (1996). From movement to thought: Anatomic substrates of the cerebellar contribution to cognitive processing. *Human Brain Mapping, 4*, 174–198.

Schnyer, D. M., Allen, J. J. B., Kaszniak, A. W., & Forster, K. I. (1999). An event-related potential examination of masked and unmasked repetition priming in Alzheimer's disease: Implications for theories of implicit memory. *Neuropsychology, 13*, 323–337.

Schwartz, T. J., Kutas, M., Butters, N., Paulsen, J. S., & Salmon, D. P. (1996). Electrophysiological insights into the nature of the semantic deficit in Alzheimer's disease. *Neuropsychologia, 34*, 827–841.

Seiss, E., & Praamstra, P. (2004). The basal ganglia and inhibitory mechanisms in response selection: Evidence from subliminal priming of motor responses in Parkinson's disease. *Brain, 127*, 330–339.

Shalgi, S., & Deouell, L.Y. (2007). Direct evidence for differential roles of temporal and frontal components of auditory change detection. *Neuropsychologia, 45*, 1878–1888.

Shibasaki, H., Barrett, G., Neshige, R., Hirata, I., & Tomoda, H. (1986). Volitional movement is not preceded by cortical slow negativity in cerebellar dentate lesion in man. *Brain Research, 368*, 361–365.

Shibasaki, H., Shima, F., & Kuroiwa, Y. (1978). Clinical studies of the movement-related cortical potential (MP) and the relationship between the dentatorubrothalamic pathway and readiness potential (RP). *Journal of Neurology, 219*, 15–25.

Slobounov, S. M., & Ray, W. J. (1998). Movement-related potentials with reference to isometric force output in discrete and repetitive tasks. *Experimental Brain Research, 123*, 461–473.

Stam, C. J., Visser, S. L., Op de Coul, A. A. W., De Sonneville, L. M. J., Schellens, R. L. L., Brunia, C. H. M., de Smet, D. S., & Gielen, G. (1993). Disturbed frontal regulation of attention in Parkinson's disease. *Brain, 116*, 1139–1158.

Stemmer, B., Segalowitz, S. J., Dywan, J., Panisset, M., & Melmed, C. (2007). The error negativity in nonmedicated and medicated patients with Parkinson's disease. *Clinical Neurophysiology, 118*, 1223–1229.

Stocco, A., Lebiere, C., & Anderson, J.R. (2010). Conditional routing of information to the cortex: a model of the basal ganglia's role in cognitive coordination. *Psychological Review, 117*, 541–574.

Swanwick, G. R. J., Rowan, M. J., Coen, R. F., Lawlor, B. A., Walsh, J. B., & Coakley, D. (1997). Electrophysiologic responses to varying mnemonic demand: A "stress test" is not diagnostic in very mild Alzheimer's disease. *Biological Psychiatry, 42*, 1073–1075.

Tachibana, H., Aragane, K., & Sugita, M. (1995). Event-related potentials in patients with cerebellar degeneration: Electrophysiological evidence for cognitive impairment. *Cognitive Brain Research, 2*, 173–180.

Tachibana, H., Kawabata, K., Tomino, Y., & Sugita, M. (1999a). Prolonged P3 latency and decreased brain perfusion in cerebellar degeneration. *Acta Neurologica Scandinavica, 100*, 310–316.

Tachibana, H., Miyata, Y., Takeda, M., Sugita, M., & Okita, T. (1999b). Event-related potentials reveal memory deficits in Parkinson's disease. *Cognitive Brain Research, 8*, 165–172.

Takeda, M., Tachibana, H., Okuda, B., Kawabata, K., & Sugita, M. (1998). Electrophysiological comparison between corticobasal degeneration and progressive supranuclear palsy. *Clinical Neurology and Neurosurgery, 100*, 94–98.

Tarkka, I. M., Massaquoi, S., & Hallett, M. (1993). Movement-related cortical potentials in patients with cerebellar degeneration. *Acta Neurologica Scandinavica, 88*, 129–135.

Timmermann, L., Gross, J., Dirks, M., Volkmann, J., Freund, H.-J., & Schnitzler, A. (2003). The cerebral oscillatory network of parkinsonian resting tremor. *Brain, 126*, 199–212.

Trillenberg, P., Verleger, R., Teetzmann, A., Wascher, E., & Wessel, K. (2004). On the role of the cerebellum in exploiting temporal contingencies: Evidence from response times and preparatory EEG potentials in patients with cerebellar atrophy. *Neuropsychologia, 42*, 754–763.

Touge, T., Werhahn, K. J., Rothwell, J. C., & Marsden, C. D. (1995). Movement-related cortical potentials preceding repetitive and random-choice hand movements in Parkinson's disease. *Annals of Neurology, 37*, 791–799.

Tsuchiya, H., Yamaguchi, S., & Kobayashi, S. (2000). Impaired novelty detection and frontal lobe dysfunction in Parkinson's disease. *Neuropsychologia, 38*, 645–654.

van Boxtel, G. (1994). *Non-motor components of slow brain potentials*. Unpublished doctoral dissertation, Katholieke Universiteit Brabant, Tilburg, the Netherlands.

Verleger, R. (1993). Valid identification of blink artefacts: Are they larger than 50 µV in EEG records? *Electroencephalography and Clinical Neurophysiology, 87*, 354–363.

Verleger, R. (2003). Event-related EEG potential research in neurological patients. In A. Zani & A. M. Proverbio (Eds.), *The cognitive electrophysiology of mind and brain*. (pp. 309–341). San Diego, CA, and London: Academic Press.

Verleger, R. (2004). Malfunctions of cortical control of movement studied with slow brain potentials in neurological patients. *Journal of Psychophysiology, 18*, 105–120.

Verleger, R., Adam, S., Rose, M., Vollmer, C., Wauschkuhn, B., & Kömpf, D. (2003). Control of hand movements after striatocapsular stroke: High-resolution temporal analysis of the function of ipsilateral activation. *Clinical Neurophysiology, 114*, 1468–1476.

Verleger, R., Hagenah, J., Weiß, M., Ewers, T., Heberlein, I., Pramstaller, P., Siebner, H. R., & Klein, C. (2010). Responsiveness to distracting stimuli, though increased in Parkinson's disease, is decreased in asymptomatic PINK1 and Parkin mutation carriers. *Neuropsychologia, 48*, 467–476.

Verleger, R., Kömpf, D., & Neukäter, W. (1992). Event-related EEG potentials in mild dementia of the Alzheimer type. *Electroencephalography and Clinical Neurophysiology, 84*, 332–343.

Verleger, R., Paehge, T., Kolev, V., Yordanova, J., & Jaśkowski, P. (2006). On the relation of movement-related potentials to the go/no-go effect on P3. *Biological Psychology, 73*, 298–313.

Verleger, R., Wascher, E., Arolt, V., Daase, C., Strohm, A., & Kömpf, D. (1999a). Slow EEG potentials (CNV and PINV) in schizophrenia: Their association to the present state and to Parkinsonian medication effects. *Clinical Neurophysiology, 110*, 1175–1192.

Verleger, R., Wascher, E., Wauschkuhn, B., Jaśkowski, P., Allouni, B., Trillenberg, P., & Wessel, K. (1999b). Consequences of altered cerebellar input for the cortical regulation of motor coordination, as reflected in EEG potentials. *Experimental Brain Research, 127*, 409–422.

Verleger, R., Wauschkuhn, B., van der Lubbe, R. H. J., Jaśkowski, P., & Trillenberg, P. (2000). Posterior and anterior contributions of hand-movement preparation to Late CNV. *Journal of Psychophysiology, 14*, 69–86.

Vidailhet, M., Stocchi, F., Rothwell, J. C., Thompson, P. D., Day, B. L., Brooks, D. J., & Marsden, C. D. (1993). The Bereitschaftspotential preceding simple foot movement and initiation of gait in Parkinson's disease. *Neurology, 43*, 1784–1788.

Vieregge, P., Verleger, R., Wascher, E., Stüven, F., & Kömpf, D. (1994). Selective attention is impaired in Parkinson's disease—event-related evidence from EEG potentials. *Cognitive Brain Research, 2*, 117–130.

Vieregge, P., Wauschkuhn, B., Heberlein, I., Hagenah, J., & Verleger, R. (1999). Selective attention is impaired in amyotrophic lateral sclerosis—a study of event-related EEG potentials. *Cognitive Brain Research, 8*, 27–35.

Vogel, E. K., & Luck, S. J. (2000). The visual N1 component as an index of a discrimination process. *Psychophysiology, 37*, 190–203.

Wang, L., Kuroiwa, Y., Kamitani, T., Li, M., Takahashi, T., Suzuki, Y., Shimamura, M., & Hasegawa, O. (2000). Visual event-related potentials in progressive supranuclear palsy, corticobasal degeneration, striatonigral degeneration, and Parkinson's disease. *Journal of Neurology, 247*, 356–363.

Wascher, E., Reinhard, M., Wauschkuhn, B., & Verleger, R. (1999). Spatial S-R compatibility with centrally presented stimuli: An event-related asymmetry study about dimensional overlap. *Journal of Cognitive Neuroscience, 11*, 214–229.

Wascher, E., Verleger, R., Vieregge, P., Jaśkowski, P., Koch, S., & Kömpf, D. (1997). Responses to cued signals in Parkinson's disease: Distinguishing between disorders of cognition and of activation. *Brain, 120*, 1355–1375.

Wessel, K., Verleger, R., Nazarenus, D., Vieregge, P., & Kömpf, D. (1994). Movement-related cortical potentials preceding sequential and goal-directed finger and arm movements in patients with cerebellar atrophy. *Electroencephalography and Clinical Neurophysiology, 92*, 331–341.

Westphal, K. P., Heinemann, H. A., Grözinger, B., Kotchoubey, B. J., Diekmann, V., Becker, W., & Kornhuber, H. H. (1998). Bereitschaftspotential in amyotrophic lateral sclerosis (ALS): Lower amplitudes in patients with hyperreflexia (spasticity). *Acta Neurologica Scandinavica, 98*, 15–21.

Winblad, B., Palmer, K., Kivipelto, M., Jelic, V., Fratiglioni, L., Wahlund, L. O., Nordberg, A., Backman, L., Albert, M., Almkvist, O., Arai, H., Basun, H., Blennow, K., de Leon, M., DeCarli, C., Erkinjuntti, T., Giacobini, E., Graff, C., Hardy, J., Jack, C., Jorm, A., Ritchie, K., van Duijn, C., Visser, P., & Petersen, R.C. (2004). Mild cognitive impairment—beyond controversies, towards a consensus: Report of the International Working Group on Mild Cognitive Impairment. *Journal of Internal Medicine, 256*, 240–246.

Wolk, D. A., Schacter, D. L., Berman, A. R., Holcomb, P. J., Daffner, K. R., & Budson, A. E. (2005). Patients with mild Alzheimer's disease attribute conceptual fluency to prior experience. *Neuropsychologia, 43*, 1662–1672.

Wolpaw, J. R., McFarland, D. J., Neat, G. W., & Forneris, C. A. (1991). An EEG-based brain–computer interface for cursor control. *Electroencephalography and Clinical Neurophysiology, 78*, 252–259.

Wright, M. J., Geffen, G. M., & Geffen, L. B. (1993). Event-related potentials associated with covert orientation of visual attention in Parkinson's disease. *Neuropsychologia, 31*, 1283–1297.

Yamaguchi, S., Tsuchiya, H., & Kobayashi, S. (1998). Visuospatial attention shift and motor responses in cerebellar disorders. *Journal of Cognitive Neuroscience, 10*, 95–107.

Yamaguchi, S., Tsuchiya, H., Yamagata, S., Toyoda, G., & Kobayashi, S. (2000). Event-related brain potentials in response to novel sounds in dementia. *Clinical Neurophysiology, 111*, 195–203.

CHAPTER

22 Homologues of Human ERP Components in Nonhuman Primates

Geoffrey F. Woodman

Abstract

This chapter chronicles the discovery of event-related potential (ERP) components in nonhuman primates. It focuses mainly on monkeys but also includes evidence from other species when it exists. The discussion generally unfolds chronologically, beginning with work from the nineteenth century and continuing up to the present. It addresses differences in the methods and tasks that have been utilized to record ERPs in various species. Such methodological differences are a necessary complication in electrophysiological studies across species. The chapter concludes by emphasizing some of our greatest needs in comparative electrophysiology and how such ERP studies have the ability to reshape what we know about ERP components and cognitive processing in humans.

Keywords: event-related potential, ERP components, nonhuman primates, monkeys, cognitive processing

Like many electrophysiologists who record the electroencephalogram (EEG) and event-related potentials (ERPs) from humans, I was a heavy user of the techniques before I became aware of the fact that EEG activity was originally observed during recordings from animals, including nonhuman primates (Caton, 1875). It was the 50-year-old studies with animals that motivated Hans Berger's discovery and naming of the EEG recorded from his son, Klaus, and other human subjects (Berger, 1929). Moreover, many users of the ERP technique may be surprised to learn just how rarely the ERP components we use to study human cognition have been studied in other model species, such as nonhuman primates. This chapter reviews what is known about ERP components from studies of nonhuman species. I conclude by pointing to some of the most glaring gaps in our knowledge and the enormous potential for ERP studies recorded from nonhuman primates.

History of the EEG

Continuous EEG was first discovered by Richard Caton (1875), a British physiologist, in his recordings of electrical activity from the surface of the heads of monkeys and other animal species. Caton (1875, 1877, 1887) recorded from electrodes placed on the skull, the dura, and the exposed cortex of monkeys, rabbits, and cats. The activity that Caton observed was just within the range of sensitivity of the galvanometer he used, and initially the voltage fluctuations seemed like just a noisy baseline before the presentation of a stimulus. Caton called the ubiquitous fluctuations in potential *feeble potentials*, which were spontaneous low-frequency variations in voltage that could be modulated by presenting stimuli to the subject.

Quoting from Caton's (1875) brief report: "Feeble currents of varying direction pass through the multiplier when the electrodes are placed on two points of the external surface" (p. 278).

Caton noted that these fluctuations in potential appeared to be related to the function of the underlying brain, with visual stimulation being more effective in modulating the fluctuations of potential than auditory or olfactory stimulation. Caton (1875) also noted that voltage fluctuations were recorded contralateral to the visual field in which a light was shone and were strongest when recording over an area that Ferrier had suggested was related to movements of the eyelids. He also reported that the ongoing fluctuations in potential were modulated by sleep, reduced by anesthesia, and increased immediately prior to death, after which changes in potential were completely absent (Brazier, 1957; Caton, 1887).

It is not clear from his writing why Caton favored the term *feeble potential* for the electrical activity he observed. Potentials generated by peripheral nerves had been shown to be orders of magnitude faster than this activity, potentially making the EEG too slow, in Caton's view, to be involved in the critical operations of the brain. Alternatively, the changes in potential were observed in the baseline periods before any stimulus was presented and therefore when the brain was in a resting state. Finally, this activity was found in all of the animal species that Caton used, and this may have suggested to him that the activity was not related to the intelligence of the organism. Regardless, coining this term for the electrical fluctuations he observed from the brain was likely a poor choice for his citation rate. Indeed, Caton's discoveries were often overlooked by other nineteenth-century electrophysiologists who repeatedly claimed to have discovered EEG and stimulus evoked potentials in many different species (e.g., Beck, 1890; Fleischl von Marxow, 1890). Realizing that his work had gone unnoticed, Caton wrote to the editor of the German journal *Centralblatt für Physiologie*, pointing out that the reports published in 1890 had neglected his initial discoveries (Brazier, 1957). However, this attempt appeared to have little effect; his work continued to be overlooked even in England (e.g., Gotch & Horsely, 1891). The discovery of EEG in animals received even less attention in the United States than by Caton's fellow Europeans; there, similar reports of EEG did not appear for over 50 years (Bartley & Newman, 1930). Fortunately for twenty-first-century electrophysiologists, one medical doctor in Germany was aware of Caton's contributions to the new field of electrophysiology.

Hans Berger was a physician studying blood flow in the brain before World War I (Haas, 2003). Unlike many of his predecessors, Berger was aware of Caton's work with nonhuman primates and other mammals. After returning to medicine at the conclusion of the war, he began similar noninvasive recordings from humans, largely from his son, Klaus. These initial observations formed the data reported in Berger's seminal work demonstrating alpha and beta wave rhythms in humans (Berger, 1929), the existence of which had been shown previously in dogs (Pravdich-Neminsky, 1913). In his preliminary publications, Berger explicitly acknowledged Caton's research. "Caton had already (1875) published experiments on the brains of dogs and apes in which bare unipolar electrodes were placed either on the surface of both hemispheres or one electrode on the cerebral cortex and the other on the surface of the skull. The currents were measured by a sensitive galvanometer. There were found distinct variations in current, which increased during sleep and with the onset of death strengthened, and after death weakened and disappeared" (Cohen of Birkenhead, 1959, p. 258). Thus, the use of EEG, and ultimately ERPs, to study different brain states was inspired by Caton's pioneering research in nonhuman species.

With Berger's reports of EEG recordings from human subjects, the field of human electrophysiology was born. However, Berger's findings were truly appreciated by readers of English language journals only after Edgar Adrian became interested in human EEGs (Walter, 1938). Here again the technique of recording EEG activity from humans would receive a boost from research with animals, if only in serving to recruit a prominent believer: Adrian. Edgar Adrian shared the Nobel Prize with Charles Sherrington in 1932 for his work recording action potentials from individual neurons in the frog and from the sensory organs of a variety of species. Adrian had also noted the low-frequency fluctuations in potential during his recordings from fish and insects, such as goldfish and water beetles (Adrian, 1932; Adrian & Buytendijk, 1931). Thus, when Berger reported similar potentials in human subjects, Adrian immediately became interested and published replications and extensions of Berger's work (Adrian & Yamagiwa, 1935).

Adrian's longest-lasting contribution to the field of human electrophysiology was probably his work recording from rabbits, as Caton had (Adrian & Matthews, 1934). The novelty of the contribution was the utilization of bipolar recordings, which have better spatial resolution than unipolar electrode recording techniques. Adrian and Matthews (1934) noted that the slow waves evident in the EEG are observed only when the active and reference electrodes are placed at a significance distance from each other (i.e., greater than 4 mm). It was with this evidence that Adrian and Matthews (1934) concluded that the slow fluctuations in potential that dominate human EEG are due to a summation of activity from networks of neurons that are generally active at the same time but not precisely in phase with each other. It is not surprising that Adrian had proposed a summation hypothesis to explain EEG, as this is similar to the conclusions he had drawn in his Nobel Prize–winning work with individual neurons. Previously, Adrian had shown that more vigorous limb movements are accompanied by higher rates of action potentials in individual neurons. These observations led Adrian to propose that neurons use modulations of firing rates of action potentials instead of transmitting electrical signals that vary in size to code information. By the early 1940s, electrophysiologists took Adrian and Matthew's summation hypothesis as a given despite Adrian's also entertaining the hypothesis that EEG was due to slower activity surrounding the dendrites of neurons (e.g., Adrian & Buytendijk, 1931). The summation hypothesis was the starting point for Kennard (1943) in a series of lesions studies with monkeys that attempted to localize the relative contributions of different structures to the observed spontaneous EEG. The logic of trying to lesion specific parts of the brain to eliminate EEG activity in animals was analogous to that of Lashley in trying to localize the reflex arc (Lashley, 1950) and proved to be just as unsuccessful (see Kennard, 1943; Kennard & Nims, 1942).

With the exception of some EEG recordings from animals during sleep studies (Desiraju, 1972; Weitzman, 1961), the post-Berger era is where monkey and human research split once again. In this case, it was the electrophysiological methods themselves that split the study of humans and nonhuman primates into distinct literatures (see additional discussion in Chapter 2, this volume). Electrophysiological study of the human brain using EEG and sensory evoked potentials gradually became widespread and then exploded in the mid-1960s with the discovery of ERP components that were sensitive to the task relevance of the stimuli and not just to the physical characteristics of the stimulation (Sutton et al., 1965; Walter et al., 1964). At the same time, continued refinement of microelectrode recording techniques from individual neurons in the brains of awake monkeys and other animals yielded richly detailed accounts of how single neurons responded to different stimuli and task contexts (Evarts & Magoun, 1957; Hubel et al., 1959). This single neuron–based unit of analysis contrasts sharply with the discovery of human ERP components that index activity related to specific cognitive operations taking place in large cell assemblies likely spanning many different areas. Thus, electrophysiological studies of humans and monkeys operated on very different levels of analysis during much of the twentieth century. However, in the mid-1980s, research began to unite the literatures again. Perhaps it is not surprising that just as the discovery of the human P3 component brought the human ERP technique into common use (Sutton et al., 1965), the modern era of monkey ERP recordings followed the discovery of a monkey homologue of the human P3 component (Arthur & Starr, 1984).

Monkey Homologues of Human ERP Components

The P3

The P3 or P300 component was one of the first human ERP components discovered that was related to the cognitive processing demands of the eliciting stimulus (see Chapter 7, this volume). Its discovery brought the field of human electrophysiology into the view of psychologists and cognitive scientists in a way that research on spontaneous EEG had not (Sutton et al., 1965). In the initial experiments, it was shown that a larger P3 was elicited by a stimulus of an infrequent category. Subsequent research showed that the P3 elicited by task-irrelevant infrequent stimuli had a more frontal distribution (i.e., the P3a) than the P3 elicited by task-relevant infrequent stimuli, which had a distribution with a parietal focus (i.e., the P3b; Knight et al., 1989). In addition, it was shown that the P3a was reduced in amplitude by frontal lesions, although the P3b was not (Knight, 1991; Knight et al., 1981). Several decades after finding the broad positive component we know as the P3b in humans, electrophysiologists working with animal subjects began to search for a similar index of cognitive processing in other species.

Starr and colleagues can be credited with discovering the P3 component first in cats (Wilder et al., 1981) and then in monkeys (Arthur & Starr, 1984). In fact, previous work had shown a P3-like potential in monkeys but had not required the monkeys to make a discriminative response, so the relevance of the stimuli for the effect could not be established (Donchin et al., 1971). Arthur and Starr (1984) trained their monkeys to perform a task in which they discriminated the frequency of tones and responded to infrequent target tones. This is precisely the same oddball paradigm in which the human P3 had initially been reported (Sutton et al., 1965) and on which a significant proportion of all human ERP experiments are based (see, e.g., Chapters 6, 7, 10, 11, 14, 15, 17, 18, 19, 20, and 21, this volume).

Arthur and Starr (1984) reported that monkeys showed a distinct positive potential following the presentation of infrequent and task-relevant tone stimuli embedded in a stream of frequent nontarget tones. The amplitude of the component was modulated systematically by the probability of the target tone (i.e., 10%, 30%, or 50% targets in a block of trials), just as the human P3b was known to behave. In addition, identical infrequent stimuli that were presented when they were not task relevant did not elicit the large positive component. One of the powerful aspects of the study of Arthur and Starr was that they recorded ERPs from humans in exactly the same task so that the waveforms could be directly compared between species (see Arthur & Starr, 1984, Figure 1).

There was a crucial difference in the methods used to record monkey ERPs from those used in human ERP recordings despite Arthur and Starr's (1984) measuring the monkey P3 using stimuli and a task that paralleled experiments with humans. The difference was the type of electrodes used to record ERPs in the two primate species. When the EEG is recorded from the scalp of humans, the small potentials produced in the brain need to pass through the brain, dura, bone, and finally skin, with very few large muscle groups interposed between the brain and the electrode. In monkeys, however, the skull is surrounded by thick layers of muscle tissue that are mostly connected to the jaw. This arrangement varies across species, but particularly in macaque monkeys—the preferred nonhuman primate model for a human—the muscle surrounding the skull leads to unacceptable amounts of muscle noise in scalp recordings. This problem is accentuated by the fact that macaques need to be reinforced with food or liquid for their behavior to continue while performing a task. This means that the muscles surrounding the skull will be active during the course of each trial as the monkeys move the lips and jaw to consume the reinforcing juice or food slurry and often during stimulus presentation as the animals anticipate the reward delivery. Arthur and Starr (1984) avoided this problem of muscle contamination by recording from screws that were implanted in the skull under general anesthesia.

Recording monkey EEG and the derived ERPs from skull screws has both advantages and disadvantages. An advantage of using screws as monkey EEG electrodes is that they can remain very well anchored to a specific location on the skull across days and even years. The disadvantage of these electrodes is that wires are typically used to provide a connection from the screws to the amplification equipment, and creating a good and stable electrical connection between an orthopedic screw and insulated metal wire on the operating table can be difficult. Even good connections can be compromised by normal activity in an animal's home cage. It might also be viewed as an advantage to record EEG from screws that extend all the way through the skull and often touch the dura when initially implanted. However, if the goal of the recordings from monkeys is to make direct comparisons to human ERPs and EEG, then this is in fact a problem. The electrical signal from the human brain passes through layers of tissue with different impedance (i.e., dura and bone) before it is recorded on the scalp, causing the signal to spread (for more information, see Luck, 2005, chap. 1). This means that by the time electrical activity is recorded from scalp electrodes on humans, the electrical fields have essentially been spatially low-pass filtered (see Nunez & Srinivansan, 2006, for a discussion of the frequency domain effects). The use of skull screws as EEG electrodes will result in signals that have not been influenced by the same factors that influence human EEG. Thus, the voltage distributions of components across the head are not directly comparable when one is trying to relate signals recorded from skull screws in monkeys to those from scalp electrodes on humans. The advantages of alternative types of monkey EEG/ERP electrodes will be discussed further below.

Following the discovery of a homologue of the human P3 component in monkeys, research focused on understanding the neural activity that gave rise to the component and the conditions under which it could be observed (Arthur & Starr, 1984; Glover et al., 1991; Javitt et al., 1992; Paller et al., 1992). One way researchers attacked the problem was to understand the role of a specific area and its contribution to the component that was observed on the surface electrodes. In a particularly interesting study, researchers lesioned the locus coeruleus, motivated by the hypothesis that this region is critical for the generation of the P3 component (Pineda et al., 1989). However, implications of the observation that the amplitude of the primate P3 was significantly reduced by lesioning the locus coeruleus may be limited by the centrality of this structure for excitation in the cortex in general. In addition, the new world monkeys used in this lesion study were very difficult to train to perform a task. As a result, they were passively processing the stimuli in this study, which evokes a P3a in humans instead of the more often studied P3b elicited by task-related stimuli. However, these findings parallel those of Kennard (1943) decades before, in which the spontaneous EEG was significantly disturbed only when the brainstem was lesioned, causing the health of the animal to deteriorate.

Another interesting, although unexpected, result of studying nonhuman ERP components was the observation of what is often called the *missing-stimulus potential* (Bullock, 2003). Early ERP studies with humans noted that when a stimulus is omitted from a regular and steady stream of stimuli, a component is elicited by the absence of the expected event. This component appears to be similar to the P3 elicited by the presentation of a rare stimulus (Simson & Ritter, 1976; Simson et al., 1977). In a surprising series of studies, Bullock and colleagues showed that this missing stimulus potential was found in essentially every organism examined, including invertebrates such as crayfish (e.g., Bullock, 2003; Ramón et al., 2001). I will discuss the implications of this type of large-scale comparative electrophysiology in greater detail in a subsequent section examining unanswered questions resulting from studies of nonhuman ERPs.

Although the discovery of a monkey P3 component made a large splash, it was not the first cognitively modulated ERP component that was discovered first in humans and subsequently found in research with monkeys. The initial report of the contingent negative variation (CNV; Walter et al., 1964) appeared 1 year before the original report of the P3 component by Sutton and colleagues (1965). Following a similar time course, a monkey ERP component similar to the human CNV appeared in the literature very shortly after the discovery of the component in humans (Borda, 1970). However, the field's evolving understanding of the CNV in humans resulted in tempered enthusiasm for the ability of monkey studies of the CNV to clarify the cognitive operations indexed by this component. Just as the interpretation of the CNV component was challenged by subsequent research with human subjects (Loveless & Sanford, 1975), the monkey CNV appeared to be less robust to modifications of the experimental paradigm than one would have hoped (see, e.g., Donchin et al., 1971).

Sensory and Perceptual Components

The first ERPs studied in humans were those elicited by the sensory processing of stimulus events (e.g., Davis, 1939). The study of nonhuman primate ERPs followed a similar path. The first reports of visually evoked potentials in monkeys appeared years before the Arthur and Starr (1984) P3 paper but with much less fanfare (Ripps & Vaughan, 1969; Van der Marel et al., 1981; Vaughan & Gross, 1969). Van der Marel and colleagues (1981, 1984) recorded ERP responses from awake macaque monkeys while they passively viewed stimuli of varying luminance and pattern complexity (e.g., gratings or checkerboards). They reported that the effects of stimulus onset and offset recorded from monkeys mirrored those from humans during passive viewing of the stimuli (Van der Marel et al., 1984). These findings agree with those of comparative anatomical studies showing that macaque monkeys and humans have very similar neuroanatomy, particularly with regard to the visual system (Kaas, 2005). Although the general pattern of visual ERPs was similar across the monkeys tested, the size of the visual ERP components varied significantly across individual animals (Van der Marel et al., 1984). This observation mirrors the human ERP literature, in which the amplitude of early visual ERPs varies significantly across individuals as well. Finally, Van der Marel et al. (1984) note that the visually evoked ERPs from the monkeys were 10–40 ms faster than similar components in humans.

The report of faster sensory ERP components in monkeys than in humans has been corroborated in subsequent studies (Lamme et al., 1992; Schroeder et al., 1991, 1992; Woodman et al., 2007). Subsequent work also showed that the task relevance of the stimuli does not modulate the latency of these components (Glover et al., 1991). Presumably, the earlier onsets across the sensory and perceptual components like the N1 and P1 are due to the smaller size of the brains of the nonhuman primates compared to human subjects. Specifically, the larger brains of humans have many more neurons and synapses, meaning that information transmitted through the human brain will have more transmission delays compared to information transmitted through the smaller macaque brain. Interestingly, a study of visually evoked potentials in great apes (i.e., gorillas and chimpanzees) suggests that the timing and morphology of the ERP components of our nearest primate relatives are even more similar to our own ERP components than those of old world monkeys like macaques (Boysen & Berntson, 1985). Size of the brain cannot be a simple scaling factor for temporal relationships among ERP components, however, because the human brain is approximately seven times larger than that of old world monkeys like macaques (Falk, 1986), while the ERP component latencies are typically only 25% shorter.

Now we return to the observation that even within species, the early sensory components differ across individuals. The waveforms from three monkeys performing a visual search task are shown in Figure 22.1 to provide a concrete and recently published example of the individual differences in monkey sensory ERPs (Woodman et al., 2007). Figure 22.1 illustrates the individual differences in amplitude of the early components and their relative speed compared to the human visual ERP components (see Chapters 4 and 11, this volume). Examination of the early sensory components evoked by these visual search arrays allows us another way to relate the observed monkey components to those from humans. The early visual components in humans (i.e., the P1 and N1 components) are modulated predictably by raw stimulus strength (Luck, 2005). Thus, as the set size of the search array increases from 2 to 4 to 8 objects, the human N1 component systematically increases in amplitude. As shown in Figure 22.1, this is precisely the pattern of results we observed in all three monkeys from an electrode approximating the location of electrode Oz in the modified 10/20 system (Jasper, 1958). In other words, as set size increased, the amplitude of the first negative ERP component increased as well. The amplitude of this N1 component at the intermediate set size of 4 objects is marked by a dashed line for reference. These waveforms also show an interesting difference between human and monkey ERPs. Whereas the human P1 component

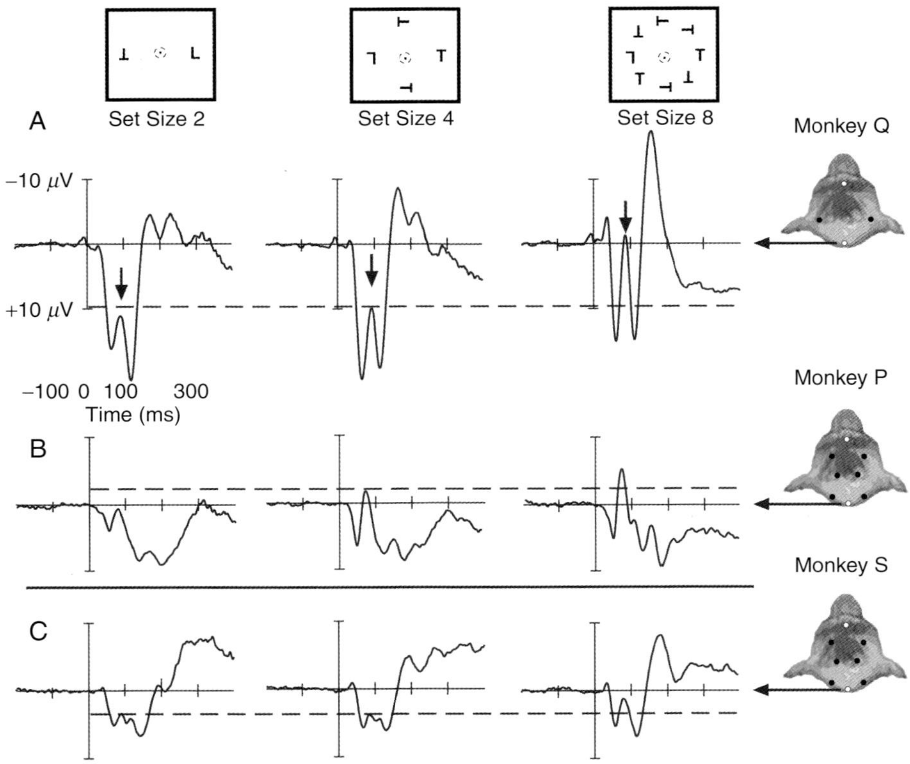

Fig. 22.1. Event-related potential waveforms recorded from a posterior midline electrode in monkey Q (A), in monkey P (B), and in monkey S (C) across set sizes 2, 4, and 8. The relevant active electrode on each monkey was equivalent to Oz, and the common frontal reference electrode is analogous to Fz. Waveforms recorded from all three monkeys show a complex of early negative-going components sensitive to the amount of sensory stimulation as a function of set size (marked by the arrows on the waveforms from monkey Q). The dashed horizontal lines mark the amplitude of the set size 4 peak of the first negativity. Adapted with permission from Woodman et al. (2007).

shows sensitivity to manipulations of the strength of sensory input, just as the N1 does, the monkey P1 does not appear to be modulated in the same manner. This demonstrates the nontrivial nature of finding homologues to the human ERP components. That is, in comparative electrophysiology, both similarities and differences between human and monkey ERPs can be found.

Studies of monkey visually evoked potentials were the first to bring multiple types of electrophysiological recordings to bear on questions of the location of component generation. The primary motivation for the Arthur and Starr (1984) paper was that by establishing the existence of a monkey P3, subsequent research using depth recordings and lesions would localize the generator or generators of the component. Using monkeys to investigate the neural origins of ERP components is an obvious advantage because this nonhuman primate model allows invasive recordings from inside the brain. Schroeder and colleagues (1991, 1992) published a series of experiments in which they collected ERP data from monkeys simultaneously with recordings of potentials across the different layers of areas such as V1. These laminar recordings not only show where potentials recorded from the surface of the scalp or skull are being generated in the cortex but also whether the candidate activity is an input to an area or is activity that arises within the area itself. This technique provides necessary evidence that an area generates electrical fields that contribute to a surface-recorded ERP component by demonstrating a polarity inversion as electrode contacts span the dipole generated in a certain brain area. That is, nearby electrodes in different layers of an area simultaneously show positive and negative potentials at the same time. For example, Schroeder and colleagues (1991) showed that the first visual ERP component appears to be generated in the supragranular layers of primary visual cortex, while the subsequent components are generated by activity in extrastriate cortical areas. Studies such as this, and those described below, in which the neural generators of specific ERP components are determined using simultaneous depth recordings in the brains of monkeys, are all too rare. The paucity of such investigations is likely due to the difficulty of these multilevel recordings and not to the richness of the dataset they provide. Nunez and Srinivasan (2006) provide an excellent discussion of how many physicists have spent their careers trying to understand principles that span spatial scales, and they point out that much more of this work is necessary for progress in understanding the signal we are recording from electrodes outside the brain.

Mismatch and Selection Negativity

The oddball paradigm typically used to investigate the P3 component yields a number of other ERP components in human subjects under different types of task demands. In the oddball paradigm, one stimulus (or stimulus class) is more frequent than the other stimulus presented in the sequence. In humans, the first difference that is observed between the waveforms elicited by frequent and infrequent stimuli occurs at around 200 ms and is known as the *mismatch negativity* (MMN; see Chapter 6, this volume). As indicated by its name, the MMN is evidenced by the waveform to the infrequent stimulus being more negative than the waveform elicited by the frequent stimuli. The MMN is elicited any time that the eliciting stimulus does not match the predominant stimuli in the sequence, even if those infrequent stimuli are not task relevant, unlike the task-related P3 (or P3b) discussed above (Näätänen, 1990; Näätänen et al., 1978; Woldorff & Hillyard, 1990; but see Woldorff et al., 1991). Thus, it appears that the MMN is a measure of the brain's recognition-100-200 ms after stimulus onset-that the current stimulus is physically different than the context in which it is presented. Using auditory presentation of stimuli, Javitt and colleagues (1992) have shown that the monkey produces a similar mismatch response to infrequent stimuli when it is not performing a task. As with the visually evoked ERP waveforms, the MMN in monkeys appears to have an earlier onset (approximately 80 ms poststimulus) than the MMN component in humans (i.e., 200 ms poststimulus). However, the greater than 100 ms discrepancy in the onset of the component between the primate species is in need of further study, as the between-species timing difference is strikingly large compared to that of other ERP components found in both species.

A slight modification of the paradigm used to elicit the MMN, in which the infrequent stimulus is also a task-relevant target stimulus, elicits a different ERP component in humans called the *selection negativity* (see Chapter 11, this volume). Whereas the MMN appears as a more negative potential for any infrequent stimulus, the selection negativity is a negative-going component elicited by infrequent task-relevant target stimuli compared to infrequent nontarget stimuli (Anllo-Vento & Hillyard, 1996; Harter et al., 1982; Hillyard & Münte, 1984; Hillyard et al., 1984). To determine whether our primate relatives share this index of attentional selection of task-relevant target information, Mehta and colleagues (2000a, 2000b) trained monkeys to perform a cross-modal attention task. The researchers then recorded from a skull screw electrode and from multicontact laminar electrodes in subcortical and visual areas of the cortex (the lateral geniculate nucleus [LGN], V1, V2, and V4). The stimuli were concurrent streams of visual and auditory oddball stimuli, and the monkeys alternated between detecting the infrequent stimuli in the visual or auditory stream. The effects of attention were determined by comparing the neural responses from the same stimuli under the condition in which they were to be ignored with the condition in which they were task relevant. That is, the ERPs elicited by a visual stimulus when the stimuli in the interleaved auditory stream were task relevant compared to the ERP response to a visual stimulus when the visual stimuli were the targets. Mehta and colleagues found that the onset of attention effects was earliest in the most downstream area studied (i.e., V4). Effects of attention were later in V2, even later in V1, and nonexistent in the LGN recordings. These findings support the view that the selection negativity originates in anterior cortical areas and that this selection signal is fed back to lower-level visual areas.

Schroeder and colleagues have used this same multisensory attention task to address fundamental questions about the brain dynamics underlying ERPs (Fu et al., 2001; Mehta et al., 2000a, 2000b; Schroeder & Foxe, 2002; Shah et al., 2004). In one of the most fundamentally important electrophysiological studies in recent years, Shah and colleagues (2004) tested the hypothesis that ERPs are not evoked by the occurrence of a discrete event, like the presentation of a visual stimulus, but instead are caused by the synchronization of ongoing EEG oscillations (e.g., Makeig et al., 2002; see also Chapters 2 and 3, this volume). Shah et al. (2004) demonstrated that local-field potential fluctuations, recorded in primary visual cortex, are generated in response to the presentation of a stimulus and are not simply the phase resetting of ongoing oscillations in the brain. In higher-order perceptual areas, specifically the inferior temporal cortex (or IT), the amplitude of the stimulus-locked waveforms was due primarily to potentials evoked by the visual stimulus, but IT also showed ongoing oscillatory activity that made a significant contribution to the time-locked ERPs. This study nicely shows how recordings of local-field potentials in the brain, also known as *intracranial EEG* (iEEG) in studies of clinical and rodent populations, can provide definitive evidence to distinguish between different models of the cortical dynamics underlying the generation of ERPs.

The N2pc

As described throughout this chapter, electrophysiologists have used three primary types of evidence to support their claims that monkeys exhibit ERP components similar to those found in humans. Studies of monkeys have shown that ERP components have relative timing that is similar to that in human studies (i.e., the components are early or late in the sequence of polarity deflections). Studies have also shown that primate ERP components are similarly sensitive to stimulus and cognitive manipulations to argue for homology between ERPs of humans and other species and have a similar distribution across the head. This section provides an example of the use of multiple criteria for establishing homology between a human ERP component and a monkey ERP component. Specifically, we discuss how the criteria of voltage distribution, timing, and sensitivity to cognitive demands were used to support the conclusion that monkeys have an ERP component related to shifting and focusing visual-spatial attention similar to that shown in humans.

In human observers, the N2pc component is a negative-going ERP waveform, typically elicited 170–200 ms after the onset of a visual search array with a posterior distribution that is contralateral with respect to where attention is deployed in the visual field (see Chapter 12, this volume). That is, the N2pc is maximal at posterior and lateral electrode locations approximately 200 ms poststimulus as attention shifts to a target or potential target item in the left or right visual field (Luck et al., 1997). Event-related potential studies of the N2pc in humans performing a visual search have been successful in revealing aspects of covert attention that cannot be observed using behavioral methods alone (Luck, 1994; Woodman & Luck, 1999, 2003a, 2003b). Of particular relevance are recent studies demonstrating that shifts of attention during a visual search can be measured using this lateralized component of human ERPs. Studies by Woodman and Luck (1999, 2003b) demonstrated that the N2pc component shifts between hemispheres as attention shifts between potential target items in visual search arrays. The N2pc has also been shown to be an index of a perceptual selection mechanism

(Luck & Hillyard, 1994b; Woodman & Luck, 2003a) that serves to suppress information from distractor objects surrounding the attended item (Luck et al., 1997). Source estimation procedures suggest that this ERP component may be generated in the human equivalent of macaque area V4 or TEO in the inferior temporal cortex (Luck & Hillyard, 1994b). Generally consistent with this possibility, a magnetoencephalographic study found that the N2pc is accompanied by a temporal lobe magnetic field that spans much of the duration of the electrical N2pc component (Hopf et al., 2000).

To determine whether monkeys exhibit a homologue of the human N2pc component, Woodman and colleagues (2007) examined the waveforms recorded from lateral-posterior electrode sites in three monkeys performing a visual search task. The visual search task required the monkeys to view an array without shifting gaze until they could make one saccade directly to the target object. This task required the monkeys to rely on covert attention to select and process the target prior to the overt eye movement because reward would rarely be obtained if the monkeys moved their eyes prior to covertly analyzing the search array. The set size of the search array varied randomly from trial to trial among 2, 4, and 8 objects. Across days, the monkeys searched for a different target object such that all stimuli served as both targets and distractors, ruling out the possibility that the lateralized effects could be entirely due to a physical stimulus confound (Woodman, 2010). The onsets of saccades were detected offline, and ERP waveforms from 20 ms preceding an eye movement were truncated. Thus, the average at each poststimulus time point was the mean of the remaining presaccadic waveforms.

The three monkeys in this study were implanted with arrays of electrodes that included posterior-lateral electrode locations, as well as parietal, central, and frontal locations. This array allowed the researchers to test for potential homologues of the human N2pc with the same contralateral and posterior distribution. Although these implanted arrays were composed of fewer electrodes than used with humans, the arrays provided greater coverage and density than is typical of nonhuman ERP recordings in which the modal number of EEG electrodes is one. The electrode array implants were constructed from Teflon-coated braided stainless steel wire and amphenol pins. During aseptic surgery, holes 1 mm deep and 1 mm in diameter were drilled into the surface of the skull, allowing the terminal end of the electrode to be tightly inserted. The use of these small electrode contacts implanted in the skull has several advantages. First, compared to the typical procedure of recording from skull screws that span the entire thickness of the bones of the skull, the electrode implants of Woodman et al. (2007) maximize the similarity of the resistive characteristics and tissue through which signals must pass in humans and in these nonhuman primate recordings. It should be noted that the skull itself is multilayered and that the different layers of bone have different conductive properties (Nunez & Srinivasan, 2006). By inserting the electrode into the most exterior 1 mm of the 3–5 mm thick skull, much of the electrical pathway was preserved across species while avoiding the tremendous noise present when recording from the scalps of the far more muscle-headed macaque monkeys. Second, as the tissue reacts to the implantation of skull screws, it is common for the bone to grow over the exposed tip of the screw penetrating the brain case. Thus, the impedance of skull screw electrodes is initially very low and can measurably change over time as additional bone layers form between the metal of the screw and the dura surrounding the brain. In contrast, the impedance of the electrodes of Woodman et al. (2007) was 2–5 kΩ at 30 Hz, which is identical to that of EEG electrodes used in human studies. The impedance of these electrodes remains stable for upwards of 5 years in healthy monkeys. Third, the electrode leads can be covered by skin that is sutured back over the skull. This allows the EEG electrodes to be minimally invasive once implanted.

Before discussing the ERP findings, it is necessary to discuss the behavioral results from the search task. As with human subjects, the monkeys' saccadic reaction times (RTs) were fastest at set size 2 and slowest at set size 8. This is shown in Figure 22.2, with the saccadic RT next to the waveform for each set size. Unlike the early sensory components, the amplitude and latency of the N2pc component are known to be related to how rapidly human observers can shift attention to a target in a search array. Specifically, a more efficient visual search is associated with larger-amplitude N2pc components due to less temporal variability when attention can be focused on the target (Luck et al., 1997; Luck & Hillyard, 1994b).

The waveforms shown in Figure 22.2 are representative of the pattern found across the three monkeys from which ERP data were recorded during the search. First, note that right visual field (VF) targets

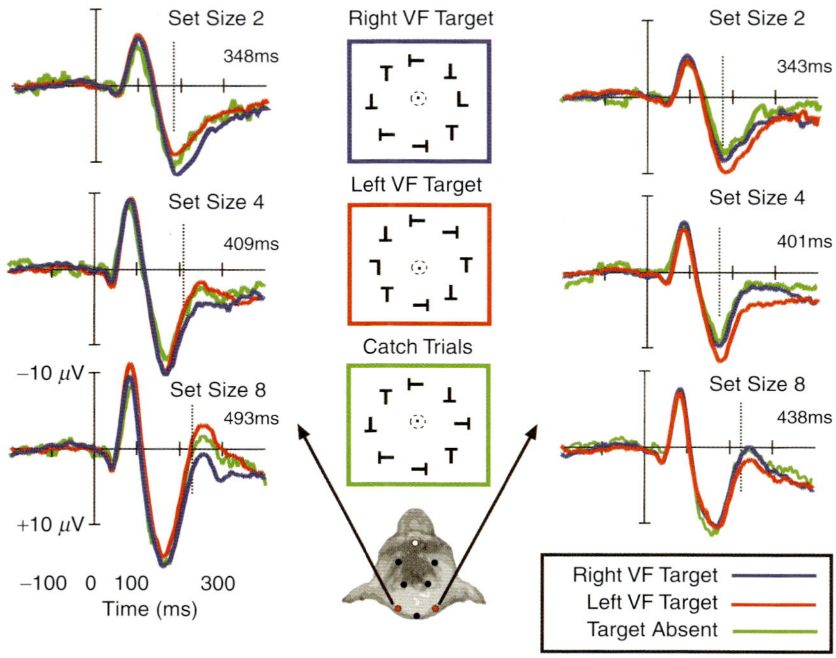

Fig. 22.2. Event-related potential waveforms recorded from the left posterior (left column) and right posterior electrodes (right column) from monkey P for right (blue traces) and left visual field (VF) targets (red traces) and target-absent trials (green traces) across set sizes 2, 4, and 8. Following a visually evoked negativity, a contralateral positivity was observed beginning ~125 ms poststimulus for lateralized targets, but not when targets appeared on the horizontal midline (data not shown), as in human observers. The amplitude of the monkey homologue of the N2pc (mN2pc) was modulated by the set size of the visual search array presented. Dashed vertical lines mark the onset of the mN2pc. The number above the waveform indicates the mean saccadic response latency for contralateral targets.

elicited a waveform that was more positive at the left hemisphere electrode site than at the right hemisphere site beginning approximately 150 ms poststimulus. The onset of this contralateral positivity is marked by the dashed line to indicate the point at which the ipsilateral and contralateral waveforms were significantly different from each other. Conversely, the waveforms elicited by left VF targets were more positive at the right hemisphere electrode than at the left hemisphere site. As discussed below, this component appears to be a contralateral positivity, whereas the human component is a contralateral negativity. Second, this contralateral positivity was sensitive to the set size of the eliciting array. As set size increased, the amplitude of the contralateral positivity decreased, the onset appeared to shift later in time, and the duration of this difference was more variable. Third, as shown in Figure 22.2, we confirmed in one monkey that on catch trials in which no target was present, the waveforms elicited by the nontarget arrays were essentially identical to those ipsilateral to the target in a search array. This parallels findings from humans shown nontarget search arrays (Luck & Hillyard, 1994a) and provides another example of using manipulations of a task in both species to determine the functional similarity of the waveforms in the two species. Moreover, these data allow us to assess whether the hemispheric difference observed in the monkeys is an ipsilateral negativity or a contralateral positivity. As shown in the green traces in Figure 22.2, the waveforms recorded on nontarget trials are essentially the same as those elicited by ipsilateral target arrays. These findings support the conclusion that shifting attention to the target location in the search array elicits a contralateral positivity. To clearly show the similarity between the macaque and human components, we ran human subjects in exactly the same visual search task as that used with the monkeys. Figure 22.3 shows the data from monkey P and the data from human J.A. for the purpose of directly comparing this attention-related component across species of primates. As you can see, the set size manipulation elicits nearly identical effects by shifting the onset and the peak amplitude of the contralateral component back in time in both the human and the monkey. Thus, this apparent macaque N2pc (or mN2pc) behaves identically to the human

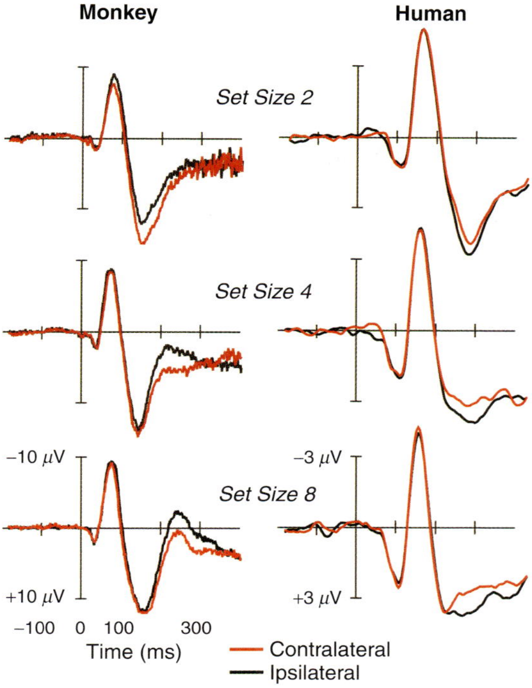

Fig. 22.3. Comparison of average ERP waveforms recorded from the posterior pair of electrodes on monkey P (left) with waveforms from electrodes OL/OR on human J.A. (right) performing the same oculomotor search task across set sizes.

component in terms of contralateral distribution and sensitivity to attentional demands of a visual search task.

The mN2pc exhibits the same anterior-posterior distribution as the human N2pc. Specifically, Figure 22.4 shows the waveforms recorded from the three pairs of posterior-to-anterior electrodes implanted in monkey P. The mN2pc is observed over the most posterior pair of electrodes, but it is noticeably and significantly reduced at the next more anterior electrodes and absent at the most anterior pair of electrodes. This distribution mirrors the posterior-to-anterior distribution of the N2pc recorded from humans performing a visual search (Luck & Hillyard, 1994a). In summary, the mN2pc recorded from all three monkeys exhibits the timing, distribution, and sensitivity to attentional demands that functionally define a monkey homologue of the human N2pc.

It is interesting to note that the human index of covert attentional deployment is a negative potential, whereas this component in the monkey is a positive potential. This was not completely unexpected due to differences in cortical folding between the two species. Because ERPs are generated by tissue that, when active, generates open electrical fields, the cortex is believed to be the principal generator of such electrical potentials (Luck, 2005; Nunez & Srinivasan, 2006). The folding of the cortical surface that contains the generating tissue will therefore determine the polarity of the observed ERP component homologue. An inversion of the cortical surface relative to the skull results in a polarity inversion of an ERP component. For example, the human C1 component is of opposite polarity when it is evoked by an upper versus a lower VF stimulus due to activation of neurons on opposite banks of the calcarine sulcus (Clark et al., 1995). Source estimation procedures suggest that the human N2pc may be generated predominantly in ventral visual areas such as V4 and IT. Whereas monkey V4 is located on a superficial gyrus, the proposed human homologue, based on functional imaging data, is in an area that has both sulci and gyri (Orban et al., 2004). It is possible that in humans the N2pc is generated in the subregion of the anatomical homologue of V4 that is folded in a sulcus. Individual differences in the folding of human cortex could invert the N2pc, and occasionally, instances have been observed in my research with humans in which contralateral positivities were found in humans as well. However, structural magnetic resonance images (MRIs) of those subjects were not available to test the hypothesis that these individuals had an anomalous pattern of folding in ventral visual cortex. Thus, the likely explanation for the polarity difference observed between human and monkey attention-related lateralizations is that the mN2pc component is generated by cortical tissue that is inverted in the macaque relative to the orientation of the functionally homologous tissue in humans.

In summary, the study by Woodman and colleagues (2007) provides an example of how comparative electrophysiological studies can use multiple types of evidence to show that an ERP component found across species is indexing the same cognitive operations during information processing. Establishing homology between ERP components found in two different species is the first step toward using an animal model to better understand the neural circuitry underlying the generation of the ERP component. Monkey models of human information processing offer the possibility of recording from structures in the brain and performing inactivation or lesions studies to determine which areas in the brain are involved in the generation of a given component. Research is currently underway to localize the neural generators of the mN2pc component using these converging operations.

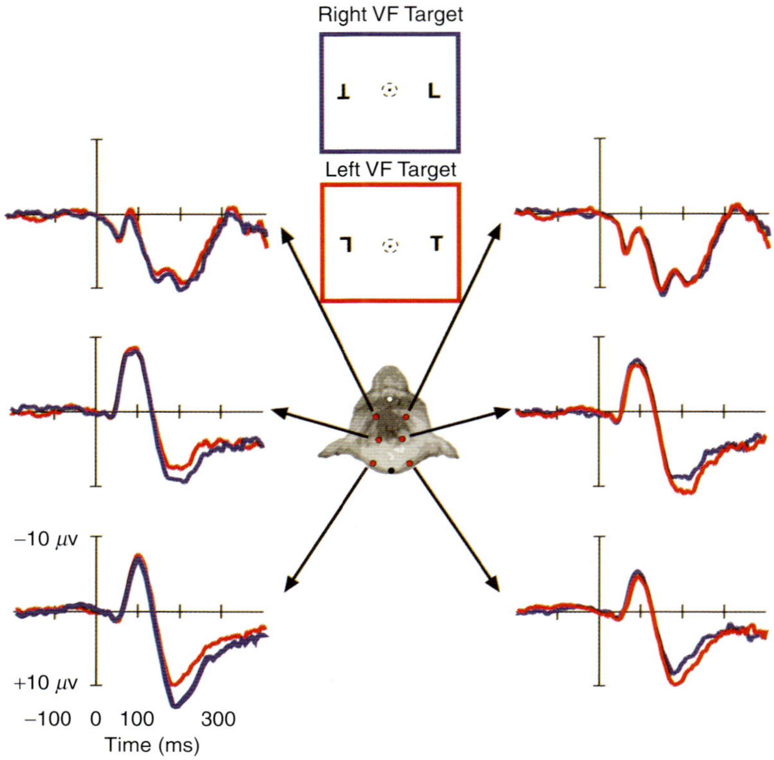

Fig. 22.4. Event-related potential waveforms recorded from monkey P for right VF targets (blue traces) and left VF targets (red traces) across all pairs of lateralized electrodes for search arrays with two items. The amplitude of the mN2pc was maximal and significant at the most posterior pair of electrodes (*ps* < .01) and decreased progressively at more anterior electrodes.

So Much to Do, So Little Known

Despite the sizable body of fundamental work examining ERPs recorded from nonhuman primates, it has not yet been established whether monkeys exhibit homologues of many of the electrophysiological indices that we use to study cognitive processes in humans. The other chapters in this book document the progress that has been made in defining the cognitive functions indexed by a large number of distinct ERP components in human subjects. However, researchers have yet to look for many of these ERP components in monkeys. This means that there is a need for many basic comparative electrophysiological studies. Obviously, the category of ERP components related to language processing and use cannot be studied in nonverbal species. However, it is possible that we can study more general semantic processing in nonhuman primate models using components that were discovered in ERP studies of language. For example, the N400 component is elicited by semantic incongruities (see Chapter 15, this volume) regardless of whether meaning is communicated through words (Kutas & Hillyard, 1980) or pictures (Nigam et al., 1992). This may be a way of studying semantic processing across species without the use of linguistic stimuli. Another limitation to consider is that the variety of ERP components that can be studied is constrained by the type of task a monkey can be trained to perform. Finally, all monkeys are overtrained on tasks relative to the modest amount of practice human subjects receive before an experiment, and we must consider whether the training of nonhuman primates renders monkey ERPs qualitatively different than human components found in the same tasks.

The neural origins of ERP components have not been definitively localized in the human brain because only rarely can potential generators of ERP components be studied intracranially in patient populations (Halgren et al., 1980; Wang et al., 2005). In addition, the temporal resolution of most imaging techniques makes them too slow to functionally localize the generators of an ERP component, which is often a brief neural event (e.g., 100 ms in duration). Interestingly, Caton's (1875) initial report of EEG in animals was presented as evidence for localization of function. Specifically, the observation that lateralized visual stimuli would elicit larger contralateral responses was taken as support for specialization of function by cortical regions (Brazier, 1957). Indeed, this was the implication of citing Ferrier's work in the initial reports of feeble potentials and stimulus-evoked fluctuations (Caton, 1875, 1887). As it remains today, the issue of localized versus distributed processing in the brain was hotly debated, with the pioneers of the fledgling field of electrophysiology viewed many of their findings as most relevant to this debate (Adrian & Matthews, 1934; Berger, 1929; Caton, 1887; Walter, 1938). Viewing EEG and ERPs as evidence for localization of function seems ironic given the limited spatial resolution of these techniques. The spatial resolution of imaging techniques with current technology is far beyond the ability of the ERP technique to localize function to regions of cortex. It seems that an area in which monkey ERPs can have the greatest impact is in our ability to have the temporal resolution of the ERP technique with the spatial resolution of depth recordings in specific brain areas.

Viewing EEG and ERP recordings from humans and nonhuman species through the lens of history brings several fundamental questions into focus. Perhaps the most basic among these lingering concerns is that we do not really understand the EEG signal in which ERPs are embedded. What is the function of the spontaneous synchrony evidenced by EEG? Why does the presentation of a stimulus cause the ongoing alpha-dominated EEG to be reduced in amplitude? This stimulus-induced alpha desychronization was one of the first observations made by Caton (1887), and yet its cause is still unknown. At the dawn of the era of ERPs in the field of electrophysiology, Donald Lindsley pointed out that despite the great enthusiasm for using ERPs to answer questions about cognition, the fundamental questions about the basic EEG signal remained unanswered (Donchin & Lindsley, 1969). Basic comparative ERP studies also still have much to discover. Bullock (2003) observed missing-stimulus potentials in every species he examined and was led to wonder what electrophysiological measures of cognitive processing are unique to humans. What is it that underlies our cognitive abilities that are so far beyond those of even our closest primate relatives? Whereas this chapter focused on the components that appear to be the same across primates, understanding what makes humans special will require us to also focus on the differences between human ERPs and those of other species of animals.

Acknowledgments
I would like to thank Jeff Schall, Jason Arita, Min-Suk Kang, Andrew Rossi, Jeremiah Cohen, Richard Heitz, Pierre Pouget, Veit Stuphorn, Mary Fuertado, Melonie Williams, Nancy Carlisle, Michelle Young, David Godlove, Braden Purcell, Robert Reinhart, Namsoo Kim, the National Eye Institute (RO1-EY019882) and the Vanderbilt Vision Research Center for their vital support of this work. Please contact for questions regarding this article (geoffrey.f.woodman@vanderbilt.edu).

References

Adrian, E. D. (1932). *The mechanism of nervous action.* Philadelphia: University of Pennsylvania Press.

Adrian, E. D., & Buytendijk, F. J. J. (1931). Potential changes in the isolated brain stem of the goldfish. *Journal of Physiology, 71*, 121–135.

Adrian, E. D., & Matthews, B. H. C. (1934). The interpretation of potential waves in the cortex. *Journal of Physiology, 81*, 440–471.

Adrian, E. D., & Yamagiwa, K. (1935). The origin of the Berger rhythm. *Brain, 58*, 323–351.

Anllo-Vento, L., & Hillyard, S. A. (1996). Selective attention to the color and direction of moving stimuli: Electrophysiological correlates of hierarchical feature selection. *Perception & Psychophysics, 58*, 191–206.

Arthur, D. L., & Starr, A. (1984). Task-relevant late positive component of the auditory event-related potential in monkeys resembles P300 in humans. *Science, 223*, 186–188.

Bartley, S. H., & Newman, E. B. (1930). Recording cerebral action currents. *Science, 71*, 587.

Beck, A. (1890). Dia strom der nervencentren. *Centralblatt für Physiologie, 4*, 572–573.

Berger, H. (1929). Ueber das Elektrenkephalogramm des Menschen. *Archives für Psychiatrie Nervenkrankheiten, 87*, 527–570.

Borda, R. P. (1970). The effect of altered drive states on the contingent negative variation (CNV) in rhesus monkeys. *Electroencephalography and Clinical Neurophysiology, 29*, 173–180.

Boysen, S. T., & Berntson, G. G. (1985). Visual evoked potentials in the great apes. *Electroencephalography and Clinical Neurophysiology, 62*, 150–153.

Brazier, M. A. B. (1957). Rise of neurophysiology in the 19th century. *Journal of Neurophysiology, 20*, 212–226.

Bullock, T. H. (2003). Have brain dynamics evolved? Should we look for unique dynamics in the sapient species? *Neural Computation, 15*, 2013–2027.

Caton, R. (1875). The electric currents of the brain. *British Medical Journal, 2*, 278.

Caton, R. (1877). Interim report on investigation of the electric currents of the brain. *British Medical Journal, 1*, Supplement 62.

Caton, R. (1887). Researches on electrical phenomena of cerebral grey matter. *Transactions of the Ninth International Medical Congress, 3*, 246–249.

Clark, V. P., Fan, S., & Hillyard, S. A. (1995). Identification of early visually evoked potential generators by retinotopic and topographic analyses. *Human Brain Mapping, 2*, 170–187.

Cohen of Birkenhead, L. (1959). Richard Caton (1842–1926); pioneer electrophysiologist. *Proceedings of the Royal Society of Medicine, 52*, 645–651.

Davis, P. A. (1939). Effects of acoustic stimuli on the waking human brain. *Journal of Neurophysiology, 2*, 494–499.

Desiraju, T. (1972). Transformations of the discharges of neurons of parietal association cortex during sleep and wakefulness in monkey. *Journal of Neurophysiology, 35*, 326–332.

Donchin, E., & Lindsley, D. B. (Eds.). (1969). *Average evoked potentials: Methods, results and evaluations*. Washington, DC: U.S. Government Printing Office.

Donchin, E., Otto, D., Gerbrandt, L. K., & Pribram, K. H. (1971). While a monkey waits: Electrocortical events recorded during the fore-period of a reaction time study. *Electroencephalography and Clinical Neurophysiology, 31*, 115–128.

Evarts, E. V., & Magoun, H. W. (1957). Some characteristics of cortical recruiting responses in unanesthetized cats. *Science, 125*(3258), 1147–1148.

Falk, D. (1986). Endocranial casts and their significance for primate brain evolution. In D. R. Swindler & J. Erwin (Eds.), *Comparative primate biology. Volume 1: Systematics, evolution and anatomy* (pp. 477–490). New York: Alan R Liss.

Fleischl von Marxow, E. (1890). Mittheilung betreffed die Physiologie der Hirnrinde. *Centralblatt für Physiologie, 4*, 537–540.

Fu, K. M., Foxe, J. J., Higgins, B. A., Murray, M. M., Javitt, D. C., & Schroeder, C. E. (2001). Shifts of intermodal selective attention produced by endogenous cues. *Journal of Cognitive Neuroscience, 12*, 145–152.

Glover, A., Ghilardi, M. F., Bodis-Wollner, I. M. O., & Mylin, L. H. (1991). Visual "cognitive" evoked potentials in the behaving monkey. *Electroencephalography and Clinical Neurophysiology, 1991*(90), 65–72.

Gotch, F., & Horsely, V. (1891). On the mammalian nervous system, its functions and their localisation determined by an electrical method. *Philosophical Transactionsof the Royal Society of London, 182B*, 267–526.

Haas, L. F. (2003). Hans Berger (1873–1941), Richard Caton (1842–1926), and electroencephalography. *Journal of Neurology, Neurosurgery and Psychiatry, 74*, 9.

Halgren, E., Squires, N. K., Wilson, C. L., Rohrbaugh, J., Babb, T. L., & Crandall, P. H. (1980). Endogenous potentials generated in the human hippocampal formation and amygdala by infrequent events. *Science, 210*, 803–805.

Harter, M. R., Aine, C., & Schroeder, C. (1982). Hemispheric differences in the neural processing of stimulus location and type: Effects of selective attention on visual evoked potentials. *Neuropsychologia, 20*, 421–438.

Hillyard, S. A., & Münte, T. F. (1984). Selective attention to color and location: An analysis with event-related brain potentials. *Perception & Psychophysics, 36*, 185–198.

Hillyard, S. A., Simpson, G. V., Woods, D. L., Van Voorhis, S., & Münte, T. (1984). Event-related brain potentials and selective attention to different modalities. In F. Reinoso-Suarez & C. Ajmone-Marsen (Eds.), *Cortical integration* (pp. 395–414). New York: Raven Press.

Hopf, J.-M., Luck, S. J., Girelli, M., Hagner, T., Mangun, G. R., Scheich, H., et al. (2000). Neural sources of focused attention in visual search. *Cerebral Cortex, 10*, 1233–1241.

Hubel, D. H., Henson, C. O., Rupert, A., & Galambos, R. (1959). Attention units in the auditory cortex. *Science, 129*(3358), 1279–1280.

Jasper, H. H. (1958). The ten-twenty electrode system of the International Federation. *Electroencephalography and Clinical Neurophysiology, 10*, 371–375.

Javitt, D. C., Schroeder, C. E., Steinshneider, M., Arezzo, J. C., & Vaughan, H. G. (1992). Demonstration of mismatch negativity in the monkey. *Electroencephalography and Clinical Neurophysiology, 83*, 87–90.

Kaas, J. H. (2005). The evolution of visual cortex in primates. In J. Kremers (Ed.), *The primate visual system* (pp 267–283). Chichester, UK: Wiley.

Kennard, M. A. (1943). Electroencephalogram of decorticate monkeys. *Journal of Neurophysiology, 6*, 233–242.

Kennard, M. A., & Nims, L. F. (1942). Effect on electroencephalogram of lesions of cerebral cortex and basal ganglia in *Macaca mulatta*. *Journal of Neurophysiology, 5*, 335–348.

Knight, R. T. (1991). Evoked potential studies of attention capacity in human frontal lobe lesions. In H. S. Levin, H. M. Eisenberg, & A. L. Benton (Eds.), *Frontal lobe function and dysfunction* (pp. 139–153). New York: Oxford University Press.

Knight, R. T., Hillyard, S. A., Woods, D. L., & Neville, H. J. (1981). The effects of frontal cortex lesions on event-related potentials during auditory selective attention. *Electroencephalography and Clinical Neurophysiology, 52*, 571–582.

Knight, R. T., Scabini, D., Woods, D. L., & Clayworth, C. C. (1989). Contributions of temporal-parietal junction to the human auditory P3. *Brain Research, 502*, 109–116.

Kutas, M., & Hillyard, S. A. (1980). Reading senseless sentences: Brain potentials reflect semantic incongruity. *Science, 207*, 203–205.

Lamme, V. A., Van Dijk, B. W., & Spekreijse, H. (1992). Texture segregation is pressed by primary visual cortex in man and monkey: Evidence from VEP experiments. *Vision Research, 32*(5), 797–807.

Lashley, K.S. (1950). In search of the engram. In Society for Experimental Biology (1950). Physiological mechanisms in animal behavior. Society's Symposium IV (Oxford, England: Academic Press), pp. 454–482.

Loveless, N. E., & Sanford, A. J. (1975). The impact of warning signal intensity on reaction time and components of the contingent negative variation. *Biological Psychology, 2*, 217–226.

Luck, S. J. (1994). Cognitive and neural mechanisms of visual search. *Current Opinion in Neurobiology, 4*, 183–188.

Luck, S. J. (2005). *An introduction to the event-related potential technique*. Cambridge, MA: MIT Press.

Luck, S. J., Girelli, M., McDermott, M. T., & Ford, M. A. (1997). Bridging the gap between monkey neurophysiology and human perception: An ambiguity resolution theory of visual selective attention. *Cognitive Psychology, 33*, 64–87.

Luck, S. J., & Hillyard, S. A. (1994a). Electrophysiological correlates of feature analysis during visual search. *Psychophysiology, 31*, 291–308.

Luck, S. J., & Hillyard, S. A. (1994b). Spatial filtering during visual search: Evidence from human electrophysiology. *Journal of Experimental Psychology: Human Perception and Performance, 20*, 1000–1014.

Makeig, S., Westerfield, M., Jung, T.-P., Enghoff, S., Townsend, J., Courchesne, E., et al. (2002). Dynamic brain sources of visual evoked responses. *Science, 25*(5555), 690–694.

Mehta, A. D., Ulbert, I., & Schroeder, C. E. (2000a). Intermodal selective attention in monkeys. I: Distribution and timing of effects across visual areas. *Cerebral Cortex, 10*(4), 343–358.

Mehta, A. D., Ulbert, I., & Schroeder, C. E. (2000b). Intermodal selective attention in monkeys. II: Physiological mechanisms of modulation. *Cerebral Cortex, 10*(4), 359–370.

Näätänen, R. (1990). The role of attention in auditory information processing as revealed by event-related potentials and other brain measures of cognitive function. *Behavior and Brain Science, 13*, 201–288.

Näätänen, R., Gaillard, A. W. K., & Mantysalo, S. (1978). Early selective-attention effect on evoked potential reinterpreted. *Acta Psychologica, 42*, 313–329.

Nigam, A., Hoffman, J. E., & Simons, R. F. (1992). N400 to semantically anomalous pictures and words. *Journal of Cognitive Neuroscience, 4*, 15–22.

Nunez, P. L., & Srinivasan, R. (2006). *Electric fields of the brain: The neurophysics of EEG*(2nd ed.). Oxford: Oxford University Press.

Orban, G. A., Van Essen, D. C., & Vanduffel, W. (2004). Comparative mapping of higher visual areas in monkeys and humans. *Trends in Cognitive Sciences, 8*(7), 315–320.

Paller, K. A., McCarthy, G., Roessler, E., Allison, T., & Wood, C. C. (1992). Potentials evoked in human and monkey medial temporal lobe during auditory and visual oddball paradigms. *Electroencephalography and Clinical Neurophysiology, 84*, 269–279.

Pineda, J. A., Foote, S. L., & Neville, H. J. (1989). Effects of locus coeruleus lesions on auditory, long-latency, event-related potentials in monkey. *Journal of Neuroscience, 9*, 81.

Pravdich-Neminsky, V. V. (1913). Ein Versuch der Registrierung der elekrischen Gehirnerscheinungen. *Centralblatt für Physiologie, 27*, 951–960.

Ramón, F., Hernández, O. H., & Bullock, T. H. (2001). Event-related potentials in an invertebrate: Crayfish emit "omitted stimulus potentials." *Journal of Experimental Biology, 204*, 4291–4300.

Ripps, H., & Vaughan, H. G. J. (1969). The spectral sensitivity of evoked potentials from the retina and cortex of nocturnal and diurnal monkeys. *Vision Research, 9*, 895–907.

Schroeder, C. E., & Foxe, J. J. (2002). The timing and laminar profile of converging input to multisensory areas of the macaque neocortex. *Cognitive Brain Research, 14*(1), 187–198.

Schroeder, C. E., Tenke, C. E., & Givre, S. J. (1992). Subcortical contributions to the surface-recorded flash-VEP in the awake macaque. *Electroencephalography and Clinical Neurophysiology, 84*, 219–231.

Schroeder, C. E., Tenke, C. E., Givre, S. J., Arezzo, J. C., & Vaughan, H. G. J. (1991). Striate cortical contribution to the surface-recorded pattern-reversal VEP in the alert monkey. *Vision Research, 31*, 1143–1157.

Shah, A. S., Bressler, S. L., Knuth, K. H., Ding, M., Mehta, A. D., Ulbert, I., et al. (2004). Neural dynamics and the fundamental mechanisms of event-related brain potentials. *Cerebral Cortex, 14*, 476–483.

Simson, R., Jr., H. G. V., & Ritter, W. (1976). The scalp topography of potentials associated with missing visual or auditory stimuli. *Electroencephalography and Clinical Neurophysiology, 40*, 33–42.

Simson, R., Vaughan, H. G., & Ritter, W. (1977). The scalp topography of potentials in auditory and visual discrimination tasks. *Electroencephalography and Clinical Neurophysiology, 42*, 528–535.

Sutton, S., Braren, M., Zubin, J., & John, E. R. (1965). Evoked potential correlates of stimulus uncertainty. *Science, 150*, 1187–1188.

Van der Marel, H., Dagnelie, G., & Spekreijse, H. (1981). Pattern evoked potentials in awake rhesus monkeys. *Investigative Ophthalmology and Visual Science, 21*, 457–466.

Van der Marel, H., Dagnelie, G., & Spekreijse, H. (1984). Subdurally recorded pattern and lunimance EPs in the alert rhesus monkey. *Electroencephalography and Clinical Neurophysiology, 57*, 354–368.

Vaughan, H. G. J., & Gross, C.G. (1969). Cortical responses to light in unanesthetized monkeys and their alteration by visual system lesion. *Experimental Brain Research, 8*, 19–36.

Walter, W. G. (1938). Critical review: The technique and application of electro-encephalography. *Journal of Neurology and Psychiatry, 1*, 359–385.

Walter, W. G., Cooper, R., Aldridge, V. J., McCallum, W. C., & Winter, A. L. (1964). Contingent negative variation: An electric sign of sensorimotor association and expectancy in the human brain. *Nature, 203*, 380–384.

Wang, C., Ulbert, I., Schomer, D. L., Marinkovic, K., & Halgren, E. (2005). Responses of human anterior cingulate cortex microdomains to error detection, conflict monitoring, stimulus-response mapping, familiarity, and orienting. *Journal of Neuroscience, 25*, 604–613.

Weitzman, E. D. (1961). A note on the EEG and eye movements during behavioral sleep in monkeys. *Electroencephalography and Clinical Neurophysiology, 13*, 790–794.

Wilder, M. B., Farley, G. R., & Starr, A. (1981). Endogenous late positive component of the evoked potential in cats corresponding to P300 in humans. *Science, 211*, 605–607.

Woldorff, M., & Hillyard, S. A. (1990). Attentional influence on the mismatch negativity (commentary on target article by R. Näätänen). *Behavioral and Brain Sciences, 13*, 258–260.

Woldorff, M. G., Hackley, S. A., & Hillyard, S. A. (1991). The effects of channel-selective attention on the mismatch negativity wave elicited by deviant tones. *Psychophysiology, 28*, 30–42.

Woodman, G. F., Kang, M.-K., Rossi, A. F., & Schall, J. D. (2007). Nonhuman primate event-related potentials indexing covert shifts of attention. *Proceedings of the National Academy of Sciences USA, 104*, 15111–15116.

Woodman, G. F. (2010). A brief introduction to the use of event-related potentials (ERPs) in studies of perception and attention. *Attention, Perception & Psychophysics, 72*, 2031–2046.

Woodman, G. F., & Luck, S. J. (1999). Electrophysiological measurement of rapid shifts of attention during visual search. *Nature, 400*, 867–869.

Woodman, G. F., & Luck, S. J. (2003a). Dissociations among attention, perception, and awareness during object-substitution masking. *Psychological Science, 14*, 605–611.

Woodman, G. F., & Luck, S. J. (2003b). Serial deployment of attention during visual search. *Journal of Experimental Psychology: Human Perception and Performance, 29*, 121–138.

INDEX

Note: Page numbers followed by "*f*" and "*t*" denote figures and tables, respectively.

D

E

M

N

O

P

R

S

W